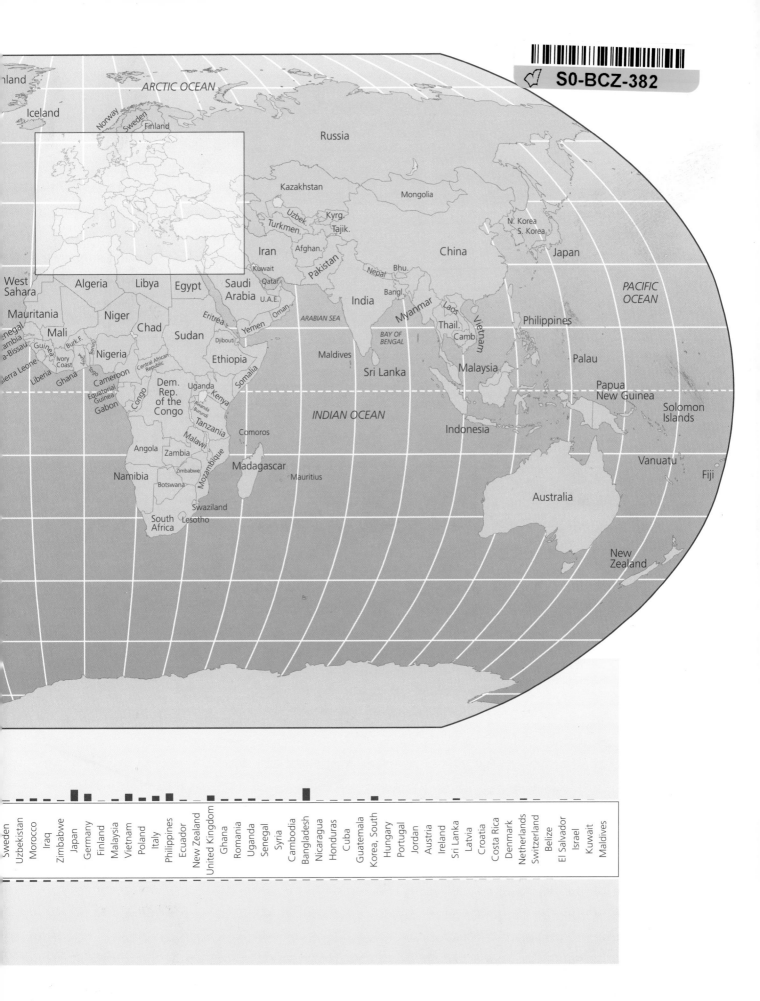

S0-BCZ-382

ARCTIC OCEAN

Iceland

Norway
Sweden
Finland

Russia

Kazakhstan

Mongolia

N. Korea
S. Korea

Uzbek.
Kyrg.
Turkmen.
Tajik.

Iran
Afghan.

China

Japan

PACIFIC
OCEAN

West
Sahara

Algeria

Libya

Egypt

Saudi
Arabia

Kuwait
Qatar
U.A.E.

Pakistan

Nepal
Bhu.

India

Bangl

Myanmar

Laos
Thail.

Vietnam
Camb.

Philippines

Mauritania

Niger

Chad

Sudan

Eritrea

Yemen
Oman

ARABIAN SEA

BAY OF
BENGAL

Palau

enegal
Mali

Burk. F.

Benin

Nigeria

Djibouti

Maldives

Sri Lanka

Malaysia

Papua
New Guinea

ambia
a-Bissau
Guinea

Ivory
Coast

Togo

Central African
Republic

Ethiopia

Somalia

Solomon
Islands

Sierra Leone
Liberia
Ghana

Cameroon
Equatorial
Guinea
Gabon

Congo

Dem.
Rep.
of the
Congo

Uganda

Rwanda
Burundi

Kenya

Indonesia

Tanzania

Comoros

INDIAN OCEAN

Angola

Malawi
Zambia

Madagascar

Mauritius

Vanuatu

Fiji

Namibia

Zimbabwe

Botswana

Mozambique

Australia

Swaziland
South
Africa

Lesotho

New
Zealand

Sweden
Uzbekistan
Morocco
Iraq
Zimbabwe
Japan
Germany
Finland
Malaysia
Vietnam
Poland
Italy
Philippines
Ecuador
New Zealand
United Kingdom
Ghana
Romania
Uganda
Senegal
Syria
Cambodia
Bangladesh
Nicaragua
Honduras
Cuba
Guatemala
Korea, South
Hungary
Portugal
Jordan
Austria
Ireland
Sri Lanka
Latvia
Croatia
Costa Rica
Denmark
Netherlands
Switzerland
Belize
El Salvador
Israel
Kuwait
Maldives

Log on.

Tune in.

Succeed.

Your steps to success.

STEP 1: Register

All you need to get started is a valid email address and the access code below. To register, simply:

1. Go to www.aw-bc.com/withgott
2. Click the appropriate book cover.
   Cover must match the textbook edition being used for your class.
3. Click "**Register**" under "**First-Time User?**"
4. Leave "**No, I Am a New User**" selected.
5. Using a coin, scratch off the silver coating below to reveal your access code.
   Do not use a knife or other sharp object, which can damage the code.
6. Enter your access code in lowercase or uppercase, without the dashes.
7. Follow the on-screen instructions to complete registration.
   During registration, you will establish a personal login name and password to use for logging into the website. You will also be sent a registration confirmation email that contains your login name and password.

Your Access Code is:

USWBE-JEHAD-SIANG-DOTED-CHOIR-MACAW

Note: If there is no silver foil covering the access code, it may already have been redeemed, and therefore may no longer be valid. In that case, you can purchase access online using a major credit card. To do so, go to www.aw-bc.com/withgott, click the cover of your textbook, click "**Buy Now**", and follow the on-screen instructions.

STEP 2: Log in

1. Go to www.aw-bc.com/withgott and click the appropriate book cover.
2. Under "**Established User?**" enter the login name and password that you created during registration. *If unsure of this information, refer to your registration confirmation email.*
3. Click "**Log In**".

STEP 3: (Optional) Join a class

Instructors have the option of creating an online class for you to use with this website. If your instructor decides to do this, you'll need to complete the following steps using the Class ID your instructor provides you. By "joining a class," you enable your instructor to view the scored results of your work on the website in his or her online gradebook.

To join a class:

1. Log into the website. For instructions, see "STEP 2: Log in."
2. Click "**Join a Class**" near the top right.
3. Enter your instructor's "**Class ID**" and then click "**Next**".
4. At the Confirm Class page you will see your instructor's name and class information. If this information is correct, click "**Next**".
5. Click "**Enter Class Now**" from the Class Confirmation page.
• *To confirm your enrollment in the class, check for your instructor and class name at the top right of the page. You will be sent a class enrollment confirmation email.*
• *As you complete activities on the website from now through the class end date, your results will post to your instructor's gradebook, in addition to appearing in your personal view of the Results Reporter.*
To log into the class later, follow the instructions under "STEP 2: Log in."

Got technical questions?

Visit http://www.aw-bc.com/techsupport/. Email technical support is available 24/7.

Join a class

Register and log in

ENVIRONMENT

THE SCIENCE BEHIND THE STORIES

SECOND EDITION

JAY WITHGOTT

SCOTT BRENNAN

PEARSON

Benjamin
Cummings

San Francisco • Boston • New York
Cape Town • Hong Kong • London • Madrid • Mexico City
Montreal • Munich • Paris • Singapore • Sydney • Tokyo • Toronto

Senior Acquisitions Editor: Chalon Bridges
Senior Project Editor: Mary Ann Murray
Executive Managing Editor: Erin Gregg
Managing Editor: Michael Early
Manufacturing Buyer: Stacy Wong
Production Supervisor: Lori Newman
Art Development: Russell Chun
Director, Media Development: Lauren Fogel
Media Producer: Ziki Dekel
Editorial Assistant: Haig MacGregor
Photo Production Manager: Travis Amos
Photo Researcher: Kristin Piljay

Marketing Manager: Jeff Hester
Production Supervisor, Media: Jennifer Mattson
Composition: TechBooks/GTS
Project Manager: Christine Knapp
Illustrations: Dragonfly Media Group
Copyeditor: Sally Peyrefitte
Proofreader: William Heckman
Design Manager: Mark Ong
Text Design: Gary Hespenheide
Cover Design: Yvo Riezebos
Cover Printer: Phoenix Color Corporation
Text Printer: Quebecor World Dubuque

Cover photograph: Phoenix Islands, Kiribati, Polynesia. Divers mapping the area find fragile table coral in Kanton Lagoon. Paul Nickelen/National Geographic/Getty Images.

Photo credits continue following the glossary.

Printed using soy-based ink. Paper is recycled containing at least 20% post-consumer waste.
ISBN 0-8053-4467-5 [Student Text Component]
ISBN 0-8053-8204-6 [Instructor]
ISBN 0-13-134642-3 [High School]

Library of Congress Cataloging-in-Publication Data
Withgott, Jay.
  Environment: the science behind the stories.–2nd ed. / Jay Withgott, Scott Brennan.
    p. cm.
  Brennan's name appears first on the previous ed.
  ISBN 0-8053-8203-8
  1. Environmental sciences. I. Brennan, Scott R. II. Title.
GE105.B74 2006
363.7—dc22   2005036344

PEARSON
Benjamin
Cummings

www.aw-bc.com   2 3 4 5 6 7 8 9 10 -QWD-08 07

About the Authors

Jay H. Withgott is a science and environmental writer with a background in scientific research and teaching. He holds degrees from Yale University, the University of Arkansas, and the University of Arizona. As a researcher, he has published scientific papers on topics in ecology, evolution, animal behavior, and conservation biology in a variety of journals including *Proceedings of the National Academy of Sciences, Proceedings of the Royal Society of London B, Evolution,* and *Animal Behavior.* He has taught university-level laboratory courses in ecology, ornithology, vertebrate diversity, anatomy, and general biology.

As a science writer, Jay has authored articles for a variety of journals and magazines including *Science, New Scientist, BioScience, Current Biology, Conservation in Practice,* and *Natural History.* He combines his scientific expertise with his past experience as a reporter and editor for daily newspapers to make science accessible and engaging for general audiences.

Jay lives with his wife, biologist Susan Masta, in Portland, Oregon, and takes every opportunity he can to explore the diverse landscapes of Oregon and the American West.

Scott Brennan has taught environmental science, ecology, resource policy, and journalism at Western Washington University and at Walla Walla Community College. He has also worked as a journalist, photographer, and consultant.

Scott has cultivated his expertise in environmental science and public policy by serving as Campaign Director of Alaskans for Responsible Mining, as Executive Conservation Fellow of the National Parks Conservation Association in Washington, D.C., and as a consultant to the U.S. Department of Defense Environmental Security Office at the Pentagon.

When not at work, Scott is likely to be found exploring the Chugach Mountains and the Bristol Bay drainages in southwest Alaska. He lives with his wife, Angela, and their dogs Raven and Hatcher, in south central Alaska's Chester Creek Watershed.

How do environmental issues affect

Integrated Central Case Studies begin each chapter and are further developed throughout the chapter text. These highlight real people and real places to bring environmental issues to life, making general concepts more understandable and interesting to learn.

149

Aggregation of zebra mussels

North America

Atlantic Ocean

Great Lakes

Pacific Ocean

Central Case: Black and White, and Spread All Over: Zebra Mussels Invade the Great Lakes

"We are seeing changes in the Great Lakes that are more rapid and more destructive than any time in the history of the Great Lakes."
—ANDY BUCHSBAUM, DIRECTOR OF THE NATIONAL WILDLIFE FEDERATION'S GREAT LAKES OFFICE

"When you tear away the bottom of the food chain, everything that is above it is going to be disrupted."
—TOM NALEPA, NATIONAL OCEANIC AND ATMOSPHERIC ADMINISTRATION RESEARCH BIOLOGIST

As if the Great Lakes hadn't been through enough already, the last thing they needed was the zebra mussel. The pollution-fouled waters of Lake Erie and the other Great Lakes shared by Canada and the United States had become gradually cleaner in the years following the Clean Water Act of 1970. As government regulation brought industrial discharges under control, people once again began to use the lakes for recreation, and populations of fish rebounded.

Then the zebra mussel arrived. Black-and-white-striped shellfish the size of a dime, zebra mussels attach to hard surfaces and open their paired shells, feeding on algae by filtering water through their gills. This mollusc, given the scientific name *Dreissena polymorpha*, is native to the Caspian Sea, Black Sea, and Azov Sea in western Asia and eastern Europe. It made its North American debut in 1988 when it was discovered in Canadian waters at Lake St. Clair, which connects Lake Erie with Lake Huron. Evidently ships arriving from Europe had discharged ballast water containing the mussels or their larvae into the Great Lakes.

Within just two years of their discovery in Lake St. Clair, zebra mussels had reached all five of the Great Lakes. The next year, these invaders entered New York's Hudson River to the east, and the Illinois River at Chicago to the west. From the Illinois River and its canals, they soon reached the Mississippi River, giving them access to a vast watershed covering 40% of the

people and places?

INVESTIGATE it! on the Withgott/Brennan Companion Website provides an additional 120 case studies beyond those presented in the text. Browse by topic or geographic region to access 100 recent articles from **The New York Times** and 20 **abc**NEWS clips that explore environmental issues that are in the news today.

Do you understand the **science** behind

The Science behind the Story highlights how scientists develop hypotheses, test predictions, and analyze and interpret data. Each *Science behind the Story* carefully walks you through the scientific process—not only *what* scientists have discovered, but *how* they discovered it. These engaging accounts help you understand "how we know what we know" about environmental issues.

The Science behind the Story

Inferring Zebra Mussels' Impacts on Fish Communities

Food webs are complicated systems, and disentangling them to infer the effects of any one species is fraught with difficulty. When zebra mussels appeared in the Great Lakes, people feared for sport fisheries, and estimated that fish population declines could cost billions of dollars. The mussels would deplete the phytoplankton and zooplankton that fish depended on, people reasoned, and many fewer fish would survive. Yet even after 15 years, there was no solid evidence of widespread harm to fish populations.

So, aquatic biologist David Strayer of the Institute of Ecosystem Studies in Millbrook, New York, joined Kathyrn Hattala and Andrew Kahnle of New York State's Department of Environmental Conservation (DEC). They mined datasets on fish populations in the Hudson River, which zebra mussels had invaded in 1991.

Strayer and others had already been studying effects of zebra mussels on aspects of the community for years. Their data showed that since the species' introduction to the Hudson:

► Biomass of phytoplankton fell 80%.
► Biomass of small zooplankton fell 76%.
► Biomass of large zooplankton fell 52%.

Zebra mussels increased filter-feeding in the community 30-fold, thereby depleting the phytoplankton and small zooplankton, and leaving all sizes of zooplankton with less phytoplankton to eat. Overall, the zooplankton and invertebrate animals of the open water that are eaten by open-water fish declined by 70%.

However, Strayer's work had also found that *benthic*, or bottom-dwelling, invertebrates in shallow water (especially in the nearshore, or *littoral*, zone) had increased by 10%, and likely much more, because the mussels' shells provide habitat structure, and their feces provide nutrients.

These contrasting trends in the benthic shallows and the open deep water led Strayer's team to hypothesize that zebra mussels would harm open-water fish that ate plankton but would help littoral-feeding fish. They predicted that after zebra mussel introduction, larvae and juveniles of six common open-water fish species would decline in number, decline in growth rate, and shift downriver toward saltier water, where mussels are absent. Conversely, they predicted that larvae and juveniles of 10 littoral fish species would increase in number, increase in growth rate, and shift upriver to regions of greatest zebra mussel density.

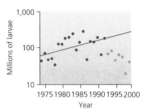

(a) American shad

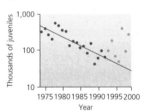

(b) Tessellated darter

Larvae of American shad (a), an open-water fish, had been increasing in abundance before zebra mussels were introduced (red points and trend line). After zebra mussel introduction, shad larvae decreased in abundance (orange points). Juveniles of the tessellated darter (b), a littoral zone fish, had been decreasing in abundance before zebra mussels were introduced (red points and trend line). After zebra mussel introduction, they increased in abundance (orange points). *Source:* Strayer, D., et al. 2004. Effects of an invasive bivalve (*Dreissena polymorpha*) on fish in the Hudson River estuary. *Can. J. Fish. Aquat. Sci.* 61: 924–941.

To test their predictions, the researchers analyzed data from three

community, removal of a keystone species will have substantial ripple effects and will alter a large portion of the food web.

Often, large-bodied secondary or tertiary consumers near the tops of food chains are considered keystone species. Top predators control populations of herbivores, which would otherwise multiply and could, through increased herbivory, greatly modify the plant community. In the United States, government bounties promoted the hunting of wolves and mountain lions, which were largely exterminated by the middle of the 20th century. In the absence of these predators, unnaturally dense deer

the news stories?

FIGURE 1.14 Indoor and outdoor air pollution contribute to millions of premature deaths each year, and environmental scientists and policymakers are working to reduce this problem in a variety of ways.

toxicologists are chronicling the impacts on people and wildlife of the many synthetic chemicals and other pollutants we emit into the environment (Chapter 14). Our most pressing pollution challenge may be to address the looming specter of global climate change (Chapter 18). Scientists have firmly concluded that human activity is altering the composition of the atmosphere and that these changes are affecting Earth's climate. Since the start of the industrial revolution, atmospheric carbon dioxide concentrations have risen by 31%, to a level not present in at least 420,000 years. This increase results from our reliance on burning fossil fuels to power our civilization. Carbon dioxide and several other gases absorb heat and warm Earth's surface, which is likely responsible for glacial melting, sea-level rise, impacts on wildlife and crops, and increased episodes of destructive weather.

The combined impact of human actions such as climate change, overharvesting, pollution, the introduction of non-native species, and particularly habitat alteration, has driven many aquatic and terrestrial species out of large parts of their ranges and toward the brink of extinction (Chapter 11). Today Earth's biological diversity, or **biodiversity,** the cumulative number and diversity of living things, is declining dramatically. Many biologists say we are already at the outset of a mass extinction event comparable to only five others documented in all of Earth's history. Biologist Edward O. Wilson has warned that the loss of biodiversity is our most serious and threatening environmental dilemma, because it is not the kind of problem that responsible human action can remedy. Rather, the extinction of species is irreversible; once a species has become extinct, it is lost forever.

Solutions to environmental problems must be global and sustainable

The nature of virtually all of these environmental issues is being changed by the set of ongoing phenomena commonly dubbed *globalization*. Our increased global interconnectedness in trade, politics, and the movement of people and of other species poses many challenging problems, but it also sets the stage for novel and effective solutions.

The most comprehensive scientific assessment of the present condition of the world's ecological systems and their ability to continue supporting our civilization was completed in 2005. In this year, over 2,000 of the world's leading environmental scientists from nearly 100 nations completed the **Millennium Ecosystem Assessment.** The four main findings of this exhaustive project are summarized in Table 1.1. The Assessment makes clear that our degradation of the world's environmental systems is having negative impacts on all of us, but that with care and diligence we can still turn many of these trends around.

Table 1.1   Main Findings of the Millennium Ecosystem Assessment

► Over the past 50 years, humans have changed ecosystems more rapidly and extensively than in any comparable period of time in human history, largely to meet rapidly growing demands for food, freshwater, timber, fiber, and fuel. This has resulted in a substantial and largely irreversible loss in the diversity of life on Earth.

► The changes made to ecosystems have contributed to substantial net gains in human well-being and economic development, but these gains have been achieved at growing costs. These costs include the degradation of ecosystems and the services they provide for us, and the exacerbation of poverty for some groups of people.

► This degradation could grow significantly worse during the first half of this century.

► The challenge of reversing the degradation of ecosystems while meeting increasing demands for their services can be partially overcome, but doing so will involve significantly changing many policies, institutions, and practices.

Adapted from *Millennium Ecosystem Assessment, Synthesis Report,* 2005.

Environmental issues change quickly, so Withgott/Brennan uses the most current data available.

References are clearly cited so you can trace the source of the information presented.

Do you know how to interpret graphs

Interpreting Graphs and Data activities at the end of each chapter give you hands-on experience in working with graphs, so you can develop the skills you'll need to understand scientific information when you see it in the news.

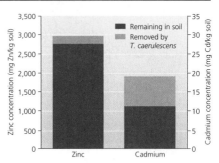

INTERPRETING GRAPHS AND DATA

In phytoremediation, plants are used to clean up soil or water contaminated by heavy metals such as lead (Pb), arsenic (As), zinc (Zn), and cadmium (Cd). For plants to absorb these metals from soil, the metals must be dissolved in soil water. For any given instance, all metal can be accounted for as either remaining bound to soil particles, being dissolved in soil water, or being stored in the plant.

In a study on the effectiveness of alpine penny-cress (*Thlaspi caerulescens*) for phytoremediation, Enzo Lombi and his colleagues grew crops of this small perennial plant for approximately one year in pots of soil from contaminated sites. They then measured the amount of zinc and cadmium in the soil and in the plants when they were harvested.

1. What were the zinc and cadmium concentrations in the soil prior to phytoremediation? What were the zinc and cadmium concentrations in the soil after one year of phytoremediation?

2. How much zinc and cadmium were removed from the soil? If the plants continued to remove zinc and cadmium from the soil at the rates shown above, approximately how long would it take to remove all the zinc and cadmium?

Removal of zinc and cadmium from contaminated soil by alpine penny-cress, *Thlaspi caerulescens*. Data from Lombi, E., et al. 2001. Phytoremediation of heavy metal-contaminated soils: natural hyperaccumulation versus chemically enhanced phytoextraction. *Journal of Environmental Quality* 30: 1919–1926.

3. Alpine penny-cress produces natural chelating agents (see "The Science behind the Story," pp. 92–93) that increase the solubility of metals in soil water. If these dissolved metals are not taken up by the plants, what may be an unintended consequence of having increased their solubility?

GRAPHit!

exercises on the Withgott/Brennan Companion Website help you to better understand how to work with and interpret graphs.

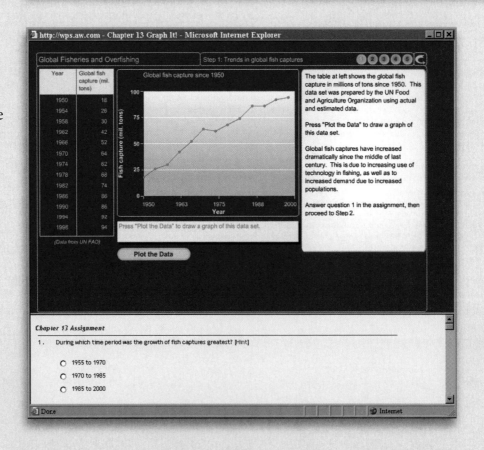

and data?

CALCULATING ECOLOGICAL FOOTPRINTS

In ecological systems, a rough rule of thumb is that when energy is transferred from plants to plant-eaters or from prey to predator, the efficiency is only about 10%. Much of this inefficiency is a consequence of the second law of thermodynamics. Another way to think of this is that eating 10 calories of plant material is the ecological equivalent of eating 1 calorie of material from an animal.

Humans are considered omnivores because we can eat both plants and animals. The choices we make about what to eat have significant ecological impacts. With this in mind, calculate the ecological energy requirements for four different diets, each of which provides a total of 2,000 dietary calories per day.

1. How many ecologically equivalent calories would it take to support you for a year on each of the four diets listed?
2. What is the relative ecological impact of including as little as 10% of your calories from animal sources (e.g., milk, dairy products, eggs, and meat)? What is the ecological impact of a strictly carnivorous diet compared with a strict vegetarian diet?
3. What percentages of the calories in your own diet do you think come from plant versus animal sources? Estimate the ecological impact of your diet, relative to a strictly vegetarian one.
4. Describe some challenges of providing food for the growing human population, especially as people in many poorer nations develop a taste for an American-style diet rich in animal protein and fat.

Diet	Source of calories	Number of calories consumed by source	Ecologically equivalent calories by source	Total ecologically equivalent calories per day
100% plant 0% animal	Plant			
	Animal			
90% plant 10% animal	Plant	1,800	1,800	3,800
	Animal	200	2,000	
50% plant 50% animal	Plant			
	Animal			
0% plant 100% animal	Plant			
	Animal			

Calculating Ecological Footprints activities at the end of each chapter let you work with numbers to evaluate the impact of actions—including your own—on a local and global scale.

Are you prepared to make **informed**

The *Viewpoints* feature in each chapter presents two opposing views on an environmental issue related to the chapter's central theme, allowing you to consider multiple sides of the story.

Reach your own conclusion on *Viewpoints* questions by accessing the *Viewpoints* link on the companion website. There you'll find questions to consider when exploring the issues further, and links to websites that support each opinion.

 VIEWPOINTS

Bioremediation

Naturally occurring microbes were put to use cleaning up beaches following the *Exxon Valdez* oil spill. **Was bioremediation a success in Prince William Sound? Can this technique address many of our society's problems with chemical contamination, or does it suffer from too many limitations?**

Bioremediation: An Effective Solution

Bioremediation was definitely a success in Prince William Sound. It was important to remove the spilled oil that reached the shorelines in an environmentally responsible way. Washing it into the sea and collecting it by skimming was the first step, but what to do about the residual oil on the beach? Bioremediation seemed a likely technology.

Oil seeps have released oil into the sea for millions of years, feeding a diverse group of microorganisms. Oil is thus very biodegradable, though unusual in that although it is rich in energy, it contains none of the inorganic nutrients required for microbial growth. The arrival of oil on a shoreline causes a dramatic increase in the number of oil-degrading microorganisms, but their growth is soon limited by the low levels of bioavailable nitrogen and phosphorus in seawater.

Our approach in Alaska was to add oleophilic (oil-adhering) and slow-release fertilizers to provide enough of these limiting nutrients over many tidal cycles so that biodegradation would be stimulated, but not so much that they would cause environmental harm. Rigorous testing, in collaboration with the State of Alaska and U.S. EPA, showed that the approach stimulated the natural rate of biodegradation some twofold to fivefold, with no detectable adverse environmental impact. Bioremediation was used on about 74 miles of shoreline—by far the largest use of this technology to date—and most shorelines were oil-free within 3 years instead of the predicted decades.

Can bioremediation address other environmental problems? Yes. There are many situations where stimulating the biodegradation of fuels, explosives, chlorinated solvents, transformer fluids, pesticides, and other substances will be an environmentally benign and effective treatment. Although the best ways of stimulating biodegradation will not be the same in all situations, working to stimulate natural processes without causing adverse effects will be an excellent way of cleaning many contaminated sites.

Roger Prince is a senior research associate at ExxonMobil's Corporate Strategic Research Laboratory in Annandale, New Jersey. He was Exxon's lead scientist in the monitoring of the successful bioremediation of shorelines oiled during the *Exxon Valdez* spill in Alaska, and he continues to work on the bioremediation of marine oil spills, including a recent multinational collaboration in the Arctic.

Biostimulation Can Sometimes Enhance Environmental Cleanup

The *Exxon Valdez* oil spill, which led to the enactment of the Oil Pollution Act of 1990, gave rise to the largest bioremediation field trial ever attempted. A research study was conducted by EPA in 1989 and 1990 to develop data to support the recommendation to go forward with a full-scale cleanup. Unfortunately, the data generated were equivocal and did not provide sufficient evidence to prove the success of the treatment. The study was equivocal because its experimental design did not provide sufficient replication or randomization of plots to allow for the calculation of experimental error, or to account for the high variability of oil contamination on the beaches.

A later field study in Delaware did provide the unequivocal evidence needed. Data generated in this and other EPA-funded field studies provided sufficient, statistically sound evidence that biostimulation of indigenous microorganisms can accelerate the disappearance of hydrocarbons at a spill site.

However, two other studies I have been involved with, both done on Canadian wetlands, have showed that biostimulation may not always be appropriate. One needs to determine the background levels of nutrients already present and make sure the affected environment is aerobic in nature. Only if nutrients are limiting in concentration, and if dissolved oxygen in the pore space is high enough to support microbial growth on the hydrocarbons, is biostimulation appropriate.

Based on various studies such as these, EPA has published two guidance documents on bioremediation on marine shorelines, freshwater wetlands, and salt marshes.

Although it is not appropriate in every case, biostimulation can play a key role in the environmental cleanup of oil spills. It is a tool to be seriously considered when contemplating how to restore a contaminated environment.

Albert D. Venosa is a senior research scientist with the U.S. Environmental Protection Agency, National Risk Management Research Laboratory, in Cincinnati, Ohio. He currently heads EPA's oil spill research program in the Office of Research and Development. He has published over 100 works in many aspects of wastewater treatment and hydrocarbon biodegradation.

> Explore this issue further by accessing **Viewpoints** at www.aw-bc.com/withgott

x

decisions on environmental issues?

Issues in environmental science often lack black-and-white answers, so critical thinking skills help you navigate the gray areas. *Weighing the Issues* questions throughout each chapter encourage you to grapple with questions about science, policy, and ethics.

Weighing the **Issues:**
Ecosystems Where You Live

Think about the area where you live. How would you describe that area's ecosystems? How do these systems interact with one another? If one ecosystem were greatly disturbed (say, if a wetland or forest were replaced by a shopping mall), what impacts might that have on nearby natural systems?

You Decide activities on the Withgott/Brennan Companion Website allow you to play the role of decision maker as you study the data, then form your own plan for saving endangered grizzlies or stopping global warming.

Preface

We live in extraordinary times. Human impact on our environment has never been so intensive or so far-reaching. The future of Earth's systems and of our society depends more critically than ever on the way we interact with the natural systems around us. Fundamental aspects of nutrient cycling, biological diversity, atmospheric composition, and climate are changing at dizzying speeds. Yet thanks to environmental science, we now understand better than ever how our planet's systems function and how we influence these systems. Understanding environmental science helps us to characterize human-induced problems and also illuminates the tremendous opportunities we have before us for effecting positive change.

The field of environmental science captures the very essence of this unique moment in history. An interdisciplinary field, environmental science integrates the natural sciences with the social sciences, studying both the workings of our planet and the workings of our own species. Environmental science draws upon the methods and findings of numerous established academic disciplines, from ecology to geology to chemistry to economics to political science to ethics. This interdisciplinary pursuit stands at the vanguard of the current need to synthesize academic disciplines and to incorporate their contributions into a big-picture understanding of the world and our place within it.

We wrote this book because we feel that the vital importance of environmental science in today's world makes it imperative to engage, educate, and inspire a broad audience of today's students—the citizens and leaders of tomorrow. We have therefore tried to implement the very best in modern teaching approaches and to clarify how the scientific process can inform human efforts. We also have aimed to maintain a balanced approach and to encourage critical thinking as we flesh out the social debate over many environmental issues. Finally, we have resolved to avoid gloom and doom and instead provide hope and solutions.

These several aims guided our crafting of the second edition of this text, as they had guided the first. Moreover, the second edition is significantly improved in many ways. We revisited every word of the text, incorporating the most current information from this fast-moving field and streamlining our presentation to make learning more straightforward. Dozens of new figures and an enhanced art style enable us to educate more effectively; the biogeochemical cycle figures in Chapter 7 are just one example. We expanded our coverage of ecology, energy, urbanization, and other topics that many instructors deem especially important or that are of growing interest. We also enhanced our focus on sustainability and on the ecological footprint concept, making them themes of the text, because of their central importance in environmental science. The new section on campus sustainability in our final chapter showcases examples of sustainable solutions that students are enacting on campuses worldwide.

Students often desire help with study skills, comprehending graphs and data, and understanding the impacts of their environmental choices in a concrete quantitative way. To address these needs, we added three new features to the end of each chapter. "Reviewing Objectives" summarizes each chapter's main points and relates them to the learning objectives presented at the opening of the chapter, enabling students to confirm that they have understood the most crucial ideas and to review concepts by turning to specified page numbers. "Interpreting Graphs and Data" uses figures from recent scientific studies to help students build quantitative and analytical skills in reading graphs and making sense of data. "Calculating Ecological Footprints" enables students to calculate the environmental impacts of their own choices and then see how individual impacts scale up to impacts at the societal level.

We have also retained the major features that made the first edition of our book unique and that are proving so successful in classrooms across North America:

▶ **Integrated Central Case Studies.** Our teaching experiences, together with feedback from colleagues across the continent, clearly reveal that telling compelling stories about real people and real places is the best way to capture students' interest. Providing narratives with concrete detail also helps teach abstract concepts, because it gives students a tangible framework with which to incorporate new ideas. Whereas many textbooks these days serve up case studies in isolated boxes, we have chosen to integrate each chapter's central case study into the main text, weaving information and elaboration throughout the chapter. In this way, we use the concrete realities of the people and places of the central case study to help illustrate the topics we cover. We are gratified that students and instructors using our first edition have consistently applauded this approach, and we hope it can help bring about a new level of effectiveness in environmental science education.

▶ **The Science behind the Story.** Our goal is not simply to present students with facts, but to engage them in the scientific process of testing and discovery. To do this, we discuss the scientific method and the social context of science in our opening chapter, and we describe hundreds of real-life studies throughout the main text. We also feature in each chapter "The Science behind the Story" boxes, which elaborate on particular studies important to the chapter topic, guiding readers through the details of the research. In this way we show not merely *what* scientists discovered, but also *how* they discovered it. Instructors using our first edition have confirmed that this feature enhances student comprehension of each chapter's material and deepens understanding of the scientific process itself—a key component of effective citizenship in today's science-driven world.

▶ **Viewpoints.** In our text we have striven to present a balanced picture of environmental issues, informed by the best science that bears upon them. Yet we all know that sometimes intelligent people can examine the same data and come to dramatically different conclusions. To ensure that students are exposed to a diversity of interpretations on key issues, we include in each chapter the *Viewpoints* feature, which consists of paired essays authored by invited experts who present different points of view on particular questions of importance. The essays provide students a taste of informed arguments directly from individuals who are actively involved in work—and debate—on environmental issues. To encourage critical thinking, we refer students to an online resource at the book's website that presents questions they can use to critically examine and discuss the ideas in these essays and that provides links to websites that support the contributors' viewpoints.

▶ **Weighing the Issues.** Because the multifaceted issues in environmental science often lack black-and white answers, students need critical-thinking skills to help navigate the gray areas at the juncture of science, policy, and ethics. We have aimed to help develop these skills with our end-of-chapter questions and with our "Weighing the Issues" feature. Several "Weighing the Issues" questions are dispersed throughout each chapter, serving as stopping points for students to absorb and reflect upon what they have read and wrestle with some of the complex dilemmas in environmental science.

▶ **An emphasis on solutions.** The complaint we most frequently hear from students in environmental science courses is that the deluge of environmental problems can seem overwhelming. In the face of so many problems, students often come to feel that there is no hope or that there is little they can personally do to make a difference. We have aimed to counter this impression by drawing out innovative solutions that have worked, are being implemented, or can be tried in the future. While we do not paint an unrealistically rosy picture of the challenges that lie ahead, we portray dilemmas as opportunities and we try to instill hope and encourage action. Indeed, for every problem that human carelessness has managed to create, human ingenuity can devise one—and likely multiple—solutions.

Environment: The Science behind the Stories has grown directly from our professional experiences in teaching, research, and writing. Jay Withgott has synthesized and presented science to a wide readership. His experience in distilling and making accessible the fruits of scientific inquiry has shaped our book's content and the presentation of its material. Scott Brennan has taught environmental science to thousands of undergraduates and has developed an intimate feeling for what works in the classroom. His knowledge and experience have shaped the pedagogical approaches we have taken in this book.

We have also been guided in our efforts by extensive input from our professional colleagues and from hundreds of instructors from around North America who have served as reviewers for our chapters and as advisors in focus group meetings arranged by Benjamin Cummings. The participation of so many learned and thoughtful experts has improved this volume in countless ways.

We sincerely hope that our efforts will come close to being worthy of the immense importance of our subject matter. We invite you, students and instructors alike, to let us know how well we have achieved our goals and where you feel we have fallen short. We are committed to continual improvement, and value your feedback. Please write the authors in care of Chalon Bridges (chalon.bridges @aw.com), Benjamin Cummings Publishing, 1301 Sansome Street, San Francisco, California, 94111.

At this most historic time to study environmental science, we are honored to serve as your guides in the quest to better understand our world and ourselves.

Jay Withgott and Scott Brennan

INSTRUCTOR SUPPLEMENTS

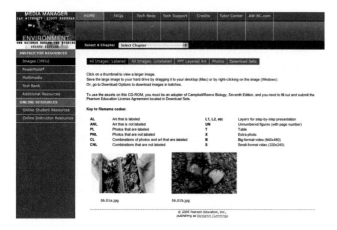

The Withgott/Brennan Media Manager
0-8053-8111-2

This powerful media package is organized chapter-by-chapter and includes all teaching resources in one convenient location. You'll find 5-minute *ABC News* Lecture Launcher videos, PowerPoint presentations, active lecture questions to facilitate class discussions (for use with or without clickers), and an image library that includes all art and tables from the text.

Instructor's Guide and Test Bank
0-8053-4468-3

This comprehensive resource provides chapter outlines, key terms, a listing of website and media resources, and teaching tips for lecture and classroom activities. A printed version of the Test Bank is conveniently included in the manual, offering hundreds of multiple-choice, short-answer, and essay questions to use on tests and quizzes. New to this edition are graphing and scenario-based questions to test students' critical-thinking abilities.

Computerized Test Bank
0-8053-8110-4

Hundreds of multiple-choice, short-answer, essay, graphing, and scenario-based questions on a cross-platform CD-ROM. Categorized by chapter objective for instructor ease in searching for question types.

Transparency Acetates
0-8053-8112-0

Includes 300 full-color acetates of all the art and tables from the text.

CourseCompass for Environment
0-8053-8198-8

This nationally hosted, easy-to-use course management tool allows professors to combine their own material with the material in the Environmental Science Place to create dynamic, online learning environments. Professors can post their syllabus, assign tutorials, customize quizzes, automatically grade them, and track the results instantly in the grade book. Go to **www.coursecompass.com**

Blackboard Premium for Environment
0-8053-8197-X

Blackboard Open Access
0-8053-8195-3

WebCT Premium for Environment
0-8053-8196-1

WebCT Open Access
0-8053-8194-5

Acknowledgments

A textbook is the product of *many* more minds and hearts than one might guess from the names on the cover. The two of us are exceedingly fortunate to be supported and guided by the tremendous staff at Benjamin Cummings and by a small army of experts in environmental science who have generously shared their time and expertise. Although we alone, as authors, bear responsibility for any inaccuracies, the strengths of this book result from the collective labor and dedication of innumerable people.

We would first like to thank our acquisitions editor, Chalon Bridges. Chalon's commitment and unremitting enthusiasm have inspired our entire team to relish the challenge of taking a successful and well-received first edition and making it still better. Moreover, her extensive interaction with instructors across North America has helped us define and refine our pedagogy and our innovative features. The approach, design, and essence of this book owe a great deal to Chalon's astute guidance and vision.

Senior project editor Mary Ann Murray worked hard every step of the way to help create this second edition, and her skill and devotion have touched every page of this book. Mary Ann's sharp eye and sound judgment improved the text in countless ways, even while she coordinated the Herculean logistics of our many features, reviews, supplements, and contributors. We are particularly grateful for her cheery optimism in the face of her authors' seemingly interminable delays.

We are excited by our book's newly enhanced art program. Senior art producer Russell Chun's ability to transform our nebulous suggestions into clear and insightful illustrations gave life and light to key points from the text. Photo researcher Kristin Piljay and photo manager Travis Amos helped provide the other half of our visual impact, reliably coming up with arresting photographic images.

We also are thrilled with two new features, "Interpreting Graphs and Data" and "Calculating Ecological Footprints." For each chapter, these were imaginatively conceived and ably authored by Jonathan Frye of McPherson College. Our thanks also go to Ned Knight of Linfield College for his insightful reviews of these features. Significant contributions from the book's first edition by April Lynch and Etienne Benson were retained in the second. We also would like to thank the authors of our Viewpoints essays, each of whom is credited along with his or her essay.

Editorial assistant Haig MacGregor was there when we needed him, coordinating reviews and a dozen other tasks. Copyeditor Sally Peyrefitte again provided thorough and meticulous examination of our text. Once the manuscript was ready, production supervisor Lori Newman saw it through to production, along with managing editors Erin Gregg and Michael Early. We thank project manager Christine Knapp and the rest of the staff at The GTS Companies for terrific work putting the whole thing together.

Finally, we would like to acknowledge the authors of our Instructor's Guide and Test Bank supplements. JodyLee Estrada Duek of Pima Community College, Debra Socci of Seminole Community College, and Steven Uyeda of Pima Community College performed extensive yet careful, quality work on a tight schedule. Kristy Manning thoroughly revised the PowerPoint slides for the main text. We also wish to thank Ziki Dekel for the development and production of the innovative media that accompanies this book.

Of course, none of this has any impact on education without the marketing staff to get the book into your hands. Marketing manager Jeff Hester dedicated his talent and enthusiasm to the book's promotion and distribution. And last but surely not least, the many field representatives who help communicate our vision and deliver our product to instructors are absolutely vital, and we deeply appreciate their work and commitment.

In the list that follows, we acknowledge the many instructors and outside experts who helped us maximize the quality and accuracy of our presentation through their chapter reviews, feature reviews, class tests, or other services. If the thoughtfulness and thoroughness of these reviewers are any indication, we feel confident that the teaching of environmental science is in excellent hands!

Lastly, Jay gives loving thanks to his wife Susan Masta, who endured the book's writing and revision with tremendous patience and sacrifice and provided support and sustenance throughout. Scott would like to thank Angela, Sean, Jonathan, Korby, Karl and Jess, and Jodi and Andy. We dedicate this book to today's students, who will shape tomorrow's world.

Jay Withgott and Scott Brennan

Reviewers

Jeffrey Albert, *Watson Institute of International Studies;* John V. Aliff, *Georgia Perimeter College;* Mary E. Allen, *Hartwick College;* Dula Amarasiriwardena, *Hampshire College;* Corey Andries, *Albuquerque Technical Vocational Institute;* David M. Armstrong, *University of Colorado–Boulder;* David L. Arnold, *Ball State University;* Joseph Arruda, *Pittsbur State University;* Thomas W. H. Backman, *Linfield College;* Marilynn Bartels, *Black Hawk College;* David Bass, *University of Central Oklahoma;* Christy Bazan, *Illinois State University;* Christopher Beals, *Volunteer State Community College;* Hans T. Beck, *Northern Illinois University;* Timothy Bell, *Chicago State University;* Terrence Bensel, *Allegheny College;* Gary Beluzo, *Holyoke Community College;* Bob Bennett, *University of Arkansas;* William B. N. Berry, *University of California, Berkeley;* Grady Price Blount, *Texas A & M University–Corpus Christi;* Marsha Bollinger, *Winthrop University;* Richard D. Bowden, Allegheny College; Frederick J. Brenner, *Grove City College;* Hugh Brown, *Ball State University;* J. Christopher Brown, *University of Kansas;* Dan Buresh, *Sitting Bull College;* John S. Campbell, *Northwestern College;* Mike Carney, *Jenks High School;* Kelly S. Cartwright, *College of Lake County;* Michelle Cawthorn, *Georgia Southern University;* Brad S. Chandler, *Palo Alto College;* Paul Chandler, *Ball State University;* David A. Charlet, *Community College Southern Nevada;* Kenneth E. Clifton, *Lewis and Clark College;* John E. Cochran, *Columbia Basin College;* Thomas L. Crisman, *University of Florida;* Jessica Crowe, *South Georgia College;* Gregory A. Dahlem, *Northern Kentucky University;* Mary E. Davis, *University of Massachusetts, Boston;* Thomas A. Davis, *Loras College;* Lola M. Deets, *Pennsylvania State University–Erie;* Roger del Moral, *University of Washington;* Craig Diamond, *Florida State University;* Darren Divine, *Community College of Southern Nevada;* Toby Dogwiler, *Winona State University;* Jeffrey Dorale, *University of Iowa;* Tracey Dosch, *Waubonsee Community College;* JodyLee Estrada Duek, *Pima Community College;* Jeffrey R. Dunk, *Humboldt State University;* Jean W. Dupon, *Menlo College;* Robert M. East, Jr., *Washington & Jefferson College;* Thomas R. Embich, *Harrisburg Area Community College;* Kenneth Engelbrecht, *Metropolitan State College of Denver;* Bill Epperly, *Robert Morris College;* Bonnie Fancher, *Switzerland County High School;* Francette Fey, *Macomb Community College;* Steve Fields, *Winthrop University;* Brad Fiero, *Pima Community College;* David G. Fisher, *Maharishi University of Management;* Linda Fitzhugh, *Gulf Coast Community College;* Laura Furlong, *Northwestern College;* Steven Frankel, *Northeastern Illinois University;* Arthur Fredeen, *University of Northern British Columbia;* Sandi B. Gardner, *Triton College;* Kristen S. Genet, *Anoka Ramsey Community College;* Marcia Gillette, *Indiana University–Kokomo;* Thad Godish, *Ball State University;* Michele Goldsmith, *Emerson College;* Amy R. Gregory, *University of Cincinnati;* Carol Griffin, *Grand Valley State University;* Judy Guinan, *Radford University;* Gian Gupta, *University of Maryland, Eastern Shore;* Mark Gustafson, *Texas Lutheran University;* Greg Haenel, *Elon University;* Grace Hanners, *Huntingtown High School;* Alton Harestad, *Simon Fraser University;* Barbara Harvey, *Kirkwood Community College;* Jill Haukos, *South Plains College;* Keith Hench, *Kirkwood Community College;* George Hinman, *Washington State University;* Joseph Hobbs, *University of Missouri;* Jason Hoeksema, *Cabrillo College;* Curtis Hollabaugh, *University of West Georgia;* David Hong, *Diamond Bar High School;* Debra Howell, *Chabot College;* April Huff, *North Seattle Community College;* Pamela Davey Huggins, *Fairmont State University;* Barbara Hunnicutt, *Seminole Community College;* Jon E. Hutchins, *Buena Vista University;* Daniel Hyke, *Alhambra High School;* Juana Ibáñez, *University of New Orleans;* Walter Illman, *University of Iowa;* Daniel Ippolito, *Anderson University;* Bonnie Jacobs, *Southern Methodist University;* Nan Jenks-Jay, *Middlebury College;* Stephen R. Johnson, *William Penn University;* Gina Johnston, *California State University, Chico;* Richard R. Jurin, *University of Northern Colorado;* Thomas M. Justice, *McLennan Community College;* Stanley S. Kabala, *Duquesne University;* Steve Kahl, *Plymouth State University;* Richard R. Keenan, *Providence Senior High School;* Dawn G. Keller, *Hawkeye Community College;* Ned J. Knight, *Linfield College;* Penelope M. Koines, *University of Maryland;* Alexander Kolovos, *University of North Carolina–Chapel Hill;* Erica Kosal, *North Carolina Wesleyan College;* Steven Kosztya, *Baldwin Wallace College;* John C. Kinworthy, *Concordia University;* Robert J. Koester, *Ball State University;* Jim Krest, *University of South Florida–South Florida;* Sushma Krishnamurthy, *Texas A&M International University;* James Kubicki, *Penn State University;* Diane M. LaCole, *Georgia Perimeter College;* Troy A. Ladine, *East Texas Baptist University;* Vic Landrum, *Washburn University;* Tom Langen, *Clarkson University;* Andrew Lapinski, *Reading Area Community College;* Kim D. B. Largen, *George Mason University;* Lissa Leege,

Georgia Southern University; John F. Looney, Jr., *University of Massachusetts–Boston;* Linda Lusby, *Acadia University;* Les M. Lynn, *Bergen Community College;* Richard A. Lutz, *Rutgers University;* Timothy F. Lyon, *Ball State University;* Sue Ellen Lyons, *Holy Cross School;* James G. March, *Washington & Jefferson College;* Blasé Maffia, *University of Miami;* Keith Malmos, *Valencia Community College East;* Anthony J.M. Marcattilio, *St. Cloud State University;* Patrick S. Market, *University of Missouri–Columbia;* Allan Matthias, *University of Arizona;* Jake McDonald, *University of New Mexico;* Dan McNally, *Bryant University;* Steven J. Meyer, *University of Wisconsin–Green Bay;* Kiran Misra, *Edinboro University of Pennsylvania;* Paul Montagna, *University of Texas–Austin;* Brian W. Moores, *Randolph-Macon College;* James T. Morris, *University of South Carolina;* Sherri Morris, *Bradley University;* Mary Murphy, *Penn State Abington;* Carla S. Murray, *Carl Sandburg College;* Richard A. Niesenbaum, *Muhlenberg College;* Mark P. Oemke, *Alma College;* Bruce Olszewski, *San Jose State University;* Nancy Ostiguy, *Penn State University;* David R. Ownby, *Stephen F. Austin State University;* Philip Parker, *University of Wisconsin–Platteville;* Brian D. Peer, *Simpson College;* Clayton Penniman, *Central Connecticut State;* Donald J. Perkey, *University of Alabama–Huntsville;* Raymond Pierotti, *University of Kansas;* Craig D. Phelps, *Rutgers University;* Frank X. Phillips, *McNeese State University;* Thomas E. Pliske, *Florida International University;* Avram G. Primack, *Miami University of Ohio;* Barbara Reynolds, *University of North Carolina–Asheville;* Samuel K. Riffell, *Mississippi State University;* Tom Robertson, *Portland Community College, Rock Creek Campus;* Mark Robson, *University of Medicine and Dentistry of New Jersey;* Angel M. Rodriguez, *Broward Community College;* Steven Rudnick, *University of Massachusetts–Boston;* Deanne Roquet, *Lake Superior College;* George E. Rough, *South Puget Sound Community College;* John Rueter, *Portland State University;* Shamili A. Sandiford; *College of Dupage;* Robert Sanford, *University of Southern Maine;* Ronald Sass, *Rice University;* Carl Schafer, *University of Connecticut;* Jeffery A. Schneider, *State University of New York–Oswego;* Mark Schwartz, *University of California–Davis;* Jennifer Scrafford, *Loyola College;* Richard Seigel, *Towson University.* Maureen Sevigny, *Oregon Institute of Technology;* Rebecca Sheesley, *University of Wisconsin–Madison;* William Shockner, *Community College of Baltimore County;* Christian V. Shorey, *University of Iowa;* Robert Sidorsky, *Northfield Mt. Hermon High School;* Cynthia Simon, *University of New England;* Michael Singer, *Wesleyan University;* Mark Smith, *Chaffey College;* Debra Socci, *Seminole Community College;* Ravi Srinivas, *University of St. Thomas;* Bruce Stallsmith, *University of Alabama–Huntsville;* Jeff Steinmetz, *Queens University of Charlotte;* Robert Strikwerda, *Indiana University–Kokomo;* Andrew Suarez, *University of*

Illinois; Keith S. Summerville, *Drake University;* Mark L. Taper, *Montana State University;* Julienne Thomas, *Robert Morris College;* Jamey Thompson, *Hudson Valley Community College;* Todd Tracy, *Northwestern College;* Frederick R. Troeh, *Iowa State University;* Virginia Turner, *Robert Morris College;* Michael Tveten, *Pima Community College;* G. Peter van Walsum, *Baylor University;* Michael Vorwerk, *Westfield State College;* Daniel W. Ward, *Waubonsee Community College;* Caryl Waggett, *Allegheny College;* Lisa Weasel, *Portland State University;* John F. Weishampel, *University of Central Florida;* James W. C. White, *University of Colorado;* Donald L. Williams, *Park University;* Ray E. Williams, *Rio Hondo College;* Dwina Willis, *Freed-Hardeman University;* James Winebrake, *Rochester Institute of Technology;* Danielle Wirth, *Des Moines Area Community College;* Marjorie Wonham, *University of Alberta;* Wes Wood, *Auburn University;* Joan G. Wright, *Truckee Meadows Community College;* Michael Wright, *Truckee Meadows Community College;* S. Rebecca Yeomans, *South Georgia College;* Lynne Zeman, *Kirkwood Community College.*

Viewpoints Essayists

Frank Ackerman, *Tufts University;* Jock Anderson, *World Bank;* Thomas M. Bonnicksen, *University of California–Davis;* Lester Brown, *Earth Island Institute;* James T. Carlton, *Williams College;* Ignacio Chapela, *University of California–Berkeley;* Timothy L. Cline, *Population Connection;* Thomas Flint, *AgFARMation;* Pete Geddes, *Foundation for Research on Economics and the Environment;* Eban Goodstein, *Lewis and Clark College;* Karl Grossman, *State University of New York–College at Old Westbury;* Adrian Herrera, *Arctic Power;* Susan Hock, *National Renewable Energy Laboratory;* Nan Jenks-Jay, *Middlebury College;* Peter Kareiva, *The Nature Conservancy;* Debra Knopman, *RAND Corporation;* Matthew Koehler, *Native Forest Network;* Michael Leech, *International Game Fish Association;* Jane Lubchenco, *Oregon State University;* David McIntosh, *Natural Resources Defense Council;* Norman Myers, *Oxford University;* Nalini M. Nadkarni, *The Evergreen State College;* Sara Nicolas, *American Rivers;* Randal O'Toole, *American Dream Coalition;* Warren Porter, *University of Wisconsin–Madison;* Daryl Prigmore, *University of Colorado–Colorado Springs;* Roger Prince, *ExxonMobil;* John Ritch, *World Nuclear Association;* Terry Roberts, *Potash & Phosphate Institute;* Lori Saldaña, *California State Assembly;* Gavin Schmidt, *NASA/Goddard Institute for Space Studies;* Jane S. Shaw, *Property and Environment Research Center;* Daniel Simberloff, *University of Tennessee;* S. Fred Singer, *Science and Environmental Policy Project;* Gary Sirota, *Coast Law Group;* Jeff Speck, *National Endowment for the Arts;* Marian K. Stanley, *American Chemistry Council;* Douglas Sylva, *Catholic Family and Human Rights Institute;* Paul H.

Templet, *Louisiana State University;* Frederick Troeh, *Iowa State University;* Indra K. Vasil, *University of Florida;* Albert D. Venosa, *U.S. Environmental Protection Agency;* Karen Wayland, *Natural Resources Defense Council;* Nathaniel Wheelwright, *Bowdoin College.*

First Edition Reviewers

Mary E. Allen, *Hartwick College;* Dulasiri Amarasiriwardena, *Hampshire College;* Gary I. Anderson, *Santa Rosa Junior College;* Joseph A. Arruda, *Pittsburg State University;* Timothy J. Bailey, *Pittsburg State University;* Stokes Baker, *University of Detroit;* David Bass, *University of Central Oklahoma;* Timothy J. Bell, *Chicago State University;* William Berry, *University of California–Berkeley;* Kristina Beuning, *University of Wisconsin–Eau Claire;* Richard Drew Bowden, *Allegheny College;* Nancy Broshot, *Linfield College;* David Brown, *California State University–Chico;* Lee Burras, *Iowa State University;* Charles E. Button, *University of Cincinnati Clermont College;* Jon Cawley, *Roanoke College;* Linda Chalker-Scott, *University of Washington;* Sudip Chattopadhyay, *San Francisco State University;* Luke W. Cole, *Center on Race, Poverty, & the Environment;* Darren Divine, *Community College of Southern Nevada;* Iver W. Duedall, *Florida Institute of Technology;* Margaret L. Edwards-Wilson, *Ferris State University;* Anne H. Ehrlich, *Stanford University;* Thomas R. Embich, *Harrisburg Area Community College;* W. F. J. Evans, *Trent University;* Jiasong Fang, *Iowa State University;* M. Siobhan Fennessy, *Kenyon College;* Linda Mueller Fitzhugh, *Gulf Coast Community College;* Doug Flournoy, *Indian Hills Community College–Ottumwa;* Johanna Foster, *Johnson County Community College;* Chris Fox, *Catonsville Community College;* Nancy Frank, *University of Wisconsin–Milwaukee;* Robert Frye, *University of Arizona;* Sandi Gardner, *Triton College;* Marcia L. Gillette, *Indiana University–Kokomo;* Jeffrey J. Gordon, *Bowling Green State University;* John G. Graveel, *Purdue University;* Cheryl Greengrove, *University of Washington;* Amy R. Gregory, *University of Cincinnati Clermont College;* Sherri Gross, *Ithaca College;* David E. Grunklee, *Hawkeye Community College;* Mark Gustafson, *Texas Lutheran University;* Daniel Guthrie, *Claremont College;* David Hacker, *New Mexico Highlands University;* Greg Haenel, *Elon University;* David Hassenzahl, *University of Nevada Las Vegas;* Joseph Hobbs, *University of Missouri–Columbia;* Catherine Hooey, *Pittsburgh State University;* Jonathan E. Hutchins, *Buena Vista University;* Juana Ibáñez, *University of New Orleans;* Walter Illman, *University of Iowa;* Gina Johnston, *California State University–Chico;* Stanley Kabala, *Duquesne University;* Carol Kearns, *University of Colorado–Boulder;* Dawn Keller, *Hawkeye Community College;* Tom Kozel, *Anderson College;* Frank T. Kuserk, *Moravian College;* William R. Lammela, *Nazareth College;* Michael T. Lares, *University of Mary;* John Latto, *University of California–Berkeley;* Joseph Luczkovich, *East Carolina University;* Jennifer Lyman, *Rocky Mountain College;* Ian R. MacDonald, *Texas A & M University;* Kenneth Mantai, *State University of New York–Fredonia;* Patrick S. Market, *University of Missouri–Columbia;* Steven R. Martin, *Humboldt State University;* John Mathwig, *College of Lake County;* Allan Matthias, *University of Arizona;* Robert Mauck, *Kenyon College;* Debbie McClinton, *Brevard Community College;* Paul McDaniel, *University of Idaho;* Gregory McIsaac, *Cornell University;* Dan McNally, *Bryant College;* Richard McNeil, *Cornell University;* Mike L. Meyer, *New Mexico Highlands University;* Patrick Michaels, *Cato Institute;* Chris Migliaccio, *Miami Dade Community College;* Kiran Misra, *Edinboro University of Pennsylvania;* Mark Mitch, *New England College;* Brian W. Moores, *Randolph-Macon College;* James T. Morris, *University of South Carolina;* Sherri Morris, *Bradley University;* William M. Murphy, *California State University–Chico;* Rao Mylavarapu, *University of Florida;* Jane Nadel-Klein, *Trinity College;* Muthena Naseri, *Moorpark College;* Michael J. Neilson, *University of Alabama–Birmingham;* Moti Nissani, *Wayne State University;* Richard B. Norgaard, *University of California–Berkeley;* Niamh O'Leary, *Wells College;* Brian O'Neill, *Brown University;* Eric Pallant, *Allegheny College;* Phillip J. Parker, *University of Wisconsin–Platteville;* Daryl Prigmore, *University of Colorado;* Loren A. Raymond, *Appalachian State University;* Barbara C. Reynolds, *University of North Carolina–Asheville;* Thomas J. Rice, *California Polytechnic State University;* Gary Ritchison, *Eastern Kentucky University;* Mark G. Robson, *University of Medicine and Dentistry of New Jersey;* Carlton Lee Rockett, *Bowling Green State University;* Armin Rosencranz, *Stanford University;* Robert E. Roth, *The Ohio State University;* Christopher T. Ruhland, *Minnesota State University;* Ronald L. Sass, *Rice University;* Richard A. Seigel, *Towson University;* Wendy E. Sera, *NDAA's National Ocean Service;* Maureen Sevigny, *Oregon Institute of Technology;* Linda Sigismondi, *University of Rio Grande;* Jeffrey Simmons, *West Virginia Wesleyan College;* Jan Simpkin, *College of Southern Idaho;* Patricia L. Smith, *Valencia Community College;* Douglas J. Spieles, *Denison University;* Bruce Stallsmith, *University of Alabama at Huntsville;* Richard J. Strange, *University of Tennessee;* Robert A. Strikwerda, *Indiana University at Kokomo;* Richard Stringer, *Harrisburg Area Community College;* Ronald Sundell, *Northern Michigan University;* Bruce Sundrud, *Harrisburg Area Community College;* Max R. Terman, *Tabor College;* Adrian Treves, *Wildlife Conservation Society;* Charles Umbanhowar, *St. Olaf College;* G. Peter van Walsum, *Baylor University;* Callie A. Vanderbilt, *San Juan College;* Elichia A. Venso, *Salisbury University;* Rob Viens, *Bellevue Community College;* Maud M. Walsh, *Louisiana State University;* Phillip L. Watson, *Ferris State University;* Richard

D. Wilk, *Union College*; James J. Winebrake, *Rochester Institute of Technology*; Jeffrey S. Wooters, *Pensacola Junior College*; Zhihong Zhang, *Chatham College.*

First Edition Class Testers
David Aborne, *University of Tennessee–Chattanooga*; Reuben Barret, *Prairie State College*; Morgan Barrows, *Saddleback College*; Henry Bart, *LaSalle University*; James Bartalome, *University of California–Berkeley*; Christy Bazan, *Illinois State University*; Richard Beckwitt, *Framingham State College*; Elizabeth Bell, *Santa Clara University*; Peter Biesmeyer, *North Country Community College*; Donna Bivans, *Pitt Community College*; Evert Brown, *Casper College*; Christina Buttington, *University of Wisconsin, Milwaukee*; Tait Chirenje, *Richard Stockton College*; Reggie Cobb, *Nash Community College*; Ann Cutter, *Randolph Community College*; Lola Deets, *Pennsylvania State University–Erie*; Ed DeGrauw, *Portland Community College*; Mrs. Dockstader, *Monroe Community College*; Dee Eggers, *University of North Carolina–Asheville*; Jane Ellis, *Presbyterian College*; Paul Fader, *Freed Hardeman University*; Joseph Fail, *Johnson C. Smith University*; Brad Fiero, *Pima Community College, West Campus*; Dane Fisher, *Pfeiffer University*; Chad Freed, *Widener University*; Sue Glenn, *Gloucester County College*; Sue Habeck, *Tacoma Community College*; Mark Hammer, *Wayne State University*;

Michael Hanson, *Bellevue Community College*; David Hassenzahl, *Oakland Community College*; Kathleen Hornberger, *Widener University*; Paul Jurena, *University of Texas–San Antonio*; Dawn Keller, *Hawkeye Community College*; David Knowles, *East Carolina University*; Erica Kosal, *Wesleyan College*; John Logue, *University of Southern Carolina Sumter*; Keith Malmos, *Valencia Community College*; Nancy Markee, *University of Nevada–Reno*; Bill Mautz, *University of New Hampshire*; Julie Meents, *Columbia College*; Mr. Getchell, *Mohawk Valley Community College*; Lori Moore, *Northwest Iowa Community College*; Elizabeth Pixley, *Monroe Community College*; John Novak, *Colgate University*; Brian Peck, *Simpson College*; Sarah Quast, *Middlesex Community College*; Roger Robbins, *East Carolina University*; Mark Schwartz, *University of California–Davis*; Julie Seiter, *University of Nevada–Las Vegas*; Brian Shmaefsky, *Kingwood College*; Diane Sklensky, *Le Moyne College*; Mark Smith, *Fullerton College*; Patricia Smith, *Valencia Community College East*; Sherilyn Smith, *Le Moyne College*; Jim Swan, *Albuquerque Technical Vocational Institute*; Amy Treonis, *Creighton University*; Darrell Watson, *The University of Mary Hardin Baylor*; Barry Welch, *San Antonio College*; Susan Whitehead, *Becker College*; Roberta Williams, *University of Nevada–Las Vegas*; Justin Williams, *Sam Houston University*; Tom Wilson, *University of Arizona.*

Brief Contents

Detailed Contents

Foundations of Environmental Science

Researcher studying
eucalyptus forest,
Australia

1

An Introduction to Environmental Science

Our island, Earth

Upon successfully completing this chapter, you will be able to:

▶ Define the term *environment*

▶ Describe natural resources and explain their importance to human life

▶ Characterize the interdisciplinary nature of environmental science

▶ Understand the scientific method and how science operates

▶ Diagnose and illustrate some of the pressures on the global environment

▶ Evaluate the concepts of sustainability and sustainable development

Our Island, Earth

Viewed from space, our home planet resembles a small blue marble suspended against a vast inky-black backdrop. Although few of us will ever get to witness that sight directly, photographs taken by astronauts convey a sense that Earth is small, isolated, and fragile. It may seem vast to us as we go about our lives on its surface, but from the astronaut's perspective it is apparent that Earth and its natural systems are not unlimited. From this perspective, it becomes clear that as our population, our technological powers, and our consumption of resources increase, so do our abilities to alter our planet and damage the very systems that keep us alive.

Our environment is the sum total of our surroundings

A photograph of Earth reveals a great deal, but it does not convey the complexity of our environment. Our **environment** (a term that comes from the French *environner,* "to surround") is more than water, land, and air; it is the sum total of our surroundings. It includes all of the **biotic factors,** or living things, with which we interact. It also includes the **abiotic factors,** or nonliving things, with which we interact. Our environment includes the continents, oceans, clouds, and ice caps you can see in the photo of Earth from space, as well as the animals, plants, forests, and farms that comprise the landscapes around us. In a more inclusive sense, it also encompasses our built environment, the structures, urban centers, and living spaces humans have created. In its most inclusive sense, our environment also includes the complex webs of scientific, ethical, political, economic, and social relationships and institutions that shape our daily lives.

From day to day, people most commonly use the term *environment* in the first, narrow sense—of a nonhuman or "natural" world apart from human society. This connotation is unfortunate, because it masks the very important fact that humans exist within the environment and are a part of nature. As one of many species of animals on Earth, we share with others the same dependence on a healthy functioning planet. The limitations of language make it all too easy to speak of "people and nature," or "human society and the environment," as though they are separate and do not interact. However, the fundamental insight of environmental science is that we are part of the natural world and that our interactions with other parts of it matter a great deal.

Environmental science explores interactions between humans and our environment

Appreciating how we interact with our environment is crucial for a well-informed view of our place in the world and for a mature awareness that we are one species among many on a planet full of life. Understanding our relationship with the environment is also vital because we are altering the very natural systems we need, in ways we do not yet fully comprehend.

We depend utterly on our environment for air, water, food, shelter, and everything else essential for living. However, our actions modify our environment, whether we intend them to or not. Many of these actions have enriched our lives, bringing us longer life spans, better health, and greater material wealth, mobility, and leisure time. However, these improvements have often degraded the natural systems that sustain us. Impacts such as air and water pollution, soil erosion, and species extinction can compromise human well-being, pose risks to human life, and threaten our ability to build a society that will survive and thrive in the long term. The elements of our environment were functioning long before the human species appeared, and we would be wise to realize that we need to keep these elements in place.

Environmental science is the study of how the natural world works, how our environment affects us, and how we affect our environment. We need to understand our interactions with our environment because such knowledge is the essential first step toward devising solutions to our most pressing environmental problems. Many environmental scientists are taking this next step, trying to apply their knowledge to develop solutions to the many environmental challenges we face.

It can be daunting to reflect on the sheer magnitude of environmental dilemmas that confront us today, but with these problems also come countless opportunities for devising creative solutions. The topics studied by environmental scientists are the most centrally important issues to our world and its future. Right now, global conditions are changing more quickly than ever. Right now, through science, we as a civilization are gaining knowledge more rapidly than ever. And right now, the window of opportunity for acting to solve problems is still open. With such bountiful challenges and opportunities, this particular moment in history is indeed an exciting time to be studying environmental science.

FIGURE 1.1 Natural resources lie along a continuum from perpetually renewable to nonrenewable. Perpetually renewable resources, such as sunlight, will always be there for us. Nonrenewable resources, such as oil and coal, exist in limited amounts that could one day be gone. Other resources, such as timber, soils, and food crops, can be renewed on intermediate time scales, if we are careful not to deplete them.

Renewable natural resources

- Sunlight
- Wind energy
- Wave energy
- Geothermal energy

- Agricultural crops
- Fresh water
- Forest products
- Soils

Nonrenewable natural resources

- Crude oil
- Natural gas
- Coal
- Copper, aluminum, and other metals

Natural resources are vital to our survival

An island by definition is finite and bounded, and its inhabitants must cope with limitations in the materials they need. On our island, Earth, human beings, like all living things, ultimately face environmental constraints. Specifically, there are limits to many of our **natural resources,** the various substances and energy sources we need to survive. Natural resources that are virtually unlimited or that are replenished over short periods are known as **renewable natural resources.** Some renewable resources, such as sunlight, wind, and wave energy, are perpetually available. Others, such as timber, food crops, water, and soil, renew themselves over months, years, or decades, if we are careful not to use them up too quickly or destructively. In contrast, resources such as mineral ores and crude oil are in finite supply and are formed much more slowly than we use them. These are known as **nonrenewable natural resources.** Once we use them up, they are no longer available.

We can view the renewability of natural resources as a continuum (Figure 1.1). Some renewable resources may turn nonrenewable if we overuse them. For example, overpumping groundwater can deplete underground aquifers and turn a lush landscape into a desert. Populations of animals and plants we harvest from the wild may be renewable if we do not overharvest them but may vanish if we do. In recent years, our consumption of natural resources has increased greatly, driven by rising affluence and the growth of the largest human population in history.

Human population growth has shaped our relationship with natural resources

For nearly all of human history, only a few million people populated Earth at any one time. Although past populations cannot be calculated precisely, Figure 1.2 gives some idea of just how recently and suddenly our population has grown beyond 6 billion people.

Two phenomena triggered remarkable increases in population size. The first was our transition from a hunter-gatherer lifestyle to an agricultural way of life. This change began to occur around 10,000 years ago and is known as the **agricultural revolution.** As people began to grow their own crops, raise domestic animals, and live sedentary lives in villages, they found it easier to meet their nutritional needs. As a result, they began to live longer and to produce more children who survived to adulthood. The second notable phenomenon, known as the **industrial revolution,** began in the mid-1700s. It entailed a shift from rural life, animal-powered agriculture, and manufacturing by craftsmen, to an urban society powered by **fossil fuels** (nonrenewable energy sources, such as oil, coal, and natural gas, produced by the decomposition and fossilization of ancient life). The industrial revolution introduced improvements in sanitation and medical technology, and it enhanced agricultural production with fossil-fuel-powered equipment and synthetic fertilizer (▸ pp. 278–282).

Thomas Malthus and population growth At the outset of the industrial revolution in England, population growth was regarded as a good thing. For parents, high birth rates meant more children to support them in old age. For society, it meant a greater pool of labor for factory work.

British economist **Thomas Malthus** (1766–1834) had a different opinion. Malthus claimed that unless population growth were controlled by laws or other social strictures, the number of people would outgrow the available food

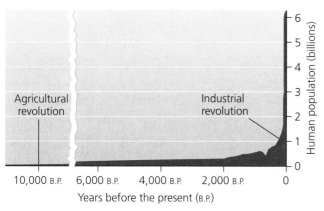

(a) World population growth

(b) Urban society

FIGURE 1.2 For almost all of human history, our population was low and relatively stable. It increased significantly as a result first of the agricultural revolution and then of the industrial revolution (**a**). Our skyrocketing population has given rise to congested urban areas, such as this city in Java, Indonesia (**b**).

supply until starvation, war, or disease arose and reduced the population (Figure 1.3). Malthus's most influential work, *An Essay on the Principle of Population,* published in 1798, argued that a growing population would eventually be checked either by limits on births or increases in deaths. If limits on births (such as abstinence and contraception) were not implemented soon enough, Malthus wrote, deaths would increase through famine, plague, and war.

Malthus's thinking was shaped by the rapid urbanization and industrialization he witnessed during the early years of the industrial revolution, but debates over his views continue today. As we will see in Chapter 8 and throughout this book, global population growth has indeed helped spawn famine, disease, and social and political conflict. However, increasing material prosperity has also helped bring down birth rates—something Malthus did not foresee.

Paul Ehrlich and the "population bomb" In our day, biologist Paul Ehrlich of Stanford University has been called a "neo-Malthusian" because he too has warned that population growth will have disastrous effects on human welfare. In his 1968 book, *The Population Bomb,* Ehrlich predicted that the rapidly increasing human population would unleash widespread famine and conflict that would consume civilization by the end of the 20th century. Like Malthus, Ehrlich argued that population was growing much faster than our ability to produce and distribute food, and he maintained that population control was the only way to prevent massive starvation and civil strife.

Although human population nearly quadrupled in the past 100 years—the fastest it has ever grown (see Figure 1.2a)—Ehrlich's predictions have not materialized on the scale he predicted. This is due, in part, to agricultural advances made in recent decades (▸ pp. 278–279). As

(a) 18th-century London, England

(b) Thomas Malthus

FIGURE 1.3 The England of Thomas Malthus's era (1766–1834), shown in this engraving (**a**), favored population growth as society industrialized. Malthus (**b**) argued that population growth could lead to disaster.

a result, Ehrlich and other neo-Malthusians have revised their predictions accordingly and now warn of a postponed, but still impending, global crisis.

Resource consumption exerts social and environmental impacts

Population growth affects resource availability and is unquestionably at the root of many environmental problems. However, the growth in consumption is also to blame. The industrial revolution enhanced the material affluence of many of the world's people by considerably increasing our consumption of natural resources and manufactured goods.

Garrett Hardin and the "tragedy of the commons"

The late Garrett Hardin of the University of California, Santa Barbara, disputed the economic theory that unfettered exercise of individual self-interest will serve the public interest. According to Hardin's best-known essay, "The Tragedy of the Commons," published in the journal *Science* in 1968, resources that are open to unregulated exploitation will eventually be depleted.

Hardin based his argument on a scenario described in a pamphlet published in 1833. In a public pasture, or "common," that is open to unregulated grazing, Hardin argued, each person who grazes animals will be motivated to increase the number of his or her animals in the pasture. Ultimately, overgrazing will cause the pasture's food production to collapse (Figure 1.4). Because no single person owns the pasture, no one has incentive to expend effort taking care of it, and everyone takes what he or she can until the resource is depleted.

Some have argued that private ownership can address this problem. Others point to cases in which people sharing a common resource have voluntarily organized and cooperated in enforcing its responsible use. Still others maintain that the dilemma justifies government regulation of the use of resources held in common by the public, from forests to clean air to clean water.

Weighing the **Issues:**
The Tragedy of the Commons

Imagine you make your living fishing for lobster. You are free to boat anywhere and set out as many traps as you like. Your harvests have been good, and nothing is stopping you from increasing the number of your traps. However, all the other lobster fishers are thinking the same thing, and the fishing grounds are getting crowded. Catches decline year by year, until one year the fishery

FIGURE 1.4 Unregulated areas that offer limited resources freely to the public are prone to be depleted by the process that Garrett Hardin dubbed "the tragedy of the commons."

crashes, leaving you and all the others with catches too meager to support your families. Some of your fellow fishers call for dividing the waters and selling access to individuals plot-by-plot. Others urge the fishers to team up, set quotas among themselves, and prevent newcomers from entering the market. Still others are imploring the government to get involved and pass laws regulating how much fishers can catch. What do you think is the best way to combat this tragedy of the commons and restore the fishery? Why?

Wackernagel, Rees, and the ecological footprint

As global affluence has increased, human society has consumed more and more of the planet's limited resources. We can quantify resource consumption using the concept of the "ecological footprint," developed in the 1990s by environmental scientists Mathis Wackernagel and William Rees. The **ecological footprint** expresses the environmental impact of an individual or population in terms of the cumulative amount of land and water required to provide the

FIGURE 1.5 The "ecological footprint" represents the total area of land and water needed to produce the resources a given person or population uses, together with the total amount of land and water needed to dispose of their waste. The footprints of the urbanized and affluent societies of today's developed nations tend to be much larger than the geographic areas these societies take up directly. Adapted from Wackernagel, M., and W. Rees. 1996. *Our ecological footprint: Reducing human impact on the Earth.* Gabriola Island, British Columbia: New Society Publishers.

raw materials the person or population consumes and to dispose of or recycle the waste the person or population produces (Figure 1.5). It measures the total amount of Earth's surface "used" by a given person or population, once all direct and indirect impacts are totaled up.

For humanity as a whole, Wackernagel and Rees have calculated that our species is using 30% more resources than are available on a sustainable basis from all the land on the planet. That is, we are depleting renewable resources 30% faster than they are being replenished—like drawing the principal out of a bank account rather than living off the interest. Furthermore, people from wealthy nations have much larger ecological footprints than do people from poorer nations. If all the world's people consumed resources at the rate of North Americans, these researchers concluded, we would need the equivalent of two additional planet Earths.

Environmental science can help us avoid mistakes made by past civilizations

It remains to be seen whether the direst predictions of Malthus, Ehrlich, and others will come to pass for today's global society, but we already have historical evidence that civilizations can crumble when pressures from population and consumption overwhelm resource availability. Easter Island is the classic case (see "The Science behind the Story," ▶ pp. 8–9), but it is not the only example. Many great civilizations have fallen after depleting resources from their environments, and each has left devastated landscapes in its wake. The Greek and Roman empires show evidence of such a trajectory, as do the Maya, the Anasazi, and other civilizations of the New World. Plato wrote of the deforestation and environmental degradation accompanying ancient Greek cities, and today further evidence is accumulating from research by archaeologists, historians, and paleoecologists who study past societies and landscapes. The arid deserts of today's Middle Eastern countries were far more vegetated when the great ancient civilizations thrived there; at that time these regions were lush enough to support the very origin of agriculture. While deforestation created deserts in temperate regions, in more tropical climates, the ancient cities of fallen civilizations became overgrown by jungle. The gigantic stone monuments of the Angkor civilization in Southeast Asia, like those of the Maya in Mexico and Central America, remained unknown to Westerners until the

The Science behind the Story

The Lesson of Easter Island

Easter Island is one of the most remote spots on the globe, located in the Pacific Ocean 3,750 km (2,325 mi) from South America and 2,250 km (1,395 mi) from the nearest inhabited island. When the first European explorers reached the island (today called Rapa Nui) in 1722, they found a barren landscape populated by fewer than 2,000 people, who lived in caves and eked out a marginal existence from a few meager crops. However, explorers also noted that the desolate island featured hundreds of gigantic statues of carved stone, evidence that a sophisticated civilization had once inhabited the island.

Historians and anthropologists long wondered how people without wheels or ropes, on an island without trees, could have moved statues 10 m (33 ft) high weighing 90 metric tons (99 tons) as far as 10 km (6.2 mi) from the quarries where they were chiseled to the coastal sites where they were erected. The explanation, scientists have discovered, lay in the fact that the island did not always lack trees, and its people were not always without rope.

Indeed, scientific research tells us that the island had once been lushly forested, with all the appeal of a South Pacific paradise, and had supported a prosperous society with a population of 6,000 to 30,000 people. Tragically, this once-flourishing

The haunting statues of Easter Island were erected by a sophisticated civilization that collapsed after depleting its resource base and devastating its island environment.

civilization overused its resources and cut down all its trees, destroying itself in a downward spiral of starvation and conflict. Today Easter Island stands as a parable and a warning for what can happen when a population grows too large and consumes too much of the limited resources that support it.

To solve the mystery of Easter Island's past, scientists have used various methods. Some, such as British scientist John Flenley, have excavated sediments from the bottom of the island's volcanic crater lakes, drilling cores deep into the mud and examining ancient grains of pollen preserved there. Because pollen grains vary from one plant species to another, scientists, by identifying specific pollen grains,

can reconstruct, layer by layer, the history of vegetation in a region through time. By analyzing pollen grains under scanning electron microscopes, Flenley and other researchers found that when Polynesian people arrived (likely between A.D. 300 and A.D. 900), the island was covered with a species of palm tree related to the Chilean wine palm, a tall and thick-trunked tree. Archaeologists located ancient palm nut casings in caves and crevices, and a geologist found carbon-lined channels in the soil that matched root channels typical of the Chilean wine palm. Furthermore, scientists deciphering the island people's script on stone tablets discerned characters etched in the form of palm trees.

By studying pollen and the remains of wood from charcoal, scientists such as French archaeologist Catherine Orliac have found that at least 21 other species of plants, many of them trees, had also been common, and are now completely gone. The island had clearly supported a diverse forest. However, starting around A.D. 750, tree populations declined and ferns and grasses became more common, according to pollen analysis from one lake site. By A.D. 950, the trees were largely gone, and around A.D. 1400 overall pollen levels plummeted, indicating a dearth of vegetation. The same sequence of events occurred about two centuries later at the other two lake sites, which were higher and more remote from village areas. Researchers first hypothesized that the forest loss was due to climate change, but evidence instead supported the hypothesis that the people had gradually denuded their own island.

The palms and other trees provided fuelwood, building material for houses and canoes, fruit to eat and fiber for clothing, and presumably, logs to move the stone statues. Several anthropologists in recent years have experimentally tested hypotheses about how the islanders moved their monoliths down from the quarries, by hiring groups of men to recreate the feat. The methods that have worked involve using numerous tree trunks as rollers or sleds, as well as great quantities of rope. The only likely source of rope on the island would have been the fibrous inner bark of the hauhau tree, a species that today is near extinction.

With the trees gone, soil would have eroded away—a phenomenon confirmed by data from the bottom of Easter Island lakes, where large quantities of sediment accumulated. Faster runoff of rainwater would have meant less fresh water available for drinking. Runoff and erosion would have degraded the islanders' agricultural land, lowering yields of crops, such as bananas, sugar cane, and sweet potatoes. Reduced agricultural production would have led to starvation and subsequent population decline.

Archaeological evidence supports the scenario of environmental degradation and civilization decline. Analysis of 6,500 bones by archaeologist David Steadman has shown that at least 6 species of land birds and 25 species of seabirds nested on Easter Island and were eaten by islanders. Today no native land birds and only 1 seabird are left. Remains from charcoal fires can be aged by radiocarbon dating (▶ pp. 94–95), and show that islanders' diets shifted over the years. Besides their crops and the island's birds, early islanders feasted on the bounty of the sea, including porpoises, fish, sharks, turtles, octopus, and shellfish. Analysis of islanders' diets in the later years indicated that little seafood was consumed. With the trees gone, the islanders could no longer build the great double-canoes their proud Polynesian ancestors had used for centuries to fish and travel among islands. Indeed, the Europeans who visited Easter Island in the 1700s observed only a few old small canoes and flimsy rafts made of reeds. As resources declined, the islanders' main domesticated food animal, the chicken, became more valuable. Archaeologists found that later islanders kept their chickens in stone fortresses with entrances designed to prevent theft. The once prosperous and peaceful civilization fell into clan warfare, as revealed by unearthed skeletons, skulls with head wounds, and artifacts of weapons made of obsidian, a hard volcanic rock.

Is the story of Easter Island as unique and isolated as the island itself, or does it hold lessons for our world today? Like the Easter Islanders, we are all stranded together on an island with limited resources. Earth may be vastly larger and richer in resources than was Easter Island, but Earth's human population is also much greater. The Easter Islanders must have seen that they were depleting their resources, but it seems that they could not stop. Whether we can learn from the history of Easter Island and act more wisely to conserve the resources on our island, Earth, is entirely up to us.

19th century, and most of these cities remain covered by rainforest. Researchers have learned enough by now, however, that scientist and author Jared Diamond in his 2005 book, *Collapse,* could synthesize this information and formulate sets of reasons why civilizations succeed and persist, or fail and collapse. Success and persistence, it turns out, depend largely on how societies interact with their environments.

Today we are confronted with news and predictions of environmental catastrophes on a regular basis, but it can be difficult to assess the reliability of such reports. It is even harder to evaluate the causes and effects of environmental change. Perhaps most difficult is to devise solutions to environmental problems. Studying environmental science will outfit you with the tools that can help you evaluate information on environmental change and think critically and creatively about possible actions to take in response. Let us examine this broad field we call environmental science, and then explore the process and methods of science in general.

The Nature of Environmental Science

Environmental scientists aim to comprehend how Earth's natural systems function, how humans are influenced by those systems, and how we are influencing those systems. In addition, many environmental scientists are motivated by a desire to develop solutions to environmental quandaries. The solutions themselves (such as new technologies, policy decisions, or resource management strategies) are applications of environmental science. However, the study of such applications and their consequences is, in turn, also part of environmental science.

People vary in their perception of environmental problems

Environmental science arose in the latter half of the 20th century as people sought to better understand environmental problems and their origins. An *environmental problem,* stated simply, is any undesirable change in the environment. However, the perception of what constitutes an undesirable change may vary from one person or group of people to another, or from one context or situation to another. A person's age, gender, class, race, nationality, employment, and educational background can all affect whether he or she considers a given environmental change to be a "problem."

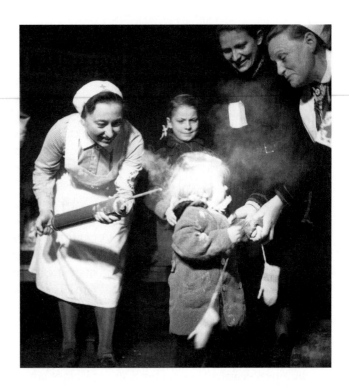

FIGURE 1.6 How a person or a society defines an environmental problem can vary with time and circumstance. In Germany in 1945, health hazards of the pesticide DDT were not yet known, so children were doused with the chemical to treat head lice. Today, knowing of its toxicity to people and wildlife, many developed nations have banned DDT. However, in some developing countries where malaria is a threat, DDT is welcomed to combat mosquitoes that transmit the disease.

For instance, today's industrial societies are more likely to view the spraying of the pesticide DDT as a problem than those societies viewed it in the 1950s, because today more is known about the health risks of pesticides (Figure 1.6). At the same time, a person living today in a malaria-infested village in Africa or India may welcome the use of DDT if it kills mosquitoes that transmit malaria, because malaria is viewed as a more immediate health threat. Thus an African and an American who have each knowledgeably assessed the pros and cons may, because of differences in their circumstances, differ in their judgment of DDT's severity as an environmental problem.

Different types of people may also vary in their awareness of problems. For example, in many cultures women are responsible for collecting water and fuelwood. As a result, they are often the first to perceive environmental degradation affecting these resources, whereas men in the same area simply might not "see" the problem. As another example, in most societies information about environmental health risks tends to reach wealthy people more readily than poor people. Thus, who you are, where you live, and what you do can have a huge effect on how you perceive your environment, how you perceive and react to change, and what impact those changes may have on how you live your life. In Chapter 2, we will examine the diversity of human values and philosophies and consider their effects on how we define environmental problems.

Environmental science provides interdisciplinary solutions

Studying and addressing environmental problems is a complex endeavor that requires expertise from many disciplines, including ecology, earth science, chemistry, biology, economics, political science, demography, ethics, and others. Environmental science is thus an **interdisciplinary** field—one that borrows techniques from numerous disciplines and brings research results from these disciplines together into a broad synthesis (Figure 1.7). Traditional established disciplines are valuable because their scholars delve deeply into topics, uncovering new knowledge and developing expertise in particular areas. Interdisciplinary fields are valuable because their practitioners take specialized knowledge from different disciplines, consolidate it, synthesize it, and make sense of it in a broad context to better serve the multifaceted interests of society.

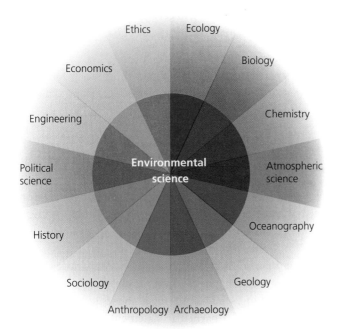

FIGURE 1.7 Environmental science is a highly interdisciplinary pursuit, involving input from many different established fields of study across the natural sciences and social sciences.

Environmental science is especially broad because it encompasses not only the **natural sciences** (disciplines that study the natural world), but also the **social sciences** (disciplines that study human interactions and institutions). The natural sciences provide us the means to gain accurate information about our environment and to interpret it reasonably. Addressing environmental problems, however, also involves weighing values and understanding human behavior, and this requires the social sciences. Most environmental science programs focus predominantly on the natural sciences as they pertain to environmental issues. In contrast, programs incorporating the social sciences heavily often prefer using the term **environmental studies** to describe their academic umbrella. Whichever approach one takes, these fields reflect many diverse perspectives and sources of knowledge.

Just as an interdisciplinary approach to studying issues can help us better understand them, an integrated approach to addressing problems can produce effective and lasting solutions. One example is the dramatic improvement in one aspect of air quality in the United States over the past few decades. Ever since automobiles were invented, lead had been added to gasoline to make cars run more smoothly, even though medical professionals knew that lead emissions from tailpipes could cause health problems, including brain damage and premature death. In 1970 air pollution was severe, and motor vehicles accounted for 78% of U.S. lead emissions. But over the following years, engineers, physicians, atmospheric scientists, and politicians all merged their knowledge and skills into a process that eventually resulted in a ban on leaded gasoline. By 1996 all gasoline sold in the United States was unleaded, and the nation's largest source of atmospheric lead emissions had been completely eliminated.

Environmental science is not the same as environmentalism

Although many environmental scientists are interested in solving problems, it would be incorrect to confuse environmental science with environmentalism, or environmental activism. They are *not* the same. Environmental science is the pursuit of knowledge about the workings of the environment and our interactions with it. **Environmentalism** is a social movement dedicated to protecting the natural world—and, by extension, humans—from undesirable changes brought about by human choices (Figure 1.8). Although environmental scientists may study many of the same issues environmentalists care about, as scientists

FIGURE 1.8 Environmental scientists and environmental activists play very different roles. Some scientists have become activists to promote particular solutions to environmental problems. However, most have not, and those who have generally try hard to keep their advocacy separate from their pursuit of objective scientific work.

they attempt to maintain an objective approach in their work. Remaining free from personal or ideological bias, and open to whatever conclusions the data demand, is a hallmark of the effective scientist. We will now proceed with a brief overview of how science works and how scientists go about this enterprise that brings our society so much valuable knowledge.

The Nature of Science

Modern scientists describe **science** (from the Latin *scire*, "to know") as a systematic process for learning about the world and testing our understanding of it. The term *science* is also commonly used to refer to the accumulated body of knowledge that arises from this dynamic process of observation, testing, and discovery.

Knowledge gained from science can be applied to address societal problems. Among the applications of science are its use in developing technology and its use in informing policy and management decisions (Figure 1.9).

These pragmatic applications in themselves are not science, but they must be informed by science in order to be effective. Many scientists are motivated simply by a desire to know how the world works, and others are motivated by the potential for developing useful applications.

Environmental science is a dynamic yet systematic way of studying the world, and it is also the body of knowledge accumulated from this process. Like science in general, environmental science informs its practical applications and often is motivated by them.

Why does science matter? The late astronomer and author Carl Sagan wrote the following in his 1995 treatise, *The Demon Haunted World: Science as a Candle in the Dark:*

> We've arranged a global civilization in which the most crucial elements—transportation, communications, and all other industries; agriculture, medicine, education, entertainment, protecting the environment; and even the key democratic institution of voting—profoundly depend on science and technology. We have also arranged things so that almost no one understands science and technology. This is a prescription for disaster. We might get away with it for a while, but sooner or later this combustible mixture of ignorance and power is going to blow up in our faces. . . . Science is an attempt, largely successful, to understand the world, to get a grip on things, to get hold of ourselves, to steer a safe course.

Sagan and many other thinkers before and since have argued that science is essential if we hope to sort fact from fiction and develop solutions to the problems—environmental and otherwise—that we face today.

Scientists test ideas by weighing evidence

How can we tell whether warnings of impending environmental catastrophes—or any other claims, for that matter— are based on scientific thinking? Scientists examine ideas about how the world works by designing tests to determine whether these ideas are supported by evidence. Ideas can be refuted by evidence but can never be absolutely proven, so, strictly speaking, scientific testing amounts to attempting to disprove ideas. If a particular statement or explanation is testable and resists repeated attempts to disprove it, scientists are likely to accept it as a useful and true explanation. Scientific inquiry thus consists of an incremental approach to the truth.

(a) Prescribed burning

(b) Methanol-powered fuel-cell car

FIGURE 1.9 Scientific knowledge can be applied in policy and management decisions and in technology. Prescribed burning, shown here in the Ouachita National Forest, Arkansas (**a**), is a management practice to restore healthy forests, and is informed by scientific research into forest ecology. Energy-efficient automobiles, like this methanol-powered fuel-cell car from Daimler–Chrysler (**b**), are technological advances made possible by materials and energy research.

The scientific method is the key element of science

Scientists generally follow a process called the **scientific method.** A technique for testing ideas with observations, it involves several assumptions and a series of interrelated steps. There is nothing mysterious about the scientific method; it is merely a formalized version of the procedure any of us might naturally take, using common sense, to resolve a question.

The scientific method is a theme with variations, however, and scientists pursue their work in many different ways. Because science is an active, creative, imaginative process, an innovative scientist may find good reason to stray from the traditional scientific method when a particular situation demands it. Moreover, scientists from different fields approach their work differently because they deal with dissimilar types of information. A natural scientist, such as a chemist, will conduct research quite differently from a social scientist, such as a sociologist. Because environmental science includes both natural and social sciences, in our discussion here we use the term *science* in its broad sense, to include both. Despite their many differences, scientists of all persuasions broadly agree on fundamental elements of the process of scientific inquiry.

The scientific method relies on the following assumptions:

▶ The universe functions in accordance with fixed natural laws that do not change from time to time or from place to place.

▶ All events arise from some cause or causes and, in turn, cause other events.

▶ We can use our senses and reasoning abilities to detect and describe natural laws that underlie the cause-and-effect relationships we observe in nature.

As practiced by individual researchers or research teams, the scientific method (Figure 1.10) typically consists of the steps outlined below.

Make observations Advances in science typically begin with the observation of some phenomenon that the scientist wishes to explain. Observations set the scientific method in motion and also function throughout the process.

Ask questions Curiosity is a fundamental human characteristic. This is evident to anyone who has observed the explorations of a young child in a new environment. Babies want to touch, taste, watch, and listen to anything that catches their attention, and as soon as they can speak,

Scientific method

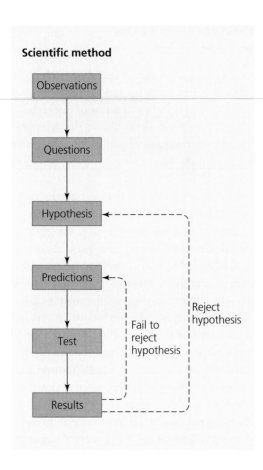

FIGURE 1.10 The scientific method is the observation-based hypothesis-testing approach that scientists use to learn how the world works. This diagram is a simplified generalization that, although useful for instructive purposes, cannot convey the true dynamic and creative nature of science. Moreover, researchers from different disciplines may pursue their work in ways that legitimately vary from this model.

they begin asking questions. Scientists, in this respect, are kids at heart. Why are certain plants or animals less common today than they once were? Why are storms becoming more severe and flooding more frequent? What is causing excessive growth of algae in local ponds? Do pesticide impacts on fish or frogs indicate that people may be affected in the same ways? All of these are questions environmental scientists have asked and attempted to answer.

Develop a hypothesis Scientists attempt to answer their questions by devising explanations that they can test. A **hypothesis** is an educated guess that explains a phenomenon or answers a scientific question. For example, a scientist investigating the question of why algae are growing excessively in local ponds might observe chemical

fertilizers being applied on farm fields nearby. The scientist might then state a hypothesis as follows: "Agricultural fertilizers running into ponds cause the amount of algae in the ponds to increase."

Make predictions The scientist next uses the hypothesis to generate **predictions,** which are specific statements that can be directly and unequivocally tested. In our algae example, a prediction might be: "If agricultural fertilizers are added to a pond, the quantity of algae in the pond will increase."

Test the predictions Predictions are tested one at a time by gathering evidence that could potentially refute the prediction and thus refute the hypothesis. The strongest form of evidence comes from experimentation. An **experiment** is an activity designed to test the validity of a hypothesis. It involves manipulating **variables,** or conditions that can change. For example, a scientist could test the hypothesis linking algal growth to fertilizer by selecting two identical ponds and adding fertilizer to one while leaving the other in its natural state. In this example, fertilizer input is an **independent variable,** a variable the scientist manipulates, whereas the quantity of algae that results is the **dependent variable,** one that depends on the fertilizer input. If the two ponds are identical except for a single independent variable (fertilizer input), then any differences that arise between the ponds can be attributed to that variable. Such an experiment is known as a **controlled experiment** because the scientist controls for the effects of all variables except the one whose effect he or she is testing. In our example, the pond left unfertilized serves as a **control,** an unmanipulated point of comparison for the manipulated **treatment** pond. Whenever possible, it is best to *replicate* one's experiment, that is, to stage multiple tests of the same comparison of control and treatment. Our scientist could perform a replicated experiment on, say, 10 pairs of ponds, adding fertilizer to one of each pair.

Experiments can establish causal relationships, showing that changes in an independent variable cause changes in a dependent variable. However, experiments are not the only way of testing a hypothesis. Sometimes a hypothesis can be convincingly addressed through **correlation,** searching for relationships among variables. Let's suppose our scientist surveys 50 ponds, 20 of which happen to be fed by fertilizer runoff from nearby farm fields and 30 of which are not. Let's also say he or she finds seven times as much algal growth in the fertilized ponds as in the unfertilized ponds. The scientist would conclude that algal growth is correlated with fertilizer input; that is, that one tends to increase along with the other. Although this type of evidence is weaker than the causal demonstration that controlled experiments can provide, sometimes it is the best approach, or the only feasible one. For example, in studying the effects of global climate change (Chapter 18), we could hardly run an experiment adding carbon dioxide to 10 treatment planets and comparing the result to 10 control planets.

Analyze and interpret results Scientists record **data,** or information, from their studies. They particularly value *quantitative* data, which is information expressed using numbers, because numbers provide precision and are easy to compare. The scientist running the fertilization experiment, for instance, might quantify the area of water surface covered by algae in each pond or might measure the dry weight of algae in a certain volume of water taken from each.

However, even with the precision that numbers provide, a scientist's results may not be clear-cut. Data from treatments and controls may vary only slightly, or different replicates may yield different results. The scientist must therefore analyze the data using statistical tests. With these mathematical methods, scientists can determine objectively and precisely the strength and reliability of patterns they find.

Some research, especially in the social sciences, involves data that is *qualitative,* or not expressible in terms of numbers. Research involving historical texts, personal interviews, surveys, detailed examination of case studies, or descriptive observation of behavior can include qualitative data on which statistical analyses may not be possible. Such studies are still scientific in the broad sense, because their data can be interpreted systematically using other accepted methods of analysis.

Weighing the Issues:
Replicates and Data Analysis

Let's say our scientist who is testing for the effects of agricultural fertilizer on algal growth uses experimental replicates, testing 10 pairs of ponds. If 5 of the 10 treatments grow more algae than controls while 5 grow less, what do you think the scientist should conclude? What if all 10 treatments grow more algae than do controls? What if 8 do? If 8 treatment ponds show 10% more growth than the control ponds they are paired with, but the remaining 2 control ponds show 300% more growth than their paired treatments, then what should

the scientist conclude? Such cases require statistical analysis so that we can judge levels of confidence to assign to our conclusions. Given possibilities like this, can you explain why scientists believe replicates are important?

If experiments refute a hypothesis, the scientist will reject it and may develop a new hypothesis to replace it. If experiments fail to reject the hypothesis, this outcome lends support to the hypothesis but does not *prove* it is correct. The scientist may choose to generate new predictions to test the hypothesis in a different way and further assess its likelihood of being true. Thus, the scientific method loops back on itself, often giving rise to repeated rounds of hypothesis-revision and new experimentation (see Figure 1.10).

If repeated tests fail to reject a particular hypothesis, evidence in favor of it accumulates, and the researcher may eventually conclude that the idea is well supported. One would ideally also want to test different potential explanations for the question of interest. For instance, our scientist might propose an additional hypothesis that algae increase in fertilized ponds because numbers of fish or invertebrate animals that eat algae decrease. It is possible, of course, that both hypotheses could be correct and that each may explain some portion of the initial observation that local ponds were experiencing algal blooms.

There are different ways to test hypotheses

An experiment in which the researcher actively chooses and manipulates the independent variable is known as a **manipulative experiment** (Figure 1.11a). A manipulative experiment provides the strongest type of evidence a scientist can obtain. In practice, however, some modes of scientific inquiry are more amenable to manipulative experimentation than others. Physics and chemistry tend to involve manipulative experiments, but many other fields deal with entities less easily manipulated than physical forces and chemical reagents. This is true of *historical sciences* such as cosmology, which deals with the history of the universe, and paleontology, which explores the history of past life. It is difficult to manipulate experimentally a star thousands of light years away, or the fossil tooth from a mastodon. Moreover, many of the most interesting questions in these fields center on the causes and consequences

FIGURE 1.11 A researcher wishing to test how temperature affects the growth of wheat might run a manipulative experiment in which wheat is grown in two identical greenhouses, one kept at 20°C (68°F) and the other kept at 25°C (77°F) **(a)**. Alternatively, the researcher might run a "natural experiment" in which he or she compares the growth of wheat in two fields at different latitudes, a cool northerly location and a warm southerly one **(b)**. Because it would be difficult to hold all variables besides temperature constant, the researcher might want to collect data on a number of northern and southern fields and correlate temperature and wheat growth using statistical methods.

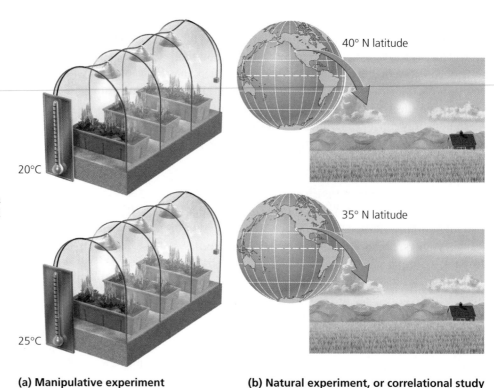

20°C

25°C

40° N latitude

35° N latitude

(a) Manipulative experiment

(b) Natural experiment, or correlational study

of particular historical events, rather than the behavior of general constants.

Disciplines that do not quite fit the so-called physics model of science sometimes rely on **natural experiments** rather than manipulative ones (Figure 1.11b). For instance, an evolutionary biologist might want to test whether animal species isolated on oceanic islands tend to evolve large body size over time. The biologist cannot run a manipulative experiment by placing animals on islands and continents and waiting long enough for evolution to do its work. However, this is exactly what nature has already done. The biologist might test the idea by comparing pairs of closely related species, in which one of each pair lives on an island and the other on a continental mainland. The experiment has in essence been conducted naturally, and it is up to the scientist to interpret the results.

In ecology, both manipulative and natural experimentation is used. The science of **ecology** deals with the distribution and abundance of organisms (living things), the interactions among them, and the interactions between organisms and their abiotic environments. When possible, ecologists try to run manipulative experiments. An ecologist wanting to measure the importance of a certain insect in pollinating the flowers of a given crop plant

might, for example, fit some flowers with a device to keep the insects out while leaving other flowers accessible, and later measure the fruit output of each group. Other questions that involve large spatial scales or long time scales may instead require natural experiments.

The social sciences generally involve less experimentation than the natural sciences, depending more on careful observation and interpretation of patterns in data. A sociologist studying how people from different cultures conceive the notion of wilderness might conduct a survey and analyze responses to its questions, looking for similarities and differences among respondents. Such analyses may be either quantitative or qualitative, depending on the nature of the data and the researchers' particular questions and approaches.

Descriptive observational studies and natural experiments can show correlation between variables, but they cannot demonstrate that one variable *causes* change in another, as manipulative experiments can. Not all variables are controlled for in a natural experiment, so a single result could give rise to several interpretations. However, correlative studies, when done well, can make for very convincing science, and they preserve the real-world complexity that manipulative experiments often sacrifice. Moreover, sometimes correlation is all we have. Because manipulations are difficult at large scales, some of the most important questions in environmental science tend to be addressed with correlative data. The large scale and complexity of many questions in environmental science also mean that few studies, manipulative or correlative, come up with neat and clean results. As such, scientists are not always able to give policymakers and society black-and-white answers to questions.

The scientific process does not stop with the scientific method

Individual researchers or teams of researchers follow the scientific method as they investigate questions that interest them. However, scientific work takes place within the context of a community of peers, and to have any impact, a researcher's work must be published and made accessible to this community. Thus, the scientific method is embedded within a larger process that takes place at the level of the scientific community as a whole (Figure 1.12).

Peer review When a researcher's work is done and the results analyzed, he or she writes up the findings and submits them to a journal for publication. Several other scientists specializing in the topic of the paper examine the manuscript, provide comments and criticism (generally anonymously), and judge whether the work merits publication in the journal. This procedure, known as **peer review,** is an essential part of the scientific process. Peer review is a valuable guard against faulty science contaminating the literature on which all scientists rely. However, because scientists are human and may have their own personal biases and agendas, politics can sometimes creep into the review process. Fortunately, just as individual scientists strive their best to remain objective in conducting their research, the scientific community does its best to ensure fair review of all work. Winston Churchill once called democracy the worst form of government, except for all the others that had been tried. The same might be said about peer review; it is an imperfect system, yet no one has come up with a better one.

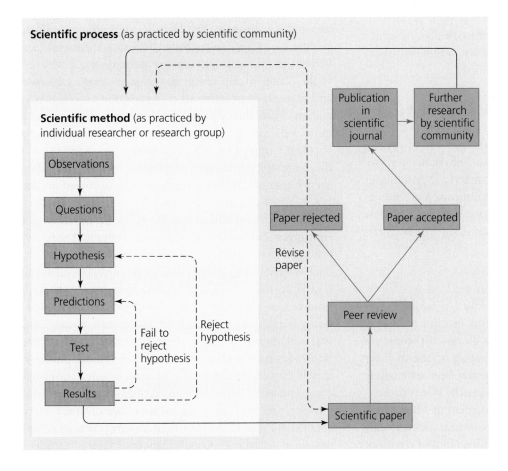

FIGURE 1.12 The scientific method (inner box) followed by individual researchers or research teams exists within the context of the overall process of science at the level of the scientific community (outer box). This process includes peer review and publication of research, acquisition of funding, and the development of theory through the cumulative work of many researchers.

Conference presentations Scientists frequently present their work at professional conferences, where they interact with colleagues and often receive informal comments on their research. When research has not yet been published, feedback from colleagues can help improve the quality of a scientist's work before it is submitted for publication.

Grants and funding Research scientists spend large portions of their time writing grant applications requesting money to fund their research from private foundations or government agencies such as the National Science Foundation. Grant applications undergo peer review just as scientific papers do, and competition for funding is often intense. Scientists' reliance on funding sources can also lead to potential conflicts of interest. A scientist who obtains data showing his or her funding source in an unfavorable light may be reluctant to publish the results for fear of losing funding—or worse yet, may be tempted to doctor the results. This situation can arise, for instance, when an industry funds research to test its products for safety or environmental impact. Most scientists do not succumb to these temptations, but some funding sources have been known to pressure their scientists for certain results. This is why as a student or informed citizen, when critically assessing a scientific study, you should always try to find out where the researchers obtained their funding.

Repeatability Sound science is based on doubt rather than certainty and on repeatability rather than one-time occurrence. Even when a hypothesis appears to explain observed phenomena, scientists are inherently wary of accepting it. The careful scientist may test a hypothesis repeatedly in various ways before submitting the findings for publication. Following publication, other scientists may attempt to reproduce the results in their own experiments and analyses.

Theories If a hypothesis survives repeated testing by numerous research teams and continues to predict experimental outcomes and observations accurately, it may potentially be incorporated into a theory. A **theory** is a widely accepted, well-tested explanation of one or more cause-and-effect relationships that has been extensively validated by a great amount of research. Whereas a hypothesis is a simple explanatory statement that may be refuted by a single experiment, a theory consolidates many related hypotheses that have been tested and have not been refuted.

Note that scientific use of the word *theory* differs from popular usage of the word. In everyday language when we say something is "just a theory," we are suggesting it is a speculative idea without much substance. Scientists, however, mean just the opposite when they use the term; to them, a theory is a conceptual framework that effectively explains a phenomenon and has undergone extensive and rigorous testing, such that confidence in it is extremely strong. For example, Darwin's theory of evolution by natural selection (▶ pp. 118–121) has been supported and elaborated by many thousands of studies over 150 years of intensive research. Such research has shown repeatedly and in great detail how plants and animals change over generations, or evolve, to express characteristics that best promote survival and reproduction. Because of its strong support and explanatory power, evolutionary theory is the central unifying principle of modern biology.

Science may go through "paradigm shifts"

It is crucial to realize that results obtained by the scientific method may sometimes later be reinterpreted to show that earlier interpretations were incorrect. Thomas Kuhn's 1962 book *The Structure of Scientific Revolutions* argued that science goes through periodic revolutions, dramatic upheavals in thought, in which one scientific **paradigm,** or dominant view, is abandoned for another. For example, before the 16th century, scientists believed that Earth was at the center of the universe, and some made elaborate and accurate measurements explaining the movements of planets from that viewpoint. Their data fit the theory quite well, yet the theory eventually was disproved by Nicolaus Copernicus, who showed that placing the sun at the center of the universe explained the planetary data even better. A similar paradigm shift occurred in the 1960s, when geologists accepted the theory of plate tectonics (▶ pp. 207–209), once evidence for the movement of continents and the action of tectonic plates had accumulated and become overwhelmingly convincing.

Understanding how science works is vital to assessing how scientific ideas and interpretations change through time as new information accrues. This process is especially relevant in environmental science, a young field that is changing rapidly as we learn vast amounts of new information, as human impacts on the planet multiply, and as lessons from the consequences of our actions become apparent. Because so much remains unstudied and undone, and because so many issues we cannot foresee are likely to arise in the future, environmental science will remain an exciting frontier for you to explore as a student and as an informed citizen throughout your life.

Sustainability and the Future of Our World

Throughout this book you will see examples of environmental scientists asking questions, developing hypotheses, conducting experiments, gathering and analyzing data, and drawing conclusions about environmental processes and the causes and consequences of environmental change. Environmental scientists who aim to understand the condition of our environment and the consequences of our impacts are studying the most centrally important issues of our time.

Population and consumption lie at the root of many environmental changes

We modify our environment in diverse ways, but the steep and sudden rise in human population has amplified nearly all of our impacts (Chapter 8). Our numbers have nearly quadrupled in the past 100 years, passing 6 billion in 1999 and 6.5 billion in 2006. We add about 78 million people to the planet each year—that's over 200,000 per day. Today, the rate of population growth is slowing, but our absolute numbers continue to increase and to shape our interactions with one another and with our environment.

Our consumption of resources has risen even faster than our population growth. The rise in affluence has been a positive development for humanity, and our conversion of the planet's natural capital has made life more pleasant for us so far. However, like rising population, rising per capita consumption amplifies the demands we make on our environment. Moreover, affluence and consumption have not grown equally for all the world's citizens. Today the 20 wealthiest nations boast 40 times the income of the 20 poorest nations—twice the gap that existed four decades earlier. The ecological footprint of the average citizen of a developed nation such as the United States is considerably larger than that of the average resident of a developing country (Figure 1.13). Within the United States, the richest fifth of people claim nearly half the income, whereas the poorest fifth receive only 5%.

We face challenges in agriculture, pollution, energy, and biodiversity

The dramatic growth in human population and consumption is due in part to our successful efforts to expand and intensify the production of food (Chapters 9 and 10).

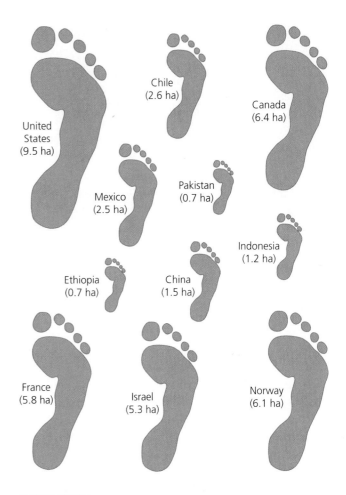

FIGURE 1.13 The citizens of some nations have larger ecological footprints than the citizens of others. U.S. residents consume more resources—and thus use more land—than residents of any other nation. Shown here are ecological footprints for average citizens of several developed and developing nations, as of 2001. Data from Global Footprint Network, 2005.

Since the agricultural revolution, new technologies have enabled us to grow increasingly more food per unit of land. These advances in agriculture must be counted as one of humanity's great achievements, but they have come at some cost. We have converted nearly half the planet's land surface for agriculture; our extensive use of chemical fertilizers and pesticides poisons organisms and alters natural systems; and erosion, climate change, and poorly managed irrigation are destroying 5–7 million hectares (ha; 12.5–17.5 million acres) of productive cropland each year.

Meanwhile, pollution from our farms, industries, households, and individual actions dirties our land, water, and air (Figure 1.14). Outdoor air pollution, indoor air pollution, and water pollution contribute to the deaths of millions of people each year (Chapters 15–17). Environmental

FIGURE 1.14 Indoor and outdoor air pollution contribute to millions of premature deaths each year, and environmental scientists and policymakers are working to reduce this problem in a variety of ways.

toxicologists are chronicling the impacts on people and wildlife of the many synthetic chemicals and other pollutants we emit into the environment (Chapter 14). Our most pressing pollution challenge may be to address the looming specter of global climate change (Chapter 18). Scientists have firmly concluded that human activity is altering the composition of the atmosphere and that these changes are affecting Earth's climate. Since the start of the industrial revolution, atmospheric carbon dioxide concentrations have risen by 31%, to a level not present in at least 420,000 years. This increase results from our reliance on burning fossil fuels to power our civilization. Carbon dioxide and several other gases absorb heat and warm Earth's surface, which is likely responsible for glacial melting, sea-level rise, impacts on wildlife and crops, and increased episodes of destructive weather.

The combined impact of human actions such as climate change, overharvesting, pollution, the introduction of non-native species, and particularly habitat alteration, has driven many aquatic and terrestrial species out of large parts of their ranges and toward the brink of extinction (Chapter 11). Today Earth's biological diversity, or **biodiversity,** the cumulative number and diversity of living things, is declining dramatically. Many biologists say we are already at the outset of a mass extinction event comparable to only five others documented in all of Earth's history. Biologist Edward O. Wilson has warned that the loss of biodiversity is our most serious and threatening environmental dilemma, because it is not the kind of problem that responsible human action can remedy. Rather, the extinction of species is irreversible; once a species has become extinct, it is lost forever.

Solutions to environmental problems must be global and sustainable

The nature of virtually all of these environmental issues is being changed by the set of ongoing phenomena commonly dubbed *globalization.* Our increased global interconnectedness in trade, politics, and the movement of people and of other species poses many challenging problems, but it also sets the stage for novel and effective solutions.

The most comprehensive scientific assessment of the present condition of the world's ecological systems and their ability to continue supporting our civilization was completed in 2005. In this year, over 2,000 of the world's leading environmental scientists from nearly 100 nations completed the **Millennium Ecosystem Assessment.** The four main findings of this exhaustive project are summarized in Table 1.1. The Assessment makes clear that our degradation of the world's environmental systems is having negative impacts on all of us, but that with care and diligence we can still turn many of these trends around.

Table 1.1 Main Findings of the Millennium Ecosystem Assessment

▶ Over the past 50 years, humans have changed ecosystems more rapidly and extensively than in any comparable period of time in human history, largely to meet rapidly growing demands for food, freshwater, timber, fiber, and fuel. This has resulted in a substantial and largely irreversible loss in the diversity of life on Earth.

▶ The changes made to ecosystems have contributed to substantial net gains in human well-being and economic development, but these gains have been achieved at growing costs. These costs include the degradation of ecosystems and the services they provide for us, and the exacerbation of poverty for some groups of people.

▶ This degradation could grow significantly worse during the first half of this century.

▶ The challenge of reversing the degradation of ecosystems while meeting increasing demands for their services can be partially overcome, but doing so will involve significantly changing many policies, institutions, and practices.

Adapted from *Millennium Ecosystem Assessment, Synthesis Report,* 2005.

FIGURE 1.15 Human activities are pushing many organisms, including the panda, toward extinction. Efforts to save endangered species and reduce biodiversity loss include many approaches, but all require that adequate areas of appropriate habitat be preserved in the wild.

Fortunately, potential solutions abound

We cannot, of course, live without exerting any impact on Earth's systems. We face trade-offs with many environmental issues, and the challenge is to develop solutions that further our quality of life while minimizing harm to the environment that supports us. Fortunately, there are many workable solutions at hand, and many more potential solutions we can achieve with further effort.

In response to agricultural problems, scientists and others have developed and promoted soil conservation, high-efficiency irrigation, and organic agriculture. Technological advances and new laws have greatly reduced the pollution emitted by industry and automobiles in wealthier countries. Although the U.S. government has resisted international efforts to rein in pollutants to halt climate change, American scientists have been at the forefront of climate change science, and other nations are beginning to address the problem, as are the governments of some U.S. states. Amid ample reasons for concern about the state of global biodiversity, advances in conservation biology are enabling scientists and policymakers in many cases to work together to protect habitat, slow extinction, and safeguard endangered species (Figure 1.15). Recycling is helping relieve our waste disposal problems, and alternative renewable energy sources are being developed to take the place of fossil fuels (Figure 1.16). These are but a few of the many solutions we will explore in the course of this book.

Are things getting better or worse?

Despite the myriad challenges we review in this book, many people maintain that the general conditions of human life and the environment are in fact getting better, not worse. A recent proponent of this view, Danish statistician Bjorn Lomborg, wrote in his book *The Skeptical Environmentalist:*

> We are not running out of energy or natural resources. There will be more and more food per head of the world's population. Fewer and fewer people are starving. In 1900 we lived for an average of 30 years; today we live for 67. . . . The air and water around us are becoming less and less polluted. Mankind's lot has actually improved in terms of practically every measurable indicator.

Furthermore, some people maintain that we will find ways to make Earth's natural resources meet all of our needs indefinitely and that human ingenuity will see us through any difficulty. Such views are sometimes characterized as *Cornucopian.* In Greek mythology, *cornucopia*—literally "horn of plenty"—is the name for a magical goat's horn that overflowed with grain, fruit, and flowers. In contrast, people who predict doom and disaster for the world because of our impact upon it have been called *Cassandras,* after the mythical princess of Troy with the gift of prophecy whose dire predictions were not believed.

At least three questions are worth asking each time you are confronted with seemingly conflicting statements from Cassandras and Cornucopians. One question is whether the impacts being debated pertain only to humans or also to other organisms and natural systems. The second question is whether the debaters are thinking in the short term or the long term. The third question is whether they are considering all costs and

FIGURE 1.16 Our dependence on fossil fuels has caused a wide array of environmental impacts. Although fossil fuels have powered our civilization since the industrial revolution, many renewable energy sources exist, such as solar energy that can be collected with panels like these. Such alternative energy sources could be further developed for sustainable use now and in the future.

benefits relevant for the question at hand, or only some. As you proceed through this book and encounter countless contentious issues, consider how one's perception of them may be influenced by these three factors.

Sustainability is a goal for the future

The primary challenge in our increasingly populated world is how to live within our planet's means, such that Earth and its resources can sustain us and the rest of Earth's biota for the foreseeable future. This is the challenge of **sustainability,** a guiding principle of modern environmental science. Sustainability means leaving our children and grandchildren a world as rich and full as the world we live in now. It means not depleting Earth's natural capital, so that after we are gone our descendants will enjoy the use of resources as we have. It means developing solutions that are able to work in the long term. Sustainability requires maintaining fully functioning ecological systems, because we cannot sustain human civilization without sustaining the natural systems that nourish it.

Sustainability is a concept you will encounter throughout this book. Our final chapter (Chapter 23) takes a wide-ranging look at emerging sustainable solutions—on college and university campuses and in the world at large.

Sustainability need not require great sacrifice of us. We will naturally always desire to enhance our quality of life, and as we will see, there are many ways we can do so while also encouraging a more sustainable lifestyle. Economists

employ the term *development* to describe the use of natural resources for economic advancement (as opposed to simple subsistence, or survival). Logging, farming, mining, and building homes and factories are all types of development, and each of them affects the environment and gives rise to changes that environmental scientists study. **Sustainable development** is the use of renewable and nonrenewable resources in a manner that satisfies our current needs without compromising future availability of resources. The United Nations defines sustainable development as development that ". . . meets the needs of the present without sacrificing the ability of future generations to meet theirs." Answering a simple question—"Can this activity continue forever?"—indicates whether a particular activity is sustainable.

Sustainability depends, in large part, on the ability of the current human population to limit its environmental impact. Doing so will require us to make an ethical commitment, while also applying information we gain from the sciences. Science can help us devise ways to limit our impact and maintain the functioning of the environmental systems on which we depend.

Conclusion

Finding effective ways of living peacefully, healthfully, and sustainably on our diverse and complex planet will require a thorough scientific understanding of both natural and social systems. Environmental science helps us understand our intricate relationship with the environment and informs our attempts to solve and prevent environmental problems.

It is important to keep in mind that identifying a problem is the first step in devising a solution to it. Many of the trends detailed in this book may cause us worry, but others give us reason to hope. One often-heard criticism of environmental science courses and textbooks is that too often they emphasize the negative. Recognizing the validity of this criticism, in this book we attempt to balance the discussion of environmental problems with a corresponding focus on potential solutions. Solving environmental problems can move us toward health, longevity, peace, and prosperity. Science in general, and environmental science in particular, can aid us in our efforts to develop balanced and workable solutions to the many environmental dilemmas we face today and to create a better world for ourselves and our children.

REVIEWING OBJECTIVES

You should now be able to:

Define the term *environment*

▶ Our environment consists of everything around us, including living and nonliving things. (p. 3)

▶ Humans are a part of the environment and are not separate from nature. (p. 3)

Describe natural resources and explain their importance to human life

▶ Resources from nature are essential to human life and civilization. (p. 4)

▶ Some resources are perpetually renewable, others are nonrenewable, and still others are renewable if we are careful not to exploit them at too fast a rate. (p. 4)

▶ Malthus and Ehrlich pointed out risks of human population growth, while Hardin and Wackernagel and Rees pioneered important concepts in resource consumption. (pp. 4–7)

Characterize the interdisciplinary nature of environmental science

▶ Environmental science uses the approaches and insights of numerous disciplines from the natural sciences and the social sciences. (p. 11)

Understand the scientific method and how science operates

▶ Science is a process of using observations to test ideas. (p. 12)

▶ The scientific method consists of a series of steps, including making observations, formulating questions, stating a hypothesis, generating predictions, testing predictions, and analyzing the results obtained from the tests. (pp. 13–15)

▶ The scientific method is not always followed strictly, and there are different ways to test questions scientifically. (pp. 15–17)

▶ Scientific research occurs within a larger process that includes peer review of work, journal publication, and interaction with colleagues. (pp. 17–18)

Diagnose and illustrate some of the pressures on the global environment

▶ Increasing human population and increasing per capita consumption exacerbate human impacts on the environment. (p. 19)

▶ Human activities such as industrial agriculture and the use of fossil fuels for energy are having diverse environmental impacts, including resource depletion, air and water pollution, habitat destruction, and the diminishment of biodiversity. (pp. 19–20)

Evaluate the concepts of sustainability and sustainable development

▶ Sustainability means living within the planet's means, such that Earth's resources can sustain us—and other species—for the foreseeable future. (pp. 20–22)

▶ Sustainable development is possible; we need not decrease our quality of life to establish sustainable lifestyles. (p. 22)

TESTING YOUR COMPREHENSION

1. What do renewable resources and nonrenewable resources have in common? How are they different? Identify two renewable and two nonrenewable resources.

2. How did the agricultural revolution affect human population size? How did the industrial revolution affect human population size? Explain your answers.

3. What is "the tragedy of the commons"? Explain how the concept might apply to an unregulated industry that is a source of water pollution.

4. What is environmental science? Name several disciplines involved in environmental science.

5. What are the two meanings of *science*? Name three applications of science.

6. Describe the scientific method. What is the typical sequence of steps?

7. Explain the difference between a manipulative experiment and a natural experiment.

8. What needs to occur before a researcher's results are published? Why is this important?

9. Give examples of three major environmental problems in the world today, along with their causes.

10. What is sustainable development?

SEEKING SOLUTIONS

1. Many resources are renewable if we use them in moderation but can become nonrenewable if we overexploit them. Order the following resources on a continuum of renewability (see Figure 1.1), from most renewable to least renewable: soils, timber, fresh water, food crops, and biodiversity. What factors influenced your decision? For each of these resources, what might constitute overexploitation, and what might constitute sustainable use?

2. Why do you think the Easter Islanders did not or could not stop themselves from stripping their island of all its trees? Do you see similarities between the history of the Easter Islanders and the modern history of our society? Why or why not?

3. What environmental problem do *you* feel most acutely yourself? Do you think there are people in the world who do not view your issue as an environmental problem? Who might they be, and why might they take a different view?

4. Name an environmental problem you would like to see solved or mitigated. Describe the scientific research you think would need to be completed so that workable solutions to this problem can be developed. Would more than science be needed?

5. If the human population were to stabilize tomorrow and never surpass 7 billion people, would that solve our environmental problems? Which types of problems might be alleviated, and which might continue to become worse?

6. Consider the historic expansion of agriculture and our ability to feed increasing numbers of people, as described in this chapter. Now ask yourself, "Are things getting better or worse?" Ask this question from four points of view: (1) from the human perspective, (2) from the perspective of other organisms, (3) from a short-term perspective, and (4) from a long-term perspective. Do your answers to this question change? If so, how?

INTERPRETING GRAPHS AND DATA

Environmental scientists study phenomena that range in size from individual molecules (Chapter 4) to the entire Earth (Chapter 7), and that occur over time periods lasting from fractions of a second to billions of years. To simultaneously and meaningfully represent data covering so many orders of magnitude, scientists have devised a variety of mathematical and graphical techniques, such as exponential notation and logarithmic scales. Below are two graphical representations *of the same data,* representing the growth of a hypothetical population from an initial size of 10 individuals at a rate of increase of approximately 2.3% per generation. The graph in part (a) uses a conventional linear scale for the population size; the graph in part (b) uses a logarithmic scale.

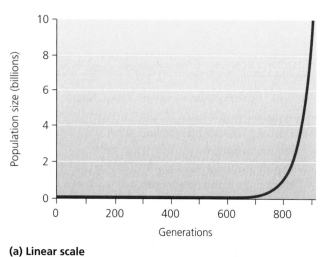

(a) Linear scale

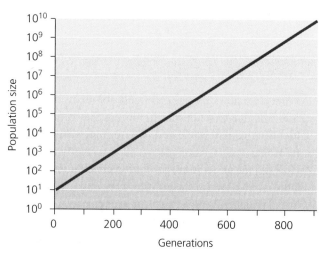

(b) Logarithmic scale

Hypothetical population growth curves, assuming an initial size of 10 and a constant rate of increase of approximately 2.3% per generation.

1. Using the graph in part (a), what would you say was the population size after 200 generations? After 400? After 600? After 800? How would you answer the same questions using the graph in part (b)? What impression does the graph in part (a) give about population change for the first 600 generations? What impression does the graph in part (b) give?

2. Compare these graphs to Figure 1.2a in the text. What does the human population appear to be doing between 10,000 B.P. and 2,000 B.P.?

3. The size of a population that is growing by a constant rate of increase will plot as a straight line on a logarithmically scaled graph like the one in part (b) above, but if the annual rate of increase changes, the line will curve. Do you think the data for the human population over the past 12,000 years would plot as a straight line on a logarithmically scaled graph? If not, when and why do you think the line would bend?

CALCULATING ECOLOGICAL FOOTPRINTS

Mathis Wackernagel and his colleagues have continued to refine the method of calculating ecological footprints—the amount of land and water required to produce the energy and natural resources we consume. In a 1999 paper, they applied their method to 52 nations that together account for 80% of the world's population and 95% of the World Domestic Product. According to their study, there are 4.9 acres available for every person in the world.

Compare the ecological footprints of each of the countries listed in the table below. Calculate their proportional relationships to the world population's average ecological footprint and to the land available globally to meet our ecological demands.

1. Why is the ecological footprint for people in Bangladesh so low?
2. Why is it so high in the United States?
3. The population of the United States is expected to grow to 349 million by 2025. What impact, if any, do you think this growth will have on the average global ecological footprint?
4. Based on the data in the table, what impacts do you think average family income has on ecological footprints?

Country	Ecological footprint (acres per person)	Proportion relative to world average footprint	Proportion relative to world land available
Bangladesh	1.2		
Colombia	4.9		1.0 (4.9/4.9)
Mexico	6.4		
Sweden	14.6		
Thailand	6.9		
United States	25.4		
World Average	6.9	1.0 (6.9/6.9)	1.4 (6.9/4.9)

Data from Wackernagel, M., et al. 1999. National natural capital accounting with the ecological footprint concept. *Ecological Economics* 29: 375–390.

Take It Further

Go to www.aw-bc.com/withgott or the student CD-ROM where you'll find:

▶ Suggested answers to end-of-chapter questions
▶ Quizzes, animations, and flashcards to help you study
▶ *Research Navigator*™ database of credible and reliable sources to assist you with your research projects

▶ **GRAPHit!** tutorials to help you master how to interpret graphs
▶ **INVESTIGATEit!** current news articles that link the topics that you study to case studies from your region to around the world

2 Environmental Ethics and Economics: Values and Choices

Kakadu National Park, Australia

Upon successfully completing this chapter, you will be able to:

▶ Characterize the influences of culture and worldview on the choices people make

▶ Outline the nature, evolution, and expansion of environmental ethics in Western cultures

▶ Describe precepts of classical and neoclassical economic theory, and summarize their implications for the environment

▶ Compare the concepts of economic growth, economic health, and sustainability

▶ Explain the fundamentals of environmental economics and ecological economics

Protestors against proposed Jabiluka uranium mine

Central Case: The Mirrar Clan Confronts the Jabiluka Uranium Mine

"For some people, what they are is not finished at the skin, but continues with the reach of the senses out into the land. If the land is summarily disfigured or reorganized, it causes them psychological pain."
—BARRY LOPEZ,
AMERICAN NATURE WRITER

"The Jabiluka uranium mine will improve the quality of the environment. The uranium resource there has been polluting the river system naturally, probably for thousands of years. . . . With the uranium resource removed and put to good use, the level of radioactivity will fall."
—MICHAEL DARBY,
AUSTRALIAN POLITICAL COMMENTATOR

The remote Kakadu region of Australia's Northern Territory is home to several groups of Australian Aborigines, native people who lived there long before the British colonization of Australia. The region also features Kakadu National Park, a World Heritage Site recognized by the United Nations for its irreplaceable natural and cultural resources. In addition, the region's land holds uranium, the naturally occurring radioactive metal valued for its use in nuclear power plants, nuclear weapons, and various medical and industrial tools. Uranium mining is a key contributor to the Australian national economy, accounting for 7% of Australia's economic output.

Many of Australia's uranium deposits occur on Aboriginal lands, giving rise to conflict between corporations seeking to develop mining operations and Aboriginal people trying to maintain their traditional culture. One such group is the Mirrar Clan, an extended family of 25 Kakadu-area Aborigines. The Mirrar have been living with the region's first uranium mine, the Ranger Mine, on their land since the Australian government approved its development in 1978.

In recent years, the Mirrar have been fighting the proposed development of a second mine, Jabiluka, nearby on their land. The Mirrar see Jabiluka as a threat to their culture and religion, which are deeply tied to the landscape. Like other Aborigines, they hold the landscape to be sacred, and they depend on its resources for their daily needs. The proposed mine site is near traditional

Mirrar hunting and gathering sites and is in the floodplain of a river that provides the clan food and water.

The Mirrar also view Jabiluka as a threat to their health and to the integrity of their environment, particularly given repeated radioactive spills at the Ranger mine. Many Mirrar fear that contaminated water could be released into area creeks and that radioactive radon gas would emanate from stored waste materials. Moreover, mindful of geological faults that exist in the area, the Mirrar worry that dams holding mine waste could fail catastrophically in an earthquake. Environmental activists worldwide have joined the Mirrar's struggle; in 1998 nearly 3,000 people traveled to the Kakadu region to protest Jabiluka.

In late 2002 their efforts appeared to have succeeded. Sir Robert Wilson, chief executive officer of the corporation holding rights to the Jabiluka ore body, announced the cancellation of mining plans at Jabiluka, citing economic factors (declining world uranium prices) and ethical factors (concerns about developing the mine without Mirrar consent). Wilson added that the company planned to rehabilitate the site and restore damage done there during exploration and assessment. However, a formal agreement was never signed, and since that time the price of uranium has risen on the world market. The corporation's plans are now in a holding pattern as it waits and hopes that the Mirrar will one day give their consent.

The Mirrar oppose mining despite the economic benefits the mining company has promised them in the form of jobs, income, development, and a higher material standard of living. The decision to bypass these economic incentives was not easy, and indeed, a number of other Aboriginal groups in the Kakadu region support mine development. In formulating their approaches to the mining proposal, the Mirrar and other Australians weighed economic, social, cultural, and philosophical questions as well as scientific ones. The story of mining and the Mirrar exemplifies some of the ways in which values, beliefs, and traditions interact with economic interests to influence the choices all of us make about how to live within our environment.

Culture, Worldview, and the Environment

The Mirrar faced difficult choices. They were offered substantial economic benefits, but they also felt that mine development ran counter to their ethical respect for their land. Such trade-offs between economic benefits and ethical concerns crop up frequently in environmental issues.

Ethics and economics involve values

As we saw in Chapter 1, environmental science examines Earth's natural systems, the ways they affect humans, and the ways humans affect them. Thus, environmental science entails a firm understanding of the natural sciences. To address environmental problems, however, it is also necessary to understand how people perceive their environment, how they relate to it philosophically and pragmatically, and how they value its elements. Ethics and economics are two quite different disciplines, but each deals with questions of what we value and how those values influence our decisions and actions. Anyone trying to address an environmental problem must try to understand not only how natural systems work, but also how values shape human behavior.

Culture and worldview influence one's perception of the environment

Almost every action we take affects our environment. Growing food requires soil, cultivation, and often irrigation. Building homes requires land, lumber, and metal. Manufacturing and fueling vehicles require metal, plastic, glass, and petroleum. From nutrition to housing to transportation, we meet our needs by altering our surroundings. Our decisions about how we manipulate and exploit our environment to meet our needs depend in part on rational assessments of costs and benefits. However, our decisions are also heavily influenced by the particular culture of which we are a part and by our particular worldview. **Culture** can be defined as the ensemble of knowledge, beliefs, values, and learned ways of life shared by a group of people. Culture, together with personal experience, influences each person's perception of the world and his or her place within it, something described as the person's **worldview.** A worldview reflects a person's (or group's) beliefs about the meaning, operation, and essence of the world (Figure 2.1).

People with different worldviews can study the same situation and review identical data yet draw dramatically different conclusions. For example, many well-meaning people have supported the Jabiluka mine while many other well-meaning people have opposed it. The officers, employees, and shareholders of the mining company, and government officials who support the mine, view uranium mining as a source of jobs, income, energy, and economic growth. They believe mining will benefit Australia in general and the Mirrar Clan in particular. Mine opponents, in contrast, foresee environmental problems, injustice, and negative social consequences. They recognize that uranium mining disturbs the landscape, pollutes air

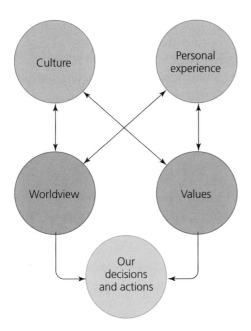

(a) Ranger mine

(b) Proposed Jabiluka mine site

FIGURE 2.1 Culture and personal experience influence a person's worldview and values, which in turn influence his or her actions and decisions. The disciplines of ethics and economics each examine (in very different ways) factors that influence our values and guide human behavior. Ethics, economics, and the behavior and preferences of individuals, together with information from the natural sciences, all inform the making of policy.

FIGURE 2.2 The Ranger mine (**a**), located on Aboriginal lands amid sacred sites, has caused enough environmental impacts to spark the Mirrar Clan's opposition to the proposed Jabiluka mine (**b**), whose preparation had begun nearby.

and water, and can expose miners to radiation, while community disruption, alcoholism, and crime frequently accompany mining booms.

Many factors shape our worldviews and perception of the environment

The traditional culture and worldview of the Mirrar Clan have played large roles in its response to the proposed Jabiluka mine. Australian Aborigines view the landscape around them as the physical embodiment of stories that express the beliefs and values central to their culture. The Australian landscape to them is a sacred text, analogous to the Bible in Christianity, the Koran in Islam, or the Torah in Judaism. Aborigines believe that spirit ancestors possessing human and animal features traveled routes called "dreaming tracks," leaving signs and lessons in the landscape. Modern Aborigines still engage in "walk-abouts," long walks that retrace the dreaming tracks. By explaining the origins of specific landscape features, dreaming-track stories assign meaning to notable landmarks and help Aborigines construct detailed mental maps of their surroundings. The stories also teach lessons concerning family relations, hunting, food gathering, and conflict resolution. Dreaming-track stories passed from one generation to the next help maintain Aboriginal culture. The Mirrar who oppose Jabiluka believe the mine would desecrate sacred sites and compromise their culture (Figure 2.2).

Weighing the Issues:
Uranium Mining in Bethlehem

Suppose a mining company discovered uranium beneath the site in Bethlehem believed to be the birthplace of Jesus—or near the Wailing Wall in Jerusalem, or the mosque at Mecca. What do you think would happen if the company announced plans to develop a mine there, assuring the public that environmental impacts would be minimal and that the mine would create jobs and stimulate economic growth? What aspects of this unlikely situation resemble that of the Mirrar case, and what factors are different? Explain your answers.

Religion is one of many factors that can shape people's worldviews and perception of the environment. A community may also share a particular view of its environment if its members have lived through similar experiences. For example, early European settlers in both Australia and North America viewed their environment as a hostile force because inclement weather, wild animals, and other natural forces frequently destroyed crops, killed livestock, and took settlers' lives. Such experiences were shared in stories and in songs and helped shape prevailing social attitudes in many frontier communities. The view of nature as a hostile force and an adversary to be overcome has passed from one generation to the next and still influences the way many North Americans and Australians view their surroundings.

A person's political ideology can also shape his or her attitude toward the environment. For instance, one's view of the proper role of government will influence whether or not one wants government to intervene in a market economy to protect environmental quality. Economic factors also sway how people perceive their environment and make decisions. An individual with a strong interest in the outcome of a decision that may result in his or her private gain or loss is said to have a *vested interest.* Mining company executives have a vested interest in a decision to open an area to mining because a new mine can increase profits, to which executive compensation is frequently tied. Likewise, a company's shareholders have a vested interest in such a decision because the value of the shares they hold increases with profits. In each case, vested interests may lead people to view a proposed mine such as Jabiluka primarily as a source of economic gain.

Throughout this book you will encounter scientific data regarding the environmental impacts of our choices (where to make our homes, how to make a living, what to wear, what to eat, how to travel, how to spend our leisure time, and so on). You will see that culture, worldviews, and values play critical roles in such choices and even can influence the interpretation of scientific data. Thus, acquiring scientific understanding is only one part of the search for solutions to environmental problems. Attention to ethics and economics helps us understand why and how we value those things we value.

Environmental Ethics

The field of **ethics** is a branch of philosophy that involves the study of good and bad, of right and wrong. The term *ethics* can also refer to the set of moral principles or values held by a person or a society. Ethicists help clarify how people judge right from wrong by elucidating the criteria, standards, or rules that people use in making these judgments. Such criteria are grounded in values—for instance, promoting human welfare, maximizing individual freedom, or minimizing pain and suffering.

People of different cultures or with different worldviews may differ in their values and thus may differ in the specific actions they consider to be right or wrong. This is why some ethicists are **relativists;** that is, they believe that ethics do and should vary with social context. However, different human societies show a remarkable extent of agreement on what moral standards are appropriate. Thus many ethicists are **universalists;** that is, they maintain that there exist objective notions of right and wrong that hold across cultures and situations. For both relativists and universalists, ethics is a *normative* or *prescriptive* pursuit; it tells us how we *ought to* behave.

Ethical standards are the criteria that help differentiate right from wrong. One classic ethical standard is *virtue,* which, as the ancient Greek philosopher Aristotle held, involves the personal achievement of moral excellence in character through reasoning and moderation. Another ethical standard is the *categorical imperative* proposed by Immanuel Kant, which roughly approximates Christianity's "golden rule": to treat others as you would prefer to be treated yourself. A third standard is the principle of *utility,* elaborated by British philosophers Jeremy Bentham and John Stuart Mill. The utilitarian principle holds that something is right when it produces the greatest practical benefits for the most people. We employ such ethical standards as tools for decision making, consciously or unconsciously, to situations in everyday life.

Environmental ethics pertains to humans and the environment

The application of ethical standards to relationships between humans and nonhuman entities is known as **environmental ethics.** This relatively new branch of ethics arose once people began to perceive environmental changes brought about by industrialization. Human interactions with the environment frequently give rise to ethical questions that can be difficult to resolve. Consider some examples:

▶ Does the present generation have an obligation to conserve resources for future generations? If so, how should this influence our decision making, and how much are we obligated to sacrifice?

▶ Are there situations that justify exposing some communities to a disproportionate share of pollution? If not, what actions are warranted in preventing this problem?

▶ Are humans justified in driving species to extinction? Are we justified in causing other changes in ecological systems? If destroying a forest would drive extinct an insect species few people have heard of but would create jobs for 10,000 people, would that action be ethically admissible? What if it were an owl species? What if only 100 jobs would be created?

We have extended ethical consideration to more entities through time

Answers to questions like those above depend partly on what ethical standard(s) a person chooses to use. They also depend on the breadth and inclusiveness of the person's domain of ethical concern. A person who feels responsibility for the welfare of insects would answer the third question very differently from a person whose domain of ethical concern ends with humans. Most of us feel moral obligations to some entities in the world, but by no means to all.

Throughout Western history, people have gradually enlarged the array of entities they feel deserve ethical consideration. The enslavement of human beings by other human beings was common in many societies until recently,

for instance. Women in the United States were not allowed to vote until 1920, and they still face lower pay for equal work. Consider, too, how little ethical consideration citizens of one nation generally extend to those of another on which their government has declared war. Human societies are only now beginning to embrace the principle that all people be granted equal ethical consideration.

Our expanding domain of ethical concern has begun to include nonhuman entities as well. Concern for the welfare of domesticated animals is evident in humane societies and in the lengths many people go to provide for their pets. Animal-rights activists voice concern for animals that are hunted, eaten, or used in laboratory testing. A great many people now accept that wild animals (at least obviously sentient animals, such as large vertebrates, with which we share similarities) merit ethical consideration. Moreover, today many environmentalists are concerned not only with certain animals but also with the well-being of whole natural communities. Some people have gone still further, suggesting that all of nature—living things and nonliving things, even rocks—should be ethically represented. The historian Roderick Nash illustrated this historical expansion of ethics in his 1989 book, *The Rights of Nature* (Figure 2.3).

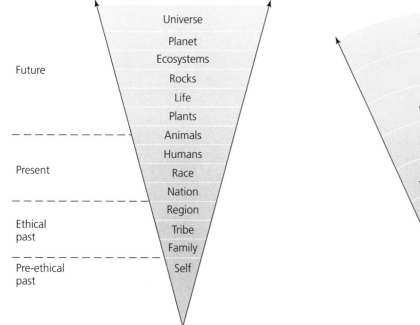

(a) The evolution of ethics

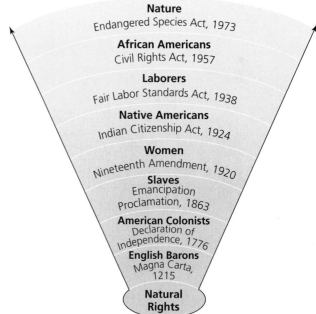

(b) The expanding concept of rights

FIGURE 2.3   Through time, people in Western cultures have broadened the scope of their ethical consideration for others. We can view ethics progressing through time in a generalized way outward from the self **(a)**. This historical expansion of ethics is reflected by key legal milestones in the expansion of rights granted by Britain and then the United States **(b)**. *Source:* Nash, R. F. 1989. *The rights of nature.* University of Wisconsin Press.

What is behind this ongoing expansion? Rising economic prosperity in Western cultures, as people gained more leisure time and became less anxious about their day-to-day survival, has helped enlarge our ethical domain. Science has also played a role. Ecology, as it has developed over the past 75 years, has made clear that all organisms are interconnected and that what affects plants, animals, and ecosystems can in turn affect humans. Evolutionary biology over the past 150 years has shown that humans are merely one species out of millions and have evolved subject to the same pressures as other organisms. Ecology and evolution have demonstrated scientifically that humans do not stand apart from nature, but rather are part of it.

For many non-Western cultures, expanded ethical domains are nothing new. Many traditional hunter-gatherer cultures have long granted nonhuman entities ethical standing. The Mirrar, who view their landscape as sacred and alive, are a case in point. However, it is worthwhile to examine Western ethical expansion because it is tied to so many of our society's beliefs and actions regarding the environment. People often simplify the continuum Nash portrayed by dividing it into three ethical perspectives: anthropocentrism, biocentrism, and ecocentrism.

Anthropocentrism **Anthropocentrism** describes a human-centered view of our relationship with the environment. An anthropocentrist denies or ignores the notion that nonhuman entities can have rights. An anthropocentrist also measures the costs and benefits of actions solely according to their impact on people (Figure 2.4). To evaluate a human action that affects the environment, an anthropocentrist might use criteria such as impacts on human health, economic costs and benefits, and aesthetic concerns. For example, if the Jabiluka mine provided a net economic benefit while doing no harm to human health and having little aesthetic impact, the anthropocentrist would conclude it was a worthwhile venture, even if it might drive some native species extinct. If protecting the area would provide spiritual, economic, or other benefits to humans now or in the future, an anthropocentrist might favor its protection. In the anthropocentric perspective, anything not providing benefit to people is considered to be of negligible value.

Biocentrism In contrast to anthropocentrism, **biocentrism** ascribes values to actions, entities, or properties on the basis of their effects on all living things or on the integrity of the biotic realm in general (see Figure 2.4). In

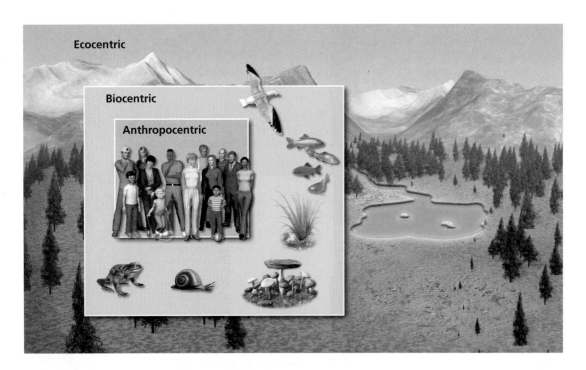

FIGURE 2.4 We can categorize people's ethical perspectives as anthropocentric, biocentric, or ecocentric. An anthropocentrist extends ethical standing only to humans and judges actions in terms of their effects on humans. A biocentrist values and considers all living things, human and otherwise. An ecocentrist extends ethical consideration to living and nonliving components of the environment. The ecocentrist also takes a holistic view of the connections among these components, valuing the larger functional systems of which they are a part.

this perspective, all life has ethical standing. A biocentrist evaluates actions in terms of their overall impact on living things, including—but not exclusively focusing on—human beings. In the case of the Jabiluka mine proposal, a biocentrist might oppose the mine if it posed a serious threat to the abundance and variety of living things in the area, even if it would create jobs, generate economic growth, and pose no threat to human health. Some biocentrists advocate equal consideration of all living things, whereas others advocate that some types of organisms should receive more than others.

Ecocentrism Ecocentrism judges actions in terms of their benefit or harm to the integrity of whole ecological systems, which consist of biotic and abiotic elements and the relationships among them (see Figure 2.4). For an ecocentrist, the well-being of an individual organism—human or otherwise—is less important than the well-being of a larger integrated ecological system. An ecocentrist might approve of an action that harmed human health, caused economic loss, or took a number of lives, if such impacts were necessary to protect an entire species, community, or ecosystem (we will study species, communities, and ecosystems in Chapters 5 through 7). Ecocentrism is a more holistic perspective than biocentrism or anthropocentrism. It not only includes a wider variety of entities, but also stresses the need to preserve the connections that tie the entities together into functional systems.

Environmental ethics has ancient roots

Environmental ethics arose as a distinct academic discipline in the early 1970s, but people have contemplated our relationship with, and possible responsibilities toward, nature for thousands of years. Ancient Aboriginal dreaming-track stories treat the environment as a source of sacred teachings, so boulders, caves, patches of lichen, and other entities are felt to have moral significance worthy of contemplation and protection. In the Western tradition, the ancient Greek philosopher Plato expressed what he considered humans' moral obligation to the environment, writing, "The land is our ancestral home and we must cherish it even more than children cherish their mother."

Some ethicists and theologians have pointed to the religious traditions of Christianity, Judaism, and Islam as sources of anthropocentric hostility toward the environment. They point out biblical passages such as, "Be fruitful and multiply, and fill the earth and subdue it; and have dominion over the fish of the sea and over the birds of the air and over every living thing that moves upon the

earth." Such wording has justified and encouraged an animosity toward nature that has characterized Western culture over the centuries, some scholars say. Others interpret sacred texts of these religions to encourage benevolent human stewardship over nature. Consider the directive, "You shall not defile the land in which you live. . . ." In fact, a 2003 poll showed that 56% of Americans supported environmental protection because they considered the environment to be "God's creation." Although people have held differing views of their ethical relationship with their environment for millennia, environmental impacts that became apparent during the industrial revolution intensified debate about our species' relationship with its environment.

The industrial revolution inspired environmental philosophers

As the industrial revolution spread in the 19th century from Great Britain throughout Europe and to North America and elsewhere, it amplified human impacts on the environment. In this period of social and economic transformation, agricultural economies became industrial ones, machines enhanced or replaced human and animal labor, and much of the rural population moved into cities. Consumption of natural resources accelerated rapidly, and pollution increased dramatically as coal combustion fueled railroads, steamships, ironworks, and factories.

Many British writers and philosophers of the time criticized the drawbacks of industrialization. Critic **John Ruskin** (1819–1900) called cities "little more than laboratories for the distillation into heaven of venomous smokes and smells." Ruskin also complained that people prized the material benefits that nature could provide but no longer appreciated its spiritual and aesthetic benefits. Motivated by similar concerns, a number of citizens' groups sprang up in 19th-century England that could be considered some of the first environmental organizations (Table 2.1).

In the United States during the 1840s, a philosophical movement called *transcendentalism* flourished, espoused in New England by the American philosophers **Ralph Waldo Emerson** and **Henry David Thoreau** and by poet **Walt Whitman.** The transcendentalists viewed nature as a direct manifestation of the divine, emphasizing the soul's oneness with nature and God. Like Ruskin, the transcendentalists objected to what they saw as their fellow citizens' obsession with material things, and through their writing they promoted their holistic view of nature. The transcendentalist worldview resembled that of the Mirrar in some respects. Both traditions identify a need to experience wild nature, and both view natural entities as symbols or messengers of some deeper truth. Although Thoreau viewed nature as

Table 2.1 Early Environmental Organizations in 19th-century Great Britain		
Organization	**Year established**	**Purpose**
Scottish Rights of Way Society	1843	Protect walking paths in and near cities
Commons Preservation Society	1865	Preserve forests and other landscapes
Society for the Protection of Ancient Buildings	1877	Protect the built environment, especially historic buildings
Selborne League	1885	Protect rare birds, plants, and landscapes
Coal Smoke Abatement Society	1898	Improve urban air quality

divine, he also observed the natural world closely and came to understand it in the manner of a scientist; he was in many ways one of the first ecologists. His book *Walden,* in which he recorded his observations and thoughts while he lived at Walden Pond away from the bustle of urban Massachusetts, remains a classic of American literature.

Conservation and preservation arose around the turn of the 20th century

One admirer of Emerson and Thoreau was **John Muir** (1838–1914), a Scottish immigrant to the United States who eventually settled in California and made the Yosemite Valley his wilderness home. Although Muir chose to live in isolation in his beloved Sierra Nevada for long stretches of time, he nonetheless became politically active and won fame as a tireless advocate for the preservation of wilderness (Figure 2.5). Muir was motivated by the rapid deforestation and environmental degradation he witnessed throughout North America and by his belief that the natural world should be treated with the same respect that cathedrals receive. Today he is associated with the **preservation** ethic, which holds that we should protect the natural environment in a pristine, unaltered state. Muir argued that nature deserved protection for its own inherent value (an ecocentrist argument), but he also maintained that nature played a large role in human happiness and fulfillment (an anthropocentrist argument). "Everybody needs beauty as well as bread," he wrote in 1912, "Places to play in and pray in, where nature may heal and give strength to body and soul alike."

Some of the same factors that motivated Muir also inspired the first professionally trained American forester, **Gifford Pinchot** (1865–1946; Figure 2.6). Both men opposed the rapid deforestation and unregulated economic development of North American lands that occurred during their lifetimes. However, Pinchot took a more anthropocentric view of how and why nature should be valued. He is today the person most closely associated with the **conservation** ethic, which holds that humans should put

FIGURE 2.5 A pioneering advocate of the preservation ethic, John Muir is also remembered for his efforts to protect the Sierra Nevada from development and for his role in founding the Sierra Club, a leading environmental organization. Here Muir (right) is shown with President Theodore Roosevelt in Yosemite National Park. After his 1903 wilderness camping trip with Muir, the president instructed his interior secretary to increase protected areas in the Sierra Nevada.

natural resources to use but also that we have a responsibility to manage them wisely. Whereas preservation aims to preserve nature for its own sake and for the aesthetic, spiritual, symbolic, and recreational benefit of people, conservation promotes the prudent, efficient, and sustainable extraction and use of natural resources for the benefit of present and future generations. The conservation ethic uses a utilitarian standard, stating that in using resources, humans should attempt to provide the greatest good to the greatest number of people for the longest time.

Pinchot and Muir came to represent different branches of the American environmental movement, and their contrasting ethical approaches often pitted them against

FIGURE 2.6 Gifford Pinchot, the first chief of what would become the U.S. Forest Service, was a leading proponent of the conservation ethic. The conservation ethic holds that humans should use natural resources, but strive to ensure the greatest good for the greatest number for the longest time.

one another on policy issues of the day. Nonetheless, they both represented reactions against a prevailing "development ethic," which holds that humans are and should be masters of nature and which promotes economic development without regard to its negative consequences. Pinchot eventually founded what would become the U.S. Forest Service and served as its chief in Theodore Roosevelt's administration. Both Pinchot and Muir left legacies that reverberate today in the different ethical approaches to environmentalism.

Weighing the Issues:
Preservation and Conservation

With which ethic do you most identify—preservation or conservation? Think of a forest or other important natural resource in your region. Give an example of a situation in which you might adopt a preservation ethic and an example of one in which you might adopt a conservation ethic. Are there conditions under which you'd follow neither, but instead adopt a "development ethic"?

Aldo Leopold's land ethic arose from the conservation and preservation ethics

As a young forester and wildlife manager, **Aldo Leopold** (1887–1949; Figure 2.7) began his career fully in the conservationist camp, having graduated from Yale Forestry

School, which Pinchot had helped found just as Roosevelt and Pinchot were advancing conservation on the national stage. As a forest manager in Arizona and New Mexico, Leopold embraced the government policy of shooting predators, such as wolves, to increase populations of deer and other game animals. At the same time, Leopold followed the development of ecological science. He eventually ceased to view certain species as "good" or "bad" and instead came to see that healthy ecological systems depend on the protection of all their interacting parts, including predators as well as prey. Drawing an analogy to mechanical maintenance, he wrote, "to keep every cog and wheel is the first precaution of intelligent tinkering."

It was more than science that pulled Leopold from an anthropocentric perspective toward a more holistic one. One day he shot a wolf, and when he reached the animal, Leopold was transfixed by "a fierce green fire dying in her eyes." The experience remained with him for the rest of his life and helped lead him to a more ecocentric ethical outlook. Years later, as a University of Wisconsin professor,

FIGURE 2.7 Aldo Leopold, a wildlife manager and pioneering environmental philosopher, articulated a new relationship between people and the environment. In his essay "The Land Ethic," he called on people to include the environment in their ethical framework.

Leopold argued that humans should view themselves and "the land" as members of the same community, and that people are obliged to treat the land in an ethical manner. In his 1949 essay "The Land Ethic," he wrote:

> All ethics so far evolved rest upon a single premise: that the individual is a member of a community of interdependent parts. . . . The land ethic simply enlarges the boundaries of the community to include soils, waters, plants, and animals, or collectively: the land. . . . A land ethic changes the role of *Homo sapiens* from conqueror of the land-community to plain member and citizen of it. . . . It implies respect for his fellow-members, and also respect for the community as such.

Leopold intended that the land ethic would help guide decision making. "A thing is right," he wrote, "when it tends to preserve the integrity, stability, and beauty of the biotic community. It is wrong when it tends otherwise." Leopold died before seeing "The Land Ethic" and his best-known book, *A Sand County Almanac*, in print, but today many view him as the most eloquent and important philosopher of environmental ethics.

Deep ecology extends environmental ethics

One philosophical perspective that goes beyond even Leopold's ecocentrism is **deep ecology,** established in the 1970s. Proponents of deep ecology describe the movement as resting on principles of "self-realization" and biocentric equality. They define self-realization as the awareness that humans are inseparable from nature and that the air we breathe, the water we drink, and the foods we consume are both products of the environment and integral parts of us. Biocentric equality is the precept that all living beings have equal value and that because we are truly inseparable from our environment, we should protect all other living things as we would protect ourselves.

Ecofeminism equates male attitudes toward nature and toward women

As deep ecology and mainstream environmentalism were extending people's ethical domains outward during the 1960s and 1970s, major social movements, such as the civil rights movement and the feminist movement, were gaining prominence. A number of feminist scholars saw parallels in human behavior toward nature and men's behavior toward women. The degradation of nature and the social oppression of women shared common roots, these scholars asserted.

Ecological feminism, or **ecofeminism,** argues that the patriarchal (male-dominated) structure of society—which traditionally grants more power and prestige to men than to women—is a root cause of both social and environmental problems. Ecofeminists hold that a worldview traditionally associated with women, which interprets the world in terms of interrelationships and cooperation, is more compatible with nature than a worldview traditionally associated with men, which interprets the world in terms of hierarchies and competition. Ecofeminists maintain that a male tendency to try to dominate and conquer what men hate, fear, or do not understand has historically been exercised against both women and the natural environment.

Environmental justice seeks equal treatment for all races and classes

Our society's domain of ethical concern has been expanding from rich to poor and from majority races and ethnic groups to minority ones. This ethical expansion involves applying a standard of fairness and equality and has given rise to the environmental justice movement. The U.S. Environmental Protection Agency (EPA) defines **environmental justice** as "the fair treatment and meaningful involvement of all people regardless of race, color, national origin, or income with respect to the development, implementation, and enforcement of environmental laws, regulations, and policies."

The environmental justice movement was fueled by the perception that poor people and minorities tend to be exposed to a greater share of pollution, hazards, and environmental degradation than are richer people and whites. A protest in the early 1980s by African Americans in Warren County, North Carolina, against a toxic waste dump in their community is widely seen as the beginning of the movement (Figure 2.8). The state had chosen to site the dump in the county with the highest percentage of African Americans, prompting Warren County residents to suspect "environmental racism." Environmental justice grew to prominence in the early 1990s as more people across North America began fighting environmental hazards in their communities. This movement—in contrast to earlier environmental movements—was made up largely of low-income people and minorities.

In 1983, a U.S. General Accounting Office (GAO) study found that three of four toxic waste landfills in the southeastern United States were located in communities where the population of minorities exceeded that of

FIGURE 2.8 Communities of poor people and people of color have suffered more than their share of environmental problems, a situation that has given rise to the environmental justice movement. The movement gained prominence with this protest of a toxic waste dump in Warren County, North Carolina.

whites, and the fourth landfill was located in a community that was 38% African American. In contrast, minorities made up only 20% of the region's population. In 1987, the United Church of Christ Commission for Racial Justice found that the percentage of minorities in areas with toxic waste sites was twice that of areas without toxic waste sites. Researchers studying air pollution, lead poisoning, pesticide exposure, and workplace hazards have found similar patterns. Today the environmental justice movement has broadened to encompass equity in transportation options, redevelopment of abandoned urban sites, worker health and safety, and access to parklands.

Weighing the Issues:
Environmental Justice

Consider the place where you grew up. Where were the factories, waste dumps, and polluting facilities located, and who lived closest to them? Who lives nearest them in the town or city that hosts your campus? Do you think the concerns of environmental justice advocates are justified? If so, what could be done to ensure that poor communities are no more polluted than wealthy ones?

The attempts of the predominantly white Australian government and uranium mining companies to open mines on traditional lands of the Mirrar (Figure 2.9) have been characterized by critics as violations of environmental

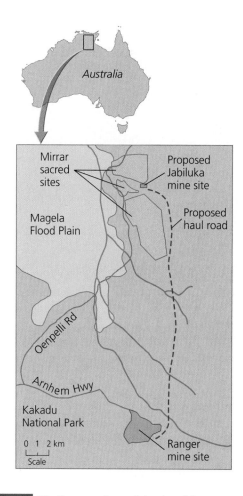

FIGURE 2.9 The Ranger mine and the site of the proposed Jabiluka mine lie amid Aboriginal lands and sacred sites.

justice. In North America, uranium mining has been a focus of environmental justice concerns as well. Native Americans of the Dene Nation, in Canada's Northwest Territories and Saskatchewan, have suffered health effects from working in uranium mines with minimal safeguards. In the southwestern United States from 1948 through the late 1960s, uranium mines employed many Native Americans, including those of the Navajo nation. Although uranium mining had already been linked to health problems and premature death, the miners had practically no awareness of radiation and its risks. The Navajo language did not even have a word for *radiation,* and for nearly two decades neither the mining industry nor the U.S. government provided information or safeguards to the miners. Many Navajo families built homes and bread-baking ovens from the abundant waste rock produced as a by-product of the mining process, not realizing it was radioactive (Figure 2.10).

Cases of lung cancer began to appear among Navajo miners in the early 1960s, but scientific studies of radiation's

FIGURE 2.10 Native Americans employed as uranium miners in Canada and the United States, such as the Navajo miner shown here, have suffered from adverse effects of mining.

effects on miners at the time excluded Native American workers. The decision to include only white miners in the studies was attributed to the researchers' desire to study a "homogeneous population." A later generation of Americans perceived this as negligence and discrimination, and their desire for justice gave rise to the Radiation Exposure Compensation Act of 1990, a federal law that compensated Navajo miners who suffered health effects from unprotected work in the mines. Such developments illustrate the interplay between changing ethical values and resultant policymaking, which we will examine further in Chapter 3. First, we will explore economics, which, like ethics, addresses people's values and widely informs policy.

Economics: Approaches and Environmental Implications

People who oppose the Jabiluka mine do so largely on the basis of ethical concerns and worries over environmental impacts. Few have challenged the mining plan on economic grounds; mine opponents generally recognize uranium as a lucrative resource that generates jobs, income, and electricity, and they do not dispute the contribution of uranium exports to the Australian economy. Support for the mine is based primarily on economic factors. Such

conflict between ethical and economic motivations is a recurrent theme in environmental issues worldwide.

Is there a trade-off between economics and the environment?

Although measures to safeguard the environment may frequently mesh well with ethical considerations, we often hear it said that environmental protection works in opposition to economic health. But is this necessarily the case? Growing numbers of economists assert that there need be no such trade-off—that in fact, environmental protection can be *good* for the economy. As we will see, the view one takes often depends on whether one thinks in the short term or the long term and whether one holds to traditional economic schools of thought or to newer ones that view human economies as coupled to the natural environment.

Economics studies the allocation of scarce resources

Like ethics, economics examines factors that guide human behavior. **Economics** is the study of how people decide to use scarce resources to provide goods and services in the face of demand for them. By this definition, environmental problems are also economic problems that can intensify as population and per capita resource consumption increase. For example, pollution may be viewed as depletion of the scarce resources of clean air, water, or soil. Indeed, the word *economics* and the word *ecology* come from the same Greek root, *oikos,* meaning "household." In its broadest context, the human "household" is Earth itself. Economists traditionally have studied the household of human society, and ecologists the broader household of all life.

Several types of economies exist today

An **economy** is a social system that converts resources into **goods,** material commodities manufactured for and bought by individuals and businesses; and **services,** work done for others as a form of business. The oldest type of economy is the **subsistence economy.** People in subsistence economies—who still comprise much of the human population—meet most or all of their daily needs directly from nature and do not purchase or trade for most of life's necessities.

A second type of economy is the **capitalist market economy.** In this system, buyers and sellers interact to determine which goods and services to produce, how much to produce, and how these should be produced and distributed. Capitalist economies are often contrasted with state socialist economies, or **centrally planned economies,** in which government determines in a top-down manner how to allocate resources. In today's world, capitalism predominates over socialism. However, a pure market economy would operate without government intervention. In reality, all capitalist market economies today, including that of the United States, have borrowed much from state socialism and are in fact hybrid systems. In modern market economies, as we will see in Chapter 3, governments typically intervene for several reasons: (1) to eliminate unfair advantages held by single buyers or sellers; (2) to provide social services, such as national defense, medical care, and education; (3) to provide "safety nets" (for the elderly, victims of natural disasters, and so on); (4) to manage the commons (▸ p. 6); and (5) to mitigate pollution.

Many societies function somewhere along a continuum between a pure subsistence economy and a capitalist market economy. For example, the Mirrar acquire essential food and water directly from their environment, but they purchase many other necessities.

Environment and economy are intricately linked

All human economies exist within the larger environment and depend on it in important ways. Economies receive inputs from the environment, process them in complex ways that enable human society to function, then discharge outputs of waste from this process into the environment. Economies are thus *open systems* (▸ p. 189) integrated with the larger environmental system of which they are a part. Earth, in turn, is a materially *closed system,* so the inputs Earth can provide to economies are ultimately limited.

Although these interactions between human economies and the nonhuman environment are readily apparent, traditional economic schools of thought have long overlooked the importance of these connections. Indeed, most conventional economists today still adhere to a worldview that largely ignores the environment (Figure 2.11a), and this worldview continues to drive most policy decisions. However, modern economists belonging to the fast-growing fields of environmental economics and ecological economics (▸ p. 45) explicitly accept that human economies are subsets of the environment and depend crucially on the environment (Figure 2.11b).

Economic activity uses resources from the environment. Natural resources (▸ p. 4) are the various substances and forces we need to survive: the sun's energy, the fresh water we drink, the trees that provide our lumber, the rocks that provide our metals, and the fossil fuels that power our machines and produce our plastics. We can think of natural resources as "goods" produced by nature. Without Earth's natural resources, there would be no human economies and, in fact, no human beings.

Environmental systems also naturally function in a manner that supports economies. Earth's ecological systems purify air and water, cycle nutrients, provide for the pollination of plants by animals, and serve as receptacles and recycling systems for the waste generated by our economic activity. Such essential services, often called **ecosystem services** (Table 2.2), support the life that makes our economic activity possible. Some ecosystem services represent the very nuts-and-bolts of our survival, and others enhance our quality of life.

While the environment enables economic activity by providing ecosystem goods and services, economic activity can affect the environment in return. When we deplete natural resources and produce too much pollution, we can degrade the ability of ecological systems to function. In fact, the Millennium Ecosystem Assessment (▸ p. 20) concluded in 2005 that 15 of 24 ecosystem services its scientists surveyed globally were being degraded or used unsustainably. The degradation of ecosystem services can in turn negatively affect economies. Currently, ecological degradation is harming poor people more than wealthy people, the Millennium Ecosystem Assessment found. As a result, restoring ecosystem services stands as a prime avenue for alleviating poverty in much of the world. These interrelationships among economic and environmental conditions have only recently become widely recognized, however. Let's briefly examine how economic thought has changed over the years, tracing the path that is now beginning to lead economies to become more compatible with natural systems.

Adam Smith and other philosophers founded classical economics

Economics shares a common intellectual heritage with ethics, and practitioners of both have long been interested in the relationship between individual action and societal well-being. Some philosophers argued that individuals acting in their own self-interest would harm society. Others believed that such behavior could benefit society, as long as the behavior was constrained by the rule of law and private property rights and operated within fairly competitive markets. The latter view was articulated by Scottish philosopher **Adam Smith** (1723–1790). Known

FIGURE 2.11 Modern environmental and ecological economists view economic activity differently from economists of the more conventional neoclassical school. Standard neoclassical economics focuses on processes of production and consumption between households and businesses (**a**), viewing the environment only as a "factor of production" that helps enable the production of goods. Environmental and ecological economists view the human economy as existing within the natural environment (**b**), receiving resources from it, discharging waste into it, and interacting with it through various ecosystem services.

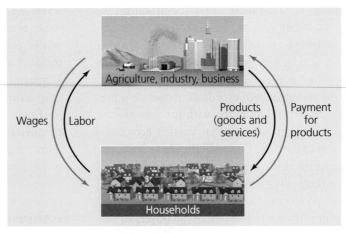

(a) Conventional view of economic activity

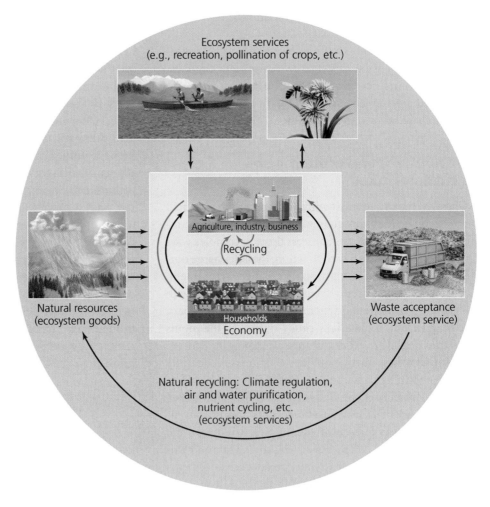

(b) Economic activity as viewed by environmental and ecological economists

Table 2.2 Ecosystem Services	
Type of ecosystem service*	**Example(s)**
Regulating atmospheric gases	Maintaining the ozone layer; balancing oxygen, carbon dioxide, and other gases
Regulating climate	Controlling global temperature and precipitation through oceanic and atmospheric currents, greenhouse gases, cloud formation, and so on
Damping impacts from disturbance	Providing storm protection, flood control, and drought recovery, mainly through vegetation structure
Regulating water flow	Providing water for agriculture, industry, transportation
Storing and retaining water	Providing water through watersheds, reservoirs, aquifers
Controlling erosion and promoting soil retention	Preventing soil loss from wind or runoff; storing silt in lakes and wetlands
Forming soil	Weathering rock; accumulating organic material
Cycling nutrients	Cycling carbon, nitrogen, phosphorus, sulfur, and other nutrients through ecosystems
Treating waste	Removing toxins, recovering nutrients, controlling pollution
Pollinating plants	Transporting floral gametes by wind or pollinating animals, enabling crops and wild plants to reproduce
Controlling populations biologically	Controlling prey with predators; controlling hosts with parasites; controlling herbivory on crops with predators and parasites
Providing habitat	Providing ecological settings in which creatures can breed, feed, rest, migrate, winter
Providing food	Producing fish, game, crops, nuts, and fruits that humans obtain by hunting, gathering, fishing, subsistence farming
Supplying raw materials	Producing lumber, fuel, metals, fodder
Furnishing genetic resources	Providing unique biological sources for medicine, materials science, genes for resistance to plant pathogens and crop pests, ornamental species (pets and horticultural plant varieties)
Providing recreational opportunities	Ecotourism, sport-fishing, hiking, birding, kayaking, other outdoor recreation
Providing cultural or noncommercial uses and goods	Aesthetic, artistic, educational, spiritual, and/or scientific values of ecosystems

*Ecosystem "goods" are here included in ecosystem services.
Adapted with permission from Costanza, R., et al. 1997. The value of the world's ecosystem services and natural capital. *Nature* 387: 253–260.

today as the father of **classical economics,** Smith believed that when people are free to pursue their own economic self-interest in a competitive marketplace, the marketplace will behave as if guided by "an invisible hand" that ensures their actions will benefit society as a whole. In his 1776 book *Inquiry into the Nature and Causes of the Wealth of Nations*, Smith wrote:

> It is not from the benevolence of the butcher, the brewer, or the baker that we expect our dinner, but from their regard to their own self-interest. [Each individual] intends only his own security, only his own gain. And he is led in this by an invisible hand to promote an end which was no part of his intention. By pursuing his own interests he frequently promotes that of society more effectually than when he really intends to.

Smith's philosophy remains a pillar of free-market thought today, and many credit it for the tremendous gains in material prosperity that industrialized nations have experienced in the past few centuries. The *laissez-faire* policies that free-market thought has spawned, however, have also been widely criticized by those who feel that market capitalism exacerbates inequalities between rich and poor and contributes to environmental degradation. Market capitalism, these critics assert, should be constrained and regulated by democratic government.

Neoclassical economics incorporates human psychology

Economists subsequently took more quantitative approaches and incorporated human psychology into their work. Modern **neoclassical economics** examines the psychological factors underlying consumer choices, explaining market prices in terms of consumer preferences for units of particular commodities. In neoclassical economic theory, buyers desire the lowest possible price, whereas sellers desire the highest possible price. This conflict between buyers and sellers results in a compromise price being reached and the "right" quantity of commodities being bought and sold. This is often phrased in terms of *supply,* the amount of a product offered for sale at a given price, and *demand,* the amount of a product people will buy at a given price if free to do so. Theoretically, when prices go up, demand drops and supply increases; and when prices fall, demand rises and supply decreases. The market automatically moves toward an equilibrium point, a price at which supply equals demand (Figure 2.12a). Similar reasoning can be applied to environmental issues, such that economists can determine "optimal" levels of resource use or pollution control (Figure 2.12b).

Cost–benefit analysis is a widespread tool

A method commonly used by neoclassical economists is **cost–benefit analysis.** In this approach, estimated costs for a proposed action are totaled up and compared to the sum of benefits estimated to result from the action. If total benefits exceed costs, the action should be pursued; if costs exceed benefits, it should not. When choosing among multiple alternative actions, the one with the greatest excess of benefits over costs should be chosen. This reasoning seems eminently logical, and because the analysis aims to be quantitative and precise, it would seem likely to result in clear recommendations for policy.

However, problems often crop up because not all costs and benefits are easily quantified, or even identified or defined. It may be quite feasible to quantify wages paid to uranium miners, the market value of uranium extracted from a mine, or the cost of measures to minimize health risks for miners. But it is difficult to assess the cost of a valued landscape being scarred by mine development, or the cost of radioactive contamination of a stream. Because some costs and benefits cannot easily be assigned monetary values—and because it is difficult to identify and agree on all costs and benefits—cost–benefit analysis is often controversial. Moreover, because economic benefits are usually more easily quantified than environmental costs, economic benefits tend to be overrepresented in traditional cost–benefit analyses. As a result, environmental advocates often feel these analyses are biased in favor of economic development and against environmental protection.

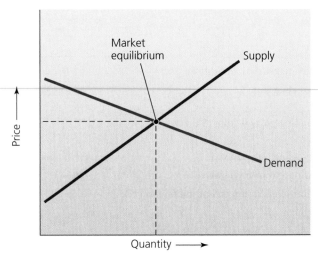

(a) Classic supply–demand curve

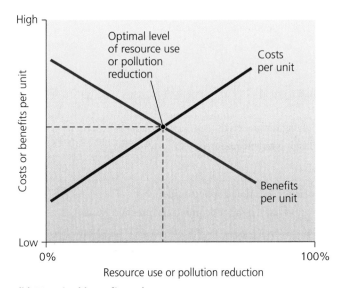

(b) Marginal benefit and cost curves

FIGURE 2.12 This basic supply-and-demand curve (**a**) illustrates the relationship between supply, demand, and market equilibrium, the "balance point" at which demand is equal to supply. We can use a similar graph (**b**) to determine an "optimal" level of resource use or pollution mitigation. In this graph, the cost per unit of resource use or pollution cleanup (blue line) rises as the resource use or pollution cleanup proceeds and it becomes expensive to extract or clean up the remaining amounts. Meanwhile, the benefits per unit of resource use or pollution cleanup (red line) decrease. The point where the lines intersect gives the optimal level.

Aspects of neoclassical economics have profound implications for the environment

Today's capitalist market systems operate largely in accord with the precepts of neoclassical economics. These systems have generated unprecedented material wealth, employment, and other desirable outcomes, but they have also contributed to environmental problems. Four fundamental assumptions of neoclassical economics have implications for the environment:

▶ Resources are infinite or substitutable.
▶ Long-term effects should be discounted.
▶ Costs and benefits are internal.
▶ Growth is good.

Resources are infinite or substitutable Neoclassical economic models generally treat workers and other resources as being either infinite or largely "substitutable and interchangeable." This implies that once we have depleted a resource—natural, human, or otherwise—we should be able to find a replacement for it. Human resources can substitute for financial resources, for instance, or manufactured resources can substitute for natural resources. Theory allows that the substituted resource may be less efficient or more costly, but some degree of substitutability is generally assumed.

Certainly it is true that many resources can be replaced; our societies have transitioned from manual labor to animal labor to steam-driven power to fossil-fuel power, and they may yet transition to renewable power sources, such as solar energy. However, Earth's material resources are ultimately limited. Nonrenewable resources such as fossil fuels can be depleted, and many renewable resources can be used up as well if we exploit them faster than they can be replenished. This is what happened to the Easter Islanders (▶ pp. 8–9) who harvested their forests faster than the forests could regrow.

Weighing the Issues:
Substitutability and the Environment

Can you think of a natural resource that might be difficult to replace with a substitute? What problems might arise from the assumption that all resources, including clean air and water, are substitutable and interchangeable? Can you think of examples that violate any of the other three assumptions of neoclassical economics? Explain your answers.

Long-term effects should be discounted Although few people would dispute that resources are *ultimately* limited, many assume that their depletion will take place so far in the future that there is no need for current generations to worry. For economists in the neoclassical tradition, an event far in the future counts much less than one in the present; in economic terminology, future effects are "discounted." In discounting, short-term costs and benefits are granted more importance than long-term costs and benefits, causing policy to play down long-term consequences of decisions we make today. Some governments and businesses use a 10% annual discount rate for decisions on resource use. This means that the long-term value of a stand of ancient trees worth $500,000 would drop by 10% each year and, after 10 years of discounting, would be worth only $174,339.22. By this logic, the more quickly the trees are cut, the more they are worth.

Costs and benefits are internal A third assumption of neoclassical economics is that all costs and benefits associated with a particular exchange of goods or services are borne by individuals engaging directly in the transaction. In other words, it is assumed that the costs and benefits of a transaction are "internal" to the transaction, experienced by the buyer and seller alone, and do not affect other members of society.

However, in many situations this is simply not the case. Pollution from a factory can harm people living nearby. In such cases, someone—often taxpayers not involved in producing the pollution—ends up paying the costs of alleviating it. Market prices do not take the social, environmental, or economic costs of this pollution into account. Costs or benefits of a transaction that involve people other than the buyer or seller are known as **externalities.** A positive externality is a benefit enjoyed by someone not involved in a transaction, and a negative externality, or **external cost,** is a cost borne by someone not involved in a transaction (Figure 2.13). Negative externalities often harm groups of people or society as a whole, while allowing certain individuals private gain. External costs commonly include the following:

▶ Human health problems
▶ Property damage
▶ Declines in desirable elements of the environment, such as fewer fish in a stream
▶ Aesthetic damage, such as that resulting from air pollution or clear-cutting
▶ Stress and anxiety experienced by people downstream or downwind from a pollution source
▶ Declining real estate values resulting from these problems

The Mirrar have experienced external costs in the form of pollution from the Ranger mine. In March 2002, the Australian Broadcasting Corporation reported that radioactive material from the Ranger mine had contaminated a stream on Mirrar land with uranium concentrations

FIGURE 2.13 An Indonesian boy wading in a polluted river suffers external costs: costs that are not borne by the buyer or seller. External costs may include water pollution, aesthetic harm, human health problems, property damage, harm to aquatic life, aesthetic degradation, declining real estate values, and other impacts.

4,000 times higher than allowed by law. According to an Aboriginal representative, this was the fourth such violation in less than three months.

By ignoring external costs, economies create a false idea of the true and complete costs of particular choices and unjustly subject people to the consequences of transactions in which they did not participate. External costs comprise one reason governments develop environmental legislation and regulations (▶ p. 63). Unfortunately, external costs are difficult to account for and eliminate. It is tough to assign a monetary value to illness, premature death, or degradation of an aesthetically or spiritually significant site.

Growth is good A fourth assumption of the neoclassical economic approach is that economic growth is required to keep employment high and maintain social order. The argument goes something like this: If the poor view the wealthy as the source of their suffering, they may revolt. Promoting economic growth can defuse this situation by creating opportunities for the poor to become wealthier themselves. By making the overall economic pie larger, everyone's slice becomes larger, even if some people still have much smaller slices than others.

The idea that economic growth is good has been encouraged over the centuries by the concept of material progress, espoused by Western cultures since the Enlightenment. The modern-day United States may represent the clearest example of the worldview that "more and bigger" is always better. We hear constantly in business news of increases in an industry's output or percentage growth in a country's economy, with increases touted as good news and decreases, stability, or even a minor drop in the *rate* of growth presented as bad news. Economic growth has become the quantitative yardstick by which progress is measured.

Is the growth paradigm good for us?

The rate of economic growth in recent decades is unprecedented in human history, and the world economy is seven times the size it was half a century ago. All measures of economic activity—trade, rates of production, amount and value of goods manufactured—are higher than they have ever been and are still increasing. This growth has brought many people much greater material wealth (although not equitably, and gaps between rich and poor remain immense).

To the extent that economic growth is a means to an end—a tool with which we can achieve greater human happiness—it can be a good thing. However, many observers today worry that growth has become an end in itself and is no longer necessarily the best tool with which to pursue happiness. Critics of the growth paradigm often note that runaway growth resembles the multiplication of cancer cells, which eventually overwhelm and destroy the organism in which they grow. These critics fear that runaway economic growth will likewise destroy the economic system on which we all depend. Resources for growth are ultimately limited, they argue, so nonstop growth is not sustainable and will fail as a long-term strategy.

Defenders of traditional economic approaches reply that Cassandras (▶ p. 22) have been saying for decades that limited resources would doom growth-oriented economies, yet most of these economies are still expanding dramatically. Thomas Malthus (▶ pp. 4–5) warned two centuries ago that increasing populations would bump into limits on resources. British classical economist David Ricardo (1772–1823) reasoned that as more high-quality farmland became occupied, profits from the lower-quality farmland that was left would progressively decrease, eventually halting growth. German philosopher Karl Marx (1818–1883) held that limits on economic growth would occur in the form of working-class revolutions against the capitalist leisure class. So why then, 150 to 200 years after such predictions were made, are we witnessing the most rapid growth of material wealth in human history?

FIGURE 2.14 Advances in technology have enabled people to push back the natural limits on growth and continue expanding their economies. Innovations in machinery for petroleum extraction, shown here, have allowed us to extract more fossil fuels than ever before and to push our economic productivity far beyond what was possible a century or two ago.

One prime reason is technological innovation. In case after case, improved technology has enabled us to push back the limits on growth. More powerful technology for extracting minerals, fossil fuels, and groundwater has expanded the amounts of these natural resources available to us (Figure 2.14). Technological developments such as automated farm machinery, fertilizers, and chemical pesticides have allowed us to grow more food per unit area of land, boosting agricultural output. Faster, more powerful machines in our factories have enabled us to translate our enhanced resource extraction and agricultural production into faster rates of manufacturing. Such increases in scale and efficiency explain why many of the dire predictions of those concerned about resource limitations have only partly come to pass.

Economists disagree on whether economic growth is sustainable

Can we conclude, then, that endless improvements in technology are possible and that we will never run into shortages of resources? At one end of the spectrum are those Cornucopians (▸ p. 21), economists, businesspeople, and policymakers who believe technology can solve everything—a philosophy that has greatly influenced economic policy in market economies over the past century.

At the other end of the spectrum, **ecological economists** argue that a couple of centuries is not a very long

period of time and that history suggests that civilizations do not, in the long run, overcome their environmental limitations. Ecological economics, which has emerged as a discipline only in the past decade or two, applies principles of ecology and systems science (Chapters 5 through 7) to the analysis of economic systems. Earth's natural systems generally operate in self-renewing cycles, not in a linear or progressive manner. Ecological economists advocate sustainability in economies and see natural systems as good models. To evaluate an economy's sustainability, ecological economists take a long-term perspective and ask, "Could we continue this activity forever and be happy with the outcome?" Most ecological economists argue that the growth paradigm will eventually fail and that if nothing is done to rein in population growth and resource consumption, depleted natural systems could plunge our economies into ruin. Many of these economists advocate economies that do not grow and do not shrink, but rather are stable. Such **steady-state economies** are intended to mirror natural ecological systems.

In the middle of the spectrum are **environmental economists.** They tend to agree with ecological economists that economies are unsustainable if population growth is not reduced and resource use is not made more efficient. Environmental economists, however, maintain that we can accomplish these changes and attain sustainability within our current economic systems. By retaining the principles of neoclassical economics but modifying them to address environmental challenges, environmental economists argue that we can keep our economies growing and that technology can continue to improve efficiency. Environmental economists were the first to develop ways to tackle the problems of external costs and discounting. In the 1950s and 1960s, a nonprofit, nonpartisan organization called Resources for the Future pioneered ways to weigh the true costs and benefits associated with resource use. These environmental economists blazed a trail, and ecological economists then went further, proposing that sustainability requires far-reaching changes leading to a steady-state economy. Thus, whereas environmental economists implemented reform, ecological economists call for revolution.

A steady-state economy is a revolutionary alternative to growth

The idea of a steady-state economy did not originate with the rise of ecological economics. Back in the 19th century, British economist **John Stuart Mill** (1806–1873) hypothesized

Gross Domestic Product versus Genuine Progress Indicator

The Science behind the Story

For decades, economists have assessed the economic robustness of a nation by calculating its **Gross Domestic Product (GDP),** the total monetary value of final goods and services produced in the country each year.

However, there are problems with using this measure of economic activity to represent a nation's economic well-being. For one, GDP does not account for the nonmarket values we discuss in this chapter. For another, GDP is not necessarily an expression of *desirable* economic activity. In fact, GDP can increase whether the economic activities driving it help or hurt the environment or society. For example, a large oil spill in a U.S. national park could increase the U.S. GDP because oil spills require cleanups, which cost money and, as a result, increase the production of goods and services. A radiation leak at a uranium mine on Aboriginal homelands would likely add to the Australian GDP because of the many monetary transactions required for cleanup and medical care.

Some economists have attempted to develop economic indicators that differentiate between desirable and undesirable economic activity. Such indicators can function as more accurate guides to nations' welfare. One alternative to the GDP is the **Genuine Progress Indicator (GPI),** introduced in 1995 by Redefining Progress, a nonprofit organization that develops economic and policy tools to promote accurate market prices and sustainability. The GPI has not yet gained widespread acceptance, but it has generated a

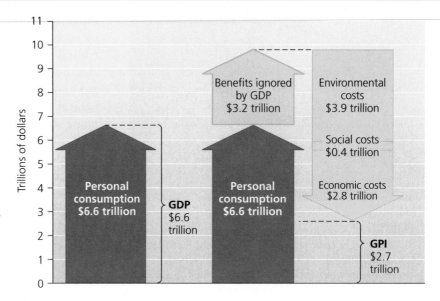

The Gross Domestic Product (GDP; red arrow) sums together all economic activity, whether good or bad. The Genuine Progress Indicator (GPI) adds to the GDP benefits such as volunteering and parenting (upward pointing gold arrow). The GPI also subtracts external environmental costs such as pollution, social costs such as divorce and crime, and economic costs such as borrowing and the gap between rich and poor (downward pointing gold arrow). Shown are values for GDP and GPI for the United States in the year 2002. Data from Venetoulis, J. and C. Cobb. 2004. *The genuine progress indicator, 1950–2002 (2004 update).* Redefining Progress.

that as resources became harder to find and extract, economic growth would slow and eventually stabilize. Economies would carry on in a state in which individuals and society subsist on steady flows of natural resources and on savings accrued during occasional productive but finite periods of growth. Such a model appears to match the economies of Australian Aborigines and many other traditional societies throughout the world before the global expansion of European culture.

Modern proponents of a steady-state global economy, such as American economist Herman Daly, are not so sanguine that a steady state will evolve on its own from a capitalist market system. Instead, most believe we will need to rethink our assumptions and fundamentally change the way we conduct economic transactions if we are to achieve a steady state.

Those resisting such notions often assume that an end to growth will mean an end to a rising quality of life. Ecological

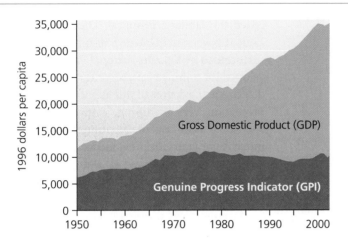

Although the GDP of the United States has increased dramatically since 1950, the GPI has risen more slowly. GPI advocates suggest that this discrepancy means that we are spending more money than ever but that our lives are not that much better. Data from Venetoulis, J. and C. Cobb. 2004. *The genuine progress indicator, 1950–2002 (2004 update)*. Redefining Progress.

great deal of discussion that has drawn attention to the weaknesses of the GDP.

To calculate GPI, economists begin with conventional economic activity and then add to it all those positive contributions to the economy that do not have to be paid for with money, such as volunteer work and parenting. They then subtract negative impacts, such as crime, pollution, gaps between rich and poor, and other detrimental social, environmental, and economic factors (see the first figure). The GPI thereby summarizes many more forms of economic activity than does the GDP, and it differentiates between economic activity that increases societal well-being and economic activity that decreases it.

Thus, whereas the GDP increases when fossil fuel use increases, the GPI declines because of the adverse environmental and social impacts of such consumption, including air and water pollution, increased road congestion and traffic accidents, and global climate change. The second figure compares changes in per capita GPI and GDP in the United States from 1950 to 2002. The country's GDP has increased greatly as a result of increased economic activity. Its GPI has also increased, but not nearly as rapidly.

The GPI is not the only alternative to the GDP. The United Nations uses a tool of its own called the Human Development Index, which is calculated by assessing a nation's standard of living, life expectancy, and education. Another alternative to the GDP is the Index of Sustainable Economic Welfare (ISEW), developed by ecological economists Herman Daly and John Cobb in 1989. The ISEW is based on income, wealth distribution, natural resource depletion, benefits associated with volunteerism, and environmental degradation. Any one of these indices, ecological economists maintain, should give a more accurate portrait of a nation's welfare than the GDP, which policymakers currently use so widely.

economists, however, argue that quality of life can continue to rise under a steady-state economy and, in fact, may be more likely to do so. Technological advances will not cease just because growth stabilizes, they argue, and neither will behavioral changes (such as greater use of recycling) that enhance sustainability. Instead, wealth and human happiness can continue to rise after economic growth has leveled off.

Attaining sustainability will certainly require the reforms that environmental economists advocate and may well require the fundamental shifts in thinking, values, and behavior that ecological economists say is necessary (see "The Science behind the Story," above). How can these goals be attained in a world whose economic policies are still largely swayed by a Cornucopian worldview that barely takes the environment into account? While keeping in mind that ecological and environmental economic approaches are still actively being developed, we will now survey a few strategies for sustainability that have been offered so far.

We can give ecosystem goods and services monetary values

As we have noted, economies receive from the environment vital resources and ecosystem services. However, any survey of environmental problems in the world today—deforestation, biodiversity loss, pollution, collapsed fisheries, climate change, and so on—makes it immediately apparent that our society often mistreats the very systems that keep it alive and healthy. Why is this? From the economist's perspective, humans exploit natural resources and systems in large part because the market assigns these entities no quantitative monetary value or, at best, assigns values that underestimate their true worth.

Think for a minute about the nature of some of these services. The aesthetic and recreational pleasure we obtain from natural landscapes, whether wildernesses or city parks, is something of real value. Yet this value is hard to quantify and appears in no traditional measures of economic worth. Or consider Earth's water cycle (▶ pp. 206–207), by which rain fills our reservoirs with drinking water, rivers give us hydropower and flush away our waste, and water evaporates, purifying itself of contaminants and readying itself to fall again as rain. This natural cycle is absolutely vital to our existence, yet because its value is not quantified, markets impose no financial penalties when we interfere with it. Ecosystem services are said to have **nonmarket values,** values not usually included in the price of a good or service (Table 2.3 and Figure 2.15). Because the market does not assign value to ecosystem services, debates such as that over the Jabiluka mine often involve comparing apples and oranges—in this case, the intangible cultural, ecological, and spiritual arguments of the Mirrar versus the hard numbers of mine proponents.

To resolve this dilemma, environmental and ecological economists have sought ways to assign values to ecosystem services. One technique, **contingent valuation,** uses surveys to determine how much people are willing to pay to protect a resource or to restore it after damage has been done. Such an exercise was conducted with a mining proposal in the Kakadu region in the early 1990s that preceded the Jabiluka proposal. The Kakadu Conservation Zone, a government-owned 50-km^2 (19-mi^2) plot of land surrounded by Kakadu National Park, was either to be developed for mining or preserved and added to the park. To determine the degree of public support for environmental protection versus mining, a government commission sponsored a contingent valuation study to determine how much Australian citizens valued keeping the Kakadu Conservation Zone preserved and undeveloped. Researchers interviewed 2,034 citizens, asking them how much money they would be willing to pay to stop mine development.

The interviewers presented two scenarios: (1) a "major impact" scenario based on predictions of environmentalists who held that mining would cause great harm, and (2) a "minor impact" scenario based on predictions of mining executives who held that development would have few downsides. After presenting both scenarios in detail, complete with photographs, the interviewers asked the respondents how much their households would pay if each scenario, in turn, was correct. Respondents on average said their households would pay $80 per year to prevent the minor-impact scenario and $143 per year to prevent the major-impact scenario. Multiplying these figures by the number of households in Australia (5.4 million at the time), the researchers found that preservation was "worth" $435 million annually to the Australian population under the minor-impact scenario, and $777 million under the major-impact scenario. Because both of these numbers exceeded the $102 million in annual economic benefits expected from mine development, the researchers concluded that preserving the land undeveloped was worth more than mining it.

Table 2.3 Values That Modern Market Economies Generally Do Not Address

Nonmarket value	Is the worth we ascribe to things . . .
Use value	that we use directly
Option value	that we do not use now but might use later
Aesthetic value	for their beauty or emotional appeal
Cultural value	that sustain or help define our culture
Scientific value	that may be the subject of scientific research
Educational value	that may teach us about ourselves and the world
Existence value	simply because they exist, even though we may never experience them directly (e.g., an endangered species in a far-off place)

(a) Existence values

(b) Use values

(c) Option values

(d) Aesthetic values

(e) Scientific values

(f) Educational values

(g) Cultural values

FIGURE 2.15 Accounting for nonmarket values such as those shown here may help us to make better environmental and economic decisions.

Because contingent valuation relies on survey questions and not actual expenditures, critics complain that in such cases people will volunteer idealistic (inflated) values rather than realistic ones, knowing that they will not actually have to pay the price they name. In part because of such concerns, the Australian government commission decided not to use the Kakadu contingent valuation study's results. (The mine was stopped, but as a result of Aboriginal opposition.)

Whereas contingent valuation measures people's *expressed* preferences, other methods aim to measure people's *revealed* preferences—preferences as revealed by data on actual behavior. For example, the amount of money, time, or effort people expend to travel to parks for recreation has been used to measure the value people place on parks. Economists have also analyzed housing prices, comparing homes with similar characteristics but different environmental settings, to infer the dollar value of landscapes, views, and peace and quiet. Another approach assigns environmental amenities value by measuring the cost required to restore natural systems that have been damaged or to mitigate harm from pollution.

In 1997 one research team reviewed studies using various valuation methods in an effort to calculate the overall global economic value of all the services that ecosystems provide (see "The Science behind the Story," ▶ pp. 50–51). The researchers came up with a figure of $33 trillion per year ($40 trillion in 2005 dollars)—greater than the combined gross domestic products of all nations in the world. A follow-up study in 2002 concluded that the economic benefits of preserving the world's remaining natural areas outweighed the benefits of exploiting them by a factor of 100 to 1.

Markets can fail

When they do not reflect the full costs and benefits of actions, markets are said to fail. **Market failure** occurs when markets do not take into account the environment's positive effects on economies (such as ecosystem services) or when they do not reflect the negative effects of economic activity on the environment or on people (external costs). Traditionally, market failure has been countered by government intervention. Governments

Calculating the Economic Value of Earth's Ecosystems

The Science behind the Story

To Robert Costanza, the problem was like an elephant in the living room, which economists had ignored for decades: Earth's ecosystems provide essential life-support services, including arable soil, waste treatment, clean water, and clean air. However, economists had failed to account for how much those services contribute economically to human welfare, as is routinely done for conventional goods and services.

So Costanza, an environmental economist at the University of Maryland, joined with 12 colleagues in 1996 at the National Center for Ecological Analysis and Synthesis in Santa Barbara, California, and combed the scientific literature. The team identified more than 100 studies that estimated the worth of such ecosystem services as water purification, greenhouse gas regulation, plant pollination, and pollution cleanup.

The studies the team reviewed had estimated values of particular ecosystem services in several ways. Methods such as contingent valuation had frequently been used because people clearly value such aspects of natural systems as biodiversity and aesthetics, even though no one pays actual money specifically for them. After poring over studies that examined the value of 17 services provided by oceans, forests, wetlands, and other ecosystems, Costanza and his colleagues synthesized the results to provide the first comprehensive quantitative estimate of the global value of ecosystem services.

To estimate the worth of the services more accurately, Costanza and his team reevaluated the data from the earlier studies using alternative valuation techniques. One method was to calculate the cost of replacing ecosystem services with technology. For example, marshes protect people from floods and filter out water pollutants. If a marsh were destroyed, the researchers would calculate the value of the services it had provided by measuring the cost of the levees and water-purification technology that would be needed to assume those tasks. The researchers then calculated the global monetary value of such wetlands by multiplying those totals by the global area occupied by the ecosystem. By calculating similar totals from coral reefs, deserts, tundra, and other ecosystems, they arrived at a global value for ecosystem services.

By their calculation, the biosphere provides at least $33 trillion ($40 trillion in 2005 dollars) worth of ecosystem services each year—greater than the gross domestic product (GDP) of all nations combined. Published in the journal *Nature* in 1997, their research paper ignited a firestorm of controversy.

Some environmental advocates and ethicists argued that it was a bad idea to put a dollar figure on priceless services such as clean air and clean water. The value of these services cannot be calculated, they held, because we would all perish if they disappeared. Environmental ethicist Timothy Weiskel of Harvard Divinity School contended that to make a commodity out of biodiversity confused "sacred space with the market place."

Some economists, meanwhile, disparaged the methods by which the researchers calculated values. Replacement costs were not a legitimate way of determining value,

can dictate limits on corporate behavior through laws and regulations. They can institute *green taxes,* which penalize environmentally harmful activities. Or, they can design economic incentives that put market mechanisms to work to promote fairness, resource conservation, and economic sustainability. We will examine legislation, regulation, green taxation, and market incentives in Chapter 3 as part of our policy discussion, but we will briefly note a few examples of alternative methods here.

Ecolabeling and permit-trading address market failure

Legislation, regulation, or taxation by government may not always effectively address market failure, so economists have contrived ways to use aspects of the market to counteract market failure. In one strategy, manufacturers of certain products are required or encouraged to designate on their labels how the products were grown, harvested, or manufactured. This method, called **ecolabeling,**

they held, because nature's services, like other economic goods and services, are worth only what people will demonstrably pay for them. Critics also argued that combining the value of ecosystem services from various small tracts of land is meaningless because people decide whether to preserve or exploit resources based on particular local considerations, not on generalized global ones.

To address economists' criticisms, Costanza joined Andrew Balmford of Cambridge University and 17 other colleagues to conduct another analysis. They compared the benefits and costs of preserving natural systems intact with those of converting wild lands for agriculture, logging, or fish farming. Again, they synthesized studies on ecosystem services, this time focusing on just five ecosystems: west African and Malaysian tropical forests, Thai mangrove swamps, Canadian wetlands, and Philippine coral reefs.

In their paper, published in the journal *Science* in 2002, the team reported that a global network of nature reserves covering 15% of Earth's land surface and 30% of the ocean would be worth between $4.4 and $5.2 trillion. This amount is about 100 times the value of the same area were it to be converted for direct exploitative human use.

That 100:1 benefit–cost ratio, they wrote, demonstrates clearly that "conservation in reserves represents a strikingly good bargain."

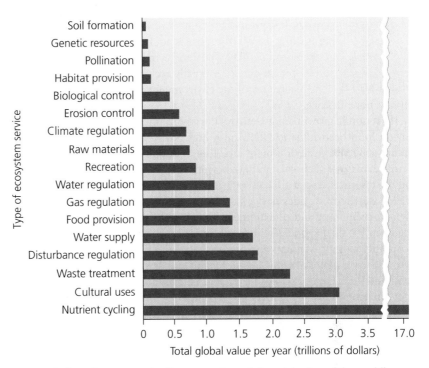

In 1997, Robert Costanza and colleagues estimated the total value of the world's ecosystem services at approximately $33 trillion. Shown are subtotals for each major class of ecosystem service. The $33 trillion figure does not include values from some ecosystems, such as deserts and tundra, for which adequate data were unavailable. Data from Costanza, R., et al. 1997. The value of the world's ecosystem services and natural capital. *Nature* 387: 253–260.

serves to tell consumers which brands use environmentally benign processes (Figure 2.16). By preferentially buying ecolabeled products, consumers can provide businesses a powerful incentive to switch to more environmentally friendly processes. The best-known example of this has been labeling cans of tuna as "dolphin-safe," indicating that the methods used to catch the tuna avoid the accidental capture of dolphins. Other examples include labeling recycled paper, organically grown foods, genetically modified foods, and lumber harvested through

sustainable forestry. We will come across such instances of ecolabeling in future chapters.

Another technique devised to counteract market failure is to create markets in permits for environmentally harmful activities. Under such a system, companies are allowed to buy and sell the right to conduct environmentally harmful activities. For instance, utilities and industries might trade permits to emit air pollution. In such a scheme, the government first determines an acceptable level of pollution, then issues permits to pollute.

Environment versus Economy?

Is environmental protection economically costly? How can we best balance the demands of economic development with the demands of environmental protection?

Economic Progress a Prerequisite for Environmental Quality

The Greek word *oikos,* meaning "household," is the root of both *economics* and *ecology.* It suggests complementarities. Protecting the environment is not free, but environmental quality and economic development are not mutually exclusive. The real enemy of the environment is poverty, not affluence.

If economic growth really destroys the environment, why do countries with large GDPs enjoy high environmental quality? The environment is cleaner in developed countries because their citizens have both the inclination and the resources to care for the environment.

Are we, as some critics assert, like the man falling from a ten-story building and concluding as he passes the second story, "so far so good"? I think not. In the long term, technological improvements and productivity gains allow us to use fewer material inputs—and to emit ever fewer pollutants—per unit of economic output. This reduces both our economic and ecological footprint.

For example, genetically modified (GM) crops and synthetic chemicals allow us greater yields on the same amount of land. As a result, the United States can afford to place 50 million acres of farmland into conservation reserves while remaining a major food exporter. The poorest countries are beneficiaries as they adopt our efficient and less environmentally damaging technologies—shortcutting the road to environmental quality.

The question we face is, in what combination and in what amounts should we seek the things we want? We value clean water and the preservation of other species. But we also value fresh produce in winter and fast and convenient transportation. Not all good things go together.

Regardless of claims, environmental quality is only one of several competing values people seek. Scarcity—the fact that virtually no resources are abundant enough to satisfy all human demands at zero cost—dictates that choices must be made among competing values. It is intellectually and ethically irresponsible to pretend away the necessity of such choices.

Pete Geddes is program director of the Foundation for Research on Economics and the Environment (FREE) and Gallatin Writers. Both are based in Bozeman, Montana.

The Trade-off Myth

Reducing pollution and protecting resources costs money. As a nation we spend over $200 billion each year, or more than 2% of GDP, on environmental protection. However, it is often contested how much environmental protection will cost in any particular case. Forecasting the costs of compliance is difficult because economists have an equally difficult time estimating future technological responses that may lower costs. For example, credible industry estimates for sulfur dioxide reduction from power plants under the 1990 Clean Air Act Amendments were eight times too high; the EPA overestimated costs by a factor of 2 to 4.

One "cost" of environmental protection that is blown out of proportion is job loss. Contrary to popular belief, there is no net "jobs–environment trade-off" in the economy, only a steady shift of jobs to cleanup work. On one hand, about 2,000 workers in the U.S. lose their jobs each year for environment-related reasons, which is less than 1/10 of 1% of all layoffs. On the other hand, as we spend more on the environment, more jobs are created. Given the industrial nature of much cleanup work, these jobs are also heavily weighted toward manufacturing and construction. Finally, and again contrary to folk wisdom, very few manufacturing plants flee the industrial countries to escape onerous environmental regulation. Plants do leave, but the overwhelming reason is the cost of labor.

What is the best way to balance costs against the benefits of environmental cleanup? Formal benefit–cost analysis is one approach, but it is of limited value when the benefits of environmental protection are highly uncertain (which is often the case), or when the costs of resource degradation are borne by a relatively small group. In these situations, the best approach is to define a health or ecological standard for cleanup and to trust democratic processes to ensure that the costs of cleanup do not rise too high.

Eban Goodstein is professor of economics at Lewis and Clark College in Portland, Oregon. He is the author of the textbook *Economics and the Environment* (John Wiley and Sons, 2004), as well as *The Trade-off Myth: Fact and Fiction about Jobs and the Environment* (Island Press, 1999).

Explore this issue further by accessing **Viewpoints** at www.aw-bc.com/withgott.

FIGURE 2.16 Ecolabeling allows businesses to promote products that have low environmental impact. Organic juices and produce are examples of ecolabeled products becoming widely available in the mass market.

Each party may emit its permitted amount, but if it is able to reduce its pollution, it receives credit for the amount it did not emit and can sell this credit to other parties that want to expand operations or are less able to control their emissions. Creating markets in permits provides companies an economic incentive to find ways to reduce emissions because the sale of permits brings in money. We will discuss **permit-trading** further in Chapter 3 (▸ p. 73). The method is not perfect. Large firms can hoard permits, deterring smaller new firms from entering the market, an action that suppresses competition. Nevertheless, permit-trading markets have shown great promise for reducing

environmental degradation while granting industries the flexibility to lessen their impacts in ways that are economically palatable to them.

Conclusion

Permit-trading, ecolabeling, and valuation of ecosystem services are some recent developments that have brought economic approaches to bear on environmental protection and resource conservation. As economics becomes more environmentally friendly, it renews some of its historic ties to ethics. Environmental ethics has expanded people's sphere of ethical consideration outward to encompass other societies and cultures, other creatures, and even nonliving entities that were formerly outside the realm of ethical concern. This ethical expansion involves the concept of distributional equity, or equal treatment, which is the aim of environmental justice. One type of distributional equity is equity among generations. Such concern by current generations for the welfare of future generations is the basis for the notion of sustainability.

Is sustainability a pragmatic pursuit for us? The answer largely depends on whether we believe that economic well-being and environmental well-being are opposed to one another or whether we accept that they can work in tandem. Equating economic well-being with economic growth, as most economists traditionally have, suggests that economic welfare entails a trade-off with environmental quality. However, if economic welfare can be enhanced in the absence of growth, we can envision economies and environmental quality benefiting from one another.

REVIEWING OBJECTIVES

You should now be able to:

Characterize the influences of culture and worldview on the choices people make

▸ A person's culture strongly influences his or her worldview. Factors such as religion and political ideology are especially influential. (pp. 28–30)

Outline the nature, evolution, and expansion of environmental ethics in Western cultures

▸ Our society's domain of ethical concern has been expanding, such that we have granted more and more entities ethical consideration. (pp. 31–32)

▸ Anthropocentrism values humans above all else, whereas biocentrism values all life and ecocentrism values ecological systems. (pp. 32–33)

▸ The preservation ethic (preserving natural systems intact) and the conservation ethic (promoting responsible long-term use of resources) have guided branches of the environmental movement during the past century. (pp. 34–35)

▸ The environmental justice movement, seeking equal treatment for people of all races and income levels, is one of several recent outgrowths of environmental ethics. (pp. 36–38)

Describe precepts of classical and neoclassical economic theory, and summarize their implications for the environment

▶ Classical economic theory proposes that individuals acting for their own economic good can benefit society as a whole. This view has provided a philosophical basis for free-market capitalism. (pp. 39–41)

▶ Neoclassical economics focuses on consumer behavior and supply and demand as forces that drive economic activity. (pp. 41–42)

▶ Several assumptions of neoclassical economic theory contribute to environmental impact. (pp. 42–44)

Compare the concepts of economic growth, economic health, and sustainability

▶ Conventional economic theory has promoted never-ending economic growth, with little regard to possible environmental impact. (p. 44)

▶ Economic growth is not necessarily required for economic well-being. (pp. 44–47, 52)

▶ In the long run, some economists believe that a steady-state economy will be necessary to achieve sustainability. (pp. 45–47)

Explain the fundamentals of environmental economics and ecological economics

▶ Environmental economists advocate reforming economic practices to promote sustainability. Key approaches are to identify external costs, assign value to nonmonetary items, and attempt to make market prices reflect real costs and benefits. (pp. 45–51)

▶ Ecological economists support these efforts and others. Many support developing a steady-state economy. (pp. 45–47)

TESTING YOUR COMPREHENSION

1. What does the study of ethics encompass? Describe the three classic ethical standards. What is environmental ethics?

2. Why in Western cultures have ethical considerations expanded to include nonhuman entities?

3. Describe the philosophical perspectives of anthropocentrism, biocentrism, and ecocentrism. How would you characterize the perspective of the Mirrar Clan?

4. Differentiate between the preservation ethic and the conservation ethic. Explain the contributions of John Muir and Gifford Pinchot in the history of environmental ethics.

5. Describe Aldo Leopold's "land ethic." How did Leopold define the "community" to which ethical standards should be applied?

6. Name four key contributions the environment makes to the economy.

7. Describe Adam Smith's metaphor of the "invisible hand." How did neoclassical economists refine classical economics?

8. Describe four ways in which critics hold that neoclassical economic approaches can negatively affect the environment.

9. Compare and contrast the views of neoclassical economists, environmental economists, and ecological economists.

10. What is contingent valuation, and what is one of its weaknesses? Describe an alternative method that addresses this weakness.

SEEKING SOLUTIONS

1. Do you feel that an introduction to environmental ethics and worldviews is an important part of a course in environmental science? Should ethics and worldviews be a component of other science courses? Explain your answers.

2. Describe your worldview as it pertains to your relationship with the environment. How do you think your culture has influenced your worldview? How do you think your personal experience has influenced it? Do you feel that you fit into any particular category discussed in this chapter? Why or why not?

3. How would you analyze the case of the Mirrar Clan and the proposed Jabiluka uranium mine from each of the following perspectives? In your description, list three questions that a person of each perspective would likely ask when attempting to decide whether the mine should be developed. Be as specific as possible, and be sure to identify similarities and differences in approaches:

▶ Preservationist
▶ Conservationist
▶ Deep ecologist

▶ Environmental justice advocate
▶ Neoclassical economist
▶ Ecological economist

4. What is a steady-state economy? Do you think this model is a practical alternative to the growth paradigm? Why or why not?

5. Do you think we should attempt to quantify and assign market values to ecosystem services and other entities that have only nonmarket values? Why or why not?

6. A manufacturing facility on a river near your home provides jobs for 200 people in your community and pays $2 million in taxes to the local government each year. Sales taxes from purchases made by plant employees and their families contribute an additional $1 million to local government coffers. However, a recent peer-reviewed study in a well-respected scientific journal revealed that the plant has been discharging large amounts of waste into the river, causing a 25% increase in cancer rates, a 30% reduction in riverfront property values, and a 75% decrease in native fish populations.

The plant owner says the facility can stay in business only because there are no regulations mandating expensive treatment of waste from the plant. If any such regulations were imposed, he says he would close the plant, lay off its employees, and relocate to a more business-friendly community.

How would you recommend resolving this situation? What further information would you want to know before making a recommendation? In arriving at your recommendation, how did you weigh the costs and benefits associated with each of the plant's impacts?

INTERPRETING GRAPHS AND DATA

As described in "The Science behind the Story" on ▶ pp. 46–47, economists use various indicators of economic well-being. One that has been used for decades is the Gross Domestic Product (GDP), the total monetary value of final goods and services produced each year. An alternative measure called the Genuine Progress Indicator (GPI) is calculated as follows:

$$GDP + (Benefits\ ignored\ by\ GDP) - (Environmental\ costs) - (Social\ and\ economic\ costs)$$

Benefits include such things as the value of parenting and volunteer work. Environmental costs include the costs of water, air and noise pollution, loss of wetlands, depletion of nonrenewable resources, and other environmental damage. Social and economic costs include investment, lending, and borrowing costs, as well as the costs of crime, family breakdown, underemployment, commuting, pollution abatement, automobile accidents, and loss of leisure time.

1. Describe economic growth as measured by GDP for the United States from 1952 to 2002. Now describe economic growth as measured by GPI over the same time period. To what factors would you attribute the growing difference between these measures?

2. For GPI to grow significantly, one or more things must happen: Either GDP must grow faster, benefits must grow faster, or social, economic, and environmental costs must shrink relative to the other terms. Which of these scenarios do you think is most likely? Which would you prefer? How do the data in the graph support your answer?

3. Even with domestic regulations for air and water pollution control, hazardous waste disposal, solid waste management, forestry practices, and species protection, environmental costs continue to increase. Why do you suppose the trend is still in that direction?

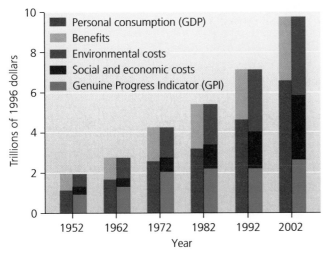

Components of GPI and GDP for the United States, 1952–2002. Data from Venetoulis, J., and C. Cobb. 2004. *The genuine progress indicator, 1950–2002 (2004 update).* Redefining Progress.

CALCULATING ECOLOGICAL FOOTPRINTS

Although the Gross Domestic Product (GDP) of the United States grew impressively between 1952 and 2002 (see the graph in the "Interpreting Graphs and Data" section), so did the U.S. population. According to the U.S. Census Bureau, the midyear population of the nation was 157,552,740 in 1952, and 288,368,698 in 2002—an 83% increase.

Estimate the values of the components of the Genuine Progress Indicator (GPI) for the United States in 1952 and 2002 from the graph, and enter your estimates into the table below. Then, using the population figures, calculate the per capita values for each component in 1952 and 2002.

Components of GPI	U.S. total in 1952 (trillions of dollars)	Per capita in 1952 (thousands of dollars)	U.S. total in 2002 (trillions of dollars)	Per capita in 2002 (thousands of dollars)
GDP	1.2	7.6	6.6	22.9
Benefits				
Environmental costs				
Social and economic costs				
GPI				

1. Consider your own life. What would you estimate is the value of the benefits in which you participate? Compare your personal estimate with the national average value in 2002.

2. What would you estimate are the values of the environmental, social, and economic costs for which you are personally responsible? How do they compare with the national average values in 2002?

3. In 2002, social and economic costs were proportionally larger relative to GDP than they were in 1952, and environmental costs were roughly the same in proportion to GDP. How would you account for these trends? What could you do to help improve these trends in your own personal accounting?

Take It Further

Go to www.aw-bc.com/withgott or the student CD-ROM where you'll find:

▶ Suggested answers to end-of-chapter questions
▶ Quizzes, animations, and flashcards to help you study
▶ *Research Navigator*™ database of credible and reliable sources to assist you with your research projects

▶ **GRAPHit!** tutorials to help you master how to interpret graphs
▶ **INVESTIGATEit!** current news articles that link the topics that you study to case studies from your region to around the world

Environmental Policy: Decision Making and Problem Solving

Border of Tijuana, Mexico (left), and San Diego, California (right)

Upon successfully completing this chapter, you will be able to:

▶ Describe environmental policy and assess its societal context

▶ Identify the institutions important to U.S. environmental policy and recognize major U.S. environmental laws

▶ Categorize the different approaches to environmental policy

▶ Delineate the steps of the environmental policy process and evaluate its effectiveness

▶ List the institutions involved with international environmental policy and describe how nations handle transboundary issues

Closed San Diego beach

Central Case: San Diego and Tijuana's Sewage Pollution Problems and Policy Solutions

"Ignorance is an evil weed, which dictators may cultivate among their dupes, but which no democracy can afford among its citizens."
—WILLIAM BEVERIDGE, BRITISH ECONOMIST

"It is the continuing policy of the Federal Government . . . to create and maintain conditions under which man and nature can exist in productive harmony and fulfill the social, economic, and other requirements of present and future generations of Americans."
—NATIONAL ENVIRONMENT POLICY ACT, 1969

On November 23, 1996, officials closed all the public beaches in San Diego, California. Stormwater runoff following heavy rains had washed pollutants into local rivers and coastal waters. This also occurred across the border in the Mexican city of Tijuana, whose aging sewer system became clogged with debris, causing raw sewage to overflow into streets and onto beaches. Such incidents, called "rogue flows,"

take place when heavy rain overwhelms the ability of sewage treatment plants to process wastewater. Rogue flows had become so common in San Diego and Tijuana that local surfers and swimmers casually referred to the initial one of each rainy season as the "first flush."

The Tijuana River winds northwestward through the arid landscape of northern Baja California, Mexico, crossing the U.S. border south of San Diego. A river's **watershed** consists of all the land from which water drains into the river, and the Tijuana River's watershed covers 4,500 km² (1,750 mi²) and is home to 2 million people of two nations. The Tijuana River watershed is a *transboundary* watershed (so named because it crosses a political boundary—in this case, a national border), with approximately 70% of its area in Mexico (Figure 3.1). On the Mexican side of the border, the river and the arroyos, or creeks, that flow into it are lined with farms, apartments, shanties, and factories, as well as leaky sewage treatment plants and toxic dump sites. Many of

FIGURE 3.1 The Tijuana River winds northwestward from Mexico into California just south of San Diego, draining 4,500 km^2 (1,750 mi^2) of land in its watershed. Pollution entering the river affects Mexican residents of the watershed and U.S. citizens on San Diego County beaches. During high-water episodes that overwhelm treatment facilities, millions of gallons of raw sewage have flushed into the river and the Pacific Ocean.

these sources release pollutants, which heavy rains wash through the arroyos into the Tijuana River and eventually onto U.S. and Mexican beaches.

Although pollution has flowed in the Tijuana River for at least 70 years, the problem has grown worse in recent years as the region's population has boomed, outstripping the capacity of sewage treatment facilities. Rogue flows have caused thousands of beach closures and pollution advisories in recent years. Garbage carried by the flows also litters the beaches. "Every day I find broken glass, balloons, or can pop-tops. I've even found hypodermic needles. It's really sad," one resident of Imperial Beach told her local newspaper in 2005.

The problem is worse on the Mexican side because most Mexican residents of the Tijuana River watershed live in poverty relative to their U.S. neighbors. In poor neighborhoods such as Loma Taurina, pollution of the river directly affects people's day-to-day lives. The rise of U.S.-owned factories, or *maquiladoras,* on the Mexican side of the border has contributed to the river's pollution, both through direct disposal of industrial waste and by attracting thousands of new workers to the already crowded region.

As impacts increased, people in the San Diego and Tijuana areas, from coastal residents to grassroots activists to businesspeople, pressed policymakers to do something. As we explore environmental policy in this chapter, we will periodically return to the Tijuana River watershed and see how citizens and policymakers together have tried to address these problems.

Environmental Policy: An Introduction

When a society reaches broad agreement that a problem exists, it may persuade its leaders to try to resolve the problem through the making of policy. **Policy** consists of a formal set of general plans and principles intended to address problems and guide decision making in specific instances. **Public policy** is policy made by governments, including those at the local, state, federal, and international levels. Public policy consists of laws, regulations, orders, incentives, and practices intended to advance societal welfare. **Environmental policy** is policy that pertains

FIGURE 3.2 Policy plays a central role in how we as a society address environmental problems. This process begins when research from the academic disciplines involved in environmental science leads to some understanding of a given environmental problem, its potential consequences, and its possible solutions. Government may use this information in formulating policy, together with ethical and economic considerations and with input from citizens and the private sector. Government policy includes laws, regulations, and market-based incentives that aim, for example, to restrict resource exploitation or ensure fairness in resource use. Along with improvements in technology and efficiency from the private sector and consumer choices exercised by citizens in the marketplace, public policy can produce lasting solutions to environmental problems.

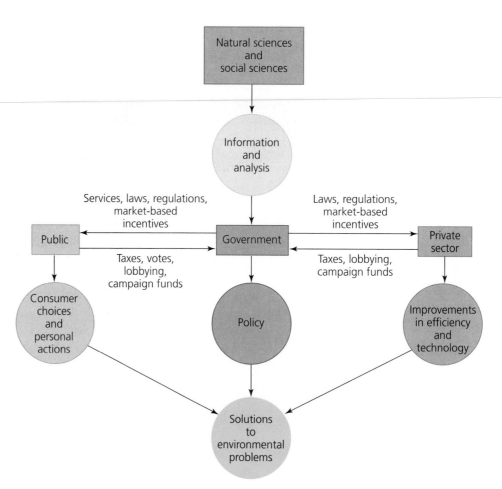

to human interactions with the environment. It generally aims to regulate resource use or reduce pollution to promote human welfare and/or protect natural systems.

Forging effective policy requires input from science, ethics, and economics. Science (Chapter 1) provides the information and analysis needed to identify and understand environmental problems and devise potential solutions to them. Ethics and economics (Chapter 2) offer criteria to assess the extent and nature of problems, and help clarify how society might like to see them addressed. Government interacts with individual citizens, organizations, and the private sector in a variety of ways to formulate policy, which is one of our major tools for devising lasting solutions to environmental problems (Figure 3.2).

Policymaking can be vital in addressing dilemmas like those of the Tijuana River watershed. Sewage-tainted rivers and beaches pose several threats. Scientific research tells us that raw sewage carries pathogens, organisms that can cause illness in humans and other animals (Table 3.1). Science also reveals how untreated sewage can radically alter conditions for aquatic and marine life, lowering concentrations of dissolved oxygen and increasing mortality for many species. Economically, pollution and beach closures cause financial harm by reducing recreation, tourism, and other economic activity associated with clean coastal areas. This is a major consideration both in Mexico and in southern California, whose beaches each year host 175 million visitors who spend over $1.5 billion. Ethically, water pollution also poses problems, because in most cases pollution from upstream users degrades water quality for downstream users. Many phenomena that come to be viewed as environmental problems share this combination of impacts—harming human health, altering ecological systems, inflicting economic damage, and creating inequities among people.

Environmental policy addresses issues of equity and resource use

People have identified environmental problems and asked government officials to develop solutions since at least the first century A.D., when Roman citizens complained to their leaders about air pollution from wood smoke. Today, our population growth and resource consumption have generated environmental problems on an unprecedented scale. Furthermore, the market capitalist economic systems of

Table 3.1 Pathogens and Health Effects Associated with Untreated Sewage	
Pathogen	**Human health effect**
Cryptosporidium	Diarrhea, vomiting, cramps, dehydration
Giardia lamblia	Diarrhea, vomiting, cramps, dehydration
Salmonella typhi	Typhoid fever
Vibrio cholerae	Cholera
Poliovirus	Poliomyelitis
Shigella spp.	Dysentery
Entamoeba histolytica	Dysentery
Hepatitis A virus	Infectious hepatitis

modern constitutional democracies are driven by incentives for short-term economic gain rather than long-term social and environmental stability. Market capitalism provides little incentive for businesses or individuals to behave in ways that minimize environmental impact or equalize costs and benefits among parties. Such *market failure* (▶ pp. 49–51) has traditionally been viewed as justification for government intervention. Thus, environmental policy aims to protect environmental quality and the natural resources people use, and also to promote equity in people's use of resources.

The tragedy of the commons Policy to protect resources held and used in common by the public is intended to safeguard these resources from depletion or degradation. As Garrett Hardin explained in his essay, "The Tragedy of the Commons" (▶ p. 6), a resource held in common that is unregulated will eventually become overused and degraded. Therefore, he argued, it is in our best interest to develop guidelines for the use of common resources. In Hardin's illustrative example of a common pasture, such guidelines might limit the number of animals each individual can graze or might require pasture users to pay to restore and manage the shared resource. These two concepts, restriction of use and active management, are central to environmental policy today.

The tragedy of the commons does not always play itself out as Hardin predicted. Some traditional societies have devised safeguards against exploitation, and in modern Western societies resource users have occasionally cooperated to prevent overexploitation. Moreover, many cases do not meet Hardin's starting assumptions; resources on public lands may not be equally accessible to everyone, but may instead be more accessible to wealthier or more established resource extraction industries. Nonetheless, the threat of overexploiting public resources is always real and has been a driving force behind much environmental policy.

Free riders Another reason to develop policy for publicly held resources is the **free rider** predicament. Let's say a community on a river suffers from water pollution that emanates from 10 different factories. The problem could in theory be solved if every factory voluntarily agreed to reduce its own pollution. However, once they all begin reducing their pollution, it becomes tempting for any one of them to stop doing so. Such a factory, by avoiding the sacrifices others are making, would in essence get a "free ride" on the efforts of others. If enough factories take a free ride, the whole effort will collapse. Because of the free rider problem, private voluntary efforts are often less effective than efforts mandated by public policy. Public policy can prevent the free rider problem and ensure that all parties sacrifice equitably by enforcing compliance with laws and regulations or by taxing parties to attain funds to pursue societal goals.

External costs Environmental policy is also developed to ensure that some parties do not use resources in ways that harm others. One way to promote fairness is by dealing with *negative externalities,* or *external costs* (▶ pp. 43–44), harmful impacts that result from market transactions but are borne by people not involved in the transactions. For example, a factory that discharges waste freely into a river may reap greater profits by avoiding paying for waste disposal or recycling. Its actions, however, impose external costs (water pollution, decreased fish populations, aesthetic degradation, or other problems) on downstream users of the river. U.S.-owned *maquiladoras* in the Tijuana River watershed dump waste that affects downstream users, such as Mexican families who use river water for washing (Figure 3.3). Likewise, the detergents these families use for washing further pollute the river, creating external costs for families farther downstream and for beachgoers in Mexico and California.

These goals of environmental policy—to protect resources against the tragedy of the commons and to promote equity by eliminating free riders and dealing with external costs—will become apparent in case after case explored in this book.

Many factors can hinder implementation of environmental policy

If the goals of environmental policy are seemingly so noble, why is it that environmental laws are so often challenged, environmental regulations frequently derided, and the ideas of environmental activists repeatedly ignored or rejected by citizens and policymakers?

FIGURE 3.3 River pollution raises many issues that have been viewed as justification for environmental policy. This woman washing clothes in the river may suffer upstream pollution from factories, and her use of detergents may cause further pollution for people living downstream.

In the United States, most environmental policy has come in the form of regulations handed down from the federal government or from state or local governments. Businesses and individuals sometimes view these regulations as overly restrictive, bureaucratic, or unresponsive to human needs. For instance, many landowners fear that zoning regulations or protections for endangered species will impose restrictions on the use of their land. Developers complain of time and money lost to bureaucracy in obtaining permits; reviews by government agencies; surveys for endangered species; and required environmental controls, monitoring, and mitigation. In the eyes of such property owners and businesspeople, environmental regulation all too often has meant inconvenience and/or economic loss.

Weighing the Issues:
Do We Really Need Environmental Policy?

Many free-market advocates maintain that environmental laws and regulations are an unnecessary government intrusion into private affairs. As you may recall from Chapter 2 (pp . 39, 41), Adam Smith argued that individuals will benefit both themselves and society by pursuing their own self-interest. Do you agree? Can you describe a situation in which an individual acting in his or her self-interest could harm society by causing an environmental problem? Can you describe how environmental policy might rectify the situation? What are some advantages and disadvantages of instituting environmental laws and regulations, versus allowing unfettered exchange of materials and services?

Another reason people sometimes resent environmental policy stems from the very nature of environmental problems, which often develop slowly and gradually. The degradation of ecosystems and public health due to human impact on the environment are long-term processes. Human behavior, however, is generally geared toward addressing short-term needs—even if they conflict with long-term needs—and this tendency is reflected in our social institutions. Businesses usually opt for short-term economic gain over long-term considerations. The news media have a short attention span traditionally based on the daily news cycle, whereby new and sudden one-time events are given more coverage than slowly developing long-term trends. Politicians most often act out of short-term interest because they depend on reelection every few years. For all these reasons, many environmental policy goals that seem admirable and that attract wide public support in theory may be obstructed in their practical implementation.

U.S. Environmental Policy: An Overview

The United States provides a good focus for understanding environmental policy in constitutional democracies worldwide. First, the United States historically pioneered innovative environmental policy. Second, U.S. policies have served as models—both of success and failure—for many other nations and international government bodies. Third, the United States exerts a great deal of influence on the affairs of other

Environmental Policy: Decision Making and Problem Solving

nations. Finally, understanding U.S. environmental policy on the federal level can enable you to better understand environmental policy at local, state, and international levels.

U.S. policy arises from the three branches of government

Environmental policy, like all U.S. policy, results from actions of the three branches of government established under the U.S. Constitution (Figure 3.4). Statutory law, or **legislation,** is created by the **legislative branch,** or Congress, which consists of the Senate and the House of Representatives. For instance, Congress passed the Tijuana River Valley Estuary and Beach Sewage Cleanup Act in 2000 to fund the treatment of sewage flowing into the United States in the Tijuana River.

Legislation is enacted (approved) or vetoed (rejected) by the president, who heads the **executive branch.** The president may also issue *executive orders,* specific legal instructions for government agencies. Once statutory laws are enacted, their implementation and enforcement is assigned to the appropriate administrative agency within the executive branch. Administrative agencies (Figure 3.5), which may be established by Congress or by presidential order, are sometimes nicknamed the "fourth branch" of government because they are the source of a great deal of policy, in the form of regulations. **Regulations** are specific rules based on the more broadly written statutory law. Besides issuing regulations, administrative agencies monitor compliance with laws and regulations and enforce them when individuals or corporations violate them. Administrative agencies have come to be the source of most U.S. environmental policy.

The **judicial branch,** or judiciary, consisting of the Supreme Court and various lower courts, is charged with interpreting law. This is necessary because of changing social factors and technologies and because Congress must write laws broadly to ensure that they apply to varied circumstances throughout the nation. Decisions rendered by the courts make up a body of law known as *case law.* Previous rulings serve as *precedents,* or legal guides, for

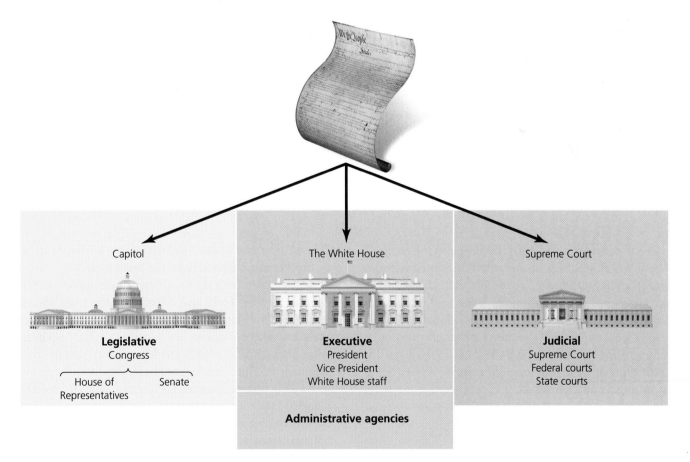

FIGURE 3.4 In the United States, federal powers are shared among the legislative, executive, and judicial branches of government. All three branches (including the administrative agencies of the executive branch) are responsible for aspects of developing and implementing environmental policy.

Federal Agencies and Departments of the Executive Branch that Affect Environmental Policy	
Environmental Protection Agency (EPA)	Department of Housing and Urban Development (HUD)
Office of Management and Budget (OMB)	Community Planning and Development
U.S. Trade Representative	Healthy Homes and Lead Hazard Control
Tennessee Valley Authority (TVA)	Policy Development and Research
Nuclear Regulatory Commission (NRC)	Department of Interior
Council on Environmental Quality (CEQ)	Bureau of Indian Affairs
Department of Treasury	Division of Energy and Mineral Resources
	Bureau of Land Management (BLM)
Community Development Financial Institutions Fund	Minerals Management Service
	National Park Service (NPS)
Department of Commerce	Office of Ethics
	Fish and Wildlife Service (FWS)
National Institute of Standards and Technology	U.S. Geological Survey (USGS)
National Oceanic and Atmospheric Administration (NOAA)	Water Resources Division
	Department of Justice
Department of Defense (DOD)	Environmental and Natural Resources Division
	Office of Legal Policy
Army Corps of Engineers	Office of Public Affairs
Department of Energy (DOE)	Department of Labor
	Mine Safety and Health Organization
Office of Environment, Safety, and Health	Department of State
Energy Efficiency and Renewable Energy Network	U.S. Agency for International Development (USAID)
Office of Civilian Radioactive Waste Management	
Office of Fossil Energy	Department of Transportation
Office of Health and Environmental Research	Office of Pipeline Safety
Office of Environmental Management	Office of Environment and Energy
Department of Health and Human Services	Department of Agriculture (USDA)
	Farm Service Agency
Public Health Service	Forest Service (USFS)
Agency for Toxic Substances and Disease Registry	Natural Resources Conservation Service (NRCS)
Centers for Disease Control and Prevention (CDC)	National Rural Development Council
Food and Drug Administration (FDA)	Food and Consumer Service

FIGURE 3.5 The many administrative agencies of the executive branch are the source of most U.S. environmental policy. *Source:* U.S. General Services Administration, Washington, D.C.

later cases, steering judicial decisions through time. The judiciary has been an important arena for environmental policy. Grassroots environmental advocates and nongovernmental organizations have used lawsuits as tools to help level the playing field with better-funded large corporations and agencies. Conversely, the courts have allowed businesses and individuals to challenge the constitutional validity of environmental laws they feel to be infringing on their rights.

State and local policy also affects environmental issues

The structure of the federal government is mirrored at the state level with governors, legislatures, judiciaries, and agencies. State laws cannot violate principles of the U.S. Constitution, and if state and federal laws are found to be in direct conflict, federal laws take precedence.

Many states with dense urban populations, such as California, New York, and Massachusetts, have strong environmental laws and well-funded environmental agencies, whereas many less-populous states, such as those of the interior West, put less priority on environmental protection. To safeguard public health in locations such as San Diego's beaches, California legislators in 1997 required state environmental health officials to set standards for testing waters for bacterial contamination. Officials issue an advisory, or warning, when bacterial concentrations in nearshore waters exceed health limits established by California law. In 1999, further legislation passed by the State Assembly mandated another state agency to develop methods for locating sources of contamination. As we proceed through our discussion of the federal government, keep in mind that important environmental policy is also created at the state and local levels.

Some constitutional amendments bear on environmental law

The U.S. Constitution laid out several principles that have come to be especially relevant to environmental policy. One of these is the clause from the Fourteenth Amendment prohibiting a state from denying "equal protection of its laws" to any person. This amendment provides the constitutional basis for the environmental justice movement (▸ pp. 36–37).

The Fifth Amendment ensures, in part, that private property shall not "be taken for public use without just compensation." Courts have interpreted this clause, known as the *takings clause*, to ban not only the literal taking of private property but also what is known as regulatory taking. A **regulatory taking** occurs when the government, by means of a law or regulation, deprives a property owner of some or all economic uses of that property. Many people cite the takings clause in arguing against environmental regulations that restrict land use and development on privately owned land. For example, some would contend that zoning regulations (▸ pp. 385–386) that prohibit a landowner from opening a hazardous waste dump in a residential neighborhood deprive the landowner of an economically valuable use of the land and, therefore, violate the Fifth Amendment.

In a landmark court case in 1992, the Supreme Court ruled that a state land use law intended to "prevent serious public harm" violated the takings clause. The case, known as *Lucas v. South Carolina Coastal Council*, involved a developer named Lucas who in 1986 purchased beachfront property in South Carolina for $975,000. In 1988, before Lucas began to build, South Carolina's legislature passed a law banning construction on eroding beaches. A state agency classified the Lucas property as an eroding beach and prohibited residential construction there. Believing that this amounted to a regulatory taking, Lucas asked a state court to overrule the new law and allow him to build homes on his property. Agreeing with Lucas, the state court ruled that he was entitled to $1.2 million in compensation. South Carolina appealed the case to the state Supreme Court, which overturned the lower court's decision. Lucas then appealed to the U.S. Supreme Court, which overturned the state Supreme Court's ruling and sided with the lower court, declaring

FIGURE 3.6 Beaches subject to erosion and hurricane damage like this one in South Carolina have given rise to environmental policy debates. Should the government be able to prevent development in areas where erosion, storms, and flooding pose risks to property and human life? If so, does such prohibition constitute a taking of private property rights, and should the property owner be compensated? These were the questions addressed in *Lucas v. South Carolina Coastal Council*.

that the state law deprived Lucas of all economically beneficial uses of his land. Today, homes stand on the land, and regulatory takings remains a contentious area of law (Figure 3.6).

Weighing the Issues:
Regulatory Takings

Imagine you have purchased land and plan to make money by clearing its forest and building condominiums on it. Now suppose an environmental group concerned about forest loss in the region points out that an endangered plant species occurs on the property and petitions the government to prevent development of the land. Use the takings clause to argue why you should be allowed to build or be financially compensated if you are not allowed to build.

Now put yourself in the shoes of the environmentalists, and make a case why the landowner should not be allowed to build nor be financially compensated.

Which argument do you feel has greater merit? Would it make a difference to you if the land in question had been zoned as developable or nondevelopable when the purchase was made? What do you think is the best way to balance private property rights with protection of the public good in cases like this?

FIGURE 3.7 Settlers such as these **(a)** took advantage of the federal government's early environmental policies, including the Homestead Act of 1862. Early mining activities on public lands were largely unregulated **(b)**. The Mineral Lands Act of 1866 required that mining could occur "subject . . . to the local customs or rules of miners," rather than in a way that would protect the environment.

(a) Settlers in Custer County, Nebraska, circa 1860

The first U.S. environmental policy addressed public land management

The laws that comprise U.S. environmental policy were created largely in three periods. Laws enacted during the first period, from the 1780s to the late 1800s, dealt primarily with the management of public lands and accompanied the westward expansion of the nation. Environmental laws of this period were intended mainly to promote settlement and the extraction and use of the West's abundant natural resources. Among these early laws were the *General Land Ordinances of 1785 and 1787,* which gave the federal government the right to manage Western lands and created a grid system for surveying them and readying them for private ownership. Between 1785 and the 1870s, the federal government promoted settlement in the West on lands it had appropriated from Native Americans by doling out as many of these lands as possible to its citizens. Western settlement provided these citizens with means to achieve prosperity, while relieving crowding in Eastern cities. It expanded the nation's geographical reach at a time when the young United States was still jostling with European powers for control of the continent. It also wholly displaced the millions of Native Americans who had inhabited these lands. U.S. environmental policy of this era reflected the public perception that Western lands were practically infinite, and inexhaustible in natural resources. The following are a few laws typical of this era:

▶ The Homestead Act of 1862 allowed any citizen to claim 65 ha (160 acres) of public land by living there for 5 years and cultivating the land or building a home, for a $16 fee (Figure 3.7a). A waiting period of only 14 months was available to those who could pay $176 for the land.

▶ The Mineral Lands Act of 1866 provided land for $5 per acre to promote mining and settlement. It allowed

(b) Nineteenth-century mining operation, Lynx Creek, Alaska

(c) Loggers felling an old-growth tree, Washington

FIGURE 3.7 *cont.* **(c)** Although the Timber Culture Act promoted tree-planting on agricultural lands, in forested regions the timber industry was allowed to clear-cut the nation's ancient trees.

mining to occur subject to local customs, with no government oversight (Figure 3.7b).

▶ The Timber Culture Act of 1873 granted 65 ha (160 acres) to any citizen promising to cultivate trees on one-quarter of that area (Figure 3.7c).

Such laws encouraged settlers, entrepreneurs, and land speculators to move west, hastening the closing of the frontier.

The second wave of U.S. environmental policy addressed impacts of the first

In the late 1800s, as the West became more populated and its resources were increasingly exploited, public perception and government policy toward natural resources began to shift. Laws of this period aimed to mitigate some of the environmental problems associated with westward expansion. In 1872 Congress designated Yellowstone as the world's first national park. In 1891 Congress, to prevent overharvesting and protect forested watersheds, passed a law authorizing the president to create "forest reserves" off-limits to logging. In 1903, President Theodore Roosevelt created the first national wildlife refuge. These acts enabled the creation, over the next few decades, of a national park system, national forest system, and national wildlife refuge system that still stand as global models (▶ pp. 364–365). These developments reflected a new understanding that the West's resources are not inexhaustible, but instead require legal protection.

Land management policies continued through the 20th century, targeting soil conservation in the Dust Bowl years (▶ p. 260) and extending through the Wilderness Act of 1964 (▶ pp. 365–366), which sought to preserve still-pristine lands "untrammeled by man, where man himself is a visitor who does not remain."

The third wave of U.S. environmental policy responded largely to pollution

Further social changes in the mid- to late 20th century gave rise to the third major period of U.S. environmental policy. In a more densely populated country driven by technology, heavy industry, and intensive resource consumption, Americans found themselves better off economically but living amid dirtier air, dirtier water, and more waste and toxic chemicals. During the 1960s and 1970s, several events triggered increased awareness of environmental problems and brought about a shift in public priorities and important changes in public policy.

A landmark event was the 1962 publication of *Silent Spring,* a book by American scientist and writer Rachel Carson (Figure 3.8). *Silent Spring* awakened the public to the negative ecological and health effects of pesticides and industrial chemicals. (The book's title refers to Carson's warning that pesticides might kill so many birds that few would be left to sing in springtime.)

FIGURE 3.8 Scientist, writer, and citizen activist Rachel Carson illuminated the problem of pollution from DDT and other pesticides in her 1962 book, *Silent Spring.*

FIGURE 3.9 In a spectacular display of the need for better control over water pollution, Ohio's Cuyahoga River caught fire several times in the 1950s and 1960s. The Cuyahoga was so polluted with oil and industrial waste that the river would burn for days at a time.

The Cuyahoga River (Figure 3.9) also did its part to raise attention to the hazards of pollution. The Cuyahoga was so polluted with oil and industrial waste that the river actually caught fire near Cleveland, Ohio, more than half a dozen times during the 1950s and 1960s. This spectacle, coupled with an enormous oil spill off the Pacific coast near Santa Barbara, California, in 1969, moved the public to prompt Congress and the president to do more to protect the environment.

Today, largely because of environmental policies enacted since the 1960s, pesticides are more strictly regulated, and the nation's air and water are considerably cleaner. The public enthusiasm for environmental protection that spurred such advances remains strong today. Polls repeatedly show that an overwhelming majority of Americans favor environmental protection. Such support is evident each year in April, when millions of people worldwide celebrate Earth Day in thousands of locally based events featuring speeches, lectures, demonstrations, hikes, bird-walks, and more. In the more than 30 years since the first Earth Day, celebrated on April 22, 1970, participation in this event has grown markedly and has spread to nearly every country in the world (Figure 3.10).

NEPA guarantees citizens input into environmental policy decisions

Besides Earth Day, two federal actions marked 1970 as the dawn of the modern era of environmental policy in the United States. On January 1, 1970, President Richard Nixon signed the **National Environmental Policy Act (NEPA)** into law (Figure 3.11). NEPA created an agency called the Council on Environmental Quality and required that an **environmental impact statement (EIS)** be prepared for any major federal action that might significantly affect en-

vironmental quality. An EIS is a report of results from detailed studies that assess the potential impacts on the environment that would likely result from development projects undertaken or funded by the federal government.

NEPA's effects have been far-reaching. The EIS process forces government agencies and any businesses that contract with them to evaluate impacts on the environment before proceeding with a new dam, highway, or building project. Although the EIS process generally does not halt such projects, it can serve as an incentive to lessen the environmental damage done by a development or activity. NEPA also grants ordinary citizens input in the policy process by requiring that environmental impact statements be made publicly available and that public comment on them be solicited and considered.

The creation of the EPA marked a shift in federal environmental policy

Six months after signing NEPA into law, Nixon issued an executive order calling for a new integrated approach to environmental policy based on the understanding that environmental problems are interrelated. "The Government's environmentally-related activities have grown up piecemeal over the years," the order stated. "The time has come to organize them rationally and systematically." Nixon's order moved elements of agencies regulating water quality, air pollution, solid waste, and other issues into a newly created agency, the **Environmental Protection Agency (EPA).** The order charged the EPA with conducting and evaluating research, monitoring environmental quality, setting standards for pollution levels, enforcing those standards, assisting the states in meeting standards and goals, and educating the public.

(a) The first Earth Day, Washington, D.C., 1970

(b) Schoolchildren celebrating Earth Day, Katmandu, Nepal, 2002

FIGURE 3.10 April 22, 1970, marked the first Earth Day celebration. **(a)** This public outpouring of support for environmental protection catalyzed the third wave of environmental policy in the United States. Three decades later, Earth Day is celebrated by millions of people across the globe **(b)**, as shown here in Nepal.

Other prominent laws followed

Ongoing public demand for a cleaner environment during this period resulted in a number of key laws that remain the linchpins of U.S. environmental policy (Figure 3.12). Today there are thousands of federal, state, and local environmental laws in the United States, and thousands more abroad. For problems like the Tijuana River's pollution, a crucial law has been the Clean Water Act of 1977. Throughout much of U.S. history, water policy was left largely to local and state governments. (An early exception was the federal Rivers and Harbors Act of 1899, which restricted dumping and discharges into navigable waters.) Prior to passage of federal legislation such as the Clean Water Act, pollution problems were subject primarily to *tort law* (law addressing harm caused to one entity by another), and individuals suffering external costs from pollution were limited to seeking redress through lawsuits. Reaction to the ruling in the case *Boomer v. Atlantic Cement Company* in 1970 effectively ended tort law's viability as a tool for preventing pollution. The court ruled that even though residents of Albany, New York, were suffering pollution from a neighborhood cement plant, the

pollution could continue because the economic costs of controlling the pollution were greater than the economic costs of the damage the pollution caused.

The flaming waters of the Cuyahoga, however, indicated to many people that tough legislation was needed. Thanks to restrictions on pollutants by the Federal Water Pollution Control Acts of 1965 and 1972, and then the Clean Water Act, U.S. waterways finally began to recover.

FIGURE 3.11 Richard Nixon served as president during the period when modern environmental policy took shape. Here he signs the National Environmental Policy Act (NEPA) into law on January 1, 1970.

FIGURE 3.12 Most major laws in modern U.S. environmental policy were enacted in the 1960s and 1970s.

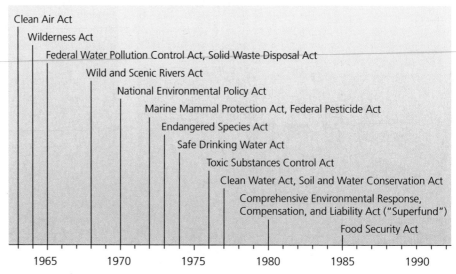

Key Environmental Protection Laws, 1963–1985

Clean Air Act
Wilderness Act
Federal Water Pollution Control Act, Solid Waste Disposal Act
Wild and Scenic Rivers Act
National Environmental Policy Act
Marine Mammal Protection Act, Federal Pesticide Act
Endangered Species Act
Safe Drinking Water Act
Toxic Substances Control Act
Clean Water Act, Soil and Water Conservation Act
Comprehensive Environmental Response, Compensation, and Liability Act ("Superfund")
Food Security Act

1965   1970   1975   1980   1985   1990

These laws regulated the discharge of wastes, especially from industry, into rivers and streams. The Clean Water Act also aimed to protect wildlife and establish a system for granting permits for the discharge of pollutants.

Weighing the Issues:
Judging Costs in Tort Law

Imagine you are the judge presiding over a lawsuit brought by citizens who live next to an industrial plant. Pollution from the plant regularly covers their homes, yards, and cars with dust and soot, making them cough and wheeze. In weighing whether to require the plant to clean up its emissions, will you consider in your ruling the financial costs the factory would need to incur? Or will you ignore economic aspects and base your ruling solely on evidence of harm to health? How easy or difficult do you think it would be to calculate and compare the costs to human health of ongoing pollution versus the costs of mandating emission controls or shutting down the plant? How could one estimate these costs?

The social context for environmental policy changes over time

Historians of environmental policy have suggested that accomplishments such as the Federal Water Pollution Control Acts and the Clean Water Act occurred in the 1960s and 1970s because three factors converged. First, evidence of environmental problems became widely and readily apparent. Second, people could visualize policies to deal with the problems. Third, the political climate was ripe, with a supportive public and leaders who were willing to act.

By the 1980s, the political climate in the United States had changed. Although public support for the goals of environmental protection remained high, many citizens and policy experts began to feel that the legislative and regulatory means used to achieve environmental policy goals too often imposed economic burdens on businesses and personal burdens on individuals. Since 1980, numerous efforts have been made at the federal level to roll back or reform environmental laws.

As the United States retreated from its leadership in environmental policy during the past two decades, other nations were increasing their political attention to environmental issues. The 1992 Earth Summit at Rio de Janeiro, Brazil, was the largest international diplomatic conference ever held, drawing representatives from 179 nations and unifying these leaders around the idea of sustainable development (▶ p. 22 and ▶ pp. 682–683). Indeed, we may now be embarking on a fourth wave of environmental policy, one focused on sustainability. This new policy approach tries to find ways to safeguard the functionality of natural systems while raising living standards for the world's poorer people (Figure 3.13). As the world's nations continue to feel the social, economic, and ecological effects of environmental degradation, environmental policy will without doubt become a more central part of governance and everyday life in all nations in the years ahead.

Approaches to Environmental Policy

Criticism of environmental legislation and regulation, and the means by which they often have been implemented, has spurred political scientists, economists, and

FIGURE 3.13 Many nations are shifting policies toward supporting sustainable development efforts, trying to increase standards of living while safeguarding the environment. Here, a woman stirs rice on a solar-powered oven at a restaurant made of recycled drink cans, showcased at the Ubunto Village at the U.N. World Summit on Sustainable Development in Johannesburg, South Africa, in 2002.

others to come up with alternative approaches for attaining environmental policy goals. The most widely developed alternatives involve the creative use of economic incentives to encourage desired outcomes, discourage undesired outcomes, and set market dynamics in motion to achieve goals in an economically efficient manner.

Top-down policy approaches are not always effective

Many environmental laws and regulations aiming to reduce pollution have simply set strict legal limits and threatened punishment for violating these limits, in what is sometimes called a **command-and-control** approach. The command-and-control approach has resulted in some major successes, as evidenced by the cleaner air and water U.S. residents enjoy today. Without doubt, our environment would be in far worse shape were it not for government regulatory intervention. However, many people have grown disenchanted with the top-down, sometimes heavy-handed nature of the command-and-control approach, and it is clear that government intervention sometimes fails every bit as badly as markets can fail.

Sometimes government actions are well intentioned but not well enough informed, so they can lead to unforeseen consequences. The predator-control policies that Aldo Leopold first supported, and later opposed, are a classic example (▸ p. 35). Killing wolves and mountain lions may have boosted deer populations for hunters and saved a few livestock for ranchers, but we now have enough ecological knowledge to recognize that it also helped cause a population explosion of deer that has altered the vegetative structure of many North American forests.

Policy can also fail if a government does not live up to its responsibilities to protect its citizens or treat them equitably. This may occur when leaders allow themselves to be unduly influenced by *interest groups*, small groups of people seeking private gain that work against the larger public interest. Finally, the command-and-control approach can fail in that it may generate opposition among citizens to government policy. If citizens view laws and regulations primarily as restrictions on their freedom, those policies will not last long in a constitutional democracy.

Subsidies are a widespread economic policy tool

Shortcomings of the command-and-control approach have led many economists to advocate the use of economic tools to compensate for market failure. One set of economic policy tools aims to encourage industries or activities that are deemed desirable. Governments may give *tax breaks* to certain types of businesses or individuals, for instance. Relieving the tax burden lowers costs for the business or individual, thus assisting the desirable industry or activity.

Probably the most widespread economic policy tool is the **subsidy,** a government giveaway of cash or publicly owned resources that is intended to encourage a particular activity. National governments commonly provide subsidies to industries they judge to benefit the nation in some way. Subsidies can be used to promote environmentally sustainable activities, but all too often they have been used to prop up unsustainable ones. Subsidies judged to be harmful to the environment and to the economy total roughly $1.45 *trillion* yearly across the globe, according to British environmental scientist Norman Myers—an amount larger than the economies of all but five nations.

The average U.S. taxpayer pays $2,000 per year in environmentally harmful subsidies, plus $2,000 more through increased prices for goods and through degradation of ecosystem services, Myers estimates. U.S. subsidies for industries and activities promoting automobile transportation alone amount to $1,700 per taxpayer per year, and the nation's heavily subsidized gasoline is generally cheaper than bottled water. The most recent Green Scissors

Report estimates that in 2003, $58 billion of U.S. taxpayers' money was budgeted for 68 environmentally harmful subsidies. The Green Scissors Report is a project of 22 nongovernmental organizations such as Friends of the Earth, Taxpayers for Common Sense, and the U.S. Public Interest Research Group. Among the subsidies highlighted in recent reports are the following:

▶ *General Mining Law of 1872.* Each year, mining companies extract $500 million to $1 billion in minerals from U.S. public lands without paying a penny in royalties to the taxpayers who own these lands. Since this law was enacted, the U.S. government has given away over $245 billion of mineral resources, and mining activities have polluted more than 40% of the watersheds in the West. The 130-year-old act still allows mining companies to buy public lands for $5 or less per acre.

▶ *Coal subsidies.* Since 1984, Congress has made $1.8 billion available to the coal industry through the Clean Coal Technology Program, paying for industry research that may lead to technologies to reduce air pollution from coal combustion. An additional $800 million per year goes to the coal industry for further research and development. These amounts dwarf subsidies granted to less-polluting renewable energy sources.

▶ *Forest Service road-building subsidies.* The U.S. Forest Service manages taxpayer-owned forests for various uses, including timber harvesting. Although many people assume that the corporations that harvest trees from public forests cover the costs of roads required to get the logs out, this is not the case. From 1992 to 1997 the Forest Service spent more than $387 million in tax dollars for road construction for timber companies (Figure 3.14). Despite legislative reform, subsidies are currently estimated at $170 million over 5 years.

Advocates of sustainable resource use have long urged governments to subsidize environmentally sustainable activities instead. To some extent, this is being done. For instance, subsidies for renewable energy sources totaled nearly $1.1 billion in the United States in 1999, according to the U.S. Department of Energy. But this amount falls short of the $2.8 billion that went to nonrenewable energy sources in that same year.

Green taxes discourage undesirable activities

Another economic policy tool—taxation—can be used to discourage undesirable activities. Taxing undesirable activities helps to "internalize" external costs by making

FIGURE 3.14 When companies cut timber in U.S. national forests, roads must be built and maintained to enable access. The costs of these roads are paid by taxpayers.

them part of the overall cost of doing business. Taxes on environmentally harmful activities and products are called **green taxes.** By taxing activities and products that cause undesirable environmental impacts, a tax becomes a tool for policy as well as simply a way to fund government.

Green taxes have yet to gain widespread support in the United States, although similar "sin taxes" on cigarettes and alcohol are tools of U.S. social policy. Taxes on pollution have been widely instituted, however, in Europe, where many nations have adopted the *polluter pays principle.* This principle specifies that the price of a good or service should include all its costs, including costs of environmental degradation that would otherwise be passed on as external costs.

Under green taxation, a factory that pollutes a waterway would pay taxes based on the amount of pollution it discharges. The idea is to give companies a financial incentive to reduce pollution, while allowing the polluter the freedom to decide how best to minimize its expenses. One polluter might choose to invest in technologies to reduce its pollution if doing so is less costly than paying the taxes. Another polluter might find abating its pollution more costly, and could choose to pay the taxes instead—funds the government might then apply toward mitigating pollution in some other way. Green taxation provides incentive for industry to lower emissions not merely to a level specified in a regulation, but to still-lower levels. However, green taxes do have disadvantages. One is that businesses will most likely pass on their tax expenses to consumers and the increased costs may affect low-income consumers disproportionately more than high-income ones.

Markets in permits can save money and produce results

A still more creative market-based approach is for government to sell or give to companies the right to pollute, by establishing markets in tradable pollution permits. After determining the overall amount of pollution it will allow an entire industry to produce, the government can issue permits to individual polluters that allow them each to emit a certain fraction of that amount. Polluters are then allowed to buy, sell, and trade these permits with other polluters. With such **marketable emissions permits,** governments create incentives for firms to reduce their pollution in a way that is compatible with market capitalism.

Suppose, for example, you are a plant owner with permits to release 10 units of pollution, but you find that you can become more efficient and release only 5 units of pollution instead. You then have a surplus of permits, which might be very valuable to some other plant that is having trouble reducing its pollution, or to one that wants to expand production. In such a case, you can sell your extra permits. Doing so meets the needs of the other plant and generates income for you while preventing any increase in the total amount of pollution. Moreover, environmental organizations can buy up surplus permits and "retire" them, thus reducing the overall amount of pollution produced.

Such a system of marketable emissions permits has been in place in the United States. It was established by 1990 amendments to the Clean Air Act, which mandated reduced emissions of sulfur dioxide (Figure 3.15), a major contributor to acidic deposition (▸ pp. 514–518). Starting in 1995, permits were issued to power plants, allowing fewer emissions gradually year by year. Los Angeles had success in the 1990s with a similar program to reduce its smog. Marketable emissions permits generally end up costing both industry and government much less than a

FIGURE 3.15 Markets in emissions permits have worked effectively in the United States to decrease the sulfur dioxide pollution that contributes to acid rain. Currently, nations and companies are developing markets in carbon emissions trading, following ratification of the Kyoto Protocol.

conventional regulatory system. Savings from the Clean Air Act permits have been estimated to add up to several billion dollars per year. Although such "cap-and-trade" programs can succeed in reducing the overall amount of pollution, they do have the drawback of allowing hotspots of pollution to occur around plants that buy permits to pollute more.

A global market in carbon emissions is currently developing as a result of the Kyoto Protocol to address climate change (▸ pp. 550, 552). Under the Protocol, nations have targets for reducing their carbon emissions from power plants, automobiles, and other sources that are driving climate change. Nations can gain carbon credits by investing in renewable energy or reforesting landscapes, as well as by reducing emissions, and any excess permits to emit carbon can be sold to other nations. European nations and major energy companies have begun such a market, and within a few years it should be clear whether this will develop into the multibillion-dollar market that many analysts predict.

Market incentives are being tried widely on the local level

At a more local scale, you may well have already taken part in transactions involving financial incentives as policy tools. Many municipalities charge residents for waste disposal according to the amount of waste they generate.

Public versus Private: In Whose Best Interest?

Transboundary pollution problems frequently give rise to debates over the best management approach—for example, public versus private, or some combination thereof. **In your view, what are the appropriate roles of the public and the private sectors in resolving the sewage pollution problems of the San Diego–Tijuana region?**

Clean, Safe Water Should Not Be a For-Profit Enterprise

The persistent problem of water pollution near the U.S.–Mexico border is partly the result of breakdowns in engineering and technology, including inadequately designed sewage treatment plants and missing or broken collection pipelines. However, it also reflects a shared history of water mismanagement and a lack of cooperative binational planning efforts.

Water is vitally important to this arid and rapidly growing region, yet over 90% is imported, used once, then dumped into the sea via sewage outfalls. Renegade sewage flows from these pipelines often pollute the river and beaches. Ultimately, to have sufficient clean water for basic health and future development, both countries will need to reconsider this pattern of use and begin conserving, reusing, and reclaiming water.

The best solutions for protecting water quality and ensuring adequate supply and sufficient treatment will be based on international cooperation and long-term planning projects. Japan is already investing in water and wastewater projects in Baja California, and the United States is beginning to help by using EPA grants to develop long-term master plans for improving water quality in the region.

However, changes in attitude and water use require support from governmental, research, and community organizations. Border residents need to understand not only their rights to the water they use, but also their role in keeping it clean and using it responsibly.

Relying on private for-profit companies to solve water problems—especially if they do not involve the community in the planning and decision-making process—can result in higher costs and fees that many families cannot afford, and will do little to change the public mindset. Public access to clean, safe water is a basic human right and is needed to protect public health. If companies fail to make a profit, who will step in to ensure that safe drinking water is available and sewage is treated?

Lori Saldaña represents District 76 in California's State Assembly. Previously she was an environmental policy researcher, writer, and community activist, specializing in water issues along the U.S.–Mexico border.

Public-Private Partnerships Mean Faster and More Efficient Environmental Change

A policy that excludes private participation in public service projects is bad policy if the government is unwilling or unable to fund project implementation.

It is a philosophical debate, not a technical question: "Should the private sector, motivated by profit, be able to participate in providing public services that are traditionally the exclusive realm of the public sector?" The private sector is more innovative, efficient, and flexible. The public-private partnership model (PPV) also provides for full disclosure, transparency, public comment, oversight, and regulatory control. The private sector can more often implement projects faster, and at a lower cost, than a lumbering government. PPV models incorporate a benefit unavailable in public sector projects: the ability to regulate by contract (as well as by statute). Desired environmental benefits are attained, while profit invigorates the market.

The Bajagua Project, a proposed PPV development model wastewater treatment plant to be built in Tijuana, Mexico, is such a project. It can be built faster and for less than any public facility. Bajagua will use technology selected by the U.S. EPA. It must comply with NEPA and all U.S., Mexican, and Californian environmental laws. Bajagua must also satisfy all contractual obligations *before* payment is due from the government, more closely emulating the "pay for services rendered" transaction model that we use every day. A policy where the public gets the environmental benefit it desires before it pays is a good policy.

In 1993, Congress capped the amount to be spent on the International Wastewater Treatment Plant. The federal agency that built the primary module in 1999 exhausted its funding and now cannot build the secondary module. Yet San Diego's beaches remain polluted, local economies remain depressed, and the federal agency remains in violation of the Clean Water Act. Clearly, PPV offers the only real possibility of success for positive environmental change.

Gary L. Sirota is a business and environmental litigator, consultant in international PPV infrastructure development, and a specialist in legislative finance and policy analysis.

Explore this issue further by accessing **Viewpoints** at www.aw-bc.com/withgott.

Other cities place taxes or disposal fees on items that require costly safe disposal, such as tires and motor oil. Still others give rebates to residents who buy water-efficient toilets and appliances, because this can cost the city less than upgrading its sewage treatment system. Likewise, power companies sometimes offer discounts to customers who buy high-efficiency lightbulbs and appliances, because doing so is cheaper for the utilities than expanding the generating capacity of their plants.

At all levels, from the local to the international, market-based incentives can reduce environmental impact while minimizing overall costs to industry and easing concerns about the intrusiveness of government regulation. Command-and-control policy is straightforward to implement, easy to monitor, and frequently works. Market-based approaches can be more complicated, but if they work, they can lessen environmental impact at a lower overall cost.

Weighing The Issues:
Environmental Policy Approaches

Imagine a factory is polluting a river near where you live. You organize your neighbors and complain to your Congresswoman, who promises to try to help solve the problem. What approach to environmental policy would you urge her to try, and why?

The Environmental Policy Process

Anyone can become involved in helping ideas become public policy. In the U.S. system, it is true that each and every person has a political voice and can make a difference. Unfortunately, it is also true that money wields influence and that some people and organizations are far more politically connected and influential than others. We will explore some of the ways people make themselves influential as we examine the main steps of the policymaking process. Our discussion pertains both to citizens at the grassroots level and to large organizations and corporations.

The environmental policy process begins when a problem is identified

The first step in the environmental policy process is to identify an environmental problem (Figure 3.16). Identifying a problem requires curiosity, observation, record keeping,

❶ Identify problem

❷ Identify specific causes of the problem

❸ Envision solution and set goals

❹ Get organized

❺ Cultivate access and influence

❻ Manage development of policy

FIGURE 3.16 Understanding the steps of the policy process is an essential element of solving environmental problems.

The Science behind the Story

Spotting Sewage by Satellite

For decades, San Diego's sewage-contaminated coast remained a stubborn mystery—until scientists turned to the sky.

Water quality along San Diego's southern beaches had begun steadily declining in the 1960s, but it was supposed to be improving for surfers and swimmers in the late 1990s. Some key sewage sources had been pinpointed along the U.S.–Mexico border, and the U.S. government had spent $260 million to build a new treatment plant in the Tijuana River Valley and clean up coastal water.

Yet bacteria levels in coastal waters remained troublesome. Between 2000 and 2003, authorities had to close various San Diego County beaches because of sewage pollution 168 times for a total of 687 beach-days. During this time period, pollution advisories were issued 920 times over a total of 5,658 beach-days. The city of Imperial Beach, near the border, saw its beach closed 161 days in 1998 because of sewage contamination. Researchers and policymakers wondered what the earlier cleanup efforts had missed. Were existing sewage treatment plants not operating as well as planned? Or had past tracking efforts overlooked some pollution sources?

In 1999, local water quality officials got together with environmental scientists at Ocean Imaging, Inc., a company based near San Diego, and came up with a hypothesis—and a plan. Undetected pollution sources likely existed along the U.S.–Mexico border, they surmised, and those sources must be large enough to cause widespread contamination. They thought such pollution flows might be traceable using an innovative type of airborne imaging technology, known as synthetic aperture radar, or SAR.

SAR was originally developed as a military tracking tool, but through a joint effort between Ocean Imaging and NASA, the technology is now used to study environmental problems. Installed aboard an airplane or satellite, SAR instruments bounce radar signals off Earth's surface and record their echoes. The echoes change in frequency, based on the material being measured. For example, clean ocean water will return one type of signal, whereas large bodies of spilled sewage or oil suspended in ocean water will usually return a different type.

By analyzing the signals received, researchers can create detailed images of the surfaces they are studying. Marine sewage flows or oil spills will often show up on SAR images as dark plumes or patches in ocean water (see the figure). SAR can function through clouds and at night, making it invaluable in tracking something as mobile and unpredictable as a sewage spill.

Using SAR to scan the coast along the U.S.–Mexico border, scientists quickly found what they had predicted—a little-known source of almost entirely raw

and an awareness of our relationship with the environment. For example, assessing the contamination of San Diego- and Tijuana-area beaches required understanding the ecological and health impacts of untreated sewage. It also required being able to detect contamination on beaches and understanding water flow dynamics among the beaches, the Pacific Ocean, and the Tijuana River watershed (see "The Science behind the Story," above).

Identifying causes of the problem is the second step in the policy process

Once an individual or group has defined a particular environmental problem, discovering specific causes of the problem is next on the agenda. A person seeking causes for pollution in the Tijuana River watershed might notice that transboundary sewage spills took on a more toxic and industrial nature during the mid-1960s, when U.S.-based companies began opening *maquiladoras* on the Mexican side of the border. Advocates of the *maquiladora* system argue that these factories provide much-needed jobs south of the border while keeping companies' costs low by paying Mexican workers far less than U.S. workers. Critics argue that the factories are waste-generating, water-guzzling polluters whose transboundary nature makes them particularly difficult to regulate.

Identifying problems and their causes requires that scientific research play a major role in the policy

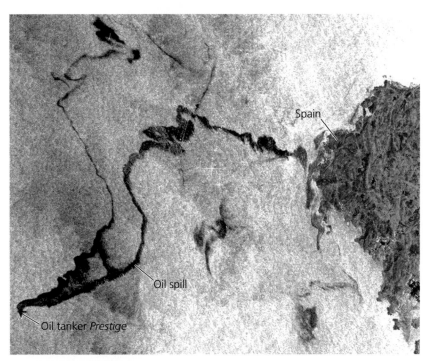

The same radar technology that has mapped San Diego's and Tijuana's sewage flows has also been used to track oil spills at sea. Here, a synthetic aperture radar image shows the sinuous trail of oil spilling from the tanker *Prestige*, which broke apart off the coast of Spain in November 2002.

sewage dumping into the surf in northern Mexico. The sewage, most likely from residential areas, was flowing through a contaminated creek at a rate of 95–133 million L (25–35 million gal) per day. Once in the ocean, the sewage flow showed up in the satellite image as a dark plume of disturbed water that, carried by currents, stretched north along the San Diego coast. Water samples taken at the same time from boats on the water confirmed that high levels of bacteria followed the plume's path.

These findings, made public in 2000, gave San Diego badly needed answers to its pollution puzzle. Everyone from environmental scientists to surfers could now "visualize what is going on," as a San Diego congressman said when the first SAR images were shown. Local policymakers used the report to lobby for more sewage tracking and cleanup efforts along the border. The SAR research gave a push to the Tijuana River Valley Estuary and Beach Sewage Cleanup Act, as well as the cross-border master plan to improve water quality in Tijuana. By unveiling an unknown source of a persistent problem, radar-based pollution tracking allowed San Diego and Tijuana to improve the cleanup of their polluted beaches.

process. Much of this work takes place in the arena of *risk assessment* (▸ pp. 422–423), in which scientists evaluate the extent and nature of problems and judge the risks that they pose to public health or environmental quality.

The third step is envisioning a solution

The better one can identify specific causes of a problem, the more effectively one will be able to envision solutions to it and argue for implementing those solutions. Science plays a role here too, through the process of *risk management* (▸ pp. 423–424), in which scientists develop strategies to minimize risk. Frequently, however, solutions will involve social or political action. In San Diego, citizen activists wanted Tijuana to enforce its own pollution laws more effectively—something that, once visualized, started to happen when San Diego city employees began training and working with their Mexican counterparts to keep hazardous wastes out of the sewage treatment system.

Getting organized is the fourth step

When it comes to gaining the ear of elected officials and influencing policy, organizations are generally more effective than lone individuals. The sole critic or crusader is easily dismissed as a crackpot or troublemaker, but a

The Science behind the Story

Assessing the Environmental Impact of Treating Transboundary Sewage

In 1990 the United States and Mexico formally agreed to construct a water treatment plant in the U.S. portion of the Tijuana River Valley. Before the South Bay International Water Treatment Plant (IWTP) could be built, however, the U.S. government was legally required by the National Environmental Policy Act (NEPA) to assess its environmental impact.

NEPA's environmental impact statement (EIS) process provides the framework for environmental impact assessment, but it is not the only law the IWTP's planners had to take into account. The federal Clean Water Act requires that harmful toxins and bacteria be removed from wastewater discharged into U.S. rivers, lakes, and oceans, and the Endangered Species Act protects species such as the Pacific pocket mouse, an inhabitant of the Tijuana River Valley.

International treaties also constrained the IWTP. For example, one 1989 agreement between the United States and Mexico required that the plant be funded by the EPA, be built on U.S. territory,

The IWTP gives sewage primary treatment, but still needs to fund facilities to provide secondary treatment. Currently, treated waste is discharged into the Pacific Ocean at the South Bay Ocean Outfall, shown here during its construction.

and treat at least 1,095 L/sec (25 million gal/day). Finally, state and local laws regulated the impact of the plant on nearby communities and on the quality of California's coastal waters. Before construction could begin, all these constraints had to be addressed by the EPA and the U.S. section of the International Boundary and Water Commission (IBWC), the organization that would own and operate the plant.

In 1991 a draft EIS for the IWTP was released for public comment. The draft provided a preliminary assessment of the plant's impact on biological and cultural resources, public health and safety, scenic and recreational values, water quality, and other environmental factors. Three years later, after extensive research and public discussion, a final EIS was released. In it, the EPA and the IBWC endorsed the plant as the

group of hundreds or thousands of individuals is not as easily dismissed. Furthermore, organizations are more effective at raising funds, which by U.S. law they are permitted to contribute to political campaigns.

As effective as large organizations can be, it is important to remember that small coalitions and even individual citizens who are motivated, informed, and organized can solve environmental problems. As renowned anthropologist Margaret Meade once remarked, "Never doubt that a small group of thoughtful, committed citizens can change the world—indeed it is the only thing that ever has." San Diego-area resident Lori Saldaña provides an

example. Concerned about the Tijuana River's pollution, Saldaña reviewed plans for the international wastewater treatment plant that the U.S. government proposed to build (see "The Science behind the Story," above). She concluded that it would merely shift pollution from the river to the ocean, where sewage would be released 5.6 km (3.5 mi) offshore. Working with her local Sierra Club chapter, Saldaña protested the plant's design and participated in a lawsuit that forced the government to conduct further studies. The EPA finally agreed to a design change, although funding for it has stalled in Congress. For her efforts, Saldaña received awards and was

"preferred alternative," and the EPA confirmed the endorsement with an official Record of Decision.

In its Record of Decision, the EPA made several choices about the IWTP's design and operation that became controversial. One choice concerned opening the plant in phases. In the first phase, expected to last 2 years, large solids and some suspended particles would be filtered out of the wastewater using an advanced primary treatment process. However, other pollutants—including toxic metals and bacteria—would remain untreated until facilities were constructed for a secondary treatment process known as activated sludge, in which microorganisms convert highly toxic waste into a less toxic substance (see ▸ pp. 458–461 for more information on wastewater treatment). As long as those facilities remained uncompleted, the plant would be releasing polluted water into the ocean, in violation of the Clean Water Act.

In response to the EIS and Record of Decision, two environmental groups sued the EPA and the IBWC. In their lawsuit, the groups argued that the EIS had failed to consider a type of secondary treatment facility known as a completely mixed aerated pond system. The suit was eventually settled out of court, but it helped spur the EPA and IBWC to conduct a supplemental EIS in which they considered seven different secondary treatment alternatives, including activated sludge and the pond system. Compared to activated sludge, the pond system would produce a smaller volume of toxic by-products and would better absorb spikes of high toxicity. Based on these considerations, the EPA eventually decided to endorse the pond system.

In the supplemental EIS, the EPA also reaffirmed its decision to release treated wastewater through the South Bay Ocean Outfall, a 5.8-km (3.6-mi) underwater tunnel with outlets 29 m (95 ft) below the ocean surface (see figure). The decision was based on a computer model of ocean currents and pollution levels off the California coast, which indicated that wastewater released through the outfall—once treated to the secondary level— would be sufficiently diluted to meet federal and state pollution standards.

Even though the formal EIS process has now largely been completed, scientific findings continue to shape the future of the IWTP. Tests conducted in 1997 and 1998 indicated that wastewater treated to the advanced primary level was still acutely toxic to fish and other marine life. The findings suggested that Tijuana's sewage would require more aggressive treatment than most wastewater generated in the United States. In 1999, local activist (and now State Assemblywoman) Lori Saldaña and oceanographer Tim Baumgartner studied water quality above the outfall. They found a noticeable decrease in water quality compared to nearby areas. Saldaña and Baumgartner posted their results on the Internet and urged policymakers to speed construction of the IWTP's secondary treatment facilities, now scheduled to be completed by 2007. The federal government is also funding a more formal monitoring process to ensure that the plant's outflow does not damage California's coastal environment.

appointed to a commission on border environmental issues by President Bill Clinton. After a decade of activism, Saldaña ran for the California State Assembly in 2004 and won. She is now the representative from California's 76th district.

Gaining access to political powerbrokers is the fifth step

The fifth step in the policy process entails gaining access to policymakers who have the clout to help enact the desired changes. People gain access and influence through lobbying, campaign contributions, and the revolving door.

Lobbying Anyone who spends time or money trying to change an elected official's mind is engaged in **lobbying**. The term was originally used to describe the activities of corporate representatives who loitered in the lobbies of Washington, D.C., establishments for opportunities to talk with members of Congress. Although anyone can lobby, it is much more difficult for an ordinary citizen than for the thousands of full-time professional lobbyists employed by the many businesses and organizations seeking a voice in Washington politics. Opponents of environmental advocacy often paint large environmental organizations, such as the National Audubon Society, as interest groups with inordinate lobbying power in Washington. But according to a 2002 report by *Fortune* magazine, these organizations are surprisingly weak. The highest-ranking environmental organization on *Fortune*'s list, the Sierra Club, ranked only at number 52. This ranking placed the Sierra Club among such groups as the Distilled Spirits Council of the United States and the National Association of Letter Carriers, far back from such heavyweights as the National Rifle Association (NRA) and the American Association of Retired Persons (AARP). Indeed, the American Petroleum Institute spends on lobbying nearly as much as the entire budgets of the top five U.S. environmental advocacy groups combined.

Campaign contributions For those of us who can't spend our time hanging around Washington lobbies, supporting a candidate's reelection efforts with money is another way to make our voices heard. Because environmental policy often regulates the activities of corporations, they have a strong interest in shaping it. Although corporations and industries may not legally make direct campaign contributions, they are allowed to establish *political action committees (PACs)* for that purpose. Generally affiliated with industries, environmental groups, and other organizations with an interest in election outcomes, PACs raise money and distribute it to political campaigns, helping like-minded candidates win elections, in hope of gaining access to those individuals once they are elected.

The revolving door Some individuals employed by government-regulated industries gain political influence when they take jobs with the very government agencies responsible for regulating their industry. Businesses also often hire former government bureaucrats who regulated their industries. Such movement of individuals between the private sector and government agencies is known as the **revolving door.** For example, the George W. Bush Administration has included a commerce secretary who was CEO of a petroleum company, a transportation secretary who worked for a leading corporation in the transportation industry, and an agriculture secretary who served on the board of the first company in the nation to market genetically engineered crops. Proponents of the revolving door system assert that corporate executives who take government jobs regulating their own industry bring with them an intimate knowledge of that industry. This inside experience makes them highly qualified and likely to benefit the nation with especially well-informed policy, proponents maintain. Critics of the system contend that taking a job regulating your former employer is a clear conflict of interest that undermines the effectiveness of the regulatory process. Regulators from the private sector will be biased toward assisting their industry, critics say, and may fail to enforce regulations, thus acting against the interests of the taxpayers who pay their salaries.

Weighing the **Issues:**
The Revolving Door

What do you think of each of the arguments for and against the revolving door system? Would the citizens of the United States be better off with or without the revolving door? How could we encourage it or discourage it?

Shepherding a solution into law is the sixth step in the policy process

Whether you're a corporate lobbyist or a grassroots activist, once your organization has the access and clout to influence policymakers, the better-known parts of the policy process come into play. Having gained access to elected officials and convinced them to hear your requests, you may be asked to prepare a bill, or draft law, that embodies the solutions you seek. Anyone can draft a bill. The hard part is finding members of the House and Senate willing to introduce the bill and shepherd it from subcommittee through full committee and on to passage by the full Congress. Lobbying and media attention intensify as the bill progresses through this process (Figure 3.17). If it passes through all of these steps, the bill may become law, but it can die in countless fashions along the way.

Of course, the policy process does not end with the enactment of legislation. Following a law's enactment, administrative agencies implement regulations. Policymakers also evaluate the policy's successes and failures and may revise the policy as necessary. Moreover, the judicial branch interprets law in response to suits in the courts, and much environmental policy has lived and died by judicial interpretation. The full policy process is long and often cumbersome, but it has resulted in a great deal of effective governance in constitutional democracies in the United States and many other nations.

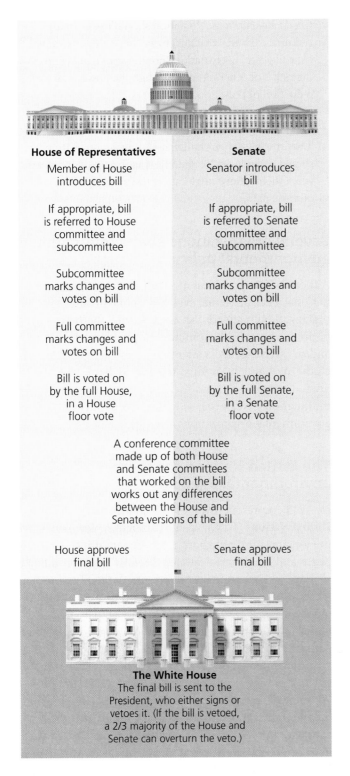

House of Representatives	Senate
Member of House introduces bill	Senator introduces bill
If appropriate, bill is referred to House committee and subcommittee	If appropriate, bill is referred to Senate committee and subcommittee
Subcommittee marks changes and votes on bill	Subcommittee marks changes and votes on bill
Full committee marks changes and votes on bill	Full committee marks changes and votes on bill
Bill is voted on by the full House, in a House floor vote	Bill is voted on by the full Senate, in a Senate floor vote

A conference committee made up of both House and Senate committees that worked on the bill works out any differences between the House and Senate versions of the bill

House approves final bill	Senate approves final bill

The White House
The final bill is sent to the President, who either signs or vetoes it. (If the bill is vetoed, a 2/3 majority of the House and Senate can overturn the veto.)

FIGURE 3.17 Before a bill becomes a law, it must clear a number of hurdles in both legislative bodies. If the bill passes the House and Senate, a conference committee must work out any differences between the House and Senate versions before the bill is sent to the president. The president may then sign or veto the bill.

International Environmental Policy

Environmental systems pay no heed to political boundaries, so environmental problems often are not restricted to the confines of particular countries. For instance, because most of the world's major rivers cross international borders, problems like those along the Tijuana River are frequently international in nature. Because U.S. law has no authority in Mexico or any other nation outside the United States, international law is vital to solving transboundary problems.

Mexico and the United States are working together to manage water

Often countries make progress on international issues not through legislation, but through creative bilateral or multilateral agreements hammered out after a lot of hard work and diplomacy. Such was the case with the successful effort to develop a long-term plan to manage drinking water and wastewater in the Tijuana metropolitan area. This master planning process, funded by the U.S. EPA under Congressional direction, began in January 2002. It involved the Comisión Estatal de Servicios Publicos de Tijuana (CESPT), the city agency that manages water and wastewater in Tijuana; the Mexican National Water Commission; the State Water Commission for Baja California; and the North American Development Bank. The resulting Tijuana Master Plan for Water and Wastewater Infrastructure aims to address a shortage of drinking water and possibilities for its reuse; water infrastructure; wastewater collection and transport; and wastewater treatment.

In 2002 the U.S. Congress also provided funding and authority for the EPA to work with CESPT to upgrade Tijuana's sewer system. The Tijuana Sewer Rehabilitation Project, known as Tijuana Sana ("Healthy Tijuana") was approved in 2001. This cooperative transboundary pollution prevention program will attempt to repair leaky sewer pipes in Tijuana. The 4-year, $43 million project will replace 131 km (81 mi), or 7.5%, of Tijuana's sewer pipes. If it succeeds, the project should reduce or eliminate the most severe sewage spills into the Tijuana River.

International law includes conventional law and customary law

Because solving transboundary dilemmas requires international cooperation, several principles of international law and a number of international organizations have arisen. Whereas U.S. law arises from the Constitution and the Bill of Rights, international environmental law is more nebulous in its origins and authorities.

International law known as **conventional law** arises from *conventions,* or *treaties,* into which nations enter. One example is the Montreal Protocol, a 1987 accord among more than 160 nations to reduce the emission of airborne chemicals that deplete the ozone layer (▶ pp. 513–514). Another example is the Kyoto Protocol to reduce fossil-fuel emissions that contribute to global climate change (▶ p. 530). This agreement took effect in February 2005 without United States participation, after a quorum of nations had ratified the accord. In 1990 Mexico and the United States signed a treaty and agreed to build an international wastewater treatment plant to handle excess sewage from Tijuana (Figure 3.18). In this case the treaty process worked well, but the results fell short of expectations. The facility reached its capacity within 3 years and then began discharging material to the ocean that did not meet safety standards established by U.S. law.

Other international law arises from long-standing practices, or customs, held in common by most cultures. This is known as **customary law.** Four principles underlie customary law as it applies to environmental issues:

FIGURE 3.18 In July of 1990, the United States and Mexico entered into a treaty and agreed to build the International Wastewater Treatment Plant (IWTP) to handle excess sewage from Tijuana. The IWTP began operating just north of the border in 1997 and treats up to 95 million L (25 million gal) of Mexican sewage each day. The IWTP has provided great benefits, but it reached its capacity within 3 years of opening.

▶ *Good neighborliness.* No nation should use its natural resources in a way that adversely affects other nations.

▶ *Due diligence.* Every nation should respect and protect the rights of other nations through the prevention and reduction of pollution.

▶ *Equitable resource use.* No nation should use more than its share of a natural resource.

▶ *The principle of information and cooperation.* Nations should provide notice and information to other nations when their actions might affect the interests and affairs of those nations.

Several organizations shape international environmental policy

Although there is no real mechanism for enforcing international environmental law, a number of international organizations regularly act to influence the behavior of nations by providing funding, applying peer pressure, and/or directing media attention. These organizations include the United Nations, the World Bank, the World Trade Organization, and the European Union and other multinational consortia, as well as a wide variety of non-governmental organizations (NGOs).

The United Nations sponsors environmental agencies

In 1945, representatives of 50 countries founded the **United Nations (U.N.).** Headquartered in New York City, this organization's purpose is "to maintain international peace and security; to develop friendly relations among nations; to cooperate in solving international economic, social, cultural and humanitarian problems and in promoting respect for human rights and fundamental freedoms; and to be a centre for harmonizing the actions of nations in attaining these ends."

The United Nations has taken an active role in shaping international environmental policy (Figure 3.19). Of several agencies within it that influence environmental policy, most notable is the United Nations Environment Programme (UNEP), created in 1972, which helps nations understand and solve environmental problems. Based in Nairobi, Kenya, its mission is sustainability, enabling countries and their citizens "to improve their quality of life without compromising that of future generations." UNEP's extensive research and outreach activities provide a wealth of information useful to policymakers and scientists throughout the world and have provided a good deal of the data cited throughout this book.

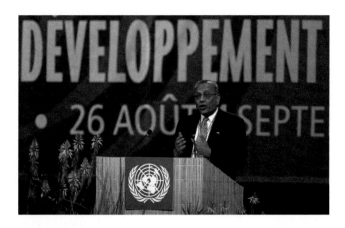

FIGURE 3.19 The United Nations is active in international environmental policymaking. For instance, it sponsored the 2002 Earth Summit in Johannesburg, South Africa.

The World Bank holds the purse strings for development

The United Nations can encourage cooperation and promote awareness of environmental problems, but the World Bank holds the purse strings. Established in 1944 and based in Washington, D.C., the **World Bank** is one of the globe's largest sources of funding for economic development. This institution can shape environmental policy through its funding of major development projects, including dams, irrigation infrastructure, and other undertakings. In 2004 the World Bank provided over $20.1 billion in loans for projects designed to benefit the poorest people in the poorest countries around the world.

Despite its admirable mission, the World Bank has frequently been criticized for funding unsustainable projects that cause more environmental problems than they solve. In 1987 the World Bank reorganized itself and established an environmental office. Since then, the Bank has given higher priority to assessing the environmental impacts of its projects. In 2001 the World Bank issued its "Environmental Strategy," a document intended as a guide to improve quality of life, promote sustainability, and protect regional and global commons. Providing for the needs of growing human populations in poor nations while minimizing damage to the environmental systems on which people depend can be a tough balancing act. Environmental scientists today agree that the concept of sustainable development must be the guiding principle for such efforts.

The European Union is active in environmental affairs

Like the World Bank and the United Nations, the **European Union (EU)** was not established primarily with environmental problem solving in mind. However, the treaty that created it held as one of its goals the promotion of solutions to environmental problems. Formed after World War II, the EU as of early 2005 contained 25 member nations. It seeks to promote Europe's unity and its economic and social progress (including environmental protection) and to "assert Europe's role in the world." The EU can sign binding treaties on behalf of its members and can enact regulations that have the same authority as national laws in each member nation. It can also issue *directives,* which are more advisory in nature. The EU's European Environment Agency works to address waste management, noise pollution, water pollution, air pollution, habitat degradation, and natural hazards. The EU also seeks to remove trade barriers among member nations. It has classified some nations' environmental regulations as barriers to trade because some northern European nations have traditionally had more stringent environmental laws that prevent the import and sale of environmentally harmful products from other member nations.

The World Trade Organization has recently attained surprising power

Based in Geneva, Switzerland, the **World Trade Organization (WTO)** was established in 1995, having grown from a 50-year-old international trade agreement. The WTO represents multinational corporations and promotes free trade by reducing obstacles to international commerce and enforcing fairness among nations in trading practices. Whereas the United Nations and the European Union have limited influence over nations' internal affairs, the WTO has real authority to impose financial penalties on nations that do not comply with its directives. These penalties can on occasion play major roles in shaping environmental policy.

Like the EU, the WTO has interpreted some national environmental laws as unfair barriers to trade. For instance, in 1995 the U.S. EPA issued regulations requiring cleaner-burning gasoline in U.S. cities, following Congress's amendments of the Clean Air Act. Brazil and Venezuela filed a complaint with the WTO, saying the new rules unfairly discriminated against the petroleum they exported to the United States, which did not burn as cleanly. The WTO agreed, ruling that even though the South American gasoline posed a threat to human health in the United States, the EPA rules represented an illegal trade barrier. The ruling forced the United States to alter its approach to regulating gasoline. Not surprisingly, critics have frequently charged that the WTO aggravates environmental problems.

- -
Weighing the Issues:
Trade Barriers and Environmental Protection

If Nation A has stricter laws for environmental protection than Nation B, and if these laws restrict the ability of Nation B to export its goods to Nation A, then by the policy of the WTO and the EU, Nation A's environmental protection laws could be overruled in the name of free trade. Do you think this is right? What if Nation A is a wealthy industrialized country and Nation B is a poor developing country that needs every economic boost it can get?

- -

Nongovernmental organizations also exert influence

A number of NGOs have grown to become international in scope and exert influence over international environmental policy (Figure 3.20). The nature of these advocacy groups is diverse. Some, such as The Nature Conservancy, focus on accomplishing conservation objectives on the ground (in its case, purchasing and managing land and habitat for rare species) without becoming politically involved. Other

FIGURE 3.20 Pursuing different visions of what makes for international environmental progress, nongovernmental organizations, such as the environmental advocacy group Greenpeace, sometimes clash with international institutions, such as the World Bank and the World Trade Organization.

groups, including Conservation International, the World Wide Fund for Nature, Greenpeace, Population Connection, and many others, attempt to shape policy directly or indirectly through research, education, lobbying, or protest. NGOs apply more funding and expertise to environmental problems, and conduct more research intended to solve them, than do many national governments.

International institutions and dynamics become more important in a globalizing world

As globalization proceeds, our world is becoming ever more interconnected. As a result, both human societies and Earth's ecological systems are being altered at unprecedented rates. Trade and technology have expanded the global reach of all societies, especially those such as the United States, which consume resources from across the world. Highly consumptive nations that import goods and resources from far and wide exert extensive impacts on the planet's environmental systems. Multinational corporations operate outside the reach of national laws and all too often have little incentive to conserve resources or conduct their business sustainably in the nations where they operate. For all these reasons, in today's globalizing world the organizations and institutions that influence international policy are becoming increasingly vital.

Conclusion

Environmental policy is a problem-solving tool that makes use of science, ethics, and economics, and requires an astute understanding of the political process. Conventional command-and-control approaches of legislation and regulation are the most common approaches to policymaking, but various innovative economic policy tools have also been developed. As we have seen in the case of the Tijuana River, environmental issues often overlap political boundaries and require international cooperation. Through the hard work of concerned citizens interacting with their government representatives, the political process eventually produced promising solutions in the Tijuana River Valley Estuary and Beach Sewage Cleanup Act (renewed in 2004) and in binational agreements and management plans. We will draw on the fundamentals of environmental policy introduced in this chapter throughout the remainder of this book. By understanding these fundamentals, you will be well equipped to develop your own creative solutions to many of the challenging problems we will encounter.

REVIEWING OBJECTIVES

You should now be able to:

Describe environmental policy and assess its societal context

▶ Policy is a tool for decision making and problem solving that makes use of information from science and values from ethics and economics. (pp. 59–60)

▶ Environmental policy is designed to protect natural resources and environmental amenities from degradation or depletion, and to promote equitable treatment of people. (pp. 60–61)

Identify the institutions important to U.S. environmental policy and recognize major U.S. environmental laws

▶ The legislative, executive, and judicial branches, together with administrative agencies, all play roles in U. S. environmental policy. (pp. 63–65)

▶ U. S. environmental policy came in three waves. The first encouraged frontier expansion and resource extraction. The second aimed to mitigate impacts of the first. The third targeted pollution and gave us many of today's major environmental laws. (pp. 66–70)

▶ Some major U.S. laws include the National Environmental Policy Act, the Clean Air Act, and the Clean Water Act. (pp. 68–70)

Categorize the different approaches to environmental policy

▶ Legislation from Congress and regulations from administrative agencies make up most federal policy.

These top-down approaches are referred to as "command-and-control." (p. 71)

▶ Economic approaches include subsidies, green taxation, and market-based permit trading. (pp. 71–73, 75)

Delineate the steps of the environmental policy process and evaluate its effectiveness

▶ The policy process entails several steps: (1) identifying the problem, (2) identifying causes of the problem, (3) envisioning solutions, (4) getting organized, (5) gaining access to power, and (6) guiding a solution into law. (pp. 75–80)

▶ In a democracy, anyone can use the policy process, although corporations and organizations with money and resources tend to have the most clout. (pp. 77–80)

List the institutions involved with international environmental policy and describe how nations handle transboundary issues

▶ Many environmental problems cross political boundaries and thus must be addressed internationally. (pp. 81–84)

▶ International policy includes conventional law (law by treaty) and customary law (law by shared traditional custom). (pp. 81–82)

▶ Institutions such as the United Nations, European Union, World Bank, World Trade Organization, and nongovernmental organizations all play roles in international policy. (pp. 82–84)

TESTING YOUR COMPREHENSION

1. Describe and critique two common justifications for environmental policy. Explain the concept of external costs, and state why it is relevant to environmental policy.
2. Outline the primary responsibilities of the legislative, executive, and judicial branches of the U.S. government. What is the "fourth branch" of the U.S. government?
3. What is meant by a *regulatory taking*?
4. Summarize the differences between the first, second, and third waves of environmental policy in U.S. history.
5. What did the National Environmental Policy Act accomplish? Briefly describe the origin and mission of the U.S. Environmental Protection Agency.

6. Differentiate between a green tax, a subsidy, a tax break, and a marketable emissions permit.
7. Describe the environmental policy process, from identification of a problem through enactment of a federal law.
8. What kinds of things can an individual citizen do to become influential in the policymaking process?
9. What is the difference between conventional law and customary law? What special difficulties do transboundary environmental problems present?
10. Why are environmental regulations sometimes considered to be unfair barriers to trade?

SEEKING SOLUTIONS

1. Do we need environmental policy? Why or why not?

2. Imagine that you live in the San Diego area and cannot safely use beaches in your neighborhood because of water pollution that originated in Mexico. Who do you think should pay to prevent the pollution of your beaches? You? The Mexican government? The state of California or the U.S. government? *Maquiladoras?* What are the pros and cons of each of these potential funding sources? Now imagine that you live in Mexico in the Tijuana River watershed and depend on its water for your drinking, washing, and the irrigation of your garden. Who do you think should pay to prevent the pollution of your water supply?

3. Compare the main approaches to environmental policy— command-and-control, tort law, and economic or market-based approaches. Can you name an advantage and disadvantage of each? Do you think any one approach is most effective? Could we do with just one approach, or does it help to have more than one?

4. Reflect on the causes for the transitions in U.S. history from one type of environmental policy to another. Now peer into the future, and think about how life might be different in 25, 50, or 100 years. What would you speculate about the environmental policy of the future? What issues might it address? Do you predict we will have more or less environmental policy?

5. Think of one environmental issue that you would like to see solved through legislation. From what you've learned about the policymaking process, how do you think you could best shepherd your ideas through the process?

6. Compare the roles of the United Nations, the European Union, the World Bank, the World Trade Organization, and nongovernmental organizations. If you could gain the support of just one of these institutions for a policy you favored, which would you choose? Why?

INTERPRETING GRAPHS AND DATA

The Clean Air Act legislation of 1970, 1977, and 1990 was designed to improve air quality in the United States by monitoring and reducing the emissions of air pollutants judged to pose threats to human health, such as carbon monoxide, nitrogen dioxide, sulfur dioxide, ozone, particulate matter, and lead (▶ pp. 507–508). The main source of lead emissions in 1970 was the exhaust of vehicles burning gasoline to which tetra-ethyl lead had been added to improve combustion. By 1985, leaded gasoline was phased out of use, although airplanes and racecars were exempted.

The 1990 amendments addressed the growing problem of urban smog by requiring the use of reformulated gas (RFG) in cities with the worst smog problems. One of the RFG requirements specifies 2% oxygen content in fuel, which has been met by adding either ethanol (▶ pp. 611–612) or methyl tert-butyl ether (MTBE). Although it burns cleanly, MTBE is water-soluble and may cause cancer, so groundwater contamination from fuel spills is a concern. Sixteen states that collectively account for 45% of U.S. consumption of MTBE have now passed legislation banning or restricting its use. The following graph shows trends in U.S. lead emissions and MTBE consumption since 1970.

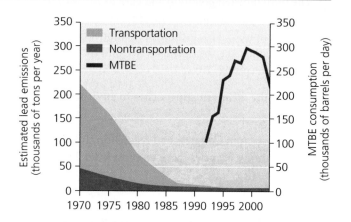

Estimated lead emissions from transportation and nontransportation sources, and consumption of MTBE in the United States. Data from U.S. Department of Transportation, Bureau of Transportation Statistics

1. Did policy resulting from Clean Air Act legislation succeed in reducing the public health risk from exposure to lead? Use data provided in the graph to support your answer.

2. In the 1990s, use of MTBE in RFG increased rapidly on the east and west coasts of the United States. Ethanol blends were used mostly in the corn-growing states of the central part of the country because ethanol is made from

corn, and so provides a market for in-state farmers while low transportation costs make it cheaper for in-state consumers. As MTBE use is being phased out in many states, use of ethanol is increasing. Name three ways in which these trends illustrate how state government actions may differ from federal government actions.

3. Do you think it is better policy to regulate air and water quality separately (e.g., under the Clean Air Act and Clean Water Act) or together under a single, comprehensive act to prevent and control pollution (as some European nations do)? How do the data in the graph support your reasoning?

CALCULATING ECOLOGICAL FOOTPRINTS

How many of us think about the destination of our waste when we flush the toilet? Some nutrient pollution from sewage generally ends up in a local waterway, but the U.S. Environmental Protection Agency establishes strict discharge standards following amendments to the U.S. Clean Water Act. One measure used is total suspended solids (TSS). To comply with federal regulations, wastewater treatment facilities in the United States must discharge water with a monthly average TSS no greater than 30 mg per liter.

   Assuming that the average toilet flush is 3 gallons and occurs 4 times per person per day, calculate the total discharge of TSS in wastewater with the maximum permissible content. Note that 1 gal = 3.7 L, 1,000 mg = 1 g, and 1,000 g = 2.2 lbs.

	TSS discharged per day	TSS discharged per month	TSS discharged per year
You	0.440 g	13.4 g	160.6 g
Your class			
Your hometown			
United States			

1. Assuming that water in an average toilet flush contains 100 g TSS, calculate the amount of TSS removed from the nation's wastewater each day, month, and year, if the EPA standard is met.
2. Do you think the standards should be stricter? What would be at least one advantage and one disadvantage of stricter standards?

Take It Further

 Go to www.aw-bc.com/withgott or the student CD-ROM where you'll find:

▶ Suggested answers to end-of-chapter questions
▶ Quizzes, animations, and flashcards to help you study
▶ *Research Navigator*™ database of credible and reliable sources to assist you with your research projects

▶ **GRAPHit!** tutorials to help you master how to interpret graphs
▶ **INVESTIGATEit!** current news articles that link the topics that you study to case studies from your region to around the world

4 From Chemistry to Energy to Life

Exxon Valdez oil tanker in Prince William Sound, Alaska

Upon successfully completing this chapter you will be able to:

▶ Explain the fundamentals of environmental chemistry and apply them to real-world situations

▶ Describe the molecular building blocks of living organisms

▶ Differentiate among the types of energy and recite the basics of energy flow

▶ Distinguish photosynthesis, respiration, and chemosynthesis, and summarize their importance to living things

▶ Itemize and evaluate the major hypotheses for the origin of life on Earth

▶ Outline our knowledge regarding early life and give supporting evidence for each major concept

Workers spraying fertilizer on an oil-coated Alaska beach

Prince William Sound, Alaska

North America

Pacific Ocean

Central Case: Bioremediation of the *Exxon Valdez* Oil Spill

"There is a dramatic difference. . . . It really cleaned the oil off the rock. It looked like someone brought in new rock."
—EPA PROGRAM MANAGER CHUCK COSTA, DESCRIBING EXPERIMENTAL BIOREMEDIATION RESULTS IN 1989

"The rush to bioremediation in Alaska was a function of the size of the problem and limited availability of options. . . . It did not turn out to be the silver bullet that many hoped it would be."
—ALASKA DEPARTMENT OF ENVIRONMENTAL CONSERVATION REPORT, 1993

On March 24, 1989, the tanker *Exxon Valdez* struck a reef in Alaska's Prince William Sound and spilled 42 million L (11 million gal) of crude oil, which eventually coated 2,100 km (1,300 mi) of Alaskan coastline. The largest oil spill in U.S. history, it killed an estimated 100,000–400,000 seabirds, 2,600–5,500 sea otters, 200–300 harbor seals, and countless fish. The oil smothered intertidal plants and animals and defiled the area's relatively pristine environment. The local economy took a nosedive as hundreds of fishermen were thrown out of work and tourism plummeted.

The massive spill was met with a massive response. Thousands of workers employed by Exxon (now ExxonMobil) and by government agencies, together with local volunteers, launched a cleanup effort of unprecedented scope. The cleanup crews corralled the oil with booms, skimmed it from the water, soaked it up with absorbent materials, and dispersed it with chemicals. They pressure-washed the beaches, removed contaminated sand with backhoes and tractors, and even tried burning the oil.

Scientists also used the opportunity to test a new cleanup strategy that enlisted nature to help take care of the mess. They stimulated naturally occurring bacteria to biodegrade, or break down, the oil. About 5% of the single-celled microbes present on Alaskan beaches feed on chemical compounds called *hydrocarbons* that are produced by the region's conifer trees. Hydrocarbons from conifers are chemically similar to the hydrocarbons that make up crude oil, so scientists predicted that the microbes might also be able to degrade oil. Scientists from the EPA and Exxon decided to put the bacteria to work in a process called **bioremediation,** the attempt to clean up pollution by enhancing natural processes of biodegradation by living organisms.

Although the bacteria were presented with an abundant new food source in the form of oil washing up on the beaches, they were not immediately able to consume it. The oil contained plenty of carbon, but not enough nitrogen and phosphorus. To remedy this imbalance of nutrients, scientists applied a fertilizing mixture containing nitrogen and phosphorus to several beaches, leaving other areas of the shore as untreated controls. The fertilizing treatment seemed to work; bacterial numbers increased, and oil residues decreased visibly. Encouraged, scientists put the program into full swing, and by the end of the year workers had treated more than 113 km (70 mi) of contaminated beach. They expanded the applications over the next 2 years.

Because there were many complicating factors, experts have interpreted the results of the study differently, and they still debate how much the treatments increased chemical breakdown of the oil. Some say degradation was sped up fivefold, whereas others think it made no difference. However, the well-publicized *Valdez* operation served as a model effort, and today bioremediation is actively researched and increasingly applied in many situations. Bioremediation has many practical limitations, but, when feasible, it can accomplish much good with a minimum of expense and environmental disturbance.

Chemistry and the Environment

The *Exxon Valdez* oil spill ignited a wide array of ecological, economic, political, and ethical concerns. Today many wildlife populations have recovered, but some have not, and pockets of oil remain throughout the region. Lawsuits against the company (some still pending) testify to the concerns of fishermen whose livelihoods were wrecked and of cleanup workers who say their health was affected. The U.S. Congress in the year following the spill passed the Oil Pollution Control Act, which required the Coast Guard and the U.S. Environmental Protection Agency (EPA) to strengthen regulations on tankers and their operators.

At the root of all these diverse impacts is the chemical makeup of the oil. It is the chemistry of crude oil that

FIGURE 4.1 The *Exxon Valdez* spill coated hundreds of thousands of seabirds with oil, impairing their ability to insulate themselves with their feathers, and bringing on fatal hypothermia. Rescue workers labored tirelessly to clean oil from those birds they could capture and treat.

causes it to gum up birds' feathers and mammals' fur, impairing their insulating abilities and bringing on hypothermia (Figure 4.1). It is oil's chemistry that causes it to float on water and accumulate on beaches. It is certain hydrocarbons from oil that, mixed in the water column or volatile in the air, are harmful to wildlife and carcinogenic to humans. Yet the chemistry of crude oil also provides the energy that powers our remarkable civilization and modern way of life—a way of life that allows us the luxury to study, reflect on, and act to address these very issues.

Examine any environmental issue, and you will likely discover chemistry playing a central role. Chemistry is crucial to understanding how gases such as carbon dioxide and methane contribute to global climate change, how pollutants such as sulfur dioxide and nitric oxide cause acid rain, and how pesticides and other artificial compounds we release into the environment affect the health of wildlife and people. Chemistry is central, too, in understanding water pollution and sewage treatment, atmospheric ozone depletion, hazardous waste and its disposal, and just about any energy issue.

Chemistry is also central to developing solutions to environmental problems. Bioremediation is one clear illustration of this, and organisms are now used to clean up pollution in a variety of situations. Hydrocarbon-consuming bacteria and fungi are used to clean up soil beneath leaky gasoline tanks that threaten drinking water supplies. Other kinds of microbes are used to degrade pesticide residues in soil. Plants such as wheat, tobacco, water hyacinth, and cattails have been employed to clean

Table 4.1 Earth's Most Abundant Chemical Elements, by Mass			
Earth's crust	**Oceans**	**Air**	**Organisms**
Oxygen (O), 49.5%	Oxygen (O), 88.3%	Nitrogen (N), 78.1%	Oxygen (O), 65.0%
Silicon (Si), 25.7%	Hydrogen (H), 11.0%	Oxygen (O), 21.0%	Carbon (C), 18.5%
Aluminum (Al), 7.4%	Chlorine (Cl), 1.9%	Argon (Ar), 0.9%	Hydrogen (H), 9.5%
Iron (Fe), 4.7%	Sodium (Na), 1.1%	Other, <0.1%	Nitrogen (N), 3.3%
Calcium (Ca), 3.6%	Magnesium (Mg), 0.1%		Calcium (Ca), 1.5%
Sodium (Na), 2.8%	Sulfur (S), 0.1%		Phosphorus (P), 1.0%
Potassium (K), 2.6%	Calcium (Ca), <0.1%		Potassium (K), 0.4%
Magnesium (Mg), 2.1%	Potassium (K), <0.1%		Sulfur (S), 0.3%
Other, 1.6%	Bromine (Br), <0.1%		Other, 0.5%

up toxic waste sites by letting them draw up heavy metals, such as lead and cadmium, through their roots (see "The Science behind the Story," ▸ pp. 92–93).

Sometimes suitable plants or bacterial cultures are introduced to a site. Sometimes naturally existing ones are fertilized, as was done at Prince William Sound. And sometimes the best way to mitigate pollution is simply to monitor naturally occurring organisms as they do their work, without disrupting the system. Bioremediation can be far less expensive, less environmentally intrusive, and more effective than conventional methods for cleaning up pollution. However, bioremediation does not always work, and it can sometimes be very slow, leave a job uncompleted, or introduce new problems. Scientists today are seeking ways to take bioremediation to the next level, by genetically engineering microbes and plants to become more efficient at the specific metabolic tasks we ask of them. Environmental chemists are excited about the countless future applications of chemistry that may help us address environmental problems.

Atoms and elements are chemical building blocks

To appreciate the complex chemistry involved in environmental science we must begin with a grasp of the fundamentals. The carbon, nitrogen, and phosphorus that played such key roles in the bioremediation of the oil spill in Prince William Sound are each elements. An **element** is a fundamental type of matter, a chemical substance with a given set of properties, which cannot be broken down into substances with other properties. Chemists currently recognize 92 elements occurring in nature, as well as more than 20 others that have been artificially created. Besides

carbon and nitrogen, elements especially abundant in living organisms include hydrogen and oxygen (Table 4.1). Each element is assigned an abbreviation, or chemical symbol. The *periodic table of the elements* (see Appendix C) summarizes information on the elements in a comprehensive and elegant way.

Elements are composed of **atoms,** the smallest components that maintain the chemical properties of the element (Figure 4.2). Every atom has a nucleus of **protons** (positively charged particles) and **neutrons** (particles lacking electric charge). The atoms of each element have a defined number of protons, called the *atomic number*. (Elemental carbon, for instance, has six protons in its nucleus; thus, its atomic number is 6.) An atom's nucleus is surrounded by negatively

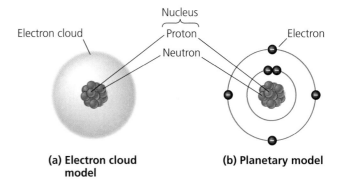

(a) Electron cloud model   **(b) Planetary model**

FIGURE 4.2 In an atom, protons and neutrons are held in the nucleus, and electrons move through space around the nucleus. Atoms, such as the carbon atom shown here, can be envisioned in different ways. In (**a**), electrons are represented by a cloud, indicating the space within which they may likely occur at any given time. In (**b**), electrons are shown orbiting the nucleus in concentric rings like planets orbit the sun. The diagram in (**a**) is more realistic, but we will use diagrams such as that in (**b**) to compare numbers of protons, neutrons, and electrons in our next figure.

Student Chemist Lets Plants Do the Dirty Work

The Science behind the Story

Bacteria that break down oil are just one example of organisms that scientists are putting to work to clean up environmental pollutants. Green plants can help, too.

When soil is contaminated with heavy metals from mining, manufacturing, oil extraction, or military facilities, the standard solution is to dig up tons of soil and pile it into a hazardous waste dump. Bulldozing so much dirt, however, can release toxic chemicals into the air and cost up to $2.5–7.5 million per hectare ($1–3 million per acre). As an alternative, scientists are developing methods of *phytoremediation*, using plants (*phyto* means "plant") to remediate, or detoxify, contaminated soils.

One researcher making advances in phytoremediation is Marc Burrell, now a senior at Rice Uni-

Rice University student Marc Burrell conducted research to find new ways to induce plants to clean up contaminated soil.

versity in Houston, Texas. While still in high school in Wisconsin, Burrell, with phytoremediation expert Peter Goldsbrough of Purdue University in Indiana and other

mentors, began researching how to coax plants to remove toxic lead from soil.

Working one summer with Greg and Maria Begonia at Jackson State University in Mississippi, Burrell ran lab experiments to test how wheat could be made to draw up lead from soil. Normally, lead is not accessible to plants, because it is tied up in compounds such as lead carbonate and lead oxide, which do not dissolve in water. But chemicals called *chelating agents* can bind to lead and make it water-soluble so plant roots can draw it up. Burrell's greenhouse experiments showed that adding the chelating agent EDTA to the soil increased wheat's uptake of lead by about 300,000 times.

Burrell hypothesized that adding an acid would enhance the effect, because an acid's hydrogen ions would help break apart lead

charged particles known as **electrons,** which balance the positive charge of the protons (Figure 4.3).

Isotopes Although all atoms of a given element contain the same number of protons, they do not necessarily contain the same number of neutrons. Atoms with differing numbers of

neutrons are referred to as **isotopes** (Figure 4.4a). Isotopes are denoted by their elemental symbol preceded by the *mass number,* or combined number of protons and neutrons in the atom. For example, ^{14}C (carbon-14) is an isotope of carbon with 8 neutrons (and 6 protons) in the nucleus rather than the normal 6 neutrons of ^{12}C (carbon-12).

FIGURE 4.3 Each chemical element has a different number of protons, neutrons, and electrons. Carbon possesses 6 of each, nitrogen 7, and phosphorus 15.

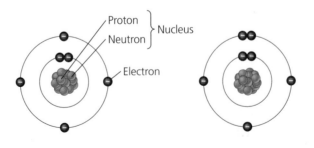
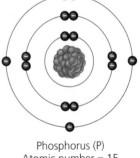

Carbon (C)
Atomic number = 6
Protons = 6
Neutrons = 6
Electrons = 6

Nitrogen (N)
Atomic number = 7
Protons = 7
Neutrons = 7
Electrons = 7

Phosphorus (P)
Atomic number = 15
Protons = 15
Neutrons = 15
Electrons = 15

compounds, freeing more lead ions to bind to EDTA. His experiments supported this hypothesis; when Burrell added acetic acid, the plants took up three times more EDTA.

Next, Burrell wanted to better understand how genes and proteins affect lead uptake. To deal with toxic metals that they take up by accident, many plants produce chelating agents called *phytochelatins* to drag the metals to vacuoles, or empty cellular sacs, where they can be stashed away without harm. Phytochelatins were thought to work with many metals, but no one had tested them with lead. Working under Heather Owen at the University of Wisconsin–Milwaukee, Burrell experimented with mustard plants engineered by geneticists who had knocked out the gene responsible for producing phytochelatins. He grew these engineered plants alongside normal

plants in lead-contaminated soil. He found that the engineered plants that could not produce phytochelatins were more susceptible to lead poisoning and died sooner. This suggested that phytochelatins do, in fact, squirrel away lead into vacuoles.

Such research results are beginning to be applied at contaminated sites. Once plants have accumulated metals, they can be harvested and put through a smelting procedure to recover the metals. Alternatively, the plants can be dried and disposed of at a hazardous waste site.

Phytoremediation is a new pursuit, and it faces some hurdles. One is time; individual plants can take up only so much of a substance, and 5 to 20 years of repeated plantings may be required to reduce a soil's metal content to an acceptable level. Another is that metals need to be in a water-soluble form. In addi-

tion, cleanup is limited to the depth of soil that plants' roots reach. Finally, plants that accumulate toxins can potentially harm insects that eat the plants, and in turn, animals that eat the insects.

Despite such obstacles, phytoremediation is catching on. At military bases in Iowa, Tennessee, and Nebraska, the U.S. Army Corps of Engineers is using vegetation in artificial wetlands to minimize contamination of groundwater by ammunition. A Virginia company, Edenspace, has used plants to extract lead from residential sites, arsenic from military and energy facilities, zinc and cadmium at EPA Superfund sites (▸ pp. 670–671), and tungsten from abandoned mines. Such efforts are at the forefront of "green chemistry" today, thanks in part to the endeavors of bright and hardworking young researchers like Marc Burrell.

Because they differ slightly in mass, isotopes of an element differ slightly in their behavior. This fact has turned out to be very useful for researchers. Scientists have been able to use isotopes to study a number of phenomena that help illuminate the history of Earth's physical environment. Researchers also have used them to study the flow of nutrients within and among organisms, and the movement of organisms from one geographic location to another (see "The Science behind the Story," ▸ pp. 94–95).

Some isotopes are radioactive and "decay," changing their chemical identity as they shed subatomic particles and emit high-energy radiation. **Radioisotopes** decay into lighter and lighter radioisotopes, until they become *stable isotopes*, isotopes that are not radioactive. Each radioisotope decays at a rate determined by that isotope's **half-life,** the amount of time it takes for one-half the atoms to give off radiation and decay. Different radioisotopes have very different half-lives, ranging from fractions of a second to billions of years. The radioisotope uranium-235 (^{235}U) is our society's source of energy for

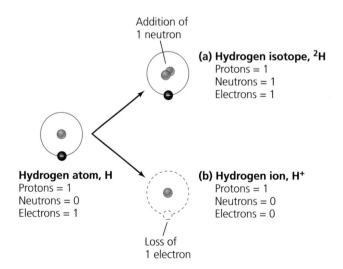

FIGURE 4.4 Atoms of an element such as hydrogen can become chemically altered to form isotopes and ions. Shown in **(a)** is an isotope of hydrogen, hydrogen-2 (^2H), or deuterium. This isotope contains one neutron, and thus it contains greater mass than a typical hydrogen atom. Shown in **(b)** is the hydrogen ion, H^+. By losing its electron, it gains a positive charge.

The Science behind the Story

How Isotopes Reveal Secrets of Earth and Life

Isotopes, those alternate versions of chemical elements, have become one of the most powerful instruments in the environmental scientist's toolkit. They have enabled scientists interested in the past to date ancient materials, reconstruct the climate of past ages, and study the lifestyles of prehistoric humans. They have allowed researchers focused on the here-and-now to work out photosynthetic pathways, measure animals' diets and health, and trace nutrient flows through organisms and ecosystems.

For researchers studying the past, *radiocarbon dating* is one informative approach. Carbon's most abundant isotope is ^{12}C, but ^{13}C and ^{14}C also occur. Carbon-14 is radioactive and occurs in organisms at the same low concentration that it occurs in the atmosphere. Once an organism dies, no new ^{14}C is incorporated into its structure, and the radioactive decay process (p. 93) gradually reduces its store of ^{14}C, converting these atoms to ^{14}N (nitrogen-14).

The decay is slow and steady enough to act as a kind of clock, so that scientists can date ancient organic materials by measuring the percent of carbon that is ^{14}C and matching this value against the

clocklike progression of decay. In this way, archaeologists and paleontologists have dated prehistoric human remains; charcoal, grain, and shells found at ancient campfires; and bones and frozen tissues of recently extinct animals, such as mammoths. Scientists can also estimate the age of a fossil by radiocarbon dating the rock, peat, or sediment that surrounds it. The most recent ice age has been dated from ^{14}C analysis of trees overrun by glacial ice sheets. Ice drilled from glaciers today can be aged by measuring the ^{14}C in air bubbles trapped during its formation.

The half-life (time it takes for one-half of a sample to decay) of ^{14}C is 5,730 years. Thus, radiocarbon dating is not useful for items over 50,000 years old, because too little ^{14}C remains to permit accurate analysis. For dating older items, other isotopes can be used. Uranium-238 (with a half-life of 4.5 billion years) has been used to age very early fossils. For dating geological formations, potassium-argon dating is useful (potassium-40 decays to argon-40). Oxygen-18 has been widely used to infer changes in climate and sea level.

Researchers interested in present-day ecology can use *stable isotopes*.

Unlike radioactive isotopes, stable isotopes occur in nature in constant ratios. For instance, nitrogen occurs as 99.63% nitrogen-14 and 0.37% nitrogen-15. Ratios of isotopes are called *isotopic signatures,* and by analyzing these signatures scientists can gain valuable information. For example, organisms tend to retain ^{15}N in their tissues but readily excrete ^{14}N. As a result, animals higher in the food chain show isotopic signatures biased toward ^{15}N, as do animals that are starving. Keith Hobson, an ecologist with Environment Canada and the University of Saskatchewan, has used nitrogen signatures to analyze the diets of seabirds and marine mammals, to show that geese fast while nesting, and to trace artificial contaminants in food chains.

Hobson and other scientists have also used stable carbon isotopes for ecological studies. Plants produce food through one of three photosynthetic pathways, and the isotopic signature of carbon in plants varies among these pathways. Grasses have higher ratios of ^{13}C to ^{12}C than oak trees do, for instance, whereas cacti have intermediate ratios. When animals eat plants, they incorporate the plants' isotopic signatures into their own tissues, and

commercial nuclear power (pp. 594–595). It decays into a series of daughter isotopes, eventually forming lead-207 (^{207}Pb), and has a half-life of about 700 million years.

Ions Atoms may also gain or lose electrons to become **ions,** electrically charged atoms or combinations of atoms (Figure 4.4b). Ions are denoted by their elemental symbol followed by their ionic charge. For instance, a common ion used

by mussels and clams to form shells is Ca^{2+}, a calcium atom that has lost two electrons, and so has a charge of positive 2.

Atoms bond to form molecules and compounds

Atoms can bond together and form **molecules,** combinations of two or more atoms. Molecules may contain one

this signal passes up the food chain. As a result, carbon isotope studies can tell ecologists what an animal has been eating. Archaeologists have used isotopic signatures in human bone to determine when ancient people switched from a hunter-gatherer diet to an agricultural one.

Carbon signature data can even tell a scientist where an animal has been. For example, nectar-feeding bats have been shown to move seasonally between communities dominated by cacti to communities dominated by trees. Such movements have been inferred for migrating warblers, for elephants hunted for ivory, and for the oceanic movements of seals and salmon.

Recently, researchers have used isotopes to track movements of birds and other animals that migrate thousands of miles. This is possible because the isotopic signature of hydrogen in rainfall varies systematically across large geographic regions. This signature gets passed from rainwater to plants, and from plants to animals, leaving a fingerprint of geographic origin in an animal's tissues. Hobson and his colleagues in 1998 used a combination of isotopic data from hydrogen and carbon to

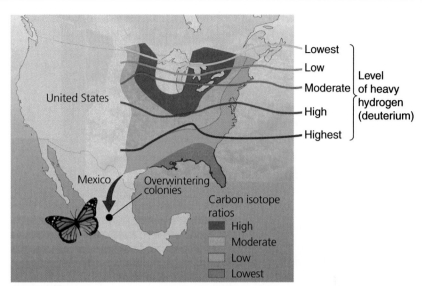

Plants in different geographic areas show different isotopic ratios for elements such as carbon and hydrogen. Caterpillars of monarch butterflies incorporate into their tissues carbon and hydrogen in the isotopic ratios present in the plants they eat. When these caterpillars metamorphose into butterflies and migrate, they carry these isotopic signals with them, providing scientists clues to their origin. Shown is a map of isotopic ratios across eastern North America produced from measurements of monarchs in the summer. The four colored bands show decreasing ratios of ^{13}C to ^{12}C from north to south. The five gray lines show increasing ratios of ^2H (heavy hydrogen or deuterium) to ^1H from north to south. By measuring carbon and hydrogen isotope ratios in monarchs wintering in Mexico, and matching the combination of these numbers against this map, researchers were able to pinpoint the geographic origin of many of the butterflies. *Source*: Wassenaar, L. I., and K. A. Hobson 1998. *Proceedings of the National Academy of Sciences of the USA* 95:15436–15439.

pinpoint the geographic origins of monarch butterflies that had migrated from throughout the United States and Canada to communal roosts in Mexico—providing important information for their conservation (see figure).

Other elements show similar standing patterns of natural variation that have not yet been used or even discovered, researchers say, so there remains much more we can learn from the use of these subtle chemical clues.

element or several. Common molecules containing only a single element include those of oxygen gas (O_2) and nitrogen gas (N_2), both of which are abundant in air. A molecule composed of atoms of two or more different elements is called a **compound.** Water is a compound; composed of two hydrogen atoms bonded to one oxygen atom, it is denoted by the chemical formula H_2O. Another compound is carbon dioxide, consisting of one car-

bon atom bonded to two oxygen atoms; its chemical formula is CO_2.

Atoms bond together because of an attraction for one another's electrons. Because the strength of this attraction varies among elements, atoms may be held together in different ways, according to whether and how they share or transfer electrons. When atoms in a molecule share electrons, they generate a **covalent bond.** For instance, two atoms of hydrogen bond to form hydrogen gas, H_2, by sharing electrons equally. Atoms in a covalent bond can also share electrons unequally, with one atom exerting a greater pull. Such is the case with water, in which oxygen attracts electrons more strongly than hydrogen, forming what are termed *polar* covalent bonds. In contrast, if the strength of attraction is unequal enough, an electron may be transferred from one atom to another. Such a transfer creates oppositely charged ions that are said to form **ionic bonds.** These associations are not considered molecules, but instead are called **ionic compounds,** or **salts.** Table salt (NaCl) contains ionic bonds between positively charged sodium ions (Na^+), each of which donated an electron, and negatively charged chloride ions (Cl^-), each of which received an electron.

Elements, molecules, and compounds can also come together in mixtures without chemically bonding. Homogenous mixtures of substances are called *solutions,* a term most often applied to liquids, but also applicable to some gases and solids. Air in the atmosphere is a solution formed of constituents such as nitrogen, oxygen, water, carbon dioxide, methane (CH_4), and ozone (O_3). Human blood, ocean water, plant sap, and metal alloys such as brass are all solutions. Crude oil at high pressure may carry natural gas in solution and often contains other substances distributed unevenly. It is a heavy liquid mixture of many kinds of molecules consisting primarily of carbon and hydrogen atoms. Its physical properties vary with its temperature, pressure, and composition.

The chemical structure of the water molecule facilitates life

Water dominates Earth's surface, covering over 70% of the globe, and its abundance is a primary reason Earth is hospitable to life. Scientists think life originated in water and stayed there for 3 billion years before moving onto land. Today every land-dwelling creature remains critically tied to water for its existence.

The water molecule's amazing capacity to support life results from its unique chemical properties. A water molecule's single oxygen atom bonds to its two hydrogen atoms at a 105-degree angle. As just mentioned, the oxygen atom attracts electrons more strongly, resulting in a polar molecule in which the oxygen end of the molecule has a partial negative charge and the hydrogen end has a partial positive charge. Because of this configuration, water molecules can adhere to one another in a special type of interaction called a *hydrogen bond,* in which the

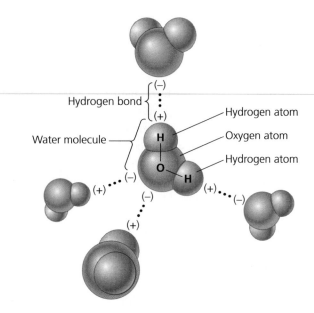

FIGURE 4.5 Water is a unique compound that has several properties crucial for life. Hydrogen bonds give water cohesion by enabling water molecules to adhere loosely to one another.

oxygen atom of one water molecule is weakly attracted to one or two hydrogen atoms of another (Figure 4.5). The weak electrical attraction of hydrogen bonding can also occur between hydrogen and certain other atoms, such as nitrogen. In water, hydrogen bonds are most stable in ice, somewhat stable in liquid water, and broken in water vapor.

These loose connections among molecules give water several properties important in supporting life and stabilizing Earth's climate:

▶ Water exhibits strong cohesion. (Think of how water holds together in drops, and how drops on a surface join together when you touch them to one another.) This cohesion facilitates the transport of chemicals, such as nutrients and waste, in plants and animals.

▶ Hydrogen bonding provides water molecules a capacity to resist temperature change. Initial heating weakens hydrogen bonds between molecules but does not speed molecular motion. As a result, water can absorb a large amount of heat with only small changes in its temperature. This quality helps stabilize systems against change, whether those systems are organisms, ponds, lakes, or climate systems.

▶ Water molecules in ice are farther apart than in liquid form (Figure 4.6a). As a result, the solid form of water is less dense than the liquid—the reverse pattern of most other compounds, which become denser as they freeze.

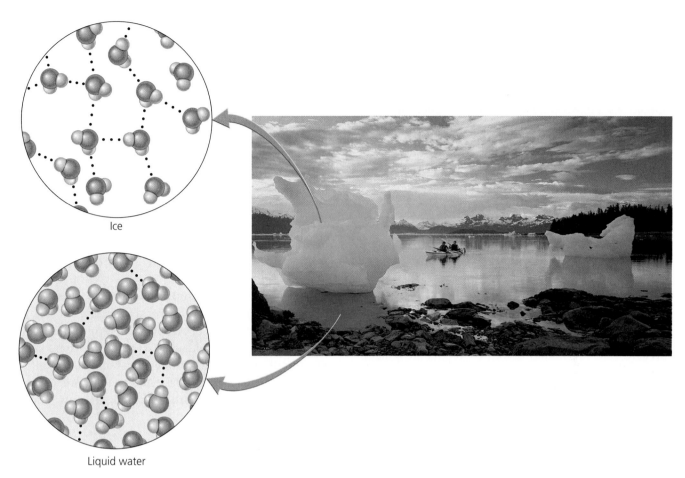

Ice

Liquid water

(a) Why ice floats on water

FIGURE 4.6 (a) Ice floats in liquid water because ice is less dense. Each molecule is connected to neighboring molecules by stable hydrogen bonds, forming a spacious crystal lattice. In liquid water, hydrogen bonds frequently break and re-form, and the molecules are closer together and less well organized. (b) Water is often called the "universal solvent" because it can dissolve so many chemicals, especially polar and ionic compounds. Seawater holds sodium and chloride ions, among others, in solution.

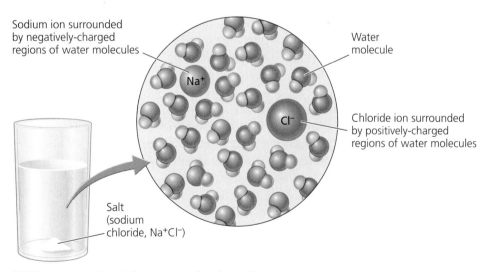

Sodium ion surrounded by negatively-charged regions of water molecules

Water molecule

Chloride ion surrounded by positively-charged regions of water molecules

Salt (sodium chloride, Na^+Cl^-)

(b) Water as a solvent; how water dissolves salt

This is why ice floats on liquid water. Floating ice has an insulating effect that can prevent water bodies from freezing solid in winter.

▶ The polar nature of water molecules allows them to bond well with other polar molecules, because the positive end of one molecule bonds readily to the negative end of another. As a result, water can hold in solution, or dissolve, many other molecules, including chemicals necessary for life (Figure 4.6b). It follows that most biologically important solutions involve water.

Water's Properties for Life

Water has several special properties that make it accommodating to life. Can you generate examples of specific ways in which each of these properties might help a particular organism, such as a fish in a pond? Can you think of ways in which any of water's properties might also bring harm to the fish?

Hydrogen ions determine acidity

In any aqueous solution, a small number of water molecules dissociate, each forming a hydrogen ion (H^+) and a hydroxide ion (OH^-). The product of hydrogen and hydroxide ion concentrations is always 10^{-14}. As the concentration of one increases, the concentration of the other decreases, and the product of their concentrations remains constant. Pure water contains equal numbers of these ions, each at a concentration of 10^{-7}, and we say that this water is neutral. Most aqueous solutions, however, contain different concentrations of these two ions. Solutions in which the H^+ concentration is greater than the OH^- concentration are **acidic,** whereas solutions in which the OH^- concentration is greater than the H^+ concentration are **basic.**

The **pH** scale (Figure 4.7) was devised to quantify the acidity or basicity of solutions. It runs from 0 to 14, because these numbers reflect the negative logarithm of the hydrogen ion concentration ("pH" comes from "potential Hydrogen"). Thus pure water has a pH of 7, because its hydrogen ion concentration is 10^{-7}. Seawater has a greater concentration of hydroxide ions, close to 10^{-6}. This means that its hydrogen ion concentration is about 10^{-8}, and thus its pH is close to 8. Solutions with pH less than 7 are acidic, those with pH greater than 7 are basic, and those with pH of 7 are neutral. Because the pH scale is logarithmic, each step on the scale represents a tenfold difference in hydrogen ion concentration. Thus, a substance with pH of 6 contains 10 times as many hydrogen ions as a substance with pH of 7, and a substance with pH of 5 contains 100 times as many hydrogen ions as one with pH of 7. Figure 4.7 shows pH for a number of common substances. Industrial air pollution has intensified the acidity of precipitation (▸ pp. 514–518), and rain in parts of the northeastern and midwestern United States now frequently dips to pH of 4 or lower.

Matter is composed of organic and inorganic compounds

Beyond their need for water, living things also depend on organic compounds, which they create and of which they are created. **Organic compounds** consist of carbon atoms (and generally hydrogen atoms) joined by covalent bonds, and may include other elements, such as nitrogen, oxygen, sulfur, and phosphorus. Carbon's unusual ability to build elaborate molecules has resulted in millions of different organic compounds that show various degrees of complexity. Because of the diversity of organic compounds and their importance in living organisms, chemists differentiate organic compounds from inorganic compounds, which lack carbon–carbon bonds.

Crude oil and petroleum products are made up of organic compounds called hydrocarbons. **Hydrocarbons** consist solely of atoms of carbon and hydrogen. The simplest hydrocarbon is methane (CH_4), the key component of natural gas; it has one carbon atom bonded to four hydrogen atoms (Figure 4.8a). Adding another carbon atom and two more hydrogen atoms gives us ethane

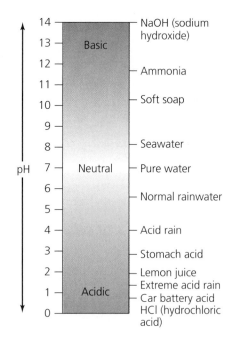

FIGURE 4.7 The pH scale measures how acidic or basic a solution is. The pH of pure water is 7, the midpoint of the scale. Acidic solutions have higher hydrogen ion concentrations and lower pH, whereas basic solutions have lower hydrogen ion concentrations and higher pH.

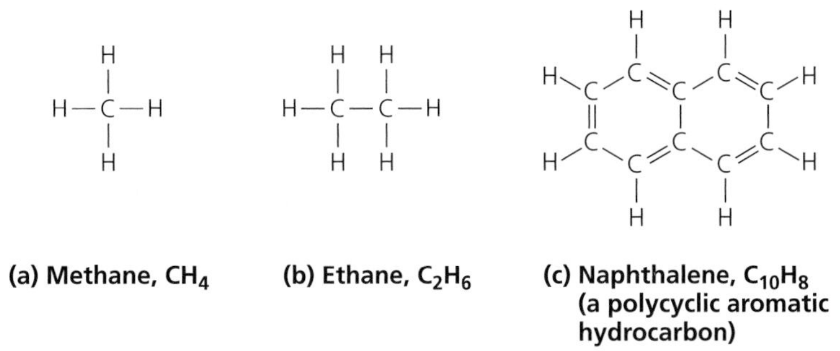

(a) Methane, CH₄ **(b) Ethane, C₂H₆** **(c) Naphthalene, C₁₀H₈ (a polycyclic aromatic hydrocarbon)**

FIGURE 4.8 Hydrocarbons are a major class of organic compound, and mixtures of them make up fossil fuels such as crude oil. The simplest hydrocarbon is methane (**a**). Many hydrocarbons consist of linear chains of carbon atoms with hydrogen atoms attached; the shortest of these is ethane (**b**). Volatile hydrocarbons with multiple rings, such as naphthalene (**c**), are called polycyclic aromatic hydrocarbons (PAHs).

(C_2H_6), the next-simplest hydrocarbon (Figure 4.8b). The smallest (and therefore lightest-weight) hydrocarbons (those consisting of four or fewer carbon atoms) exist in a gaseous state at normal temperatures and pressures. Larger (therefore heavier) hydrocarbons are liquids, and those consisting of over 20 carbon atoms long are normally solids. Some hydrocarbons from petroleum are known to pose health hazards to wildlife and people. For example, polycyclic aromatic hydrocarbons, or PAHs (Figure 4.8c), which are volatile molecules with a structure of multiple carbon rings, can evaporate from spilled oil and gasoline and can mix with water. The eggs and young of fish and other aquatic creatures are often most at risk. PAHs also occur in particulate form in various combustion products, including cigarette smoke, wood smoke, and charred meat.

Bacteria used in the bioremediation of petroleum spills do not actually consume the entire hydrocarbon molecules they attack. Rather, the bacteria, facilitated by oxygen, degrade complex hydrocarbon structures into simpler ones, or into their simplest components, hydrogen and carbon. Often this involves sequences of many chemical reactions. For example, bacterial degradation of the PAH naphthalene shown in Figure 4.8c involves 11 steps, during which atoms are added, removed, and rearranged, eventually producing the simpler products pyruvate and acetaldehyde (Figure 4.9).

Macromolecules are building blocks of life

Just as the carbon atoms in hydrocarbons may be strung together in chains, other organic compounds can sometimes combine to form long chains of repeated molecules. Some of these chains, called **polymers,** play key roles as building blocks of life. Three types of polymers are essential to life: proteins, nucleic acids, and carbohydrates. Lipids are not considered polymers but are also fundamental to life. These four types of molecules are referred to as **macromolecules** because of their large size.

Proteins Amino acids are organic molecules consisting of a central carbon linked to a hydrogen atom, an acidic carboxyl group (—COOH), a basic amine group (—NH₂), and an organic side chain unique to each type of amino acid (Figure 4.10a). Organisms combine up to 20 different types of amino acids into long chains to build **proteins** (Figure 4.10b). A protein's identity is determined by its particular sequence of amino acids and by the shape the protein molecule assumes as it folds. Protein molecules typically have highly convoluted shapes, with certain parts of the chain exposed and others hidden inside the folds (Figure 4.10c). A protein's folding pattern affects its function, because the position of each chemical group helps determine how it interacts with cell surfaces and with other molecules.

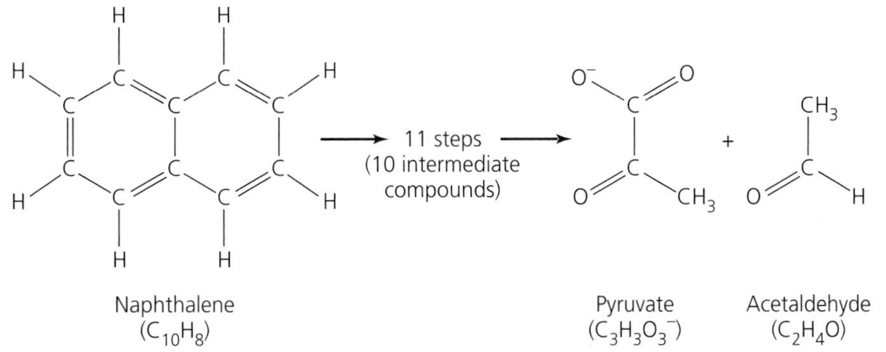

Naphthalene ($C_{10}H_8$) → 11 steps (10 intermediate compounds) → Pyruvate ($C_3H_3O_3^-$) + Acetaldehyde (C_2H_4O)

FIGURE 4.9 Bacterial degradation of naphthalene involves 11 steps and results in the simpler organic compounds pyruvate and acetaldehyde. The many chemicals involved in this complicated process are omitted from this simplified diagram.

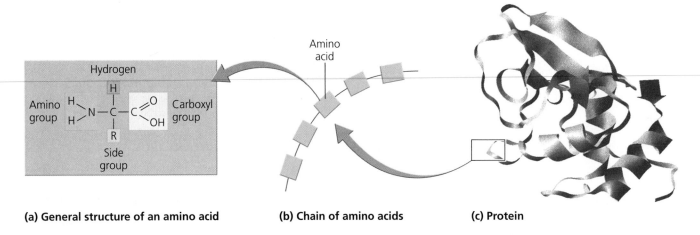

(a) General structure of an amino acid   **(b) Chain of amino acids**   **(c) Protein**

FIGURE 4.10 Proteins are polymers vital for life. They are made up of long chains of amino acids **(a, b)** and fold up into complex convoluted shapes **(c)** that help determine their functions.

Proteins serve many functions. Some help produce tissues and provide structural support for the organism. For example, animals use proteins to generate skin, hair, muscles, and tendons. Some proteins help store energy, and others transport substances. Some function as components of the immune system, defending the organism against foreign attackers. Still others act as hormones, molecules that serve as chemical messengers within an organism. Finally, proteins can serve as enzymes, molecules that catalyze, or promote, certain chemical reactions. For example, bacteria used for bioremediation use specialized enzymes to break down hydrocarbons, just as we use enzymes to digest our food.

Nucleic acids Protein production is directed by **nucleic acids.** The two nucleic acids—**deoxyribonucleic acid (DNA)** and **ribonucleic acid (RNA)**—carry the hereditary information for organisms and are responsible for passing traits from parents to offspring. Nucleic acids are composed of series of nucleotides, each of which contains a sugar molecule, a phosphate group, and a nitrogenous base (Figure 4.11a). DNA includes four types of nucleotides, each with a different nitrogenous base: adenine (A), guanine (G), cytosine (C), and thymine (T). RNA is similar to DNA in structure, except that its sugar group is ribose (instead of deoxyribose), thymine is replaced by uracil (U), and RNA is generally single-stranded whereas DNA is double-stranded. Within DNA or RNA, nucleotides are linked together to form extremely long chains, with a sugar and phosphate backbone and nitrogenous base pairs. Adenine (A) pairs with thymine (T), and cytosine (C) pairs with guanine (G). In DNA, the paired base chains can be pictured as rungs of a ladder, with the ladder twisted into a spiral,

giving the entire molecule a shape called a double helix (Figure 4.11b).

In the process of *transcription,* the hereditary information in the nucleotide sequence of DNA is rewritten to a molecule of RNA. Then, during the process of *translation,* RNA directs the order in which amino acids assemble to build proteins (Figure 4.12). Proteins go on to influence the structure and maintenance of the organism. Genetic information from DNA is passed from one generation to another as the strands replicate during cell division

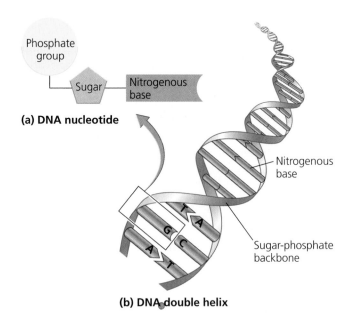

(a) DNA nucleotide

(b) DNA double helix

FIGURE 4.11 DNA is the molecule that carries genetic information from parent to offspring across the generations. The information is coded in the sequence of nucleotides **(a)**, small molecules that pair together like rungs of a ladder that twist into the shape of a double helix **(b)**.

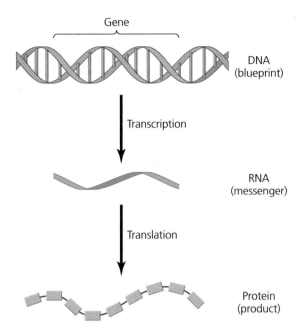

FIGURE 4.12 DNA serves as the blueprint for synthesizing the proteins that help build and maintain organisms. The genetic information in DNA (functional regions of which are called *genes)* instructs the creation of RNA through a process called *transcription.* Messenger RNA then directs the synthesis of proteins through a process called *translation.*

and egg or sperm formation. Regions of DNA coding for particular proteins that perform particular functions are called **genes.** In most organisms, the *genome*—the set of all an organism's genes—is divided into chromosomes. Different types of organisms have different numbers of genes and chromosomes. Most bacteria have a single circular chromosome, for instance, whereas humans have 46 linear ones.

Carbohydrates **Carbohydrates** constitute a third class of biologically vital polymer. These organic compounds consist of atoms of carbon, hydrogen, and oxygen (Figure 4.13). Simple carbohydrates, called sugars or monosaccharides, have structures, or skeletons, that are three to seven carbon atoms long, and formulas that are some multiple of CH_2O. Glucose ($C_6H_{12}O_6$) is one of the most common and important sugars, providing energy that fuels plant and animal cells. Glucose also serves as a building block for complex carbohydrates, or polysaccharides. Plants use starch, a glucose-based polysaccharide, to store energy, and animals eat plants to acquire starch.

In addition, both plants and animals use complex carbohydrates to build structure. Insects and crustaceans form hard shells from the carbohydrate chitin. Cellulose, the most abundant organic compound on Earth, is a complex carbohydrate found in the cell walls of leaves, bark, stems, and roots. Cellulose, like starch, is composed of glucose molecules, bound together in a different way.

Lipids A fourth type of macromolecule includes a chemically diverse group of compounds that are classified together because they do not dissolve in water. These **lipids** include fats, phospholipids, waxes, and steroids:

▶ *Fats and oils* are convenient forms of energy storage, especially for mobile animals. Their hydrocarbon structures somewhat resemble gasoline, a similarity echoed in their function: to effectively store energy and release it when burned.

▶ *Phospholipids* are similar to fats but consist of one water-repellant side and one water-attracting side. This characteristic allows them, when arranged in a double

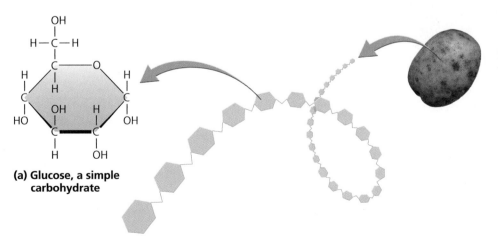

(a) Glucose, a simple carbohydrate

(b) Starch, a polysaccharide

FIGURE 4.13 The monosaccharide glucose (**a**) is the simplest and most abundant carbohydrate and is a vital energy source for organisms. Linked glucose molecules form starch (**b**), an important polysaccharide.

layer, to make up the primary component of animal cell membranes.

▶ *Waxes* are lipids that are digestible by some but not all organisms, and they can play structural roles (for instance, beeswax in bees' hives).

▶ *Steroids* are used in animal cell membranes and in production of hormones, including the sex hormones estrogen and androgen, vital to sexual maturation.

Organisms use cells to compartmentalize macromolecules

All living things are composed of **cells,** the most basic unit of organismal organization. Organisms range in complexity from single-celled bacteria to plants and animals that contain millions of cells. Cells vary greatly in size, shape, and function.

Biologists classify organisms into two groups based on the structure of their cells. **Eukaryotes** include plants, animals, fungi, and protists. The cells of eukaryotes (Figure 4.14a) consist of an outer membrane of lipids and an inner fluid-filled chamber containing **organelles,** internal structures that perform specific functions. These internal structures include (among others) ribosomes, which are organelles that synthesize proteins, and mitochondria, where the last step in the extraction of energy from sugars and fats occurs. Eukaryotes also have within each of their cells a membrane-enclosed nucleus that houses DNA. Eukaryotic organisms generally have many cells.

Prokaryotic organisms are much simpler. Prokaryotes are generally single-celled, and their cells lack membrane-bound organelles and a nucleus (Figure 4.14b). All bacteria are prokaryotes, as are the lesser-known microorganisms called archaea. Bacteria are diverse and are ubiquitous in the environment, and of course they do far more than attack oil spills. Many types of bacteria perform functions

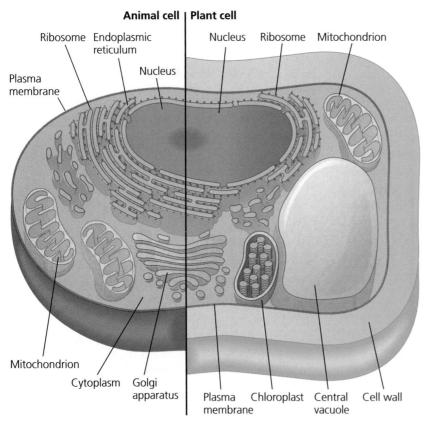

(a) Eukaryotic cell

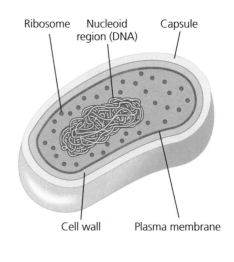

(b) Prokaryotic cell

FIGURE 4.14 Cells are the smallest unit of life that can function independently. Eukaryotic cells **(a)** contain organelles, such as mitochondria and chloroplasts, as well as a membrane-enclosed nucleus that contains DNA. Plant cells (right) have rigid cell walls of cellulose, whereas animal cells (left) have more flexible cell membranes. Prokaryotic cells **(b)** are simpler, lacking membrane-bound organelles and an enclosed nucleus.

Hierarchy of Matter within Organisms		
	Organism	An individual living thing
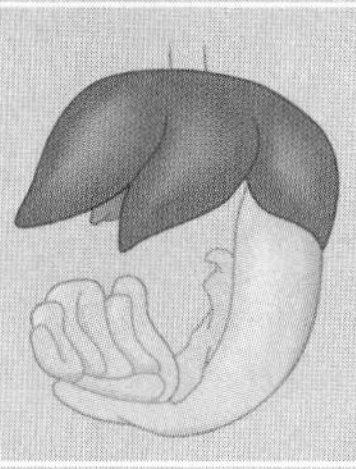	Organ system	An integrated system of organs whose action is coordinated for a particular function
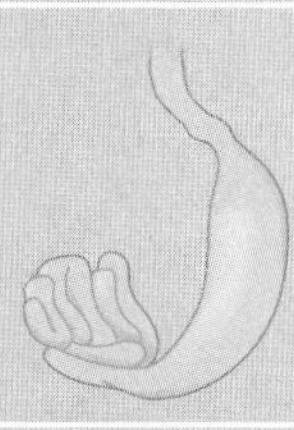	Organ	A structure in an organism composed of several types of tissues and specialized for some particular function
	Tissue	A group of cells with common structure and function
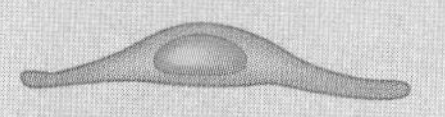	Cell	The smallest unit of living matter able to function independently, enclosed in a semi-permeable membrane
	Organelle	A structure inside a eukaryotic cell that performs a particular function
	Macro-molecule	An organic compound large in size and important for life (includes proteins, nucleic acids, carbohydrates, and lipids)
	Molecule	A combination of two or more atoms chemically bonded together
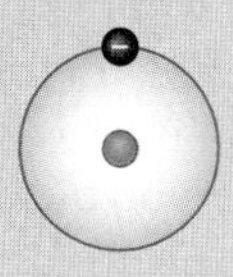	Atom	The smallest component of an element that maintains the element's chemical properties

FIGURE 4.15 Within an organism, we can view matter as being organized in a hierarchy of levels. Atoms (with their protons, neutrons, and electrons) form the base of the hierarchy, and they build the next level, molecules. Macromolecules make up portions of cells, including (in eukaryotes) cell organelles. Cells of similar types function together in tissues, tissues make up organs, and organs make up organ systems, which in turn collectively comprise the organism.

vital to human life—for instance, aiding in digestion and preventing the buildup of harmful wastes.

In eukaryotes, cells specialize in different roles and are organized into collections of cells performing the same function, called *tissues*. Tissues make up *organs*, and organisms are composed of *organ systems*. We have now completed a (very quick!) review of the hierarchy in which matter is organized in living things on Earth (Figure 4.15). Over the next three chapters, we will explore the levels of this hierarchy above the organismal level, as we study the science of ecology. But first we will examine energy, something that underlies every process in environmental science.

Energy Fundamentals

Creating and maintaining organized complexity, whether of a cell or an organism or an ecological system, requires energy. Energy is needed to power the geological forces that shape our planet, to organize matter into complex forms such as biological polymers, to build and maintain cellular structure, and to power the interactions that take place among species. Indeed, energy is somehow involved in nearly every biological, chemical, and physical event.

But what, exactly, is energy? An intangible phenomenon, **energy** is that which can change the position, physical composition, or temperature of matter. A sparrow in flight expends energy to propel its body through the air. When the sparrow lays an egg, its body uses energy to create the calcium-based eggshell and color it with pigment. The sparrow sitting on its nest transfers energy from its body in heating the developing chicks inside its eggs. Some of the most dramatic releases of energy in nature do not involve living things; think of volcanoes erupting or tornadoes sweeping across the plains.

Scientists differentiate between two types of energy: **potential energy,** energy of position; and **kinetic energy,** energy of motion. Consider river water held behind a dam. By preventing water from moving downstream, the dam causes the water to accumulate potential energy. When the dam gates are opened, the potential energy is

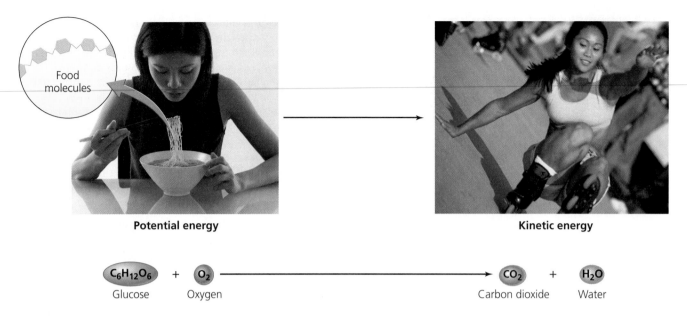

Potential energy

Kinetic energy

$C_6H_{12}O_6$ + O_2 → CO_2 + H_2O

Glucose   Oxygen   Carbon dioxide   Water

FIGURE 4.16 Energy is released when potential energy is converted to kinetic energy. Potential energy stored in sugars, such as glucose, in the food we eat (**a**), combined with oxygen, becomes kinetic energy when we exercise (**b**), releasing carbon dioxide and water as by-products.

converted to kinetic energy, in the form of water's motion as it rushes downstream.

Such energy transfers take place at the atomic level every time a chemical bond is broken or formed. **Chemical energy** is potential energy held in the bonds between atoms. Bonds differ in their amounts of chemical energy, depending on the atoms they hold together. Converting a molecule with high-energy bonds (such as the carbon–carbon bonds of petroleum products) into molecules with lower-energy bonds (such as the bonds in water or carbon dioxide) releases energy by changing potential energy into kinetic energy and produces motion, action, or heat. Just as our automobile engines split the hydrocarbons of gasoline to release chemical energy and generate movement, our bodies split glucose molecules in our food for the same purpose (Figure 4.16).

Energy is always conserved . . .

Although energy can change from one form to another, it cannot be created or lost. The total energy in the universe remains constant and thus is said to be conserved. Scientists have dubbed this principle the **first law of thermodynamics.** The potential energy of the water behind a dam will equal the kinetic energy of its eventual movement down the riverbed. Similarly, burning converts the potential energy in a log of firewood to an equal amount of energy produced as heat and light. We obtain energy

from the food we eat, which we expend in exercise, put toward the body's maintenance, or store as fat. We do not somehow create additional energy or end up with less than the food gives us. Any individual system can temporarily increase or decrease in energy, but the total amount in the universe remains constant.

. . . But energy changes in quality

Although the first law of thermodynamics requires that the overall amount of energy be conserved in any process of energy transfer, the **second law of thermodynamics** states that the nature of energy will change from a more-ordered state to a less-ordered state, if no force counteracts this tendency. The degree of disorder in a substance, system, or process is called **entropy,** and the second law of thermodynamics holds that systems tend to move toward increasing entropy. For instance, after death every organism undergoes decomposition and loses its structure. A log of firewood—the highly organized and structurally complex product of many years of slow tree growth—transforms in the campfire to a residue of carbon ash, smoke, and gases such as carbon dioxide and water vapor, as well as the light and the heat of the flame (Figure 4.17). With the help of oxygen, the complex biological polymers making up the wood are converted into a disorganized assortment of rudimentary molecules and heat and light energy.

The nature of any given energy source helps determine how easily humans can harness it. Sources such as

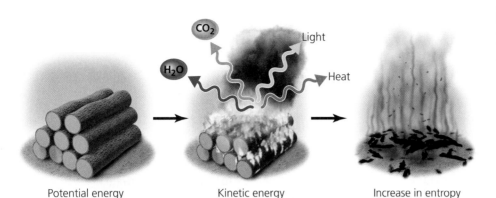

Potential energy
(stored in the molecular
bonds of wood)

Kinetic energy
(released as heat and light)

Increase in entropy

FIGURE 4.17 The burning of firewood demonstrates energy transfer leading from a more-ordered to a less-ordered state. This increase in entropy reflects the second law of thermodynamics.

petroleum products and high-voltage electricity contain concentrated energy that is easily released. It is relatively easy for us to gain large amounts of energy efficiently from these high-quality sources. In contrast, sunlight and the heat stored in ocean water are considered low-quality energy sources. Each and every day the world's oceans absorb heat energy from the sun equivalent to that of 250 billion barrels of oil—more than 3,000 times as much as our global society uses in a year. But because this energy is spread out across such vast spaces, it is diffuse and difficult to harness.

Weighing the Issues:
Energy Quality and Energy Policy

Contrast the ease of harnessing high-quality energy, such as that of petroleum, with the ease of harnessing low-quality energy, such as that of heat from the oceans. How do you think this difference has affected our society's energy policy and energy sources through the years?

In every transfer of energy, some portion usable to us is lost. The inefficiency of some of the most common energy conversions that power our society can be surprising. When we burn gasoline in an automobile engine, only about 16% of the energy released is used to power the automobile's movement. The rest of the energy is converted to heat. Incandescent light bulbs are worse; only 5% of their energy is converted to the light that we use them for, while the rest escapes as heat. Viewed in this context, the 15% efficiency of much current solar energy technology does not look bad at all.

Although the second law of thermodynamics specifies that systems tend to move toward disorder, the order of an object or system can be increased through the input of

additional energy from outside the system. This is precisely what living organisms do. Organisms maintain their structure and function by consuming energy. They represent a constant struggle to maintain order and combat the natural tendency toward disorder.

Light energy from the sun powers most living systems

The energy that powers Earth's organisms and that flows through ecological systems comes primarily from the sun. The sun releases radiation from large portions of the electromagnetic spectrum, although our atmosphere filters much of this out, and we can see only some of this radiation as visible light (Figure 4.18). Most of the sun's energy is reflected, or else absorbed and re-emitted, by the atmosphere, land, or water. Solar energy drives our weather and climate patterns, including winds and ocean currents. A small amount (less than 1% of the total) powers plant growth, and a still smaller amount flows from plants into the organisms that eat them and the organisms that decompose dead organic matter. A minuscule amount of energy, relatively speaking, is eventually deposited below ground in the form of the chemical bonds in fossil fuels.

The sun's light energy is used directly by some organisms to produce their own food. Such organisms, called **autotrophs** or primary **producers,** include green plants, algae, and cyanobacteria. (Cyanobacteria are a type of bacteria named for their characteristic blue-green, or cyan, color.) Autotrophs turn light energy from the sun into chemical energy in a process called photosynthesis (Figure 4.19). In **photosynthesis,** sunlight powers a series of chemical reactions that convert carbon dioxide and water into sugars, transforming low-quality energy from the sun into high-quality energy the

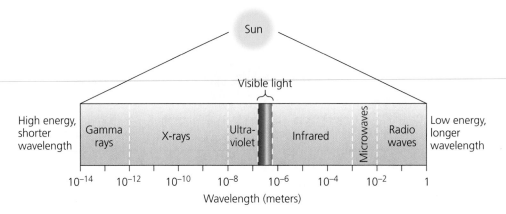

FIGURE 4.18 The sun emits radiation from many portions of the electromagnetic spectrum, and visible light makes up only a small proportion of this energy. Some radiation that reaches our planet is reflected back, some is absorbed by air, land, and water, and a small amount powers photosynthesis.

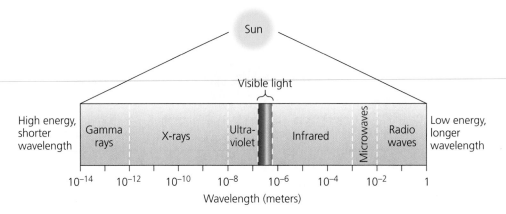

organism can use. It is an example of moving toward a state of lower entropy, and as such it requires a substantial input of outside energy.

Photosynthesis produces food for plants and animals

Photosynthesis occurs within cell organelles called *chloroplasts*, where the light-absorbing pigment *chlorophyll* (which is what makes plants green) uses solar energy to initiate a series of chemical reactions called *light-dependent reactions*. During these reactions, water molecules are split, and they react to form hydrogen ions (H^+) and molecular oxygen (O_2), thus creating the oxygen that we breathe. The light-dependent reactions also produce small, high-energy molecules that are used to fuel a set of *light-independent reactions*. In these reactions, carbon atoms from carbon dioxide are linked together to manufacture sugars. Photosynthesis is a complex process, but the overall reaction can be summarized with the following equation:

$$6CO_2 + 12H_2O \longrightarrow C_6H_{12}O_6 + 6O_2 + 6H_2O$$
$$+ \text{energy from the sun} \qquad \text{(sugar)}$$

The numbers preceding each molecular formula indicate how many of each molecule are involved in the reaction. Note that the sum of the numbers on each side of the equation for each element are equal; that is, there are 6 C, 24 H, and 24 O on each side. This illustrates how chemical equations are balanced, with each atom recycled and matter conserved. No atoms are lost; they are simply rearranged among molecules. Note also that water appears on both sides of the equation. The reason is that for every 12 water molecules that are input and dissociated in the process, 6 water molecules are newly created. We can streamline the photosynthesis equation by showing only the net loss of 6 water molecules:

$$6CO_2 + 6H_2O \longrightarrow C_6H_{12}O_6 + 6O_2$$
$$+ \text{energy from the sun} \qquad \text{(sugar)}$$

Thus in photosynthesis, water, carbon dioxide, and light energy from the sun are transformed to produce sugar (glucose) and oxygen. To accomplish this, green plants draw up water from the ground through their roots, suck in carbon dioxide from the air through their

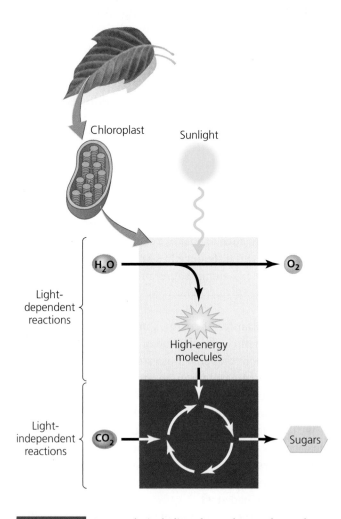

FIGURE 4.19 Autotrophs including plants, algae, and cyanobacteria use sunlight to convert carbon dioxide and water into sugars and oxygen in photosynthesis. Autotrophs provide themselves and the many heterotrophs that eat them with energy for life.

leaves, and harness sunlight. With these ingredients, they create sugars for their growth and maintenance, and release oxygen as a by-product. Animals, in turn, depend on these two outputs of photosynthesis. Animals survive by eating plants, or by eating animals that have eaten plants, and by taking in oxygen. In fact, it is thought that animals appeared on Earth's surface only after the planet's atmosphere had been supplied with oxygen by the earliest autotrophs, the cyanobacteria.

Cellular respiration releases chemical energy

The chemical energy created by photosynthesis can later be used by organisms in a process called **cellular respiration.** To release the chemical energy of glucose, cells use the reactivity of oxygen to convert glucose back into its original starting materials, water and carbon dioxide. The energy released during this process is used to form chemical bonds or to perform other tasks within cells. The net equation for cellular respiration is the exact opposite of that for photosynthesis:

$$C_6H_{12}O_6 \text{ (sugar)} + 6O_2 \rightarrow 6CO_2 + 6H_2O + \text{energy}$$

However, the energy gained per glucose molecule in respiration is only two-thirds of the energy input per glucose molecule in photosynthesis—a prime example of the second law of thermodynamics in action. The extraction of energy from glucose through respiration occurs in the autotrophs that created the glucose and also in the animals that gain glucose by consuming autotrophs. Organisms that consume autotrophs are called **consumers,** or **heterotrophs** (*hetero* means "different"), and include most animals, as well as the fungi and microbes that decompose organic matter. In most ecological systems, plants, algae, or cyanobacteria form the base of a food chain through which energy passes to heterotrophs (▶ pp. 156–158).

Geothermal energy also powers Earth's systems

Although the sun is Earth's primary power source, it is not the only one. A minor additional source is the gravitational pull of the moon, which causes ocean tides. This low-quality energy is weak and diffuse in comparison to solar energy. A more significant additional energy source is radiation emanating from inside Earth, powered by radioactivity. When we think of radioactivity, nuclear power plants and atomic weapons may come to mind, but radioactivity is a natural phenomenon. As discussed earlier (▶ p. 93), it consists of the gradual release of high-energy

FIGURE 4.20 This geyser in the Black Rock Desert of Nevada propels scalding water into the air, powered by geothermal energy from deep below ground. The bright colors of the rocks are from colonies of bacteria that thrive in the hot mineral-laden water.

rays or particles by radioisotopes as their nuclei decay. Radiation from radioisotopes deep inside Earth heats the inside of the planet, and this heat gradually makes its way to the surface. There it drives plate tectonics (▶ pp. 207–209), heats magma that erupts from volcanoes, and warms groundwater, which in some regions of the planet shoots out of the ground in the form of geysers (Figure 4.20). Called **geothermal energy,** this heating from deep within the planet is now being harnessed for commercial power (▶ pp. 633–635).

Long before humans came along, however, geothermal energy was powering other biological communities. On the floor of the ocean, jets of geothermally heated water—essentially underwater geysers—gush into the icy-cold depths. One of the amazing scientific discoveries of recent decades was the realization that these *hydrothermal vents* can host entire communities of organisms that thrive in the extreme high-temperature, high-pressure conditions. Gigantic clams, immense tubeworms, and odd mussels, shrimps, crabs, and fish all flourish in the seemingly hostile environment near scalding water that shoots out of tall chimneys of encrusted minerals (Figure 4.21).

These locations are so deep underwater that they completely lack sunlight, so their communities cannot fuel themselves through photosynthesis. Instead, bacteria in deep-sea vents use the chemical-bond energy of hydrogen sulfide (H_2S) to transform inorganic carbon into organic carbon compounds in a process called **chemosynthesis:**

$$6CO_2 + 6H_2O + 3H_2S \rightarrow C_6H_{12}O_6 \text{ (sugar)} + 3H_2SO_4$$

(a) Hydrothermal vent

(b) Giant tubeworms

FIGURE 4.21 Hydrothermal vents on the ocean floor (**a**) send spouts of hot mineral-rich water into the cold blackness of the deep sea. Amazingly, specialized biological communities thrive in these unusual conditions. Odd creatures such as these giant tubeworms (**b**) survive thanks to bacteria that produce food from hydrogen sulfide.

There are many different types of chemosynthesis, but note how this particular reaction for chemosynthesis closely resembles the photosynthesis reaction. These two processes use different energy sources, but each uses water and carbon dioxide to produce sugar and a by-product, and each produces potential energy that is later released during respiration. Energy from chemosynthesis passes through the deep-sea-vent animal community as heterotrophs such as clams, mussels, and shrimp gain nutrition from chemoautotrophic bacteria. Hydrothermal vent communities excited scientists not only because they were novel and unexpected, but also because some researchers believe they may help us understand how life itself originated.

The Origin of Life

How and where life originated is one of the most centrally important—and intensely debated—questions in modern science. In searching for the answer, scientists have learned a great deal about the history of life on Earth and about what early Earth was like.

Early Earth was a very different place

Earth formed about 4.5 billion years ago in the same way as the other planets of our solar system; dispersed bits of material whirling through space around our sun were

FIGURE 4.22 The young Earth on which life originated was a very different place from our planet today. Microbial life first evolved amid sulfur-spewing volcanoes, intense ultraviolet radiation, frequent extraterrestrial impacts, and an atmosphere containing ammonia.

drawn by gravity into one another, coalescing into a series of spheres. For several hundred million years after the planets formed, there remained enough stray material in the solar system that Earth and the other young planets were regularly bombarded by large chunks of debris in the form of asteroids, meteorites, and comets. The largest impacts were probably so explosive that they vaporized our planet's newly formed oceans. Add to this the severe volcanic and tectonic activity and the intense ultraviolet radiation, and it is clear that early Earth was a pretty hostile place (Figure 4.22). Any life that got under way during this "bombardment stage" might easily have been killed off. Only after most debris was cleared from the solar system was life able to gain a foothold and keep it.

Earth's early atmosphere was probably very different from our atmosphere today. For instance, oxygen was largely lacking until photosynthesizing microbes started producing it. Whereas today's atmosphere is dominated by nitrogen and oxygen (see Table 4.1), many scientists in the 20th century thought Earth's early atmosphere contained large amounts of hydrogen, ammonia (NH_3), methane, and water vapor. More recently, opinion has shifted, and most scientists now infer that the atmosphere contained less hydrogen but more carbon dioxide, nitrogen, and carbon monoxide (CO) than previously thought.

Several hypotheses have been proposed to explain life's origin

Most biochemists interested in life's origin think that life must have begun when inorganic chemicals linked themselves into small molecules and formed organic compounds. Some of these compounds gained the ability to replicate, or reproduce themselves, whereas others found ways to group together into proto-cells. There is much debate and ongoing research, however, on the details of this process, especially concerning the location of the first chemical reactions and the energy source(s) that powered them.

Primordial soup: The heterotrophic hypothesis In the 1930s, scientists J. B. S. Haldane and Aleksandr Oparin independently advanced the idea that life evolved from a primordial soup of simple inorganic chemicals—carbon dioxide, oxygen, and nitrogen—dissolved in the ocean's surface waters or tidal shallows. They suggested how simple amino acids might have formed under these conditions and how more complex organic compounds could have eventually followed, including simple ribonucleic acids that could replicate themselves. This hypothesis is termed *heterotrophic* because it proposes that the first life forms used organic compounds from their environment as an energy source.

This hypothesis has traditionally been favored, and lab experiments have provided evidence that the proposed process can work. In 1953, biochemists Stanley Miller and Harold Urey passed electricity through a mixture of water vapor, hydrogen, ammonia, and methane, which was believed at that time to represent the early atmosphere, and were readily able to produce impressive amounts of organic compounds, including amino acids. Subsequent experiments confirmed Miller's and Urey's findings, but now that most scientists believe early atmospheric conditions were different, this hypothesis seems somewhat less likely to represent what actually happened.

Seeds from space: The extraterrestrial hypothesis A modification of the heterotrophic hypothesis proposes that early chemical reactions on Earth may have received help from outer space. In the early 1900s, Swedish chemist Svante Arrhenius proposed that microbes from space might have traveled on meteorites that crashed to Earth, seeding our planet with life. Scientists largely rejected this idea, also called the *panspermia hypothesis,* believing that even if amino acids or bacteria were to exist in space, they could not be transported to Earth because the high temperatures that comets and meteors attain as they enter our atmosphere should destroy all biological compounds.

However, the Murchison meteorite, which fell in Australia in 1969, was found to contain many amino acids, suggesting that amino acids within rock can survive impact. Since then, experiments simulating impact conditions have shown that organic compounds and some bacteria can withstand a surprising amount of abuse. Furthermore, planetary scientists have shown that large asteroid impacts on one planet (such as Mars) can throw up so much material that some eventually may make its way to other planets (such as Earth). And recent astrobiology research has made a case that comets have brought large amounts of water, and possibly organic compounds, to Earth throughout its history. As a result of such findings, long-distance travel of microbes through space and into our atmosphere now seems more plausible than previously thought.

Life from the depths: The chemoautotrophic hypothesis In the 1970s and 1980s, several scientists, among them Jack Corliss, a discoverer of deep-sea hydrothermal vents, proposed that life may have emanated from the deep sea. The chemoautotrophic hypothesis proposes that life originated at scalding-hot deep-sea vent systems, where sulfur was abundant. In this scenario, the first organisms were chemoautotrophs, creating their own food from hydrogen sulfide.

Genetic analysis of the relationships of present-day organisms (p. 111) suggests that some of the most ancient ancestors of today's life forms lived in extremely hot, wet environments. Additionally, the extreme heat of hydrothermal vents could act to speed up chemical reactions that link atoms together into long molecules, a necessary early step in life's formation. Scientists have shown experimentally that it is possible to form amino acids and begin a chain of steps that might potentially lead to the formation of life under high-temperature, high-pressure conditions similar to those of hydrothermal vents.

Weighing the **issues:**
Hypotheses on Life's Origin

Which lines of evidence in the debate over the origin of life strike you as the most convincing, and why? Which strike you as the least convincing, and why? Can you think of any further scientific research that could be done to address the question of how life originated?

Self-replication and cell formation were crucial steps

Whether they formed in deep-sea vents, tidal pools, or comet craters, early organic polymers had to develop the ability to self-replicate before life as we know it could commence. This involves a chicken-or-egg paradox. To replicate biological polymers, nucleic acids, typically DNA, are needed—but to create DNA, enzymes, a type of biological polymer, are required. In order to get one, you would seem to need the other. This is why scientists think the first self-replicating molecule may have been RNA, because RNA can act as both enzyme and information carrier. Early RNA molecules may have acted as their own enzymes, replicating without the help of other polymers. Over time, scientists hypothesize, the replicating process of early RNA was modified, and DNA, which is more stable, came to dominate as the information carrier in modern cells.

The formation of cells must have been another key step in life's origin. Cells have semipermeable membranes and/or walls, which allow them to maintain an internal environment different from their surroundings. Chance aggregations of polymers may have formed cell-like structures, and structures that happened to contain RNA may have replicated. Such cell-like structures have formed spontaneously in lab experiments, have been able to divide, and—if enzymes are provided—have been able to absorb substances from their surroundings, as cells do.

The fossil record has taught us much about life's history

The earliest evidence of life on Earth comes from rocks about 3.5 billion years old. Although these earliest traces are controversial, there is ample evidence that simple forms of life, such as single-celled bacteria, were present on Earth well over 3 billion years ago. Remains of these microscopic life forms (or their chemical by-products) have been preserved just as have later, much larger, creatures, such as dinosaurs—by fossilization.

As organisms die, some are buried by sediment. Under the right conditions, the hard parts of their bodies—such as bones, shells, and teeth—may be preserved as the sediments are compressed into rock (▸ pp. 195–196). Minerals replace the organic material, leaving behind a **fossil,** an imprint in stone of the dead organism (Figure 4.23). In thousands of locations around the world, geological processes over millions of years have buried sedimentary rock layers and later brought them to the surface, revealing assemblages of fossils representing plants and animals from different time periods. The cumulative body of fossils worldwide is known as the **fossil record.** Paleontologists study the fossil record to infer the history of past life on Earth.

The fossil record clearly shows that:

▸ The species living today are but a tiny fraction of all the species that ever lived; the vast majority of Earth's species are long extinct.

FIGURE 4.23 The fossil record has helped reveal the history of life on Earth. The numerous fossils of trilobites suggest that these animals, now extinct, were abundant in the oceans from roughly 540 million to 250 million years ago.

▸ Earlier types of organisms changed, or evolved, into later ones.

▸ The number of species existing at any one time has increased through history.

▸ There have been several episodes of *mass extinction,* or simultaneous loss of great numbers of species (▸ pp. 127–128).

▸ Many organisms present early in history were smaller and simpler than modern organisms.

The fossil record also tells us that for most of life's history, microbes, like the bacteria that consume hydrocarbons or the cyanobacteria that produce oxygen, were the only life on Earth. It was not until about 600 million years ago that large and complex organisms such as animals, land plants, and fungi appeared.

The crude oil with which we began this chapter is itself a kind of fossil. Plant and animal matter can be preserved as they sink to the seafloor and are buried in the absence of oxygen; eventually they become compressed and turn into the amorphous mixes of hydrocarbons we call fossil

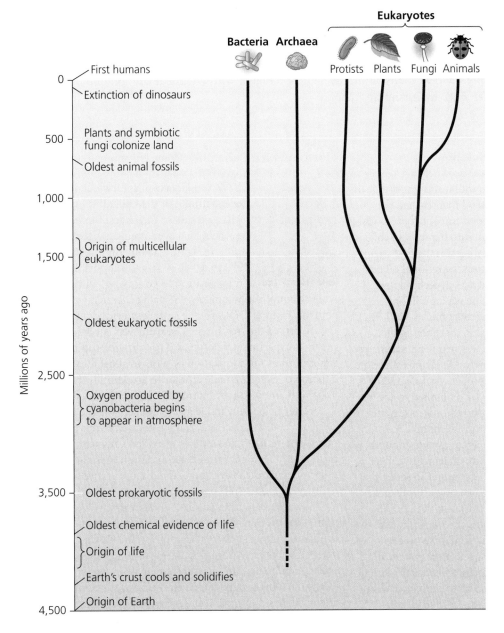

Eukaryotes

Bacteria Archaea Protists Plants Fungi Animals

Millions of years ago

0 — First humans
Extinction of dinosaurs

500 — Plants and symbiotic fungi colonize land
Oldest animal fossils

1,000 —

1,500 — Origin of multicellular eukaryotes

Oldest eukaryotic fossils

2,500 —

Oxygen produced by cyanobacteria begins to appear in atmosphere

3,500 — Oldest prokaryotic fossils

Oldest chemical evidence of life

Origin of life

Earth's crust cools and solidifies

Origin of Earth

4,500 —

FIGURE 4.24 Microbes have lived on the planet for most of its 4.5-billion-year history. Large complex eukaryotes such as vertebrates, in contrast, are relative newcomers. The fossil record and the analysis of present-day organisms and their genes allow scientists to reconstruct evolutionary relationships among organisms and to build a "tree of life." As you progress upward from the trunk of the tree to the tips of its branches, you are moving forward in time. Each fork denotes the divergence of major groups of organisms, each group of which in this greatly simplified diagram includes many thousands of species. The tree of life as understood by scientists today consists of three main groups—the bacteria, the recently discovered archaea, and the diverse eukaryotes. Protists and ancestral eukaryotes are poorly known, and their future study may well produce discoveries that further revise our understanding of life's history.

fuels (▸ p. 559). Coal, oil, and natural gas are the fossil fuels we use to power our civilization. When we drive a car, ignite a stove, or flick on a light switch, we are using energy from life buried millions of years ago.

Present-day organisms and their genes also help us decipher life's history

Besides fossils, biologists also use present-day organisms to infer how evolution proceeded in the past. By comparing the genes and/or the external characteristics of organisms, scientists can create branching trees, similar to family genealogies, that show the relationships among organisms and thus their history of divergence through

time. As you follow such a tree from its trunk to the tips of its branches, you proceed forward through time, tracing the history of life. A major advance was made in recent years as scientists discovered an entire new domain of life, the archaea, single-celled prokaryotes that are genetically very different from bacteria. Today most biologists view the tree of life as a three-pronged edifice consisting of the bacteria, the archaea, and the eukaryotes (Figure 4.24). We will examine the flowering of the diversity of life on our planet in Chapter 5 and further in Chapter 11. The relationships among organisms, and those between organisms and their environments, form the basis for ecology, a discipline of primary importance to environmental science.

Bioremediation

Naturally occurring microbes were put to use cleaning up beaches following the *Exxon Valdez* oil spill. **Was bioremediation a success in Prince William Sound? Can this technique address many of our society's problems with chemical contamination, or does it suffer from too many limitations?**

Bioremediation: An Effective Solution

Bioremediation was definitely a success in Prince William Sound. It was important to remove the spilled oil that reached the shorelines in an environmentally responsible way. Washing it into the sea and collecting it by skimming was the first step, but what to do about the residual oil on the beach? Bioremediation seemed a likely technology.

Oil seeps have released oil into the sea for millions of years, feeding a diverse group of microorganisms. Oil is thus very biodegradable, though unusual in that although it is rich in energy, it contains none of the inorganic nutrients required for microbial growth. The arrival of oil on a shoreline causes a dramatic increase in the number of oil-degrading microorganisms, but their growth is soon limited by the low levels of bioavailable nitrogen and phosphorus in seawater.

Our approach in Alaska was to add oleophilic (oil-adhering) and slow-release fertilizers to provide enough of these limiting nutrients over many tidal cycles so that biodegradation would be stimulated, but not so much that they would cause environmental harm. Rigorous testing, in collaboration with the State of Alaska and U.S. EPA, showed that the approach stimulated the natural rate of biodegradation some twofold to fivefold, with no detectable adverse environmental impact. Bioremediation was used on about 74 miles of shoreline—by far the largest use of this technology to date—and most shorelines were oil-free within 3 years instead of the predicted decades.

Can bioremediation address other environmental problems? Yes. There are many situations where stimulating the biodegradation of fuels, explosives, chlorinated solvents, transformer fluids, pesticides, and other substances will be an environmentally benign and effective treatment. Although the best ways of stimulating biodegradation will not be the same in all situations, working to stimulate natural processes without causing adverse effects will be an excellent way of cleaning many contaminated sites.

Roger Prince is a senior research associate at ExxonMobil's Corporate Strategic Research Laboratory in Annandale, New Jersey. He was Exxon's lead scientist in the monitoring of the successful bioremediation of shorelines oiled during the *Exxon Valdez* spill in Alaska, and he continues to work on the bioremediation of marine oil spills, including a recent multinational collaboration in the Arctic.

Biostimulation Can Sometimes Enhance Environmental Cleanup

The *Exxon Valdez* oil spill, which led to the enactment of the Oil Pollution Act of 1990, gave rise to the largest bioremediation field trial ever attempted. A research study was conducted by EPA in 1989 and 1990 to develop data to support the recommendation to go forward with a full-scale cleanup. Unfortunately, the data generated were equivocal and did not provide sufficient evidence to prove the success of the treatment. The study was equivocal because its experimental design did not provide sufficient replication or randomization of plots to allow for the calculation of experimental error, or to account for the high variability of oil contamination on the beaches.

A later field study in Delaware did provide the unequivocal evidence needed. Data generated in this and other EPA-funded field studies provided sufficient, statistically sound evidence that biostimulation of indigenous microorganisms can accelerate the disappearance of hydrocarbons at a spill site.

However, two other studies I have been involved with, both done on Canadian wetlands, have showed that biostimulation may not always be appropriate. One needs to determine the background levels of nutrients already present and make sure the affected environment is aerobic in nature. Only if nutrients are limiting in concentration, and if dissolved oxygen in the pore space is high enough to support microbial growth on the hydrocarbons, is biostimulation appropriate.

Based on various studies such as these, EPA has published two guidance documents on bioremediation on marine shorelines, freshwater wetlands, and salt marshes.

Although it is not appropriate in every case, biostimulation can play a key role in the environmental cleanup of oil spills. It is a tool to be seriously considered when contemplating how to restore a contaminated environment.

Albert D. Venosa is a senior research scientist with the U.S. Environmental Protection Agency, National Risk Management Research Laboratory, in Cincinnati, Ohio. He currently heads EPA's oil spill research program in the Office of Research and Development. He has published over 100 works in many aspects of wastewater treatment and hydrocarbon biodegradation.

 Explore this issue further by accessing **Viewpoints** at www.aw-bc.com/withgott.

Conclusion

Carbon-based life has flourished on Earth for over 3 billion years, stemming from an origin that scientists are eagerly attempting to understand. Deciphering how life originated depends in part on understanding energy, energy flow, and chemistry. Knowledge in these areas also enhances our understanding of how present-day organisms interact with one another, how they relate to their nonliving environment, and how environmental systems function. Energy and chemistry are in some way tied to nearly every significant process involved in environmental science.

Chemistry can also be a tool for finding solutions to environmental problems. Cleaning up chemical pollution through bioremediation with microbes or plants is just one example. Knowledge of chemistry can be a powerful ally, whether you are interested in analyzing agricultural practices, managing water resources, reforming energy policy, conducting toxicological studies, or finding ways to mitigate global climate change.

REVIEWING OBJECTIVES

You should now be able to:

Explain the fundamentals of environmental chemistry and apply them to real-world situations

▶ Understanding chemistry provides a powerful tool for developing solutions to many environmental problems. (pp. 90–91)
▶ Atoms form molecules, and changes at the atomic level can result in alternative forms of elements, such as ions and isotopes. (pp. 91–96)
▶ Characteristics of the water molecule help facilitate life. (pp. 96–98)
▶ Living things depend on organic compounds, which are carbon-based. (pp. 98–99)

Describe the molecular building blocks of living organisms

▶ Macromolecules, including proteins, nucleic acids, carbohydrates, and lipids, are key building blocks of life. (pp. 99–102)
▶ Organisms use cells to compartmentalize macromolecules. (pp. 102–103)

Differentiate among the types of energy and recite the basics of energy flow

▶ Energy can change in its nature between potential and kinetic energy. Chemical energy is potential energy in the bonds between atoms. (pp. 103–104)
▶ The total amount of energy in the universe is conserved; it cannot be created or lost. (p. 104)
▶ Systems tend to increase in entropy, or disorder, unless energy is added to build or maintain order and complexity. (pp. 104–105)
▶ Earth's systems are powered by radiation from the sun and by geothermal heating from the planet's core. (pp. 105–108)

Distinguish photosynthesis, respiration, and chemosynthesis, and summarize their importance to living things

▶ In photosynthesis, autotrophs use carbon dioxide, water, and solar energy to produce the sugars they need, as well as oxygen. (pp. 105–107)
▶ In respiration, organisms extract energy from sugars by converting them in the presence of oxygen into carbon dioxide and water. (p. 107)
▶ In chemosynthesis, specialized autotrophs use carbon dioxide, water, and chemical energy from minerals to produce the sugars they need. (pp. 107–108)

Itemize and evaluate the major hypotheses for the origin of life on Earth

▶ The heterotrophic hypothesis proposes that life arose from chemical reactions in surface or shallow waters of the ocean. (p. 109)
▶ The panspermia hypothesis proposes that substances needed for life's origin on Earth arrived from space. (p. 109)
▶ The chemoautotrophic hypothesis proposes that life arose from chemical reactions near deep-sea hydrothermal vents. (p. 109)

Outline our knowledge regarding early life, and give supporting evidence for each major concept

▶ The fossil record has revealed many patterns in the history of life, including that species evolve, most species are extinct, and species numbers on Earth have increased. (pp. 110–111)
▶ By comparing modern-day organisms scientists can infer genetic relationships among them and understand their evolutionary history. (p. 111)

TESTING YOUR COMPREHENSION

1. What are the basic building blocks of matter? Provide examples using chemicals common in living organisms.
2. Name four ways in which the chemical nature of the water molecule facilitates life.
3. What are the three classes of biological polymer, and what are their functions?
4. Describe the two major forms of energy, and give examples of each.
5. State the first law of thermodynamics, and describe some of its implications.

6. What is the second law of thermodynamics, and how might it affect our interactions with the environment?
7. What are the two major sources of energy that power Earth's environmental systems?
8. What substances are produced by photosynthesis? By cellular respiration? By chemosynthesis?
9. Compare and contrast three competing hypotheses for the origin of life.
10. Name three things scientists have learned from the fossil record.

SEEKING SOLUTIONS

1. Under what types of conditions might bioremediation be a successful strategy, and when might it not be?
2. Can you think of an example of an environmental problem not mentioned in this chapter that a good knowledge of chemistry could help us solve?
3. Describe an example of energy transformation from one form to another that is not mentioned in this chapter.
4. Give three examples of ways in which the input of energy can resist the tendency toward disorder that the second law of thermodynamics describes.

5. Referring to the chemical reactions for photosynthesis and respiration, provide an argument for why increasing amounts of carbon dioxide in the atmosphere due to global climate change might potentially increase amounts of oxygen in the atmosphere. Give an argument for why it might potentially decrease amounts of atmospheric oxygen. What would you need to know to determine which of these two outcomes might occur?
6. The debate over the origin of life is heated and ongoing. For each of the main hypotheses, what evidence would convince you of its validity over the others?

INTERPRETING GRAPHS AND DATA

In phytoremediation, plants are used to clean up soil or water contaminated by heavy metals such as lead (Pb), arsenic (As), zinc (Zn), and cadmium (Cd). For plants to absorb these metals from soil, the metals must be dissolved in soil water. For any given instance, all metal can be accounted for as either remaining bound to soil particles, being dissolved in soil water, or being stored in the plant.

In a study on the effectiveness of alpine penny-cress (*Thlaspi caerulescens*) for phytoremediation, Enzo Lombi and his colleagues grew crops of this small perennial plant for approximately one year in pots of soil from contaminated sites. They then measured the amount of zinc and cadmium in the soil and in the plants when they were harvested.

1. What were the zinc and cadmium concentrations in the soil prior to phytoremediation? What were the zinc and cadmium concentrations in the soil after one year of phytoremediation?

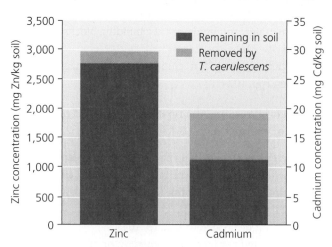

Removal of zinc and cadmium from contaminated soil by alpine penny-cress, *Thlaspi caerulescens*. Data from Lombi, E., et al. 2001. Phytoremediation of heavy metal-contaminated soils: natural hyperaccumulation versus chemically enhanced phytoextraction. *Journal of Environmental Quality* 30: 1919–1926.

2. How much zinc and cadmium were removed from the soil? If the plants continued to remove zinc and cadmium from the soil at the rates shown above, approximately how long would it take to remove all the zinc and cadmium?

3. Alpine penny-cress produces natural chelating agents (see "The Science behind the Story," pp. 92–93) that increase the solubility of metals in soil water. If these dissolved metals are not taken up by the plants, what may be an unintended consequence of having increased their solubility?

CALCULATING ECOLOGICAL FOOTPRINTS

In ecological systems, a rough rule of thumb is that when energy is transferred from plants to plant-eaters or from prey to predator, the efficiency is only about 10%. Much of this inefficiency is a consequence of the second law of thermodynamics. Another way to think of this is that eating 10 calories of plant material is the ecological equivalent of eating 1 calorie of material from an animal.

Humans are considered omnivores because we can eat both plants and animals. The choices we make about what to eat have significant ecological impacts. With this in mind, calculate the ecological energy requirements for four different diets, each of which provides a total of 2,000 dietary calories per day.

Diet	Source of calories	Number of calories consumed by source	Ecologically equivalent calories by source	Total ecologically equivalent calories per day
100% plant 0% animal	Plant			
	Animal			
90% plant 10% animal	Plant	1,800	1,800	3,800
	Animal	200	2,000	
50% plant 50% animal	Plant			
	Animal			
0% plant 100% animal	Plant			
	Animal			

1. How many ecologically equivalent calories would it take to support you for a year on each of the four diets listed?

2. What is the relative ecological impact of including as little as 10% of your calories from animal sources (e.g., milk, dairy products, eggs, and meat)? What is the ecological impact of a strictly carnivorous diet compared with a strict vegetarian diet?

3. What percentages of the calories in your own diet do you think come from plant versus animal sources? Estimate the ecological impact of your diet, relative to a strictly vegetarian one.

4. Describe some challenges of providing food for the growing human population, especially as people in many poorer nations develop a taste for an American-style diet rich in animal protein and fat.

Take It Further

Go to www.aw-bc.com/withgott or the student CD-ROM where you'll find:

▶ Suggested answers to end-of-chapter questions
▶ Quizzes, animations, and flashcards to help you study
▶ *Research Navigator*™ database of credible and reliable sources to assist you with your research projects

▶ **GRAPHit!** tutorials to help you master how to interpret graphs
▶ **INVESTIGATEit!** current news articles that link the topics that you study to case studies from your region to around the world

5 Evolution, Biodiversity, and Population Ecology

Monteverde cloud forest, Costa Rica

Upon successfully completing this chapter, you will be able to:

▶ Explain the process of natural selection, and cite evidence for this process

▶ Describe the ways in which evolution results in biodiversity

▶ Discuss reasons for species extinction and mass extinction events

▶ List the levels of ecological organization

▶ Outline the characteristics of populations that help predict population growth

▶ Assess logistic growth, carrying capacity, limiting factors, and other fundamental concepts of population ecology

▶ Identify efforts and challenges involved in the conservation of biodiversity

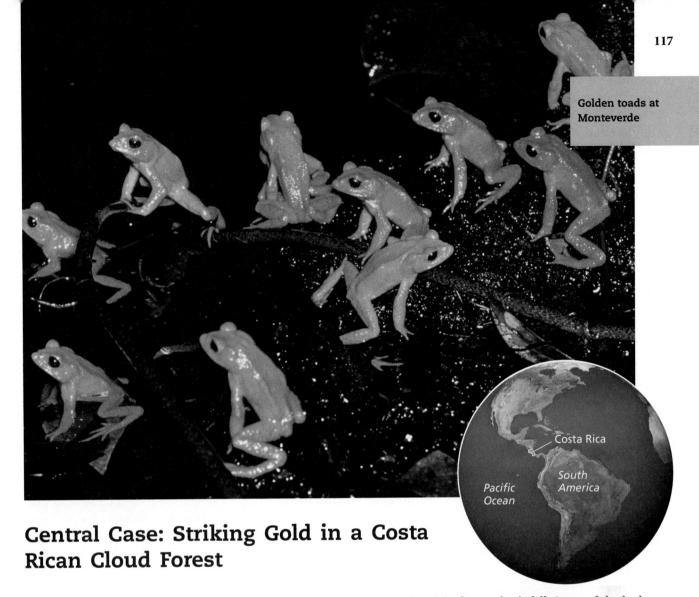

Golden toads at Monteverde

Costa Rica

South America

Pacific Ocean

Central Case: Striking Gold in a Costa Rican Cloud Forest

> "I must confess that my initial response when I saw them was one of disbelief and suspicion that someone had dipped the examples in enamel paint."
> —Dr. Jay M. Savage, describing the golden toad in 1966

> "What a terrible feeling to realize that within my own lifetime, a species of such unusual beauty, one that I had discovered, should disappear from our planet."
> —Dr. Jay M. Savage, describing the golden toad in 1998

During a 1963 visit to Central America, biologist Jay M. Savage heard rumors of a previously undocumented toad living in Costa Rica's mountainous Monteverde region. The elusive amphibian, according to local residents, was best known for its color: a brilliant golden yellow-orange. Savage was told the toad was hard to find because it appeared only during the early part of the region's rainy season.

Monteverde means "green mountain" in Spanish, and the name couldn't be more appropriate. The village of Monteverde sits beneath the verdant slopes of the Cordillera de Tilarán, mountains that receive over

400 cm (157 in) of annual rainfall. Some of the lush forests above Monteverde, which begin around 1,600 m (5,249 ft, just under a mile high), are known as *lower montane rainforests*. They are also known as *cloud forests* because much of the moisture they receive arrives in the form of low-moving clouds that blow inland from the Caribbean Sea. Monteverde's cloud forest was not fully explored at the time of Savage's first visit, and researchers who had been there described the area as pristine, with a rich bounty of ferns, liverworts, mosses, clinging vines, orchids, and other organisms that thrive in cool, misty environments. Savage knew that such conditions create ideal habitat for many toads and other amphibians.

In May of 1964, Savage organized an expedition into the muddy mountains above Monteverde to try to document the existence of the previously unknown toad species in its natural habitat. Late in the afternoon of May 14, he and his colleagues found what they were looking for. Approaching the mountain's crest, they spotted bright orange patches on the forest's black floor.

In one area that was only 5 m (16.4 ft) in diameter, they counted 200 golden toads.

The discovery received international attention, making a celebrity of the tiny toad—which Savage named *Bufo periglenes* (literally, "the brilliant toad")—and making a travel destination of its mountain home. At the time, no one knew that the Monteverde ecosystem was about to be transformed. No one foresaw that the oceans and atmosphere would begin warming due to global climate change (Chapter 18) and cause Monteverde's moisture-bearing clouds to rise, drying the forest. No one could guess that this newly discovered species of toad would become extinct in less than 25 years.

Evolution as the Wellspring of Earth's Biodiversity

The golden toad was new to science, and countless species still await discovery, but scientists understand quite well how the world became populated with the remarkable diversity of organisms we see today. We know that the process of biological evolution has brought us from a stark planet inhabited solely by microbes to a lush world of 1.5 million (and likely millions more) species (Figure 5.1).

Perceiving how organisms adapt to their environments and change over time is crucial for understanding the history of life. Understanding evolution is also vital for appreciating ecology, a central component of environmental science. Evolutionary processes are relevant to many aspects of environmental science, including pesticide resistance, agriculture, medicine, and environmental health.

The term *evolution* in the broad sense means change over time, but scientists most often use the term to refer specifically to biological evolution. Biological **evolution** consists of genetic change in organisms across generations. This genetic change often leads to modifications in the appearance, functioning, or behavior of organisms through time. Biological evolution results from random genetic changes, and may proceed randomly or be directed by natural selection.

Natural selection is the process by which traits that enhance survival and reproduction are passed on more frequently to future generations than those that do not, altering the genetic makeup of populations through time. The theory of evolution by natural selection is one of the best-supported and most illuminating concepts in all of science, yet it has remained socially controversial among some nonscientists who fear it may threaten their religious beliefs. Although scientists sometimes disagree about the specific mechanisms thought to drive evolution in particular cases, or about the time scales on which it takes place, this routine scientific debate should not be equated with the socially driven opposition of some nonscientists. From a scientific standpoint, evolutionary theory is indispensable, because it is the foundation of modern biology.

(a) Resplendent quetzal

(b) Puffball mushroom

(c) Harlequin frog

(d) Scutellerid bug

FIGURE 5.1 Much of our planet's biological diversity resides in tropical rainforests. Monteverde's cloud-forest community includes organisms such as this **(a)** resplendent quetzal *(Pharomachrus mocinno)*, **(b)** puffball mushroom *(Calostoma cinnabarina)*, **(c)** harlequin frog *(Atelopus varius)*, and **(d)** scutellerid bug *(Pachycoris torridus)*.

Natural selection shapes organisms and diversity

In 1858, **Charles Darwin** and **Alfred Russell Wallace** each proposed the concept of natural selection as a mechanism for evolution and as a way to explain the great variety of living things. Darwin and Wallace were exceptionally keen naturalists from England who had studied plants and animals in such exotic locales as the Galapagos Islands and the Malay Archipelago. Both men recognized that organisms face a constant struggle to gain sufficient resources to survive and reproduce. Both were influenced by the writings of Thomas Malthus (▸ pp. 4–5), who feared that human population growth would outstrip resource availability and lead to widespread death and social upheaval. Darwin and Wallace observed that organisms produce more offspring than can possibly survive, and they realized that some offspring may be more likely than others to survive and reproduce. Furthermore, they recognized that whichever characteristics give certain individuals an advantage in surviving and reproducing might be inherited by their offspring. These characteristics, Darwin and Wallace realized, would tend to become more prevalent in the population in future generations.

Natural selection is a simple concept that offers an astonishingly powerful explanation for patterns apparent in nature. The idea of natural selection follows logically from a few straightforward premises (Table 5.1). One is that individuals of the same species vary in their characteristics. Although not known in Darwin and Wallace's time, we now know that variation is due to differences in genes, the environments within which genes are expressed, and the interactions between genes and environment. As a result of this variation, some individuals within a species will happen to be better suited to their environment than

others and thus will be able to survive longer and/or reproduce more.

Many characteristics are passed from parent to offspring through the genes, and a parent that is long-lived, robust, and produces many offspring will pass on genes to more offspring than a weaker, shorter-lived individual that produces only a few offspring. In the next generation, therefore, the genes of better-adapted individuals will be more prevalent than those of less well-adapted individuals. From one generation to another through time, species will evolve to possess characteristics that lead to better and better success in a given environment. A trait that promotes success is called an **adaptive trait,** or an **adaptation.** A trait that reduces success is *maladaptive.*

Natural selection acts on genetic variation

For an organism to pass a trait along to future generations—that is, for the trait to be *heritable*—genes in the organism's DNA (▸ pp. 100–101) must code for the trait. Accidental alterations that arise during DNA replication give rise to genetic variation among individuals. In an organism's lifetime, its DNA will be copied millions of times by millions of cells. In all this copying and recopying, sometimes a mistake is made. Accidental changes in DNA, called **mutations,** can range in magnitude from the addition, deletion, or substitution of single nucleotides (▸ p. 100) to the insertion or deletion of large sections of DNA. If a mutation occurs in a sperm or egg cell, it may be passed on to the next generation. Although most mutations have little effect, some can be deadly, whereas others can be beneficial. Those that are not lethal provide the genetic variation on which natural selection acts.

Sexual reproduction also generates variation. In sexual organisms, genetic material is mixed, or recombined, so that a portion of each parent's genome is included in the genome of the offspring. This process of *recombination* produces novel combinations of genes, generating variation among individuals.

Genetic variation can lead to variation in organismal-level traits. We can visualize how traits vary using distribution graphs, which help us see that selection can alter the characteristics of organisms through time in three main ways (Figure 5.2). Selection that drives a feature in one direction rather than another—for example, toward larger or smaller, faster or slower—is called *directional selection.* In contrast, *stabilizing selection* produces intermediate traits, in essence preserving the status quo. Under *disruptive selection,* traits diverge from their starting condition in two or more directions.

Table 5.1	**The Logic of Natural Selection**

▸ Organisms produce more offspring than can survive

▸ Individuals vary in their characteristics

▸ Many characteristics are inherited by offspring from parents

Therefore,

    ▸ Some individuals will be better suited to their environment than other individuals

    ▸ By producing more offspring and/or offspring of higher quality, better-suited individuals transmit more genes to future generations than poorly suited individuals

    ▸ Future generations will contain more genes, and thus more characteristics, of the better-suited individuals; as a result, characteristics evolve across generations through time

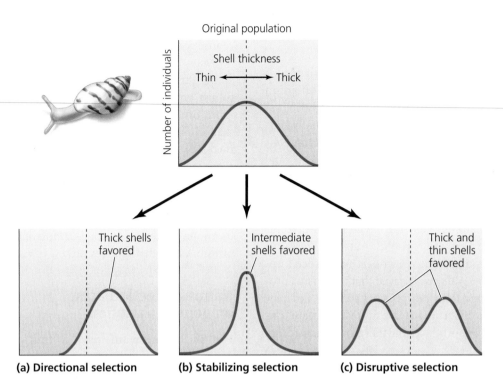

FIGURE 5.2 Selection can act in three ways. Consider snails living in tropical cloud forest, and assume that we begin with a population of snails with shells of different thicknesses (top graph). Because shells protect snails against predators, snails with thick shells may be favored over those with thin shells, through *directional selection* (**a**). Alternatively, suppose that a shell that is too thin breaks easily, whereas a shell that is too thick wastes resources that are better used for feeding or reproduction. In such as case, *stabilizing selection* (**b**) could act to favor snails with shells that are neither too thick nor too thin. Under *disruptive selection* (**c**), extreme traits are favored. For example, perhaps thin-shelled snails are so resource-efficient that they can outreproduce intermediate-shelled snails, whereas thick-shelled snails are so well protected from predators that they also outreproduce intermediate-shelled snails. In such a case, each of the "extreme" strategies works more effectively than a compromise between the two, and natural selection increases the relative numbers of thin- *and* thick-shelled snails, while reducing the number of intermediate-shelled ones.

An organism's environment determines what pressures natural selection will exert on the organism. However, environments change, and organisms may move to new places and encounter new conditions. In either case, a trait that is adaptive in one location or season may prove maladaptive in another. Golden toads that had adapted to the moist conditions of Monteverde's cloud forest would not have survived in drier forests, and apparently did not persist after Monteverde's climate became drier starting 25 years ago. Differences in environmental conditions in time and space make adaptation a moving target for organisms.

In all these ways, variable genes and variable environments interact as organisms engage in a perpetual process of adapting to the changing conditions around them. During this process, natural selection does not simply weed out unfit individuals. It also helps to elaborate and diversify traits that in the long term may help lead to the formation of new species and whole new types of organisms. In this way, natural selection has helped bring about the wondrous flowering of life on our planet.

Evidence of natural selection is all around us

The results of natural selection are all around us, visible in every adaptation of every organism (Figure 5.3). In addition, countless lab experiments (mostly with fast-reproducing organisms, such as bacteria and fruit flies) have demonstrated rapid evolution of traits. The evidence for selection that may be most familiar to us is that which Darwin himself cited prominently in his work 150 years ago: our breeding of domesticated animals. In our dogs, our cats, and our livestock, we have conducted selection under our own direction. We have chosen animals with traits we like and bred them together, while not breeding those with variants we do not like. Through such selective

Kauai ʻAkialoa
Hemignathus procerus

Grosbeak Finch
Psittirostra kona

ʻAkiapolaʻau
Hemignathus wilsoni

Palila
Psittirostra bailleui

ʻŌʻū
Psittirostra psittacea

ʻAkepa
Loxops coccinea

ʻAkikiki
Loxops maculata bairdi

Maui Parrotbill
Pseudonestor xanthophrys

ʻAmakihi
Loxops virens

FIGURE 5.3 Natural selection has produced tremendous diversity among organisms in the wild. In the group of birds known as Hawaiian honeycreepers, closely related species have adapted to different food resources, habitats, or ways of life, as indicated by the diversity in their plumage colors and the shapes of their bills. Such a burst of species formation due to natural selection is known as an *adaptive radiation.*

breeding, we have been able to exaggerate particular traits we prefer. Consider the great diversity of dog breeds (Figure 5.4a), all of which comprise variations on a single species. From Great Dane to Chihuahua, they can interbreed freely and produce viable offspring, yet breeders maintain striking differences between them by allowing only like individuals to breed with like. This process of selection conducted under human direction is termed **artificial selection.**

Artificial selection has also given us the many crop plants we depend on for food, all of which were domesticated from wild ancestors and carefully bred over years, centuries, or millennia (Figure 5.4b). Through selective breeding, we have created corn with larger sweeter kernels; wheat and rice with larger and more numerous grains; and apples, pears, and oranges with better taste.

We have diversified single types into many, for instance, breeding variants of the plant *Brassica oleracea* to create broccoli, cauliflower, cabbage, and Brussels sprouts. Our entire agricultural system is based on artificial selection.

Weighing the Issues:
Artificial Selection

Consider some of the pets and farm animals that humans have domesticated through artificial selection, such as horses, dogs, and cats. In what ways do these domesticated animals differ from their wild relatives? What kinds of characteristics do people prefer in pets and farm animals? Do the evolved traits of the domesticated animals match these preferences?

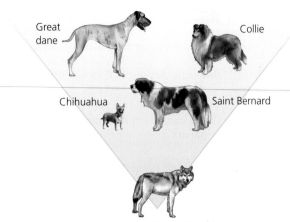

(a) Ancestral wolf and derived dog breeds

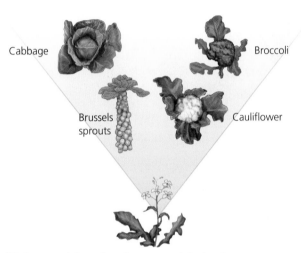

(b) Ancestral *Brassica oleracea* and derived crops

FIGURE 5.4 Selection imposed by humans (selective breeding, or artificial selection) has resulted in the numerous breeds of dogs. By starting with the gray wolf *(Canis lupus)* as the ancestral wild species, and by breeding like with like and selecting for the traits we prefer, we have evolved breeds as different as Great Danes and Chihuahuas **(a).** By this same process we have created the immense variety of crop plants we depend on for sustenance. Cabbage, Brussels sprouts, broccoli, and cauliflower were all evolved from a single ancestral species, *Brassica oleracea* **(b).**

Evolution generates biological diversity

When Charles Darwin wrote about the wonders of a world full of diverse animals and plants, he conjured up the vision of a "tangled bank" of vegetation harboring all kinds of creatures. Such a vision fits well with the arching vines, dripping leaves, and mossy slopes of the tropical cloud forest of Monteverde. Indeed, tropical forests worldwide teem with life and harbor immense biological diversity (see Figure 5.1).

Biological diversity, or **biodiversity** for short, refers to the sum total of all organisms in an area, taking into account the diversity of species, their genes, their populations, and their communities. A **species** is a particular type of organism or, more precisely, a population or group of populations whose members share certain characteristics and can freely breed with one another and produce fertile offspring. A **population** is a group of individuals of a particular species that live in the same area. We have already discussed genes (▶p. 101), and we will introduce communities shortly (▶p. 130; Chapter 6).

Scientists have described between 1.5 million and 1.8 million species, but many more remain undiscovered or unnamed. Estimates for the total number of species in the world range up to 100 million, with many of them thought to occur in tropical forests. In this light, the discovery of a new toad species in Costa Rica in 1964 seems far less surprising. Although Costa Rica covers a tiny fraction (0.01%) of Earth's surface area, it is home to 5–6% of all species known to scientists. And of the 500,000 species scientists estimate exist in the country, only 87,000 (17.4%) have been inventoried and described.

Tropical rainforests such as Costa Rica's, however, are by no means the only places rich in biodiversity. Step outside anywhere on Earth, even in a major city, and you will find numerous species within easy reach. They may not always be large and conspicuous like Yellowstone's bears or Africa's elephants, but they will be there. Plants poke up from cracks in asphalt in every city in the world, and even Antarctic ice harbors microbes. In a handful of backyard soil there may exist an entire miniature world of life, including several insect species, several types of mites, a millipede or two, many nematode worms, a few plant seeds, countless fungi, and millions upon millions of bacteria. We will examine Earth's biodiversity in detail in Chapter 11.

Speciation produces new types of organisms

How did Earth come to have so many species? Whether there are 1.5 million or 100 million, such large numbers require scientific explanation. The process by which new species are generated is termed **speciation.** Speciation can occur in a number of ways, but most biologists consider the main mode of species formation to be *allopatric speciation,* species formation due to the physical separation of populations over some geographic distance. To understand allopatric speciation, begin by picturing a population of organisms. Individuals within the population

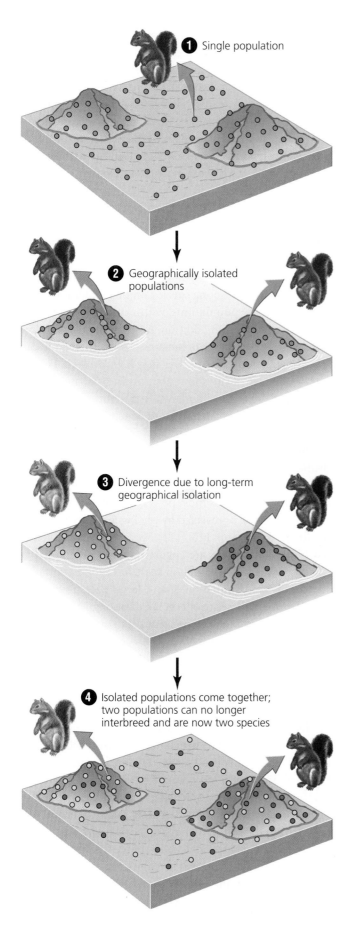

① Single population

② Geographically isolated populations

③ Divergence due to long-term geographical isolation

④ Isolated populations come together; two populations can no longer interbreed and are now two species

possess many similarities that unify them as a species, because they are able to reproduce with one another and share genetic information. However, if the population is broken up into two or more populations that become isolated from one another, individuals from one population cannot reproduce with individuals from the others.

When a mutation arises in the DNA of an organism in one of these isolated populations, it cannot spread to the other populations. Over time, each population will independently accumulate its own set of mutations. Eventually, the populations may diverge, or grow different enough, that their members can no longer mate with one another. Perhaps they no longer recognize one another as being the same species because they have diverged so much in appearance or behavior. Once this has happened, there is no going back; the two populations cannot interbreed, and they have embarked on their own independent evolutionary trajectories as separate species (Figure 5.5). The populations will continue diverging in their characteristics as chance mutations accumulate that confer traits causing the populations to become different in random ways. If environmental conditions happen to be different for the two populations, then natural selection may accelerate the divergence. Through the speciation process, single species can generate multiple species, each of which can in turn generate more.

Populations can be separated in many ways

The long-term geographic isolation of populations that can lead to allopatric speciation can occur in various ways (Table 5.2). Glacial ice sheets may move across continents during ice ages and split populations in two. Major rivers may change course and do the same. Mountain ranges may rise and divide regions and their organisms. Drying climate may partially evaporate lakes, subdividing them into multiple smaller bodies of water. Warming or cooling temperatures may cause whole plant communities to move northward or southward, or upslope or downslope, creating new patterns of plant and animal distribution. Regardless of the mechanism of separation, in

FIGURE 5.5 Allopatric speciation has generated much of Earth's diversity. In this process, some geographical barrier splits a population. In this diagram, two mountaintops (1) are turned into islands by rising sea level (2), isolating populations of squirrels. Each isolated population accumulates its own independent set of genetic changes over time, until individuals become genetically distinct and unable to breed with individuals from the other population (3). The two populations now represent separate species and will remain so even if the geographical barrier is removed and the new species intermix (4).

Table 5.2	Mechanisms of Population Isolation That Can Give Rise to Allopatric Speciation

▶ Glacial ice sheets advance

▶ Mountain chains are uplifted

▶ Major rivers change course

▶ Sea level rises, creating islands (see Figure 5.5)

▶ Climate warms, pushing vegetation up mountain slopes and fragmenting it

▶ Climate dries, dividing large single lakes into multiple smaller lakes

▶ Ocean current patterns shift

▶ Islands are formed in the sea by volcanism

order for speciation to occur, populations must remain isolated for a long time, generally thousands of generations.

If the geological or climatic process that has isolated populations reverses itself—if the glacier recedes, or the river returns to its old course, or warm temperatures turn cool again—then the populations can come back together. This is the moment of truth for speciation. If the populations have not diverged enough, their members will begin interbreeding and reestablish gene flow, mixing those mutations that each population accrued while isolated. However, if the populations have diverged sufficiently, they will not interbreed, and two species will have been formed, each fated to continue on its own evolutionary path.

Although allopatric speciation has long been considered the main mode of species formation, speciation appears to occur in other ways as well. *Sympatric speciation* occurs when species form from populations that become reproductively isolated within the same geographic area. For example, populations of some insects may become isolated by feeding and mating exclusively on different types of plants. Or they may mate during different seasons, isolating themselves in time rather than space. In some plants, speciation apparently has occurred as a result of hybridization between species. In others, it seems to have resulted from mutations that changed the numbers of chromosomes, creating plants that could not mate with plants with the original number of chromosomes. Garnering solid evidence for speciation mechanisms is difficult, so biologists still actively debate the relative prevalence of each of these modes of speciation.

Life's diversification results from numerous speciation events

Repeated speciation events have generated complex patterns of diversity at levels above the species level. Such patterns are studied by evolutionary biologists, who examine how groups of organisms arose and how they evolved the characteristics they show. For instance, how did we end up with plants as different as mosses, palm trees, daisies, and redwoods? Why do fish swim, snakes slither, and sparrows sing? How and why did the ability to fly evolve independently in birds, bats, and insects? To address such questions, one needs to know how the major groups diverged from one another, and this pattern ultimately results from the history of individual speciation events.

We saw in Chapter 4 how the history of divergence can be represented in a treelike diagram (Figure 4.24, ▶ p. 111). Such branching diagrams, called cladograms, or **phylogenetic trees,** illustrate scientists' hypotheses as to how divergence took place (Figure 5.6). Phylogenetic trees can show relationships among species, among major groups of species, among populations within a species, or even among individuals. In addition, by mapping traits onto a tree according to which organisms possess them, one can trace how the traits themselves may have evolved. For instance, the tree of life shows that birds, bats, and insects are distantly related, with many other flightless groups between them. So, it is far simpler to conclude that the three groups evolved flight independently than to conclude that the many flightless groups all lost an ancestral ability to fly. Because phylogenetic trees help biologists make such inferences about so many traits, they have become one of the modern biologist's most powerful tools.

Life's history, as revealed by phylogenetic trees and by the fossil record (▶ pp. 110–111), is complex indeed, but a few big-picture trends are apparent. As we mentioned in Chapter 4, life in its 3.5 billion years has evolved complex structures from simple ones, and large sizes from small ones. However, these are only generalizations. Many organisms have evolved to become simpler or smaller when natural selection favored it. Many very complex life forms have disappeared (Figure 5.7), and it is easy to argue that Earth still belongs to the bacteria and other microbes, some of them little changed over eons.

Even fans of microbes, however, must marvel at some of the exquisite adaptations that animals, plants, and fungi have evolved: The heart that beats so reliably for an animal's entire lifetime that we take it for granted. The complex organ system of which the heart is a part. The stunning plumage of a peacock in full display. The ability of each and every plant on the planet to lift water and nutrients from the soil, gather light from the sun, and turn it into food. The staggering diversity of beetles and other insects. The human brain and its ability to reason. All these and more have resulted from the process of evolution as it has generated new species and whole new branches on the tree of life.

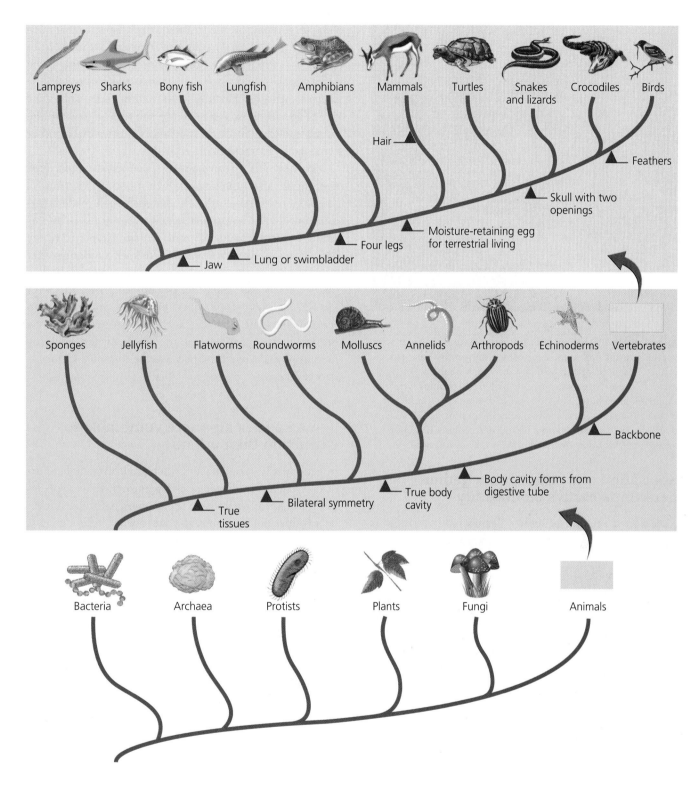

FIGURE 5.6 Phylogenetic trees show the history of life's divergence. Similar to family genealogies, these trees illustrate relationships among groups of organisms, as inferred from the study of similarities and differences among present-day creatures. The diagram here is a greatly simplified representation of relationships among a few major groups—one small portion of the huge and complex "tree of life." Each branch results from a speciation event, and time proceeds upward from bottom to top. By mapping traits onto phylogenetic trees, biologists can study how traits have evolved over time. In this diagram, several major traits are mapped, using triangular arrows indicating the point at which they originated. For instance, all vertebrates "above" the point at which jaws are indicated have jaws, whereas lampreys diverged before jaws originated and thus lack them.

FIGURE 5.7 Life has not always progressed from simple to complex during evolution. Many complex organisms have gone extinct, taking their designs and innovations with them. For example, the strange creatures portrayed in this painting were found fossilized in the Burgess Shale of the Canadian Rockies in British Columbia. They lived in marine environments 530 million years ago and vanished without leaving descendants. Painting by Marella J. Sibbick. *Source:* The National History Museum, London.

Speciation and extinction together determine earth's biodiversity

Although speciation generates Earth's biodiversity, it is only one part of the equation—for, as you will recall from Chapter 4 (▶p. 110), the vast majority of species that once lived are now gone. The disappearance of a species from Earth is called **extinction.** From studying the fossil record, paleontologists calculate that the average time a species spends on Earth is 1–10 million years. The number of species in existence at any one time is equal to the number added through speciation minus the number removed by extinction.

Extinction is a natural process, but human impact can profoundly affect the rate at which it occurs (Figure 5.8). The apparent extinction of the golden toad made headlines worldwide, but unfortunately it was not such an unusual occurrence. As we will see in Chapter 11, the biological diversity that makes Earth such a unique planet is being lost at an astounding rate. This loss affects humans directly, because other organisms provide us with life's necessities—food, fiber, medicine, and ecosystem services (▶p. 39). Species extinction brought about by human impact may well be the single biggest environmental problem we face, because the loss of a species is irreversible.

Some species are more vulnerable to extinction than others

In general, extinction occurs when environmental conditions change rapidly or severely enough that a species cannot adapt genetically to the change; natural selection simply does not have enough time to work. All manner of environmental events can cause extinction, from climate

FIGURE 5.8 Until 10,000 years ago, the North American continent teemed with a variety of large mammals, including mammoths, camels, giant ground sloths, lions, saber-toothed cats, and various types of horses, antelope, bears, and others. Nearly all of this megafauna went extinct suddenly about the time that humans first arrived on the continent. Similar extinctions occurred in other areas simultaneously with human arrival, suggesting to many scientists that overhunting or other human impacts were responsible. *Source:* National Museum of Natural History, Smithsonian Institution.

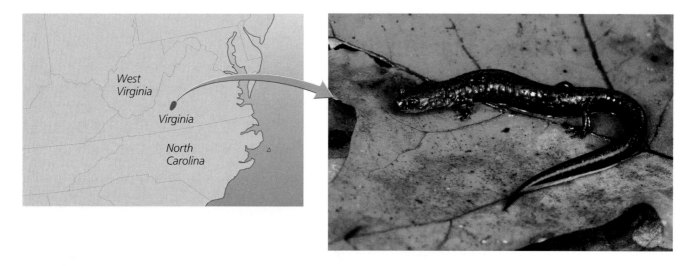

FIGURE 5.9 Forty salamander species in the United States are restricted in range to areas the size of a typical county, and some of these live atop single mountains. Such small range sizes leave these creatures vulnerable to extinction if severe local environmental changes occur. The Peaks of Otter salamander *(Plethodon hubrichti)* pictured here lives on only a few mountains in Virginia's Blue Ridge Mountains.

change to the rise and fall of sea level, to the arrival of new harmful species, to severe weather events such as extended droughts. In general, small populations and species narrowly specialized on some particular resource or way of life are most vulnerable to extinction from environmental change.

The golden toad was a prime example of a vulnerable species. It was **endemic** to the Monteverde cloud forest, meaning that it occurred nowhere else on the planet. Endemic species face relatively high risks of extinction because all their members belong to a single, sometimes small, population. At the time of its discovery, the golden toad was known from only a 4-km² (988-acre) area of Monteverde. It also required very specific conditions to breed successfully. During the spring at Monteverde, water collects in shallow pools within the network of roots that span the cloud forest's floor. The golden toad gathered to breed in these root-bound reservoirs, and it was here that Jay Savage and his companions collected their specimens in 1964. Monteverde provided ideal habitat for the golden toad, but the minuscule extent of that habitat meant that any environmental stresses that deprived the toad of the resources it needed to survive might doom the entire world population of the species.

In the United States, a number of amphibians are limited to very small ranges and thus are vulnerable to extinction. The Yosemite toad is restricted to a small region of the Sierra Nevada in California, the Houston toad occupies just a few areas of Texas woodland, and the Florida bog frog lives in a tiny region of Florida wetland. Fully 40 salamander species in the United States are restricted to areas the size of a typical county, and some of these live atop single mountains (Figure 5.9).

Earth has seen several episodes of mass extinction

Most extinction occurs gradually, one species at a time. The rate at which this type of extinction occurs is referred to as the *background extinction rate.* However, Earth has seen five events of staggering proportions that killed off massive numbers of species at once. These episodes, called **mass extinction events,** have occurred at widely spaced intervals in Earth history and have wiped out half to 95% of our planet's species each time. The best-known mass extinction occurred 65 million years ago and brought an end to the dinosaurs (although birds are modern representatives of dinosaurs). Evidence suggests that the impact of a gigantic asteroid caused this event, called the Cretaceous-Tertiary, or K-T, event (see "The Science behind the Story," ▶pp. 128–129).

The K-T event, as massive as it was, was moderate compared to the mass extinction at the end of the Permian period 250 million years ago (see Appendix D for geologic periods). Paleontologists estimate that 75–95% of all species on Earth may have perished during this event. Precisely what caused the event scientists don't know. The evidence for extraterrestrial impact is much weaker than it is for the K-T event, and other ideas

The K-T Mass Extinction

The Science behind the Story

On five occasions in the history of life, huge numbers of species went extinct in a geologic instant. The last mass extinction occurred 65 million years ago, at the dividing line between the Cretaceous and Tertiary periods, called the *K-T boundary*. About 70% of the species then living, including the dinosaurs, disappeared.

When he first started working at Bottaccione Gorge in Italy, American geologist Walter Alvarez had no idea he would soon help discover what killed off the dinosaurs. Alvarez was developing a new method to determine the age of sedimentary rocks, and he'd chosen Bottaccione Gorge because it formed an ideal geological archive. Its 400-m (1,300-ft) walls are stacked like layer cake with beds of rose-colored limestone that formed between 100 million and 50 million years ago from dust that had settled to the bottom of an ancient sea. While analyzing these layers, Alvarez noticed a band of reddish clay 1 cm (0.4 in.) thick sandwiched between

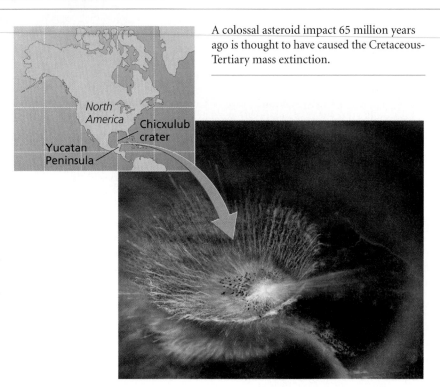

A colossal asteroid impact 65 million years ago is thought to have caused the Cretaceous-Tertiary mass extinction.

two layers of limestone. The older layer just below it was packed with fossils of globotruncana, a sand-sized animal that lived in the late Cretaceous period. The newer layer just above it contained just a few scattered fossils of a cousin of globotruncana, typical of sedimen-

tary rock formed in the early Tertiary period. What Alvarez found interesting was that the intermediate clay layer, which had formed just as dinosaurs were going extinct, had no fossils at all.

To see how quickly the K-T mass extinction event occurred,

abound. The hypothesis with the most support so far is that massive volcanism threw into the atmosphere a global blanket of soot and sulfur, smothering the planet, reducing sunlight, and inducing severe climate gyrations.

The sixth mass extinction is upon us

Many biologists have concluded that Earth is currently entering its sixth mass extinction event—and that we are the cause. The changes to Earth's natural systems set in motion by human population growth, development, and resource depletion have driven many species extinct and

are threatening countless more. The alteration and outright destruction of natural habitats, the hunting and harvesting of species, and the introduction of invasive species from one place to another where they can harm native species—these processes and many more have combined to threaten Earth's biodiversity.

When we look around us, it may not appear as though a human version of an asteroid impact is taking place, but we cannot judge such things on a human timescale. On the geological timescale, extinction over 100 years or over 10,000 years appears every bit as instantaneous as extinction over a few days.

Walter and his father, physicist Luis Alvarez, analyzed the intermediate clay layer to see how long it had taken to form, using the rare metal iridium as a kind of clock. Almost all of the iridium on Earth's surface comes from dust from tiny meteorites (shooting stars) that burn up in the atmosphere. Because the same amount of meteorite dust rains down each year, the Alvarezes measured iridium levels in the clay layer to determine how many years' worth of iridium had accumulated.

Iridium levels in the limestone were typical for sedimentary rocks, about 0.3 parts per billion. In the clay layer, however, the Alvarezes were surprised to find levels 30 times higher. To make sure the finding was not unique to Bottaccione Gorge, they checked the K-T clay layer at a Danish sea cliff. It had 160 times more iridium than the surrounding rock. The Alvarezes hypothesized that the excess iridium had come from a massive asteroid that smashed into Earth, causing a global environmental catastrophe that had wiped out the dinosaurs.

To convince themselves and the science community that an asteroid impact did in fact cause the K-T event, the Alvarezes had to rule out other possible explanations. For example, the extra iridium could have come from seawater. Calculations, however, proved that seawater did not contain enough iridium to account for the high levels in the clay layers.

An asteroid 10 km (6.2 mi) wide strikes Earth, on average, every 100 million years. Such an impact would unleash an explosion 1,000 times more forceful than the 1883 eruption of the Indonesian volcano Krakatau, which scattered so much dust around the world that sunsets were intense for 2 years afterward. An asteroid impact 65 million years ago, they suggested, kicked up enough soot to blot out the sun for several years. This would have inhibited photosynthesis, causing plants to die off, food webs to collapse, and most animals, including dinosaurs, to die of starvation. Only a few smaller animals survived, feeding on rotting vegetation. When sunlight returned, plants sprouted from dormant seeds, and a long recovery began.

Published in the journal *Science* in 1980, the Alvarezes' explanation was immediately attacked by other geologists, who claimed that spectacular volcanic eruptions more likely explained the high iridium levels in Bottaccione Gorge. But throughout the 1980s, scientists kept finding evidence supporting the asteroid-impact hypothesis. Iridium-enriched clay turned up at K-T layers around the world, as did bits of minerals called shocked quartz and stishovite, which form only under the extreme pressure of thermonuclear explosions and asteroid impacts. The Alvarezes' hypothesis became widely accepted after 1991, once scientists pinpointed a 65-million-year-old crater of the expected size in the ocean off the coast of Mexico. Today, scientists broadly agree that an asteroid impact 65 million years ago caused our planet's most recent mass extinction.

Weighing the Issues:
Should We Care about Extinction?

Although many scientists say biodiversity loss is our biggest environmental problem, some critics say we should not be concerned about biodiversity loss or a new mass extinction. What do you think can account for such a disparity of viewpoints? Thinking back to our discussion of ethics and economics in Chapter 2, can you elaborate reasons why we should be concerned about loss of biodiversity and extinction of species?

Levels of Ecological Organization

The extinction of species, their generation through speciation, and other evolutionary mechanisms and patterns have substantial influence on ecology. Moreover, it's often said that ecology provides the stage on which the play of evolution unfolds. The two, it's clear, are tightly intertwined in many ways. As we discussed in Chapter 1, ecology is the study of interactions among organisms and between organisms and their environments.

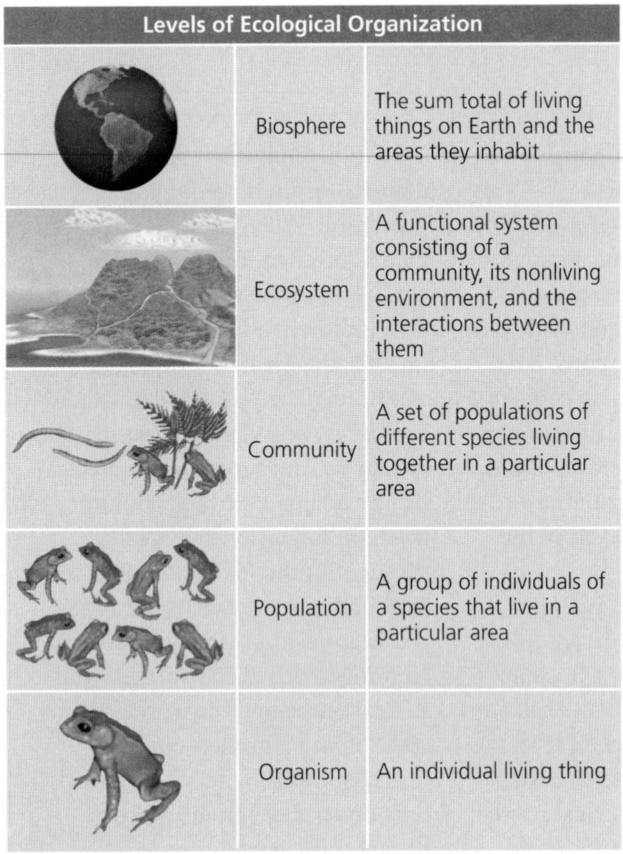

Levels of Ecological Organization		
	Biosphere	The sum total of living things on Earth and the areas they inhabit
	Ecosystem	A functional system consisting of a community, its nonliving environment, and the interactions between them
	Community	A set of populations of different species living together in a particular area
	Population	A group of individuals of a species that live in a particular area
	Organism	An individual living thing

FIGURE 5.10 Life exists in a hierarchy of levels. Ecology includes the study of the organismal, population, community, and ecosystem levels and, increasingly, the level of the biosphere. Levels below the organismal level were illustrated in Figure 4.15, ▶p. 103.

Ecology is studied at several levels

Life occurs in a hierarchy of levels. The atoms, molecules, and cells we reviewed in Chapter 4 represent the lowest levels in this hierarchy (see Figure 4.15, ▶p. 103). Aggregations of cells of particular types form tissues, and tissues form organs, all housed within an individual living organism. Ecologists study relationships on the higher levels of this hierarchy (Figure 5.10), namely on the organismal, population, community, and ecosystem levels. **Communities** are made up of multiple interacting species that live in the same area. A population of golden toads, a population of resplendent quetzals, populations of ferns and mosses, together with all of the other interacting plant, animal, fungal, and microbial populations in the Monteverde cloud forest, would be considered a community. **Ecosystems** encompass communities and the abiotic (nonliving) material and forces with which their members interact. Monteverde's

cloud-forest ecosystem consists of the community plus the air, water, soil, nutrients, and energy the community's organisms use.

At the organismal level, the science of ecology describes relationships between organisms and their physical environments. It helps us understand, for example, what aspects of the golden toad's environment were important to it, and why. **Population ecology** investigates the quantitative dynamics of how individuals within a species interact with one another. It helps us understand why populations of some species (such as the golden toad) decline while populations of others (such as ourselves) increase. **Community ecology** focuses on interactions among species, from one-to-one interactions to complex interrelationships involving entire communities. In the case of Monteverde, it allows us to study how the golden toad and many other species of its cloud-forest community interact. Finally, **ecosystem ecology** reveals patterns, such as energy and nutrient flow, by studying living and nonliving components of systems in conjunction. As we will see, changing climate has had a strong influence on the organisms of Monteverde's cloud-forest ecosystem.

As improving technologies allow scientists to learn more about the complex operations of natural systems on a global scale, ecologists are increasingly expanding their horizons beyond ecosystems to the biosphere as a whole. In this chapter we explore ecology up through the population level. In Chapter 6 we examine the community level, and in Chapter 7 we explore the ecosystem and biosphere levels.

Habitat, niche, and degree of specialization are important in organismal ecology

On the organismal level, each organism relates to its environment in ways that tend to maximize its survival and reproduction. One key relationship involves the specific environment in which an organism lives—its **habitat.** A species' habitat consists of both living and nonliving elements—of rock, soil, leaf litter, and humidity, as well as the other organisms around it. The golden toad lived in a habitat of cloud forest— more specifically, on the moist forest floor, using seasonal pools for breeding and burrows for shelter. The plants known as epiphytes use other plants as habitat; they grow on trees for physical support, obtaining water from the air and nutrients from organic debris that collects among their leaves. Epiphytes thrive in cloud forests because they require a habitat with high humidity, and Monteverde hosts more than 330 species of epiphytes, mostly ferns, orchids, and bromeliads (pineapple relatives). By collecting pools of rainwater and pockets of leaf litter, epiphytes create habitat for many other organisms, including many invertebrates and even frogs that lay their eggs in the rainwater pools.

Habitats are scale-dependent. A tiny soil mite may perceive its habitat as a mere square meter of soil. A vulture, in contrast, may view its habitat in terms of miles upon miles of hills and valleys that it easily traverses by air.

Each organism thrives in certain habitats and not in others, leading to nonrandom patterns of **habitat use.** Mobile organisms actively select habitats in which to live from among the range of options they encounter, a process called *habitat selection.* In the case of plants and sessile animals, whose progeny disperse passively, patterns of habitat use result from success in some habitats and failure in others. The criteria by which organisms favor some habitats over others can vary greatly. The soil mite may judge available habitats in terms of the chemistry, moisture, and compactness of the soil and the percentage and type of organic matter. The vulture may ignore not only soil but also topography and vegetation, focusing solely on the abundance of dead animals in the area that it scavenges for food. Every species judges habitats differently because every species has different needs.

Habitat selection is important in environmental science because the availability and quality of habitat are crucial to an organism's well-being. Indeed, because habitats provide everything an organism needs, including nutrition, shelter, breeding sites, and mates, the organism's very survival depends on the availability of suitable habitats. Often this engenders conflict with people who want to alter or develop a habitat for their own purposes.

Another way in which an organism relates to its environment is through its niche. A species' **niche** reflects its use of resources and its functional role in a community. This includes its habitat use, its consumption of certain foods, its role in the flow of energy and matter, and its interactions with other organisms. The niche is a multidimensional concept, a kind of summary of everything an organism does. We will examine the niche concept further in Chapter 6 (▸pp. 151–152).

Organisms vary in the breadth of their niche. Species with narrow breadth, and thus very specific requirements, are said to be **specialists.** Those with broad tolerances, able to use a wide array of habitats or resources, are **generalists.** For example, in a study of eight Costa Rican bird species that feed from epiphytes, ornithologist T. Scott Sillett found that four were generalists. The other four were specialists on the insect resources the epiphytes provided and spent more than 75% of their foraging efforts feeding from ephiphytes. Specialist and generalist strategies each have advantages and disadvantages. Specialists can be successful over evolutionary time by being extremely good at the things they do, but they are vulnerable when conditions change and threaten the habitat or resource on which they have specialized. Generalists meet with success by being able to live in many different places and weather variable conditions, but they may not thrive in any one situation to the degree that a specialist does. An organism's habitat, niche, and degree of specialization each reflect the adaptations of the species and are products of natural selection.

Population Ecology

Individuals of the same species inhabiting a particular area make up a population. Species may consist of multiple populations that are geographically isolated from one another. This is the case with a species characteristic of Monteverde—the resplendent quetzal *(Pharomachrus mocinno),* considered one of the world's most spectacular birds (see Figure 5.1a). Although it ranges from southernmost Mexico to Panama, the resplendent quetzal lives only in high-elevation tropical forest and is absent from low-elevation areas. Moreover, human development has destroyed much of its forest habitat. Thus, the species today exists in many separate populations scattered across Central America.

In contrast, humans have become more mobile than any other species and have spread into nearly every corner of the planet. As a result, it is difficult to define a distinct human population on anything less than the global scale. Some would maintain that in the ecological sense of the word, all 6.5 billion of us comprise one population.

Populations exhibit characteristics that help predict their dynamics

Whether one is considering humans or quetzals or golden toads, all populations show characteristics that help population ecologists predict the future dynamics of that population. Attributes such as density, distribution, sex ratio, age structure, and birth and death rates all help the ecologist understand how a population may grow or decline. The ability to predict growth or decline is especially useful in monitoring and managing threatened and endangered species (▸pp. 331–335). It is also vital in applying to human populations (Chapter 8). Understanding human population dynamics, their causes, and their consequences is one of the central elements of environmental

(b) 19th-century lithograph of pigeon hunting in Iowa

(a) Passenger pigeon

FIGURE 5.11   The passenger pigeon (**a**) was once North America's most numerous bird, and its flocks literally darkened the skies when millions of birds passed overhead (**b**). However, human cutting of forests and hunting drove the species to extinction within a few decades.

science and one of the prime challenges for our society today.

Population size Expressed as the number of individual organisms present at a given time, **population size** may increase, decrease, undergo cyclical change, or remain the same over time. Extinctions are generally preceded by population declines. As late as 1987, scientists documented a golden toad population at Monteverde in excess of 1,500 individuals, but in 1988 and 1989 scientists sighted only a single toad. By 1990, the species had disappeared.

The passenger pigeon *(Ectopistes migratorius),* also now extinct, illustrates the extremes of population size (Figure 5.11). It was once the most abundant bird in North America; flocks of passenger pigeons literally darkened the skies. In the early 1800s, ornithologist Alexander Wilson wrote of watching a flock of 2 billion birds 390 km (240 mi) long that took 5 hours to fly over and sounded like a tornado. Passenger pigeons nested in gigantic colonies in the forests of the upper Midwest and southern Canada. Once people began cutting the forests, however, the birds' great concentrations made them easy targets for market hunters, who gunned down thousands at a time and shipped them to market by the wagonload.

By the end of the 19th century, the passenger pigeon population had declined to such a low number that they could not form the large colonies they apparently needed to breed effectively. In 1914, the last passenger pigeon on Earth died in the Cincinnati Zoo, bringing the continent's most numerous bird species to extinction within just a few decades.

Population density The flocks and breeding colonies of passenger pigeons showed high population density, another attribute that ecologists assess to better understand populations. **Population density** describes the number of individuals within a population per unit area. For instance, the 1,500 golden toads counted in 1987 within 4 km^2 (988 acres) indicated a density of 375 toads/km^2. In general, larger organisms have lower population densities because they require more resources to survive.

High population density can make it easier for organisms to group together and find mates, but it can also lead to conflict in the form of competition if space, food, or mates are in limited supply. Overcrowded organisms may also become more vulnerable to the predators that feed on them, and close contact among individuals can increase the transmission of infectious disease. For these reasons, organisms sometimes leave an area when densities become too high. In contrast, at low population densities, organisms benefit from more space and resources but may find it harder to locate mates and companions.

Overcrowding at high population densities is thought to have doomed Monteverde's harlequin frog (*Atelopus varius*; see Figure 5.1c), an amphibian that disappeared from the cloud forest at the same time as the golden toad. The harlequin frog is a habitat specialist, favoring "splash zones," areas alongside rivers and streams that receive spray from waterfalls and rapids. As Monteverde's climate grew warmer and drier in the 1980s and 1990s, water flow decreased, and many streams dried up. Splash zones grew smaller and fewer, and harlequin frogs were forced to cluster together in what remained of the splash-zone habitat. Researchers J. Alan Pounds and Martha Crump recorded frog population densities up to 4.4 times higher than normal, with more than 2 frogs per meter (3.3 ft) of stream. Such overcrowding likely made the frogs more vulnerable to disease transmission, predator attack, and assault from parasitic flies. From their field research, during which Pounds and Crump witnessed 40 frogs dead or dying, the researchers concluded that these factors led to the harlequin frog's disappearance from Monteverde.

Thankfully, a new population of harlequin frogs was found in 2003 on a private reserve elsewhere in Costa Rica, so there is still hope that the species may survive. The frog was rediscovered by University of Delaware student Justin Yeager, who was doing field research during his study abroad trip in Costa Rica that summer.

Population distribution It was not simply the harlequin frog's density, but also its distribution in space that led to its demise at Monteverde. **Population distribution,** or **population dispersion,** describes the spatial arrangement of organisms within an area. Ecologists define three distribution types: random, uniform, and clumped (Figure 5.12). In a *random distribution,* individuals are located haphazardly in space in no particular pattern. This type of distribution can occur when the resources an organism needs are found throughout an area and other organisms do not strongly influence where members of a population settle.

A *uniform distribution* is one in which individuals are evenly spaced. This can occur when individuals hold territories or otherwise compete for space. For instance, in a desert where there is little water, each plant may need a certain amount of space for its roots to gather adequate moisture. As a result, each individual plant may be equidistant from others.

(a) Random

(b) Uniform

(c) Clumped

FIGURE 5.12 Individuals in a population can be spatially distributed over a landscape in three fundamental ways. In a random distribution (**a**), organisms are dispersed at random through the environment. In a uniform distribution (**b**), individuals are spaced evenly, at equal distances from one another. Territoriality can result in such a pattern. In a clumped distribution (**c**), individuals occur in patches, concentrated more heavily in some areas than in others. Habitat selection or flocking to avoid predators can result in such a pattern.

In a *clumped distribution,* the pattern most common in nature, organisms arrange themselves according to the availability of the resources they need to survive. Many desert plants grow in patches around isolated springs or along arroyos that flow with water after rainstorms. During their mating season, golden toads were found clumped at seasonal breeding pools. Humans, too, exhibit clumped distribution; people frequently aggregate together in urban centers. Clumped distributions often indicate habitat selection. Distributions can depend on the scale at which one measures them. At very large scales, all organisms show clumped or patchy distributions, because some parts of the total area they inhabit are bound to be more hospitable than others.

Sex ratios For organisms that reproduce sexually and have distinct male and female individuals, the sex ratio of a population can help determine whether it will increase or decrease in size over time. A population's **sex ratio** is its proportion of males to females. In monogamous species (in which each sex takes a single mate), a 50/50 sex ratio maximizes population growth, whereas an unbalanced ratio leaves many individuals of one sex without mates.

Age structure Populations most often consist of individuals of different ages. **Age distribution,** or **age structure,** describes the relative numbers of organisms of each age within a population. Like sex ratio, age distribution can have strong effects on rates of population growth or decline. A population made up mostly of individuals past reproductive age will tend to decline over time. In contrast, a population with many individuals of reproductive age or soon to be of reproductive age is likely to increase. A population with an even age distribution will likely remain stable as births keep pace with deaths.

Age structure diagrams, often called *age pyramids,* are visual tools scientists use to show the age structure of populations (Figure 5.13). The width of each horizontal bar represents the relative size of each age class. A pyramid with a wide base has a relatively large age class that has not yet reached its reproductive stage, indicating a population much more capable of rapid growth. In this respect, a wide base of an age pyramid is like an oversized engine on a rocket—the bigger the booster, the faster the increase. We will examine age pyramids further in Chapter 8 (▸pp. 224–226) in reference to human populations.

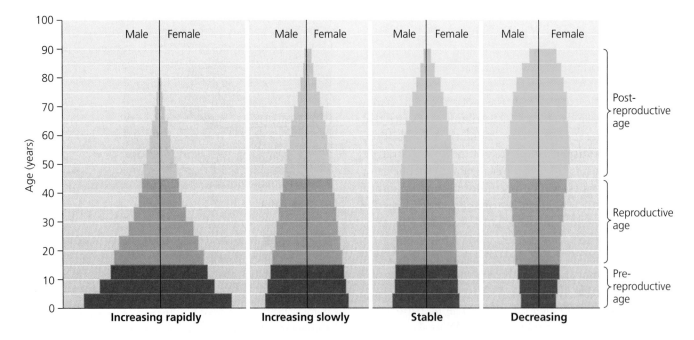

FIGURE 5.13 Age structure diagrams show the relative frequencies of individuals of different age classes in a population. In this example for humans, populations heavily weighted toward young age classes (at left) grow most quickly, whereas those weighted heavily toward old age classes (at right) decline.

Birth and death rates All the preceding factors can influence the rates at which individuals within a population are born and die. A convenient way to express birth and death rates is to measure the number of births and deaths per 1,000 individuals for a given time period. Such a rate is termed a *crude birth rate* or *crude death rate*.

Just as individuals of different ages have different abilities to reproduce, individuals of different ages show different probabilities of dying. For instance, people are more likely to die at old ages than young ages; if you were to follow 1,000 10-year-olds and 1,000 80-year-olds for a year, you would find that at year's end more 80-year-olds had died than 10-year-olds. However, this pattern does not hold for all organisms. Amphibians such as the golden toad produce large numbers of young, which suffer high death rates. For a toad, death is less likely (and survival more likely) at an older age than at a very young age. To show how the likelihood of death can vary with age, ecologists use graphs called **survivorship curves** (Figure 5.14). There are three fundamental types of survivorship curves. Humans, with higher death rates at older ages, show a *type I* survivorship curve. Toads, with highest death rates at young ages, show a *type III* survivorship curve. A *type II* survivorship curve is intermediate and indicates equal

rates of death at all ages. Many birds are thought to show type II curves.

Populations may grow, shrink, or remain stable

Now that we have outlined some key attributes of populations, we are ready to take a quantitative view of population change by examining some simple mathematical concepts used by population ecologists and *demographers* (those who study human populations). Population growth, or decline, is determined by four factors:

1. Births within the population *(natality)*
2. Deaths within the population *(mortality)*
3. **Immigration** (arrival of individuals from outside the population)
4. **Emigration** (departure of individuals from the population)

To understand how a population changes, we measure its **growth rate,** which can be calculated as the crude birth rate plus the immigration rate, minus the crude death rate plus the emigration rate, each expressed as the number per 1,000 individuals per year:

(Crude birth rate + immigration rate) − (Crude death rate + emigration rate) = Growth rate

The resulting number tells us the net change in a population's size per 1,000 individuals. For example, a population with a crude birth rate of 18 per 1,000, a crude death rate of 10 per 1,000, an immigration rate of 5 per 1,000, and an emigration rate of 7 per 1,000 would have a growth rate of 6 per 1,000:

$$(18/1,000 + 5/1,000) − (10/1,000 + 7/1,000) = 6/1,000$$

Thus, a population of 1,000 in one year will reach 1,006 in the next. If the population is 1,000,000, it will reach 1,006,000 the next year. These population increases are often expressed as percentages, which we can calculate using the formula:

Growth rate × 100%

Thus, a growth rate of 6/1,000 would be expressed as:

$$6/1,000 × 100\% = 0.6\%$$

By measuring population growth in terms of percentages, scientists can compare increases and decreases in species that have far different population sizes. They can also project changes that will occur in the population over longer periods, much like you might calculate the amount of interest your savings account will earn over time.

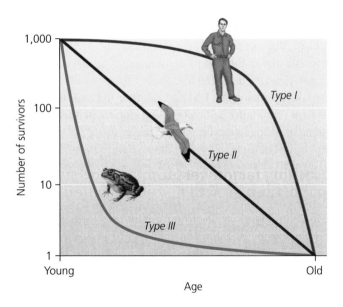

FIGURE 5.14 In a type I survivorship curve, survival rates are high when organisms are young and decrease sharply when organisms are old. In a type II survivorship curve, survival rates are equivalent regardless of an organism's age. In a type III survivorship curve, most mortality takes place at young ages, and survival rates are greater at older ages. Some examples include humans (type I), birds (type II), and amphibians (type III).

Table 5.3	Exponential Growth in a Savings Account with 5% Annual Compound Interest
Age (in years)	**Principal**
0 (birth)	$1,000
10	$1,629
20	$2,653
30	$4,322
40	$7,040
50	$11,467
60	$18,679
70	$30,426
80	$49,561

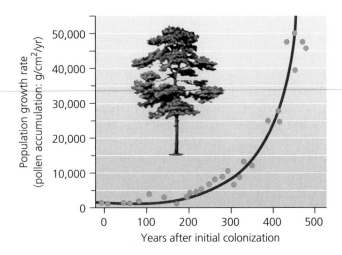

FIGURE 5.15 Although no species can maintain exponential growth indefinitely, some may grow exponentially for a time when colonizing an unoccupied environment or exploiting an unused resource. Scientists have used pollen records to determine that the Scots pine *(Pinus sylvestris)* increased exponentially after the retreat of glaciers following the last ice age around 9,500 years ago. Go to **GRAPHit!** at www.aw-bc.com/withgott or on the student CD-ROM. Data from Bennett, K. D. 1983. Postglacial population expansion of forest trees in Norfolk, U.K. *Nature* 303: 164–167.

Unregulated populations increase by exponential growth

When a population, or anything else, increases by a fixed percentage each year, it is said to undergo **exponential growth.** A savings account is a familiar frame of reference for describing exponential growth. If at the time of your birth your parents had invested $1,000 in a savings account earning 5% interest compounded each year, you would have only $1,629 by age 10, and $2,653 by age 20, but you would have over $30,000 when you turn 70. If you could wait just 10 years more, that figure would rise to nearly $50,000 (Table 5.3). Only $629 was added during your first decade, but approximately $19,000 was added during the decade between ages 70 and 80. The reason is that a fixed percentage of a small number makes for a small increase, but that same percentage of a large number produces a large increase. Thus, as savings accounts (or populations) become larger, each incremental increase likewise gets larger. Such acceleration is a characteristic of exponential growth.

We can visualize changes in population size by using population growth curves. The J-shaped curve in Figure 5.15 shows exponential increase. As Thomas Malthus realized, populations of all organisms increase exponentially unless they meet constraints. Each organism reproduces by a certain amount, and as populations get larger, more individuals reproduce by that amount. If there are no external limits on growth, ecologists theoretically expect exponential growth.

Exponential growth usually occurs in nature when a population is small and environmental conditions are ideal for the organism in question. Most often, these conditions occur when organisms are introduced to a new en-

vironment. Mold growing on a piece of bread or fruit, or bacteria colonizing a recently dead animal, are cases in point. But species of any size may show exponential growth under the right conditions. A population of the Scots pine, *Pinus sylvestris,* grew exponentially when it began colonizing the British Isles after the end of the last ice age (see Figure 5.15). Receding glaciers had left conditions ideal for its exponential expansion.

Limiting factors restrain population growth

However, exponential growth rarely lasts long. If even a single species in Earth's history had increased exponentially for very many generations, it would have blanketed the planet's surface, and nothing else could have survived. Instead, every population eventually is constrained by **limiting factors,** physical, chemical, and biological characteristics of the environment that restrain population growth. The interaction of these factors determines the **carrying capacity,** the maximum population size of a species that a given environment can sustain.

Ecologists use the curve in Figure 5.16 to show how an initial exponential increase is slowed and finally brought to a standstill by limiting factors. Called the **logistic**

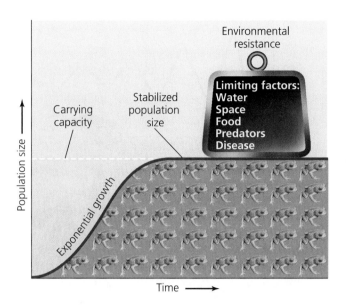

FIGURE 5.16 The logistic growth curve shows how population size may increase rapidly at first, then grow more slowly, and finally stabilize at a carrying capacity. Carrying capacity is determined both by the biotic potential of the organism and by various external limiting factors, collectively termed *environmental resistance.*

growth curve, it rises sharply at first but then begins to level off as the effects of limiting factors become stronger. Eventually the force of these factors—which taken together are termed *environmental resistance*—stabilizes the population size at its carrying capacity.

The logistic curve is a simplified model, and real populations can behave differently. Some may cycle indefinitely above and below the carrying capacity. Some may show cycles that become less extreme and approach the carrying capacity. Others may overshoot the carrying capacity and then crash, fated either for extinction or recovery (Figure 5.17).

Many factors contribute to environmental resistance and influence a population's growth rate and carrying capacity. Space is one factor that limits the number of individuals a given environment can support; if there is no physical room for additional individuals, they are unlikely to survive. Other limiting factors for animals in a terrestrial environment include the availability of food, water, mates, shelter, and suitable breeding sites; temperature extremes; prevalence of disease; and abundance of predators. Plants are often limited by amounts of sunlight and moisture and the type of soil chemistry, in addition to disease and attack from plant-eating animals. In aquatic systems, limiting factors include salinity, sunlight, temperature, dissolved oxygen, fertilizers, and pollutants.

Sometimes one limiting factor may outweigh all others and restrict population growth. For example, scientists hypothesize that Monteverde's population of golden toads had plenty of space, food, and shelter, but that it lacked adequate moisture. If moisture were the primary limiting factor, then increasing moisture would have increased the carrying capacity of the habitat for the toads. Indeed, to determine limiting factors, ecologists often conduct experiments in which they increase or decrease a hypothesized limiting factor to observe its effects on population size. Unfortunately in the case of the golden toad, such experiments could not be conducted before its disappearance.

Carrying capacities can change

Because limiting factors can be numerous, and because environments are complex and ever-changing, carrying capacity can vary constantly. The human species illustrates another reason that carrying capacity is not necessarily a fixed entity. Although all organisms are subject to environmental resistance, they may be capable of altering their environment to reduce this resistance. Our own species has proved particularly effective at this. When our ancestors began to build shelters and use fire for heating and cooking, they reduced the environmental resistance of areas with cold climates and were able to expand into new territory. As limiting factors are overcome (either through the development of new technologies or through natural environmental change), the carrying capacity for a species increases. We humans have managed so far to increase the planet's carrying capacity for ourselves, but unfortunately for the golden toad, the limiting factors on its population growth seemed to exert ever-increasing pressure during the late 1980s.

Weighing the Issues:
Carrying Capacity and Human Population Growth

As we saw in Chapter 1, the global human population has risen from fewer than 1 billion 200 years ago to 6.5 billion today, and we have far exceeded our historic carrying capacity. What factors increased Earth's carrying capacity for people? Do you think there are limiting factors for the human population? What might they be? Do you think we can keep raising our carrying capacity in the future? Might Earth's carrying capacity for us decrease?

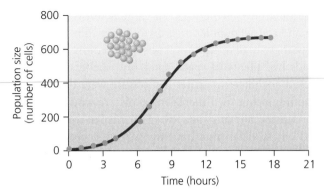

(a) Yeast cells, *Saccharomyces cerevisiae*

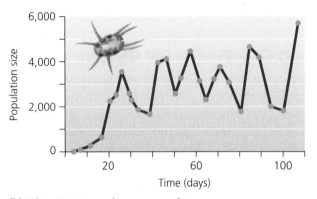

(b) Mite, *Eotetranychus sexmaculatus*

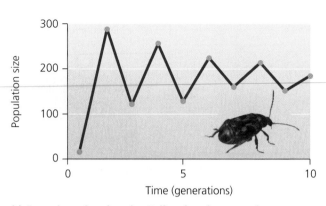

(c) Stored-product beetle, *Callosobruchus maculatus*

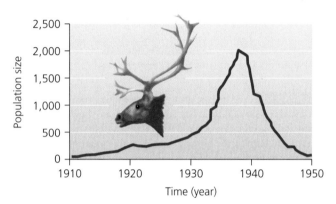

(d) St. Paul reindeer, *Rangifer tarandus*

FIGURE 5.17 Population growth in nature often departs from the stereotypical logistic growth curve, and it can do so in several fundamental ways. Yeast cells from an early lab experiment show logistic growth **(a)** that closely matches the theoretical model. Some organisms, like the mite shown here, show cycles in which population fluctuates indefinitely above and below the carrying capacity **(b).** Population oscillations can also dampen, lessening in intensity and eventually stabilizing at carrying capacity **(c),** as in a lab experiment with the stored-product beetle. Populations that rise too fast and deplete resources may crash just as suddenly **(d),** like the population of reindeer introduced to the Bering Sea island of St. Paul. Data from Pearl, R. 1927. The growth of populations. *Quarterly Review of Biology* 2: 532–548, (a); Huffaker, C. B. 1958. Experimental studies on predation: Dispersion factors and predator-prey oscillations, *Hilgardia* 27: 343–383, (b); Utida, S. 1967. Damped oscillation of population density at equilibrium, *Researches on Population Ecology* 9: 1–9, (c); Scheffer, V. C. 1951. Rise and fall of a reindeer herd, *Scientific Monthly* 73: 356–362, (d).

The influence of some factors depends on population density

Just as carrying capacity is not a fixed entity, the influence of limiting factors can vary with changing conditions. In particular, the density of a population can increase or decrease the impact of certain factors on that population. Recall that high population density can help organisms find mates but can also increase competition and the risk of predation and disease. Such factors are said to be **density-dependent** factors, because their influence waxes and wanes according to population density. The logistic growth curve in Figure 5.16 represents the effects of density dependence. The more

population size rises, the more environmental resistance kicks in.

Density-independent factors are limiting factors whose influence is not affected by population density. Temperature extremes and catastrophic events such as floods, fires, and landslides are examples of density-independent factors, because they can eliminate large numbers of individuals without regard to their density.

Biotic potential and reproductive strategies vary from species to species

Limiting factors from an organism's environment provide only half the story of population regulation. The other half comes from the attributes of the organism itself. For example, organisms differ in their *biotic potential,* or ability to produce offspring. A fish with a short gestation period that lays thousands of eggs at a time has high biotic potential, whereas a whale with a long gestation period that gives birth to a single calf at a time has low biotic potential. The interaction between an organism's biotic potential and the environmental resistance to its population growth helps determine the fate of its population.

Giraffes, elephants, humans, and other large animals with low biotic potential produce a relatively small number of offspring and take a long time to gestate and raise each of their young. Species that take this approach to reproduction compensate by devoting large amounts of energy and resources to caring for and protecting the relatively few offspring they produce during their lifetimes. Such species are said to be **K-selected.** K-selected species are so named because their populations tend to stabilize over time at or near their carrying capacity, and *K* is an abbreviation for carrying capacity. Because their populations stay close to carrying capacity, these organisms must be good competitors, able to hold their own in a crowded world. Thus in these species, natural selection favors individuals that invest in producing offspring of high quality that can be good competitors.

In contrast, species that are **r-selected** focus on quantity, not quality. Species considered to be r-selected have high biotic potential and devote their energy and resources to producing as many offspring as possible in a relatively short time. Their offspring do not require parental care after birth, so r-strategists simply leave their survival to chance. The abbreviation *r* denotes the rate at which a population increases in the absence of limiting factors. Populations of r-selected species fluctuate greatly, such that they are often well below carrying capacity. This is why natural selection in these species favors traits that lead to rapid population growth. Many fish, plants, frogs, insects, and others are r-selected. The golden toad is one example. Each adult female laid 200–400 eggs, and its tadpoles spent 5 weeks unsupervised in the breeding pools metamorphosing into adults.

Table 5.4 summarizes stereotypical traits of r-selected and K-selected species. However, it is important to note that these are two extremes on a continuum and that most species fall somewhere between these endpoints. Moreover, some organisms show combinations of traits that do not clearly correspond to a place on the continuum. A redwood tree *(Sequoia sempervirens),* for instance, is large and long-lived, yet it produces many small seeds and offers no parental care.

Table 5.4 Traits of r-selected and K-selected species

r-selected species	K-selected species
Small size	Large size
Fast development	Slow development
Short-lived	Long-lived
Reproduction early in life	Reproduction later in life
Many small offspring	Few large offspring
Fast population growth rate	Slow population growth rate
No parental care	Parental care
Weak competitive ability	Strong competitive ability
Variable population size, often well below carrying capacity	Constant population size, close to carrying capacity
Variable and unpredictable mortality	More constant and predictable mortality

Populations of K-selected species are generally regulated by density-dependent factors such as disease, predation, and food limitation. In contrast, density-independent factors tend to regulate populations of r-selected species, whose success or failure is often determined by large-scale environmental change. Many r-selected species frequently experience large swings in population size, such as rapid increases during the breeding season and rapid declines soon after, when unfit and unlucky young are removed from the population. For this reason, scientists often have difficulty determining whether steep population declines are a part of natural cycles or a sign of serious trouble. For years, scientists debated the golden toad's apparent extinction. Now that it has failed to reappear for over 15 years, most agree that the toad's population crash was not part of a normal, repeating cycle.

Changes in populations influence the composition of communities

In the late 1980s, the golden toad and the harlequin frog were the most diligently studied species affected by changing environmental conditions in the Costa Rican cloud forest. However, once scientists began looking at populations of other species at Monteverde, they began to notice more troubling changes. By the early 1990s, not only had golden toads, harlequin frogs, and other organisms been pushed from their cloud-forest habitat into apparent extinction, but many species from lower, drier habitats had also begun to appear at Monteverde. These immigrants included species tolerant of drier conditions,

The Science behind the Story

Climate Change and Its Effects on Monteverde

Soon after the golden toad's disappearance, scientists began to investigate the potential role of climate change in driving cloud-forest species toward extinction. They had noted that the period from July 1986 to June 1987 was the driest on record in Monteverde, with unusually high temperatures and record-low stream flows. These conditions had caused the golden toad's breeding pools to dry up shortly after they filled in the spring of 1987, likely killing nearly all of the eggs and tadpoles present in the pools.

Scientists began reviewing reams of weather data and eventually found that the number of dry days and dry periods each winter in the Monteverde region had increased between 1973 and 1998. Biologists knew that such local climate trends were bad news for amphibians like the golden toad and harlequin frog. Because amphibians breathe and absorb moisture through their skin, they are susceptible to dry conditions, high temperatures, acid rain, and pollutants concentrated by reduced water levels. Based on these facts, herpetologists J. Alan Pounds and Martha Crump in 1994 hypothesized that hot, dry conditions were to blame for increased adult mortality and breeding problems among golden toads and other amphibians.

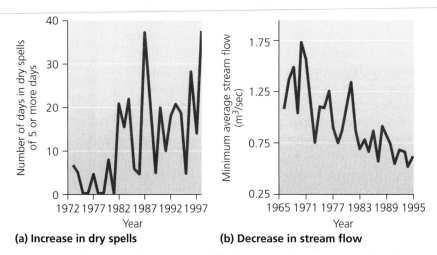

(a) Increase in dry spells

(b) Decrease in stream flow

Warming and drying trends in Monteverde's climate may have contributed to the region's amphibian declines. Evidence gathered over 25 years shows **(a)** an increase in the annual number of dry days and **(b)** a decrease in the amount of annual stream flow. Data from Pounds, J. A., et al. 1999. Biological response to climate change on a tropical mountain. *Nature* 398: 611–615.

Throughout this period, scientists worldwide were realizing that the oceans and atmosphere were warming because of human release of carbon dioxide and other gases into the atmosphere. Global climate change (Chapter 18), experts were learning, could produce varying effects on climate at regional and local levels. With this in mind, Pounds and others concerned about Monteverde's changing conditions reviewed the scientific literature on ocean and atmospheric science to analyze the effects on Monteverde's local climate of warming patterns in the ocean regions around Costa Rica.

By 1997 these researchers had determined that Monteverde's cloud forest was becoming drier because the clouds that had given the forest its name and much of its moisture now passed by at higher elevations, where they were no longer in contact with the trees. The primary factor determining the clouds' altitude, the researchers found, is nearby ocean temperatures; as ocean temperatures increase, clouds pass over Monteverde at higher elevations. Once the cloud forest's water supply was pushed upward, out of reach of the mountaintops, the cloud forest began to dry out.

such as blue-crowned motmots *(Momotus momota)* and brown jays *(Cyanocorax morio)*. By the year 2000, 15 dry-forest species had moved into the cloud forest and begun to breed. Meanwhile, population sizes of several cloud-forest bird species had declined. After 1987, 20 of 50 frog species vanished from one part of Monteverde, and ecologists later reported more disappearances, including

those of two lizards native to the cloud forest. Scientists hypothesized that the warming, drying trends that researchers were documenting (see "The Science behind the Story," above) were causing population fluctuations and unleashing changes in the composition of the community.

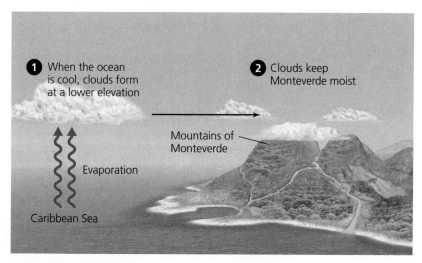

(a) Cool ocean conditions

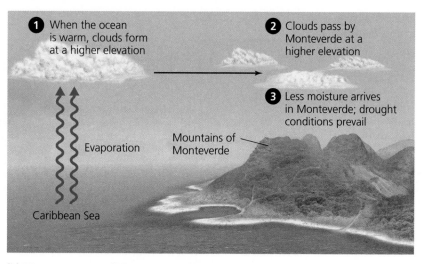

(b) Warm ocean conditions

Monteverde's cloud forest gets its name and life-giving moisture from clouds that sweep inland from the oceans. **(a)** When ocean temperatures are cool, the clouds keep Monteverde moist. **(b)** Warmer ocean conditions resulting from global climate change cause clouds to form at higher elevations and pass over the mountains, drying the cloud forest.

In a 1999 paper in the journal *Nature,* Pounds and two colleagues reported these findings. Their conclusion—that broad-scale climate modification was causing local changes at the species, population, and community levels— explained a great number of events occurring at Monteverde. Rising cloud levels and decreasing moisture could explain not only the disappearance of the golden toad and harlequin frog, but also the concurrent population crashes in 1987 and subsequent disappearance of 20 species of frogs and toads from the Monteverde region. Amphibians that survived underwent population crashes in each of the region's three driest years.

Pounds and his co-workers further described "a constellation of demographic changes that have altered communities of birds, reptiles and amphibians" in the area as likely additional consequences of this shift in moisture availability. As these mountaintop forests dried out, dry-tolerant species crept in, and moisture-dependent species were stranded at the mountaintops by a rising tide of dryness. Although organisms may in general be driven from one area to another by changing environmental conditions, if a species has nowhere to go, then extinction may result.

The Conservation of Biodiversity

Changes in populations and communities have been taking place naturally as long as life has existed, but today human development, resource extraction, and population pressure are speeding the rate of change and altering the types of change. The ways we modify our environment cannot be fully understood in a scientific vacuum, however. The actions that threaten biodiversity have complex social, economic, and political roots, and environmental scientists appreciate that we must understand these aspects if we are to develop solutions.

Fortunately, there are things people can do to forestall population declines of species threatened with extinction. Millions of people around the world are already taking action to safeguard the biodiversity and ecological and evolutionary processes that make Earth such a unique place (Chapter 11). Costa Ricans have been confronting the challenges to their nation's biodiversity, and their actions so far show what even a small country of modest means can do.

Social and economic factors affect species and communities

Many of the threats to Costa Rica's species and ecological communities result from past economic and social forces whose influences are still evident. European immigrants and their descendants viewed Costa Rica's lush forests as an obstacle to agricultural development, and timber companies saw them simply as a source of wood products. Costa Rica's leading agricultural products have long included beef and bananas, whose production and cultivation require extensive environmental modification. Between 1945 and 1995, the country's population grew from 860,000 to 3.34 million, and the percentage of land devoted to pasture increased from 12% to 33%. With much of the formerly forested land converted to agriculture, the proportion of the country covered by forest decreased from 80% to 25%. In 1991, Costa Rica was losing its forests faster than any other country in the world— nearly 140 ha (350 acres) per day. As a result, populations of innumerable species were declining, and some were becoming endangered (Figure 5.18). As had occurred in the history of the United States, few people foresaw the need to conserve biological resources until it became clear that they were being rapidly lost.

FIGURE 5.18 Costa Rica is home to a number of species classified as globally threatened or endangered. The golden-cheeked warbler, *Dendroica chrysoparia* (a), winters in Central America but breeds in the Texas hill country, where its habitat is being rapidly lost to housing development. The green sea turtle, *Chelonia mydas* (b), is widely distributed throughout the world's oceans and lays eggs on beaches in Costa Rica and elsewhere, but it has undergone steep population declines. The red-backed squirrel monkey, *Saimiri oerstedii* (c), is endemic to a tiny area in Costa Rica and is judged vulnerable to forest loss because of its small geographic range. These vertebrate species receive attention from scientists and the media and are the focus of recovery efforts, but many more plants, insects, and other less-celebrated species are declining in number while most of us go about our days unaware of their existence.

Costa Rica took steps to protect its environment

During the 1950s a group of Quakers, Christian pacifists who opposed the U.S. military draft, emigrated from Alabama to Costa Rica and founded the village of Monteverde.

(a) Golden-cheeked warbler

(b) Green sea turtle

(c) Red-backed squirrel monkey

FIGURE 5.19 Costa Rica has protected a wide array of its diverse natural areas. This protection has stimulated the nation's economy through ecotourism. Here, visitors experience a walkway through the forest canopy in one of the nation's parks.

The Quakers relied on milk and cheese for much of their economic activity, but they also set aside one-third of their land for conservation. The Quakers' efforts, along with contributions from international conservation organizations, provided the beginnings of what is today the Monteverde Cloud Forest Biological Reserve. This privately managed 10,500-ha (26,000-acre) reserve was established in 1972 to protect the forest and its populations of 2,500 plant species, 400 bird species, 500 butterfly species, 100 mammal species, and 120 reptile and amphibian species, including the golden toad.

In 1970, the Costa Rican government and international representatives came together to create the country's first national parks and protected areas. The first parks centered on areas of spectacular scenery, such as the Poas Volcano National Park. Santa Rosa National Park encompassed valuable tropical dry forest, Tortuguero National Park contained essential nesting beaches for the green turtle (*Chelonia mydas;* see Figure 5.18b), and Cahuita National Park was meant to protect a prominent coral reef system. Initially the government gave the parks little real support. According to Costa Rican conservationist Mario Boza, in their early years the parks were granted only five guards, one vehicle, and no funding.

Today government support is greater. Fully 12% of the nation's area is contained in national parks, and a further 16% is devoted to other types of wildlife and conservation reserves. Costa Ricans, along with international biologists, are working to protect endangered species and recover their populations. Costa Rica and its citizens are now reaping the benefits of their conservation efforts—not only ecological benefits, but also economic ones. Because of its parks and its reputation for conservation, tourists from around the world now visit Costa Rica, a phenomenon called **ecotourism** (Figure 5.19). The ecotourism industry draws more than 1 million visitors to Costa Rica each year, provides thousands of jobs to Costa Ricans, and is a major contributor to the country's economy. Today's Costa Rican economy is fueled in large part by commerce and tourism, whose contributions (40%) outweigh those of industry (22%) and agriculture (13%) combined.

Weighing the **Issues:**
How Best to Conserve Biodiversity?

Most people view national parks and ecotourism as excellent ways to help keep ecological systems intact. Yet the golden toad went extinct despite living within a reserve established to protect it, and climate change does not pay attention to park boundaries. What lessons can we take from this about the conservation of biodiversity?

It remains to be seen how effectively ecotourism can help preserve natural systems in Costa Rica in the long term. As forests outside the parks disappear, the parks are beginning to suffer from illegal hunting and timber extraction. Conservationists like Boza say the parks are still underprotected and underfunded. Ecotourism will likely need to generate still more money to preserve habitat, protect endangered species, and restore altered communities to their former condition. Restoration is being carried out in Costa Rica's Guanacaste Province, for instance, where scientists are restoring dry tropical forest from grazed pasture. Restoration of ecological communities is one phenomenon we will examine in our next chapter, as we move from populations to communities.

Conclusion

The golden toad and other organisms of the Monteverde cloud forest have helped illuminate the fundamentals of evolution and population ecology that are integral to environmental science. The evolutionary processes of natural selection, speciation, and extinction help determine Earth's biodiversity. Understanding how ecological processes work at the population level is crucial to protecting biodiversity threatened by the mass extinction event that many biologists maintain is already under way.

Conservation at Monteverde

What lessons, if any, can we learn from ecological changes that have occurred in the Monteverde region since the golden toad's discovery in 1964?

Lessons from the Green Mountain: Changes in Ecological and Human Communities

Monteverde—the "Green Mountain"—is one of the best-studied cloud forests in the world. Over the years, researchers have documented negative effects of human activities on the diverse biota. These include forest fragmentation and its isolation of plant and animal populations; lengthened dry seasons due to regional deforestation; and the upward shift of lower-elevation animal and plant populations due to global climate change.

One complex aspect of increasing human presence in Monteverde is ecotourism. The annual influx of nearly 80,000 visitors has had negative effects on the human community. Consumerism has increased. Television has replaced square-dancing. Cell telephones have replaced the single community party line used in previous decades. Guides rely on having cars rather than feet to show visitors around.

However, the presence of outside visitors who are deeply interested in nature has also affected Monteverde in positive ways. Some Monteverde families have abandoned dairy farming as their source of income, and have been able to turn to natural history guiding or managing small hotels. This has led to the reforestation of pastures, and the coalescence of formerly fragmented forest patches.

In addition, each visitor to Monteverde becomes a potential conservation activist. By seeing the unique wildlife and habitats—and learning about the causes that threaten it—visitors can take political and economic action to promote conservation when they return home. Grass roots conservation and educational organizations such as the Monteverde Conservation League and the Cloud Forest School have done much to promote conservation. These organizations rely on the contributions of ecotourists who have been touched by their experiences of seeing a quetzal or flowering orchid, and who become moved to preserve the habitat of the region.

Nalini M. Nadkarni is a member of the faculty at The Evergreen State College in Olympia, Washington. Her research focuses on the ecological interactions that occur in tropical and temperate rainforest canopies. The co-founder and president of the International Canopy Network, she is co-author of the book *Monteverde: Ecology and Conservation of a Tropical Cloud Forest.*

Conservation Successes in the Shadow of the Golden Toad

In the late 1970s, the world began to discover the pristine montane forests and peaceful farming community of Monteverde, Costa Rica. Overnight, it seemed, ecotourism boomed, new houses, hotels, and restaurants sprung up, and tens of thousands of visitors arrived annually, eager to see resplendent quetzals and epiphyte-laden cloud forests. But a decade later, equally suddenly, the golden toad, a stunning amphibian found nowhere else on Earth, vanished. Shortly afterward, 19 other species of frogs and salamanders—40% of all the amphibian species that were present when I began my dissertation research there in 1979—disappeared from the area. Fortunately, none but the golden toad was endemic to Monteverde.

Yet, surprisingly, there have also been impressive achievements in conservation at Monteverde over this same time period. In 1979, poaching of large mammals and birds was commonplace. Now the very people who hunted scarce animals with rifles use binoculars instead as they lead natural history tours. Tapirs and guans are more common today than they have been for more than half a century. As the Monteverde Cloud Forest Preserve has expanded tenfold, clearings on the Atlantic slope have reverted to lush forest. The Guacimal River, formerly rancid because of waste dumped by the community dairy plant, is much cleaner now.

Thus, although ecotourism can have a negative effect on local populations of some plants and animals, it can also have enduring positive impacts on conservation—not just locally by increasing economic opportunities and incentives for land preservation, but also globally by educating the public about environmental values. Monteverde has inspired visitors to appreciate tropical forests and, through that experience, their own natural heritage—and that may help protect threatened habitats worldwide.

Nathaniel Wheelwright is Professor of Biology at Bowdoin College in Brunswick, Maine, director of the Bowdoin Scientific Station on Kent Island, New Brunswick, and co-editor of *Monteverde: Ecology and Conservation of a Tropical Cloud Forest.*

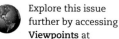 Explore this issue further by accessing **Viewpoints** at www.aw-bc.com/withgott.

REVIEWING OBJECTIVES

You should now be able to:

Explain the process of natural selection and cite evidence for this process

▶ Because organisms produce excess young, individuals vary in their traits, and many traits are inherited, some individuals will prove better at surviving and reproducing. Their genes will be passed on and become more prominent in future generations. (pp. 118–119)

▶ Mutations and recombination provide the genetic variation for natural selection. (pp. 119–120)

▶ We have produced our pets, farm animals, and crop plants through artificial selection. (pp. 120–122)

Describe the ways in which evolution results in biodiversity

▶ Natural selection can act as a diversifying force as organisms adapt to their environments in myriad ways. (pp. 120–122, 124)

▶ Speciation (by geographic isolation and other means) produces new species. (pp. 122–124)

▶ Once they have diverged, lineages continue diverging, a process represented in a phylogenetic tree. (pp. 124–125)

Discuss reasons for species extinction and mass extinction events

▶ Extinction often occurs when species that are highly specialized or that have small populations encounter rapid environmental change. (pp. 126–127)

▶ Earth's life has experienced five known episodes of mass extinction, due to an asteroid impact and possibly volcanism and other factors. (pp. 127–129)

List the levels of ecological organization

▶ Ecologists study phenomena on the organismal, population, community, and ecosystem levels—and, increasingly, on the biosphere level. (pp. 129–130)

Outline the characteristics of populations that help predict population growth

▶ Populations are characterized by population size, population density, population distribution, sex ratio, age structure, and birth and death rates. (pp. 131–135)

▶ Immigration and emigration, as well as birth and death rates, determine how a population will grow or decline. (p. 135)

Assess logistic growth, carrying capacity, limiting factors, and other fundamental concepts of population ecology

▶ Populations unrestrained by limiting factors will undergo exponential growth until they meet environmental resistance. (p. 136)

▶ Logistic growth describes the effects of density dependence; exponential growth slows as population size increases, and population size levels off at a carrying capacity. (pp. 136–138)

▶ K-selection and r-selection describe theoretical extremes in how organisms can allocate growth and reproduction. (pp. 138–139)

Identify efforts and challenges involved in the conservation of biodiversity

▶ Social and economic factors influence our impacts on natural systems. (pp. 141–144)

▶ Extensive efforts to protect and restore species and habitats will be needed to prevent further erosion of biodiversity. (pp. 141–144)

TESTING YOUR COMPREHENSION

1. Explain the premises and logic that supports the concept of natural selection.
2. How does allopatric speciation occur?
3. Name two examples of evidence for natural selection.
4. Name three organisms that have gone extinct, and give a probable reason for each extinction.
5. What is the difference between a species and a population? Between a population and a community?
6. Contrast the concepts of habitat and niche.
7. List and describe each of the five major population characteristics discussed in this chapter. Explain how each shapes population dynamics.
8. Could any species undergo exponential growth forever? Explain your answer.
9. Describe how limiting factors relate to carrying capacity.
10. Explain the difference between K-selected species and r-selected species. Can you think of examples of each that were not mentioned in the chapter?

SEEKING SOLUTIONS

1. In what ways has artificial selection changed people's quality of life? Give examples. Can you imagine a way in which artificial selection could be used to improve our quality of life further? Can you imagine a way it could be used to lessen our environmental impact?

2. What types of species are most vulnerable to extinction, and what kinds of factors threaten them? Can you think of any species in your region that are threatened with extinction today? What reasons lie behind their endangerment?

3. Do you think the human species can continue raising its global carrying capacity? How so, or why not? Do you think we *should* try to keep raising our carrying capacity? Why or why not?

4. Describe the evidence suggesting that changes in temperature and precipitation led to the extinction of the golden toad and to population crashes for other amphibians at Monteverde. What do you think could be done to help make future such declines less likely?

5. What are the advantages of ecotourism for a country like Costa Rica? Can you think of any disadvantages? What would you recommend that Costa Rica do to prevent the loss of its biodiversity?

6. As Monteverde changed and some species disappeared, scientists reported that others moved in from lower, drier areas. If this is true, should we be concerned about the extinction of the golden toad and disappearance of other species from Monteverde? Explain your answer.

INTERPRETING GRAPHS AND DATA

Amphibians are sensitive biological indicators of climate change because their reproduction and survival are so closely tied to water. One way in which drier conditions may affect amphibians is by reducing the depth of the pools of water in which their eggs develop. Shallower pools offer less protection from UV-B (ultraviolet) radiation, which some scientists maintain may kill embryos directly or make them more susceptible to disease.

Herpetologist Joseph Kiesecker and colleagues conducted a field study of the relationships among water depth, UV-B radiation, and survivorship of western toad *(Bufo boreas)* embryos in the Pacific Northwest. In manipulative experiments, the researchers placed toad embryos in mesh enclosures at three different depths of water. The researchers placed protective filters that blocked all UV-B radiation over some of these embryos, while leaving other embryos unprotected without the filters. Some of the study's results are presented in the accompanying graph.

1. If the UV-B radiation at the surface has an intensity of 0.27 watts/m^2, approximately what is its intensity at depths of 10 cm, 50 cm, and 100 cm?

2. Approximately how much did survival rates at the 10-cm depth differ between the protected and unprotected treatments? Why do you think survival rates differed significantly at the 10-cm depth but not at the other depths?

3. What do you think would be the effect of drier-than-average years on the western toad population, if the

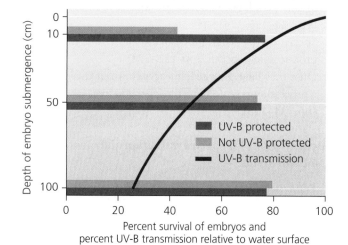

Embryo survivorship in western toads *(Bufo boreas)* at different water depths and UV-B light intensities. Red bars indicate embryos protected under a filter that blocked UV-B light; orange bars indicate unprotected embryos. The blue line indicates the amount of UV-B light reaching different depths in the water column, expressed as a percentage of the UV-B radiation at the water surface. Data from Kiesecker, J. M., et al. 2001. Complex causes of amphibian population declines. *Nature* 410: 681–684.

average depth of pools available for toad spawning dropped? How do the data above address your hypothesis? Do they support cause-and-effect relationships among water depth, UV-B exposure, disease, and toad mortality?

CALCULATING ECOLOGICAL FOOTPRINTS

In 2004, coffee consumption in the United States topped 2.7 billion pounds (out of 14.8 billion pounds produced globally). Next to petroleum, coffee is the most valuable commodity on the world market, and the United States is its leading importer. Most coffee is produced in large tropical plantations, where coffee is the only tree species and is grown in full sun. However, approximately 2% of coffee is produced in small groves where coffee trees and other species are intermingled. These *shade-grown* coffee forests maintain greater habitat diversity for tropical rainforest wildlife. Given the information above, estimate the coffee consumption rates in the table below.

	Population	Pounds of coffee per day	Pounds of coffee per year
You (or the average American)	1	0.025	9
Your class			
Your hometown			
Your state			
United States			

Data from O'Brien, T. G. and M. F. Kinnaird. 2003. Caffeine and conservation. *Science* 300: 587; and International Coffee Organization.

1. What percentage of global coffee production is consumed in the United States? If only shade-grown coffee were consumed in the United States, how much would shade-grown production need to increase to meet that demand?

2. How much extra would you be willing to pay for a pound of shade-grown coffee, if you knew that your money would help to prevent habitat loss or extinction for animals such as Sumatran tigers, rhinoceroses, and the many songbirds that migrate between Latin America and North America each year?

3. If everyone in the United States were willing to pay as much extra per pound for shade-grown coffee as you are, how much additional money would that provide for conservation of biodiversity in the tropics each year?

Take It Further

 Go to www.aw-bc.com/withgott or the student CD-ROM where you'll find:

▶ Suggested answers to end-of-chapter questions
▶ Quizzes, animations, and flashcards to help you study
▶ *Research Navigator*™ database of credible and reliable sources to assist you with your research projects

▶ **GRAPHit!** tutorials to help you master how to interpret graphs
▶ **INVESTIGATEit!** current news articles that link the topics that you study to case studies from your region to around the world

6 Species Interactions and Community Ecology

Marsh community along the shore of Lake Ontario

Upon successfully completing this chapter, you will be able to:

▶ Compare and contrast the major types of species interactions

▶ Characterize feeding relationships and energy flow, using them to construct trophic levels and food webs

▶ Distinguish characteristics of a keystone species

▶ Characterize the process of succession and the debate over the nature of communities

▶ Perceive and predict the potential impacts of invasive species in communities

▶ Explain the goals and methods of ecological restoration

▶ Describe and illustrate the terrestrial biomes of the world

Aggregation of zebra mussels

Central Case: Black and White, and Spread All Over: Zebra Mussels Invade the Great Lakes

"We are seeing changes in the Great Lakes that are more rapid and more destructive than any time in the history of the Great Lakes."
—ANDY BUCHSBAUM, DIRECTOR OF THE NATIONAL WILDLIFE FEDERATION'S GREAT LAKES OFFICE

"When you tear away the bottom of the food chain, everything that is above it is going to be disrupted."
—TOM NALEPA, NATIONAL OCEANIC AND ATMOSPHERIC ADMINISTRATION RESEARCH BIOLOGIST

As if the Great Lakes hadn't been through enough already, the last thing they needed was the zebra mussel. The pollution-fouled waters of Lake Erie and the other Great Lakes shared by Canada and the United States had become gradually cleaner in the years following the Clean Water Act of 1970. As government regulation brought industrial discharges under control, people once again began to use the lakes for recreation, and populations of fish rebounded.

Then the zebra mussel arrived. Black-and-white-striped shellfish the size of a dime, zebra mussels attach to hard surfaces and open their paired shells, feeding on algae by filtering water through their gills. This mollusc, given the scientific name *Dreissena polymorpha,* is native to the Caspian Sea, Black Sea, and Azov Sea in western Asia and eastern Europe. It made its North American debut in 1988 when it was discovered in Canadian waters at Lake St. Clair, which connects Lake Erie with Lake Huron. Evidently ships arriving from Europe had discharged ballast water containing the mussels or their larvae into the Great Lakes.

Within just two years of their discovery in Lake St. Clair, zebra mussels had reached all five of the Great Lakes. The next year, these invaders entered New York's Hudson River to the east, and the Illinois River at Chicago to the west. From the Illinois River and its canals, they soon reached the Mississippi River, giving them access to a vast watershed covering 40% of the

United States. By 1992, zebra mussels had reached the Ohio, Arkansas, and Tennessee Rivers, and by 1994 they had colonized waters in 19 U.S. states and two Canadian provinces.

How could a mussel spread so quickly? The zebra mussel's larval stage is well adapted for long-distance dispersal. Its tiny larvae drift freely for several weeks, traveling as far as the currents take them. Adults that attach themselves to boats and ships may be transported from one place to another, even to small isolated lakes and ponds well away from major rivers. Moreover, in North America the mussels encountered none of the particular species of predators, competitors, and parasites that had evolved to limit their population growth in the Old World.

But why all the fuss? Zebra mussels are best known for clogging up water intake pipes at factories, power plants, municipal water supplies, and wastewater treatment facilities (Figure 6.1a). At one Michigan power plant, workers counted 700,000 mussels per square meter of pipe surface. Great densities of these organisms can damage boat engines, degrade docks, foul fishing gear, and sink buoys that ships use for navigation. Through such impacts, it is estimated that zebra mussels cost the U.S. economy hundreds of millions of dollars each year.

Zebra mussels also have severe impacts on the ecological systems they invade. They eat primarily phytoplankton, microscopic algae that drift in open water. Because each mussel filters a liter or more of water every day, they consume so much phytoplankton that they can deplete populations. Phytoplankton is the foundation of the Great Lakes food web, so its depletion is bad news for zooplankton, the tiny aquatic animals that eat phytoplankton—and for the fish that eat both. Water bodies with zebra mussels have fewer zooplankton and open-water fish than water bodies without them, researchers are finding. Zebra mussels also interfere with native molluscs, suffocating them by attaching to their shells (Figure 6.1b).

However, zebra mussels also benefit some bottom-feeding invertebrates and fish. By filtering algae and organic matter from open water and depositing nutrients in their feces, they shift the community's nutrient balance to the bottom and benefit the species that feed there. Once they have cleared the water, sunlight penetrates more deeply, spurring the growth of large-leafed underwater plants and algae. Such changes have further ripple effects throughout the community that scientists are only beginning to understand.

(a) Clogging a pipe

(b) Suffocating native clams

FIGURE 6.1 Zebra mussels clog up water intake pipes **(a)** of power plants and industrial facilities. They also starve and suffocate native clams by adhering to their shells en masse and sealing them shut **(b)**.

Species Interactions

By interacting with many species in a variety of ways, zebra mussels have set in motion an array of changes in the ecological communities they have invaded. Interactions among species are the threads in the fabric of communities, holding them together and determining their nature. Ecologists have organized species interactions into several fundamental categories. Most prominent are competition, predation, parasitism, herbivory, and mutualism. Table 6.1 summarizes the positive and negative impacts of each type of interaction for each participant.

Table 6.1 Effects of Species Interactions on Their Participants		
Type of interaction	**Effect on species 1**	**Effect on species 2**
Mutualism	+	+
Commensalism	+	0
Predation, parasitism, herbivory	+	−
Neutralism	0	0
Amensalism	−	0
Competition	−	−

"+" denotes a positive effect; "−" denotes a negative effect; "0" denotes no effect.

Competition can occur when resources are limited

When multiple organisms seek the same limited resource, their relationship is said to be one of **competition.** Competing organisms do not usually fight with one another directly and physically. Competition is generally more subtle, involving the consequences of one organism's ability to match or outdo others in procuring resources. The resources for which organisms compete can include just about anything an organism might need to survive, including food, water, space, shelter, mates, sunlight, and more. Competitive interactions can take place among members of the same species (*intraspecific competition*) or among members of two or more different species (*interspecific competition*).

We have already discussed intraspecific competition in Chapter 5, without naming it as such. Recall that density dependence (▸ p. 138) limits the growth of a population; individuals of the same species compete with one another for limited resources, such that competition is more acute when there are more individuals per unit area (denser populations). Thus, intraspecific competition is really a population-level phenomenon. Interspecific competition, however, can have substantial effects on the composition of communities.

Interspecific competition can give rise to different types of outcomes. If one species is a very effective competitor, it may exclude another species from resource use entirely. This outcome, called *competitive exclusion,* occurred in Lake St. Clair and western Lake Erie as the zebra mussel outcompeted a native mussel species.

Alternatively, if neither competing species fully excludes the other, the species may live side by side at a certain ratio of population sizes. This result, called *species coexistence,* may produce a stable point of equilibrium, in which the population size of each remains fairly constant through time.

Coexisting species that use the same resources tend to adjust to their competitors to minimize competition with them. Individuals can do this by changing their behavior so as to use only a portion of the total array of resources they are capable of using. In such cases, individuals are not fulfilling their entire *niche*, or ecological role (▸ p. 131). The full niche of a species is called its **fundamental niche** (Figure 6.2a). An individual that plays only part of its role because of competition or other species interactions is said to be displaying a **realized niche** (Figure 6.2b), the portion of its fundamental niche that is actually filled, or realized.

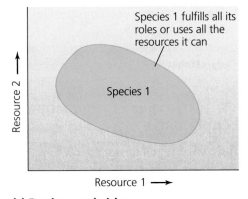

(a) Fundamental niche

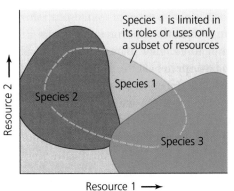

(b) Realized niche

FIGURE 6.2 An organism facing competition may be forced to play a lesser ecological role or use fewer resources than it would in the absence of its competitor. With no competitors, an organism can exploit its full fundamental niche (**a**). But when competitors restrict what an organism can do or what resources it can use, the organism is limited to a realized niche (**b**), which covers only a subset of its fundamental niche. In considering niches, ecologists have traditionally focused on competition, but they now recognize that other species interactions also are influential.

FIGURE 6.3 When species compete, they tend to partition resources, each specializing on a slightly different resource or way of attaining a shared resource. A number of types of birds—including the woodpeckers, creeper, and nuthatch shown here—feed on insects from tree trunks, but they use different portions of the trunk, seeking different foods in different ways.

White-breasted nuthatch climbs downward along trunk looking for insects

Yellow-bellied sapsucker drills rows of holes in trunk and consumes sap and insects that get stuck in sap

Pileated woodpecker digs deeply into wood to find large insects

Brown creeper climbs upward along trunk looking for tiny insects

Species make similar adjustments over evolutionary time. They adapt to competition by evolving to use slightly different resources or to use their shared resources in different ways. If two bird species eat the same type of seeds, one might come to specialize on larger seeds and the other to specialize on smaller seeds. Or one bird might become more active in the morning and the other more active in the evening, thus avoiding direct interference. This process is called **resource partitioning,** because the species divide, or partition, the resource they use in common by specializing in different ways (Figure 6.3). Resource partitioning can lead to *character displacement,* which occurs when competing species evolve physical characteristics that reflect their reliance on the portion of the resource they use. By becoming more different from one another, two species reduce their competition. Through natural selection (▶pp. 118–121), birds that specialize on larger seeds may evolve larger bills that enable them to make best use of the resource, whereas birds specializing on smaller seeds may evolve smaller bills. This is precisely what extensive research has revealed about the finches from the Galapagos Islands that were first described by Charles Darwin.

Several types of interactions are exploitative

In competitive interactions, each participant has a negative effect on other participants, because each takes resources the others could have used. This is reflected in the two minus signs shown for competition in Table 6.1. In other types of interactions, some participants benefit while others are harmed (note the +/− interactions in Table 6.1). We can think of interactions in which one member exploits another for its own gain as exploitative interactions, or *exploitation.* Such interactions include predation, parasitism, herbivory, and related concepts, as outlined below.

FIGURE 6.4 Predator-prey interactions have ecological and evolutionary consequences for both prey and predator. Here, a Halloween snake *(Pliocercus euryzonus)* devours a frog in the Monteverde cloud forest we studied in Chapter 5.

Predators kill and consume prey

Every living thing needs to procure food, and for most animals, that means eating other living organisms. **Predation** is the process by which individuals of one species, a *predator,* hunt, capture, kill, and consume individuals of another species, its *prey* (Figure 6.4). Along with competition, predation has traditionally been viewed as one of the primary organizing forces in community ecology. Interactions between predators and prey structure the food webs that we will examine shortly, and they influence community composition by helping determine the relative abundance of predators and prey.

Zebra mussel predation on phytoplankton has reduced phytoplankton populations by up to 90%, according to many studies in the Great Lakes and Hudson River. Zebra mussels also consume the smaller types of zooplankton. This predation, combined with the competition mentioned above, has caused zooplankton population sizes and biomass to decline by up to 70% in Lake Erie

and the Hudson River since zebra mussels arrived. Meanwhile, the mussels do not eat some cyanobacteria (▸p. 105), so concentrations of these cyanobacteria rise in lakes with zebra mussels. Most predators are also prey, however, and zebra mussels have become a food source for a number of North American species since their introduction. These include diving ducks, muskrats, crayfish, flounder, sturgeon, eels, and several types of fish with grinding teeth, such as carp and freshwater drum.

Predation can sometimes drive population dynamics by causing cycles in population sizes. An increase in the population size of prey creates more food for predators, which may survive and reproduce more effectively as a result. As the predator population rises, additional predation drives down the population of prey. Fewer prey in turn causes some predators to starve, so that the predator population declines. This allows the prey population to begin rising again, starting the cycle anew. Most natural systems involve so many factors that such cycles don't last long, but in some cases we see extended cycles (Figure 6.5).

Predation also has evolutionary ramifications. Individual predators that are more adept at capturing prey will likely live longer, healthier lives and be better able to provide for their offspring than will less adept individuals. Thus, natural selection on individuals within a predator species leads to the evolution of adaptations that make them better hunters. Prey face an even stronger selective pressure—the risk of immediate death. For this reason, predation pressure has caused organisms to evolve an elaborate array of defenses against being eaten (Figure 6.6).

Parasites exploit living hosts

Organisms can exploit other organisms without killing them. **Parasitism** is a relationship in which one organism, the *parasite,* depends on another, the *host,* for nourishment or some other benefit while simultaneously doing

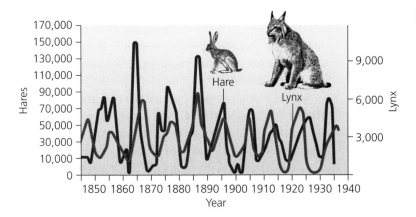

FIGURE 6.5 Predator-prey systems sometimes show paired cycles, in which increases and decreases in one organism apparently drive increases and decreases in the other. Although such cycles are predicted by theory and are seen in lab experiments, they are very difficult to document conclusively in natural systems. Data from MacLulich, D.A. 1937. Fluctuation in the numbers of varying hare *(Lepus americanus). Univ. Toronto Stud. Biol. Ser. 43,* Toronto: University of Toronto Press.

(a) Cryptic coloration

(b) Warning coloration

(c) Mimicry

FIGURE 6.6 Natural selection to avoid predation has resulted in many fabulous adaptations. Some prey hide from predators by *crypsis,* or camouflage, such as this gecko on tree bark (**a**). Other prey are brightly colored to warn predators that they are toxic or distasteful, such as this poison dart frog (**b**). Still others fool predators with mimicry. Some, like walking sticks imitating twigs, mimic for crypsis. Others mimic toxic, distasteful, or dangerous organisms, like this caterpillar (**c**); when it is disturbed, the caterpillar swells and curves its tail end and shows eyespots, to look like a snake's head.

the host harm. Unlike predation, parasitism usually does not result in an organism's immediate death, although it sometimes contributes to the host's eventual death.

Many parasites live in close contact with their hosts. These parasites include disease pathogens, such as the protists that cause malaria and dysentery, as well as animals, such as tapeworms, that live in the digestive tracts of their hosts. Other parasites live on the exterior of their hosts, such as the ticks that attach themselves to their hosts' skin, and the sea lamprey *(Petromyzon marinus),* another invader of the Great Lakes (Figure 6.7). Sea

FIGURE 6.7 Parasites harm their host organism in some way. With its suction-like mouth and rasping tongue, the sea lamprey *(Petromyzon marinus)* attaches itself to fish and sucks their blood for days or weeks, sometimes killing the fish. Sea lampreys wreaked havoc on Great Lakes fisheries after entering the lakes through human-built canals.

lampreys are tube-shaped vertebrates that grasp the bodies of fish with a suction-cup mouth and a rasping tongue, sucking blood from the fish for days or weeks. Sea lampreys invaded the Great Lakes from the Atlantic Ocean after people dug canals to connect the lakes for shipping, and the lampreys soon devastated economically important fisheries of chubs, lake herring, whitefish, and lake trout. Since the 1950s, U.S. and Canadian fisheries managers have reduced lamprey populations by applying chemicals that selectively kill lamprey larvae.

Other types of parasites are free-living and come into contact with their hosts only infrequently. For example, the cuckoos of Eurasia and the cowbirds of the Americas parasitize other birds by laying eggs in their nests and letting the host bird raise the parasite's young.

Some parasites cause little harm, but others may kill their hosts. Many insects parasitize other insects, often killing them in the process, and are called *parasitoids.* Various species of parasitoid wasps lay eggs on caterpillars. When the eggs hatch, the wasp larvae burrow into the caterpillar's tissues and slowly consume them. The wasp larvae metamorphose into adults and fly from the body of the dying caterpillar.

Just as predators and prey evolve in response to one another, so do parasites and hosts, in a process termed *coevolution.* Hosts and parasites can become locked in a duel of escalating adaptations, a situation sometimes referred to as an *evolutionary arms race.* Like rival nations racing to stay ahead of one another in military technology, host and parasite may repeatedly evolve new responses to the other's latest advance. In the long run, though, it may not be in a

FIGURE 6.8 Herbivory is a common way to make a living. The world holds many thousands, and perhaps millions, of species of plant-eating insects, such as this larva (caterpillar) of the death's head hawk moth *(Acherontia atropos)* from western Europe.

parasite's best interest to become too harmful to its host. Instead, a parasite might leave more offspring in the next generation—and thus be favored by natural selection—if it allows its host to live a longer time, or even to thrive.

Herbivores exploit plants

One of the most common types of exploitation is **herbivory,** which occurs when animals feed on the tissues of plants. Insects that feed on plants are the most widespread type of herbivore; just about every plant in the world is attacked by some type of insect (Figure 6.8). In most cases, herbivory does not kill a plant outright, but may affect its growth and reproduction.

Like animal prey, plants have evolved a wide array of defenses against the animals that feed on them. Many plants produce chemicals that are toxic or distasteful to herbivores. Others arm themselves with thorns, spines, or irritating hairs. In response, herbivores may evolve ways to overcome these defenses, and the plant and the animal may embark on an evolutionary arms race.

Some plants go a step further and recruit certain animals as allies to assist in their defense. Many such plants encourage ants to take up residence by providing thorns or swelled stems for the ants to nest in or nectar-bearing structures for the ants to feed from. These ants protect the plant in return by attacking other insects that land or crawl on it. Other plants respond to herbivory by releasing volatile chemicals when they are bitten or pierced. The airborne chemicals attract predatory insects that may attack the herbivore. Such cooperative strategies, trading defense for food, are examples of our next type of species interaction, mutualism.

Mutualists help one another

Mutualism is a relationship in which two or more species benefit from interaction with one another. Generally each partner provides some resource or service that the other needs.

Many mutualistic relationships—like many parasitic relationships—occur between organisms that live in close physical contact. (Indeed, biologists hypothesize that many mutualistic associations evolved from parasitic ones.) Such physically close association is called **symbiosis.** Thousands of terrestrial plant species depend on mutualisms with fungi; plant roots and some fungi together form symbiotic associations called mycorrhizae. In these symbioses, the plant provides energy and protection to the fungus, while the fungus assists the plant in absorbing nutrients from the soil. In the ocean, coral polyps, the tiny animals that build coral reefs, share beneficial arrangements with algae known as zooxanthellae. The coral provide housing and nutrients for the algae in exchange for a steady supply of food—90% of their nutritional requirements.

You, too, are part of a symbiotic association. Your digestive tract is filled with microbes that help you digest food—microbes for which you are providing a place to live. Indeed, we may owe our very existence to symbiotic mutualisms. It is now widely accepted that the eukaryotic cell (▶p. 102) originated after certain prokaryotic cells engulfed other prokaryotic cells and established mutualistic symbioses. Scientists have inferred that some of the engulfed cells eventually evolved into cell organelles.

Not all mutualists live in close proximity. One of the most important mutualisms in environmental science involves free-living organisms that may encounter each other only once in their lifetimes. This is *pollination* (Figure 6.9), an interaction of key significance to agriculture and our food supply (▶pp. 285–286). Bees, birds, bats, and other creatures transfer pollen (male sex cells) from one flower to another, fertilizing eggs that become embryos within seeds. Most

FIGURE 6.9 In mutualism, organisms of different species benefit one another. An important mutualistic interaction for environmental science is pollination. This hummingbird visits flowers to gather nectar and in the process transfers pollen between flowers, helping the plant reproduce. Pollination is of key importance to agriculture, ensuring the reproduction of many crop plants.

pollinating animals visit flowers for their nectar, a reward the plant uses to entice them. The pollinators receive food, and the plants are pollinated and reproduce. Various types of bees alone pollinate 73% of our crops, one expert has estimated—from soybeans to potatoes to tomatoes to beans to cabbage to oranges.

Some interactions have no effect on some participants

Two other types of species interaction get far less attention. **Amensalism** is a relationship in which one organism is harmed and the other is unaffected. In **commensalism**, one species benefits and the other is unaffected. Amensalism has been difficult to pin down, because it is hard to prove that the organism doing the harm is not in fact besting a competitor for a resource. For instance, some plants release poisonous chemicals that harm nearby plants (a phenomenon called *allelopathy*), and some experts have suggested that this is an example of amensalism. However, allelopathy can also be viewed as one plant investing in chemicals to outcompete others for space.

One association commonly cited as an example of commensalism is when the conditions created by one plant happen to make it easier for another plant to establish and grow. For instance, palo verde trees in the Sonoran Desert create shade and leaf litter that allow the soil beneath them to hold moisture longer, creating an area that is cooler and moister than the surrounding sunbaked ground. Young plants find it easier to germinate and grow in these conditions, so seedling cacti and other desert plants generally grow up directly beneath "nurse" trees such as palo verde. This phenomenon, called *facilitation*, influences the structure and composition of communities and how they change through time.

Ecological Communities

In Chapter 5 we defined a *community* as a group of populations of organisms that live in the same place at the same time. The members of a community interact with one another in the ways described above, and the direct interactions among species often have indirect effects that ripple outward to affect other community members. The strength of interactions also varies, and together species interactions determine the species composition, structure, and function of communities. *Community ecologists* are interested in what species coexist, how they relate to one another, how communities change through time, and why these patterns exist.

Energy passes among trophic levels

The interactions among members of a community are many and varied, but some of the most important involve who eats whom. As we saw in Chapter 4 (▸pp. 105–106), the energy that drives such interactions in most systems comes ultimately from the sun via photosynthesis. As organisms feed on one another, this energy moves through the community, from one rank in the feeding hierarchy, or **trophic level,** to another (Figure 6.10).

Producers Producers, or autotrophs ("self-feeders"), comprise the first trophic level, as we saw in Chapter 4 (▸pp. 105–108). Terrestrial green plants, cyanobacteria, and algae capture solar energy and use photosynthesis to produce sugars. The chemosynthetic bacteria of hot springs and deep-sea hydrothermal vents use geothermal energy in a similar way to produce food.

Consumers Organisms that consume producers are known as *primary consumers* and comprise the second

Aquatic examples Terrestrial examples

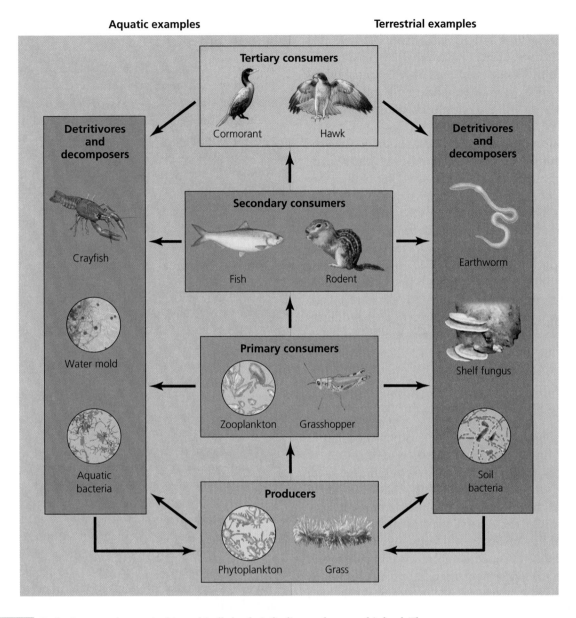

FIGURE 6.10 Ecologists organize species hierarchically by their feeding rank, or trophic level. The diagram shows aquatic (left) and terrestrial (right) examples at each level. Arrows indicate the direction of energy flow. Producers produce food by photosynthesis, primary consumers (herbivores) feed on producers, secondary consumers eat primary consumers, and tertiary consumers eat secondary consumers. Communities can have more or fewer trophic levels than in this example. Detritivores and decomposers feed on nonliving organic matter and the remains of dead organisms from all trophic levels, and they "close the loop" by returning nutrients to the soil or the water column for use by producers.

trophic level. Grazing animals, such as deer and grasshoppers, are primary consumers. The third trophic level consists of *secondary consumers,* which prey on primary consumers. Wolves that prey on deer are considered secondary consumers, as are rodents and birds that prey on grasshoppers. Predators that feed at even higher trophic levels are known as *tertiary consumers.* Examples of tertiary consumers include hawks and owls that eat rodents that have eaten grasshoppers. Note that most primary consumers are *herbivores* because they consume plants, whereas secondary and tertiary consumers are *carnivores* because they eat animals. Animals that eat both plant and animal food are referred to as *omnivores.*

Detritivores and decomposers *Detritivores* and *decomposers* consume nonliving organic matter. Detritivores, such as millipedes and soil insects, scavenge the waste products or the dead bodies of other community members. Decomposers, such as fungi and bacteria, break down leaf litter and other nonliving matter further into simpler constituents that can then be taken up and used by plants. These organisms play an essential role as the community's recyclers, making nutrients from organic matter available for reuse by living members of the community.

In Great Lakes communities, phytoplankton are the main producers, floating freely and photosynthesizing with sunlight that penetrates the upper layer of the water. Zooplankton are primary consumers, feeding on the phytoplankton. Phytoplankton-eating fish are primary consumers, and zooplankton-eating fish are secondary consumers. At higher trophic levels are tertiary consumers such as larger fish and birds that feed on plankton-eating fish. Zebra mussels, by eating both phytoplankton and zooplankton, function on multiple trophic levels. When any of these organisms dies and sinks to the bottom, detritivores scavenge its tissues and microbial decomposers recycle its nutrients.

Energy, biomass, and numbers decrease at higher trophic levels

At each trophic level, most of the energy that organisms use is lost through respiration. Only a small amount of the energy is transferred to the next trophic level through predation, herbivory, or parasitism. The first trophic level (producers) contains a large amount of energy, but the second (primary consumers) contains less energy—only that amount gained from consuming producers. The third trophic level (secondary consumers) contains still less energy, and higher trophic levels (tertiary consumers) contain the least. A general rule of thumb is that each trophic level contains just 10% of the energy of the trophic level below it, although the actual proportion can vary greatly.

This pattern, which can be visualized as a pyramid, generally also holds for the numbers of organisms at each trophic level. Generally, fewer organisms exist at higher trophic levels than at lower trophic levels. A grasshopper eats many plants in its lifetime, a rodent eats many grasshoppers, and a hawk eats many rodents. Thus, for every hawk in a community there must be many rodents, still more grasshoppers, and an immense number of plants. Because the difference in numbers of organisms among trophic levels tends to be large, the same pyramid-like relationship also often holds true for biomass. Even though rodents are larger than grasshoppers, and hawks larger than rodents, the sheer number of prey relative to the predators means that prey biomass will likely be greater overall.

Food webs show feeding relationships and energy flow

As energy is transferred from species on one trophic level to species on other trophic levels, it is said to pass up a *food chain.* Plant, grasshopper, rodent, and hawk make up a food chain, a linear series of feeding relationships. Thinking in terms of food chains is conceptually useful, but in reality ecological systems are far more complex than simple linear chains. A more accurate representation of the feeding relationships in a community is a **food web,** a visual map of feeding relationships and energy flow, showing the many paths by which energy passes among organisms as they consume one another.

Figure 6.11 shows a food web from a temperate deciduous forest of eastern North America. Like virtually all diagrams of ecological systems, it is greatly simplified, leaving out the vast majority of species and interactions that occur. Note, however, that even within this simplified diagram, we can pick out a number of different food chains involving different sets of species.

A Great Lakes food web would involve the phytoplankton and cyanobacteria that photosynthesize near the water's surface, the zooplankton that eat them, the fish that eat all these, the larger fish that eat the smaller fish, and the lampreys that parasitize the fish. It would include a number of native mussels and clams and, since 1988, the zebra mussel that is crowding them out. It would include diving ducks that used to feed on native bivalves and now are preying on zebra mussels. This food web would also show that an array of bottom-dwelling invertebrates feed from the refuse of zebra mussels. These waste products promote bacterial growth and disease pathogens that harm native bivalves, but they also provide nutrients that nourish crayfish and many smaller *benthic* (bottom-dwelling) invertebrate animals. Finally, the food web would include underwater plants and macroscopic algae, whose growth is promoted by zebra mussels. The mussels clarify the water by filtering out phytoplankton, and sunlight penetrates more deeply into the water column, spurring photosynthesis and plant growth. Overall, zebra mussels alter this food web essentially by shifting productivity from the open-water regions to the benthic and *littoral* (nearshore) regions.

The many direct interactions in this food web also create indirect interactions. For example, zebra mussels affect fish indirectly, helping benthic and littoral fishes and making life harder for open-water fishes (see "The Science behind the Story," ▶ pp. 160–161).

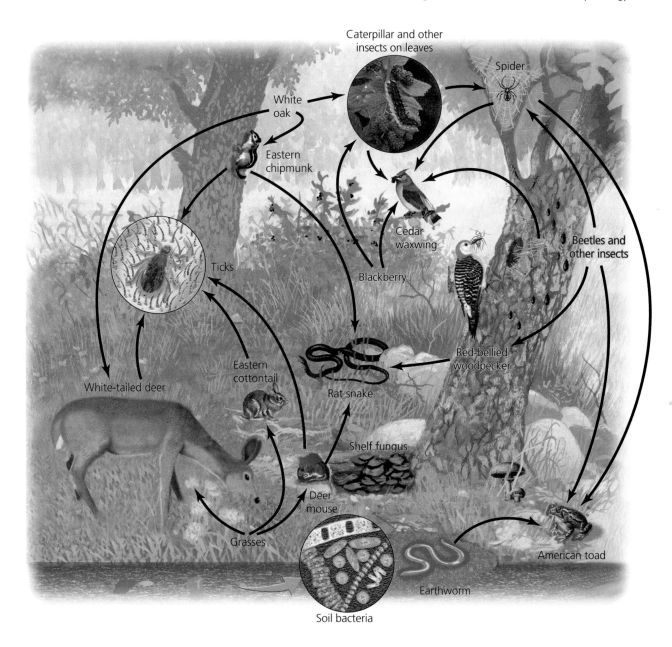

FIGURE 6.11 Food webs are conceptual representations of feeding relationships in a community. This food web pertains to eastern North America's temperate deciduous forest and includes organisms on several trophic levels. In a food web diagram, arrows are drawn from one organism to another to indicate the direction of energy flow as a result of predation, parasitism, or herbivory. For example, an arrow leads from the grass to the cottontail rabbit to indicate that cottontails consume grasses. The arrow from the cottontail to the tick indicates that parasitic ticks derive nourishment from cottontails. Communities include so many species and are complex enough, however, that most food web diagrams are bound to be gross simplifications.

Some organisms play bigger roles in communities than others

"Some animals are more equal than others," George Orwell wrote in his 1945 book *Animal Farm.* Although Orwell was making wry sociopolitical commentary, his remark hints at a truth in ecology. In communities, ecologists have found, some species exert greater influence than do others. A species that has particularly strong or far-reaching impact is often called a **keystone species.** A keystone is the wedge-shaped stone at the top of an arch that is vital for holding the structure together; remove the keystone, and the arch will collapse. In an ecological

Inferring Zebra Mussels' Impacts on Fish Communities

The Science behind the Story

Food webs are complicated systems, and disentangling them to infer the effects of any one species is fraught with difficulty. When zebra mussels appeared in the Great Lakes, people feared for sport fisheries, and estimated that fish population declines could cost billions of dollars. The mussels would deplete the phytoplankton and zooplankton that fish depended on, people reasoned, and many fewer fish would survive. Yet even after 15 years, there was no solid evidence of widespread harm to fish populations.

So, aquatic biologist David Strayer of the Institute of Ecosystem Studies in Millbrook, New York, joined Kathryn Hattala and Andrew Kahnle of New York State's Department of Environmental Conservation (DEC). They mined datasets on fish populations in the Hudson River, which zebra mussels had invaded in 1991.

Strayer and others had already been studying effects of zebra mussels on aspects of the community for years. Their data showed that since the species' introduction to the Hudson:

▶ Biomass of phytoplankton fell 80%.

▶ Biomass of small zooplankton fell 76%.

▶ Biomass of large zooplankton fell 52%.

Zebra mussels increased filter-feeding in the community 30-fold, thereby depleting the phytoplankton and small zooplankton, and leaving all sizes of zooplankton with less phytoplankton to eat. Overall, the zooplankton and invertebrate animals of the open water that are eaten by open-water fish declined by 70%.

However, Strayer's work had also found that *benthic,* or bottom-dwelling, invertebrates in shallow water (especially in the nearshore, or *littoral,* zone) had increased by 10%, and likely much more, because the mussels' shells provide habitat structure, and their feces provide nutrients.

These contrasting trends in the benthic shallows and the open deep water led Strayer's team to hypothesize that zebra mussels would harm open-water fish that ate plankton but would help littoral-feeding fish. They predicted that after zebra mussel introduction, larvae and juveniles of six common open-water fish species would decline in number, decline in growth rate, and shift downriver toward saltier water, where mussels are absent. Conversely, they predicted that larvae and juveniles of 10 littoral fish species would increase in number, increase in growth rate, and shift upriver to regions of greatest zebra mussel density.

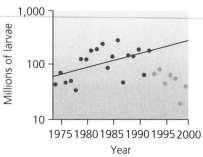

(a) American shad

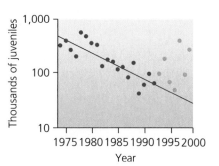

(b) Tessellated darter

Larvae of American shad **(a),** an open-water fish, had been increasing in abundance before zebra mussels were introduced (red points and trend line). After zebra mussel introduction, shad larvae decreased in abundance (orange points). Juveniles of the tessellated darter **(b),** a littoral zone fish, had been decreasing in abundance before zebra mussels were introduced (red points and trend line). After zebra mussel introduction, they increased in abundance (orange points). *Source:* Strayer, D., et al. 2004. Effects of an invasive bivalve (*Dreissena polymorpha*) on fish in the Hudson River estuary. *Can. J. Fish. Aquat. Sci.* 61: 924–941.

To test their predictions, the researchers analyzed data from three

community, removal of a keystone species will have substantial ripple effects and will alter a large portion of the food web.

Often, large-bodied secondary or tertiary consumers near the tops of food chains are considered keystone species. Top predators control populations of herbivores,

which would otherwise multiply and could, through increased herbivory, greatly modify the plant community. In the United States, government bounties promoted the hunting of wolves and mountain lions, which were largely exterminated by the middle of the 20th century. In the absence of these predators, unnaturally dense deer

types of fish surveys carried out over 26 years spanning periods before and after the zebra mussel's arrival. One data set came from surveys conducted from 1985 to 1999 by the DEC. The other two came from surveys conducted from 1974 to 1999 by biologists hired by electric utilities, which in New York are required to monitor fish populations in return for using the Hudson's water for cooling at their power plants.

The researchers compared values for abundance, growth, and distribution for young fish before 1991 with those after 1991. The results supported their predictions. Larvae and juveniles of open-water fish, such as American shad, blueback herring, and alewife, tended to decline in abundance in the years after zebra mussel introduction (first figure, part (a)). Those of littoral fish, such as tessellated darter, bluegill, and largemouth bass, tended to increase (first figure, part (b)). Growth rates showed the same trend: Open-water fish grew more slowly after zebra mussel introduction, whereas littoral fish grew more quickly. In terms of distribution in the 248-km (154-mi) stretch of river studied, open-water fish shifted downstream toward areas with fewer zebra mussels, whereas littoral fish shifted upstream

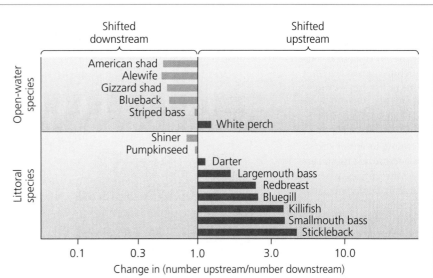

Young of open-water fish, such as American shad, blueback herring, and alewife, tended to shift downstream toward areas with fewer zebra mussels in the years following zebra mussel arrival. Young of littoral fish, such as killifish, bluegill, and largemouth bass, tended to shift upstream toward areas with more zebra mussels. *Source:* Strayer, D., et al. 2004. Effects of an invasive bivalve (*Dreissena polymorpha*) on fish in the Hudson River estuary. *Can. J. Fish. Aquat. Sci.* 61: 924–941.

toward areas with more zebra mussels (second figure). The results were published in 2004 in the *Canadian Journal of Fisheries and Aquatic Sciences.*

Overall, the results supported the hypothesis that the fish community would respond to changes in its food resources caused by zebra mussels. The results are correlative, and correlation does not prove causation. However, previous attempts to address these questions experimentally were so limited in time and scale that they could not

reflect true effects in natural systems.

Research such as this helps illuminate the often obscure connections and impacts that particular species interactions have on communities as a whole. In this case, the research may also help fisheries managers predict changes in commercially and recreationally important fish populations. With this knowledge, biologists may be able to manage fisheries more effectively in the Hudson and other areas invaded by zebra mussels.

populations have overgrazed forest-floor vegetation and eliminated whole cohorts of tree seedlings, causing major changes in forest structure.

The removal of top predators in the United States was an uncontrolled large-scale experiment with unintended consequences. But ecologists have verified the keystone

species concept in controlled experiments. Classic work by marine biologist Robert Paine established that the predatory starfish *Pisaster ochraceus* has great influence on the community composition of intertidal organisms on the Pacific coast of North America. When *Pisaster* is present in this community, species diversity is high, with several

FIGURE 6.12 A keystone is the wedge-shaped stone at the top of an arch that holds its structure together (**a**). A keystone species, such as the sea otter, is one that exerts great influence on a community's composition and structure (**b**). Sea otters consume sea urchins that eat kelp in marine nearshore environments of the Pacific. When otters are present, they keep urchin numbers down, which allows lush underwater forests of kelp to grow and provide habitat for many other species. When otters are absent, urchin populations increase and the kelp is devoured, destroying habitat and depressing species diversity. See "The Science behind the Story," ▶pp. 164–165.

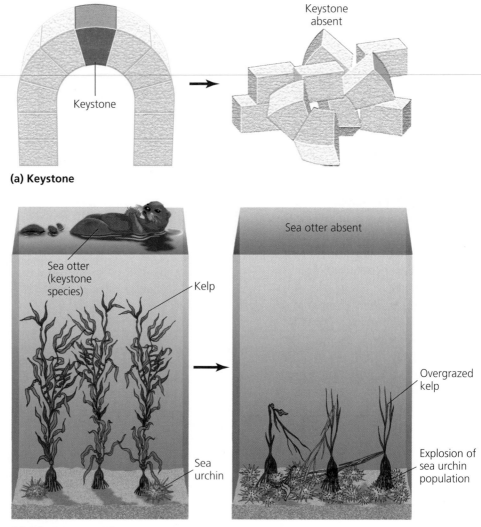

(a) Keystone

(b) A keystone species

types of barnacles, mussels, and algae. When *Pisaster* is removed, the mussels it preys on become numerous and displace other species, suppressing species diversity.

Animals at high trophic levels, such as wolves, starfish, and sea otters (Figure 6.12), are most often seen as keystone species. But other species attain keystone status as "ecosystem engineers" by physically modifying the environment shared by community members. Beavers build dams and turn streams into ponds, flooding acres of dry land and turning them to swamp. Prairie dogs dig burrows that aerate the soil and serve as homes for other animals. Less conspicuous organisms and those toward the bottoms of food chains can potentially be viewed as keystone species, too. Remove the fungi that decompose dead matter, or the insects that control plant growth, or the phytoplankton that are the base of the marine food chain, and a community may change very rapidly indeed. Because there are usually more species at lower trophic levels, however, it is less

likely that any one of them alone might have wide influence; if one species is removed, other species that remain may be able to perform many of its functions.

Identifying keystone species is no simple task, and there is no cut-and-dried definition of the term to help us. Community dynamics are complex, species interactions differ in their strength, and the strength of species interactions can vary through time and space. "The Science behind the Story" (▶pp. 164–165) gives an idea of the surprises that are sometimes in store for ecologists studying these interactions.

Weighing the **Issues:**
Keystone Species and Conservation

Imagine the government is funding a development project in your town and is gathering citizen input on three options. The environmental impact statement (EIS) (▶p. 68) states that option 1 would likely result in the

Oaks, hardwoods

Pines

Shrubs, poplar trees

Shrubs, seedlings

Grasses, herbs, forbs

Time ⟶

FIGURE 6.13 Secondary succession in a terrestrial setting occurs after a disturbance, such as fire, landslides, or farming, removes most vegetation from an area. Here is shown a typical series of changes in a plant community of eastern North America following the abandonment of a farmed field.

extermination of bobcats, a tertiary consumer in the community. Option 2, the EIS predicts, would probably kill off a species of pocket mouse, a primary consumer that is common in the community. Option 3 would likely eliminate a species of lupine, a plant that covers a large percentage of the ground in the present community.

You are a citizen desiring minimal change in the natural community so that your children grow up in an area like the one in which you grew up. What questions would you ask of an ecologist about the bobcat, pocket mouse, and lupine so that you could decide which might most likely be a keystone species? If you instead had to provide input without further information, what would you advise the government?

Communities respond to disturbance in different ways

The removal of a keystone species and the spread of an invasive species are just two of many types of disturbance that can modify the composition, structure, or function of an ecological community. Over time, any given community may experience natural disturbances ranging from gradual phenomena such as climate change to sudden events such as hurricanes, floods, or avalanches.

Communities are dynamic systems and may respond to disturbance in several ways. A community that resists change and remains stable despite disturbance is said to show **resistance** to the disturbance. Alternatively, a community may show **resilience,** meaning that it changes in response to disturbance but later returns to its original state. Or, a community may be modified by disturbance permanently and may never return to its original state.

Succession follows severe disturbance

If a disturbance is severe enough to eliminate all or most of the species in a community, the affected site will undergo a somewhat predictable series of changes that ecologists call **succession.** In the traditional view of this process, ecologists described two types of succession. **Primary succession** follows a disturbance so severe that no vegetation or soil life remains from the community that occupied the site. In primary succession, a biotic community is built essentially from scratch. In contrast, **secondary succession** begins when a disturbance dramatically alters an existing community but does not destroy all living things or all organic matter in the soil. In secondary succession (Figure 6.13), vestiges of the previous community remain, and these building blocks help shape the process.

At terrestrial sites, primary succession takes place after a bare expanse of rock, sand, or sediment becomes newly exposed to the atmosphere. This can occur when glaciers retreat, lakes dry up, or volcanic lava flows spread across the landscape. Species that arrive first and colonize the new substrate are referred to as **pioneer species.** Pioneer species are well adapted for colonization, having traits such as spores or seeds that can travel long distances. The pioneers best suited to colonizing bare rock are the mutualistic aggregates of fungi and algae known as *lichens.* Lichens succeed because their algal component provides food and energy via photosynthesis while the fungal component takes a firm hold on rock and captures the moisture that both organisms need to survive. As lichens grow, they secrete acids that break down the rock surface. The resulting waste material forms the beginnings of soil, and once soil begins to form, small plants, insects, and worms find the rocky outcrops more hospitable. As new organisms arrive,

Otters, Urchins, Kelp, and a Whale of a Chain Reaction

The Science behind the Story

Ecologists required years of careful study to comprehend the relationship among sea otters, sea urchins, and kelp forests diagrammed in Figure 6.12. And even after they thought they understood it, some surprises were in store.

Sea otters live in coastal waters of the Pacific Ocean. These mammals float on their backs amid the waves, feasting on sea urchins that they pry from the bottom and bring to the surface. Once abundant, sea otters were hunted nearly to extinction for their fur. Their protection by international treaty in 1911 allowed their numbers to regrow. Otters returned to high densities in some regions (such as off Alaska and the Aleutian Islands), but failed to return at all in others.

Biologists noted that regions with otters hosted dense "forests" of kelp. Kelp is a brown alga (seaweed) that anchors to the seafloor and reaches toward sunlit surface waters, growing up to 60 m (200 ft) high. Kelp forests provide complex physical structure in which diverse communities of fish and invertebrates find shelter and food.

In regions without sea otters, scientists found kelp forests absent. Urchins in such areas become so numerous that they may eat every last bit of kelp, creating empty seafloors called "urchin barrens" that are relatively devoid of life.

Ecologists observed that once areas with urchin barrens were recolonized by sea otters, urchin numbers declined and kelp forests returned. Through comparative research of this kind, ecologists determined that otters were largely responsible for the presence of the kelp forest community, simply by keeping urchin numbers in check through predation. This research—mostly by James Estes of the University of California at Santa Cruz and his colleagues—established sea otters as a prime example of a keystone species.

But the story did not end there. In the 1990s, otter populations dropped precipitously in areas off Alaska and the Aleutians. No one knew why. So, Estes and his co-workers placed radio tags on Aleutian otters and studied them at sea. Their first hypothesis was that fertility rates had dropped, but the radio-tracking observations showed that females were raising pups without problem. They then tested a second hypothesis, that otters were simply moving to other locations. But the radio tracking showed no unusual dispersal, and the population declines were taking place evenly over large areas. They were left with one viable hypothesis: increased mortality.

Biologists were finding no evidence of disease or starvation, however. Then one day in 1991, Estes's team witnessed something never seen before. They watched as a sea otter was killed and eaten by an orca, or killer whale. These striking black-and-white predators grow up to 10 m (32 ft) long, hunt in groups, and had always attacked larger prey. A sea otter to them is a mere snack. Yet over the following years, Estes's team saw nine more cases of orca predation on otters. Could killer whales be killing off the otters?

The researchers calculated the hours they had spent studying otters at sea before and after 1991 to assess the likelihood that they had simply missed seeing predation by orcas earlier. Their statistical analysis convinced them the rate of attacks had indeed increased. Then they compared a bay where otters were vulnerable to orcas with a lagoon where they were protected. Otter numbers in the lagoon remained stable over 4 years, whereas those in the bay dropped by 76%. Radio tracking showed no movement between these locations.

Finally, using data on otter birth rates, death rates, and population age structure (▶p. 134), they estimated that to account for the otter decline, 6,788 orca attacks per year would have had to occur in their study area. This produced an *expected* rate of observed attacks that matched their *actual* number of observed attacks well. These lines of evidence led the researchers to propose that predation by orcas was eliminating otters.

As otters declined in the Aleutians, urchins increased, and kelp

---

they provide more nutrients and habitat for future arrivals. As time passes, larger plants establish themselves, the amount of vegetation increases, and species diversity rises.

Secondary succession on land begins when a fire, a hurricane, logging, or farming removes much of the bi-

otic community. Consider a farmed field in eastern North America that has been abandoned. In the first few years after farming ends, the site will be colonized by pioneer species of grasses, herbs, and forbs that were already in the vicinity and that disperse effectively. As

density fell dramatically (see the figure). These changes supported the idea that otters were keystone species, but now it seemed that one keystone species was being controlled by another.

Why had orcas suddenly started eating otters? A possible answer came in 2003, after Alan Springer, Estes, and others determined that sea otters were only the latest in a series of population crashes in the northern Pacific. Harbor seals had declined by more than 90% in the late 1970s and 1980s. Fur seals had fallen by 60% since the 1970s. Sea lions crashed by 80% in the 1980s. And preceding these declines were the collapses of the great whales—gray whales, blue whales, humpback whales, and others.

Historically, most orcas specialized on eating great whales, which they would kill in groups, like a wolf pack taking down an elk. But industrial whaling by ships from Japan, Russia, and other nations since the 1950s caused populations of great whales to plummet by 99% between 1965 and 1973 in the northern Pacific. Once human hunting decimated the great whales, the orcas had to turn elsewhere. Springer, Estes, and their colleagues argued that they had shifted their diet to smaller, less-favored seals and sea lions. When their predation had depleted those populations, the chain reaction continued, and they turned to smaller prey still—sea otters.

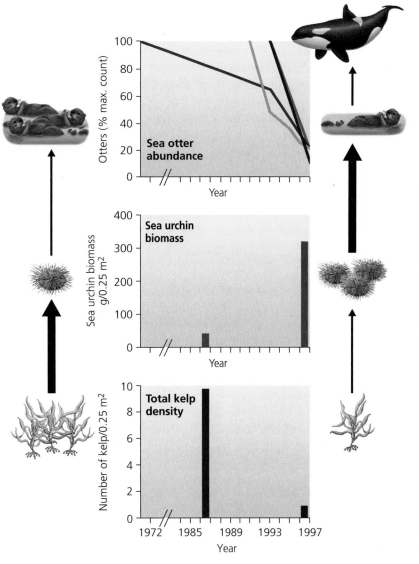

Before the 1990s (left side of figure), otters kept urchin numbers in check, allowing kelp forests to grow. By the end of the 1990s (right side of figure), orcas (killer whales), deprived of their usual food sources, were eating otters. This set off a chain reaction across several trophic levels: Otters decreased, sea urchins increased, and kelp decreased. The lines in the top graph indicate trends in otter populations from four different Aleutian islands. Width of arrows indicates strength of interaction. *Source:* Estes, J., et al. 1998. Killer whale predation on sea otters linking oceanic and nearshore ecosystems. *Science* 282: 473–476.

time passes, shrubs and fast-growing trees such as aspens rise from the field. Pine trees subsequently rise above the aspens and shrubs, forming a pine-dominated forest. This pine forest develops an understory of hardwood trees, because pine seedlings do not grow well under mature pines, whereas some hardwood seedlings do. Eventually the hardwoods outgrow the pines, creating a hardwood forest (see Figure 6.13).

Succession also occurs in aquatic systems. A lake or pond that originates as nothing but water on a lifeless

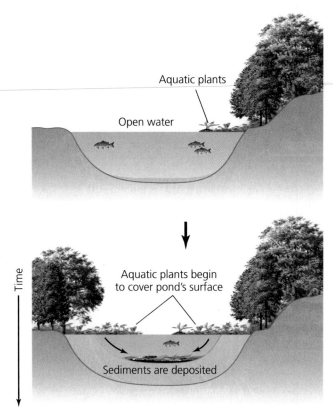

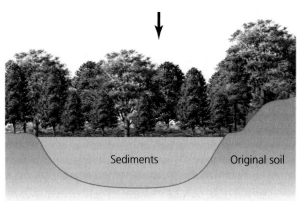

FIGURE 6.14 Primary aquatic succession occurs when plant growth gradually fills in a pond or lake and converts an aquatic system to a wet meadow and ultimately to a terrestrial system. Increased nutrient input can accelerate this process.

substrate begins to undergo succession as it is colonized by algae, microbes, plants, and zooplankton. As these organisms grow, reproduce, and die, the water body slowly fills with organic matter. The lake or pond acquires further organic matter and sediments from the water it receives from rivers, streams, and surface runoff. Eventually, the water body fills in and undergoes a gradual transition to a terrestrial system (Figure 6.14).

In this traditional view of succession that we have described, the transitions between stages of succession

eventually lead to a *climax community,* which remains in place, with little modification, until some disturbance restarts succession. Early ecologists felt that each region had its own characteristic climax community, determined by the region's climate.

Today, ecologists recognize that succession is far more variable and less predictable than originally thought. The trajectory of succession can vary greatly according to chance factors, such as which particular species happen to gain an early foothold. The stages of succession blur into one another and vary from place to place, and some stages may sometimes be skipped completely. In addition, climax communities are not predetermined solely by climate, but may vary with other conditions from one time or place to another. Once a climax community is disturbed and succession is set in motion, there is no guarantee that the community will ever return to that climax state. Many communities disturbed by human impact have not returned to their former conditions. This is the case with vast areas of the Middle East that once were fertile enough to support productive farming but now are deserts.

How cohesive are communities?

Ecologists who have studied how communities change in response to disturbance have conceptualized communities in different ways. Early in the 20th century, botanist Frederick Clements promoted the view that communities are cohesive entities whose members remain associated over time and space. Communities, he argued, are discrete units with integrated parts, much like organisms. Clements's view implied that the many varied members of a community share similar limiting factors and evolutionary histories.

Henry Gleason disagreed. Gleason, also a botanist, maintained that each species responds independently to its own limiting factors and that species can join or leave communities without greatly altering their composition. Communities, Gleason argued, are not cohesive units, but temporary associations of individual species that can reassemble into different combinations.

Today ecologists side largely with Gleason, although most see validity in aspects of both men's ideas. Indeed, many ecologists still find it useful to refer to communities by names that highlight certain key plants (such as "oak-hickory forest," "tallgrass prairie," and "pine-bluestem community"). Ecologists find labeling communities as though they were cohesive units to be a pragmatic tool, even though they know the associations could change radically decades or centuries hence.

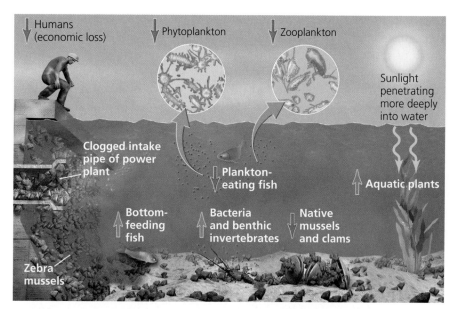

(a) Impacts of zebra mussels on members of a Great Lakes nearshore community

FIGURE 6.15 The zebra mussel is a prime example of a biological invader that has modified an ecological community. By filtering phytoplankton and small zooplankton from open water, it generates a number of impacts on other species, both negative (red downward arrows) and positive (green upward arrows) **(a)**. This map **(b)** shows the range of the zebra mussel in the United States as of 2005. In less than two decades it has spread from the Great Lakes east to Vermont and Connecticut; west to Nebraska and Kansas; and south to Louisiana and Mississippi (shaded states and red dots), and it has been transported by people to other states as well (blue dots). *Source:* (b) U.S. Geological Survey.

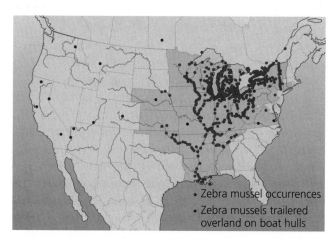

(b) Occurrence of zebra mussels in the United States, 2005

Invasive species pose new threats to community stability

Traditional concepts of community cohesion and of succession involve sets of organisms understood to be native to an area. But what if a new organism arrives from elsewhere? And what if this non-native organism turns *invasive,* spreading widely and becoming dominant in a community? Such **invasive species** can potentially alter a community substantially and are one of the central ecological forces in today's world.

Most often, invasive species are non-native species that people have introduced, intentionally or by accident, from elsewhere in the world. Species become invasive when limiting factors that regulate their population growth are removed. Many ecologists have suggested that

plants and animals brought to one area from another may leave their predators, parasites, and competitors behind and be freed from these constraints on their population growth. If there happen to be few organisms in the new environment that can act as predators, parasites, or competitors, the introduced species can do very well. As it proliferates, it may exert diverse influences on its fellow community members (Figure 6.15).

An example is the chestnut blight, an Asian fungus that killed nearly every mature American chestnut *(Castanea dentata),* the dominant tree species of many forests of eastern North America, in the quarter-century preceding 1930. Asian trees had evolved defenses against the fungus over long millennia of coevolution, but the American chestnut had not. A different fungus caused Dutch elm disease, destroying most of the American elms *(Ulmus americana)* that once gracefully lined the streets of many U.S. cities. Grasses introduced in the American West by ranchers have overrun entire regions, pushing out native vegetation. Fish introduced into streams for sport compete with and exclude native fish. Hundreds of island-dwelling animals and plants worldwide have been driven extinct by the goats, pigs, and rats introduced by human colonists. We will examine more examples in our discussion of biodiversity in Chapter 11 (▶pp. 320–322). The impact of invasive species on native species and ecological communities is severe already, and it is growing year by year with the increasing mobility of humans and the globalization of our society.

Our global trade helped spread zebra mussels, which were unintentionally transported in the ballast water of cargo ships. To maintain stability at sea, ships take water into their hulls as they begin their voyage and discharge that water at their destination. Decades of unregulated exchange of ballast water have ferried hundreds of species across the oceans.

In North America, zebra mussels—and the media attention they generated—helped put invasive species on the map as a major environmental and economic problem. Scientific research into introduced species proliferated, and many ecologists came to view invasive species as the second-greatest threat to species and natural systems, behind only habitat destruction. In 1990 the U.S. Congress passed the Nonindigenous Aquatic Nuisance Prevention and Control Act, which became the National Invasive Species Act of 1996. Among other things, this law directed the Coast Guard to ensure that ships dump their freshwater ballast at sea and exchange it with saltwater before entering the Great Lakes.

Since then, funding has become widely available for the control and eradication of invasive species. Managers have been trying a wide variety of techniques to control the spread of zebra mussels—removing them manually, applying toxic chemicals, drying them out, depriving them of oxygen, introducing predators and diseases, and stressing them with heat, sound, electricity, carbon dioxide, and ultraviolet light. However, most of these are localized and short-term fixes that are not capable of making a dent in the huge populations at large in the environment. In case after case, managers are finding that controlling and eradicating invasive species are so difficult and expensive that preventive measures (such as ballast water regulations) represent a much better investment.

Weighing the **Issues:**
Are Invasive Species All Bad?

Some ethicists have questioned the notion that all invasive species should automatically be considered bad. If we introduce a non-native species to a community and it greatly modifies the community, do you think that is a bad thing? What if it drives another species extinct? What if the invasive species arrived on its own, rather than through human intervention? What ethical standard(s) (▸ p. 30) would you apply to determine whether an invasive species should be battled or accepted?

Altered communities can be restored to their former condition

Invasive species are adding to the tremendous transformations that humans have already forced on natural landscapes and communities through habitat alteration, deforestation, hunting of keystone species, pollution, and other activities.

With so much of Earth's landscape altered by human impact, it is impossible to find areas that are truly pristine. This realization has given rise to the conservation effort known as **ecological restoration.** The practice of ecological restoration is informed by the science of **restoration ecology.** Restoration ecologists research the historical conditions of ecological communities as they existed before our industrialized civilization altered them. They then try to devise ways to restore some of these areas to an earlier condition, often to a natural "presettlement" condition.

For instance, in the United States nearly every last scrap of tallgrass prairie that once covered the eastern Great Plains and parts of the Midwest was converted to agriculture in the 19th century. Now a number of efforts are underway to restore small patches of prairie by planting native prairie plants, weeding out invaders and competitors, and introducing controlled fire to mimic the fires that historically maintained this community. Illinois boasts several of the largest prairie restoration projects so far. The Morton Arboretum in Lisle has a 40-ha (100-acre) prairie, and one at the Forest Glen Preserve includes more than 120 species of native prairie plants. A 184-ha (455-acre) area inside the massive ring of the Fermilab nuclear accelerator in Batavia is being restored, and plans are underway for a larger project at an old U.S. Army ammunition facility near Joliet.

Perhaps the world's largest restoration project is the ongoing effort to restore the Florida Everglades. The Everglades, a 7,500-km^2 (4,700-mi^2) ecosystem of marshes and seasonally flooded grasslands, has been drying out for decades because the water that feeds it has been managed for flood control and overdrawn for irrigation and development. Populations of wading birds have dropped 90–95%, and economically important fisheries have suffered greatly as a result. The 30-year, $7.8-billion restoration project intends to restore water by undoing damming and diversions of 1,600 km (1,000 mi) of canals, 1,150 km (720 mi) of levees, and 200 water control structures.

One of the most intriguing new restoration efforts is the drive to restore the Mesopotamian marshes between the Tigris and Euphrates Rivers in Iraq, formerly one of the world's greatest wetlands. The government of Saddam Hussein diverted water from the marshes in an effort to debilitate the minority peoples who had lived here for thousands of years. Following the U.S. occupation of Iraq, ecologists from many nations want to help restore the marshes and give their human residents a place to live again in their traditional lifestyle. Whether the funds, resources, and access are granted to allow this project to succeed remains to be seen. Regardless, the more our population grows and development spreads, the more ecological restoration will become a vital conservation strategy for the future.

2. Wolves were hunted nearly to extinction in the 1930s and were only reintroduced to Yellowstone in 1995. How, would you suspect, has their reintroduction affected scavenger populations since then? Why?

3. What effect would you predict that continued shorter, warmer winters would have on scavenger populations? Why? Are the predicted effects of wolf reintroduction and of climate change compounded, or do they tend to cancel one another out?

CALCULATING ECOLOGICAL FOOTPRINTS

Species appearing in a new area are generally called "invasive" if they increase markedly in population and also increase in their impacts on the biotic communities and landscapes around them. By these measures, are human beings an invasive species? The table below shows human population and per capita energy consumption for the United States across the time period of roughly two generations, from 1950 to 1975, and from 1975 to 2000. Total energy consumption gives us a very rough measure of total environmental impact. Calculate total energy consumption in the table by multiplying the population and per capita consumption values.

1. In percentage terms, how much did total energy consumption, as calculated here, increase between 1950 and 1975? Between 1975 and 2000?

2. What effects on biotic communities do you think this increase in total energy consumption has had? Speculate on overall effects, and give several specific known or likely examples.

3. After considering these data, do you consider yourself to be a member of an invasive species? Why or why not?

Year	Human population (U.S., in millions)	Per capita energy consumption (U.S., in million BTU)	Total energy consumption (U.S., in quadrillion BTU)
1950	151	229	
1975	216	334	
2000	281	352	

Data source: U.S. Census Bureau and U.S. Energy Information Administration.

Take It Further

 Go to www.aw-bc.com/withgott or the student CD-ROM where you'll find:

▶ Suggested answers to end-of-chapter questions
▶ Quizzes, animations, and flashcards to help you study
▶ *Research Navigator*™ database of credible and reliable sources to assist you with your research projects

▶ **GRAPHit!** tutorials to help you master how to interpret graphs
▶ **INVESTIGATEit!** current news articles that link the topics that you study to case studies from your region to around the world

7 Environmental Systems and Ecosystem Ecology

The Mississippi River as it enters the Gulf of Mexico

Upon successfully completing this chapter, you will be able to:

▶ Describe the nature of environmental systems

▶ Define ecosystems and evaluate how living and nonliving entities interact in ecosystem-level ecology

▶ Compare and contrast how carbon, phosphorus, nitrogen, and water cycle through the environment

▶ Explain how plate tectonics and the rock cycle shape the earth beneath our feet

Fisherman in the Gulf of Mexico's "dead zone"

United States

Mississippi River

Mexico

Gulf of Mexico

Central Case: The Gulf of Mexico's "Dead Zone"

"In nature there is no 'above' or 'below,' and there are no hierarchies. There are only networks nesting within other networks."
—FRITJOF CAPRA, THEORETICAL PHYSICIST

"Let's say you put Saran Wrap over south Louisiana and suck the oxygen out. Where would all the people go?"
—NANCY RABALAIS, BIOLOGIST FOR THE LOUISIANA UNIVERSITIES MARINE CONSORTIUM

Louisiana fishermen have long hauled in more seafood than those of any other U.S. state except Alaska. Each year they have plied the rich waters of the northern Gulf of Mexico and have sent nearly 600 million kg (1.3 billion lb) of shrimp, fish, and shellfish to our dinner tables. Then in 2005, Hurricane Katrina and Hurricane Rita pummeled the Gulf Coast and left Louisiana's fisheries in ruin. Boats, docks, marinas, and fueling stations were destroyed, and thousands were suddenly out of work. Today, fishermen are struggling to reestablish their livelihoods as the industry continues to rebuild itself.

But for years before the hurricanes hit, fishing had become increasingly difficult. In the words of longtime Louisiana fisherman Johnny Glover, it was "getting harder and harder to make a living." The reason? Each year billions of organisms were suffocating in the Gulf's "dead zone," a region of water so depleted of oxygen that marine organisms are killed or driven away.

The low concentrations of dissolved oxygen in the bottom waters of this region represent a condition called **hypoxia** (see "The Science behind the Story," ▶ pp. 202–203 and Figure 7.5, ▶ p. 189). Aquatic animals obtain oxygen by respiring through their gills, and, like us, these animals will asphyxiate if deprived of oxygen. Fully oxygenated water contains up to 10 parts per million (ppm) of oxygen, but when concentrations drop below 2 ppm, creatures that can leave an affected area will do so. Below 1.5 ppm, most marine organisms die. In the Gulf's hypoxic zone, oxygen concentrations frequently drop well below these levels.

The dead zone appears each spring and grows through the summer and fall, starting near the mouths of the Mississippi and Atchafalaya Rivers off the Louisiana coast. In 2002 the dead zone reached a record 22,000 km^2 (8,500 mi^2)—an area larger than New Jersey. Shrimp boats came up with nets nearly empty. One shrimper derided his meager catch as "cat food." Others, ironically, said they hoped a hurricane would strike and stir some oxygen into the Gulf's stagnant waters.

What's starving these waters of oxygen? Scientists studying the dead zone have identified modern Midwestern farm practices and other human impacts hundreds of kilometers away. The Gulf, they say, is being over-enriched by nitrogen and phosphorus flushed down the Mississippi River. This nutrient pollution comes from fertilizers used on farms far upstream in the Mississippi River basin, as well as from other sources, including urban runoff, industrial discharges, atmospheric deposition from fossil fuel combustion, and municipal sewage outflow.

The U.S. government has acted on these findings, proposing that farmers in states such as Ohio, Iowa, and Illinois cut down on fertilizer use. Farmers' advocates protest that farmers are being singled out while urban pollution sources are being ignored. Meanwhile, coastal dead zones have appeared in 150 other areas throughout the world, from Chesapeake Bay to Oregon to Denmark to the Black Sea. The story of how scientists have determined the causes of these dead zones involves understanding environmental systems and the often complex behavior they exhibit.

Earth's Environmental Systems

Our planet's environment consists of complex networks of interlinked systems. In the realm of community ecology (Chapter 6), these systems include the ecological webs of relationships among species. At the ecosystem level, they include the interaction of living species with the nonliving entities around them. Earth's systems also include cycles that guide the flow of key chemical elements and compounds that support life, regulate climate, and control other aspects of Earth's functioning. We depend on these systems for our very survival.

Assessing questions holistically by taking a "systems approach" is helpful in environmental science, in which so many issues are multifaceted and complex. Such an approach poses a challenge, however, because systems often show behavior that is difficult to understand and predict. The scientific method operates best when researchers isolate and manipulate small parts of complex systems, focusing in depth on manageable components. However,

environmental scientists increasingly are accepting the challenge of studying systems broadly and are beginning to find solutions to problems such as the Gulf of Mexico's hypoxic zone.

Systems show several defining properties

A **system** is a network of relationships among parts, elements, or components that interact with and influence one another through the exchange of energy, matter, or information. Systems receive inputs of energy, matter, or information, process these inputs, and produce outputs. Energy inputs to Earth's environmental systems include solar radiation as well as heat released by geothermal activity, organismal metabolism, and human activities such as fossil fuel combustion. Information inputs can come in the form of sensory cues from visual, olfactory (chemical), magnetic, or thermal signals. Inputs of matter occur when chemicals or physical material moves among systems, such as when seeds are dispersed long distances, migratory animals deposit waste far from where they consumed food, or plants convert carbon in the air to living tissue by photosynthesis. As a system, the Gulf of Mexico receives inputs of freshwater, sediments, nutrients, and pollutants from the Mississippi and other rivers. Shrimpers harvest some of the Gulf system's output: matter and energy in the form of shrimp. This output subsequently becomes input to the human economic system and to the digestive systems of the many people who consume the shrimp. Even greater than the output of shrimp is that of menhaden, a fish that supplies oil for many consumer products and is used to feed farmed animals.

Sometimes a system's output can serve as input to that same system, a circular process described as a **feedback loop.** Feedback loops are of two types, negative and positive. In a **negative feedback loop** (Figure 7.1a), output that results from a system moving in one direction acts as input that moves the system in the other direction. Input and output essentially neutralize one another's effects, stabilizing the system. A thermostat, for instance, stabilizes a room's temperature by turning the furnace on when the room gets cold and shutting it off when the room gets hot. Similarly, negative feedback regulates our body temperature. If we get too hot, our sweat glands pump out moisture that evaporates to cool us down, or we may move from sun to shade. If we get too cold, we shiver, creating heat, or we move into the sun or put on more clothing. Another example of negative feedback is a predator-prey system in which predator and prey populations rise and fall in response to one another (see Figure 6.5, ▶p. 153). Most systems in nature involve negative feedback

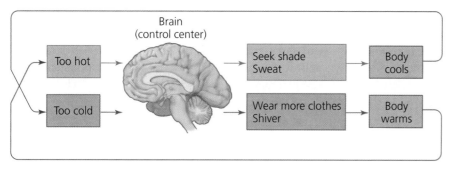

(a) Negative feedback

Negative feedback loops **(a)** exert a stabilizing influence on systems and are common in nature. The human body's response to heat and cold involves a negative feedback loop. Positive feedback loops **(b)** have a destabilizing effect on systems and push them toward extremes. Rare in nature, they are common in natural systems altered by human impact. The clearing of forested land, for instance, can lead to a runaway process of soil erosion.

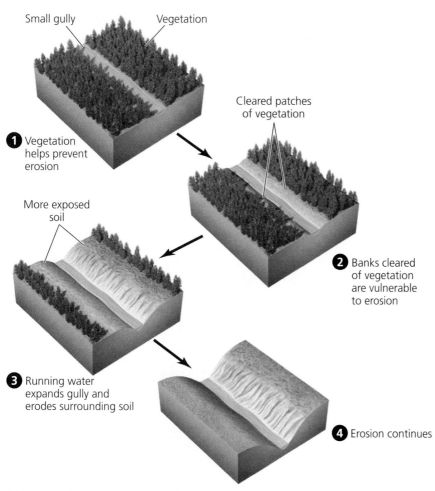

(b) Positive feedback

loops. Negative feedback loops enhance stability, and in the long run, only those systems that are stable will persist.

Positive feedback loops have the opposite effect. Rather than stabilizing a system, they drive it further toward one extreme or another. Exponential growth in human population (▶ pp. 218–219) provides an example. The more people are born, the more there are to give birth to further people; increased output leads to increased input, leading to further increased output. Another example is the spread of cancer; as cells multiply out of control, the process is self-accelerating. Positive feedback can also occur with the process of erosion, the removal of soil by water or wind (▶ pp. 257–258). Once vegetation has been cleared to expose soil, erosion may become progressively more severe if the forces of water or wind surpass the rate of vegetative regrowth. Water flowing through an eroded gully may expand the gully and lead to further erosion (Figure 7.1b). Positive feedback can alter a system substantially. Positive feedback loops are rare in nature, but they are common in natural systems altered by human impact.

The inputs and outputs of complex natural systems usually occur simultaneously, keeping the system constantly active. Earth's climate system, for instance, does not ever stop. When processes within a system move in opposing directions at equivalent rates so that their effects balance out, the process is said to be in **dynamic equilibrium.** Processes in dynamic equilibrium can contribute to **homeostasis,** the tendency of a system to maintain constant or stable internal conditions. When homeostasis exists, organisms and other systems can keep their internal conditions within a range that allows them to function. Homeostatic systems are often thought of as being in a steady state; however, the steady state itself may change slowly over time while the system maintains its ability to stabilize conditions internally. For instance, organisms grow and mature. Similarly, Earth has experienced a gradual increase in atmospheric oxygen over its history (▶p. 108), yet life has adapted, and Earth remains, by most definitions, a homeostatic system.

Often it is difficult to understand systems fully by focusing on their individual components because systems can show **emergent properties,** characteristics not evident in the components alone. Stating that systems possess emergent properties is a lot like saying, "The whole is more than the sum of its parts." For example, if you were to reduce a tree to its component parts (leaves, branches, trunk, bark, roots, fruit, and so on) you would not be able to predict the whole tree's emergent properties, which include the role the tree plays as habitat for birds, insects, parasitic vines, and other organisms (Figure 7.2). You could analyze the tree's chloroplasts (photosynthetic cell organelles), diagram its branch structure, and evaluate its fruit's nutritional content, but you would still be unable to understand the tree as habitat, as part of a forest landscape, or as a reservoir for carbon storage.

Systems seldom have well-defined boundaries, so deciding where one system ends and another begins can be difficult. Consider a desktop computer system. It is

FIGURE 7.2 A system's emergent properties are not evident when we break the system down into its component parts. For example, a tree serves as wildlife habitat and plays roles in forest ecology and global climate regulation, but you would not know that from considering the tree only as a collection of leaves, branches, and chloroplasts. If we try to understand systems solely by breaking them into component parts, we will miss much of what makes them important.

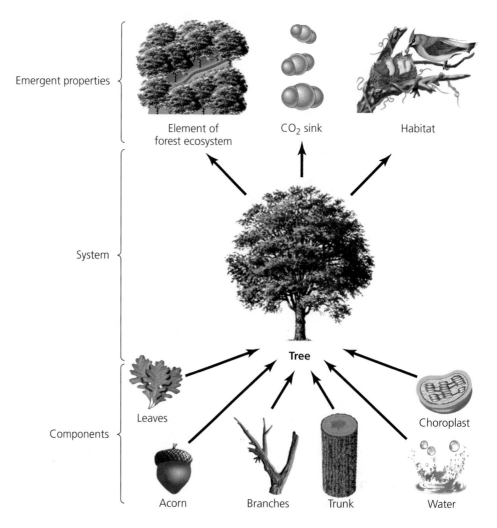

Emergent properties

Element of forest ecosystem

CO_2 sink

Habitat

System

Tree

Components

Leaves

Acorn

Branches

Trunk

Choroplast

Water

certainly a network of parts that interact and exchange energy and information, but what are its boundaries? Is the system what arrives in a packing crate and sits on top of your desk? Or does it include the network you connect it to at school, home, or work? What about the energy grid you plug it into, with its distant power plants and transmission lines? And what of the Internet? Browsing the Web, you are drawing in digitized text, light, and sound from around the world. No matter how we attempt to isolate or define a system, we soon see that it has many connections to systems larger and smaller than itself. Systems overlap, and one may be contained within others, so where we draw boundaries may depend on the spatial or temporal scale on which we choose to focus.

Scientists will often treat a system as if it is a **closed system,** one that is isolated and self-contained, for the purpose of simplifying some problem with which they are grappling. However, no matter how closed a system might seem, if we look closely enough or wait long enough, we will detect interactions with other systems. Thus, when viewed in context, all systems are **open systems,** exchanging energy, matter, and information with other systems. Our planet is clearly an open system in regard to energy because it receives sunlight continuously and emanates heat into space. In regard to physical matter, Earth is largely a closed system, but comets, asteroids, and meteors do introduce small amounts of material every day.

Understanding the dead zone requires considering Mississippi River and Gulf of Mexico systems together

The Gulf of Mexico and the Mississippi River are systems that interact with one another. On a map, the Mississippi River appears as a branched and braided network of water channels. But where are this system's boundaries? You might argue that the Mississippi consists primarily of water, originates in Minnesota, and ends in the Gulf of Mexico near New Orleans. But what about the rivers that feed it and the farms, cities, and forests that line its banks (Figure 7.3)? Major rivers such as the Missouri, Arkansas, and Ohio flow into the Mississippi. Hundreds of smaller tributaries drain vast expanses of farmland, woodland, fields, cities, towns, and industrial areas before their water joins the Mississippi's. These waterways carry with them millions of tons of sediment, hundreds of species of plants and animals, and numerous pollutants.

For an environmental scientist interested in runoff and the flow of water, sediment, or pollutants, it may make best sense to view the Mississippi River's watershed as a system. As with the Tijuana River in Chapter 3, one must consider the entire area of land a river drains to comprehend and solve problems of river pollution. However, for a scientist interested in the Gulf of Mexico's dead zone, it may make best sense to view the Mississippi River watershed together with the Gulf as the

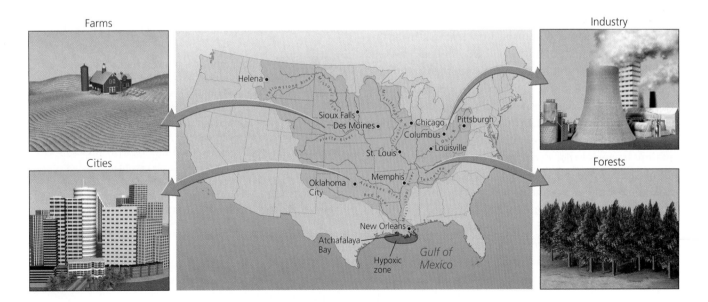

FIGURE 7.3 The Mississippi River system is the largest in North America. The river's watershed encompasses 3.2 million km^2 (1.2 million mi^2), or 41% of the area of the lower 48 U.S. states. The river carries water, sediment, and pollutants from a variety of sources downriver to the Gulf of Mexico, where nutrient pollution has given rise to a hypoxic zone.

system of interest because their interaction is central to the problem. In environmental science, one's delineation of a system can and should depend on the questions one is addressing.

The reason for the Gulf of Mexico's dangerously low levels of oxygen, scientists have concluded, is abnormally high levels of nutrients such as nitrogen and phosphorus. The excess nutrients originate with sources in the Mississippi River watershed, particularly nitrogen- and phosphorus-rich fertilizers applied to crops. Inorganic nitrogen fertilizers used on Midwestern farms account for roughly 30% of total nitrogen contributions to the river, an amount equal to that from natural soil decomposition. The remainder comes from various sources, including animal manure, nitrogen-fixing crops, sewage treatment facilities, street runoff, and industrial and automobile emissions. Altogether, agricultural sources are thought to contribute 74% of the nitrate and 65% of the total nitrogen carried in the river. Much of the nitrate originates in the upper portions of the watershed (Figure 7.4a) from farms growing corn and soybeans in Iowa, Illinois, Indiana, Minnesota, and Ohio. Nitrogen fertilizer input to farmland in the Mississippi River watershed has increased dramatically since 1950 (Figure 7.4b). Since 1980, the Mississippi River and the Atchafalaya River (which drains a third of the Mississippi's water through a second delta) have pumped about 1 million metric tons of nitrate into the Gulf of Mexico each year, three times as much as during the 1960s.

The enhanced nitrogen input to the Gulf boosts the growth of *phytoplankton* (▶ p. 150), microscopic photosynthetic algae, protists, and cyanobacteria that drift near the surface. Phytoplankton ordinarily are limited in their growth by scarcity of nutrients such as nitrogen and phosphorus. As phytoplankton flourish at the surface and as zooplankton (▶ p. 150) consume them, more dead phytoplankton and waste products of phytoplankton and zooplankton drift to the bottom, providing food for organisms (mainly bacteria) that decompose them. The result is a population explosion of bacteria. These decomposers consume enough oxygen to cause oxygen concentrations in bottom waters to plummet, suffocating shrimp and fish that live at the bottom, and creating the dead zone. The fresh water from the river remains naturally stratified in a layer at the surface that mixes only very slowly with the denser salty ocean water, so that oxygenated surface water does not make its way down to the bottom-dwelling life that needs it. This process of nutrient over-enrichment, blooms of algae, increased production of organic matter, and subsequent ecosystem degradation is known as **eutrophication** (Figure 7.5). As we will see in

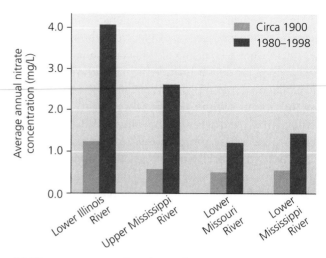

(a) Nitrate concentrations in portions of the Mississippi River watershed

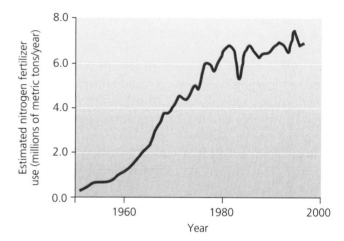

(b) Nitrogen use in the Mississippi-Atchafalaya River basin

FIGURE 7.4 Concentrations of nitrogen in the Mississippi and its tributaries rose during the 20th century, especially in the upper portion of the watershed (**a**). Orange bars show average concentrations from circa 1900 (based on more recent analysis of sediments), and red bars show average concentrations measured from 1980 to 1998. Usage of nitrogen-based fertilizer in the Mississippi-Atchafalaya River basin skyrocketed after 1950 (**b**). About 15% of nitrogen from fertilizer applied to farms eventually reaches rivers that flow to the Gulf. Data from National Science and Technology Council: Committee on Environment and Natural Resources. May 2000. Hypoxia: An integrated assessment in the northern Gulf of Mexico.

Chapters 15 and 16, eutrophication can take place in both freshwater and saltwater systems.

Moderate amounts of additional nutrients may increase the productivity of fisheries, but at higher concentrations, this fertilizing effect is offset by hypoxia, and fishery yields decline.

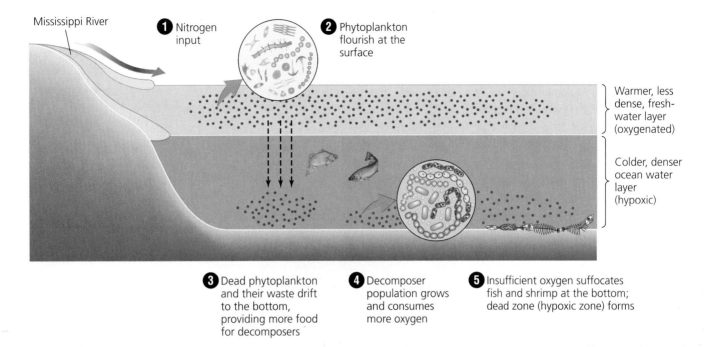

Mississippi River

1 Nitrogen input

2 Phytoplankton flourish at the surface

Warmer, less dense, fresh-water layer (oxygenated)

Colder, denser ocean water layer (hypoxic)

3 Dead phytoplankton and their waste drift to the bottom, providing more food for decomposers

4 Decomposer population grows and consumes more oxygen

5 Insufficient oxygen suffocates fish and shrimp at the bottom; dead zone (hypoxic zone) forms

FIGURE 7.5 Excess nitrogen causes eutrophication in coastal marine systems such as the Gulf of Mexico. Coupled with stratification (layering) of water, eutrophication can severely deplete dissolved oxygen. Nitrogen from river water (1) boosts growth of phytoplankton (2), which die and are decomposed at the bottom by bacteria (3). Stability of the surface layer prevents deeper water from absorbing oxygen to replace that consumed by decomposers (4), and the oxygen depletion suffocates or drives away bottom-dwelling marine life (5). This process gives rise to hypoxic zones like that of the Gulf of Mexico.

Environmental systems may be perceived in various ways

There are many ways to delineate natural systems, and your choice will depend on the particular issues in which you are interested. Categorizing environmental systems can help us make Earth's dazzling complexity comprehensible to the human brain and accessible to problem solving. However, a fundamental insight of environmental science is that systems overlap and interact.

For instance, scientists sometimes divide Earth's components into structural spheres. The **lithosphere** is the rock and sediment beneath our feet, in the planet's uppermost layers. The **atmosphere** is composed of the air surrounding our planet. The **hydrosphere** encompasses all water—salt or fresh, liquid, ice, or vapor—in surface bodies, underground, and in the atmosphere. The **biosphere** consists of the total of all the planet's living organisms and the abiotic (nonliving) portions of the environment with which they interact. We will examine the workings of various environmental systems within these structural spheres in more detail in later chapters.

Although these categories can be useful, their boundaries overlap, so the systems interact. Picture a robin plucking

an earthworm from the ground after a rain. You are witnessing an organism (the robin) consuming another organism (the earthworm) by removing it from part of the lithosphere (soil) that the earthworm had been modifying—all this made possible because rain (from the hydrosphere) recently wet the ground. The robin might then fly through the air (the atmosphere) to a tree (an organism), in the process respiring (combining oxygen from the atmosphere with glucose from an organism, and adding water to the hydrosphere and carbon dioxide and heat to the atmosphere). Finally, the bird might defecate, adding nutrients from an organism to the lithosphere below. The study of such interactions among living and nonliving things is a key part of ecology at the ecosystem level. As scientists become more inclined to approach systems holistically, ecology at the ecosystem level and beyond has become increasingly important.

Ecosystems

An **ecosystem** consists of all organisms and nonliving entities that occur and interact in a particular area at the same time. The ecosystem concept builds on the idea of

The Science behind the Story

Biosphere 2

In September 1991, eight people and nearly 3,800 species of plants and animals were sealed within Biosphere 2, a collection of airtight, interconnected domes spanning more than 1.2 hectares (3.0 acres) in the Arizona desert (see figure). The goal of the biospherians, as they were known, was to survive for two years within a self-contained ecosystem that might someday be used to colonize other planets, while also learning about environmental processes on Earth. Only nine months later, however, oxygen levels in Biosphere 2's artificial atmosphere began to drop at an alarming rate. Within 18 months, the biospherians were literally gasping for breath.

The near-failure of Biosphere 2's life-support system was not due to a lack of data. Scattered throughout Biosphere 2's ocean, rainforest, savanna, and desert biomes were more than 1,000 sensors that tracked day-to-day changes in oxygen, carbon dioxide, temperature, pH, and other environmental variables. Nonetheless, despite a sophisticated

Within 18 months of being sealed off from Earth's atmosphere, Biosphere 2 had oxygen levels of less than 14%—close to the lowest levels at which humans can survive.

computer monitoring system and the help of external advisers, the biospherians, only some of whom were scientists, were unable to locate the missing oxygen. In desperation, they called on Wallace S. Broecker, a geochemist at Columbia University's Lamont-Doherty Earth Observatory.

Broecker and graduate student Jeff Severinghaus quickly ruled out the possibility that the biospherians themselves were consuming the oxygen. The amount disappearing—roughly 450 kg (1,000 lb) of oxygen per month—was far larger than they, or any of the project's large animals,

the biological community (Chapter 6), but ecosystems include abiotic components as well as biotic ones. In ecosystems, energy flows and matter cycles among these living and nonliving components.

Ecosystems are systems of interacting living and nonliving entities

The idea of ecosystems originated early last century with people such as British ecologist Arthur Tansley, who saw that biological entities are tightly intertwined with chemical and physical entities. Tansley and others felt that there was so much interaction and feedback between organisms and their abiotic environments that it made most sense to view living and nonliving elements together. For instance,

the input of moisture from clouds plays a key role in the Monteverde cloud forest ecosystem we visited in Chapter 5. The flow of water, sediment, and nutrients from the Mississippi River plays a key role in the nearshore ecosystem of the northern Gulf of Mexico. And as we saw in our discussion of biomes in Chapter 6 (▶ pp. 170–177), abiotic factors such as temperature, precipitation, latitude, and elevation have substantial influence over which species and communities exist in a given locality.

Ecologists soon began analyzing ecosystems as an engineer might analyze the operation of a machine. In this view, ecosystems are systems that receive inputs of energy, process and transform that energy while cycling matter internally, and produce a variety of outputs (such as heat, water flow, and animal waste products) that can move

could be using. So, they focused their investigation on Biosphere 2's 30,000 tons of soil, which had an extraordinarily high percentage— nearly 30%—of organic matter. They determined that microbes in the soil were converting unexpectedly large amounts of oxygen to carbon dioxide. Still, one question remained: If the microbes were responsible for the severe decrease in oxygen, then CO_2 levels in the atmosphere should have skyrocketed, when instead the opposite had occurred. Broecker and Severinghaus suspected that the missing carbon dioxide was being stored in the soil itself.

On close examination, however, that proved not to be the case, so the researchers turned their attention to another possible culprit: the exposed concrete supporting the building's glass and metal shell. In Earth's atmosphere, they knew, concrete can react with carbon dioxide to form a solid substance called calcium carbonate. Usually, the reaction takes place only in a thin outer layer of exposed concrete, which is what the designers of

Biosphere 2 had expected. Concentrations of CO_2 in Biosphere 2's atmosphere, however, ranged 3–10 times higher than Earth's, and the researchers found that carbon dioxide, converted to calcium carbonate, had been deposited up to 15 cm (6 in.) deep in the concrete walls.

The most effective solution would have been to replace all 30,000 tons of soil, but that would have cost millions of dollars and set the project back several years. Instead, Biosphere 2's management team settled on two stopgap measures. First, to address the urgent need for oxygen, they injected more than 23 tons of pure oxygen gas. Then, to minimize calcium carbonate deposits, they covered the exposed concrete with a layer of paint. With the most pressing problem solved, at least temporarily, the biospherians were able to focus on other issues, such as invasive species. Aquatic fire worms were preying on coral, hundred-foot-long morning glory vines were overwhelming the rainforest, and crazy ants—an aggressive species

unintentionally included in the project—had spread throughout Biosphere 2 and driven most of its other insect species extinct.

Nonscientific problems also plagued the project. By the time the biospherians emerged from their two-year seclusion in September 1993, the project had become embroiled in accusations of mismanagement and scientific fraud. In 1994, its financial backer, Texas oil billionaire Edward P. Bass, dismissed the project's management team and invited a panel of independent scientists to assess its future.

In 1996, Columbia University began the difficult process of transforming Biosphere 2 into a center for research and education on global climate change. Despite some successes, the project remained troubled, and Columbia eventually backed out. Whatever the future of Biosphere 2 may hold, the lesson of its first mission is clear: Creating self-contained ecosystems from scratch is no easy task. We're better off preserving the ones we have.

---

into other ecosystems. Energy flows in one direction through ecosystems; most arrives as radiation from the sun, powers the system, and exits in the form of heat. Matter, in contrast, is generally recycled within ecosystems. We saw in Chapter 6 (▶pp. 156–159) how energy and matter are passed from one organism to another through food web relationships, from producers to primary consumers to secondary consumers. Matter is recycled because when organisms die and decay their nutrients remain in the system. In contrast, most energy that organisms gain is lost during their lifetimes through respiration.

Ecosystems can be conceptualized at different scales. An ecosystem can be as small as an ephemeral puddle of water where brine shrimp and tadpoles feed on algae and detritus with mad abandon as the pool dries up. Or an

ecosystem might be as large as a lake or a forest. For some purposes, scientists even view the entire biosphere as a single all-encompassing ecosystem (see The Science behind the Story, above). The term is most often used, however, to refer to systems of moderate geographic extent that are somewhat self-contained. For example, Monteverde's mountaintop system is delineated by drier forests and cleared agricultural land on its lower slopes.

Landscape ecologists study geographic areas with multiple ecosystems

Ecosystems are open systems, and those that physically abut one another may interact extensively, no matter how distinctly different they may appear. For instance, a pond ecosystem is very different from a forest ecosystem that surrounds it, but the two interact. Salamanders that develop in the pond live their adult lives under logs on the forest floor until returning to the pond to breed. Rainwater that nourishes forest plants may eventually make its way to the pond, carrying with it nutrients from the forest's leaf litter. Similarly, coastal dunes, the ocean, and the lagoon or salt marsh between them all interact, as do forests and prairie where they converge. Areas where ecosystems meet may consist of transitional zones called **ecotones,** in which elements of each ecosystem mix.

Because of this mixing, sometimes ecologists find it useful to view these systems at a larger landscape scale that focuses on geographic areas that include multiple ecosystems. For instance, if you are interested in large mammals, such as black bears, that move seasonally from mountains to valleys or move between mountain ranges, you had better consider the landscape scale that includes all these habitats. Such a broad-scale approach, often called **landscape ecology,** is important in studying birds that migrate between continents, or fish, such as salmon, that move between marine and freshwater ecosystems. Some conservation groups, such as the Nature Conservancy, have begun applying the landscape ecology approach widely in their land management strategies.

Weighing the Issues:
Ecosystems Where You Live

Think about the area where you live. How would you describe that area's ecosystems? How do these systems interact with one another? If one ecosystem were greatly disturbed (say, if a wetland or forest were replaced by a shopping mall), what impacts might that have on nearby natural systems?

Energy is converted to biomass

Regardless of scale, the energy flow in most ecosystems begins with radiation from the sun. In Chapter 4 (▶pp. 105–107) we explored how autotrophs, such as green plants and phytoplankton, use photosynthesis to capture the sun's energy and produce food. We then saw in Chapter 6 (▶pp. 156–159) how organisms at higher trophic levels consume organisms at lower trophic levels, transferring energy.

As autotrophs convert solar energy to the energy of chemical bonds in sugars, they perform *primary production*. Specifically, the assimilation of energy by autotrophs is termed **gross primary production.** Autotrophs must use a portion of this production to power their own

metabolism by respiration (▶ p. 107). The energy that remains after respiration and is used to generate biomass ecologists call **net primary production.** Thus, net primary production equals gross primary production minus respiration. Net primary production can be measured by the energy or the organic matter stored by autotrophs after they have metabolized enough for their own maintenance.

Another way to think of net primary production is that it represents the energy or biomass available for consumption by heterotrophs. Plant matter not eaten by herbivores becomes fodder for detritivores and decomposers. Heterotrophs use the energy they gain from plants for their own metabolism, growth, and reproduction. The total biomass that heterotrophs generate by consuming autotrophs is termed *secondary production*.

Ecosystems vary in the rate at which autotrophs convert energy to biomass. The rate at which production occurs is termed **productivity,** and ecosystems whose producers convert solar energy to biomass rapidly are said to have high **net primary productivity.** Freshwater wetlands, tropical forests, and coral reefs and algal beds tend to have the highest net primary productivities, whereas deserts, tundra, and open ocean tend to have the lowest (Figure 7.6a). Variation among ecosystems and among biomes in net primary productivity results in geographic patterns of variation across the globe (Figure 7.6b). In terrestrial ecosystems, net primary productivity tends to increase with temperature and precipitation. In aquatic ecosystems, net primary productivity tends to rise with light and the availability of nutrients.

Nutrients can limit ecosystem productivity

Nutrients are elements and compounds that organisms consume and require for survival. Organisms need several dozen naturally occurring chemical elements to survive. Elements and compounds required in relatively large amounts are called *macronutrients* and include such nutrients as nitrogen, carbon, and phosphorus. Nutrients needed in small amounts are called *micronutrients*.

Nutrients stimulate production by plants, and the lack of nutrients can limit production. The availability of nitrogen or phosphorus frequently is a limiting factor (▶ p. 136) for plant growth. When these nutrients are added to a system, plants show the greatest response to whichever nutrient has been in shortest supply. Phosphorus tends to be limiting in freshwater systems, and nitrogen in marine systems. Thus the Gulf of Mexico's hypoxic zone is thought to result primarily from excess nitrogen, whereas ponds and lakes in the Mississippi

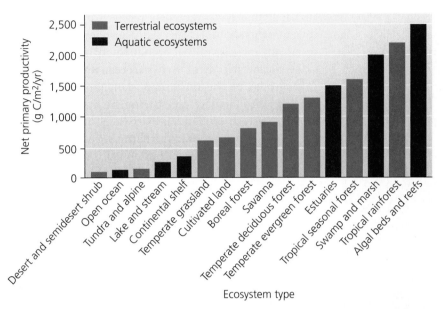

(a) Net primary productivity for major ecosystem types

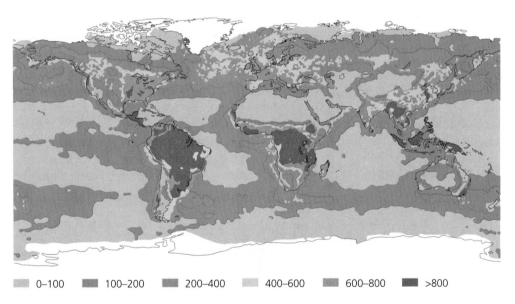

| 0–100 | 100–200 | 200–400 | 400–600 | 600–800 | >800 |

(b) Global map of net primary productivity

FIGURE 7.6 (a) Freshwater wetlands, tropical forests, coral reefs, and algal beds show high net primary productivities on average, whereas deserts, tundra, and the open ocean show low values. (b) On land, net primary production varies geographically with temperature and precipitation, which also influence the locations of biomes (▸ pp. 170–178). In the world's oceans, net primary production is highest around the margins of continents, where nutrients (of both natural and human origin) run off from land. Data in (a) from Whittaker, R. H. 1975. *Communities and ecosystems*, 2nd ed., New York: MacMillan. Map in (b) from satellite data presented by Field, C. B., et al. 1998. Primary production of the biosphere: Integrating terrestrial and oceanic components. *Science* 281: 237–240.

River Valley tend to suffer eutrophication when they contain too much phosphorus.

Canadian ecologist David Schindler and others demonstrated the effects of phosphorus on freshwater systems in the 1970s by experimentally manipulating entire lakes. In one experiment, the researchers divided a 16-ha (40-acre) lake in Ontario in half with a plastic barrier. To one half the researchers added carbon, nitrate, and phosphate; to the other they added only carbon and nitrate. Soon after the experiment began, they witnessed a

FIGURE 7.7 A portion of this lake in Ontario was experimentally treated with the addition of phosphate. This treated portion experienced an immediate, dramatic, and prolonged algal bloom, visible in the opaque water in the topmost part of this photo. Photo by David W. Schindler.

dramatic increase in algae in the half of the lake that received phosphate, whereas the other half hosted algal levels typical for lakes in the region (Figure 7.7). This difference held until shortly after they stopped fertilizing seven years later, when algae decreased to normal levels in the half that had previously received phosphate. Such experiments showed clearly that phosphorus addition can markedly increase primary productivity in lakes.

Similar experiments in coastal ocean waters have shown nitrogen to be the more important limiting factor for primary productivity. In experiments in the 1980s and 1990s, Swedish ecologist Edna Granéli took samples of ocean water from sites in the Baltic Sea and added phosphate, nitrate, or nothing. Chlorophyll and phytoplankton increased greatly in the flasks with nitrate, whereas those with phosphate did not differ from the controls. Seawater experiments in Long Island Sound by other researchers have shown similar results. For open ocean waters far from shore, research has shown that iron is a highly effective nutrient.

Because nutrients run off from land in the Baltic, Long Island Sound, the Gulf of Mexico, and worldwide, primary productivity in the oceans tends to be greatest in nearshore waters, which receive nutrient runoff from land, and lowest in open ocean areas far from land (see Figure 7.6b). Satellite imaging technology that reveals phytoplankton densities has given scientists an improved view of productivity at regional and global scales, and has helped them track blooms of algae that contribute to coastal hypoxic zones (Figure 7.8).

The number of known dead zones is increasing globally, with about 150 documented so far. Most are located off the coasts of Europe and the eastern United States. Specific causes vary from place to place, but most result from rising nutrient pollution from farms, cities, and industry.

FIGURE 7.8 Satellite images like this one have helped scientists track runoff and phytoplankton blooms. Shown is the Louisiana coast on December 30, 1997. Greenish brown at the top is land, and deep blue at the bottom right is water of the Gulf of Mexico. Along the coast, two light brown plumes of sediment are visible expanding into the Gulf's waters from the mouths of the Mississippi and Atchafalaya Rivers. The shades of turquoise represent phytoplankton blooms that result from nutrient inputs and that foster the dead zone. Image from the SeaWiFS Project, NASA/Goddard Space Flight Center, and ORBIMAGE.

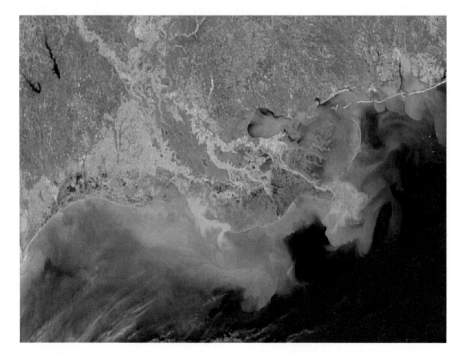

In North America, Chesapeake Bay may be the most severely affected area besides the Gulf of Mexico. Decades of pollution and human impact here have devastated fisheries and greatly altered the bay's ecology. Good news of a sort comes from the Black Sea, which borders Ukraine, Russia, Turkey, and eastern Europe. This immense inland sea had long suffered one of the world's worst hypoxic zones, and many of its valuable fisheries and seagrass beds were destroyed. Then in the 1990s, after the Soviet Union collapsed, industrial agriculture in the region declined drastically. With fewer fertilizers running off into the sea, this water body began to recover, and today fish and mussel beds are reviving. However, agricultural collapse is not a strategy anyone would choose to alleviate hypoxia. Rather, scientists are proposing a variety of innovative and economically acceptable ways to reduce nutrient runoff.

Biogeochemical Cycles

Just as nitrogen and phosphorus from Minnesota soybean fields end up in Gulf shrimp on our dinner plates, all nutrients move through the environment in complex and fascinating ways. Whereas energy enters an ecosystem from the sun, flows from one organism to another, and is dissipated to the atmosphere as heat, the physical matter of an ecosystem is circulated over and over again.

Nutrients circulate through ecosystems in biogeochemical cycles

Nutrients move through ecosystems in **nutrient cycles** or **biogeochemical cycles.** In these cycles they travel through the atmosphere, hydrosphere, and lithosphere, and from one organism to another, in dynamic equilibrium. A carbon atom in your fingernail today might have helped comprise the muscle of a cow a year earlier, may have resided in a blade of grass a month before that, and may have been part of a dinosaur's tooth 100 million years ago. After we die, the nutrients in our bodies will spread widely through the environment, eventually being incorporated by an untold number of organisms far into the future.

Nutrients move from one *pool*, or *reservoir*, to another, remaining for varying amounts of time *(residence time)* in each. The dinosaur, the grass, the cow, and you are each reservoirs for carbon atoms. The movement of nutrients among pools is termed *flux*, and the rates of flux between any given pair of pools can change over time. Human activity has influenced certain flux rates. For instance, it has increased the flux of nitrogen from the atmosphere to

pools on the Earth's surface and has shifted the flux of carbon in the opposite direction. Reservoirs that release more nutrients than they accept are called *sources*, and reservoirs accepting more nutrients than they are releasing are called *sinks*. As we discuss biogeochemical cycles, think about how they involve negative feedback loops that promote dynamic equilibrium, and about how some human actions can generate positive feedback loops.

The carbon cycle circulates a vital organic nutrient

As the definitive component of organic molecules (▸ pp. 98–99), carbon is an ingredient in carbohydrates, fats, and proteins, and in the bones, cartilage, and shells of all living things. From fossil fuels to DNA, from plastics to pharmaceuticals, carbon (C) atoms are everywhere. The **carbon cycle** describes the routes that carbon atoms take through the environment (Figure 7.9).

Photosynthesis, respiration, and food webs Producers, including terrestrial and aquatic plants, algae, and cyanobacteria, pull carbon dioxide out of the atmosphere and out of surface water to use in photosynthesis. Photosynthesis (▸ pp. 105–107) breaks the bonds in carbon dioxide (CO_2) and water (H_2O) to produce oxygen (O_2) and carbohydrates (e.g., glucose, $C_6H_{12}O_6$). Autotrophs use some of the carbohydrates to fuel their own respiration, thereby releasing some of the carbon back into the atmosphere and oceans as CO_2. When producers are eaten by primary consumers, who in turn are eaten by secondary and tertiary consumers, more carbohydrates are broken down in respiration, producing carbon dioxide and water. The same process occurs when decomposers consume waste and dead organic matter. Respiration from all these organisms releases carbon back into the atmosphere and oceans.

All organisms use carbon for structural growth, so a portion of the carbon an organism takes in becomes incorporated into its tissues. The abundance of plants and the fact that they take in so much carbon dioxide for photosynthesis makes plants a major reservoir for carbon. Because CO_2 is a greenhouse gas of primary concern (▸ pp. 530–532), much research on global climate change has been directed toward measuring the amount of CO_2 that plants tie up. Scientists are working toward understanding exactly how much this portion of the carbon cycle influences Earth's climate.

Sediment storage of carbon As organisms die, their remains may settle in sediments in ocean basins

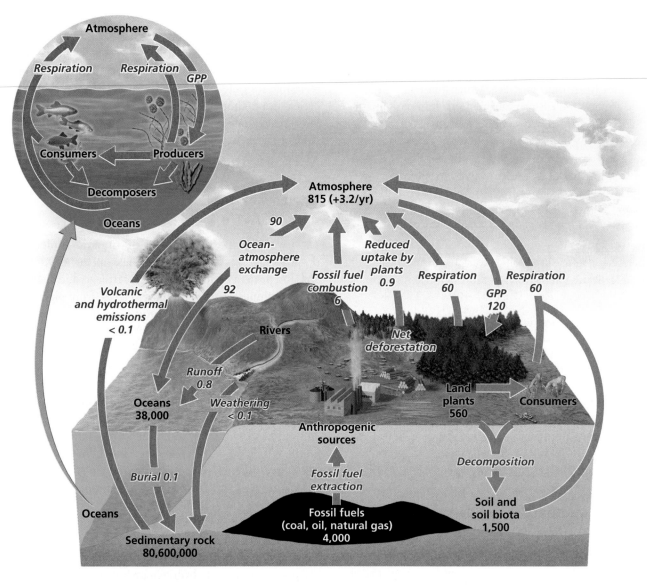

FIGURE 7.9 The carbon cycle summarizes the many routes that carbon atoms take as they move through the environment. Gray arrows represent fluxes among reservoirs, or pools, for carbon. In the carbon cycle, plants use carbon dioxide from the atmosphere for photosynthesis (gross primary production, or "GPP" in the figure). Carbon dioxide is returned to the atmosphere through respiration by plants, their consumers, and decomposers. The oceans sequester carbon in their water and in deep sediments. The vast majority of the planet's carbon is stored in sedimentary rock. In the figure, pool names are printed in plain bold type, and numbers in plain bold type represent pool sizes, expressed in units of 10^{15} g C. Processes printed in italic bold red type give rise to fluxes, printed in italic bold red type and expressed in units of 10^{15} g C per year. Data from Schlesinger, W. H. 1997. *Biogeochemistry: An analysis of global change.* 2nd ed. London: Academic Press.

or in freshwater wetlands. As layers of sediment accumulate, older layers are buried more deeply, experiencing high pressure over long periods of time. These conditions can convert soft tissues into fossil fuels—coal, oil, and natural gas—and shells and skeletons into sedimentary rock, such as limestone. Sedimentary rock comprises the largest single reservoir in the carbon cycle. Although any given carbon atom spends a relatively short time in the atmosphere, carbon trapped in sedimentary rock may reside there for hundreds of millions of years.

Carbon trapped in sediments and fossil fuel deposits may eventually be released into the oceans or atmosphere by geological processes such as uplift, erosion, and volcanic eruptions. It also reenters the atmosphere when we extract and burn fossil fuels.

The oceans The world's oceans are the second-largest reservoir in the carbon cycle. They absorb carbon-containing compounds from the atmosphere, from terrestrial runoff, from undersea volcanoes, and from the waste products and detritus of marine organisms. Some carbon atoms absorbed by the oceans—in the form of carbon dioxide, carbonate ions (CO_3^{2-}), and bicarbonate ions (HCO_3^-)—combine with calcium ions (Ca^{2+}) to form calcium carbonate ($CaCO_3$), an essential ingredient in the skeletons and shells of microscopic marine organisms. As these organisms die, their calcium carbonate shells sink to the ocean floor and begin to form sedimentary rock. The rates at which the oceans absorb and release carbon depend on many factors, including temperature and the numbers of marine organisms converting CO_2 into carbohydrates and carbonates.

Humans are shifting carbon from the lithosphere to the atmosphere

By mining fossil fuel deposits, we are essentially removing carbon from an underground reservoir with a residence time of millions of years. By combusting fossil fuels in our automobiles, homes, and industries, we release carbon dioxide and greatly increase the flux of carbon from the ground to the air. Since the mid-18th century, our fossil fuel combustion has added 246 billion metric tons (271 billion tons) of carbon to the atmosphere. Meanwhile, the movement of CO_2 from the atmosphere back to the hydrosphere, lithosphere, and biosphere has not kept pace.

In addition, cutting down forests and burning fields removes carbon from the pool of vegetation and releases it to the air. And if less vegetation is left on the surface, there are fewer plants to draw CO_2 back out of the atmosphere.

As a result, scientists estimate that today's atmospheric carbon dioxide reservoir is the largest that Earth has experienced in the past 420,000 years, and perhaps in the past 20 million years. The ongoing flux of carbon out of the fossil-fuel reservoir and into the atmosphere is a driving force behind global climate change (Chapter 18).

The phosphorus cycle involves mainly lithosphere and ocean

The element phosphorus (P) is a key component of DNA, RNA, and cell membranes (▶pp. 100–102). Organisms also use phosphorus to build two important biochemical compounds, adenosine diphosphate (ADP) and adenosine triphosphate (ATP). Cells use ADP and ATP to transfer and convert energy from one form to another during metabolism, and these compounds play roles in processes involving DNA and RNA. Although phosphorus is vital for life in these ways, the amount of phosphorus in organisms is dwarfed by the vast amounts in rocks, soil, sediments, and the oceans. Unlike the carbon and nitrogen cycles, the **phosphorus cycle** (Figure 7.10) has no appreciable atmospheric component besides the transport of tiny amounts of windblown dust and seaspray.

Geology and phosphorus availability Most of Earth's phosphorus is contained within rocks and is released only by weathering (▶pp. 251–252), an extremely slow process. The weathering of rocks releases phosphate (PO_4^{3-}) ions into water. Phosphates dissolved in lakes or in the oceans precipitate into solid form, settle to the bottom, and reenter the lithosphere's phosphorus reservoir in sediments. Because most phosphorus is bound up in rock and only slowly released, environmental concentrations of phosphorus available to organisms tend to be very low. This relative rarity explains why phosphorus is frequently a limiting factor for plant growth and why an artificial influx of phosphorus can produce immediate and dramatic effects.

Food webs Plants can take up phosphorus through their roots only when phosphate is dissolved in water. Primary consumers acquire phosphorus from water and plants and pass it on to secondary and tertiary consumers. Consumers also pass phosphorus to the soil through the excretion of waste. Decomposers break down phosphorus-rich organisms and their wastes and, in so doing, return phosphorus to the soil.

We affect the phosphorus cycle

Humans influence the phosphorus cycle in several ways. We mine rocks containing phosphorus to extract this nutrient for the inorganic fertilizers we use on crops and lawns. Treated and untreated sewage discharge also tends to be rich in phosphates. Those phosphates that run off into waterways can boost algal growth and cause eutrophication, leading to murkier waters and altering the structure and function of aquatic ecosystems. Phosphates are also present in detergents, and one way each of us can reduce phosphorus input into the environment is to purchase phosphate-free detergents.

The nitrogen cycle involves specialized bacteria

Nitrogen (N) makes up 78% of our atmosphere by mass, and is the sixth most abundant element on Earth. It is an essential ingredient in the proteins, DNA, and RNA that

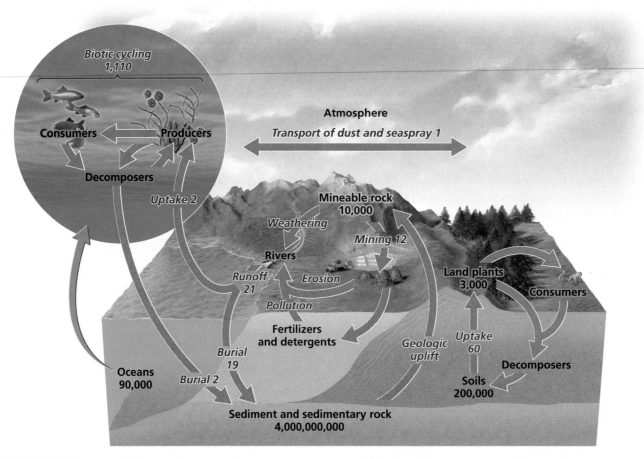

FIGURE 7.10   The phosphorus cycle summarizes the many routes that phosphorus atoms take as they move through the environment. Gray arrows represent fluxes among reservoirs, or pools, for phosphorus. Most phosphorus resides underground in rock and sediment, but the phosphorus cycle moves this element through the soil, the oceans, and freshwater and terrestrial ecosystems. Rocks containing phosphorus are uplifted geologically and weathered away in this slow process, and small amounts of phosphorus cycle through food webs, where this nutrient is often a limiting factor for plant growth. In the figure, pool names are printed in plain bold type, and numbers in plain bold type represent pool sizes, expressed in units of 10^{12} g P. Processes printed in italic bold red type give rise to fluxes, printed in italic bold red type and expressed in units of 10^{12} g P per year. Data from Schlesinger, W. H. 1997. *Biogeochemistry: An analysis of global change.* 2nd ed. London: Academic Press.

build our bodies and, like phosphorus, is an essential nutrient for plant growth. Thus the **nitrogen cycle** (Figure 7.11) is of vital importance to us and to all other organisms. Despite its abundance in the air, nitrogen gas (N_2) is chemically inert and cannot cycle out of the atmosphere and into living organisms without assistance from lightning, highly specialized bacteria, or human intervention. For this reason, the element is relatively scarce in the lithosphere and hydrosphere and in organisms. However, once nitrogen undergoes the right kind of chemical change, it becomes biologically active and available to the organisms that need it, and it can act as a potent fertilizer.

Its scarcity makes biologically active nitrogen a limiting factor for plant growth.

Nitrogen fixation   To become biologically available, inert nitrogen gas (N_2) must be "fixed," or combined with hydrogen in nature to form ammonia (NH_3), whose water-soluble ions of ammonium (NH_4^+) can be taken up by plants. **Nitrogen fixation** can be accomplished in two ways: by the intense energy of lightning strikes, or when air in the top layer of soil comes in contact with particular types of **nitrogen-fixing bacteria.** These bacteria live in a mutualistic relationship (▶ pp. 155–156) with many types

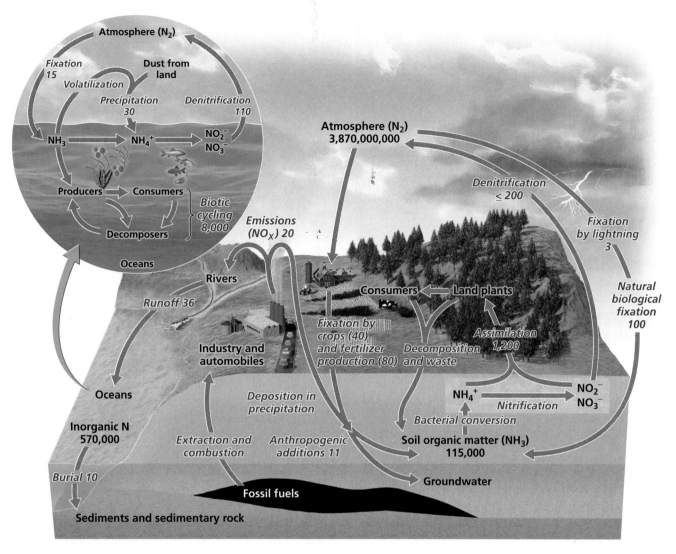

FIGURE 7.11 The nitrogen cycle summarizes the many routes that nitrogen atoms take as they move through the environment. Gray arrows represent fluxes among reservoirs, or pools, for nitrogen. In the nitrogen cycle, specialized bacteria play key roles in "fixing" atmospheric nitrogen and converting it to chemical forms that plants can use, while other types of bacteria convert nitrogen compounds back to the atmospheric gas N_2. In the oceans, inorganic nitrogen is buried in sediments while nitrogen compounds are cycled through food webs as they are on land. In the figure, pool names are printed in plain bold type, and numbers in plain bold type represent pool sizes, expressed in units of 10^{12} g N. Processes printed in italic bold red type give rise to fluxes, printed in italic bold red type and expressed in units of 10^{12} g N per year. Data from Schlesinger, W. H. 1997. *Biogeochemistry: An analysis of global change.* 2nd ed. London: Academic Press.

of plants, including soybeans and other legumes, providing them nutrients by converting nitrogen to a usable form. As we will see in Chapter 9, farmers have long nourished their soils by planting crops that host nitrogen-fixing bacteria among their roots (Figure 7.12).

Nitrification and denitrification Other types of specialized bacteria then perform a process known as **nitrification.** In this process, ammonium ions are first converted

into nitrite ions (NO_2^-), then into nitrate ions (NO_3^-). Plants can take up these ions, which also become available after atmospheric deposition on soils or in water or after application of nitrate-based fertilizer.

Animals obtain the nitrogen they need by consuming plants or other animals. Recall from Chapter 4 ("The Science behind the Story," ▶ pp. 94–95) how scientists use stable isotopes of nitrogen to study the trophic level and nutritional condition of animals. Decomposers obtain

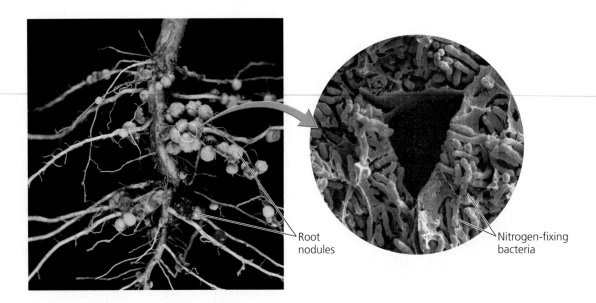

FIGURE 7.12 Specialized bacteria live in nodules on the roots of this legume plant. In the process of nitrogen fixation, the bacteria convert nitrogen to a form that the plant can take up into its roots.

nitrogen from dead and decaying plant and animal matter and from the urine and feces of animals. Once decomposers process the nitrogen-rich compounds they take in, they release ammonium ions, making these available to nitrifying bacteria to convert again to nitrates and nitrites.

The next step in the nitrogen cycle occurs when **denitrifying bacteria** convert nitrates in soil or water to gaseous nitrogen via a multistep process. Denitrification thereby completes the cycle by releasing nitrogen back into the atmosphere as a gas.

Humans have greatly influenced the nitrogen cycle

The impacts of excess nitrogen from agriculture and other human activities in the Mississippi River watershed have become painfully evident to shrimpers and scientists with an interest in the Gulf of Mexico (see "The Science behind the Story" on ▶pp. 202–203). But hypoxia in the Gulf and other coastal locations around the world is hardly the only problem resulting from human manipulation of the nitrogen cycle.

Historically, nitrogen fixation was a *bottleneck*, a step that limited the flux of nitrogen out of the atmosphere. This changed when the research of two German chemists enabled us to fix nitrogen on an industrial scale. Fritz Haber worked in the German army's chemical weapons program during World War I. Shortly before the war, Haber found a way to combine nitrogen and hydrogen gases to synthesize ammonia, a key ingredient in modern

explosives and agricultural fertilizers. Several years later, Carl Bosch built on Haber's work and devised methods to produce ammonia on an industrial scale. The work of these two scientists enabled people to overcome the limits on productivity long imposed by nitrogen scarcity in nature. The widespread application of their findings has enhanced agriculture and thereby contributed to the enormous increase in human population over the past 90 years. Farmers, golf course managers, and homeowners have all taken advantage of the fertilizers made possible by the **Haber-Bosch process.** These developments have also led to a dramatic alteration of the nitrogen cycle, however. Today, using the Haber-Bosch process, our species is fixing at least as much nitrogen artificially as is being fixed naturally. We have effectively doubled the natural rate of nitrogen fixation on Earth (Figure 7.13).

By fixing atmospheric nitrogen, we increase its flux out of the atmosphere and into other reservoirs. In addition, we have affected fluxes in other parts of the cycle. When we burn forests and fields, we force nitrogen out of soils and vegetation and into the atmosphere. When we burn fossil fuels, we increase the rate at which nitric oxide (NO) enters the atmosphere and reacts to form nitrogen dioxide (NO_2). This compound is a precursor to nitric acid (HNO_3), a key component of acid precipitation (▶pp. 514–518). We introduce another nitrogen-containing gas, nitrous oxide (N_2O), by allowing anaerobic bacteria to break down the tremendous volume of animal waste produced in agricultural feedlots (▶pp. 296–297). We have also accelerated the introduction of nitrogen-rich compounds into terrestrial

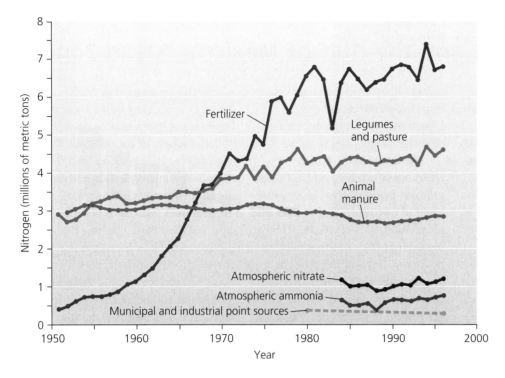

FIGURE 7.13 In the past half century, human inputs of nitrogen into the environment have greatly increased, such that today fully half of the nitrogen entering the environment is of human origin. Agricultural fertilizer has for the past 35 years been the leading source of all nitrogen inputs, natural and artificial. Data from National Science and Technology Council: Committee on Environment and Natural Resources. May 2000. *Hypoxia: An integrated assessment in the northern Gulf of Mexico.*

and aquatic systems by destroying wetlands and cultivating more legume crops that host nitrogen-fixing bacteria in their roots.

These activities increase amounts of nitrogen available to aquatic plants and algae, boosting their growth. Algal populations soon outstrip the availability of other required nutrients and begin to die and decompose. As in the Gulf of Mexico, this large-scale decomposition can rob other aquatic organisms of oxygen, leading to shellfish die-offs and other significant impacts on ecosystems. Increased nitrate pollution, such as that found in groundwater in the Mississippi River basin, can also lead to human health effects.

In 1997, a team of scientists led by Peter Vitousek of Stanford University summarized the dramatic changes humans have caused in the global nitrogen cycle. Each of these alterations can reverberate, through feedback loops, to produce unintended outcomes and unpredictable results. Human alterations of the nitrogen cycle, according to the Vitousek team's report, have:

▶ Doubled the rate that fixed nitrogen enters terrestrial ecosystems (and the rate is still increasing).

▶ Increased atmospheric concentrations of the greenhouse gas N_2O and of other oxides of nitrogen that produce smog.

▶ Depleted essential nutrients, such as calcium and potassium, from soils, because fertilizer helps flush them out.

▶ Acidified surface water and soils.

▶ Greatly increased transfer of nitrogen from rivers to oceans.

▶ Encouraged plant growth, causing more carbon to be stored within terrestrial ecosystems.

▶ Reduced biological diversity, especially plants adapted to low nitrogen concentrations.

▶ Changed the composition and function of estuaries and coastal ecosystems.

▶ Harmed many coastal marine fisheries.

Weighing the **Issues:**
Nitrogen Pollution and Its Financial Impacts

Most nitrate that enters the Gulf of Mexico originates from farms and other sources in the upper Midwest, yet many of its negative impacts are borne by downstream users, such as fishermen and shrimpers along the Gulf Coast. Who do you believe should be responsible for addressing this problem? Should environmental policies on this particular issue be developed and enforced by state governments, the federal government, both, or neither? Explain the reasons for your answer.

In 1998, the U.S. Congress passed the Harmful Algal Bloom and Hypoxia Research and Control Act. This law called for an "integrated assessment" of hypoxia in the northern Gulf to address the extent, nature, and causes of the dead zone, as well as its ecological and economic

The Science behind the Story

Hypoxia and the Gulf of Mexico's "Dead Zone"

She was prone to seasickness, but Nancy Rabalais cared too much about the Gulf of Mexico to let that stop her. Leaning over the side of an open boat idling miles from shore, she hauled a water sample aboard—and helped launch the long effort to breathe life back into the Gulf's "dead zone."

Since that expedition in 1985, Rabalais, her colleague and husband Eugene Turner, and fellow scientists at the Louisiana Universities Marine Consortium (LUMCON) and Louisiana State University have made great progress in unraveling the mysteries of the region's hypoxia—and in getting it on the political radar screen.

A few scientists had noticed abnormally low oxygen levels in parts of the Gulf in the 1970s. But such observations gained urgency in the 1980s as fishing trawlers had to head farther and farther offshore to find anything to catch. Such early signs of trouble led to bigger questions for Rabalais and her colleagues: How widespread was the hypoxia? What was causing it? How much worse might it get?

Rabalais and other researchers started tracking oxygen levels at nine sites in the Gulf every month, and continued those measurements for five years. At dozens of other spots near the shore and in deep water, they took less frequent oxygen readings. For some of this work, the researchers have relied on mobile oxygen probes. Sensors, as they are lowered into the water, measure oxygen levels and send continuous readings back to a shipboard

computer. Further data have come from fixed, submerged oxygen meters that continuously measure dissolved oxygen and store the data.

The team also collected hundreds of coastal and Gulf water samples, using lab tests to measure levels of nitrogen, salt, bacteria, and phytoplankton. LUMCON scientists logged hundreds of miles in their boats, regularly monitoring more than 70 sites in the Gulf. They also donned scuba gear to view firsthand the condition of shrimp, fish, and other sea life. Such a range of long-term data allowed the scientists to build a "map" of the dead zone, tracking its location and effects.

In 1991, Rabalais made that map public, earning immediate headlines. That year, her group mapped the size of the zone at more than 10,000 km^2 (about 4,000 mi^2). Bottom-dwelling shrimp were stretching out of their burrows, straining for oxygen. Many fish had fled. The bottom waters, infused with sulfur from bacterial decomposition, smelled of rotten eggs.

The group's years of continuous tracking also explained the dead zone's predictable emergence. As rivers rose each spring (and as fertilizers were applied in the Midwestern farm states), oxygen would start to disappear in the northern Gulf. The hypoxia would last through the summer or fall, until seasonal storms mixed oxygen into hypoxic areas. Over time, monitoring linked the dead zone's size to the volume of river flow and its nutrient load; the 1993 flooding of the Mississippi created a zone much larger than the year before.

Conversely, a drought in 2000 brought lower river flows, lower nutrient loads, and a smaller dead zone.

The source of the problem, Rabalais said, lay back on land. The Mississippi and Atchafalaya Rivers draining into the Gulf were polluted from runoff, and that pollution spurred algal blooms whose decomposition snuffed out oxygen in wide stretches of ocean water. The rivers were carrying high concentrations of nitrates and other chemicals from agricultural fertilizers and other sources. As outlined in this chapter, excess nitrogen sets off a chain reaction in marine waters that eventually drains oxygen in deeper waters.

Many Midwestern farming advocates and some scientists, such as Derek Winstanley, chief of the Illinois State Water Survey, challenged the findings. They argued that the Mississippi naturally carries high loads of nitrogen from the rich prairie soil and that the Rabalais team had not ruled out upwelling in the Gulf as a source of nutrients. But sediment analyses showed that Mississippi River mud was much lower in nitrates early in the century, and Rabalais and Turner found that silica residue from phytoplankton blooms had increased in Gulf sediments between 1970 and 1989, paralleling rising nitrogen levels. In 2000, the federal integrative assessment involving dozens of scientists laid the blame for the dead zone on nutrients from fertilizers and other sources.

Then in 2004, while representatives of farmers and fishermen bickered over political fixes, EPA water quality scientist Howard Marshall suggested that phosphorus

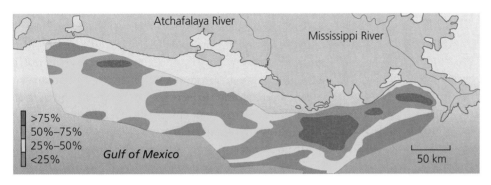

(a) Frequency of hypoxia, 1985–1999

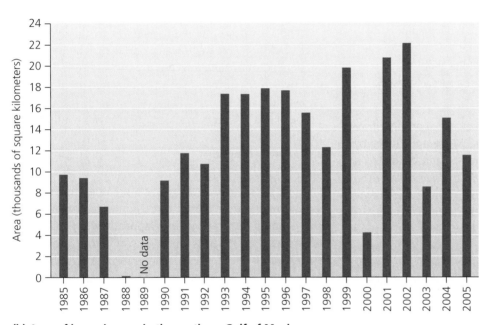

(b) Area of hypoxic zone in the northern Gulf of Mexico

Some parts of the Gulf of Mexico suffer from hypoxia more frequently than others **(a).** Areas in red have experienced oxygen concentrations below 2 ppm in more than 75% of surveys; those in orange experienced hypoxia in 50–75% of surveys; those in yellow were affected in 25–50% of surveys; and those in green were affected in less than 25% of surveys. The size of the Gulf's hypoxic zone varies **(b)** as a result of several factors. These include floods, which increase its size by bringing additional runoff (as with the Mississippi River floods of 1993), and tropical storms, which decrease its size by mixing oxygen-rich water into the dead zone (as in 2003). Between 1985 and 2005, the hypoxic zone averaged 12,700 km^2 in size and ranged up to 22,000 km^2. Data in (a) from Rabalais, N., et al. 2002. Beyond science into policy: Gulf of Mexico hypoxia and the Mississippi. *BioScience* 52: 129–142. Data in (b) from Nancy Rabalais, LUMCON.

pollution from industry and sewage treatment might instead be at fault. His reasoning: the ratio of nitrogen to phosphorus in the Gulf was so overwhelmingly biased toward nitrogen that phosphorus had become the limiting factor. Because phytoplankton need both, they now had more nitrogen than they could use but not enough phosphorus. Thus, Marshall suggested, we'd be best off reducing phosphorus if we want to alleviate the dead zone.

Other scientists are giving this idea a mixed reception, and many have proposed that nitrogen and phosphorus should be managed jointly. Further suggestions have come from still more research. One is that the federally mandated 30% reduction in nitrogen in the river will not be enough. Another is that large-scale restoration of wetlands along the river and at the river's delta would best filter pollutants before they reached the Gulf. All

this research is guiding a federal plan to reduce farm runoff, clean up the Mississippi, restore wetlands, and shrink the Gulf's dead zone—and it has also led to a better understanding of hypoxic zones around the world.

"What people do 800 miles away from the Gulf of Mexico directly affects the Gulf of Mexico," Rabalais said when she accepted a major environmental award in 1999. "It's hard for many people to realize that."

impacts. The assessment report published two years later also outlined potential solutions to the problem, along with estimates of the social and economic costs associated with these solutions. The report proposed that the federal government work with Gulf Coast and Midwestern communities to:

- Reduce nitrogen fertilizer use on Midwestern farms.
- Change the timing of fertilizer application to minimize rainy-season runoff.
- Use alternative crops.
- Manage nitrogen-rich livestock manure more effectively.
- Restore nitrogen-absorbing wetlands in the Mississippi River basin.
- Use artificial wetlands to filter farm runoff.
- Install better nitrogen-removing technologies in sewage treatment plants.
- Restore frequently flooded lands to reduce runoff.
- Restore wetland ecosystems near the Mississippi River's mouth to enhance nitrogen-absorbing ability.
- Continue evaluating which of these approaches work and which do not.

A combined state and federal task force proposed reducing nitrogen flowing down the Mississippi River by 30% by 2015, although some scientists later estimated it will require cuts of 50% to decrease the size of the dead zone. Farmers' advocates argued that farmers were being unfairly singled out, when many other artificial and natural sources contribute to the problem. They also argued that severe restrictions on fertilizer use would hurt farmers economically and decrease crop yields. As people debated the science and the proposed remedies, Congress in 2003 reauthorized the Harmful Algal Bloom and Hypoxia Research and Control Act for an additional five years, promising funding for further research and the development of solutions.

Many scientists and farmers are now searching for innovative solutions that would alleviate pollution while not hurting agriculture. One proposal offers farmers insurance and economic incentives for not using excess fertilizer. Another program is testing new farming strategies to see whether any can maintain yields while decreasing fertilizer use. A third approach involves planting cover crops in the off-season to reduce runoff from bare fields. Yet another encourages farmers to maintain artificial wetlands on their lands that serve as natural buffers against pollution. Wetland plants host denitrifying bacteria that convert nitrates to nitrogen gas, so wetlands can effectively clean up a large amount of nitrogen pollution. Many Midwestern farmers have already taken part in one or more of these strategies.

The hydrologic cycle influences all other cycles

Water is so integral to life that we frequently take it for granted. Water is the essential medium for all manner of biochemical reactions (▸ pp. 96–98). It plays key roles in nearly every environmental system, including each of the nutrient cycles we have just discussed. Water carries nutrients and sediments from the continents to the oceans via rivers, streams, and surface runoff, and it distributes sediments onward in ocean currents. Increasingly, water also distributes artificial pollutants. The water cycle, or **hydrologic cycle** (Figure 7.14), summarizes how water—in liquid, gaseous, and solid forms—flows through our environment. Our brief introduction to the hydrologic cycle here sets the stage for our more in-depth discussion of freshwater and marine systems in Chapters 15 and 16.

The oceans are the main reservoir in the hydrologic cycle, holding 97% of all water on Earth. The freshwater we depend on for our survival accounts for less than 3%, and two-thirds of this small amount is tied up in glaciers, snowfields, and ice caps. Thus, considerably less than 1% of the planet's water is in a form that we can readily use—groundwater, surface freshwater, and rain from atmospheric water vapor.

Evaporation and transpiration Water moves from surface reservoirs—oceans, lakes, ponds, rivers, and moist soil—into the atmosphere by **evaporation,** the conversion of a liquid to gaseous form. Warm temperatures and strong winds speed rates of evaporation. A greater degree of exposure has the same effect; an area logged of its forest or converted to agriculture or residential use will lose water more readily than a comparable area that remains vegetated. Water also enters the atmosphere by **transpiration,** the release of water vapor by plants through their leaves. Transpiration and evaporation act as natural processes of distillation, effectively creating pure water by filtering out minerals carried in solution.

Precipitation, runoff, and surface water Water returns from the atmosphere to Earth's surface as **precipitation** when water vapor condenses and falls as rain or snow. Precipitation may be taken up by plants and used by animals, but much of it flows as **runoff** into streams, rivers, lakes, ponds, and oceans. Amounts of precipitation vary greatly from region to region globally, helping give rise to the variety of biomes.

Groundwater Some precipitation and surface water soaks down through soil and rock to recharge under-

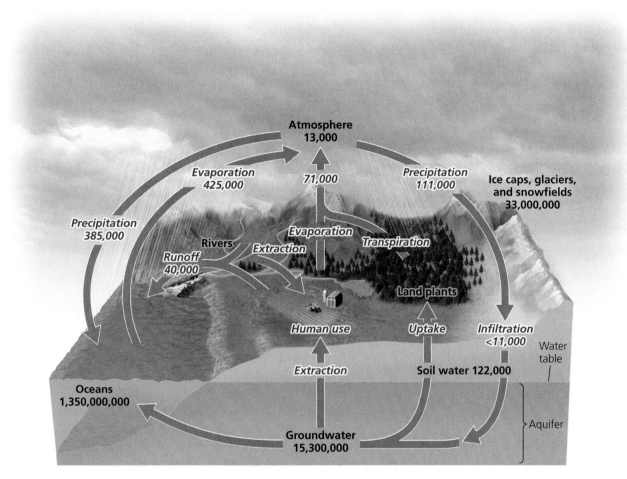

FIGURE 7.14 The hydrologic cycle summarizes the many routes that water molecules take as they move through the environment. Gray arrows represent fluxes among reservoirs, or pools, for water. The hydrologic cycle is a system unto itself but also plays key roles in other biogeochemical cycles. Oceans hold 97% of our planet's water, while most freshwater resides in groundwater and icecaps. Water vapor in the atmosphere condenses and falls to the surface as precipitation, then evaporates from land and transpires from plants to return to the atmosphere. Water flows downhill into rivers, eventually reaching the oceans. In the figure, pool names are printed in plain bold type, and numbers in plain bold type represent pool sizes, expressed in units of km^3. Processes printed in italic bold red type give rise to fluxes, printed in italic bold red type and expressed in units of km^3 per year. Data from Schlesinger, W. H. 1997. *Biogeochemistry: An analysis of global change,* 2nd ed. London: Academic Press.

ground water reservoirs known as **aquifers.** Aquifers are spongelike regions of rock and soil that hold **groundwater,** water found underground beneath layers of soil. The upper limit of groundwater held in an aquifer is referred to as the **water table.** Aquifers may hold groundwater for long periods of time, so the water may be quite ancient. In some cases groundwater can take hundreds or even thousands of years to fully recharge after being depleted. Groundwater becomes exposed to the air where the water table reaches the surface, and the exposed water can run off toward the ocean or evaporate into the atmosphere.

Our impacts on the hydrologic cycle are extensive

Human activity has affected every aspect of the water cycle. By damming rivers to create reservoirs, we have increased evaporation and, in some cases, infiltration of surface water into aquifers. By altering Earth's surface and its vegetation, we have increased surface runoff and erosion. By spreading water on agricultural fields, we have depleted rivers, lakes, and streams, and have increased evaporation. By removing forests and other vegetation, we have reduced transpiration and have lowered water tables. And by emitting into the atmosphere pollutants that dissolve in water droplets, we have changed the chemical nature of precipitation, in effect sabotaging the natural distillation process that evaporation and transpiration provide. Perhaps most threatening to our future, we have drawn groundwater to the surface for drinking, irrigation, and industrial use, and have thereby begun to deplete groundwater resources (Figure 7.15). Water shortages have already given rise to numerous conflicts worldwide, from the Middle East to the American West (▶p. 444).

Weighing the **Issues:**
Your Water

Are you aware of any evidence of water shortages or conflict over water use in your region? What about pollution and water quality? In what ways might these issues affect the hydrologic cycle in your area?

Geological Systems: How Earth Works

Biogeochemical cycles are not the planet's only cyclical environmental systems. Physical processes of geology determine Earth's landscape and form the foundation for the biotic patterns that overlay the landscape.

The rock cycle is a fundamental environmental system

We tend to think of rock as pretty solid stuff; this is clear when we say something is "hard as rock." Yet in the long run, over geological time, rocks do change. Rocks and the minerals that comprise them are heated, melted, cooled, broken down, and reassembled in a very slow process called the **rock cycle** (Figure 7.15). The type of rock in a given region helps determine soil chemistry and thereby influences the biotic components of the region's ecosystems.

Igneous rock All rocks can melt. At high enough temperatures, rock will enter a molten, liquid state called **magma.** If magma is released from the lithosphere (as in a volcanic eruption), it may flow or spatter across Earth's surface as **lava.** Rock that forms when magma cools is called **igneous** (from the Latin *ignis,* meaning "fire") **rock** (Figure 7.15a). Igneous rock comes in several different types, because there are different ways in which magma

can solidify. Magma that cools slowly while it is well below Earth's surface is known as *intrusive* igneous rock. Half Dome and many other famous rock formations at Yosemite National Park in California were formed in this way and later exposed above the surface. Granite is the best-known type of intrusive rock. A slow cooling process allows minerals of different types to segregate from one another and aggregate with minerals of their own type, forming the crystals that give granite its multicolored, coarse-grained appearance. In contrast, when magma is ejected from a volcano, it cools quickly, so minerals have little time to differentiate into clusters. This kind of igneous rock is called *extrusive* igneous rock, and its most common representative is basalt.

Sedimentary rock All rock weathers away with time. The relentless forces of wind, water, freezing, and thawing eat away at rocks, stripping off one tiny grain (or large chunk) at a time. Particles of rock blown by wind or washed away by water finally come to rest somewhere, where they help to form **sediments.** These eroded remains of rocks usually are deposited very slowly, but floods can accelerate the process. The floods that sweep down the Mississippi River deposit sediments and nutrients at the river's mouth, building up its delta. They also spread them across the floodplain along the river's banks, where soils are enriched as a result. Sediments collect downhill, downstream, or downwind from their sources and are deposited in layers. These layers accumulate over time, causing the weight and pressure of overlying layers to increase. **Sedimentary rock** (Figure 7.15b) is formed when dissolved minerals seep through sediment layers and act as a kind of glue, crystallizing and binding sediment particles together. The formation of rock through these processes of compaction, binding, and crystallization is termed **lithification.** The fossils of organisms—and the fossil fuels we use for energy—are created by similar processes of physical compaction and chemical transformation (▶pp. 561–562).

Like igneous rock, the several types of sedimentary rock are classified by the way they form and the size of particles they contain. Rock such as limestone and rock salt forms by chemical means when rocks dissolve and their components crystallize to form new rock. A second type of sedimentary rock forms when layers of sediment are compressed and physically bonded to one another. Examples include conglomerate, made up of large particles that give it the appearance of nougat; sandstone, made of cemented sand particles; and shale, composed of still smaller mud particles.

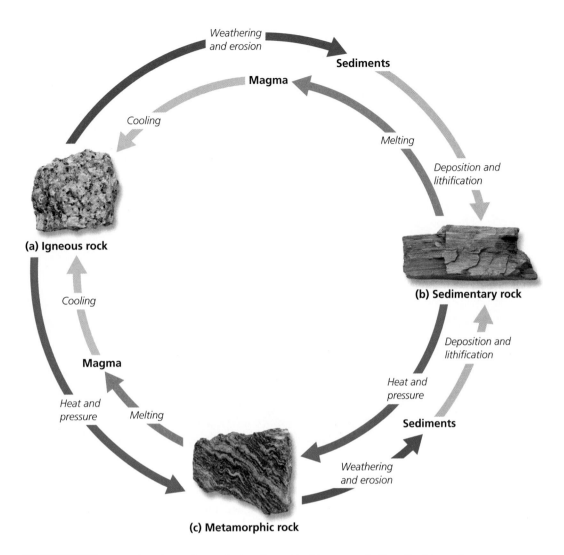

FIGURE 7.15 The rock cycle is a slow cycle that shapes Earth's crust and affects Earth's nutrient cycles. Igneous rock (**a**) is formed when rock melts to form magma, and the magma then cools. Sedimentary rock (**b**) is formed when rock is weathered and eroded, and the resulting sediments are compressed to form new rock. Metamorphic rock (**c**) is formed when rock is subjected to intense heat and pressure underground. Through these several processes (shown by differently colored arrows in the figure), each type of rock can be converted into either of the other two types.

Metamorphic rock Geological forces may bend, uplift, compress, or stretch rock. When great heat or pressure is exerted on rock, the rock may change its form, becoming **metamorphic** (from the Greek for "changed form" or "changed shape") **rock** (Figure 7.15c). The forces that metamorphose rock occur at temperatures lower than the rock's melting point but high enough to reshape crystals within the rock and change its appearance and physical properties. Metamorphic rock may resemble the rock from which it was created or may be quite different. Common types of metamorphic rock include marble, formed when limestone is heated and pressurized, strengthening its structure; and slate, formed when shale is heated and metamorphosed.

The changes that occur as rocks are altered from one type to another may proceed in any direction. Understanding the transition of rocks from one stage in the rock cycle to another enables us to appreciate more clearly the formation and conservation of soils, mineral resources, fossil fuels, and other natural resources.

Plate tectonics shapes Earth's geography

The rock cycle takes place within the broader context of **plate tectonics,** a process that underlies earthquakes and volcanoes and that determines the geography of the Earth's surface. Earth's surface consists of a lightweight

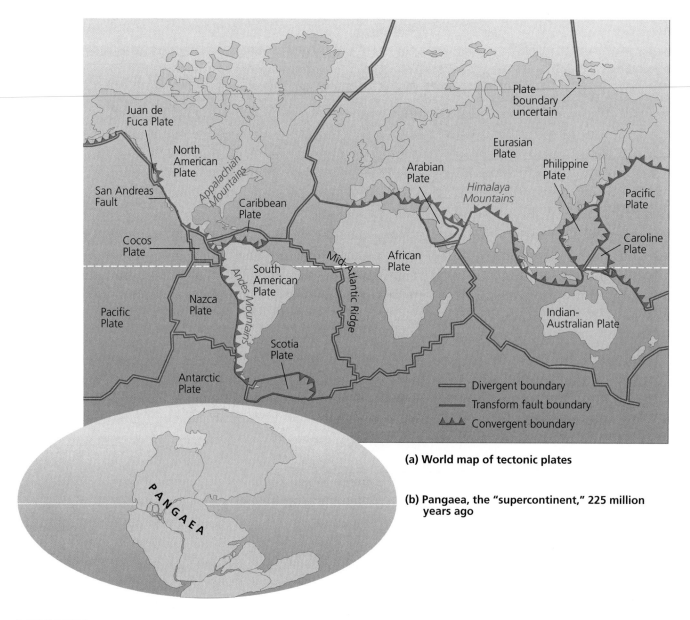

(a) World map of tectonic plates

(b) Pangaea, the "supercontinent," 225 million years ago

FIGURE 7.16   Earth's crust consists of roughly 15 major plates (**a**) that move through time by the process of plate tectonics. Today's continents were joined together in the "supercontinent" Pangaea (**b**) about 225 million years ago.

thin **crust** of rock floating atop a malleable **mantle,** which in turn surrounds a molten heavy **core** made mostly of iron. Earth's internal heat drives convection currents that flow in loops in the mantle, pushing the mantle's soft rock cyclically upward (as it warms) and downward (as it cools), like a gigantic conveyor belt. As the mantle material moves, it drags large plates of crust along its surface edge.

Earth's surface consists of about 15 major tectonic plates, most including some combination of ocean and continent (Figure 7.16a). Imagine peeling an orange and putting the pieces of peel back onto the fruit; the ragged pieces of peel are like the plates of crust riding atop Earth's

surface. These plates move at rates of roughly 2–15 cm (1–6 in.) per year. This movement has influenced Earth's climate and life's evolution throughout our planet's history as the continents combined, separated, and recombined in various configurations. By studying ancient rock formations throughout the world, geologists have determined that at least twice, all landmasses were joined together in a supercontinent scientists have dubbed *Pangaea* (Figure 7.16b).

At **divergent plate boundaries,** magma surging upward to the surface divides plates and pushes them apart, creating new crust as it cools and spreads (Figure 7.17a). A

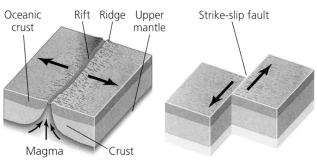

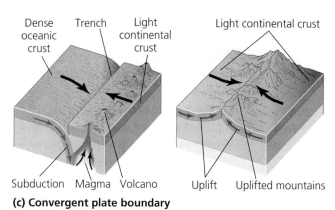

(a) Divergent plate boundary   (b) Transform plate boundary   (c) Convergent plate boundary

FIGURE 7.17 Different types of boundaries between tectonic plates result in different geologic processes. At a divergent plate boundary, such as a mid-ocean ridge on the seafloor (**a**), magma extrudes from beneath the crust, and the two plates move gradually away from the boundary in the manner of conveyor belts. At a transform plate boundary (**b**), two plates slide alongside one another, creating friction that leads to earthquakes. Where plates collide at a convergent plate boundary (**c**), one plate may be subducted beneath another, leading to volcanism, or both plates may be uplifted, leading to the formation of mountain ranges.

prime example is the Mid-Atlantic Ridge, part of a 74,000-km (46,000-mi) system of magmatic extrusion cutting across the seafloor. Plates expanding outward from divergent plate boundaries at mid-ocean ridges bump against other plates, creating different types of plate boundaries.

When two plates meet, they may slip and grind alongside one another, forming a **transform plate boundary** (Figure 7.17b). The friction at these boundaries spawns earthquakes along slipstrike faults. This is the case with the Pacific Plate and the North American Plate, which rub against each other along California's San Andreas Fault. Southern California is slowly inching its way toward northern California along this fault, and Los Angeles will eventually reach San Francisco.

When plates collide at **convergent plate boundaries,** either of two consequences may result (Figure 7.17c). First, one plate of crust may slide beneath another in a process called **subduction.** The subducted crust is heated as it dives into the mantle, and it may send up magma that erupts through the surface in volcanoes. Mount Saint Helens in Washington, which erupted violently in 1980 and renewed its activity in 2004, is fueled by magma from subduction. When denser ocean crust slides beneath lighter continental crust, volcanic mountain ranges are formed that parallel coastlines. Examples are the Cascades, which include Mount Saint Helens, and South America's Andes Mountains, where the Nazca Plate slides below the South American Plate. When one plate of oceanic crust is subducted beneath another, the resultant volcanism may form arcs of islands, such as Japan and the Aleutians. In

addition, deep trenches may be created, such as the Mariana Trench, the planet's deepest abyss.

Alternatively, two colliding plates of continental crust may slowly lift material from both plates. The Himalayas, the world's highest mountains, are the result of the Indian-Australian Plate's collision with the Eurasian Plate 40–50 million years ago, and these mountains are still being uplifted today. The Appalachian Mountains of the eastern United States, once the world's highest mountains themselves, resulted from a much earlier collision with the edge of what is today Africa.

Weighing the issues:
Geology and Topography

Think about the topography in your region. Can you infer what geological processes produced it? Now recall the most recent earthquake or volcano in the news. With what type of plate boundary do you suspect this event was associated?

Amazingly, this environmental system of such fundamental importance was totally unknown to us just half a century ago. Our civilization was sending humans to the moon by the time our geologists were explaining the movement of land under our very feet. It is humbling to reflect on this; what other fundamental systems might we not yet appreciate or understand while our technology—and our ability to affect Earth's processes—continues racing ahead?

The Dead Zone

Scientific research has indicated that nitrogen fertilizers from the Mississippi River watershed are contributing to hypoxia in the Gulf of Mexico. **Do you agree this is occurring? Why or why not? If so, what steps should be taken to solve the problem?**

Evidence Not Conclusive

Scientific evidence that nitrogen (N) fertilizer is polluting the northern Gulf of Mexico is not conclusive. Hypoxia in the Gulf has been recognized since 1935, long before fertilizer use became widespread in the 1960s. Nitrogen fertilizer use in the Mississippi River basin has remained relatively stable in the last two decades, but the size of the hypoxic zone has fluctuated markedly, especially since 1988.

According to the U.S. Geological Survey, the annual discharge of N from the Mississippi River has tripled in the last 30 years, with most of the increase occurring from 1970 to 1983. However, since 1980, river N discharge has changed very little, whereas N fertilizer use has grown by almost 10%. From 1980 to 1999, N fertilizer sales in the Mississippi River basin explained 8% of the variation in river nitrate-N discharge, whereas the Mississippi River's average annual flow explained nearly 80% of its variation.

Recent data suggest that the molar ratio of inorganic N to inorganic phosphorus (P) is the principal indicator of phytoplankton blooms that cause hypoxia, and that the N:P ratio in freshwater entering the Gulf is high enough that any increase in inorganic P loading will increase the hypoxic zone. The blame is shifting from N to P fertilizer. However, P fertilizer sales in the Mississippi River Basin have declined by 17% since 1980.

Can hypoxia be blamed on sales of N or P fertilizer? Numerous nutrient sources contribute to Gulf loading. Atmospheric deposition, decomposition of crop residue and soil organic matter, legumes, animal manure, municipal sewage sludge and effluent, and composted household wastes all contribute nutrients to the Gulf.

Hypoxia results from a complex interaction of chemical, biological, and physical factors. Fertilizer is a potential pollutant if used improperly, but used correctly, it increases food production and helps protect the environment.

Terry L. Roberts is vice president of the Potash & Phosphate Institute (PPI) and vice president of the Foundation for Agronomic Research (FAR), located in Norcross, Georgia. Dr. Roberts is a Certified Crop Adviser and a Fellow of the American Society of Agronomy.

Act Now to Save These Resources

The springtime area of low-oxygen (anoxic) water in the Gulf of Mexico, known as the dead zone, is driven by a massive influx of nutrients into a system no longer able to process them. Eutrophication begins when nutrients from farmlands in the floodplain states wash to the sea. These nutrients (nitrogen fertilizers) now present in the water lead to plankton blooms, which in turn reduce dissolved oxygen in the water and eventually kill fish.

Taking a system view is slightly more complicated, but understanding the system is important for the most effective long-term management. Before people built levees all along the delta, the Mississippi River flooded each spring, and the waters of the river covered the extensive wetlands. This important renewal process deposited sediment on the wetlands to build up the soil base while the plants of the wetlands made use of the nutrient pollutants in the water. The result was cleaner water, richer wetlands, and a sustained environment. Levees now prevent the flooding, the dead zone emerges, and the wetlands are lost as they sink below sea level. Rises in sea level speed the loss.

Loss of wetlands is serious. The wetlands are the base of the fisheries of the Gulf of Mexico, and their loss is irreversible. Saving the wetlands and reversing the dead zone requires a twofold approach. First, reduce the amount of fertilizer so that it is used more efficiently by plants and so that less enters streams. This has the added benefit of saving money and reducing energy consumption (making fertilizers is energy intensive). Second, reinstate the flooding of the wetlands.

Should we wait to act? No. We know enough now to design strategies that can sustain these resources, and new information is unlikely to change what we know. The precautionary principle, which environmental managers use, says that even if information is imperfect, it is important to act before the resource is lost entirely and while any possible cost of error is small and manageable. We need to act now to save these resources.

Paul Templet is a professor at the Institute for Environmental Studies at Louisiana State University. He organized the first Earth Day at LSU in 1970 and served as the secretary of the Louisiana Department of Environmental Quality from 1988 to 1992.

 Explore this issue further by accessing **Viewpoints** at www.aw-bc.com/withgott.

Conclusion

Earth hosts many interacting systems, and the way one perceives them depends on the questions in which one is interested. Physical systems and processes such as the hydrologic cycle, the rock cycle, and plate tectonics lay the groundwork for the ways in which life spreads itself across the planet. Life interacts with its abiotic environment in ecosystems, systems through which energy flows and materials are recycled. Understanding the biogeochemical cycles that describe the movement of nutrients within and among ecosystems is crucial, because human activities are causing significant changes in the ways those cycles operate.

Thinking in terms of systems is important in understanding how Earth works, so that we may learn how to avoid disrupting its processes and how to mitigate any disruptions we cause. By studying the environment from a systems perspective and by integrating scientific findings with the policy process, people who care about the Mississippi River and the Gulf of Mexico are working today to solve their pressing problems. Their model is one that we can adapt to many other issues in environmental science.

We might also consider adopting other models more generally. Think again about unperturbed ecosystems, their use of renewable solar energy, their recycling of nutrients, and the extent to which they exhibit dynamic equilibrium and involve negative feedback loops. The environmental systems we see on Earth today are those that have survived the test of time. Our industrialized civilization is young in comparison. Might we not be able to take a few lessons about sustainability from a careful look at the natural systems of our planet?

REVIEWING OBJECTIVES

You should now be able to:

Describe the nature of environmental systems

▶ Systems are networks of interacting components that generally involve feedback loops, show dynamic equilibrium, and result in emergent properties. (pp. 184–187)

▶ Because Earth's natural systems are so complex, environmental scientists often take a holistic approach to studying environmental systems. (pp. 184, 187–188)

▶ Because environmental systems interact and overlap, one's delineation of systems depends on the questions in which one is interested. (p. 189)

Define ecosystems and evaluate how living and nonliving entities interact in ecosystem-level ecology

▶ Ecosystems consist of all organisms and nonliving entities that occur and interact in a particular area at the same time. (pp. 189–190)

▶ Energy flows in one direction through ecosystems, whereas matter is recycled. (p. 191)

▶ Energy is converted to biomass, and ecosystems vary in their productivity. (pp. 192–193)

▶ Input of nutrients can boost productivity, but an excess of nutrients can alter ecosystems in ways that cause severe ecological and economic consequences. (pp. 192–195)

Compare and contrast how carbon, phosphorus, nitrogen, and water cycle through the environment

▶ Most carbon is contained in sedimentary rock. Substantial amounts also occur in the oceans and in soil. Carbon flux between organisms and the atmosphere occurs via photosynthesis and respiration. (pp. 195–197)

▶ Phosphorus, like carbon, is most abundant in sedimentary rock, with substantial amounts in soil and the oceans. Phosphorus has no appreciable atmospheric pool, however. It is a key nutrient for plant growth. (pp. 197–198)

▶ Nitrogen, like phosphorus, is a vital nutrient for plant growth. Most nitrogen is in the atmosphere, however, and must be "fixed" by specialized bacteria or lightning before plants can use it. (pp. 197–200)

▶ Water moves widely through the environment in the hydrologic cycle. (pp. 204–205)

▶ Humans are causing substantial impacts to Earth's biogeochemical cycles. These impacts include shifting carbon from fossil fuel reservoirs into the atmosphere, shifting nitrogen from the atmosphere to the planet's surface, and depleting groundwater supplies, among many others. (pp. 197, 200–206)

Explain how plate tectonics and the rock cycle shape the earth beneath our feet

▶ Matter is cycled within the lithosphere, and rocks transform from one type to another. (pp. 206–207)

▶ Plate tectonics is a fundamental system that produces earthquakes and volcanoes and guides Earth's physical geography. (pp. 207–209)

TESTING YOUR COMPREHENSION

1. Which type of feedback loop is most common in nature, and which more commonly results from human action? How might the emergence of a positive feedback loop affect a system in homeostasis?
2. Describe how hypoxic conditions can develop in coastal marine ecosystems such as the northern Gulf of Mexico.
3. What is the difference between an ecosystem and a community?
4. Describe the typical movement of energy through an ecosystem. Describe the typical movement of matter through an ecosystem.
5. What role do each of the following play in the carbon cycle?
   ▶ Cars
   ▶ Photosynthesis
   ▶ The oceans
   ▶ Earth's crust
6. Describe the difference between the function performed by nitrogen-fixing bacteria and that performed by denitrifying bacteria.
7. How has human activity altered the carbon cycle? The phosphorus cycle? The nitrogen cycle? To what environmental problems have these changes given rise?
8. What is the difference between evaporation and transpiration? Give examples of how the hydrologic cycle interacts with the carbon, phosphorus, and nitrogen cycles.
9. Name the three main types of rocks, and describe how each type may be converted to the others via the rock cycle.
10. How does plate tectonics account for mountains? For volcanoes? For earthquakes? Why do you think it took so long for scientists to discover such a fundamental environmental system as plate tectonics?

SEEKING SOLUTIONS

1. In this chapter we discussed how system boundaries can be difficult to determine. Can you think of a truly closed system whose boundaries are easily defined? If so, try to describe such a system and its boundaries.
2. Consider the ecosystem(s) that surround(s) your campus. How do some of the principles from our discussion on ecosystems apply to the ecosystem(s) around your campus?
3. A simple change in the flux rate between just two reservoirs in a single nutrient cycle can potentially have major consequences for ecosystems and, indeed, for the globe. Explain how this can be, using one example from the carbon cycle and one example from the nitrogen cycle.
4. How do you think we might solve the problem of eutrophication in the Gulf of Mexico? Assess several possible solutions, your reasons for believing they might work, and the likely hurdles we might face. Explain who should be responsible for implementing solutions, and why.
5. Imagine that you are a shrimper on the Louisiana coast and that your income is decreasing year by year because the dead zone is making it harder and harder to catch shrimp. One day your senator comes to town, and you have a one-minute audience with her. What steps would you urge her to take in Washington, D.C., to try to help alleviate the dead zone and bring back the shrimp fishery?
6. Now imagine that you are an Iowa farmer and that you have learned that the federal government is insisting that you use 30% less fertilizer on your crops each year. You know that in good growing years you could do without that fertilizer, and you'd be glad not to have to pay for it. But in bad growing years, you need the fertilizer to ensure a harvest so that you can continue making a living. And you must apply the fertilizer each spring before you know whether it will be a good or bad year. What would you tell your senator when she comes to town?

INTERPRETING GRAPHS AND DATA

Scientists are debating what effects global climate change (Chapter 18) may have on nutrient cycles. As soil becomes warmer, especially at far northern latitudes, nutrients in the soil should become more available to plants, stimulating plant growth. One hypothesis is that more carbon will end up stored in the soil as a result, because plants will pull carbon from the atmosphere and transfer it to the soil reservoir as they shed leaves or die.[1] Under this hypothesis, increased flux of carbon from the atmosphere to the soil would act as negative feedback counteracting climate warming, because less carbon in the atmosphere would lead to less warming.

To test whether the carbon flux actually changes in this way when nutrients are made more available in a tundra ecosystem, researchers are conducting a long-term study in Alaska.[2] For 20 years, they have added fertilizer to treatment plots while leaving control plots unfertilized. Recently, they estimated amounts of carbon by measuring biomass aboveground and belowground in both sets of plots. Aboveground biomass consists of living plant material, whereas belowground biomass consists mostly of nonliving organic material stored in the soil and not yet decomposed. Some of the research team's results are presented in the graph below.

1. Calculate the sizes of the aboveground, belowground, and total carbon pools (in g C/m^2) for the Control and Fertilized treatment groups.
2. What do the aboveground data indicate about the effect of fertilizer on plant growth? What do the belowground data indicate about the effect of fertilizer on organic material stored in the soil?
3. Do the data support the hypothesis that the net effect of increased nutrient availability will be to remove carbon from the atmosphere and store it in the soil? Using Figure 7.9 as a reference, can you suggest a different hypothesis that might explain the data better? Based on this data, would you predict that the warming of tundra soil will decrease the atmospheric concentration of CO_2 and act as negative feedback to climate change, or increase it and act as positive feedback?

[1]Hobbie, S. E., et al. 2002. A synthesis: The role of nutrients as constraints on carbon balances in boreal and arctic regions. *Plant Soil* 242: 163–170.
[2]Mack, M. C., et al. 2004. Ecosystem carbon storage in arctic tundra reduced by long-term nitrogen fertilization. *Nature* 431: 440–443.

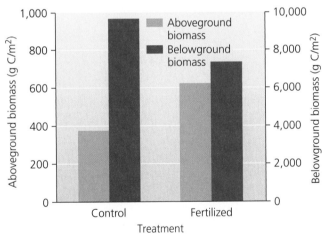

Effects of 20 years of fertilization (10 g N and 5 g P/m^2/yr) on C pools in tundra near Toolik Lake, Alaska. Differences are statistically significant for the aboveground, belowground, and total carbon pools.

CALCULATING ECOLOGICAL FOOTPRINTS

In the United States, a common dream is to own your own home, surrounded by a weed-free, green lawn. Nationwide there are about 20 million acres of lawn grass, making it our largest single crop!

Assuming that all of the populations indicated in the following table have lawns and will fertilize them at a typical fertilizer application rate of 45 lb per acre, calculate the total amount of nitrogen that will be applied to their lawns. When estimating the number of lawns, assume that the typical household includes three people.

Fertilizer application	Number of lawns	Pounds of nitrogen
To your 1/3-acre lawn	1	15
To the lawns of your classmates		
To the lawns of all your schoolmates		
To all the lawns in your hometown		
To all the lawns in your state		
To all the lawns in the United States	60,000,000	

1. Where does all of this nitrogen go? Where does it come from?
2. What other environmental impacts are caused by fertilizer production, transport, and application?

Take It Further

 Go to www.aw-bc.com/withgott or the student CD-ROM where you'll find:

▶ Suggested answers to end-of-chapter questions
▶ Quizzes, animations, and flashcards to help you study
▶ *Research Navigator*™ database of credible and reliable sources to assist you with your research projects

▶ **GRAPHit!** tutorials to help you master how to interpret graphs
▶ **INVESTIGATE it!** current news articles that link the topics that you study to case studies from your region to around the world

PART TWO

Environmental Issues and the Search for Solutions

Canal Street, New Orleans, after Hurricane Katrina

8 Human Population

Hong Kong, China

Upon successfully completing this chapter, you will be able to:

▶ Assess the scope of human population growth

▶ Evaluate how human population, affluence, and technology affect the environment

▶ Explain and apply the fundamentals of demography

▶ Outline and assess the concept of demographic transition

▶ Describe how wealth and poverty, the status of women, and family planning programs affect population growth

▶ Characterize the dimensions of the HIV/AIDS epidemic

Central Case:
China's One-Child Policy

"Population growth is analogous to a plague of locusts. What we have on this earth today is a plague of people."
—TED TURNER,
MEDIA MAGNATE AND
SUPPORTER OF THE UNITED
NATIONS POPULATION FUND

"There is no population problem."
—SHELDON RICHMAN,
SENIOR EDITOR,
CATO INSTITUTE

The People's Republic of China is the world's most populous nation, home to one-fifth of the 6.5 billion people living on Earth at the start of 2006. When Mao Zedong founded the country's current regime 57 years earlier, roughly 540 million people lived in a mostly rural, war-torn, impoverished nation. Mao believed population growth was desirable, and under his leadership China grew and changed. By 1970, improvements in food production, food distribution, and public health allowed China's population to swell to approximately 790 million people. At that time, the average Chinese woman gave birth to 5.8 children in her lifetime.

Unfortunately, the country's burgeoning population and its industrial and agricultural development were eroding the nation's soils, depleting its water, leveling its forests, and polluting its air. Chinese leaders realized that the nation might not be able to feed its people if their numbers grew much larger. They saw that continued population growth could exhaust resources and threaten the stability and economic progress of Chinese society. The government decided to institute a population-control program that precluded large numbers of Chinese couples from having more than one child.

The program began with education and outreach efforts encouraging people to marry later and have fewer children. Along with these efforts, the Chinese government increased the accessibility of contraceptives and abortion. By 1975, China's annual population growth rate had dropped from 2.8% to 1.8%. To further decrease birthrates, in 1979 the government took the more drastic step of instituting a system of rewards and punishments to enforce a one-child limit. One-child families received better access to schools, medical care, housing, and government jobs, and mothers with only one child were given longer maternity leaves. Families with more than one child, meanwhile, were subjected to social scorn and ridicule, employment discrimination, and monetary fines. In some cases, the fines exceeded half the offending couple's annual income.

In enforcing these policies, China has, in effect, been conducting one of the largest and most controversial social experiments in history. In purely quantitative terms, the experiment has been a major success; the nation's growth rate is now down to 0.6%, making it easier for the country to deal with its many social, economic, and environmental challenges. However, China's population control policies have also produced unintended consequences, such as widespread killing of female infants, an unbalanced sex ratio, and a black-market trade in teen-aged girls. Moreover, the policies have elicited intense criticism from those who oppose government intrusion into personal reproductive choices.

China embarked on its policy because its leaders felt it necessary. As other nations become more and more crowded, might their governments also feel forced to turn to drastic policies that restrict individual freedoms? In this chapter, we examine human population dynamics worldwide, consider their causes, and assess their consequences for the environment and our society.

Human Population Growth: Baby 6 Billion and Beyond

While China was working to slow its population growth and speed its economic growth, on the other side of the Eurasian continent, a milestone was reached in 1999. On the morning of October 12 of that year, the first cries of a newborn baby in Sarajevo, Bosnia-Herzegovina, marked the arrival of the six-billionth human being on our planet (Figure 8.1). At least that was how the milestone was symbolically marked by the United Nations, which monitors human population growth, among other global trends.

Just how much is 6 billion? We often have trouble conceptualizing the scale of huge numbers like a billion. Keep in mind that a billion is 1,000 times greater than a million. If you were to count once each second without ever sleeping, it would take over 30 years to reach a billion. In order to put a billion miles on your car, you would need to drive from New York to Los Angeles more than 350,000 times.

FIGURE 8.1 U.N. Secretary-General Kofi Annan recognized the newborn son of Fatima Nevic and her husband, Jasminko, as our six-billionth neighbor. Although it is impossible to know the precise moment—or day, week, or even month—the world's population reached 6 billion, U.N. population experts pinpointed October 12, 1999, as the best approximation to make the symbolic declaration. Many observers interpreted the selection of a child born in war-ravaged Sarajevo as a harbinger of the hard times that could face future generations as population grows and competition for scarce resources increases.

The human population is growing nearly as fast as ever

As we saw in Chapter 1 (▶ pp. 4–5), the human population has been growing at a tremendous rate. Our population has doubled just since 1964 and is growing by roughly 78 million people annually (nearly 2.5 people every *second*). This is the equivalent of adding all the people of California, Texas, and New York to the world each year. It took until after 1800, virtually all of human history, for our population to reach 1 billion. Yet we reached 2 billion by 1930, and 3 billion in just 30 more years, in 1960. Our population added its next billion in just 15 years (1975), its next billion in a mere 12 years (1987), and its most recent billion in another 12 years (Figure 8.2). Think about when you were born and how many people have been added to the planet just since that time. This unprecedented growth means that today's generations are in circumstances that previous generations never experienced. Our grandparents never had to deal with the number of people that crowd our planet today.

How and why has our growth accelerated? We saw in Chapter 5 (▶ p. 136) how exponential growth—the increase in a quantity by a *fixed percentage* per unit time—accelerates the absolute increase of population size over time, just as compound interest accrues in a savings

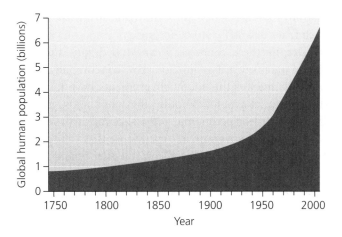

FIGURE 8.2 The global human population has grown exponentially, rising from less than 1 billion in 1800 to over 6.5 billion today. Data from U.S. Bureau of the Census.

account. The reason, you will recall, is that a given percentage of a large number is a greater quantity than the same percentage of a small number. Thus, even if the growth rate remains steady, population size will increase by greater increments with each successive generation.

In fact, our growth rate has not remained steady. Instead, for much of the 20th century, the growth rate of the human population actually rose from year to year. It peaked at 2.1% during the 1960s and has declined to 1.2% since then. Although 1.2% may sound small, exponential growth endows small numbers with large consequences. For instance, a hypothetical population starting with 1 man and 1 woman that grows at 1.2% gives rise to a population of 2,939 after 40 generations and 112,695 after 60 generations. In today's world, rates of annual growth vary greatly from region to region. Figure 8.3 maps this variation.

At a 2.1% annual growth rate, a population doubles in size in only 33 years. For low rates of increase, we can estimate doubling times with a handy rule-of-thumb. Just take the number 70, and divide it by the annual percentage growth rate: 70 ÷ 2.1 = 33.3. Had China not instituted its one-child policy—that is, had its growth rate remained unchecked at 2.8%—it would have taken only 25 years to double in size. Had population growth continued at this rate, China's population would have surpassed 2 billion people in 2004.

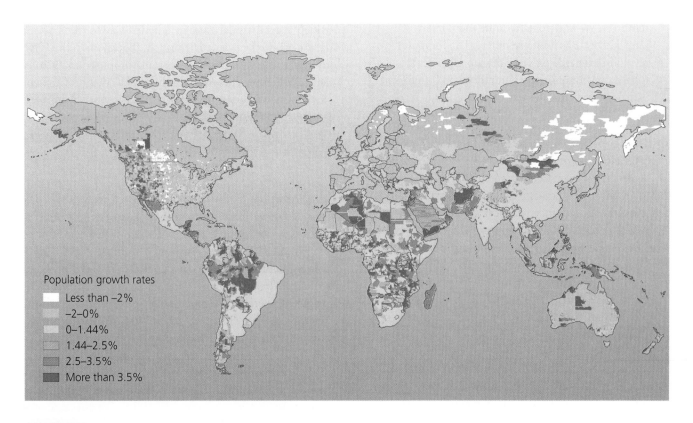

FIGURE 8.3 A map of population growth rates from the period 1990–1995 shows great variation from place to place. Population is growing fastest in tropical regions and in some desert and rainforest areas that have historically been sparsely populated. Data from Center for International Earth Science Information Network (CIESIN), Columbia University; and Harrison, P., and F. Pearce. 2000. *AAAS atlas of population and environment.* Berkeley, CA: University of California Press.

Is population growth really a "problem"?

Our ongoing population growth has resulted largely from technological innovations, improved sanitation, better medical care, increased agricultural output, and other factors that have led to a decline in death rates, particularly a drop in rates of infant mortality. Birth rates have not declined as much, so births have outpaced deaths for many years now. Thus, the so-called population problem actually arises from a very good thing—our ability to keep more of our fellow humans alive longer.

Indeed, just as the mainstream view in the day of Thomas Malthus (▸pp. 4–5) held that population increase was a good thing, there are many people today who argue that population growth poses no problems. Under the Cornucopian view that many economists hold, resource depletion due to population increases is not a problem if new resources can be found to replace depleted resources (▸p. 43). Libertarian writer Sheldon Richman expressed this view at the time the six-billionth baby was born:

> The idea of carrying capacity doesn't apply to the human world because humans aren't passive with respect to their environment. Human beings create resources. We find potential stuff and human intelligence turns it into resources. The computer revolution is based on sand; human intelligence turned that common stuff into the main component [silicon] of an amazing technology.

In contrast to Richman's point of view, environmental scientists recognize that few resources are actually created by humans and that not all resources can be replaced once they are depleted. For example, once species have gone extinct, we cannot replicate their exact function in ecosystems, or know what medicines or other practical applications we might have obtained from them, or regain the educational and aesthetic value of observing them. Another irreplaceable resource is land, that is, space in which to live; we cannot expand Earth like a balloon to increase its surface area.

Even if resource substitution could hypothetically enable population growth to continue indefinitely, could we maintain the *quality* of life that we would desire for ourselves and our descendants? Surely some of today's resources are bound to be easier or cheaper to use, and less environmentally destructive to harvest or mine, than any resources that can replace them. Replacing such resources might make our lives more difficult or less pleasant. In any case, unless resource availability keeps pace with population growth, the average person in the future will have less space in which to live, less food to eat, and less material wealth than the average person does today. Thus population increases are indeed a problem if they create stress on resources, social systems, or the natural environment, such that our quality of life declines.

Despite these considerations—and despite the fact that in today's world population growth is correlated with poverty, not wealth—many governments have found it difficult to let go of the notion that population growth increases a nation's economic, political, or military strength. Many national governments, even those that view global population increase as a problem, still offer financial and social incentives that encourage their own citizens to produce more children. Governments of countries currently experiencing population declines (such as many in Europe) feel especially uneasy. According to the Population Reference Bureau, more than 3 of every 5 European national governments now take the view that their birth rates are too low, and none state that theirs is too high. However, outside Europe, 56% of national governments feel their birth rates are too high, and only 8% feel they are too low.

Population is one of several factors that affect the environment

The extent to which population increase can be considered a problem involves more than just numbers of people. One widely used formula gives us a handy way to think about factors that affect the environment. Nicknamed the **IPAT model,** it is a variation of a formula proposed in 1974 by Paul Ehrlich (▸pp. 5–6) and John Holdren, a professor of environmental policy at Harvard University. The IPAT model represents how our total impact (I) on the environment results from the interaction among population (P), affluence (A), and technology (T):

$$I = P \times A \times T$$

Increased population intensifies impact on the environment as more individuals take up space, use natural resources, and generate waste. Increased affluence magnifies environmental impact through the greater per capita resource consumption that generally has accompanied enhanced wealth. Changes in technology may either decrease or increase human impact on the environment. Technology that enhances our abilities to exploit minerals, fossil fuels, old-growth forests, or ocean fisheries generally increases impact, but technology to reduce smokestack emissions, harness renewable energy, or improve manufacturing efficiency can decrease impact.

We might also add a sensitivity factor (S) to the equation to denote how sensitive a given environment is to human pressures:

$$I = P \times A \times T \times S$$

For instance, the arid lands of western China are more sensitive to human disturbance than the moist regions of southeastern China. Plants grow more slowly in the arid west, making deforestation and soil degradation more likely. Thus, adding an additional person to western China should have more environmental impact than adding one to southeastern China.

We could refine the IPAT equation further by adding terms for the effects of social institutions, such as education, laws and their enforcement, stable and cohesive societies, and ethical standards that promote environmental well-being. Factors like these all affect how population, affluence, and technology translate into environmental impact.

Impact can be thought of in various ways, but it can generally be boiled down to either pollution or resource consumption. Pollution became a problem in the modern world once our population grew large enough that we produced great quantities. The depletion of resources by larger and hungrier populations has been a focus of scientists and philosophers since Malthus's time. Recall how on Easter Island (▸ pp. 8–9), islanders brought down their own civilization by depleting their most important limited resource, trees. History offers other cases in which resource depletion helped end civilizations, from the

Mayans to the Mesopotamians. Some environmental scientists have predicted similar problems for our global society in the near future if we do not manage to embark on a path toward sustainability (Figure 8.4).

However, as we noted in Chapter 1, Malthus and his "neo-Malthusian" followers have not yet seen their direst predictions come true. The reason is that we have developed technology—the T in the IPAT equation—time and again to alleviate our strain on resources and allow us to further expand our population. For instance, we have employed technological advances to increase global agricultural production faster than our population has risen (▸ pp. 278–279).

Modern-day China shows how all elements of the IPAT formula can combine to cause tremendous environmental impact in very little time. The world's fastest-growing economy over the past two decades, China is "demonstrating what happens when large numbers of poor people rapidly become more affluent," in the words of Earth Policy Institute president Lester Brown. While millions of Chinese are increasing their material wealth and their consumption of resources, the country is battling unprecedented environmental challenges brought

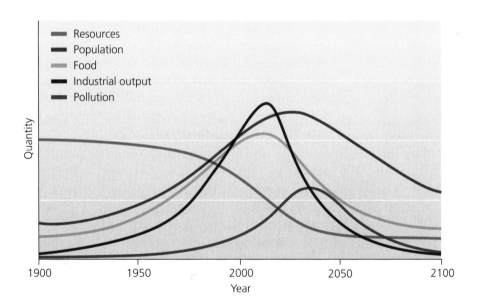

FIGURE 8.4 Environmental scientists Donella Meadows, Jorgen Randers, and Dennis Meadows used computer simulations to generate a series of projections of trends in human population, resource availability, food production, industrial output, and pollution. Their projections for trends over the coming century are based on data from the past century and current scientific understanding of the environment's biophysical limits. Shown here is their projection for a world in which "society proceeds in a traditional manner without any major deviation from the policies pursued during most of the twentieth century." In this projection, population and production increase until declining nonrenewable resources make further growth impossible, causing population and production to decline rather suddenly. The researchers also ran their simulations with different parameters to examine possible alternative futures. Under a scenario with policies aimed at sustainability, population leveled off at 8 billion, production and resource availability leveled off at medium-high levels, and pollution declined to low levels. Data from Meadows, D., et al. 2004. *Limits to growth: The 30-year update.* White River Junction, VT: Chelsea Green Publishing.

about by its pell-mell economic development. Intensive agriculture has expanded westward out of the country's historic moist rice-growing areas, causing farmland to erode and literally blow away, much like the Dust Bowl tragedy that befell the U.S. agricultural heartland in the 1930s (▶ p. 260). China has overpumped many of its aquifers and has drawn so much water for irrigation from the Yellow River that the once-mighty waterway now dries up in many stretches. Although China has been reducing its air pollution from industry and charcoal-burning homes, the country faces new urban pollution and congestion threats from rapidly increasing numbers of automobiles. As the world's developing countries try to attain the level of material prosperity that industrialized nations enjoy, China is a window on what much of the rest of the world could soon become.

Weighing the **Issues:**
Population Growth and Reproductive Freedom

It is often suggested that if human population growth remains unchecked, everyone will eventually suffer a poorer quality of life. Would you be willing to make this sacrifice if it meant that people in other countries (such as China) could avoid government-imposed limitations on their reproductive freedom? If your own government ever implemented a strict reproductive policy, how would you feel? Would you rather have the government limit your reproductive freedom or your consumption?

Demography

As we have seen, it is a fallacy to think of people as being somehow outside nature. Humans exist within their environment as one species out of many. As such, all the principles of population ecology we outlined in Chapter 5 that apply to toads, frogs, and passenger pigeons apply to humans as well. Environmental factors set limits on our population growth, and the environment has a carrying capacity (▶ pp. 136–137) for our species, just as it does for every other.

We happen to be a particularly successful organism, however—one that has repeatedly raised its carrying capacity by developing technology to overcome the natural limits on its growth. We did so with the agricultural and the industrial revolutions (▶ p. 4) and likely before that with our invention of tools (Figure 8.5).

Environmental scientists who have tried to pin a number to the human carrying capacity have come up with wildly differing estimates. The most rigorous estimates range from 1–2 billion people living prosperously in a healthy environment to 33 billion living in extreme poverty in a degraded world of intensive cultivation without natural areas. As our population climbs toward 7 billion and beyond, we may yet continue to find ways to raise our carrying capacity. Given our knowledge of population ecology, however, we have no reason to presume that human numbers can go on growing indefinitely. Indeed, as we have seen (see Figure 5.17d, ▶ p. 138), populations that exceed their carrying capacity can crash.

FIGURE 8.5 Tool making, the advent of agriculture, and industrialization each allowed our species to raise its global carrying capacity. The logarithmic scale of the axes makes it easier to visualize this pattern. Data from Goudie, A. 2000. *The human impact.* Cambridge, MA: MIT Press.

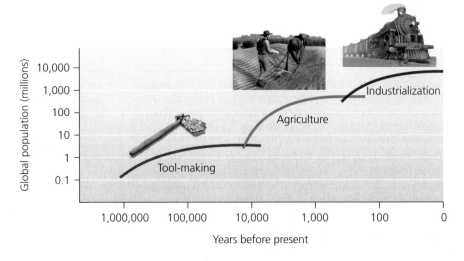

Demography is the study of human population

The application of population ecology principles to the study of statistical change in human populations is the focus of the social science of **demography.** The field of demography developed along with and partly preceding population ecology, and the disciplines have influenced and borrowed from one another. Data gathered by demographers help us understand how differences in population characteristics and related phenomena (for instance, decisions about reproduction) affect human communities and their environments. Demographers study population size, density, distribution, age structure, sex ratio, and rates of birth, death, immigration, and emigration of humans, just as population ecologists study these characteristics in other organisms. Each of these characteristics is useful for predicting population dynamics and potential environmental impacts.

Population size The global human population of more than 6.5 billion consists of well over 200 nations with populations ranging from China's 1.3 billion, India's 1.1 billion, and the 300 million of the United States down to a number of island nations with populations below 100,000 (Figure 8.6). The size that our global population will eventually reach remains to be seen (Figure 8.7). However, population size alone—the absolute number of individuals—doesn't tell the whole story. Rather, a population's environmental impact depends on its density, distribution, and composition (as well as on affluence, technology, and other factors outlined earlier).

Population density and distribution People are distributed very unevenly over the globe. In ecological terms, our distribution is clumped (▸ pp. 133–134) at all spatial scales. At the largest scales (Figure 8.8), population density is high in regions with temperate, subtropical, and tropical climates, such as China, Europe, Mexico, southern Africa, and India. Population density is low in regions with extreme-climate biomes, such as desert, deep rainforest, and tundra. Dense along seacoasts and rivers, human population is less dense at locations far from water. At intermediate scales, we cluster together in cities and suburbs and are spread more sparsely across rural areas. At small scales, we cluster in certain neighborhoods and in individual households.

This uneven distribution means that certain areas bear far more environmental impact than others. Just as the Yellow River has experienced intense pressure from millions of Chinese farmers, the world's other major rivers, from the Nile to the Danube to the Ganges to the

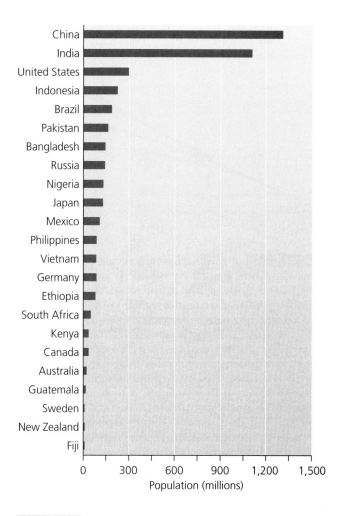

FIGURE 8.6 The world's nations range in human population from several thousand (on some South Pacific islands) up to China's 1.3 billion. Shown here are the 2005 populations for the world's most populous 15 countries, followed by a selection of other countries. Data from Population Reference Bureau. 2005. *2005 world population data sheet.*

Mississippi, have all received more than their share of human impact. The urban way of life entails the packaging and transport of goods, intensive fossil fuel consumption, and hotspots of pollution. However, people's concentration in cities relieves pressure on ecosystems in less-populated areas by releasing some of them from direct human development (▸ p. 395).

At the same time, areas with low population density are often vulnerable to environmental impacts, because the reason they have low populations in the first place is that they are sensitive and cannot support many people (a high S value in our revised IPAT model). Deserts, for instance, are easily affected by development that commandeers a substantial share of available water. Grasslands can be turned to deserts if they are farmed too intensively, as has happened across vast stretches of the Sahel region

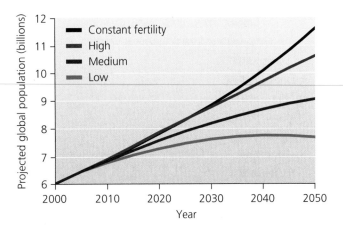

FIGURE 8.7 The United Nations predicts trajectories of world population growth, presenting its estimates in several scenarios based on different assumptions of fertility rates. In this 2004 projection, population is estimated to reach 11.7 billion in the year 2050 if fertility rates remain constant at 2004 levels (top line in graph). However, U.N. demographers expect fertility rates to continue falling, so they arrived at a best guess (*medium* scenario) of 9.1 billion for the human population in 2050. In the *high* scenario, if women on average have 0.5 child more than in the medium scenario, population will reach 10.6 billion in 2050. In the *low* scenario, if women have 0.5 child less than in the medium scenario, the world will contain 7.7 billion people in 2050. Data from United Nations Population Division. 2004. *World population prospects: The 2004 revision.*

bordering Africa's Sahara Desert, in the Middle East, and in parts of China and the United States.

Age structure Data on the age structure or age distribution of human populations are especially valuable to demographers trying to predict future dynamics of populations. As we saw in Chapter 5 (▶p. 134), large proportions of individuals in young age groups portend a great deal of reproduction and, thus, rapid population growth. Examine age pyramids for the nations of Canada and Madagascar (Figure 8.9). Not surprisingly, it is Madagascar that has the greater population growth rate. In fact, its annual growth rate, 2.7%, is 9 times that of Canada's 0.3%.

By causing dramatic reductions in the number of children born since 1970, China virtually guaranteed that its population age structure would change. Indeed, in 1995 the median age in China was 27; by 2030 it will be 39. In 1997 there were 125 children under age 5 for every 100 people 65 or older in China, but by 2030 there will be only 32. The number of people older than 65 will rise from 100 million in 2005 to 236 million in 2030 (Figure 8.10). This dramatic shift in age structure will challenge China's economy, health care systems, families, and military forces because fewer working-age people will be available to support social programs that assist the increasing number of older people. However, the shift in age structure also reduces the proportion of dependent children. The reduced

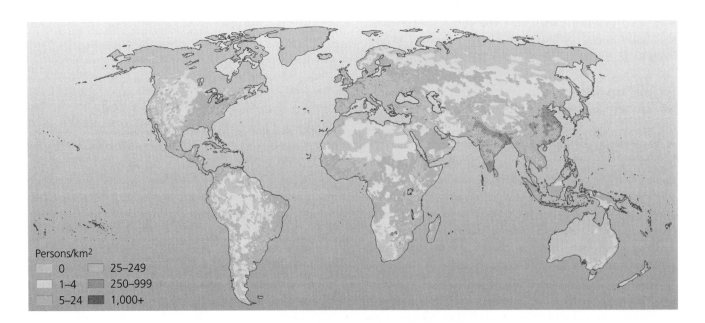

FIGURE 8.8 Human population density varies tremendously from one region to another. Arctic and desert regions have the lowest population densities, whereas areas of India, Bangladesh, and eastern China have the densest populations. Data are for 2000, from Center for International Earth Science Information Network (CIESIN), Columbia University; and Centro Internacional de Agricultura Tropical (CIAT), 2004.

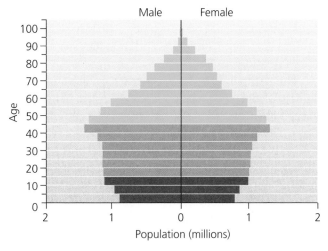

(a) Age pyramid of Canada in 2005

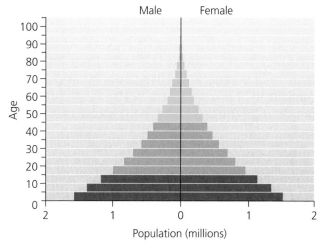

(b) Age pyramid of Madagascar in 2005

FIGURE 8.9   Canada (**a**) shows a balanced age structure, with relatively even numbers of individuals in various age classes. Madagascar (**b**) shows an age distribution heavily weighted toward young people. Madagascar's population growth rate is 9 times that of Canada's. Go to **GRAPHIT!** at www.aw-bc.com/withgott or on the student CD-ROM. Data from U.N. Population Division.

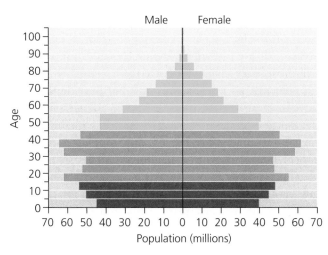

(a) Age pyramid of China in 2005

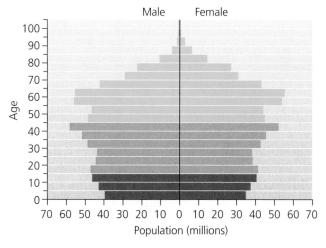

(b) Projected age pyramid of China in 2030

FIGURE 8.10   As China's population ages, older people will outnumber the young. Age pyramids show the predicted graying of the Chinese population between 2005 (**a**) and 2030 (**b**). Today's children may, as working-age adults (**c**), face pressures to support greater numbers of older citizens than has any previous generation. Data from U.N. Population Division.

(c) Young female factory workers in Hong Kong

number of young adults may mean a decrease in the crime rate. Moreover, older people are often productive members of society, contributing volunteer activities and services to their children and grandchildren. Clearly, in terms of both benefits and drawbacks, life in China will continue to be profoundly affected by the particular approach its government has taken to population control.

Weighing the Issues:
China's Reproductive Policy

Consider the benefits as well as the problems associated with a reproductive policy such as China's. Do you think a government should be able to enforce strict penalties for citizens who fail to abide by such a policy? If you disagree with China's policy, what alternatives can you suggest for dealing with the resource demands of a quickly growing population?

This pattern of aging in the population is occurring in many countries, including the United States (Figure 8.11). Older populations will present new challenges for many nations, as increasing numbers of older people require the care and financial assistance of relatively fewer working-age citizens.

Sex ratios The ratio of males to females also can affect population dynamics. Imagine two islands, one populated by 99 men and 1 woman and the other by 50 men and 50 women. Where would we be likely to see the greatest population increase over time? Of course, the island with an equal number of men and women would have a greater number of potential mothers and thus a greater potential for population growth.

The naturally occurring sex ratio in human populations at birth features a slight preponderance of males; for every 100 female infants born, 105 to 106 male infants are born. This phenomenon may be an evolutionary adaptation to the fact that males are slightly more prone to death during any given year of life. It usually ensures that the ratio of men to women is approximately equal at the time people reach reproductive age. Thus, a slightly uneven sex ratio at birth may be beneficial. However, a greatly distorted ratio can lead to problems.

In recent years, demographers have witnessed an unsettling trend in China: The ratio of newborn boys to girls has become strongly skewed. In the 2000 census, 120 boys were reported born for every 100 girls. Some provinces reported sex ratios as high as 138 boys for every 100 girls. A leading hypothesis for these unusual sex ratios is that many parents, having learned the sex of their fetuses by ultrasound, are selectively aborting female fetuses. Traditionally, Chinese culture has valued sons because they can carry on the family name, assist with farm labor in rural areas, and care for aging parents. Daughters, in contrast, will most likely marry and leave their parents, as the culture dictates. As a result, they will not provide the same benefits to their parents as will sons. Sociologists hold that this cultural gender preference, combined with the government's one-child policy, has led some couples to selectively abort female fetuses or to abandon or kill female infants.

China's skewed sex ratio may have the effect of further lowering population growth rates. However, it has proved tragic for the "missing girls." It is also beginning to have the undesirable social consequence of leaving many Chinese men single. This, in turn, has resulted in a grim new phenomenon. In parts of rural China, teen-aged girls are being kidnapped and sold to families in other parts of the country as brides for single men.

Population growth depends on rates of birth, death, immigration, and emigration

Just as they do for other organisms, rates of birth, death, immigration, and emigration help determine whether a human population grows, shrinks, or remains stable. The formula for measuring population growth that we used in Chapter 5 (▶ p. 135) also pertains to humans: birth and immigration add individuals to a population, whereas death and emigration remove individuals. Technological advances have led to a dramatic decline in

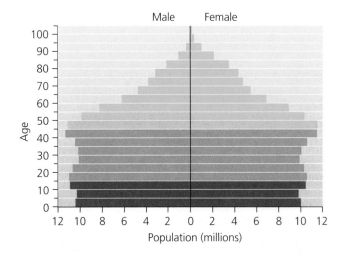

FIGURE 8.11   The "baby boom" is visible in the 2005 age pyramid for the United States, in the age brackets between 40 and 50. In future years the nation will experience an aging population as baby-boomers grow older. Data from U.N. Population Division.

human death rates, widening the gap between birth rates and death rates and resulting in the global human population expansion.

In today's ever-more-crowded world, immigration and emigration are playing increasingly large roles. Refugees, people forced to flee their home country or region, have become more numerous in recent decades as a result of war, civil strife, and environmental degradation. The United Nations puts the number of refugees who flee to escape poor environmental conditions at 25 million per year and possibly many more. Often the movement of refugees causes environmental problems in the receiving region as these desperate victims try to eke out an existence with no livelihood and with no cultural or economic attachment to the land or incentive to conserve its resources. The millions who fled Rwanda following the genocide there in the mid-1990s, for example, inadvertently destroyed large areas of forest while trying to obtain fuelwood, food, and shelter to stay alive once they reached the Democratic Republic of Congo (Figure 8.12).

For most of the past 2,000 years, China's population has been relatively stable. The first significant increases resulted from enhanced agricultural production and a powerful government during the Qing, or Manchu, Dynasty in the 1800s. Population growth began to outstrip food supplies by the mid-1850s, and quality of life for the average Chinese peasant began to decline. From the mid-1800s, an era of increased European intervention in China, until 1949, China's population grew very slowly, at about 0.3% per year. This slow population growth was due, in part, to food shortages and political instability. As we have seen, population growth rates rose again follow-

Table 8.1 Trends in China's Population Growth				
Measure	1950	1970	1990	2005
Total fertility rate	5.8	5.8	2.2	1.6
Rate of natural population increase (% per year)	1.9	2.6	1.4	0.6
Doubling time (years)	37	27	49	117

Data from China Population Information and Research Center, 2005; and Population Reference Bureau. 2005. *2005 World population data sheet.*

ing Mao's establishment of the People's Republic, and they have declined since the establishment of the one-child policy (Table 8.1).

Since 1970, growth rates in many countries have been declining, even without population control policies, and the global growth rate has declined (Figure 8.13). This decline has come about, in part, from a steep drop in birth rates.

A population's total fertility rate influences population growth

One key statistic demographers calculate to examine a population's potential for growth is the **total fertility rate (TFR),** or the average number of children born per female member of a population during her lifetime. **Replacement fertility** is the TFR that keeps the size of a population stable. For humans, replacement fertility is equal to a TFR of 2.1. When the TFR drops below 2.1, population size, in the absence of immigration, will shrink.

FIGURE 8.12 The flight of refugees from Rwanda into the Democratic Republic of Congo in 1994 following the Rwandan genocide caused tremendous hardship for the refugees and tremendous stress on the environment into which they moved.

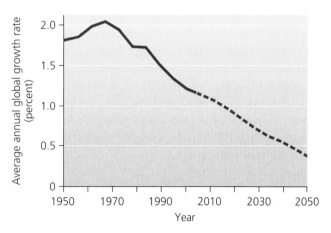

FIGURE 8.13 The annual growth rate of the global human population peaked in the 1960s and has been declining since then. The dashed line indicates projected future trends. Data from United Nations Population Division. 2004. *World population prospects: The 2004 revision.*

Table 8.2	Total Fertility Rates for Major Continental Regions
Region	**Total fertility rate (TFR)**
Africa	5.1
Latin America and the Caribbean	2.6
Asia	2.5
Oceania	2.1
North America	2.0
Europe	1.4

Data from Population Reference Bureau. 2005. *2005 World population data sheet.*

Various factors influence TFR and have acted to drive it downward in many countries in recent years. Historically, people tended to conceive many children, which helped ensure that at least some would survive, but lower infant mortality rates have made this less necessary. Increasing urbanization has also driven TFR down; whereas rural families need children to contribute to farm labor, in urban areas children are usually excluded from the labor market, are required to go to school, and impose economic costs on their families. If a government provides some form of social security, as most do these days, parents need fewer children to support them in their old age when they can no longer work. Finally, with greater education and changing roles in society, women tend to shift into the labor force, putting less emphasis on child rearing.

All these factors have come together in Europe, where TFR has dropped from 2.6 to 1.4 in the past half-century. Every European nation now has a fertility rate below the replacement level, and populations are declining in 18 of 43 European nations. In 2005, Europe's overall annual **natural rate of population change** (change due to birth and death rates alone, excluding migration) was –0.1%. Worldwide by 2005, a total of 71 countries had fallen below the replacement fertility of 2.1. These countries made up roughly 45% of the world's population and included China (with a TFR of 1.6). Table 8.2 shows the TFRs of major continental regions.

Weighing the Issues:
Consequences of Low Fertility?

In the United States, Canada, and every European nation, the total fertility rate has now dipped below the replacement fertility rate. What economic or social consequences do you think might result from below-replacement fertility rates?

Some nations have experienced a change called the demographic transition

Many nations that have lowered their birth rates and TFRs have been going through a similar set of interrelated changes. In countries with good sanitation, good health care, and reliable food supplies, more people than ever before are living long lives. As a result, over the past 50 years the life expectancy for the average person has increased from 46 to 67 years as the global crude death rate has dropped from 20 deaths per 1,000 people to 9 deaths per 1,000 people. Strictly speaking, **life expectancy** is the average number of years that an individual in a particular age group is likely to continue to live, but often people use this term to refer to the average number of years a person can expect to live from birth. Much of the increase in life expectancy is due to reduced rates of infant mortality. Societies going through these changes are mostly the ones that have undergone urbanization and industrialization and have been able to generate personal wealth for their citizens.

To make sense of these trends, demographers developed a concept called the **demographic transition.** This is a model of economic and cultural change proposed in the 1940s and 1950s by demographer Frank Notestein and elaborated on by others to explain the declining death rates and birth rates that have occurred in Western nations as they became industrialized. Notestein believed nations moved from a stable pre-industrial state of high birth and death rates to a stable post-industrial state of low birth and death rates. Industrialization, he proposed, caused these rates to fall naturally by first decreasing mortality and then lessening the need for large families. Parents would thereafter choose to invest in quality of life rather than quantity of children. Because death rates fall before birth rates fall, a period of net population growth results. Thus, under the demographic transition model, population growth is seen as a temporary phenomenon that occurs as societies move from one condition to another.

The pre-industrial stage Notestein's demographic model describing the population impacts of industrialization proceeds in several stages (Figure 8.14). The first is the **pre-industrial stage,** characterized by conditions that have defined most of human history. In pre-industrial societies, both death rates and birth rates are high. Death rates are high because disease is widespread, medical care rudimentary, and food supplies unreliable and difficult to obtain. Birth rates are high because people must compensate for high mortality rates in infants and young children by having several children. In this stage, children are

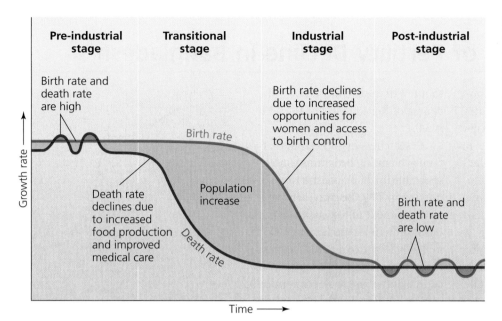

FIGURE 8.14 The demographic transition is an idealized process that has taken some populations from a pre-industrial state of high birth rates and high death rates to a post-industrial state of low birth rates and low death rates. In this diagram, the wide green area between the two curves illustrates the gap between birth and death rates that causes rapid population growth during the middle portion of this process. Data from Kent, M. M. and K. A. Crews. 1990. *World population: Fundamentals of growth.* Population Reference Bureau.

valuable as additional workers who can help meet a family's basic needs. Populations within the pre-industrial stage are not likely to experience much growth, which is why the human population was relatively stable from Neolithic times until the industrial revolution.

Industrialization and falling death rates Industrialization initiates the second stage of the demographic transition, known as the **transitional stage.** This transition from the pre-industrial stage to the industrial stage is generally characterized by declining death rates due to increased food production and improved medical care. Birth rates in the transitional stage remain high, however, because people have not yet grown used to the new economic and social conditions. As a result, population growth surges.

The industrial stage and falling birth rates The third stage in the demographic transition is the **industrial stage.** Industrialization increases opportunities for employment outside the home, particularly for women. Children become less valuable, in economic terms, because they do not help meet family food needs as they did in the pre-industrial stage. If couples are aware of this, and if they have access to birth control, they may choose to have fewer children. Birth rates fall, closing the gap with death rates and reducing the rate of population growth.

The post-industrial stage In the final stage, the **post-industrial stage,** both birth and death rates have fallen to low and stable levels. Population sizes stabilize or decline slightly. The society enjoys the fruits of industrialization without the threat of runaway population growth.

Is the demographic transition a universal process?

The demographic transition has occurred in many European countries, the United States, Canada, Japan, and several other developed nations over the past 200 to 300 years. Nonetheless, it is a model that may not apply to all developing nations as they industrialize now and in the future. Some social scientists doubt that it will apply; they point out that population dynamics may be different for developing nations that adopt the Western world's industrial model rather than devising their own. Some demographers assert that the transition will fail in cultures that place greater value on childbirth or grant women fewer freedoms.

Moreover, natural scientists warn that there are not enough resources in the world to enable all countries to attain the standard of living that developed countries now enjoy. It has been estimated that for all nations to enjoy the quality of life that United States citizens enjoy, we would need the natural resources of two more planet Earths. Whether developing nations, which include the vast majority of the planet's people, pass through the demographic transition as developed nations have is one of the most important and far-reaching questions for the future of our civilization and Earth's environment.

Population and Society

Demographic transition theory links the statistical study of human populations with various societal factors that influence, and are influenced by, population dynamics. Let's now examine a few of these major societal factors more closely.

Causes of Fertility Decline in Bangladesh

The Science behind the Story

Research in developing countries indicates that poverty and overpopulation can create a vicious cycle, in which poverty encourages high fertility and high fertility obstructs economic development. Are there policy steps that such countries can take to bring down fertility rates? Scientific analysis of family-planning programs in the South Asian nation of Bangladesh suggests that the answer is yes.

Bangladesh is one of the poorest, most densely populated countries on the planet. Its 145 million people live in an area about the size of Wisconsin, and 45% of them live below the poverty line. With few natural resources and 1,000 people per km^2 (over 2,500/mi^2—more than twice the population density of New Jersey), limiting population growth is critically important. As Bangladeshi president Ziaur Rahman declared in 1976, "If we cannot do something about population, nothing else that we accomplish will matter much."

Fortunately, Bangladesh has made striking progress in controlling population growth in the past three decades. Despite stagnant economic development, low literacy rates, poor health care, and limited rights for women, the nation's total fertility rate (TFR) has dropped markedly. In the 1970s, the average woman in Bangladesh gave birth to more than six children over the course of her life. Today, the TFR is 3.0.

Researchers hypothesized that family-planning programs were responsible for Bangladesh's rapid reduction in TFR. Because conducting an experiment to test such a hypothesis is difficult, some researchers took advantage of a natural experiment. By comparing Bangladesh to countries that are socioeconomically similar but have had less success in lowering TFR, such as Pakistan, researchers concluded that Bangladesh succeeded because of aggressive, well-funded outreach efforts that were sensitive to the values of its traditional society.

However, because no two countries are identical, it is difficult to draw firm conclusions from such broad-scale studies. This is why the Matlab Family Planning and Health Services Project, in the isolated rural area of Matlab, Bangladesh, has become one of the best-known experiments in family planning in developing countries. The Matlab Project was an intensive outreach program run collaboratively by the Bangladeshi government and international aid organizations. Each household in the project area received biweekly visits from local women offering counseling, education, and free contraceptives. Compared to a similar government-run program in a nearby area, the Matlab

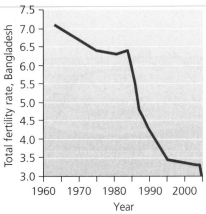

Total fertility rate has declined markedly in Bangladesh in the past 40 years, in part because of the enhanced availability of contraceptives. However, TFR has leveled off in recent years, suggesting that other societal changes are needed to lower it further. Data from Bangladesh Bureau of Statistics; Bangladesh Fertility Survey; The Global Reproductive Health Forum at Harvard; International Centre for Diarrhoeal Disease Research, Bangladesh; National Family Planning and Fertility Survey; United States Agency for International Development.

Project featured more training, more services, and more frequent visits. In both areas, a highly organized health surveillance system gave researchers detailed information about births, deaths, and health-related behaviors such as contraceptive use. The result was an experiment comparing the Matlab Project with the government-run area.

When Matlab Project director James Phillips and his colleagues

Women's empowerment greatly affects population growth rates

Many demographers had long believed that fertility rates were influenced largely by degrees of wealth or poverty.

However, affluence alone cannot determine TFR, because a number of developing countries now have fertility rates lower than that of the United States. Instead, recent research is highlighting factors pertaining to the social empowerment of women. Drops in TFR have

reviewed a decade's worth of data in 1988, they found that fertility rates had declined in both areas. The decline appeared to be due almost entirely to a rise in contraceptive use, because other factors—such as the average age of marriage—remained the same. Phillips and his colleagues also found that the declines had been significantly greater in the Matlab area than in the government-run area. These findings suggested that high-intensity outreach efforts can affect fertility rates even in the absence of significant improvements in women's status, education, or economic development.

But why exactly was the outreach program successful? One hypothesis was that visits from health care workers had helped convince local women that small families are desirable. However, in 1999, Mary Arends-Kuenning, a graduate student in economics at the University of Michigan, and her colleagues reported that there was no relationship between women's perception of the ideal family size and the number of visits made by outreach workers, either in Matlab or nearby comparison areas. Ideal family size declined equally in all areas. Instead of creating new demand for birth control, the Matlab Project appears to have helped women convert an already-existing desire for fewer children into behaviors, such as

In the Matlab Project, Bangladeshi households received visits from local women offering counseling, education, and free contraceptives.

contraceptive use, that reduce fertility.

Bangladesh's ability to rein in fertility rates despite unfavorable social and economic conditions bodes well for impoverished nations facing explosive population growth. However, significant challenges remain. Since the 1990s, Bangladesh's TFR has appeared to level off at slightly more than 3 children per woman. If rates fail to decline further, the country's population could double to 290 million—nearly the size of today's U.S. population—

within 30 years. Scientific research has helped illuminate the impact of family-planning programs on fertility, but further reductions may require fundamental social, political, and economic changes that are difficult to implement in traditional, resource-strapped countries such as Bangladesh. Nonetheless, the scientific evidence collected at Matlab since the 1970s has played an important role in informing population control efforts in Bangladesh and elsewhere.

been most noticeable in countries where women have gained improved access to contraceptives and education, particularly family-planning education (see "The Science behind the Story," above; also see Figure 8.15 and Figure 8.16).

In 2005, 53% of married women worldwide (ages 15–49) reported using some modern method of contraception to plan or prevent pregnancy. China, at 86%, had the highest rate of contraceptive use of any nation. Six western European nations showed rates of contraceptive

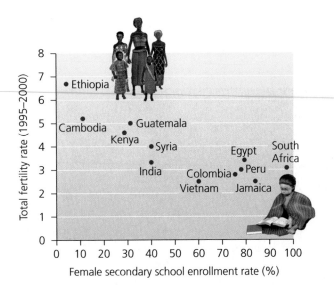

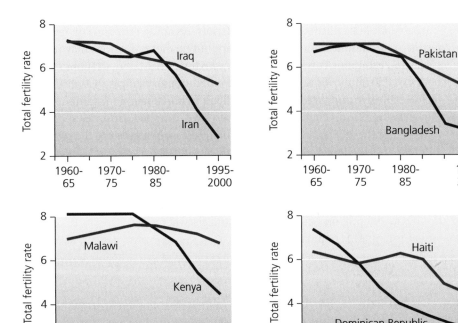

FIGURE 8.15 Increasing female literacy is strongly associated with reduced birth rates in many nations. Data from McDonald, M., and D. Nierenberg. 2003. Linking population, women, and biodiversity. *State of the World 2003*. Washington, D.C.: Worldwatch Institute.

use above 70%, as did Costa Rica, Cuba, New Zealand, Canada, Brazil, and Thailand (the U.S. rate was 68%). At the other end of the spectrum, 23 African nations had rates below 10%. These low rates of contraceptive use contribute to high fertility rates in sub-Saharan Africa, where the region's TFR is 5.6 children per woman. By comparison, in Asia, where the TFR in 1950 was 5.9, today it is 2.5—in part a legacy of the population control policies of China and some other Asian countries.

These data clearly indicate that in societies where women have little power, substantial numbers of pregnancies are unintended. Unfortunately, many women still lack the information and personal freedom of choice to allow them to make their own decisions about when to have children and how many to have. Today, many social scientists and policymakers recognize that for population growth to slow and stabilize, women need to achieve equal power with men in societies worldwide. Studies show that in societies in which women are freer to decide whether and when to have children, fertility rates have fallen, and the resulting children are better cared for, healthier, and better educated.

Unfortunately, we are still a long way from achieving gender equality. Over two-thirds of the world's people who cannot read, and 60% of those living in poverty, are women. Violence against women remains shockingly common. In many societies, by tradition men restrict women's decision-making abilities, including decisions as to how many children they will bear. The gap between the power held by men and the power held by women is just as obvious at the highest levels of government. Worldwide, only 13% of elected government officials in national legislatures are women (Figure 8.17). The United States

FIGURE 8.16 Data from four pairs of neighboring countries demonstrate the effectiveness of family planning in reducing fertility rates. In each case, the nation that invested in family planning and (in some cases) made other reproductive rights, education, and health care more available to women (blue lines) reduced its total fertility rate (TFR) far more dramatically than its neighbor (red lines). Data from U.N. Population Division; and Harrison, P., and F. Pearce. 2000. *AAAS atlas of population and environment* Berkeley, CA: University of California Press.

| Table 8.3 Per Capita Wealth, with Rates of Fertility, Population Growth, and Contraceptive Use, for Selected Nations | | | | | |
Nation	Per Capita GNI PPP (U.S. $)*	Population increase (% per year)	Children born per woman (TFR)	Population density (per mi²)	Infant mortality (per 1,000)	Percentage of couples using birth control
Ethiopia	810	2.5	5.9	182	100	6
Niger	830	3.4	8.0	29	153	4
Haiti	1,680	1.9	4.7	774	80	22
Cameroon	2,090	2.3	5.0	89	74	13
Pakistan	2,160	2.4	4.8	528	85	20
India	3,100	1.7	3.0	869	60	43
Nicaragua	3,300	2.7	3.8	115	36	66
Syria	3,550	2.7	3.7	257	22	35
China	5,530	0.6	1.6	353	27	86
Brazil	8,020	1.4	2.4	56	27	70
Mexico	9,590	1.9	2.6	142	25	59
Czech Republic	18,400	−0.1	1.2	335	4	58
Spain	25,070	0.1	1.3	223	4	53
United Kingdom	31,460	0.2	1.7	635	5	79
Japan	30,040	0.1	1.3	876	3	48
Canada	30,660	0.3	1.5	8	5	73
United States	39,710	0.6	2.0	80	7	68

*GNI PPP is "gross national income in purchasing power parity," a measure that standardizes income and makes it comparable among nations, by converting income to "international" dollars using a conversion factor. International dollars indicate the amount of goods and services one could buy in the United States with a given amount of money. Data from Population Reference Bureau. 2005. *2005 World population data sheet.*

the Bush administration has withheld funds, pointing out that U.S. law prohibits funding any organization that "supports or participates in the management of a program of coercive abortion or involuntary sterilization," and claiming that the Chinese government has been implicated in both these activities. Many nations and organizations criticized the U.S. decision, and the European Union offered additional funding to UNFPA to offset the loss of U.S. contributions. What do you think of the U.S. decision? Should the United States fund family planning efforts in other nations? What conditions, if any, should it place on the use of such funds?

Poverty is strongly correlated with population growth

The alleviation of poverty, one target of the Cairo conference, has been linked to population because poorer societies tend to show higher population growth rates than do wealthier societies. This pattern is consistent with demographic transition theory. Note in Table 8.3 how poorer nations tend to have higher fertility and growth rates,

along with higher birth and infant mortality rates and lower rates of contraceptive use.

Trends such as these have affected the distribution of people on the planet. In 1960, 70% of all people lived in developing nations. By 2005, 81% of the world's population was living in these countries. Moreover, fully 98% of the next billion people to be added to the global population will be born in these poor, less developed regions (Figure 8.18). This is unfortunate from a social standpoint, because these people will be added to the countries that are least able to provide for them. It is also unfortunate from an environmental standpoint, because poverty often results in environmental degradation. People dependent on agriculture in an area of poor farmland, for instance, may need to try to farm even if doing so degrades the soil and is not sustainable. This is largely why Africa's once-productive Sahel region, like many regions of western China, is turning to desert (Figure 8.19). Poverty also drives the hunting of many large mammals in Africa's forests, including the great apes that are now disappearing as local settlers and miners kill them for their "bush meat."

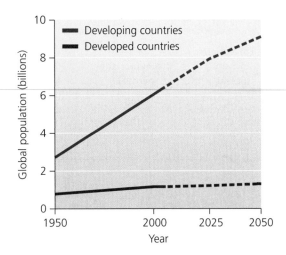

FIGURE 8.18 Nearly 98% of the next 1 billion people added to Earth's human population will reside in the less developed, poorer parts of the world. The dashed line indicates projected future trends. Data from U.N. Population Division; and Harrison, P., and F. Pearce. 2000. *AAAS atlas of population and environment.* 2000. Berkeley, CA: University of California Press.

Consumption from affluence creates environmental impact

Poverty can lead people into environmentally destructive behavior, but wealth can produce even more severe and far-reaching environmental impacts. The affluence that characterizes a society such as the United States, Japan, or the Netherlands is built on massive and unprecedented levels of resource consumption. Much of this chapter has dealt with numbers of people rather than on the amount of resources each member of the population consumes or the amount of waste each member produces. The environmental impact of human activities, however, depends not only on the number of people involved but also on the way those people live. Recall the A for affluence in the IPAT equation. Patterns of affluence and consumption are spread unevenly across the world, and affluent societies generally consume resources from other societies as well as from their own.

In Chapter 1 (▸pp. 6–7, 19, and 25), we introduced the concept of the *ecological footprint,* the cumulative amount of Earth's surface area required to provide the raw materials a person or population consumes and to dispose of or recycle the waste that they produce. Individuals from affluent societies leave a considerably larger per capita ecological footprint (see Figure 1.13, ▸p. 19). This fact should remind us that the "population problem" does not lie entirely with the developing world. Just as population is rising, so is consumption, and environmental scientists have calculated that we are already living beyond the planet's means to support us sustainably. One recent

FIGURE 8.19 In the semi-arid Sahel region of Africa, where population is increasing beyond the land's ability to handle it, dependence on grazing agriculture has led to environmental degradation.

analysis concluded that humanity's global ecological footprint surpassed Earth's capacity to support us in 1987 and that our species is now living more than 20% beyond its means (Figure 8.20).

The wealth gap and population growth contribute to violent conflict

The stark contrast between affluent and poor societies in today's world is, of course, the cause of social as well as

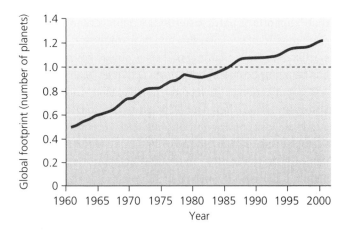

FIGURE 8.20 The global ecological footprint of the human population is 2.5 times larger than it was in 1961 and now exceeds what Earth can bear in the long run, scientists have calculated. The estimate shown here indicates that we have already overshot our carrying capacity by at least 20%; that is, we are using renewable natural resources 20% faster then they are being replenished. Data from WWF-World Wide Fund for Nature, 2004. *Living planet report.* Gland, Switzerland: WWF.

(a) A family living in the United States

(b) A family living in Egypt

FIGURE 8.21 A typical U.S. family **(a)** may own a large house, keep numerous material possessions, and have enough money to afford luxuries such as vacation travel. A typical family in a developing nation such as Egypt **(b)** may live in a small, sparsely furnished dwelling with few material possessions and little money or time for luxuries. The ubiquity of television sets, even among poor families of the developing world, means that the world's poor see representations (both real and exaggerated) of wealth in the United States as depicted on American TV shows. Many sociologists hold that this has increased the poor's awareness of the global wealth gap and has spurred aspirations for consumption among the poor of developing nations.

environmental stress. Over half the world's people live below the internationally defined poverty line of U.S. $2 per day. The richest one-fifth of the world's people possesses over 80 times the income of the poorest one-fifth (Figure 8.21). The richest one-fifth also uses 86% of the world's resources. That leaves only 14% of global resources—energy, food, water, and other essentials—for the remaining four-fifths of the world's population to share. As the gap between rich and poor grows wider and

as the sheer numbers of those living in poverty continue to increase, it seems reasonable to predict increasing tensions between the "haves" and the "have-nots." This is why the inequitable distribution of wealth is one of the key factors the U.S. Departments of Defense and State take into account when assessing the potential for armed conflict around the world, whether it be conventional warfare or terrorism.

HIV/AIDS is a major influence on populations in parts of the world

The rising material wealth and falling fertility rates of many industrialized nations today is slowing population growth in accordance with the demographic transition model. Some other nations, however, are not following Notestein's script. Instead, in these countries mortality is beginning to increase, presenting a scenario more akin to Malthus's fears. This is especially the case in countries where the HIV/AIDS epidemic has taken hold (Figure 8.22). African nations are being hit hardest. Of the 38 million people around the world infected with HIV/AIDS as of 2004, 25 million live in the nations of sub-Saharan Africa. The low rate of use of contraceptives, which contributes to this region's high fertility rate, also fuels the expansion of AIDS. One in every 13 people aged 15 to 49 in sub-Saharan Africa is infected with HIV, and for southern African nations, the figure is more than one in five.

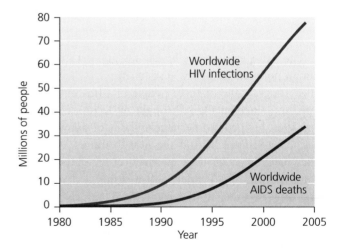

FIGURE 8.22 AIDS cases are increasing rapidly in much of the world. As of 2004, total cumulative HIV infections since 1980 were estimated at nearly 78 million, and 34 million people are estimated to have died from the disease so far. Data from UNAIDS; and *Vital signs 2005*. Washington, D.C.: Worldwatch Institute.

AIDS Resistance Genes and the Black Death: An Unexpected Connection?

The Science behind the Story

When HIV was first identified as the cause of AIDS in the early 1980s, some scientists predicted that a cure for the disease would be found within a few years. In the two decades that followed, however, progress toward a cure or vaccine remained slow, even as the number infected grew to more that 60 million. Thus, the discovery in 1996 of a genetic mutation that appeared to confer resistance to AIDS was an important breakthrough. Still more surprising was a possible connection between AIDS and the Black Death, a disease that decimated Europe in the mid-14th century.

Beginning in 1984, geneticists Stephen J. O'Brien, Michael Dean, and colleagues at the National Cancer Institute (NCI) began searching for individuals who were naturally resistant to HIV. They gathered genetic samples from people with and without the virus,

trying to identify mutations unique to people who had been exposed to HIV but remained uninfected.

Until the mid-1990s, the search was largely unsuccessful. Then in 1995, NCI researchers identified molecules called chemokines that carry signals from one part of the immune system to another. Other researchers discovered that HIV uses the immune system's own chemokine receptors to penetrate the cell membrane. Together, these advances revealed how HIV gains entry into human hosts. They also gave O'Brien, Dean, and other researchers a promising place to look for resistance genes.

The search quickly produced exciting results. In 1996, three separate groups reported the discovery of a mutation conferring resistance to HIV. The mutation—a deletion of 32 base-pairs of DNA—affected the chemokine receptor CCR5, which is expressed on the surface of macrophages, the first cells HIV

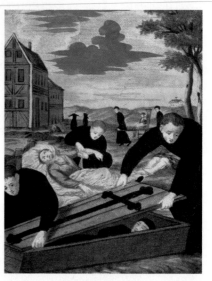

The Black Death killed one of every three Europeans in the 14th century but may have conferred on the descendants of its survivors some resistance to AIDS.

attacks when it enters a new host. People with two copies of the mutation appeared to be immune to HIV. People with one copy developed AIDS, but slowly.

The AIDS epidemic is having the greatest impact on human populations of any disease since the Black Death killed roughly one of three people in 14th-century Europe, and since smallpox brought by Europeans to the New World wiped out perhaps millions of native people (see "The Science behind the Story," above). As AIDS takes roughly 6,000 lives in Africa every day, the epidemic is unleashing a variety of demographic changes. Infant mortality in sub-Saharan Africa has risen to 96 deaths out of 1,000 live births—14 times the rate in the developed world. The high numbers of infant deaths and premature deaths of young adults has caused life expectancy in parts of southern Africa to fall from a high of close to 59 years in the early 1990s back down to less than 40 years, where it stood in the early 1950s. AIDS is also leaving behind millions of orphans. As of 2002, 14 million

children under the age of 15 worldwide had lost one or both parents to the disease.

Weighing the Issues:
HIV/AIDS and Population

What sorts of problems would you predict might occur in the surviving population after a major disease such as AIDS kills a high percentage of the population?

Severe demographic changes have social, political, and economic repercussions

Beyond its demographic effects, AIDS is having immediate social, economic, and political consequences. Everywhere in sub-Saharan Africa, AIDS is undermining the

Surprisingly, the mutation was unevenly distributed across human populations. One research team, led by Belgian scientists Michael Samson and Marc Parmentier, found that about 9% of the Caucasians they tested had at least one copy of the mutation, but none of the Japanese or Africans they tested had the mutation. Other groups found similar results, and it soon became clear that the mutation was present only in Europeans and their descendants.

A key step toward answering why the mutation was unique to Europeans occurred when O'Brien and Dean determined when the mutation had first arisen. To do so, they analyzed the mutation's association with distinctive genetic variants, or alleles, on the chromosome. In each generation, the association, or linkage, between the mutation and its neighbors is weakened by recombination, the chromosomal reshuffling that takes place during reproduction. Scientists can thus use the strength of the linkage to identify the mutation's approximate age. When O'Brien and Dean used this technique, they found that the CCR5 mutation was about 700 years old.

That put the mutation's origin at about the time when the Black Death, which most scientists have identified as bubonic plague, was ravaging Europe. Like the AIDS virus, the bacterium responsible for bubonic plague *(Yersinia pestis)* initially attacks macrophages, using them as a refuge before spreading to the rest of the body. Thus a mutation in CCR5 could potentially have conferred resistance to bubonic plague, just as it now confers resistance to AIDS. Under the strong selection pressure of the Black Death, which killed a third of Europe's population, the prevalence of the mutation could have risen greatly. Thus, the researchers suggested, the plague had one positive legacy: it offered the descendants of Europeans who survived it some degree of immunity to AIDS.

As with any exciting hypothesis, other researchers set out to confirm or refute it by testing the idea in different ways. This time, two labs taking different approaches cast doubt on the idea. Joan Mecsas and colleagues at Stanford University experimented with lab mice and found that mice with the CCR5 mutation were *not* protected against bubonic plague.

Shortly before the Mecsas team published its results in February 2004, two researchers from University of California–Berkeley conducted a modeling study that considered the population dynamics of diseases, and found that plague was not the most likely candidate for causing natural selection for the CCR5 mutation. Instead, they suggested that smallpox—another disease that killed millions in the Middle Ages—played this role.

ability of developing countries to make the transition to modern technologies because it is removing many of the youngest and most productive members of society. For example, in 1999 Zambia lost 600 teachers to AIDS, and only 300 new teachers graduated to replace them. In Rwanda, more than one in three college-educated residents of the city of Kigali are infected with the virus. South Africa loses an estimated $7 billion per year to declines in its labor force as AIDS patients fill the nation's hospitals (Figure 8.23). The loss of productive household members to AIDS causes families and communities to break down as income and food production decline while medical expenses and debt skyrocket.

These problems are hitting many countries at a time when their governments are already experiencing what has been called *demographic fatigue.* Demographically

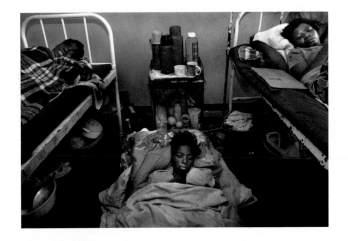

FIGURE 8.23 AIDS patients occupy 60% of South Africa's hospital beds. By 2010, AIDS in Africa may orphan an estimated 40 million children.

fatigued governments face overwhelming challenges related to population growth, including educating and finding jobs for their swelling ranks of young people. With the added stress of HIV/AIDS, these governments face so many demands that they are stretched beyond their capabilities to address problems. As a result, the problems grow worse, and citizens lose faith in their governments' abilities to help them.

If nations in sub-Saharan Africa—and other regions where the disease is spreading fast, such as India and southeast Asia—do not take aggressive steps soon, and if the rest of the world does not step in to help, these countries could fail to advance through the demographic transition. Instead, their rising death rates could push birth rates back up, potentially causing these countries to fall back to the pre-industrial stage of the demographic transition model. Such an outcome would lead to greater population growth while economic and social conditions worsen. It would be a profoundly negative outcome, both for human welfare and for the well-being of the environment.

If one of humanity's goals is to generate a high standard of living and quality of life for all the world's people, then developing nations must find ways to reduce their population growth. However, those of us living in the industrialized world must also be willing to reduce our consumption. Earth does not hold enough resources to sustain all 6.5 billion of us at the current North American standard of living, nor can we go out and find extra planets; so, we must make the best of the one place that supports us all.

Conclusion

Today, several years after welcoming its six-billionth member, the human population is larger than at any time in the past. Our growing population, as well as our growing consumption, affects the environment and our ability to meet the needs of all the world's people. Approximately 90% of children born today are likely to live their lives in conditions far less healthy and prosperous than most of us in the industrialized world are accustomed to.

However, there are at least two major reasons to be encouraged. First, although global population is still rising, the *rate* of growth has decreased nearly everywhere, and some countries are even seeing population declines. Most developed nations have passed through the demographic transition, showing that it is possible to lower death rates while stabilizing population and creating more prosperous societies. A second reason to feel encouraged is the progress in expanding rights for women worldwide. Although there is still a long way to go, women are slowly being treated more fairly, receiving better education, obtaining more economic independence, and gaining more ability to control their reproductive decisions. Aside from the clear ethical progress these developments entail, they are helping slow population growth.

Human population cannot continue to rise forever. The question is how it will stop rising: through the gentle and benign process of the demographic transition, through restrictive governmental intervention such as China's one-child policy, or through the miserable Malthusian checks of disease and social conflict caused by overcrowding and competition for scarce resources. Moreover, sustainability demands a further challenge—that we stabilize our population size in time to avoid destroying the natural systems that support our economies and societies. We are indeed a special species. We are the only one to come to such dominance as to change fundamentally so much of Earth's landscape, and even its climate system. We are also the only species with the intelligence needed to turn around an increase in our own numbers before we destroy the very systems on which we depend.

REVIEWING OBJECTIVES

You should now be able to:

Assess the scope of human population growth

▶ Our global population of 6.5 billion people adds about 78 million people per year (2.5 people every second). (p. 218)

▶ Our growth rate peaked at 2.1% in the 1960s and now stands at 1.2%. Growth rates vary among regions of the world. (p. 219)

▶ Rising population is a problem to the extent that it depletes resources, intensifies pollution, stresses social systems, or degrades ecosystems, such that the natural environment or our quality of life decline. (p. 220)

Evaluate how human population, affluence, and technology affect the environment

▶ The IPAT model summarizes how environmental impact (I) results from interactions among population size (P), affluence (A), and technology (T). (pp. 220–221)

▶ Rising population and rising affluence (leading to greater consumption) each increase environmental impact. Technological advances have frequently exacerbated environmental degradation, but they can also help mitigate our impact. (pp. 220–222)

Explain and apply the fundamentals of demography

▶ Demography applies principles of population ecology to the statistical study of human populations. (p. 223)

▶ Demographers study size, density, distribution, age structure, and sex ratios of populations, as well as rates of birth, death, immigration, and emigration. (pp. 223–227)

▶ Total fertility rate (TFR) contributes greatly to change in a population's size. (pp. 227–228)

Outline and assess the concept of demographic transition

▶ The demographic transition model explains why population growth has slowed in industrialized nations. Industrialization and urbanization have reduced the economic need for children, while education and the empowerment of women have decreased unwanted pregnancies. Parents in developed nations choose to invest in quality of life rather than quantity of children. (pp. 228–229)

▶ The demographic transition may or may not proceed to completion in all of today's developing nations. Whether it does is of immense importance for the quest for sustainability. (p. 229)

Describe how wealth and poverty, the status of women, and family planning programs affect population growth

▶ When women are empowered and achieve equality with men, fertility rates fall, and children tend to be better cared for, healthier, and better educated. (pp. 230–233)

▶ Family-planning programs and reproductive education have successfully reduced population growth in many nations. (pp. 230–233)

▶ Poorer societies tend to have higher population growth rates than do wealthier societies. (p. 235)

▶ The high consumption rates of affluent societies may make their ecological impact greater than that of poorer nations with larger populations. (pp. 236–237)

Characterize the dimensions of the HIV/AIDS epidemic

▶ About 38 million people worldwide are infected with HIV/AIDS, of which 25 million live in sub-Saharan Africa. (pp. 237–238)

▶ Epidemics that claim large numbers of young and productive members of society influence population dynamics and can have severe social and political ramifications. (pp. 238–240)

TESTING YOUR COMPREHENSION

1. What is the approximate current human global population? How many people are being added to the population each day?

2. Why has the human population continued to grow in spite of environmental limitations?

3. Contrast the views of environmental scientists with those of the libertarian writer Sheldon Richman and similar-thinking economists over whether population growth is a problem. Why does Richman think the concept of carrying capacity does not apply to human populations?

4. Explain the IPAT model. How can technology either increase or decrease environmental impact? Provide at least two examples.

5. What characteristics and measures do demographers use to study human populations? Which of these help determine the impact of human population on the environment?

6. What is the total fertility rate (TFR)? Can you explain why the replacement fertility for humans is approximately 2.1? How is Europe's TFR affecting its natural rate of population change?

7. Why have fertility rates fallen in many countries?

8. In the demographic transition model, why is the pre-industrial stage characterized by high birth and death rates, and the industrial stage by falling birth and death rates?

9. How does the demographic transition model explain the increase in population growth rates in recent centuries? How does it explain the decrease in population growth rates in recent decades?

10. Why do poorer societies have higher population growth rates than wealthier societies? How does poverty affect the environment? How does affluence affect the environment?

SEEKING SOLUTIONS

1. China's reduction in birth rates is leading to significant change in the nation's age structure. Review Figure 8.10, which portrays the projected change. You can see that the population is growing older, based on the top-heavy age pyramid for the year 2030. What sorts of effects might this ultimately have on Chinese society? Explain your answer.

2. The World Bank estimates that half the world's people survive on less than the equivalent of two dollars per day. What effect would you expect this situation to have on the political stability of the world? Explain your answer.

3. Apply the IPAT model to the example of China provided in the chapter. How do population, affluence, technology, and ecological sensitivity affect China's environment? Now consider your own country, or your own state. How do population, affluence, technology, and ecological sensitivity affect your environment? How can we regulate the relationship between population and its effects on the environment?

4. Do you think that all of today's developing nations will complete the demographic transition and come to enjoy a permanent state of low birth and death rates? Why or why not? What steps might we as a global society take to help ensure that they do? Now think about developed nations like the United States and Canada. Do you think these nations will continue to lower and stabilize their birth and death rates in a state of prosperity? What factors might affect whether they do so?

5. Imagine that India's prime minister puts you in charge of that nation's population policy. India has a population growth rate of 1.7% per year, a TFR of 3.0, a 43% rate of contraceptive use, and a population that is 72% rural. What policy steps would you recommend, and why?

6. Now imagine that you have been tapped to design population policy for Germany. Germany is losing population at an annual rate of 0.1%, has a TFR of 1.3, a 72% rate of contraceptive use, and a population that is 88% urban. What policy steps would you recommend, and why?

INTERPRETING GRAPHS AND DATA

Below are graphed data representing the economic condition of the world's population. The *y* axis indicates the per capita income for each country or region expressed as purchasing power in U.S. dollars (termed *gross national income* *in purchasing power parity,* or *GNI PPP*; see Table 8.3). The *x* axis indicates the cumulative percentage of the world population whose per capita GNI PPP is equal to or greater than that country's or region's per capita GNI PPP. The horizontal dotted line indicates the global average per capita GNI PPP.

1. What percentage of the world population lives at or below the global average per capita GNI PPP? What percentage lives at or below one half of the global average per capita GNI PPP? What percentage lives at or above twice the global average per capita GNI PPP?

2. Given a global average per capita GNI PPP of $8,540 and a world population of 6,477,000,000 people (as of mid-2005), what is the total global GNI PPP? What would the global GNI PPP be if everyone lived at the level of affluence of the United States?

3. How do you personally resolve the ethical conflict between the desirable goal of raising the standard of living of the billions of desperately poor people in the world and the likelihood that increasing their affluence (A in the equation I = PAT) will have a negative impact on the environment?

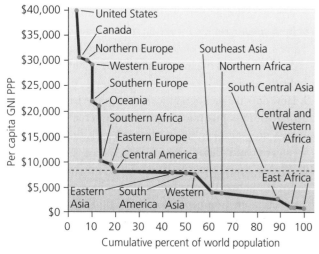

Percent of world population at various income levels. Data source: Population Reference Bureau. 2005. *World population data sheet 2005.*

CALCULATING ECOLOGICAL FOOTPRINTS

The equation I = PAT (Impact = Population × Affluence × Technology) suggests that a population's size and affluence are not the only determinants of its ecological impact; its technological choices also have an effect. Technologies can be either efficient or wasteful. One way of gauging the relative value of T is to calculate a per capita value of I/A (equivalent to I divided by A divided by P). The table presents per capita values of I (estimated ecological footprints) and A (income). Calculate the relative values of T by completing the blank column.

Country	Impact (ecological footprint, in acres per capita)	Affluence (per capita income, in GNI PPP)	Technology (I/A) (footprint per $1,000 income)
Bangladesh	1.2	$1,980	0.61
Colombia	4.9	$6,820	
Mexico	6.4	$9,590	
Sweden	14.6	$29,770	
Thailand	6.9	$8,020	
United States	25.4	$39,710	
World average	6.9	$8,540	

Data sources: Population Reference Bureau. 2005. *World population data sheet 2005*; and Wackernagel, M., et al. 1999. National natural capital accounting with the ecological footprint concept. *Ecological Economics* 29: 375–390.

1. If the world average value of T were decreased (improved) to that of the United States, what per capita GNI PPP could be supported at the current average per capita ecological footprint of 6.9 acres?

2. What value of T would enable the world's population to live at its current affluence within the 4.9 acres per capita that Wackernagel et al. estimate are available? Do you think this is achievable?

3. Which country's technological choices would you choose to study if you were interested in learning how to maximize your standard of living while minimizing your ecological impact? Using the value of T for this country and the mid-2005 world population of 6,477,000,000, calculate the following:

   (a) The number of people the world could support at the current per capita impact of 6.9 acres and affluence of $8,540

   (b) The number of people the world could support sustainably on the available 4.9 acres per capita and at an affluence of $8,540

   (c) The per capita GNI PPP that the world's current population could achieve on 6.9 acres per capita

   (d) The per capita GNI PPP that the world's current population could achieve on 4.9 acres per capita

---

Take It Further

Go to www.aw-bc.com/withgott or the student CD-ROM where you'll find:

▶ Suggested answers to end-of-chapter questions

▶ Quizzes, animations, and flashcards to help you study

▶ *Research Navigator*™ database of credible and reliable sources to assist you with your research projects

▶ **GRAPHit!** tutorials to help you master how to interpret graphs

▶ **INVESTIGATEit!** current news articles that link the topics that you study to case studies from your region to around the world

9 Soil and Agriculture

Wheat fields in Paraná, Brazil

Upon successfully completing this chapter, you will be able to:

▶ Explain the importance of soils to agriculture, and describe the impacts of agriculture on soils

▶ Outline major historical developments in agriculture

▶ Delineate the fundamentals of soil science, including soil formation and the properties of soil

▶ State the causes and predict the consequences of soil erosion and soil degradation

▶ Recite the history and explain the principles of soil conservation

No-till farm in Ceara, Brazil

Atlantic Ocean

Pacific Ocean

Brazil

Paraná

Santa Catarina

Rio Grande do Sul

Central Case: No-Till Agriculture in Southern Brazil

"The nation that destroys its soil destroys itself."
— U.S. President Franklin D. Roosevelt

"There are two spiritual dangers in not owning a farm. One is the danger of supposing that breakfast comes from the grocery, and the other that heat comes from the furnace."
— Conservationist and philosopher Aldo Leopold

In southernmost Brazil, hundreds of thousands of people make their living farming. The warm climate and rich soils of this region's rolling highlands and coastal plains have historically made for bountiful harvests. However, repeated cycles of plowing and planting over many decades diminished the productivity of the soil. More and more topsoil—the valuable surface layer of soil richest in organic matter and nutrients—was being eroded away by water and wind. Meanwhile, the synthetic fertilizers used to restore nutrients were polluting area waterways. Yields were falling, and by 1990 farmers were looking for help.

As a result, many of southern Brazil's farmers abandoned the conventional practice of tilling the soil after harvests. In its place, they turned to *no-tillage* farming, otherwise known as *zero-tillage, no-till,* or in Brazil, *plantio direto.* Turning the earth by tilling (plowing, disking, harrowing, or chiseling) aerates the soil and works weeds and old crop residue into the soil to nourish it. Tilling, however, also leaves the surface bare of vegetation for a period of time, during which erosion by wind and water can remove precious topsoil. Although tilling historically boosted the productivity of agriculture in Europe, many experts now think it is less appropriate for soils in subtropical regions such as southern Brazil. The reason is that the heavy rainfall of tropical and subtropical regions results in greater rates of erosion, causing tilled soils to lose organic matter and nutrients.

Working with agricultural scientists and government extension agents, southern Brazil's farmers began leaving crop residues on their fields after harvesting, and planting "cover crops" to keep soil protected during periods when they weren't raising a commercial crop.

When they went to plant the next crop, they merely cut a thin, shallow groove into the soil surface, dropped in seeds, and covered them. They did not invert the soil as they had when tilling, and the soil stayed covered with plants or their residues at all times, reducing erosion by 90%.

With less soil eroding away, and more organic material being added to it, the soil held more water and was better able to support crops. The improved soil quality meant better plant growth and greater crop production. In the state of Santa Catarina, maize yields per hectare increased by 47% between 1991 and 1999, wheat yields rose by 82%, and soybean yields by 83%, according to local farmers, extension agents, and international scientists. In the states of Paraná and Rio Grande do Sul, maize yields were up 67% over 10 years, and soybean yields were up 68%.

Besides boosting yields, no-till farming methods reduced farmers' costs, because farmers now used less labor and less fuel. No-till agriculture spread quickly in the region as farmers saw their neighbors' successes and traded information through "Friends of the Land" clubs organized on local, municipal, regional, and statewide levels. In Paraná and Rio Grande do Sul, the area being farmed with no-till methods shot up from 700,000 ha (1.7 million acres) in 1990 to 10.5 million ha (25.9 million acres) in 1999, when it involved 200,000 farmers. In Santa Catarina, where farms are generally smaller, over 100,000 farmers now apply no-till methods to 880,000 ha (2.2 million acres) of farmland. No-till farming is now spreading northward into Brazil's tropical regions and to other parts of Latin America.

By enhancing soil conditions and reducing erosion, no-till techniques have benefited southern Brazil's society and environment as well; its air, waterways, and ecosystems are less polluted. Similar effects are being felt in parts of the United States and elsewhere in the world where no-till and reduced-tillage methods are increasingly being applied.

Reduced tillage is certainly not a panacea for all areas of the world. In general, tropical areas benefit more than temperate regions, because erosion is greater in the tropics and hot weather can overheat tilled soil. The benefits and drawbacks of different tillage approaches vary with location, soil characteristics, and type of crop. In regions suitable for reduced tillage, proponents say these approaches can help make agriculture sustainable. We will need sustainable agriculture if we are to feed the world's human population while protecting the natural environment, including the soils that vitally support our production of food.

Soil: The Foundation for Feeding a Growing Population

As the human population has increased, so have the amounts of land and resources we devote to agriculture, which currently covers 38% of Earth's land surface. We can define **agriculture** as the practice of raising crops and livestock for human use and consumption. We obtain most of our food and fiber from **cropland,** land used to raise plants for human use, and **rangeland** or pasture, land used for grazing livestock.

Healthy soil is vital for agriculture, for forestry (Chapter 12), and for the functioning of Earth's natural systems. **Soil** is not merely lifeless dirt; it is a complex plant-supporting system consisting of disintegrated rock, organic matter, water, gases, nutrients, and microorganisms. Each of these components can be altered by the way we treat soil. Productive soil is a renewable resource, but if we abuse it through careless or uninformed practices, we can greatly reduce its productivity.

As population and consumption increase, soils are being degraded

If we are to feed the world's rising human population, we will need to change our diet patterns or increase agricultural production—and do so sustainably, without degrading the environment and reducing its ability to support agriculture. However, we cannot simply keep expanding agriculture into new areas, because land suitable and available for farming is running out. Instead, we must find ways to improve the efficiency of food production in areas that are already in agricultural use.

Today many lands unsuitable for farming are being farmed, causing considerable environmental damage. Mismanaged agriculture has turned grasslands into deserts and has removed ecologically precious forests. It has extracted nutrients from soils and added them to water bodies, harming both systems. It has diminished biodiversity; encouraged invasive species; and polluted soil, air, and water with toxic chemicals. Poor agricultural practices have allowed countless tons of fertile soil to be blown and washed away.

As our planet gains over 70 million people each year, we lose 5–7 million ha (12–17 million acres) of productive cropland annually. Throughout the world, especially in drier regions, it has gotten more difficult to raise crops and graze livestock as soils have become eroded and

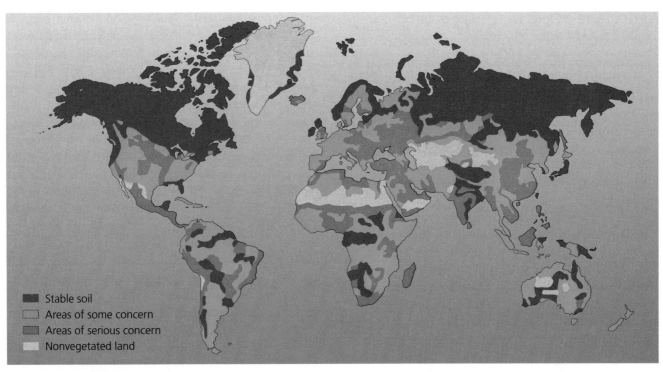

(a) World soil conditions

Stable soil
Areas of some concern
Areas of serious concern
Nonvegetated land

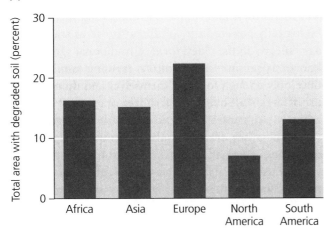

(b) Soil degradation by continent

FIGURE 9.1 Soils are becoming degraded in many areas worldwide **(a)**. Europe currently has a higher proportion of degraded land than other continents **(b)** because of its long history of intensive agriculture, but degradation is rising quickly in developing countries in Africa and Asia. Go to **GRAPHit!** at www.aw-bc.com/withgott or on the student CD–ROM. Data from International Soil Reference and Information Centre (ISRIC) and United Nations Environment Programme (UNEP), 1996. *Human–induced soil degradation.* Rome: ISRIC, UNEP, and U.N. Food and Agriculture Organization (FAO) (a); UNEP 2002. *Global environmental outlook 3.* London: UNEP and Earthscan Publ. (b).

degraded (Figure 9.1). Soil degradation around the globe has resulted from roughly equal parts forest removal, cropland agriculture, and overgrazing of livestock (Figure 9.2).

Soil degradation has direct impacts on agricultural production. It is estimated that degradation over the past 50 years has reduced potential rates of global grain production by 13% on cropland and 4% on rangeland. By the middle of the 21st century, there will likely be 3 billion more mouths to feed. For these reasons, it is imperative that we learn to farm in sustainable ways that are gentler on the land and that maintain the integrity of soil.

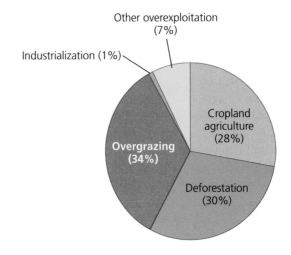

FIGURE 9.2 The great majority of the world's soil degradation results from cropland agriculture, overgrazing by livestock, and deforestation. Data from Wali, M. K. et al., 1999. Assessing terrestrial ecosystem sustainability: Usefulness of regional carbon and nitrogen models. *Nature and Resources* 35: 21–33.

Agriculture began to appear around 10,000 years ago

Agriculture is a relatively new approach for meeting our nutritional needs; on the scale of human history, there was no such thing as a farm until very recently. During most of our species' 160,000-year existence, we were hunter-gatherers, depending on wild plants and animals. Then about 10,000 years ago, as the climate warmed following a period of glaciation, people in some cultures began to raise plants from seed and to domesticate animals.

How and why did agriculture begin? The most plausible hypothesis is that it began as hunter-gatherers brought back to their encampments wild fruits, grains, and nuts. Some of these foods fell to the ground, were thrown away, or were eaten and survived passage through the digestive system. The plants that grew from these seeds near human encampments likely produced fruits that were on average larger and tastier than those in the wild, because they sprang from seeds of fruits selected by people because they were especially large and delicious. As these plants bred with others nearby that shared their characteristics, they gave rise to subsequent generations of plants with large and flavorful fruits. Eventually, people realized that they could guide this selective process through conscious effort, and our ancestors began intentionally planting seeds from the plants whose produce was most desirable. This is, of course, artificial selection at work (▸pp. 120–122). This practice of selective breeding continues to the present day and has produced the many hundreds of crops we enjoy, all of which are artificially selected versions of wild plants. People followed the same process of selective breeding with animals, creating livestock from wild species.

Once our ancestors learned to cultivate crops and raise animals, they began to settle in more permanent camps and villages, often near water sources. Agriculture and a sedentary lifestyle likely reinforced one another in a positive feedback cycle (▸p. 185). The need to harvest crops kept people sedentary, and once they were sedentary, it made sense to plant more crops. Population increase resulted from these developments and further promoted them. Moreover, the ability to grow excess farm produce enabled some people to leave farming and live off the food that others produced. This led to the development of professional specialties, commerce, technology, densely populated urban centers, social stratification, and politically powerful elites. For better or worse, the advent of agriculture eventually brought us the civilization we have today.

Archaeological and paleoecological evidence suggests that agriculture was invented independently by different cultures in at least five areas of the world and possibly 10 or more (Figure 9.3). The earliest widely accepted archaeological evidence for plant domestication is from the "Fertile Crescent" region of the Middle East about 10,500 years ago, and the earliest evidence for animal domestication also is from that region, just 500 years later. Crop remains have been dated using radiocarbon dating (▸p. 94) and similar methods. Wheat and barley originated in the Fertile Crescent, as did rye, peas, lentils, onions, garlic, carrots, grapes, and other food plants familiar to us today. The people of this region also domesticated goats and sheep. Meanwhile, in China, domestication began as early as 9,500 years ago, leading eventually to the rice, millet, and pigs we know today. Agriculture in Africa (coffee, yams, sorghum, and more) and the Americas (corn, beans, squash, potatoes, llamas, and more) developed later in several areas, 4,500–7,000 years ago.

For most of these thousands of years, the work of cultivating, harvesting, storing, and distributing crops was performed by human and animal muscle power, along with hand tools and simple machines (Figure 9.4). This biologically powered agriculture is known as **traditional agriculture.** In the oldest form of traditional agriculture, known as *subsistence agriculture,* farming families produce only enough food for themselves and do not make use of large-scale irrigation, fertilizer, or teams of laboring animals. In contrast, *intensive traditional agriculture* sometimes uses draft animals and employs significant quantities of irrigation water and fertilizer, but it stops short of using fossil fuels. This type of agriculture aims to produce food for the farming family, as well as excess food to sell in the market.

Industrialized agriculture is newer still

The industrial revolution introduced large-scale mechanization and fossil fuel combustion to agriculture just as it did to industry, enabling farmers to replace horses and oxen with faster and more powerful means of cultivating, harvesting, transporting, and processing crops. Other advances facilitated irrigation and fertilizing, while the invention of chemical pesticides reduced competition from weeds and herbivory by insects and other crop pests. To be efficient, however, **industrialized agriculture** demands that vast fields be planted with single types of crops. The uniform planting of a single crop, termed **monoculture,** is distinct from the *polyculture* approach of much traditional agriculture, such as Native American

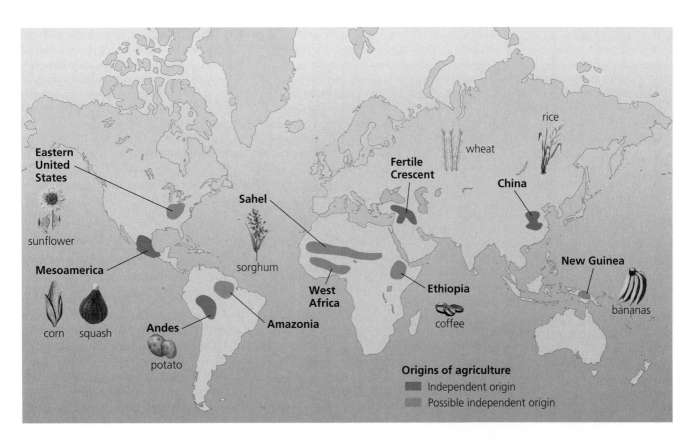

FIGURE 9.3 Agriculture appears to have originated independently in multiple locations throughout the world, as different cultures domesticated certain plants and animals from wild species living in their environments. This depiction summarizes conclusions from diverse sources of research on evidence for early agriculture. Areas where people are thought to have independently invented agriculture are colored green. (China may represent two independent origins.) Areas colored blue represent regions where people either invented agriculture independently or obtained the idea from cultures of other regions. A few of the many crop plants domesticated in each region are shown. Data from syntheses in Diamond, J., 1997. *Guns, germs, and steel.* New York: W.W. Norton; and Goudie, A. 2000. *The human impact,* 5th ed. Cambridge, MA: MIT Press.

FIGURE 9.4 Hunting and gathering was the predominant human lifestyle until the onset of agriculture and sedentary living, which centered around farms, villages, and cities, beginning nearly 10,000 years ago. Over the millennia, societies practicing traditional agriculture gradually replaced hunter-gatherer cultures. Only within the past century has industrialized agriculture spread, replacing much traditional agriculture.

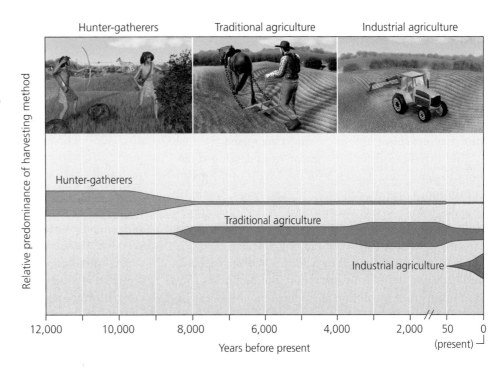

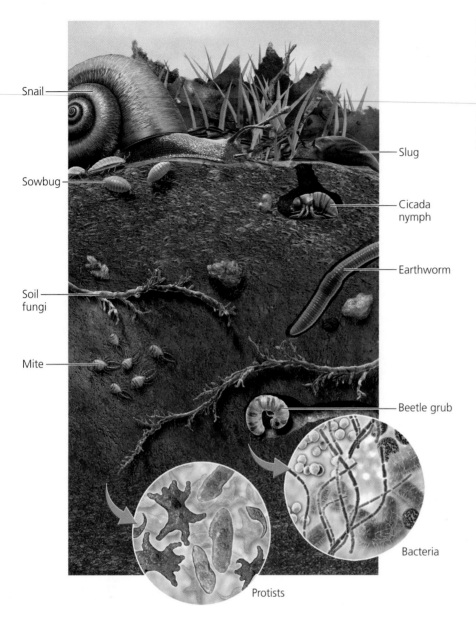

Snail

Sowbug

Soil fungi

Mite

Protists

Slug

Cicada nymph

Earthworm

Beetle grub

Bacteria

FIGURE 9.5 Soil is a complex mixture of organic and inorganic components and is full of living organisms whose actions help keep it fertile. In fact, entire ecosystems exist in soil. Most soil organisms, from bacteria to fungi to insects to earthworms, decompose organic matter. Many, such as earthworms, also help to aerate the soil.

farming systems that mixed maize, beans, squash, and peppers in the same fields. Today, industrialized agriculture occupies about 25% of the world's cropland.

Industrialized agriculture spread from developed nations to developing nations with the advent of the **green revolution,** a phenomenon we will explore in Chapter 10 (▶pp. 279–282). Beginning around 1950, the green revolution introduced new technology, crop varieties, and farming practices to the developing world. These advances dramatically increased yields per acre of cropland, and helped millions avoid starvation. But despite its successes, the green revolution is exacting a high price. The intensive cultivation of farmland is creating new problems and exacerbating old ones. Many of these problems pertain to the integrity of soil, which is the very foundation of our terrestrial food supply.

Soil as a System

We generally overlook the startling complexity of soils. We tend to equate the word *soil* with the word *dirt,* which connotes something useless or undesirable. Soil, however, is much more. It is not merely loose material derived from rock; it also contains a large biotic component, is molded by life, and is capable of supporting plant growth (Figure 9.5).

By volume, soil consists very roughly of half mineral matter and up to 5% organic matter. The rest consists of pore space taken up by air or water. The organic matter in soil includes living and dead microorganisms as well as decaying material derived from plants and animals. Most of us tend to think of soil as inert and lifeless, but a single teaspoon of soil can contain 100 million bacteria, 500,000 fungi, 100,000 algae, and 50,000 protists. Soil also provides habitat for earthworms, insects, mites, millipedes, centipedes, nematodes, sow bugs, and other invertebrates, as well as burrowing mammals, amphibians, and reptiles. The composition and quality of a region's soil can have as much influence on the region's ecosystems as do the climate, latitude, and elevation. In fact, because soil is composed of living and nonliving components that interact in complex ways, soil itself meets the definition of an ecosystem (▶pp. 189–191).

Soil formation is slow and complex

The formation of soil plays a key role in terrestrial primary succession (▶p. 163), which begins when the lithosphere's parent material is exposed to the effects of the atmosphere, hydrosphere, and biosphere. **Parent material** is the base geological material in a particular location. It can include lava or volcanic ash; rock or sediment deposited by glaciers; wind-blown dunes; sediments deposited by rivers, in lakes, or in the ocean; or **bedrock,** the continuous mass of solid rock that makes up Earth's crust.

The processes most responsible for soil formation are weathering, erosion, and the deposition and decomposition of organic matter. **Weathering** describes the physical, chemical, and biological processes that break down rocks and minerals, turning large particles into smaller particles (Figure 9.6).

Physical or *mechanical weathering* breaks rocks down without triggering a chemical change in the parent material. Wind and rain are two main forces of physical weathering. Daily and seasonal temperature variation aids their action by causing the thermal expansion and contraction of parent material. Areas with extreme temperature fluctuations experience rapid rates of physical weathering. Water freezing and expanding in cracks in rock also causes physical weathering.

Chemical weathering results when water or other substances chemically interact with parent material. Warm, wet conditions usually accelerate chemical weathering.

Biological weathering occurs when living things break down parent material by physical or chemical means. For instance, lichens initiate primary terrestrial succession by producing acid, which chemically weathers rock. A tree

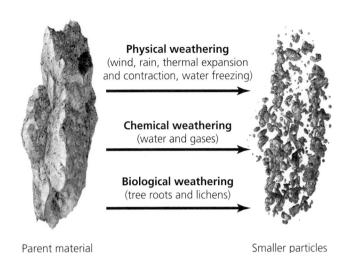

Physical weathering
(wind, rain, thermal expansion and contraction, water freezing)

Chemical weathering
(water and gases)

Biological weathering
(tree roots and lichens)

Parent material (rock)

Smaller particles of parent material

FIGURE 9.6 The weathering of parent material is the first step in soil formation. Rock is broken down into finer particles by physical, chemical, or biological means.

may accelerate weathering through the physical action of its roots as they grow and rub against rock. It may also accelerate weathering chemically through the decomposition of its leaves and branches or with chemicals it releases from its roots.

Weathering produces fine particles, and is the first step in soil formation. Another process often involved is **erosion,** the movement of soil from one area to another. Erosion may sometimes help form soil in one locality by depositing material it has depleted from another. Erosion is particularly prevalent when soil is denuded of vegetation, leaving the surface exposed to water and wind that may wash or blow it away. Although erosion can sometimes help build new soil in the long term, on the timescale of human lifetimes and for the natural systems on which we depend, erosion is generally perceived as a destructive process that reduces the amount of life that a given area of land can support.

Biological activity contributes to soil formation through the deposition, decomposition, and accumulation of organic matter. As plants, animals, and microbes die or deposit waste, this material is incorporated into the substrate, mixing with minerals. The deciduous trees of temperate forests, for example, drop their leaves each fall, making leaf litter available to the detritivores and decomposers (▶pp. 157–158) that break it down and incorporate its nutrients into the soil. In decomposition, complex organic molecules are broken down into simpler ones, including those that plants can take up through their roots. Partial decomposition of organic matter creates *humus,* a

Table 9.1	Five Factors That Influence Soil Formation
Factor	**Effects**
Climate	Soil forms faster in warm, wet climates. Heat speeds chemical reactions and accelerates weathering, decomposition, and biological growth. Moisture is required for many biological processes and can speed weathering.
Organisms	Earthworms and other burrowing animals mix and aerate soil, add organic matter, and facilitate microbial decomposition. Plants add organic matter and affect a soil's composition and structure.
Topographical relief	Hills and valleys affect exposure to sun, wind, and water, and they influence where and how soil moves. Steeper slopes result in more runoff and erosion and in less leaching, accumulation of organic matter, and differentiation of soil layers.
Parent material	Chemical and physical attributes of the parent material influence properties of the resulting soil.
Time	Soil formation takes decades, centuries, or millennia. The four factors above change over time, so the soil we see today may be the result of multiple sets of factors.

Adapted from: Jenny, H. 1941. *Factors of soil formation: A system of quantitative pedology.* New York: McGraw-Hill, Inc. Reprinted 1994 by Dover Publications, Mineola, New York.

dark, spongy, crumbly mass of material made up of complex organic compounds. Soils with high humus content hold moisture well and are productive for plant life.

Weathering, erosion, the accumulation and transformation of organic matter, and other processes that contribute to soil formation are all influenced by outside factors. Soil scientists cite five primary factors that influence the formation of soil (Table 9.1).

Weighing the Issues:
Earth's Soil Resources

It can take 500 to 1,000 years to produce 1 inch of natural topsoil. Is soil a renewable resource? How do you think soil's long renewal time should influence its management? What types of practices encourage the formation of new topsoil?

A soil profile consists of distinct layers known as horizons

Once weathering has produced an abundance of small particles between the parent material and the atmosphere, then wind, water, and organisms begin to move and sort them. Eventually, distinct layers develop. Each layer of soil is known as a **horizon,** and the cross-section as a whole, from surface to bedrock, is known as a **soil profile.**

The simplest way to categorize soil horizons is to recognize A, B, and C horizons corresponding to topsoil, subsoil, and parent material. However, soil scientists often

find it useful to subdivide the layers more finely, by their characteristics and the processes that take place within them. For our purposes we will discuss six major horizons, known as the O, A, E, B, C, and R horizons (Figure 9.7). Soils from different locations vary, and few soil profiles contain all six of these horizons, but any given soil contains at least some of them. Generally, the degree of weathering and the concentration of organic matter decrease as one moves downward in the soil profile.

Many soil profiles include an uppermost layer consisting mostly of organic matter, such as decomposing branches, leaves, and animal waste. This thin layer is designated the **O horizon** (O for *organic*) or litter layer. Just below the O horizon lies the **A horizon,** consisting of inorganic mineral components such as weathered substrate, with organic matter and humus from above mixed in. The A horizon is often referred to as **topsoil,** that portion of the soil that is most nutritive for plants and therefore most vital to ecosystems and agriculture. Topsoil takes its loose texture and dark coloration from its humus content. The O and A horizons are home to most of the countless organisms that give life to soil.

Beneath the A horizon in some soils lies the **E horizon.** E refers to *eluviation,* meaning loss, and the E horizon is characterized by the loss of some minerals and organic matter through leaching. **Leaching** is the process whereby solid particles suspended or dissolved in liquid are transported to another location. Generally in soils, the solvent is water, and leaching carries minerals downward. Soil that undergoes leaching is a bit like coffee grounds in a drip filter. When it rains, water infiltrates the soil (just as it infiltrates coffee grounds), dissolves

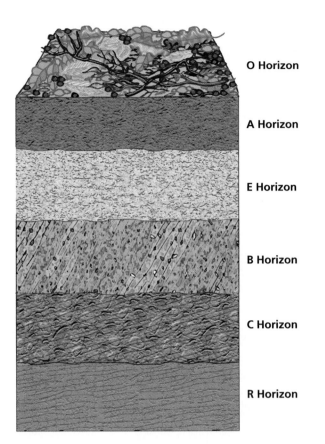

O Horizon

A Horizon

E Horizon

B Horizon

C Horizon

R Horizon

FIGURE 9.7 Mature soil consists of layers, or horizons, that have different compositions and characteristics. The number and depth of horizons vary from place to place and from soil type to soil type, producing different soil profiles. In general, organic matter and the degree of weathering decrease as one moves downward in a soil profile. The O horizon consists mostly of organic matter deposited by organisms. The A horizon, or topsoil, consists of some organic material mixed with mineral components. Minerals tend to leach out of the E horizon down into the B horizon. The C horizon consists largely of weathered parent material, which may overlie an R horizon of pure parent material.

some of its components, and carries them downward into the deeper horizons. Minerals commonly leached from the E horizon include iron, aluminum, and silicate clay. In some soils, minerals may be leached so rapidly that plants are deprived of nutrients. Minerals that leach rapidly from soils may be carried into groundwater and can pose human health threats when the water is extracted.

Minerals leaching from the A and E horizons move into the layer beneath them, the **B horizon,** or subsoil. This horizon collects and accumulates minerals from above. Often called the *illuviation horizon, zone of accumulation,* or *zone of deposition,* the B horizon contains a greater concentration of minerals and organic acids leached from above than does the E horizon.

The **C horizon,** if present, is located below the B horizon and consists of parent material unaltered or only slightly altered by the processes of soil formation. It therefore contains rock particles that are larger and less weathered than the layers above. The C horizon sits directly above the **R horizon,** or parent material.

Soil can be characterized by color, texture, structure, and pH

The six horizons presented above depict an idealized, "typical" soil, but soils display great variety. U.S. soil scientists classify soils into 12 major groups, based largely on the processes thought to form them. Within these 12 "orders," there are dozens of "suborders," hundreds of "great groups," and thousands of soils belonging to lower categories, all arranged in a hierarchical system. Scientists classify soils using properties such as color, texture, structure, and pH.

Soil color The color of soil (Figure 9.8) can indicate soil composition and sometimes soil fertility. Black or dark brown soils are usually rich in organic matter, whereas a

Clay

Peat

Chalk

Silt

FIGURE 9.8 The color of soil may vary drastically from one location to another. A soil's composition affects its color. For instance, soils high in organic matter tend to be dark brown or black.

FIGURE 9.9 The texture of soil depends on its mix of particle sizes. Using the triangular diagram shown, scientists classify soil texture according to the relative proportions of sand, silt, and clay. After measuring the percentage of each type of particle size in a soil sample, a scientist can trace the appropriate white lines extending inward from each side of the triangle to determine what type of soil texture that particular combination of values creates. Loam is generally the best for plant growth, although some types of plants grow better in other textures of soil.

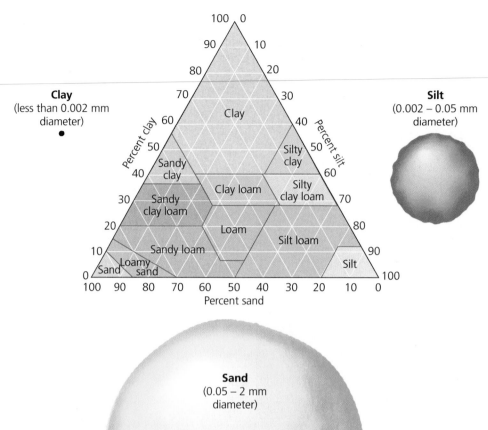

pale gray to white color often indicates leaching or low organic content. This color variation occurs among soil horizons in any given location and also among soils from different geographic locations. Long before modern analytical tests of soil content were developed, the color of topsoil provided farmers and ranchers with information about a region's potential to support crops and provide forage for livestock.

Soil texture Soil texture is determined by the size of particles and is the basis on which the United States Department of Agriculture (USDA) assigns soils to one of three general categories (Figure 9.9). **Clay** consists of particles less than 0.002 mm in diameter, **silt** of particles 0.002–0.05 mm, and **sand** of particles 0.05–2 mm. Sand grains, as any beachgoer knows, are large enough to see individually and do not adhere to one another. Clay particles, in contrast, readily adhere to one another and give clay a sticky feeling when moist. Soil with a relatively even mixture of the three particle sizes is known as **loam.**

For the farmer, soil texture influences a soil's "workability," its relative ease or difficulty of cultivation. Soil texture also influences *soil porosity,* a measure of the size of spaces between particles. In general, the finer the particles, the smaller the spaces between them. The smaller the

spaces, the harder it is for water and air to travel through the soil, slowing infiltration and reducing the amount of oxygen available to soil biota. Conversely, soils with large particles allow water to pass through (and beyond the reach of plants' roots) too quickly. Thus, crops planted in sandy soils require frequent irrigation. For this reason, silty soils with medium-sized pores, or loamy soils with mixtures of pore sizes, are generally best for plant growth and crop agriculture.

Soil structure Soil structure is a measure of the "clumpiness" of soil. Some degree of structure encourages soil productivity, and biological activity helps promote this structure. However, soil clumps that are too large can discourage plant roots from establishing if soil particles are compacted too tightly together. Repeated tilling can compact soil and make it less able to absorb water. When farmers repeatedly till the same field at the same depth, they may end up forming *plowpan,* a hard layer that resists the infiltration of water and the penetration of roots.

Soil pH The degree of acidity or alkalinity (▸p. 98) influences a soil's ability to support plant growth. Plants can die in soils that are too acidic or alkaline, but moderate variation can influence the availability of nutrients for

FIGURE 9.10 In tropical forested areas, the traditional form of farming is *swidden* agriculture, as seen here in Surinam. In this practice, forest is cut, the plot is farmed for one to a few years, and the farmer then moves on to clear another plot, leaving the first to regrow into forest. This frequent movement is necessary because tropical soils are nutrient-poor, with nearly all nutrients held in the vegetation. Burning the cut vegetation adds nutrients to the soil, which is why this practice is often called "slash-and-burn" agriculture. At low population densities, this form of farming had little large-scale impact on forests, but at today's high population densities, it is a leading cause of deforestation.

plants' roots. During leaching, for instance, acids from organic matter may remove some nutrients from the sites of exchange between plant roots and soil particles, and water carries these nutrients deeper.

Regional differences in soil traits can affect agriculture

The characteristics of soil and soil profiles can vary from place to place. One example that bears on agriculture is the difference between soils of tropical rainforests and those of temperate grasslands. Although rainforest ecosystems have high primary productivity (▶pp. 192–193), most of their nutrients are tied up in plant tissues and not in the soil. The soil of Amazonian rainforest in northern Brazil is in fact much less productive than the soil of grassland in Kansas.

To understand how this can be, consider the main differences between the two regions: temperature and rainfall. The enormous amount of rain that falls in the Amazon readily leaches minerals and nutrients out of the topsoil and E horizon. Those not captured by plants are taken quickly down to the water table, out of reach of most plants' roots. High temperatures speed the decomposition of leaf litter and the uptake of nutrients by plants, so amounts of humus remain small, and the topsoil layer remains thin.

Thus when forest is cleared for farming, cultivation quickly depletes the soil's fertility. This is why the traditional form of agriculture in tropical forested areas is *swidden* agriculture, in which the farmer cultivates a plot for one to

a few years and then moves on to clear another plot, leaving the first to grow back to forest (Figure 9.10). This method may work well at low population densities, but with today's high human populations, soils may not be allowed enough time to regenerate. As a result, intensive agriculture has ruined the soils and forests of many tropical areas.

In temperate grassland areas such as the Kansas prairie, in contrast, rainfall is low enough that leaching is reduced and nutrients remain high in the soil profile, within reach of plants' roots. Plants take up nutrients and then return them to the topsoil when they die; this cycle maintains the soil's fertility. The thick, rich topsoil of temperate grasslands can be farmed repeatedly with minimal loss of fertility if proper farming techniques are used. However, growing and harvesting crops without returning adequate organic matter to the soil gradually depletes organic material, and leaving soil exposed to the elements increases erosion of topsoil. It is such consequences that farmers in southern Brazil, the U.S. Midwest, and other locations have sought to forestall through the use of reduced tillage.

Soil Degradation: Problems and Solutions

Scientists' studies of soil and the practical experience of farmers have shown that the most desirable soil for agriculture is a loamy mixture with a pH close to neutral that is workable and capable of holding nutrients. Many soils

The Science behind the Story

Measuring Erosion

Can a hedge of grass help stop soil erosion?

Grass hedges are widely used, especially in the tropics, to trap eroding soil by slowing runoff from rain. But Jerry Ritchie, a researcher with the U.S. Department of Agriculture, wanted to measure how well they actually work. Going beyond the visible signs of soil loss, he used techniques ranging from simple measuring pins to complex radiation detectors to document erosion and calculate just how much soil has moved.

In the 1990s, Ritchie and his team of researchers began by measuring erosion around hedges planted near a set of gullies in Maryland. They relied on cheap, simple tools known as erosion pins (see the figure), which were developed in the 1960s and 1970s by scientists working for the U.N. Food and Agriculture Organization. Erosion pins are spikes that can be made from almost anything, includ-

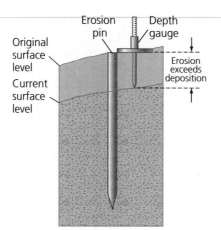

As soil erodes around an erosion pin, more of the pin is exposed, enabling the soil scientist to measure the amount of soil loss.

ing bamboo stakes or pieces of plastic pipe. The pins, each cut to a uniform length, are driven into the soil until their tops are level with the ground's surface. Over time, if soil in the area is eroding, the soil surface will recede, and the erosion pins will be increasingly exposed.

By using many pins over a wide area and averaging their readings, scientists can determine an overall erosion rate for the area.

Researchers often dig expansive holes, called catchpits, nearby. These pits are lined with plastic and serve as collection sites for eroding soil. Researchers measure the volume of soil that accumulates in the catchpits and compare that data with the extent of exposure on erosion pins. In the Maryland experiment, Ritchie used such techniques and found that 1–2 cm (0.4–0.8 in.) of soil was accumulating upslope from the hedges per year, indicating that the grass was trapping soil as it moved.

Erosion pins and catchpits work well in one spot but are impractical over a large region. To get the best evidence of erosion on a wide scale, scientists have turned to measuring a modern-day leftover rarely considered an environmental benefit— nuclear fallout from atomic

---

deviate from this ideal and prevent land from being arable or limit the productivity of arable land. Increasingly, limits to productivity are being set by human impact that has degraded many once-excellent soils. Common problems affecting soil productivity include erosion, desertification, salinization, waterlogging, nutrient depletion, structural breakdown, and pollution.

Erosion can degrade ecosystems and agriculture

Erosion, as we have noted, is the removal of material from one place and its transport toward another by the action of wind or water. *Deposition* is the arrival of eroded material at its new location. Erosion and deposition are natural processes that in the long run can help create soil. Flowing water can deposit eroded sediment in river valleys and deltas,

producing rich and productive soils. This is why floodplains are excellent for farming and why flood-control measures can decrease the productivity of agriculture in the long run.

However, erosion often becomes a problem locally for ecosystems and agriculture because it nearly always takes place much more quickly than soil is formed. Furthermore, erosion tends to remove topsoil, the most valuable soil layer for living things. People have increased the vulnerability of fertile lands to erosion through three widespread practices:

1. Overcultivating fields through poor planning or excessive plowing, disking, or harrowing
2. Overgrazing rangelands with more livestock than the land can support
3. Clearing forested areas on steep slopes or with large clear-cuts (▶pp. 355–357)

weapons testing. In 1945, the United States exploded the world's first nuclear bomb, and since then more than 2,000 nuclear devices have been tested by the United States, the former Soviet Union, and other nations. Nuclear weapons testing has spread radioactive material through the atmosphere worldwide, and fallout from the atmosphere has covered Earth's surface with minuscule but measurable amounts of nuclear debris.

Fallout includes cesium-137, a radioactive isotope of the element cesium. As discussed in Chapter 4 (▶pp. 94–95), isotopes of chemical elements are often tracked and measured by environmental scientists. Cesium-137, a product of nuclear fuel and weapons reactions, has a half-life of 30 years—a duration that enables soil scientists to use the isotope as a universal environmental tracer for erosion and sediment deposits. Soil tends to absorb cesium-137 quickly and

evenly, so if soil in an area hasn't moved or been heavily disturbed, testing will show fairly uniform levels of the isotope. But if such measurements are not uniform—with some areas showing lower concentrations of cesium-137 and others showing higher concentrations—then erosion may be at work. Ritchie decided to use cesium-137 tests on the area around the hedges and compare the results against his physical soil measurements.

In studies involving cesium-137, soil samples are tested using a gamma spectrometer. This device measures gamma rays, which serve as unique signatures of energy emitted by chemical elements in soil or rock. As they are emitted, gamma rays show up as sharp emission lines on a spectrum. The energy represented in these emissions reflects which elements are present, and the intensity of the lines reveals the concentration of the element. Thus it is possible to calculate the

amount of an isotope such as cesium-137 in a test sample. In erosion tests, each sample from the study area is measured for cesium-137, and those levels are compared to baseline levels for the region. By pinpointing places in the study area with lower or higher levels of accumulated cesium, scientists can detect where soil has moved and how much has shifted.

In Maryland, the radioactive testing helped Ritchie determine that hedges may offer only partial help against erosion. Although his team's physical measurements of soil accumulation showed soil being deposited near the hedges, the cesium-137 tests revealed that the area around the hedges had nonetheless undergone a net loss of soil over a period of four decades. Grass hedges can help, Ritchie wrote when releasing his findings for the Federal Agricultural Research Service in 2000, but they "should not be seen as a panacea."

Erosion can be gradual and hard to detect. For example, an erosion rate of 12 tons/ha (5 tons/acre) removes only a penny's thickness of soil. In many parts of the world, scientists, farmers, and extension agents are measuring erosion rates in hopes of identifying areas in danger of serious degradation before they become too badly damaged (see "The Science behind the Story," above).

Soil erodes by several mechanisms

Grasslands, forests, and other plant communities protect soil from wind and water erosion. Vegetation breaks the wind and slows water flow, while plant roots hold soil in place and take up water. Removing plant cover will nearly always accelerate erosion. Several types of erosion can occur, including wind erosion and four principal kinds of water erosion (Figure 9.11).

Splash erosion (Figure 9.11a) occurs when rain striking the soil surface breaks aggregates into smaller sizes. Soil particles are released and fill in gaps between the remaining clumps, decreasing a soil's ability to absorb water. In *sheet erosion,* or overland flow (Figure 9.11b), surface water flows downhill, washing topsoil away in relatively uniform layers. *Rill erosion* (Figure 9.11c) takes place when water runs along small furrows on the surface of the topsoil, gradually deepening and widening the furrows into rills, or small channels. Rills can merge to form larger and larger channels and eventually gullies. *Gully erosion* (Figure 9.11d) is least common but causes the most dramatic and visible changes in the landscape.

Research indicates that rill erosion has the greatest potential to move topsoil, followed by sheet erosion and splash erosion, respectively. All types of water erosion—particularly gully erosion—are more likely to occur where

(a) Splash erosion

(b) Sheet erosion

(c) Rill erosion

(d) Gully erosion

FIGURE 9.11 The erosion of soil by water can be classified into at least four categories. Splash erosion (**a**) occurs as raindrops strike the ground with enough force to dislodge small amounts of soil. Sheet erosion (**b**) results when thin layers of water traverse broad expanses of sloping land. Rill erosion (**c**) leaves small pathways along the surface where water has carried topsoil away. Gully erosion (**d**) cuts deep into soil, leaving large gullies that can expand as erosion proceeds.

slopes are steeper. In general, steeper slopes, greater precipitation intensities, and sparser vegetative cover all lead to greater water erosion.

One study conducted in the early 1990s determined that at erosion rates typical for the United States, U.S. croplands lose about 2.5 cm (1 in.) of topsoil every 15–30 years, reducing corn yields by 4.7–8.7% and wheat yields by 2.2–9.5%. According to U.S. government figures, erosion rates in the United States declined from 9.1 tons/ha (3.7 tons/acre) in 1982 to 5.9 tons/ha (2.4 tons/acre) in 2001, thanks to soil conservation measures discussed below. Yet in spite of these measures, U.S. farmlands still lose 6 tons of soil for every ton of grain harvested.

Soil erosion is a global problem

Erosion has become a major problem in many areas of the world, including Australia, sub-Saharan Africa, central Asia, India, the Middle East, and parts of South America, Central America, Europe, and the United States. In total, more than 19 billion ha (47 billion acres) of the world's croplands suffer from erosion and other forms of soil degradation resulting from human activities. Between 1957 and 1990, China lost as much arable farmland as exists in Denmark, France, Germany, and the Netherlands combined. In Kazakhstan, central Asia's largest nation, industrial cropland agriculture imposed on land better suited for grazing caused tens of millions of hectares to be degraded by wind erosion. For Africa, projections indicate that soil degradation over the next 40 years could reduce crop yields by half. Couple these declines in soil quality and crop yields with the rapid population growth occurring in many of these areas, and we begin to see why some observers describe the future of agriculture as a crisis situation.

In today's world, humans are the primary cause of erosion, and we have accelerated it to unnaturally high rates. A 2004 study by geologist Bruce Wilkinson analyzed prehistoric erosion rates from the geologic record and compared these with modern rates. Wilkinson concluded that humans are over 10 times more influential at moving soil than are all other natural processes on the surface of the planet combined.

Arid land may lose productivity by desertification

Much of the world's population lives and farms in arid environments, where **desertification** is a concern. This term describes a loss of more than 10% productivity due to erosion, soil compaction, forest removal, overgrazing, drought, salinization, climate change, depletion of water sources, and other factors. Severe desertification can result in the expansion of desert areas or creation of new ones in areas that once supported fertile land. This process has occurred in many areas of the Middle East that have been inhabited, farmed, and grazed for long periods of time. To appreciate the cumulative impact of centuries of traditional agriculture, we need only look at the present desertified state of that portion of the Middle East where agriculture originated, nicknamed the "Fertile Crescent." These arid lands—in present-day Iraq, Syria, Turkey, Lebanon, and Israel—are not so fertile anymore.

Arid and semiarid lands are prone to desertification because their precipitation is too meager to meet the demand for water from growing human populations. According to

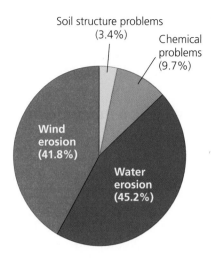

FIGURE 9.12 Soil degradation on drylands is due primarily to erosion by wind and water. Data from United Nations Environment Programme. 2002. *Tackling land degradation and desertification.* Washington and Rome: Global Environment Facility and International Fund for Agricultural Development.

the United Nations Environment Programme (UNEP), 40% of Earth's land surface can be classified as drylands, arid areas that are particularly subject to degradation. Declines of soil quality in these areas have endangered the food supply or well-being of more than 1 billion people around the world. Of the affected lands, most degradation results from wind and water erosion (Figure 9.12).

It has been estimated that desertification affects fully one-third of the planet's land area, impinging on people in 110 countries. Desertification cost the world's people at least $300–600 billion in income just in the period 1978–1991, UNEP estimates. China alone loses $6.5 billion annually from desertification. In its western reaches, desert areas are expanding and joining one another because of overgrazing from over 400 million goats, sheep, and cattle. In the Sistan Basin along the border of Iran and Afghanistan, an oasis that supported a million livestock recently turned barren in just 5 years, and windblown sand buried over 100 villages. In Africa, the continent's most populous nation, Nigeria, loses an amount of land equal to half the state of Delaware each year to the expanding Sahara Desert. In Kenya, overgrazing and deforestation fueled by rapid population growth has left 80% of its land vulnerable to desertification. In a positive feedback cycle, the soil degradation forces ranchers to crowd onto more marginal land and farmers to reduce fallow periods, both of which further exacerbate soil degradation.

As a result of desertification, in recent years gigantic dust storms from denuded land in China have blown across the Pacific Ocean to North America, and dust storms from Africa's Sahara Desert have blown across the

Atlantic Ocean to the Caribbean Sea. Such massive dust storms occurred in the United States during the Dust Bowl days of the early 20th century, when desertification shook American agriculture and society to their very roots.

similar to the silicosis that afflicts coal miners exposed to high concentrations of coal dust. Large numbers of farmers were forced off their land, and many who remained had to rely on government assistance programs to survive.

The Dust Bowl was a monumental event in the United States

Prior to large-scale cultivation of the southern Great Plains of the United States, native prairie grasses of this temperate grassland region held erosion-prone soils in place. In the late 19th and early 20th centuries, however, many homesteading settlers arrived in Oklahoma, Texas, Kansas, New Mexico, and Colorado with hopes of making a living there as farmers. Between 1879 and 1929, cultivated area in the region soared from around 5 million ha (12 million acres) to 40 million ha (100 million acres). Farmers grew abundant wheat, and ranchers grazed many thousands of cattle, sometimes expanding onto unsuitable land. Both types of agriculture contributed to erosion by removing native grasses and breaking down soil structure.

During the early 1930s, a drought in the region exacerbated the ongoing human impacts on the soil. The region's strong winds began to carry away millions of tons of topsoil. Dust storms traveled up to 2,000 km (1,250 mi), blackening rain and snow as far away as New York and Vermont. Some areas in the affected states lost as much as 10 cm (4 in.) of topsoil in a few short years (Figure 9.13). The affected region in the Great Plains became known as the **Dust Bowl,** a term now also used for the historical event itself. The "black blizzards" of the Dust Bowl destroyed livelihoods and caused many people to suffer a type of chronic lung irritation and degradation known as dust pneumonia,

The Soil Conservation Service pioneered measures to slow soil degradation

In response to the devastation in the Dust Bowl, the U.S. government, along with state and local governments, increased its support of research into soil conservation measures. The U.S. Congress also passed the Soil Conservation Act of 1935. This act described soil erosion as a threat to the nation's well-being and established the Soil Conservation Service (SCS) to address the problem. The new agency worked closely with farmers to develop conservation plans for individual farms, following several aims and principles:

▶ Assess the land's resources, its problems, and opportunities for conservation.
▶ Draw on science to prepare an integrated plan for the property.
▶ Work closely with land users to ensure that conservation plans harmonize with the users' objectives.
▶ Implement conservation measures on individual properties to contribute to the overall quality of life in the watershed or region.

The early teams that the SCS formed to combat erosion typically included soil scientists, forestry experts, engineers, economists, and biologists. These teams were among the earliest examples of interdisciplinary approaches to environmental problem solving. The first director of the SCS, Hugh Hammond Bennett, was an innovator and

FIGURE 9.13 Drought combined with poor agricultural practices brought devastation and despair to millions of U.S. farmers in the 1930s, especially in the Dust Bowl region of the southern Great Plains. The tragedy spurred the development of soil conservation practices that have since been put into place in the United States and around the world.

evangelist for soil conservation. Under his leadership, the agency promoted soil-conservation practices through county-based **conservation districts.** These districts operate with federal direction, authorization, and funding, but they are organized by state law. The districts implement soil conservation programs locally and aim to empower local residents to plan and set priorities in their home areas. In 1994 the SCS was renamed the *Natural Resources Conservation Service,* and its responsibilities were expanded to include water quality protection and pollution control.

The SCS served as a model for similar efforts elsewhere in the world (Figure 9.14). Southern Brazil's no-till movement came about through local grass-roots organization by farmers, with the help of agronomists and government extension agents who provided them information and resources. In this model of collaboration between local farmers and trained experts, 8,000 Friends of the Land clubs now exist in Paraná and Rio Grande do Sul, and 7,700 in Santa Catarina. Many of these groups are delineated by the boundaries of the more than 3,000 small-scale watersheds *(microbacias)* in which they farm.

No-till agriculture as practiced in southern Brazil is one of many approaches to soil conservation. Hugh Hammond Bennett advocated a complex approach, combining techniques such as crop rotation, contour farming, strip-cropping, terracing, grazing management, and reforestation, as well as wildlife management. Such measures have been widely applied in many places around the world.

Farmers can protect soil against degradation in various ways

Several farming techniques can reduce the impacts of conventional cultivation on soils (Figure 9.15). Some of these have been promoted by the SCS since the Dust Bowl. Some, like no-till farming in Brazil, are finding popularity more recently. Others have been practiced by certain cultures for centuries.

Crop rotation The practice of alternating the kind of crop grown in a particular field from one season or year to the next is **crop rotation** (Figure 9.15a). Rotating crops can return nutrients to the soil, break cycles of disease associated with continuous cropping, and minimize the erosion that can come from letting fields lie fallow. Many U.S. farmers rotate their fields between wheat or corn and soybeans from one year to the next. Soybeans are legumes, plants that have specialized bacteria on their roots that can fix nitrogen (▶ pp. 198–200). Soybeans revitalize soil that the previous crop had partially depleted of nutrients. Crop rotation also reduces insect pests; if an insect is adapted to feed and lay eggs on one particular

FIGURE 9.14 Government agricultural extension agents assist farmers by providing information on the newest research and techniques that can help them farm productively while minimizing damage to the land. Such specialists have helped U.S. farmers since the Dust Bowl and now assist farmers worldwide. Here, an extension agent from Colombia's Instituto Colombiano Agropecuario inspects yuca plants grown by farmer Pedro Gomez on a farm in Valle del Cauca.

crop, planting a different crop will leave its offspring with nothing to eat.

In a practice similar to crop rotation, southern Brazil's farmers plant "cover crops" designed to prevent erosion and forestall nitrogen loss from leaching during times of the year when the main crops are not growing. Santa Catarina's extension agents have worked with farmers to test over 60 species as cover crops, including both legumes and other plants, such as oats and turnips.

Contour farming Water running down a hillside can easily carry soil away, particularly if there is too little vegetative cover to hold the soil in place. Thus, sloped agricultural land is especially vulnerable to erosion. Several methods have been developed for farming on slopes. **Contour farming**

(a) Crop rotation

(b) Contour farming

(c) Intercropping

(d) Terracing

(e) Shelterbelts

(f) No-till farming cover crop

FIGURE 9.15 The world's farmers have adopted various strategies to conserve soil. Rotating crops such as soybeans and corn (**a**) helps restore soil nutrients and reduce impacts of crop pests. Contour farming (**b**) reduces erosion on hillsides. Intercropping (**c**) can reduce soil loss while maintaining soil fertility. Terracing (**d**) minimizes erosion in steep mountainous areas. Shelterbelts (**e**) protect against wind erosion. In (**f**), corn grows up from amid the remnants of a "cover crop" used in no-till agriculture.

(Figure 9.15b) consists of plowing furrows sideways across a hillside, perpendicular to its slope, to help prevent formation of rills and gullies. The technique is so named because the furrows follow the natural contours of the land. In contour farming, the downhill side of each furrow acts as a small dam that slows runoff and catches soil before it is carried away. Contour farming is most effective on gradually sloping land with crops that grow well in rows. Extension agents from Santa Catarina in Brazil have helped those farmers who still plow to switch to contour plowing and to plant barriers of grass along their contours.

Intercropping Farmers may also gain protection against erosion by **intercropping,** planting different types of crops in alternating bands or other spatially mixed arrangements (Figure 9.15c). Intercropping helps slow erosion by providing more complete ground cover than does a single crop. Like crop rotation, intercropping offers the additional benefits of reducing vulnerability to insect and disease incidence, and, when a nitrogen-fixing legume is one of the crops, of replenishing the soil. Some southern Brazilian farmers intercrop food crops with cover crops. The cover crops are physically mixed with primary crops, which include maize, soybeans, wheat, onions, cassava, grapes, tomatoes, tobacco, and orchard fruit.

Terracing On extremely steep terrain, terracing (Figure 9.15d) is the most effective method for preventing erosion. Terraces are level platforms, sometimes with raised edges, that are cut into steep hillsides to contain water from irrigation and precipitation. **Terracing** transforms slopes into series of steps like a staircase, enabling farmers to cultivate hilly land without losing huge amounts of soil to water erosion. Terracing is common in ruggedly mountainous regions, such as the foothills of the Himalayas and the Andes, and has been used for centuries by farming communities in such areas. Terracing is labor-intensive to establish but in the long term is likely the only sustainable way to farm in mountainous terrain.

Shelterbelts A widespread technique to reduce erosion from wind is to establish **shelterbelts** or *windbreaks* (Figure 9.15e). These are rows of trees or other tall, perennial plants that are planted along the edges of fields to slow the wind. Shelterbelts have been widely planted across the U.S. Great Plains, where fast-growing species such as poplars are often used. Shelterbelts have also been combined with intercropping in a practice known as *agroforestry,* or *alley cropping.* In this approach, fields planted in rows of mixed crops are surrounded by or interspersed with rows of trees that provide fruit, wood,

Table 9.2 No-Till Farming in Brazil

Benefits of no-till farming

- ▶ Conserves biodiversity in soil and in terrestrial and aquatic ecosystems
- ▶ Produces sustainable, high crop yields
- ▶ Heightens environmental awareness among farmers
- ▶ Provides shelter and winter food for animals
- ▶ Reduces irrigation demands by 10–20%
- ▶ Crop residues act as a sink for carbon (1 metric ton/ha)
- ▶ Reduces fossil fuel use by 40–70%
- ▶ Enhances food security by increasing drought resistance
- ▶ Reduces erosion by 90%

Other benefits arising from the reduction in erosion

- ▶ Reduces silt deposition in reservoirs
- ▶ Reduces water pollution from chemicals
- ▶ Increases groundwater recharge and lessens flooding
- ▶ Increases sustained crop yields and lowers food prices
- ▶ Lowers costs of treating drinking water
- ▶ Reduces costs of maintaining dirt roads
- ▶ Eliminates dust storms in towns and cities
- ▶ Increases efficiency in use of fertilizer and machinery

Modified from Shaxson, T. F. 1999. The roots of sustainability: Concepts and practice: Zero tillage in Brazil, *ABLH Newsletter ENABLE; World Association for Soil and Water Conservation (WASWC) Newsletter.*

or protection from wind. Such methods have been used in India, Africa, and in Brazil, where coffee growers near a national conservation area have established farming systems combining farming and forestry.

Reduced tillage To plant using the zero-tillage method (Figure 9.15f), a tractor pulls a "no-till drill" that cuts long furrows through the O horizon of dead weeds and crop residue and the upper levels of the A horizon. The device drops seeds into the furrow and closes the furrow over the seeds. Often a localized dose of fertilizer is added to the soil along with the seed. Reduced-tillage agriculture disturbs the soil surface more than no-tillage does, but less than conventional cultivation does. By increasing organic matter and soil biota while reducing erosion, no-till and reduced tillage farming can build soil up, restore it, and improve it. Based on the Brazilian experience, proponents of no-till farming have claimed that the practice offers a number of benefits (Table 9.2).

No-till and reduced tillage methods were pioneered in the United States and United Kingdom, where no-till is still rare but where reduced tillage has been slowly spreading for decades. Today nearly half of U.S. acreage is farmed with reduced-tillage methods. As the appeal of no-till farming spread in Brazil, it also spread in neighboring Argentina and Paraguay. In Argentina, the area under no-till farming exploded from 100,000 ha (247,000 acres) in 1990 to 7.3 million ha (18.0 million acres) in 1999, covering 30% of all arable land in the country. The results there parallel those in Brazil: increased crop yields, reduced erosion, enhanced soils, and a healthier environment. Maize yields grew by 37% and soybean yields by 11%, while costs to farmers fell by 40–57%. Erosion, pesticide use, and water pollution declined. As in Brazil, the techniques spread largely because of the actions of farmers themselves and their national no-till farmers' organization.

Critics of no-till and reduced-tillage farming in the United States have noted that these techniques often require substantial use of chemical herbicides (because weeds are not physically removed from fields) and synthetic fertilizer (because other plants take up a significant portion of the soil's nutrients). In many industrialized countries, this has indeed been the case. Proponents of the Brazilian program, however, assert that it does not always need to be so. Southern Brazil's farmers have departed somewhat from the industrialized model by relying more heavily on *green manures* (dead plants as fertilizer) and by rotating fields with cover crops, including nitrogen-fixing legumes. The manures and legumes nourish the soil, and cover crops also reduce weeds by taking up space the weeds might occupy. Critics maintain, however, that green manures are generally not practical for large-scale intensive agriculture. Certainly, reduced tillage methods work well in some areas but not in others, and they work better with some crops than with others. Farmers will do best by educating themselves on the options and doing what is best for their particular crops on their own land.

The methods we have described to combat soil degradation can be used in combination. When they have been, the results have sometimes been dramatic. One town in the Guatemalan highlands that established shelterbelts, crop rotation, and cover crops with the help of a U.S.-based nonprofit organization improved its corn production from 0.4 tons/ha (1.0 tons/acre) in 1972 to 2.5 tons/ha (6.2 tons/acre) in 1979. It went on to improve production to 4.5 tons/ha (11.1 tons/acre) by 1994.

Weighing the Issues:
How Would You Farm?

You are a farmer owning land on both sides of a steep ridge. You want to plant a sun-loving crop on the sunny, but very windy, south slope of the ridge and a crop that needs a great deal of irrigation on the north slope. What

FIGURE 9.16 Vast swathes of countryside in western China have been planted with fast-growing poplar trees. These "reforestation" efforts do not create ecologically functional forests—the plantations are too biologically simple—but they do greatly slow soil erosion.

type of farming techniques might maximize conservation of your soil? What other factors might you want to know about before you decide to commit to one or more methods?

Erosion-control practices protect and restore plant cover

Farming methods to control erosion make use of the general principle that maximizing vegetative cover will protect soils, and this principle has been applied widely beyond farming. It is common throughout the developed world to stabilize eroding banks along creeks and roadsides by planting plants to anchor the soil. In areas with severe and widespread erosion, some nations have planted vast plantations of fast-growing trees. China has embarked on the world's largest tree-planting program to slow its soil loss (Figure 9.16). Although such "reforestation" efforts do help slow erosion, they do not at the same time produce ecologically functioning forests, because tree species are selected only for their fast growth and are planted in monocultures.

Irrigation has boosted productivity but has also caused long-term soil problems

Erosion is not the only threat to the health and integrity of soils. Soil degradation can result from other factors as well, such as impacts caused by our application of water to crops. The artificial provision of water to support agriculture is known as **irrigation.** Some crops, such as rice and cotton, require large amounts of water, whereas others, such as beans and wheat, require relatively little. Other factors influencing the amount of water required for growth include the rate of evaporation, as determined by climate, and the soil's ability to hold water and make it available to plant roots. If the climate is too dry or too much water evaporates or runs off before it can be absorbed into the soil, crops may require irrigation. By irrigating crops, people have managed to turn previously dry and unproductive regions into fertile farmland. Seventy percent of all freshwater withdrawn by people is used for irrigation. Irrigated acreage has increased dramatically around the world, reaching 276 million ha (683 million acres) in 2002, greater than the entire area of Mexico and Central America. We will examine irrigation further in Chapter 15 (▸pp. 440–441).

If some water is good for plants and soil, it might seem that more must be better. But this is not necessarily the case; there is indeed such a thing as too much water. Overirrigation in poorly drained areas can cause or exacerbate certain soil problems. Soils too saturated with water may become waterlogged. When **waterlogging** occurs, the water table is raised to the point that water bathes plant roots, depriving them of access to gases and essentially suffocating them. If it lasts long enough, waterlogging can damage or kill plants.

An even more frequent problem is **salinization,** the buildup of salts in surface soil layers. In dryland areas where precipitation is minimal and evaporation rates are high, water evaporating from the soil's A horizon may pull water from lower horizons upward by capillary action. As this water rises through the soil, it carries dissolved salts, and when it evaporates at the surface, those salts precipitate and are left at the surface. Irrigation in arid areas generally hastens salinization, because it provides repeated doses of moderate amounts of water, which dissolve salts in the soil and gradually raise them to the surface. Moreover, because irrigation water often contains some dissolved salt in the first place, irrigation introduces new sources of salt to the soil. Overirrigation and waterlogging can worsen salinization problems, and in many areas of farmland, soil is turning whitish with encrusted salt. Salinization now inhibits agricultural production on one-fifth of all irrigated cropland globally, costing more than $11 billion annually.

Salinization is easier to prevent than to correct

The remedies for mitigating salinization once it has occurred are more expensive and difficult to implement than the techniques for preventing it in the first place. The best way to prevent salinization is to avoid planting crops that require a great deal of water in areas that are prone to

the problem. A second way is to irrigate with water that is as low as possible in salt content. A third way is to irrigate efficiently, supplying no more water than the crop requires, thus minimizing the amount of water that evaporates and hence the amount of salt that accumulates in the topsoil. Currently, irrigation efficiency worldwide is low; only 43% of the water applied actually gets used by plants. Drip irrigation systems (Figure 9.17) that target water directly to plants are one solution to the problem. These systems allow more control over where water is aimed and waste far less water. Once considered expensive to install, they are becoming cheaper, such that more farmers in developing countries will be able to afford them.

(a) Conventional irrigation

(b) Drip irrigation

FIGURE 9.17 Currently, less than half the water we apply in irrigation actually gets taken up by plants. Conventional methods that lose a great deal of water to evaporation (**a**) are now being replaced by more efficient ones in which water is more precisely targeted to plants. In drip irrigation systems, such as this one watering grape vines in California (**b**), hoses are arranged so that water drips from holes in the hoses directly onto the plants that need the water.

If salinization has occurred, one potential way to mitigate it would be to stop irrigating and wait for rain to flush salts from the soil. However, this solution is unrealistic because salinization generally becomes a problem in dryland areas where precipitation is never adequate to flush soils. A better option may be to plant salt-tolerant plants, such as barley, that can be used as food or pasture. A third option is to bring in large quantities of less-saline water with which to flush the soil. However, using too much water may cause waterlogging. As is the case with many environmental problems, preventing salinization is easier than correcting it after the fact.

FIGURE 9.18 Farmers often add nutrients to soils with fertilizers. Organic fertilizers such as manure and vegetation may be used, or synthetically manufactured chemicals may be applied to supply nitrogen, phosphorus, and other nutrients, as this North Dakota farmer is doing.

Weighing the Issues:
Measuring and Regulating Soil Quality

The U.S. EPA has adopted measures of air quality and water quality and has set legal standards for allowable levels of various pollutants in air and water. Could such standards be developed for soil quality? If so, what properties should be measured to inform such standards? Should such standards be developed? Why or why not?

Agricultural fertilizers boost crop yields but can be over-applied

Salinization is not the only source of chemical damage to soil. Overapplying fertilizers can also chemically damage soils. Plants grow through photosynthesis, requiring sunlight, water, and carbon dioxide, but they also require nitrogen, phosphorus, and potassium, as well as smaller amounts of over a dozen other nutrients. Plants remove these nutrients from soil as they grow, and leaching likewise removes nutrients. If agricultural soils come to contain too few nutrients, crop yields decline. Therefore, a great deal of effort has aimed to enhance nutrient-limited soils by adding **fertilizer,** any of various substances that contain essential nutrients (Figure 9.18).

There are two main types of fertilizers. **Inorganic fertilizers** are mined or synthetically manufactured mineral supplements. **Organic fertilizers** consist of natural materials (largely the remains or wastes of organisms) and include animal manure; crop residues; fresh vegetation, known as *green manure;* and *compost,* a mixture produced when decomposers break down organic matter, including food and crop waste, in a controlled environment. Organic fertilizers can provide some benefits that inorganic fertilizers cannot. The proper use of compost improves soil structure, nutrient retention, and water-retaining capacity, helping to prevent erosion. As a form of recycling, composting reduces the amount of waste consigned to landfills

and incinerators (▶p. 658). However, organic fertilizers are no panacea. For instance, manure, when applied in amounts needed to supply sufficient nitrogen for a crop, may introduce excess phosphorus that can run off into waterways. Inorganic fertilizers are generally more susceptible than are organic fertilizers to leaching and runoff, and they are somewhat more likely to cause unintended off-site impacts. Inorganic and organic fertilizer use is growing globally (Figure 9.19). Unfortunately, its mismanagement is causing increasingly severe pollution problems.

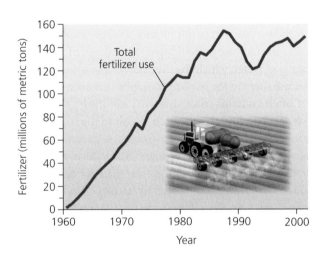

FIGURE 9.19 Use of synthetic fertilizers has risen sharply over the past half-century and now stands at over 140 million metric tons. (The temporary drop during the early 1990s was due to economic decline in countries of the former Soviet Union following that nation's dissolution.) Data from Food and Agriculture Organization of the United Nations (FAO); and Worldwatch Institute, 2001. *Vital signs 2001.*

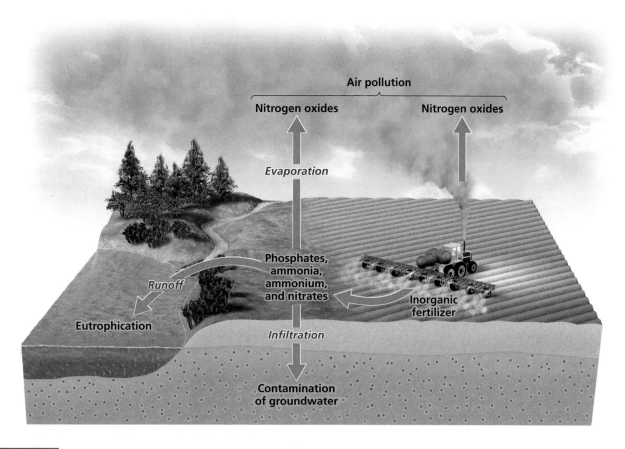

FIGURE 9.20 The overapplication of inorganic (or organic) fertilizers can have effects beyond the farm field, because nutrients that are not taken up by plants may end up in other places. Nitrates can leach into groundwater, where they can pose a threat to human health in drinking water. Phosphates and some nitrogen compounds can run off into surface waterways and alter the ecology of streams, rivers, ponds, and lakes through eutrophication. Some compounds like nitrogen oxides can even enter and pollute the air. Anthropogenic inputs of nitrogen have greatly modified the nitrogen cycle (▶pp. 197–199), and now account for one-half the total nitrogen flux on Earth.

Applying substantial amounts of fertilizer to crop-lands has impacts far beyond the boundaries of the fields (Figure 9.20). We saw in Chapter 7 one impact of excess fertilizer use. Nitrogen and phosphorus runoff from farms and other sources in the Mississippi River basin each year spurs phytoplankton blooms in the Gulf of Mexico and creates an oxygen-depleted "dead zone" that kills fish and shrimp. Such eutrophication occurs at many river mouths, lakes, and ponds throughout the world. Moreover, nitrates readily leach through soil and contaminate groundwater, and components of some nitrogen fertilizers can even volatilize (evaporate) into the air. Through these processes, unnatural amounts of nitrates and phosphates spread through ecosystems and pose human health risks, including cancer and methemoglobinemia, or blue-baby disease, which can asphyxiate and kill infants. The U.S. Environmental Protection Agency (EPA) has determined that nitrate concentrations in excess of 10 mg/L for adults and 5 mg/L for infants in drinking water are unsafe, yet many sources around the world exceed even the looser standard of 50 mg/L set by the World Health Organization.

Grazing practices and policies can contribute to soil degradation

We have focused in this chapter largely on the cultivation of crops as a source of impacts on soils and ecosystems, but raising livestock also has such impacts. When sheep, goats, cattle, or other livestock graze on the open range, they feed primarily on grasses. As long as livestock populations do not exceed a range's carrying capacity (▶pp. 136–137) and do not consume grasses faster than grasses can be replaced, grazing may be sustainable. However, when too many animals eat too much of the plant cover, impeding plant regrowth and

FIGURE 9.21 When grazing by livestock exceeds the carrying capacity of rangelands and their soil, overgrazing can set in motion a series of consequences and positive feedback loops that degrade soils and grassland ecosystems.

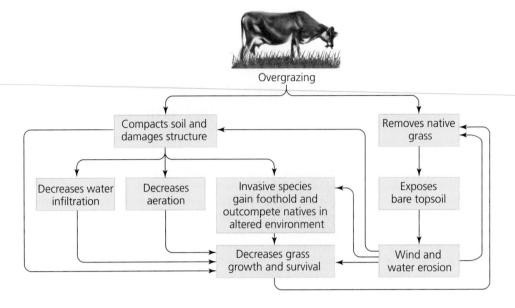

preventing the replacement of biomass, the result is **overgrazing.**

Rangeland scientists have shown that overgrazing causes a number of impacts, some of which give rise to positive feedback cycles that exacerbate damage to soils, natural communities, and the land's productivity for grazing (Figure 9.21). When livestock remove too much of an area's plant cover, more soil surface is exposed and made vulnerable to erosion. Soil erosion makes it difficult for vegetation to regrow, perpetuating the lack of cover and giving rise to more erosion. Moreover, non-native weedy plants may invade denuded soils (Figure 9.22). These invasive plants are usually less palatable to livestock and can outcompete native vegetation in the new, modified environment, further decreasing native plant cover.

In addition, overgrazing can compact soils and alter their structure. Soil compaction makes it harder for water to infiltrate, harder for soils to be aerated, harder for plants' roots to expand, and harder for roots to conduct cellular respiration (▶ p. 107). All of these effects further decrease the growth and survival of native plants.

Overgrazing is a serious problem worldwide. As a cause of soil degradation, it is equal to cropland agriculture, and it is a greater cause of desertification. Humans keep a total of 3.3 billion cattle, sheep, and goats. Rangeland classified as degraded now adds up to 680 million ha (1.7 billion acres), five times the area of U.S. cropland, although some estimates put the number as high as 2.4 billion ha (5.9 billion acres), fully 70% of the world's rangeland area. Rangeland degradation is estimated to cost $23.3 billion per year. Grazing exceeds the sustainable

supply of grass in India by 30% and in parts of China by up to 50%. To relieve pressure on rangelands, both nations are now beginning to feed crop residues to livestock.

Range managers in the United States do their best to assess the carrying capacity of rangelands and inform livestock owners of these limits, so that herds are rotated from site to site as needed to conserve grass cover and soil integrity. Managers also can establish and enforce limits on grazing on publicly owned land when necessary. U.S.

FIGURE 9.22 The effects of overgrazing can be striking, as shown in this photo along a fence line separating a grazed plot (right) from an ungrazed plot (left) in the Konza Prairie Reserve in Kansas. The overgrazed plot has lost much of its native grass and has been invaded by weedy plants that compete more effectively in conditions of degraded soil.

Soil Conservation

Productive farming depends on fertile soil, but our farming practices have all too often eroded and degraded soil. **How much erosion constitutes a problem? How can we best protect the condition of soil while we farm?**

Land Policy Is Necessary for Soil Conservation

Claiming that soil erosion is a major problem in the more-developed parts of the world is environmental nonsense. Pierre Crosson of Resources for the Future, Washington, D.C., has looked carefully at the U.S. data over many years and concluded that soil erosion does *not* constitute a significant problem for contemporary agriculture and its sustainability.

There are surely many cases of local erosive effects, but these usually amount to small quantities of soil shifting around the agricultural landscape. Erosion can be beneficial in some cases, such as in the creation of new farmland in Bangladesh, which formed when materials washed down from Nepal and other parts of the Himalayas.

Soil conservation is a problem that must be approached from all sides. The most important factor will be to secure legally protected property rights for resource users. Farmers must have the right to own their own land resource. Ethiopia is usually regarded as one of the most eroded countries in the world. The state owns all farmland and has a history of moving people against their will to different areas. If a farmer thinks he will not be able to continue managing his farm for the next several years, he will not invest his scarce resources in erosion-reducing practices such as terracing, stone building, maintaining vegetative contour strips, and reducing grazing pressure. For resource custody, and hence the treatment of soil, land policy really matters.

Land policy also affects public investment in infrastructure such as rural roads, which reduce the cost of farmers' transactions. Farmers must haul supplies such as fertilizer to their farms and must transport their product to market. Inadequate access to markets can significantly influence the incentives to employ resource-consuming soil conservation practices. Thus, social science has key roles to play along with physical and biological sciences in addressing soil management.

Jock R. Anderson joined the World Bank in its Agriculture and Natural Resources Department, where he served variously as adviser of agricultural technology policy and, more recently, as adviser of strategy and policy for agriculture and rural development. He is a fellow of the Australian Institute of Agricultural Science and of the American Agricultural Economics Association.

Sustainable Soil Productivity Requires Good Land Stewards

How much soil erosion is tolerable? The standard answer to this question ranges from 1 to 5 tons of soil per acre annually, depending on soil depth and other characteristics that affect sustainability. However, this answer is admittedly subjective and debatable.

Some say these values are too low. They argue that land has been used for centuries with higher rates of erosion, and it is difficult or impossible to reduce erosion on some land without drastic reductions in high-value crops. Others say the standard tolerable erosion values are too high. They argue that the average annual rate of soil formation is less than 1 ton per acre, so a greater erosion rate will inevitably make the soil shallower and less productive. Most agree that excessive erosion is undesirable and a major cause of land abandonment.

Tillage was fundamental to the development of civilization, but it also exposed the soil to erosion—a problem that became more serious as population increased. Tillage can now be reduced or even eliminated by using modern no-till equipment that can cut through crop residues and plant seeds in a narrow slot. Weeds, insects, and plant diseases are controlled by integrated pest management using crop rotations, biological methods, and chemicals. We can now grow crops with much less erosion and still produce good yields.

Conserving water and planting drought-tolerant crops in semi-arid climates is another important advance in areas where erosion due to summer fallow has been common. These practices are effective ways to conserve soil.

The most important requirement for sustainable productivity is a fervent desire to be good land stewards—people who use the best soil conservation practices and are alert for detecting problem spots.

Frederick R. Troeh grew up in Idaho, surveyed soils for the Soil Conservation Service, earned his Ph.D. from Cornell University, and taught soil science and researched for Iowa State University. He has worked internationally in Uruguay, Argentina, and Morocco, and he is lead author of the textbooks *Soils and Soil Fertility* and *Soil and Water Conservation*.

 Explore this issue further by accessing **Viewpoints** at www.aw-bc.com/withgott.

Overgrazing and Fire Suppression in the Malpai Borderlands

The Science behind the Story

In the high desert of southern Arizona and New Mexico, scientists trying to heal the scars left by decades of cattle ranching and overgrazing found they had to contend with a creature even more damaging than a hungry steer: Smokey the Bear.

Wildfires might seem a natural enemy of grasslands, but in the Malpai Borderlands, researchers realized that people's efforts to suppress fire had done far more harm. Before large numbers of Anglo settlers and ranchers arrived more than a century ago, this rocky stretch of arid Western landscape thrived under an ecological cycle common to many grasslands. Scrubby trees such as mesquite grew near creeks. Hardy grasses such as black grama covered the drier plains. Periodic wildfires, usually sparked by lightning, burned back shrubs and trees and kept grasslands open. Native grazing animals, such as deer, rabbits, and bighorn sheep, ate the grasses, but rarely ate enough to deplete the range. Fed by seasonal rains, new grasses sprouted without being overeaten or crowded out by larger plants.

To restore native grasslands, the Malpai Borderlands Group reinstated fire as a natural landscape process, conducting controlled burns on over 100,000 ha since 1994. Monitoring indicates that restoring fire has improved ecological conditions in the region.

By the early 1990s, however, those grasslands were increasingly scarce. Ranchers had brought large cattle herds to the area in the late 1800s. The cows chewed through vast expanses of grass, trampled soil, and scattered mesquite seed into areas where grasses had dominated. Ranchers fought wildfires to keep their herds safe, and the federal government joined in the firefighting efforts. Gradually, the Malpai's ranching families found themselves struggling to feed their cattle and make a living from the land. Decades of photos taken by ranchers showed how the area's soil had eroded, and how trees and brush had overgrown the grass. The ranchers knew their cattle were part of the problem, but they also suspected that firefighting efforts might be to blame.

In 1993, a group of ranchers launched an innovative plan. They formed the Malpai Borderlands

ranchers have traditionally had little incentive to conserve rangelands because most grazing has taken place on public lands leased from the government, not on lands privately owned by ranchers. In addition, the U.S. government has heavily subsidized grazing. As a result, overgrazing has been extensive and has caused many environmental problems in the American West. Today increasing numbers of ranchers are working cooperatively with government agencies, environmental scientists, and even environmental advocates to find ways to ranch more sustainably and safeguard the health of the land (see "The Science behind the Story," above).

Forestry, too, has impacts on soil

Farming and grazing are agricultural practices that help feed human populations, that depend on healthy soils, and that affect the conditions of those soils. Forestry, the

Group, designating about 325,000 ha (800,000 acres) of land as an area for protection and study. Through study of the region, ranchers joined government agencies, environmentalists, and scientists to bring back grasses, restore native animal species, and return periodic fires to the landscape (see the figure).

The group's research efforts have centered on the Gray Ranch, a 121,000-ha (300,000-acre) parcel in the heart of the borderlands. Scientists led by researcher Charles Curtin created study sites by dividing pastureland on the ranch into four study areas of about 890 ha (2,200 acres) each. Each area is further divided into four different "treatments," or areas with varying land management techniques:

▶ In Treatment 1, fire is not suppressed, and grazing by cattle and native animals is permitted.

▶ In Treatment 2, fire is not suppressed, and grazing is not allowed.

▶ In Treatment 3, animals can graze, but fire is suppressed.

▶ In Treatment 4, fire is suppressed, and grazing is not permitted.

Treatments 1 and 3, which allow grazing, also feature small fenced-off areas that prevent animals from eating grass. These exclosures allow scientists to make precise side-by-side comparisons of how grazing affects grasses. Scientists measure rainfall in each area and monitor soils for degradation and erosion. Teams of wildlife and vegetation specialists monitor each treatment for the distribution and abundance of birds, insects, animals, and vegetation.

By comparing areas where fire is suppressed to those where fire can burn, researchers have documented how the suppression of fire leads to more brush and trees and less grass. When fire burns an area, woody plants such as mesquite are damaged and their seed production disrupted, and the flames are often followed by a return of grass. Such benefits follow both natural fires and carefully monitored, deliberately set controlled burns.

Cattle don't do heavy damage to grass if they are managed carefully, the researchers have found. The scientists have helped ranchers develop a cycle of grassbanking, in which herds are allowed to graze on shared plots of land while other areas recover. And ranchers must work with the weather. Scientists have found that controlled burns or grassbanking should track with cycles of rain and drought to bring back grass.

Because of such research, controlled burns are now a regular part of the Malpai landscape. More than 100,000 ha (250,000 acres) have been burned since 1994. Natural fires are often allowed to run their course with little or no intervention. Damaged areas have been reseeded with native grasses. Scientists increasingly believe that ranching, if managed properly, can help bring damaged grazing areas back to life. "We cannot assume rangelands will recover on their own," Curtin wrote in a recent study on the Malpai Borderlands. "Conservation of grazed lands requires restoring and sustaining natural processes."

cultivation of trees, is a similar practice that we will discuss in Chapter 12. Forestry practices can have substantial impacts on soils, just as farming and ranching can. As with farming and ranching techniques, forestry practices have been modified over the years to try to minimize damage to soils. Some practices, such as clear-cutting—the removal of all trees from an area at once—can lead to severe erosion. This is particularly the case on steep slopes (Figure 9.23). Alternative timbering methods that remove fewer trees over longer periods of time are more successful in minimizing erosion.

A number of U.S. and international programs promote soil conservation

In recent years, the U.S. Congress has enacted a number of laws promoting soil conservation. The Food Security Act of 1985 required farmers to adopt soil conservation

FIGURE 9.23 Deforestation, discussed further in Chapter 12, can be a major cause of erosion, particularly when trees are clear-cut for timber on steep slopes.

plans and practices as a prerequisite for receiving price supports and other government benefits. The Conservation Reserve Program, also enacted in 1985, pays farmers to stop cultivating highly erodible cropland and instead place it in conservation reserves planted with grasses and trees. The USDA estimates that for an annual cost of $1 billion, this program saves 700 million metric tons (771 million tons) of topsoil each year. Besides reducing erosion, the Conservation Reserve Program has generated income for farmers and has provided habitat for native wildlife. In 1996, Congress extended the program by passing the Federal Agricultural Improvement and Reform Act. Also known as the "Freedom to Farm Act," this law aimed to reduce subsidies and government

influence over many farm products. It also created the Environmental Quality Incentive Program and the Natural Resources Conservation Foundation to promote and pay for the adoption of conservation practices in agriculture. In 1998, the USDA initiated the Low-Input Sustainable Agriculture Program to provide funding for individual farmers to develop and practice sustainable agriculture.

Internationally, the United Nations promotes soil conservation and sustainable agriculture through a variety of programs of its Food and Agriculture Organization (FAO). The FAO's Farmer-Centered Agricultural Resource Management Program (FAR) is a project undertaken in partnership by China, Thailand, Vietnam, Indonesia, Sri Lanka, Nepal, the Philippines, and India to support innovative approaches to resource management and sustainable agriculture. This program studies agricultural success stories and tries to help other farmers duplicate the successful efforts. Rather than following a top-down, government-controlled approach, the FAR program calls on the creativity of local communities to educate and encourage farmers throughout Asia to conserve soils and secure their food supply.

Conclusion

Many of the policies enacted and the practices developed to combat soil degradation in the United States and worldwide have been quite successful, particularly in reducing the erosion of topsoil. Despite these successes, however, soil is still being degraded at a rate that calls into question the sustainability of industrial agriculture.

Our species has enjoyed a 10,000-year history with agriculture, yet despite all we have learned about soil degradation and conservation, many challenges remain. It is clear that even the best-conceived soil conservation programs require research, education, funding, and commitment from both farmers and governments if they are to fulfill their potential. In light of continued population growth, we will likely need better technology and wider adoption of soil conservation techniques to avoid an eventual food crisis. Increasingly, it seems relevant to consider whether the universal adoption of Aldo Leopold's land ethic (▶ pp. 35–36) will also be required if we are to feed the 9 billion people expected to crowd our planet in mid-century.

REVIEWING OBJECTIVES

You should now be able to:

Explain the importance of soils to agriculture, and describe the impacts of agriculture on soils

▶ Successful agriculture requires healthy soil. (p. 246)
▶ As the human population grows and consumption increases, pressures from agriculture are degrading Earth's soil, and we are losing 5–7 million ha (12–17 million acres) of productive cropland annually. (pp. 246–247)

Outline major historical developments in agriculture

▶ Beginning about 10,000 years ago, people began breeding crop plants and domesticating animals. (p. 248)
▶ Domestication took place through the process of selective breeding, or artificial selection. (p. 248)
▶ Agriculture is thought to have originated multiple times independently in different cultures across the world. (pp. 248–249)
▶ Industrial agriculture is gradually replacing traditional agriculture, which largely replaced hunting and gathering. (pp. 248–250)

Delineate the fundamentals of soil science, including soil formation and the properties of soil

▶ Soil includes diverse biotic communities that decompose organic matter. (pp. 250–251)
▶ Climate, organisms, relief, parent material, and time are factors influencing soil formation. (pp. 251–252)
▶ Soil profiles consist of distinct horizons with characteristic properties. (pp. 252–253)
▶ Soil can be characterized by color, texture, structure, and pH. (pp. 253–255)
▶ Soil properties affect the potential for plant growth and agriculture in any given location. (p. 255)

State the causes and predict the consequences of soil erosion and soil degradation

▶ Some agricultural practices have resulted in high rates of erosion across the world, lowering crop yields. (pp. 256–258)
▶ Desertification affects a large portion of the world's soils, especially those in arid regions. (pp. 259–260)
▶ Overirrigation can cause salinization and waterlogging, which lower crop yields and are difficult to mitigate. (pp. 264–266)
▶ Overapplying fertilizers can cause pollution problems that affect ecosystems and human health. (pp. 266–267)
▶ Overgrazing can cause soil degradation on grasslands, as well as diverse impacts to native ecosystems. (pp. 267–270)
▶ Careless forestry practices, such as deforesting steep slopes, are a major cause of erosion. (pp. 270–271)

Recite the history and explain the principles of soil conservation

▶ The Dust Bowl in the United States and similar events elsewhere have encouraged scientists and farmers to develop ways of better protecting and conserving topsoil. (pp. 260–261)
▶ Farming techniques such as crop rotation, contour farming, intercropping, terracing, shelterbelts, and reduced tillage enable farmers to reduce soil erosion and boost crop yields. (pp. 261–264)
▶ In the United States and across the world, governments are devising innovative policies and programs to deal with the problems of soil degradation. (pp. 271–272)

TESTING YOUR COMPREHENSION

1. How did the practices of selective breeding and human agriculture begin roughly 10,000 years ago? Summarize the influence of agriculture on the development and organization of human communities.
2. Describe the methods used in traditional agriculture, and contrast subsistence agriculture with intensive traditional agriculture. What makes industrialized agriculture different from traditional agriculture?
3. What processes are most responsible for the formation of soil? Describe the three types of weathering that may contribute to the process of soil formation.
4. Name the five primary factors thought to influence soil formation, and describe one effect of each.
5. How are soil horizons created? What is the general pattern of distribution of organic matter in a typical soil profile?
6. Why is erosion generally considered a destructive process? Name three human activities that can promote

soil erosion. Describe four kinds of soil erosion by water. What factors affect the intensity of water erosion?

7. List innovations in soil conservation introduced by Hugh Hammond Bennett, first director of the SCS. What other farming techniques can help reduce the risk of erosion due to conventional cultivation methods?

8. How does terracing effectively turn very steep and mountainous areas into arable land? Explain the

method of no-till farming. Why does this method reduce soil erosion?

9. How do fertilizers boost crop growth? How can large amounts of fertilizer added to soil also end up in water supplies and the atmosphere?

10. What policies can be linked to the practice of overgrazing? Describe the effects of overgrazing on soil. What conditions characterize sustainable grazing practices?

SEEKING SOLUTIONS

1. How do you think a farmer can best help to conserve soil? How do you think a scientist can best help to conserve soil? How do you think a national government can best help to conserve soil?

2. How and why might actual soils differ from the idealized six-horizon soil profile presented in the chapter? How might departures from the idealized profile indicate the impact of human activities? Provide at least three examples.

3. What method of farming would you choose to employ on a gradual slope with the threat of natural erosion? What kinds of plants might you use to prevent erosion, and why?

4. Discuss how the methods of no-till or reduced tillage farming, as described in this chapter, can enhance soil quality. What drawbacks or negative effects might no-till or reduced-tillage practices have on soil quality, and how might these be prevented?

5. Discuss how methods of locally based sustainable agriculture described in this chapter are promoting the science of soil conservation. In reference to Aldo Leopold's land ethic (▶ pp. 35–36), how are humans and the soil members of the same community?

6. Imagine that you are the head of an international granting agency that assists farmers with soil conservation and sustainable agriculture. You have $10 million to disburse. Your agency's staff has decided that the funding should go to (1) farmers in an arid area of Africa prone to salinization, (2) farmers in a fast-growing area of Indonesia where swidden agriculture is practiced, (3) farmers in southern Brazil practicing no-till agriculture, and (4) farmers in a dryland area of Mongolia undergoing desertification. What types of projects would you recommend funding in each of these areas, how would you apportion your funding among them, and why?

INTERPRETING GRAPHS AND DATA

Kishor Atreya and his colleagues at Kathmandu University in Nepal conducted a field experiment to test the effects of reduced tillage versus conventional tillage on erosion and nutrient loss in the Himalayan Mountains in central Nepal. The region in which they worked has extremely steep terrain (with an average slope of 18%), and receives over 138 cm (55 in.) of rain per year, with 90% of it coming during the monsoon season from May to September. Atreya's team measured the amounts of soil, organic carbon, and nitrogen lost from the research plots (which were unterraced) over the course of a year. Some of their results are presented in the graph.

1. Under the conditions of the study reported above, how much soil, organic carbon, and nitrogen would be saved annually in fields with reduced tillage relative to fields with conventional tillage? Express your answers both in absolute units and as percentages.

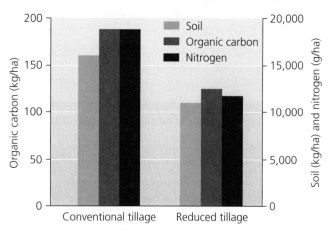

Annual soil and nutrient losses in plots under conventional and reduced tillage systems. All reduced tillage values are significantly different than their conventional tillage counterparts. Data from Atreya, K., et al. 2005. Applications of reduced tillage in hills of central Nepal. *Soil & Tillage Research*, in press.

2. Given that annual crop yields in the study plots were approximately 4 metric tons/ha, what is the ratio of soil lost to crop yield under conventional tillage? Under reduced tillage?

3. Is reduced tillage a sustainable management practice for Nepalese farmers? If so, what data from the study above would you cite in support of your answer? If not, or if you cannot say, then what concerns raised by the data above would still need to be addressed, or what additional data would be needed, to answer the question?

CALCULATING ECOLOGICAL FOOTPRINTS

As you learned in this chapter, rates of soil loss due to erosion can be high. Even in the United States, approximately 6 pounds of topsoil are lost for every 1 pound of grain harvested. Erosion rates vary greatly with soil type, topography, tillage method, and crop type. For simplicity let us assume that the 6:1 ratio applies to all plant crops and that a typical diet includes 1 pound of plant material or its derived products (sugar, for example) per day. In the first two columns of the table, calculate the annual topsoil losses associated with growing this food for you and for other groups, assuming the same diet.

	Plant products consumed (lb)	Soil loss at 6:1 ratio (lb)	Soil loss at 4:1 ratio (lb)	Reduced soil loss at 4:1 relative to 6:1 ratio (lb)
You	365	2,190	1,460	730
Your class				
Your state				
United States				

1. Improved soil conservation measures reduced erosion by approximately one-third from 1982 to 1997. If additional measures were again able to reduce the current rate of soil loss by a third, the ratio of soil lost to grain harvested would fall from 6:1 to 4:1. Calculate the soil losses associated with food production at a 4:1 ratio, and record your answers in the third column of the table.

2. Calculate the amount of topsoil hypothetically saved by the additional conservation measures in Question 1, and record your answers in the fourth column of the table.

3. Define a "sustainable" rate of soil loss. Describe how you might determine if a given farm was practicing sustainable use of soil.

Take It Further

 Go to www.aw-bc.com/withgott or the student CD-ROM where you'll find:

▶ Suggested answers to end-of-chapter questions
▶ Quizzes, animations, and flashcards to help you study
▶ *Research Navigator*™ database of credible and reliable sources to assist you with your research projects

▶ **GRAPHit!** tutorials to help you master how to interpret graphs
▶ **INVESTIGATEit!** current news articles that link the topics that you study to case studies from your region to around the world

10 Agriculture, Biotechnology, and the Future of Food

Organic vegetable farm in Whatcom County, Washington

Upon successfully completing this chapter, you will be able to:

▶ Explain the challenge of feeding a growing human population

▶ Identify the goals, methods, and environmental impacts of the "green revolution"

▶ Categorize the strategies of pest management

▶ Discuss the importance of pollination

▶ Describe the science behind genetically modified food

▶ Evaluate controversies and the debate over genetically modified food

▶ Ascertain approaches for preserving crop diversity

▶ Assess feedlot agriculture for livestock and poultry

▶ Weigh approaches in aquaculture

▶ Evaluate sustainable agriculture

Farmer in maize field in Oaxaca, Mexico

Central Case: Possible Transgenic Maize in Oaxaca, Mexico

"Worrying about starving future generations won't feed them. Food biotechnology will.... At Monsanto, we now believe food biotechnology is a better way forward."
—ADVERTISING CAMPAIGN OF THE MONSANTO COMPANY

"Industrial agriculture has not produced more food. It has destroyed diverse sources of food, and it has stolen food from other species ... using huge quantities of fossil fuels and water and toxic chemicals in the process."
—VANDANA SHIVA, DIRECTOR OF THE RESEARCH FOUNDATION FOR SCIENCE, TECHNOLOGY, AND NATURAL RESOURCE POLICY, INDIA

Corn is a staple grain of the world's food supply. We can trace its ancestry back roughly 5,500 years, when people in the highland valleys of what is now the state of Oaxaca in southern Mexico first domesticated that region's wild maize plants. The corn we eat today arose from some of the many varieties that evolved from the early selective crop breeding conducted by the people of this region.

Today Oaxaca remains a world center of biodiversity for maize, with many native varieties, or *cultivars*, growing in the rich, well-watered soil. Preserving such varieties of crops in their ancestral homelands is important for securing the future of our food supply, scientists maintain, because these varieties serve as reservoirs of genetic diversity—reservoirs we may need to draw on to sustain or advance our agriculture.

Thus, it caused global consternation when, in 2001, Mexican government scientists conducting routine genetic tests of Oaxacan farmers' maize announced that they had turned up DNA that matched genes from genetically modified (GM) corn. GM corn was widely grown in the United States, but Mexico had banned its cultivation in 1998. Corn is one of many crops that scientists have genetically engineered to express desirable traits such as large size, fast growth, and resistance to insect pests. To genetically engineer crops, scientists extract genes from the DNA of one organism and transfer them into the DNA of another. The aim is to improve crop performance and feed the world's hungry, but many people worry that the **transgenes** from these **transgenic** plants could cause unpredictable harm. One concern is that transgenic crops might crossbreed with native crops and thereby "contaminate" the genetic makeup of native crops.

Two researchers at the University of California at Berkeley, Ignacio Chapela and his postdoctoral associate David Quist, shared these concerns, and they ventured to Oaxaca to test samples of native maize. Their analyses seemed to confirm the government scientists' findings, revealing what they argued were traces of DNA from genetically engineered corn in the genes of native maize plants. They also suggested that the invading genes had split up and spread throughout the maize genome. Quist and Chapela published their findings in the scientific journal *Nature* in November 2001. Activists opposed to GM food trumpeted the news and urged a ban on imports of transgenic crops from producer countries such as the United States into developing nations. The agrobiotech industry defended the safety of its crops and questioned the validity of the research—as did many of Quist and Chapela's peers. Responding to criticisms from researchers, *Nature* took the unprecedented step of stating that Quist and Chapela's paper should never have been published—a move that created a firestorm of controversy (see "The Science behind the Story," ▸ pp. 292–293).

Since that time, further research by the Mexican government has confirmed that transgenes exist in Mexican maize. These findings have not yet (as of late 2005) been published in a scientific journal, but they were accepted by a special commission of experts convened under the North American Free Trade Agreement (NAFTA) to study the issue. In November 2004 the commission reported that in 11 communities, 3–13% of maize contained transgenes, and in four other localities, 20–60% of maize contained transgenes. It also reported that 37% of maize grains distributed by the government food distribution agency, Diconsa, were transgenic. The commission reasoned that corn imported from the United States was the source. U.S. corn shipments contain an undifferentiated mix of GM and non-GM grain. Once the transgenes were inside the country, they would have easily spread by wind pollination and by interbreeding with native cultivars of maize.

A harder question to answer is how genetically modified crops may affect people and the environment. In this chapter, we take a wide-ranging view of the ways people have devised to increase agricultural output, the environmental effects of these efforts, and the food of the future.

The Race to Feed the World

Although human population growth has slowed, we can still expect our numbers to swell to 9 billion by the middle of this century. For every two people living today, there will be three in 2050. Feeding 50% more mouths half a century from now while protecting the integrity of soil, water, and ecosystems will require sustainable agriculture. This could involve approaches as diverse as organic farming and the genetically modified crops that are eliciting so much controversy in Oaxaca and elsewhere.

Agricultural production has outpaced population increase so far

Over the past half century, our ability to produce food has grown even faster than global population (Figure 10.1). However, largely because of political obstacles and inefficiencies in distribution, today 850 million people in developing countries do not have enough to eat. Every 5 seconds, somewhere in the world, a child starves to death. Agricultural scientists and policymakers pursue a goal of **food security,** the guarantee of an adequate, reliable, and available food supply to all people at all times. Making a food supply sustainable depends on maintaining healthy soil, water, and biodiversity. As we saw in Chapter 9, careless expansion of agriculture can have devastating effects on the environment and the long-term ability of the world's soils to continue supporting crops and livestock.

Starting in the 1960s, a number of scientists including Paul Ehrlich (▸ pp. 5–6) predicted widespread starvation and a catastrophic failure of agricultural systems, arguing that the human population could not continue to grow without outstripping its food supply. However, the human population has continued to increase well past their predictions. Although it is tragic that 850 million people are hungry today, this number is smaller than the 960 million who lacked reliable and sufficient food in 1970. In percentage terms, we have reduced hunger by half, from 26% of the population in 1970 to 13% today.

We have achieved these advances in part by increasing our ability to produce food (see Figure 10.1). We have increased food production by devoting more energy (especially fossil fuel energy) to agriculture; by planting and harvesting more frequently; by increasing the use of irrigation, fertilizer, and pesticides; by increasing the amount of cultivated land; and by developing (through crossbreeding and genetic engineering) more productive crop and livestock varieties.

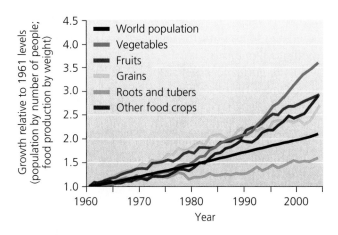

FIGURE 10.1 Global agricultural food production rose by over two-and-a-half times in the past four decades, growing at a faster rate than world population. Production of all types of foods, particularly vegetables, increased from 1961 to 2004. Trend lines show cumulative increases relative to 1961 levels. Data from Food and Agriculture Organization of the United Nations.

However, with grain crops, the world's staple foods, we are producing less food per person each year. World grain production per person peaked in 1985 and has since slowly fallen (Figure 10.2). Moreover, the world's soils are in decline, and nearly all the planet's arable land has already been claimed. Simply because agricultural production has outpaced population growth so far, there is no guarantee that it will continue to do so.

We face undernourishment, overnutrition, and malnourishment

Although many people lack access to adequate food, others are affluent enough to consume more than is healthy. People who are **undernourished,** receiving less than 90% of their daily caloric needs, live mostly in the developing world. Meanwhile, in the developed world, many people suffer from **overnutrition,** receiving too many calories each day. In the United States, where food is available in abundance and people tend to lead sedentary lives with little exercise, more than three out of five adults are technically overweight, and over one out of four are obese. For most people who are undernourished, the reasons are economic. One-fifth of the world's people live on less than $1 per day, and over half live on less than $2 per day, the World Bank estimates. Hunger is a problem even in the United States, where the U.S. Department of Agriculture (USDA) has classified 31 million Americans as "food insecure," lacking the income required to procure sufficient food at all times.

Just as the *quantity* of food a person eats is important for health, so is the *quality* of food. **Malnutrition,** a

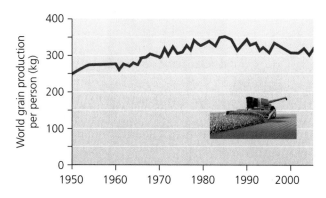

FIGURE 10.2 World per capita production of grain rose until 1985. Since then it has declined slightly as grain production has grown more slowly than global population. Some have pointed to this trend as a sign that our agriculture is no longer able to cope with the deteriorating soil and environmental conditions it has caused. Others propose that the decline is due to two factors: (1) populations in developed nations, where people may be near their limit of grain consumption already, have stabilized; and (2) population growth has been greatest in developing nations, where people consume far less than in developed nations. As a result, a larger share of the world's people lives in societies in which people consume less, thus dragging down the global average demand for grain production. Data from UN FAO and U.S. Department of Agriculture.

shortage of nutrients the body needs, including a complete complement of vitamins and minerals, can occur in both undernourished and overnourished individuals. Malnutrition can lead to disease (Figure 10.3). When people eat a high-starch diet but not enough protein or essential amino acids (▶pp. 99–100), then *kwashiorkor* results. Children who have recently stopped breast-feeding are most at risk for developing kwashiorkor, which causes bloating of the abdomen, deterioration and discoloration of hair, mental disability, immune suppression, developmental delays, anemia, and reduced growth. Protein deficiency together with a lack of calories can lead to *marasmus*, which causes wasting or shriveling among millions of children in the developing world.

The "green revolution" led to dramatic increases in agricultural production

The desire for greater quantity and quality of food for the growing human population led in the mid- and late 20th century to the **green revolution** (first introduced in Chapter 9, ▶p. 250). Realizing that farmers could not go on indefinitely cultivating more and more land to increase crop output, agricultural scientists created methods and technologies to increase crop output per unit area of existing cultivated land. Industrialized nations had been

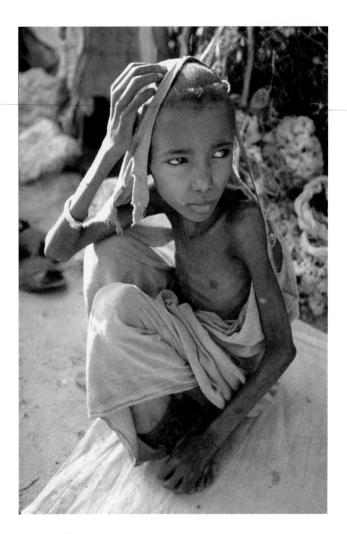

FIGURE 10.4 Norman Borlaug holds examples of the wheat variety he bred that helped launch the green revolution. The high-yielding disease-resistant wheat helped increase agricultural productivity in many developing countries.

FIGURE 10.3 Millions of children, including this child in Somalia, suffer from forms of malnutrition, such as kwashiorkor and marasmus.

dramatically increasing their per-area yields; the average hectare of U.S. cornfield during the 20th century, for instance, upped its corn output fivefold. Many people saw such growth in production and efficiency as key to ending starvation in developing nations.

The transfer of technology to the developing world that marked the green revolution began in the 1940s, when U.S. agricultural scientist Norman Borlaug introduced Mexico's farmers to a specially bred type of wheat (Figure 10.4). This strain of wheat produced large seed heads, was short in stature to resist wind, was resistant to diseases, and produced high yields. Within two decades of planting and harvesting this specially bred crop, Mexico tripled its wheat production and began exporting wheat. The stunning success of this program inspired others. Borlaug— who won the Nobel Peace Prize for his work—took his wheat to India and Pakistan and helped transform agriculture there. Soon many developing countries were

increasing their crop yields using selectively bred strains of wheat, rice, corn, and other crops from developed nations. Some varieties yielded three or four times as much per acre as did their predecessors.

The green revolution has caused the environment both benefit and harm

Along with the new grains, developing nations imported the methods of industrialized agriculture. They began applying large amounts of synthetic fertilizers and chemical pesticides on their fields, irrigating crops with generous amounts of water, and using heavy equipment powered by fossil fuels. From 1900 to 2000, humans expanded the world's total cultivated area by 33%, yet increased energy inputs into agriculture by 80 *times*.

This high-input agriculture succeeded dramatically in allowing farmers to harvest more corn, wheat, rice, and soybeans from each hectare of land. Intensive agriculture saved millions in India from starvation in the 1970s and eventually turned that nation into a net exporter of grain (Figure 10.5). However, it had mixed effects on the environment. On the positive side, the intensified use of already-cultivated land reduced pressures to convert additional natural lands for new cultivation. Between 1961 and 2002, food production rose 150% and population rose 100%, while area converted for agriculture increased only 10% (see "Interpreting Graphs and Data," ▶ pp. 306–307). For this reason, the green revolution prevented some degree

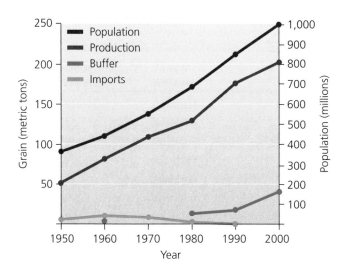

FIGURE 10.5 Thanks to green revolution technology, India's grain production has kept pace with its rapid population growth. This has enabled India to stop importing grain and to maintain extra grain in reserve as a buffer against food shortages.

Weighing the Issues:
The Green Revolution and Population

In the 1960s, India's population was growing at an unprecedented rate, and its traditional agriculture was not producing enough food to support the growth. By adopting green revolution agriculture, India sidestepped mass starvation. In the years since intensifying its agriculture, India has added several hundred million more people and continues to suffer widespread poverty and hunger. Norman Borlaug called his green revolution methods "a temporary success in man's war against hunger and deprivation," something to give us breathing room in which to deal with what he called "the Population Monster."

Do you think we can call the green revolution a success? Do you think the green revolution has solved problems, or delayed our resolution of problems, or created new ones? How sustainable are green revolution approaches? Consider our discussion of demographic transition theory from Chapter 8 (▸pp. 228–229). Have we been dealing with the "Population Monster" during the breathing room that the green revolution has bought for us?

of deforestation and habitat conversion in many countries at the very time those countries were experiencing their fastest population growth rates. In this sense, the green revolution was a boon for the preservation of biodiversity and natural ecosystems. However, the intensive use of water, fossil fuels, and chemical fertilizers and pesticides had extensive negative impacts on the environment in terms of pollution, salinization, and desertification (Chapter 9).

One key aspect of green revolution techniques has had negative consequences for biodiversity and mixed consequences for crop yields. The planting of crops in *monocultures*, large expanses of single crop types (Figure 10.6a), has made planting and harvesting more

(a) Wheat monoculture

(b) Armyworm

FIGURE 10.6 Most agricultural production in industrialized countries comes from monocultures—large stands of single types of crop plant, such as this wheat field in Washington (**a**). Clustering crop types in uniform fields on large scales greatly improves the efficiency of planting and harvesting, but it also decreases biodiversity and makes crops susceptible to outbreaks of pests that specialize on particular crops. Armyworms (**b**) are major agricultural pests whose outbreaks can substantially reduce crop yields. The caterpillars of these moths defoliate a wide variety of crops, including wheat, corn, cotton, alfalfa, and beets.

efficient and has thereby increased output. However, monocultural planting has reduced biodiversity over huge areas, because many fewer wild organisms are able to live in monocultures than in native habitats or in traditional small-scale polycultures. Moreover, when all plants in a field are genetically similar, as in monocultures, all will be equally susceptible to viral diseases, fungal pathogens, or insect pests that can spread quickly from plant to plant (Figure 10.6b). For this reason, monocultures bring significant risks of catastrophic failure.

Monocultures have also contributed to a narrowing of the human diet. Globally, 90% of the food we consume now comes from only 15 crop species and eight livestock species—a drastic reduction in diversity from earlier times. The nutritional dangers of such dietary restriction have been alleviated by the fact that expanded global trade has provided many people access to a wider diversity of foods from around the world. However, this effect has benefited wealthy people far more than poor people. One reason farmers and scientists were so concerned about transgenic contamination of Oaxaca's native maize is that Oaxacan maize varieties serve as a valuable source of genetic variation in a world where so much variation is being lost to monocultural practices.

Pests and Pollinators

Throughout the history of agriculture, the insects, fungi, viruses, rats, and weeds that eat or compete with our crop plants have taken advantage of the ways we cluster food plants into agricultural fields. These organisms, in making a living for themselves, cut crop yields and make it harder for farmers to make a living. As just one example of thousands, various species of moth caterpillars known as armyworms (see Figure 10.6b) lower yields of everything from beets to sorghum to millet to canola to pasture grasses. Pests and weeds have always posed problems for traditional agriculture, and they pose an even greater threat to monocultures, where a pest adapted to specialize on the crop can easily move from one individual plant to many others of the same type.

What humans term a *pest* is any organism that damages crops that are valuable to us. What we term a *weed* is any plant that competes with our crops. These are subjective categories that we define entirely by our own economic interests. There is nothing inherently malevolent in the behavior of a pest or a weed. These organisms are sim-

Table 10.1 Most Commonly Used Pesticides* in Agriculture in the United States		
Active ingredient†	**Type of pesticide**	**Millions of kg applied per year**
Glyphosate	Herbicide	39–41
Atrazine	Herbicide	34–36
Metam sodium	Fumigant	26–28
Acetochlor	Herbicide	14–16
2,4-D	Herbicide	13–15
Malathion	Insecticide	9–11
Methyl bromide	Fumigant	9–11
Dichloropropene	Fumigant	9–11
Metolachlor-s	Herbicide	9–11
Metolachlor	Herbicide	7–10

*Includes only "conventional pesticides" used in agriculture. Does not include many other types of pesticides, such as disinfectants and wood preservatives.
†Includes only active ingredients, not ingredients such as oil, sulfur, and sulfuric acid.
Data from Kiely, T., et al. 2004. *Pesticides industry sales and usage: 2000 and 2001 market estimates.* Washington, DC: U.S. Environmental Protection Agency.

ply trying to survive and reproduce. From the viewpoint of an insect that happens to be adapted to feed on corn, grapes, or apples, a grain field, vineyard, or orchard represents an endless buffet.

Many thousands of chemical pesticides have been developed

To prevent pest outbreaks and to limit competition with weeds, people have developed thousands of artificial chemicals to kill insects (*insecticides*), plants (*herbicides*), and fungi (*fungicides*). Such poisons that target pest organisms are collectively termed **pesticides.** Table 10.1 shows the 10 most widely used pesticides in U.S. agriculture. All told, roughly 400 million kg (900 million lb) of active ingredients from conventional pesticides are applied in the United States each year. Three-quarters of this total is applied on agricultural land. The remainder is used by industry and municipalities (13%) and applied by homeowners to homes, lawns, and gardens (11%). Since 1960, pesticide use has risen fourfold worldwide. Usage in the United States and the rest of the industrialized world has leveled off in the past two decades, but it continues to increase in the developing world. Today more than $32 billion is expended annually on pesticides, with one-third

① Outbreak of pests on crops

② Application of pesticide

③ All pests except a few with innate resistance are killed

④ Survivors breed and produce a pesticide-resistant population

⑤ Pesticide is applied again

⑥ Pesticide has little effect and new, more toxic pesticides must be developed

FIGURE 10.7 Through the process of natural selection, crop pests frequently evolve resistance to the poisons we apply to kill them. This simplified diagram shows that when a pesticide is applied to an outbreak of insect pests, it may kill virtually all individuals except those few with an innate immunity to the poison. Those surviving individuals may found a population with genes for resistance to the poison. Future applications of the pesticide may then be ineffective, forcing us to develop a more potent poison or an alternative means of pest control.

of that total spent in the United States. We will address the health consequences of synthetic pesticides for humans and other organisms in Chapter 14.

Pests evolve resistance to pesticides

Despite the toxicity of these chemicals (▸ pp. 406–407), their usefulness tends to decline with time as pests evolve resistance to them. Recall from our discussion of natural selection (▸ pp. 118–121) that organisms within populations vary in their traits. Because most insects and microbes occur in huge numbers, it is likely that a small fraction of individuals may by chance have genes that confer some degree of immunity to a given pesticide. Even if a pesticide application kills 99.99% of the insects in a field, 1 in 10,000 survives. If an insect survives by being genetically resistant to a pesticide, and if it mates with other resistant individuals of the same species, the insect

population may grow. This new population will consist of individuals that are genetically resistant to the pesticide. As a result, pesticide applications will cease to be effective (Figure 10.7).

In many cases, industrial chemists are caught up in an "evolutionary arms race" with the pests they battle, racing to increase or retarget the toxicity of their chemicals while the armies of pests evolve ever-stronger resistance to their efforts. The number of species known to have evolved resistance to pesticides has grown over the decades. As of 2000, there were nearly 2,700 known cases of resistance by 540 species to over 300 pesticides. Several insects, such as the green peach aphid, Colorado potato beetle, and diamondback moth, have evolved resistance to multiple insecticides. Resistant pests can take a significant economic toll on crops. As just one example, gummy stem blight destroyed two-thirds of Texas's melon crop in 1997 after the blight evolved resistance to the pesticide Benlate.

Biological control pits one organism against another

Because of pesticide resistance and the health risks of some synthetic chemicals, agricultural scientists increasingly battle pests and weeds with organisms that eat or infect them. This strategy, called **biological control,** or **biocontrol** for short, operates on the principle that "the enemy of one's enemy is one's friend." For example, parasitoid wasps (▶p. 154) are natural enemies of many caterpillars. These wasps lay eggs on a caterpillar, and the larvae that hatch from the eggs feed on the caterpillar, eventually killing it. Parasitoid wasps have been used as biocontrol agents in many situations. Some such efforts have succeeded at pest control and have led to steep reductions in chemical pesticide use.

One classic case of successful biological control is the introduction of the cactus moth, *Cactoblastis cactorum,* from Argentina to Australia in the 1920s to control invasive prickly pear cactus that was overrunning rangeland (Figure 10.8). Within just a few years, the moth managed to free millions of hectares of rangeland from the cactus.

A widespread modern biocontrol effort has been the use of ***Bacillus thuringiensis* (Bt),** a naturally occurring soil bacterium that produces a protein that kills many caterpillars and the larvae of some flies and beetles. Farmers have used the natural pesticidal activity of this bacterium to their advantage by spraying spores of this bacterium on their crops. If used correctly, Bt can protect crops from pest-related losses.

Biological control agents themselves may become pests

In most cases, biological control involves introducing an animal or microbe into a foreign ecosystem, often on another continent. Such relocation helps ensure that the target pest has not already evolved ways to deal with the biocontrol agent, but it also introduces risks. In some cases, biocontrol has produced unintended consequences once the biocontrol agent became invasive and began affecting nontarget organisms. Following the cactus moth's success in Australia, for example, it was introduced in other countries to control prickly pear. Moths introduced to Caribbean islands spread to Florida on their own and are now eating their way through rare native cacti in Florida and spreading to other states. If these moths reach Mexico and the southwestern United States, they could decimate many native and economically important species of prickly pear there.

(a) Before cactus moth introduction

(b) After cactus moth introduction

FIGURE 10.8 In one of the classic cases of biocontrol, larvae of the cactus moth, *Cactoblastis cactorum,* were used to clear non-native prickly pear cactus from millions of hectares of rangeland in Queensland, Australia. These photos from the 1920s show an Australian ranch before **(a)** and after **(b)** introduction of the moth.

One recent study revealed the extent to which some biocontrol agents in Hawaii have missed their targets. Wasps and flies have been introduced to control agricultural pests in Hawaii at least 122 times over the past century, and biologists Laurie Henneman and Jane Memmott suspected that some of these might be adversely affecting native Hawaiian caterpillars that were not pests. They sampled parasitoid wasp larvae from 2,000 caterpillars of various species in a remote mountain swamp far from farmland. In this area, designated as a wilderness preserve, they found that fully 83% of the parasitoids were biocontrol agents that had been intended to combat lowland agricultural pests.

Scientists debate the relative benefits and risks of biocontrol measures. If biocontrol works as planned, it can be a permanent solution that requires no further maintenance and is environmentally benign. However, if the agent has nontarget effects, the harm done may also become permanent, because removing the agent from the system once it is established is far more difficult than simply stopping a chemical pesticide application. One noted skeptic of biocontrol, ecologist Daniel Simberloff (▶p. 169, and ▶pp. 332–333), has remarked that biocontrol "should be used with our eyes wide open—and as a last resort." However, two British scientists reviewing cases as of the year 2000 concluded that only a small percentage of efforts have resulted in demonstrable nontarget effects, and perhaps less than 10% of these effects were substantial. Because of concerns about unintended impacts, researchers now study biocontrol proposals carefully before putting them into action, and government regulators must approve these efforts. However, there will never be a sure-fire way of knowing in advance whether a given biocontrol program will work as planned.

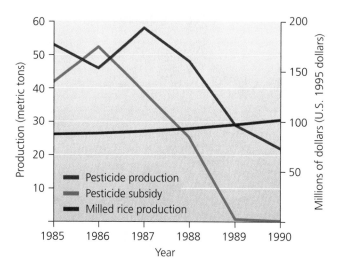

FIGURE 10.9 The Indonesian government threw its weight behind integrated pest management starting in 1986. Within just a few years, pesticide production and pesticide imports were down drastically, pesticide subsidies were phased out, and yields of rice increased slightly.

Integrated pest management combines biocontrol and chemical methods

As it became clear that both chemical and biocontrol approaches have their drawbacks, many agricultural scientists and farmers developed a more sophisticated strategy, trying to combine the best attributes of each approach. In **integrated pest management (IPM),** numerous techniques are integrated to achieve long-term suppression of pests, including biocontrol, use of chemicals, close monitoring of populations, habitat alteration, crop rotation, transgenic crops, alternative tillage methods, and mechanical pest removal. IPM is broadly enough defined that it encompasses a wide variety of strategies.

In recent decades, IPM has become popular in many parts of the world. Indonesia stands as an exemplary case (Figure 10.9). The nation had subsidized pesticide use heavily for years, but its scientists came to understand that pesticides were actually making pest problems worse. They were killing the natural enemies of the brown planthopper, which began to devastate rice fields as its populations exploded. Concluding that pesticide subsidies were costing money, causing pollution, and apparently decreasing yields, the Indonesian government in 1986 banned the importation of 57 pesticides, slashed pesticide subsidies, and encouraged IPM. Within 4 years, pesticide production fell to below half its 1986 level, imports fell to one-third, and subsidies were phased out (saving $179 million annually). Rice yields rose 13%.

We depend on insects to pollinate crops

Managing insect pests is such a major issue in agriculture that many people fall into a habit of thinking of all insects as somehow bad or threatening. But in fact, most insects are harmless to agriculture, and some are absolutely essential. The insects that pollinate agricultural crops are one of the most vital, yet least understood and least appreciated, factors in cropland agriculture. Pollinators are the unsung heroes of agriculture.

Pollination is the process by which male sex cells of a plant (pollen) fertilize female sex cells of a plant; it is the botanical version of sexual intercourse. Without pollination, no plants could reproduce sexually, and no plant species would persist for long. Plants such as ferns, conifer trees, and grasses achieve pollination by the wind. Millions of minuscule pollen grains are blown long distances, and by chance a small number land on the female parts of other plants of their species. The many kinds of plants that sport showy flowers, however, typically are pollinated by animals, such as hummingbirds, bats, and insects (Figure 10.10). Flowers are, in fact, evolutionary adaptations that function to attract pollinators. The sugary nectar and protein-rich pollen in flowers serve as rewards to lure these sexual intermediaries, and the sweet smells and bright colors of flowers are signals to advertise these rewards.

Although our staple grain crops are derived from grasses and are wind-pollinated, many other crops

FIGURE 10.10 Many agricultural crops depend on insects or other animals to pollinate them. Our food supply, therefore, depends partly on conservation of these vital organisms. These apple blossoms are being visited by a European honeybee. Flowers use colors and sweet smells to advertise nectar and pollen, enticements that attract pollinators.

FIGURE 10.11 European honeybees are widely used to pollinate crop plants, and beekeepers transport hives of bees to crops when it is time for flowers to be pollinated. However, honeybees have recently suffered devastating epidemics of parasitism, making it increasingly important for us to conserve native species of pollinators.

depend on insects for pollination. The most complete survey to date, by tropical bee biologist Dave Roubik, documented 800 species of cultivated plants that rely on bees and other insects for pollination. An estimated 73% of cultivars are pollinated, at least in part, by bees; 19% by flies; 5% by wasps; 5% by beetles; and 4% by moths and butterflies. In addition, bats pollinate 6.5% and birds 4%. As one of many examples, alfalfa has long been a major cash crop in the U.S. Great Basin states and is pollinated mostly by native alkali bees that live in the soil as larvae. In the 1940s to 1960s, farmers began plowing the land and increasing pesticide use in an effort to boost yields. These measures killed vast numbers of the soil-dwelling bees, and alfalfa production declined. By 1990, only 15% of U.S. alfalfa-growing lands were inhabited by alkali bees and native leaf-cutter bees. Despite their decline in numbers, bees and their pollination services help these lands produce fully 85% of the U.S. alfalfa crop.

Conservation of pollinators is vital

Preserving the biodiversity of native pollinators is especially important today because the domesticated workhorse of pollination, the European honeybee *(Apis apis)*, is being devastated by parasites. North American farmers regularly hire beekeepers to bring colonies of this introduced honeybee to their fields when it is time to pollinate

crops (Figure 10.11). In recent years, certain parasitic mites have swept through honeybee populations, decimating hives and pushing many beekeepers toward financial ruin. Moreover, research indicates that honeybees are sometimes less effective pollinators than many native species, and often outcompete them, keeping the native species away from the plants.

Farmers and homeowners alike can help maintain populations of pollinating insects by reducing or eliminating pesticide use. All insect pollinators, including honeybees, are vulnerable to the vast arsenal of insecticides that modern industrial agriculture applies to crops and that many homeowners apply to lawns and gardens. Some insecticides are designed to specifically target certain types of insects, but many are not. Without full and detailed information on the effects of pesticides, farmers and homeowners trying to control the "bad" bugs that threaten the plants they value all too often kill the "good" insects as well.

Homeowners, even in the middle of a large city, can encourage populations of pollinating insects by planting gardens of flowering plants that nourish pollinating insects and by providing nesting sites for bees. These can be simple contraptions made of wood that have holes and plastic straws. And by allowing noncrop flowering plants (such as clover) to grow around the edges of their fields, farmers can maintain a diverse community of insects—some of which will pollinate their crops.

Genetic Modification of Food

The green revolution enabled us to feed a greater number and proportion of the world's people, but relentless population growth demands still more. A new set of potential solutions began to arise in the 1980s and 1990s as advances in genetics enabled scientists to directly alter the genes of organisms, including crop plants and livestock. The genetic modification of organisms that provide us food holds promise for increasing nutrition and the efficiency of agriculture while lessening the impacts of agriculture on the planet's environmental systems. However, genetic modification may also pose risks that are not yet well understood, which has given rise to protest around the globe from consumer advocates, small farmers, opponents of big business, and environmental activists. Because genetically modified foods have generated so much emotion and controversy, it is vital at the outset to clear up the terminology and clarify exactly what the techniques involve.

Genetic modification of organisms depends on recombinant DNA

The genetic modification of crops and livestock is one type of genetic engineering. **Genetic engineering** is any process whereby scientists directly manipulate an organism's genetic material in the lab, by adding, deleting, or changing segments of its DNA. **Genetically modified (GM) organisms** are organisms that have been genetically engineered using a technique called recombinant DNA technology. **Recombinant DNA** is DNA that has been patched together from the DNA of multiple organisms. In this process, scientists break up DNA from multiple organisms and then splice segments together, trying to place genes that produce certain proteins and code for certain desirable traits (such as rapid growth, disease and pest resistance, or higher nutritional content) into the genomes of organisms lacking those traits.

Recombinant DNA technology was developed in the 1970s by scientists studying the bacterium *Escherichia coli*. As shown in Figure 10.12, scientists first isolate plasmids, small, circular DNA molecules, from a bacterial culture. At the same time, DNA containing a gene of interest is removed from the cells of another organism. Scientists insert the gene of interest into the plasmid to form recombinant DNA. This recombinant DNA enters new bacteria, which then reproduce, generating many copies of the desired gene.

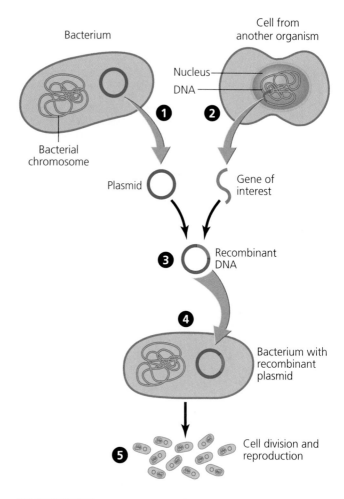

FIGURE 10.12 The creation of recombinant DNA is a key to genetically modifying organisms. In this process, a gene of interest is excised from the DNA of one type of organism and is inserted into a stretch of bacterial DNA called a plasmid. The plasmid is then introduced into cells of the organism to be modified. If all goes as planned, the new gene will be expressed in the GM organism as a desirable trait, such as rapid growth or high nutritional content in a food crop.

When scientists use recombinant DNA technology to develop new varieties of crops, they can often introduce the recombinant DNA directly into a plant cell and regenerate an entire plant from that single cell. Some plants, including many grains, are not receptive to plasmids, in which case scientists may use a "gene gun" to shoot DNA directly into plant cells. An organism that contains DNA from another species is called a *transgenic* organism, and the genes that have moved between them are called *transgenes*. The creation of transgenic organisms is one type of **biotechnology,** the material application of biological science to create products derived from organisms. Recombinant DNA and other types of biotechnology have helped us develop medicines, clean up pollution,

Several Notable Examples of Genetically Modified Food Technology	
Food	Development
Golden rice 	Millions of people in the developing world get too little vitamin A in their diets, causing diarrhea, blindness, immune suppression, and even death. The problem is worst with children in east Asia, where the staple grain, white rice, contains no vitamin A. Researchers took genes from plants that produce vitamin A and spliced the genes into rice DNA to create more-nutritious "golden rice" (the vitamin precursor gives it a golden color). Critics charged that biotech companies hyped their product, which contains only small amounts of the nutrient and may not be the best way to combat vitamin A deficiency. India's foremost critic of GM food, Vandana Shiva, charged that "vitamin A rice is a hoax... a very effective strategy for corporate takeover of rice production, using the public sector as a Trojan horse." Backers of the technology counter that the nutritive value can be further improved and could enhance the health of millions of people.
Flavr Savr tomato 	By reversing the function of a normal tomato gene, the Calgene Corporation created the Flavr Savr tomato, which Calgene maintained would ripen longer on the vine, taste better, stay firm during shipping, and last longer in the produce department. The U.S. Food and Drug Administration approved the Flavr Savr tomato for sale in the United States in 1994. Calgene stopped selling the Flavr Savr in 1996, however, for several reasons, including problems with the technique and public safety concerns.
Ice-minus strawberries 	University of California–Berkeley researcher Steven Lindow removed a gene that facilitated the formation of ice crystals from the DNA of a particular bacterium, *Pseudomonas syringae*. The modified, frost-resistant bacteria could then serve as a kind of antifreeze when sprayed on the surface of frost-sensitive crops such as strawberries. The multiplying bacteria would coat the berries, protecting them from frost damage. However, early news coverage of this technique showed scientists spraying plants while wearing face masks and protective clothing, an image that caused public alarm.
Bt crops 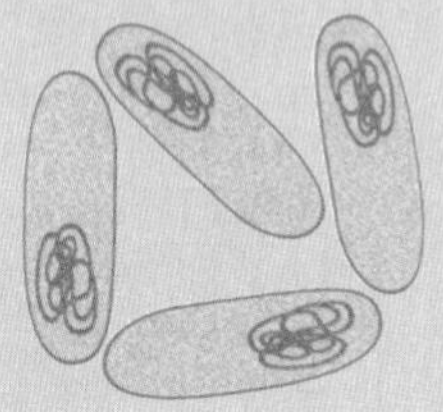	By equipping plants with the ability to produce their own pesticides, scientists hoped to boost crop yields by reducing losses to insects. By the late 1980s, scientists working with *Bacillus thuringiensis* (Bt) had pinpointed the genes responsible for producing that bacterium's toxic effects on insects, and had managed to insert the genes into the DNA of crops. The USDA and EPA approved Bt versions of 18 crops for field testing, from apples to broccoli to cranberries. Corn and cotton are the most widely planted Bt crops today. Proponents say Bt crops reduce the need for chemical pesticides. However, critics worry that the continuous presence of Bt in the environment will induce insects to evolve resistance to the toxins and that Bt crops might cause allergic reactions in humans. Another concern is that the crops may harm nontarget species. A 1999 study reported that pollen from Bt corn can kill the larvae of monarch butterflies, a nontarget species, when corn pollen drifts onto milkweed plants monarchs eat. Another study that year showed the Bt toxin could leach from corn roots and poison the soil.

FIGURE 10.13 The early development of genetically modified foods has been marked by a number of cases in which these products ran into trouble in the marketplace or were opposed by activists. A selection of these cases serves to illustrate some of the issues that proponents and opponents of GM foods have being debating.

understand the causes of cancer and other diseases, dissolve blood clots after heart attacks, and make better beer and cheese. Figure 10.13 details several of the most notable developments in GM foods. These examples and the stories behind them illustrate both the promises and pitfalls of food biotechnology.

Genetic engineering is like, and unlike, traditional agricultural breeding

The genetic alteration of plants and animals by humans is nothing new; we have been influencing the genetic makeup of our livestock and crop plants for thousands of years. As we saw in Chapter 9 (▶p. 248), our ancestors altered the gene pools of our domesticated plants and animals through selective breeding by preferentially mating individuals with favored traits so that offspring would inherit those traits. Early farmers selected plants and animals that grew faster, were more resistant to disease and drought, and produced large amounts of fruit, grain, or meat.

Proponents of GM crops often stress this continuity with our past and say there is little reason to expect that today's GM food will be any less safe than selectively bred

Several Notable Examples of Genetically Modified Food Technology	
Food	**Development**
StarLink corn 	StarLink corn, a variety of Bt corn, had been approved and used in the United States for animal feed but not for human consumption. In 2000, StarLink corn DNA was discovered in taco shells and other corn products, causing fears that the corn might cause allergic reactions. No such health effects have been confirmed, but the corn's French manufacturer, Aventis CropScience, chose to withdraw the product from the market. Although StarLink corn was grown on only a tiny portion of U.S. farmland, its transgene apparently spread widely to other corn through cross-pollination. This episode cost U.S. taxpayers, because the U.S. government spent $20 million to purchase contaminated corn and remove it from the food supply.
Sunflowers and superweeds 	Sunflowers have also been engineered to express the Bt toxin. Research on Bt sunflowers suggests that their transgenes might spread to other plants and turn them into vigorous weeds that compete with the crop. This is most likely to happen with crops like squash, canola, and sunflowers that can breed with their wild relatives. In 2002, Ohio State University researcher Alison Snow and colleagues bred wild sunflowers with Bt sunflowers and found that hybrids with the Bt gene produced more seeds and suffered less herbivory than hybrids without it. They concluded that if Bt sunflowers were planted commercially, the Bt gene would spread into wild sunflowers, potentially turning them into superweeds. Researcher Norman Ellstrand of the University of California–Riverside, meanwhile, had found that transgenes from radishes were transferred to wild relatives 1 km (0.6 mi) away and that hybrids produced more seeds, so the gene could be expected to spread in wild populations. He found the same results with sorghum and its weedy relative, johnsongrass. Such results suggest that transgenic crops can potentially create superweeds that can compete with crops and harm nontarget organisms.
Roundup Ready crops 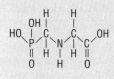	The Monsanto Company manufactures a widely used herbicide called Roundup. Roundup kills weeds, but kills crops too, so farmers must apply it carefully. Thus, Monsanto engineered Roundup Ready crops, including soybeans, corn, cotton, and canola, that are immune to the effects of its herbicide. With these variants, farmers can spray Roundup on their fields without killing their crops, in theory making the farmer's life easier. Of course, this also creates an incentive for farmers to use Monsanto's Roundup herbicide rather than a competing brand. Unfortunately, Roundup is not completely benign; its active ingredient, glyphosate, is the third-leading cause of illness for California farm workers. It also harms nitrogen-fixing bacteria and desirable fungi in soils that are essential for crop production. Biotech proponents have argued that GM crops are good for the environment because they reduce pesticide use. This may often be the case, but some studies have shown that farmers apply more herbicide when they use Roundup Ready crops.
Terminator seeds	In the late 1990s the USDA worked with Delta and Pine Land Company to engineer a line of crop plants that can kill their own seeds. This so-called "terminator" technology would ensure that farmers buy seeds from seed companies every year rather than planting seeds saved from the previous year's harvest. Because GM crops require a great deal of research and development, seed companies reason that they need to charge farmers annually for seeds in order to recoup their investment. Critics worried that pollen from terminator plants might fertilize normal plants, damaging the crops of farmers who save seeds from year to year. Some nations, like India and Zimbabwe, banned terminator seeds. These countries saw the efforts of biotech seed companies to sell them terminator seeds as a ploy to make poor farmers dependent on multinational corporations for seeds. In the face of this opposition, in 1999 agrobiotech companies Monsanto and AstraZeneca announced that they would not bring their terminator technologies to market.

food. Dan Glickman, head of the USDA from 1995 to 2001, remarked:

> Biotechnology's been around almost since the beginning of time. It's cavemen saving seeds of a high-yielding plant. It's Gregor Mendel, the father of genetics, cross-pollinating his garden peas. It's a diabetic's insulin, and the enzymes in your yogurt.... Without exception, the biotech products on our shelves have proven safe.

However, as biotech critics are quick to point out, the techniques geneticists use to create GM organisms do differ from traditional selective breeding in several ways. For one, selective breeding generally mixes genes of individuals of the same species, whereas with recombinant DNA technology, scientists mix genes of different ones, even species as different as viruses and crops, or spiders and goats. For another, selective breeding deals with whole organisms living in the field, whereas genetic engineering involves lab experiments dealing with genetic material apart from the organism. And whereas traditional breeding selects from among combinations of genes that come together on their own, genetic engineering creates the novel combinations directly. Thus, traditional breeding changes organisms through the process of selection, whereas genetic engineering is more akin to the process of mutation.

Biotechnology is transforming the products around us

In just three decades, GM foods have gone from science fiction to big business. As recombinant DNA technology first developed in the 1970s, scientists debated among themselves whether the new methods were safe. They collectively regulated and monitored their own research until most scientists were satisfied that reassembling genes in bacteria did not create dangerous superbacteria. Once the scientific community declared itself confident that the technique was safe in the 1980s, industry leaped at the chance to develop hundreds of applications, from improved medicines (such as hepatitis B vaccine and insulin for diabetes) to designer plants and animals. Traits engineered into crops, such as built-in pest resistance and herbicide resistance, made it efficient, and in some cases more economical, for large-scale commercial farmers to do their jobs. As a result, sales of GM seeds to these farmers in the United States and other countries rose quickly.

Today well over two-thirds of the U.S. harvests of soybeans, corn, and cotton consists of genetically modified strains. Worldwide, more than half the soybean harvest is transgenic, as is one of every four cotton plants, one of every five canola plants, and one of every six corn plants. Globally in 2004, it was estimated that GM crops grew on 81 million ha (200 million acres) of farmland, an area the size of Kansas, Nebraska, and California combined.

Of the 17 nations growing GM crops in 2004, five (the United States, Argentina, Canada, Brazil, and China) accounted for 96% of GM crops, with the United States alone accounting for three-fifths of the global total (Figure 10.14). Because these nations are major food exporters, much of the produce on the world market is transgenic for crops such as soybeans, corn, cotton, and canola. The global area planted in GM crops has jumped by more than 10% annually every year since 1996, and over half the world's people now live in nations in which GM crops are grown. The market value of GM crops in 2004 was estimated at $4.7 billion.

What are the impacts of GM crops?

As GM crops were adopted, as research proceeded, and as biotech business expanded, many citizens, scientists, and policymakers became concerned. Some feared the new foods might be dangerous for people to eat. Others were concerned that transgenes might escape and pollute ecosystems and damage nontarget organisms. Still others worried that pests would evolve resistance to the supercrops and become "superpests" or that transgenes would

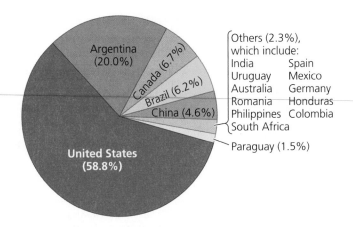

FIGURE 10.14 At 58.8% of the global total, the United States leads the world in land area dedicated to genetically modified crops. Following far behind the United States is Argentina at 20.0%. Data from International Service for the Acquisition of Agri-Biotech Application (ISAAA), 2004.

be transferred from crops to other plants and turn them into "superweeds." Some, like Quist and Chapela, worried that transgenes might ruin the integrity of native ancestral races of crops.

Because the technology is new and its large-scale introduction into the environment is newer still, there remains a lot scientists don't know about how transgenic crops behave in the field. Certainly, millions of Americans eat GM foods every day without outwardly obvious signs of harm, and evidence for negative ecological effects is limited so far. However, it is still too early to dismiss the concerns discussed above without further scientific research. Therefore, critics argue that we should proceed with caution, adopting the **precautionary principle,** the idea that one should not undertake a new action until the ramifications of that action are well understood.

The best efforts so far to test the ramifications of GM crops produced results in 2003 through 2005. The British government, in considering whether to allow the planting of GM crops, commissioned three large-scale studies. The first study, on economics, found that GM crops could produce long-term financial benefits for Britain, although short-term benefits would be minor. The second study addressed health risks and found little to no evidence of harm to human health, but noted that effects on wildlife and ecosystems should be tested before crops are approved. The third study (involving 19 researchers, 200 sites, and $8 million in funding) tested effects on bird and invertebrate populations from four GM crops modified for herbicide resistance. Results showed that fields of GM beets and GM spring oilseed rape supported less biodiversity than fields of their non-GM counterparts.

Fields of GM maize supported more, however, and fields of winter oilseed rape showed mixed results. Policymakers had hoped that the biodiversity study would end the debate, but the science showed that the impacts of GM crops are complex.

Debate over GM foods involves more than science

Much more than science has been involved in the debate over GM foods. Ethical issues have played a large role. For many people, the idea of "tinkering" with the food supply seems dangerous or morally wrong. Even though our agricultural produce is the highly artificial product of thousands of years of selective breeding, people tend to think of food as natural. Furthermore, because every person relies on food for survival and cannot choose *not* to eat, the genetic modification of dietary staples such as corn, wheat, and rice essentially forces people to consume GM products or to go to special effort to avoid them.

The perceived lack of control over one's own food has driven widespread concern about domination of the global food supply by a few large businesses. Gigantic agrobiotech companies, among them Monsanto, Syngenta, Bayer CropScience, Dow, DuPont, and BASF, create GM technologies. Many activists say these multinational corporations threaten the independence and well-being of the small farmer. This perceived loss of democratic local control is a driving force in the opposition to GM foods, especially in Europe and the developing world. Critics of biotechnology also voice concern that much of the research into the safety of GM organisms is funded, overseen, or conducted by the corporations that stand to profit if their transgenic crops are approved for human consumption, animal feed, or ingredients in other products.

So far, GM crops have not lived up to their promise of feeding the world's hungry. Nearly all commercially available GM crops have been engineered to express either pesticidal properties (e.g., Bt crops) or herbicide tolerance. Often, these GM crops are tolerant to herbicides that the same company manufactures and profits from (e.g., Monsanto's Roundup Ready crops). Crops with traits that might benefit poor small-scale farmers of developing countries (such as increased nutrition, drought tolerance, and salinity tolerance) have not been widely commercialized, perhaps because corporations have less economic incentive to do so. Whereas the green revolution was a largely public venture, the "gene revolution" promised by GM crops has been largely driven by market considerations of companies selling proprietary products.

FIGURE 10.15 Saskatchewan farmer Percy Schmeiser was accused by the Monsanto Company of planting its patented Roundup Ready canola in his field without a contract with the company. Schmeiser said his non-GM plants were contaminated with Monsanto's transgenes from neighboring farms. Monsanto won the David-and-Goliath case in Canada's Supreme Court, but Schmeiser became a hero to small farmers and anti-GM food activists worldwide.

When the U.S.-based Monsanto Company began developing GM products in the mid-1980s, it foresaw public anxiety and worked hard to inform, reassure, and work with environmental and consumer advocates, whom the company feared would otherwise oppose the technology. Monsanto even lobbied the U.S. government to regulate the industry so the public would feel safer about it. These efforts were undermined, however, when the company's first GM product to market, a growth hormone to spur milk production in cows, alarmed consumers concerned about children's health. Then, when the company went through a leadership change, its new head changed tactics and pushed new products aggressively without first reaching out to opponents. Opposition built, and the company lost the public's trust, especially in Europe and in the developing world.

In Canada, Monsanto has been engaged in a high-publicity battle with a third-generation Saskatchewan farmer named Percy Schmeiser (Figure 10.15). Schmeiser maintained that pollen from Monsanto's Roundup Ready canola (see Figure 10.13) used by his neighbors blew onto his land and pollinated his non-GM canola. Schmeiser had never purchased Monsanto's patented

Transgenic Contamination of Native Maize?

David Quist and Ignacio Chapela's *Nature* paper reporting DNA from genetically engineered corn in native Mexican maize was controversial from the moment it was published. The idea that a transgene had entered native maize in remote areas of Oaxaca was provocative enough. But their claim that the transgene also had been jumping throughout the genome came under particular fire.

Soon after publication, several geneticists pointed out flaws in the methods Quist and Chapela used to determine the location of the transgene. Interpreting results of an experimental technique known as inverse polymerase chain reaction (i-PCR), Quist and Chapela had reported that the transgene was surrounded by essentially random sequences of DNA. Critics argued that i-PCR was unreliable and that

Ignacio Chapela (left) and David Quist (right) ignited a firestorm of controversy with their scientific paper reporting transgenic corn DNA in Mexican maize.

Quist and Chapela had used insufficient controls. Their results, the critics suggested, could have arisen from similarities between the transgene and stretches of maize DNA.

Several geneticists also attacked Quist and Chapela's more fundamental claim that transgenes had been integrated into the genome of native maize. Matthew Metz of the

University of Washington and Johannes Fütterer of the Institute of Plant Sciences in Zürich, Switzerland, argued that the low levels of transgenic DNA detected indicated, at most, first-generation hybrids between local cultivars and transgenic varieties. Other critics, such as Paul Christou, editor of the journal *Transgenic Research,* noted that i-PCR was a highly sensitive technique and that all of Quist and Chapela's findings could be unreliable if laboratory practices had been careless.

On April 11, 2002, *Nature* published two letters from scientists critical of the study, along with an editorial note concluding that "the evidence available is not sufficient to justify the publication of the original paper." Quist and Chapela responded by acknowledging that some of their initial findings, particularly those based on the i-PCR technique, were probably invalid. But they also

seed and said he did not want the crossbreeding. Monsanto investigators took seed samples from his plants and charged him with violating Canada's law that makes it illegal for farmers to reuse patented seed or grow the seed without a contract with the company. Monsanto sued Schmeiser, and the court sided with Monsanto, ordering the farmer to pay the corporation roughly $238,000. Schmeiser appealed the case to the Supreme Court of Canada, which in 2004 ruled by a 5-4 vote that Schmeiser had violated Monsanto's patent (although it spared the 74-year-old farmer from having to make payments to the company). Schmeiser received wide public support, a government committee called for revising the patent law, and the National Farmers Union of Canada called for a moratorium on GM food. Meanwhile, Monsanto continues to demand that small farmers in Canada and the United States heed the Canadian and U.S. patent laws.

Weighing the Issues:
Early Hurdles for GM Foods

As the vignettes in Figure 10.13 illustrate, a number of GM foods have run into difficulties. Do you think this reveals an underlying problem with the approach, or are these simply examples of unavoidable glitches that are bound to occur during the early development of any new technology? Can you take any lessons from the stories in Figure 10.13? Do you expect that debate between proponents and opponents of GM organisms will subside in the future, or not?

Given such developments, the future of GM foods seems likely to hinge on social, economic, legal, and political factors as well as scientific ones. European consumers have expressed widespread unease about possible

presented new analyses to support their fundamental claim and pointed to a Mexican government study that also found high rates of transgenic contamination. However, this did little to convince skeptics, especially because the Mexican government's results remained unpublished.

The correspondence published in *Nature* was only part of a broader debate that took place on the editorial pages of journals such as *Nature Biotechnology* and *Transgenic Research,* in newspapers and magazines, over the Internet, and within the halls of academe. The debate sometimes turned personal. GM supporters pointed out that Chapela and Quist had long opposed transgenic crops and biotechnology corporations. Chapela had spoken out against his university's plan to enter into a $25 million partnership with the biotechnology firm Novartis (now Syngenta),

sparking a debate that created divisions among Berkeley faculty. GM opponents countered that Quist and Chapela's most vocal critics received funding from biotechnology corporations. As the controversy progressed, it became clear that few participants in the debate could be considered entirely objective. Even *Nature,* whose impartiality is critical to its reputation as a first-tier science journal, was engaged in commercial partnerships with biotechnology corporations.

Chapela was denied tenure by Berkeley's administration, even though his department and dean had recommended him. Alleging that he had received unfair treatment, Chapela waged a 3-year fight to win tenure and gained a following that included 230 academics who signed a letter of support. In May 2005, a new chancellor awarded him tenure.

The twists and turns of the still-unresolved debate illustrate that there can be more to the scientific process than merely the scientific method. Almost overlooked, however, was the fact that nearly all researchers agreed that gene flow between transgenic corn and Mexico's native landraces was bound to happen at some point.

Researchers anxious about transgenic crops emphasize that such gene flow could decrease genetic diversity and negatively affect agriculture and the environment. Those in favor of transgenic foods argue that gene flow is natural and that the genetic diversity of Mexican maize has persisted despite thousands of years of cross-pollination. The paucity of adequate scientific data—and the high economic and environmental stakes surrounding Mexico's ban on transgenic corn—go a long way toward explaining why the debate has been so heated.

risks of GM technologies, whereas U.S. consumers have largely accepted the GM crops approved by U.S. agencies. Opposition in nations of the European Union resulted in a de facto moratorium on GM foods from 1998 to 2003, blocking the importation of hundreds of millions of dollars in U.S. agricultural products. This prompted the United States to bring a case before the World Trade Organization in 2003, complaining that Europe's resistance was hindering free trade. Europeans now demand that GM foods be labeled and criticize the United States for not joining 100 other nations in signing the Cartagena Protocol on Biosafety, a treaty that lays out guidelines for open information about exported crops.

Transnational spats between Europe and the United States will surely affect the future direction of agriculture, but the world's developing nations could exert the most influence in the end. Recent decisions by the governments

of India and Brazil to approve GM crops (following long and divisive debate) are already adding greatly to the world's transgenic agriculture. Moreover, China is aggressively expanding its use of transgenic crops.

A counterexample is Zambia, one of several African nations that refused U.S. food aid meant to relieve starvation during a drought in late 2002. The governments of these nations worried that their farmers would plant some of the GM corn seed meant to be eaten and that GM corn would thereby establish itself in their countries. They viewed this as undesirable because African economies depend on exporting food to Europe, which has put severe restrictions on GM food. In the end, Zambia's neighbors accepted the grain after it had been milled (so none could be planted), but Zambia held out. Citing health and environmental risks, uncertain science, and the precautionary principle, the Zambian government declined the aid, despite the fact that 2–3 million of its people were at risk

Genetically Modified Foods

Proponents of GM foods say these products can alleviate hunger and malnutrition while posing no known threats to human health or the environment. Opponents say that these foods help only the large corporations that sell them and that they do pose risks to human health, wild organisms, and ecosystems. **What do you think? Should we encourage the continued development of GM foods? If so, what, if any, restrictions should we put on their dissemination?**

A Global Experiment Without Controls

Genetic engineering, specifically transgenesis, gives us the unprecedented capacity to move DNA. In so doing, this technology breaches boundaries established through millions of years of evolution. As such, we should expect fundamental alterations in ecosystems with the release of transgenic crops, fish, insects, microbes, and so forth into uncontrolled areas. These alterations are similar in nature to those caused by the introduction of exotic species into new environments. Both processes are unpredictable and could have serious consequences.

Yet because of political and short-term economic imperatives, releases of transgenic organisms have continued unabated for at least a decade. Science has barely started to imagine the ecological and evolutionary consequences of releasing transgenic crops. Not only are we experiencing a global experiment without controls, we don't have the tools to document it. Serious research, although extremely scarce, has already confirmed some of the theoretical fears concerning transgenesis aired by scientists a quarter century ago.

Today, we have cataclysmic world hunger paired with food surpluses. The claim that transgenesis can solve this problem is merely a diversion tailored to conceal how transgenesis manipulates the biosphere. Molecular biology might one day become part of the solution to world hunger, but it is certainly not the science most relevant to address the problem today.

What checks should be placed on the release of transgenic organisms? Every check. Through the unaccountable releases so far, we have seen enough, and possibly caused enough, environmental insult for me to say today that we should stop. We need to take stock of the consequences of transgenesis and continue researching under strictly regulated conditions.

Ignacio H. Chapela is associate professor (microbial ecology) in the Department of Environmental Science, Policy, and Management at the University of California, Berkeley. He helped found the Mycological Facility: Oaxaca, Mexico, where he serves as scientific director.

The United States Should Begin a Phased Deregulation of Biotech Crops

During the past two decades the international scientific community, biotechnology industry, and regulatory agencies in many countries have accumulated and critically evaluated a wealth of information about the production and use of biotech crops and products. Biotech crops have been planted since 1996 on more than 1 billion acres of farmland in nearly 20 countries. More than 1 billion humans and hundreds of millions of farm animals have consumed biotech foods and products. Yet there is not a single instance in which biotech crops and foods have been shown to cause illness in humans or animals or to damage the environment.

In spite of this exemplary safety record, a small but well-organized, well-financed, and vocal antibiotechnology lobby has alleged that biotech crops and products are unsafe for humans and a danger to the environment, demanding a moratorium or outright ban on biotech crops. The rhetoric of the antibiotechnology groups is alarming, confusing, and frightening to the public, but it is devoid of any substance, because they have never provided any credible scientific evidence to support their allegations.

Any further delay in combining the power of biotechnology with conventional breeding will seriously endanger future food security, political and economic stability, and the environment. Plant biotechnology is still the best hope for meeting the food needs of the ever-growing world population. Biotech crops are already helping to conserve valuable natural resources, reduce the use of harmful agro-chemicals, produce more nutritious foods, and promote economic development.

Twenty years ago, the United States set the precedent by developing regulations for the development and use of biotech crops. Now, as the world leader in plant biotechnology, it is imperative that it lead again by phasing out these redundant regulations in an organized and responsible manner.

Indra K. Vasil is graduate research professor emeritus at the University of Florida (Gainesville, Florida). His research focuses on the biotechnology of cereal crops, and he has been recognized as one of the world's most highly cited authors in the plant and animal sciences.

Explore this issue further by accessing **Viewpoints** at www.aw-bc.com/withgott.

FIGURE 10.16 Debate over GM foods reached a dramatic climax in Zambia in 2002, when the government refused U.S. shipments of GM corn that were intended to relieve starvation due to drought. Here a Zambian mother with her child waits in a line for food assistance.

of starvation (Figure 10.16). Intense debate followed within the country and around the world. Eventually the United Nations delivered non-GM grain, and in April 2003 the Zambian government announced a plan to coordinate a comprehensive long-term policy on GM foods.

The Zambian experience demonstrates some of the ethical, economic, and political dilemmas modern nations face. The corporate manufacturers of GM crops naturally aim to maximize their profits, but they also aim to develop products that can boost yields, increase food security, and reduce hunger. Although industry, activists, policymakers, and scientists all agree that hunger and malnutrition are problems and that agriculture should be made environmentally safer, they often disagree about the solutions to these dilemmas and the risks that each proposed solution presents.

Preserving Crop Diversity

As the excitement over Quist and Chapela's findings in Oaxaca demonstrate, one concern many people harbor about transgenic crops is that transgenes might move, by pollination, into local native races of crop plants.

Crop diversity provides insurance against failure

Preserving the integrity of native variants gives us a bulwark against commercial crop failure. The regions where crops first were domesticated generally remain important repositories of crop biodiversity. Although modern industrial agriculture relies on a small number of plant types, its foundation lies in the diverse varieties that still exist in places like Oaxaca. These varieties contain genes that, through conventional crossbreeding or genetic engineering, might confer resistance to disease, pests, inbreeding, and other pressures that challenge modern agriculture. Monocultures essentially place all our eggs in one basket, such that any single catastrophic cause could potentially wipe out entire crops. Having available the wild relatives of crop plants or the domesticated varieties, or cultivars, of crop plants gives us the genetic diversity that may include ready-made solutions to unforeseen problems. Because accidental interbreeding can decrease the diversity of local variants, many scientists argue that we need to protect areas like Oaxaca. For this reason, the Mexican government helped create the Sierra de Manantlan Biosphere Reserve around an area harboring the localized plant thought to be the direct ancestor of maize. For this reason, too, it imposed a national moratorium in 1998 on the planting of transgenic corn.

We have lost a great deal of genetic diversity in our crop plants already. The number of wheat varieties in China is estimated to have dropped from 10,000 in 1949 to 1,000 by the 1970s, and Mexico's famed maize varieties now number only 30% of what was extant in the 1930s. In the United States, many fruit and vegetable crops have decreased in diversity by 90% in less than a century. A primary cause of this loss of diversity is that market forces have discouraged diversity in the appearance of fruits and vegetables. Commercial food processors prefer items to be similar in size and shape, for convenience. Consumers, for their part, have shown preferences for uniform, standardized food products over the years. Now that local organic agriculture is growing in affluent societies, however, consumer preferences for diversity are increasing.

Seed banks are living museums for seeds

Protecting areas with high crop diversity is one way to preserve genetic assets for our agricultural systems. Another is to collect and store seeds from crop varieties and periodically plant and harvest them to maintain a diversity of cultivars. This is the work of **seed banks** or **gene banks,** institutions that preserve seed types as a kind of living museum of genetic diversity (Figure 10.17). In total, these facilities hold roughly 6 million seed samples, keeping them in cold, dry conditions to encourage long-term viability. The $300 million in global funding for these facilities is not adequate for proper storage and for the labor of growing out the seed periodically to renew

(a) Traditional food plants of the Desert Southwest

(b) Pollination by hand

FIGURE 10.17 Seed banks preserve genetic diversity of traditional crop plants. Native Seeds/SEARCH of Tucson, Arizona, preserves seeds of food plants important to traditional diets of Native Americans of Arizona, New Mexico, and northwestern Mexico. Beans, chiles, squashes, gourds, maize, cotton, and lentils are all in its collections, as well as less-known plants such as amaranth, lemon basil, and devil's claw **(a).** Traditional foods such as mesquite flour, prickly pear pads, chia seeds, tepary beans, and cholla cactus buds help fight the diabetes that Native Americans frequently suffer from having adopted a Western diet. At the farm where seeds are grown, care is taken to pollinate varieties by hand **(b)** to protect their genetic distinctiveness.

the stocks. Therefore, it is questionable how many of these 6 million seeds are actually preserved. Major efforts include large seed banks such as the U.S. National Seed Storage Laboratory at Colorado State University, the Royal Botanic Garden's Millennium Seed Bank in Britain, Seed Savers Exchange in Iowa, and the Wheat and Maize Improvement Center (CIMMYT) in Mexico.

Feedlot Agriculture: Livestock and Poultry

Food from cropland agriculture makes up a large portion of the human diet, but most people also eat animal products. People don't *need* to eat meat or other animal products to live full, active, healthy lives, but for most people it is difficult to obtain a balanced diet without incorporating animal products. Most of us do eat animal products, and this choice has significant environmental, social, agricultural, and economic impacts.

Consumption of animal products is growing

As wealth and global commerce have increased, so has our consumption of meat, milk, eggs, and other animal products (Figure 10.18). The world population of domesticated animals raised for food rose from 7.3 billion animals to 20.6 billion animals between 1961 and 2000. Most of these animals are chickens. Global meat production has increased fivefold since 1950, and per capita meat consumption has nearly doubled. The most-eaten meat per unit weight is pork.

High consumption has led to feedlot agriculture

In traditional agriculture, livestock were kept by farming families near their homes or were grazed on open grasslands by nomadic herders or sedentary ranchers. These traditions have survived, but the advent of industrial agriculture has

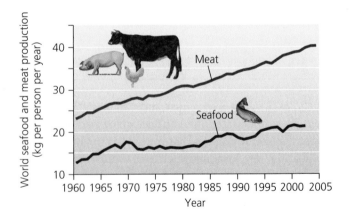

FIGURE 10.18 Per capita consumption of meat from farm animals has increased steadily worldwide over the past few decades, as has per capita consumption of seafood (marine and freshwater, harvested and farmed). Data from Food and Agriculture Organization of the United Nations.

FIGURE 10.19 These chickens at a Pennsylvania factory farm are housed several to a cage and have been "debeaked," the tips of their beaks cut off to prevent them from pecking one another. The hens cannot leave the cages and essentially spend their lives eating, defecating, and laying eggs, which roll down slanted floors to collection trays. The largest U.S. chicken farms house hundreds of thousands of individuals.

added a new method. **Feedlots,** also known as *factory farms* or *concentrated animal feeding operations (CAFOs),* are essentially huge warehouses or pens designed to deliver energy-rich food to animals living at extremely high densities (Figure 10.19). Today over half of the world's pork and poultry come from feedlots, as does much of its beef.

Feedlot operations allow for greater production of food and are probably necessary for a country with a level of meat consumption like that of the United States. Feedlots have one overarching benefit for environmental quality: Taking cattle, sheep, goats, and other livestock off the land and concentrating them in feedlots reduces the impact they would otherwise exert on large portions of the landscape. In Chapter 9 (▸pp. 267–268, and 270) we saw how overgrazing can degrade soils and vegetation and that hundreds of millions of hectares of land are considered overgrazed. Animals that are densely concentrated in feedlots will not contribute to overgrazing and soil degradation.

Of course, feedlots are not without impact, and many environmental advocates have attacked them for their contributions to water and air pollution. Waste from feedlots can emit strong odors and can pollute surface water and groundwater, because livestock produce prodigious

amounts of feces and urine. One dairy cow can produce about 20,400 kg (44,975 lb) of waste in a single year. Greeley, Colorado, is home to North America's largest meatpacking plant and two adjacent feedlots, all of which are owned by the agribusiness firm ConAgra. Each feedlot has room for 100,000 cattle that are fed surplus grain and injected with anabolic steroids to stimulate growth. During its stay at the feedlot, a typical steer will eat 1,360 kg (3,000 lb) of grain, gain 180 kg (400 lb) in body weight, and generate 23 kg (50 lb) of manure each day. The amount of manure that 200,000 such animals generate exceeds the amount of waste produced by all the human residents of Atlanta, St. Louis, Boston, and Denver combined. Poor waste containment practices at feedlots in North Carolina, Maryland, and other states have been linked to outbreaks of disease, including virulent strains of *Pfiesteria,* a microbe that poisons fish. The crowded and dirty conditions under which animals are often kept necessitates heavy use of antibiotics to control disease. These chemicals can be transferred up the food chain, and their overuse can cause microbes to evolve resistance to them.

Feedlot impacts can be minimized when properly managed, and both the EPA and the states regulate U.S. feedlots. Most feedlot manure is applied to farm fields as fertilizer, reducing the need for chemical fertilizers. Manure in liquid form can be injected into the ground where plants need it, and farmers can conduct tests to determine amounts that are appropriate to apply.

Weighing the Issues:
Feedlots and Animal Rights

Animal rights activists decry factory farming because they say it mistreats animals. Chickens, pigs, and cattle are kept crowded together in small pens their entire lives, fattened up, and slaughtered. Chickens are often "debeaked." Do you think animal rights concerns should be given weight as we determine how best to raise our food? Do you think these are as important as the environmental issues? Are conditions at feedlots a good reason for being vegetarian?

Our food choices are also energy choices

What we choose to eat has ramifications for how we use energy and the land that supports agriculture. Recall our discussions of thermodynamics and trophic levels (▸pp. 104–105 and ▸pp. 156–157). Every time energy moves from one trophic level to the next, as much as 90% is lost. For example, if we feed grain to a cow and then eat beef from the cow, we lose a great deal of the grain's energy to the cow's digestion and metabolism. Energy is used up

Feed input

Produce output (edible weight)

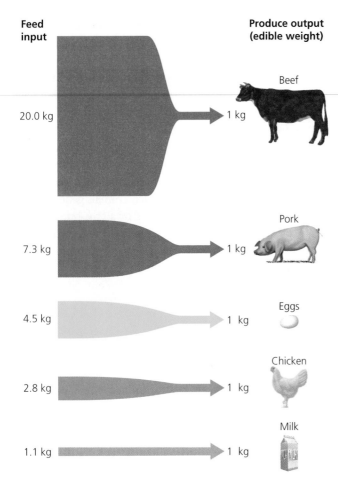

FIGURE 10.20 Different animal food products require different amounts of input of animal feed. Chickens must be fed 2.8 kg of feed for each 1 kg of resulting chicken meat, for instance, whereas 20 kg of feed must be provided to cattle to produce 1 kg of beef. Go to **GRAPHit!** at www.aw-bc.com/withgott or on the student CD-ROM. Data from Smil, V. 2001. *Feeding the world: A challenge for the twenty-first century.* Cambridge, MA: MIT Press.

when the cow converts the grain to tissue as it grows, and as the cow uses its muscle mass on a daily basis to maintain itself. For this reason, eating meat is far less energy-efficient than relying on a vegetarian diet. The lower in the food chain from which we take our food sources, the greater the proportion of the sun's energy we put to use as food, and the more people Earth can support.

Some animals convert grain feed into milk, eggs, or meat more efficiently than others (Figure 10.20). Scientists have calculated relative energy-conversion efficiencies for different types of animals. Such energy efficiencies have ramifications for land use—land and water are required to raise food for the animals, and some animals require more than others. Figure 10.21 shows the area of land and weight of water required to produce 1 kg (2.2 lb) of food protein for milk, eggs, chicken, pork, and beef.

Producing eggs and chicken meat requires the least space and water, whereas producing beef requires the most. Such differences make clear that when we choose what to eat, we are also indirectly choosing how to make use of resources such as land and water.

In 1900 we fed about 10% of global grain production to animals. In 1950 this number had reached 20%, and by the beginning of the 21st century we were feeding 45% of global grain production to animals. Although much of the grain fed to animals is not of a quality suitable for human consumption, the resources required to grow it could have instead been applied toward growing food for people. One partial solution is to feed livestock crop residues, plant matter such as stems and stalks that we would not consume anyway, and this is increasingly being done.

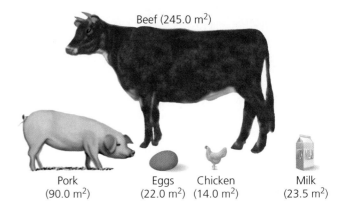

(a) Land required to produce 1 kg of protein

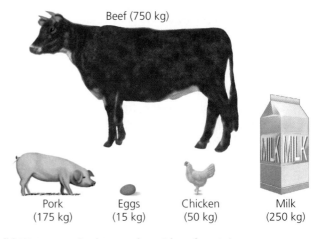

(b) Water required to produce 1 kg of protein

FIGURE 10.21 Producing different types of animal products requires different amounts of land and water. Raising cattle for beef requires by far the most land and water of all animal products. Go to **GRAPHit!** at www.aw-bc.com/withgott or on the student CD–ROM. Data from Smil, V. 2001. *Feeding the world: A challenge for the twenty-first century,* Cambridge, MA: MIT Press.

Aquaculture

In addition to plants grown in croplands and animals raised on rangelands and in feedlots, we rely on aquatic organisms for food. Wild fish populations are plummeting throughout the world's oceans as increased demand and new technologies have led us to overharvest most marine fisheries (▶ pp. 482–485). This means that raising fish and shellfish on "fish farms" may be the only way to meet the growing demand for these foods.

We call the raising of aquatic organisms for food in controlled environments **aquaculture.** Many aquatic species are grown in open water in large, floating net-pens. Others are raised in land-based ponds or holding tanks. People pursue both freshwater and marine aquaculture. Aquaculture is the fastest-growing type of food production; in the past 20 years, global output has increased sevenfold (Figure 10.22). Aquaculture today provides a third of the world's fish for human consumption, is most common in Asia, and involves over 220 species. Some, such as carp, are grown for local consumption, whereas others, such as salmon and shrimp, are exported to affluent countries.

Aquaculture brings a number of benefits

When conducted on a small scale by families or villages, as in China and much of the developing world, aquaculture helps ensure people a reliable protein source. This type of small-scale aquaculture can be sustainable, and it is compatible with other activities. For instance, uneaten fish scraps make excellent fertilizers for crops. Aquaculture on larger scales can help improve a region's or nation's food security by increasing overall amounts of fish available. Aquaculture on any scale also has the benefit of reducing fishing pressure on overharvested and declining wild stocks. Reducing fishing pressure also reduces the *by-catch* (▶ p. 485; the unintended catch of nontarget organisms) that results from commercial fishing. Furthermore, aquaculture relies far less on fossil fuels than do fishing vessels and provides a safer work environment than does commercial fishing. Fish farming can also be remarkably energy-efficient, producing as much as 10 times more fish per unit area than is harvested from oceanic waters on the continental shelf and up to 1,000 times as much as is harvested from the open ocean.

Aquaculture has negative impacts

Along with its benefits, aquaculture has disadvantages. Dense concentrations of farmed animals can increase

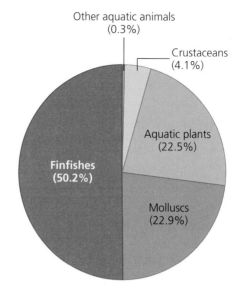

(a) **World aquaculture production by groups**

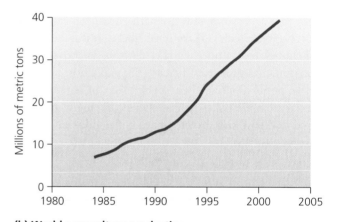

(b) **World aquaculture production**

FIGURE 10.22 Aquaculture involves a wide diversity of marine and freshwater organisms (**a**). Global production of meat (all species other than plants) from aquaculture has risen steeply in the past two decades (**b**). Data in (a) from FAO 2004. *The state of world fisheries and aquaculture, 2004;* and (b) FAO *Aquaculture production statistics, 1984–1993* and *Fishery statistics: Aquaculture production.*

the incidence of disease, which reduces food security, necessitates antibiotic treatment, and results in additional expense. A virus outbreak wiped out half a billion dollars in shrimp in Ecuador in 1999, for instance. Aquaculture can also produce prodigious amounts of waste, from the farmed organisms and from the feed that goes uneaten and decomposes in the water column. Farmed fish often are fed grain, and as we have discussed, growing grain to feed animals that we then eat reduces the energy efficiency of food production and consumption. In other cases, farmed fish are fed fish meal made from wild ocean fish such as herring and anchovies, whose harvest may place additional stress on wild fish populations.

If farmed aquatic organisms escape into ecosystems where they are not native, they may spread disease to native stocks or may outcompete native organisms for food or habitat. The possibility of competition also arises when the farmed animals have been genetically modified. Like the transgenic corn that has influenced Mexican maize, transgenic fish have become a part of the food production system in recent years. Genetic engineering of Pacific salmon has produced transgenic fish that weigh up to 11 times more than nontransgenic ones. Transgenic Atlantic salmon raised in Scotland have been engineered to grow to 5–50 times the normal size for their species (Figure 10.23). GM fish such as these may outcompete their non-GM wild cousins while also spreading disease to them. They may also interbreed with native and hatchery-raised fish and weaken already troubled stocks. Researchers have concluded that under certain circumstances, escaped transgenic salmon may increase the extinction risk that native populations of their species face, in part because the larger male fish (such as those carrying a gene for rapid and excessive growth) have better odds of mating successfully.

FIGURE 10.23 Efforts to genetically modify important food fish have resulted in the creation of transgenic salmon (top), which can be considerably larger than wild salmon of the same species.

Sustainable Agriculture

Industrialized agriculture involves many adverse environmental impacts, from the degradation of soils (Chapter 9) to reliance on fossil fuels (Chapter 19) to problems arising from pesticide use, genetic modification, and intensive feedlot and aquaculture operations. Although many of these developments in intensive commercial agriculture have alleviated some environmental pressures, they have often exacerbated others. Industrial agriculture in some form seems necessary to feed our planet's 6.5 billion people, but many feel we will be better off in the long run by practicing less-intensive methods of raising animals and crops.

Farmers and researchers have made great advances toward sustainable agriculture in recent years. **Sustainable agriculture** is agriculture that does not deplete soils faster than they form. It is farming and ranching that does not reduce the amount of healthy soil, clean water, and genetic diversity essential to long-term crop and livestock production. It is, simply, agriculture that can be practiced in the same way far into the future. For example, the no-till agriculture practiced in southern Brazil that we examined in Chapter 9 appears to fit the notion of sustainable agriculture, as does the traditional Chinese practice of carp aquaculture in small ponds. Sustainable agriculture is closely related to *low-input agriculture,* agriculture that uses smaller amounts of pesticides, fertilizers, growth hormones, water, and fossil fuel energy than are currently used in industrial agriculture. Food-growing practices that use no synthetic fertilizers, insecticides, fungicides, or herbicides—but instead rely on biological approaches such as composting and biocontrol—are termed **organic agriculture.**

Organic agriculture is on the increase

Citizens, government officials, farmers, and agricultural industry representatives have debated the meaning of the word *organic* for many years. Experimental organic gardens began to appear in the United States in the 1940s, but it was decades before the U.S. government developed a clear definition of what it meant to be organic. In 1990, Congress passed the Organic Food Production Act. This law established national standards for organic products and facilitated the sale of organic food. As required by this act, the USDA in 2000 issued criteria by which crops and livestock could be officially certified as organic (Table 10.2). California legislation passed the same year established stricter state guidelines for labeling foods organic, and Washington and Texas followed with still

Table 10.2	USDA Criteria for Certifying Crops and Livestock as Organic

For crops to be considered organic

▶ The land where they are grown must be free of prohibited substances for at least 3 years.

▶ They must not be genetically engineered.

▶ They must not be treated with ionizing radiation (a means of eliminating bacteria in packaged food).

▶ The use of sewage sludge is prohibited.

▶ They must be produced without fertilizer containing synthetic ingredients.

▶ Fertility and crop nutrients must be achieved through crop rotations, cover crops, animal and crop wastes, or synthetic materials approved by the National Organic Standards Board.

▶ Use of most conventional pesticides is prohibited.

▶ Use of organic seeds and other planting stock is preferred.

▶ Crop pests, weeds, and diseases should be controlled through physical, mechanical, and biological management practices or with synthetic substances approved by the National Organic Standards Board.

For livestock to be considered organic

▶ Mammals must be raised under organic management from the last third of gestation; poultry, no later than the second day of life.

▶ Producers must feed livestock 100% organic agricultural feed; however, vitamin and mineral supplements are allowed.

▶ Producers of existing dairy herds must provide 80% organically produced feed for 9 months, followed by 3 months of 100% organically produced feed.

▶ Use of hormones or antibiotics is prohibited, although vaccines are permitted.

▶ Animals must have access to the outdoors.

Data from The National Organic Program. 2002. *Organic production and handling standards.* U.S. Department of Agriculture.

stricter guidelines. Today 17 U.S. states have laws spelling out standards for organic products.

Weighing the **Issues:**
Do You Want Your Food Labeled?

The USDA issues labels to certify that produce claiming to be organic has met the government's organic standards. Increasingly, critics of GM products want them to be labeled as well. Given that 70% of processed food currently contains GM ingredients, labeling would cause added—and many people think, unnecessary—costs. But the European Union currently labels such foods. Do you want your food to be labeled? Would you choose among foods based on whether they are organic or genetically modified? Do you feel your food choices have environmental impacts, good or bad? Is purchasing power an effective way to make your views heard?

Long viewed as a small niche market, the market for organic foods is on the increase. Although it accounts for only 1% of food expenditures in the United States and Canada, sales of organic products increased 20% annually from 1989 to 2002, when global sales of organic products reached $25 billion. In 2001, 3–5% (close to $10 billion) of Europe's food market was organic. Although 3–5% may not seem like much, organic agriculture expanded by a factor of 35 between 1985 and 2001 in Europe, representing an annual growth rate of 30%.

Production is increasing along with demand. Although organic agriculture takes up less than 1% of cultivated land worldwide (24 million ha [59 million acres] in 2004), this area is rapidly expanding. In the United States and Canada, the amount of land used in organic agriculture has recently increased 15–20% each year. As of 2001, 550,000 ha (1.36 million acres) of U.S. farmland and 1 million ha (2.47 million acres) of Canadian farmland were in organic production. Today farmers in more than 130 nations practice organic farming commercially to some extent.

Two motivating forces have fueled these trends. Many consumers favor organic products because of concern that consuming produce grown with the use of pesticides may pose risks to their health. Consumers also buy organic produce out of a desire to improve environmental quality by reducing chemical pollution and soil degradation (see "The Science behind the Story," ▶p. 302). Many other consumers, however, will not buy organic produce because it usually is more expensive and often looks less uniform and aesthetically appealing in the supermarket aisle compared to the standard produce of high-input agriculture. Overall, enough consumers are willing to pay more for organic meat, fruit, and vegetables that businesses are making such foods more widely available. In early 2000, one of Britain's largest supermarket brands announced that it would sell only organic food—and that the new organic products would cost their customers no more than had nonorganic products. In addition to food products, many textile makers (among them The Gap, Levi's, and Patagonia) are increasing their use of organic cotton.

Government initiatives have also spurred the growth of organic farming. For example, several million hectares of land has undergone conversion from conventional to

The Science behind the Story

Organic Farming

Fields of wheat and potatoes, some grown organically and some cultivated with the synthetic chemicals favored by industrialized agriculture, stand side by side on an experimental farm in Switzerland. Although conventionally farmed fields receive up to 50% more fertilizer, they produce only 20% more food than organically farmed fields. How are organic fields able to produce decent yields without synthetic agricultural chemicals? The answer, scientists have found, lies in the soil.

Researchers have long documented that organic farming is better for the environment, but farmers have long found that synthetic chemicals help them fight pests and increase crop yields. Although organic farming might put fewer synthetic chemicals into the air and water, some have contended that it will never contribute much to the world's food supply.

To address concerns about crop yields, Swiss researchers at the Research Institute of Organic Agriculture have been comparing organic and conventional fields since 1978, using a series of growing areas that feature four different farming systems. One group of plots mirrors conventional farms, in which large amounts of chemical pesticides, herbicides, and fertilizer are applied to soil and plants. Another set of fields is treated with a mixed approach of conventional and organic practices, including chemical additives, synthetic sprays, and livestock manure as fertilizer. Organic plots use only manure, mechanical weeding machines, and plant extracts to control pests. A fourth group of

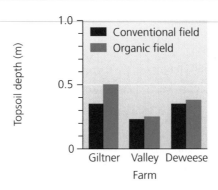

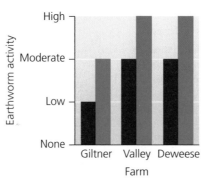

Researchers in Nebraska and North Dakota demonstrated that organic farming at three sites increased topsoil depth moderately and activity of earthworms dramatically. Data from Liebig, M. A., and J. W. Doran, 1999. Impact of organic production practices on soil quality indicators, *Journal of Environmental Quality* 28: 1601–1609.

plots follows similar organic practices but also uses extra natural boosts, such as adding herbal extracts to compost. The two organic plots receive about 35–50% less fertilizer than the conventional fields and 97% fewer pesticides.

Over more than 20 years of monitoring, the organic fields yielded 80% of what the conventional fields produced, researchers reported in the journal *Science* in 2002. Organic crops of winter wheat yielded about 90% of the conventional wheat crop yield. Organic potato crops averaged about 68% of

the conventional potato yields. The comparatively low potato yield was due to nutrient deficiency and a fungus-caused potato blight.

Scientists have hypothesized that organic farms keep their yields high because organic agricultural practices better conserve soil quality, keeping soil fertile over the long term. In one study that appears to back this hypothesis, U.S. researchers compared five pairs of organic and conventional farms in the mid-1990s in North Dakota and Nebraska. After extracting soil samples from each of the farms, the researchers analyzed the soil for water-holding ability, microbial biomass, and nutrients such as carbon and nitrogen. Organic farming, they found, produced soils that contained more naturally occurring nutrients, held greater quantities of water, and had higher concentrations of microbial life than conventionally farmed soil. Organic farms also had deeper nutrient-rich topsoil and greater earthworm activity—all signs of soil healthy enough to produce impressive crops without help from synthetic chemicals.

Scientists at the Swiss research project have found similar signs of soil fertility on their organic research plots. Increasingly, researchers are concluding that organically managed soil supports a more diverse range of microbial and plant life, which translates into increased biodiversity, self-sustaining fields, and strong crop yields. Such findings may be pivotal as large growers increasingly debate whether to turn to organic farming.

organic farming in Europe since the European Union adopted a policy in 1993 to support farmers financially during the first years of conversion. The United States offers no such support, which may be why the U.S. organic market lags behind that of Europe. Such support is important, because conversion often means a temporary loss in income for farmers. More and more studies, however, suggest that reduced inputs and higher market prices can, in the long run, make organic farming more profitable for the farmer than conventional methods.

Locally supported agriculture is growing

Increasing numbers of farmers and consumers are also supporting local small-scale agriculture. Farmers' markets (Figure 10.24) are becoming more numerous throughout North America as consumers rediscover the joys of fresh, locally grown produce. The average food product sold in U.S. supermarkets travels at least 2,300 km (1,400 mi) between the farm and the shelf, and supermarket produce is often chemically treated to preserve freshness and color. At farmers' markets, consumers can buy fresh produce in season from local farmers and often have a wide choice of organic items and unique local varieties.

Some consumers are even partnering with local farmers in a phenomenon called *community-supported agriculture*. In this practice, consumers pay farmers in advance for a share of their yield, usually in the form of weekly deliveries of produce. Consumers get fresh seasonal produce, while farmers get a guaranteed income stream up front to invest in their crops—an alternative to taking out loans and being at the mercy of the weather. As of 2005, about 1,700 U.S. farms were supplying 340,000 families per week in community-supported agriculture programs.

Cuba has embraced organic agriculture

Perhaps no other nation has implemented local organic farming to the extent that Cuba has. Long a close ally of the former Soviet Union, Cuba suffered economic and agricultural upheaval following the Soviet Union's dissolution. In 1989, as the USSR was breaking up, Cuba lost 75% of its total imports, 53% of its oil imports, and 80% of its fertilizer and pesticide imports. Faced with such losses, Cuba's farmers had little choice but to "go organic."

Because far less oil was available to fuel Cuba's transportation system, farmers began growing food closer to cities and even within them. By 1998 the Cuban government's Urban Agriculture Department had encouraged the development of more than 8,000 gardens in the capital city of Havana (Figure 10.25). Over 30,000 people, including farmers, government workers, and private citizens, worked in these gardens, which covered 30% of the

FIGURE 10.24 Farmers' markets, like this one in San Francisco, have become more widespread as consumers have rediscovered the benefits of buying fresh, locally grown produce.

FIGURE 10.25 Organic gardening takes place within the city limits of Havana, Cuba, out of necessity. With little money to pay for the large amounts of fertilizers and pesticides required for industrialized agriculture, Cubans get much of their food from local agriculture without these inputs.

city's available land. Cuba has also taken steps to compensate for the loss of fossil fuels, fertilizers, and pesticides by, for example, using oxen instead of tractors, using integrated pest management, encouraging people to live outside urban areas and to remain involved in agriculture, and establishing centers to breed organisms for biological pest control.

Cuba's agriculture likely requires more human labor per unit output than do intensive commercial farms of developed nations, and Cuba's economic and agricultural policies are guided by tight top-down control in a rigid state socialist system. Nevertheless, Cuba's low-input farming has produced some positive achievements. The practices have led to the complete control of the sweet-potato borer, a significant pest insect, and in the 1996–1997 growing season the Cuban people produced record yields for 10 crops. Although Cuba's move toward organic agriculture was involuntary, its response to its economic and agricultural crisis illustrates how other nations might, by choice, begin to farm in ways that rely less on enormous inputs of fossil fuels and synthetic chemicals.

Organic and sustainable agriculture will likely need to play a large role in our future

Organic agriculture succeeds in part because it alleviates many problems introduced by high-input agriculture, even while passing up many of the benefits. For instance, although in many cases more insect pests attack organic crops because of the lack of chemical pesticides, biocontrol methods can often keep these pests in check. Moreover, the lack of synthetic chemicals maintains soil quality and encourages helpful pollinating insects. In the end, consumer choice will determine the future of organic agriculture. Falling prices and wider availability suggest that organic agriculture will continue to increase. In addition, sustainable agriculture, whether organic or not, will sooner or later need to become the rule rather than the exception.

Conclusion

Many of the intensive commercial agricultural practices we have discussed have substantial negative environmental impacts. At the same time, it is important to realize that many aspects of industrialized agriculture have had positive environmental effects by relieving certain pressures on land or resources. Whether Earth's natural systems would be under more pressure from 6.5 billion people practicing traditional agriculture or from 6.5 billion people living under the industrialized agriculture model is a very complicated question.

What is certain is that if our planet is to support 9 billion people by mid-century without further degradation of the soil, water, pollinators, and other ecosystem services that support our food production, we must find ways to shift to sustainable agriculture. Approaches such as biological pest control, organic agriculture, pollinator conservation, preservation of native crop diversity, sustainable aquaculture, and likely some degree of careful and responsible genetic modification of food may all be parts of the game plan we will need to set in motion. What remains to be seen is the extent to which individuals, governments, and corporations will be able to put their own interests and agendas in perspective to work together toward a sustainable future.

REVIEWING OBJECTIVES

You should now be able to:

Explain the challenge of feeding a growing human population

▶ Our food production has outpaced the growth of our population, yet there are still 850 million hungry people in the world. (pp. 278–293)

Identify the goals, methods, and environmental impacts of the "green revolution"

▶ The goal of the green revolution was to increase agricultural productivity per unit area of land to feed the world's hungry without further degrading natural lands. (pp. 279–280)

▶ Agricultural scientists used selective breeding to develop strains of crops that grew quickly, were more nutritious, or were resistant to disease or drought. (p. 280)

▶ The expanded use of fossil fuels and chemical fertilizers and pesticides has increased pollution. However, the increased efficiency of production has reduced the amount of natural land converted for farming. (pp. 280–281)

Categorize the strategies of pest management

▶ Most "pests" and "weeds" are killed with synthetic chemicals that also can pollute the environment and pose health hazards. (pp. 282–283)

▶ Pests tend to evolve resistance to chemical pesticides, forcing chemists to design ever more toxic poisons. (p. 283)

▶ Natural enemies of pests can be employed against them in the practice of biological control. (pp. 284–285)

▶ Integrated pest management includes a combination of techniques, and attempts to minimize use of synthetic chemicals. (p. 285)

Discuss the importance of pollination

▶ Insects and other organisms are essential for ensuring the reproduction of many of our crop plants. (pp. 285–286)

▶ Conservation of native pollinating insects is vitally important to our food supply. (p. 286)

Describe the science behind genetically modified food

▶ Genetic modification depends on the technology of recombinant DNA. Genes containing desirable traits are moved from one type of organism into another. (pp. 287–288)

▶ Modification through genetic engineering is both like and unlike traditional selective breeding. (pp. 288–289)

▶ GM crops may have ecological impacts, including the spread of transgenes, the creation of "superweeds," and indirect impacts on biodiversity. More research is needed to determine how widespread or severe these impacts may be. (pp. 288–291)

Evaluate controversies and the debate over genetically modified food

▶ Little evidence exists so far for human health impacts from GM foods, but anxiety over health impacts inspires wide opposition to GM foods. (p. 290–291)

▶ Many people have ethical qualms about altering the food we eat through genetic engineering. (p. 291)

▶ Opponents of GM foods view multinational biotechnology corporations as a threat to the independence of small farmers. (pp. 291–292)

Ascertain approaches for preserving crop diversity

▶ Protecting regions of diversity of native crop varieties, such as Oaxaca, can provide insurance against failure of major commercial crops. (p. 295)

▶ Seed banks preserve rare and local varieties of seed, acting as storehouses for genetic diversity. (pp. 295–296)

Assess feedlot agriculture for livestock and poultry

▶ Increased consumption of animal products has driven the development of high-density feedlots. (pp. 296–297)

▶ Feedlots create tremendous amounts of waste and other environmental impacts, but they also relieve pressure on lands that could otherwise be overgrazed. (pp. 297–298)

Weigh approaches in aquaculture

▶ Aquaculture provides economic benefits and food security, can relieve pressures on wild fish stocks, and can be sustainable. (p. 299)

▶ Aquaculture also creates pollution, habitat loss, and other environmental impacts. (pp. 299–300)

Evaluate sustainable agriculture

▶ Organic agriculture has fewer environmental impacts than industrial agriculture. It is a small part of the market but is growing rapidly. (pp. 300–304)

▶ Locally supported agriculture, as shown by farmers' markets and community-supported agriculture, is also growing. (pp. 303–304)

TESTING YOUR COMPREHENSION

1. What kinds of techniques have people employed to increase agricultural food production? How did agricultural scientist Norman Borlaug help inaugurate the green revolution?
2. Explain how pesticide resistance occurs.
3. Explain the concept of biocontrol. List several components of a system of integrated pest management (IPM).
4. About how many and what types of cultivated plants are known to rely on insects for pollination? Why is it important to preserve the biodiversity of native pollinators?
5. What is recombinant DNA? How is a transgenic organism created? How is genetic engineering different from traditional agricultural breeding? How is it similar?
6. Describe several reasons why many people support the development of genetically modified organisms, and name several uses of such organisms that have been developed so far.
7. Describe the scientific concerns of those opposed to genetically modified crops. Describe some of the other concerns.
8. Name several positive and negative environmental effects of feedlot operations. Why is beef an inefficient food from the perspective of energy consumption?
9. What are some economic benefits of aquaculture? What are some negative environmental impacts?
10. What are the objectives of sustainable agriculture? What factors are causing organic agriculture to expand?

SEEKING SOLUTIONS

1. Assess several ways in which high-input agriculture can be beneficial for the environment and several ways in which it can be detrimental to the environment. Now suggest several ways in which we might modify industrial agriculture to lessen its environmental impact.
2. What factors make for an effective biological control strategy of pest management? What risks are involved in biocontrol? If you had to decide whether to use biocontrol against a particular pest, what questions would you want to have answered before you decide?
3. From what you have learned in this chapter about the staple crop corn, how would you choose to farm corn, if you had to do so for a living? Would you choose to grow genetically modified corn? How would you manage pests and weeds? Would you grow corn for people, livestock, or both? Think of the various ways corn is grown, purchased, and valued in different places—such as the United States, Europe, Oaxaca, and Zambia—as you formulate your answer.
4. Those who view GM foods as solutions to world hunger and pesticide overuse often want to speed their development and approval. Others adhere to the precautionary principle and want extensive testing for health and environmental safety. How much caution do you think is warranted before a new GM crop is introduced?
5. Imagine it is your job to make the regulatory decision as to whether to allow the planting of a new genetically modified strain of cabbage that produces its own pesticide and has twice the vitamin content of regular cabbage. What questions would you ask of scientists before deciding whether to approve the new crop? What scientific data would you want to see, and how much would be enough? Would you also consult nonscientists or take ethical, economic, and social factors into consideration?
6. Cuba adopted low-input organic agriculture out of necessity. If the country were to become economically prosperous once more, do you think Cubans would maintain this form of agriculture, or do you think they would turn to intensive, high-input farming instead? What path do you think they should pursue, and why?

INTERPRETING GRAPHS AND DATA

In the year 2000, over 80 million metric tons of nitrogen fertilizer was used in producing food for the world's 6 billion people. Food production, use of nitrogen fertilizers, and world population all had grown over the preceding 40 years, but at somewhat different rates. Food production grew slightly faster than population while relatively little additional land was converted to agricultural use during this time. Fertilizer use grew most rapidly.

1. Express the year 2002 values of the four graphed indices as percentages of the value of each index in 1961.
2. Calculate the ratio of the food production index to the nitrogen fertilizer use index in 1961 and in 2002. What does comparing these two ratios tell you about how the efficiency of nitrogen use in agriculture has changed? Is this an example of the law of diminishing returns?

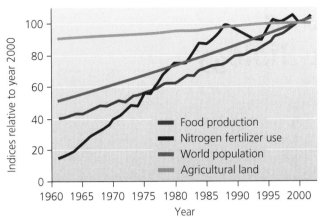

Global food production, nitrogen fertilizer use, human population, and land converted to agriculture, 1961–2002, relative to 2000 levels (2000 = 100). Data from Food and Agriculture Organization of the United Nations.

3. As world population has grown, so has the demand for food, yet little additional land has been devoted to food production. Calculate the ratio of the agricultural land index to the population index for 1961 and 2002. What does comparing these two ratios tell you about how the per capita demand on agricultural land has changed over the years? To what factors can you attribute this change?

CALCULATING ECOLOGICAL FOOTPRINTS

As food production became more industrialized during the 20th century, several trends emerged. One trend, documented in this chapter, was a loss in the number of varieties of crops grown. A second trend was the increasing amount of energy expended to store food and ship it to market. In the United States today, food travels an average of 1,400 miles from the field to your table. The price you pay for the food covers the cost of this long-distance transportation, which in 2004 was approximately one dollar per ton per mile. Assuming that the average person eats 2 pounds of food per day, calculate the food transportation costs for each category in the table below.

Consumer	Daily Cost	Annual Cost
You	$1.40	$511
Your class		
Your hometown		
Your state		
United States		

1. What specific challenges to environmental sustainability are imposed by a food production and distribution system that relies on long-range transportation to bring food to market?

2. A study by Pirog and Benjamin* noted that locally produced food traveled only 50 miles or so to market, thus saving 96% of the transportation costs. Locally grown foods may be fresher and cause less environmental impact as they are brought to market, but what are the disadvantages to you as a consumer in relying on local food production? Do you think the advantages outweigh those disadvantages?

3. What has happened to gasoline prices recently? Would future increases in the price of gas affect your answers to the preceding questions?

*Data from Pirog, R., and A. Benjamin. 2003. Checking the food odometer: Comparing food miles for local versus conventional produce sales to Iowa institutions. Ames, IA: Leopold Center for Sustainable Agriculture, Iowa State University.

Take It Further

Go to www.aw-bc.com/withgott or the student CD-ROM where you'll find:

▶ Suggested answers to end-of-chapter questions
▶ Quizzes, animations, and flashcards to help you study
▶ *Research Navigator*™ database of credible and reliable sources to assist you with your research projects

▶ **GRAPHit!** tutorials to help you master how to interpret graphs
▶ **INVESTIGATEit!** current news articles that link the topics that you study to case studies from your region to around the world

11 Biodiversity and Conservation Biology

The Sikhote-Alin Mountains meet the Pacific Ocean

Upon successfully completing this chapter, you will be able to:

▶ Characterize the scope of biodiversity on Earth

▶ Describe ways to measure biodiversity

▶ Contrast background extinction rates and periods of mass extinction

▶ Evaluate the primary causes of biodiversity loss

▶ Specify the benefits of biodiversity

▶ Assess conservation biology and its practice

▶ Explain island biogeography theory and its application to conservation biology

▶ Compare and contrast traditional and more innovative biodiversity conservation efforts

Amur River
Russia
Mongolia
China
Sikhote-Alin Mountains
India

Central Case: Saving the Siberian Tiger

"Future generations would be truly saddened that this century had so little fore-sight, so little compassion, such lack of generosity of spirit for the future that it would eliminate one of the most dramatic and beautiful animals this world has ever seen."
—GEORGE SCHALLER,
WILDLIFE BIOLOGIST,
ON THE TIGER

"Except in pockets of ignorance and malice, there is no longer an ideological war between conserva-tionists and developers. Both share the perception that health and prosperity decline in a deteriorating environment. They also understand that useful products cannot be harvested from extinct species."
—EDWARD O. WILSON,
HARVARD UNIVERSITY
BIODIVERSITY EXPERT

Historically, tigers roamed widely across Asia from Turkey to northeast Russia to Indonesia. Within the past 200 years, however, people have driven the majestic striped cats from most of their historic range. Today, tigers are exceedingly rare and are creeping toward extinction.

Of the tigers that still survive, those of the subspecies known as the Siberian tiger are the largest cats in the world. Males reach 3.66 m (12 ft) in length and weigh up to 363 kg (800 lb). Also named Amur tigers for the watershed they occupied along the Amur River, which divides Siberian Russia from Manchurian China, these cats now find their last refuge in the forests of the remote Sikhote-Alin Mountains of the Russian Far East.

For thousands of years the Siberian tiger coexisted with the region's native people and held a prominent place in native language and lore. These people referred to the tiger as "Old Man" or "Grandfather" and equated it with royalty or viewed it as a guardian of the mountains and forests. Indigenous people of the region rarely killed a tiger unless it had preyed on a person.

The Russians who moved into the region and exerted control in the early 20th century had no such cultural traditions. They hunted tigers for sport and hides, and some Russians reported killing as many as 10 tigers in a single hunt. In addition, poachers began killing tigers to sell their body parts to China and other Asian countries, where they are used in traditional medicine and as aphrodisiacs. Meanwhile, road building, logging, and agriculture began to fragment tiger habitat and provide easy access for well-armed hunters. The tiger population dipped to perhaps 20–30 animals.

International conservation groups began to get involved, working with Russian biologists to try to save the dwindling tiger population. One such group was the Hornocker Wildlife Institute, now part of the Wildlife Conservation Society. In 1991 the group helped launch the Siberian Tiger Project, devoted to studying the tiger and its habitat. The team put together a plan to protect the tiger, began educating people regarding the tiger's importance and value, and worked closely with those who live in proximity to the big cats.

Thanks to such efforts by conservation biologists, today Siberian tigers in the wild number roughly 330–370, and about 600 more survive in zoos and captive breeding programs around the world. The outlook for the species' survival still looks challenging, but many people are trying to save these endangered animals. It is one of many efforts around the world today to stem the loss of our planet's priceless biological diversity.

Our Planet of Life

Growing human population and resource consumption are putting ever-greater pressure on the flora and fauna of the planet, from tigers to tiger beetles. We are diminishing Earth's diversity of life, the very quality that makes our planet so special. In Chapter 5 we introduced the concept of **biological diversity,** or **biodiversity,** as the sum total of all organisms in an area, taking into account the diversity of species, their genes, their populations, and their communities. In this chapter we will refine this definition and examine current biodiversity trends and their relevance to our lives. We will then explore science-based solutions to biodiversity loss.

Biodiversity encompasses several levels

Biodiversity is a concept as multifaceted as life itself, and definitions of the term are plentiful. As sociologist of science David Takacs explains in his 1996 book, *The Idea of Biodiversity,* different biologists employ different working definitions according to their own aims, interests, and values. Nonetheless, there is broad agreement that the concept applies across several major levels in the organization of life (Figure 11.1). The level that is easiest to visualize and most commonly used is species diversity.

Ecosystem diversity

Species diversity

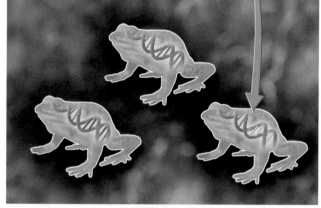

Genetic diversity

FIGURE 11.1 The concept of biodiversity encompasses several levels in the hierarchy of life. Species diversity (middle frame of figure) refers to the number or variety of species. Genetic diversity (bottom frame) refers to variation in DNA composition among individuals within a species. Ecosystem diversity (top frame) and related concepts refer to variety at levels above the species level, such as ecosystems, communities, habitats, or landscapes.

Species diversity As you recall from Chapter 5 (▸p. 122), a *species* is a distinct type of organism, a set of individuals that uniquely share certain characteristics and can breed with one another and produce fertile offspring. Biologists may use somewhat differing criteria to delineate species boundaries; some emphasize characteristics shared because of common ancestry, whereas others emphasize ability to interbreed. In practice, however, scientists broadly agree on species identities. We can express **species diversity** in terms of the number or variety of species in the world or in a particular region. One component of species diversity is *species richness,* the number of species. Another is *evenness* or *relative abundance,* the extent to which numbers of individuals of different species are equal or skewed.

As we saw in Chapter 5 (▸pp. 122–126), speciation generates new species, adding to species richness, whereas extinction decreases species richness. Although immigration, emigration, and local extinction may increase or decrease species richness locally, only speciation and extinction change it globally.

Taxonomists, the scientists who classify species, use an organism's physical appearance and genetic makeup to determine its species. Taxonomists also group species by their similarity into a hierarchy of categories meant to reflect evolutionary relationships. Related species are grouped together into *genera* (singular, *genus*), related genera are grouped into families, and so on (Figure 11.2). Every species is given a two-part Latin or Latinized scientific name denoting its genus and species. The tiger, *Panthera tigris,* differs from the world's other species of large cats such as the jaguar *(Panthera onca),* the leopard *(Panthera pardus),* and the African lion *(Panthera leo).* These four species are closely related in evolutionary terms, as indicated by the genus name they share, *Panthera.* They are more distantly related to cats in other genera such as the cheetah *(Acinonyx jubatus)* and the bobcat *(Felis rufus),* although all cats are classified together in the family Felidae.

Biodiversity exists below the species level in the form of *subspecies,* populations of a species that occur in different geographic areas and differ from one another in some characteristics. Subspecies are formed by the same processes that drive speciation, but result when divergence does not proceed far enough to create separate species. Scientists denote subspecies with a third part of the scientific name. The Siberian tiger, *Panthera tigris altaica,* is one of five subspecies of tiger still surviving (Figure 11.3). Tiger subspecies differ in color, coat thickness, stripe patterns, and size. For example, *Panthera tigris altaica* is 5–10 cm (2–4 in.) taller at the shoulder than the Bengal tiger *(Panthera tigris tigris)* of India and Nepal, and it has a thicker coat and larger paws.

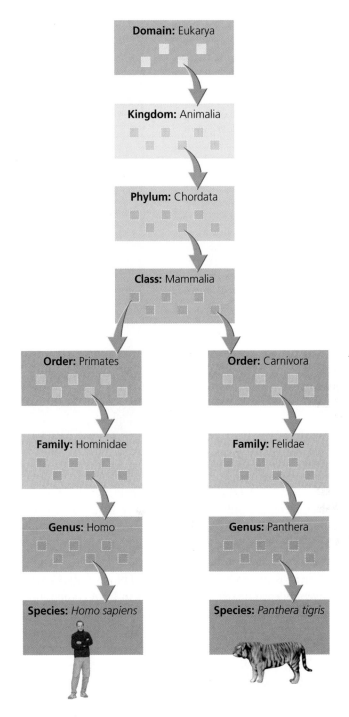

FIGURE 11.2 Taxonomists classify organisms using a hierarchical system meant to reflect evolutionary relationships. Species that are similar in their appearance, behavior, and genetics (because of recent common ancestry) are placed in the same genus. Organisms of similar genera are placed within the same family. Families are placed within orders, orders within classes, classes within phyla, phyla within kingdoms, and kingdoms within domains. For instance, humans *(Homo sapiens,* a species in the genus *Homo)* and tigers *(Panthera tigris,* a species in the genus *Panthera)* are both within the class Mammalia. However, the differences between our two species, which have evolved over millions of years, are great enough that we are placed in different orders and families.

FIGURE 11.3 Three of the eight subspecies of tiger became extinct during the 20th century. The Bali, Javan, and Caspian tigers are extinct. Today only the Siberian (Amur), Bengal, Indochina, Sumatran, and South China tigers persist, and the Chinese government estimates that less than 30 individuals of the South China tiger remain. Deforestation, hunting, and other pressures from people have caused tigers of all subspecies to disappear from most of the geographic range they historically occupied. This map contrasts the ranges of the eight subspecies in the years 1800 (orange) and 2000 (red). Data from the Tiger Information Center.

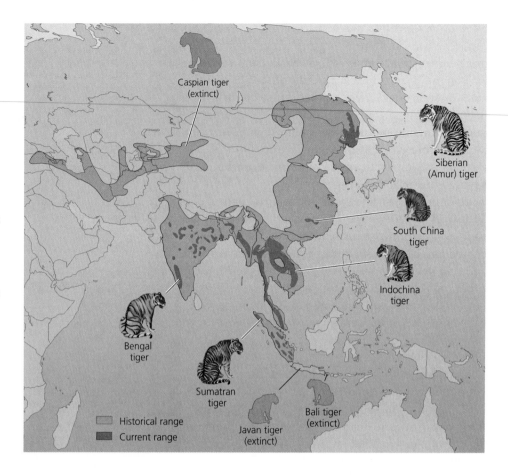

Genetic diversity Scientists designate subspecies when they recognize substantial genetically based differences among individuals from different populations of a species. However, all species consist of individuals that vary genetically from one another to some degree, and this genetic diversity is an important component of biodiversity. **Genetic diversity** encompasses the differences in DNA composition among individuals within species and populations.

Genetic diversity provides the raw material for adaptation to local conditions. A diversity of genes for coat thickness in tigers allowed natural selection to favor genes for thin coats of fur in Bengal tigers living in warm regions, and genes for thick coats of fur for Siberian tigers living in cold regions. In the long term, populations with more genetic diversity may stand better chances of persisting, because their variation better enables them to cope with environmental change. Populations with little genetic diversity are vulnerable to environmental change for which they are not genetically prepared. Populations with depressed genetic diversity may also be more vulnerable to disease and may suffer *inbreeding depression,* which occurs when genetically similar parents mate and produce weak or defective offspring. Scientists have sounded warnings over low genetic diversity in species

that have dropped to low population sizes in the past, including cheetahs, bison, and elephant seals, but the full consequences of reduced diversity in these species remain to be seen. Diminishing genetic diversity in our crop plants also is a prime concern to humanity, as we saw in Chapter 10 (▶pp. 295–296).

Ecosystem diversity Biodiversity also encompasses levels above the species level. *Ecosystem diversity* refers to the number and variety of ecosystems, but biologists may also refer to the diversity of biotic community types or habitats within some specified area. If the area is large, scientists may also consider the geographic arrangement of habitats, communities, or ecosystems at the landscape level, including the sizes, shapes, and interconnectedness of patches of these entities. Under any of these concepts, a seashore of rocky and sandy beaches, forested cliffs, offshore coral reefs, and ocean waters would hold far more biodiversity than the same acreage of a monocultural cornfield. A mountain slope whose vegetation changes from desert to hardwood forest to coniferous forest to alpine meadow would hold more biodiversity than an area the same size consisting of only desert, forest, or meadow.

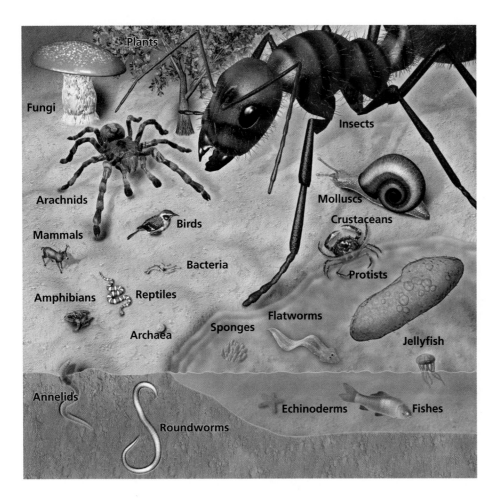

Plants

Fungi

Insects

Arachnids

Molluscs

Crustaceans

Mammals

Birds

Bacteria

Protists

Amphibians

Reptiles

Flatworms

Sponges

Archaea

Jellyfish

Annelids

Echinoderms

Fishes

Roundworms

FIGURE 11.4 This illustration shows organisms scaled in size to the number of species known so far from each major taxonomic group, giving a visual sense of the disparity in species richness among groups. However, because most species are not yet discovered or described, some groups (such as bacteria, archaea, insects, nematodes, protists, fungi, and others) may contain far more species than we now know of. Data from Groombridge, B., and M. D. Jenkins. 2002. *Global biodiversity: Earth's living resources in the 21st century*. UNEP-World Conservation Monitoring Centre. Cambridge, U.K.: Hoechst Foundation.

Measuring biodiversity is not easy

Coming up with precise quantitative measurements to express a region's biodiversity is difficult. This is partly why scientists often express biodiversity in terms of its most easily measured component, species diversity, and in particular, species richness. Species richness is a good gauge for overall biodiversity, but we still are profoundly ignorant of the number of species that exist worldwide. So far, scientists have identified and described 1.7–2.0 million species of plants, animals, and microorganisms. However, estimates for the total number that actually exist range from 3 million to 100 million, with our best educated guesses ranging from 5 million to 30 million.

Species are not evenly distributed among taxonomic groups. In terms of number of species, insects show a staggering predominance over all other forms of life (Figures 11.4 and 11.5). Within insects, about 40% are beetles. Beetles outnumber all noninsect animals and all plants. No wonder the 20th-century British biologist J. B. S. Haldane famously quipped that God must have had "an inordinate fondness for beetles."

Our knowledge of species numbers is incomplete for several reasons. First, some areas of Earth remain little

explored. We have barely sampled the ocean depths, hydrothermal vents (▸pp. 107–108), or the tree canopies and soils of tropical forests. Second, many species are tiny and easily overlooked. These inconspicuous organisms include bacteria, nematodes (roundworms), fungi, protists, and soil-dwelling arthropods. Third, many organisms are so difficult to identify that ones thought to be identical sometimes turn out, once biologists look more closely, to be multiple species. This is frequently the case with microbes, fungi, and small insects, but also sometimes with organisms as large as birds, trees, and whales.

Smithsonian Institution entomologist Terry Erwin pioneered one method of estimating species numbers. In 1982, Erwin's crews fogged rainforest trees in Central America with clouds of insecticide and then collected insects, spiders, and other arthropods as they died and fell from the treetops. Erwin concluded that 163 beetle species specialized on the tree species *Luehea seemannii*. If this were typical, he figured, then the world's 50,000 tropical tree species would hold 8,150,000 beetle species and— since beetles represent 40% of all arthropods—20 million arthropod species. If canopies hold two-thirds of all arthropods, then arthropod species in tropical forests alone would number 30 million. Many assumptions were

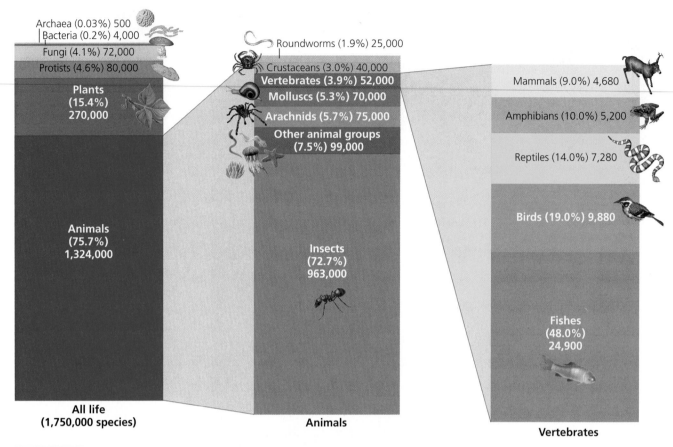

FIGURE 11.5 In the left portion of the figure, we see that three-quarters of known species are animals. The central portion subdivides animals, revealing that nearly three-quarters of animals are insects and that vertebrates comprise only 3.9% of animals. Among vertebrates (right portion of figure), nearly half are fishes, and mammals comprise only 9%. As noted, most species are not yet discovered or described, so some groups may contain far more species than we now know of. Data from Groombridge, B., and M.D. Jenkins. 2002. *Global biodiversity: Earth's living resources in the 21st century.* UNEP-World Conservation Monitoring Centre. Cambridge, U.K.: Hoechst Foundation.

involved in this calculation, and several follow-up studies have revised Erwin's estimate downward.

Biodiversity is unevenly distributed

Numbers of species tell only part of the story of Earth's biodiversity. Living things are distributed across our planet unevenly, and scientists have long sought to explain the distributional patterns they see. For example, as we have noted, some groups of organisms include only one or a few species, whereas other groups contain many. Some groups have given rise to many species in a relatively short period of time through the process of adaptive radiation (see Figure 5.3, ▶p. 121). Species diversity also varies according to biome. Tropical dry forests and rainforests tend to support more species than tundra and boreal forests, for instance. The variation in diversity by biome is related to one of the planet's most striking patterns of

species diversity: the fact that species richness generally increases as one approaches the equator (Figure 11.6). This pattern of variation with latitude, called the *latitudinal gradient,* has been one of the most obvious patterns in ecology, but it also has been one of the most difficult ones for scientists to explain.

Hypotheses abound for the cause of the latitudinal gradient in species richness, but it seems likely that plant productivity and climate stability play key roles in the phenomenon (Figure 11.7). Greater amounts of solar energy, heat, and humidity at tropical latitudes lead to more plant growth, making areas nearer the equator more productive and able to support larger numbers of animals. In addition, the relatively stable climates of equatorial regions—their similar temperatures and rainfall from day to day and season to season—help ensure that single species won't dominate ecosystems, but that instead numerous species can coexist. Whereas varying environmental

FIGURE 11.6 For many types of organisms, number of species per unit area tends to increase as one moves toward the equator. This trend, the latitudinal gradient in species richness, is one of the most readily apparent—yet least understood—patterns in ecology. One example is bird species in North and Central America: In any one spot in arctic Canada and Alaska, 30 to 100 species can be counted; in areas of Costa Rica and Panama, the number rises to over 600. Adapted from Cook, R. E. 1969. Variation in species density in North American birds. *Systematic Zoology* 18: 63–84.

FIGURE 11.7 Ecologists have offered many hypotheses for the latitudinal gradient in species richness, and one set of ideas is summarized here. The variable climates (across days, seasons, and years) of polar and temperate latitudes favor organisms that can survive a wide range of conditions. Such generalist species have expansive niches; they can do many things well enough to survive, and they spread over large areas. In tropical latitudes, the abundant solar energy, heat, and humidity induce greater plant growth, which supports more organisms. The stable climates of equatorial regions favor specialist species, which have restricted niches but do certain things very well. Together these factors are thought to promote greater species richness in the tropics.

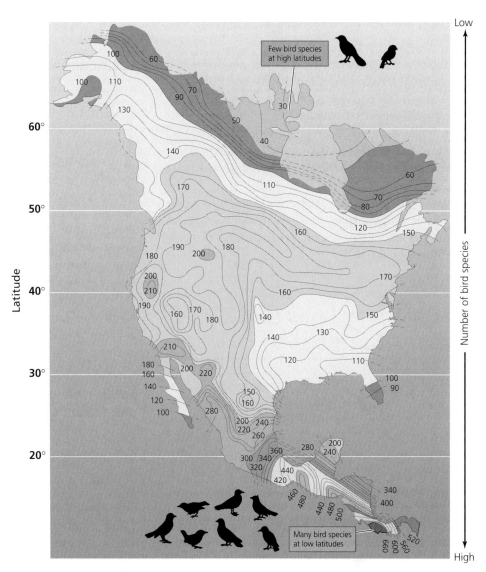

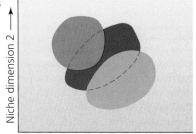

Temperate and polar latitudes
- Variable climate favors fewer species, and species that are widespread generalists.

Tropical latitudes
- Greater solar energy, heat, and humidity promote more plant growth to support more organisms. Stable climate favors specialist species. Together these encourage greater diversity of species.

conditions favor generalists—species that can deal with a wide range of circumstances but that do no single thing very well—stable conditions favor organisms with specialized niches that do particular things very well. In addition, polar and temperate regions may be relatively lacking in species because glaciation events repeatedly forced organisms out of these regions and toward more tropical latitudes.

We will discuss further geographic patterns in biodiversity later in this chapter, when we explore solutions to the ongoing loss of global biodiversity that our planet is currently experiencing.

Biodiversity Loss and Species Extinction

Biodiversity at all levels is being lost to human impact, most irretrievably in the extinction of species. Once vanished, a species can never return. *Extinction* (►p. 126) occurs when the last member of a species dies and the species ceases to exist, as apparently was the case with Monteverde's golden toad. The disappearance of a particular population from a given area, but not the entire species globally, is referred to as **extirpation.** The tiger has been extirpated from most of its historic range, but it is not yet extinct. Although a species that is extirpated from one place may still exist in others, extirpation is an erosive process that can, over time, lead to extinction.

Extinction is a natural process

Extirpation and extinction occur naturally. If organisms did not naturally go extinct, we would be up to our ears in dinosaurs, trilobites, ammonites, and the millions of other types of creatures that vanished from Earth long before humans appeared. Paleontologists estimate that roughly 99% of all species that have ever lived are now extinct. This means that the wealth of species on our planet today comprises only about 1% of all species that ever lived. Most extinctions preceding the appearance of humans have occurred one by one for independent reasons, at a rate that paleontologists refer to as the **background rate of extinction.** For example, the fossil record indicates that for mammals and marine animals, one species out of 1,000 would typically become extinct every 1,000 to 10,000 years. This translates to an annual rate of one extinction per 1 to 10 million species.

Earth has experienced five previous mass extinction episodes

Extinction rates have risen far above this background rate during several mass extinction events in Earth's history. In the past 440 million years, our planet has experienced five major episodes of **mass extinction** (Figure 11.8). Each of these events has eliminated more than one-fifth of life's families and at least half its species (Table 11.1). The most severe episode occurred at the end of the Permian period, 248 million years ago, when close to 54% of all families, 90% of all species, and 95% of marine species went extinct.

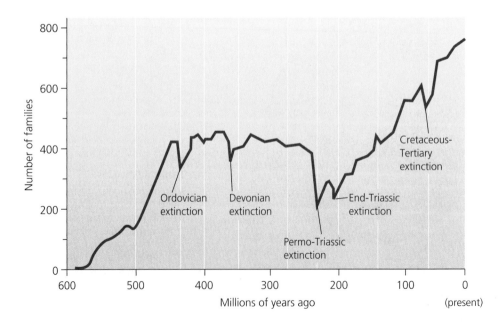

FIGURE 11.8 The fossil record shows evidence of five episodes of mass extinction during the past half-billion years of Earth history. At the end of the Ordovician, Devonian, Permian, Triassic, and Cretaceous periods, 50–95% of the world's species appear to have gone extinct. Each time, biodiversity later rebounded to equal or higher levels, but the rebound required millions of years in each case. Data from Raup, D. M., and J. J. Sepkoski. 1982. Mass extinctions in the marine fossil record. *Science* 215: 1501–1503.

Table 11.1	Mass Extinctions			
Event	Date (millions of years ago)	Cause	Types of life most affected	Percentage of life depleted
Ordovician	440 mya	Unknown	Marine organisms; terrestrial record is unknown	>20% of families
Devonian	370 mya	Unknown	Marine organisms; terrestrial record is unknown	>20% of families
Permo-Triassic	250 mya	Possibly volcanism	Marine organisms; terrestrial record is less known	>50% of families; 80–95% of species
End-Triassic	202 mya	Unknown	Marine organisms; terrestrial record is less known	20% of families; 50% of genera
Cretaceous-Tertiary	65 mya	Asteroid impact	Marine and terrestrial organisms, including dinosaurs	15% of families; >50% of species
Current	Beginning 0.01 mya	Human impact, through habitat destruction and other means	Large animals, specialized organisms, island organisms, and organisms hunted or harvested by humans	Ongoing

The best-known episode occurred at the end of the Cretaceous period, 65 million years ago, when an apparent asteroid impact brought an end to the dinosaurs and many other groups (▶pp. 128–129). In addition, there is evidence for further mass extinctions in the Cambrian period and earlier, more than half a billion years ago.

If current trends continue, the modern era, known as the Quaternary period, may see the extinction of more than half of all species. Although similar in scale to previous mass extinctions, today's ongoing mass extinction is different in two primary respects. First, humans are causing it. Second, humans will suffer as a result of it.

Humans set the sixth mass extinction in motion years ago

We have recorded many instances of human-induced species extinction over the past few hundred years. Sailors documented the extinction of the dodo on the Indian Ocean island of Mauritius in the 17th century, and we still have a few of the dodo's body parts in museums. Among North American birds in the past two centuries, we have driven into extinction the Carolina parakeet, great auk, Labrador duck, and passenger pigeon (▶p. 132), and probably the Bachman's warbler and Eskimo curlew. Several more species, including the whooping crane, California condor, Kirtland's warbler, and the ivory-billed woodpecker, recently rediscovered in the wooded swamps of Arkansas, teeter on the brink of extinction.

However, species extinctions caused by humans precede written history. Indeed, people may have been hunting species to extinction for thousands of years. Archaeological evidence shows that in case after case, a wave of extinctions followed close on the heels of human arrival on islands and continents (Figure 11.9). After Polynesians reached Hawaii, half its birds went extinct. Birds, mammals, and reptiles vanished following human arrival on many other oceanic islands, including large island masses such as New Zealand and Madagascar. The pattern appears to hold for at least two continents, as well. Dozens of species of large vertebrates died off in Australia after Aborigines arrived roughly 50,000 years ago, and North America lost 33 genera of large mammals after people arrived on the continent at least 10,000 years ago.

Current extinction rates are much higher than normal

Today, species loss is accelerating as our population growth and resource consumption put increasing strain on habitats and wildlife. A decade ago, 1,500 of the world's leading scientists reported to the United Nations in their Global Biodiversity Assessment that in the preceding 400 years, 484 animal and 654 plant species were known to have become extinct, and that more than 30,000 plant and animal species currently faced extinction. In 2005, scientists with the Millennium Ecosystem Assessment (▶p. 20) calculated that the current global extinction rate is 100 to 1,000 times greater than the background rate. Moreover, they projected that the rate would increase tenfold or more in future decades.

To keep track of the current status of endangered species, the World Conservation Union maintains the

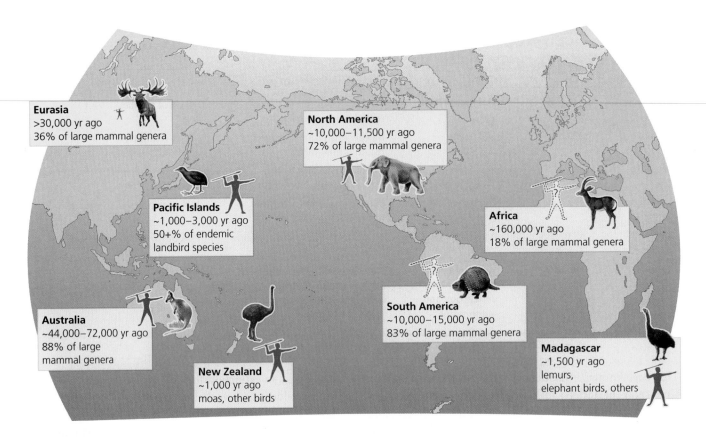

Eurasia
>30,000 yr ago
36% of large mammal genera

North America
~10,000–11,500 yr ago
72% of large mammal genera

Pacific Islands
~1,000–3,000 yr ago
50+% of endemic
landbird species

Africa
~160,000 yr ago
18% of large mammal genera

Australia
~44,000–72,000 yr ago
88% of large
mammal genera

South America
~10,000–15,000 yr ago
83% of large mammal genera

Madagascar
~1,500 yr ago
lemurs,
elephant birds, others

New Zealand
~1,000 yr ago
moas, other birds

FIGURE 11.9 This map shows for each region the time of human arrival and the extent of the recent extinction wave. Illustrated are representative extinct megafauna from each region. The human hunter icons are sized according to the degree of evidence that human hunting was a cause of extinctions; larger icons indicate more certainty that humans (as opposed to climate change or other forces) were the cause. Data for South America and Africa are so far too sparse to be conclusive, and future archaeological and paleontological research could well alter these interpretations. Adapted from Barnosky, A. D., et al. 2004. Assessing the causes of late Pleistocene extinctions on the continents. *Science* 306: 70–75; and Wilson, E. O. 1992. *The diversity of life.* Cambridge, MA: Belknap Press.

Red List, an updated list of species facing high risks of extinction. The 2004 Red List reported that 23% (1,101) of mammal species and 12% (1,213) of bird species are threatened with extinction. Among other major groups (for which assessments are not fully complete), estimates of the percentage of species threatened ranged from 31% to 86%. Since 1970, at least 58 fish species, 9 bird species, and 1 mammal species have become extinct, and in the United States alone over the past 500 years, 236 animals and 17 plants are confirmed to have gone extinct. For all of these figures, the *actual* numbers of species extinct and threatened, like the actual number of total species in the world, are doubtless greater than the *known* numbers.

Among the 1,101 mammals facing possible extinction on the Red List is the tiger, which despite—or perhaps because of—its tremendous size and reputation as a fierce predator, is one of the most endangered large animals on

the planet. In 1950, eight tiger subspecies existed (see Figure 11.3). Today, three are extinct. The Bali tiger, *Panthera tigris balica,* went extinct in the 1940s; the Caspian tiger, *Panthera tigris virgata,* during the 1970s; and the Javan tiger, *Panthera tigris sondaica,* during the 1980s.

Biodiversity loss involves more than extinction

Statistics on extinction tell only part of the story of biodiversity loss. The larger part of the story is the decline in population sizes of many organisms. Declines in numbers are accompanied by shrinkage of species' geographic ranges. Thus, many species today are less numerous and occupy less area than they once did. These patterns mean that genetic diversity and ecosystem diversity, as well as species diversity, are being lost. To measure and quantify

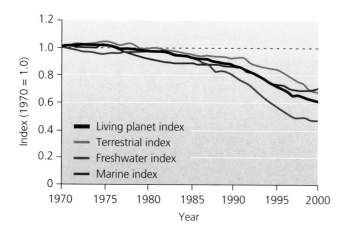

FIGURE 11.10 The Living Planet Index serves as an indicator of the state of global biodiversity. Index values summarize population trends for 1,145 species. Between 1970 and 2000, the Living Planet Index fell by roughly 40%. The indices for terrestrial and marine species fell 30%, and the index for freshwater species fell 50%. Data from World Wide Fund for Nature and U.N. Environment Programme. 2004. *The Living Planet Report, 2004.* Gland, Switzerland: WWF.

this degradation, scientists at the World Wildlife Fund and the United Nations Environment Programme (UNEP) developed a metric called the *Living Planet Index.* This index summarizes trends in the populations of 555 terrestrial species, 323 freshwater species, and 267 marine species that are well enough monitored to provide reliable data. Between 1970 and 2000, the Living Planet Index fell by roughly 40% (Figure 11.10).

There are several major causes of biodiversity loss

Reasons for the decline of any given species are often multifaceted and complex, so they can be difficult to determine. The current precipitous decline in populations of amphibians throughout the world provides an example. Frogs, toads, and salamanders worldwide are decreasing drastically in abundance. Several have already gone extinct, and scientists are struggling to explain why. Recent studies have implicated a wide array of factors, and most scientists now suspect that such factors may be interacting synergistically (see "The Science behind the Story," ▶pp. 322–323).

Nonetheless, overall, scientists have identified four primary causes of population decline and species extinction: habitat alteration, invasive species, pollution, and overharvesting. Global climate change (Chapter 18) now threatens to become the fifth. Each of these factors is exacerbated by human population growth and by our increase in per capita consumption of resources.

Habitat alteration Nearly every human activity can alter the habitat of the organisms around us. Farming replaces diverse natural communities with simplified ones of only one or a few plant species. Grazing modifies the structure and species composition of grasslands. Either type of agriculture can lead to desertification. Clearing forests removes the food, shelter, and other resources that forest-dwelling organisms need to survive. Hydroelectric dams turn rivers into reservoirs upstream and thereby affect water conditions and floodplain communities downstream. Urbanization and suburban sprawl supplant diverse natural communities with simplified human-made ones, driving many species from their homes.

Such changes in habitat generally have negative effects. Organisms are already adapted to the habitats in which they live, so any change is likely to render the habitat less suitable for them. Of course, human-induced habitat change may benefit some species. Animals such as starlings, house sparrows, pigeons, and gray squirrels do very well in urban and suburban environments and benefit from our modification of natural habitats. However, the species that benefit are relatively few; for every species that gains, more lose. Furthermore, the species that do well in our midst tend to be weedy, cosmopolitan generalists that are in little danger of disappearing any time soon.

Habitat alteration is by far the greatest cause of biodiversity loss today. It is the primary source of population declines for 83% of threatened mammals and 85% of threatened birds, according to UNEP data. As just one example of thousands, the prairies native to North America's Great Plains have been almost entirely converted to agriculture. The area of prairie habitat has been reduced by more than 99%. As a result, grassland bird populations have declined by an estimated 82–99%. Many grassland species have been extirpated from large areas, and the two species of prairie chickens still persisting in pockets of the Great Plains could soon go extinct. Habitat destruction has occurred widely in nearly every biome. Estimates by UNEP in 2002 reported that 45% of Earth's forests, 50% of its mangrove ecosystems, and 10% of its coral reefs had been destroyed by recent human activity.

Invasive species Human introduction of non-native species to new environments, where some may become invasive (Figure 11.11), has also pushed native species toward extinction. Some introductions have been accidental. Examples include aquatic organisms (such as zebra mussels; Chapter 6) transported among continents in the ballast water of ships, animals that have escaped from the pet trade, and the weed seeds that cling to our socks as we travel from place to place. Other introductions have been

Invasive Species			
Species	**Native to...**	**Invasive in...**	**Effects**
Mosquito fish (*Gambusia affinis*) 	North America	Africa, Asia, Europe, and Australia	Introduced to control mosquito populations, the mosquito fish outcompetes native fish, eats their eggs, and does no better than native species in controlling mosquitoes.
Zebra mussels (*Dreissenna polymorpha*) 	Caspian Sea	Freshwater ecosystems including the Great Lakes of Canada and the United States	Zebra mussels (Chapter 6) most likely made their way from their home by traveling in ballast water taken on by cargo ships. They compete with native species and clog water treatment facilities and power plant cooling systems.
Kudzu (*Pueraria montana*) 	Japan	Southeastern United States	A vine that can grow 30 m (100 ft) in a single season. The U.S. Soil Conservation Service introduced kudzu in the 1930s to help control erosion. Adaptable and extraordinarily fast-growing, kudzu has taken over thousands of hectares of forests, fields, and roadsides in the southeastern United States.
Asian long-horned beetles (*Anoplophora glabripennis*) 	Asia	United States	Having first arrived in the United States in imported lumber in the 1990s, these beetles burrow into hardwood trees and interfere with the trees' ability to absorb and process water and nutrients. They may wipe out the majority of hardwood trees in an area. Several U.S. cities, including Chicago in 1999 and Seattle in 2002, have cleared thousands of trees after detecting these invaders.
Rosy wolfsnail (*Euglandina rosea*) 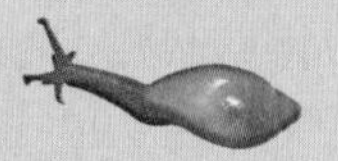	Southeastern United States and Latin America	Hawaii	In the 1950s, well-meaning scientists introduced the rosy wolfsnail to Hawaii to prey upon and reduce the population of another invasive species, the giant African land snail (*Achatina fulica*), which had been introduced early in the 20th century as an ornamental garden animal. Within a few decades, however, the carnivorous rosy wolfsnail had instead driven more than half of Hawaii's native species of banded tree snails to extinction.
Cane toad (*Bufo marinus*) 	Southern United States to tropical South America	Northern Australia and other locations	Since being introduced 70 years ago to control insects in sugarcane fields, the cane toad has wreaked havoc across northern Australia (and other locations). The skin of this tropical American toad can kill its predators, and the cane toad outcompetes native amphibians.
Bullfrog (*Rana catesbiana*) 	Eastern North America	Western North America	The bullfrog is contributing to amphibian and reptile declines in western North America. Bullfrog tadpoles grow large and can outcompete and prey on other tadpoles, but need to grow a long time in permanent water to do so. Historically most water bodies in the arid West dried up part of the year, making it impossible for bullfrogs to live there, but artificial impoundments—dams, farm ponds, canals—gave the bullfrogs bases from which they could spread.

FIGURE 11.11 Invasive species are species that thrive in areas where they are introduced, outcompeting, preying on, or otherwise harming native species. Of the many thousands of invasive species, this chart shows a few of the best known.

intentional. People have brought with them food crops, domesticated animals, and other organisms as they colonized new places, generally unaware of the ecological consequences that could result. Species native to islands are especially vulnerable to disruption from introduced species because the native species have been in isolation for so long with relatively few parasites, predators, and competitors. As a result, they have not evolved the

Invasive Species			
Species	Native to...	Invasive in...	Effects
Gypsy moth (*Lymantria dispar*)	Eurasia	Northeastern United States	In the 1860s, a scientist introduced the gypsy moth to Massachusetts in the mistaken belief that it might be bred with others to produce a commercial-quality silk. The gypsy moth failed to start a silk industry, and instead spread through the northeastern United States and beyond, where its outbreaks defoliate trees over large regions every few years.
European starling (*Sturnus vulgaris*)	Europe	North America	The bird was first introduced to New York City in the late 19th century by Shakespeare devotees intent on bringing every bird mentioned in Shakespeare's plays to the new continent. It only took 75 years for the birds to spread to the Pacific coast, Alaska, and Mexico, becoming one of the most abundant birds on the continent. Starlings are thought to outcompete native birds for nest sites.
Indian mongoose (*Herpestes auropunctatus*)	Southeast Asia	Hawaii	Rats that had invaded the Hawaiian islands from ships in the 17th century were damaging sugarcane fields, so in 1883 the Indian mongoose was introduced to control rat populations. Unfortunately, the rats were active at night and the mongooses fed during the day, so the plan didn't work. Instead mongooses began preying on native species like ground-nesting seabirds and the now-endangered Nene or Hawaiian goose (*Branta sandvicensis*).
A green alga (*Caulerpa taxifolia*)	Tropical oceans and seas	Mediterranean Sea	Dubbed the "killer algae," *Caulerpa taxifolia* has spread along the coasts of several Mediterranean countries since it apparently escaped from Monaco's aquarium in 1984. Creeping underwater over the sand and mud like a green shag carpet, it crowds out other plants, is inedible to most animals, and tangles boat propellers. It has been the focus of intense eradication efforts since arriving recently in Australia and California.
Cheatgrass (*Bromus tectorum*)	Eurasia	Western United States	In just 30 years after its introduction to Washington state in the 1890s, cheatgrass has spread across much of the western United States. Its secret: fire. Its thick patches that choke out other plants and use up the soil's nitrogen burn readily. Fire kills many of the native plants, but not cheatgrass, which grows back even stronger amid the lack of competition.
Brown tree snake (*Boiga irregularis*)	Southeast Asia	Guam	Nearly all native forest bird species on the South Pacific island of Guam have disappeared. The culprit is the brown tree snake. The snakes were likely brought to the island inadvertantly as stowaways in cargo bays of military planes in World War II. Guam's birds had not evolved with tree snakes, and so had no defenses against the snake's nighttime predation. The snakes also cause numerous power outages each year on Guam and have spread to other islands where they are repeating their ecological devastation. The arrival of this snake is the greatest fear of conservation biologists in Hawaii.

defenses necessary to resist invaders that are better adapted to these pressures.

Most organisms introduced to new areas perish, but the few types that survive may do very well, especially if they find themselves without the predators and parasites that attacked them back home or without the competitors that had limited their access to resources. Once released from the limiting factors of predation, parasitism, and

Amphibian Diversity and Amphibian Declines

Amphibians illustrate the two most salient aspects of Earth's biodiversity today. Scientists are discovering more and more species while more and more populations and species are vanishing.

New species of most classes of vertebrates are discovered at a rate of only one or a few per year, but the number of known amphibian species (which include frogs, salamanders, and others)—about 5,800 as of 2005—has jumped nearly 42% just since 1985.

At the same time, however, over 200 species are in steep decline. Researchers feel that they may be naming some species just before they go extinct and losing others before they are even discovered. At least 32 species of frogs, toads, and salamanders studied just years or decades ago, including the golden toad (Chapter 5), are now altogether gone.

These losses are especially worrying because amphibians are widely regarded as "biological indicators" that indicate whether an ecosystem is in good shape or is degraded. Amphibians rely on both

The odd-looking purplish frog *Nasikabatrachus sahyadrensis*, of India, is one of many new amphibian species recently discovered.

aquatic and terrestrial environments and may breathe and absorb water through their skin, so they are sensitive to pollution and other environmental stresses. The link between amphibians and environmental quality suggests that studying the reasons for their declines can tell us much about the state of our environment.

In Sri Lanka and other countries, scientific scrutiny and improved technology have revealed amphibian "hot spots." In the 1990s, an international team of scientists

set out to determine whether Sri Lanka, a large tropical island off the coast of India, held more than the 40 frog species that were already known. Researcher Madhava Meegaskumbura and his team combed through trees, rivers, ponds, and leaf litter for 8 years, collecting more than 1,400 frogs at 300 study sites. The scientists analyzed the frogs' physical appearance, habitat use, and vocalizations. They also examined the frogs' genes by obtaining sequences of nucleotides in several regions of their DNA. They then compared these genetic, physical, and behavioral characteristics to those of known species of frogs.

The team found that the DNA from many of their frogs didn't match that of known species. And they found that many of their frogs looked different, sounded different, or behaved differently from known species. Clearly, they had discovered new species of frogs unknown to science. Some of these novel species live on rocks and sport leg fringes and markings that help disguise them as clumps of moss. Others are tree frogs that lay their eggs in baskets they construct. In all, more

competition, an introduced species may increase rapidly, spread, and displace native species. Moreover, invasive species cause billions of dollars in economic damage each year.

Pollution Pollution can negatively affect organisms in many ways. Air pollution (Chapter 17) can degrade forest ecosystems. Water pollution (Chapter 15) can adversely affect fish and amphibians. Agricultural runoff (including

fertilizers, pesticides, and sediments; Chapters 7, 9, and 10) can harm many terrestrial and aquatic species. Heavy metals, PCBs, endocrine-disrupting compounds, and various other toxic chemicals can poison people and wildlife (Chapter 14), and the effects of oil and chemical spills on wildlife are dramatic and well known. However, although pollution is a substantial threat, it tends to be less significant than public perception holds it to be. The damage to wildlife and ecosystems caused by pollution can be severe,

than 100 new species of amphibians were discovered—all on an island only slightly larger than the state of West Virginia! When reported in the journal *Science* in 2002, the study caught the attention of conservation biologists worldwide.

Such promising discoveries, however, come against a backdrop of distressing declines. Observed numbers of amphibians are down around the globe. Scientists are racing to pin down the causes, and have found evidence for causes as varied as habitat destruction, chemical pollution, disease, invasive species, and climate change. Most worrisome is that many populations are vanishing even when no direct damage, such as habitat loss, is apparent. In some cases, researchers surmise that a combination of factors may be at work.

In one study, researchers Rick Relyea and Nathan Mills presented young frogs with two common dangers—pesticides and predators—to see how the mix affected their survival. The team collected 10 pairs of tree frogs from a wildlife area in Missouri and placed their eggs in clean water. When tad-poles emerged from the eggs, groups of 10 were each put in different tubs of water. Some tubs contained pure water, others contained varying levels of the pesticide carbaryl, and others contained the harmless solvent acetone as a control. To some of each of these three types of tubs, the researchers added a hungry predator—a young salamander. The salamander was caged and couldn't reach the tadpoles, but the tadpoles were aware of its presence. In a series of experiments, over up to 16 days, researchers watched to see how many tadpoles survived the different combinations of stress factors.

Their results, published in the *Proceedings of the National Academy of Sciences* in 2001, revealed that tadpoles that withstood one type of stress might not survive two. As expected, all tadpoles in clean water with no predators survived, and all tadpoles exposed to high concentrations of carbaryl died within several days, regardless of predator presence. But when carbaryl levels were lower, the presence of the salamander made a noticeable difference. In one trial, about 75% of tadpoles survived the pesticide if no predator was present, but in the presence of the salamander, survival rates dropped to about 25%. Thus, when both stresses were present (a condition likely in the tadpoles' natural habitat), death rates increased by two to four times.

One year later, a study published in the same journal by herpetologist Joseph Kiesecker found similar results with pathogens and pesticides. His field and lab experiments revealed that wood frogs were more vulnerable to parasitic infections that cause limb deformities when they were exposed to water containing pesticides.

As scientists learn more about how such factors combine to threaten amphibians, they are gaining a clearer picture of how the fate of these creatures may foreshadow the future for other organisms. "Amphibians have been around for 300 million years. They're tough, and yet they're checking out all around us," says David Wake, a biologist at the University of California at Berkeley, who was among the first to note the creatures' decline. "We really do see amphibians as biodiversity bellwethers."

but it tends to be less than the damage caused by habitat alteration or invasive species.

Overharvesting For most species, a high intensity of hunting or harvesting by humans will not *in itself* pose a threat of extinction, but for some species it can. The Siberian tiger is one such species. Large in size, few in number, long-lived, and raising few young in its lifetime—a classic K-selected species (▶p. 139)—the Siberian tiger is just the type of animal to be vulnerable to population reduction by hunting. The advent of Russian hunting nearly drove the animal extinct, whereas decreased hunting during and after World War II contributed to a population increase. By the mid-1980s, the Siberian tiger population was likely up to 250 individuals. The political freedom that came with the Soviet Union's breakup in 1989, however, brought with it a freedom to harvest Siberia's natural resources, the tiger included, without

FIGURE 11.12 Body parts from tigers have long been used as medicines or aphrodisiacs in some traditional Asian cultures. Hunters and poachers have illegally killed countless tigers through the years to satisfy market demand for these items. Here a street vendor in northern China displays tiger penises and other body parts for sale.

regulations or rules. This coincided with an economic expansion in many Asian countries, where tiger penises are traditionally used to try to boost human sexual performance and where tiger bones, claws, whiskers, and other body parts are used to treat a wide variety of maladies (Figure 11.12). Thus, the early 1990s brought a boom in poaching (poachers killed at least 180 Siberian tigers between 1991 and 1996), as well as a dramatic increase in logging of the Korean pine forests on which the tigers and their prey depend.

Over the past century, hunting has led to steep declines in the populations of many other K-selected animals. The Atlantic gray whale has gone extinct, and several other whales remain threatened or endangered. Gorillas and other primates that are killed for their meat may be facing extinction soon. Thousands of sharks are killed each year simply for their fins, which are used in soup. Today the oceans contain only 10% of the large animals they once did (▶ p. 483).

Climate change The preceding four types of human impacts affect biodiversity in discrete places and times. In contrast, our manipulation of Earth's climate system (Chapter 18) is beginning to have global impacts on biodiversity. As we will explore in Chapter 18, our emissions of carbon dioxide and other "greenhouse gases" that trap heat in the atmosphere are causing average temperatures to warm worldwide, modifying global weather patterns and increasing the frequency of extreme weather events. Scientists foresee that these effects, together termed *global climate change,* will accelerate and become more severe in the years ahead until we find ways to reduce our emissions from fossil fuels.

Climate change is beginning to exert effects on plants and animals. Extreme weather events such as droughts put increased stress on populations, and warming temperatures are forcing species to move toward the poles and higher in altitude. Some species will be able to adapt, but others will not. Consider the cloud-forest fauna from Monteverde that we examined in Chapter 5. Mountaintop organisms cannot move further upslope to escape warming temperatures, so they will likely perish. Trees may not be able to move poleward fast enough. Animals and plants may find themselves among different communities of prey, predators, and parasites to which they are not adapted. The impacts of climate change will likely play a large role in shaping the future world that we and our children will inhabit.

All five of these avenues are influenced by human population growth and rising per capita consumption. More people and more consumption mean more habitat alteration, more invasive species, more pollution, more overharvesting, and more climate change. Growth in population and growth in consumption are the ultimate reasons behind the proximate threats to biodiversity.

Benefits of Biodiversity

Scientists worldwide are presenting us with data that confirm what any naturalist who has watched the habitat change in his or her hometown already knows: From amphibians to tigers, biodiversity is being lost rapidly and visibly within our lifetimes. This suggests the question, "Does it matter?" There are many ways to answer this question, but we can begin by considering the ways that biodiversity benefits people. Scientists have offered a number of tangible, pragmatic reasons for preserving biodiversity, showing how biodiversity directly or indirectly supports human society. In addition, many people feel that organisms have an intrinsic right to exist and that ethical and aesthetic dimensions to biodiversity preservation cannot be ignored.

Biodiversity provides ecosystem services free of charge

Contrary to popular opinion, some things in life can indeed be free, as long as we choose to protect the living systems that provide them. Intact forests provide clean air and buffer hydrologic systems against flooding and drought. Native crop varieties provide insurance against disease and drought. Abundant wildlife can attract tourists and boost the economies of developing nations. Intact ecosystems provide these and other valuable processes, known as *ecosystem services* (▸pp. 39–41, 48–51), for all of us, free of charge.

Maintaining these ecosystem services is one clear benefit of protecting biodiversity. According to UNEP, biodiversity:

▸ Provides food, fuel, and fiber
▸ Provides shelter and building materials
▸ Purifies air and water
▸ Detoxifies and decomposes wastes
▸ Stabilizes and moderates Earth's climate
▸ Moderates floods, droughts, wind, and temperature extremes
▸ Generates and renews soil fertility and cycles nutrients
▸ Pollinates plants, including many crops
▸ Controls pests and diseases
▸ Maintains genetic resources as inputs to crop varieties, livestock breeds, and medicines
▸ Provides cultural and aesthetic benefits
▸ Gives us the means to adapt to change

Organisms and ecosystems support a vast number of vital processes that humans could not replicate or would need to pay for if nature did not provide them. As we saw in Chapter 2, the annual value of just 17 of these ecosystem services may be in the neighborhood of $16–54 trillion per year.

Biodiversity helps maintain ecosystem function

Even if functioning ecosystems are important, however, does biodiversity really help them maintain their function? Ecologists have found that the answer appears to be

yes. Research has demonstrated that high levels of biodiversity tend to increase the *stability* of communities and ecosystems. Research has also found that high biodiversity tends to increase the *resilience* of ecological systems— their ability to weather disturbance, bounce back from stresses, or adapt to change. Most of this research has dealt with species diversity, but new work is finding similar effects for genetic diversity. Thus, a decrease in biodiversity could diminish a natural system's ability to function and to provide services to our society.

What about the extinction of selected species, however? Skeptics have asked whether the loss of a few endangered species will really make much difference in an ecosystem's ability to function. Ecological research suggests that the answer to this question depends on which species are removed. Removing a species that can be functionally replaced by others may make little difference. Recall, however, from Chapter 6 our discussion of keystone species (▸pp. 159–162). Like the keystone that holds together an arch, a keystone species is one whose removal results in significant changes in an ecological system. If a keystone species is extirpated or driven extinct, other species may disappear or experience significant population changes as a result.

Often top predators, such as tigers, are considered keystone species. A single top predator may prey on many other carnivores, each of which may prey on many herbivores, each of which may consume many plants. Thus the removal of a single individual at the top of a food chain can have impacts that multiply as they cascade down the food chain. Moreover, top predators such as tigers, wolves, and grizzly bears are among the species most vulnerable to human impact. Large animals are frequently hunted, and also need large areas of habitat, making them susceptible to habitat loss and fragmentation. In addition, top predators are vulnerable to the buildup of toxic pollutants in their tissues through the process of biomagnification (▸p. 416).

Top predators are not the only species that exert far-reaching influence over their ecosystems. The influence of other species, including "ecosystem engineers" such as ants and earthworms, can be equally significant. Ecosystems are complex, and it is difficult to predict which particular species may be important. Thus, many people prefer to apply the precautionary principle in the spirit of Aldo Leopold (▸pp. 35–36), who advised, "To keep every cog and wheel is the first precaution of intelligent tinkering."

Biodiversity enhances food security

Biodiversity benefits our agriculture, as well. As our discussion of native landraces of corn in Oaxaca, Mexico, in Chapter 10 showed, genetic diversity within crop species and their ancestors is enormously valuable. In 1995, Turkey's wheat crops received at least $50 billion worth of disease resistance from wild wheat strains. California's barley crops annually receive $160 million in disease resistance benefits from Ethiopian strains of barley. During the 1970s a researcher discovered a maize species in Mexico known as *Zea diploperennis*. This maize is highly resistant to disease, and it is a perennial, meaning it will grow back year after year without being replanted. At the time of its discovery, its entire range was limited to a 10-ha (25-acre) plot of land in the mountains of the Mexican state of Jalisco.

Other potentially important food crops await utilization (Figure 11.13). The babassu palm *(Orbignya phalerata)* of the Amazon produces more vegetable oil than any other plant. The serendipity berry *(Dioscoreophyllum cumminsii)* produces a sweetener that is 3,000 times sweeter than table sugar. Several species of salt-tolerant grasses and trees are so hardy that farmers can irrigate them with saltwater. These same plants also produce animal feed, a substitute for conventional vegetable oil, and other economically important products. Such species could be immeasurably beneficial to areas undergoing soil salinization due to poorly managed irrigation (▶pp. 265–266).

Biodiversity provides drugs and medicines

People have made medicines from plants for centuries, and many of today's widely used drugs were discovered by studying chemical compounds present in wild plants, animals, and microbes (Figure 11.14). Each year pharmaceutical products owing their origin to wild species generate up to $150 billion in sales.

It can truly be argued that every species that goes extinct represents one lost opportunity to find a cure for cancer or AIDS. The rosy periwinkle *(Catharanthus roseus)* produces compounds that treat Hodgkin's disease and a particularly deadly form of leukemia. Had this native plant of Madagascar become extinct prior to its discovery by medical researchers, two deadly diseases would have claimed far more victims than they have to date. In Australia, where the government has placed high priority on research into products from rare and endangered species, a rare species of cork, *Duboisia leichhardtii*, now provides hyoscine, a compound that physicians use to treat cancer, stomach disorders, and motion sickness. Another Australian plant, *Tylophora*, provides a drug that treats lymphoid leukemia. Researchers are now exploring the potential of the compound prostaglandin E2

Food Security and Biodiversity: Potential new food sources		
Species	Native to...	Potential uses and benefits
Amaranths (three species of *Amaranthus*) 	Tropical and Andean America	Grain and leafy vegetable; livestock feed; rapid growth, drought resistant
Buriti palm (*Mauritia flexuosa*) 	Amazon lowlands	"Tree of life" to Amerindians; vitamin-rich fruit; pith as source for bread; palm heart from shoots
Maca (*Lepidium meyenii*) 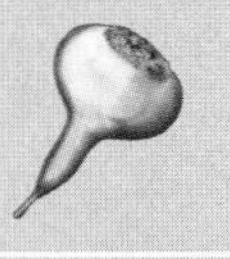	Andes Mountains	Cold-resistant root vegetable resembling radish, with distinctive flavor; near extinction
Tree tomato (*Cyphomandra betacea*) 	South America	Elongated fruit with sweet taste
Babirusa (*Babyrousa babyrussa*) 	Indonesia: Moluccas and Sulawesi	A deep-forest pig; thrives on vegetation high in cellulose and hence less dependent on grain
Capybara (*Hydrochoeris hydrochoeris*) 	South America	World's largest rodent; meat esteemed; easily ranched in open habitats near water
Vicuna (*Lama vicugna*) 	Central Andes	Threatened species related to llama; valuable source of meat, fur, and hides; can be profitably ranched
Chachalacas (*Ortalis*, many species) 	South and Central America	Birds, potentially tropical chickens; thrive in dense populations; adaptable to human habitations; fast-growing
Sand grouse (*Pterocles*, many species) 	Deserts of Africa and Asia	Pigeon-like birds adapted to harshest deserts; domestication a possibility

FIGURE 11.13 By protecting biodiversity, we can enhance food security. The wild species shown here are a tiny fraction of the many plants and animals that could someday supplement our food supply. Adapted from Wilson, E. O. 1992. *The diversity of life.* Cambridge, MA: Belknap Press.

Medicines and Biodiversity: Natural sources of pharmaceuticals		
Plant	**Drug**	**Medical application**
Pineapple (*Ananas comosus*) 	Bromelain	Controls tissue inflammation
Autumn crocus (*Colchicum autumnale*) 	Colchicine	Anticancer agent
Yellow cinchona (*Cinchona ledgeriana*) 	Quinine	Antimalarial
Common thyme (*Thymus vulgaris*) 	Thymol	Cures fungal infection
Pacific yew (*Taxus brevifolia*) 	Taxol	Anticancer (especially ovarian cancer)
Velvet bean (*Mucuna deeringiana*) 	L-Dopa	Parkinson's disease suppressant
Common foxglove (*Digitalis purpurea*) 	Digitoxin	Cardiac stimulant

FIGURE 11.14 By protecting biodiversity, we can enhance our ability to treat illness. Shown here are just a few of the plants that have so far been found to provide chemical compounds of medical benefit. Adapted from Wilson, E. O. 1992. *The diversity of life.* Cambridge, MA: Belknap Press.

in treating gastric ulcers. This compound was first discovered in two frog species unique to the rainforest of Queensland, Australia. Scientists believe that both species are now extinct.

Weighing the Issues:
Bioprospecting in Costa Rica

Bioprospectors search for organisms that can provide new drugs, foods, or other valuable products. Scientists working for pharmaceutical companies, for instance, scour biodiversity-rich countries for potential drugs and medicines. Many have been criticized for harvesting indigenous species to create commercial products that do not benefit the country of origin. To make sure it would not lose the benefits of its own biodiversity, Costa Rica reached an agreement with the Merck pharmaceutical company in 1991. The nonprofit National Biodiversity Institute of Costa Rica (INBio) allowed Merck to evaluate a limited number of Costa Rica's species for their commercial potential in return for $1 million, plus equipment and training for Costa Rican scientists.

Do you think that both sides win in this agreement? What if Merck discovers a compound that could be turned into a billion-dollar drug? Does this provide a good model for other countries? For other companies?

Biodiversity provides economic benefits through tourism and recreation

Besides providing for our food and health, biodiversity can represent a direct source of income through tourism, particularly for developing countries in the tropics that have impressive species diversity. As we saw in Chapter 5 with Costa Rica, many people like to travel to experience protected natural areas, and in so doing they create economic opportunity for residents living near those natural areas. Visitors spend money at local businesses, hire local people as guides, and support the parks that employ local residents. Ecotourism thus can bring jobs and income to areas that otherwise might be poverty-stricken.

Ecotourism has become a vital source of income for nations such as Costa Rica, with its rainforests; Australia, with its Great Barrier Reef; Belize, with its reefs, caves, and rainforests; and Kenya and Tanzania, with their savanna wildlife. The United States, too, benefits from ecotourism; its national parks draw millions of visitors domestically and from around the world. Ecotourism serves as a powerful financial incentive for nations, states, and local communities to preserve natural areas and reduce impacts on the landscape and on native species.

As ecotourism increases in popularity, however, critics have warned that too many visitors to natural areas can degrade the outdoor experience and disturb wildlife. Anyone who has been to Yosemite, the Grand Canyon, or the Great Smokies on a crowded summer weekend can attest to this. Ecotourism's effects on species living in parks and reserves are much debated, and likely they vary enormously from one case to the next. As ecotourism continues to increase, so will debate over its costs and benefits for local communities and for biodiversity.

People value and seek out connections with nature

Not all of the benefits of biodiversity to humans can be expressed in the hard numbers of economics or the day-to-day practicalities of food and medicine. Some scientists and philosophers argue that there is a deeper

FIGURE 11.15 Edward O. Wilson is the world's most recognized authority on biodiversity and its conservation and has inspired many people who study our planet's life. A Harvard professor and world-renowned expert on ants, Wilson has written over 20 books and has won two Pulitzer prizes. His books *The Diversity of Life* and *The Future of Life* address the value of biodiversity and its outlook for the future.

importance to biodiversity. E. O. Wilson (Figure 11.15) has described a phenomenon he calls **biophilia,** "the connections that human beings subconsciously seek with the rest of life." Wilson and others have cited as evidence of biophilia our affinity for parks and wildlife, our keeping of pets, the high value of real estate with a view of natural landscapes, and our interest—despite being far removed from a hunter-gatherer lifestyle—in hiking, bird-watching, fishing, hunting, backpacking, and similar outdoor pursuits.

In a 2005 book, writer Richard Louv adds that as today's children are increasingly deprived of outdoor experiences and direct contact with wild organisms, they suffer what he calls "nature-deficit disorder." Although it is not a medical condition, this alienation from biodiversity and the natural environment, Louv argues, may damage childhood development and lie behind many of the emotional and physical problems young people in developed nations face today.

Weighing the Issues:
Biophilia and Nature-Deficit Disorder

What do you think of the concepts of biophilia and "nature-deficit disorder"? Have you ever felt a connection to other living things that you couldn't explain in scientific or economic terms? Do you think that an affinity for other living things is innately human? How could you determine whether or not most people in your community feel this way?

Do we have ethical obligations toward other species?

If Wilson, Louv, and others are right, then biophilia not only may affect ecotourism and real estate prices, but also may influence our ethics. When Maurice Hornocker and his associates first established the Siberian Tiger Project, he wrote: "Saving the most magnificent of all the cat species and one of the most endangered should be a global responsibility. . . . If they aren't worthy of saving, then what are we all about? What is worth saving?"

On one hand, we humans are part of nature, and like any other animal we need to use resources and consume other organisms to survive. In that sense, there is nothing immoral about our doing so. On the other hand, we have conscious reasoning ability and are able to control our actions and make conscious decisions. Our ethical sense has developed from this intelligence and ability to choose. As our society's sphere of ethical consideration has widened over time, and as more of us take up biocentric or ecocentric worldviews (▶pp. 32–33), more people have come to feel that other organisms have intrinsic value and an inherent right to exist.

Despite our ethical convictions, however, and despite biodiversity's many benefits—from the pragmatic and economic to the philosophical and spiritual—the future of biodiversity is far from secure. Even our protected areas and national parks are not big enough or protected well enough to ensure that biodiversity is fully safeguarded within their borders. The search for solutions to today's biodiversity crisis is an exciting and active one, and scientists are playing a leading role in developing innovative approaches to maintaining the diversity of life on Earth.

Conservation Biology: The Search for Solutions

Today, more and more scientists and citizens perceive a need to do something to stem the loss of biodiversity. In his 1994 autobiography, *Naturalist,* E. O. Wilson wrote:

> When the [20th] century began, people still thought of the planet as infinite in its bounty. The highest mountains were still unclimbed, the ocean depths never visited, and vast wildernesses stretched across the equatorial continents. . . . In one lifetime exploding human populations have reduced wildernesses to threatened nature reserves. Ecosystems and species are vanishing at the fastest rate in 65 million years. Troubled by what we have wrought, we have begun to turn in our role from local conqueror to global steward.

Conservation biology arose in response to biodiversity loss

The urge to act as responsible stewards of natural systems, and to use science as a tool in that endeavor, helped spark the rise of conservation biology. **Conservation biology** is a scientific discipline devoted to understanding the factors, forces, and processes that influence the loss, protection, and restoration of biological diversity. It arose as biologists became increasingly alarmed at the degradation of the natural systems they had spent their lives studying.

Conservation biologists choose questions and pursue research with the aim of developing solutions to such problems as habitat degradation and species loss. Conservation biology is thus an applied and goal-oriented science, with implicit values and ethical standards. This perceived element of advocacy sparked some criticism of conservation biology in its early years. However, as scientists have come to recognize the scope of human impact on the planet, more of them have directed their work to address environmental problems. Today conservation biology is a thriving pursuit that is central to environmental science and to achieving a sustainable society.

Conservation biologists work at multiple levels

Conservation biologists integrate an understanding of evolution and extinction with ecology and the dynamic nature of environmental systems. They use field data, lab data, theory, and experiments to study the impacts of humans on other organisms. They also attempt to design, test, and implement ways to mitigate human impact.

These researchers address the challenges facing biological diversity at all levels, from genetic diversity to species diversity to ecosystem diversity. At the genetic level, *conservation geneticists* study genetic attributes of organisms, generally to infer the status of their populations. If two populations of a species are found to be genetically distinct enough to be considered subspecies, they may have different ecological needs and may require different types of management. In addition, as a population dwindles, genetic variation is lost from the gene pool. Conservation geneticists ask how small a population can become and how much genetic variation it can lose before running into problems such as inbreeding depression. By determining a *minimum viable population size* for a given population, conservation geneticists and population biologists provide wildlife managers with an indication of how vital it may be to increase the population.

Problems for populations and subspecies spell problems for species, because declines and local extirpation generally precede range-wide endangerment and extinction. As we will soon see, it is at the species level that much of the funding and resources for conservation biology exist. However, many efforts also revolve around habitats, communities, and ecosystems.

Island biogeography theory is a key component of conservation biology

Safeguarding habitat for species and conserving communities and ecosystems requires thinking and working at the landscape level. One key conceptual tool for doing so is the **equilibrium theory of island biogeography.** This theory, introduced by E. O. Wilson and ecologist Robert MacArthur in 1963, explains how species come to be distributed among oceanic islands. Since then, researchers have also applied it to "habitat islands"— patches of one habitat type isolated within "seas" of others. The Sikhote-Alin Mountains, last refuge of the Siberian tiger, are a habitat island, isolated from other mountains by deforested regions, a seacoast, and populated lowlands.

Island biogeography theory predicts the number of species on an island based on the island's size and its distance from the nearest mainland. The number of species on an island results from a balance between the number added by immigration and the number lost through extinction (or more precisely, extirpation from the particular island). Immigration and extinction are ongoing

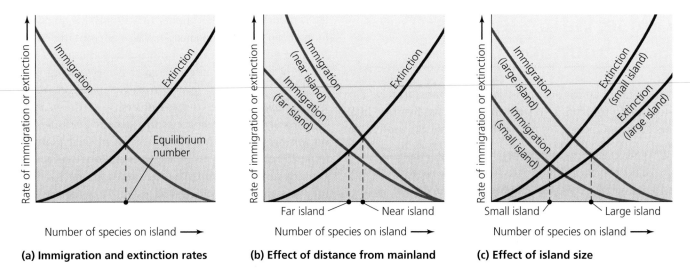

(a) Immigration and extinction rates   (b) Effect of distance from mainland   (c) Effect of island size

FIGURE 11.16 Island biogeography theory explains species richness on islands as a function of immigration and extinction rates interacting with island size and distance from the mainland. In (a), immigration rates are highest for islands with few species (because each new immigrant species represents a high proportion of the island's total species). Extinction rates are greatest on islands with many species (because interspecific competition for resources keeps population sizes small). These two trends set the stage for examining the effects of island size and distance. In (b), islands closer to a mainland tend to hold more species, because immigration rates are higher, whereas extinction rates are not affected by distance. In (c), immigration is greater on large islands because large islands are easier for dispersing organisms to discover, while extinction rates are lower on large islands, because more area allows populations to grow larger. Together these trends give large islands a higher number of species at equilibrium.

dynamic processes, so the balance between them represents an equilibrium state (Figure 11.16a).

The farther an island is located from a continent, the fewer species tend to find and colonize it. Thus, remote islands host fewer species because of lower immigration rates (Figure 11.16b). Large islands have higher immigration rates because they present fatter targets for wandering or dispersing organisms to encounter. Large islands have lower extinction rates because more space allows for larger populations, which are less vulnerable to dropping to zero by chance. Together, these trends give large islands more species at equilibrium than small islands (Figure 11.16c). Very roughly, the number of species on an island is expected to double as island size increases tenfold. This effect can be illustrated with *species-area curves* (Figure 11.17). Large islands also tend to contain more species because they generally possess more habitats than smaller islands, providing suitable environments for a wider variety of arriving species.

These theoretical patterns have been widely supported by empirical data from the study of species on islands (see "The Science behind the Story," ▸ pp. 332–333). The patterns hold up for terrestrial habitat islands, such as forests

fragmented by logging and road building (Figure 11.18). Small islands of forest lose their diversity fastest, starting with large species that were few in number to begin with. In a landscape of fragmented habitat, species requiring the habitat will gradually disappear, winking out from one fragment after another over time. Fragmentation of forests and other habitats constitutes one of the prime threats to biodiversity. In response to habitat fragmentation, conservation biologists have designed landscape-level strategies to try to optimize the arrangement of areas to be preserved. We will examine a few of these strategies in our discussion of parks and preserves in Chapter 12 (▸ pp. 364–369).

Weighing the Issues:
Fragmentation and Biodiversity

Suppose a critic of conservation tells you that human development increases biodiversity, pointing out that when a forest is fragmented, new habitats, such as grassy lots and gardens, may be introduced to an area and allow additional species to live there. How would you respond?

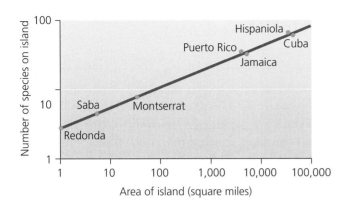

FIGURE 11.17 The larger the island, the greater the number of species—a prediction borne out by data from around the world. By plotting the number of amphibians and reptile species on Caribbean islands as a function of the areas of these islands, the species-area curve shows that species richness increases with area. The increase is not linear, but logarithmic; note the scales of the axes. Go to **GRAPHit!** at www.aw-bc.com/withgott or on the student CD-ROM. Data from MacArthur, R. H., and E. O. Wilson. 1967. *The theory of island biogeography.* Princeton University Press.

Should endangered species be the focus of conservation efforts?

The primary legislation for protecting biodiversity in the United States is the **Endangered Species Act (ESA).** Passed in 1973, the ESA forbids the government and private citizens from taking actions (such as developing land) that destroy endangered species or their habitats. The ESA also forbids trade in products made from endangered species. The aim is to prevent extinctions, stabilize declining populations, and enable populations to recover. As of 2005, there were 1,264 species in the United States listed as "endangered" or as "threatened," the latter status considered one notch less severe than endangered.

The ESA has had a number of notable successes. Following the ban on the pesticide DDT and years of intensive effort by wildlife managers, the peregrine falcon, brown pelican, bald eagle, and other birds have recovered and have been removed from the endangered list (▶p. 416). Intensive management programs with other species, such as the red-cockaded woodpecker, have held formerly declining populations steady in the face of continued pressure on habitat. In fact, roughly 40% of declining populations have been held stable.

This success comes despite the fact that the U.S. Fish and Wildlife Service and the National Marine Fisheries Service, the agencies responsible for upholding the ESA, have been perennially underfunded for the job. These agencies have faced repeated budgetary shortfalls for en-

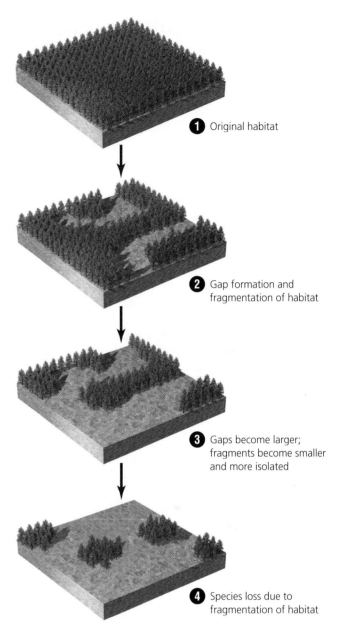

FIGURE 11.18 Forest clearing, farming, road building, and other types of human land use and development can fragment natural habitats. Habitat fragmentation usually begins when gaps are created within a natural habitat. As development proceeds, these gaps expand, join together, and eventually dominate the landscape, stranding islands of habitat in their midst. As habitat becomes fragmented, fewer populations can persist, and numbers of species in the fragments decrease with time.

dangered species protection. Moreover, efforts to reauthorize the ESA have faced stiff opposition in the U.S. Congress since the 1990s. As this book went to press, Republican Congressman Richard Pombo was leading efforts to weaken the ESA, including stripping it of its ability to safeguard habitat.

The Science behind the Story

Testing and Applying Island Biogeography Theory

The researchers who first experimentally tested the equilibrium theory of island biogeography and applied it to conservation biology leaned on their own resourcefulness, along with some plastic, some pesticides, and a small station wagon.

Robert MacArthur and Edward O. Wilson had originally developed island biogeography theory by using observational data from oceanic islands, correlating numbers of species found on islands with island size and distance between landmasses. Yet as of 1966, no one had tested its precepts in the field with a manipulative experiment. Wilson decided to remedy that.

Wilson began looking for islands in the United States where he could run an experimental test: to remove all animal life from islands and then observe and measure recolonization. To be suitable, the islands would need to be small,

contain few forms of life, and be situated close to the mainland to ensure an influx of immigrating species. Wilson found his research sites off the tip of Florida: six small mangrove islands 11–18 m (36–59 ft) in diameter, home only to trees and a few dozen species of insects, spiders, centipedes, and other arthropods.

Daniel Simberloff, Wilson's graduate student at the time, painstakingly counted each island's arthropods, breaking up bark and poking under branches to find every mite, midge, and millipede. Then with the help of professional exterminators, Wilson and Simberloff wrapped each island in a plastic tarpaulin and gassed the interior with the pesticide methyl bromide. After removing the tarpaulins, Simberloff checked to make sure no creatures were left alive. The researchers then waited to see how life on the islands would return.

Over the next 2 years, Simberloff scrambled up trees and turned over leaves, looking for newly arrived organisms. His monitoring showed that life recovered on most islands within a year, regaining about the same number of species and total number of arthropods the islands had sheltered originally. Larger islands once again became home to a greater number of species than smaller islands. Outlying islands recovered more slowly and reached lower species diversities than did islands near the mainland. These results provided the first evidence from a manipulative experiment for the predictions of island biogeography theory.

Published in the journal *Ecology* in 1969 and 1970, Simberloff and Wilson's research gave new empirical rigor to a set of ideas that was increasingly helping scientists understand geographic patterns of biodiversity. Their research also fueled a question of

Daniel Simberloff and E. O. Wilson used mangrove islands in the Florida Keys to test island biogeography theory.

pressing concern for conservation biology: Could island biogeography theory also be applied to isolated "islands" of habitat on continents? At the University of Michigan, biology graduate student William Newmark set out to address this question.

Newmark had learned that many North American national parks kept records that documented sightings of wildlife over the course of the parks' existence. He surmised that by examining these historical records, he could infer which species had vanished from parks, which were new arrivals, and roughly when these changes occurred. The parks, increasingly surrounded by development, were islands of natural habitat isolated by farms, roads, towns, and cities, so Newmark hypothesized that island biogeography theory would apply.

In 1983, Newmark drove his Toyota station wagon to 24 parks in the western United States and Canada. At each park, he studied the wildlife sighting records, focusing on larger mammals such as bear, lynx, and river otter (but not species, such as wolves, that had been deliberately eradicated by hunting). Newmark found a few species missing from many parks, and they added up to a troubling total. Forty-two species had disappeared in all, and not as a result of direct human action. The red fox and river otter had vanished from Sequoia and Kings Canyon National Parks, for example, and the white-tailed jackrabbit and spotted skunk

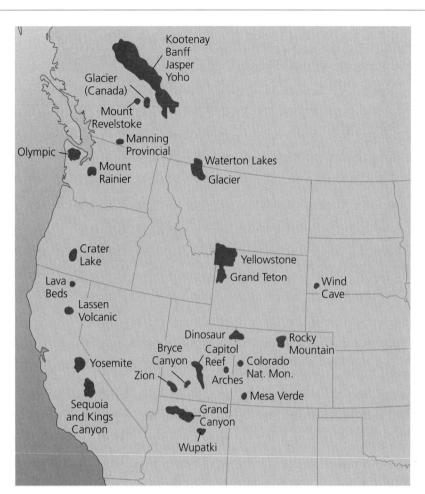

William Newmark visualized national parks of western North America as islands of natural habitat in a sea of development. His data showed that mammal species were disappearing from these recently created terrestrial "islands," in accordance with island biogeography theory. Adapted from Quammen, D. 1997. *The song of the dodo.* New York: Simon & Schuster.

no longer lived in Bryce Canyon National Park. As theory predicted, the smallest parks showed the greatest number of losses, and the largest parks retained a greater number of species.

These species disappeared because the parks, Newmark concluded, were too small to sustain their populations in the long term.

Moreover, the parks had become island habitats too isolated to be recolonized by new arrivals. Newmark's findings, published in the journal *Nature* in 1987, placed island biogeography theory squarely on the mainland. Today, this theory helps inform national park policy and biodiversity conservation plans around the world.

FIGURE 11.19 In efforts to save the California condor *(Gymnogyps californianus)* from extinction, biologists have raised hundreds of chicks in captivity with the help of hand puppets designed to look and feel like the heads of adult condors. Using these puppets, biologists feed the growing chicks in an enclosure and shield them from all contact with humans, so that when the chick is grown it does not feel an attachment to people.

Polls repeatedly show that most Americans support the idea of protecting endangered species. Yet some oppose and resent provisions of the ESA. Many opponents feel that the ESA places more value on the life of an endangered organism than it does on the livelihood of a person. This was a common perception in the Pacific Northwest in the 1990s, when protection for the northern spotted owl slowed logging in old-growth rainforest and caused many loggers to fear for their jobs. Resentment toward the ESA also comes from landowners worried that federal officials will restrict the use of private land if threatened or endangered species are found on it. This has led in many cases to a practice described as "Shoot, shovel, and shut up," among landowners who want to conceal the presence of such species on their land.

ESA supporters maintain that such fears are overblown, pointing out that the ESA has stopped few development projects. Moreover, a number of provisions of the ESA and its amendments promote cooperation with landowners. *Habitat conservation plans* and *safe harbor agreements* allow landowners to harm species in some ways if they voluntarily improve habitat for the species in others.

Debate over the U.S. law has influenced other nations' approaches to species protection. When Canada enacted its long-awaited endangered species law in 2002, the *Species at Risk Act (SARA),* the Canadian government was careful to stress cooperation with landowners and provincial governments, rather than presenting the law as a decree from the national government. Canada's environment

minister, David Anderson, wanted none of the hostility the U.S. act had unleashed. Environmentalists and many scientists, however, protested that SARA was too weak and failed to protect habitat adequately. Today a number of nations have laws protecting species. In Russia, the government issued Decree 795 in 1995, creating a Siberian tiger conservation program and declaring the tiger one of the nation's most important natural and national treasures. Time will tell how effective this decree will be.

Captive breeding, reintroduction, and cloning are single-species approaches

In the effort to save threatened and endangered species, biologists are going to impressive lengths. Zoos and botanical gardens have become centers for the **captive breeding** of these species, so that individuals can be raised and reintroduced into the wild. One example is the program to save the California condor, North America's largest bird (Figure 11.19). Condors were persecuted in the early 20th century, collided with electrical wires, and succumbed to lead poisoning from scavenging carcasses of animals killed with lead shot. By 1982, only 22 condors remained, and biologists decided to take all the birds into captivity, in hopes of boosting their numbers and then releasing them. The ongoing program is succeeding. So far, over 100 of the 250 birds raised in captivity have been released into the wild at sites in California and Arizona, where a few pairs have begun nesting.

Other reintroduction programs have been more controversial. Reintroducing wolves to Yellowstone National Park has proven popular with the public, but reintroducing them to sites in Arizona and New Mexico has met stiff resistance from ranchers who fear the wolves will attack their livestock. The program is making slow headway; several of the wolves have been shot.

The newest idea for saving species from extinction is to create more individuals by cloning them. In this technique, DNA from an endangered species is inserted into a cultured egg without a nucleus, and the egg is implanted into a closely related species that can act as a surrogate mother. So far two Eurasian mammals have been cloned in this way. With future genetic technology, some scientists even talk of recreating extinct species from DNA recovered from preserved body parts. However, even if cloning can succeed from a technical standpoint, most biologists agree that such efforts are not an adequate response to biodiversity loss. Without ample habitat and protection in the wild, having cloned animals in a zoo does little good.

Some species act as "umbrellas" for protecting habitat and communities

Protecting habitat and conserving communities, ecosystems, and landscapes are the goals of many conservation biologists. Often, particular species are essentially used as tools to conserve communities and ecosystems. This is because the ESA provides legal justification and resources for species conservation, but no such law exists for communities or ecosystems. Large species that roam great distances, such as the Siberian tiger, require large areas of habitat. Meeting the habitat needs of these so-called *umbrella species* automatically helps meet those of thousands of less charismatic animals, plants, and fungi that would never elicit as much public interest.

Environmental advocacy organizations have found that using large and charismatic vertebrates as spearheads for biodiversity conservation has been an effective strategy. This approach of promoting particular *flagship species* is evident in the longtime symbol of the World Wide Fund for Nature (World Wildlife Fund in North America), the panda. The panda is a large endangered animal requiring sizeable stands of undisturbed bamboo forest. Its lovable appearance has made it a favorite with the public—and an effective tool for soliciting funding for conservation efforts that protect far more than just the panda. At the same time, many conservation organizations today are moving beyond the single-species approach. The Nature Conservancy, for instance, has in recent years focused more on whole communities and landscapes. The most ambitious effort may be the Wildlands Project, a group proposing to restore huge amounts of North America's land to its presettlement state.

Weighing the **Issues:**
Single-Species Conservation?

What would you say are some advantages of focusing on conserving single species, versus trying to conserve broader communities, ecosystems, or landscapes? What might be some of the disadvantages? Which do you think is the better approach, or should we use both?

International conservation efforts include widely signed treaties

At the international level, biodiversity protection has been pursued in a variety of ways. Most effective so far have been several treaties facilitated by the United Nations. The 1973 **Convention on International Trade in Endangered Species of Wild Fauna and Flora (CITES)** protects endangered species by banning the international transport of their body parts. When nations enforce it, CITES can protect the tiger and other rare species whose body parts are traded internationally.

In 1992, leaders of many nations agreed to the **Convention on Biological Diversity.** This treaty embodies three goals: to conserve biodiversity, to use biodiversity in a sustainable manner, and to ensure the fair distribution of biodiversity's benefits. The Convention addresses a number of topics, including the following:

▶ Providing incentives for biodiversity conservation
▶ Managing access to and use of genetic resources
▶ Transferring technology, including biotechnology
▶ Promoting scientific cooperation
▶ Assessing the effects of human actions on biodiversity
▶ Promoting biodiversity education and awareness
▶ Providing funding for critical activities
▶ Encouraging every nation to report regularly on their biodiversity conservation efforts

Since the treaty was proposed, UNEP has identified a number of accomplishments, such as ensuring that Ugandan people share in the economic benefits of wildlife preserves, increasing global markets for "shade-grown" coffee and other crops grown without removing forests, and replacing pesticide-intensive farming practices with sustainable ones in some rice-producing Asian nations. As of 2005, 188 nations had become parties to the

FIGURE 11.20 The golden lion tamarin *(Leontopithecus rosalia)*, a species endemic to Brazil's Atlantic rainforest, is one of the world's most endangered primates. Captive breeding programs have produced roughly 500 individuals in zoos, but the tamarin's habitat is fast disappearing.

Convention on Biological Diversity. Those choosing *not* to do so include Iraq, Somalia, the Vatican, and the United States. This decision is just one example of why the U.S. government is no longer widely regarded as a leader in biodiversity conservation efforts.

Biodiversity hotspots pinpoint areas of high diversity

One international approach oriented around geographic regions, rather than single species, has been the effort to map **biodiversity hotspots.** The concept of biodiversity hotspots was introduced in 1988 by British ecologist Norman Myers (▶p. 338) as a way to prioritize regions that are most important globally for biodiversity conservation. A hotspot is an area that supports an especially great number of species that are **endemic** to the area, that is, found nowhere else in the world (Figure 11.20). To qualify as a hotspot, a location must harbor at least 1,500 endemic plant species, or 0.5% of the world total. In addition, a

hotspot must have already lost 70% of its habitat as a result of human impact and be in danger of losing more.

The nonprofit group Conservation International maintains a list of 34 biodiversity hotspots (Figure 11.21). The ecosystems of these areas together once covered 15.7% of the planet's land surface, but today, because of habitat loss, cover only 2.3%. This small amount of land is the exclusive home for 50% of the world's plant species and 42% of all terrestrial vertebrate species. The hotspot concept gives incentive to focus on these areas of endemism, where the greatest number of unique species can be protected with the least amount of effort.

Community-based conservation is increasingly popular

Taking a global perspective and prioritizing optimal locations to set aside as parks and reserves makes good sense. However, setting aside land for preservation affects the people that live in and near these areas. In past decades, many conservationists from developed nations, in their zeal to preserve ecosystems in other nations, too often neglected the needs of people in the areas they wanted to protect. Many developing nations came to view this international environmentalism as a kind of neocolonialism.

Today this has largely changed, and many conservation biologists actively engage local people in efforts to protect land and wildlife in their own backyards, in an approach sometimes called **community-based conservation.** Setting aside land for preservation deprives local people of access to natural resources, but it can also guarantee that these resources will not be used up or sold to foreign corporations and can instead be sustainably managed. Moreover, parks and reserves draw ecotourism (▶p. 143), which can support local economies.

In the small Central American country of Belize, conservation biologist Robert Horwich and his Wisconsin-based group Community Conservation, Inc., have helped start a number of community-based conservation projects. The Community Baboon Sanctuary consists of tracts of riparian forest that farmers have agreed to leave intact, to serve as homes and traveling corridors for the black howler monkey, a centerpiece of ecotourism. The fact that the reserve uses the local nickname for the monkey signals respect for residents, and today a local women's cooperative is running the project. A museum was built, and residents receive income for guiding and housing visiting researchers and tourists. Some other projects have not turned out as well, however. Nearby on the Belizean coast, efforts to create a locally run reserve for the manatee have

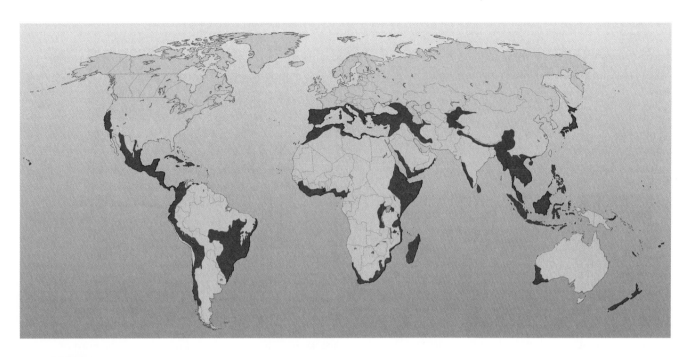

FIGURE 11.21 Some areas of the world possess exceptionally high numbers of species found nowhere else. Some conservation biologists have suggested prioritizing habitat preservation in these areas, dubbed *biodiversity hotspots*. Shown in red are the 34 biodiversity hotspots mapped by Conservation International in 2005. Data from Conservation International, 2005.

struggled because of shortfalls in funding. Community-based conservation has not always been successful, but in a world of increasing human population, locally based management that meets people's needs sustainably will likely be essential.

Innovative economic strategies are being employed

As conservation moves from single-species approaches to the hotspot approach to community-based conservation, innovative economic strategies are also being attempted. One strategy is the *debt-for-nature swap.* In such a swap, a conservation organization raises money and offers to pay off a portion of a developing country's international debt in exchange for a promise by the country to set aside reserves, fund environmental education, and better manage protected areas.

A newer strategy that Conservation International has pioneered is the *conservation concession.* Nations often sell concessions to foreign multinational corporations, allowing them to extract resources from the nation's land. A nation can, for instance, earn money by selling to an international logging company the right to log its forests. Conservation International has stepped in and paid nations for concessions for conservation rather than resource extraction. The nation gets the money *and* keeps its natural resources intact. The South American country of Surinam, which still has extensive areas of pristine rainforest, entered into such an agreement and has virtually halted logging while pulling in $15 million. It remains to be seen how large a role such strategies will play in the future protection of biodiversity.

Conclusion

The erosion of biological diversity on our planet threatens to result in a mass extinction event equivalent to the mass extinctions of the geological past. Human-induced habitat alteration, invasive species, pollution, and overharvesting of biotic resources are the primary causes of biodiversity loss. This loss matters, because human society could not function without biodiversity's pragmatic benefits. As a result, conservation biologists are rising to the challenge of conducting science aimed at saving endangered species, preserving their habitats, restoring populations, and keeping natural ecosystems intact. The innovative strategies of these scientists hold promise to slow the erosion of biodiversity that threatens life on Earth.

Biodiversity

Biodiversity is being lost worldwide at an accelerating pace. **How should we respond to biodiversity loss? What solutions should we seek, and what strategies should we prioritize?**

Mainstreaming Conservation through Ecosystem Services

Parks and nature reserves are the jewels of conservation, but they are not, and never will be, enough. Even the most ambitious conservation plans aspire to having no more than 20–30% of the world's lands protected as nature reserves, and the reality is that closer to 5–10% of land is currently under protection. If we put a fence around our parks to keep humans out, our nature preserves will still end up deteriorating because of how we treat the remaining 70–90% of Earth. Nature cannot be sequestered in some small portion of the globe while the bulk of land and water are left at the mercy of humans. Conservation will never be attainable unless humans learn to live in, work in, exploit, and harvest nature in a manner that does not ravage biodiversity.

An additional 3 billion people are expected in the next 50 years, bringing the total population to 9 billion. Many of the most impoverished people in the world depend on nature for their livelihood. All of us depend on nature for flood control, protection against storm damage, and other regulating services. To protect biodiversity and meet human needs, we must align biodiversity protection with these ecosystem services. As we make clear the links between ecosystem services and biodiversity protection, institutions and business should increasingly be willing to pay for nature's protection.

The disastrous tsunami of 2004 did minimal damage in coastal areas sheltered by natural mangrove forests. If only more of the Asia-Pacific region had been willing to invest in the insurance provided by mangroves, we would have saved human lives and biodiversity. The solution to biodiversity loss lies in clearly recognizing and valuing ecosystem services and in setting priorities that focus simultaneously on biodiversity and ecosystem services.

Peter Kareiva is lead scientist for The Nature Conservancy and an adjunct professor at the University of California at Santa Barbara and at Santa Clara University. Dr. Kareiva has also served as director of the Division for Conservation Biology at NOAA Fisheries, at the Northwest Fisheries Science Center, Seattle. His research has concerned organisms as diverse as whales, owls, Antarctic seabirds, ladybug beetles, butterflies, wildebeest, and genetically engineered microbes and crops.

Parks: The Best Way to Protect Biodiversity?

Although protected areas cover 12% of Earth's land surface, we need many more of them. Consider 34 "biodiversity hotspots" containing the last habitats of 50% of Earth's vascular plant species and 42% of vertebrate species (excluding fish). Once covering 16% of Earth's land surface, their habitats have since lost 86% of their expanse and now cover just 2.3%. If we could safeguard these relatively small areas, we could reduce the number of extinct species by at least one-third. Protection of the hotspots (parks, reserves, and so on) would cost roughly $1.5 billion per year for 5 years—just one-seventh of all conservation funding worldwide. What a massive need and massive opportunity for protected areas!

In the tropics, however, where most biodiversity is found and where it is most threatened, no island is an island, so to speak. The Kruger Park in South Africa is drying out because rivers arising outside the park are losing their waters to ranches and other development works. The park is also being overtaken by acid rain from South Africa's main industrial region upwind.

Moreover, global warming will shift temperature bands, and consequently vegetation communities, away from the equator and toward the poles. Many plants and animals in Hawaii will have little place to go but into the sea, as will those in the southern tip of Africa, northern Philippines, and dozens of other unfortunate locales. Even if they were to be turned into giant parks and perfectly protected on the ground, they would still be vulnerable to global warming—half of which is caused by carbon dioxide emissions from fossil fuels.

Biodiversity enthusiasts, here's a key question for you: Which country has less than one-twentieth of the world's population but causes one-quarter of the world's carbon dioxide emissions?

Norman Myers is a Fellow of Oxford University and a member of the U.S. National Academy of Sciences. He works as an independent scientist, advising the United Nations, the World Bank, the World Conservation Union, the World Wildlife Fund, and dozens of other conservation bodies around the world.

Explore this issue further by accessing **Viewpoints** at www.aw-bc.com/withgott.

REVIEWING OBJECTIVES

You should now be able to:

Characterize the scope of biodiversity on Earth

▶ Biodiversity can be thought of at three levels, commonly called species diversity, genetic diversity, and ecosystem diversity. (pp. 310–312)

▶ Roughly 1.7–2.0 million species have been described so far, but scientists agree that the world holds millions more. (pp. 313–314)

▶ Some taxonomic groups (such as insects) hold far more diversity than others. (pp. 313–314)

▶ Diversity is unevenly spread across different habitats and areas of the world. (pp. 314–316)

Describe ways to measure biodiversity

▶ Global estimates of biodiversity are based on extrapolations from scientific assessments in local areas and certain taxonomic groups. (pp. 313–314)

Contrast background extinction rates and periods of mass extinction

▶ Species have gone extinct at a background rate of roughly one species per 1 to 10 million species each year. Most species that have ever lived are now extinct. (p. 316)

▶ Earth's life has experienced five mass extinction events in the past 440 million years. (pp. 316–317)

▶ Human impact is presently causing the beginnings of a sixth mass extinction. (pp. 317–319)

Evaluate the primary causes of biodiversity loss

▶ Habitat alteration is the main cause of current biodiversity loss. Invasive species, pollution, and overharvesting are also important causes. Climate change threatens to become a major cause very soon. (pp. 319–324)

Specify the benefits of biodiversity

▶ Biodiversity is vital for functioning ecosystems and the services they provide us. (p. 325)

▶ Wild species are sources of food, medicine, and economic development. (pp. 325–327)

▶ Many people feel humans have a psychological need to connect with the natural world. (pp. 327–328)

Assess conservation biology and its practice

▶ Conservation biology is an applied science that studies biodiversity loss and seeks ways to protect and restore biodiversity at all its levels. (p. 329)

Explain island biogeography theory and its application to conservation biology

▶ Island biogeography theory explains how size and distance influence the number of species occurring on islands. (pp. 329–330)

▶ The theory applies to terrestrial islands of habitat in fragmented landscapes. (pp. 330–331)

Compare and contrast traditional and more innovative biodiversity conservation efforts

▶ Most conservation efforts and laws so far have focused on threatened and endangered species. Efforts include captive breeding and reintroduction programs. (pp. 331, 334–335)

▶ Species that are charismatic and well known are often used as tools to conserve habitats and ecosystems. Increasingly, landscape-level conservation is being pursued in its own right. (p. 335)

▶ International conservation approaches include treaties, biodiversity hotspots, community-based conservation, debt-for-nature swaps, and conservation concessions. (pp. 335–337)

TESTING YOUR COMPREHENSION

1. What is biodiversity? List and describe three levels of biodiversity.
2. What are the five primary causes of biodiversity loss? Can you give a specific example of each?
3. List and describe five invasive species and the adverse effects they have had.
4. Define the term *ecosystem services*. Give five examples of ecosystem services that humans would have a hard time replacing if their natural sources were eliminated.
5. What is the relationship between biodiversity and food security? Between biodiversity and pharmaceuticals? Give three examples of potential benefits of biodiversity conservation for food security and medicine.
6. Describe four reasons why people suggest biodiversity conservation is important.
7. What is the difference between an umbrella species and a keystone species? Could one species be both an umbrella species and a keystone species?
8. Explain the theory of island biogeography. Use the example of the Siberian tiger to describe how this theory can be applied to fragmented terrestrial landscapes.

9. Name two successful accomplishments of the U.S. Endangered Species Act. Now name two reasons some people have criticized it.

10. What is a biodiversity hotspot? Describe community-based conservation.

SEEKING SOLUTIONS

1. In one of the quotes that open this chapter, biologist E. O. Wilson argues that "except in pockets of ignorance and malice, there is no longer an ideological war between conservationists and developers. Both share the perception that health and prosperity decline in a deteriorating environment." Do you agree or disagree? How do people in your community view biodiversity?

2. Many arguments have been advanced for the importance of preserving biodiversity. Which argument do you think is most compelling, and why? Which argument do you think is least compelling, and why?

3. Some people argue that we shouldn't worry about endangered species because extinction has always occurred. How would you respond to this view?

4. Imagine that you are an influential legislator in a country that has no endangered species act and that you want to introduce legislation to protect your country's vanishing biodiversity. Consider the U.S. Endangered Species Act and the Canadian Species At Risk Act, as well as international efforts such as CITES and the Convention on Biological Diversity. What strategies would you write into your legislation? How would your law be similar to and different from the U.S., Canadian, and international efforts?

5. Environmental advocates from developed nations who want to preserve biodiversity globally have long argued for setting aside land in biodiversity-rich regions of developing nations. Many leaders of developing nations have responded by accusing the advocates of neocolonialism. "Your nations attained their prosperity and power by overexploiting their environments decades or centuries ago," these leaders asked, "so why should we now sacrifice our own development by setting aside our land and resources?" What would you say to these leaders of developing countries? What would you say to the environmental advocates? Do you see ways that both preservation and development goals might be reached?

6. Compare the biodiversity hotspot approach to the approach of community-based conservation. What are the advantages and disadvantages of each? Can we—and should we—follow both approaches?

INTERPRETING GRAPHS AND DATA

Habitat alteration is the primary cause of present-day biodiversity loss. Of all human activities, the one that has resulted in the most habitat alteration is agriculture. Between 1850 and 2000, 95% of the native grasslands of the Midwestern United States were converted to agricultural use. As a result, conventional farming practices replaced diverse natural communities with greatly simplified ones. The vast monocultures of industrialized agriculture produce bountiful harvests, but at substantial costs in lost ecosystem services.

Data from a recent study reviewing the scientific literature on the effects of organic farming versus conventional farming practices on biodiversity are shown in the graph.

1. How many studies showed a positive effect of organic farming on biodiversity, relative to conventional farming? How many studies reported a negative effect? How many studies reported no effect?

2. For which group or groups of organisms is evidence of positive effects the strongest? Reference the numbers to support your choice(s).

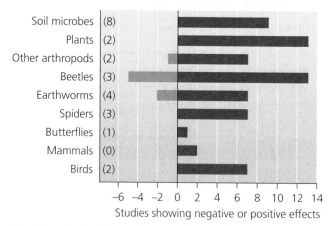

Numbers of scientific studies reporting negative or positive effects on biodiversity of organic agriculture versus conventional farming practices. In parentheses are numbers of studies reporting no effect. Data from Hole, D., et al. 2005. Does organic farming benefit biodiversity? *Biological Conservation* 122: 113–130.

3. Recall the ecosystem services provided by biodiversity (▶ p. 325). What services do the groups you chose in Question 2 provide?

CALCULATING ECOLOGICAL FOOTPRINTS

Of the five major causes of biodiversity loss discussed in this chapter, habitat alteration arguably has the greatest impact. In their 1996 book introducing the ecological footprint concept, authors Mathis Wackernagel and William Rees present a consumption/land-use matrix for an average North American. Each cell in the matrix lists the number of hectares of land of that type required to provide for the different categories of a person's consumption (food, housing, transportation, consumer goods, and services). Of the 4.27 hectares required to support this average person, 0.59 hectares are forest, with most (0.40 hectares) being used to meet the housing demand. Using this information, calculate the missing values in the table.

	Hectares of forest used for housing	Total forest hectares used
You	0.40	0.59
Your class		
Your state		
United States		

Data from Wackernagel, M., and W. Rees. 1996. *Our ecological footprint: reducing human impact on the earth.* British Columbia, Canada: New Society Publishers.

1. Approximately two-thirds of the forests' productivity is consumed for housing. To what use(s) would you speculate that most of the other third is put?
2. If the harvesting of forest products exceeds the sustainable harvest rate, what will be the likely consequence for the forest?
3. What will be the impact of deforestation, or of the loss of old-growth forests and their replacement with plantations of young trees, on the species diversity of the forest community? In your answer, discuss the possibilities of both extirpation and extinction.

Take It Further

Go to www.aw-bc.com/withgott or the student CD-ROM where you'll find:

▶ Suggested answers to end-of-chapter questions
▶ Quizzes, animations, and flashcards to help you study
▶ *Research Navigator*™ database of credible and reliable sources to assist you with your research projects

▶ **GRAPH**it! tutorials to help you master how to interpret graphs
▶ **INVESTIGATE**it! current news articles that link the topics that you study to case studies from your region to around the world

12 Resource Management, Forestry, Land Use, and Protected Areas

Clearcuts on timber industry land adjacent to Gifford Pinchot National Forest, Washington

Upon successfully completing this chapter, you will be able to:

▶ Identify the principles, goals, and approaches of resource management

▶ Summarize the ecological roles and economic contributions of forests, and outline the history and scale of forest loss

▶ Explain the fundamentals of forest management, and describe the major methods of harvesting timber

▶ Analyze the scale and impacts of agricultural land use

▶ Identify major federal land management agencies and the lands they manage

▶ Recognize types of parks and reserves, and evaluate issues involved in their design

Anti-logging
protest at
Clayoquot Sound

Central Case: Battling Over the Last Big Trees at Clayoquot Sound

"What we have is nothing less than an ecological Holocaust occurring right now in British Columbia."
—MARK WAREING, WESTERN CANADA WILDERNESS COMMITTEE, 1990

"Clear-cutting . . . may be either desirable or undesirable, acceptable or unacceptable, according to the type of forest and the management objectives for the forest."
—DR. HAMISH KIMMINS, UNIVERSITY OF BRITISH COLUMBIA, 1992

It was the largest act of civil disobedience in Canadian history, and it played out along a seacoast of stunning majestic beauty, at the foot of some of the world's biggest trees. Protestors blocked logging trucks, preventing them from entering stands of ancient temperate rainforest, the forest that had once carpeted most of Vancouver Island, British Columbia.

The activists were opposing **clear-cutting**, the logging practice that removes all trees from an area. Most of Canada's old-growth temperate rainforest had already been cut, and the forests of Clayoquot Sound on the western coast of Vancouver Island were among the largest undisturbed stands of temperate rainforest left on the planet.

The protestors chanted slogans, sang songs, and chained themselves to trees. The loggers complained that the protestors were keeping them from doing their jobs and making a living. In the end, 850 of the 12,000 protestors were arrested in full view of the global media, and this remote, mist-enshrouded land of cedars and hemlocks became ground zero in the global debate over how we manage forests.

That was in 1993. Timber from old-growth forests had powered British Columbia's economy since the province's early days. Historically, one in five jobs in British Columbia depended on its $13 billion timber industry, and many small towns would have gone under without it. By 1993, however, the timber industry was cutting thousands of jobs a year because of mechanization, and the looming depletion of old growth threatened to slow the industry.

Meanwhile, half a world away in Great Britain, the environmental group Greenpeace was busy trying to convince British companies and customers to boycott forest products made from British Columbia trees that had been clear-cut by multinational timber company MacMillan Bloedel. Soon two British corporations cancelled contracts worth several million dollars with the beleaguered company, and British Columbia's premier found himself touring European nations trying to persuade them not to boycott his province's main export.

Then in 1995, the provincial government called for an end to clear-cutting at Clayoquot Sound, after its appointed scientific panel of experts submitted a new forestry plan for the region. The plan recommended reducing harvests, retaining 15–70% of old-growth trees in each stand, decreasing the logging road network, designating forest reserve areas, and managing riparian zones. Two years later the provincial government reversed many of these regulations on logging and a new premier pronounced forest activists "enemies of British Columbia," but the progressive management objectives for Clayoquot Sound have largely survived to the present day.

The antagonists called a truce and struck a deal; wilderness advocates and MacMillan Bloedel agreed to logging of old growth in limited areas, using more environmentally friendly practices. In 1998, Native people of the region formed a timber company, Iisaak, in agreement with MacMillan Bloedel's successor, Weyerhaeuser, and began logging in a more environmentally sensitive manner.

Meanwhile, leaving most of the trees standing had accomplished just what forest advocates had predicted: People from all over the world—1 million each year— were now visiting Clayoquot Sound for its natural beauty, and kayaking and whale-watching in its waters. Ecotourism (along with fishing and aquaculture) surpassed logging as the driver of local economies. The United Nations designated the site as an international biosphere reserve, encouraging land protection and sustainable development. The trees appeared to be worth more left standing than cut down.

Tensions continue today, however. Another timber company, Interfor, is harvesting areas near park and biosphere reserve boundaries. Local forest advocates worry that the provincial government's new Working Forest Policy will increase logging in the region. The town of Tofino has petitioned the province to exempt Clayoquot Sound's forests from the new policy so that pristine valleys can be preserved and the town's ecotourism economy can be maintained. As long as our demand for lumber, paper, and forest products keeps increasing, pressures will keep building on the remaining forests on Vancouver Island and around the world.

Resource Management

Debates over forest resources epitomize the broader questions of how to manage natural resources in general. We need to manage the resources we take from the natural world because many of them are limited. In Chapter 1, we saw how some resources, such as fossil fuels, are nonrenewable on human time scales, whereas other resources, such as the sun's energy, are perpetually renewable. Between these extremes lie resources such as timber, which are renewable if they are not exploited too rapidly or carelessly. **Resource management** is the practice of harvesting potentially renewable resources in ways that do not deplete them. Resource managers are guided in their decision making by available research in the natural sciences, as well as by political, economic, and social factors.

A key question in managing resources is whether to focus narrowly on the resource of interest or to look more broadly at the environmental system of which the resource is a part. Taking a broader view can often help avoid damaging the system and can thereby help sustain the availability of the resource in the long term.

Several natural resources are vital to us

Besides timber, several other types of natural resources are vital to our civilization. These include soils, freshwater, wildlife and fisheries, rangeland, and minerals (Figure 12.1). All natural resources also serve functions in the ecosystems of which they are a part.

Soils Soil resources, particularly topsoil, are of direct importance to us because they support the plants we grow for food and fiber and thus play a central role in agriculture. As we saw in Chapter 9 (▸ pp. 261–264), certain farm practices, such as terracing and use of windbreaks, can help guard against loss of topsoil to erosion, and other farm practices, such as planting nitrogen-fixing crops, can help maintain soil quality. Healthy soils also support forests and other natural communities, serving as a site for decomposition and a reservoir for nutrients.

Freshwater Each of us depends directly on freshwater, so ensuring a dependable supply of drinking water is a life-or-death issue. Freshwater also is necessary for agriculture; indeed, we use most freshwater not for drinking but for irrigating crops. In addition, waterways and wetlands are crucial for wildlife and properly functioning ecosystems, so those who manage water resources try to maintain supplies for all these reasons. Water managers also try to protect the

FIGURE 12.1 Our civilization depends on a number of natural resources, including timber, soils, freshwater, wildlife and fisheries, rangeland, and minerals. Minerals are a nonrenewable resource that is mined, but the others can be kept renewable through responsible resource management.

quality of these supplies by guarding against pollution. We will examine freshwater resources in depth in Chapter 15.

Wildlife and fisheries Humans have long hunted animals for food. Managers regulate the hunting of game animals, such as deer and quail, for food and sport in an effort to maintain populations of these animals at desired levels. As populations of many other terrestrial species decline from habitat loss and other causes (▶ pp. 319–324), management for nongame species is becoming increasingly important. Marine animals are also harvested, and despite extensive management of fisheries, many stocks of fish and shellfish have declined throughout the world's oceans from overharvesting, as we will see in Chapter 16. Besides their use as food, living organisms provide humans with many other benefits, from materials to medicines to aesthetic appreciation (▶ pp. 325–328). Organisms are also, of course, vital components of the ecosystems that sustain our world.

Rangeland Most of our food from animals today comes not from wild animals but from domesticated ones that are farmed. Most cattle in North America are raised in crowded feedlots, but they have traditionally been raised by grazing on open grassland. As we saw in Chapter 9 (▶ pp. 267–268, 270), grazing can be sustainable, but overgrazing results in damage to soils, waterways, and vegetative communities. Range managers are responsible for regulating ranching on public lands, and they advise ranchers on sustainable grazing practices on private lands.

Minerals Our civilization depends on numerous minerals, and until we achieve a closed-loop economy (▶ pp. 664–665), we will rely on the mining of minerals. Iron is mined and processed to make steel. Copper is used in pipes, electrical wires, and a variety of other applications. Aluminum is extracted via bauxite ore and used in packaging and other end products. Lead is used in batteries, to shield medical patients from radiation, and in many other ways. Zinc, tungsten, phosphate, uranium, gold, silver—the list goes on and on.

Although we rely on these mineral resources, we do not manage their extraction as we do with the aforementioned resources. Like fossil fuels, minerals are nonrenewable resources that are mined rather than harvested. Therefore, the mining industry has no built-in incentive to conserve. Instead, it benefits by extracting as much as it can as fast as it can and then, once extraction has become too inefficient to be profitable, moving on to new sites.

The process of mining also has historically degraded surrounding environments, sometimes on massive scales. Mining may directly remove vegetation from large areas, cause erosion, and produce acidic runoff that poisons area waterways. In addition, the smelting of metals following mining can create severe air pollution. We touch on some of the environmental and social costs of coal mining in Chapter 19 (▶ pp. 576–577), but these issues exist with all types of mining. Improved technology and more environmentally sensitive extraction procedures can help minimize environmental impacts. Such advances come as a result of public pressure, government legislation, or economic savings. Industries extracting potentially renewable resources, in contrast, have an added incentive to reduce ecological impacts: doing so may help sustain their ability to harvest resources—and this is where resource management comes in.

Managers have tried to achieve maximum sustainable yield

One guiding principle in resource management has been **maximum sustainable yield.** The aim is to achieve the maximum amount of resource extraction possible, without depleting the resource from one harvest to the next. At first this goal may sound ideal, but in reality it may sometimes harm the ecosystems from which the resource is derived. If those ecosystems cease to function effectively as a result, then in time this may decrease the availability of the resource.

For instance, recall the logistic growth curve (Figure 5.16, ▶ p. 137), which shows that a population grows most quickly when it is at an intermediate size. A fisheries manager aiming for maximum sustainable yield will therefore prefer to keep fish populations at intermediate levels so that they rebound quickly after each harvest. Doing so should

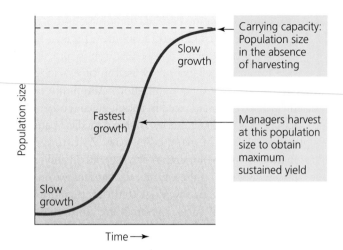

FIGURE 12.2 Using the concept of maximum sustainable yield, resource managers attempt to maximize the amount of resource harvested, as long as the harvest is sustainable in perpetuity. In the case of a wildlife population or fisheries stock that grows according to a logistic growth curve, managers aim to keep the population at an intermediate level, well below the carrying capacity, because populations grow fastest at intermediate sizes.

result in the greatest amount of fish harvested over time, while the population sustains itself (Figure 12.2). However, this management approach keeps the fish population at only about half its carrying capacity (▸ p. 136)—well below the level it would attain in the absence of fishing. Reducing one population in this way will likely have effects on other species and on the food web dynamics of the community. From an ecological point of view, management for maximum sustainable yield may thereby set in motion complex and significant ecological changes.

In forest management, maximum sustainable yield argues for cutting trees shortly after they have gone through their fastest stage of growth, and trees often grow most quickly at intermediate ages. Thus, trees may be cut long before they have grown as large as they would in the absence of harvesting. Although this practice may maximize timber production over time, it can cause drastic changes in the ecology of a forest by eliminating habitat for species that depend on mature trees.

Today many managers pursue ecosystem-based management

Because of these dilemmas, increasing numbers of managers today espouse ecosystem-based management. **Ecosystem-based management** attempts to manage the harvesting of resources in ways that minimize impact on the ecosystems and ecological processes that provide the resource. The plan proposed in 1995 by the Scientific Panel for Sustainable Forest Practices on Clayoquot Sound and approved

by British Columbia's government was essentially a plan for ecosystem-based management. By carefully managing ecologically important areas such as riparian corridors, by considering patterns at the landscape level, and by affording protection to some forested areas, the plan aimed to allow continued timber harvesting at reduced levels while preserving the functional integrity of the ecosystem.

Although ecosystem-based management has gained a great deal of support in recent years, it is challenging for managers to determine how best to implement this type of management. Ecosystems are complex, and our understanding of how they operate is limited. Thus, ecosystem-based management has often come to mean different things to different people.

Adaptive management evolves and improves

Some management actions will succeed, and some will fail. A wise manager will try new approaches if old ones are not effective. In recent years, many resource managers have taken this a step further by implementing adaptive management. **Adaptive management** involves systematically testing different management approaches with the aim of improving methods as time goes on. It calls for changing practices midstream if necessary, as managers learn which work best. This approach represents a true fusion of science and management, because hypotheses about how best to manage resources are explicitly tested.

Adaptive management can be time-consuming and complicated. It entails monitoring the results of one's practices and continually adjusting them as needed, based on what is learned. It has posed a challenge for many managers, because those who adopt new approaches must often overcome inertia and resistance to change from proponents of established practices. Adaptive management is beginning to be used in forestry (see "The Science behind the Story," ▸ pp. 348–349). The management of timber and other forest resources is a clear and representative example of resource management.

Forest Management

Forests cover much of Earth's land surface, provide habitat for countless organisms, and play key roles in our planet's biogeochemical cycles (▸ pp. 195–206). Forests have also long provided humanity with wood for fuel, construction, paper production, and more. Foresters, those professionals who manage forests through the practice of **forestry,** today must balance the central importance

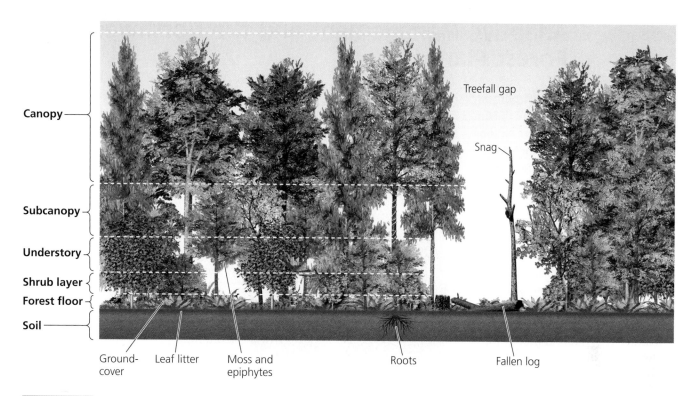

Canopy

Subcanopy

Understory

Shrub layer

Forest floor

Soil

Treefall gap

Snag

Ground-cover Leaf litter Moss and epiphytes Roots Fallen log

FIGURE 12.3 Mature forests are complex ecosystems. In this cross-section of a generalized mature forest, the crowns of the largest trees form the canopy, and smaller trees beneath them form the shaded subcanopy and understory. Shrubs and groundcover grow just above the forest floor, which may be covered in leaf litter rich with invertebrate animals. Vines, mosses, lichens, and epiphytes cover portions of trees and the forest floor. Snags (standing dead trees), whose wood can be easily hollowed out, provide food and nesting and roosting sites for woodpeckers and other animals. Fallen logs nourish the soil and young plants and provide habitat for countless invertebrates as the logs decompose. Treefall gaps caused by fallen trees let light through the canopy and create small openings in the forest, allowing early successional plants to grow in patches within the mature forest.

of forests as ecosystems with civilization's demand for wood products.

Forests are ecologically valuable

Most of the world's forests occur as boreal forest, the biome (▸ pp. 170, 176–177) that stretches across much of Canada, Russia, and Scandinavia, or as tropical rainforest, the biome that occurs in South and Central America, equatorial Africa, and Indonesia and southeast Asia. Temperate forests cover less area globally, in part because so many have already been cleared by people.

Because of their structural complexity and their ability to provide many niches for organisms, forests comprise some of the richest ecosystems for biodiversity (Figure 12.3). Trees furnish food and shelter for an immense diversity of vertebrate and invertebrate animals. Countless insects, birds, mammals, and other organisms subsist on the leaves, fruits, and seeds that trees produce.

Some animals are adapted for living in dense treetop canopies. Here beetles, caterpillars, and other leaf-eating insects abound, providing food for birds such as tanagers and warblers, while arboreal mammals from squirrels to sloths to monkeys consume fruit and leaves. Other animals specialize on the subcanopies of trees, and still others utilize the bark, branches, and trunks. Cavities in trunks provide nest and shelter sites for a wide variety of vertebrates. Dead and dying trees are valuable for many species; these snags are decayed by insects that are eaten by woodpeckers and other animals.

Understory shrubs and groundcover plants give a forest structural complexity and provide habitat for still more organisms. Moreover, the leaves, stems, and roots of forest plants are colonized by an extensive array of fungi and microbes, in both parasitic and mutualistic relationships (▸ pp. 153–155). And much of a forest's diversity resides in the forest floor, where the soil is generally nourished by leaf litter. As we saw in Chapter 11, myriad

Adaptive Management and the Northwest Forest Plan

The Science behind the Story

While environmental advocates and the timber industry were battling one another at Clayoquot Sound, similar debates were occurring in the United States, particularly in the Pacific Northwest. Here loggers were closing in on some of the nation's last stands of old-growth conifers, and preservationists were staging protests to protect the stands. Environmentalists also were winning lawsuits to force the U.S. Forest Service and other agencies to enforce provisions to protect the northern spotted owl (and thus its habitat, old-growth forest) under the Endangered Species Act (▶ pp. 331–333). As a result, logging was being significantly restricted.

In 1993, the Clinton administration waded into the impasse and helped direct the development of a new plan to manage the forests of western Washington, Oregon, and northwestern California so that consensus could be reached and logging could continue with adequate protections for species and ecosystems. The resulting **Northwest Forest Plan,** approved in February 1994, used

Interstate 90 at Snoqualmie Pass, Washington, is a barrier to many animal species traveling through the forested areas on either side of it.

science extensively to guide management.

Research conducted over the past decade since the plan's inception has fallen into several categories: (1) wildlife conservation and population viability, (2) aquatic conservation, (3) socioeconomic research, (4) ecological processes and function, (5) spatial and landscape studies, (6) tree growth studies, (7) adaptive management concepts, and (8) adaptive management areas.

The Northwest Forest Plan represented one of the first large-scale applications of adaptive management. Ten adaptive management areas (AMAs) were established throughout the region. These areas ranged from 37,000 to 200,000 ha (92,000 to 500,000 acres), and were located in regions where reduced timber harvests were affecting local economies.

Several studies were carried out at one AMA, Snoqualmie Pass in Washington's Cascade Mountains (see figure). Some of this research found that understanding plant associations at the landscape scale could help predict disturbance from fire, insects, and disease. Another study addressed habitat fragmentation and its effects on wildlife. Land ownership at Snoqualmie Pass in 1994 was like a checkerboard, with alternating blocks of land owned by various government agencies, timber companies, and individuals. Landscape ecologists thought the area might

soil organisms help decompose plant material and cycle nutrients.

In general, forests with a greater diversity of plants host a greater diversity of organisms overall. And, in general, fully mature forests, such as the undisturbed old-growth forests remaining at Clayoquot Sound, contain more biodiversity than younger forests, because older forests contain more structural diversity and thus more microhabitats and resources for more species.

The complex systems we call forests provide all manner of vital ecosystem services (▶ pp. 39–41). Forest vegetation acts to stabilize soil and prevent erosion. Forest plants also help regulate the hydrologic cycle, slowing runoff, lessening flooding, and purifying water as they take it in from the soil and release it to the atmosphere. Forest plants also store carbon, release oxygen, and moderate climate. By performing such ecological functions, forests are indispensable for our survival and help make our planet's environment what it is.

Forest products are economically valued

Forests also provide people with economically valuable wood products. For millennia, wood from forests has fueled our fires, keeping people warm and well fed. It has

serve as a vital north–south corridor for wildlife to travel, so land was transacted to try to consolidate ownership to form corridors of publicly managed land. But an outstanding question was whether wildlife would be able to cross Interstate 90, which runs east–west over the pass.

Forest Service researchers led by Peter Singleton and John Lehmkuhl spent 2 years studying the movements of wildlife in relation to a 48-km (30-mi) stretch of I-90 to find out how much of a barrier the highway represented.

First, they used GIS methods (see the next "Science behind the Story" on ▸ pp. 352–353) to predict how accessible areas were to animal species with different dispersal abilities and habitat preferences.

They also set up cameras that took pictures of animals that passed in front of heat and motion detectors. From data at 115 stations, they found no difference between sites near and far from the interstate. However, they did find that some animals, for example, bobcats, skunks, hares, and bears, appeared near certain portions of the highway more than other portions.

Third, the researchers mapped all road-killed animals discovered along 86 km (54 mi) of the interstate by Washington Department of Transportation employees. They found that the 450 deer and elk killed in the 2-year period were concentrated along four particular stretches of interstate.

Fourth, they surveyed the interstate in winter for tracks left in snow by animals crossing the highway. Twenty-three of 37 crossings by coyotes, bobcats, and raccoons took place in a single 2-mile stretch of road.

Fifth, they monitored underpasses and overpasses, as well as drainage culverts crossing beneath the interstate, to determine how animals were using these structures. Using cameras and plates of soot or sand that recorded footprints, the researchers detected 15 mammal species using culverts to cross the interstate, with deer mice the most frequent crossers. Different species used different types of culverts, but dry ones 0.4–1.1 m in diameter were regularly used.

With these data, Singleton and Lehmkuhl could pinpoint locations preferred by wildlife and could recommend that opportunities for safely crossing the interstate be provided at those locations. In particular, they wrote, providing dry drainage culverts near areas of mature forest could help promote the ecological functions that small mammals provide, such as dispersing plant seeds and fungal spores. They also urged that habitat be restored and maintained on the landscape level in patterns that facilitated animal movement.

Key for adaptive management is that data continue to be collected, to assess whether actions that are taken succeed or fail. In this respect, the AMAs under the Northwest Forest Plan have not yet lived up to their potential. Doing adaptive management right requires time, money, and a rare combination of flexibility and commitment. However, science and management under the Northwest Forest Plan is ongoing, and the Plan's proponents are committed to making it work.

housed people, keeping us sheltered. It built the ships that carried people and cultures from one continent to another. It allowed us to produce paper, the medium of the first information revolution.

In recent decades, industrial harvesting has allowed the extraction of more timber than ever before, supplying all these needs of a rapidly growing human population and its expanding economy. The exploitation of forest resources has been instrumental in helping our society achieve the standard of living we enjoy today. Indeed, without industrial timber harvesting, you would not be reading this book.

Most commercial logging today takes place in Canada, Russia, and other nations that hold large expanses of boreal forest, and in tropical countries with large amounts of rainforest, such as Brazil. In the United States, most logging takes place on land both private and public, primarily in the conifer forests of the West and the pine plantations of the South.

Demand for wood has led to deforestation

We all depend in some way on wood, from the subsistence herder in Nepal cutting trees for firewood to the

FIGURE 12.4 Nations in Africa and in Latin America are experiencing rapid deforestation as they attempt to develop, extract resources, and provide new agricultural land for their growing populations. In parts of Europe and North America, meanwhile, forested area is slowly increasing as some former farmed areas are abandoned and allowed to grow back into forest. Go to **GRAPHit!** at www.aw-bc.com/withgott or on the student CD-ROM. Data from United Nations Environmental Programme. 2002. *Global environmental outlook 3.* London: Earthscan Publications.

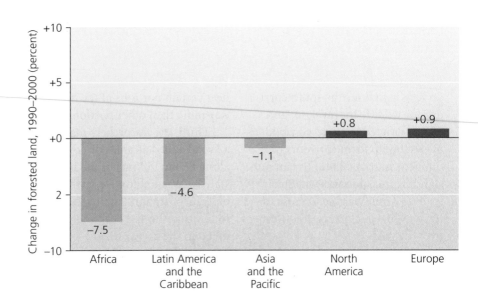

American student consuming reams upon reams of paper in the course of getting a degree. For such reasons, people have cleared forests for millennia to exploit timber resources. Still more forest has been cleared to make way for agriculture.

Deforestation, the clearing and loss of forests, has altered the landscapes and ecosystems of much of the planet. It has caused soil degradation, population declines, and species extinctions, and, as we saw with Easter Island (▶ pp. 8–9), has in some cases helped bring whole civilizations to ruin. Continued deforestation threatens dire ecological and economic consequences. Impacts are greatest in tropical areas, because of the potentially massive loss of biodiversity, and in arid regions, because of the vulnerability to desertification (▶ pp. 259–260). In addition, deforestation adds carbon dioxide (CO_2) to the atmosphere: CO_2 is released when plant matter is burned or decomposed, and thereafter less vegetation remains to soak up CO_2. Deforestation is thereby one contributor to global climate change (▶ p. 530).

All continents have experienced deforestation, but forests are being felled at the fastest rates today in the tropical rainforests of Latin America and Africa (Figure 12.4). Developing countries in these regions are striving to expand areas of settlement for their burgeoning populations and to boost their economies by extracting natural resources and selling them abroad. Meanwhile, areas of Europe and eastern North America are slowly gaining forest cover as they recover from severe deforestation of past decades and centuries. In total, depending on one's definition, one-fifth to one-third of Earth's land area is currently covered by forest (see "The Science behind the Story," ▶ pp. 352–353).

The growth of the United States and Canada was fed by deforestation

Deforestation for timber and farmland propelled the growth of the United States and Canada throughout their phenomenal expansion westward across the North American continent over the past 400 years. The vast deciduous forests of the East were virtually stripped of their trees by the mid-19th century, making way for countless small farms. Timber from these forests built the cities of the Atlantic seaboard. Later, cities such as Chicago were constructed with timber felled in the vast pine and hardwood forests of Wisconsin and Michigan. As a farming economy shifted to an industrial one, wood was used to stoke the furnaces of industry. Logging operations moved south to the Ozarks of Missouri and Arkansas, while the pine woodlands of the South were logged and converted to pine plantations. Once most mature trees were removed from these areas, timber companies moved west, cutting the continent's biggest trees in the Rocky Mountains, the Sierra Nevada, the Cascade Mountains, and the Pacific Coast ranges (Figure 12.5).

By the early 20th century, very little virgin timber was left in the lower 48 U.S. states (Figure 12.6). Today, the largest oaks and maples found in eastern North America, and even most redwoods of the California coast, are merely **second-growth** trees, trees that have sprouted and grown to partial maturity after old-growth timber has been cut. The size of the gargantuan trees they replaced can be seen in the enormous stumps that remain in the more recently logged areas of the Pacific coast. The scarcity of old-growth trees on the North American continent

FIGURE 12.5 A mule team drags logs of Douglas fir from a clear-cut in Mason County, Washington, in 1901. Early timber harvesting practices in North America caused significant environmental impacts and removed virtually all the virgin timber from one region after another.

(a) Areas of natural forest, 1620

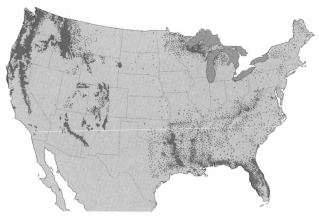

(b) Areas of natural forest, 1920

FIGURE 12.6 When Europeans first were colonizing North America, the entire eastern half of the continent and substantial portions of the western half were covered in forest (**a**). By the early 20th century, the vast majority of this forest had been cut, replaced mostly by agriculture and other human land uses (**b**). Since that time, most of the remaining original forest has been cut, but much of the landscape has also become reforested with second-growth forest. *Source:* Williams, M. 1989. *Americans and their forests.* Cambridge: Cambridge University Press. As adapted by Goudie, A., 2000. *The human impact.* Cambridge, MA: MIT Press.

today explains the passion with which environmental advocates have fought for the preservation of ancient forests in areas such as Clayoquot Sound.

The fortunes of loggers have risen and fallen with the availability of big trees. As each region was denuded of timber, the industry declined there and the timber companies moved on, while local loggers lost their jobs. If the remaining ancient trees of North America—most in British Columbia and Alaska—are cut, many U.S. and Canadian loggers will likely be out of jobs once again. Meanwhile, their employers will move on to nations of the developing world, as many already have.

Using Geographic Information Systems to Survey Earth's Forests

One map is washed in splashes of green, showing broad stretches of largely undisturbed forest. Another outlines political boundaries. A third map marks human population centers, denoting their density with different colors. A fourth lays out parks, reserves, and other land protected from development. On its own, each map shows only one physical characteristic of Earth. But layered together through the use of a powerful technology, these maps reveal something entirely different: the condition of the world's last intact forests.

Used in a groundbreaking study published by the United Nations Environment Programme (UNEP) in 2001, these maps, in conjunction with a *geographic information system (GIS),* have become an essential tool for environmental planners and resource managers around the world.

A GIS consists of software that builds layered maps, using different sets of data to compile various views of landscapes (see figure). Environmental researchers most often use data on the status and uses of natural resources. In the case of the

UNEP study, researchers wanted a comprehensive look at global forests—where they are, which countries govern them, which forests face the most pressure from encroaching human settlements, and which are most protected. The researchers especially wanted to know the status of "closed" forests: forests whose canopies cover more than 40% of their area. Conducting such a survey on the ground, going from forest to forest and country to country, would have been neither practical nor accurate. Synthesizing the findings from such research into a comprehensive paper map would have been almost impossible. With a GIS, however, researchers pull data such as satellite imagery and computerized maps into an all-inclusive view of the area they are studying.

Some of the researchers' most critical data are gathered via a sensor, an advanced very high resolution radiometer (AVHRR), orbiting the Earth by satellite. The rotating mirror, telescope, and internal electronics of the AVHRR measure wavelengths of energy rising from Earth's surface. Different types of substrate release different

wavelengths, which the AVHRR translates into color images. Different colors in the satellite imagery thus denote whether an area is covered by water, rock, houses, or vegetation.

For more than a decade, scientists have taken advantage of AVHHR to produce the Normalized Difference Vegetation Index (NDVI). NDVI researchers survey forests and plants and gauge changes to vegetation. Using the NDVI's catalog of worldwide vegetation images, UNEP researchers loaded forest data into their GIS. They then added three other sets of electronic data into the GIS: the distribution of the world's population, global political boundaries, and land protected against development.

Once compiled, the data layers offered UNEP researchers an unprecedented global view of forests. About 20% of Earth's surface remains covered by closed forest, the study indicated. Moreover, the study confirmed previous findings that forests in densely populated countries, such as India and Indonesia, are under pressure from expanding human settlement. Such

Deforestation is proceeding rapidly in many developing nations

Uncut tropical forests still remain in many developing countries, and these nations are in the position the United States and Canada faced a century or two ago: having a vast frontier that they can develop for human use. Today's advanced technology, however, has allowed these countries to exploit their resources and push back their frontiers even faster than occurred in North America. As a result, deforestation is rapid in places such as Brazil and Indonesia.

Developing nations are often desperate enough for economic development that they impose few or no restrictions on logging. Often their timber is extracted by foreign multinational corporations, which have paid fees to the developing nation's government for a *concession,* or right to extract the resource. In such cases, the foreign corporation has little or no incentive to manage forest resources sustainably. Many of the short-term economic benefits are reaped not by local residents but by the corporations that log the timber and export it elsewhere.

countries require stronger and immediate conservation efforts, researchers determined. The team also found that more than 80% of the world's intact forests were concentrated in just 15 countries and that 88% of those forests were sparsely inhabited by people. These findings could prove critical in starting conservation efforts in regions at risk, to help prevent further forest degradation.

GIS is not foolproof; GIS maps are only as good as the data that go into them. Some of the UNEP data—those on population size and protected areas—may not be reliable, researchers warn, because the countries providing the information may not always keep accurate records. Some satellite images used in the UNEP study were ambiguous, as was the case with the maps indicating forest cover; the team couldn't be sure every tree farm or plantation had been removed from that set of data before loading into the GIS.

However, the UNEP scientists stressed that without GIS such a major forest assessment would never have been possible. With it, forest planners can focus conservation priorities on areas with the best prospects for continued existence.

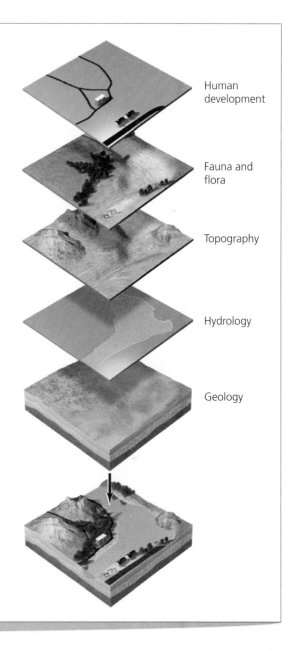

Human development

Fauna and flora

Topography

Hydrology

Geology

Geographic information systems (GIS) allow the layering of different types of data on natural landscape features and human land uses so as to produce maps integrating this information. GIS can then be used to explore informative correlations between these data sets.

Local people may or may not receive temporary employment from the corporation, but once the timber is harvested they no longer have the forest and the ecosystem services it once provided.

In Sarawak, the Malaysian portion of the island of Borneo, foreign corporations that were granted logging concessions have deforested several million hectares of tropical rainforest since 1963. The clearing of this forest—one of the world's richest, hosting such organisms as orangutans and the world's largest flower, *Rafflesia arnoldii*—has

had direct impacts on the 22 tribes of people who live as hunter-gatherers in Sarawak's rainforest. The Malaysian government did not consult the tribes about the logging, which decreased the wild game on which these people depended. Oil palm agriculture was established afterward, leading to pesticide and fertilizer runoff that killed fish in local streams. The tribes protested peacefully and finally began blockading logging roads. The government, which at first jailed them, now is negotiating, but it insists on converting the tribes to a farming way of life.

FIGURE 12.7 Federal agencies own and manage well over 250 million ha (600 million acres) of land in the United States, particularly in the western states. These include national forests, national parks, national wildlife refuges, Native American reservations, and Bureau of Land Management lands. Data from United States Geological Survey.

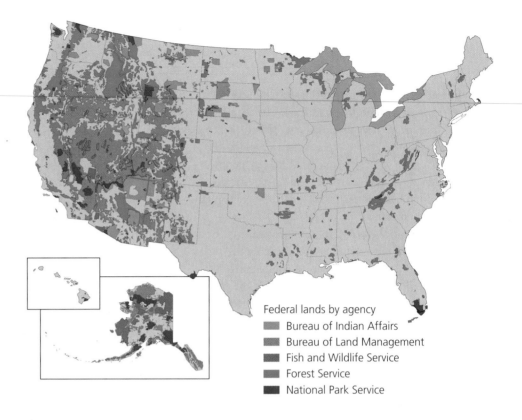

Federal lands by agency
- Bureau of Indian Affairs
- Bureau of Land Management
- Fish and Wildlife Service
- Forest Service
- National Park Service

Weighing the **Issues:**
Logging Here or There

Imagine you are an environmental activist protesting a logging operation that is cutting old-growth trees near your hometown. Now let's say you know that if the protest is successful, the company will move to a developing country and cut its virgin timber instead. Would you still protest the logging in your hometown? Would you pursue any other approaches?

Fear of a "timber famine" spurred establishment of national forests

In the United States, the depletion of the eastern forests and widespread fear of a "timber famine" spurred the formation of a system of forest reserves: public lands set aside to grow trees, produce timber, protect watersheds, and serve as insurance against future scarcities of lumber. Today the U.S. **national forest** system consists of 77 million ha (191 million acres) spread across all but a few states (Figure 12.7). The system, managed by the U.S. Forest Service, covers over 8% of the nation's land area.

The U.S. Forest Service was established in 1905 under the leadership of Gifford Pinchot (▶pp. 34–35). Pinchot

and others developed the concepts of resource management and conservation during the Progressive Era, a time of social reform when people urged that science and education be applied to public policy to improve society. In line with Pinchot's conservation ethic, the Forest Service aimed to manage the forests for "the greatest good of the greatest number in the long run." Pinchot believed the nation should extract and use resources from its public lands, so timber harvesting was, from the start, a goal behind establishing the national forests. But conservation meant planting trees as well as harvesting them, and the Forest Service intended to pursue wise management of timber resources.

Management goals are similar for the Canadian Forest Service and the provincial forestry ministries. Of Canada's 310 million ha (765 million acres) of forested land, 77% belongs to the provinces, and only 16% is federally owned and 7% privately owned. About one-third of the nation's forests are in British Columbia, and 38% are in Quebec and Ontario.

Timber is extracted from public and private lands

Timber is extracted from publicly held forests in the United States and Canada not by the governments of these nations, but by private timber companies. In the United States, Forest Service employees plan and manage timber sales and build roads to provide access for logging companies, which sell the timber they harvest for profit.

However, most timber harvesting in the United States these days takes place on private land, including land owned by timber companies. In 2001, the most recent year for which good data are available, timber companies extracted 31.7 million m^3 (340 million ft^3) of live timber from national forests. Although this is a large amount, it is considerably less than the amount cut from private lands and other public forests (Figure 12.8). Timber harvesting declined on national forests during the 1980s and 1990s, and in 2001 tree regrowth outpaced tree removal on these lands by nearly 12 to 1. Overall, timber production in the United States and other developed countries has remained roughly stable for the past 40 years. Meanwhile, it has more than doubled in developing countries.

The equivalent rates of growth and removal on private lands in Figure 12.8 reflect attempts by timber companies to manage their resources in accordance with the maximum sustainable yield approach, so that they can obtain maximal profits over many years for their owners and investors. On public lands, rates of growth and removal reflect not only economic forces but social and political ones as well, and these have changed over time. From the U.S. national forests, private timber extraction began to increase in the 1950s as the country experienced a postwar economic boom, consumption of paper products rose, and the population expanded into newly built suburban homes. In more recent decades, harvests from national forests decreased as economic trends shifted, public concern over clear-cutting grew, and forest management philosophy evolved.

Plantation forestry has grown

Today the North American timber industry focuses most on maximizing production from tree plantations in the Northwest and the South. These plantations feature stands of fast-growing tree species, and are single-species monocultures (▸ pp. 281–282). Because all trees in a given stand are planted at the same time, the stands are **even-aged**, with all trees the same age (Figure 12.9). Stands are cut after a certain number of years (called the *rotation time*), and the land is replanted with seedlings. Most ecologists and foresters view these plantations more as crop agriculture than as ecologically functional forests. Because there are few tree species and little variation in tree age, plantations do not offer many forest organisms the habitat they need. However, some harvesting methods aim to maintain **uneven-aged** stands, where a mix of ages (and often a mix of tree species) makes the stand more similar to a natural forest.

Timber is harvested by several methods

When they harvest trees, timber companies use any of several methods. From the 1950s through the 1970s, many timber harvests were conducted using the clear-cutting method, in which all trees in an area are cut, leaving only stumps. Clear-cutting is generally the most cost-efficient method in the short term, but it has the greatest impacts

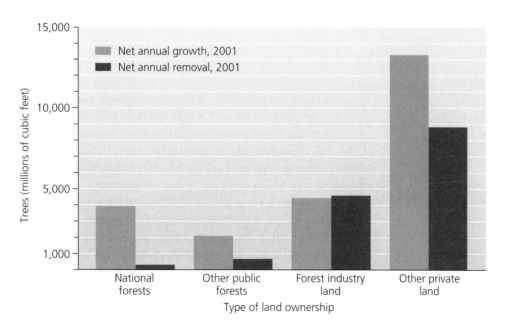

FIGURE 12.8 Forest Service data indicate that in the United States, trees (measured in cubic feet of wood biomass) are growing at a faster rate than they are being removed. The exception is on land privately owned by the timber industry, where extraction is narrowly outpacing growth. Go to **GRAPHIT!** at www.aw-bc.com/withgott or on the student CD-ROM. Data are for 2001, from United States Forest Service, 2001.

FIGURE 12.9 Even-aged management is practiced on tree plantations where all trees are of equal age, as seen in the stand in the foreground that is regrowing after clear-cutting. In uneven-aged management, harvests are designed so as to maintain a mix of tree ages, as seen in the more mature forest in the background. The increased structural diversity of uneven-aged stands provides superior habitat for most wild species and makes these stands more akin to ecologically functional forests.

on forest ecosystems (Figure 12.10). In the best-case scenario, clear-cutting may mimic natural disturbance events such as fires, tornadoes, or windstorms that knock down trees across large areas. In the worst-case scenario, entire communities of organisms are destroyed or displaced, soil erodes, and the penetration of sunlight to ground level changes microclimatic conditions such that new types of plants replace those that dominated the native forest. Essentially, an artificially driven process of succession (▶ p. 163) is set in motion, in which the resulting

FIGURE 12.10 Clear-cutting is the most cost-efficient method for timber companies, but it can have severe ecological consequences, including soil erosion and species turnover. Although certain species do use clear-cuts as they regrow, most people find these areas aesthetically unappealing, and public reaction to clear-cutting has driven changes in forestry methods.

climax community may turn out to be quite different from the original climax community.

Widespread clear-cutting occurred across North America at a time when public awareness of environmental problems was blossoming. The combination produced public outrage toward the timber industry and public forest managers. Eventually the industry integrated other harvesting methods (Figure 12.11). Clear-cutting (Figure 12.11a) is still widely practiced, but other methods involve cutting some trees and leaving some standing. In the *seed-tree* approach (Figure 12.11b), small numbers of mature and vigorous seed-producing trees are left standing so that they can reseed the logged area. In the *shelterwood* approach (also Figure 12.11b), small numbers of mature trees are left in place to provide shelter for seedlings as they grow. These three methods all lead to even-aged stands of trees.

Selection systems, in contrast, allow uneven-aged stand management. In selection systems (Figure 12.11c), only some trees in a forest are cut at any one time. The stand's overall rotation time may be the same as in an even-aged approach, because multiple harvests are made, but the stand remains mostly intact between harvests. Selection systems include single-tree selection, in which widely spaced trees are cut one at a time, and group selection, in which small patches of trees are cut.

It was a form of selection harvesting that MacMillan Bloedel and other timber companies pursued at Clayoquot Sound, after old-growth advocates applied pressure and the scientific panel published its guidelines. Not wanting to bring a complete end to logging when so many local people depended on the industry for work, these activists and scientists instead promoted what they considered a

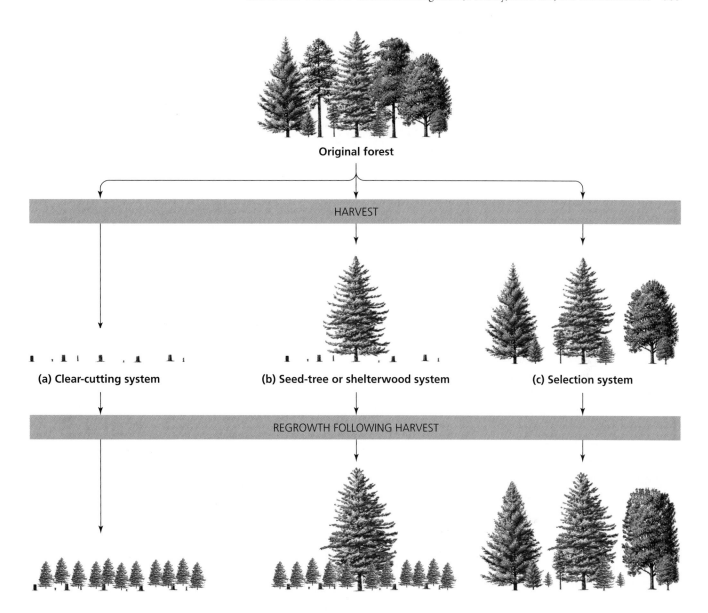

FIGURE 12.11 Foresters and timber companies have devised various methods of harvesting timber from forests. In clear-cutting **(a)**, all trees in an area are cut, extracting a great deal of timber inexpensively but leaving a vastly altered landscape. In seed-tree systems and shelterwood systems **(b)**, small numbers of large trees are left in clearcuts, to help re-seed the area or provide shelter for growing seedlings. In selection systems **(c)**, a minority of trees is removed at any one time, while most are left standing. These latter methods involve less environmental impact than clear-cutting, but all methods can cause significant changes to the structure and function of natural forest communities.

more environmentally friendly method of timber removal. However, selection systems are by no means ecologically harmless. Moving trucks and machinery over an extensive network of roads and trails to access individual trees compacts the soil and disturbs much of the forest floor. Selection methods are also unpopular with timber companies because they are expensive, and loggers dislike them because they are more dangerous than clear-cutting.

All methods of logging result in habitat disturbance, which invariably affects the plants and animals inhabiting an area. All methods change forest structure and composition. Most methods increase soil erosion, leading to siltation of waterways, which can degrade habitat and affect drinking water quality. Most methods also speed runoff, sometimes causing flooding. In extreme cases, as when steep hillsides are clear-cut, landslides can result.

Public forests may be managed for recreation and ecosystems

In recent decades, increased awareness of these problems has prompted many citizens to protest the way public forests are managed in the United States and Canada. These citizens have urged that the national, state, and provincial forests be managed for recreation, wildlife, and ecosystem integrity, rather than for timber. They want forests managed as ecologically functional entities, not as cropland for trees.

Critics of the U.S. Forest Service have also protested the fact that taxpayers' money is used to subsidize the extraction of publicly held resources by private corporations. Scientists who have analyzed government subsidies have concluded that the U.S. Forest Service loses at least $100 million of taxpayers' money each year by selling timber well below its costs for marketing and administering the harvest and for building access roads. Subsidies also inflate harvest levels beyond what would occur in a free market.

In one sense, the U.S. Forest Service has long had a policy of attending to interests beside timber production. For the past half century, forest management has nominally been guided by the policy of **multiple use,** meaning that the national forests were to be managed for recreation, wildlife habitat, mineral extraction, and various other uses. In reality, however, timber production was most often the primary use.

In 1976 the U.S. Congress passed the **National Forest Management Act,** which mandated that plans for renewable resource management be drawn up for every national forest. These plans were to be explicitly based on the concepts of multiple use and sustained yield, and they were to be subject to broad public participation under the National Environmental Policy Act (▸ p. 68). Guidelines specified that these plans:

▸ Consider both economic and environmental factors.
▸ Provide for diversity of plant and animal communities and preserve the regional diversity of tree species.
▸ Ensure research and monitoring of management practices.
▸ Permit increases in harvest levels only if sustainable.
▸ Ensure that timber is harvested only where soils and wetlands will not be irreversibly damaged, lands can be restocked quickly, and economic return alone does not guide the choice of harvest method.
▸ Ensure that logging is conducted only where impacts have been assessed; cuts are shaped to the terrain; maximum size limits are established; and "cuts are carried out in a manner consistent with the protection of soil, watershed, fish, wildlife, recreation, and aesthetic resources, and the regeneration of the timber resource."

Over the years following passage of the National Forest Management Act, the U.S. Forest Service began responding to the increasing public demand that national forests be managed for uses other than timber. It developed new programs to manage wildlife, nongame animals, and endangered species. It pushed for ecosystem-based management and even ran extensive programs of ecological restoration, attempting to recover whole plant and animal communities that had been lost or degraded. Moreover, timber harvesting methods were brought more in line with ecosystem-based management goals. A set of approaches dubbed **new forestry** called for timber cuts that came closer to mimicking natural disturbances. For instance, "sloppy clear-cuts" that leave a variety of trees standing were intended to mimic the changes a forest might experience if hit by a severe windstorm.

In late 2004, however, the George W. Bush administration issued new regulations that bucked these trends. These new rules freed local forest managers from requirements imposed by the National Forest Management Act, granting them more flexibility in managing forests, but loosening environmental protections and restricting public oversight.

Then in 2005, the Bush administration repealed the Clinton administration's roadless rule, by which 23.7 million ha (58.5 million acres)—31% of national forest land and 2% of total U.S. land—were in 2001 put off limits to further road construction or maintenance. Although the roadless rule had been supported by a record 4.2 million public comments, the Bush administration overturned it and required state governors to petition the federal government if they want to keep areas in their states roadless. The states of California, Oregon, and New Mexico responded by suing the federal government, asking that the roadless rule be reinstated.

Fire policy has also stirred controversy

Some ecosystem management efforts, ironically, run counter to the U.S. Forest Service's best-known symbol, Smokey Bear. The cartoon bear wearing a ranger's hat who advises us to fight forest fires is widely recognized, but unfortunately Smokey's message is badly outdated, and many scientists assert that it has done great harm to American forests.

For over a century, the Forest Service and other land management agencies suppressed fire whenever and wherever it broke out. Yet ecological research now clearly shows that many ecosystems depend on fire. Certain plants have seeds that germinate only in response to fire, and researchers studying tree rings have documented that many ecosystems historically experienced frequent fire. Burn marks in a tree's rings reveal past fires, giving scientists

Managing Forests

The U.S. Forest Service attempts to manage the national forests for multiple uses, including timber production, wildlife habitat, and recreation. **How well is the agency doing in balancing its multiple goals? What changes, if any, do you think are needed in the way the national forests are managed?**

Restoring America's National Forests

The U.S. Forest Service completed the 20th century with a celebrated and criticized record. Its challenge is that laws direct the Forest Service to provide society with often conflicting values from the national forests, including wood, water, wildlife, fish, forage, and recreation.

The Forest Service also must "preserve and protect" forests as required in the Organic Act of 1897. Therefore, conservation guides management because it integrates the need to use and protect national forests.

For much of the last century, the Forest Service protected forests by putting out fires, without realizing that frequent, gentle fires kept forests thinned and healthy. Then they stopped timber harvests and active management on many national forests, so forests grew even thicker and less healthy.

Today's forests are 10 to 20 times denser than historic forests. Consequently, between the years 2000 and 2003, unnatural wildfires destroyed 9.7 million ha (24 million acres) of forest, many human lives, and thousands of homes. Insects devoured millions more hectares of unhealthy forest.

Restoration should guide the Forest Service in the 21st century. Restoration means using active management to restore ecologically and economically sustainable forests that represent natural historic landscapes. Restored forests provide diverse public values, from scenery and recreation to lumber and safety.

However, we don't have the billions of tax dollars needed to manage our forests. That means we must form a partnership with the private sector to make restoration economically feasible.

We have a choice. Adopt a "hands-off" policy and let the harsh indifference of unnatural wildfires and mindless insects determine the fate of national forests, or shape their destiny through restoration forestry.

Thomas M. Bonnicksen is a historian of North American forests and originator of the concept of restoration forestry. He is professor emeritus of forest science at Texas A&M University, visiting professor at the University of California at Davis, and the author of *America's Ancient Forests* (John Wiley, 2000.)

Protect Our Forests from Logging

Most Americans are shocked when they learn that over the past century the Forest Service's management has largely emphasized logging, road building, and other forms of resource extraction.

Unfortunately, because the Forest Service's budgets are still tied to logging and resource extraction—not to forest protection and restoration—the public's clean water, wildlife habitat, wildlands, and recreational opportunities continue to be sacrificed. Just consider these facts:

▶ There are 716,500 km (445,000 mi) of roads on national forests—enough to circle Earth 18 times.
▶ An estimated 50% of riparian areas on national forests require restoration because of impacts from logging, road building, grazing, mining, and offroad vehicles.
▶ Taxpayers spend over $1 billion annually subsidizing private logging companies to cut down national forests—all to supply less than 2% of our nation's wood products.
▶ Although less than 5% of America's ancient, old-growth forests remain, these heritage forests continue to be logged, with over 160 km^2 (100 mi^2) of the public's ancient forests currently on the chopping block in the Northwest alone.

Fully protecting and restoring our national forests will take a heroic effort. The first step in the restoration process is to prevent further ecological degradation by protecting our national forests from logging and other forms of resource extraction. Next, we need to redirect taxpayer subsidies toward ecologically based restoration projects—such as road removal and watershed restoration—with the goal of restoring natural processes and reestablishing fully functioning ecosystems.

Only once this happens will we see the Forest Service's management of national forests in step with the desires of an American public that wants to see our forests protected and restored.

Matthew Koehler is executive director of the Native Forest Network and a co-founder of the National Forest Protection Alliance, a national network of 130 grassroots forest protection organizations working to protect and restore national forests.

Explore this issue further by accessing **Viewpoints** at www.aw-bc.com/withgott.

FIGURE 12.12 Forest fires are natural phenomena to which many plants are well adapted and which maintain many ecosystems. The suppression of fires by humans over the past century has led to a buildup of leaf litter and young trees, which serve as fuel to increase the severity of fires when they do occur. As a result, catastrophic forest fires (such as this one in Yellowstone National Park in 1988) have become more common in recent years. These unnaturally severe fires can do great damage to ecosystems and human communities. The best solution, most fire ecologists agree, is to forego suppressing natural fires as much as possible and to institute controlled burning to reduce fuel loads and restore forest ecosystems.

an accurate history of fire events extending back hundreds or even thousands of years. Researchers have found that North America's grasslands and open pine woodlands burned regularly. Ecosystems dependent on fire are adversely affected by its suppression; pine woodlands become cluttered with hardwood understory that ordinarily would be cleared away by fire, for instance, and animal diversity and abundance decline in such cluttered habitats.

In the long term, fire suppression can lead to catastrophic fires that truly do damage forests, destroy human property, and threaten human lives. Fire suppression allows limbs, logs, sticks, and leaf litter to accumulate on the forest floor over the years, effectively producing kindling for a catastrophic fire. Such fuel buildup helped cause the 1988 fires in Yellowstone National Park (Figure 12.12), the 2003 fires in southern California, the 2003 fires in British Columbia, and thousands of other fires across the continent. Fire suppression and fuel buildup have made catastrophic fires significantly greater problems than they were in the past. At the same time, increasing residential development on the edges of forested land is placing more homes in fire-prone situations.

To reduce fuel load and improve the health and safety of forests, the Forest Service and other land management agencies have in recent years been burning areas of forest under carefully controlled conditions. These **prescribed burns** or **controlled burns** have worked effectively, but have been implemented on only a relatively small amount of land. And every once in a while, a prescribed burn may

get out of control, as happened in 2000 when homes and government labs were destroyed at Los Alamos, New Mexico. All too often, these worthy efforts have been impeded by public misunderstanding and by interference from politicians who have not taken time to understand the science behind the approach.

In the wake of the 2003 California fires, the U.S. Congress, intending to make forests less fire-prone, passed the Bush administration's Healthy Forests Restoration Act. Although this legislation encourages some prescribed burning, it primarily promotes the physical removal of small trees, underbrush, and dead trees by timber companies. The removal of dead trees, or snags, following a natural disturbance is called **salvage logging.** From an economic standpoint, salvage logging may seem to make good sense. However, ecologically, snags have immense value; the insects that decay them provide food for wildlife, and many birds, mammals, and reptiles depend on holes in snags for nesting and roosting sites. Conducting timber removal operations on recently burned land can also cause severe erosion and soil damage. Many scientists and environmental advocates have criticized the Healthy Forests Restoration Act, saying it increases commercial logging in national forests while doing little to reduce catastrophic fires near populated areas. By streamlining procedures for timber removal on public lands, the law also decreases oversight and public participation in enforcing environmental regulations, critics contend.

How to Handle Fire?

A century of fire suppression has left millions of hectares of forested lands in North America in danger of catastrophic wildfires. Yet we will probably never have adequate resources to conduct careful prescribed burning over all these lands. Can you suggest any possible solutions that might help protect people's homes near forests while improving the ecological condition of some forested lands?

Sustainable forestry is gaining ground

Any company can claim that its timber harvesting practices are sustainable, but how is the purchaser of wood products to know whether they really are? In the last several years, a consumer movement has grown that is making informed consumer choice possible. Several organizations now examine the practices of timber companies and offer **sustainable forestry certification** to products produced using methods they consider sustainable (Figure 12.13).

Organizations such as the International Organization for Standardization (ISO), the Sustainable Forestry Initiative (SFI) program, and the Forest Stewardship Council (FSC) have varying standards for certification. Consumers can look for the logos of these organizations on forest products they purchase. The FSC is widely perceived to have the strictest certification standards. In 2001, Iisaak, the Native-run timber company at Clayoquot Sound, became the first tree farm license holder in British Columbia to receive FSC certification.

Consumer demand for sustainable wood has been great enough that several major retail businesses have announced that they will sell only sustainable wood. Home Depot in 2002 began selling only FSC-certified lumber, and the company says it is doing its best to keep prices as low as possible. B&Q, a major British retailer similar to Home Depot, also switched to sustainable wood, and the company's head said he was "taken aback" by the favorable public response. The decisions of such retailers are influencing the logging practices of many timber companies. In British Columbia, 70% of the province's annual harvest now is certified or meets ISO requirements.

Sustainable forestry is more costly for the timber industry, but if certification standards can be kept adequately strong, then consumer choice in the marketplace can be a powerful driver for good forestry practices for the future.

FIGURE 12.13 A Brazilian woodcutter taking inventory marks timber harvested from a forest certified for sustainable management in Amazonian Brazil. A consumer movement centered on independent certification of sustainable wood products is allowing consumer choice to promote sustainable forestry practices.

Agricultural Land Use

Having replaced many forests, agriculture now covers more of the planet's surface than does forest. Thirty-eight percent of Earth's terrestrial surface is devoted to agriculture—more than the area of North America and Africa combined. Of this land, 26% supports pasture, and 12% consists of crops and arable land. Agriculture is the most widespread type of human land use, and causes tremendous impacts on land and ecosystems. Although agricultural methods such as organic farming and no-till farming can be sustainable, the majority of the world's cropland hosts intensive traditional agriculture and monocultural industrial agriculture, involving heavy use of fertilizers, pesticides, and irrigation, and often producing soil erosion, salinization, and desertification (Chapters 9 and 10).

In theory, the marketplace should discourage people from farming with intensive methods that degrade land they own if such practices are not profitable. But agriculture in many countries is supported by massive subsidies. Governments of 30 developed nations handed out $311 billion in farm subsidies in 2001, averaging $12,000 per farmer. Roughly one-fifth of the income of the average U.S. or Canadian farmer comes from subsidies. Proponents of such subsidies stress that the vagaries of weather make profits and losses from farming unpredictable from year to year. To persist in the long term,

FIGURE 12.14 Cropland agriculture exerts dramatic effects over a large portion of Earth's landscape. The huge green circles visible during any airplane trip over the Great Plains are created by center-pivot irrigation. Immense sprinklers pivot around a central point, watering a circular area. Shown are fields in eastern Oregon.

these proponents say, an agricultural system needs some way to compensate farmers for bad years. Opponents of subsidies argue that farmers can buy insurance to protect themselves against crop failures and that subsidizing environmentally destructive agricultural practices is unsustainable (Figure 12.14).

Wetlands have been drained for farming

Many of today's crops grow on the sites of former wetlands (▸ pp. 435–436)—swamps, marshes, bogs, and river floodplains—that were drained and filled in (Figure 12.15). Throughout recent history, governments have encouraged laborious efforts to drain wetlands. To promote settlement and farming, the United States passed a series of laws known as the Swamp Lands Acts in 1849, 1850, and 1860, which encouraged draining wetlands for agriculture. The government transferred over 24 million ha (60 million acres) of wetlands to state ownership (and eventually to private hands) to stimulate drainage, conversion, and flood control.

In the Mississippi River valley, the Midwest, and a handful of states from Florida to Oregon, these transfers eradicated malaria (because mosquito vectors breed in wetlands) and created over 10 million ha (25 million acres) of new farmland. A U.S. Department of Agriculture (USDA) program in 1940 provided monetary aid and technical assistance to farmers draining wetlands on their property, resulting in the conversion of almost 23 million ha (57 million acres).

Today, less than half the original wetlands in the lower 48 U.S. states and southern Canada remain. However, many people now have a new view of wetlands. Rather than viewing them as worthless swamps, science has made clear that they are valuable ecosystems. This scientific knowledge, along with a preservation ethic, has induced policymakers to develop regulations to safeguard remaining wetlands. However, because of loopholes, differing state laws, development pressures—and even debate over the legal definition of wetlands—many of these valuable ecosystems are still being lost.

Financial incentives are also being used as a tool to influence agricultural land use in the United States. The Conservation Reserve Program begun in 1985 was a landmark initiative that provided farmers with a different kind of subsidy—it paid them to take highly erodible lands out of production and instead encouraged them to make the areas more habitable for wildlife. Now, under the Wetland Reserve Program, the U.S. government is offering subsidies to landowners who refrain from developing wetland areas.

Weighing the **issues:**
Subsidies, Soil, and Wetlands

Do you think that subsidy programs such as the Conservation Reserve Program and the Wetland Reserve Program are a good use of taxpayers' money? Are financial incentives such as these better tools than government regulation for promoting certain land use goals?

FIGURE 12.15 Most of North America's wetlands have been drained and filled, and the land converted to agricultural use. The northern Great Plains region of Canada and the United States was pockmarked with thousands of "prairie potholes," water-filled depressions that served as nesting sites for most of the continent's waterfowl. Today many of these wetlands have been lost; shown are farmlands encroaching on prairie potholes in North Dakota.

Livestock graze one-fourth of Earth's land

Cropland agriculture uses less than half the land taken up by livestock grazing, which covers a quarter of the world's land surface. Human use of rangeland, however, does not necessarily exclude its use by wildlife or its continued functioning as a grassland ecosystem. Grazing can be sustainable if done carefully and at low intensity. In the American West, ranching proponents claim that cattle are merely taking the place of the vast herds of bison that once roamed the plains. Indeed, most of the world's grasslands have historically been home to large herds of grass-eating mammals, and grasses have adapted to herbivory. Nonetheless, poorly managed grazing, as we saw in Chapter 9, can have adverse impacts on soil and grassland ecosystems.

Most U.S. rangelands are federally owned and managed by the **Bureau of Land Management (BLM).** The BLM is the nation's single largest landowner; its 106 million ha (261 million acres) are spread across 12 western states (see Figure 12.7). Ranchers are allowed to graze cattle on BLM lands for inexpensive fees, a practice that many public lands advocates say encourages overgrazing. Thus ranchers and environmentalists have traditionally been at loggerheads. In the past several years, however, ranchers and environmentalists have been finding common ground, teaming up to preserve ranchland against what each of them views in common as a threat—the encroaching housing developments of suburban sprawl

(▸ pp. 378–379). Although developers often pay high prices for ranchland, many ranchers do not want to see the loss of the wide open spaces and the ranching lifestyle that they cherish.

Land use in the American West might have been better managed

Land uses such as grazing, farming, and timber harvesting need not have strongly adverse impacts. It is not these activities per se that cause environmental problems, but rather the overexploitation of resources beyond what ecosystems can handle. In the American West, a great deal of damage was done to the land by poor farming practices, overgrazing, and attempts to farm arid lands that were more suitable for grazing or preservation.

Most land to the west of the 100th meridian, the longitudinal line slicing through the Great Plains from Manitoba south through Texas, receives less than 50 cm (20 in.) of rain per year, making it too arid for unirrigated agriculture. One man in U.S. history recognized this fact and attempted to reorient policy so the West could be settled in a way that allowed farmers to succeed. John Wesley Powell, an extraordinary individual who explored the raging Colorado River by boat despite having lost an arm in the Civil War, undertook vast surveys of the Western lands for the U.S. government in the late 19th century (Figure 12.16). A pioneer in calling for government

FIGURE 12.16 John Wesley Powell, Civil War hero and 19th-century Western explorer, tried to shift U.S. land use policy to take account of the aridity of Western lands.

agencies to base their policies on science, Powell maintained that lands beyond the 100th meridian were too dry to support farming on the 65-ha (160-acre) plots the government parceled out through the Homestead Act (▶ p. 66). Plots in the West, he said, would have to be 16 times as large and would require irrigation. Moreover, to prevent individuals or corporations from monopolizing scarce water resources, he urged the government to organize farmers into cooperative irrigation districts, with each district encompassing a watershed.

Powell's ideas, based on science and close study of the land, were too revolutionary for the entrenched political interests and prevailing misconceptions of his time, which held that the West was a utopia for frontier settlement. The ideas in Powell's 1878 *Report on the Lands of the Arid Region of the United States* were, for the most part, never implemented. Instead, existing land use policies contributed to failures such as the Dust Bowl of the 1930s (▶ p. 260).

For agriculture and forestry alike, debates continue today over how best to use land and manage resources. Resource extraction from public lands in the United States and Canada has helped propel the economies of these countries. But as resources dwindle, as forests and soils are degraded, and as the landscape fills with more people, the arguments for conservation of resources—for their sustainable use—have grown stronger. Also growing stronger is the argument for preservation of land—setting aside tracts of relatively undisturbed land intended to remain forever undeveloped.

Parks and Reserves

Preservation has been part of the American psyche ever since John Muir rallied support for saving scenic lands in the Sierras (▶ p. 34). For ethical reasons as well as pragmatic ecological and economic ones, U.S. citizens and many other people worldwide have chosen to set aside tracts of land in perpetuity to be preserved and protected from development.

Why have we created parks and reserves?

What specifically in this mix of ethics, ecology, and economics has driven so many cultures to refrain voluntarily from exploiting land for material resources? The historian Alfred Runte has cited four traditional reasons that parks and protected areas have been established:

1. Enormous, beautiful, or unusual features such as the Grand Canyon, Mount Rainier, or Yosemite Valley inspire people to protect them—an impulse termed *monumentalism* (Figure 12.17).
2. Protected areas offer recreational value to tourists, hikers, fishers, hunters, and others.
3. Protected areas offer utilitarian benefits. For example, undeveloped watersheds provide cities with clean drinking water and a buffer against floods.
4. Parks make use of sites lacking economically valuable material resources; land that holds little monetary value is easy to set aside because no one wants to buy it.

To these four traditional reasons, a fifth has been added in recent years: the preservation of biodiversity. As we saw in Chapter 11, human impact alters habitats and has led to countless population declines and species extinctions. A park or reserve is widely viewed as a kind of Noah's Ark, an island of habitat that can, scientists hope, maintain species that might otherwise disappear.

Federal parks and reserves began in the United States

The striking scenery of the American West impelled the U.S. government to create the world's first **national parks,** publicly held lands protected from resource extraction and development but open to nature appreciation and various forms of recreation. In 1872, Yellowstone National Park was established as "a public park or pleasuring-ground for the benefit and enjoyment of the people." Yosemite, General Grant, Sequoia, and Mount Rainier National Parks followed after 1890. The Antiquities Act of 1906 gave the president authority to declare selected

FIGURE 12.17 The awe-inspiring beauty of some regions of the western United States was one reason for the establishment of national parks. Images of scenic vistas such as this one of Bridal Veil Falls in Yosemite National Park, portrayed by the landscape painter Albert Bierstadt, have inspired millions of people from North America and abroad to visit these parks.

public lands as national monuments, which can be an interim step to national park status. Presidents from Theodore Roosevelt to Bill Clinton have used this authority to expand the nation's system of protected lands.

The National Park Service (NPS) was created in 1916 to administer the growing system of parks and monuments, which today numbers 388 sites totaling 32 million ha (79 million acres) and includes national historic sites, national recreation areas, national wild and scenic rivers, and other types of areas (see Figure 12.7). This most widely used park system in the world received 277 million reported recreation visits in 2004—about as many visits as there are U.S. residents. The high visitation rates signal the success of the park system, but they also create overcrowded conditions at some parks. Many observers have therefore suggested that the parks' popularity indicates a pressing need to expand the system.

Many other nations now have national park systems. Canada's system covers 26.5 million ha (65.5 million acres) and receives 16 million visits yearly. At Clayoquot Sound, Pacific Rim National Park Reserve is a protected area designated for future national park status. The Clayoquot Sound region also encompasses several provincial parks. Provincial parks in Canada cover more area than national parks. In the United States, state parks are numerous and tend to be more oriented toward recreation than are national parks.

Another type of federal protected area in the United States is the **national wildlife refuge.** The system of national wildlife refuges, begun in 1903 by President Theodore Roosevelt, now totals 37 million ha (91 million acres) spread over 541 sites (see Figure 12.7). The U.S. Fish and Wildlife Service administers the refuges with management ranging "from preservation to active manipulation of habitats and populations." Indeed, these refuges not only serve as havens for wildlife, but also in many cases encourage hunting, fishing, wildlife observation, photography, environmental education, and other public uses. Some wildlife advocates find it ironic that hunting is allowed at many refuges, but hunters have long been in the forefront of the conservation movement and have traditionally supplied the bulk of funding for land acquisition and habitat management for the refuges. Many refuges are managed for waterfowl, but the FWS increasingly considers nongame species as well as game species. The FWS manages at the habitat and ecosystem levels, engaging in ecological restoration of marshes and grasslands, for example.

Wilderness areas have been established on various federal lands

In response to the public's desire for undeveloped areas of land, in 1964 the U.S. Congress passed the Wilderness Act, which allowed areas of existing federal lands to be designated as **wilderness areas.** These areas are off-limits to development of any kind, but they are open to public recreation such as hiking, nature study, and other activities that have minimal impact on the land. Congress declared that wilderness areas were necessary "to assure that an increasing population, accompanied by expanding settlement and growing mechanization, does not occupy and modify all areas within the United States and its possessions, leaving no lands designated for preservation and protection in their natural condition." Wilderness areas have been established within portions of national forests, national parks, national wildlife refuges, and BLM land, and they are overseen by the agencies that

FIGURE 12.18 Wilderness areas were designated on various federally managed lands in the United States following the 1964 Wilderness Act. These include areas little disturbed by human activities, such as the Selway-Bitterroot Wilderness in Idaho, shown here.

administer those areas (Figure 12.18). Some preexisting extractive land uses, such as grazing and mining, were "grandfathered in," or allowed to continue, within wilderness areas as a political compromise so the act could be passed.

Not everyone supports land set-asides

The restriction of activities in wilderness areas has helped generate opposition to U.S. land protection policies. Sources of such opposition include the governments of some western states, where large portions of land are federally owned. When those states came into existence, the federal government retained ownership of much of the acreage inside their borders. Idaho, Oregon, and Utah own less than 50% of the land within their borders, and in Nevada 80% of the land is federally owned. Western state governments have traditionally sought to obtain land from the federal government and encourage resource extraction and development on it.

The drive to extract more resources, secure local control of lands, and expand recreational access to public lands is epitomized by the **wise-use movement,** a loose confederation of individuals and groups that coalesced in the 1980s and 1990s in response to the increasing success of environmental advocacy. Wise-use advocates are dedicated to protecting private property rights; opposing government regulation; transferring federal lands to state,

local, or private hands; and promoting more motorized recreation on public lands. Wise-use advocates include farmers, ranchers, trappers, and mineral prospectors at the grassroots level who live off the land, as well as groups representing the large corporations of industries that extract timber, mineral, and fossil fuel resources.

Debate between mainstream environmental groups and wise-use spokespeople has been vitriolic. Each side claims to represent the will of the people and paints the other as the oppressive establishment. Wise-use advocates have played key roles in ongoing debates over national park policy, such as whether recreational activities that disturb wildlife, such as snowmobiles and jet-skis, should be allowed. Under the Bush administration, wilderness protection policies have been weakened, and federal agencies have generally shifted policies and enforcement away from preservation and conservation, and toward recreation and extractive uses.

Nonfederal entities also protect land

Efforts to set aside land—and the debates over such efforts—at the federal level are paralleled at regional and local levels. Each U.S. state and Canadian province has agencies that manage resources on state or provincial lands, as do many counties and municipalities. As just one example, New York State in the 19th century created Adirondack State Park in a mountainous area whose streams converge to form the headwaters of the Hudson River, which flows south past Albany to New York City. Seeing the need for river water to power industries, keep canals filled, and provide drinking water, the state set the land aside, a farsighted decision that has paid dividends through the years.

Private nonprofit groups also preserve land. **Land trusts** are local or regional organizations that preserve lands valued by their members. In most cases, land trusts purchase land outright with the aim of preserving it in its natural condition. The Nature Conservancy can be considered the world's largest land trust, but smaller ones are springing up throughout North America. By one estimate, there are 900 local and regional land trusts in the United States that together own 177,000 ha (437,000 acres) and have helped preserve an additional 930,000 ha (2.3 million acres), including well-known scenic areas such as Big Sur on the California coast, Jackson Hole in Wyoming, and Maine's Mount Desert Island.

Parks and reserves are increasing internationally

Many nations have established national park systems and are benefiting from ecotourism as a result—from Costa Rica (Chapter 5) to Ecuador to Thailand to Tanzania. The total worldwide area in protected parks and reserves increased more than fourfold from 1970 to 2000, and in 2003 the world's 38,536 protected areas covered 1.3 billion ha (3.2 billion acres), or 9.6% of the planet's land area. However, parks in developing countries do not always receive the funding they need to manage resources, provide for recreation, and protect wildlife from poaching and timber from logging. Thus many of the world's protected areas are merely *paper parks*—protected on paper but not in reality.

Some types of protected areas fall under national sovereignty but are designated or partly managed internationally by the United Nations. *World heritage sites* are an example; currently over 560 sites across 125 countries are listed for their cultural value and nearly 150 for their natural value. One such site is Australia's Kakadu National Park, discussed in Chapter 2. Another is the mountain gorilla reserve shared by three African countries. The gorilla reserve, which integrates national parklands of Rwanda, Uganda, and the Democratic Republic of Congo, is also an example of a *transboundary park,* an area of protected land overlapping national borders. Transboundary parks can be quite large, and they account for 10% of protected areas worldwide, involving 113 countries. A North American example is Waterton-Glacier National Parks on the

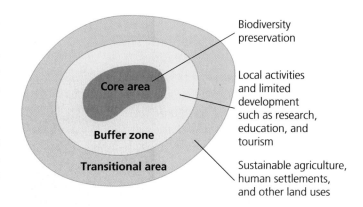

FIGURE 12.19 Biosphere reserves are international efforts that couple preservation with sustainable development to benefit local residents. Each reserve includes a core area that preserves biodiversity, a buffer zone that allows limited development, and a transition zone that permits various uses.

Canadian–U.S. border. Some transboundary reserves function as *peace parks,* helping ease tensions by acting as buffers between nations that have quarreled over boundary disputes. This is the case with Peru and Ecuador as well as Costa Rica and Panama, and many people hope that peace parks can also help resolve conflicts between Israel and its neighbors.

Biosphere reserves are tracts of land with exceptional biodiversity that couple preservation with sustainable development to benefit local people. They are designated by UNESCO (the United Nations Educational, Scientific, and Cultural Organization) following application by local stakeholders. Each biosphere reserve consists of (1) a core area that preserves biodiversity, (2) a buffer zone that allows local activities and limited development that do not hinder the core area's function, and (3) an outer transition zone in which agriculture, human settlement, and other land uses can be pursued in a sustainable way (Figure 12.19).

Clayoquot Sound was designated as Canada's 12th biosphere reserve in 2000, in an attempt to help build cooperation among environmentalists, timber companies, Native peoples, and local residents and businesses. The core area consists of provincial parks and Pacific Rim National Park Reserve. Environmentalists hoped the designation would help promote stronger land preservation efforts. Local residents supported it because outside money was being offered for local development efforts. The timber industry did not stand in the way once it was clear that harvesting operations would not be affected. The designation has brought Clayoquot Sound more international attention, but it has not created new protected areas and has not altered land use policies.

(a) Mount Hood National Forest, Oregon

(b) Wood thrush

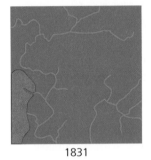

1831

1882

1902

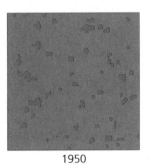

1950

(c) Fragmentation of wooded area (green) in Cadiz Township, Wisconsin

FIGURE 12.20 As human populations have grown and human impacts have increased, most large expanses of natural habitat have become fragmented into smaller disconnected areas. Forest fragmentation from timber harvesting, for example, is evident on the Mount Hood National Forest, Oregon **(a)**. Fragmentation has significant impacts on forest-dwelling species such as the wood thrush, *Hylocichla mustelina* **(b),** a migrant songbird of eastern North America. In forest fragments, wood thrush nests are parasitized by cowbirds that thrive in open country and edge habitats. Forest fragmentation has been extreme in the eastern and midwestern United States; shown in **(c)** are historical changes in forested area in Cadiz Township, Wisconsin, between 1831 and 1950. *Source for (c):* Curtis, J. T. 1956. The modification of mid-latitude grasslands and forests by man. In Thomas, W. L. Jr., ed., *Man's role in changing the face of the earth.* Chicago: Univ. of Chicago Press.

The design of parks and reserves has consequences for biodiversity

Often it is not outright destruction of habitat that threatens species, but rather the fragmentation of habitat (▶ pp. 330–331). Expanding agriculture, spreading cities, highways, logging, and many other impacts have chopped up large contiguous expanses of habitat into small disconnected ones (Figure 12.20a and 12.20c). When this happens, many species suffer. Bears, mountain lions, and other animals that need large ranges in which to roam may disappear. Bird species that thrive in the interior of forests may fail to reproduce when forced near the edge of a fragment (Figure 12.20b). Their nests often are attacked by predators and parasites that favor open habitats surrounding the fragment or that travel along habitat edges. Avian ecologists judge forest fragmentation to be a main reason why populations of many songbirds of eastern North America are declining.

Because habitat fragmentation is such a central issue in biodiversity conservation, and because there are limits on how much land can be set aside, conservation biologists have argued heatedly about whether it is better to make reserves large in size and few in number, or many in number but small in size. Nicknamed the **SLOSS dilemma,** for "single large or several small," this debate is ongoing and complex, but it seems clear that large species that roam great distances, such as the Siberian tiger (Chapter 11), benefit more from the "single large" approach to reserve design. In contrast, creatures such as insects that live as larvae in small areas may do just fine in a number of small isolated reserves, if they can disperse as adults by flying from one reserve to another.

A related issue is whether **corridors** of protected land are important for allowing animals to travel between islands of protected habitat. In theory, connections between fragments provide animals access to more habitat and help enable gene flow to maintain populations in the long term. Many land management agencies and environmental groups try, when possible, to join new reserves to existing reserves for these reasons. It is clear that we will need to think on the landscape level if we are to preserve a great deal of our natural heritage.

Conclusion

Managing natural resources is necessary for resources such as timber, which can be either responsibly and sustainably managed or carelessly exploited and overharvested. The United States, Canada, and other nations have established various federal and regional agencies to oversee and manage publicly held land and the natural resources that are extracted from public land.

Forest management in North America reflects trends in land and resource management in general. Early emphasis on resource extraction evolved into policies on sustained yield and multiple use, a shift that occurred as land and resource availability declined and as the public became more aware of environmental degradation. Public forests today are managed not only for timber production, but also for recreation, wildlife habitat, and ecosystem integrity.

Meanwhile, public support for preservation of natural lands has resulted in parks, wilderness areas, and other reserves, both in North America and abroad. These trends are positive ones, because the preservation and conservation of land and resources is essential if we wish our society to be sustainable and to thrive in the future.

REVIEWING OBJECTIVES

You should now be able to:

Identify the principles, goals, and approaches of resource management

▶ Resource management enables us to sustain natural resources that are renewable if we are careful not to deplete them. (pp. 344–345)
▶ Resource managers have increasingly focused not only on extraction, but also on sustaining the ecological systems that make resources available. (pp. 344–346)
▶ Resource managers have long managed for maximum sustainable yield and are beginning to implement ecosystem-based management and adaptive management. (pp. 345–346)

Summarize the ecological roles and economic contributions of forests, and outline the history and scale of forest loss

▶ Forests provide us economically important timber, but also support biodiversity and contribute ecosystem services. (pp. 347–349)
▶ Developed nations deforested much of their land as settlement, farming, and industrialization proceeded. Today deforestation is taking place most rapidly in developing nations. (pp. 350–354)

Explain the fundamentals of forest management, and describe the major methods of harvesting timber

▶ The U.S. national forests were established to conserve timber for the nation and allow for its sustainable extraction. (p. 354)
▶ Harvesting methods include clear-cutting and other even-aged techniques, as well as selection strategies that maintain uneven-aged stands that more closely resemble natural forest. (pp. 355–357)
▶ Foresters have responded to public demand by beginning to manage for recreation, wildlife habitat, and ecosystem integrity, as well as timber production. (p. 358)
▶ Fire policy has been politically controversial, but scientists agree that we need to take steps to reverse the impacts of a century of fire suppression. (pp. 358, 360)
▶ Certification of sustainable forest products is allowing consumer choice in the marketplace to influence forestry techniques. (p. 361)

Analyze the scale and impacts of agricultural land use

▶ Agriculture has contributed greatly to deforestation, and has had enormous impacts on landscapes and ecosystems worldwide. (pp. 361–364)

Identify major federal land management agencies and the lands they manage

▶ The U.S. Forest Service, National Park Service, Fish and Wildlife Service, and Bureau of Land Management manage U.S. national forests, national parks, national wildlife refuges, and BLM land. (pp. 354, 364–366)

Recognize types of parks and reserves, and evaluate issues involved in their design

▶ Public demand for preservation and recreation has led to the creation of parks, reserves, and wilderness areas in North America and across the world. (pp. 364, 367)

▶ Biosphere reserves are one of several types of internationally managed protected lands. (p. 367)

▶ Because habitat fragmentation threatens wildlife, and landscape patterns matter, conservation biologists are working on how best to design parks and reserves. (pp. 368–369)

TESTING YOUR COMPREHENSION

1. How do minerals differ from timber when it comes to resource management?
2. Compare and contrast maximum sustainable yield, adaptive management, and ecosystem management. Why may pursuing maximum sustainable yield sometimes conflict with what is ecologically desirable?
3. Name several major causes of deforestation. Where is deforestation most severe today?
4. Describe the major methods of timber harvesting.
5. What are some ecological effects of logging? What has been the U.S. Forest Service's response to public concern over the ecological effects of logging?
6. Are forest fires a bad thing? Explain your answer.

7. Approximately what percentage of Earth's land is used for agriculture? What policies have caused conversion of wetlands for agriculture in the United States?
8. Name five reasons that parks and reserves have been created. Why did the U.S. Congress determine in 1964 that wilderness areas were necessary? How do these areas differ from national parks and national wildlife refuges?
9. Why do some people in the United States oppose federal land protection?
10. Roughly what percentage of Earth's land is protected? What types of protected areas have been established in countries outside the United States?

SEEKING SOLUTIONS

1. Do you think maximum sustained yield represents an appropriate policy for resource managers to follow?
2. Consider the economic importance of timber and the ecological roles that forests play. How would you manage the public forests of your country, if you were in charge?
3. People in developed countries are fond of warning people in developing countries to stop destroying rainforest. People of developing countries often respond that this is hypocritical, because the developed nations became wealthy by deforesting their land and exploiting its resources in the past. What would you say to the president of a developing nation, such as Brazil, that is seeking to clear much of its forest?
4. Can you think of a land use conflict that has occurred in your region? How was it resolved, or is it unresolved?

5. What are some ecological effects of agricultural subsidies? Propose arguments for and against subsidies from an ecological point of view.
6. Imagine you have just been elected mayor of a town on Clayoquot Sound. A timber company that employs 20% of your town's residents wants to log a hillside above the town, and the provincial government is supportive of the harvest. But owners of ecotourism businesses that run whale-watching excursions and rent kayaks to out-of-town visitors are complaining that the logging would destroy the area's aesthetic appeal and devastate their businesses—and these businesses provide 40% of the tax base for your town. Greenpeace is organizing a demonstration in your town soon, and news reporters are beginning to call your office, asking what you will do. How will you proceed?

INTERPRETING GRAPHS AND DATA

The invention of the moveable-type printing press by Johannes Gutenberg in 1450 stimulated a demand for paper that has only increased as the world population has grown. The 20th-century invention of the xerographic printing process used in photocopiers and laser printers has accelerated our demand for paper, with most of the raw fiber for paper production coming from wood pulp from forest trees.

1. Approximately how many millions of tons of paper and paperboard were consumed worldwide in 1970? 1980? 1990? 2000?
2. By what percentage did worldwide consumption of paper and paperboard increase from 1970 to 1980? From 1980 to 1990? From 1990 to 2000?
3. If consumption continues to increase at current rates, what do you predict the worldwide consumption of paper and paperboard will be in 2010?

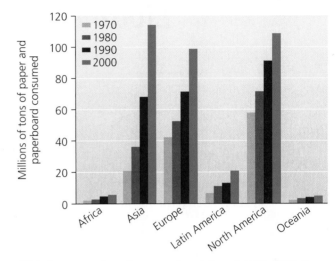

Global consumption of paper and paperboard, 1970–2000. Data from the Food and Agriculture Organization of the United Nations.

CALCULATING ECOLOGICAL FOOTPRINTS

	Population (millions)*	Total paper consumed in 2000 (millions of tons)	Per capita paper consumed in 2000 (pounds)
Africa	840	6	14
Asia	3,766		
Europe	728		
Latin America	531		
North America	319		
Oceania	32		
World	6,216		~114

Data source: Population Reference Bureau.

How much paper do you think you use? You may be surprised to learn that the average North American uses over 300 kg (660 lb) of paper and paperboard per year. Using the estimates of paper and paperboard consumption for each region of the world for the year 2000—as shown in the table in the "Interpreting Graphs and Data" section—calculate the per capita consumption of paper and paperboard for each region of the world using the population data in the table.

1. How much paper would North Americans save each year if we consumed paper at the rate of Europeans?
2. How much paper would be consumed if everyone in the world used as much paper as the average European? As the average North American?
3. Why do you think people in other regions consume less paper, per capita, than North Americans? Name three things you could do to reduce your paper consumption.

Take It Further

Go to www.aw-bc.com/withgott or the student CD-ROM where you'll find:

▶ Suggested answers to end-of-chapter questions
▶ Quizzes, animations, and flashcards to help you study
▶ *Research Navigator*™ database of credible and reliable sources to assist you with your research projects

▶ **GRAPH**it! tutorials to help you master how to interpret graphs
▶ **INVESTIGATE**it! current news articles that link the topics that you study to case studies from your region to around the world

13 Urbanization and Creating Livable Cities

Intersection near Shibuya station, Tokyo, Japan

Upon successfully completing this chapter, you will be able to:

▶ Describe the scale of urbanization

▶ Assess urban and suburban sprawl

▶ Outline city and regional planning and land use strategies

▶ Evaluate transportation options

▶ Describe the roles of urban parks

▶ Analyze environmental impacts and advantages of urban centers

▶ Assess the pursuit of sustainable cities

Pedestrians and joggers use Portland, Oregon's Tom McCall Waterfront Park

Portland
North America
Atlantic Ocean
Pacific Ocean

Central Case: Managing Growth in Portland, Oregon

"Sagebrush subdivisions, coastal condomania, and the ravenous rampage of suburbia in the Willamette Valley all threaten to mock Oregon's status as the environmental model for the nation."
—OREGON GOVERNOR TOM MCCALL, 1973

"We have planning boards. We have zoning regulations. We have urban growth boundaries and 'smart growth' and sprawl conferences. And we still have sprawl."
—ENVIRONMENTAL SCIENTIST DONELLA MEADOWS, 1999

With the fighting words above, Oregon governor Tom McCall challenged his state's legislature in 1973 to take action against runaway urbanization, which many Oregon residents feared would ruin the communities and landscapes they had come to love. McCall echoed the growing concerns of state residents that farms, forests, and open space were being gobbled up for development, including housing for people moving in from California and elsewhere. Foreseeing a future of subdivisions, strip malls, and traffic jams engulfing the pastoral Willamette Valley,

Oregon acted. With Senate Bill 100, the state legislature in 1973 passed a sweeping land use law that would become the focus of acclaim, criticism, and careful study for years afterward by other states and communities trying to manage their own urban and suburban growth.

Oregon's law required every city and county to draw up a comprehensive land use plan, in line with statewide guidelines that had gained popular support from the state's electorate. As part of each land use plan, each metropolitan area had to establish an **urban growth boundary (UGB)**, a line on a map intended to separate areas desired to be urban from areas desired to remain rural. Development for housing, commerce, and industry would be encouraged within these urban growth boundaries, but severely restricted beyond them. The intent was to revitalize city centers, prevent suburban sprawl, and protect farmland, forests, and open landscapes around the edges of urbanized areas.

At the time, Oregon was taking many pioneering environmental steps. The legislature passed the world's first bottle bill (▶ pp. 660, 662), while policymakers and citizen volunteers enhanced state parks and began cleaning up the Willamette River. In Portland, the state's largest city, voters turned down a proposal to construct a new freeway, then chose to rip out an old highway along the riverfront and replace it with a public park. Portland-area residents also established a new regional entity to help plan how land would be apportioned in their region. The Metropolitan Service District, or Metro, represents 25 municipalities and three counties.

Metro adopted the Portland-area urban growth boundary in 1979 and has tried to focus growth on existing urban centers and to build communities where people can walk or take mass transit between home, work, and shopping. These policies have largely worked as intended; Portland's downtown and older neighborhoods have thrived, regional urban centers are becoming denser and more community oriented, mass transit has expanded, and development has been limited on land beyond the UGB. Portland began attracting international attention for its "livability."

To many Portlanders today, the UGB remains the key to maintaining quality of life in city and countryside alike. In the view of its critics, however, the "Great Wall of Portland" is an elitist and intrusive government regulatory tool. Ironically, the Portland area's successes may one day prove its undoing. A continuing influx of people has meant rapid development and rising housing prices. Still, most citizens have supported Oregon's land use rules for the past 30 years, and the system has survived three state referenda and many legal challenges.

In November 2004, however, Oregon voters approved a ballot measure that threatens to eviscerate the very land use reforms they had backed for three decades. Ballot Measure 37 requires the state to compensate landowners if government regulation has decreased the value of their land. For example, regulations prevent landowners outside UGBs from subdividing their lots and selling them for housing development. Under Ballot Measure 37, the state now has to pay these landowners to make up for theoretically lost income, or else allow them to ignore the regulations. Because the state does not have enough money to pay such claims, the measure could effectively gut Oregon's zoning, planning, and land use rules.

A county judge struck down the measure in late 2005, and its fate now seems destined to be decided in the courts. Events in Oregon over the next few years could tell us much about how our cities and landscapes may change in the future.

Our Urbanizing World

We live at a turning point. Beginning about the year 2007, for the first time in human history, more people will live in urban areas than in rural areas. This shift from the countryside into towns and cities, or **urbanization,** is arguably the single greatest change our society has undergone since its transition from a nomadic hunter-gatherer lifestyle to a sedentary agricultural one.

Industrialization has driven the move to urban centers

Agricultural harvests that produced surplus food freed a proportion of citizens from farm life and allowed the rise of specialized manufacturing professions, class structure, political hierarchies, and urban centers (►p. 248). Technological innovations spawned by the industrial revolution (►p. 4) created jobs and opportunities in urban centers for people no longer needed on farms. Industrialization and urbanization bred further technological advances that increased production efficiencies, both on the farm and in the city. This process of positive feedback continues today.

Worldwide, the proportion of population that is urban rose from 30% half a century ago to 48% today. Since 1950, the world's urban population has quadrupled, growing faster than population overall. Between 1975 and 2000, the global urban population increased by 2.53% each year, while the rural population rose only by 0.92% annually. From 2000 to 2030, the United Nations projects that the urban population will grow by 1.83% annually, while the rural population will decline by 0.03% each year.

Trends differ between developed and developing nations, however. In developed nations such as the United States and Canada, urbanization has slowed, because roughly three of every four people already live in cities, towns, and **suburbs,** the smaller communities that ring cities. In 1850, the U.S. Census Bureau classified only 15% of U.S. citizens as urban dwellers. That percentage passed 50% shortly before 1920 and now stands at 80%. Most U.S. urban dwellers reside in suburbs; fully 50% of the U.S. population today is suburban.

In contrast, today's developing nations, where many people still reside on farms, are urbanizing rapidly (Figure 13.1). In nations such as China, India, and Nigeria, rural people are streaming to cities in search of jobs and urban lifestyles. U.N. demographers estimate that virtually all the world's population growth over the next 25 years will be absorbed by urban areas of developing nations.

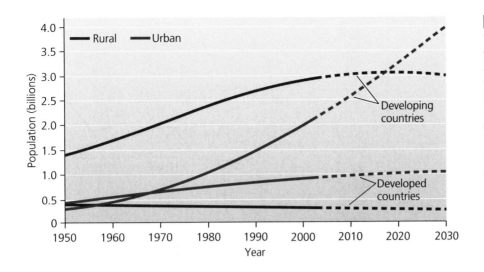

FIGURE 13.1 In developing countries today, urban populations are growing quickly, whereas rural populations are leveling off and may soon begin to decline. Developed countries are already largely urbanized, so in these countries urban populations are growing more slowly, whereas rural populations are falling. Solid lines in the graph indicate past data, and dashed lines indicate projections of future trends. Data from United Nations Population Division. 2004. *World urbanization prospects: The 2003 revision.* New York: UNPD.

Today's urban centers are unprecedented in scale

Cities in themselves are nothing novel. Urban centers where population—and political power—is concentrated have been part of human culture for several thousand years. Ancient Mediterranean civilizations, the great Chinese dynasties, and the Mayan and Incan empires all featured sophisticated and powerful urban centers.

What is new is the sheer scale of today's metropolitan areas. Human population growth (Chapter 8) has placed greater numbers of people in towns and cities than ever before. Today, 46 cities hold more than 5 million people, and 20 cities are home to over 10 million residents (Table 13.1). The metropolitan area of the world's most populous city, Tokyo, Japan, is home to 35 million people. North America's largest metropolises, Mexico City and New York City, each hold about 18 million. However, less than 5% of urban dwellers live in cities of greater than 10 million. Rather, most of them live in smaller cities, such as Portland, Omaha, Winnipeg, Raleigh, Austin, and their still-smaller suburbs.

Urban growth has often been rapid

Urban populations are growing for two reasons: (1) more people are moving from farms to cities than are moving from cities to farms, and (2) the human population overall is growing. Most cities are influenced by both of these trends, but the particular reasons for growth (or decline) vary from one city to another. Portland got its start in the mid-19th century as pioneers arriving by the Oregon Trail settled where the Willamette River flowed into the Columbia River. Situated at the juncture of these two major rivers, Portland had a strategic advantage in trade. The city grew as

it shipped farm products from the fertile Willamette Valley overseas and accepted products shipped in from other North American ports and from Asia. Like many U.S. cities, Portland's population growth stalled in the 1950s to 1970s, as crowding and deteriorating economic conditions caused

Table 13.1 Metropolitan Areas with 10 Million Inhabitants or More, as of 2003

City, Country	Millions of people
Tokyo, Japan	35.0
Mexico City, Mexico	18.7
New York, United States	18.3
Sao Paulo, Brazil	17.9
Mumbai (Bombay), India	17.4
Delhi, India	14.1
Calcutta, India	13.8
Buenos Aires, Argentina	13.0
Shanghai, China	12.8
Jakarta, Indonesia	12.3
Los Angeles, United States	12.0
Dhaka, Bangladesh	11.6
Osaka-Kobe, Japan	11.2
Rio de Janeiro, Brazil	11.2
Karachi, Pakistan	11.1
Beijing, China	10.8
Cairo, Egypt	10.8
Moscow, Russian Federation	10.5
Metro Manila, Philippines	10.4
Lagos, Nigeria	10.1

Source: United Nations Population Division. 2004. *World urbanization prospects: The 2003 revision.* New York: UNPD.

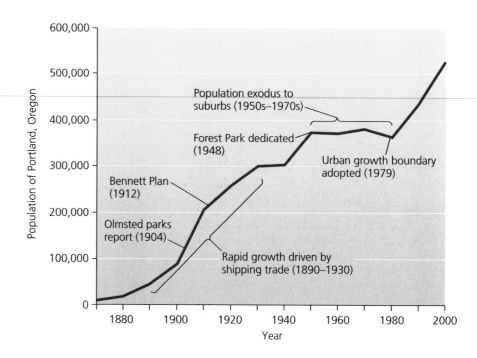

FIGURE 13.2 Once Portland established its position as a strategically located port, international shipping trade provided jobs and boosted its economy, and the city's population grew rapidly. City residents began leaving for the suburbs in the 1950s to 1970s, but policies designed to make the city center more attractive then restarted the city's growth.

many city residents to move outward into the growing suburbs. Unlike many U.S. cities, however, policies undertaken to improve the city center's attractiveness helped restart Portland's growth (Figure 13.2).

Some American cities have grown faster than Portland. For instance, Chicago was incorporated in 1833 with a population of 350, but this frontier town grew with extraordinary speed as railroads funneled through it the resources of the vast lands to the west on their way to the cities and consumers to the east. Chicago became a center for grain processing, livestock slaughtering, meatpacking, and much else. With ample employment opportunities, the city's population soared to 500,000 in 1880, 1.1 million in 1890, and 3.5 million during World War II. In recent years, many cities in the southern and western United States have gone through similar growth spurts, as people (particularly retirees) from northern and eastern states have moved south and west in search of warmer weather or more space. Between 1990 and 2000, for instance, the population of the Atlanta metropolitan area grew by 39%, while that of the Phoenix region grew by 45%, and that of the Las Vegas metropolitan area grew by 83%.

Internationally, most fast-growing cities today are in the developing world, because industrialization is decreasing the need for farm labor and is increasing commerce and jobs in cities. Sadly, another reason is because wars, conflict, and ecological degradation are driving millions of people out of the countryside and into cities. Cities like Mumbai (Bombay), India; Lagos, Nigeria; and Cairo, Egypt are growing in population even faster than American cities did. All too often, they are doing so without the economic growth to match their population growth. As a

result, many of these cities are facing overcrowding, pollution, and poverty. Nearly three of every four governments of developing nations have by now enacted policies to discourage the movement of people from the countryside into cities.

Various factors influence the geography of urban areas

Real estate agents have long used the saying, "Location, location, location," to stress how much a home's whereabouts determines its value—and location is vitally important for urban centers, as well. Successful cities tend to be located in places that give them economic advantages. Think of any major city, and chances are it's situated along a major river, seacoast, railroad, or highway—some corridor for trade that has driven economic growth. Environmental variables such as climate, topography, and the configuration of waterways go a long way toward determining whether a small settlement will become a large city. Many well-located cities, like Portland and Chicago, have acted as linchpins in trading networks, funneling in resources from agricultural regions, processing them and manufacturing products, and shipping those products to other markets.

In fact, all cities, from ancient times to the present day, have supported themselves by drawing in resources from outlying rural areas through trade, persuasion, or conquest. In turn, cities have historically influenced how people use land in surrounding areas (Figure 13.3). Although city life and country life may seem very different, cities and the rural regions surrounding them have always been linked by tight economic relationships.

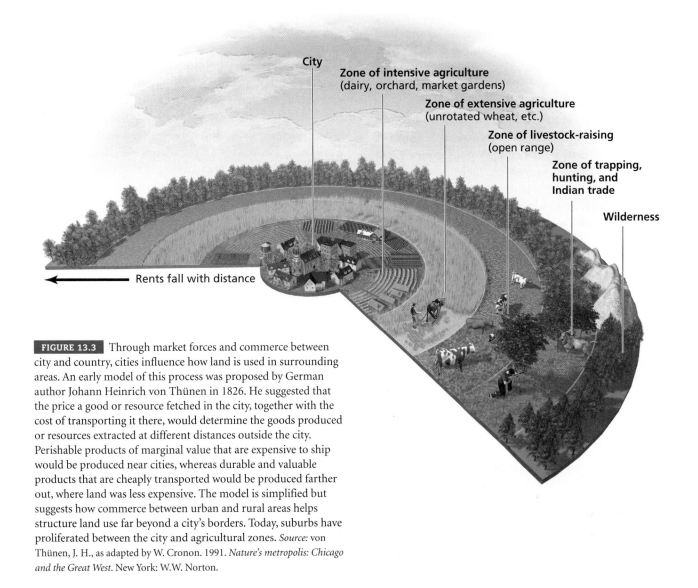

City

Zone of intensive agriculture
(dairy, orchard, market gardens)

Zone of extensive agriculture
(unrotated wheat, etc.)

Zone of livestock-raising
(open range)

**Zone of trapping,
hunting, and
Indian trade**

Wilderness

← Rents fall with distance

FIGURE 13.3 Through market forces and commerce between city and country, cities influence how land is used in surrounding areas. An early model of this process was proposed by German author Johann Heinrich von Thünen in 1826. He suggested that the price a good or resource fetched in the city, together with the cost of transporting it there, would determine the goods produced or resources extracted at different distances outside the city. Perishable products of marginal value that are expensive to ship would be produced near cities, whereas durable and valuable products that are cheaply transported would be produced farther out, where land was less expensive. The model is simplified but suggests how commerce between urban and rural areas helps structure land use far beyond a city's borders. Today, suburbs have proliferated between the city and agricultural zones. *Source:* von Thünen, J. H., as adapted by W. Cronon. 1991. *Nature's metropolis: Chicago and the Great West.* New York: W.W. Norton.

Weighing the Issues:
What Made Your City?

Consider the town or city in which you live, or the major urban center located nearest you. Why do you think it developed into an urban area? What physical, social, or environmental factors may have aided its growth?

Spatial patterns of urbanization can change, however, with changing times. Today several factors are enabling population centers to decentralize in developed nations. For one thing, people now are globally interconnected to an unprecedented degree. Being located on a river or seacoast is no longer as vital to a city's success in our age of global commerce, jet travel, diplomacy, television, cell phones, and the Internet. Globalization has connected distant societies, and businesses and individuals can more easily communicate from locations away from major city centers. Moreover, fossil fuels have enabled the outward spread of cities. By easing long-distance transport, fossil fuels have made it easier to commute into and out of cities and to import and export resources, goods, and waste. Such factors have enabled a shift of population from cities to suburbs, particularly in the United States and Canada.

People have moved to suburbs

By the mid-20th century, many cities in the United States, Canada, and other developed nations had accumulated more people than these cities had jobs to offer. Unemployment rose, and crowded inner-city areas began to suffer increasing poverty and crime. As inner cities declined economically from the 1950s onward, many affluent city dwellers chose to move outward to the cleaner, less crowded, and more parklike suburban communities beginning to surround the cities. These people were pursuing more space, better economic opportunities, cheaper real estate, less crime, and better schools for their children. The exodus to the suburbs further hastened the economic decline of central cities. Chicago's population declined to 80% of its peak because so many residents moved to its suburbs. Philadelphia's population fell to 76% of its peak, and Detroit's to just 55%.

In most ways, suburbs have delivered the qualities people sought in them. The wide spacing of houses, with each house on its own plot of land, gives families room and privacy. However, by allotting more space to each person, suburban growth has spread human impact across the landscape. Natural areas have disappeared as housing developments are constructed. We have built extensive road networks to ease travel, but suburbanites now find themselves needing to drive everywhere. They commute longer distances to work and spend more time commuting in more congested traffic. The expanding rings of suburbs surrounding cities have grown larger than the cities themselves, and towns are running into one another. These aspects of suburban growth have inspired a new term: *sprawl.*

Sprawl

The term *sprawl* has become laden with meanings and connotes different things to different people. To some, sprawl is aesthetically ugly, environmentally harmful, and economically inefficient. To others, it is the collective outgrowth of reasonable individual desires and decisions in a world of growing human population. We can begin our discussion by giving **sprawl** a simple nonjudgmental definition: the spread of low-density urban or suburban development outward from an urban center.

Today's urban areas spread outward

As urban and suburban areas have grown in population, they have also grown spatially. This growth is obvious from maps and satellite images of rapidly spreading cities such as Las Vegas (Figure 13.4). Houses and roads supplant over 1 million ha (2.5 million acres) of U.S. rural land each year—over 2,700 ha (6,700 acres) every day.

Because suburban growth entails allotting more space per person than in cities, in most cases this outward spatial growth across the landscape has outpaced the growth in

(a) Las Vegas, Nevada, 1972

(b) Las Vegas, Nevada, 2002

FIGURE 13.4 Satellite images show the type of rapid urban and suburban expansion that many people have dubbed *sprawl.* Las Vegas, Nevada, is currently one of the fastest-growing cities in North America. Between 1972 **(a)** and 2002 **(b)**, the population increased more than fivefold, and the developed area has risen more than threefold.

(a) Uncentered commercial strip development

(b) Low-density single-use development

(c) Scattered, or leapfrog, development

(d) Sparse street network

FIGURE 13.5 Several standard approaches to development can result in sprawl. In uncentered commercial strip development (**a**), businesses are arrayed in a long strip along a roadway, and no attempt is made to create a centralized community with easy access for consumers. In low-density, single-use residential development (**b**), homes are located on large lots in residential tracts far away from commercial amenities. In scattered or leapfrog development (**c**), developments are created at great distances from a city center and are not integrated. In developments with a sparse street network (**d**), roads are far enough apart that moderate-sized areas go undeveloped, but not far enough apart for these areas to function as natural areas or sites for recreation. All these development approaches necessitate frequent automobile use.

numbers of people. In fact, many researchers define *sprawl* as the physical spread of development at a rate greater than the rate of population growth. For instance, Phoenix grew from 105,000 residents spread over 44 km² (17 mi²) in 1950 to 1.3 million residents spread over 1,200 km² (470 mi²) in 2002; its land area grew 27 times larger, while the population grew 12 times larger. Between 1950 and 1990, the population of 58 major U.S. metropolitan areas rose by 80%, but the land area they covered rose by 305%. Even in 11 metro areas where population declined between 1970 and 1990, the amount of land covered increased.

Several types of development approaches can lead to sprawl (Figure 13.5). As a result, Chicago's metropolitan area now consists of more than 9 million people spread over 23,000 km² (9,000 mi²)—an area 40 times the size of the city proper. Each person in the suburban region takes up an average of 11 times as much space as do residents of the city proper. As for Portland, today its growing suburbs hold as many people as does the city proper.

Sprawl has several causes

There are two main components of sprawl. One is human population growth—there are simply more people alive each year. The other is per capita land consumption—each person takes up more land. The amount of sprawl is a function of the number of people added to an area times the amount of land the average person occupies.

A study of the 100 major metropolitan areas of the United States between 1970 and 1990 found that, on average, each of these two factors contributes about equally to sprawl. Cities varied, however, in which factor was more important. The Los Angeles metro area increased in population density by 9% between 1970 and 1990, becoming the nation's most densely populated metro area. Increasing density should be a good recipe for preventing sprawl. Yet L.A. still grew in size by a whopping 1,021 km^2 (394 mi^2). This spatial growth, despite the increase in density, clearly resulted from an overwhelming influx of new people.

Detroit provides a contrasting example. The Detroit metro area lost 7% of its population between 1970 and 1990, yet it expanded in area by 28%. Clearly population growth was not the issue here; rather, sprawl was caused solely by increased per capita land consumption.

We discussed reasons for human population growth in general in Chapter 8. As for the increase in per capita land consumption, there are numerous reasons. Technologies such as telecommunications and the Internet have fostered movement away from city centers. Because of these technologies, many businesses no longer need the infrastructure a major city provides, and many workers have greater flexibility to live wherever they desire. The primary reasons for greater per capita land consumption, however, are that most people simply like having some space and privacy and dislike congestion. Furthermore, in the consumption-oriented American lifestyle that promotes bigger houses, bigger cars, and bigger TVs, having more space to house one's possessions becomes important. Unless there are overriding economic or social disadvantages, most people prefer living in a cleaner, more spacious, more affluent community.

Economists, politicians, and city boosters have almost universally encouraged the unbridled spatial expansion of cities and suburbs. The conventional assumption has been that growth is good and that attracting business, industry, and residents will unfailingly increase a community's economic well-being, political power, and cultural influence. Today, however, this assumption is increasingly being challenged. As the negative effects of sprawl on citizens' lifestyles accumulate, growing numbers of people have begun to question the mantra that all growth is good.

What is wrong with sprawl?

Sprawl means different things to different people. To some, the word evokes strip malls, homogenous commer-

cial development, and tracts of cookie-cutter houses encroaching on farmland and ranchland. It may suggest traffic jams, destruction of wildlife habitat, and loss of natural land around cities. However, for other people, sprawl represents the collective result of choices made by millions of well-meaning individuals trying to make a better life for themselves and their families. In this view, those who decry sprawl are elitist, show disdain for the popular will, and fail to appreciate the good things about suburban life. Let us try, then, to leave the emotional debate aside and assess the impacts of sprawl (see "The Science behind the Story," ▶ pp. 382–383).

Transportation Most studies show that sprawl constrains transportation options, essentially forcing people to drive cars. These constraints include the need to own a vehicle and to drive it most places, the need to drive greater distances or to spend more time in vehicles, a lack of mass transit options, and more traffic accidents. Across the United States, during the 1980s and 1990s the average length of work trips rose by 36%, and total vehicle miles driven increased at three times the rate of population growth. An automobile-oriented culture also increases dependence on nonrenewable petroleum, with the attendant economic and environmental consequences (▶ pp. 575–580).

Pollution Sprawl's effects on transportation give rise to increased pollution. Carbon dioxide emissions from vehicles exacerbate global climate change (Chapter 18) while nitrogen- and sulfur-containing air pollutants contribute to tropospheric ozone, urban smog, and acid precipitation (▶ pp. 514–518). Waterways are polluted by substances such as motor oil and road salt from roads and parking lots. Runoff of polluted water from paved areas is estimated to be about 16 times greater than from naturally vegetated areas. Such air and water pollution has been shown to degrade natural environments and pose risks to human health.

Health Besides the health impacts of pollution, some recent studies suggest that sprawl promotes physical inactivity because driving cars largely takes the place of walking during daily errands. This research has suggested that physical inactivity has increased obesity and high blood pressure, which can in turn lead to other ailments. A 2003 study found that people from the most-sprawling U.S. counties weigh 2.7 kg (6 lb) more for their height than people from the least-sprawling U.S.

Suburban Sprawl

Do you see problems with urban and suburban sprawl? **If so, what are the best solutions? What strategies should we pursue in making our communities more livable?**

The Myth of Urban Sprawl

Russians say Americans have no real problems, so they make them up. Urban sprawl is one of those made-up problems. Yet the proposed remedy to sprawl— sometimes called "smart growth," though it is anything but smart—will cause far more problems than it solves.

The U.S. Department of Agriculture says that urban development does not threaten American farm productivity. Nor is it a threat to rural open space: All of the cities, suburbs, and towns in the United States occupy less than 4% of the nation's land area.

University of Southern California planning professor Peter Gordon points out that low-density development is a remedy for, not the cause of, congestion, air pollution, and many other problems. Studies claiming that suburbs cause obesity, crime, and other social problems are little more than junk science, being based on inadequate data with little statistical significance and usually confusing cause and effect.

So-called smart growth says more people should live in high-density, mixed-use developments. These developments can be attractive to some people, mainly young adults with no children. But the market for them is limited. Most Americans still find a single-family home with a large yard to be their American dream.

Attempting to impose smart growth on more people has many unfortunate effects. Because it doesn't significantly reduce the miles people drive, it increases congestion; and because cars pollute more in stop-and-go traffic, it increases air pollution. Smart growth makes housing unaffordable to low- and even middle-income families, and makes neighborhoods more vulnerable to crime.

Instead of attempting to impose their lifestyle preferences on others, city officials should simply ensure that people pay the full costs of whatever lifestyle they prefer. Once that happens, people can be free to choose to live in high densities or low, and to drive, walk, bicycle, or ride transit.

Randal O'Toole is an economist and the director of the American Dream Coalition, which seeks to solve urban problems without reducing people's personal freedom. He is the author of *The Vanishing Automobile and Other Urban Myths: How Smart Growth Will Harm American Cities.* He has taught environmental economics at Yale, the University of Calfornia at Berkeley, and Utah State University.

The Real Problem with Sprawl

The most visible problem associated with sprawl is one of livability—sprawl's effect on our quality of life. However, the bigger but less visible problem is sprawl's contribution to global warming.

Until the mid-20th century, every village, town, and city in the world was made up of mixed-use, pedestrian-friendly neighborhoods. But after World War II, the time-tested *neighborhood* was discarded in favor of an untested invention, now known as *sprawl.*

Sprawl is based on two simple premises: first, that each land use be separated from every other; and second, that these now distant land uses be connected by a massive automotive infrastructure. In sprawl, walking not only serves limited purposes but also can often be dangerous. For this reason, a typical suburban household generates more than 12 one-way car trips per day.

The ecological impact is profound. Motor vehicles, the lifeblood of sprawl, are the single greatest contributor to global warming. Over the past 60 years, we have created a built environment that requires most adults to own a car and drive it every day. Unless we quickly make the change to nonpolluting vehicles, global warming will remain the strongest argument against sprawl.

But what if cars did not pollute; would sprawl then offer a satisfactory solution? This is where quality of life enters the picture. In a society in which cars are a prerequisite to social and economic viability, those who can't drive— one-third of the population—become second-class citizens. And because everyone who can drive must drive, we spend inordinate amounts of time stuck in traffic, time that would be better spent in less stressful pursuits. The frustration with this situation—the hours that we spend trying to reconnect our artificially disassociated lives—is the reason why most people who argue against sprawl cite concerns over quality of life.

Jeff Speck is director of design at the National Endowment for the Arts. Prior to joining the Endowment in 2003, he spent 10 years as Director of Town Planning at the firm of Duany Plater-Zyberk and Co., Architects and Town Planners (DPZ). With Andres Duany and Elizabeth Plater-Zyberk, he is the co-author of the book *Suburban Nation: The Rise of Sprawl and the Decline of the American Dream,* published in March 2000 by North Point/Farrar Straus Giroux.

Explore this issue further by accessing **Viewpoints** at www.aw-bc.com/withgott.

Measuring the Impacts of Sprawl

The Science behind the Story

Critics of sprawl have blamed it for so many societal ills that it can make a person feel guilty just for being born in the suburbs or shopping at a mall. But what does scientific research tell us are the actual consequences of sprawl?

When Reid Ewing of Rutgers University and his team set out to measure the impacts of sprawl, they discovered that researchers studying sprawl have been hard-pressed even to agree on a definition of the term or on how to measure it. Surveying the literature, Ewing's team found that researchers using different criteria ranked cities in very different ways. For instance, in most studies Los Angeles was deemed more sprawling than Portland, but in some, Portland was judged to suffer worse sprawl than L.A.

So Ewing, Rolf Pendall of Cornell University, and Don Chen of the nonprofit group Smart Growth America tried to define *sprawl* in terms as simple as possible, without mixing sprawl's consequences into the definition. They decided that sprawl occurs when the spread of development across the landscape far outpaces population growth.

Ewing, Pendall, and Chen then devised four criteria by which to rank 83 of the largest U.S. metropolitan areas in terms of sprawl. Sprawling cities would show:

Traffic congestion is one of the most recognized impacts of sprawl.

▶ Low residential density
▶ Distant separation of homes, employment, shopping, and schools
▶ Lack of "centeredness," that is, lack of activity in community centers and downtown areas
▶ Street networks that make many streets hard to access

For each criterion, Ewing's team measured multiple factors—22 variables in all. They then devised a way to analyze the variables and arrive at a cumulative index of sprawl. Finally, they obtained data from municipalities throughout the country.

Ewing's team's rankings showed that the most sprawling area in the nation was the Riverside–San Bernardino, California, region, which has expanded quickly in recent decades. The area with the least sprawl was New York City,

whose historically dense population and vibrant neighborhoods kept it geographically compact, relative to its number of inhabitants. The 10 most- and least-sprawling areas are listed in the accompanying tables.

With their sprawl scores in hand, the researchers next correlated those scores with a number of transportation variables. They found that people in the 10 most-sprawling metros owned more cars (180 per 100 households) than people in the 10 least-sprawling metros (162 per 100 households). They also found that residents of the most-sprawling metros drove an average of 43 km (27 mi) per day, whereas those of the least-sprawling metros drove only 34 km (21 mi). In addition, people in the most-sprawling metros used public transit far less and suffered 67% more traffic fatalities than those in the least-sprawling metros.

Strikingly, the study found no significant difference in commute time from home to work for people of sprawling versus less-sprawling metro areas. Critics of sprawl have long blamed sprawl for traffic congestion and commute delays. Advocates of suburban spread have argued that more streets ease commutes and that regions can sprawl their way out of congestion. The Ewing team's results seem to suggest that each side may have a point and that the effects may cancel one another out.

counties. Over 25% of people from the most-sprawling counties showed hypertension (high blood pressure), whereas fewer than 23% of people from the least-sprawling U.S. counties showed this condition.

Land use   The spread of low-density development over large areas of land means that more land is developed while less is left as forests, fields, farmland, or ranchland. Of the estimated 1 million ha (2.5 million acres) of U.S.

Because vehicle emissions cause air pollution, the researchers also measured levels of tropospheric ozone (▶ p. 507). Sprawling areas had worse air pollution, they found; ozone levels were 40% higher in the most-sprawling metros.

None of these results prove that sprawl *causes* these impacts, because statistical correlation alone does not imply causation. However, taken together, the results suggest that spatial patterns of development may influence people's options, impacts, and behavior relative to transportation. The results were published in 2003 in the *Transportation Research Record*, and in a 2002 report published by Smart Growth America.

Ewing and his colleagues are continuing to examine other impacts of sprawl. For instance, a study in 2004 went beyond ozone to look at other health correlates of sprawl. Because other studies have found that residents of sprawling areas depend more on cars and walk less, the researchers hypothesized that they would find more obesity and poorer health in people living in sprawling areas. That is exactly what they found: People from sprawling metros are on average heavier for their height and show increased instances of high blood pressure (although not more diabetes or cardiovascular disease).

The 10 Most-Sprawling American Urban Areas

Rank	Metropolitan Region
1	Riverside–San Bernardino, CA
2	Greensboro–Winston-Salem–High Point, NC
3	Raleigh–Durham, NC
4	Atlanta, GA
5	Greenville–Spartanburg, SC
6	West Palm Beach–Boca Raton–Delray Beach, FL
7	Bridgeport–Stamford–Norwalk–Danbury, CT
8	Knoxville, TN
9	Oxnard–Ventura, CA
10	Fort Worth–Arlington, TX

1 = most-sprawling, 10 = less-sprawling.

The 10 Least-Sprawling American Urban Areas

Rank	Metropolitan Region
83	New York, NY
82	Jersey City, NJ
81	Providence–Pawtucket–Woonsocket, RI
80	San Francisco, CA
79	Honolulu, HI
78	Omaha, NE–IA
77	Boston–Lawrence–Salem–Lowell–Brockton, MA
76	Portland, OR
75	Miami–Hialeah, FL
74	New Orleans, LA

83 = least-sprawling, 74 = more-sprawling. *Source:* Ewing, R., et al. 2002. *Measuring sprawl and its impact.* Washington, D.C.: Smart Growth America.

As more studies on the impacts of sprawl accumulate, we will have a better idea of the benefits and costs of our choices in urban design.

land converted each year, roughly 60% is agricultural land and 40% is forest. These types of open space provide vital resource production, aesthetic beauty, habitat for wildlife, cleansing of water, places for recreation, and many other ecosystem services (▶ p. 39, 41). Sprawl generally diminishes all these amenities.

Economics Sprawl drains tax dollars from existing communities and funnels them into infrastructure for new development on the fringes of those communities. Money that could be spent maintaining and improving downtown centers is instead spent on extending the road system, water and sewer system, electricity grid, and telephone lines to distant developments, and extending police and fire service, schools, and libraries. For instance, one study calculated that sprawling development at Virginia Beach, Virginia, would require 81% more in infrastructure costs and would drain 3.7 times more from the community's general fund each year than compact urban development. Advocates for sprawling development argue that taxes on new development eventually pay back the investment made in infrastructure, but studies have found that in most cases taxpayers continue to subsidize new development, especially if municipalities do not pass on infrastructure costs to developers.

Weighing the **Issues:**
Sprawl Near You

Is there sprawl in the area where you live? Are you bothered by it, or not? Has development in your area had any of the impacts described above? Do you think your city or town should use its resources to encourage outward growth?

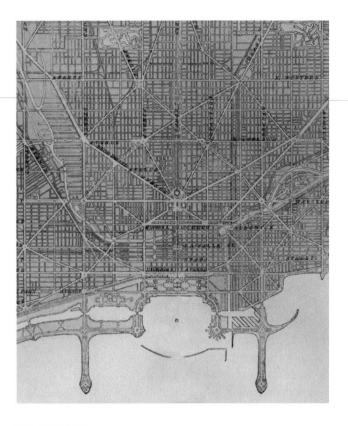

FIGURE 13.6 Daniel Burnham's 1909 *Plan of Chicago* included parks and greenways, efficient transportation routes, and increased access to the lakefront for city residents. The plan has served as a model for city planners ever since.

Creating Livable Cities

Architects, planners, developers, and policymakers across North America today are trying to restore the vitality of city centers and respond to the challenges that suburban sprawl presents. But efforts to design and restore cities are as old as cities themselves.

City and regional planning are means for creating livable urban areas

City planning is the professional pursuit that attempts to design cities so as to maximize their efficiency, functionality, and beauty. City planners advise policymakers on where different types of development should be allowed, transportation needs, public parks, and other matters.

City planning in North America came into its own in the early 20th century. Landscape architect Daniel Burnham's 1909 *Plan of Chicago* (Figure 13.6) represented the first thorough plan for an American city, and it was largely implemented over the following years and decades. This plan expanded city parks and playgrounds, improved neighborhood living conditions, streamlined traffic systems, and cleared industry and railroads from the shore of Lake Michigan to provide public access to the lake.

Portland gained its own comprehensive plan just 3 years later. Edward Bennett's *Greater Portland Plan* recommended rebuilding the harbor; dredging the river channel; constructing new docks, bridges, tunnels, and a waterfront railroad; superimposing wide radial boulevards on the old city street grid; establishing civic centers downtown; and greatly expanding the number of parks. Voters approved the plan by a 2:1 margin, but they defeated a bond issue that would have paid for park development. As the century progressed, several other major planning efforts were conducted, and some ideas, such as establishing a downtown public square, came to fruition.

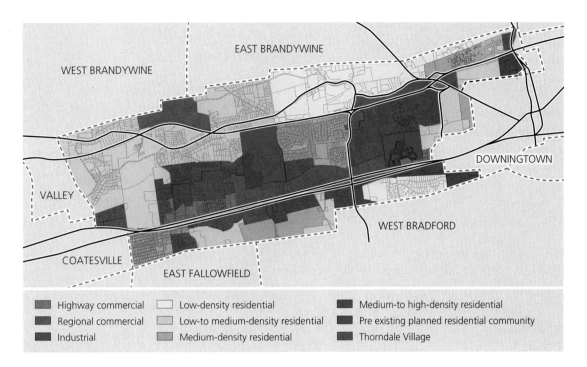

FIGURE 13.7 By zoning areas of a city for different uses, planners can guide how an urban area develops. Zoning puts restrictions on what private landowners can do with their land, but it is intended to maximize prosperity, efficiency, and quality of life for the community. This zoning map for Caln Township, Pennsylvania, shows several patterns common to modern zoning practice. Public and institutional uses are clustered together in a downtown area. Industrial uses are clustered together, away from most residential areas. Commercial uses are clustered along major roadways, and residential zones generally are higher in density toward the center of town.

City planning grew in importance throughout the 20th century as urban populations expanded, inner cities decayed, and wealthier residents fled to the suburbs. In today's world of sprawling metropolitan areas, **regional planning** has become just as important. Regional planners deal with the same issues as city planners, but they work on broader geographic scales and must coordinate their work with multiple municipal governments. In some places, regional planning has been institutionalized in formal governmental bodies; the Portland area's Metro is the epitome of such a regional planning entity.

Weighing the **Issues:**
Your Urban Area

Think of your favorite parts of the city you know best. What aspects do you like about them? What do you dislike about some of your least favorite parts of the city? What could this city do to improve the quality of life for its inhabitants?

Zoning is a key tool for planning

One tool planners use is **zoning,** the practice of classifying areas for different types of development and land use (Figure 13.7). Industrial plants may be kept out of districts zoned for residential use to preserve the cleanliness and tranquility of residential neighborhoods, for instance. The specification of zones for different types of development gives planners a powerful means of guiding what gets built where. Zoning can restrict areas to just one use, as is often done with suburban residential tracts in so-called bedroom communities. Or, zoning can allow the type of mixed use—residential and commercial, for instance—that some planners say can reinvigorate urban neighborhoods. Zoning also gives home buyers and business owners security; they know in advance what types of development can and cannot be located nearby.

Zoning involves government restriction of the use of private land and represents a top-down constraint on personal property rights. For this reason, some people consider zoning a regulatory taking (▶ p. 65) that violates individual freedoms. Others defend zoning, saying that government has a proper role in setting certain limitations on property rights for the good of the community. Similar debates arise with endangered species management (▸pp. 331–334) and other environmental issues.

Oregon voters sided with private property rights when in 2004 they passed Ballot Measure 37, which shackled government's ability to enforce zoning regulations with landowners who had owned their land before the regulations were enacted. However, opponents of the measure insisted that the state's voters had not understood the complicated terms of the proposal. They predicted that many will change their minds once they begin seeing new development they do not wish to occur. For the most part, people have supported zoning over the years because the common good it produces for communities is widely felt to outweigh the restrictions on private use.

Weighing the Issues:
Zoning and Development

Imagine you own a 10-acre parcel of land that you want to sell for housing development—but the local zoning board rezones the land so as to prohibit the development. How would you respond?

Now imagine that you live next to someone else's undeveloped 10-acre parcel, and you enjoy the privacy it provides—but the local zoning board rezones the land so that it can be developed into a dense housing subdivision. How would you respond?

What factors do you think members of a zoning board should take into consideration when deciding how to zone or rezone land in a community?

Urban growth boundaries have become popular

Planners intended Oregon's urban growth boundaries to limit sprawl by containing future growth largely within existing urbanized areas. The UGBs aimed to revitalize downtowns; protect farms, forests, and their industries; and ensure urban dwellers some access to open space near cities. Among Willamette Valley towns, the long-term goal was to head off the growth of a potential megalopolis stretching from Eugene to Seattle (Figure 13.8).

Since Oregon began its experiment, a number of other states, regions, and cities have adopted UGBs—from Boulder, Colorado, to Lancaster, Pennsylvania, to many California communities. In their own ways, all the UGBs aim to concentrate development, prevent sprawl, and pre-

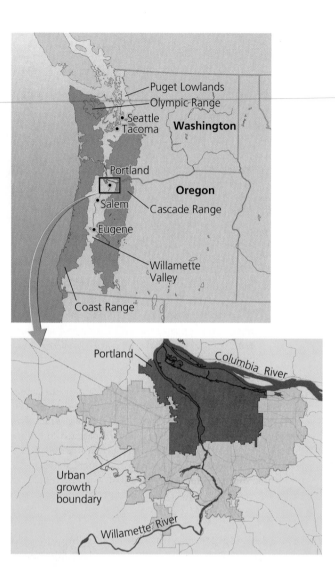

FIGURE 13.8 Oregon's urban growth boundaries (UGBs) were a response to fears that suburban sprawl might one day stretch in a great megalopolis from Eugene, Oregon, up the Willamette Valley to Portland, and up the Puget Lowlands to Seattle, Washington. The Portland area's 956-km^2 (369-mi^2) UGB encompasses Portland (dark gray) and portions of 24 other communities and three counties (light gray). Its jagged edge separates areas planners have earmarked for urban development from areas they have chosen to protect from urban development.

serve working farms, orchards, ranches, and forests. UGBs unquestionably help maintain farms and natural areas. UGBs also appear to reduce the amounts municipalities have to pay for infrastructure, compared to sprawl. The best estimate nationally is that UGBs save taxpayers about 20% on infrastructure costs. However, UGBs also seem to increase housing prices within their boundaries, although precisely how much is difficult to assess; the best estimate for Portland is that its UGB adds roughly $10,000 to the average home price.

Although housing is becoming less affordable, in most other ways the Portland-area UGB (see Figure 13.8) is working as intended. It has lowered prices for land outside the UGB while increasing prices within it. It has restricted development outside the UGB. It has increased the density of new housing inside the UGB by over 50% as homes are built on smaller lots and as multistory apartments fulfill a vision of "building up, not out." Downtown employment rose by 73% between 1970 and 1995 as businesses and residents alike invested in the central city. And Portland has been able to absorb considerable immigration while avoiding rampant sprawl. However, urbanized area still did increase by 101 km^2 (39 mi^2) in the decade after the UGB was established, because 146,000 people were added to the population. This fact suggests that relentless population growth may thwart even the best antisprawl efforts.

When Oregon's lawmakers passed Senate Bill 100, they understood that the UGBs would likely need to be enlarged from time to time to accommodate population growth. Indeed, Metro has enlarged the Portland-area UGB three dozen times, expanding it by over 10% between 1998 and 2004. Population projections for the region suggest there will be pressure for still more expansion. Many other locations that have instituted UGBs have chosen to expand them later, as well. For instance, a group of 48 local governments from cities, towns, and counties along the Colorado Front Range in 1997 agreed to a UGB that allowed expansion of the urbanized area to 1,130 km^2 (700 mi^2) by the year 2020, a 30% increase in size. Soon, however, this limit was raised three times, finally to 1,369 km^2 (850 mi^2) by the year 2030. Given the reality of repeated UGB expansions, it must be asked whether UGBs will truly limit sprawl in the long run.

"Smart growth" aims to counter sprawl

As more people have begun to feel negative effects of sprawl on their everyday lives, efforts to control growth have sprung up throughout North America. Oregon's Senate Bill 100 was one of the first, and since then dozens of states, regions, and cities have adopted similar land use policies. Urban growth boundaries and many other ideas from these policies have coalesced under the concept of **smart growth**.

Proponents of smart growth want communities to manage their growth in ways that maintain or improve quality of life for residents. To them, this means guiding the rate, placement, and style of development such that it serves the environment, the economy, and the commu-

Table 13.2   Ten Principles of "Smart Growth"
▶ Mix land uses
▶ Take advantage of compact building design
▶ Create a range of housing opportunities and choices
▶ Create walkable neighborhoods
▶ Foster distinctive, attractive communities with a strong sense of place
▶ Preserve open space, farmland, natural beauty, and critical environmental areas
▶ Strengthen and direct development towards existing communities
▶ Provide a variety of transportation choices
▶ Make development decisions predictable, fair, and cost-effective
▶ Encourage community and stakeholder collaboration in development decisions

Source: U.S. Environmental Protection Agency, 2005.

nity. Smart growth tries to promote healthy neighborhoods and communities, jobs and economic development, transportation options, and environmental quality. It aims to counter sprawl and to rejuvenate the older existing communities that so often are drained and impoverished by sprawl. Smart growth often means "building up, not out"—focusing development and economic investment in existing urban centers and favoring multistory shop-houses and high-rises over one-story homes spreading outward. Ten principles of smart growth are listed in Table 13.2.

People have pursued the goals of smart growth in various ways. Several hundred referenda and ballot measures in dozens of U.S. states in the past decade have proposed various solutions to slow sprawl and rechannel development into more community-friendly paths. Many of these ballot measures have won, and a number of experiments are going on right now across North America.

The "new urbanism" is now in vogue

Much of what Portland and cities like it have been trying to accomplish is compatible with a school of thought called the **new urbanism**, currently in vogue among many architects, planners, and developers. This approach seeks to design neighborhoods on a walkable scale, with homes, businesses, schools, and other amenities all close together for convenience. The aim is to create functional neighborhoods in which most of a family's needs can be met close to home without the use of a car. Green spaces, trees, a mix of architectural styles,

FIGURE 13.9 This plaza at Mizner Park, Boca Raton, Florida, is part of a planned community built in the style of the "new urbanism." Homes, schools, and businesses are mixed close together in a centered neighborhood so that most amenities are within walking distance.

and creative street layouts add to the visual interest and pleasantness of new urbanist developments (Figure 13.9). In these ways, these developments mimic the traditional urban neighborhoods that existed until the advent of suburbs.

Over 600 communities in the new urbanist style across North America are in various stages of planning or construction, and many are now complete. Seaside, Florida, was the first, and others have followed, such as Kentlands in Gaithersburg, Maryland; Addison Circle in Addison, Texas; Mashpee Commons in Mashpee, Massachusetts; Harbor Town in Memphis, Tennessee; Celebration in Orlando, Florida; and Orenco Station, west of Portland.

New urbanist neighborhoods are generally connected to public transit systems. In a related approach called *transit-oriented development,* compact communities in the new urbanist style are arrayed around the stops on a major rail transit line, enabling people to travel most places they need to go by train and foot alone. Several lines of the Washington, D.C. Metro system have been successfully developed in this manner.

Transportation options are vital to livable cities

A key ingredient in any planner's recipe for improving the quality of urban life is making multiple transportation options available to citizens. These options include public buses, trains and subways, and light rail (smaller rail systems powered by overhead electric wires). As long as an

urban center is large enough to support the infrastructure necessary, these mass transit options are cheaper, more energy-efficient, and cleaner than roadways choked with cars (Figure 13.10). They also ease traffic congestion by carrying passengers who would otherwise be driving cars. An average Portland bus is estimated to keep about 250 cars off the road each day. Compared to mass transit rail systems, road networks take up more space, and cars emit more pollution. The fuel and productivity lost to traffic jams have been estimated to cost the U.S. economy $74 billion each year.

The nation's most-used train systems have been the extensive heavy rail systems in America's largest cities, such as New York's subways, Washington, D.C.'s Metro, the T in Boston, and the San Francisco Bay area's BART, each of which carries more than one-fourth of each city's daily commuters. In Canada, rail systems in Montreal and Toronto enjoy similarly heavy use. Major cities internationally from Moscow to Beijing to Paris to Tokyo also have large and heavily utilized subway systems. Some cities with severe traffic problems, such as Bangkok, Thailand, and Athens, Greece, have recently opened new rail systems that carry hundreds of thousands of commuters a day. Light rail use is increasing in Europe, and U.S. ridership is now rising faster than is the rate of new car drivers. Most countries have bus systems that are far more accessible to citizens than are those of the United States.

One city famous for its efficient transportation system is Curitiba, Brazil. Faced with a heavy influx of immigrants from outlying farms in the 1970s, visionary city leaders led by Mayor Jaime Lerner decided to pursue an

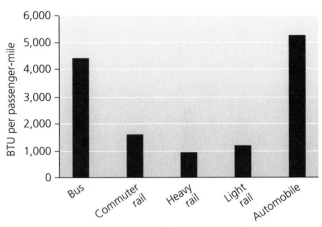

(a) Energy consumption for different modes of transit

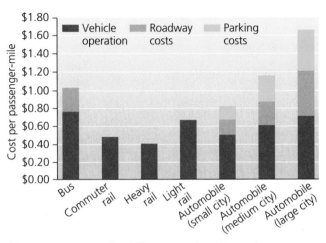

(b) Operating costs for different modes of transit

FIGURE 13.10 Rail transit tends to be cheaper and more energy-efficient than road-based transportation. Rail transit consumes far less energy per passenger mile (**a**) than bus or automobile transit. Rail transit involves fewer costs per passenger mile (**b**) than bus or automobile transit. *Source:* Litman, T. 2005. *Rail transit in America: A comprehensive evaluation of benefits.* Victoria, B.C.: Victoria Transport Policy Institute.

(a) MAX light rail train

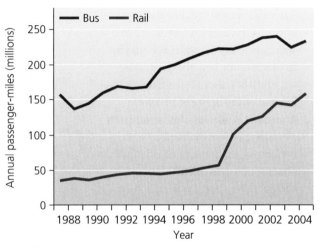

(b) Portland transit ridership trends

FIGURE 13.11 Portland's light rail system (**a**) is one component of an urban planning strategy that has helped make it one of North America's most livable cities. In Portland, bus ridership has been increasing slowly, whereas ridership on the MAX light rail system is growing more quickly (**b**).

aggressive planning process so that they could direct growth rather than being overwhelmed by it. They reconfigured Curitiba's road system to maximize the efficiency of a large fleet of public buses. Today this metropolis of 2.5 million people has an outstanding bus system that is used each day by three-quarters of the population. The 340 bus routes, 250 terminals, and 1,900 buses accompany measures to encourage bicycles and pedestrians. All of this has resulted in a steep drop in car usage, despite the city's rapidly growing population.

Portland's well-organized bus system carries 63 million riders per year. In 1986 Portland introduced a light rail system called MAX, and Metro policy encouraged the development of self-sufficient neighborhood communi-

ties in the new urbanist style along the rail lines. Light rail ridership has steadily increased as the system has expanded (Figure 13.11).

Establishing mass transit is not always easy, however. Planners and zoning boards face many constraints, and one of the biggest is the road-based transportation system that encourages individuals to drive their own automobiles. Once a road system has been developed and businesses and homes are built alongside roads, it can be difficult and expensive to replace or complement the road system with a mass transit system. In addition, modes of mass transit differ in their effectiveness (see "The Science behind the Story", ► pp. 390–391), depending on city size, size of the transit system, and other factors.

The Science Behind the Story

Assessing Benefits of Rail Transit

Most urban planning experts see benefits in public mass transit and in making transit systems readily accessible to citizens. But building a mass transit system—and then maintaining and expanding it—can be a hugely expensive undertaking. Thus, planners and policymakers greatly value quantitative information on the costs and benefits of different types of transit systems and on how extensive a given transit system should be to produce the benefits they desire.

For this reason, researcher Todd Litman of the nonprofit Victoria Transport Policy Institute conducted a comprehensive evaluation of the benefits of rail transit in the United States, published in 2005 by the Victoria Transport Policy Institute, with support from the American Public Transport Association.

Litman first divided American cities into three categories. "Bus-only" cities have bus service but no rail service. "Small-rail" cities have rail systems serving fewer than 12% of daily commuters; these include 16 cities ranging from Atlanta to St. Louis to Denver to Salt Lake City. "Large-rail" cities feature rail service

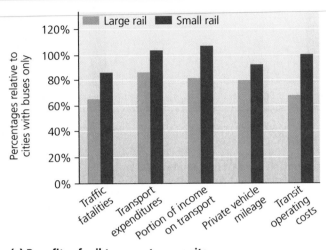

(a) Benefits of rail transport per capita

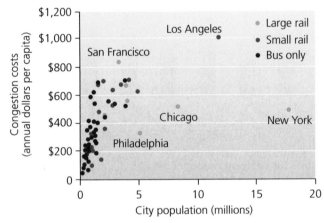

(b) Transport mode, congestion costs, and city size

Large-rail cities outperform bus-only cities in transport-oriented variables, but the benefits of rail (a) are not so clear-cut for small-rail cities. Per capita congestion costs increase with city size, except in large-rail cities (b). *Adapted from: Litman, T. 2005, Rail transit in America: A comprehensive evaluation of benefits.* Victoria, B.C.: Victoria Transport Policy Institute.

Fortunately, installing mass transit systems is not the only way to make urban transportation run more efficiently. Governments can also raise fuel taxes, tax inefficient modes of transport, reward carpoolers with carpool lanes, encourage bicycle use and bus ridership, and charge trucks for road damage. Overall, they can choose to minimize investment in infrastructure that encourages sprawl, and stimulate investment in renewed urban centers.

Parks and open space are key elements of livable cities

City dwellers often desire some sense of escape from the noise, commotion, and stress of urban life. Natural lands, public parks, and open space provide greenery, scenic beauty, freedom of movement, and places for recreation. These lands also keep ecological processes functioning by regulating climate, producing oxygen, filtering air and water pollutants, and providing habitat for wildlife. The

as a major component of the transportation system; these include New York, Washington, D.C., Boston, San Francisco, Chicago, Philadelphia, and Baltimore, whose rail systems serve 12–48% of daily commuters.

Litman compared these three groups of cities for a number of variables. Several key results are summarized in part (a) in the accompanying figure. Compared with bus-only cities, large-rail cities had 36% fewer per capita traffic deaths each year. Residents of large-rail cities drove 21% fewer miles yearly than those of bus-only cities. Large-rail city residents also saved money on transportation, spending 14% less than bus-only city residents on transportation—equivalent to annual savings of $448. Transportation costs took up only 12% of their household budgets, compared to 14.9% of those of bus-only city residents.

The analysis also revealed benefits of large-rail systems for municipalities running the systems. Per passenger mile, large-rail cities paid only 42 cents in operating costs, versus 63 cents paid by bus-only cities. In addition, large-rail cities recovered far more of their costs than bus-only cities (38% cost recovery versus 24%).

Small-rail cities were intermediate in traffic deaths and vehicle mileage, as might be expected. However, they performed no better than bus-only cities in terms of operating costs per passenger mile or transit cost recovery. And interestingly, residents of small-rail cities fared slightly worse than residents of bus-only cities in annual transport expenditures and proportion of income spent on transportation. These results suggest that not all benefits of rail transit begin to accrue with small systems. Rather, for many factors, a rail system has to be large enough or accommodate enough riders to gain the economy of scale needed to provide benefits.

A similar pattern was found for annual per capita costs due to traffic congestion (part (b) in the figure). For bus-only and small-rail cities, congestion costs increased with city size. But for large-rail cities, congestion costs did not vary with size, and were lower than comparably sized cities.

Litman went on to quantify and assess benefits and costs of rail transit overall. He found that each year governments spend $12.5 billion on rail transit systems that they do not get back in revenues from fares. That means governments are subsidizing rail transit by $140 per resident of cities with rail systems.

Although these numbers are considerable, they are surpassed by the monetary benefits of rail systems. Each year, rail systems are estimated to save $19.4 billion in congestion costs, $22.6 billion in consumer transportation costs, $8.0 billion in roadway costs, $12.1 billion in parking costs, and $5.6 billion in accident costs, for a total of $67.7 billion in total annual savings. And that figure does not even account for indirect savings due to enhanced environmental quality.

A number of studies have critiqued rail transit and portrayed it as ineffective. The Litman study indicates that although not all rail systems are cost-effective, rail systems become more beneficial as they become larger and carry a greater proportion of a city's commuters.

animals and plants of urban parks and natural lands also serve to satisfy biophilia (▸p. 328), our natural affinity for contact with other organisms.

Protecting natural lands and establishing public parks become more important as human societies become more urbanized, because many urban dwellers come to feel increasingly isolated and disconnected from nature. In the wake of urbanization and sprawl, people of every industrialized society in the world today have, to some degree, chosen to set aside land in public parks.

City parks were widely established at the turn of the last century

At the turn of the 20th century in urban America, civic improvement was garnering interest and support as politicians and citizens alike yearned for ways to make their crowded and dirty cities more livable. In the late 1800s, public parks began to be established in eastern U.S. cities, using aesthetic ideals borrowed from European parks, gardens, and royal hunting grounds. The

FIGURE 13.12 City parks were developed in many urban areas in the late 19th century to provide citizens aesthetic pleasure, recreation, and relief from the stresses of the city. Manhattan's Central Park, shown here, was one of the first.

lawns, shaded groves, curved pathways, and pastoral vistas we see today in many American city parks and cemeteries originated with these European ideals, as interpreted by the leading American landscape architect, Frederick Law Olmsted. Olmsted designed New York's Central Park in 1853 and a host of urban park systems afterwards (Figure 13.12). Olmsted and his followers sought to create landscapes that provided tranquility, greenery, and an escape from pollution and stress.

Two sometimes conflicting goals motivated the establishment and design of early city parks. On the one hand, the parks were meant to be "pleasure grounds" for the wealthy, who helped support their establishment financially and who would ride the parks' winding roadways in carriages. On the other hand, the parks were meant to alleviate congestion and allow some escape for the many poverty-stricken immigrants who lived in America's cities at the time. These park users were more interested in active recreation, such as ballgames, than in carriage rides. At times the aesthetic interests of the educated elite and the recreational interests of the laboring class came into conflict—a friction that survives today in debates over recreation in city and national parks.

East Coast cities, such as New York, developed parks early on, but cities further west were not far behind. In Chicago, civic boosters striving to overcome the city's reputation as a rough-and-ready frontier town of slaughterhouses and meatpacking plants were glad to have Daniel Burnham design parks and playgrounds and turn a lakeside

swamp into stately Jackson Park. Burnham's *Plan of Chicago* also included a program for a greenbelt of forest preserves circling the city's outskirts (Figure 13.13). Thanks to these lands, Chicago-area residents today have places nearby to hike, fish, bird-watch, study nature, or just relax and escape urban noise and commotion. Further west, in San Francisco, William Hammond Hall transformed 2,500 ha (1,000 acres) of the peninsula's natural landscape of dunes into a verdant playground of lawns, trees, gardens, and sports fields. Golden Gate Park remains today one of the world's foremost city parks.

Portland's quest for urban parks began in 1900, when city leaders created a parks commission and then hired Frederick Law Olmsted's son, John Olmsted, to design a citywide park system. His 1904 plan recommended acquiring land to ring the city generously with parks, but no action was taken. Both Olmsted's plan and Edward Bennett's city plan recommended that the large forested ridge on the northwest side of the city be made a public park. The city instead planned to have it developed, but in the Depression, economic troubles, coupled with a few landslides, caused

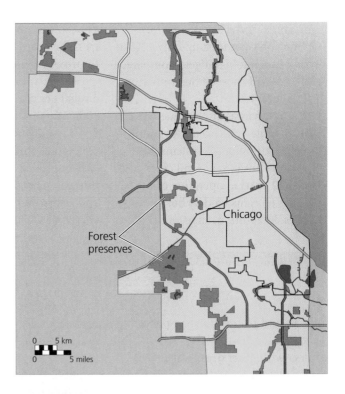

FIGURE 13.13 The forest preserves of Cook County, DuPage County, Lake County, and Will County, Illinois, wind through the suburbs surrounding the city of Chicago. This regional system features 40,000 ha (100,000 acres) of woodlands, fields, marshes, and prairie wildflowers, comprising the largest holding of locally owned public conservation land in the United States.

FIGURE 13.14 Urban community gardens like this one in Seattle provide city residents with a place to grow vegetables and also serve as greenspaces that beautify cities.

developers and investors to forfeit their holdings on the ridge to the city. Finally, citizen pressure resulted in the park's creation in 1948. At 11 km (7 mi) long, Forest Park is the largest city park in the United States.

Smaller public spaces are also important

Large city parks are a key component of a healthy urban environment, but even small spaces can make a big difference. Playgrounds are of immense value for providing places where children can be active outdoors and interact with their peers. Another type of urban space, community gardens, allows people to grow their own vegetables and flowers in a neighborhood setting (Figure 13.14). Portland, Seattle, Baltimore, Boston, and many other cities feature thriving community gardens.

Still another type of greenspace is the corridor or strip of land that connects parks or neighborhoods, often called a *greenway*. Greenways, often located along rivers, streams, or canals, may provide access to networks of

walking trails. They can also protect water quality, boost property values, and serve as corridors for the movement of birds and wildlife. The Rails-to-Trails Conservancy has spearheaded the conversion of abandoned railroad rights-of-way into trails for walking, jogging, and biking. To date, nearly 24,000 km (15,000 mi) of 1,200 rail lines have been converted across North America.

Besides creating new types of urban spaces, many cities are working to enhance the "naturalness" of their parks, through ecological restoration (▶ p. 168), the practice of restoring native communities. In Portland parks, volunteer teams remove English ivy, an invasive plant that covers trees and smothers native plants on the forest floor. At some Chicago-area forest preserves, scientists and volunteers use prescribed burns (▶ p. 360) to restore patches of prairie native to the region. In San Francisco's Presidio, certain areas are being restored to the native dune communities that were wholly displaced by urban development.

Weighing the Issues:
Nature in the City

Do you feel it is important that urban residents have access to parks and natural areas within their cities and towns? Why or why not? What advantages strike you as most important? What types of parks or natural areas do you think are most helpful, and why?

Urban Sustainability

Urbanization and the ways that cities and suburbs function have immense consequences for the natural environment. Urban centers exert both positive and negative environmental impacts, and these impacts depend strongly on how we utilize resources, produce goods, transport materials, and deal with waste. Proponents of urban sustainability seek ways to increase the good that comes from living in urban centers, minimize the bad, and make cities sustainable.

Urban resource consumption brings a mix of environmental impacts

Most of us might guess that urban living involves greater consumption of resources than rural living and thus has a greater environmental impact. However, the picture is not so simple; instead, urbanization brings a complex mix of consequences.

Resource sinks Cities and towns are sinks for resources, having to import from widespread sources beyond their borders nearly everything they need to feed, clothe, and house their inhabitants. Urban and suburban areas are utterly dependent on large expanses of agricultural land to supply food and other crops and on undeveloped land to provide natural resources such as water, timber, metal ores, and mined fuels. Urban centers also need areas of natural land to provide ecosystem services, including purification of water and air, nutrient cycling, and waste treatment (▸ pp. 39–41). Major cities such as New York, Boston, San Francisco, and Los Angeles depend for their day-to-day survival on the water they pump in from faraway watersheds (Figure 13.15). As cities have grown, and as the material wealth of most societies has risen, the inexorable pull of resources from the countryside to the cities has become greater and greater. And as urban areas extend their reach, it becomes increasingly hard for urban residents isolated from natural lands to have a tangible sense of the environmental impacts of their choices.

The long-distance transportation of resources and goods requires a great deal of fossil fuel use, which has significant environmental impacts (Chapter 19). For this reason, the centralization of resource use that urbanization entails may seem a bad thing for the environment. However, imagine that all the world's 3 billion urban residents were instead spread evenly across the landscape. What would the transportation requirements be, then, to move all those resources and goods around to all those people? A world without cities would likely require *more* transportation to provide people the same level of access to resources and goods.

Efficiency Once resources have arrived at an urban center where people are densely concentrated, cities should be able to minimize per capita consumption by maximizing the efficiency of resource use and delivery of goods and services. For instance, providing electricity from a power plant for numerous urban houses close together is more efficient than providing electricity to far-flung homes in the countryside. The density of cities also facilitates the provision of many other social services that improve quality of life, including medical services, education, water and sewer systems, waste disposal, and public transportation.

More consumption Because cities draw resources from afar, their ecological footprints are much greater than their actual land areas. For instance, urban scholar Herbert Girardet calculated that the ecological footprint of London, England, extends 125 times larger than the city's actual

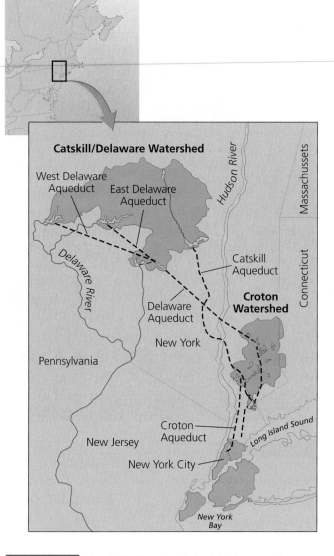

FIGURE 13.15 New York City pipes in its drinking water from upstate reservoirs in the Croton and Catskill/Delaware Watersheds. The city has gone to lengths to acquire, protect, and manage watershed land to minimize pollution of these water sources. When New York City was confronted in 1989 with an order by the Environmental Protection Agency to build a $6 billion filtration plant to protect its citizens against waterborne disease, the city opted instead to purchase and better protect watershed land, for a fraction of the cost.

area. By another estimate, cities take up only 2% of the world's land surface but consume over 75% of its resources. However, the ecological footprint concept is most meaningful when used on a per capita basis. So, in asking whether urbanization causes increased resource consumption, we must ask whether the average urban dweller has a larger footprint than the average rural dweller. The answer is yes, but urban and suburban residents also tend to be wealthier than rural residents, and wealth correlates best with resource consumption. Thus, although urban citizens tend to consume more than rural ones, the reason could be simply that they tend to be wealthier.

Urban centers preserve land

Although the ecological footprints of urban areas are large, because people are packed densely together in cities, more land outside cities is left open and undeveloped. This is, indeed, the idea behind urban growth boundaries. If cities did not exist, and if instead all 6.5 billion of us were evenly spread across the planet's land area, we would have much less room for agriculture, let alone wilderness, biodiversity, or privacy. There would be no large blocks of land left uninhabited by people or of habitat unfragmented for wildlife. The fact that half the human population is concentrated in discrete locations helps allow room for natural ecosystems to maintain themselves, continue functioning, and provide the ecosystem services on which all of us, urban and rural, depend.

Urban centers suffer and export pollution

Just as cities import resources, they export wastes. By exporting their waste, either passively through pollution or actively through trade, urban centers transfer the costs of their activities to other regions—and mask the costs from their own residents. Citizens of Toronto may not recognize that pollution from coal-fired power plants in their region exacerbates acid precipitation hundreds of miles to the east. Citizens of New York may not realize how much garbage their city produces if it is shipped to other states or nations for disposal.

However, not all waste and pollution leaves the city. Heavy metals, industrial compounds, and chemicals from manufactured products accumulate in soil and water. Airborne pollutants cause photochemical smog, industrial smog, and acid precipitation. Fossil fuel combustion releases carbon dioxide and other pollutants, leading to climate change. Urban residents also suffer noise pollution and light pollution. *Noise pollution* consists of undesired ambient sound. Excess noise degrades one's surroundings aesthetically, can induce stress, and at intense levels (such as with prolonged exposure to leafblowing, lawnmowing, and jackhammering) can harm hearing. *Light pollution* describes the way that city lights obscure the night sky, impairing the visibility of stars.

These various forms of pollution and the health threats it poses are not evenly shared among urban residents. Those who receive the brunt of the pollution are often those who are too poor to live in cleaner areas. Environmental justice concerns (▶pp. 36–38) center on the fact

that a disproportionate number of people living near, downstream from, or downwind from factories, power plants, and other polluting facilities are people who are poor and, often, people of racial minorities.

Urban centers foster innovation

One of the greatest impacts of urbanization on environmental quality is also one of the most indirect and intangible. Cities promote a flourishing cultural life and, by mixing together diverse people and influences, spark innovation and creativity. The urban environment can promote education and scientific research, and cities have long been viewed as engines of technological and artistic inventiveness. This inventiveness can lead to solutions to societal problems, including ways to reduce the environmental impacts we have just discussed. For instance, research into renewable energy sources is helping us develop ways to replace fossil fuels. Technological advances have helped us reduce pollution. Wealthy and educated urban populations provide markets for low-impact goods, such as organic produce. Recycling programs help reduce the solid waste stream. Environmental education (Figure 13.16) is helping people choose their own ways to live cleaner, healthier, lower-impact lives. All these phenomena are outgrowths of the education, innovation, science, and technology that are part of urban culture.

Some seek sustainability for cities

Improving quality of life for people while also improving environmental quality—and making these gains permanent—is at the heart of the quest for sustainability. Unfortunately, many critics have noted that the model of a city that imports all its resources and exports all its wastes is not sustainable. Modern cities have a linear, one-way metabolism. Such linear models of production and consumption tend to destabilize environmental systems. Proponents of sustainability for cities urge us to reject the paradigm of an infinite quest for growth and, instead, stress the need to develop circular systems, akin to systems found in nature. Natural systems recycle materials and use renewable sources of energy. For these reasons, they are able to persist indefinitely.

Researchers in the field of **urban ecology** hold that cities can be viewed explicitly as ecosystems and that the fundamentals of ecosystem ecology and systems science (Chapter 7) apply to urban areas. Only by accepting that ecological principles apply to cities, they say, will we have hope of making them sustainable. Major urban ecology projects are ongoing in Baltimore and Phoenix. Researchers

FIGURE 13.16   Urban and suburban children take part in an environmental education program in a Chicago-area forest preserve. Although natural land is rare in urbanized areas, the education and innovation that urbanization often promotes can lead to solutions that reduce environmental impact.

funded by the National Science Foundation's Long Term Ecological Research (LTER) program are studying these two cities explicitly as ecosystems. The researchers are examining such topics as nutrient cycling of carbon and nitrogen, biodiversity patterns, and air and water quality. They are also exploring ways in which humans perceive their environment and react to environmental health threats.

To help cities maintain their standard of living while reducing their environmental impacts, urban sustainability advocates suggest that cities:

▶ Maximize efficient use of resources
▶ Recycle as much as possible
▶ Develop environmentally friendly technologies
▶ Account fully for external costs
▶ Offer tax incentives to encourage sustainable practices
▶ Use locally produced resources
▶ Use organic waste and wastewater to restore soil fertility
▶ Encourage urban agriculture

Increasing numbers of cities are adopting one or more of these strategies. For instance, urban agriculture is a growing pursuit in many urban areas, from Portland to Cuba (▶pp. 303–304) to Japan. Singapore produces all its meat and 25% of its vegetable needs within its city limits. In Berlin, Germany, 80,000 people grow food in community gardens, and 16,000 more are on waiting lists.

In general, experts advise that developed countries should invest in resource-efficient technologies to reduce their impacts and enhance their economies, whereas developing countries should invest in basic infrastructure to improve health and living conditions. Curitiba, Brazil shows the kind of success that can result when a city invests in well-planned infrastructure. Besides the highly effective bus transportation network described earlier, the city provides recycling, environmental education, job training for the poor, and free health care. Surveys show that its citizens are unusually happy and better off economically than people living in other Brazilian cities. Successes in places like Curitiba suggest that cities need not be unsustainable. Indeed, because they affect the environment in some positive ways and have the potential for efficient resource use, cities can and should be a key element in achieving environmental progress.

Conclusion

As half the human population has shifted from rural to urban lifestyles, the nature of our impact on the environment has changed. As urban and suburban dwellers, our impacts are less direct but often more far-reaching. Resources must be delivered to us over long distances, requiring the use of still more resources. Limiting the waste of those resources by making our urban areas more sustainable will be vital for the future. Fortunately, the innovative cultural environment that cities foster has helped us develop solutions to alleviate impact and promote sustainability.

Part of seeking urban sustainability lies in making urban areas better places to live. One key component of these efforts involves expanding transportation options to relieve congestion and make cities run more efficiently. Another lies in ensuring access to adequate park lands and greenspaces near and within our urban centers, to keep us from becoming wholly isolated from nature. Accomplishments in city and regional planning have made many American cities more livable than they once were, and we should be encouraged about such progress. Proponents of smart growth and the new urbanism believe they have solutions to the challenges posed by urban and suburban sprawl, although free-market theorists maintain that if people desire suburban lifestyles, governments should not stand in the way of sprawl. Continuing experimentation in cities from Portland, Oregon, to Curitiba, Brazil, will help us determine how best to ensure that urban growth improves our quality of life and does not degrade the quality of our environment.

REVIEWING OBJECTIVES

You should now be able to:

Describe the scale of urbanization

▶ The world's population is becoming predominantly urban. (p. 374)

▶ The shift from rural to urban living is driven largely by industrialization and is proceeding fastest now in the developing world. (pp. 374–375)

▶ Most of the world's metropolises are in developing nations, and nearly all future population growth will be in cities of the developing world. (pp. 375–376)

▶ The geography of urban areas is changing as cities decentralize and suburbs grow and expand. (pp. 376–378)

Assess urban and suburban sprawl

▶ Urban and suburban expansion has resulted in sprawl, that is, growth that covers large areas of land with relatively low density development. Both population growth and increased per capita land usage contribute to sprawl. (pp. 378–380)

▶ Sprawl has resulted from the home-buying choices of individuals who preferred suburbs to cities. (p. 380)

▶ Sprawl may lead to negative impacts involving transportation, pollution, health, land use, natural habitat, and economics. (pp. 380, 382–384)

Outline city and regional planning and land use strategies

▶ City and regional planning and zoning are key tools for improving the quality of urban life. (pp. 384–386)

▶ "Smart growth," urban growth boundaries, and the "new urbanism" attempt to recreate compact and vibrant urban spaces. (pp. 386–388)

Evaluate transportation options

▶ Mass transit systems can enhance the efficiency of urban areas, but bus and train systems of different sizes bring different benefits. (pp. 388–391)

Describe the roles of urban parks

▶ Urban park lands are vital for active recreation, soothing the stress of urban life, and keeping people in touch with natural areas. (pp. 390–393)

Analyze environmental impacts and advantages of urban centers

▶ Cities are resource sinks, and urban dwellers have high per capita resource consumption. However, cities also allow natural lands to be preserved. (pp. 394–395)

▶ Urban centers can maximize efficiency and help foster innovation that can lead to solutions for environmental problems. (p. 395)

Assess the pursuit of sustainable cities

▶ The linear mode of consumption and production is unsustainable, and more circular modes will be needed to create sustainable cities. (pp. 395–396)

TESTING YOUR COMPREHENSION

1. What factors lie behind the shift of population from rural areas to urban areas? What types of cities and countries are experiencing the fastest urban growth today, and why?

2. Why have so many city dwellers in the United States, Canada, and other nations moved into suburbs?

3. Give two definitions of *sprawl*. Describe five negative impacts that have been suggested to result from sprawl.

4. What are city planning and regional planning? Contrast planning with zoning. Give examples of some of the suggestions made by early planners, such as Daniel Burnham and Edward Bennett.

5. How are some people trying to prevent or slow sprawl? Describe some key elements of "smart growth." What effects, positive and negative, do urban growth boundaries tend to have?

6. Describe several apparent benefits of rail transit systems. What is a potential drawback?

7. How are city parks thought to make urban areas more livable? What types of smaller spaces in cities can serve some of the functions of parks?

8. Why do urban dwellers tend to consume more resources per capita than rural dwellers?

9. Describe the connection between urban ecology and sustainable cities.

10. Name two positive effects of urban centers on the natural environment.

SEEKING SOLUTIONS

1. Assess the reasons why urban populations are rising and rural populations are stable or falling. Do you think these trends will continue in the future, or might they change for some reason?

2. Evaluate the causes of the spread of suburbs and of the environmental, social, and economic impacts of sprawl. Overall, do you think the spread of urban and suburban development that many people label *sprawl* is predominantly a good thing or a bad thing? Do you think it is inevitable? Give reasons for your answers.

3. Would you personally want to live in a neighborhood developed in the style of the new urbanism? Would you like to live in a city or region with an urban growth boundary? Why or why not?

4. All things considered, do you feel that cities are a positive thing or a negative thing for environmental quality? How much do you feel we may be able to improve the sustainability of our urban areas?

5. Which is more ecologically sustainable: living in a high-rise apartment in a big city, or living on a 40-acre ranch abutting a national forest? Why?

6. You may soon be faced with a decision about where to live. You can live in the midst of a densely populated city; or, in a suburb where you have more space, but where development and sprawl may soon surround you for many miles; or, in a rural area with plenty of space but few cultural amenities. Where would you choose to live? Why?

INTERPRETING GRAPHS AND DATA

In the following graph, urban population density is used as an indicator of sprawl, and carbon emissions per capita provide some measure of the environmental impact of the transportation system or preferences for each of the cities represented.

1. Describe the relationship between urban density and carbon emissions, as shown in the graph.

2. Assuming that the standard of living is similar in these cities, to what might you attribute the relationship described in your answer to question 1?

3. If zoning ordinances successfully slowed urban sprawl and resulted in a doubling of urban population density in a city like Houston, Texas, what do you predict the change would be in carbon emissions per capita?

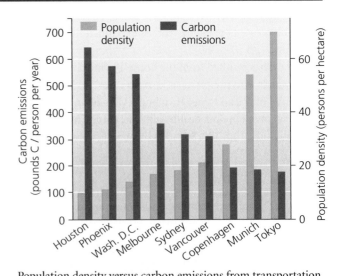

Population density versus carbon emissions from transportation in 1990. Data from Kenworthy, J., et al.,1999. *An international sourcebook of automobile dependence in cities,* Boulder, CO: University Press of Colorado, as cited by Sheehan, M. O. 2002. *What will it take to halt sprawl?* Washington D.C.: Worldwatch Institute.

CALCULATING ECOLOGICAL FOOTPRINTS

One way of altering your ecological footprint is to consider transportation alternatives. Each gallon of gasoline is converted to approximately 20 lb of carbon dioxide (CO_2) during combustion, and this CO_2 is then released into the atmosphere. Typical amounts of CO_2 released for each person per mile, through various forms of transportation, are listed in the table below, assuming typical fuel efficiencies.

For an average person who travels 12,000 miles per year, calculate the cumulative impact of choosing the alternatives to driving alone, and record your results in the table. Clearly, it is unlikely that any of us will walk or bicycle 12,000 miles per year or only travel in vanpools of eight people. In the last two columns, estimate how much of the 12,000 miles you think you might actually travel by each method, and then calculate the CO_2 emissions that you would be responsible for generating over the course of a year.

	CO_2 per person per mile	CO_2 per person per year	CO_2 emission reduction	Your estimated mileage per year	Your CO_2 emissions per year
Automobile (driver only)	0.825 lb	9,900 lb	0		
Automobile (2 persons)	0.413 lb				
Automobile (4 persons)	0.206 lb				
Vanpool (8 persons)	0.103 lb				
Bus	0.261 lb				
Walking	0.082 lb				
Bicycle	0.049 lb				
				Total = 12,000	

Take It Further

Go to www.aw-bc.com/withgott or the student CD-ROM where you'll find:

▶ Suggested answers to end-of-chapter questions
▶ Quizzes, animations, and flashcards to help you study
▶ *Research Navigator*™ database of credible and reliable sources to assist you with your research projects

▶ **GRAPHit!** tutorials to help you master how to interpret graphs
▶ **INVESTIGATEit!** current news articles that link the topics that you study to case studies from your region to around the world

14 Environmental Health and Toxicology

Alligator hatchling from Lake Apopka, Florida

Upon successfully completing this chapter, you will be able to:

▶ Identify the major types of environmental health hazards and explain the goals of environmental health

▶ Describe the types, abundance, distribution, and movement of synthetic and natural toxicants in the environment

▶ Discuss the study of hazards and their effects, including case histories, epidemiology, animal testing, and dose-response analysis

▶ Assess risk assessment and risk management

▶ Compare philosophical approaches to risk

▶ Describe policy and regulation in the United States and internationally

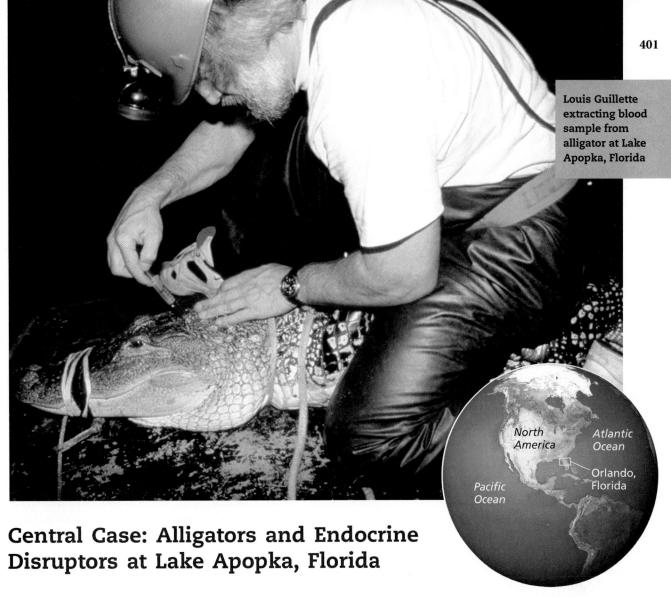

Louis Guillette extracting blood sample from alligator at Lake Apopka, Florida

North America

Atlantic Ocean

Orlando, Florida

Pacific Ocean

Central Case: Alligators and Endocrine Disruptors at Lake Apopka, Florida

"Over the past 50 years we have all been unwitting participants in a vast, uncontrolled, worldwide chemistry experiment involving the oceans, air, soils, plants, animals, and human beings."

—UNITED NATIONS
ENVIRONMENT PROGRAMME,
IN A GUIDE TO THE
STOCKHOLM CONVENTION
ON PERSISTENT ORGANIC
POLLUTANTS

"Every man in this room is half the man his grandfather was."

—REPRODUCTIVE BIOLOGIST
LOUIS GUILLETTE TO A U.S.
CONGRESSIONAL COMMITTEE
IN 1995, REGARDING
DECLINING SPERM COUNTS IN
MEN WORLDWIDE OVER THE
PREVIOUS HALF-CENTURY

the male sex hormone testosterone, and female hatchlings showed greatly elevated levels of the female sex hormone estrogen. The young animals had abnormal gonads, too. Testes of the males produced sperm at a premature age, and ovaries of the females contained multiple eggs per follicle instead of the expected one egg per follicle.

As Guillette and his co-workers tested and rejected hypotheses for the reproductive abnormalities one by one, they grew to suspect that environmental contaminants might be somehow responsible. Lake Apopka had suffered a major spill of the pesticides dicofol and DDT in 1980, and yearly surveys thereafter showed a precipitous decline in the number of juvenile alligators in the lake. In addition, the lake received high levels of chemical runoff from agriculture and was experiencing eutrophication (▶pp. 188–189 and ▶p. 453) from nutrient input from fertilizers.

The puzzle pieces fell together when a colleague of Guillette's told him that recent research was showing that some environmental contaminants mimic

When biologist Louis Guillette began studying the reproductive biology of alligators in Florida lakes in 1985, he soon discovered that not all was well. Alligators from one lake in particular, Lake Apopka northwest of Orlando, showed a number of bizarre reproductive problems.

Females were having trouble producing viable eggs. Male hatchling alligators had severely depressed levels of

hormones and interfere with the functioning of animal endocrine (hormone) systems. At very low doses, these endocrine disruptors can affect animals, mimicking estrogen and feminizing males. Guillette realized that his alligator observations were similar to results from tests in which rodents were exposed to estrogen during embryonic development. He formulated the hypothesis that certain chemical contaminants in Lake Apopka were disrupting the endocrine systems of alligators during development in the egg.

To test his hypothesis, Guillette and his co-workers compared alligators from heavily polluted Lake Apopka with those from cleaner lakes nearby. They found that Lake Apopka alligators had abnormally low hatching rates in the years after the pesticide spill and, even as hatching rates recovered in the 1990s, continued to show aberrant hormone levels and bizarre gonad abnormalities. In addition, the penises of Lake Apopka male alligators were 25% smaller than those of male alligators from surrounding lakes.

Similar problems began cropping up in other lakes that experienced runoff of chemical pesticides. In the lab, researchers found that several contaminants detected in alligator eggs and young could bind to receptors for estrogen and exist in concentrations great enough to cause sex reversal of male embryos. One chemical in particular, atrazine—the most widely used herbicide in the United States—appeared to disrupt hormones by inducing production of aromatase, an enzyme that converts testosterone to estrogen. In 2003, Guillette reported preliminary findings that nitrate from fertilizer runoff may also act as an endocrine disruptor; when nitrate concentrations in lakes are above the standard for drinking water, juvenile male alligators have smaller penises and 50% lower testosterone levels.

Guillette's results have raised concern not only for alligator health but also for human health. Because alligators and humans share many of the same hormones, many scientists suspect that chemical contaminants could be affecting people, just as they have affected alligators.

Environmental Health

The study of impacts of human-made chemicals on wildlife and people is one aspect of the broad field of environmental health. The study and practice of **environmental health** assesses environmental factors that influence human health and quality of life. Those environmental factors include wholly natural aspects of the environment over which we have little or no control, as well as anthropogenic (human-caused) factors that spread through the environment and can affect people who do not cause them. Practitioners of environmental health seek to prevent adverse effects on human health and on the ecological systems that are essential to environmental quality and human well-being.

Environmental hazards can be physical, chemical, biological, or cultural

There are innumerable environmental health threats, or hazards, in the world around us. For ease of understanding, we can categorize them into four main types: physical, biological, chemical, and cultural. For each of these four types of hazards, there is some amount of risk that we cannot avoid—but there is also some amount of risk that we *can* avoid by taking precautions. Much of environmental health consists of taking steps to minimize the risks of encountering hazards and to minimize the impacts of hazards when we do encounter them.

Physical hazards Some *physical* processes that occur naturally in our environment can pose health hazards. These include discrete events such as earthquakes, volcanic eruptions, fires, floods, blizzards, landslides, hurricanes, and droughts. They also include ongoing natural phenomena, such as ultraviolet (UV) radiation from sunlight (Figure 14.1a). At excessive exposure, UV radiation damages DNA and has been tied to skin cancer, cataracts, and immune suppression in humans. We can do little to predict the timing of a natural disaster such as an earthquake, and nothing to prevent one. However, scientists can map geologic faults to determine areas at risk of earthquakes, engineers can design buildings in ways that help them resist damage, and citizens and governments can take steps to prepare for the aftermath of a severe quake.

Some common practices have increased our vulnerability to certain physical hazards. Deforesting slopes makes landslides more likely, for instance, and diking rivers can make flooding more likely in some areas while reducing flooding in others. We can reduce risk from such hazards by improving our forestry and flood control practices and by choosing not to build in areas prone to floods, landslides, fires, and coastal waves. For hazards such as exposure to UV light, we can reduce our exposure and risk by using clothing and sunscreen to shield our skin from intense sunlight.

(a) Physical hazard

(b) Chemical hazard

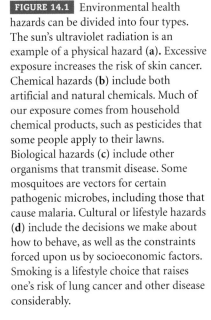

FIGURE 14.1 Environmental health hazards can be divided into four types. The sun's ultraviolet radiation is an example of a physical hazard (**a**). Excessive exposure increases the risk of skin cancer. Chemical hazards (**b**) include both artificial and natural chemicals. Much of our exposure comes from household chemical products, such as pesticides that some people apply to their lawns. Biological hazards (**c**) include other organisms that transmit disease. Some mosquitoes are vectors for certain pathogenic microbes, including those that cause malaria. Cultural or lifestyle hazards (**d**) include the decisions we make about how to behave, as well as the constraints forced upon us by socioeconomic factors. Smoking is a lifestyle choice that raises one's risk of lung cancer and other disease considerably.

(c) Biological hazard

(d) Cultural hazard

Chemical hazards *Chemical* hazards include many of the synthetic chemicals that our society produces, such as disinfectants, pesticides (Figure 14.1b), and the compounds that contributed to reproductive problems for the alligators at Lake Apopka. Chemicals produced naturally by organisms also can be hazardous. Some hazardous chemicals are so widespread they are impossible to avoid, but with concerted effort, we may be able to reduce our exposure. Following our overview of environmental health, much of our chapter will focus on chemical health hazards and the ways people study and regulate them.

Biological hazards *Biological* hazards result from ecological interactions among organisms (Figure 14.1c). When we become sick from a virus, bacterial infection, or other pathogen, we are suffering parasitism by other species that are simply fulfilling their ecological roles. This is **infectious disease,** also called *communicable* or *transmissible disease.* Infectious diseases such as malaria, cholera, tuberculosis, and influenza (flu) all are considered environmental health hazards. As with physical and chemical hazards, it is impossible for us to avoid risk from biological agents completely, but we can take steps to reduce the likelihood of infection.

Cultural hazards Hazards that result from the place we live, our socioeconomic status, our occupation, or our behavioral choices can be thought of as *cultural* or *lifestyle* hazards. For instance, choosing to smoke cigarettes, or living or working with people who do, can greatly increase our risk of lung cancer (Figure 14.1d). Choosing to smoke is a personal behavioral decision, but exposure to secondhand smoke in the home or workplace may not be under one's control and may be influenced by socioeconomic constraints. Much the same might be said for drug use, diet and nutrition, crime, and mode of transportation. As advocates of environmental justice (▸ pp. 36–38) argue, such health factors as living in proximity to toxic waste sites or working unprotected with pesticides are often correlated with socioeconomic deprivation.

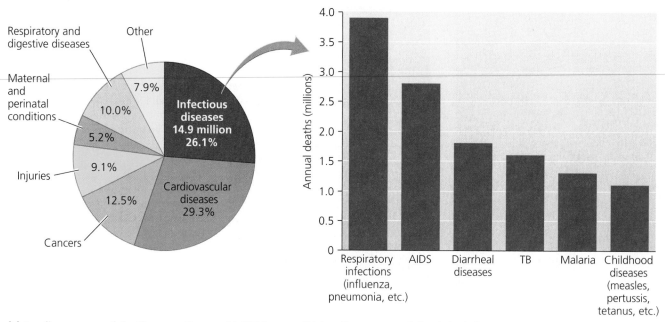

(a) **Leading causes of death across the world, 2004**

(b) **Leading causes of death by infectious disease, 2004**

FIGURE 14.2 Infectious diseases are the second-leading cause of death worldwide, accounting for over one-quarter of all deaths per year (**a**). Six types of diseases—respiratory infections, AIDS, diarrhea, tuberculosis (TB), malaria, and childhood diseases such as measles—account for 80% of all deaths from infectious disease (**b**). Data from World Health Organization, 2004.

Disease is a major focus of environmental health

Among the hazards people face, disease stands preeminent. Despite all our technological advances, we still find ourselves battling disease, which causes the vast majority of human deaths worldwide (Figure 14.2a). Many major killers such as cancer, heart disease, and respiratory disorders have genetic bases but are also influenced by environmental factors. For instance, whether a person develops asthma is influenced not only by the genes, but also by environmental conditions. Pollutants from fossil fuel combustion worsen asthma, and children raised on farms suffer less asthma than children raised in cities, studies have shown. Malnutrition (▸pp. 279–280) can foster a wide variety of illnesses, as can poverty and poor hygiene. Moreover, lifestyle choices can affect risks of acquiring some noninfectious diseases: Smoking can lead to lung cancer, and lack of exercise to heart disease, for example.

Infectious diseases account for 26% of deaths that occur worldwide each year—nearly 15 million people (Figure 14.2b). Some pathogenic microbes attack us directly, and sometimes infection occurs through a *vector,* an organism that transfers the pathogen to the host. Infectious diseases account for close to half of all deaths in developing countries, but for very few deaths in developed

nations. This discrepancy is due to differences in hygiene conditions and access to medicine, which are tightly correlated with wealth. Public health efforts have lessened the impact of infectious disease in developed nations and even have eradicated some diseases. Nevertheless, other diseases, among them tuberculosis, acquired immunodeficiency syndrome (AIDS), and West Nile virus, are increasing (Figure 14.3).

To predict and prevent infectious disease, environmental health experts deal with the often complicated interrelationships among technology, land use, and ecology. One of the world's leading infectious diseases, malaria (which takes nearly 1.3 million lives each year), provides an example. The microscopic protist that causes malaria depends on mosquitoes as a vector. The protist can sexually reproduce only within a mosquito, and it is the mosquito that injects the protist into a human or other host. Thus the primary mode of malaria control has been to use insecticides such as DDT to kill mosquitoes. Large-scale eradication projects involving insecticide use and draining of wetlands have removed malaria from large areas of the temperate world where it used to occur, such as throughout the southern United States. However, types of land use that create pools of standing water in formerly well-drained areas can boost mosquito populations and allow malaria to reinvade an area.

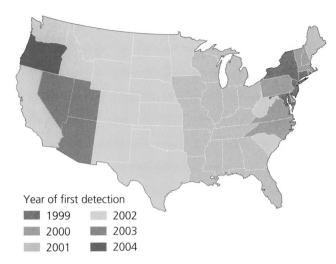

Year of first detection
- 1999
- 2000
- 2001
- 2002
- 2003
- 2004

(a) Spread of West Nile virus

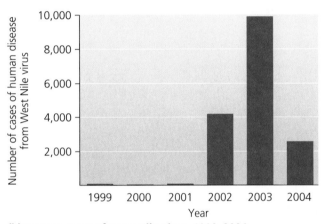

(b) Human cases of West Nile virus, 1999–2004

FIGURE 14.3 Modern public health efforts have greatly diminished the burden of disease in developed nations, but some infectious diseases are currently on the increase. One example is West Nile virus, an Old World virus that has spread rapidly through the Western Hemisphere since it was first detected in the New York City area in 1999. Some types of birds are most affected by this mosquito-transmitted virus, but over 15,000 human disease cases, including over 500 deaths, had been recorded in the United States through the end of 2004. In the map **(a)**, U.S. states are color-coded to show the year that West Nile virus was first detected within their borders. The disease spread westward across the country in less than 5 years. The graph **(b)** shows the number of cases reported nationwide in each year. Data from Centers for Disease Control and Prevention, 2005.

Environmental health hazards exist indoors as well as outdoors

Outdoor hazards are generally more familiar to us, but we spend most of our lives indoors. Therefore, we must consider the spaces inside our homes and workplaces to be part of our environment and, as such, the sources of

Table 14.1 Selected Environmental Hazards
Air
▶ Smoking and secondhand smoke
▶ Chemicals from automotive exhaust
▶ Chemicals from industrial pollution
▶ Tropospheric ozone
▶ Pesticide drift
▶ Dust and particulate matter
Water
▶ Pesticide and herbicide runoff
▶ Nitrates and fertilizer runoff
▶ Mercury, arsenic, and other heavy metals in groundwater and surface water
Food
▶ Natural toxins
▶ Pesticide and herbicide residues
Indoors
▶ Asbestos
▶ Radon
▶ Lead in paint and pipes
▶ Toxicants in plastics and consumer products

potential environmental hazards (Table 14.1). Radon is one indoor hazard (▶pp. 522–523). Radon is a highly toxic radioactive gas that is colorless and undetectable without specialized kits. Radon seeps up from the ground in areas with certain types of bedrock and can build up inside basements and homes with poor air circulation. The U.S. Environmental Protection Agency (EPA) estimates that slightly less than 1 person in 1,000 may contract lung cancer as a result of a lifetime of radon exposure at average levels for U.S. homes.

Lead poisoning represents another indoor health hazard. When ingested, lead, a heavy metal, can cause damage to the brain, liver, kidney, and stomach; learning problems and behavioral abnormalities; anemia; hearing loss; and even death. It has been suggested that the downfall of ancient Rome was caused in part by chronic lead poisoning. The elites of the power structure regularly drank wine sweetened with mixtures prepared in leaden vessels. Several historical personalities, such as the composer Beethoven, suffered maladies that some historians have attributed to lead poisoning. Today lead poisoning can result from drinking water that has passed through the lead pipes common in older houses. Even in newer pipes, lead solder was widely used into the 1980s and is still sold in stores. Lead is perhaps most dangerous, however, to children through its presence in paint. Until 1978, most paints contained lead, and interiors in most houses were painted with lead-based paint. Babies and young children often take an interest in peeling paint from walls and may ingest or inhale some of it. Today lead poisoning is thought to affect 3–4 million young children, fully one in six children under age 6.

Another indoor hazard is asbestos. There are several types of *asbestos,* each of which is a mineral that forms long, thin microscopic fibers. This structure allows asbestos to insulate heat, muffle sound, and resist fire. Because of these qualities, asbestos was used widely as insulation in buildings, as well as in many products. Unfortunately, its fibrous structure also makes asbestos dangerous when inhaled. When lodged in lung tissue, asbestos induces the body to produce acid to combat it. The acid scars the lung tissue but does little to dislodge or dissolve the asbestos. Consequently, within a few decades the scarred lungs may cease to function, a disorder called *asbestosis.* Asbestos can also cause types of lung cancer. Because of these risks, asbestos has been removed from many schools and offices (Figure 14.4). However, many people have argued that the dangers of exposure from asbestos removal exceed the dangers of leaving it in place.

Some indoor hazards likely have not yet come to light. One example of a recently recognized hazard is a group of chemicals known as polybrominated diphenyl ethers (PBDEs). These compounds provide fire-retardant properties and are used in a diverse array of consumer products, including computers, televisions, plastics, and furniture. They appear to be released during production and disposal of products and also to evaporate at very slow rates throughout the lifetime of products. These chemicals persist and accumulate in living tissue, and their abundance in the environment and in people in the United States is doubling every few years.

Like the chemicals affecting Louis Guillette's alligators, PBDEs appear to be endocrine disruptors; lab testing with animals shows them to affect thyroid hormones. Animal testing also shows limited evidence that PBDEs may affect brain and nervous system development and might possibly cause cancer. Concern about PBDEs rose after a study showed concentrations in the breast milk of Swedish mothers increasing exponentially from 1972 to 1997. The European Union decided in 2003 to ban PBDEs, and industries in Europe had already begun phasing them out. As a result, concentrations in breast milk of European mothers have fallen substantially. In the United States, however, there has so far been little movement to address the issue.

FIGURE 14.4 Asbestos was widely used in building insulation and other products. A cause of lung cancer and asbestosis, the substance has now been removed from many buildings in which it was used. Its removal poses risks as well, however, and removal workers must wear protective clothing and respirators.

Toxicology is the study of poisonous substances

Studying the health effects of chemical agents suspected to be harmful, such as PBDEs, is the focus of the field of **toxicology,** the science that examines the effects of poisonous substances on humans and other organisms. Toxicologists assess and compare substances to determine their *toxicity,* the degree of harm a chemical substance can inflict. The concept of toxicity among chemical hazards is analogous to that of *pathogenicity* or *virulence* of the biological hazards that spread infectious disease. Just as types of microbes differ in their ability to cause disease, chemical hazards differ in their capacity to endanger us. However, any chemical substance may exert negative effects if it is ingested in great enough quantities or if exposure is extensive enough. Conversely, a toxic agent, or **toxicant,** in a minute enough quantity may pose no health risk at all. These facts are often summarized in the catchphrase, "The dose makes the poison." In other words, a substance's toxicity depends not only on its chemical identity, but also on its quantity.

During the past century, our ability to produce new chemicals has expanded, concentrations of chemical contaminants in the environment have increased, and public concern for health and the environment have grown. These trends have driven the rise of **environmental toxicology,** which deals specifically with toxic substances that come from or are discharged into the environment. Environmental toxicology includes the study of health effects on humans, other animals, and ecosystems, and it represents one approach within the broader scope of environmental health.

Toxicologists generally focus on human health, using other organisms as models and test subjects. In environmental toxicology, animals are also studied out of concern for their welfare and because—like canaries in a coal mine—animals can serve as indicators of health threats that could soon affect humans. For example, the reproductive abnormalities of Lake Apopka alligators seem analogous to those seen in some Taiwanese boys. Studies showed that these boys were born to mothers who used cooking oil contaminated with toxicants called polychlorinated biphenyls (PCBs), which are by-products of chemicals used in transformers and other electrical equipment.

As we review the sometimes poisonous effects of human-made chemicals throughout this chapter, it is important to keep in mind that artificially produced chemicals have played a crucial role in giving us the standard of living we enjoy today. These chemicals have helped create the industrial agriculture that produces our food, the medical advances that protect our health and prolong our lives, and many of the modern materials and conveniences we use every day. It is important to remember these benefits as we examine some of the unfortunate side effects of these advances and as we search for better alternatives.

Toxic Agents in the Environment

The environment contains countless natural chemical substances that may pose health risks. These substances include oil oozing naturally from the ground; radon gas seeping up from bedrock; and toxic chemicals stored or manufactured in the tissues of living organisms—for example, *toxins* that plants use to ward off herbivores and

FIGURE 14.5 Synthetic chemicals, such as those in household products, are everywhere around us in our everyday lives. Some of these compounds may potentially pose environmental or human health risks.

toxins that insects use to defend themselves from predators. In addition, we are exposed to many synthetic (artificial, or human-made) chemicals.

Synthetic chemicals are ubiquitous in our environment

Synthetic chemicals are all around us in our daily lives (Figure 14.5). Thousands of synthetic chemicals have been manufactured (Table 14.2), and many have found their way into soil, air, and water. A 2002 study by the U.S. Geological Survey found that 80% of U.S. streams contain at least trace amounts of 82 wastewater contaminants, including antibiotics, detergents, drugs, steroids, plasticizers, disinfectants, solvents, perfumes, and other substances. The pesticides we use to kill insects and weeds (▶pp. 282–283) on farms, lawns, and golf courses are some of the most widespread synthetic chemicals. As a result of all this exposure, every one of us carries traces of numerous industrial chemicals in our bodies.

Table 14.2	Estimated Numbers of Chemicals in Commercial Substances during the 1990s
Type of chemical	**Estimated number**
Chemicals in commerce	100,000
Industrial chemicals	72,000
New chemicals introduced per year	2,000
Pesticides (21,000 products)	600
Food additives	8,700
Cosmetic ingredients (40,000 products)	7,500
Human pharmaceuticals	3,300

Data from Harrison, P., and F. Pearce. 2000. *AAAS Atlas of Population and Environment.* Berkeley, CA: University of California Press.

This should not *necessarily* be cause for alarm. Not all synthetic chemicals pose health risks, and relatively few are known with certainty to be toxicants. However, of the roughly 100,000 synthetic chemicals on the market today, very few have been thoroughly tested for harmful effects. For the vast majority, we simply do not know what effects, if any, they may have.

Why are there so many synthetic chemicals around us? Let's consider pesticides and herbicides such as those that Louis Guillette found were disrupting sex hormones and feminizing male alligator embryos. Our many pesticides and herbicides were made possible by advances in chemistry during and following World War II. Mechanization of production and the growth of large corporations enabled immense quantities of synthetic chemicals to be produced.

As material prosperity grew in Westernized nations in the decades following the war, people began using pesticides not only for agriculture, but also to improve the look of their lawns and golf courses and to fight termites, ants, and other insects inside their homes and offices. Pesticides were viewed as means toward a better quality of life.

It was not until the 1960s that people began to learn about the risks of exposure to pesticides. Many people began to speak up against indiscriminate pesticide use. Indeed, concern over pesticides was a catalyst that helped spur the environmental movement in the United States. The key event was the publication of Rachel Carson's 1962 book *Silent Spring* (▸ p. 67), which brought the pesticide dichloro-diphenyl-trichloroethane (DDT) to the attention of the public.

Silent Spring began the public debate over synthetic chemicals

Rachel Carson was a naturalist, author, and government scientist. In *Silent Spring,* she brought together a diverse collection of scientific studies, medical case histories, and other data that no one had previously synthesized and presented to the general public. Her message was that DDT in particular and artificial pesticides in general were hazardous to people's health, the health of wildlife, and the well-being of ecosystems. Carson wrote at a time when large amounts of pesticides virtually untested for health effects were indiscriminately sprayed over residential neighborhoods and public areas, on the assumption that the chemicals would do no harm to people (Figure 14.6). Most

FIGURE 14.6 Before the 1960s, the environmental and health effects of potent pesticides such as DDT were not widely studied or publicly known. Public areas such as parks, neighborhoods, and beaches were regularly sprayed for insect control without safeguards against excessive human exposure. Here children on a Long Island, New York, beach are fogged with DDT from a pesticide spray machine being tested in 1945.

consumers had no idea that the store-bought chemicals they used in their houses, gardens, and crops might be toxic.

Although challenged vigorously by spokespeople for the chemical industry, who attempted to discredit both the author's science and her personal reputation, Carson's book was a best-seller. Carson suffered from cancer as she finished *Silent Spring,* and she lived only briefly after its publication. However, the book helped generate significant social change in views and actions toward the environment. The use of DDT was banned in the United States in 1973 and has been banned in a number of other nations.

The United States still manufactures and exports DDT to countries that do use it, however. Many developing countries with tropical climates use DDT to control human disease vectors, such as mosquitoes that transmit malaria. In these countries, malaria represents a far greater health threat than do the toxic effects of the pesticide.

Weighing the Issues:
The Circle of Poison

It has been called the "circle of poison." Although the United States has banned the use of DDT, U.S. companies still manufacture and export the compound to many developing nations. Thus, it is possible that pesticide-laden food can be imported back into the United States. How do you feel about this? Is it unethical for one country to sell to others a substance that it has deemed toxic? Are there factors or circumstances that might change the view you take?

Toxicants come in several different types

Toxicants, whether they are natural or synthetic, can be classified into different types based on their particular effects on health. The best-known are **carcinogens,** which are chemicals or types of radiation that cause cancer. In cancer, malignant cells grow uncontrollably, creating tumors, damaging the body's functioning, and often leading to death. In our society today, the greatest number of cancer cases is thought to result from carcinogens contained in cigarette smoke. Carcinogens can be difficult to identify because there may be a long lag time between exposure to the agent and the detectable onset of cancer. Historically, much toxicological work focused on carcinogens. Now, however, we know that toxicants can produce many different types of effects, so scientists have many more endpoints, or health impacts, to look for.

FIGURE 14.7 The drug thalidomide turned out to be a potent teratogen. It was banned in the 1960s, but not before causing thousands of birth defects in babies born to mothers who took the product to relieve nausea during pregnancy. Butch Lumpkin was an exceptional "thalidomide baby" who learned to overcome his short arms and deformed fingers, becoming a professional tennis instructor in Georgia.

Mutagens are chemicals that cause mutations in the DNA of organisms (▸pp. 100–101). Although most mutations have little or no effect, some can lead to severe problems, including cancer and other disorders. If mutations occur in an individual's sperm or egg cells, then the individual's offspring suffer the effects.

Chemicals that cause harm to the unborn are called **teratogens.** Teratogens that affect the development of human embryos in the womb can cause birth defects. One example involves the drug thalidomide, developed in the 1950s as a sleeping pill and to prevent nausea during pregnancy. Tragically, the drug turned out to be a powerful teratogen, and its use caused birth defects in thousands of babies (Figure 14.7). Even a single dose during pregnancy could result in limb deformities and organ defects. Thalidomide was banned in the 1960s once scientists recognized its connection with birth defects. Ironically, today the drug shows promise in treating a wide range of diseases, including Alzheimer's disease, AIDS, and various types of cancer.

The human immune system protects our bodies from disease. Some toxicants weaken the immune system, reducing the body's ability to defend itself against bacteria, viruses, allergy-causing agents, and other attackers. Others, called **allergens,** overactivate the immune system, causing an immune response when one is not necessary. One hypothesis for the increase in asthma in recent years is that allergenic synthetic chemicals are more prevalent in our environment.

Still other chemical toxicants, **neurotoxins,** assault the nervous system. Neurotoxins include various heavy metals such as lead, mercury, and cadmium, as well as pesticides and some chemical weapons developed for use in war. A famous case of neurotoxin poisoning occurred in Japan, where a chemical factory dumped mercury waste into Minamata Bay between the 1930s and 1960s. Thousands of people in and around the town on the bay were poisoned by eating fish contaminated with the mercury. First the town's cats began convulsing and dying, and then people began to show odd symptoms, including slurred speech, loss of muscle control, sudden fits of laughter, and in some cases death. The company and the government eventually paid out about $5,000 in compensation to each affected resident.

Most recently, scientists have recognized the importance of **endocrine disruptors,** toxicants that interfere with the *endocrine system,* or hormone system. The endocrine system consists of chemical messengers *(hormones)* that travel though the body. Sent through the bloodstream at extremely low concentrations, these messenger molecules have many vital functions. They stimulate growth, development, and sexual maturity, and they regulate brain function, appetite, sexual drive, and many other aspects of our physiology and behavior. Hormone-disrupting toxicants can affect an animal's endocrine system by blocking the action of hormones or accelerating their breakdown. Many endocrine disruptors are so similar to some hormones in their molecular structure and chemistry that they "mimic" the hormone by interacting with receptor molecules just as the actual hormone would (Figure 14.8). One common type of endocrine disruption involves the feminization of male animals, as Louis Guillette found with alligators, and as other studies have indicated with fish, frogs, and other organisms. Feminization may be widespread because a number of chemicals appear to mimic the female sex hormone estrogen and bind to estrogen receptors.

Endocrine disruption may be widespread

Scientists first noted endocrine-disrupting effects as far back as the 1960s with the pesticide DDT. However, the idea that synthetic chemicals might be altering the hormones of animals was not widely appreciated until the

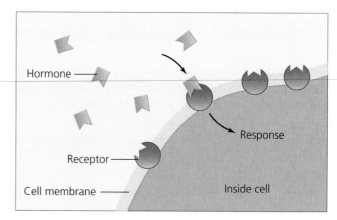

(a) Normal hormone binding

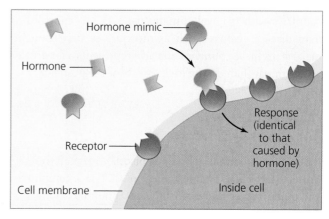

(b) Hormone mimicry

FIGURE 14.8 Many endocrine-disrupting substances act by mimicking the structure of hormone molecules. Like a key similar enough to fit into another key's lock, the hormone mimic binds to a cellular receptor for the hormone, causing the cell to react as though it had encountered the hormone.

1996 publication of the book *Our Stolen Future,* by Theo Colburn, Dianne Dumanoski, and J. P. Myers. Colburn, a scientist with the World Wildlife Fund, had spent years bringing toxicologists, medical doctors, and wildlife biologists together to share their findings. She and her co-authors then presented the data and stories in *Our Stolen Future* (and on a Web page that provides updates on new research). Like *Silent Spring,* this book integrated scientific work from various fields and presented a unified picture that shocked many readers—and brought criticism from some scientists and from the chemical industry.

One line of research that has sparked debate is that of scientist Tyrone Hayes of the University of California at Berkeley, who studies frogs and has found in them gonadal abnormalities similar to those of Guillette's alligators. As

FIGURE 14.9 According to research by Tyrone Hayes of the University of California at Berkeley, frogs may suffer reproductive abnormalities from exposure to the best-selling herbicide atrazine. Industry-backed scientists have disputed the findings.

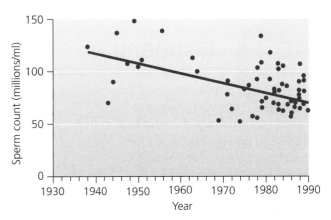

(a) Declining sperm count in humans, based on 61 studies

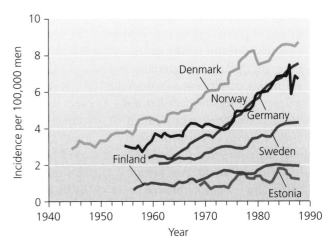

(b) Increasing incidence of testicular cancer

FIGURE 14.10 Research in 1992 synthesized the results of 61 studies that had reported sperm counts in men from various localities since 1938. The data were highly variable but showed a significant decrease in human sperm counts over time **(a)**. Many scientists have hypothesized that this decrease may result from exposure to endocrine-disrupting chemicals in the environment. Some have also hypothesized that endocrine disruptors could be behind an increased incidence of testicular cancer **(b)**. (a) Data from Carlsen, E., et al. 1992. Evidence for decreasing quality of semen during the last 50 years. *British Medical Journal* 305: 609–613, as adapted by Toppari, J., et al. 1996. Male reproductive health and environmental xenoestrogens. *Environmental Health Perspectives* 104(Suppl 4): 741–803. (b) Adami, H.O., et al. 1994. Testicular cancer in nine northern European countries. *International Journal of Cancer* 50: 33–38.

with the Lake Apopka alligators, Hayes has attributed the frog abnormalities to atrazine (Figure 14.9). In lab experiments, male frogs raised in water containing very low doses of the herbicide became feminized and hermaphroditic, developing both testes and ovaries. In field studies, leopard frogs across North America showed similar problems in areas of atrazine usage. Hayes found these effects at atrazine concentrations well below EPA guidelines for human health and at concentrations lower than commonly exist in U.S. drinking water. Thus, he suggests, people may be at risk.

To date, endocrine effects have been found most widely in nonhuman animals, but many scientists attribute the striking drop in sperm counts among men worldwide to endocrine disruptors (Figure 14.10a). In the most noted study, Danish researchers reported in 1992 that the number and motility of sperm in men's semen had declined by 50% since 1938. The research involved a review of 61 studies that included 15,000 men from 20 nations on six continents. Subsequent studies by other researchers—including some who set out to disprove the

findings—have largely confirmed the results using other methods and other populations. Although there is tremendous geographic variation that remains unexplained, it seems that human sperm counts have declined notably in many parts of the world. Some researchers have also voiced concerns about rising rates of testicular cancer (Figure 14.10b), undescended testicles, and genital birth

The Science behind the Story

Bisphenol-A

Can a compound in everyday plastic products damage the most basic biological processes necessary for healthy pregnancies and births? A scientist in Ohio has found evidence that it does—all because one of her lab assistants reached for the wrong soap.

At a laboratory at Case Western Reserve University in August 1998, geneticist Patricia Hunt was making a routine check of her female lab mice, which included extracting and examining developing eggs from the ovaries. The results made her wonder what had gone wrong. About 40% of the eggs showed problems with their chromosomes, and 12% had irregular amounts of genetic material, a condition that can cause serious reproductive problems in mice and people alike.

Detective work revealed that a lab assistant had mistakenly washed the lab's plastic mouse cages and water bottles with an especially harsh soap, A-33, manufactured by Airkem Professional

Dose and Meiosis Abnormalities in Mice Exposed to Bisphenol-A	
BPA dose*	Chromosomal problems during cell division
Control: 0 mg/g	2/115 (1.7%)
20 ng/g (0.00002 mg)	10/172 (5.8%)
40 ng/g (0.00004 mg)	19/255 (7.5%)
100 ng/g (0.0001 mg)	5/46 (10.9%)

*Oral Reference Dose (or safe intake level) established by the U.S. Environmental Protection Agency is 0.05 mg/kg of body weight per day. The European Commission's Scientific Committee on Food's Tolerable Daily Intake is 0.01 mg/kg of body weight per day (recently revised down from U.S. EPA level of 0.05). Data from Hunt, P. A., et al. 2003. Bisphenol A exposure causes meiotic aneuploidy in the female mouse. *Current Biology* 13: 546–553.

Products. The soap damaged the cages so badly that parts of them seemed to have melted. The cages were made from a type of plastic called polycarbonate, which contains the chemical bisphenol-A (BPA). BPA is used to manufacture thousands of plastic products, from baby bottles to food containers to auto parts. The plastics industry uses over 900 million kg (2 billion lb) of BPA every year, according to industry estimates.

The chemical, however, is an estrogen mimic and a long-suspected endocrine disruptor. Some studies link BPA to reproductive abnormalities in mice, such as low sperm counts and early sexual development. Other studies indicate that BPA can leach out of plastic products into water and food. One such study conducted by researchers for the magazine *Consumer Reports,* published in 1999, found that BPA seeped out of the plastic walls of heated baby bottles into infant formula. The plastics industry insisted that BPA was safe and fought back with its own studies. U.S. health

defects in men. Other scientists have proposed that the rise in breast cancer rates (one in eight U.S. women today develops breast cancer) may also be due to hormone disruption, because an excess of estrogen appears to feed tumor development in older women.

Endocrine disruptors can affect more than just the reproductive system. Some impair the brain and nervous system. North American studies have shown neurological problems associated with PCB contamination. In one study in Michigan, mothers who ate Great Lakes fish contaminated with PCBs had babies with lower birth weights and smaller heads, compared to mothers who did not eat fish. These babies grew into children who showed weak and jerky reflexes and tested poorly in intelligence tests.

Endocrine disruption research has generated debate

Much of the research into hormone disruption has brought about strident debate. This is partly because a great deal of scientific uncertainty is inherent in any young and developing field. Another reason is that negative findings about chemicals pose an economic threat to the manufacturers of those chemicals. For instance, Tyrone Hayes's work has met with fierce criticism from scientists associated with atrazine's manufacturer, which stands to lose many millions of dollars if its top-selling herbicide were to be banned or restricted in the United States.

Indeed, our society has invested heavily in some chemicals that now are suspect. One example is bisphenol-A (see

officials, caught in a scientific stalemate, declined to strengthen regulations on BPA.

Hunt, however, thought the chemical might be adversely affecting her mice. The mice in Hunt's lab were suddenly showing dramatic problems during meiosis, the division of chromosomes during the formation of eggs. The 40% of mouse eggs with meiotic problems represented a huge jump from the 1–2% rate of such defects seen before the cage-washing incident. The 12% with irregular amounts of genetic material, a dangerous cell division error called aneuploidy, would likely lead to nonviable pregnancies or birth defects. In humans, aneuploidy is the main cause of miscarriage and Down syndrome.

Deciding to recreate the BPA exposure experimentally, Hunt instructed researchers in her lab to intentionally wash polycarbonate cages and water bottles using varying levels of A-33. They then compared mice kept in damaged

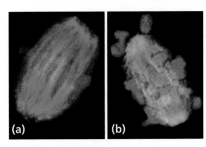

During normal cell division, chromosomes align properly (**a**). Exposure to bisphenol-A, however, causes abnormal cell division (**b**), where chromosomes scatter and are distributed improperly and unevenly between daughter cells.

cages with plastic water bottles to mice kept in undamaged cages with glass water bottles. The developing eggs of mice exposed to BPA through the deliberately damaged plastic showed levels of meiotic problems similar to those in the original A-33 incident, whereas the eggs of mice in the control cages were normal. In an additional round of tests, three sets of female mice were given daily oral doses of BPA over 3, 5, and 7 days, and the

same meiotic abnormalities were observed, although at lower levels. The mice given BPA for 7 days were most severely affected.

Published in 2003 in the journal *Current Biology,* Hunt's findings quickly set off a new wave of concern over the safety of BPA. Hunt wrote, "We have observed meiotic defects in mice at exposure levels close to or even below those considered 'safe' for humans. Clearly, the possibility that BPA exposure increases the likelihood of genetically abnormal offspring is too serious to be dismissed without extensive further study."

Although Hunt's research focused on mice, and not humans, the findings were disturbing because sex cells of mice and humans divide and function in similar ways. Regulators in Europe have recently lowered the level of BPA intake they consider safe. In the United States, some scientists are now pushing for more research and renewed federal scrutiny into the safety of BPA.

"The Science behind the Story," above). An estrogen mimic, bisphenol-A occurs in a great variety of plastic products we use daily, from baby bottles to food cans to eating utensils to auto parts. It has even been used in a cavity-preventing coating for children's teeth. The chemical leaches out of many of these products into water and food, and recent evidence ties it to birth defects in lab mice. The plastics industry vehemently protests that the chemical is safe, however, and points to other research backing its contention.

Research results with bisphenol-A and mice, and those with atrazine and frogs, each showed effects at extremely low levels of the chemical. This is also the case with research on other known or purported endocrine disruptors. The likely reason is that the endocrine system is geared to respond to minute concentrations of substances (normally,

hormones in the bloodstream). Because the endocrine system responds to minuscule amounts of chemicals, it may be especially vulnerable to effects from environmental contaminants that are dispersed and diluted through the environment and that reach our bodies in very low concentrations.

Toxicants may concentrate in surface water or groundwater

Toxicants are not evenly distributed in the environment, and they move about in specific ways (Figure 14.11). For instance, water, in the form of runoff, often carries toxicants from large areas of land and concentrates them in small volumes of surface water. The U.S. Geological Survey findings regarding surface waterways (▸ p. 407)

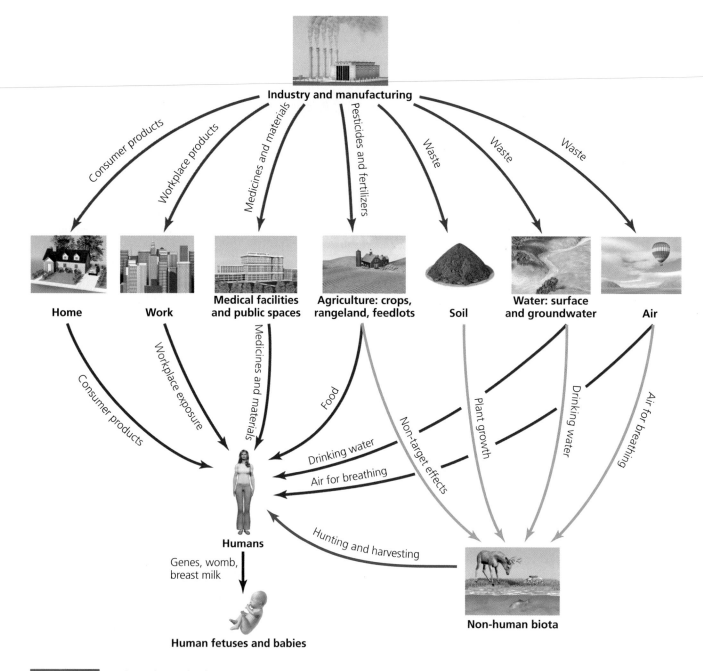

FIGURE 14.11 Synthetic chemicals take many routes in traveling through the environment. Although humans take in only a tiny proportion of these compounds, and although many compounds are harmless, humans—particularly babies—receive small amounts of toxicants from many sources.

demonstrated this concentrating effect. If chemicals can persist in soil, they can leach down into groundwater and contaminate drinking water supplies.

Many chemicals are soluble in water, and these are often the ones that are most accessible to organisms, entering organisms' tissues through drinking or absorption. For this reason, aquatic animals such as fish, frogs, and stream invertebrates are especially good indicators of pollution. When aquatic organisms become sick, we can take it as an early warning that something is amiss. We can then focus research to identify the problem, ideally before it harms humans or the ecological systems that support us. This is why findings that show impacts of low concentrations of pesticides on frogs, fish, and invertebrates have caused such a stir. Many scientists see such findings as a warning that humans could be next, because the pesticides that wash into streams and rivers also flow and seep into the water we drink, and drift through the air we breathe.

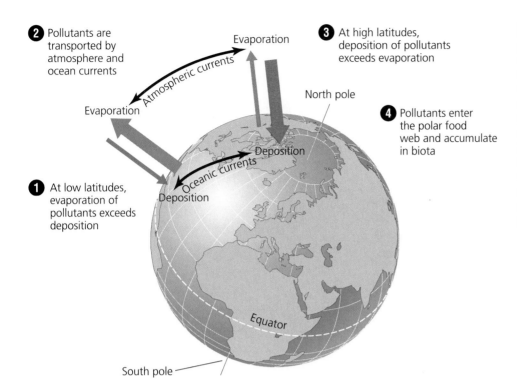

2 Pollutants are transported by atmosphere and ocean currents

Evaporation

Evaporation

Atmospheric currents

3 At high latitudes, deposition of pollutants exceeds evaporation

North pole

4 Pollutants enter the polar food web and accumulate in biota

Deposition

Oceanic currents

Deposition

1 At low latitudes, evaporation of pollutants exceeds deposition

Equator

South pole

FIGURE 14.12 In a process called *global distillation,* pollutants that evaporate and rise high into the atmosphere at lower latitudes, or are deposited in the ocean, are carried preferentially toward the poles by atmospheric currents of air and oceanic currents of water. For this reason, polar organisms take in more than their share of toxicants, despite the fact that relatively few synthetic chemicals are manufactured or used near the poles.

Airborne toxicants can travel widely

Because many chemical substances can be transported by air (Chapter 17), the toxicological effects of chemical use can occur far from the site of direct chemical use. For instance, airborne transport of pesticides is sometimes termed *pesticide drift.* The Central Valley of California is widely considered the most productive agricultural region in the world. But because it is a naturally arid area, food production depends on intensive use of irrigation, fertilizers, and pesticides. Roughly 143 million kg (315 million lb) of pesticide active ingredients are used in California each year, most of these in the Central Valley. The region's frequent winds often blow the airborne spray—and dust particles containing pesticide residue—for long distances. Families living in towns in the Central Valley have suffered health impacts, and activists for farmworkers maintain that hundreds of thousands of the state's residents are at risk. In the mountains of the Sierra Nevada, research has associated pesticide drift from the Central Valley with population declines in four species of frogs.

Synthetic chemical contaminants are ubiquitous worldwide. Despite being manufactured and applied mainly in the temperate and tropical zones, contaminants appear in substantial quantities in the tissues of Arctic polar bears, Antarctic penguins, and people living in Greenland. Scientists can travel to the most remote and seemingly pristine alpine lakes in British Columbia and find them contaminated with foreign toxicants, such as

PCBs. The surprisingly high concentrations in polar regions result from patterns of global atmospheric circulation that move airborne chemicals systematically toward the poles (Figure 14.12).

Some toxicants persist for a long time

Once toxic agents arrive somewhere, they may degrade quickly and become harmless, or they may remain unaltered and persist for many months, years, or decades. The rate at which chemicals degrade depends on factors such as temperature, moisture, and sun exposure, and on how these factors interact with the chemistry of the toxicant. Toxicants that persist in the environment have the greatest potential to harm many organisms over long periods of time. A major reason people have been so concerned about toxic chemicals such as DDT and PCBs is that they have long persistence times. In contrast, the Bt toxin used in biocontrol and in genetically modified (GM) crops (▶ p. 284; Figure 10.13, ▶ p. 288) has a very short persistence time. The herbicide atrazine is variable in its persistence, depending on environmental conditions. There are large numbers of persistent synthetic chemicals because many are designed to be persistent. Those used in plastics, for instance, are used precisely because they resist breakdown.

Most toxicants eventually degrade into simpler compounds called *breakdown products.* Often these are less harmful than the original substance, but sometimes they are just as toxic as the original chemical, or more so. For instance, DDT breaks down into DDE, a highly persistent and toxic compound in its own right. Atrazine produces a large number of breakdown products whose effects have not been fully studied.

Toxicants may accumulate and move up the food chain

Toxicants that organisms absorb, breathe, or consume are subjected to the organism's metabolic processes. Some will be quickly excreted, and others will be degraded into harmless breakdown products. However, some will not be broken down but instead will remain in the body. Toxicants that are fat-soluble or oil-soluble (often organic compounds such as DDT and DDE) are absorbed and stored in fatty tissues. Others, such as methyl mercury, may be stored in muscle tissue. Such toxicants may build up in an animal in a process termed **bioaccumulation.**

Toxicants that bioaccumulate in the tissues of one organism may be transferred to other organisms as predators consume prey. When one organism consumes another, it takes in any stored toxicants and stores them itself, along with the toxicants it has received from eating many other prey. Thus with each step up the food chain, from producer to primary consumer to secondary consumer and so on, concentrations of toxicants can be greatly magnified. This process, called **biomagnification,** occurred most famously with DDT. Top predators, such as birds of prey, ended up with high concentrations of the pesticide because concentrations were magnified as DDT moved from water to algae to plankton to small fish to bigger fish and finally to fish-eating birds (Figure 14.13).

Biomagnification caused populations of many North American birds of prey to decline precipitously from the 1950s to the 1970s. The peregrine falcon was almost totally wiped out in the eastern United States, and the bald eagle, the U.S. national bird, was virtually eliminated from the lower 48 states. The brown pelican vanished from its Atlantic Coast range, remaining only in Florida, and the osprey and other hawks saw substantial population declines. Eventually scientists determined that DDT was causing these birds' eggshells to grow thinner, so that eggs were breaking while in the nest.

Although all these birds' populations have rebounded since the U.S. ban on DDT, such scenarios are by no means a thing of the past. The polar bears of Svalbard Island in Arctic Norway are at the top of the food chain and feed on seals that have biomagnified toxicants. Despite their remote Arctic location, Svalbard Island's polar bears show some of the highest levels of PCB contamination of any

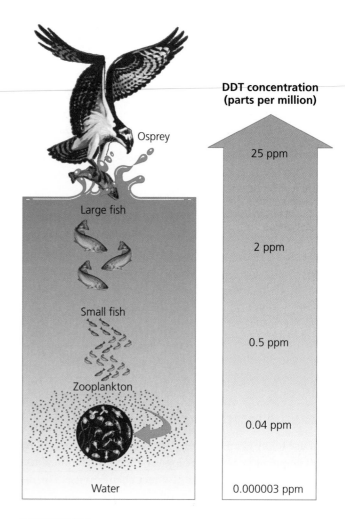

FIGURE 14.13 Fat-soluble compounds such as DDT bioaccumulate in the tissues of organisms. As animals at higher trophic levels eat organisms lower on the food chain, their load of toxicants passes up to each consumer. DDT moves from zooplankton through various types of fish, finally becoming highly concentrated in fish-eating birds such as ospreys.

wild animals tested, as a result of biomagnification and the process of global distillation shown in Figure 14.12. The contaminants are likely responsible for the immune suppression, hormone disruption, and high cub mortality that the bears seem to be suffering. Cubs that survive receive PCBs in their mothers' milk, so that contamination persists and accumulates between generations.

Not all toxicants are synthetic

Although we have focused on synthetic chemicals thus far, it is vital to recognize that chemical toxicants also exist naturally in the environment around us and in the foods we eat. We have good reason as citizens and consumers to insist on being informed about risks synthetic chemicals may pose, but it is a mistake to assume that all artificial chemicals are unhealthy and that all natural chemicals are healthy. In fact, the plants and animals we eat contain many chemicals that can cause us harm. Recall that plants produce toxins to ward off animals that eat them. In domesticating crop plants, we have selected for strains with reduced toxin content, but we have not eliminated these dangers. Furthermore, when we consume animal meat, we take in toxins the animals have ingested from plants or animals they have eaten.

Scientists have actively debated just how much risk natural toxicants pose. Biochemist Bruce Ames of the University of California at Berkeley maintains that the amounts of synthetic chemicals in our food from pesticide residues are dwarfed by the quantities of natural toxicants. Ames also holds that the natural defenses against toxins that our bodies have evolved are largely effective against synthetic toxicants. Therefore, he reasons, synthetic chemicals are a relatively minor worry. Moreover, he contends, if fear of pesticide residues or increased cost to remove them causes people to eat fewer fruits and vegetables, then people will be at greater risk for disease, because a balanced diet that includes fruits and vegetables is important for health.

A respected senior scientist, Ames was a hero of environmentalists for his early work. The *Ames test,* a bacterial assay for mutagens, allows easy screening of suspected toxins. Today Ames is a target of criticism from many environmental advocates. His critics say that natural toxicants are usually more readily metabolized and excreted by the body than synthetic ones, that synthetic toxicants persist and accumulate in the environment, and that synthetic chemicals expose people (such as farmworkers and factory workers) to risks in ways other than the ingestion of food. What is clear is that more research is required in this area.

Weighing the Issues:
Natural and Synthetic Estrogen Mimics

Skeptics of the idea that synthetic chemicals act as endocrine disruptors point out that humans are exposed to phytoestrogens, natural estrogens from plants, which are ubiquitous in the environment. However, phytoestrogens are easily broken down by the body and are not stored in tissue. Do you think there is a distinction between synthetic and natural estrogen-mimicking chemicals? Should manufacturers of synthetic chemicals that mimic estrogen be held responsible for health impacts, given that natural chemicals can also mimic estrogen?

Studying Effects of Hazards

Determining health effects of particular environmental hazards is a challenging job, particularly because any given person or organism likely has a complex history of exposure to many hazards throughout life. Scientists rely on several different methods, including correlative surveys and manipulative experiments (▸pp. 15–17).

Wildlife studies use careful observations in the field and the lab

When scientists were zeroing in on the impacts of DDT, one key piece of evidence came from museum collections of wild birds' eggs from the decades before synthetic pesticides were manufactured. Eggs from museum collections had measurably thicker shells than the eggs scientists were studying in the field from present-day birds. Scientists have pieced together the puzzle of toxicant effects on alligators by taking measurements from animals in the wild, then doing controlled experiments in the lab to test hypotheses. With frogs and atrazine, scientists first measured toxicological effects in lab experiments, then sought to demonstrate correlations with herbicide usage in the wild.

Often the study of wildlife advances in the wake of some conspicuous mortality event. Off the California coast in 1998–2001, populations of sea otters fell noticeably, and many dead otters washed ashore. Field biologists documented the population decline, and specialists went to work in the lab performing autopsies to determine causes of death. The most common cause of death was found to be infection with the protist parasite *Toxoplasma,* which killed otters directly and also made them vulnerable to shark attack. *Toxoplasma* occurs in the feces of cats, so scientists hypothesized that sewage runoff containing waste from litter boxes was entering the ocean from urban areas and infecting the otters.

Human studies rely on case histories, epidemiology, and animal testing

In studies of human health, as in wildlife studies, we have gained much knowledge by studying sickened individuals directly. Medical professionals have long treated victims of poisonings, for instance, so the effects of common poisons are well known. Autopsies help us understand what constitutes a lethal dose. This process of observation and analysis of individual patients is known as a *case history* approach. Case histories have advanced our understanding of human illness, but they do not always help us infer

The Science behind the Story

Pesticides and Child Development in Mexico's Yaqui Valley

With spindly arms and big, round eyes, one set of pictures shows the sorts of stick figures drawn by young children everywhere. Next to them is another group of drawings, mostly disconnected squiggles and lines, resembling nothing. Both sets of pictures are intended to depict people. The main difference identified between the two groups of young artists: long-term pesticide exposure.

Children's drawings are not a typical tool of toxicology, but Elizabeth Guillette, an anthropologist married to Louis Guillette, wanted to try new methods. Guillette and her colleagues were interested in the effects of pesticides on children. They devised tests to measure childhood development based on techniques from anthropology and medicine. Searching for a study site, Guillette found the Yaqui Valley region of northwestern Mexico.

The Yaqui Valley is farming country, worked for generations by the indigenous group that gives the region its name. Synthetic pesticides arrived in the area in the 1940s. Some Yaqui embraced the agricultural innovations, spraying their farms in the valley to increase their yields. Yaqui farmers in the surrounding foothills, however, generally chose to bypass the chemicals and to continue following more traditional farming practices. Although

4-year-olds   5-year-olds

Drawings by children in the foothills

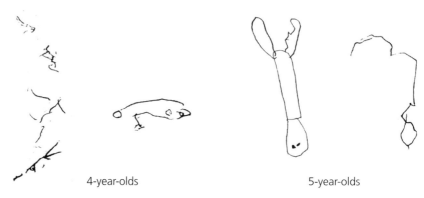

4-year-olds   5-year-olds

Drawings by children in the valley

Elizabeth Guillette's study in Mexico's Yaqui Valley offers a startling example of apparent neurological effects of pesticide poisoning. Young children from foothills areas where pesticides were not commonly used drew recognizable figures of people. Children the same age from valley areas where pesticides were used heavily in industrialized agriculture could draw only scribbles. Adapted from Guillette, E. A., et al. 1998. *Environmental Health Perspectives* 106: 347–353.

differing in farming techniques, Yaqui in the valley and foothills continued to share the same culture, diet, education system, income levels, and family structure.

At the time of the study, in 1994, valley farmers planted crops twice a year, applying pesticides up to 45 times from planting to harvest. A previous study conducted in the valley in 1990, focusing on areas with the largest farms, had indicated high levels of multiple pesticides in the breast milk of mothers and in the

the effects of hazards people rarely come into contact with, of newly manufactured compounds, or of chemicals existing at low environmental concentrations and exerting minor long-term effects. Case histories also tell us little about probability and risk, such as how many extra deaths we might expect in a population due to a particular cause.

For such situations, which are common in environmental toxicology, epidemiological studies are necessary. **Epidemiological studies** involve large-scale comparisons among groups of people, usually contrasting a group known to have been exposed to some hazard and a group that has not. Epidemiologists track the fate of all people in the study, generally for a long period of time (often

umbilical cord blood of newborn babies. In contrast, foothill families avoided chemical pesticides in their gardens and homes.

To understand how pesticide exposure affects childhood development, Guillette and four fellow researchers studied 50 preschoolers aged four to five, of whom 33 were from the valley and 17 from the foothills. Each child underwent a half-hour exam, during which researchers showed a red balloon, promising to give the balloon later as a gift, and using the promise to evaluate long-term memory. Each child was then put through a series of physical and mental tests:

▶ Catching a ball from distances of up to 3 m (10 ft) away, to test overall coordination

▶ Jumping in place for as long as possible, to assess endurance

▶ Drawing a picture of a person, as a measure of perception

▶ Repeating a short string of numbers, to test short-term memory

▶ Dropping raisins into a bottle cap from a height of about 13 cm (5 in.), to gauge fine motor skills

The researchers also measured each child's height and weight but, because of lack of time and money, stopped short of taking blood or tissue samples to check for pesticides or other toxins. When all tests were completed, each child was asked what he or she had been promised and received a red balloon.

Although the two groups of children were not significantly different in height and weight, they differed markedly in other areas of development. Valley children were far behind the foothill children developmentally in coordination, physical endurance, long-term memory, and fine motor skills:

▶ From a distance of 3 m (10 ft), valley children had great difficulty catching the ball.

▶ Valley children could jump for an average of 52 seconds, compared to 88 seconds for foothill children.

▶ Most valley children missed the bottle cap when dropping their raisins, whereas foothill children dropped them into the caps far more often.

▶ Each group did fairly well repeating numbers, but valley children showed poor long-term memory. At the end of the test, all but one of the foothill children remembered that they had been promised a balloon, and 59% remembered it was red. However, of the valley children only 27% remembered the color of the balloon, only 55% remembered they'd be getting a balloon, and 18% were unable to remember anything about the balloon.

It was the children's drawings, however, that exhibited the most dramatic difference between valley and foothill children (see figure). The researchers determined each drawing could earn 5 points, with 1 point each for a recognizable feature: head, body, arms, legs, and facial features. The foothill children drew pictures that looked like people, averaging about 4.5 points per drawing. The valley children, in contrast, averaged 1.6 points per drawing; their scribbles resembled little that looked like a person. By the standards of developmental medicine, the four- and five-year-old valley children drew at the level of a two-year-old.

Some scientists greeted Guillette's study skeptically, pointing out that its sample size was too small to be meaningful. Others said that factors the researchers missed, such as different parenting styles or unknown health problems, could be to blame. Prominent toxicologists argued that without blood or tissue tests on the children, the study results couldn't be tied to agricultural chemicals. Regardless of these criticisms, Guillette maintains that her findings show that nontraditional study methods are a valid way to track the effects of environmental toxins and that pesticides present a complex long-term risk to human growth and health.

years or decades) and measure the rate at which deaths, cancers, or other health problems occur in each group. The epidemiologist then analyzes the data, looking for observable differences between the groups, and statistically tests hypotheses accounting for differences. When a group exposed to a hazard shows a significantly greater degree of harm, it suggests that the hazard may be responsible. This process is akin to a natural experiment (▶pp. 16–17), in which the experimenter takes advantage of the presence of groups of subjects made possible by some event that has already occurred. A slightly different type of natural experiment was conducted by anthropologist Elizabeth Guillette (see "The Science behind the Story," above).

The advantages of epidemiological studies are their realism and their ability to yield relatively accurate predictions about risk. The drawbacks include the need to wait a long time for results and the inability to address future effects of new products just coming to market. In addition, participants in epidemiological studies encounter many factors that affect their health besides the one under study. Epidemiological studies measure a statistical association between a toxicant and an effect, but they do not confirm that the toxicant *causes* the effect.

Manipulative experiments are needed to truly nail down causation. However, subjecting people to massive doses of toxicants in a lab experiment would clearly be unethical. So researchers have traditionally used other animals as subjects to test toxicity. Foremost among these animal models have been laboratory strains of rats, mice, and other mammals. Because of our shared evolutionary history with these mammals, their bodies function similarly to ours. Yet we are also different, of course. The extent to which results from animal lab tests apply to humans is always somewhat uncertain, and it can be expected to vary from one study to the next. Although some people feel the use of rats and mice for testing is unethical, animal testing enables scientific advances that would be impossible or far more difficult otherwise. However, new techniques (with human cell cultures, bacteria, or tissue from chicken eggs) are being devised that may one day replace some live-animal testing.

Dose-response analysis is a mainstay of toxicology

The standard method of testing with lab animals in toxicology is dose-response analysis. Scientists quantify the toxicity of a given substance by measuring how much effect a toxicant produces at different doses or how many animals are affected by different doses of the toxic agent. The *dose* is the amount of toxicant the test animal receives, and the *response* is the type or magnitude of negative effects the animal exhibits as a result. The response is generally quantified by measuring the proportion of animals exhibiting negative effects. The data are plotted on a graph, with dose on the *x* axis and response on the *y* axis (Figure 14.14a). The resulting curve is called a **dose-response curve.**

Once they have plotted a dose-response curve, toxicologists can calculate a convenient shorthand gauge of a substance's toxicity: the amount of toxicant it takes to kill half the population of study animals used. This lethal dose for 50% of individuals is termed the **LD_{50}.** A high LD_{50} indicates low toxicity, and a low LD_{50} indicates high

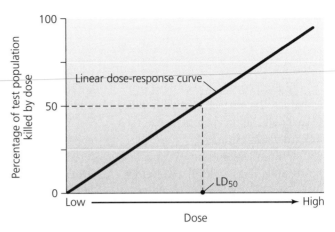

(a) Linear dose-response curve

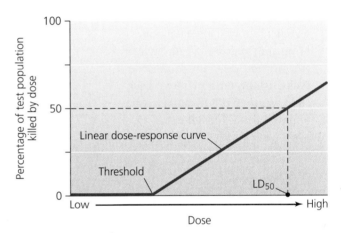

(b) Dose-response curve with threshold

FIGURE 14.14 In a classic linear dose-response curve (**a**), the percentage of animals killed or otherwise affected by the substance rises with the dose. The point at which 50% of the animals are killed is labeled the lethal-dose-50, or LD_{50}. For some toxic agents, a threshold dose (**b**) exists, below which doses have no measurable effect. Go to **GRAPHit!** at www.aw-bc.com/withgott or on the student CD-ROM.

toxicity. Of course, the experimenter may instead be interested in nonlethal health effects. A researcher often will want to document the level of toxicant at which 50% of the population of test animals is affected in whatever way is of interest in the study (for instance, what level of toxicant causes 50% of lab mice to lose their hair?). Such a level is called the effective-dose-50%, or **ED_{50}.** Chemicals with an especially low LD_{50} or ED_{50} are assumed likely to be poisonous to humans, whereas those with an extremely high LD_{50} or ED_{50} are thought likely to be safe.

Sometimes responses occur only above a certain dose. Such a **threshold** dose (Figure 14.14b) might be expected if the body's organs can fully metabolize or excrete a toxicant at low doses but become overwhelmed at higher concentrations. It might also occur if cells can repair damage to their DNA only up to a certain point.

Scientists generally give lab animals much higher doses relative to body mass than humans would receive in the environment. This is so that the response is great enough to be measured, and so that differences between the effects of small and large doses are evident. Data from a range of doses help give shape to the dose-response curve. Once the data from animal tests are plotted, scientists generally extrapolate downward to estimate the effect of still-lower doses on a hypothetically large population of animals. This way, they can come up with an estimate of, say, what dose causes cancer in 1 mouse in 1 million. A second extrapolation is then required to estimate the effect on humans, with our greater body mass. Because these two extrapolations go beyond the actual data obtained, they introduce uncertainty into the interpretation of what doses are acceptable for humans. As a result, to be on the safe side, regulatory agencies set standards for maximum allowable levels of toxicants that are well below the minimum toxicity levels estimated from lab studies.

Knowing the shape of dose-response curves is crucial if one is planning to extrapolate from them to predict responses at doses below those that have been tested. And sometimes responses *decrease* with increased dose. Toxicologists are finding that some dose-response curves are U-shaped or J-shaped, or shaped like an inverted U. Such counterintuitive curves often appear to apply to endocrine disruptors. This likely occurs because the hormone system is geared to function with extremely low concentrations of hormones and is thus vulnerable to disruption by toxicants at extremely low concentrations. Inverted dose-response curves present a challenge for policymakers attempting to set safe environmental levels for toxicants. If many chemicals behave in these ways, we may have underestimated the dangers of these compounds, because many chemicals exist in very low concentrations over wide areas.

Individuals vary in their responses to hazards

Different individuals may respond quite differently to identical exposures to hazards. These differences can be genetically based, or can be due to a person's current condition. People in poorer health are often more sensitive to biological and chemical hazards. Sensitivity also can vary with sex, age, and weight. Because of their smaller size and rapidly developing organ systems, fetuses, infants, and young children tend to be much more sensitive to toxicants than are adults. The degree of this difference was not appreciated for many years. Regulatory agencies such as the EPA traditionally set standards for adults and extrapolated downward for infants and children. However, they have subsequently found that in many cases their linear extrapolations did not lower standards enough to protect babies adequately. Many critics today contend that despite improvements, regulatory agencies still do not account explicitly enough for risks to fetuses, infants, and children.

The type of exposure can affect the response

The risk posed by a hazard often varies according to whether a person experiences high exposure for short periods of time, known as **acute exposure**, or lower exposure over long periods of time, known as **chronic exposure**. Incidences of acute exposure are easier to recognize, because they often stem from discrete events, such as accidental ingestion, an oil spill, a chemical spill, or a nuclear accident. Lab tests and LD_{50} values generally reflect acute toxicity effects. However, chronic exposure is more common—and more difficult to detect and diagnose. Chronic exposure often affects organs gradually, as when smoking causes lung cancer, or when alcohol abuse induces liver or kidney damage. Pesticide residues on food or low levels of arsenic in drinking water (▸ pp. 456–457) also pose chronic risk. Because of the long time periods involved, relationships between cause and effect may not be readily apparent.

Mixes may be more than the sum of their parts

It is difficult enough to determine the impact of a single hazard on an organism, but the task becomes astronomically more difficult when multiple hazards interact. For instance, chemical substances, when mixed, may act in concert in ways that cannot be predicted from the effects of each in isolation. Mixed toxicants may sum each other's effects, cancel out each other's effects, or multiply each other's effects. Whole new types of impacts may arise when toxicants are mixed together. Such interactive impacts—those that are more than or different from the simple sum of their constituent effects—are called **synergistic effects.**

Examples of synergistic effects with toxic substances are numerous. With Florida's alligators, lab experiments have indicated that DDE can either help cause or inhibit sex reversal, depending on the presence of other chemicals. Mice exposed to a mixture of nitrate, atrazine, and aldicarb have been found to show immune, hormone, and nervous-system effects that were not evident from exposure to each of these chemicals alone. Wood frogs in the wild are increasingly suffering limb deformities, apparently the result of being parasitized by trematode flatworms. However, being near an agricultural field with pesticide runoff increases the rate of parasitic infection, because, as lab studies have shown, pesticides suppress the frog's immune response, making the amphibian more vulnerable to parasites.

Traditionally, environmental health has tackled effects of single hazards one at a time. In toxicology, the complex experimental designs required to test interactions, and the sheer number of chemical combinations, have meant that single-substance tests have received priority. This approach is changing, but scientists in environmental health and toxicology will never be able to test all possible combinations. There are simply too many hazards in the environment.

Risk Assessment and Risk Management

Policy decisions to ban chemicals or restrict their use are not made casually, but generally follow years of rigorous testing for toxicity. Likewise, strategies for combating disease and other environmental health threats are often based on extensive research. Policy and management decisions also reach beyond the scientific results on health to incorporate considerations about economics and ethics. And all too often, they are influenced by political pressure from powerful interests. The steps between the collection and interpretation of scientific data and the formulation of policy involve assessing and managing risk.

Risk is expressed in terms of probability

Exposure to an environmental health threat does not invariably produce some given effect. Rather, it causes some probability of harm, some statistical chance that damage will result. To understand the impact of an environmental health threat, a scientist must know more than just its identity and strength. He or she must also know the chance that an organism will encounter it, the frequency with which the organism may encounter it, the amount of substance or degree of threat to which the organism is exposed, and the organism's sensitivity to the threat. Such factors help determine the overall risk posed. Risk can be measured in terms of *probability*, a quantitative description of the likelihood of a certain outcome. The mathematical probability that some harmful outcome (for instance, injury, death, environmental damage, or economic loss) will result from a given action, event, or substance expresses the **risk** posed by that phenomenon.

Our perception of risk may not match reality

Every action we take and every decision we make involves some element of risk, some (generally small) probability that things will go wrong. Some actions and decisions are more risky than others, and we try our best in everyday life to behave in ways that minimize risk. Interestingly, our perceptions of risk do not always match statistical reality (Figure 14.15). People often worry unduly about negligibly small risks but happily engage in other activities that pose high risks. For instance, most people perceive flying in an airplane as a riskier activity than driving a car, but driving a car is statistically far more dangerous.

Psychologists agree that this difference between risk perception and reality stems from the fact that we feel more at risk when we are not controlling a situation and more safe when we are "at the wheel"—regardless of the actual risk involved. When we drive a car, we feel we are in control, even though statistics show we are at greater risk than as a passenger in an airplane. This psychology can account for people's great fear of nuclear power, toxic waste, and pesticide residues on foods—environmental hazards that are invisible or little understood, and whose presence in their lives is largely outside their personal control. In contrast, people are more ready to accept and ignore the risks of smoking cigarettes, overeating, and not exercising, all voluntary activities statistically shown to pose far greater risks to health.

Risk assessment analyzes risk quantitatively

The quantitative measurement of risk and the comparison of risks involved in different activities or substances together are termed **risk assessment.** Risk assessment is a way of identifying and outlining problems. In environmental health, it helps ascertain which substances and activities pose health threats to people or wildlife and which are largely safe. Assessing risk for a chemical substance

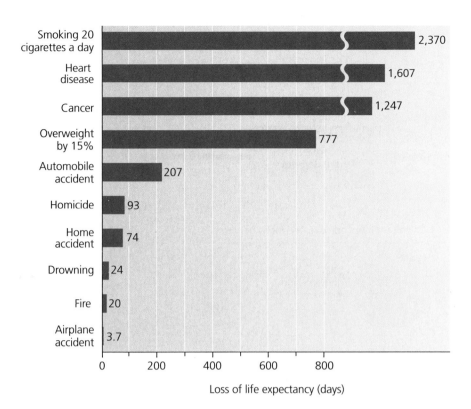

FIGURE 14.15 Our perceptions of risk do not always match the reality of risk. Listed here are several leading causes of death in the United States, along with a measure of the risk each poses. Risk is measured in days of lost life expectancy, or the number of days of life lost by people suffering the hazard, spread across the entire population—a measure commonly used by insurance companies. By this measure, one common source of anxiety, airplane accidents, poses 20 times less risk than home accidents, over 50 times less risk than auto accidents, and over 200 times less risk than being overweight. Data from Cohen, B. 1991. Catalog of risks extended and updated. *Health Physics* 61: 317–335.

involves several steps. The first steps involve the scientific study of toxicity outlined above—determining whether a given substance has toxic effects and, through dose-response analysis, measuring how effects on an organism vary with the degree of toxicant exposure. Subsequent steps involve assessing the individual's or population's likely extent of exposure to the substance, including the frequency of contact, the concentrations likely encountered, and the length of time the substance is expected to be encountered. Risk assessment studies are often performed by scientists associated with the industries that manufacture toxicants, which in many people's minds may undermine the objectivity of the process.

Weighing the Issues:
A Nationwide Health Tracking System?

When 16 children were diagnosed with leukemia in the small town of Fallon, Nevada, it raised alarm. The identification of such "cancer clusters" has prompted many U.S. scientists and policymakers to back the notion of creating a national environmental health tracking system. Some states already track environmental hazards and illnesses, but comparable statistics are not compiled everywhere. It would take an estimated $1 billion to integrate these disparate sets of information and to maintain records.

Do you think the potential benefits are worth it? How might this affect epidemiological studies? How might it affect risk assessments? Could it speed the identification of causes of poisonings or disease outbreaks? How might finding the causes of such events help to mitigate them or to prevent them in the future?

Risk management combines science and other social factors

Accurate risk assessment is a vital step toward effective **risk management,** which consists of decisions and strategies to minimize risk (Figure 14.16). In most developed nations, risk management is handled largely by federal agencies, such as the EPA, the Centers for Disease Control and Prevention (CDC), and the Food and Drug Administration (FDA) in the United States. In risk management, scientific assessments of risk are considered in light of economic, social, and political needs and values. The costs and benefits of addressing risk in various ways are assessed with regard to both scientific and nonscientific concerns. Decisions whether to reduce or eliminate risk are then made.

In environmental health and toxicology, comparing costs and benefits (▸ p. 42) is often difficult because the benefits are often economic and the costs often pertain to health. Moreover, economic benefits are generally known,

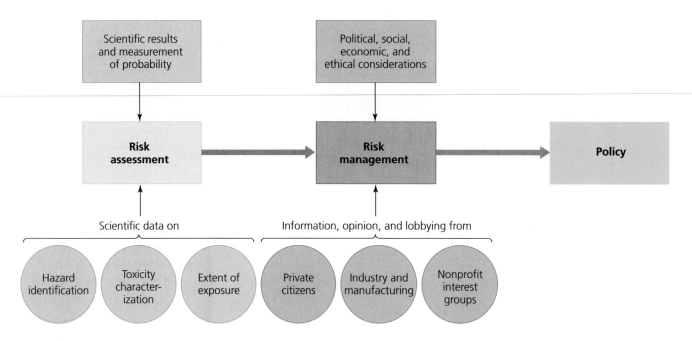

FIGURE 14.16 Risks must be assessed before policy steps can be taken to minimize them. Thus the first step in addressing the risk of an environmental hazard is risk assessment, a process of quantifying the risk of the hazard and comparing it to other risks. Once science identifies and measures risks, then risk management can proceed. In this process, economic, political, social, and ethical issues are considered in light of the scientific data from risk assessment. The consideration of all these types of information is designed to result in policy decisions that minimize the risk of the environmental hazard.

easily quantified, and of a discrete and stable amount, whereas health risks are hard-to-measure probabilities, often involving a very small percentage of people likely to suffer greatly and a large majority likely to experience little effect. When a government agency bans a pesticide, it generally means measurable economic loss for the manufacturer and the farmer, whereas the benefits accrue less predictably over the long term to some percentage of factory workers, farmers, and the general public. Because of the lack of equivalence in the way costs and benefits are measured, risk management frequently tends to stir up debate.

Philosophical and Policy Approaches

Because we do not know a substance's toxicity until we measure and test it, and because there are so many untested chemicals and combinations, science will never eliminate the many uncertainties that accompany risk assessment. In such a world of uncertainty, there are two basic philosophical approaches to categorizing substances as safe or dangerous (Figure 14.17).

Two approaches exist for determining safety

One approach is to assume that substances are harmless until shown to be harmful. We might nickname this the *innocent-until-proven-guilty approach*. Because thoroughly testing every existing substance (and combination of substances) for its effects is a hopelessly long, complicated, and expensive pursuit, the innocent-until-proven-guilty approach has the benefit of not slowing down technological innovation and economic advancement. However, it has the disadvantage of putting into wide use some substances that may later turn out to be dangerous.

The other approach is to assume that substances are harmful until they are shown to be harmless. This approach follows the precautionary principle first discussed in Chapter 10 in regard to genetically modified foods (▸p. 290). This more cautious approach should enable us to identify troublesome toxicants before they are released into the environment, but it may also significantly impede the pace of technological and economic advance.

These two approaches are actually two ends of a continuum of possible approaches. The two endpoints differ

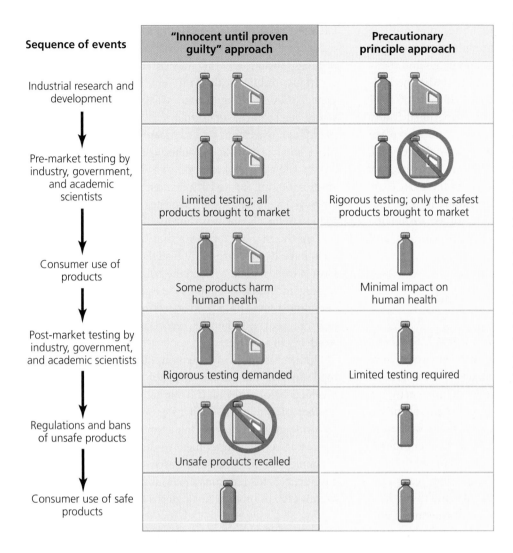

Sequence of events	"Innocent until proven guilty" approach	Precautionary principle approach
Industrial research and development		
Pre-market testing by industry, government, and academic scientists	Limited testing; all products brought to market	Rigorous testing; only the safest products brought to market
Consumer use of products	Some products harm human health	Minimal impact on human health
Post-market testing by industry, government, and academic scientists	Rigorous testing demanded	Limited testing required
Regulations and bans of unsafe products	Unsafe products recalled	
Consumer use of safe products		

FIGURE 14.17 Testing a new chemical compound or product for toxicity rarely gives a black-or-white answer, and many tests must be run before a substance's properties are well understood. Given such uncertainty, two main approaches can be taken to introducing new substances to the market. In one approach, substances are innocent until proven guilty; they are brought to market relatively quickly after limited testing. Products reach consumers more quickly, but some fraction of them may cause harm to some fraction of people. In the other approach, the precautionary principle is adopted, and substances are brought to market cautiously, only after extensive testing. Products that reach the market should be safe, but many perfectly safe products will be delayed in reaching consumers.

mainly in where they lay the burden of proof—specifically, whether product manufacturers are required to prove safety or whether government, scientists, or citizens are required to prove danger.

Weighing the **Issues:**
The Precautionary Principle

Given the substantial costs of testing chemicals for safety and the increasing concerns about their spread through the environment, should proof of safety be required by government prior to a chemical's release into the environment? Should the burden of proof fall to the company that stands to make a profit from a product's release? How do you think adopting the precautionary principle would affect the number of chemicals on the market?

Philosophical approaches are reflected in policy

One's philosophical approach has immediate and far-reaching impact on policy decisions, directly affecting what materials are allowed into our environment. Most nations follow a blend of the two approaches, but there is marked variation among countries. At the present time, European nations are to a great extent following the precautionary principle regarding the regulation of synthetic chemicals, whereas the United States is not. Although industry frequently complains that government regulation is cumbersome, environmental and consumer advocates criticize U.S. policies for largely following the innocent-until-proven-guilty approach. For instance, compounds involved in cosmetics require no FDA review or approval before being sold to the public.

In the United States, several federal agencies apportion responsibility for tracking and regulating synthetic chemicals. The FDA, under the Food, Drug, and Cosmetic Act of 1938 and its subsequent amendments, regulates foods and food additives, cosmetics, and drugs and medical devices. The EPA regulates pesticides under the Federal Insecticide, Fungicide, and Rodenticide Act of 1947 (FIFRA) and its amendments. The Occupational Safety and Health Administration (OSHA) regulates workplace hazards under a 1970 act. Several other agencies regulate other substances. Synthetic chemicals not covered by other laws are regulated by the EPA under the 1976 Toxic Substances Control Act (TSCA). Examining in more detail the policy process for pesticides and for toxic substances will provide an idea of how government and industry interact.

The EPA regulates pesticides and other substances

FIFRA was enacted as the post–World War II U.S. chemical industry was expanding, but before the environmental activism of the 1960s and 1970s that gave rise to so many environmental laws. As such, FIFRA was primarily intended not to protect public health or the environment, but to assure consumers that products actually worked as their manufacturers claimed. Subsequent amendments shifted the focus somewhat toward protecting health and safety and charged the EPA with "registering" each new pesticide that manufacturers propose to bring to market.

The registration process involves risk assessment and risk management. The EPA first asks the pesticide manufacturer to provide information, including the results of safety assessments the company has performed according to EPA guidelines. The EPA examines the company's research and all other relevant scientific research. It examines the product's ingredients and how the product will be used and evaluates whether the chemical poses risks to humans, other organisms, or water or air quality. The EPA then approves, denies, or sets limits on the chemical's sale and use. It also must approve the language used on the product's label.

Because the registration process takes economic considerations into account, critics say it allows hazardous chemicals to be approved if the economic benefits outweigh the hazards. Here the challenges of weighing intangible risks involving human health and environmental quality against the tangible and quantitative numbers of economics become apparent.

TSCA directed the EPA to monitor some 75,000 industrial chemicals manufactured in or imported into the United States, including PCBs and other compounds involved in plastics. The act gave the agency power to regulate these chemicals and ban them if they are found to pose excessive risk. TSCA was the first law to require screening of these substances before they entered the marketplace.

Many public health and environmental advocates view TSCA as being far too weak. They note that the screening required of industry is minimal and that to mandate more extensive and meaningful testing, the EPA must show proof of the chemical's toxicity. In other words, the agency is trapped in a Catch-22: To push for studies looking for toxicity, it must have proof of toxicity already. The result, these advocates say, is that most synthetic chemicals are not thoroughly tested before being put on the market. Of those that fall under TSCA, only 10% have been thoroughly tested for toxicity; only 2% have been screened for carcinogenicity, mutagenicity, or teratogenicity; fewer than 1% are government-regulated; and almost none have been tested for endocrine, nervous, or immune system damage, according to the U.S. National Academy of Sciences.

Industry's critics say chemical manufacturers should be made to bear the burden of proof for the safety of their products before they hit the market. Industry's supporters say that safety advocates will never be satisfied that industry has done enough. They say that mandating more toxicological research will hamper the introduction of products that consumers want, increase the price of products as research costs are passed on to consumers, and cause U.S. companies to move to nations where standards are more lax.

Toxicants are regulated internationally

In April 2003, European Union (EU) commissioners proposed legislation that would require chemical manufacturers to test and register 30,000 chemicals already in use and that would impose restrictions on 1,500 chemicals already considered hazardous. In announcing the policy, EU environment commissioner Margot Wallström said, "It is high time to place the responsibility where it belongs, with industry." Industry was not pleased and estimated that the law would cost it $7–8 billion over a decade. At the same time, some people on all sides agreed that the new policy would have the positive effect of spurring research into safer products and creating new markets for them. As of late 2005, the proposal was slowly making its way toward becoming law.

Action regarding chemical toxicants has also been taken in the form of international treaties. The Stockholm Convention on Persistent Organic Pollutants (POPs), introduced in 2001, appears on its way to ratification. POPs are toxic chemicals that persist in the environment, bioaccumulate in the food chain, and often can travel long

VIEWPOINTS

Chemical Product Testing: Industry or Government?

The testing of chemical products for safety can take the so-called innocent-until-proven-guilty or precautionary approaches. **Should manufacturers be held responsible for comprehensive testing of new chemical products before they are introduced to the public? What would be some of the advantages and disadvantages? How extensive a role should government play in the testing process?**

Testing Must Ensure Public Health

Like most things in life, the controversy over product testing arises because both approaches have valid advantages and disadvantages.

Allowing industry to follow the innocent-until-proven-guilty approach, with limited testing, reduces development costs for new chemical products and may lead to greater economic activity. If industry were required to comprehensively test chemical product safety before introduction to the public, chemical industry profits could fall and result in job loss and costly new product development. Consumer prices might rise to cover these costs.

However, comprehensive testing would lower the number of chemicals that adversely affect biological species.

Our definition of *innocent* is often too narrow. In the innocent-until-proven-guilty approach, *what the consumer actually buys is never tested,* because only ultra-pure active ingredients are tested. Surfactants and organic soaps ("other ingredients") are added to improve the active ingredients' lipid or water solubility, and these other ingredients are frequently very active biologically. Also, production contaminants are not tested and registered. Therefore, a so-called innocent product can cause cancer and reproductive defects.

The assumption of a linear dose response is also coming under increased scrutiny, especially at very low physiological doses, where hormonal, immune, and neurological processes respond. At much higher pharmacological doses, where toxicity testing is typically done, the responses of physiological systems to the same chemical can be very different.

Given the inherent inadequacies of the testing process and the uncertainty of the economic impacts, both government and industry should share the responsibility of testing to ensure public safety.

Warren Porter is a toxicologist and physiological ecologist at the University of Wisconsin–Madison. He evaluates the connections among climate, animal energetics, and behavior using statistical experimental design.

An Industry Perspective

Chemical manufacturers already take an active role in testing new chemicals. This process is part of current EPA regulations. Additionally, government often tests chemicals to elucidate either hazard or exposure. Generation of these data by government adds to the body of knowledge generated by industry.

Chemical risk depends on two factors: hazard (toxicity) and exposure. To evaluate a chemical, manufacturers typically start by conducting screening-level toxicological and environmental studies and proceed to more or higher-tier studies as warranted. There is no single comprehensive testing program that is appropriate for all industrial chemicals.

The Toxic Substances Control Act requires almost all new commercial substances to undergo Premanufacture Notification (PMN) review, and to describe this preliminary process as an innocent-until-proven-guilty approach is an oversimplification. When the EPA reviews a PMN, it considers the physical and chemical properties of the substance, structural similarity to other compounds of known toxicity, and potential for human exposure and environmental release. If there is no evidence of harm from preliminary testing, longer-term or more specialized testing may not be conducted. In some cases, the EPA may require additional testing to determine whether the chemical poses an unreasonable risk to human health or the environment. If the EPA finds that risk can be addressed by reducing exposure, it may enter into a binding agreement with the manufacturer to require exposure reduction activities, rather than additional laboratory testing.

Manufacturers often voluntarily conduct new studies to support the continued safe use of their chemicals. The 150 member companies of the American Chemistry Council represent about 90% of U.S. chemical production. These companies are committed to Responsible Care®, under which chemical manufacturers, as good stewards of their products, continue to test as new data and methodologies become available. There is a role for both government and industry in chemical testing, and it is important that the EPA and manufacturers work together in evaluating chemicals to improve health, safety, and the environment.

Marian K. Stanley is senior director for the American Chemistry Council, which she joined in 1990. She is responsible for managing chemical-specific issue groups in the Council's self-funded CHEMSTAR Department.

Explore this issue further by accessing **Viewpoints** at www.aw-bc.com/withgott.

Table 14.3 The "Dirty Dozen" Persistent Organic Pollutants (POPs) Targeted by the Stockholm Convention

Toxicant	Type	Description
Aldrin	Pesticide	Kills termites, grasshoppers, corn rootworm, and other soil insects
Chlordane	Pesticide	Kills termites and is a broad-spectrum insecticide on various crops
DDT	Pesticide	Widely used in the past to protect against malaria, typhus, and other insect-spread diseases; continues to be applied in several countries to control malaria
Dieldrin	Pesticide	Controls termites and textile pests; also used against insect-borne diseases and insects in agricultural soil
Dioxins	Unintentional by-product	Produced by incomplete combustion and in chemical manufacturing; released in some kinds of metal recycling, pulp and paper bleaching, automobile exhaust, tobacco smoke, and wood and coal smoke
Endrin	Pesticide	Kills insects on cotton and grains; also used against rodents
Furans	Unintentional by-product	Result from the same processes that release dioxins; also are found in commercial mixtures of PCBs
Heptachlor	Pesticide	Kills soil insects and termites, cotton insects, grasshoppers, and malaria-carrying mosquitoes
Hexachlorobenzene	Fungicide; unintentional by-product	Kills fungi that affect crops; released during chemical manufacture and from processes that give rise to dioxins and furans
Mirex	Pesticide	Combats ants and termites; also is a fire retardant in plastics, rubber, and electrical goods
PCBs	Industrial chemical	Used in industry as heat-exchange fluids, in electrical transformers and capacitors, and as additives in paint, carbonless copy paper, sealants, and plastics
Toxaphene	Pesticide	Kills insects on cotton, grains, fruits, nuts, and vegetables; kills ticks and mites on livestock

Data from United Nations Environment Programme (UNEP), 2001.

distances. The PCBs and other contaminants found in polar bears are a prime example. Because these contaminants so often cross international boundaries, an international treaty seemed the best way of dealing fairly with such transboundary pollution. The Stockholm Convention aims first to end the use and release of 12 of the POPs shown to be most dangerous, a group nicknamed the "dirty dozen" (Table 14.3). It sets guidelines for phasing out these chemicals and encourages transition to safer alternatives.

Conclusion

International agreements such as the Stockholm Convention represent a hopeful sign that governments will act to protect the world's people, wildlife, and ecosystems from toxic chemicals and other environmental hazards. At the same time, solutions can often come more easily when they do not arise from government regulation alone. To many

minds, consumer choice, exercised through the market, may be the best way to influence industry's decision making. Consumers of products can make decisions that influence industry when they have full information from scientific research regarding the risks involved. Once scientific results are in, a society's philosophical approach to risk management will determine what policy decisions are made.

Whether the burden of proof is laid at the door of industry or of government, it is important to realize that we will never attain complete scientific knowledge of any risk. At some point we must choose whether or not to act on the information available. Synthetic chemicals have brought us innumerable modern conveniences, a larger food supply, and medical advances that save and extend human lives. Human society would be very different without them. Yet a safer and happier future, one that safeguards the well-being of both humans and the environment, depends on knowing the risks that some hazards pose and on having means in place to phase out harmful substances and replace them with safer ones.

REVIEWING OBJECTIVES

You should now be able to:

Identify the major types of environmental health hazards and explain the goals of environmental health

▶ Environmental health seeks to assess and mitigate environmental factors that adversely affect human health and ecological systems. (p. 402)
▶ Environmental health threats include physical, chemical, biological, and cultural hazards. (pp. 402–403)
▶ Disease is a major focus of environmental health. (pp. 404–405)
▶ Environmental hazards exist indoors as well as outdoors. (pp. 405–406)
▶ Toxicology is the study of poisonous substances. (pp. 406–407)

Describe the types, abundance, distribution, and movement of synthetic and natural toxicants in the environment

▶ Many thousands of potentially toxic substances exist all around us in varying degrees. (pp. 407–408)
▶ Toxicants may be of human or natural origin. They include carcinogens, mutagens, teratogens, allergens, neurotoxins, and endocrine disruptors. (pp. 409–413)
▶ Toxicants may enter and move through surface and groundwater reservoirs, or they may travel long distances through the atmosphere. (pp. 413–415)
▶ Some chemicals break down very slowly and thus persist in the environment. (pp. 415–416)
▶ Some organic poisons bioaccumulate and may move up the food chain, poisoning consumers on high trophic levels through the process of biomagnification. (p. 416)

Discuss the study of hazards and their effects, including case histories, epidemiology, animal testing, and dose-response analysis

▶ In case histories, researchers study health problems in individual people. (pp. 417–418)

▶ Epidemiology involves gathering data from large groups of people over long periods of time and comparing groups with and without exposure to the environmental health threat being assessed. (pp. 418–420)
▶ In dose-response analysis, scientists measure the response of test animals to various doses of the suspected toxicant. (pp. 420–421)
▶ Toxicity or strength of response may be influenced by the dose or amount of exposure, the nature of exposure (acute or chronic), individual variation among subjects, and synergistic interactions with other hazards. (pp. 421–422)

Assess risk assessment and risk management

▶ Risk assessment involves measuring risk quantitatively and comparing risks involved in different activities or substances. (pp. 422–423)
▶ Risk management integrates science with political, social, and economic concerns, in order to design strategies to minimize risk. (pp. 423–424)

Compare philosophical approaches to risk

▶ An innocent-until-proven-guilty approach assumes that a substance is not harmful unless it is shown to be so. (pp. 424–425)
▶ A precautionary approach entails assuming that a substance may be harmful unless proven otherwise. (pp. 424–425)

Describe policy and regulation in the United States and internationally

▶ The EPA, CDC, FDA, and OSHA are responsible for regulating environmental health threats under U.S policy. (pp. 425–426)
▶ European nations take a more precautionary approach than does the United States when it comes to testing chemical products. (pp. 426, 428)

TESTING YOUR COMPREHENSION

1. What four major types of health hazards does research in the field of environmental health encompass?
2. In what way is disease the greatest hazard that humans face? What kinds of interrelationships must environmental health experts study to learn about how diseases affect human health?

3. Where does most exposure to lead, asbestos, radon, and PBDEs occur? How has each of these exposure problems been addressed?
4. When did concern over the effects of pesticides start to grow in the United States? Describe the argument presented by Rachel Carson in *Silent Spring*. What policy

resulted from the book's publication? Is DDT still used?

5. List and describe the six types or general categories of toxicants described in this chapter.

6. How do toxicants travel through the environment, and where are they most likely to be found? What are the life spans of toxic agents? Describe the processes of bioaccumulation and biomagnification.

7. What are epidemiological studies, and how are they most often conducted?

8. Why are animals used in laboratory experiments in toxicology? Explain the dose-response curve. Why are high LD_{50} and ED_{50} levels considered safe and low LD_{50} and ED_{50} levels considered unsafe for humans?

9. What factors may affect an individual's response to a toxic substance? Why is chronic exposure to toxic agents often more difficult to measure and diagnose than acute exposure? What are synergistic effects, and why are they difficult to measure and diagnose?

10. How do scientists identify and assess risks from substances or activities that may pose health threats?

SEEKING SOLUTIONS

1. Describe some environmental hazards that you think you may be living with indoors. How do you think you may have been affected by indoor or outdoor environmental hazards in the past? What philosophical approach do you plan to take in dealing with these toxicants in your own life?

2. Why is it that research on endocrine disruption has spurred so much debate? What steps do you think could be taken to help establish more consensus among scientists, industry, regulators, policymakers, and the public?

3. Name some naturally occurring substances that can act as toxic agents. Explain the arguments of Bruce Ames and those of his critics regarding the prevalence and effects of natural toxicants.

4. Do you feel that laboratory-bred animals should be used in experiments in toxicology? Why or why not?

5. Discuss ways that we may cope with the uncertainty of risk assessment for synthetic chemicals in environmental health. Can you think of alternatives to taking one of the two philosophical approaches discussed in the chapter? Should these approaches apply to natural toxicants as well?

6. Describe what you have learned from this chapter regarding the policies of the United States and the European Union toward the study and management of the risks of synthetic chemicals. Which do you believe is better, the policies of the United States or the European Union, and why?

INTERPRETING GRAPHS AND DATA

To minimize their exposure to ultraviolet (UV) radiation and thus their risk of skin cancer, people have increased use of sunscreen lotions in recent decades. Recently, however, research has shown that chemicals in sunscreens may themselves pose some risk to human health. The compounds most commonly used as UV protectants are fat-soluble, environmentally persistent, and prone to bioaccumulation. Moreover, they exhibit estrogenic effects in laboratory rats (see Schlumpf et al., 2001). Although the benefits and risks of sunscreen use are not yet well understood, a hypothetical trade-off between the risk factors of UV exposure and sunscreen use illustrates the balancing act known as risk management.

1. What dosage of applied sunscreen on the graph corresponds to the greatest risk due to UV exposure? What dosage corresponds to the greatest risk due to

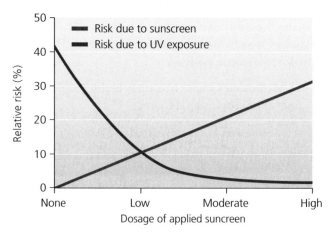

Hypothetical risk distributions for individuals using an estrogenic sunscreen to prevent skin cancer. Reference: Schlumpf, M., et al. 2001. *In vitro* and *in vivo* estrogenicity of UV screens. *Environmental Health Perspectives* 109: 239–244.

chemicals in the sunscreen? Which of these two points on the graph is associated with the greater risk?

2. What dosage of applied sunscreen on the graph corresponds to the least risk due to UV exposure? What dosage corresponds to the least risk due to chemicals in the sunscreen? Which of these two points is associated with the greater risk?

3. The total risk to the individual is the sum of the two individual risks. What point on the graph corresponds to the greatest total risk? What sunscreen dosage corresponds to the least total risk? Based on the data shown here, how much sunscreen would you choose to apply the next time you go to the beach? Is there any other information you'd like to know before you change the way you use sunscreen? Can you think of any other cases that illustrate this sort of trade-off between dose-dependent risk factors?

CALCULATING ECOLOGICAL FOOTPRINTS

In 2001, the population of the United States was approximately 285 million, and the world's population totaled 6.16 billion. In that same year, pesticide use in the United States was approximately 1.20 billion pounds of active ingredient, and world pesticide use totaled 5.05 billion pounds of active ingredient. Pesticides include hundreds of chemicals used as insecticides, fungicides, herbicides, rodenticides, repellants, and disinfectants. They are used by farmers, governments, industries, and individuals. In the table, calculate your share of pesticide use as a U.S. citizen in 2001 and the amount used by (or on behalf of) the average citizen of the world.

	Annual pesticide use (pounds of active ingredient)
You	4.21
Your class	
Your state	
United States	
World (total)	
World (per capita)	

1. What is the ratio of your annual pesticide use to the world's per capita average? Refer back to the "Calculating Ecological Footprints" question for Chapter 1 (▶ p. 25), and find the ecological footprints of the average U.S. citizen and the average world citizen. Compare the ratio of pesticide usage with the ratio of the overall ecological footprints. How would you explain the difference?

2. Does the figure for per capita pesticide use for you as a U.S. citizen seem reasonable for you personally? Why or why not? Do you find this figure alarming or of little concern? What else would you like to know to assess the risk associated with this level of pesticide use?

Take It Further

Go to www.aw-bc.com/withgott or the student CD-ROM where you'll find:

▶ Suggested answers to end-of-chapter questions
▶ Quizzes, animations, and flashcards to help you study
▶ *Research Navigator*™ database of credible and reliable sources to assist you with your research projects

▶ **GRAPHit!** tutorials to help you master how to interpret graphs
▶ **INVESTIGATEit!** current news articles that link the topics that you study to case studies from your region to around the world

Hoover Dam on the Colorado River

Upon successfully completing this chapter, you will be able to:

▶ Explain the importance of water and the hydrologic cycle to ecosystems, human health, and economic pursuits

▶ Delineate freshwater distribution on Earth

▶ Describe major types of freshwater ecosystems

▶ Discuss how we use water and alter freshwater systems

▶ Assess problems of water supply and propose solutions to address freshwater depletion

▶ Assess problems of water quality and propose solutions to address water pollution

▶ Explain how wastewater is treated

Colorado River
delta

Pacific
Ocean

Colorado
River

Los
Angeles

North
America

San
Diego

Central Case: Plumbing the Colorado River

"We've gone from being
assured that we lived in
this magical place where
the rules of water didn't
apply to [a] wake-up call
about the fact that we do
live in the California desert.
People have lived in this
false water utopia."
—BUFORD CRITES, CITY
   COUNCILOR, PALM DESERT
   CITY, CALIFORNIA

"Water promises to be to
the 21st century what oil
was to the 20th century: the
precious commodity that
determines the wealth of
nations."
—FORTUNE MAGAZINE,
   MAY 2000

As the clock struck midnight on New Year's Eve,
millions of Californians toasted the arrival of 2003
with champagne. But some people in the state
that night had another liquid on their minds: water. Their
fears were borne out the next day when the U.S. government followed through on its threat to cut off 15% of the
water that California takes from the Colorado River.

Water is the lifeline for any civilization in an arid
environment. Without generous supplies of freshwater

delivered from elsewhere, southern California society
as we know it could simply not exist. In ordering the
New Year's Day cutoff, U.S. Interior Secretary Gale
Norton was simply holding up her end of a deal that an
irrigation district in California had scuttled. It may seem
bizarre that a 3–2 vote of one county irrigation district
could block enough water for 1.6 million households in
Los Angeles and San Diego, but it was just the latest
episode in the colorful history of California water politics
and the battles among seven states jockeying for rights
to what was once the West's wildest river.

The Colorado River begins in the high peaks of the
Rocky Mountains, charges through the Grand Canyon,
crosses the border into Mexico, and empties into the Gulf
of California, draining 637,000 km^2 (246,000 mi^2) of
southwestern North America. Its raging waters have
chiseled through thousands of feet of bedrock, creating
the Grand Canyon and leaving extraordinary scenery
along its 2,330-km (1,450-mi) length. Today, however, only
a small amount of water—often none at all—reaches the
river's mouth. Instead, the waters of the Colorado River

irrigate 7% of U.S. cropland, quench the thirst of over 20 million people, keep hundreds of golf courses green in the desert, and fill the swimming pools and fountains of Las Vegas casinos. The Colorado provides vital water to the rapidly growing metropolitan areas of the arid U.S. Southwest—Phoenix, Tucson, Las Vegas, San Diego, Los Angeles, and many others. The massive dams built across the river provide flood control and recreation, produce 12 billion kilowatt-hours of electricity from hydroelectric power each year, and provide irrigation that makes agriculture possible in this arid region.

For 80 years, the seven states along the Colorado have divided the river's water among themselves, guided by the Colorado River Compact they signed in 1922, which apportioned water to each state. California had long been permitted to exceed its allotment because Colorado, Wyoming, Utah, Nevada, New Mexico, and Arizona were not using all of their allotted portions. With the populations of these states booming, however, Interior Secretary Bruce Babbitt in 2000 pressured California to reduce its withdrawals by roughly 15% over 15 years.

California worked hard to get its agricultural districts, which controlled most water distribution in the state, to

agree. At the last minute, however, the Imperial Irrigation District backed out of the agreement. The New Year's deadline passed, and 2003 saw the federal cutoff implemented. After 10 months of bickering and negotiation, the deal was patched up, and the cutoff was ended. Southern California's residents were able to continue living—at least for a little while longer—their mirage in the desert.

Freshwater Systems

"Water, water, everywhere, nor any drop to drink." The well-known line from Coleridge's poem *The Rime of the Ancient Mariner* describes the situation on our planet quite well. Water may seem abundant to us, but water that we can drink is actually quite rare and limited (Figure 15.1). Roughly 97.5% of Earth's water resides in the oceans and is too salty to drink or use to water crops. Only 2.5% is considered **freshwater,** water that is relatively pure, with few dissolved salts. Because most freshwater is tied up in glaciers, icecaps, and underground aquifers, just over 1 part in 10,000 of Earth's water is easily accessible for human use.

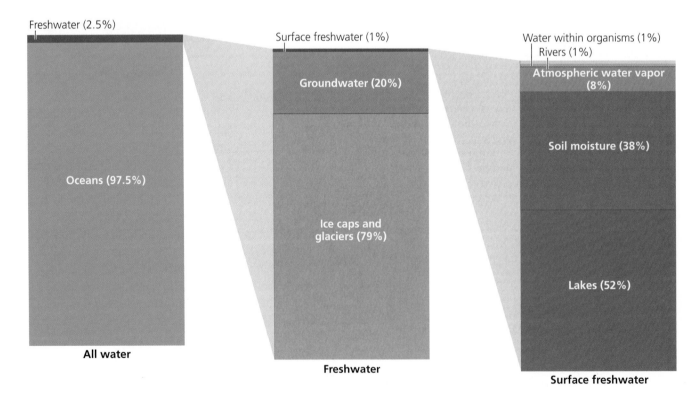

FIGURE 15.1 Only 2.5% of Earth's water is freshwater. Of that 2.5%, most is tied up in glaciers and ice caps. Of the 1% that is surface water, most is in lakes and soil moisture. Data from United Nations Environment Programme (UNEP) and World Resources Institute.

FIGURE 15.2 Rivers and streams flow downhill, shaping landscapes, as shown by an oxbow of this meandering river in Colorado.

Water is constantly moving among the reservoirs specified in Figure 15.1 via the *hydrologic cycle* (Figure 7.14, ▶ pp. 204–205). As water moves, it redistributes heat, erodes mountain ranges, builds river deltas, maintains organisms and ecosystems, shapes civilizations, and gives rise to political conflicts. Let's first examine the portions of the hydrologic cycle that are most conspicuous to us—surface water bodies—and take stock of the ecological systems they support.

Rivers and streams wind through landscapes

Water from rain, snowmelt, or springs runs downhill and converges in small channels, which join to form streams, creeks, or brooks. These watercourses join one another and eventually merge into rivers, whose water eventually reaches the ocean (or sometimes ends in a landlocked water body). As we learned in Chapter 3, a smaller river flowing into a larger one is a *tributary,* and the area of land drained by a river and all its tributaries is that river's *watershed.*

Rivers shape the landscape through which they run. The force of water rounding a river's bend gradually eats away at the outer shore, eroding soil from the bank. Meanwhile, sediment is deposited along the inside of the bend, where water currents are weaker. In this way, over time, river bends become exaggerated in shape (Figure 15.2). Eventually, a bend may become such an extreme loop (called an *oxbow*) that water erodes a shortcut from one end of the loop to the other, pursuing a direct course.

The bend is cut off, and remains as an isolated, U-shaped water body called an *oxbow lake.*

Over thousands or millions of years, a river may shift from one course to another, back and forth over a large area, carving out a flat valley. Areas nearest a river's course that are flooded periodically are said to be within the river's **floodplain.** Frequent deposition of silt from flooding makes floodplain soils especially fertile. As a result, agriculture thrives in floodplains, and *riparian* (riverside) forests are productive and species-rich.

Diverse ecological communities exist in the water of rivers and streams. Algae and detritus support many types of invertebrates, from water beetles to crayfish. Insects as diverse as dragonflies, mayflies, and mosquitoes develop as larvae in streams and rivers before maturing into adults that take to the air. Fish consume aquatic insects, and birds such as kingfishers, herons, and ospreys dine on fish. Many amphibians spend their larval stages in streams, and some live their entire lives in streams. Salmon migrate from oceans up rivers and streams to spawn.

Wetlands include marshes, swamps, and bogs

Systems that combine elements of freshwater and dry land are enormously rich and productive. Often lumped under the term *wetlands,* such areas include different types of systems. Freshwater marshes (Figure 15.3) feature water shallow enough to allow plants to grow from the bottom and to rise above the water surface. Cattails and bulrushes

FIGURE 15.3 Shallow water bodies with ample vegetation are called wetlands, and include swamps, bogs, and marshes such as this one in Botswana, Africa.

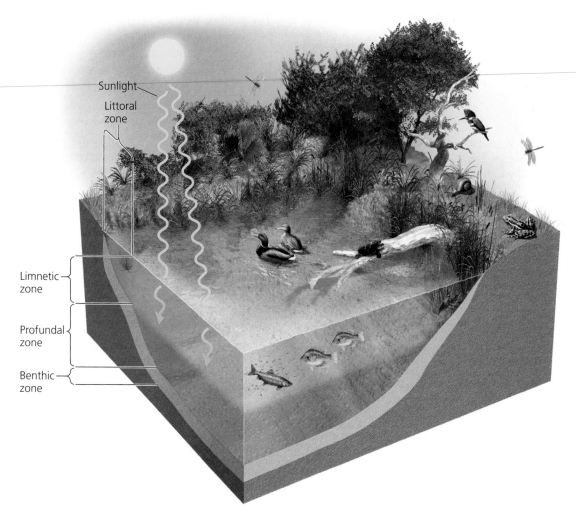

FIGURE 15.4 Lakes and ponds are open, still bodies of water consisting of different zones. Emergent plants grow around the shoreline in the littoral zone. The limnetic zone is the layer of open, sunlit water where photosynthesis takes place. Sunlight does not reach the deeper profundal zone. The benthic zone, which is the bottom of the water body, often is muddy, rich in detritus and nutrients, and low in oxygen.

are plants typical of North American marshes. Swamps also consist of shallow water rich in vegetation, but they occur in forested areas. The cypress swamps of the southeastern United States, where cypress trees grow in standing water, are an example. Swamps are also created when beavers build dams across streams with limbs from trees they have cut, flooding wooded areas upstream. Bogs are ponds thoroughly covered with thick floating mats of vegetation, and can represent a stage in aquatic succession.

All these types of wetlands are extremely valuable as habitat for wildlife. They also provide ecosystem services by slowing runoff, reducing flooding, recharging aquifers, and filtering pollutants. Wetlands have been extensively drained and filled by people, largely for agriculture (▸ pp. 362–363). It is estimated that southern Canada and the United States have lost well over half their wetlands since European colonization.

Lakes and ponds are ecologically diverse systems

Lakes and ponds are bodies of open standing water. Their depth varies greatly, and the physical conditions and types of life within them vary with depth and the distance from shore. As a result, scientists have described several zones typical of lakes and ponds (Figure 15.4).

The region ringing the edge of a water body is named the **littoral zone.** Here the water is shallow enough that aquatic plants grow from the mud and reach above the water's surface. The nutrients and productive plant growth of the littoral zone make it rich in invertebrates—such as insect larvae, snails, and crayfish—on which fish, birds, turtles, and amphibians feed. Many invertebrates live in the mud on the bottom, feeding on detritus or preying on one another. The bottom layer of a lake or pond is the **benthic zone.** The benthic zone extends along the bottom of the entire water body, from shore to the deepest point.

In the open portion of a lake or pond, away from shore, sunlight penetrates shallow waters of the **limnetic zone.** Because light enables photosynthesis and plant growth, the limnetic zone supports phytoplankton, which in turn support zooplankton, both of which are eaten by fish. Within the limnetic zone, sunlight intensity (and therefore water temperature) decreases with depth. The water's turbidity affects the depth of this zone; water that is clear allows sunlight to penetrate deeply, whereas turbid water does not. Below the limnetic zone is the **profundal zone,** the volume of open water that sunlight does not reach. This zone lacks plant life and thus is lower in dissolved oxygen and supports fewer animals. Aquatic animals rely on dissolved oxygen, and its concentration depends on the amount released by photosynthesis and the amount removed by animal and microbial respiration, among other factors.

Ponds and lakes change over time naturally, as streams and runoff bring them sediment and nutrients. *Oligotrophic* lakes and ponds, which have low-nutrient and high-oxygen conditions, may slowly give way to the high-nutrient, low-oxygen conditions of *eutrophic* water bodies. Eventually, water bodies may fill in completely by the process of aquatic succession (Figure 6.14, ▶ p. 166). As lakes or ponds change over time, species of fish, plants, and invertebrates adapted to oligotrophic conditions may give way to those that thrive under eutrophic conditions.

Some lakes are so large that they differ substantially in their characteristics from small lakes. These large lakes are sometimes known as inland seas. North America's Great Lakes are prime examples. Because they hold so much water, most of their biota is adapted to open water. Major fish species of the Great Lakes include lake sturgeon, lake whitefish, northern pike, alewife, bass, walleye, and perch. Lake Baikal in Asia is the world's deepest lake, at 1,637 m (5,370 ft) deep, and the Caspian Sea is the world's largest freshwater body, at 371,000 km^2 (143,000 mi^2).

Groundwater plays key roles in the hydrologic cycle

It is easy for us to understand surface water systems because we witness them all the time, but groundwater and its functions in the hydrologic cycle are more difficult to visualize (Figure 15.5). Any precipitation reaching Earth's land surface that does not evaporate, flow into waterways, or get taken up by organisms infiltrates the surface. Most percolates downward through the soil to become groundwater, which makes up one-fifth of Earth's freshwater supply and plays a key role in meeting human water needs.

Groundwater is contained within **aquifers,** porous, spongelike formations of rock, sand, or gravel that hold

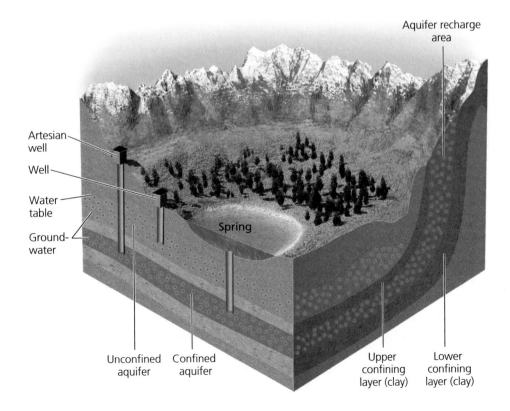

FIGURE 15.5 Groundwater may occur in unconfined aquifers above impermeable layers or in confined aquifers under pressure between impermeable layers. Water may rise naturally to the surface at springs and through the wells we dig. Artesian wells tap into confined aquifers to mine water under pressure.

Aquifer recharge area

Artesian well

Well

Water table

Ground-water

Spring

Unconfined aquifer

Confined aquifer

Upper confining layer (clay)

Lower confining layer (clay)

water. An aquifer's upper layer, or zone of aeration, contains pore spaces partly filled with water. In the lower layer, or zone of saturation, the spaces are completely filled with water. The boundary between these two zones is the **water table.** Picture a sponge resting partly submerged in a tray of water; the lower part of the sponge is completely saturated, whereas the upper portion may be moist but contains plenty of air in its pores. Any area where water infiltrates Earth's surface and reaches an aquifer below is known as an *aquifer recharge zone.*

There are two broad categories of aquifers. A **confined aquifer,** or **artesian aquifer,** exists when a water-bearing porous layer of rock, sand, or gravel is trapped between upper and lower layers of less permeable substrate (often clay). In such a situation, the water is under great pressure. In contrast, an **unconfined aquifer** has no such upper layer to confine it. Thus its water is under considerably less pressure than that of a confined aquifer and it can be readily recharged by surface water.

Just as surface water becomes groundwater by infiltration and percolation, groundwater becomes surface water through springs (and human-drilled wells), sometimes keeping streams flowing when surface conditions are otherwise dry. Groundwater flows downhill and from areas of high pressure to areas of low pressure. A typical rate of groundwater flow might be about 1 m (3 ft) per day. Because of this slow movement, groundwater may remain in an aquifer for a long time. In fact, groundwater can be ancient. The average age of groundwater has been estimated at 1,400 years, and some is tens of thousands of years old. Nonetheless, volumes of groundwater are large enough that each day in the United States alone, aquifers release 1.9 trillion L (492 billion gal) of groundwater into bodies of surface water—nearly as much as the daily flow of the Mississippi River. The world's largest known aquifer is the Ogallala Aquifer, which underlies the Great Plains of the United States (Figure 15.6). It spans 453,000 km² (176,700 mi²), is 370 m (1,200 ft) deep at its thickest point, and has a water-holding capacity of 3,700 km³ (881 mi³).

Water is unequally distributed across Earth's surface

The Great Plains and its farmlands boast the massive Ogallala Aquifer as a source of freshwater, but many other areas are not so endowed. Different regions possess vastly different amounts of groundwater, surface water, and precipitation. Precipitation ranges from about 1,200 cm (470 in.) per year at Mount Waialeale on the Hawaiian island of Kauai to virtually zero in Chile's Atacama Desert.

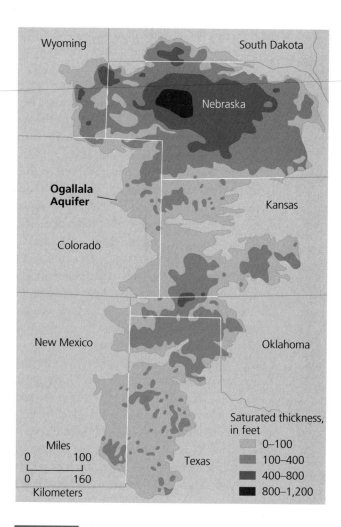

FIGURE 15.6 The Ogallala Aquifer is the world's largest aquifer, and it held 3,700 km³ (881 mi³) of water before pumping began. This aquifer underlies 453,000 km² (175,000 mi³) of the Great Plains beneath eight U.S. states from South Dakota to Texas. Overpumping for irrigation is currently reducing the volume and extent of this aquifer.

People are not distributed across the globe in accordance with water availability. Many areas with high population density are water-poor (Figure 15.7), leading to inequalities in per capita water resources among and within nations. For example, Canada has 20 times more water for each of its citizens than does China. The Amazon River carries 15% of the world's runoff, but its watershed holds less than half a percent of the world's human population. Nations that hold back water in transboundary rivers exacerbate such natural inequities. Many densely populated nations, such as Pakistan, Iran, and Egypt, face serious water shortages. Asia possesses the most water of any continent but has the least water available per person, whereas Australia, with the least amount of water, boasts the most water available per person. Because of this mismatched distribution of water and

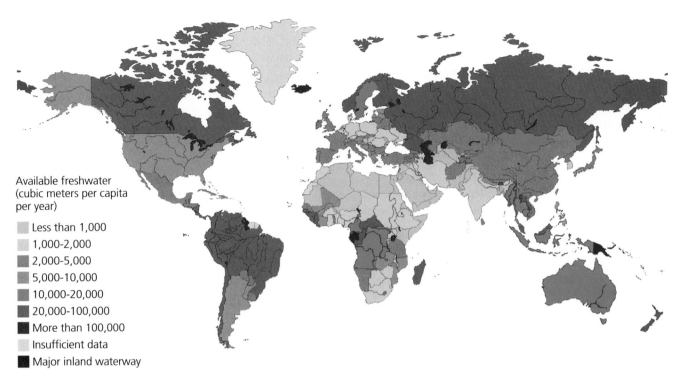

**Available freshwater
(cubic meters per capita
per year)**

- Less than 1,000
- 1,000-2,000
- 2,000-5,000
- 5,000-10,000
- 10,000-20,000
- 20,000-100,000
- More than 100,000
- Insufficient data
- Major inland waterway

(a) Water-stressed nations: Available freshwater resources, 2000

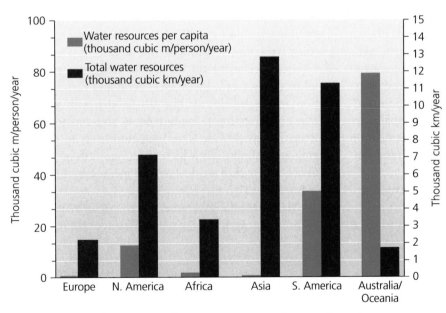

(b) Water-stressed nations: Total water resources and per capita water resources

FIGURE 15.7 Nations vary tremendously in the amount of freshwater per capita available to their citizens (**a**). For example, Iceland, Papua New Guinea, Gabon, and Guyana have more than 100 times as much water as do many Middle Eastern and North African countries. There is almost as much variation among continents, which show great imbalances between total water resources and per capita water resources (**b**). Heavily populated Asia has tremendous total water resources but extremely low amounts per capita, for example, whereas Australia and the oceanic island nations have little total water but high amounts per capita because of their low populations. Go to **GRAPHit!** at www.aw-bc.com/withgott or on the student CD-ROM. Data from (a) UNEP and World Resources Institute, as presented by Harrison, P., and F. Pearce. 2000. *AAAS atlas of population and the environment.* Berkeley, CA: University of California Press. (b) U.N. Sustainable Development Programme, 2002.

human population, one challenge for human societies has always been to transport freshwater from its source to where it is needed. In nearly every modern country, such transport is vital to equalizing access among people.

Freshwater is distributed unevenly in time as well as space. India's monsoon season brings concentrated storms in which half of a region's annual rain may fall in a few hours. Northwest China receives three-fifths of its annual precipitation during 3 months when crops do not need it. Uneven distribution of water across time is one reason people have erected dams to store water, so that it may be distributed when needed.

How We Use Water

In our attempts to harness freshwater sources for countless purposes and pursuits, we have achieved impressive engineering accomplishments. In so doing, we also have altered many environmental systems. It is estimated that 60% of the world's largest 227 rivers (and 77% of those in North America and Europe) have been strongly or moderately affected by artificial dams, canals, and diversions.

We are also using too much water. Data indicate that at present our freshwater consumption in much of the world is unsustainable, and we are depleting many sources of surface water and groundwater. At least 1.7 billion people live in regions of water scarcity (with less than 1,000 m^3 (35,000 ft^3) of water per person per year), and this number is expected to grow to at least 2.4 billion by 2025.

Water supplies our households, agriculture, and industry

Every one of us uses water at home for drinking, cooking, and cleaning. Farmers and ranchers use water to irrigate crops and water livestock. We use water in most manufacturing and industrial processes. The proportions of each of these three types of use—residential/municipal, agricultural, and industrial—vary dramatically among nations (Figure 15.8). Nations with arid climates tend to use more freshwater for agriculture, and heavily industrialized nations use a great deal for industry. Globally, we spend about 70% of our annual freshwater allotment on agriculture. Industry accounts for roughly 20%, and residential and municipal uses for only 10%.

Freshwater use can be consumptive or nonconsumptive. In **consumptive use,** water is removed from a particular aquifer or surface water body and is not returned. A large portion of agricultural irrigation and of many industrial and residential uses is consumptive. **Nonconsumptive use** of water does not remove, or only temporarily removes, water from an aquifer or surface water body. Using water to generate electricity at hydroelectric dams is an example of nonconsumptive use; water is taken in, passed through dam machinery to turn turbines, and released on the downstream side.

Inefficient irrigation wastes water

The green revolution (▶ pp. 250, 280–281) required significant increases in irrigation, and 60% more water is withdrawn for irrigation today than in 1960. During this period, the amount of land under irrigation has roughly doubled (Figure 15.9). Expansion of irrigated agriculture has kept pace with population growth; irrigated area per capita has remained stable for at least four decades at around 460 m^2 (4,900 ft^2).

Irrigation can more than double crop yields by allowing farmers to control the application of water when and where it is needed. The world's 274 million ha (677 million acres) of irrigated cropland make up only 18% of world farmland but yield fully 40% of world agricultural produce, including 60% of the global grain crop. Still, most irrigation remains highly inefficient. Only about 45% of the freshwater we use for irrigation actually is taken up by crops. Inefficient "flood and furrow" irrigation, in which fields are liberally flooded with water that may evaporate from standing pools, accounts for 90% of irrigation worldwide. Overirrigation leads to waterlogging and salinization, which affect one-fifth of farmland today and reduce world farming income by $11 billion (▶ pp. 264–265).

Many national governments have subsidized irrigation to promote agricultural self-sufficiency. Unfortunately,

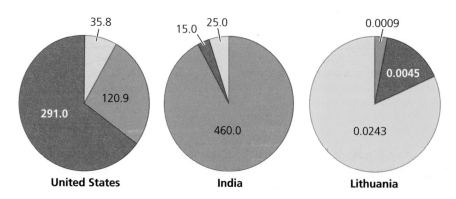

Water consumption by type of use (billion cubic meters)
■ Industry ■ Agriculture ■ Domestic

FIGURE 15.8 Nations apportion their freshwater consumption differently. Industry consumes most water used in the United States, agriculture uses the most in India, and most water in Lithuania goes toward domestic use. Data from World Bank, as presented by: Harrison, P., and F. Pearce. 2000. *AAAS atlas of population and the environment.* Berkeley, CA: University of California Press.

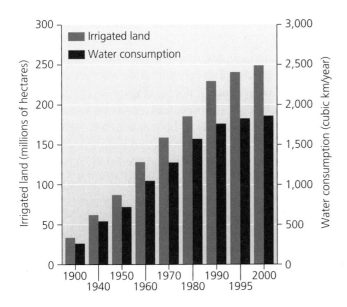

FIGURE 15.9 Throughout the 20th century, overall global water consumption rose in tandem with the area of land irrigated for agriculture. Data from United Nations Sustainable Development Commission on Freshwater, 2002.

inefficient irrigation methods in arid areas such as the Middle East are using up huge amounts of groundwater for little gain. Worldwide, roughly 15–35% of water withdrawals for irrigation are thought to be unsustainable. Figure 15.10 shows areas where agriculture is demanding more freshwater than can be sustainably supplied. In these areas, *water mining*—withdrawing water faster than it can be replenished—is taking place; aquifers are being depleted or surface water is being piped in from other regions.

We are depleting groundwater

Groundwater is more easily depleted than surface water because most aquifers recharge very slowly. One-third of Earth's human population—including 99% of the rural population of the United States—relies on groundwater for its needs. To obtain groundwater, many people in rural areas of developing countries walk long distances to haul water from local wells. In developed countries, most homes in rural areas have electrically powered wells.

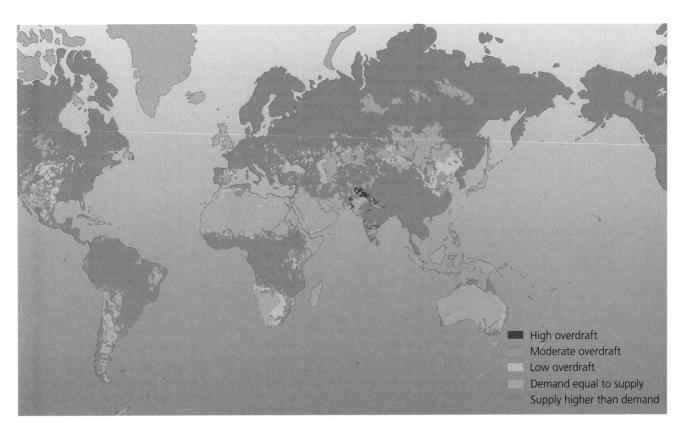

High overdraft
Moderate overdraft
Low overdraft
Demand equal to supply
Supply higher than demand

FIGURE 15.10 Globally, 15–35% of water withdrawals are estimated to be unsustainable. Mapped are regions where freshwater use for agriculture exceeds the available supply, requiring groundwater depletion or diversion of water from other regions. "High overdraft" equals more than 1 km^3/yr; "moderate overdraft" equals 0.1–1 km^3/yr; and "low overdraft" equals 0–0.1 km^3/yr. In areas of high overdraft, water tables are being drawn down 1.6 m/yr or more; in low overdraft areas, water tables are falling by less than 0.1 m/yr. Data from Millennium Ecosystem Assessment, 2005.

Homes and businesses in urban and suburban areas receive water through complex networks of pipes and pumps generally maintained by municipal governments.

If we compare an aquifer to a bank account, we are making more withdrawals than deposits, and the balance is shrinking. Globally, over the past 60 years we have been withdrawing groundwater in amounts that increase 2.5–3% annually, greater than the rate of population growth. Today we are extracting 160 km³ (5.65 trillion ft³) more water each year than is finding its way back into the ground. This degree of overpumping is equal to the quantity of water needed to produce 10% of the world grain supply. As aquifers are depleted, water tables drop. Groundwater becomes more difficult and expensive to extract, and eventually it may run out. In parts of Mexico, India, China, and other nations in Asia and the Middle East, water tables are falling 1–3 m (3–10 ft) per year. In the United States, by the late 1990s overpumping had drawn the Ogallala Aquifer down by 325 km³ (11.5 trillion ft³). This volume is equal to the yearly flow of 18 Colorado Rivers.

Overpumping of groundwater in coastal areas can cause salt water to intrude into aquifers, making water undrinkable. This has occurred widely in Middle Eastern nations and in localities as varied as Florida, Turkey, and Bangkok.

As aquifers lose water, their substrate can become weaker and less capable of supporting overlying strata, and the land surface above may subside. For this reason, cities from Venice to Bangkok to Beijing are slowly sinking. Mexico City's downtown has sunk over 10 m (33 ft) since the time of Spanish arrival; streets are buckled, old buildings lean at angles, and underground pipes break so often that 30% of the system's water is lost to leaks. Sometimes subsidence can occur suddenly in the form of **sinkholes,** areas where the ground gives way with little warning, occasionally swallowing people's homes (Figure 15.11). Once the ground subsides, soil becomes compacted, losing the porosity that enabled it to hold water. Recharging a depleted aquifer may thereafter become more difficult. It has been estimated that compacted aquifers under California's Central Valley have lost storage capacity equal to that of 40% of the state's artificial surface reservoirs.

Falling water tables can do vast ecological harm. Permanent wetlands exist where water tables are high enough to reach the surface, so when water tables drop, wetland ecosystems dry up. In Jordan, the Azraq Oasis covered 7,500 ha (18,500 acres) and enabled migratory birds and other animals to find water in the desert. The water table beneath this oasis dropped from 2.5 m (8.2 ft) to 7 m (23 ft) during the 1980s because of increased well use by the city of Amman. As a result, the oasis dried up altogether during the 1990s. Today international donors are collaborating with the Jordanian government to try to find alternate sources of water and restore this oasis.

We divert—and deplete—surface water to suit our needs

In areas near surface water, people have found it far easier to make use of rivers, streams, lakes, and ponds, diverting water from these sources to farm fields, homes, and cities.

FIGURE 15.11 When too much groundwater is withdrawn too quickly, the land above it may collapse in sinkholes, sometimes bringing buildings down with it.

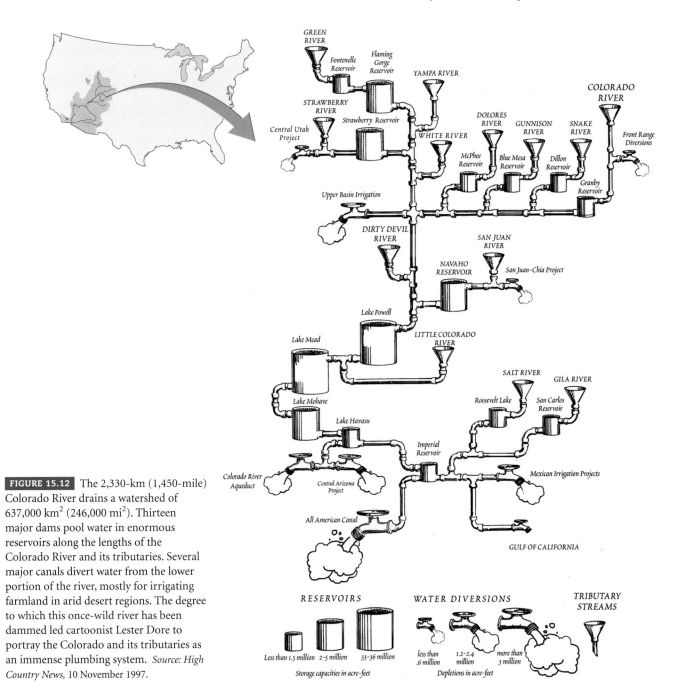

FIGURE 15.12 The 2,330-km (1,450-mile) Colorado River drains a watershed of 637,000 km² (246,000 mi²). Thirteen major dams pool water in enormous reservoirs along the lengths of the Colorado River and its tributaries. Several major canals divert water from the lower portion of the river, mostly for irrigating farmland in arid desert regions. The degree to which this once-wild river has been dammed led cartoonist Lester Dore to portray the Colorado and its tributaries as an immense plumbing system. *Source: High Country News,* 10 November 1997.

The Colorado River's water is heavily diverted and utilized (Figure 15.12). Early on in its course, some Colorado River water is piped through a mountain tunnel and down the Rockies' eastern slope to supply the city of Denver. More is removed for Las Vegas and other cities and for farmland as the water proceeds downriver. When it reaches Parker Dam on the California-Arizona state line, large amounts are diverted into the Colorado River Aqueduct, which brings water to millions in the Los Angeles and San Diego areas via a long open-air canal. From Parker Dam, Arizona also draws water, transporting it in the large canals of the Central Arizona Project. Further south at Imperial Dam, water is diverted into the Coachella and All-American Canals, destined for agriculture, mostly in California's Imperial Valley. To make this desert bloom, Imperial Valley farmers soak the soil with subsidized water for which they pay one penny per 795 L (210 gal).

What water is left of the Colorado River after all the diversions comprises just a trickle making its way to the Gulf of California over the sediments of the once-rich delta (Figure 15.13). On some days, the water does not reach the Gulf at all.

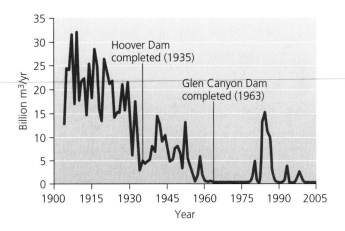

FIGURE 15.13 Flow at the mouth of the Colorado River has greatly decreased over the past century as a result of withdrawals, mostly for agriculture. The river now often runs dry at its mouth. Data from Postel, S., 2005. *Liquid assets: The critical need to safeguard freshwater ecosystems.* Worldwatch Paper 170. Washington, D.C.: Worldwatch Institute.

The Colorado's plight is not unique. Several hundred miles to the east, the Rio Grande also frequently runs dry, the victim of overextraction by both Mexican and U.S. farmers in times of drought. The story is even worse for China's Yellow River, so the Chinese government wants to build a massive aqueduct to supplement its flow with water from the Yangtze River. Even the river that has nurtured human civilization as long as any other, the Nile, now peters out before reaching its mouth.

Nowhere are the effects of surface water depletion so evident as at the Aral Sea. Once the fourth-largest lake on Earth, just larger than Lake Huron, it has lost four-fifths of its volume in 40 years and could soon disappear altogether (Figure 15.14). This dying inland sea, on the border of present-day Uzbekistan and Kazakhstan, is the victim of irrigation practices. The former Soviet Union instituted large-scale cotton farming in this region by flooding the dry land with water from the two rivers leading into the Aral Sea. For a few decades this action boosted Soviet cotton production, but it led the Aral Sea to shrink, while the irrigated soil became waterlogged and salinized. Today 60,000 fishing jobs are gone, winds blow pesticide-laden dust up from the dry lakebed, and what cotton grows on the blighted soil cannot bring the regional economy back. Scientists are struggling to save the Aral Sea, whose ecosystems have been seriously damaged.

Weighing the Issues:
The Klamath Crisis

In 2001, angry farmers of Klamath County, Oregon, disobeyed a federal order to divert irrigation waters downstream to save two endangered species of fish.

During that bone-dry year, there simply wasn't enough water to irrigate farmers' fields and also keep endangered salmon and suckerfish alive. The government had long ago begun allocating more water than was sustainable in the region. Although the homesteaders enjoyed federal incentives to settle and farm the region at the turn of the century, their children are now subject to a conflicting federal mandate—the Endangered Species Act. Can you think of solutions to the Klamath crisis? What would you do to reconcile these needs?

Will we see a future of water wars?

Freshwater depletion leads to shortages, and resource scarcity can lead to conflict. We have only to look to the Colorado River to see evidence of this. In 1933 the governor of Arizona, foreseeing that California's water diversion from the Colorado might endanger Arizona's future allotment, sent the state's National Guard to threaten the construction of Parker Dam. After a long standoff, the U.S. interior secretary halted the project to avoid hostilities while the issue was mediated in court. Arizona won the court case, but California got the dam authorized by Congress, and Arizona chose not to tackle the U.S. Army troops sent to protect the dam's construction.

Many predict that water's role in regional conflicts will increase as human population continues to grow in water-poor areas. A total of 261 major rivers, whose watersheds cover 45% of the world's land area, cross national borders, and transboundary disagreements are common. The World Water Commission's chairman, Ismail Serageldin, has remarked that "the wars of the twenty-first century will be fought over water."

On the positive side, many nations have cooperated with neighbors to resolve water disputes. India has struck cooperative agreements over management of transboundary rivers with Pakistan, Bangladesh, Bhutan, and Nepal. In Europe, international conventions have been signed by multiple nations along the Rhine and the Danube rivers. Such progress gives reason to hope that future water wars will be few and far between.

Dikes and levees are meant to control floods

We have applied a great deal of engineering brainpower and muscle to transport water from place to place, but we have also expended vast efforts to keep water in place. Flood prevention ranks high among reasons we control the movement of freshwater. People have always been attracted to riverbanks for their water supply and for the flat topography

(a) Ships stranded by the Aral Sea's fast-receding waters

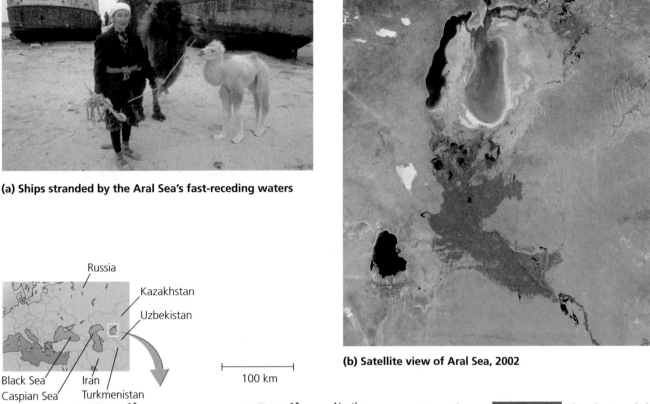

(b) Satellite view of Aral Sea, 2002

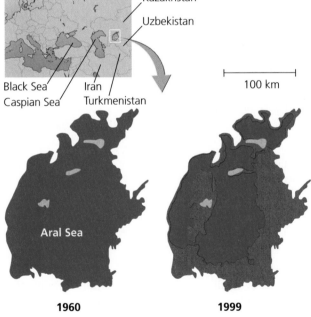

(c) The shrinking Aral Sea, then and now

FIGURE 15.14 Ships lie stranded in the sand (**a**) because the waters of Central Asia's Aral Sea have receded so far and so quickly (**b**). The Aral Sea was once the world's fourth-largest lake. However, it has been shrinking for the past four decades (**c**) and could disappear completely in the near future. The primary cause has been overwithdrawal of water to irrigate cotton crops. Map adapted from *New Scientist* 179(2404): 9

and fertile soil of floodplains. Flooding is a normal, natural process due to snowmelt or heavy rain, and, as we have seen, floodwaters spread nutrient-rich sediments over large areas, benefiting both natural systems and human agriculture.

In the short term, however, floods can do tremendous damage to the farms, homes, and property of people who choose to live in floodplains. To protect against floods, individuals and governments have built *dikes* and *levees*

(long raised mounds of earth) along the banks of rivers to hold rising water in main channels. Many dikes are small and locally built, but in the United States the Army Corps of Engineers has constructed thousands of miles of massive levees along the banks of major waterways (those that failed in New Orleans after Hurricane Katrina are examples). Although these structures prevent flooding at most times and places, they can sometimes exacerbate flooding

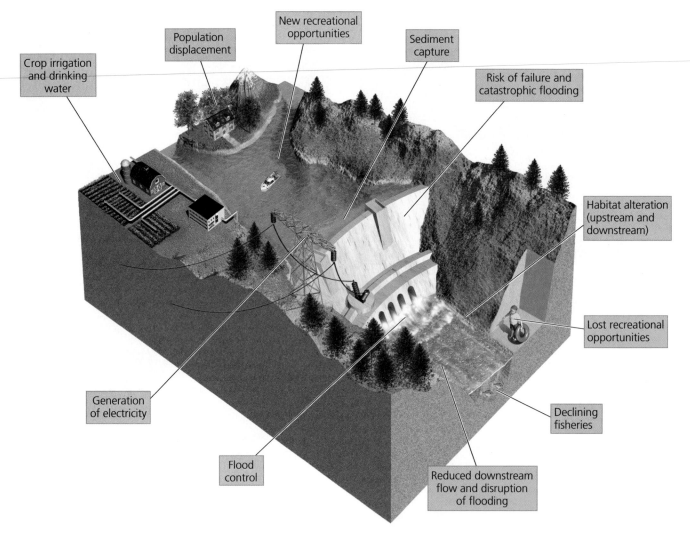

FIGURE 15.15 Damming rivers has many diverse consequences for people and the environment. The generation of clean and renewable electricity is one of several major benefits (green boxes) of hydroelectric dams. Habitat alteration is one of several negative impacts (red boxes).

because they force water to stay in channels and accumulate, building up enormous energy and leading to occasional catastrophic overflow events.

We have erected thousands of dams

A **dam** is any obstruction placed in a river or stream to block the flow of water so that water can be stored in a reservoir. We build dams to prevent floods, provide drinking water, facilitate irrigation, and generate electricity (Figure 15.15; Table 15.1). Power generation with hydroelectric dams is discussed in Chapter 20 (▶ pp. 612–615). Worldwide, more than 45,000 large dams (greater than 15 m, or 49 ft, high) have been erected

across rivers in over 140 nations. Additionally, tens of thousands of smaller dams have been built. Virtually the only major rivers in the world that remain undammed and free-flowing run through the tundra and taiga of Canada, Alaska, and Russia, and in remote regions of Latin America and Africa.

The largest of our dams are some of the greatest engineering feats humans have produced. The two behemoths of the Colorado River, Hoover Dam and Glen Canyon Dam, stand 221 m (726 ft) and 216 m (710 ft) high, respectively, stretch for 379 m (1,244 ft) and 476 m (1,560 ft) across, and consist of 6.6 and 7.3 million tons of concrete and steel. Hoover Dam holds back 35.2 km³ (8.4 mi³) of water in a reservoir that stretches 177 km (110 mi) long

Table 15.1 Major Benefits and Costs of Dams	
Benefits	**Costs**
▶ **Power generation.** Hydroelectric dams (▶ pp. 612–614) provide a great deal of inexpensive electricity to help run our economies and power our modern conveniences. Some nations gain nearly all their electricity from hydropower.	▶ **Habitat alteration.** Reservoirs flood riparian habitats and displace or kill riparian species. Dams modify rivers downstream, making them tranquil and low in dissolved oxygen, so that many fish adapted to fast-flowing rivers cannot survive. Shallow warm water downstream from a dam is periodically flushed with cold water from the reservoir, stressing or killing both cold- and warm-water species.
▶ **Emissions reduction.** Hydropower provides electricity without producing emissions that pollute the air. By replacing fossil fuel combustion as an electricity source, hydropower reduces air pollution, climate change, and their health and environmental consequences.	▶ **Decline in fisheries.** Fish that migrate up rivers to spawn encounter dams as a barrier. Although "fish ladders" have been built at many dams to allow passage, most fish do not make it. Population declines of salmon species have devastated fishing economies.
▶ **Crop irrigation.** Reservoir water can be withdrawn for irrigating crops. Irrigation is vital for agriculture in arid areas. By storing large amounts of water, reservoirs can release irrigation water when farmers most need it and can buffer regions against drought.	▶ **Population displacement.** Reservoirs generally flood fertile farmland, and have flooded many human settlements, even large modern cities. An estimated 40–80 million people globally have been displaced by dam projects over the past half century.
▶ **Drinking water.** Many reservoirs store water for municipal drinking water supplies. Such water is generally plentiful, reliable, and clean, provided that watershed lands draining into the reservoir are not developed or polluted.	▶ **Sediment capture.** As fast-flowing rivers give way to slow-flowing reservoirs, sediment falls through the water column and settles, trapped behind dams. Downstream floodplains and estuaries are no longer nourished, and reservoirs fill with silt.
▶ **Flood control.** Dams can prevent floods by storing seasonal surges, such as those following snowmelt or heavy rain. Water can then be released gradually to even out flows downstream. Flood prevention saves lives and prevents property damage.	▶ **Disruption of flooding.** In the long term, flooding is a valuable ecosystem service. Floods create productive farmland by depositing rich sediment. Without flooding, topsoil is lost and farmland deteriorates.
▶ **Shipping.** By replacing rocky river beds and rapids with deep placid pools, dams enable ships to transport goods over longer distances.	▶ **Risk of failure.** There is always a small risk that a dam could fail, causing massive property damage, ecological damage, and loss of life.
▶ **New recreational opportunities.** People can fish from boats and use personal watercraft on reservoirs in regions where such recreation was not possible before.	▶ **Lost recreational opportunities.** When a wild river is dammed, recreational opportunities such as tubing, whitewater rafting, and kayaking are lost.

and reaches 152 m (500 ft) deep. Glen Canyon Dam holds back 33.3 km³ (7.9 mi³) of water in a reservoir that stretches 300 km (186 mi) long and reaches 171 m (560 ft) deep. Together these reservoirs store four times more water than flows down the river in an entire year.

China's Three Gorges Dam is the world's largest

The complex mix of benefits and costs that dams produce is exemplified by the world's largest dam project. The Three Gorges Dam on China's Yangtze River, 186 m (610 ft) high and 2 km (1.3 mi) wide, was completed in 2003 (Figure 15.16a). When filled in 2009, the resulting reservoir should be 616 km (385 mi) long, as long as Lake Superior. The reservoir will hold over 38 trillion L (10 trillion gal) of water. It will generate hydroelectric power, enable boats and barges to travel farther upstream, and

provide flood control. The power generation may be enough to replace dozens of large coal or nuclear plants.

One of the costs of the Three Gorges Dam, aside from its $25 billion construction price tag, is that its reservoir is flooding 22 cities and the homes of 1.13 million people, requiring the largest resettlement project in China's history (Figure 15.16b). The reservoir behind the dam is also inundating archaeological sites 10,000 years old and submerging productive farmlands and wildlife habitat. Moreover, the reservoir will slow the river's flow so that suspended sediment will settle and begin to fill the reservoir as soon as it is completed. Many scientists worry that the Yangtze's many pollutants will be trapped in the reservoir, making the water undrinkable. Indeed, high levels of bacteria were found in water as it began building up behind the dam. The Chinese government plans to sink $5 billion into building hundreds of sewage treatment and waste disposal facilities.

Dam Removal

Dams bring us many benefits, but also exert ecological and social impacts. **Have some dams outlived their usefulness, and if so, should we dismantle them?**

Dams for Today and Tomorrow

Our need for dams today is greater than ever because water is a finite resource. Over the past century, global water use has increased at twice the rate of population growth. Currently, world population is expected to grow by 50%, to 9 billion people in total, by 2050. Dams and reservoirs address the needs of a growing world by efficiently storing and regulating water for multiple uses. Our world relies on the benefits they bring—drinking water, flood control, power generation, irrigation, and recreation.

Ninety percent of the dams in the United States are small, local projects that lack controversy. Consider what dams do every day for millions of people. Along the Mississippi, 70% of America's grain exports are barged to the Gulf of Mexico. Dams support 55 million irrigated acres of crop and pasture land (mostly in the arid West). Dams and reservoirs carry water to millions of people via canals and aqueducts. And dams help communities avoid billions of dollars in flood damage.

With dams, however, come environmental concerns such as fish passage, changes to water quality, and altered habitats. The challenge for any community is to balance the economic, environmental, and social considerations in utilizing dams. Today's choices often reflect the values, needs, wealth, and options of different communities and countries. It should be no surprise that in a world where 1.7 billion people are without electricity, hydropower is being developed in 80 countries.

Our challenge is to make decisions that embrace what research, sound science, technological innovation, and engineering prowess offers. We can embrace these things while also staying true to our historic and evolving cultural, environmental, and economic values. We owe it to future generations to make thoughtful, responsible policy choices about dams that affect not only our way of life, but also theirs.

Thomas Flint is a fifth-generation farmer, actively farming in Grant County, Washington. He was elected to the Grant County Public Utility District board of commissioners in 2000, is founder and director of the public education effort known as AgFARMation, is a grassroots activist, and holds director's positions on the Black Sands Irrigation District and the Columbia Basin Development League.

The Case for Dam Removal

The dams currently in existence in the United States were built to provide a variety of services, including flood control, water supply, and hydropower, which runs mills and generates electricity. Although many dams continue to provide a useful service, large numbers are considered obsolete, providing no direct economic, safety, or social function. For example, many mill dams continue to stand across streams and rivers 100 to 200 years after the mill they powered went out of operation or was torn down. These dams should be considered for removal.

Regardless of size, all dams harm riparian environments. Dams block the free flow of water down a stream corridor and create a pool, or impoundment, behind them—an artificial lake in the middle of a stream community. Impounded waters often divide into layers by temperature and depth, with heated waters in the upper layer and oxygen-poor cooler water in the lower layer. The macroinvertebrates that fish depend on for food cannot survive under these lake conditions. Carp and non-native lake fish that can survive in hotter and oxygen-poor waters often displace trout and other cold-water stream species.

Dams block the movement of migratory fish and other aquatic species, preventing them from reaching upstream areas to feed, spawn, and successfully reproduce. Dams also block river sediments that would normally travel downstream and replenish beaches or gravel stream bottoms, where most macroinvertebrates live and where fish spawn.

Rivers are dynamic systems. They move within floodplains, exchanging nutrients, sediments, and interacting on many levels. When dams interrupt that exchange, river functions are impaired, and the fish and wildlife dependent on free-flowing river systems do not thrive as well. Once a dam has outlived its utility, it makes great sense to restore the river back to its original condition.

Sara Nicholas is associate director of dam programs for American Rivers and works out of their mid-Atlantic office in Harrisburg, Pennsylvania. She has a master of science degree in environmental science from the Yale School of Forestry.

Explore this issue further by accessing **Viewpoints** at www.aw-bc.com/withgott.

(a) The Three Gorges Dam in Yichang, China

(b) Displaced people in Sichuan Province, China

FIGURE 15.16 China's Three Gorges Dam, completed in 2003, is the world's largest dam **(a)**. Well over a million people were displaced and whole cities were leveled for its construction, as shown here in Sichuan Province **(b)**. The reservoir began filling in 2003 and will continue filling for several years.

Some dams are now being removed

People who feel that the costs of some dams have outweighed their benefits have been pushing for such dams to be dismantled. By removing dams and letting rivers flow free, these people say, we can restore riparian ecosystems, reestablish economically valuable fisheries, and revive river recreation such as fly-fishing and rafting.

Increasingly, private dam owners and the Federal Energy Regulatory Commission (FERC), the U.S. government agency charged with renewing licenses for dams, have agreed. Roughly 500 dams have been removed in the United States, nearly 200 of them in the past decade. One reason is that many aging dams are in need of costly repairs or have outlived their economic usefulness.

The drive to remove dams first gathered steam in 1999 with the dismantling of the Edwards Dam on Maine's Kennebec River, which resulted from the first FERC determination that the environmental benefits of removing a dam outweighed the economic benefits of relicensing it. Within just a year after the 7.3-m (24-ft) high, 279-m (917-ft) long dam was removed, large numbers of 10 species of migratory fish, including salmon, sturgeon, shad, herring, alewife, and bass, ventured upstream and began using the 27-km (17-mi) stretch of river above the dam site. Some property owners along the former reservoir who had opposed the dam's removal had a change of heart once they saw the healthy and vibrant river that now ran past their property. More small dams—and perhaps some large ones—will come down in years ahead as over 500 FERC licenses come up for renewal in the next decade.

Solutions to Freshwater Depletion

Ensuring adequate quantities of freshwater is an endeavor as old as our species. Technological advances will not eliminate this challenge. Human population growth, expansion of irrigated agriculture, and industrial development doubled our global annual freshwater use between 1960 and 2000. We now use an amount equal to 10% of total global runoff. The hydrologic cycle makes freshwater a renewable resource, but if our usage exceeds what a lake, river, or aquifer can provide, we must either reduce our use, find another water source, or be prepared to run out of water.

Solutions can address supply or demand

To address freshwater depletion, we can aim either to increase supply or to reduce demand. Strategies for reducing demand include conservation and efficiency measures. Lowering demand is more difficult politically in the short term but may be necessary in the long term. In the developing world, international aid agencies are increasingly funding demand-based solutions over supply-based solutions, because demand-based solutions offer better economic returns and cause less ecological and social damage.

Strategies most often used to increase supply in a particular area have involved transporting water through pipes and aqueducts from areas where it is more plentiful or accessible. In some instances this has been done by mutual agreement, but in other instances water-poor regions have forcibly appropriated water from communities too weak to keep it for themselves. For instance, Los Angeles built itself up with water it appropriated from the Owens Valley, Mono Lake, and other rural and less-inhabited regions of California. Transporting water from place to place is not a just solution if it harms the region that loses the water. Another strategy is to develop new technologies to find or "make" more water.

FIGURE 15.17 Kuwaiti engineers walk along water intake pipes for a desalination plant on the Persian Gulf.

Weighing the Issues:
Reaching for Water

In 1941, the burgeoning metropolis of Los Angeles needed water and decided to divert streams feeding into Mono Lake, over 565 km (350 mi) away in northern California. As the lake level fell 14 m (45 ft) in 40 years, salt concentrations doubled and aquatic communities suffered. Other desert cities—such as Las Vegas, Phoenix, and Denver—are expected to double in population in coming decades. Where will they go for water to sustain their communities? What challenges might they face in trying to pipe in water from distant sources? How will people living in these source areas be affected? How else could these cities meet their future water needs?

Desalination "makes" more water

The best-known technological approach to generate freshwater is **desalination,** or *desalinization,* the removal of salt from seawater or other water of marginal quality. One method of desalination mimics the hydrologic cycle by hastening evaporation from allotments of ocean water with heat and then condensing the vapor—essentially distilling freshwater. Another method involves forcing water through membranes to filter out salts; the most common process of this type is called reverse osmosis.

Over 7,500 desalination facilities are operating worldwide, most in the arid Middle East (Figure 15.17) and some in small island nations that lack groundwater. The largest plant, in Saudi Arabia, produces 485 million L (128 million gal) of freshwater every day. However, desalination is currently quite expensive. In 1992 the world's largest reverse osmosis plant was completed along the Colorado River near Yuma, Arizona. It was intended to re-

duce the salinity of irrigation runoff reentering the river, but it proved too expensive to operate and closed after only 8 months.

Agricultural demand can be reduced

Although supply-side solutions are worth pursuing, there is a need to reduce demand as well. Because most water use is for agriculture, it makes sense to look first to agriculture for ways to decrease demand. Farmers can improve efficiency by lining irrigation canals to prevent leaks, leveling fields to minimize runoff, and adopting efficient irrigation methods. Techniques to increase irrigation efficiency include low-pressure spray irrigation, which sprays water downward toward plants, and drip irrigation systems, which target individual plants and introduce water directly onto the soil (see Figure 9.17, ▶ p. 265). Both methods reduce water lost to evaporation and surface runoff. Low-pressure precision sprinklers in Texas have efficiencies of 80–95% and have resulted in water savings of 25–37%. Drip irrigation, which has efficiencies as high as 90%, could cut water use in half while raising yields by 20–90% and giving developing-world farmers $3 billion in extra annual income, it has been estimated. Such improvements in Jordan more than doubled crop yields from 1973 to 1986.

Choosing crops to match the land and climate in which they are being farmed can save huge amounts of water. Currently, crops that require a great deal of water, such as cotton, rice, and alfalfa, are often planted in arid areas with government-subsidized irrigation. As a result, the true cost of water is not part of the costs of growing the crop. Eliminating subsidies and growing crops in climates with adequate rainfall could greatly reduce water use in many parts of the world. Finally, selective breeding (▸ pp. 120–122; 248) and genetic modification (▸ pp. 287–295) can result in some crop varieties that require less water.

We can lessen residential, municipal, and industrial water use in many ways

We can reduce our household water use by installing low-flow faucets, showerheads, washing machines, and toilets. Automatic dishwashers, studies show, use less water than does washing dishes by hand. If our homes have lawns, it is best to water them at night, when water loss from evaporation is minimal. Better yet, replacing water-intensive lawns with native plants adapted to your region's natural precipitation patterns saves the most water. *Xeriscaping*, landscaping using plants adapted to arid conditions, has become a popular approach in much of the U.S. Southwest.

Industry and municipalities can take water-saving steps as well. Manufacturers have shifted to processes that use less water and in doing so have reduced their costs. Something as seemingly obvious as finding and patching leaks in pipes has saved some cities and companies large amounts of water—and money—once they have invested in the search. Boston and its suburbs reduced water demand by 31% between 1987 and 2004 by patching leaks, retrofitting homes with efficient plumbing, auditing industry, and promoting conservation to the public. This wildly successful program enabled Massachusetts to avoid an unpopular $500-million river diversion scheme.

Another water conservation practice is recycling municipal wastewater for irrigation and industrial uses, as some cities, among them St. Petersburg, Florida, are doing. In yet another approach, governments in Arizona and in England are capturing excess surface runoff during their rainy seasons and pumping it into aquifers.

Various economic approaches to water conservation are being debated

Many economists have suggested market-based strategies for achieving sustainability in water use, ending government subsidies of inefficient practices, and letting water become a commodity whose price reflects the true costs of its extraction. Others worry that making water a fully priced commodity would make it less available to the world's poor and increase the gap between rich and poor. Because industrial use of water can be 70 times more profitable than agricultural use, market forces alone could favor uses that would benefit wealthy and industrialized people, companies, and nations at the expense of the poor and less industrialized.

Similar concerns surround another potential solution, the privatization of water supplies. During the 1990s, some public water systems were partially or wholly privatized, and their construction, maintenance, management, or ownership transferred to private companies. This was done in hope of increasing the systems' efficiency, but many people worry that firms have little incentive to allow equitable access to water for rich and poor alike. Already in some developing countries, rural residents without access to public water supplies, who are forced to buy water from private vendors, end up paying on average 12 times more than those connected to public supplies.

Other experiences indicate that decentralization of control over water, from the national level to the local level, may help conserve water. In Mexico, the effectiveness of irrigation systems improved dramatically once they were transferred from public ownership to the control of 386 local water user associations.

Regardless of how demand is addressed, the ongoing shift from supply-side to demand-side solutions is beginning to pay dividends. In Europe, a new focus on demand (through government mandates and public education) has decreased public water consumption, and industries are becoming more efficient in their water use. The United States decreased its total water consumption by 10% from 1980 to 1995, thanks to conservation measures, even while its population grew 16%.

Freshwater Pollution and Its Control

The quantity and distribution of freshwater poses one set of environmental and social challenges. Safeguarding the *quality* of water involves another collection of environmental and human health dilemmas. To be safe for human consumption and other organisms, water must be relatively free of disease-causing organisms and toxic substances.

Although developed nations have made admirable advances in cleaning up water pollution over the past few decades, the World Commission on Water in 1999 concluded that over half the world's major rivers are "seriously depleted and polluted, degrading and poisoning the surrounding ecosystems, threatening the health and livelihood of people who depend on them." The largely invisible pollution of groundwater, meanwhile, has been termed a "covert crisis." Groundwater and surface water are polluted by numerous anthropogenic sources, including untreated sewage, agricultural runoff, and petroleum and chemical spills.

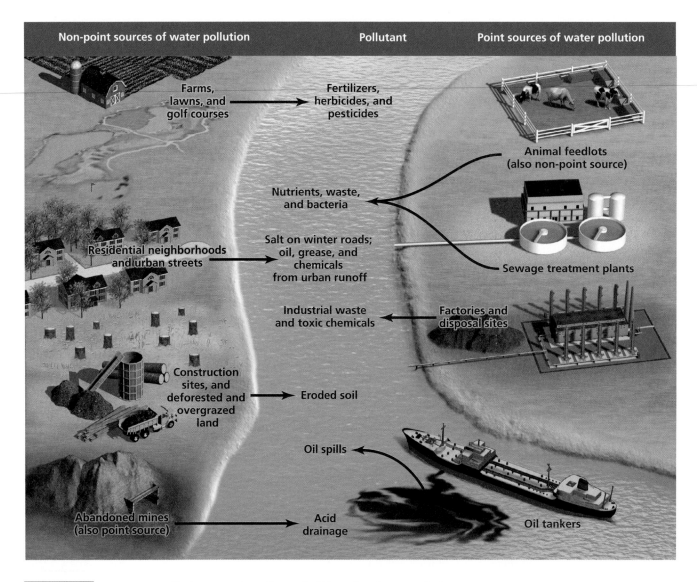

Non-point sources of water pollution   **Pollutant**   **Point sources of water pollution**

Farms, lawns, and golf courses → Fertilizers, herbicides, and pesticides

Animal feedlots (also non-point source)

Nutrients, waste, and bacteria

Sewage treatment plants

Residential neighborhoods and urban streets → Salt on winter roads; oil, grease, and chemicals from urban runoff

Industrial waste and toxic chemicals ← Factories and disposal sites

Construction sites, and deforested and overgrazed land → Eroded soil

Oil spills

Abandoned mines (also point source) → Acid drainage

Oil tankers

FIGURE 15.18 Point-source pollution comes from discrete facilities or locations, usually from single outflow pipes. Non-point-source pollution (such as runoff from streets, residential neighborhoods, lawns, and farms) originates from numerous sources spread over large areas.

Water pollution comes from point sources and from diffuse non-point sources

The term **pollution** describes any matter or energy released into the environment, whether from human activity or natural sources, that causes undesirable impacts on the health and well-being of humans or other organisms. Pollution can be physical, chemical, or biological, and can affect water, air, or soil.

Water pollution can be emitted from **point sources**—discrete locations, such as a factory or sewer pipe. Alternatively, it can consist of **non-point-source** pollution arising from multiple cumulative inputs over larger areas, such as

farms, city streets, and residential neighborhoods (Figure 15.18). The U.S. Clean Water Act, by targeting industrial discharges, has addressed point-source pollution with some success, such that non-point-source pollution has a greater impact on water quality in the United States today. Many common activities give rise to non-point-source water pollution, such as applying fertilizers and pesticides to lawns, applying salt to roads in winter, and changing automobile oil.

Water pollution comes in many forms and can cause diverse impacts on aquatic ecosystems and human health. We can categorize pollution into several types, including nutrient pollution, biological pollution by disease-causing organisms, toxic chemical pollution, physical pollution by sediment, and thermal pollution.

Nutrient pollution We saw in Chapter 7 how nutrient pollution from fertilizers and other sources can lead to eutrophication and hypoxia in coastal marine areas, as exemplified by the Gulf of Mexico's dead zone (see Fig. 7.5, ▶ p. 189). Eutrophication proceeds in a similar fashion in freshwater systems, where phosphorus is usually the nutrient that spurs growth (▶ pp. 192–194). When excess phosphorus enters surface waters, it fertilizes algae and aquatic plants, boosting their growth rates and populations. Although such growth provides oxygen and food for other organisms, algae can cover the water's surface, depriving deeper-water plants of sunlight. As algae die off, they provide food for decomposing bacteria. Decomposition requires oxygen, so the increased bacterial activity drives down levels of dissolved oxygen. These levels can drop too low to support fish and shellfish, leading to dramatic changes in aquatic ecosystems.

Eutrophication (Figure 15.19) is a natural process, but excess nutrient input from runoff from farms, golf courses, lawns, and sewage can dramatically increase the rate at which it occurs. We can reduce nutrient pollution by treating wastewater, reducing fertilizer application, and planting vegetation to increase nutrient uptake.

Pathogens and waterborne diseases Many disease-causing organisms (pathogenic viruses, protists, and bacteria) survive in surface water. Some enter inadequately treated drinking water supplies when these become contaminated with human or animal waste. Specialists monitoring water quality can tell when water has been contaminated by waste when they detect concentrations of fecal coliform bacteria, which live in the intestinal tracts of people and other vertebrates. These bacteria are usually not pathogenic themselves, but rather serve as indicators of fecal contamination, which may mean that the water holds other pathogens that can cause ailments such as giardiasis, typhoid, or hepatitis A.

Biological pollution by pathogens causes more human health problems than any other type of water pollution. A study of global water supply and sanitation issues by the World Health Organization (WHO) and the United Nations Children's Fund (UNICEF) in 2000 showed that despite advances in many parts of the world, major problems still existed. On the positive side, 4.9 billion people (82% of the population) had access to safe water as a result of some form of improvement in their water

(a) Oligotrophic water body

(b) Eutrophic water body

FIGURE 15.19 An oligotrophic water body **(a)** with clear water and low nutrient content may eventually become a eutrophic water body **(b)** with abundant vegetation and high nutrient content. Pollution of freshwater bodies by excess nutrients accelerates the process of eutrophication.

supply—an increase from 4.1 billion (79% of the population) in 1990. However, over 1.1 billion people were still without safe water supplies. In addition, 2.4 billion people had no sewer or sanitation facilities. Most of these people were Asians and Africans, and four-fifths of the people without sanitation lived in rural areas. These conditions contribute to widespread health impacts and 5 million deaths per year.

We have developed strategies for reducing the risks that water borne pathogens pose. Treating sewage constitutes one approach. Another is using chemical or other means to disinfect drinking water. Finally, hygienic measures can fight waterborne disease. Among these are public education to encourage personal hygiene and government enforcement of regulations to ensure the cleanliness of food production, processing, and distribution.

Toxic chemicals Our waterways have become polluted with toxic organic substances of our own making, including pesticides, petroleum products, and other synthetic chemicals. Many of these can poison animals and plants, alter aquatic ecosystems, and cause a wide array of human health problems, including cancer. In addition, toxic metals such as arsenic, lead, and mercury, and acids from acid precipitation and from acid drainage from mining sites (▸ p. 576), also cause negative impacts on human health and the environment. Health impacts of toxic chemicals are discussed in Chapter 14.

Legislating and enforcing more stringent regulations of industry can help reduce releases of these toxic inorganic chemicals. Better yet, we can modify our industrial processes to rely less on these substances.

Sediment Although floods build fertile farmland, sediment that rivers transport can also impair aquatic ecosystems. Mining, clear-cutting, land clearing for real estate development, and careless cultivation of farm fields all expose soil to wind and water erosion (▸ pp. 256–259). Although some water bodies, such as the Colorado River and China's Yellow River, are naturally sediment-rich, many others are not. When a clear-water river receives a heavy influx of eroded sediment, aquatic habitat can change dramatically, and fish adapted to clear-water environments may not be able to handle the change. We can reduce sediment pollution by more thoughtfully managing farms and forests and by avoiding large-scale disturbance of vegetation.

Heat and cold Because water's ability to hold dissolved oxygen decreases as temperature rises, some aquatic organisms may not survive temperature increases in surface waters. Many human activities can increase water temperatures. When water is withdrawn from a river or stream and used to cool industrial facilities, it transfers heat energy from the facility back into the surface water to which it is returned. People also raise surface water temperatures by removing streamside vegetation that shades water.

Too little heat can also cause problems. On the Colorado and many other dammed rivers, water at the bottoms of reservoirs is much colder than water at the surface. If dam operators release water from the depths of a reservoir, downstream water temperatures become much lower. In the Colorado River system, these low water temperatures have favored cold-loving invasive trout over an endangered native species of suckerfish.

Scientists use several indicators of water quality

Most forms of water pollution are not very visible to the human eye, so scientists and technicians measure certain physical, chemical, and biological properties of water to characterize its quality. Biological properties include the presence of fecal coliform bacteria and other disease-causing organisms, as discussed above. In addition, algae and aquatic invertebrates are commonly used as biological indicators of water quality.

Chemical properties include nutrient concentrations, pH (▸ p. 98), taste and odor, and hardness. Hard water contains high concentrations of calcium and magnesium ions, prevents soap from lathering, and leaves chalky deposits behind when heated or boiled. Another chemical characteristic is dissolved oxygen content. Dissolved oxygen is an indicator of aquatic ecosystem health because surface waters low in dissolved oxygen are less capable of supporting aquatic life.

Among physical characteristics, turbidity measures the density of suspended particles in a water sample. Fast-moving rivers that cut through arid or eroded landscapes, such as the Colorado and Yellow rivers, carry a great deal of sediment and are turbid and muddy-looking as a result. Water color can reveal particular substances present in a water sample. Some forest streams run the color of iced tea because of chemicals called tannins that occur naturally in decomposing leaf litter. Finally, temperature can be used to assess water quality. High temperatures can interfere with some biological processes, and warmer water holds less dissolved oxygen.

Groundwater pollution is a serious problem

Most efforts at pollution control have focused on surface water bodies. Yet increasingly, groundwater sources once assumed to be pristine have been contaminated by pollution from industrial and agricultural practices. Groundwater pollution is largely hidden from view and is extremely difficult to monitor; it can be out-of-sight, out-of-mind for decades until widespread contamination of drinking supplies is discovered.

Groundwater pollution can also be more difficult to manage than surface water pollution. Rivers flush their pollutants fairly quickly, but groundwater retains its contaminants until they decompose, which in the case of persistent pollutants can be many years or decades. The long-lived pesticide DDT, for instance, is still found widely in U.S. aquifers even though it was banned 35 years ago. Moreover, chemicals are broken down much more slowly in aquifers than in surface water or soils. Groundwater generally contains less dissolved oxygen, microbes, minerals, and organic matter, so decomposition is slower. For instance, concentrations of the herbicide alachlor are reduced by half after 20 days in soil but in groundwater this takes almost 4 years.

There are many sources of groundwater pollution

Various chemicals that are toxic at high concentrations, including aluminum, fluoride, nitrates, and sulfates, occur naturally in groundwater. The poisoning of Bangladesh's wells by arsenic is one case of natural contamination (see "The Science behind the Story," ▸ pp. 456–457). However, groundwater pollution from human activity is widespread. Industrial, agricultural, and urban wastes—from heavy metals to petroleum products to industrial solvents to pesticides—can leach through soil and seep into aquifers. Pathogens and other pollutants can enter groundwater through improperly designed wells. Contamination even results from the intentional pumping of liquid hazardous waste below ground (▸ pp. 668–670).

Leakage from underground storage tanks is another contributor to groundwater pollution. These include septic tanks, tanks of industrial chemicals, and tanks of oil and gas. In the United States, the Environmental Protection Agency (EPA) has embarked on a nationwide cleanup program to unearth and repair leaky tanks before they do further damage to soil and groundwater quality (Figure 15.20). After more than a decade of work, the EPA in 2005 had confirmed leaks from 449,000 tanks, had initiated cleanups on 416,000 of them, and had completed cleanups of 324,000. Intercepting carcinogenic or otherwise toxic pollutants such as chlorinated solvents and gasoline before they reach aquifers is vital because once an aquifer is contaminated, it is extremely difficult to remediate.

Agriculture also contributes to groundwater pollution. Nitrate from fertilizers has leached into groundwater in Canada and in 49 U.S. states. Pesticides were detected in over half of the shallow aquifer sites tested in the United

FIGURE 15.20 Leaky underground storage tanks are a major source of groundwater pollution. Under an EPA program, hundreds of thousands of these tanks are being unearthed and repaired.

States in the mid-1990s, although generally below the standards set by the EPA for drinking water. Nitrate in drinking water has been linked to cancers, miscarriages, and "blue-baby" syndrome, in which infants suffocate. Agriculture can also contribute pathogens; in 2000, the groundwater supply of Walkerton, Ontario, became contaminated with the bacterium *Escherichia coli*, or *E. coli*. Two thousand people became ill, and seven died.

Manufacturing industries and military sites have been heavy polluters through the years. Although legislation has forced them to reduce discharges, groundwater can bear a toxic legacy long afterwards. During World War II, the U.S. Army operated the world's largest facility to produce trinitrotoluene (TNT) near St. Louis. Nitroaromatic by-products seeped into the drinking water for miles around. The site was high on the list to be cleaned up under the 1980 Superfund legislation (▸ pp. 670–671). At another Superfund site, the Hanford Nuclear Reservation in Washington State, radioactive waste has seeped into groundwater, some of it with a half-life of a quarter-million years.

Legislative and regulatory efforts have helped reduce pollution

As numerous as our freshwater pollution problems may seem, it is important to remember that many of them were worse a few decades ago. Citizen activism and government response during the 1960s and 1970s in the

Arsenic in the Waters of Bangladesh

The Science behind the Story

In the 1970s, UNICEF, with the help of environmental scientists at the British Geological Survey, launched a campaign to improve access to freshwater in Bangladesh. By digging thousands of small artesian wells, the designers of the program hoped to reduce Bangladeshis' dependence on disease-ridden surface waters. In the mid-1990s, however, scientists began to suspect that the wells dug to improve Bangladeshis' health were contaminated with arsenic, a poison that, if ingested frequently, can cause serious skin disorders and other illnesses, including cancer.

A medical doctor sounded the first alarm. In 1983, dermatologist K. C. Saha of the School of Tropical Medicine in Calcutta, India, saw the first of many patients from West Bengal, an area of India just west of Bangladesh, who showed signs of arsenic poisoning. Through a process of elimination, contaminated well water was identified as the likely cause of the poisoning. The hypothesis was confirmed by

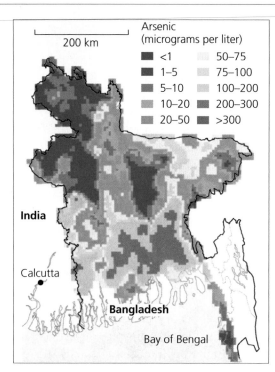

In a modern tragedy, thousands of wells dug for drinking water in Bangladesh at the urging of international aid workers turned out to be laced with arsenic. This map shows that the highest concentrations of arsenic—100–300 or more micrograms per liter of water—are found in the southern portion of the country. *Source:* Kinniburgh, D. G., and P. L. Smedley, eds. 2001. Arsenic contamination of groundwater in Bangladesh. Department of Public Health Engineering Bangladesh, British Geological Survey Report.

groundwater testing and by the work of epidemiologists, among them Dipankar Chakraborti of Calcutta's Jadavpur University.

However, it was not until the late 1990s that large-scale testing of Bangladesh's wells began. By 2001, when the British Geological Survey and the government of Bangladesh published their final report, 3,524 wells had been tested. Of the shallow wells, those less than 150 m (490 ft) deep, 46% exceeded the World Health Organization's maximum recommended level of 10 µg/L (1.33×10^{-6} oz/gal) of arsenic. Extrapolating across all of Bangladesh, the scientists estimated that as many as 2.5 million wells serving 57 million people were contaminated. As the

United States resulted in legislation such as the Federal Water Pollution Control Act of 1972 (later amended and renamed the Clean Water Act in 1977). These acts made it illegal to discharge pollution from a point source without a permit, set standards for industrial wastewater, set standards for contaminant levels in surface waters, and funded construction of sewage treatment plants. Thanks to such legislation, point-source pollution in the United States has been reduced, and rivers and lakes are cleaner than they have been in decades.

Other developed nations have also reduced pollution. In Japan, Singapore, China, and South Korea, legislation, regulation, enforcement, and investment in wastewater treatment have brought striking water quality improvements. However, non-point-source pollution, eutrophication, and acid precipitation remain major challenges.

The Great Lakes of Canada and the United States represent a success story in fighting water pollution. In the 1970s these lakes, which hold 18% of the world's surface freshwater, were badly polluted with wastewater, fertilizers, and toxic chemicals. Algal blooms occurred along beaches, and Lake Erie was pronounced dead. Today, efforts of the Canadian and U.S. governments have paid

figure shows, arsenic contamination is most prevalent in southern Bangladesh, but localized hot spots are found in northern regions of the country.

Scientists have not yet reached consensus on the chemical processes by which Bangladesh's shallow aquifers became contaminated. All agree that the arsenic is of natural origin; what remains unclear is how the low levels of arsenic naturally present in soils were dissolved in the aquifers in elemental and highly toxic form. One initial explanation, suggested by Chakraborti and his colleagues, placed most of the blame on agricultural irrigation. By drawing large amounts of water out of aquifers during Bangladesh's dry season, they argued, irrigation had permitted oxygen to enter the aquifers and prompted the release of arsenic from pyrite, a common mineral.

Other scientists contend that pyrite oxidation cannot explain most cases of arsenic contamination. In a 1998 paper in the journal *Nature,* Ross Nickson of the British Geological Survey and his colleagues suggested that arsenic was being released from iron oxides carried into Bangladesh by the Ganges River. They pointed to results of a hydrochemical survey that measured the chemical composition of aquifers throughout Bangladesh. Contrary to the predictions of the pyrite oxidation hypothesis, the survey found that arsenic concentrations tended to increase with aquifer depth and to be inversely correlated with concentrations of sulfur, a component of pyrite. Nickson and his colleagues concluded that highly reducing chemical conditions created by buried organic matter, such as peat, had probably leached arsenic from iron oxides over thousands of years.

Recently, Massachusetts Institute of Technology hydrologist Charles Harvey and colleagues suggested that irrigation may contribute to the arsenic problem after all, but not because of pyrite oxidation. In a 2002 paper in the journal *Science,* they described an experiment in which more than a dozen wells were dug near the capital city of Dhaka. Contrary to the pyrite oxidation hypothesis, and in agreement with Nickson and his colleagues, they found little evidence of a connection between sulfur or oxygen and arsenic.

However, they also found that they could increase arsenic concentrations by injecting organic matter, such as molasses, into their experimental wells. In the process of being metabolized by microbes, the molasses appeared to be freeing arsenic from iron oxides. A similar process might take place naturally, Harvey's team argued, when runoff from rice paddies, ponds, and rivers recharges aquifers that have been depleted by heavy pumping for irrigation. In support of this hypothesis, they found that much of the carbon in the shallow wells was of recent origin. Other scientists, however, have found arsenic in much older waters. This finding suggests that Bangladesh's arsenic problem may be caused by multiple hydrological and geological factors.

off. According to Environment Canada, releases of seven toxic chemicals have been reduced by 71%, municipal phosphorus has been decreased by 80%, and chlorinated pollutants from paper mills are down by 82%. Levels of PCBs and DDE are down by 78% and 91%. Bird populations are rebounding, and Lake Erie is now home to the world's largest walleye fishery. The Great Lakes' troubles are by no means over—sediment pollution is still heavy, PCBs and mercury still settle on the lakes from the air, and fish are not always safe to eat. However, the progress so far shows how conditions can improve when citizens push their governments to take action.

Drinking water is treated before it reaches your tap

Technological advances have also improved our ability to control pollution and mitigate its impacts. The treatment of drinking water and wastewater are widespread and successful mainstream practices in developed nations today. The U.S. EPA sets standards for over 80 drinking water contaminants, which local governments and private water suppliers are obligated to meet. Before being sent to your tap, water from a reservoir or aquifer is treated with chemicals to remove particulate matter; passed through filters of sand, gravel, and charcoal; and/or disinfected with small amounts of an agent such as chlorine.

It is better to prevent pollution than to mitigate it after it occurs

In many cases, solutions to pollution will need to involve prevention, not simply treatment and cleanup. One prominent expert has said that addressing groundwater pollution will require a complete overhaul of the way we live and dispose of waste, not just an "end-of-pipe" approach of piecemeal cleanups. Preventing pollution in the first place would seem to be the best strategy when one considers the other options for dealing with groundwater contamination:

▶ Filtering groundwater before distributing it can be expensive; facilities in the U.S. Midwest by one estimate spend $400 million annually just to remove the herbicide atrazine from water supplies.

▶ Pumping water out of an aquifer, treating it, then injecting it back in, repeatedly, takes an impracticably long time. Most Superfund sites with contaminated groundwater are using this method, but work has begun on just over 1% of all sites with heavily polluted groundwater, and the eventual cleanup bill has been estimated at $1 trillion.

▶ Restricting pollutants on lands above selected aquifers would alleviate contamination in those aquifers—but could simply shift pollution elsewhere.

Although a sea change in our attitudes, behavior, or technology may be necessary to clean up our water completely, there are many things ordinary people can do to help minimize freshwater pollution. One is to exercise the power of consumer choice in the marketplace by purchasing phosphorus-free detergents and other "environmentally friendly" products. Another is to become involved in protecting local waterways. Locally based "riverwatch" groups or watershed associations enlist volunteers to collect data and help state and federal agencies safeguard the health of rivers and other water bodies. Such programs are emerging in many countries throughout the world as citizens and policymakers increasingly demand clean water.

Wastewater and Its Treatment

Wastewater refers to water that has been used by people in some way. It includes water carrying sewage; water from showers, sinks, washing machines, and dishwashers;

water used in manufacturing or cleaning processes by businesses and industries; and stormwater runoff.

Although natural systems can process moderate amounts of wastewater, the large and concentrated amounts generated by our densely populated areas can harm ecosystems and pose threats to human health. Thus, attempts are now widely made to treat wastewater before releasing it into the environment.

Municipal wastewater treatment involves several steps

In rural areas, **septic systems** are the most popular method of wastewater disposal. In a septic system, wastewater runs from the house to an underground septic tank, inside which solids and oils separate from water. The clarified water proceeds downhill to a drain field of perforated pipes laid horizontally in gravel-filled trenches underground. The wastewater they emit is decomposed there by microbes. Periodically, solid waste needs to be pumped from the septic tank and taken to a landfill.

In more densely populated areas, municipal sewer systems carry wastewater from homes and businesses to centralized treatment locations. There, pollutants in wastewater are removed by physical, chemical, and biological means (Figure 15.21).

Primary treatment, the physical removal of contaminants in settling tanks or clarifiers, generally removes about 60% of suspended solids from wastewater. Wastewater then proceeds to **secondary treatment,** in which water is aerated by being stirred up, and aerobic bacteria degrade organic pollutants. Roughly 90% of suspended solids are generally removed after secondary treatment.

FIGURE 15.21 Shown here is a generalized process from a modern, environmentally sensitive wastewater treatment facility. Wastewater initially passes through screens to remove large debris and into grit tanks to let grit settle (1). It then enters tanks called primary clarifiers (2), in which solids settle to the bottom and oils and greases float to the top for removal. Clarified water then proceeds to aeration basins (3) that oxygenate the water to encourage decomposition by aerobic bacteria. Water then passes into secondary clarifier tanks (4) for removal of further solids and oils. Next, the water may be purified by chemical treatment with chlorine, passage through carbon filters, and/or exposure to ultraviolet light (5). The treated water (called effluent) may then be piped into natural water bodies, used for urban irrigation, flowed through an artificial wetland, or used to recharge groundwater. In addition, most treatment facilities control odor in the early steps and use anaerobic bacteria to digest sludge removed from the wastewater. Sludge from digesters may be sent to farm fields as fertilizer, and gas from digestion may be used to generate electric power.

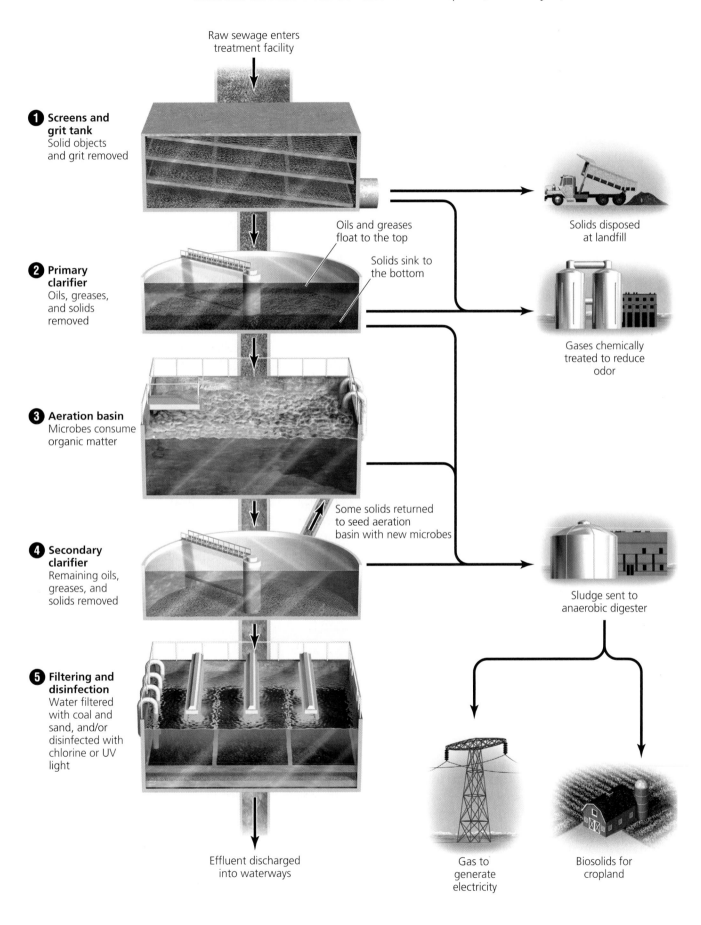

Raw sewage enters
treatment facility

1 **Screens and
grit tank**
Solid objects
and grit removed

2 **Primary
clarifier**
Oils, greases,
and solids
removed

Oils and greases
float to the top

Solids sink to
the bottom

Solids disposed
at landfill

Gases chemically
treated to reduce
odor

3 **Aeration basin**
Microbes consume
organic matter

4 **Secondary
clarifier**
Remaining oils,
greases, and
solids removed

Some solids returned
to seed aeration
basin with new microbes

Sludge sent to
anaerobic digester

5 **Filtering and
disinfection**
Water filtered
with coal and
sand, and/or
disinfected with
chlorine or UV
light

Effluent discharged
into waterways

Gas to
generate
electricity

Biosolids for
cropland

The Science behind the Story

Using Nature to Treat Our Wastewater

On a stretch of northern California's scenic Redwood Coast, a beautiful marsh hugs the waterfront lining the town of Arcata. Bulrushes, cattails, and wildflowers wave in the breeze. Locals come to jog or walk their dogs. Thousands of birds forage for food in the shallow waters. Few visitors would guess the whole thing is built on sewage.

About 450 km (280 mi) up the coast from San Francisco, the Arcata Marsh and Wildlife Sanctuary is a 121-ha (300-acre) site where science and nature merge to clean up pollution. The marsh is a human-engineered wetland created to filter wastewater. Aquatic plants and microbes perform secondary wastewater treatment, cleaning partially treated sewage water.

Arcata, a town of 16,000 people that sits on Humboldt Bay, built its treatment wetland after requirements of the Federal Water Pollution Control Act took effect in 1974. Arcata's first plan to meet the new requirements called for building a larger treatment plant that would cost more than $50 million and pipe wastewater far out into the ocean. Local residents opposed the plan, pushing for less expensive options more in step with the local environment. Two environmental scientists from nearby Humboldt State University, Robert Gearheart and George Allen, came forward, saying wetlands might provide the treatment Arcata needed.

Gearheart's and Allen's Marsh Pilot Project began in 1979 with 10 small constructed wetlands that together handled 10% of Arcata's wastewater. After 2 years, the pilot study revealed that wetlands could clean wastewater and also restore coastal wetland habitat. Arcata's civic leaders voted to use marshland to solve their sewage treatment problem. Wetland construction started in 1981, and the Arcata marsh system opened in 1986.

Treatment begins with heavy filtering at a conventional wastewater plant. To remove nitrogen compounds, pathogens, and suspended solids that initial sewage treatment fails to catch, Gearheart and Allen engineered a system of plants and microbes in the 20 ha (49 acres) of aeration ponds, where oxygen-breathing bacteria break down sewage, and in the six treatment and enhancement marshes, which receive the water for further filtering. The marshes are lined with a mix of aquatic plants, including duckweed, pennywort, and a native species, hardstem bulrush. The roots and stems of these plants form a dense network of underwater vegetation, hosting a wide variety of single-celled microbes and fungi. An especially productive zone lies in the plants' rhizome network, where tiny root hairs form a thick web that shelters algae, bacteria, and other microscopic sewage-treaters.

Some of the roots and microbes in this network trap suspended solids, holding them until the plants can use them as fertilizer. Other microbes break down nitrogen compounds such as nitrate through denitrification (▶ pp. 199–200). Still other microbes consume coliform bacteria, which may also die when exposed to sunlight in the marsh's

Finally, the clarified water is treated with chlorine, and sometimes ultraviolet light, to kill bacteria. Most often, the treated water, or effluent, is piped into rivers or the ocean following primary and secondary treatment. Sometimes, however, "reclaimed" water is used for lawns and golf courses, for irrigation, or for industrial purposes such as cooling water in power plants.

The solid material that is removed as water is purified throughout the treatment process is termed *sludge*. Sludge is sent to digesting vats, where microorganisms decompose much of the matter. The result, a wet solution of "biosolids," is then dried and either disposed of in a landfill, incinerated, or used as fertilizer on cropland. Each year about 6 million dry tons of sludge are generated in the United States.

Weighing the Issues:
Sludge on the Farm

It is estimated that anywhere from 38% to over half of the sewage sludge, or biosolids, produced each year is used as fertilizer on agricultural lands. This practice makes productive use of the sludge, increases crop output, and conserves landfill space, but many people have voiced concern over accumulation of toxic metals, proliferation of dangerous pathogens, and odors. Do you feel this practice represents an efficient use of resources or an unnecessary risk? What further information would you want to know to inform your decision?

shallow waters. Scientists who monitor the marshes know the cleaning system is working when the roots and stems of marsh plants are slimy and slippery: evidence that cleansing microbes are growing and feeding and that the rhizome network is trapping suspended solids. After 2 months of treatment, the wastewater is released into Humboldt Bay.

The marshes are monitored carefully to avoid overwhelming the wetland's cleaning abilities. Too much nitrogen can lead to algal blooms. If initial sewage treatment does not catch contaminants such as lead, these toxins can accumulate and harm microbes or wildlife. Dissolved oxygen, temperature, and nitrogen content are measured regularly, and marsh plants are thinned every few years to keep them from choking out other forms of life in the wetland.

The system's low maintenance needs and simplicity have won awards and saved money. The cost of creating the marshes was approximately $7 million, far less than that

The Arcata Marsh and Wildlife Sanctuary is the site of an artificially constructed wetland that helps treat this northern Californian city's wastewater.

of the conventional sewage plant that was first proposed. The marsh treatment system has also brought the Arcata waterfront back to life. More than 100,000 people visit the marsh each year, and more than 250 species of birds have been observed there. Other cities have used the

Arcata Marsh and Wildlife Sanctuary as a model for their own wastewater treatment wetlands, and the marsh is a site for ongoing research on wetlands and solutions to water pollution problems.

Artificial wetlands can aid treatment

Natural wetlands already perform the ecosystem service of water purification, and wastewater treatment engineers are now manipulating wetlands and even constructing wetlands *de novo* to employ them as tools to cleanse wastewater. At the same time, these wetland areas serve as havens for wildlife and areas for human recreation.

A project in Arcata, California, was one of the first such attempts (see "The Science behind the Story," above). A more recent example is Sweetwater Wetlands in Tucson, Arizona. The artificial marshes constructed here in 1996 comprise a wetland oasis for birds and wildlife in this desert region while helping to recharge a depleted aquifer.

The practice of treating wastewater with artificial wetlands is growing fast; today over 500 artificially constructed or restored wetlands in the United States are performing this service.

Conclusion

Citizen action, government legislation and regulation, new technologies, economic incentives, and public education are all enabling us to confront what will surely be one of the great environmental challenges of the new century:

ensuring adequate quantity and quality of freshwater for ourselves and for the planet's ecosystems.

Accessible freshwater comprises only a minuscule percentage of the hydrosphere, but we have grown used to taking it for granted. With our expanding population and increasing water usage, we are approaching conditions of widespread scarcity. Water depletion and water pollution are already taking a toll on the health, economies, and societies of the developing world, and they are beginning to do so in arid areas of the developed world. There is reason to hope that we may yet attain sustainability in our water usage, however. Potential solutions are numerous, and the issue is too important to ignore.

REVIEWING OBJECTIVES

You should now be able to:

Explain the importance of water and the hydrologic cycle to ecosystems, human health, and economic pursuits

▶ We depend utterly on drinkable water, and a properly functioning hydrologic cycle is vital to maintaining ecosystems and our civilization. (pp. 434–435)

Delineate freshwater distribution on Earth

▶ Of all the water on Earth, only about 1% is readily available for our use. (p. 434)

▶ Water availability varies in space and time, and regions vary greatly in the amounts they possess. (pp. 438–439)

Describe major types of freshwater ecosystems

▶ The main types of freshwater ecosystems include rivers and streams, wetlands, and lakes and ponds. (pp. 435–437)

Discuss how we use water and alter freshwater systems

▶ We use water for agriculture, industry, and residential use. The ratio of these uses varies among societies, but globally most is used for agriculture. (pp. 440–441)

▶ We pump water from aquifers and surface water bodies, sometimes at unsustainable rates. (pp. 441–442)

▶ We divert the flow of water with canals and irrigation ditches and attempt to control floods with dikes and levees. (pp. 442–446)

▶ Most of the world's rivers are dammed. Dams bring a diverse set of benefits and costs. Increasingly, people are calling for the removal of some dams. (pp. 446–449)

Assess problems of water supply and propose solutions to address freshwater depletion

▶ Water tables are dropping worldwide from unsustainable groundwater extraction. Surface water extraction has caused rivers to run dry and some water bodies to shrink. (pp. 441–445)

▶ Because of inequalities of water distribution amid shrinking supplies, we can expect heightened political tensions over water in the future. (p. 444)

▶ Solutions to expand supply, such as desalination, are worth pursuing, but not to the exclusion of finding ways to decrease demand. (pp. 449–450)

▶ Solutions to reduce demand include technology, approaches, and consumer products that increase efficiency in agriculture, industry, and the home. (pp. 450–451)

Assess problems of water quality and propose solutions to address water pollution

▶ Water pollution stems from point sources and non-point sources. (p. 452)

▶ Water pollutants include excessive nutrients, microbial pathogens, toxic chemicals, sediments, and thermal pollution. (pp. 453–454)

▶ Scientists who monitor water quality use biological, chemical, and physical indicators. (pp. 454–455)

▶ Groundwater pollution can be more persistent than surface water pollution. (pp. 455–456)

▶ Legislation and regulation have improved water quality in developed nations in recent decades. (pp. 456–457)

▶ Prevention of water pollution is better than mitigation. (p. 458)

Explain how wastewater is treated

▶ Septic systems are used to treat wastewater in rural areas. (p. 458)

▶ Wastewater is treated physically, biologically, and chemically in a series of steps at municipal wastewater treatment facilities. (pp. 458–460)

▶ Artificial wetlands can enhance wastewater treatment while restoring habitat for wildlife. (pp. 460–461)

TESTING YOUR COMPREHENSION

1. Define *groundwater*. What role does groundwater play in the hydrologic cycle?
2. Why are sources of freshwater unreliable for some people and plentiful for others?
3. Describe three benefits of damming rivers, and three costs. What particular environmental, health, and social concerns has China's Three Gorges Dam and its reservoir raised?
4. Why do the Colorado, Rio Grande, Nile, and Yellow rivers now slow to a trickle or run dry before reaching their deltas?
5. Why are water tables dropping around the world? What are some environmental costs of falling water tables?
6. Name three major types of water pollutants, and provide an example of each. List three properties of water that scientists use to determine water quality.
7. Why do many scientists consider groundwater pollution a greater problem than surface water pollution?
8. What are some anthropogenic (human) sources of groundwater pollution?
9. Describe how drinking water is treated. How does a septic system work?
10. Describe and explain the major steps in the process of wastewater treatment. How can artificial wetlands aid such treatment?

SEEKING SOLUTIONS

1. Discuss possible strategies for equalizing distribution of water throughout the world. Consider supply and transport issues. Have our methods of drawing, distributing, and storing water changed very much throughout history? How is the scale of our efforts affecting the availability of water supplies?
2. How can agricultural demand for water be decreased? Describe some ways that we can reduce household water use. How can industrial uses of water be reduced?
3. Have the provisions of the Clean Water Act been effective? Discuss some of the methods we can adopt, in addition to "end-of-pipe" solutions, to prevent contamination and ensure "water security."
4. How might desalination technology help "make" more water? Describe two methods of desalination. Where is this technology being used?
5. Your state's governor has put you in charge of water policy for the state. The aquifer beneath your state has been overpumped, and many wells have gone dry

already. Agricultural production last year decreased for the first time in a generation, and farmers are clamoring for you to do something. Meanwhile, the state's largest city is growing so fast that more water is needed for the burgeoning urban population. What policies would you consider to restore your state's water supply? Would you try to take steps to increase supply, to decrease demand, or both? Explain why you would choose such policies.
6. Having solved the water depletion problem in your state, your next task is to deal with pollution of the groundwater that provides your state's drinking water supply. Recent studies have shown that one-third of the state's groundwater has levels of pollutants that violate EPA standards for human health. The federal government is threatening enforcement, and citizens are fearful for their safety. What steps would you consider taking to safeguard the quality of your state's groundwater supply, and why?

INTERPRETING GRAPHS AND DATA

Close to 75% of the freshwater used by people is used in agriculture, and about 1 of every 14 people live where water is scarce, according to a review by hydrologist J. S. Wallace. By the year 2050, scientists project that two-thirds of the world's population will live in water-scarce areas, including most of Africa, the Middle East, India, and China. How much water is required to feed over 6 billion people a basic dietary requirement of 2,700 calories per day? The answer depends on the efficiency with which we use water in agricultural production and on the type of diet we consume.

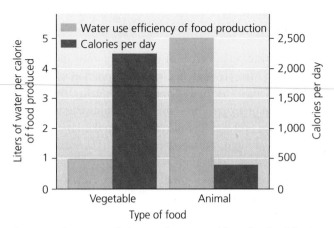

Amount of water needed to produce vegetable and animal food, and global average calories per day consumed of vegetable and animal food. Data from Wallace, J. S. 2000. Increasing agricultural water use efficiency to meet future food production. *Agriculture, Ecosystems and Environment* 82: 105–119.

1. How many liters of water are needed to produce 2,300 calories of vegetable food? How many liters of water are needed to produce 400 calories of animal food? How many liters of water are needed daily to provide this diet? Annually?

2. How many liters of water would be saved daily, compared to the diet in the graph, if the 2,700 calories were provided entirely by vegetables? Annually?

3. Reflect on the quote at the beginning of this chapter: "Water promises to be to the 21st century what oil was to the 20th century: the precious commodity that determines the wealth of nations." How do you think the demographic pressure on the water supply could affect world trade, particularly trade of agricultural products? Do you think it could affect prospects for peace and stability in and among nations? How so?

CALCULATING ECOLOGICAL FOOTPRINTS

In the United States, the EPA estimates that household water use averages 750 liters per person per day. One of the single greatest personal uses of water is for showering. Standard showerheads dispense 15 liters of water per minute, but so-called low-flow showerheads dispense only 9 liters per minute. Given an average daily shower time of 10 minutes, calculate the amounts of water used and saved over the course of a year with standard versus low-flow showerheads, and record your results in the table below.

1. What percentage of personal water consumption would you calculate is used for showering?

2. How much additional water would you be able to save by shortening your average shower time from 10 minutes to 8 minutes? To 5 minutes?

3. Can you think of any factors that are not being considered in this scenario of water savings? Explain.

	Annual water use with standard showerheads (liters)	Annual water use with low-flow showerheads (liters)	Annual water savings with low-flow showerheads (liters)
You	54,750	32,850	21,900
Your class			
Your state			
United States			

Data from U.S. EPA, 1995. *Cleaner water through conservation: Chapter 1—How we use water in the United States.* EPA 841-B-95-002.

Take It Further

Go to www.aw-bc.com/withgott or the student CD-ROM where you'll find:

▶ Suggested answers to end-of-chapter questions

▶ Quizzes, animations, and flashcards to help you study

▶ *Research Navigator*™ database of credible and reliable sources to assist you with your research projects

▶ **GRAPHit!** tutorials to help you master how to interpret graphs

▶ **INVESTIGATEit!** current news articles that link the topics that you study to case studies from your region to around the world

The Oceans: Natural Systems, Human Use, and Marine Conservation

Portion of the
Florida Keys

Upon successfully completing this chapter you will be able to:

▶ Identify physical, geographical, chemical, and biological aspects of the marine environment

▶ Describe major types of marine ecosystems

▶ Outline historic and current human uses of marine resources

▶ Assess human impacts on marine environments

▶ Review the current state of ocean fisheries and reasons for their decline

▶ Evaluate marine protected areas and reserves as innovative solutions

466

Coral reef in the
Florida Keys
National Marine
Sanctuary

North
America

Atlantic
Ocean

Florida Keys

Central Case: Seeding the Seas with Marine Reserves

"I saw the fisheries spiral
down. There's less fish
and they're smaller. It's
not anything like it was
15 years ago."
—DON DEMARIA,
COMMERCIAL FISHERMAN,
FLORIDA KEYS

"We have no water-quality
problems. We have no
problems with our reef. The
sanctuary was shoved
down our throats."
—BETTYE CHAPLIN, FLORIDA
KEYS REALTOR AND
CO-FOUNDER OF THE
CONCH COALITION

Stretching southwest from the southern tip of
Florida for 320 km (200 mi), the string of islands
known as the Florida Keys hosts some of North
America's richest marine and coastal ecosystems. These
islands boast the world's third-largest barrier coral reef,
as well as sea grass meadows, coastal mangrove forests,
and estuaries. As more people have come to enjoy this
remarkable natural environment, tourism in the Keys has
grown to 3 million visitors each year.

As tourism was increasing, however, so was human
impact. Overfishing, trash and sewage dumping, boat
groundings, and careless anchoring were damaging the

region's ecosystems while its waters received inputs of
pesticides, oil, and heavy metals from roads, residential
areas, and farms on the islands. Runoff rich in
sediments, fertilizer, and nutrients from leaky septic
systems were damaging sea grass beds and causing
plankton blooms and eutrophication in Florida Bay,
between the Keys and the mainland. All these impacts
were harming the coral, living animals whose skeletons
give structure to coral reefs, which are home to so much
biodiversity. From 1966 to 2000, biologists documented
declines in the diversity of living coral and the area
covered by coral at two-thirds of the sites they
monitored.

These impacts on water quality, sea grass, and coral
reefs combined with overfishing to depress fish stocks.
Since the 1970s, scientists and fishers alike had reported
that the Keys' fish and lobster populations were
declining and that the remaining fish and lobsters were
smaller. Scientists monitoring the Keys' fish populations
concluded that reef fish were severely overexploited by
commercial and recreational fishing; a report held that

13 of 16 grouper populations, 7 of 13 snapper populations, and 2 of 5 grunt populations had been overfished.

To protect the area's natural and cultural resources, Congress in 1989 established the Florida Keys National Marine Sanctuary (Figure 16.1). This sanctuary incorporates several previously established protected areas and today safeguards 9,800 km^2 (3,800 mi^2) of marine habitat, including 530 km^2 (205 mi^2) of coral reef, and over 6,000 plant, fish, and invertebrate species. Oil exploration, mining, dumping, and large ships are banned. Fishing is allowed throughout most of the sanctuary, but 24 smaller areas recently zoned as reserves protect 65% of the sanctuary's shallow reef habitat and prohibit all harvesting of natural resources.

Although these 24 reserves amount to only 6% of the sanctuary's area, they have been a magnet for controversy. After the sanctuary was established, a group of Keys residents formed the Conch Coalition to protest the sanctuary and the proposal for no-fishing reserves within it. (A conch is a large marine snail that is a popular food item in the Keys.) The Conch Coalition, which eventually claimed 3,000 supporters, fought the proposal in public meetings and brought a lawsuit. Some Conch Coalition supporters hanged and burned in effigy

the sanctuary superintendent and another sanctuary proponent during a protest. Today, however, as fish populations have begun to increase both inside and outside the reserves, most Keys residents have come to support the sanctuary and its no-fishing reserves.

Sanctuaries and reserves are types of *marine protected areas,* portions of ocean that are protected from some human activities. Whereas national parks and other types of protected areas (▶ pp. 364–369) have existed on land for over a century, the oceanic equivalent is quite new. "Most people think it's common sense on land to have areas where we don't hunt," James Bohnsack, a Florida-based National Marine Fisheries Service researcher, explained. "We're now trying to create natural water areas—to see the buffalo roam, so to speak. It's a major change of thinking that protects the ecosystem and biodiversity, but it also protects the fishery."

Oceanography

The oceans cover the vast majority of our planet's surface, and understanding them is crucial for understanding how our planet's systems work. The oceans influence global climate, teem with biodiversity, facilitate transportation

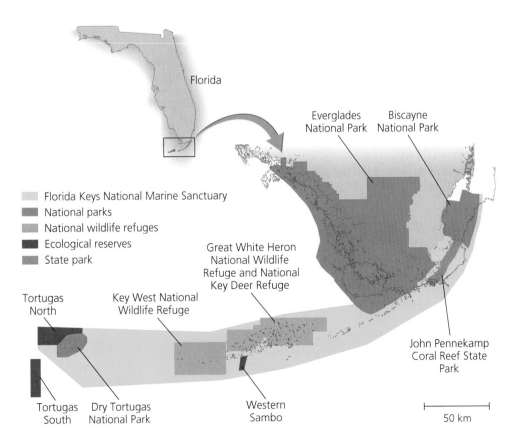

FIGURE 16.1 The Florida Keys National Marine Sanctuary includes over 9,800 km^2 (3,800 mi^2) of mangrove islands, coral reefs, and shallow waters, from the southeastern tip of Florida westward to the distant keys called the Tortugas. The sanctuary is adjacent to or encompasses three national parks, one state park, two national wildlife refuges, and three marine ecological reserves.

Florida

Everglades National Park
Biscayne National Park

Florida Keys National Marine Sanctuary
National parks
National wildlife refuges
Ecological reserves
State park

Great White Heron National Wildlife Refuge and National Key Deer Refuge

Tortugas North
Key West National Wildlife Refuge

John Pennekamp Coral Reef State Park

Tortugas South
Dry Tortugas National Park
Western Sambo

50 km

FIGURE 16.2 The world's oceans are connected in a single vast body of water but are given different names for convenience. The Pacific Ocean is the largest and, like the Atlantic and Indian Oceans, includes both tropical and temperate waters. The smaller Arctic and Antarctic Oceans include the waters in the north and south polar regions, respectively. Many smaller bodies of water are named as seas or gulfs; a selected few are shown here.

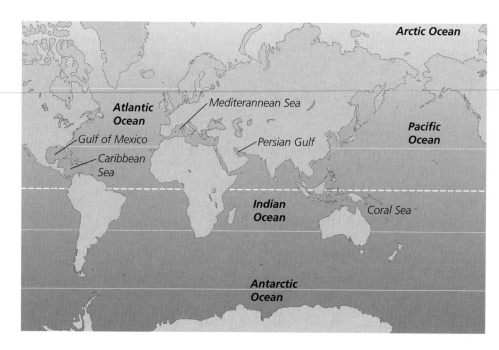

and commerce, and provide us many resources. Even landlocked areas far from salt water are affected by the oceans and the ways we interact with them. The oceans provide fish for people to eat in Iowa, supply oil to power cars in Ontario, and influence the weather in Nebraska. The study of the physics, chemistry, biology, and geography of the oceans is called **oceanography.**

Oceans cover most of Earth's surface

Although we generally speak of the world's oceans (Figure 16.2) in the plural, giving each major basin a name—Pacific, Atlantic, Indian, Arctic, and Antarctic—all these oceans are connected, comprising a single vast body of water. This one "world ocean" covers 71% of our planet's surface and contains 97.2% of its surface water. The oceans take up most of the hydrosphere, influence the atmosphere, interact with the lithosphere, and encompass a large portion of the biosphere, including at least 250,000 species. The world's oceans touch and are touched by virtually every environmental system and every human endeavor.

The oceans contain more than water

Ocean water contains approximately 96.5% H_2O by mass; most of the remainder consists of ions from dissolved salts (Figure 16.3). Ocean water is salty primarily because ocean basins are the final repositories for water that runs off the land. Rivers carry sediment and dissolved salts from the continents into the ocean, as do winds. Evaporation

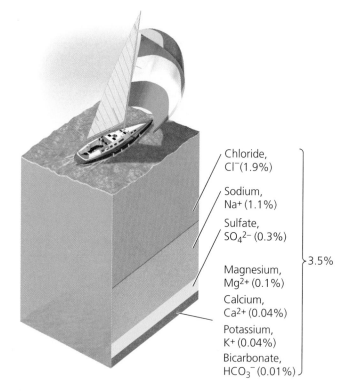

Chloride, Cl^- (1.9%)

Sodium, Na^+ (1.1%)

Sulfate, SO_4^{2-} (0.3%)

Magnesium, Mg^{2+} (0.1%)

Calcium, Ca^{2+} (0.04%)

Potassium, K^+ (0.04%)

Bicarbonate, HCO_3^- (0.01%)

3.5%

FIGURE 16.3 Ocean water consists of 3.5% salt, by mass. Most of this salt is NaCl in solution, so sodium and chloride ions are abundant. A number of other ions and trace elements are also present.

from the ocean surface then removes pure water, leaving a higher concentration of salts. If we were able to evaporate all the water from the oceans, the world's ocean basins would be covered with a layer of dried salt 63 m (207 ft) thick.

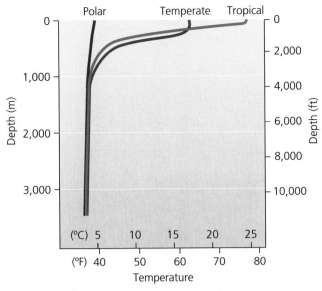

(a) Temperature profiles at polar, temperate, and tropical latitudes

FIGURE 16.4 Ocean water varies in temperature and density with depth. Water temperatures **(a)** near the surface are warmer because of daily heating by the sun, and become rapidly colder with depth over the top 1,000 m (3,300 ft). This temperature differential is greatest in the tropics because of intense solar heating and is least in the polar regions. Deep water at all latitudes is equivalent in temperature. Density decreases unevenly with depth, giving rise to several distinct zones **(b)**. Waters of the surface zone are well mixed and roughly equivalent in density, whereas density decreases rapidly with depth in the pycnocline. Waters of the deep zone resist mixing and are largely unaffected by sunlight, winds, and storms.

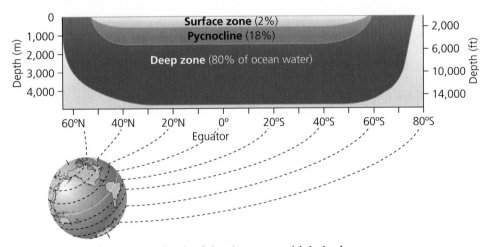

(b) Variation in the average depth of density zones with latitude

The salinity of ocean water generally ranges from 33 to 37 parts per thousand (ppt), varying from place to place because of differences in evaporation, precipitation, and freshwater runoff from land and glaciers. Salinity near the equator is low because this region has a great deal of precipitation, which is relatively salt-free. In contrast, at latitudes roughly 30–35 degrees north and south, evaporation exceeds precipitation, making the surface salinity of the oceans higher than average.

Besides the dissolved salts shown in Figure 16.3, nutrients such as nitrogen and phosphorus occur in seawater in trace amounts (well under 1 part per million) and play essential roles in nutrient cycling in marine ecosystems. Another aspect of ocean chemistry is dissolved gas content, particularly the dissolved oxygen on which many marine animals depend. Oxygen concentrations are high-est in the upper layer of the ocean, reaching 13 ml/L of water. Roughly 36% of the gas dissolved in seawater is oxygen, which is produced by photosynthetic plants, bacteria, and phytoplankton, and by diffusion from the atmosphere.

Ocean water is vertically structured

Surface waters in tropical regions receive more solar radiation and therefore are warmer than surface waters in temperate or polar regions. In all regions, however, temperature declines with depth (Figure 16.4a). Water density by definition increases as salinity rises and as temperature falls. These relationships give rise to different layers of water; heavier (colder and saltier) water sinks, and lighter (warmer and less salty) water remains nearer the surface

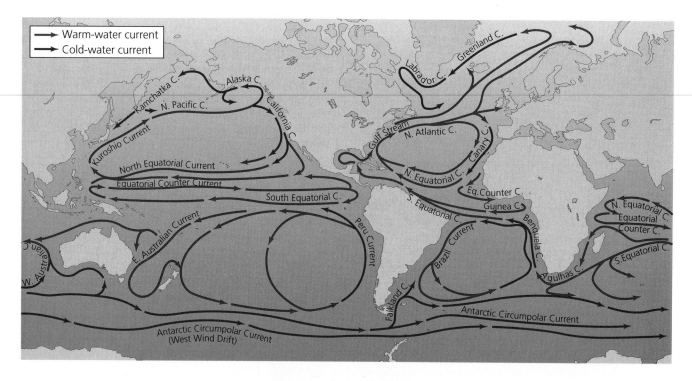

FIGURE 16.5 The upper waters of the oceans move in currents, which are long-lasting and predictable global patterns of water movement. Warm- and cold-water currents interact with the planet's climate system and have been used by people to navigate the oceans for centuries. *Source:* Garrison, T. S. 1999. *Oceanography,* 3rd ed. Belmont, CA: Wadsworth.

(Figure 16.4b). Waters of the surface zone are heated by sunlight each day and are stirred by wind such that they are of similar density throughout, down to a depth of approximately 150 m (490 ft). Below the zone of surface water lies the *pycnocline,* a region in which density increases rapidly with depth. The pycnocline contains about 18% of ocean water by volume, compared to the surface zone's 2%. The remaining 80% resides in the deep zone beneath the pycnocline. The dense water in this zone is sluggish and not affected by winds and storms, sunlight, and daily temperature fluctuations.

Despite the daily heating and cooling of surface waters, ocean temperatures are much more stable than temperatures on land. Midlatitude oceans experience yearly temperature variation of only around 10 °C (18 °F), and tropical and polar oceans are still more stable. The reason for this stability is that water has a very high *heat capacity,* a measure of the heat required to increase temperature by a given amount. It takes much more heat energy to increase the temperature of water than it does to increase the temperature of air by the same amount. High heat capacity enables the oceans to absorb a tremendous amount of heat from the atmosphere. In fact, the heat content of the entire atmosphere is equal to that of just the top 2.6 m (8.5 ft) of the oceans. By absorbing heat and later releas-

ing it to the atmosphere, the oceans help shape Earth's climate (Chapter 18). Also influencing climate is the ocean's surface circulation, a system of currents that move in the pycnocline and the surface zone.

Ocean water flows horizontally in currents

Far from being a static pool of water, Earth's ocean is composed of vast riverlike flows (Figure 16.5) driven by density differences, heating and cooling, gravity, and wind. These surface **currents** move in the upper 400 m (1,300 ft) of water, horizontally and for great distances. These long-lasting patterns influence global climate and play key roles in the phenomena known as El Niño and La Niña (▸ pp. 534, 536–537). They also have been crucial in navigation and human history; currents helped carry Polynesians to Easter Island, Darwin to the Galapagos, and Europeans to the New World. Currents transport heat, nutrients, pollution, and the larvae of many marine species.

Some currents are very slow. Others, like the Gulf Stream, are rapid and powerful. From the Gulf of Mexico, the Gulf Stream moves eastward along the southern edge of the Florida Keys, then northward past Miami at a rate of 160 km per day (nearly 2 m/sec, or over 4.1 mi/hr). An

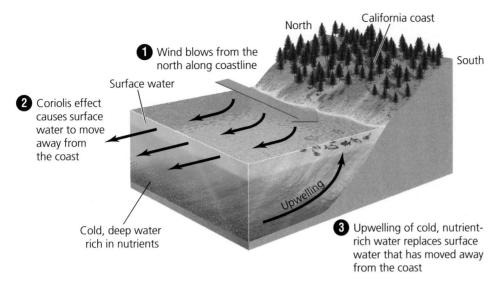

1 Wind blows from the north along coastline

North

California coast

South

Surface water

2 Coriolis effect causes surface water to move away from the coast

Upwelling

Cold, deep water rich in nutrients

3 Upwelling of cold, nutrient-rich water replaces surface water that has moved away from the coast

FIGURE 16.6 Upwelling is the movement of bottom waters upward. This type of vertical current often brings nutrients up to the surface, creating rich areas for marine life. For example, north winds blow along the California coastline, while the Coriolis effect draws wind and water away from the coast. Water is then drawn up from the bottom to replace the water that moves away from shore.

average width of 70 km (43 mi) across, the Gulf Stream continues across the North Atlantic, bringing warm water to Europe and moderating that continent's climate, which otherwise would be much colder.

Vertical movement of water affects marine ecosystems

Surface winds and heating also create vertical currents in seawater. **Upwelling,** the vertical flow of cold, deep water toward the surface, occurs in areas where horizontal currents diverge, or flow away from one another. Because up-welled water is rich in nutrients from the bottom, upwellings are often sites of high primary productivity (▸ pp. 192–193) and lucrative fisheries. Upwellings also occur where strong winds blow away from or parallel to coastlines (Figure 16.6). An example is the California coast, where north winds and the Coriolis effect (▸ pp. 504–505) move surface waters away from shore, raising nutrient-rich water from below and creating a biologically rich region. The cold water also chills the air along the coast, giving San Francisco its famous fog and cool summers.

In areas where surface currents converge, or come together, surface water sinks, a process called **downwelling.** Downwelling transports warm water rich in dissolved gases, providing an influx of oxygen for deep-water life. Vertical currents also occur in the deep zone, where differences in water density can lead to rising and falling convection currents, such as those seen in molten rock (▸ pp. 208–209) and in air (▸ pp. 501–502). The North Atlantic Deep Water (▸ p. 534) is an example of such circulation that has far-reaching effects on global climate.

Weighing the Issues:
Why Understand Ocean Currents?

Mapping ocean currents is crucial to understanding where and how the larvae of many fish and marine invertebrates become distributed from place to place. Currents help determine where larvae will settle and mature into adults. Why do you think this information might be important for people to know? What else can be carried by currents? Describe another reason that an understanding of ocean currents can be helpful and important.

Seafloor topography can be rugged and complex

Although oceans are depicted on most maps and globes as smooth, blue swaths, parts of the ocean floor are just as complex as the terrestrial portion of the lithosphere. Underwater volcanoes shoot forth enough magma to build islands above sea level, such as the Hawaiian Islands. Steep canyons similar in scale to Arizona's Grand Canyon lie just offshore of some continents. The deepest spot in the oceans—the Mariana Trench, located in the South Pacific near Guam—is deeper than Mount Everest is high, by over 2.1 km (1.3 mi). Our planet's longest mountain range is under water—the Mid-Atlantic Ridge runs the length of the Atlantic Ocean (Figure 16.7).

We can gain an understanding of the major types of underwater geographic features by examining a stylized map that reflects *bathymetry* (the study of ocean depths)

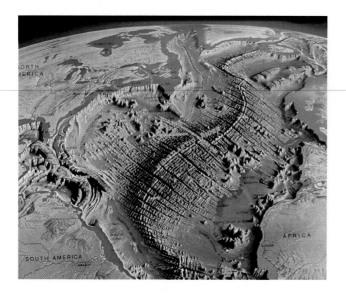

FIGURE 16.7 The seafloor can be every bit as rugged as continental topography. The spreading margin between tectonic plates at the Mid-Atlantic Ridge gives rise to a vast underwater volcanic mountain chain, cross-hatched by immense perpendicular breaks in the oceanic crust.

and topography (physical geography, or the shape and arrangement of landforms) (Figure 16.8). In bathymetric profile, gently sloping **continental shelves** underlie the shallow waters bordering the continents. Continental shelves vary in width from 100 m (330 ft) to 1,300 km (800 mi), averaging 70 km (43 mi) wide, with an average

slope of 1.9 m/km (10 ft/mi). These shelves drop off with relative suddenness at the *shelf-slope break.* The *continental slope* angles somewhat steeply downward, connecting the continental shelf to the deep ocean basin below.

Most of the seafloor is flat, but volcanic peaks that rise above the ocean floor provide physical structure for marine animals and are frequently the site of productive fishing grounds. Some island chains, such as the Florida Keys, are formed by reef development and lie atop the continental shelf. Others, such as the Aleutian Islands curving across the North Pacific from Alaska toward Russia, are volcanic in origin, with peaks that rise above sea level. The Aleutians are also the site of a deep trench that, like the Mariana Trench, formed at a convergent tectonic plate boundary, where one slab of crust dives under another in the process of subduction (▸ p. 209).

Oceanic zones differ greatly, and some support more life than others. The uppermost 10 m (33 ft) of ocean water absorbs 80% of the solar energy that reaches its surface. For this reason, nearly all of the oceans' primary productivity occurs in the well-lit top layer, or *photic zone.* Generally, the warm, shallow waters of continental shelves are most biologically productive and support the greatest species diversity. Biological oceanographers, or marine biologists, tend to classify marine habitats and ecosystems into two types. Those occurring between the ocean's surface and floor are **pelagic,** whereas those that occur on the ocean floor are **benthic.** Each of these major areas contains several vertical zones (Figure 16.9).

FIGURE 16.8 A stylized bathymetric profile shows key geologic features of the submarine environment. Shallow regions of water exist around the edges of continents over the continental shelf, which drops off at the shelf-slope break. The relatively steep dropoff called the continental slope gives way to the more gradual continental rise, all of which are underlain by sediments from the continents. Vast areas of seafloor are flat abyssal plain. Seafloor spreading occurs at oceanic ridges, and oceanic crust is subducted in trenches. Volcanic activity along trenches often gives rise to island chains such as the Aleutian Islands. Features on the left side of this diagram are more characteristic of the Atlantic Ocean, and features on the right side of the diagram are more characteristic of the Pacific Ocean. Adapted from Thurman, H. V. 1990. *Essentials of oceanography,* 4th ed. New York: Macmillan.

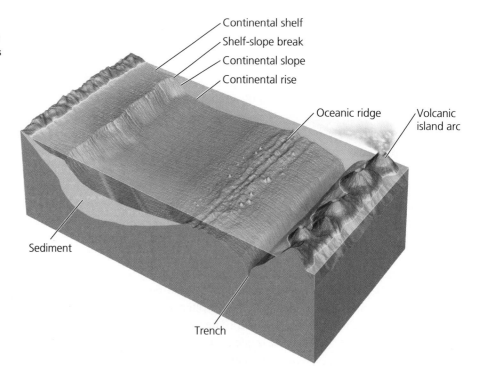

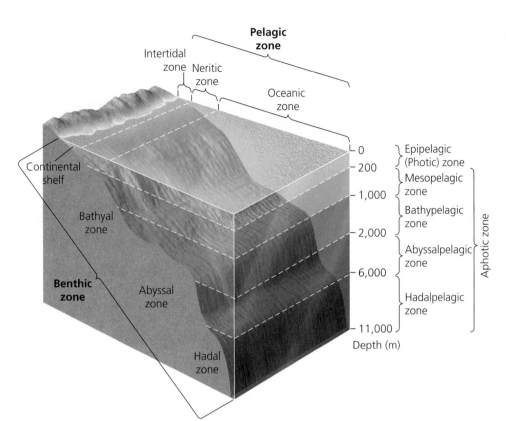

FIGURE 16.9 Oceanographers classify marine habitats into benthic (seafloor) and pelagic (open-water) categories and subdivide these major categories into zones based on depth. Pelagic waters extend from the epipelagic zone at and near the surface down to the hadalpelagic zone at depths below 6,000 m (19,700 ft). Near-surface waters that receive adequate light for photosynthesis are also often termed the *photic zone,* and waters above the continental shelves are also said to be in the *neritic zone.* Benthic zones range from the intertidal zone where the ocean meets the land, to the continental shelves, and down to the bathyal, abyssal, and hadal zones.

Marine Ecosystems

With their variation in topography, temperature, salinity, nutrients, and sunlight, marine environments feature a variety of ecosystems. Most marine ecosystems are powered by solar energy, with sunlight driving photosynthesis by phytoplankton in the photic zone. Yet even the darkest ocean depths host life.

Open-ocean ecosystems vary in their biological diversity

Biological diversity in pelagic areas of the open ocean is highly variable in its distribution. Plant productivity and animal life near the surface are concentrated in regions of nutrient-rich upwelling. Microscopic phytoplankton constitute the base of the marine food chain in the pelagic zone. These photosynthetic algae, protists, and cyanobacteria feed zooplankton, which in turn become food for fish, jellyfish, whales, and other free-swimming animals (Figure 16.10). Predators at higher trophic levels include larger fish, sea turtles, and sharks. In addition, many bird species feed at the surface of the open ocean, returning periodically to nesting sites on islands and coastlines.

In recent years biologists have been learning more about animals of the deep ocean, although tantalizing questions remain and many organisms are not yet discovered. In deep-water ecosystems, animals have adapted to

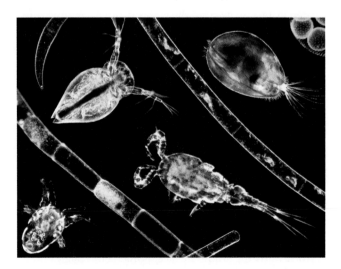

FIGURE 16.10 The uppermost reaches of ocean water contain billions upon billions of phytoplankton—tiny photosynthetic algae, protists, and bacteria that form the base of the marine food chain—as well as zooplankton, tiny animals and protists that dine on phytoplankton and comprise the next trophic level.

FIGURE 16.11 Life is scarce in the dark depths of the deep ocean, but the creatures that do live there often appear bizarre to us. The anglerfish lures prey toward its mouth with a bioluminescent (glowing) organ that protrudes from the front of its head.

FIGURE 16.12 Tall brown algae known as *kelp* grow from the floor of the continental shelf and provide structure with kelp forests, a key marine ecosystem. Numerous fish and other creatures eat kelp or find refuge among its fronds.

deal with extreme water pressures and to live in the dark without food from plants. Some of these often bizarre-looking creatures scavenge carcasses or detritus (organic particles) that fall from above. Others are predators, and still others attain food from symbiotic mutualistic (▸ pp. 155–156) bacteria. Some species carry bacteria that produce light chemically by bioluminescence (Figure 16.11).

Finally, as we explored in Chapter 4 (▸ pp. 106–107), some ecosystems form around hydrothermal vents, where heated water spurts from the seafloor, often carrying minerals that precipitate to form large rocky structures. Tubeworms, shrimp, and other creatures in these recently discovered systems use symbiotic bacteria to derive their energy ultimately from chemicals in the heated water rather than from sunlight. They manage to thrive within the amazingly narrow zones between scalding-hot and icy cold water.

Kelp forests harbor many organisms in temperate waters

Large brown algae, or **kelp** (often nicknamed *seaweed),* grow from the floor of continental shelves, reaching upward toward the sunlit surface. Some kelp reaches 60 m (200 ft) in height and can grow 45 cm (18 in.) in a single day. Dense stands of kelp form underwater forests on the continental shelves in many temperate waters (Figure 16.12). Kelp forests, with their complex structure, supply shelter and food for invertebrates and fish, which in

turn provide food for higher-trophic-level predators, such as seals and sharks. Indeed, kelp forests were the setting for our discussion of keystone species in Chapter 6 (▸ pp. 159–165). Recall that sea otters control sea urchin populations, and when otters disappear, urchins overgraze the kelp, destroying the forests and creating "urchin barrens" in their place. Kelp forests also absorb wave energy and protect shorelines from erosion. People in Asian cultures eat some types of kelp, and kelp provides compounds known as alginates, which serve as thickeners in a wide range of consumer products, from cosmetics to paints to paper to soaps.

Coral reefs are treasure troves of biodiversity

In subtropical and tropical waters, kelp forests give way to coral reefs. A reef is an outcrop of rock, sand, or other material that rises near the surface of a relatively shallow body of salt water. A **coral reef** is a mass of calcium carbonate composed of the skeletons of tiny colonial marine organisms known as *corals.* A coral reef may occur as an extension of a shoreline, along a *barrier island* paralleling a shoreline, or as an *atoll,* a ring around a sunken island.

Corals are tiny invertebrate animals related to sea anemones and jellyfish. Some corals have hard external skeletons encompassing a soft hollow body, whereas others have flexible internal skeletons. Corals are sessile (stationary), attached to rock or existing reef and capturing

(a) Elkhorn coral

(c) Partially bleached staghorn coral

(b) Moray eel hiding behind seafan

FIGURE 16.13 Different species of reef-dwelling corals give rise to a variety of distinct structures on reefs in the Florida Keys and elsewhere. One example is elkhorn coral **(a).** Fish take refuge among corals, but reef crevices also hide predators, such as moray eels **(b),** that ambush them. Coral reefs face multiple environmental stresses from natural and human impacts, and many corals have died as a result of coral bleaching, in which corals lose their zooxanthellae. Such bleaching is evident in the whitened portions of this antler-shaped staghorn coral **(c).**

passing food with stinging tentacles. They also derive nourishment from symbiotic algae, known as *zooxanthellae,* that inhabit their bodies and produce food through photosynthesis. Most corals are colonial, and the colorful surface of a coral reef is made up of thousands or millions of densely packed individuals. As corals die, their skeletons remain part of the reef while new corals grow atop them, increasing the size of the reef.

The reefs of the Florida Keys National Marine Sanctuary host many types of coral (Figure 16.13a). Coral reefs absorb wave energy and protect the shore from damage. They also provide essential habitat for many of the Keys' 6,000 species of marine organisms. Globally, coral reefs host as much biodiversity as any other type of ecosystem. The likely reason is that coral reefs provide complex physical structure (and thus many habitats) in shallow nearshore waters, which are regions of high primary productivity. Besides the staggering diversity of anemones, sponges, hydroids, tubeworms, and other sessile invertebrates, innumerable molluscs, flatworms, starfish, and urchins patrol the reefs, and

thousands of fish species find food and shelter in reef nooks and crannies. Larger predators, such as grouper and moray eels (Figure 16.13b), feed on the smaller fish.

Coral reefs are experiencing worldwide declines, however. Many reefs have undergone "coral bleaching," a process that occurs when zooxanthellae leave the coral, depriving it of nutrition (Figure 16.13c). Corals lacking zooxanthellae lose color and frequently die, leaving behind ghostly white patches in the reef. Coral bleaching is thought to result from increased sea surface temperatures associated with global climate change, from the influx of pollutants, from unknown natural causes, or from some combination of these factors. Nutrient pollution in coastal waters also promotes the growth of algae, which are blanketing reefs in the Florida Keys and many other regions. Coral reefs also sustain damage when divers stun fish with cyanide to capture them for food or for the pet trade, a common practice in waters of Indonesia and the Philippines. It is estimated that for each fish caught in this way, cyanide poisoning destroys one square meter of reef.

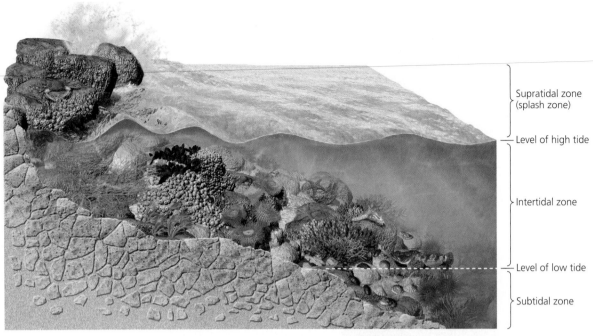

(a) Tidal zones

Supratidal zone
(splash zone)

Level of high tide

Intertidal zone

Level of low tide

Subtidal zone

(b) Tidepools at low tide

FIGURE 16.14 The rocky intertidal zone is the region along a rocky shoreline between the lowest and highest reaches of the tides **(a).** This is an ecosystem rich in biodiversity, typically containing large invertebrates such as seastars (starfish), barnacles, crabs, sea anemones, corals, bryozoans, snails, limpets, chitons, mussels, nudibranchs (sea slugs), and sea urchins. Fish swim in tidal pools **(b),** and many types of algae cover the rocks. Areas higher on the shoreline are exposed to the air more frequently and for longer periods, so those organisms that can tolerate exposure best specialize in the upper intertidal zone. The lower intertidal zone is less frequently exposed and for shorter periods, so organisms less able to tolerate exposure thrive here.

Weighing the Issues:
The Coral Crisis

Some scientists are exploring new technologies to enhance coral growth and the abundance of reef-inhabiting organisms, including chemical cues to attract larvae and special wavelengths of light to attract fish. Do you think such technology could offer adequate solutions to the global decline of coral reefs? Why or why not? What other solutions can you suggest to address the problems facing coral reefs? What challenges might each of these solutions encounter?

A few coral species thrive in waters outside the tropics and build reefs on the ocean floor at depths of 200–500 m (650–1,650 ft). These little-known reefs, which occur in cold-water areas off the coasts of Norway, Spain, the British Isles, and elsewhere, are only now beginning to be studied by scientists. Already, however, many have been badly damaged by trawling, the fishing practice in which heavy nets are dragged over the seafloor (▸ p. 485). Norway and other countries are now beginning to protect some of these deep-water reefs.

Intertidal zones undergo constant change

Where the ocean meets the land, **intertidal,** or **littoral,** ecosystems (Figure 16.14) lie along shorelines between the farthest reach of the highest tide and the lowest reach of the lowest tide. **Tides** are the periodic rising and falling of the ocean's height at a given location, caused by the gravitational pull of the moon and sun. High and low tides occur roughly 6 hours apart, although three overlapping tidal cycles make the timing and height of tides complex. Intertidal organisms spend part of each day submerged in water, part of the day dry and exposed to the air and sun, and part of the day being lashed and beaten by waves. Subject to tremendous extremes in temperature, moisture, sun exposure, and salinity, these creatures must also protect themselves from marine predators at high tide and terrestrial predators at low tide.

The intertidal environment is a tough place to make a living, but it is home to a remarkable diversity of organisms. Rocky shorelines can be full of life among the crevices, which provide shelter and pools of water (tide pools) during low tides. Sessile animals such as anemones, mussels, and barnacles live attached to rocks, filter-feeding on plankton in the water that washes over them. Urchins, sea slugs, chitons, and limpets eat intertidal algae or scrape food from the rocks. Starfish creep slowly along, preying on the filter-feeders and herbivores at high tide. Crabs clamber around the rocks, scavenging detritus. At low tide, birds, raccoons, and other land animals come by and dine on exposed animals.

The rocky intertidal zone is so diverse because environmental conditions, including temperature, salinity, and the presence or absence of water, change dramatically from the highest to the lowest reaches. This environmental variation gives rise to conspicuous bands formed by dominant organisms as they array themselves according to their habitat needs. Sandy intertidal areas, such as those of the Florida Keys, host less biodiversity, yet plenty of organisms burrow into the sand at low tide to await the return of high tide, when they emerge to feed.

Salt marshes cover large areas of coastline in temperate areas

Along many of the world's coastlines at temperate latitudes, **salt marshes** occur where the tides wash over gently sloping sandy or silty substrates. Rising and falling tides flow into and out of channels called *tidal creeks* and at highest tide spill over onto elevated marsh flats (Figure 16.15). Marsh flats grow thick with grasses, rushes, shrubs, and other herbaceous plants. Grasses such as those in the genera *Spartina* and *Distichlis* comprise the dominant vegetation in most salt marshes. Salt marshes boast very high primary productivity and provide critical habitat for shorebirds, waterfowl, and the adults and young of many commercially important fish and shellfish species. In many parts of the world, people have altered salt marshes

FIGURE 16.15 Salt marshes occur in temperate intertidal zones where the substrate is muddy, allowing salt-adapted grasses to grow. Tidal waters generally flow through marshes in channels called *tidal creeks,* amid flat areas called *benches,* sometimes partially submerging the grasses. In this salt marsh in Delaware, people have cut linear ditches to drain shallow areas to control mosquito populations.

to make way for coastal shipping, industrial facilities, farming, and other development.

Mangrove forests line coastlines in the tropics and subtropics

In tropical and subtropical latitudes, mangrove forests replace salt marshes along gently sloping sandy and silty coasts. **Mangroves** are trees with unique types of roots, some of which curve upward like snorkels to attain oxygen lacking in the mud, and some of which curve downward, serving as stilts to support the tree in changing water levels (Figure 16.16). Fish, shellfish, crabs, snakes, worms, and other organisms thrive among the root networks, and birds find habitat for feeding and nesting in the dense foliage of these coastal forests. Mangroves also provide materials that people use for food, medicine, tools, and construction. Mangroves just reach the southern edge of the United States, and the Florida Keys National Marine Sanctuary contains tens of thousands of acres forested by three species of mangrove, each adapted to specialize in a different tidal zone.

In south Florida and elsewhere, mangrove forests have been destroyed as people have converted coastal areas for residential, recreational, and commercial uses. It is estimated that people have eliminated half the world's mangrove forests and that mangrove forest area continues to decline by 2–8% per year. When mangroves are

FIGURE 16.16 Mangrove forests are important ecosystems along tropical and subtropical coastlines throughout the world. Mangrove trees, such as these at Lizard Island, Australia, show specialized adaptations for growing in salt water and provide habitat for many types of fish, birds, crabs, and other animals.

removed, coastal areas lose the ability to slow runoff, filter pollutants, and retain soil. As a result, offshore systems such as eelgrass beds and coral reefs are more readily degraded. Only about 1% of the world's remaining mangroves benefit from some sort of protection against development.

Freshwater meets salt water in estuaries

Many salt marshes and mangrove forests occur in or near **estuaries,** water bodies where rivers flow into the ocean, mixing freshwater with salt water. Biologically productive ecosystems, estuaries experience significant fluctuations in salinity as tidal currents and freshwater runoff vary daily and seasonally. For shorebirds and for many commercially important shellfish species, estuaries provide critical habitat. For anadromous fishes (those, like salmon, that spawn in freshwater and mature in salt water), estuaries provide a transitional zone where young fish make the passage from freshwater to salt water.

Estuaries around the world have been affected by urban and coastal development, water pollution, habitat alteration, and overfishing. The estuary of Florida Bay, where freshwater from the Everglades system mixes with salt water, has suffered pollution and a reduction in freshwater input due to irrigation and fertilizer use by sugarcane farmers, housing development, septic tank leakage, and other human impacts. Coastal ecosystems have borne the brunt of human impact because two-thirds of Earth's people choose to live within 160 km (100 mi) of the ocean.

Human Use and Impact

Our species has a long history of interacting with the oceans. We have long traveled across their waters, clustered our settlements along coastlines, and been fascinated by the beauty, power, and vastness of the seas. We have also left our mark upon them by exploiting oceans for their resources and polluting them with our waste.

Oceans provide transportation routes

The oceans have provided transportation routes for thousands of years and continue to provide affordable means of moving people and products over vast distances. Ocean shipping has accelerated the global reach of some cultures and has promoted interaction among long-isolated peoples. It has had substantial impacts on the environment as well. The thousands of ships plying the world's oceans today carry everything from cod to cargo containers to crude oil. Ships transport ballast water as well, which, when discharged at ports of destination, may transplant aquatic organisms picked up at ports of departure. Some of these species—such as the zebra mussel (Chapter 6)—establish and become invasive in their new homes.

We extract energy and minerals

We use the oceans as sources of commercially valuable energy. By the 1980s about 25% of our production of crude oil and natural gas came from exploitation of deposits

FIGURE 16.17 Crude oil and natural gas from beneath the seafloor are some of the economically valuable resources that we take from the oceans. Offshore petroleum drilling at platforms such as these off Santa Barbara, California, creates some of the many human impacts on the marine environment.

beneath the seafloor (Figure 16.17). According to recent estimates, offshore areas may contain as much as 2 trillion barrels of oil, roughly half the amount known to exist underground on land. The exploitation of oil and gas deposits in the Gulf of Mexico has triggered debate in recent years. The risk of an oil spill in the Florida Keys National Marine Sanctuary concerns many sanctuary supporters.

Ocean sediments also contain a novel potential source of fossil fuel energy. **Methane hydrate** is an ice-like solid consisting of molecules of methane (CH_4, the main component of natural gas) embedded in a crystal lattice of water molecules. Methane hydrates are stable at temperature and pressure conditions found in many sediments on the Arctic seafloor and the continental shelves. The U.S. Geological Survey estimates that the world's deposits of methane hydrates may hold twice as much carbon as all known deposits of oil, coal, and natural gas combined. Some people hope that methane hydrates can be developed as an energy source to power our civilization through the 21st century and beyond. However, a great deal of research remains before scientists and engineers can be sure of how to extract these energy sources safely. Destabilizing a methane hydrate deposit could lead to a catastrophic release of gas, which could cause a massive landslide and tsunami (tidal wave). This would also release huge amounts of methane, a potent greenhouse gas, into the atmosphere, exacerbating global climate change.

The oceans also hold potential for providing renewable energy sources that do not emit greenhouse gases. Engineers have developed ways of harnessing energy from waves, tides, and the heat of ocean water (▶ pp. 635–637).

These promising energy sources are awaiting further research, development, and investment.

We extract minerals from the ocean floor, as well. By using large vacuum-cleaner-like hydraulic dredges, miners collect sand and gravel from beneath the sea. Also extracted are sulfur from salt deposits in the Gulf of Mexico and phosphorite from offshore areas near the California coast and elsewhere. Other valuable minerals found on or beneath the seafloor include calcium carbonate (used in making cement), silica (used as fire-resistant insulation and in manufacturing glass), and rich deposits of copper, zinc, silver, and gold ore. Many minerals are found concentrated in manganese nodules, small ball-shaped accretions that are scattered across parts of the ocean floor. It is estimated that over 1.5 trillion tons of manganese nodules exist in the Pacific Ocean alone and that their reserves of metal exceed all terrestrial reserves. The logistical difficulty of mining them, however, has kept their extraction uneconomical so far.

Marine pollution threatens resources

People have long made the oceans a sink for waste and pollution. Even into the mid-20th century, it was common for coastal cities in the United States to dump trash and untreated sewage along their shores. Fort Bragg, a bustling town on the northern California coast, boasts of its Glass Beach, an area where beachcombers can collect sea glass, the colorful surf-polished glass sometimes found on beaches after storms. Glass Beach is in fact the site of the former town dump, and besides well-polished glass, the perceptive visitor may also spot old batteries, rusting car frames, and all other manner of trash protruding from the bluffs above the beach.

Oil, plastic, industrial chemicals, and excess nutrients all eventually make their way from land into the oceans. Raw sewage and trash from cruise ships and abandoned fishing gear from fishing boats add to the input. The scope of trash in the sea can be gauged by the amount picked up each September by volunteers who trek beaches in the Ocean Conservancy's annual International Coastal Cleanup. In this nonprofit organization's 2004 cleanup, 300,000 people from 88 nations picked up 3.5 million kg (7.7 million lb) of trash from 17,700 km (11,000 mi) of shoreline.

Nets and plastic debris endanger marine life

Plastic bags and bottles, fishing nets, gloves, fishing line, buckets, floats, abandoned cargo, and much else that people transport on the sea or deposit into it can harm marine organisms. Because most plastic is not biodegradable, it can drift for decades before washing up on beaches. Marine mammals, seabirds, fish, and sea turtles may mistake floating plastic debris for food and can die as a result of ingesting material they cannot digest or expel. Fishing nets that are lost or intentionally discarded can continue snaring animals for decades.

Of 115 marine mammal species, 49 are known to have eaten or become entangled in marine debris, and 111 of 312 species of seabirds are known to ingest plastic. Sea turtles of five species in the Gulf of Mexico have died from consuming or contacting marine debris. Marine debris harms people, as well. A survey of fishers off the Oregon coast indicated that more than half had encountered equipment damage or other problems from plastic debris, and debris has caused over $50 million dollars in insurance payments.

Oil pollution comes not only from massive spills

Major oil spills, such as the *Exxon Valdez* spill in Prince William Sound, Alaska (Chapter 4), make headlines and cause serious environmental problems. Yet it is important to put such accidents into perspective. The majority of oil pollution in the oceans comes not from large spills in a few particular locations, but from the accumulation of innumerable widely spread small sources, including leakage from small boats and runoff from human activities on land. In addition, the amount of petroleum spilled into the oceans in recent years is equaled by the amount that seeps into the water from naturally occurring seafloor deposits (Figure 16.18a).

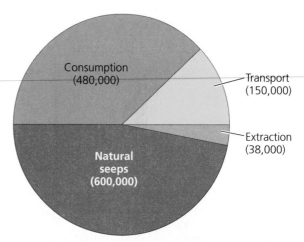

(a) Sources of petroleum input into oceans (metric tons)

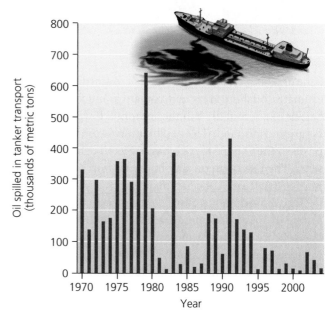

(b) Quantity of petroleum spilled from tankers, 1970–2004

FIGURE 16.18 Of the 1.3 million metric tons of petroleum spilled into the world's oceans each year, nearly half is from natural seeps (**a**). Petroleum consumption by people accounts for 38% of total input, and this includes numerous diffuse non-point sources, especially runoff from rivers and coastal communities and leakage from two-stroke engines. Spills during petroleum transport account for 12%, and leakage during petroleum extraction accounts for 3%. Less oil is being spilled into ocean waters today in large tanker spills, thanks in part to regulations on the oil shipping industry and improved spill response techniques (**b**). The figure shows cumulative quantities of oil spilled worldwide from nonmilitary spills over 7 metric tons. Data from: (a) National Research Council. 2003. *Oil in the sea III. Inputs, fates, and effects.* Washington, DC: National Academies Press. (b) International Tanker Owners Pollution Federation Ltd. 2005. *Oil tanker spill statistics: 2004.* London: ITOPF.

Nonetheless, minimizing the amount of oil we release into coastal waters is important, because petroleum pollution is detrimental to the marine environment and the human economies that draw sustenance from that environment. Petroleum can physically coat and kill marine organisms, and ingested chemical components in petroleum can poison marine life. In response to headline-grabbing oil spills, governments around the world have begun to implement more stringent safety standards for tankers, such as requiring industry to pay for tugboat escorts in sensitive and hazardous coastal waters, and to develop prevention and response plans for major oil spills.

The U.S. Oil Pollution Act of 1990 created a $1 billion prevention and cleanup fund and required that by 2015 all oil tankers in U.S. waters be equipped with double hulls as a precaution against puncture. The oil industry has resisted many such safeguards, and today the ship that oiled Prince William Sound is still plying the world's oceans, renamed the *Sea River Mediterranean,* and still featuring only a single hull. However, over the past three decades, the amount of oil spilled in U.S. and global waters has decreased (Figure 16.18b), in part because of an increased emphasis on spill prevention and response.

Excess nutrients cause algal blooms

Pollution from fertilizer runoff or other nutrient inputs can have dire effects on marine ecosystems, as we saw with the Gulf of Mexico's dead zone in Chapter 7. The release of excess nutrients into surface waters can spur unusually high growth rates and population densities of phytoplankton, causing eutrophication in freshwater and saltwater systems. Such problems have occurred in the Florida Keys, leading the U.S. Environmental Protection Agency (EPA) in 2001 to prohibit the discharge of sewage and other waste from boats into the waters of the Florida Keys National Marine Sanctuary.

Excessive nutrient concentrations sometimes give rise to population explosions among several species of marine algae that produce powerful toxins that attack the nervous systems of vertebrates. Blooms of these algae are known as **harmful algal blooms.** Some algal species produce reddish pigments that discolor surface waters, and blooms of these species are nicknamed **red tides** (Figure 16.19). Harmful algal blooms can cause illness and death among zooplankton, birds, fish, marine mammals, and humans as their toxins are passed up the food chain. They also cause economic loss for communities dependent on fishing or beach tourism. Reducing nutrient runoff into coastal waters can lessen the risk of these outbreaks, and health impacts can be minimized by monitoring to prevent human consumption of affected organisms.

As severe as the impacts of marine pollution can be, however, most marine scientists concur that the more worrisome dilemma is overharvesting. Unfortunately, the old cliché that "there are always more fish in the sea" appears not to be true; the oceans today have been overfished, and many fishing stocks have been largely depleted.

(a) Dinoflagellate (*Gymnodinium*)

FIGURE 16.19 In a harmful algal bloom, certain types of algae multiply to great densities in surface waters, producing toxins that can bioaccumulate and harm organisms. Red tides are a type of harmful algal bloom in which the algae, such as dinoflagellates of the genus *Gymnodinium* (**a**), produce pigment that turns the water red (**b**).

(b) Red tide, Gulf of Carpentaria, Australia

Emptying the Oceans

The oceans and their biological resources have provided for human needs for thousands of years, but today we are placing unprecedented pressure on marine resources. Half the world's marine fish populations are fully exploited, meaning that we cannot harvest them more intensively without depleting them, according to a 2004 U.N. Food and Agriculture Organization (FAO) report. An additional 25% of marine fish populations are overexploited and already being driven toward extinction, the FAO reported. Thus only one-quarter of the world's marine fish populations can yield more than they are already yielding without being driven into decline. Total global fisheries catch, after decades of increases, leveled off after about 1988 (Figure 16.20).

As our population grows, we will become even more dependent on the oceans' bounty. Existing fishing practices are not sustainable given present consumption rates, many scientists and fisheries managers have concluded. This makes it vital, they say, that we take immediate steps to modify our priorities and improve our use of science in fisheries management.

Overfishing is nothing new

People have harvested fish, shellfish, turtles, seals, and other animals from the oceans for millennia. Archaeological evidence from ancient coastal communities reveals shellfish-rich diets, and many sites around the world include vast middens, or piles, of discarded oyster and clam shells. Although much of this harvesting may have been sustainable, paleoecologists are learning that the depletion of marine animals by humans did not begin with to-

day's industrialized fishing fleets. Rather, overfishing took a toll on a variety of marine species beginning centuries or millennia ago. It then accelerated during the colonial period of European expansion and intensified further in the 20th century.

A recent synthesis of historical evidence by marine biologist Jeremy Jackson and others revealed that ancient overharvesting likely affected ecosystems in astounding ways we only partially understand today. Several large animals, including the Caribbean monk seal, Steller's sea cow, and Atlantic gray whale, were hunted to extinction long enough ago that scientists never studied them or the ecological roles they played. Overharvesting of the vast oyster beds of Chesapeake Bay led to the collapse of its oyster fishery in the late 19th century. Eutrophication and hypoxia similar to that of the Gulf of Mexico (Chapter 7) have resulted, because there are no longer oysters to filter algae and bacteria from the water.

Florida Bay, according to Jackson and his colleagues, suffers today from the results of overhunting of green sea turtles in past centuries. The once-abundant turtles ate sea grass (often called turtle grass) and likely kept it cropped low, like a lawn. But with today's turtle population a tiny fraction of what it once was, sea grass grows thickly, dies, and rots, giving rise to disease like sea grass wasting disease, which ravaged Florida Bay in the 1980s.

A better-known case of historical overharvesting is the near-extinction of many species of whales. This resulted from commercial whaling that began centuries ago and was curtailed only in 1986. Although overfishing has a long history, the industrialized methods, new technology, and global reach of today's commercial fleets are making our impacts much more rapid and far-reaching than in past centuries.

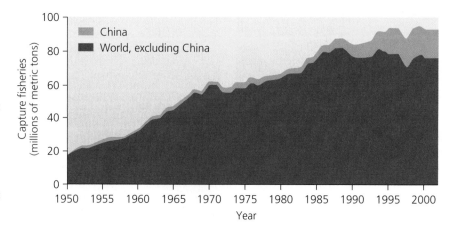

FIGURE 16.20 The total global fisheries catch has increased over the past half-century, but in recent years growth has stalled, and many fear that a global catch decline is imminent if conservation measures are not taken soon. The figure shows trends with and without China's data (see "The Science behind the Story," ▶ pp. 486–487). With China's data, global catch has leveled off since the mid-1990s. Without China's data, catch has decreased slightly since 1988. Go to GRAPHit! at www.aw-bc.com/withgott or on the student CD-ROM. Data from U.N. Food and Agricultural Organization (FAO). 2004. *World review of fisheries and aquaculture.* Rome: FAO.

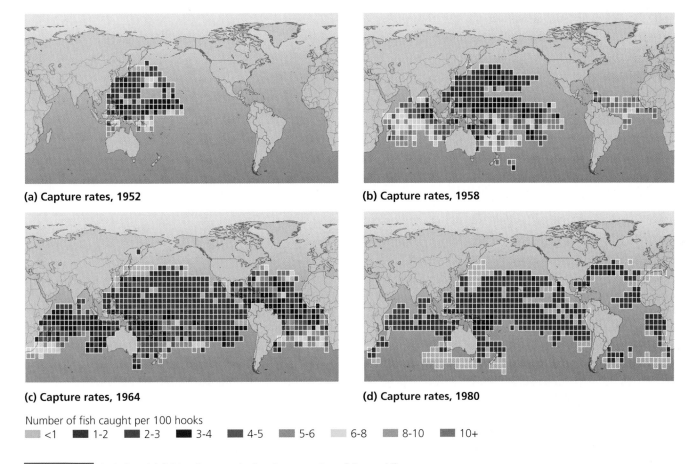

(a) Capture rates, 1952

(b) Capture rates, 1958

(c) Capture rates, 1964

(d) Capture rates, 1980

Number of fish caught per 100 hooks

<1 ■ 1-2 ■ 2-3 ■ 3-4 ■ 4-5 ■ 5-6 ■ 6-8 ■ 8-10 ■ 10+

FIGURE 16.21 As industrial fishing fleets reached each new region of the world's oceans, capture rates of large predatory fishes were initially high and then within a decade declined markedly. In the figure, reds and oranges signify high capture rates, and maroons and purples signify low capture rates. High capture rates in the southwestern Pacific in 1952 (**a**) gave way to low ones in later years. Excellent fishing success in the tropical Atlantic and Indian Oceans in 1958 (**b**) had turned mediocre by 1964 (**c**) and poor by 1980 (**d**). High capture rates in the north and south Atlantic in 1964 (**c**) gave way to low capture rates there in 1980 (**d**). Data from Myers, R. A. and B. Worm. 2003. *Nature* 423: 280–283.

Modern fishing fleets deplete marine life rapidly

Today's industrialized fishing fleets can deplete marine populations quickly. In a 2003 study, fisheries biologists Ransom Myers and Boris Worm analyzed fisheries data from FAO archives, looking for changes in the catch rate of fish in various regions of ocean since they were first exploited by industrialized fishing. For one region after another, they found the same pattern: Catch rates dropped precipitously, with 90% of large-bodied fish and sharks eliminated within only a decade (Figure 16.21). Following that, populations stabilized at 10% of their former levels. This means, Myers and Worm concluded, that the oceans today contain only one-tenth of the large-bodied animals they once did. It also means that declines happened so fast in most regions of the world that scientists never knew the original abundance of these animals.

Many fisheries are collapsing today

The proportion of marine fish stocks that are overfished increased tenfold between 1950 and 1990. Many fisheries have collapsed, and others are in danger of collapsing. These collapses are ecologically devastating and also take a severe economic toll on human communities that depend on fishing.

A prime example of fishery collapse took place in the 1990s, affecting groundfish in the North Atlantic off the Canadian and U.S. coasts. The term *groundfish* refers to various species that live in benthic habitats, such as Atlantic cod, haddock, halibut, and flounder. These fish are major food sources that powered fishing economies in Newfoundland, Labrador, the Maritime Provinces, and the New England states for close to 400 years. Yet fishing pressure became so intense that most stocks collapsed in recent years, bringing fishing

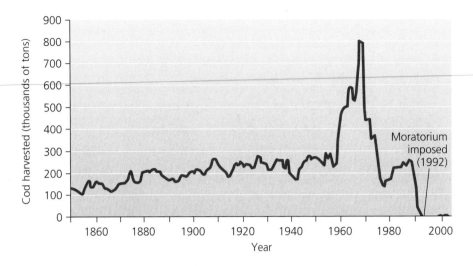

FIGURE 16.22 In the North Atlantic off the coast of Newfoundland, commercial catches of Atlantic cod increased with intensified fishing in the 1960s and 1970s. The fishery subsequently crashed, and moratoria imposed in 1992 and 2003 have not brought it back. Data from Millennium Ecosystem Assessment, 2005.

economies down with them (Figure 16.22). With Canada's cod stocks down by 99% and showing no sign of recovery, the Canadian government in 1992 ordered a complete ban on cod fishing in the Grand Banks region off Newfoundland and Labrador. The moratorium was partially lifted in 1998, but catches declined, so it was reimposed in 2003. To soften the economic blow, the government offered $50 million to affected fishers of the region.

The experience on the U.S. side of the border is showing that such bans can sometimes help restore depleted fisheries. When the groundfish fisheries of Georges Bank in the Gulf of Maine collapsed in the mid-1990s, three areas totaling 17,000 km² (6,600 mi²) were closed to fishing. The closures worked; five years later, haddock, flounder, and yellowtail were recovering, and scallops rebounded strongly, attaining sizes 9–14 times as large as before the closures. Fishers began having better luck, especially just outside the closed regions. Unfortunately, cod have not recovered. Because adult cod eat fish that prey on young cod, some ecologists hypothesize that once the adult cod were depleted, intensified predation on young cod prevented the population from recovering.

Fisheries declines are masked by several factors

Although industrialized fishing has depleted fish stocks in region after region, the overall global catch has remained roughly stable for two decades (see Figure 16.20). You might wonder how this could be. The seeming stability of the total global catch can be explained by several factors that mask population declines. One is that fishing fleets have been traveling longer distances to reach less-fished

portions of the ocean. They also have been fishing in deeper waters; average depth of catches was 150 m (495 ft) in 1970 and 250 m (820 ft) in 2000. Moreover, fishing fleets have been spending more time fishing and have been setting out more nets and lines—expending increasing effort just to catch the same number of fish.

Improved technology also helps explain high catches despite declining stocks. Today's Japanese, European, Canadian, and U.S. fleets can reach almost any spot on the globe with boats that can attain speeds of 80 kph (50 mph). They have access to an array of technologies that militaries have developed for spying and for chasing enemy submarines, including advanced sonar mapping equipment, satellite navigation, and thermal sensing systems. Some fleets rely on aerial spotters to find schools of commercially valuable fish, such as bluefin tuna.

Finally, another cause of misleading stability in global catch numbers is that not all data supplied to international monitoring agencies may be accurate (see "The Science behind the Story," ▶ pp. 486–487).

We are "fishing down the food chain"

Overall figures on total global catch tell only part of the story, because they do not include information on the species, age, and size of fish harvested. Careful analyses of fisheries data have revealed in case after case that as fishing increases, the size and age of fish caught decline. In addition, as particular species become too rare to fish profitably, fleets begin targeting other species that are in greater abundance. Generally this means shifting from large, desirable species to smaller, less desirable ones. Fleets have time and again depleted popular food fish such as cod and snapper and shifted their emphasis to species that were previously of lower value. Because this often entails catching species at lower trophic levels, this phenomenon has been termed "fishing down the food chain."

Some fishing practices kill nontarget animals and damage ecosystems

Some fishing practices catch more than just the species they target. **By-catch** refers to the capture of animals not meant to be caught, and it accounts for the deaths of many thousands of fish, sharks, marine mammals, and birds each year.

Boats that drag huge *driftnets* through the water (Figure 16.23a) capture substantial numbers of dolphins, seals, and sea turtles, as well as countless nontarget fish. Most of these end up drowning (mammals and turtles need to surface to breathe) or dying from air exposure on deck (fish breathe through gills in the water). Many nations have banned or restricted driftnetting because of excessive by-catch. The widespread death of dolphins in driftnets also motivated consumer efforts to label tuna as "dolphin-safe" if its capture uses methods designed to avoid dolphin by-catch. Such measures helped reduce dolphin deaths from an estimated 133,000 per year in 1986 to less than 2,000 per year since 1998.

Similar by-catch problems exist with *longline fishing* (Figure 16.23b), which involves dragging extremely long lines with baited hooks spaced along their lengths. Longline fishing kills turtles, sharks, and many albatrosses, magnificent seabirds with wingspans up to 3.6 m (12 ft). It is estimated that 300,000 seabirds of various species die each year from longline fishing.

Other fishing practices can directly damage entire ecosystems. *Bottom-trawling* (Figure 16.23c) involves dragging weighted nets over the floor of the continental shelf to catch such benthic organisms as scallops and groundfish. Trawling crushes many organisms in its path and leaves long swaths of damaged sea bottom. It is especially destructive to structurally complex areas, such as reefs, that provide shelter and habitat for many animals. Only in recent years has underwater photography begun to reveal the extent of structural and ecological disturbance done by trawling.

Consumer choice can influence fishing practices

To most of us, marine fishing practices may seem a distant phenomenon over which we have no control. Yet by exercising careful choice when we buy seafood, consumers can influence the ways fisheries function. Purchasing ecolabeled seafood products such as dolphin-safe tuna is one

(a) Driftnetting

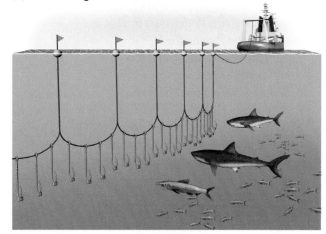

(b) Longlining

(c) Bottom-trawling

FIGURE 16.23 Commercial fishing fleets use several methods of capture. In driftnetting (**a**), huge nets are dragged through the open water to capture schools of fish. In longlining (**b**), lines with numerous baited hooks are pulled through the open water. In bottom-trawling (**c**), weighted nets are dragged along the floor of the continental shelf. All methods result in large amounts of by-catch, or capture of nontarget animals. Bottom-trawling can also result in severe structural damage to reefs and benthic habitats.

The Science behind the Story

China's Fisheries Data

China is responsible for a larger share of the world's fisheries catch than any other nation, according to the United Nations Food and Agriculture Organization (FAO). Thus, China's catch data has a major impact on the FAO's attempts to assess the health of global fisheries. In 2001, two fisheries scientists published a paper suggesting that China had exaggerated its catch data by as much as 100% during the 1990s. The inflated catch data had led to complacency among the world's fisheries managers and policymakers, these scientists argued, because it led the FAO to overestimate the true amount of fish left in the ocean.

The authors of the paper, University of British Columbia researchers Reg Watson and Daniel Pauly, had become suspicious of China's data for several reasons. First, China's coastal fisheries had been overexploited for decades, yet reported catches continued to increase. Second, China's "catch per unit effort" remained unchanged from 1980 to 1995, even though the abundance of fish had decreased. Finally, China's catch statistics suggested that its coastal waters were far more productive than ecologi-

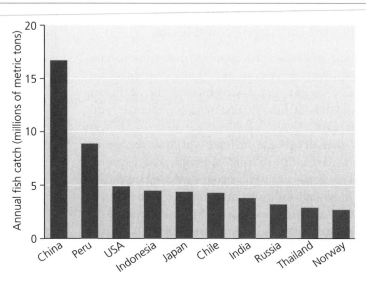

China's reported fish catches in recent years have dwarfed those of other nations, but a recent study asserted that Chinese fisheries managers had been falsely inflating their catches. Data from U.N. Food and Agriculture Organization, 2004.

cally similar areas fished by other nations.

In the November 29, 2001, issue of the journal *Nature,* Watson and Pauly reported the results of their study testing the hypothesis that China's apparent productivity was unrealistically high. Using a number of different databases, including FAO catch data collected since the 1950s, they built a statistical model to predict global fisheries catch. They divided the world's oceans into approximately 180,000 "cells," each

spanning half a degree longitudinally and latitudinally. The cells were then characterized by factors that affect catch size, such as depth, primary productivity, fishing rights, and species distribution. The researchers then used these factors to predict total annual catch for each cell.

For most cells, the catch predicted by the model was similar to the catch reported by fleets that fished those waters. In large swaths of China's coastal waters, however, predicted catches were far smaller than

---

way to exercise choice, but in most cases consumers have no readily available information about how their seafood was caught. Thus, several nonprofit organizations have recently devised concise guides to help consumers make choices. These guides differentiate fish and shellfish that are overfished or whose capture is ecologically damaging from those that are harvested more sustainably. For instance, the Monterey Bay Aquarium provides a wealth of this information on its Web site.

Marine Conservation Biology

Because we bear responsibility and stand to lose a great deal if valuable ecological systems collapse, marine scientists have been working to develop solutions to the problems that threaten the oceans. Many have begun by taking a hard look at the strategies used traditionally in fisheries management.

reported catches. For instance, the model predicted that in 1999 China's total catch would be 5.5 million metric tons, whereas the figure reported by Chinese fisheries authorities was 10.1 million metric tons.

When Watson and Pauly proceeded to analyze global fisheries trends using the model's predictions rather than China's official numbers, they found that the total annual global catch had decreased by 360,000 metric tons since 1988, instead of increasing by 330,000 metric tons, as reported by the FAO. Their finding suggested that the global catch, rather than remaining stable through the 1990s, actually had begun to decline in the 1980s.

Fisheries managers and policymakers depend on the FAO for information about the state of the world's fisheries. By using China's apparently inflated numbers, Watson later wrote, the FAO had encouraged a global "mopping-up operation" in which the last of the world's fish were being decimated.

The Chinese government and the FAO both criticized the study. China's reported catch was "basically correct," said Yang Jian, director-general of the Chinese Bureau of Fisheries. Watson and

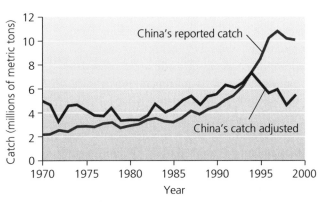

Fisheries biologists used models to estimate China's fish harvests, and found estimated amounts to be much lower than reported amounts. Data from Watson, R., and D. Pauly. 2001. Systematic distortions in world fisheries catch trends. *Nature* 414: 536–538.

Pauly's model, he suggested, failed to take into account unique aspects of China's fisheries, such as its large catch of crabs and jellyfish. Yang also noted that China had already begun to address the problems of overfishing and overreporting that did exist. In 1998, for instance, the Chinese government had promoted a "zero-growth policy" that capped China's total catch at 1998 levels.

For its part, the FAO noted that it was already treating China's statistics with caution. In its recent publications, global fisheries catches were being reported with and without China's data (see Figure 16.20). The FAO had also held several meetings

with Chinese officials to discuss ways of reducing overreporting. Moreover, the FAO suggested, nothing in its reports had encouraged complacency. Although it had indeed reported that the global catch remained stable through the 1990s, it also noted that many individual fisheries had declined or collapsed and that others were being aggressively overfished.

Assigning blame for the state of the world's fisheries is not the most productive response to Watson and Pauly's findings, says fisheries scientist Andy Rosenberg of the University of New Hampshire. "This is a global problem, not a case of a few bad actors."

Fisheries management has been based on maximum sustainable yield

For decades, fisheries management has sought to use scientific assessments to ensure sustainable harvests. Historically, fisheries managers have studied fish population biology and used that knowledge to regulate the timing of harvests, the techniques used to catch fish, and the scale of harvests. The goal was to allow for maximal harvests of

particular populations while keeping fish available for the future—the concept of *maximum sustainable yield* (▶ pp. 345–346). If data indicated that current yields were unsustainable, managers might limit the number or total mass of that fish species that could be harvested or restrict the type of gear fishers could use.

Despite such efforts, a number of fish stocks have plummeted, and many marine scientists and managers now feel it is time to rethink fisheries management. One key change these reformers suggest is to shift the focus away from the individual fish species and toward viewing marine resources as elements of larger ecological systems. This means considering the effects of fishing practices on habitat quality, on interspecific interactions, and on other factors that may have indirect or long-term effects on populations. One key aspect of such an *ecosystem-based approach* is to set aside areas of ocean where systems can function without human interference.

We can protect areas in the ocean

Large numbers of **marine protected areas (MPAs)** have been established, most of them along the coastlines of developed countries. The United States now contains nearly 300 federally managed MPAs. However, despite the name, marine protected areas do not necessarily protect their natural resources, because nearly all MPAs allow fishing or other extractive activities. As a recent report from an environmental advocacy group put it, even national marine sanctuaries "are dredged, trawled, mowed for kelp, crisscrossed with oil pipelines and fiber-optic cables, and swept through with fishing nets."

Because of the lack of true refuges from fishing pressure, many scientists—and some fishers—now want to establish areas where no fishing is allowed. Such "no-take" areas have come to be called **marine reserves.** Designed to preserve entire ecosystems intact without human interference, marine reserves are also intended to improve fisheries. Scientists have argued that marine reserves can act as production factories for fish for surrounding areas, because fish larvae produced inside reserves will disperse outside and stock other parts of the ocean. By serving both purposes, proponents argue, marine reserves are a win-win proposition for environmentalists and fishers alike.

Weighing the **Issues:**
Preservation on Land and at Sea

Almost 4% of U.S. land area is designated as wilderness, yet far less than 1% of coastal waters are protected in reserves. Why do you think it is taking so long for the preservation ethic to make the leap to the oceans?

Marine reserves have met forceful opposition

Many fishers don't like the idea of no-take reserves, however. Nearly every marine reserve proposed has met with pockets of intense opposition from people and businesses who use the area for fishing or recreation. Opposition comes from industrial fishing fleets that fish commercially as well as from individuals who fish recreationally. Both types of fishers are concerned that marine reserves will simply put more areas off-limits to fishing.

In the Florida Keys, property rights advocates who had opposed the sanctuary protested the 1998 establishment of reserve zones where fishing was to be prohibited. They rallied citizen opposition in local newspapers and filed a $27 million lawsuit against the federal government. In other parts of the world, such protests have become violent. Fishermen in the Galapagos Islands have rioted, looted, and destroyed the administration building at Galapagos National Park to protest fishing restrictions.

However, people in the Florida Keys have come to see that the reserves have not greatly threatened their access to areas. They have also witnessed improvements in the marine life around them. Surveys just 3 years after establishment of the no-take zones showed increases in the size and number of spiny lobsters and in the populations of three of four major reef fish. Keys fishers are catching more fish, and fishing revenues are up. As a result, many opponents of the reserve system have become supporters. Reserve workers estimated in 2000 that 70% of Keys residents supported the no-take reserves and that over 50% would support establishing more of them.

Reserves can work for both fish and fishers

In the past decade, data synthesized from marine reserves around the world have been indicating that reserves *do* work as win-win solutions that benefit ecosystems, fish populations, and fishing economies. In 2001, 161 prominent marine scientists signed a "consensus statement" summarizing the effects of marine reserves. Besides boosting fish biomass, total catch, and record-sized fish, the report stated, marine reserves yield several benefits. Within reserve boundaries, they

▶ Produce rapid and long-term increases in abundance, diversity, and productivity of marine organisms.
▶ Decrease mortality and habitat destruction.
▶ Lessen the likelihood of extirpation of species.

Outside reserve boundaries, marine reserves

▶ Can create a "spillover effect" when individuals of protected species spread outside reserves.
▶ Allow larvae of species protected within reserves to "seed the seas" outside reserves.

VIEWPOINTS

Marine Reserves

Can marine reserves or other forms of "no-fishing" zones help us solve problems facing the oceans today? Why or why not?

"No-Fishing" Zones Do Not Prevent Overfishing

If you close off part of the ocean from all forms of "take," there will obviously be more fish in the protected area. That is just common sense. However, a major problem with our oceans is severe overfishing. Fish are being killed faster than they can replenish themselves. There is a simple way to improve this situation—stop killing so many fish.

"No-fishing" zones don't accomplish this. What they do is to shift the fishing pressure to another area. Anything short of reducing the number of fish being killed, such as marine reserves, not only won't solve the problem, but also may give a false impression that something is being done, thus delaying action to actually solve the problem.

There are many traditional fishery management tools that can be used to prevent overfishing. These tools include bag limits, size limits, slot limits, closed seasons, protected species, and catch-and-release-only species. For commercial fishermen there are quotas, gear restrictions, trip limits, and limited entry, plus all the management tools listed above for recreational anglers.

A major problem for fishery managers has been politics. Commercial fishing interests have very effective lobbyists who have been instrumental in preventing or delaying needed management restrictions. If lobbying fails, they often challenge regulations through the court system. The result is management regulations that are not restrictive enough. Marine reserves will not solve this problem.

In cases where remote, pristine areas are meant to be preserved, perhaps marine reserves allowing only catch and release could be effective. Examples of this are the remote reefs in northern Hawaii and some snapper spawning areas far west of Key West, Florida. However, "no-fishing" zones by themselves will not rebuild fisheries.

Michael Leech joined the International Game Fish Association (IGFA) in 1983 and became president of IGFA in 1992. Under his leadership, the IGFA recently constructed their permanent world headquarters in Dania Beach, Florida. The IGFA Fishing Hall of Fame and Museum not only is a tourist attraction but also serves as the world center for most fishing-related information.

Marine Reserves Restore Ecosytems

Marine reserves are a powerful tool for protecting and restoring marine ecosystems. They are successful because they protect not only species but also habitats. In the past, there were innumerable naturally recurring marine reserves around the oceans—places that were too far from land, too deep, or too rocky to fish. Modern technology systematically eliminated those reserves, and today we protect far less than 1% of the oceans in established reserves.

Recent scientific studies have demonstrated that reserves provide a clear benefit to conservation of marine organisms: Biomass, density, individual size, and diversity all increase inside reserves. It is especially important to protect large marine organisms, which produce disproportionately more offspring than smaller ones. A 37-cm (14.6-in.) vermilion rockfish produces 150,000 young, whereas a 60-cm (23.6-in.) one produces 1.7 million young! Allowing individuals to get big and fat is very valuable.

Modeling results indicate that reserves substantially benefit fisheries, from both spillover (fish spilling from the reserve) and export (larvae produced inside the reserve and transported away by currents). However, not all species will recover immediately when a reserve is established. Species that grow slowly and reproduce late will need longer to colonize and recover.

Networks of marine reserves, connected by the movement of organisms, are likely to provide the best combination of conservation and fishery benefit. A network provides protection for a large total area and a long perimeter over which organisms can escape to reseed adjacent areas. Networks of reserves are not a panacea—they need to be coupled with good fishery management and pollution control—but marine reserves are one of the most promising tools available for solving the problems plaguing ocean environments today.

Jane Lubchenco is a professor of marine biology and zoology at Oregon State University. She works with policymakers, business leaders, private foundations, religious leaders, other scientists, governmental and nongovernmental organizations, and students to help figure out how to make a transition to sustainability. She is president of the International Council for Science and co-founded the Aldo Leopold Leadership Program, PISCO (Partnership for Interdisciplinary Studies of Coastal Oceans), and COMPASS (Communication Partnership for Science and the Sea).

Explore this issue further by accessing **Viewpoints** at www.aw-bc.com/withgott.

The Science behind the Story

Do Marine Reserves Work?

In November 2001, a team of fisheries scientists published a paper in the journal *Science,* providing some of the first clear evidence that marine reserves can benefit nearby fisheries. The team, led by York University researcher Callum Roberts, focused on reserves off the coasts of Florida and the Caribbean island of St. Lucia.

Following the establishment in 1995 of the Soufrière Marine Management Area (SMMA), a network of reserves intended to help restore St. Lucia's severely depleted coral reef fishery, Roberts and his colleague, Julie Hawkins, conducted annual visual surveys of fish abundance in the reserves and nearby areas. Within 3 years, they found that the biomass of five commercially important families of fish— surgeonfishes, parrot fishes, groupers, grunts, and snappers— had tripled inside the reserves and doubled outside them. Roberts and Hawkins also interviewed local

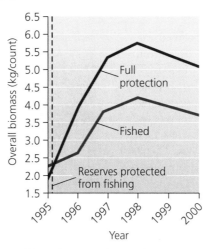

Established in 1995, the Soufrière Marine Management Area (SMMA), along the coast of St. Lucia, had a rapid impact. By 1998, fish biomass within the five reserves tripled and in adjacent, fished areas, it doubled. Data from Roberts, C., et al. 2001. Effects of marine reserves on adjacent fisheries. *Science* 294: 1920–1923.

fishers and found that those with large traps were catching 46% more fish per trip in 2000–2001 than they had in 1995–1996 and that fishers with small traps were catching 90%

more. Roberts and his colleagues concluded that "in 5 years, reserves have led to improvement in the SMMA fishery, despite the 35% decrease in area of fishing grounds."

Roberts and his coworkers also studied the oldest fully protected marine reserve in the United States, the Merritt Island National Wildlife Refuge (MINWR), established in 1962 as a buffer around what is today the Kennedy Space Center on Cape Canaveral, Florida. In a previous study, Darlene Johnson and James Bohnsack of the National Oceanic and Atmospheric Administration and Nicholas Funicelli of the United States Geological Service had found that the reserve contained more and larger fish than did nearby unprotected areas. This team also found that some of the reserve's fish appeared to be migrating to nearby fishing areas.

Bohnsack, Roberts, and their colleagues corroborated the evidence for migration by analyzing trophy records from the International

The consensus statement was backed up by research into reserves in the Caribbean and Florida (see "The Science behind the Story," above) and others worldwide. At Apo Island in the Philippines, biomass of large predators increased eightfold inside a marine reserve, and outside the reserve fishing improved. At two coral reef sites in Kenya, commercially fished and keystone species were up to 10 times more abundant in the protected area as in the fished area. At Leigh Marine Reserve in New Zealand, snapper increased 40-fold, and spiny lobsters were increasing by 5–11% yearly. Spillover from this reserve improved fishing and ecotourism, and—as in

Florida— local residents who once opposed the reserve now support it.

The review of data from existing marine reserves as of 2001 revealed that just 1–2 years after their establishment, marine reserves

▶ Increased densities of organisms on average by 91%.
▶ Increased biomass of organisms on average by 192%.
▶ Increased average size of organisms by 31%.
▶ Increased species diversity by 23%.

From data like these and considerations of socioeconomic factors, the scientific consensus statement concluded that

Game Fish Association. They found that the proportion of Florida's record-sized fish caught near Merritt Island increased significantly after 1962. Nine years after the refuge was established, for instance, the number of spotted sea trout records from the Merritt Island area jumped dramatically. Bohnsack, Roberts, and their colleagues hypothesized that the reserve was providing a protected zone in which fish could grow to trophy size before migrating to nearby areas, where they were caught by recreational fishers.

Not everyone saw the St. Lucia and Merritt Island cases as proof that marine reserves could rescue depleted fisheries. In February 2002, several alternative interpretations were published as letters in *Science.* Mark Tupper, a fisheries scientist at the University of Guam, suggested that the St. Lucia results were relevant only to coral reef fisheries in developing nations, whereas Florida's boost in fish populations

was due primarily to limits on recreational fishing. Karl Wickstrom, editor-in-chief of *Florida Sportsman* magazine, suggested that the increase in trophy fish near MINWR was caused by commercial fishing regulations and changes in how trophies were recorded and promoted. And Ray Hilborn, a fisheries scientist at the University of Washington, challenged the study's scientific methods. In the St. Lucia case, he pointed out, there had been no control condition.

In response, Roberts and his colleagues reaffirmed the validity of their results while acknowledging some limitations. They agreed with Tupper that marine reserves are not always effective and often need to be complemented by other management tools, such as size limits. "We agree that inadequately protected reserves are useless," they wrote, "but our study shows that well-enforced reserves can be extremely effective and can play a critical role in achieving sustainable fisheries."

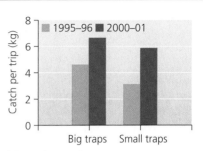

(a) Catch per trip

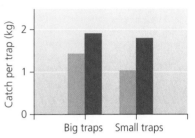

(b) Catch per trap

Roberts and his colleagues studied biomass of fish caught at the SMMA over two 5-month periods in 1995–1996 and 2000–2001. For fishers with big traps, catch increased by 46%, and for those with small traps, it increased by 90%. Per trap, catch increased by 36% for big traps and by 80% for small traps. Data from Roberts, C., et al. 2001. Effects of marine reserves on adjacent fisheries. *Science* 294: 1920–1923.

existing scientific information justifies the immediate application of fully protected marine reserves as a central management tool.

How should reserves be designed?

If marine reserves work in principle, the question becomes how best to design reserves and arrange them into networks. Scientists today are asking how large reserves need to be, how many there need to be, and where they need to be placed. Involving fishers directly in the planning process is crucial for coming up with answers to such

questions. Of the 40 studies that have estimated how much area of the ocean should be protected in no-take reserves, estimates range from 10% to 65%, with most falling between 20% and 50%. Other studies are modeling how to optimize the size and spacing of individual reserves so that ecosystems are protected, fisheries are sustained, and people are not overly excluded from marine areas (Figure 16.24). If marine reserves are designed strategically to take advantage of ocean currents, many scientists say, then they may well seed the seas and help lead us toward solutions to one of our most pressing environmental problems.

Effects of Different–Sized Marine Reserves				
	Consequences for conservation	Consequences for small fisheries	Consequences for commercial fisheries	Overall consequences
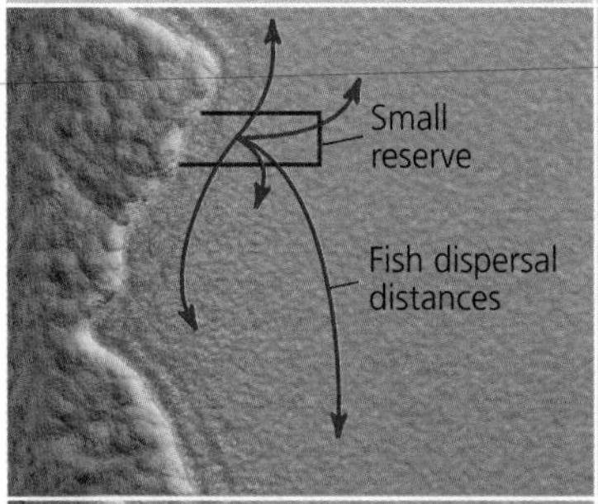 Small reserve / Fish dispersal distances	Reserve is not self-sustaining; most species are lost.	High periphery-to-area ratio makes many fish available to fishers outside reserve. This is unsustainable, however, because small reserves offer too little protection for fish populations.	Provides little or no boost to fish populations outside reserve, but does not appreciably decrease area of fishing grounds.	Too many fish wander out of reserve, yet fisheries experience little gain. Many populations are not sustained. Reserve fails to meet any stakeholder goals.
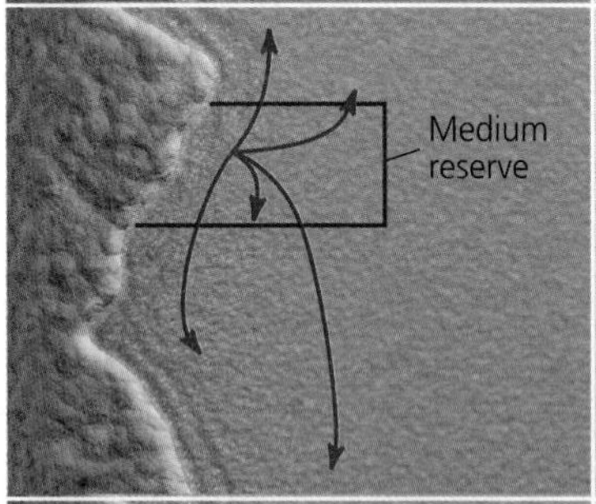 Medium reserve	Reserve is moderately self-sustaining; some species are lost.	Intermediate periphery-to-area ratio makes some fish available to fishers outside reserve, while offering a large enough area of refuge to protect other fish, thus sustaining many populations.	Provides a significant boost to fish populations outside reserve, with only a moderate decrease in area of fishing grounds.	Conservation is effective inside reserve, and fisheries gain fish while not losing too much fishing area. Good balance of benefits for all stakeholders.
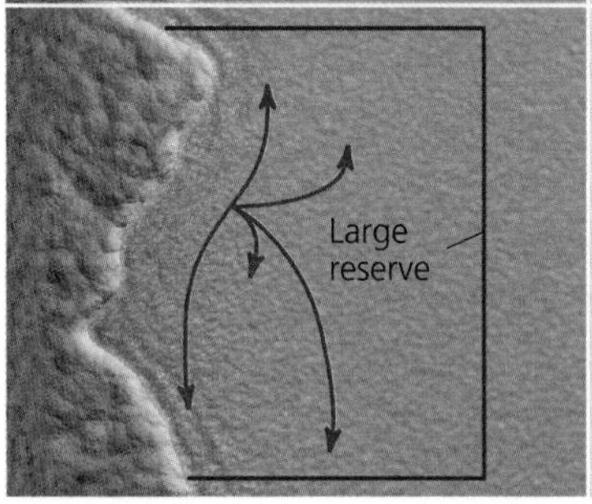 Large reserve	Reserve is completely self-sustaining; all species are retained.	Low periphery-to-area ratio makes relatively few fish available to fishers outside reserve; the "spillover" for fisheries is small in relation to the area protected in the reserve.	Provides relatively little boost to fish populations in regions around reserve, and severely decreases area of fishing grounds.	Preservation is effective inside reserve, but too few fish spill over into fishing grounds that have been much reduced in area. Outcome not acceptable to fisheries stakeholders.

FIGURE 16.24 Marine reserves of different sizes may have varying effects on ecological communities and fisheries. Young and adult fish and shellfish of different species can disperse different distances, as indicated by the red arrows in the figure. A small reserve may fail to protect animals because too many disperse out of the reserve. A large reserve may protect fish and shellfish very well but will provide relatively less "spillover" into areas where people can legally fish. Thus medium-sized reserves may offer the best hope of preserving species and ecological communities while also providing adequate fish to fishermen and human communities. Determining the actual size of such reserves requires a great deal of data on dispersal distances, water currents, community composition, animal behavior, and human economies. *Source:* Halpern, B. S., and R. R. Warner. 2003. Matching marine reserve design to reserve objectives. *Proceedings of the Royal Society of London B: Biological sciences* 270: 1871–1878.

Conclusion

Oceans cover most of our planet and contain diverse topography and ecosystems, some of which we are only now beginning to explore and understand. We are learning more about the oceans and coastal environments while we are intensifying our use of their resources. In so doing, we are coming to understand better how to use these resources without depleting them or causing undue ecological harm. In the Florida Keys and hundreds of other areas, scientists are demonstrating that setting aside protected areas of the ocean can serve to maintain natural systems and also to enhance fisheries. As historical studies reveal more information on how much biodiversity our oceans formerly contained and have now lost, we may increasingly look beyond simply making fisheries stable and instead consider restoring the ecological systems that used to flourish in our waters.

REVIEWING OBJECTIVES

You should now be able to:

Identify physical, geographical, chemical, and biological aspects of the marine environment

▶ Oceans cover 71% of Earth's surface and contain over 97% of its surface water. (p. 468)

▶ Ocean water contains 96.5% H_2O by mass and various dissolved salts. (p. 468–469)

▶ Colder, saltier water is denser and sinks. Water temperatures vary with latitude, and temperature variation is greater in surface layers. (pp. 469–470)

▶ Persistent currents move horizontally through the oceans, driven by density differences, sunlight, and wind. (pp. 470–471)

▶ Vertical water movement includes upwelling and downwelling, which affect the distribution of nutrients and life. (p. 471)

▶ Seafloor topography can be complex and rugged. (pp. 471–472)

Describe major types of marine ecosystems

▶ Major types of marine and coastal ecosystems include pelagic and deep-water open ocean systems, kelp forest, coral reefs, intertidal zones, salt marshes, mangrove forests, and estuaries. (pp. 473–478)

▶ Many of these systems are highly productive and rich in biodiversity. Many also suffer heavy impacts from human influence. (pp. 473–478)

Outline historic and current human uses of marine resources

▶ For millennia, people have drawn resources from the oceans and used ocean waters for transportation. (p. 478)

▶ Today we extract energy and minerals from the oceans, as well as using them for transportation. (pp. 478–479)

Assess human impacts on marine environments

▶ People pollute ocean waters with trash, untreated sewage, petroleum spills, plastic that harms marine life, and nutrient pollution that leads to harmful algal blooms. (pp. 479–481)

▶ Overharvesting is perhaps the major human impact on marine systems. (p. 482)

Review the current state of ocean fisheries and reasons for their decline

▶ Half the world's marine fish populations are fully exploited, 25% are already overexploited, and 25% can yield more without declining. (p. 482)

▶ People began depleting marine resources long ago, but impacts have intensified in recent decades. (p. 483)

▶ Global fish catches have stopped growing since the late 1980s, despite increased fishing effort and improved technologies. (p. 482, 484)

▶ Today's oceans hold only one-tenth the number of large animals that they did before the advent of industrialized commercial fishing. (pp. 483–484)

▶ As fishing intensity increases, the fish available become smaller. (pp. 484–485)

▶ Fishing practices such as driftnetting, longline fishing, and trawling capture nontarget organisms, called bycatch. (p. 485)

▶ Traditional fisheries management has not stopped declines, so many scientists feel that ecosystem-based management is needed. (pp. 487–488)

Evaluate marine protected areas and reserves as innovative solutions

▶ We have established fewer protected areas in the oceans than we have on land and most marine protected areas allow many extractive activities. (p. 488)

▶ No-take marine reserves can protect ecosystems while also boosting fish populations and making fisheries sustainable. (pp. 488–491)

TESTING YOUR COMPREHENSION

1. What proportion of Earth's surface do oceans cover? About how much salt does ocean water contain? How are water density, salinity, and temperature related in each layer of ocean water?

2. What factors drive the system of ocean currents? In what directions do ocean currents move, and how do such movements affect conditions for life in the oceans?

3. Where in the oceans are productive areas of biological activity likely to be found?

4. Describe three kinds of ecosystems found near coastal areas and the kinds of life they support.

5. Why are coral reefs biologically valuable? What are some possible ways in which they are being degraded by human impact? What is causing the disappearance of mangrove forests and salt marshes?

6. Describe three major forms of pollution in the oceans and the consequences of each.

7. Provide an example of how overfishing can lead to ecological damage and fishery collapse.

8. Explain the conclusion of the Myers and Worm study of 2003 (see Figure 16.21).

9. Name three industrial fishing practices that create bycatch and harm marine life, and explain how they do so.

10. How does a marine protected area differ from a marine reserve? Why do many fishers oppose marine reserves? Explain why many scientists say no-take reserves will be good for fishers.

SEEKING SOLUTIONS

1. What benefits do you derive from the oceans? How does your behavior affect the oceans? Give specific examples.

2. We have been able to reduce the amount of oil we spill into the oceans, but petroleum-based products such as plastic continue to litter our oceans and shorelines. Discuss some ways that we can reduce this threat to the marine environment.

3. What factors account for the trends in global fish capture over the past 20 years?

4. Consider what you know about biological productivity in the oceans, about the scientific data on marine reserves, and about the social and political issues surrounding the establishment of marine reserves. What ocean regions do you think it would be particularly appropriate to establish as marine reserves? Why?

5. Why does the 2001 scientific consensus statement argue for networks of reserves to be established? What role

should science and scientists play in this kind of decision making? Discuss how you would engage a group of fishers, environmentalists, and scientists in a reserve-planning process.

6. You are mayor of a coastal town where some residents are employed as commercial fishers and others make a living serving ecotourists who come to snorkel and scuba-dive at the nearby coral reef. In recent years, several fish stocks have crashed, and ecotourism is dropping off as fish disappear from the increasingly degraded reef. Scientists are urging you to help establish a marine reserve around portions of the reef, but most commercial and recreational fishers are opposed to this idea. What steps would you take to restore your community's economy and environment?

INTERPRETING GRAPHS AND DATA

The accompanying graph presents trends in the status of ocean fisheries. Fully exploited fisheries (green line) are ones currently producing their maximum sustainable yield. Moderately exploited and underexploited fisheries (blue line) are ones that could be more heavily fished than they are and could produce larger, yet sustainable, yields. Overexploited, depleted, and recovering fisheries (red line) are ones that have been overfished.

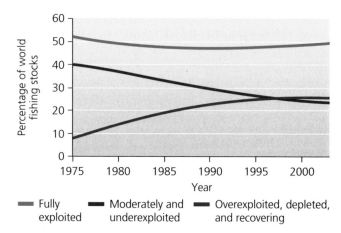

Global trends in the state of ocean fishing stocks from 1975 to 2003. Data from U.N. Food and Agriculture Organization (FAO), Fisheries Department. 2004. *The state of world fisheries and aquaculture: 2004.*

1. What is the sum of the percentages for any given year in the figure above? Explain why this is so.
2. Describe the trend from 1975 to 2003 in the percentage of fisheries categorized as overexploited, depleted, or recovering. Describe the trend from 1975 to 2003 in the percentage of fisheries categorized as moderately exploited or underexploited.
3. Based on the data in the graph, predict the likely trend in overall ocean fisheries production over the next 25 years. Explain your answer.

CALCULATING ECOLOGICAL FOOTPRINTS

The relationship between the ecological goods and services used by individuals and the amount of *land* area needed to provide those goods and services is relatively well developed. People also use goods and services from Earth's oceans, where the concept of *area* is less useful. It is clear, however, that our removal of fish from the oceans has an impact, or an ecological footprint, on remaining fish populations.

The table shows data on the mean annual per capita consumption from ocean fisheries for North America, China, and the world as a whole. Using the data provided, calculate the amount of fish each consumer group would consume at the annual per capita consumption rates for each of these three regions, and record your results in the table.

	Annual per capita consumption rate		
Consumer group	North America 21.6 kg	China 27.7 kg	World 16.2 kg
You			
Your class			
Your state			
United States	6.48×10^9 kg	8.31×10^9 kg	4.86×10^9 kg
World			

Data from U.N. Food and Agriculture Organization (FAO), Fisheries Department. 2004. *The state of world fisheries and aquaculture: 2004.* Data are for 2002, the most recent year for which comparative data are available.

1. Calculate the ratio of North America's per capita fish consumption rate to that of the world. Compare this ratio to the ratio of the per capita ecological footprints for the United States, Canada, and Mexico (see Figure 1.13, ▶ p. 19) versus the world average footprint of 2.2 ha/person/year. Can you account for similarities and differences between these ratios?
2. The population of China has grown at an annual rate of 1.1% since 1987, while over the same period fish consumption in China has grown at an annual rate of 8.9%. Speculate on the reasons behind China's rapidly increasing consumption of fish.
3. What ecological concerns do the combined trends of human population growth and increasing per capita fish consumption raise for you? What role might you play in contributing to these concerns or to their solutions?

Take It Further

Go to www.aw-bc.com/withgott or the student CD-ROM where you'll find:

▶ Suggested answers to end-of-chapter questions
▶ Quizzes, animations, and flashcards to help you study
▶ *Research Navigator*™ database of credible and reliable sources to assist you with your research projects

▶ **GRAPHit!** tutorials to help you master how to interpret graphs
▶ **INVESTIGATEit!** current news articles that link the topics that you study to case studies from your region to around the world

17 Atmospheric Science and Air Pollution

View of Earth's
atmosphere from
space

Upon successfully completing this chapter, you will be able to:

▶ Describe the composition, structure, and function of Earth's atmosphere

▶ Outline the scope of outdoor air pollution and assess potential solutions

▶ Explain stratospheric ozone depletion and identify steps taken to address it

▶ Define acidic deposition and illustrate its consequences

▶ Characterize the scope of indoor air pollution and assess potential solutions

London • Europe
Atlantic Ocean
Africa

Central Case: The 1952 "Killer Smog" of London

"You had this swirling, like somebody had set a load of car tires on fire."
—STAN CRIBB, EYEWITNESS TO THE DECEMBER 1952 "GREAT SMOKE" OF LONDON

"The modern field of environmental health owes much to the tragedy that befell Greater London some 50 years ago."
—DEVRA L. DAVIS, MICHELLE L. BELL, AND TONY FLETCHER, IN AN EDITORIAL FOR THE JOURNAL ENVIRONMENTAL HEALTH PERSPECTIVES

December 5, 1952, was a particularly cold Friday in London, and many residents stoked their coal stoves to keep the chill away. Few people took notice when thick, foul-smelling smog, a fog polluted with smoke and chemical fumes, first settled over the city. After all, London had been famous for its "pea souper" fogs for well over a century.

But the smog that December weekend was particularly thick; the city's air quality was ten times worse than usual for that year. Visibility was so poor that pedestrians could not see across the street.

Transportation came to a standstill, and roads became clogged with abandoned cars. Schools closed, cattle asphyxiated, and an opera was halted because the audience could not see the stage.

A wind finally relieved Londoners of the miserable smog on Tuesday, December 9. By that time, however, authorities estimated that over 4,000 people had died, mostly from lung ailments such as bronchitis that were induced or aggravated by the pollution. A 2002 study by American researchers estimated that the actual death toll, including delayed cases that appeared over the next 2 months and were considered unrelated at the time, may have been as high as 12,000.

The "killer smog" of 1952 was remarkable, but hardly unique. Similar, although less severe, phenomena had occurred in London as early as 1813 and again in 1873, 1880, 1891, and 1948. Other such events have taken lives in Pennsylvania, New York, Mexico, and Malaysia. London's killer smog, together with other severe smog events, helped change the way the public viewed air pollution. Before the 1950s, most people considered urban smog a necessary burden. Today, we recognize the importance of clean air and view air pollution as an environmental challenge that can be overcome.

We have overcome much already; declines in air pollution represent some of the biggest successes of environmental policy. These declines have been due largely to limits on emissions brought about through legislation, such as the British Clean Air Acts of 1956 and 1968 and the U.S. Clean Air Acts of 1970 and 1990. Today, the air in many U.S. cities is cleaner, and the concentration of airborne particles around London averages one-tenth that of the 1950s.

However, much remains to be done. In 2002, a London governmental body estimated that vehicle emissions contribute to the premature deaths of 380 city residents each year. Furthermore, many cities in developing nations that have increasing populations, older technologies, and lax legislation and enforcement face conditions similar to those of 1950s London. For example, in 1995 air pollution in Delhi, India, was measured at 1.3 times London's average for the year 1952, and air pollution in Lanzhou, China, was measured at 2.7 times London's 1952 level. To understand what caused London's killer smog and how we can prevent recurrences of such events, we must examine both natural atmospheric conditions and human-made pollutants.

Atmospheric Science

Every breath we take reaffirms our connection to the **atmosphere,** the thin layer of gases that surrounds Earth. We live at the bottom of this layer, which provides us oxygen, absorbs hazardous solar radiation, burns up incoming meteors, transports and recycles water and nutrients, and moderates climate.

The atmosphere consists of roughly 78% nitrogen gas (N_2) and 21% oxygen gas (O_2). The remaining 1% is composed of argon gas (Ar) and minute concentrations of several other *permanent gases* that remain at stable concentrations. The atmosphere also contains a number of *variable gases* that vary in concentration from time to time or place to place, as a result of natural processes or human activities on Earth's surface (Figure 17.1).

Over Earth's long history, the atmosphere's chemical composition has changed. Oxygen gas began to build up in an atmosphere dominated by carbon dioxide (CO_2), nitrogen, carbon monoxide (CO), and hydrogen (H_2) about 2.7 billion years ago, with the emergence of

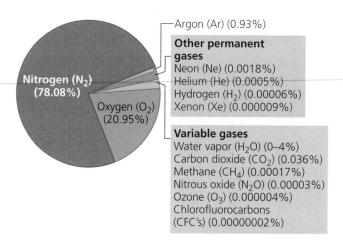

FIGURE 17.1 Earth's atmosphere consists mostly of nitrogen, secondarily of oxygen, and lastly of a mix of gases at dilute concentrations. Permanent gases are fixed in concentration. Variable gases vary in concentration as a result of either natural processes or human activities. Data from Ahrens, D.C. 1998. *Essentials of meteorology*, 2nd ed. New York: Wadsworth.

autotrophic microbes that emitted oxygen as a by-product of photosynthesis (▶ p. 108). Just as these early organisms had a substantial impact on the Earth's atmosphere long ago, human activity is altering the quantities of some atmospheric gases today, such as carbon dioxide, methane (CH_4), and ozone (O_3). In this chapter and in Chapter 18, we will explore the atmospheric changes brought about by artificial pollutants, but we must first begin with an overview of Earth's atmosphere.

The atmosphere consists of several layers

The atmosphere that stretches so high above us and seems so vast is actually just a thin coating about 1/100th of Earth's diameter, like the fuzzy skin of a peach. This coating consists of four layers whose boundaries are not visible to the human eye, but which atmospheric scientists recognize by measuring differences in temperature, density, and composition (Figure 17.2).

The bottommost layer, the **troposphere,** blankets Earth's surface and provides organisms the air they need to live. The movement of air within the troposphere is also largely responsible for the planet's weather. Although it is thin (averaging 11 km [7 mi] high) relative to the atmosphere's other layers, the troposphere contains three-quarters of the atmosphere's mass, because air is denser near Earth's surface. On average, tropospheric air temperature declines by about 6 °C for each

FIGURE 17.2 Some aspects of the atmosphere change with altitude across its four layers. Temperature drops with altitude in the troposphere, rises with altitude in the stratosphere, drops in the mesosphere, then rises again in the thermosphere. The tropopause separates the troposphere from the stratosphere. Ozone reaches a peak in a portion of the stratosphere, giving rise to the term *ozone layer.* Adapted from Jacobson, M. Z. 2002. *Atmospheric pollution: History, science, and regulation.* Cambridge: Cambridge University Press; Parson, E. A. 2003. *Protecting the ozone layer: Science and strategy.* Oxford: Oxford University Press.

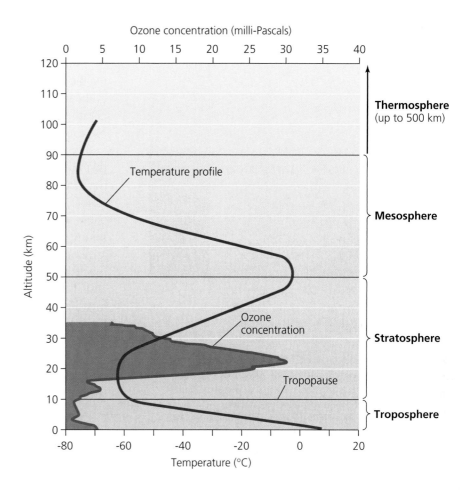

kilometer in altitude (or 3.5 °F per 1,000 ft), dropping to roughly −52 °C (−62 °F) at its highest point. At the top of the troposphere, however, temperatures cease to decline with altitude, marking a boundary called the *tropopause.* The tropopause acts like a cap, limiting mixing between the troposphere and the atmospheric layer above it, the stratosphere.

The **stratosphere** extends 11–50 km (7–31 mi) above sea level. Similar in composition to the troposphere, the stratosphere is 1,000 times drier and less dense. Its gases experience little vertical mixing, so once substances (including pollutants) enter it, they tend to remain for a long time. The stratosphere attains a maximum temperature of −3 °C (27 °F) at its highest altitude and becomes cooler in its lower reaches. The reason is that ozone and oxygen absorb and scatter the sun's ultraviolet (UV) radiation (▶p. 106), so that much of the UV radiation penetrating the upper stratosphere fails to reach the lower stratosphere. Most of the atmosphere's minute amount of ozone is concentrated in a portion of the stratosphere roughly

17–30 km (10–19 mi) above sea level, a region that has come to be called Earth's **ozone layer.** The ozone layer greatly reduces the amount of UV radiation that reaches Earth's surface. Because UV light can damage living tissue and induce DNA mutations, the ozone layer's protective effects are vital for life on Earth.

Above the stratosphere lies the *mesosphere,* which extends 50–90 km (31–56 mi) above sea level. Air pressure is extremely low here, and temperatures decrease with altitude, reaching their lowest point (−90 °C, or −130 °F) at the top of the mesosphere. From here, the *thermosphere* extends upward to an altitude of 500 km (300 mi). In the thermosphere, molecules are so few and far between that they collide only rarely. As a result, heavier molecules (such as nitrogen and oxygen) sink, and light ones (such as hydrogen and helium) end up near the top of the thermosphere. This stands in contrast to the atmosphere's lower three layers, where frequent collisions among molecules keep the chemical composition (see Figure 17.1) relatively constant throughout.

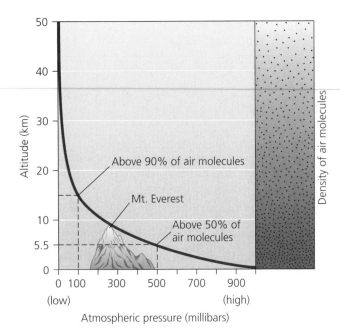

FIGURE 17.3   As one climbs higher through the atmosphere, gas molecules become less densely packed. As density decreases, so does atmospheric pressure. Because the vast majority of air molecules lie low in the atmosphere, one needs to be only 5.5 km (3.4 mi) high to be above half the planet's air molecules. Adapted from Ahrens, D. C. 1998. *Essentials of meteorology,* 2nd ed. New York: Wadsworth.

Atmospheric properties include temperature, pressure, and humidity

Although the lower atmosphere is stable in its composition, it is dynamic in its movement; air moves within it as a result of differences in the physical properties of air masses. Among these properties are pressure and density, relative humidity, and temperature. Gravity pulls gas molecules toward Earth's surface, causing air to be most dense near the surface and less so as altitude increases. **Atmospheric pressure,** which measures the force per unit area produced by a column of air, also decreases with altitude, because at higher altitudes fewer molecules are pulled down by gravity (Figure 17.3). At sea level, atmospheric pressure is 14.7 lb/in.² or 1,013 millibar (mb). Mountain climbers trekking to Mount Everest, the world's highest mountain, can look up and view their destination from Kala Patthar, a nearby peak, at roughly 5.5 km (18,000 ft). At this altitude, pressure is 500 mb, and half the atmosphere's air molecules are above the climber and half are below. A climber who reaches Everest's peak (8.85 km [29,035 ft]), where the "thin air" is just over 300 mb, stands above two-thirds of the molecules in the atmosphere. When we fly on a commercial jet airline, at a typical cruising altitude of 11 km (36,000 ft), we are above roughly 80% of the atmosphere's molecules.

Another important property of air is **relative humidity,** the ratio of water vapor a given volume of air contains to the maximum amount it *could* contain at a given temperature. Relative humidity varies considerably from place to place and time to time. Average daily relative humidity in June in Phoenix, Arizona, is only 31% (meaning that the air contains just less than a third of the water vapor it possibly can at its temperature), whereas on the tropical island of Guam, relative humidity rarely drops below 88%. People are sensitive to changes in relative humidity because we perspire to cool our bodies. When relative humidity is high, the air is already holding nearly as much water vapor as it can, so sweat evaporates slowly and the body cannot cool itself efficiently. This is why high humidity makes it feel hotter than it really is. Low humidity speeds evaporation and makes it feel cooler than it really is.

The temperature of air also varies with location and time. At the global scale, temperature varies over Earth's surface because the sun's rays strike some areas more directly than others. At more local scales, temperature varies because of topography, plant cover, proximity of land to water, and many other factors.

Solar energy heats the atmosphere, helps create seasons, and causes air to circulate

Energy from the sun influences temperatures by heating air in the atmosphere. Solar energy also drives the atmosphere's air movement, helps create seasons, and influences both weather and climate. An enormous amount of solar energy continuously bombards the upper atmosphere—over 1,000 watts/m², many thousands of times greater than the total output of electricity generated by human society. Of that solar energy, about 70% is absorbed by the atmosphere and planetary surface, while the rest is reflected back into space (see Figure 18.1, ▸ p. 531).

The spatial relationship between Earth and the sun determines the amount of solar radiation that strikes each point of Earth's surface. Sunlight is most intense when it shines directly overhead and meets the planet's surface at a perpendicular angle. At this angle, sunlight passes through a minimum of energy-absorbing atmosphere, and Earth's surface receives a maximum of solar energy per unit surface area. Conversely, solar energy that approaches Earth's surface at an oblique angle loses intensity as it traverses a longer distance through the atmosphere, and it is less concentrated when it reaches the surface.

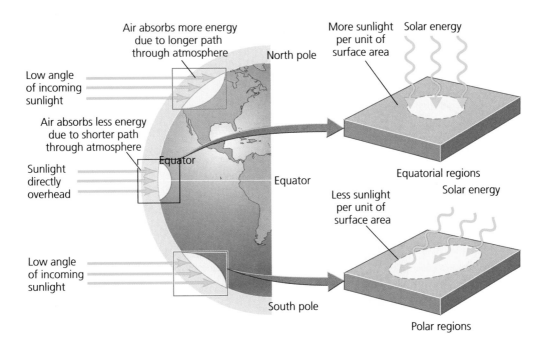

FIGURE 17.4 Because of Earth's curvature, polar regions receive on average less solar energy than equatorial regions. One reason is that sunlight gets spread over a larger area when striking the surface at an angle, a phenomenon that increases with distance from the equator. Another reason is that as sunlight approaches at a lower angle near the poles, it must traverse a longer distance through the atmosphere before reaching the surface, during which more energy is absorbed or reflected. These patterns represent year-round averages; the latitude at which radiation approaches the surface perpendicularly varies with the seasons (see Figure 17.5).

This is why, on average, solar radiation intensity is highest near the equator and weakest near the poles (Figure 17.4).

Because Earth is tilted on its axis (an imaginary line connecting the poles, running perpendicular to the equator) by about 23.5 degrees, the Northern and Southern Hemispheres each tilt toward the sun for half the year, resulting in the change in seasons (Figure 17.5). Regions near the equator are largely unaffected by this tilt; they experience about 12 hours each of sunlight and darkness every day throughout the year. Near the poles, however, the effect is strong, and seasonality is pronounced.

Land and surface water absorb solar energy, radiating some heat and causing some water to evaporate. Air near Earth's surface therefore tends to be warmer and moister than air at higher altitudes. These differences set into motion a process of **convective circulation** (Figure 17.6). Warm air, being less dense, rises and creates vertical currents. As air rises into regions of lower atmospheric pressure, it expands and cools. Once the air cools, it descends and becomes denser, replacing warm air that is rising. The air picks up heat and moisture near ground level and prepares to rise again, continuing the process. Similar convective circulation patterns occur within ocean

waters (▶p. 471), in columns of magma beneath Earth's surface (▶p. 208), and even in a simmering pot of soup. Convective circulation influences both weather and climate.

The atmosphere drives weather and climate

In everyday speech, we often use the terms *weather* and *climate* interchangeably. However, these words have very distinct meanings. Both concepts involve the physical properties of the troposphere, such as temperature, pressure, humidity, cloudiness, and wind. However, **weather** specifies atmospheric conditions over short time periods, typically hours or days, and within relatively small geographic areas. **Climate,** in contrast, describes the pattern of atmospheric conditions found across large geographic regions over long periods of time, typically seasons, years, or millennia. Mark Twain once noted the distinction between climate and weather by saying, "Climate is what we expect, weather is what we get." For example, even very dry climates (such as that of the desert around Phoenix) occasionally have wet weather.

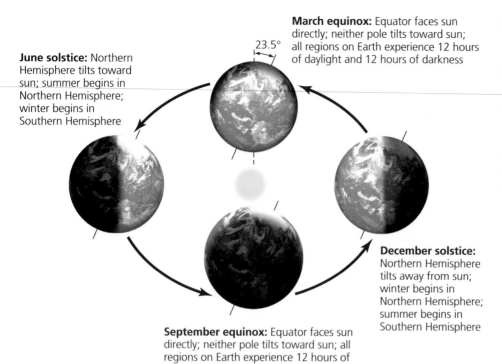

March equinox: Equator faces sun directly; neither pole tilts toward sun; all regions on Earth experience 12 hours of daylight and 12 hours of darkness

23.5°

June solstice: Northern Hemisphere tilts toward sun; summer begins in Northern Hemisphere; winter begins in Southern Hemisphere

December solstice: Northern Hemisphere tilts away from sun; winter begins in Northern Hemisphere; summer begins in Southern Hemisphere

September equinox: Equator faces sun directly; neither pole tilts toward sun; all regions on Earth experience 12 hours of daylight and 12 hours of darkness

FIGURE 17.5 The seasons occur because Earth is tilted on its axis by 23.5 degrees. As Earth revolves around the sun, the Northern Hemisphere tilts toward the sun for one half of the year, and the Southern Hemisphere tilts toward the sun for the other half of the year. In each hemisphere, summer occurs during the period in which the hemisphere receives the most solar energy because of its tilt toward the sun.

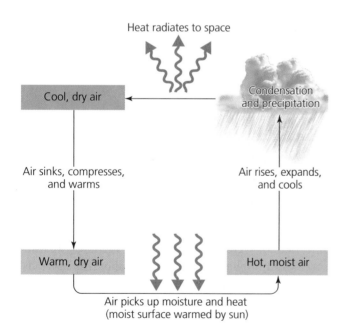

Heat radiates to space

Cool, dry air

Air sinks, compresses, and warms

Warm, dry air

Condensation and precipitation

Air rises, expands, and cools

Hot, moist air

Air picks up moisture and heat (moist surface warmed by sun)

FIGURE 17.6 Weather is driven in part by the convective circulation of air in the atmosphere. Air being heated near Earth's surface picks up moisture and rises. Once aloft, this air cools, and moisture condenses, forming clouds and precipitation. Cool, drying air begins to descend, compressing and warming in the process. Warm dry air near the surface begins the cycle anew.

Weather is produced by interacting air masses

Many changes in weather occur when air masses with different physical properties meet. The boundary between air masses that differ in temperature and moisture (and therefore density) is called a *front*. The boundary along which a mass of warmer, moister air replaces a mass of colder, drier air is termed a **warm front.** Some of the warm, moist air behind a warm front rises over the cold air mass and then cools and condenses to form clouds that may produce light rain. A **cold front** is the boundary along which a colder, drier air mass displaces a warmer, moister air mass. The colder air, being denser, tends to wedge beneath the warmer air. The warmer air rises, then cools and expands to form clouds that can produce thunderstorms. Once a cold front passes through, the sky usually clears, and the temperature and humidity drop (Figure 17.7).

Adjacent air masses may also differ in atmospheric pressure. An air mass with relatively high atmospheric pressure, or a **high-pressure system,** contains air that moves outward away from the center of high pressure as it descends. High-pressure systems typically bring fair weather. In a **low-pressure system,** air moves toward the low atmospheric pressure at the center of the system and spirals upward. The air expands and cools, and clouds and precipitation often result.

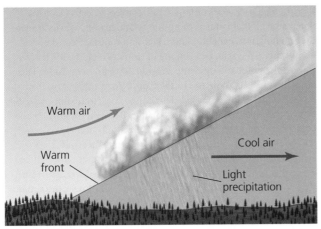

(a) Warm front

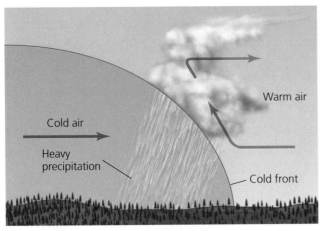

(b) Cold front

FIGURE 17.7 When a warm front approaches, warmer air rises over cooler air, causing light to moderate precipitation as moisture in the warmer air condenses (**a**). When a cold front approaches, colder air pushes beneath warmer air, and the warmer air rises, causing condensation and resulting in heavy precipitation (**b**).

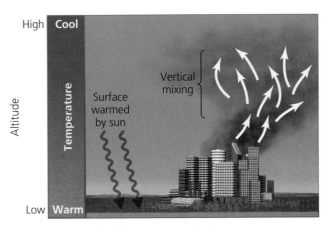

(a) Normal conditions

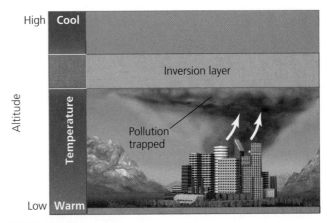

(b) Thermal inversion

FIGURE 17.8 A thermal inversion is a natural atmospheric occurrence that can exacerbate air pollution locally. Under normal conditions (**a**), tropospheric temperature decreases with altitude, and air of different altitudes mixes somewhat freely, dispersing most pollutants upward and outward from their sources. During a thermal inversion (**b**), cool air remains near the ground underneath an "inversion layer" of warmer air. No mixing occurs, and pollutants are trapped within the cool layer near the surface.

Under most conditions, air in the troposphere decreases in temperature as altitude increases. Because warm air rises, this causes vertical mixing. Occasionally, however, a layer of relatively cool air occurs beneath a layer of warmer air. This departure from the normal temperature profile is known as a **temperature inversion,** or **thermal inversion** (Figure 17.8). The band of air in which temperature rises with altitude is called an **inversion layer** (because the normal direction of temperature change is inverted). The cooler air at the bottom of the inversion layer is denser than the warmer air at the top of the inversion layer, so it resists vertical mixing and remains stable. Thermal inversions can occur in different ways, sometimes involving cool air at ground level and sometimes producing an inversion layer higher above the

ground. One common type of inversion occurs in mountain valleys where slopes block morning sunlight, keeping ground-level air within the valley shaded and cool.

Whereas vertical mixing normally allows ground-level air pollution to be diluted upward, thermal inversions trap pollutants near the ground. It was a thermal inversion that sparked London's killer smog. A high-pressure system settled over the city, acting like a cap on the air pollution and keeping it in place. Although London's 1952 smog event was the worst single such event recorded, inversions regularly cause smog buildup in many areas worldwide. This is a problem particularly in large metropolitan areas in hilly terrain, such as Los Angeles; Mexico City; Seoul, Korea; and Rio de Janeiro and São Paulo, Brazil.

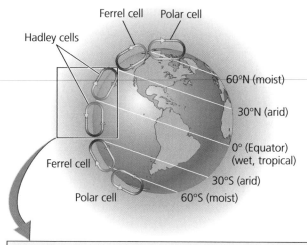

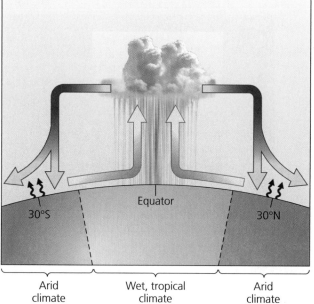

(a) Convection currents

FIGURE 17.9 A series of large-scale convective cells **(a)** helps determine global patterns of humidity and aridity. Warm air near the equator rises, expands, and cools, and moisture condenses, giving rise to a warm, wet climate in tropical regions. Air travels toward the poles and descends around 30 degrees latitude. This air, which lost its moisture in the tropics, causes regions around 30 degrees latitude to be arid. This convective circulation, a Hadley cell, occurs on both sides of the equator. Between roughly 30 and 60 degrees latitude north and south, Ferrel cells occur; and between 60 and 90 degrees latitude, polar cells occur. As a result, air rises around 60 degrees latitude, creating a moist climate, and falls around 90 degrees, creating a dry climate. Global wind currents **(b)** show latitudinal patterns as well. Trade winds near the equator blow westward, while westerlies between 30 and 60 degrees latitude blow eastward.

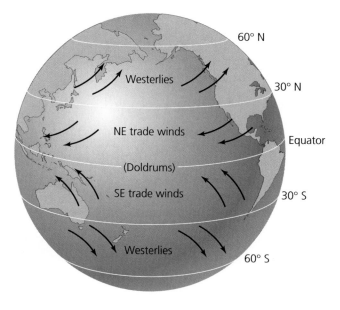

(b) Global wind patterns

Global climate patterns result from large-scale circulation systems

At larger geographic scales, convective air currents contribute to climatic patterns that are maintained over long periods of time (Figure 17.9a). Near the equator, solar radiation sets in motion a pair of convective cells known as **Hadley cells.** Here, where sunlight is most intense, surface air warms, rises, and expands. As it does so, it releases moisture, producing the heavy rainfall that gives rise to tropical rainforests near the equator. After releasing much of its moisture, this air diverges and moves in currents heading northward and southward. The air in these currents cools and descends back to Earth at about 30 degrees latitude north and south. Because the descending air has low relative humidity, the regions around 30 degrees lati-

tude are quite arid, giving rise to deserts. Two pairs of similar but less intense convective cells, called **Ferrel cells** and **polar cells,** lift air and create precipitation around 60 degrees latitude north and south and cause air to descend at around 30 degrees latitude and in the polar regions.

These three pairs of cells account for the latitudinal distribution of moisture across Earth's surface: wet, tropical climates near the equator; arid climates near 30 degrees latitude; somewhat moist regions near 60 degrees latitude; and dry, arctic conditions near the poles. These patterns, combined with temperature variation, help explain why biomes tend to be arrayed in latitudinal bands (Figure 6.16, ▸p. 170).

The Hadley, Ferrel, and polar cells interact with Earth's rotation to produce the global wind patterns shown in Figure 17.9b. As Earth rotates on its axis, locations on the equator spin faster than locations near the poles. As a result, the north-south air currents of the convective cells appear to be deflected from a straight path, as some portions of the globe move beneath them more quickly than others. This apparent deflection is called the **Coriolis effect,** and it results in the curving global wind patterns evident in Figure 17.9b. Near the equator lies a region with few latitudinal winds known as the *doldrums.* Between the equator and 30 degrees latitude lie the *trade winds,* which blow from east to west. From 30 to 60 degrees latitude are the *westerlies,* which originate from the west and blow east.

People used these global circulation patterns for centuries to facilitate ocean travel by wind-powered sailing ships. Moreover, the atmosphere interacts with the oceans to affect weather, climate, and the distribution of biomes. For instance, winds and convective circulation in ocean water together maintain ocean currents (▸ pp. 470–471), and trade winds weaken periodically, leading to El Niño conditions (▸ pp. 534, 536–537). The atmosphere's interactions with other systems of the planet can be complex, but even a basic understanding of how the atmosphere functions can help us comprehend how our pollution of the atmosphere can affect ecological systems, economies, and human health.

Outdoor Air Pollution

Throughout human history, we have made the atmosphere a dumping ground for our airborne wastes. Whether from primitive wood fires or modern coal-burning power plants, people have generated significant quantities of **air pollutants,** gases and particulate material added to the atmosphere that can affect climate or harm people or other organisms. **Air pollution** refers to the release of air pollutants. In recent decades, government policy and improved technologies have helped us diminish *outdoor air pollution* (often called *ambient air pollution*) substantially in countries of the developed world. However, outdoor air pollution remains a problem, particularly in developing nations and in urban areas worldwide.

Natural sources can pollute

When we think of outdoor air pollution, we tend to envision smokestacks belching black smoke from industrial plants. However, natural processes produce a great deal of the world's air pollution. Some of these natural impacts can be exacerbated by human activity and land-use policies.

Winds sweeping over arid terrain can send huge amounts of dust aloft. In 2001, strong westerlies lifted soil from deserts in Mongolia and China. The dust blanketed Chinese towns, spread to Japan and Korea, traveled eastward across the Pacific Ocean to the United States, then crossed the Atlantic and left evidence atop the French Alps. Every year, hundreds of millions of tons of dust are blown westward by trade winds across the Atlantic Ocean from northern Africa to the Americas (Figure 17.10a). Fungal and bacterial spores carried along with the dust have been linked to die-offs in Caribbean coral reef systems. Although dust storms are natural, the immense scale of these events results from nonsustainable farming and grazing practices that strip vegetation from the soil and promote wind erosion. Continental-scale dust storms took place in the United States in the 1930s, when soil from the drought-stricken Dust Bowl states blew eastward to the Atlantic (▸ p. 260).

Volcanic eruptions release large quantities of particulate matter, as well as sulfur dioxide and other gases, into the troposphere. Major eruptions may blow matter into the stratosphere, where it can remain for months or years. The 1980 eruption of Mount Saint Helens in Washington produced 1.1 billion m^3 (1.4 billion yd^3) of dust that circled the planet in 15 days (Figure 17.10b). The massive 1883 eruption on the Indonesian island of Krakatau blew enough dust into the atmosphere to cause a 1°C drop in global temperature (and produce gorgeous sunsets throughout the world).

The burning of vegetation also pollutes the atmosphere with soot and gases. Over 60 million ha (150 million acres) of forest and grassland burn in a typical year (Figure 17.10c). Fires occur naturally, but many today result from "slash-and-burn" forest clearing for farming and grazing in the tropics (▸ p. 255). In 1997, a severe drought brought on by the 20th century's strongest El Niño event caused fires in Indonesia to rage out of control. Their smoke sickened 20 million Indonesians, caused cargo ships to collide, and brought about a plane crash in Sumatra. More than 170 million metric tons of carbon monoxide were released from these fires. These, along with tens of thousands of fires in drought-plagued Mexico, Central America, and Africa, released more carbon monoxide into the atmosphere during 1997–1998 than did the worldwide burning of fossil fuels.

Human activities create various types of outdoor air pollution

Human activity can exacerbate the severity of natural air pollution and can also introduce new sources of air pollution.

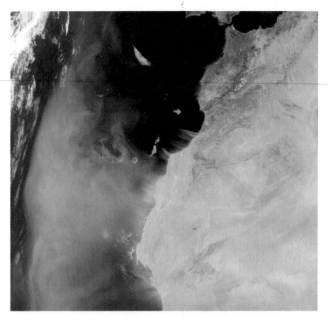

(a) Dust storm from Africa to the Americas

(b) Mount Saint Helens eruption, 1980

(c) Natural fire in California

FIGURE 17.10 Massive dust storms, such as this one blowing across the Atlantic Ocean from Africa to the Americas (**a**), are one type of natural air pollution. Volcanoes are another, as shown by Mount Saint Helens (**b**), which erupted in Washington State in 1980. A third cause is natural fires in forests and grasslands (**c**).

may be stationary or may emit pollutants while in motion, as do automobiles, aircraft, ships, locomotives, construction equipment, and lawn mowers.

Once pollutants are in the atmosphere at sufficient concentrations, they may do harm directly or they may induce chemical reactions that produce harmful compounds. **Primary pollutants,** such as soot and carbon monoxide, are pollutants emitted into the troposphere in a form that can be directly harmful or that can react to form harmful substances. **Secondary pollutants** are harmful substances produced when primary pollutants interact or react with constituents of the atmosphere.

Arguably the greatest human-induced air pollution problem today is our emission of greenhouse gases that contribute to global climate change. We will address this issue separately in Chapter 18.

Clean Air Act legislation has addressed pollution in the United States

To address air pollution in the United States, Congress has passed a series of laws, beginning with the Air Pollution Control Act of 1955. The Clean Air Act of 1963 funded research into pollution control and encouraged emissions standards for automobiles and stationary point sources, such as industrial plants. Subsequent amendments

As with water pollution, air pollution can emanate from *point sources* or *non-point sources* (▶ p. 452). A point source describes a specific spot—such as a factory's smokestacks—where large quantities of pollutants are discharged. In contrast, *non-point sources* are more diffuse, often consisting of many small sources (such as millions of automobiles). In 1952 London, coal-burning power plants acted as point sources contributing to the killer smog, while millions of home fireplaces together comprised a potent non-point source. Pollution sources

expanded the legislation's scope and established a nationwide air quality monitoring system.

In 1970, Congress thoroughly revised the law in what came to be known as the **Clean Air Act of 1970.** This legislation set stricter standards for air quality, imposed limits on emissions from new stationary and mobile sources, provided new funds for pollution-control research, and enabled citizens to sue parties violating the standards. Once some of these goals came to be viewed as too ambitious, amendments in 1977 altered some standards and extended some deadlines for compliance.

The **Clean Air Act of 1990** sought to strengthen regulations pertaining to air quality standards, auto emissions, toxic air pollution, acidic deposition, and stratospheric ozone depletion. It also introduced an emissions trading program (▶p. 73) for sulfur dioxide. Beginning in 1995, businesses and utilities were allocated permits for emitting this pollutant, and could then buy, sell, or trade these allowances with one another. Each year the overall amount of allowed pollution was decreased. This market-based incentive program has proven successful, and has spawned similar programs at state and regional levels and for other pollutants.

As a result of Clean Air Act legislation, the U.S. Environmental Protection Agency (EPA) sets nationwide standards for emissions of pollutants and concentrations of pollutants in ambient air throughout the nation. However, it is largely up to the states to monitor air quality and develop, implement, and enforce regulations within their boundaries. States submit implementation plans to the EPA for approval, and if a state's plans are not adequate, the EPA can take over enforcement in that state.

The EPA sets standards for "criteria pollutants"

The EPA and the states focus on six **criteria pollutants,** pollutants judged to pose especially great threats to human health—carbon monoxide (CO), sulfur dioxide (SO_2), nitrogen dioxide (NO_2), tropospheric ozone (O_3), particulate matter, and lead (Pb). For these, the EPA has established *national ambient air quality standards (NAAQS),* which are maximum allowable concentrations of these pollutants in ambient outdoor air. Through risk assessment procedures (▶pp. 422–423), these six pollutants were selected on the basis of criteria relating to human health.

Carbon monoxide Carbon monoxide is a colorless, odorless gas produced primarily by the incomplete combustion of fuels. In the United States in 2004, 87.2 million tons of CO were released, making it the most abundant air pollutant by mass. Vehicles account for about 62% of these emissions, but other sources include lawn and garden equipment (10%), forest fires (6%), open burning of industrial waste (3%), and residential wood burning (2%). Carbon monoxide poses risk to humans and other animals, even in small concentrations. It can bind irreversibly to hemoglobin in red blood cells, preventing the hemoglobin from binding with oxygen. U.S. emissions of CO have decreased in recent decades largely because of cleaner-burning motor vehicle engines.

Sulfur dioxide Like CO, sulfur dioxide is a colorless gas. Of the 15.2 million metric tons of SO_2 released in the United States in 2004, about 70% resulted from the combustion of coal for electricity generation and industry. During combustion, elemental sulfur (S) in coal reacts with oxygen gas (O_2) to form SO_2. Once in the atmosphere, SO_2 may react to form sulfur trioxide (SO_3) and sulfuric acid (H_2SO_4), which may then fall back to Earth in the form of acid precipitation.

Nitrogen dioxide Nitrogen dioxide is a highly reactive, foul-smelling reddish brown gas that contributes to smog and acid precipitation. Along with nitric oxide (NO), NO_2 belongs to a family of compounds called nitrogen oxides (NO_x). Nitrogen oxides result when atmospheric nitrogen and oxygen react at the high temperatures created by combustion engines. Of the 18.8 million tons of nitrogen oxides released in the United States in 2004, over half resulted from combustion in vehicle engines. Electrical utility and industrial combustion accounted for most of the rest.

Tropospheric ozone Although ozone in the stratosphere shields organisms from the dangers of UV radiation, O_3 from human activity forms and accumulates low in the troposphere and acts as a pollutant. In the troposphere, this colorless gas results from the interaction of sunlight, heat, nitrogen oxides, and volatile carbon-containing chemicals. Ozone is therefore categorized as a secondary pollutant. A major component of smog, O_3 can pose health risks as a result of its instability as a molecule; this triplet of oxygen atoms will readily release one of its threesome, leaving a molecule of oxygen gas and a free oxygen atom. The free oxygen atom may then participate in reactions that can injure living tissues and cause respiratory problems. Although concentrations fell by 11–18% (depending on how they were measured) in the United States from 1982 to 2001, tropospheric O_3 is the pollutant that most frequently exceeds the EPA standard.

Particulate matter Particulate matter is composed of solid or liquid particles small enough to be suspended in the atmosphere. Particulate matter includes primary pollutants such as dust and soot, as well as secondary pollutants such as sulfates and nitrates. Particulate matter can damage respiratory tissues when inhaled. Most particulate matter in the atmosphere is wind-blown dust (60%), but 2.5 million tons of particulate matter was released by human activities in the United States in 2004. Along with sulfur dioxide, it was largely the emission of particulate matter from industrial and residential coal-burning sources that produced London's 1952 killer smog and the deaths resulting from that episode.

Lead Lead is a heavy metal that enters the atmosphere as a particulate pollutant. The lead-containing compounds tetraethyl lead and tetramethyl lead, when added to gasoline, improve engine performance. However, the exhaust from leaded gasoline emits lead into the atmosphere, from which it can be inhaled or can be deposited on land and water. Lead can enter the food chain, accumulate within body tissues, and cause central nervous system malfunction, mental retardation among children, and a variety of other ailments. Once people recognized the dangers of lead, leaded gasoline was phased out in the United States and other industrialized nations, and U.S. lead emissions plummeted 93% from 1980 to 1990. Since then, lead emissions have remained steady and low. Today most lead emitted in the United States comes from industrial metal smelting. However, many developing nations continue to add lead to gasoline and experience significant lead pollution.

Agencies monitor pollutants that affect air quality

State and local agencies also monitor, calculate, and report to the EPA emissions of major pollutants that affect ambient concentrations of the six criteria pollutants. These include the four criteria pollutants that are primary pollutants (carbon monoxide, sulfur dioxide, particulate matter, and lead), as well as all nitrogen oxides (because NO reacts readily in the atmosphere to form NO_2, which is both a primary and secondary pollutant). Tropospheric ozone is a secondary pollutant only, so there are no emissions to monitor. Instead the EPA monitors emissions of volatile organic compounds, which can react to produce ozone and other secondary pollutants.

Volatile organic compounds (VOCs) are carbon-containing chemicals used in industrial processes such as dry-cleaning and manufacturing. One group of VOCs consists of hydrocarbons (▶ pp. 98–99) such as methane (CH_4, the primary component of natural gas), propane

(C_3H_8, used as a portable fuel), butane (C_4H_{10}, found in cigarette lighters), and octane (C_8H_{18}, a component of gasoline). Human activity accounts for about half the VOC emissions in the United States, and the remainder comes from natural sources. For example, plants produce isoprene (C_5H_8) and terpene ($C_{10}H_{15}$), and animals produce methane. The largest sources of anthropogenic VOC emissions include industrial use of solvents (28%) and vehicle emissions (27%).

Air pollution has decreased markedly since 1970

Since the Clean Air Act of 1970, emissions of each of the six monitored pollutants have decreased, and total emissions of the six together have declined by 54% (Figure 17.11a).

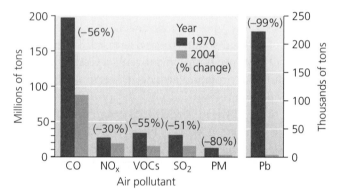

(a) Declines in six major pollutants

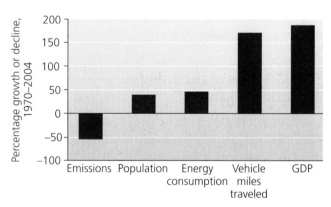

(b) Trends in major indicators

FIGURE 17.11 The EPA tracks emissions of several major pollutants into ambient air. Shown are emissions of four of the six "criteria pollutants," along with nitrogen oxides and volatile organic compounds. Each of these pollutants has shown substantial declines since 1970, and emissions from all six together have declined by 54% **(a)**. This decrease in emissions has occurred despite increases in U.S. population, energy consumption, vehicle miles traveled, and gross domestic product **(b)**. Go to **GRAPHit!** at www.aw-bc.com/withgott or on the student CD-ROM. Data from U.S. EPA, 2004.

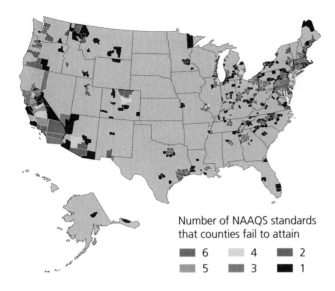

FIGURE 17.12 Nearly half of all Americans live in counties that periodically fail to meet the EPA's national ambient air quality standards (NAAQS) for at least one criteria pollutant. This map shows counties that failed to attain the standards for one (blue) through six (red) of the six criteria pollutants. Data are for April 2005, from U.S. EPA.

Number of NAAQS standards that counties fail to attain

6 4 2
5 3 1

This has occurred despite substantial increases in the nation's population, energy consumption, miles traveled by vehicle, and gross domestic product (Figure 17.11b). Because of the success in reducing emissions, air quality in the United States has improved markedly. As just one indicator, EPA data show that the percentage of days U.S. citizens were exposed to unhealthy air dropped from 10% in 1988 to 3% in 2001.

The reduction of outdoor air pollutant levels since 1970 represents one of the greatest accomplishments in safeguarding human health and environmental quality in the United States. However, the nation still has plenty of room to improve. Outdoor air pollution remains a problem, and in recent years, EPA monitoring has found that nearly half of all Americans live in counties where at least one of the six criteria pollutants periodically reaches unhealthy levels (Figure 17.12).

Weighing the Issues:
Your County's Air Quality

Locate where you live on the map in Figure 17.12. What is the status of your county's air quality, and how does your county compare to the rest of the nation? What factors do you think account for the quality of its air? Can you propose any solutions for reducing air pollution in your county?

Toxic substances are also major pollutants

Other chemicals known to cause serious health or environmental problems are classified as **toxic air pollutants.** These include substances known to cause cancer, reproductive defects, or neurological, developmental, immune system, or respiratory problems. Also included are substances that cause substantial ecological harm by affecting the health of nonhuman animals and plants. Some toxic air pollutants are produced naturally. For example, hydrogen sulfide gas (H_2S) gives the mud of swamps and bogs the odor of rotten eggs. However, most toxic air pollutants are produced by human activities, such as metal smelting, sewage treatment, and industrial processes. The 1990 Clean Air Act identifies 188 different toxic air pollutants, ranging from the heavy metal mercury (from coal-burning power plant emissions and other sources) to VOCs such as benzene (a component of gasoline) and methylene chloride (found in paint stripper). Among the 188 pollutants are 21 from mobile sources (such as diesel exhaust) and 33 "urban hazardous" pollutants judged to pose the greatest health risks in urban areas.

State and federal agencies do not monitor toxic air pollutants as extensively as they do the six criteria pollutants, but so far 300 monitoring sites are operating, and coverage is improving. The EPA estimates that because of Clean Air Act regulations, from 1990 to 1999 emissions of toxic air pollutants decreased by 30%.

Burning fossil fuels produces industrial smog

In response to the increasing incidence of fogs polluted by the smoke of Britain's industrial revolution, a British scientist coined the term *smog* long before the 1952 event in London. Today the term is used worldwide to describe unhealthy mixtures of air pollutants that often form over urban areas. The smog that enveloped London in 1952 was what we would today call **industrial smog,** or gray-air smog. When coal or oil is burned, some portion is completely combusted, forming CO_2; some is partially combusted, producing CO; and some remains unburned and is released as soot, or particles of carbon. Moreover, coal contains varying amounts of contaminants, including mercury and sulfur. Sulfur reacts with oxygen to form sulfur dioxide, which can undergo a series of reactions to form sulfuric acid and ammonium sulfate (Figure 17.13a). These chemicals and others produced by further reactions, along with soot, are the main components of industrial smog, and give the smog its characteristic gray color.

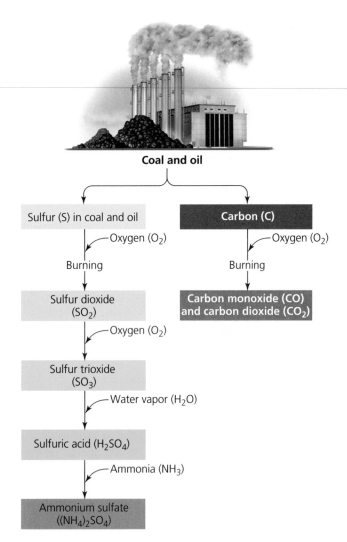

Coal and oil

Sulfur (S) in coal and oil → Oxygen (O_2) → Burning → Sulfur dioxide (SO_2) → Oxygen (O_2) → Sulfur trioxide (SO_3) → Water vapor (H_2O) → Sulfuric acid (H_2SO_4) → Ammonia (NH_3) → Ammonium sulfate ((NH_4)$_2SO_4$)

Carbon (C) → Oxygen (O_2) → Burning → Carbon monoxide (CO) and carbon dioxide (CO_2)

(a) Burning sulfur-rich oil or coal without adequate pollution control technologies

(b) Donora, Pennsylvania, at midday in the 1948 smog event

FIGURE 17.13 Emissions from the combustion of coal and oil in manufacturing plants and utilities without pollution control technologies can create industrial smog. Industrial smog consists primarily of sulfur dioxide and particulate matter, as well as carbon monoxide and carbon dioxide from the carbon component of fossil fuels. Sulfur contaminants in fossil fuels when combusted create the sulfur dioxide, which in the presence of other chemicals in the atmosphere can produce several other sulfur compounds **(a)**. Under certain weather conditions, industrial smog can blanket whole towns or regions, as it did in Donora, Pennsylvania, shown here in the daytime during its deadly 1948 smog episode **(b)**.

Industrial smog is far less common today in developed nations than it was 50–100 years ago. In the wake of the 1952 London episode and others, the governments of most developed nations began regulating industrial emissions to minimize the external costs (▶pp. 43–44) they impose on citizens. However, in industrializing regions such as China, India, and Eastern Europe, heavy reliance on coal burning (both by industry and by citizens heating and cooking in their homes), combined with lax air pollution controls, produces industrial smog that poses significant health risks in many urban areas.

Although coal combustion supplies the chemical constituents for industrial smog, weather also plays a role, as it did in London in 1952. A similar event occurred 4 years earlier in Donora, Pennsylvania. A thermal inversion trapped smog containing particulate matter emissions from a steel and wire factory. Twenty-one people were killed, and over 6,000 people—nearly half the town—

became ill (Figure 17.13b). In Donora's killer smog, air near the ground cooled during the night. Normally, morning sunlight warms the land and air, causing air to rise. However, because Donora is located in hilly terrain, too little sun reached the valley floor to warm and disperse the cold air. The resulting thermal inversion kept a pall of smog over the town long enough to impair visibility and cause serious health problems. Hilly topography such as Donora's is a factor in the air pollution of many other cities where surrounding mountains trap air and create inversions. This is true for the Los Angeles basin, which has long symbolized chronic smog problems in American popular culture. Modern-day Los Angeles, however, suffers from a different type of smog, one called photochemical smog.

Photochemical smog is produced by a complex series of atmospheric reactions

A photochemical process is one whose activation requires light. **Photochemical smog,** or brown-air smog, is formed through light-driven chemical reactions of primary pollutants and normal atmospheric compounds that produce a mix of over 100 different chemicals, tropospheric ozone

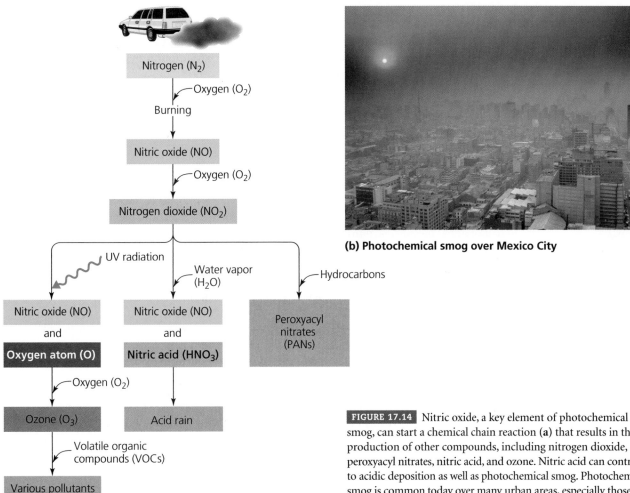

(a) Formation of photochemical smog

(b) Photochemical smog over Mexico City

FIGURE 17.14 Nitric oxide, a key element of photochemical smog, can start a chemical chain reaction (**a**) that results in the production of other compounds, including nitrogen dioxide, peroxyacyl nitrates, nitric acid, and ozone. Nitric acid can contribute to acidic deposition as well as photochemical smog. Photochemical smog is common today over many urban areas, especially those with hilly topography or frequent inversion layers. Mexico City (**b**) is one city that frequently experiences photochemical smog.

often being the most abundant among them (Figure 17.14a). High levels of NO_2 cause photochemical smog to form a brownish haze over cities (Figure 17.14b). Hot, sunny, windless days in urban areas provide perfect conditions for the formation of photochemical smog. Exhaust from morning traffic releases large amounts of NO and VOCs into a city's air. Sunlight then promotes the production of ozone and other constituents of photochemical smog. Levels of photochemical pollutants in urban areas typically peak in mid-afternoon and at sufficient levels can irritate people's eyes, noses, and throats. Air pollutants called *peroxyacyl nitrates,* created by the reaction of NO_2 with hydrocarbons, can induce further reactions that damage living tissues in animals and plants.

Photochemical smog afflicts many major cities, especially those with topography and weather conditions that promote it. In Athens, Greece, site of the 2004 Olympics, the problem had been bad enough that the city government provided incentives to replace aging automobiles. It also mandated that cars with odd-numbered license plates be driven only on odd-numbered days, and those with even-numbered plates only on even-numbered days. According to Greek officials, smog has been reduced by 30% since 1990 as a result.

Synthetic chemicals deplete stratospheric ozone

A pollutant at low altitudes, ozone is a highly beneficial gas at altitudes centering around 25 km (15 mi) in the lower stratosphere, where it is concentrated in the so-called *ozone layer.* Here, concentrations of ozone are only about 12 parts per million. However, ozone molecules are so effective at absorbing incoming ultraviolet radiation from the sun that this concentration is adequate to protect life on Earth's surface from the damaging effects of UV radiation.

In the 1960s, atmospheric scientists began wondering why their measurements of ozone were lower than theoretical models predicted. Researchers speculating that natural or artificial chemicals were depleting ozone finally

Air Pollution

VIEWPOINTS

Despite improvements in air quality, air pollution leading to photochemical smog remains a health and environmental concern in and around many major cities. **What is needed to reduce this type of pollution?**

A Policy Portfolio Approach to Fighting Smog

Federal and state regulations to control emissions from factories, power plants, and cars have reduced smog. Further progress will require more effective and affordable means to reduce smog "precursors," nitrogen oxides and volatile organic compounds, from old and new sources. Simply tightening the same regulations used in the past and applying them nationwide to the same sources is unlikely to significantly reduce smog for three reasons: control costs will be high and the marginal gains low for many "old" stationary sources already being regulated; emissions from many smaller sources, such as sport-utility vehicles, trucks, lawnmowers, and motorboats, are a growing concern; and the right mix of reductions in nitrogen oxide and volatile organic compound emissions depends on local circumstances.

There is no silver bullet. Regional air quality regulators will need to apply several different policy approaches, called a policy portfolio, to achieve further reductions. For large stationary sources, emissions trading is an economically attractive supplement to current technology-based regulations. Emissions trading works as follows: First, government regulators set a regional limit on nitrogen oxides. Then, regulators assign or auction to emissions sources only as many allowances (for example, one allowance equals one ton of nitrogen oxide emissions) as the overall limit (or "cap") permits. Lastly, emissions sources could take one of three actions: reduce their own emissions to stay within their allowances; sell or retire any excess allowances; or buy others' excess allowances to meet their own obligations. They would choose according to market prices, with the cheapest sources of reductions being made available first.

Small mobile sources that run on diesel fuel and gasoline are also a problem. For now, the federal government still needs to press manufacturers to build cleaner-burning vehicles and fuel makers to make cleaner fuels, as the EPA is doing. Meanwhile, regional regulators need to be given the flexibility and enforcement tools to tailor their efforts to highly variable regional climate and atmospheric conditions.

Debra Knopman is vice president and director of the Infrastructure, Safety, and Environment division of the RAND Corporation. This statement is based primarily on work she did at the Progressive Policy Institute, where she was director of the Center for Innovation and the Environment from 1995 to 2001.

To Reduce Smog, States Must Take Up the Slack

Nearly half the people in this country—136 million individuals—live in areas where smog levels exceed the Environmental Protection Agency's health standard. Unfortunately for the state governments that are responsible for those areas, a substantial amount of the pollution that forms smog in their cities and valleys originates from power plants in up-wind states.

In May 2005, EPA issued a regulation to address smog-forming pollution that crosses state lines. The rule will require states in the eastern half of the country to tighten limits on emissions of nitrogen oxides from their power plants.

The prescribed limits are too weak, however, to eliminate all of the interstate smog transport, much less clean up all of the smog. Without additional measures, 20 million people will live in areas exceeding EPA's smog standard in 2015—5 years after the latest smog cleanup deadline in the Clean Air Act.

To protect their citizens from asthma attacks, hospitalizations, and missed schooldays and workdays, state governments must pick up the slack. That will require new urban planning measures to reduce vehicle miles traveled, because tailpipe emissions contribute to smog. It will also require states to impose new emissions limits—more stringent than the federally prescribed ones—on their own power plants, oil refineries, and factories.

In an attempt to avoid those new limits, large polluters in 2004 and again in 2005 convinced congressional allies to attach to the federal energy bill a provision that would have extended the Clean Air Act's smog cleanup deadlines by a decade. Fortunately, the measure was stripped from the bill after state officials made clear they were not interested in ignoring the problem. That was a good start. Now there is a lot of work to do.

David McIntosh is a staff attorney at the Natural Resources Defense Council in Washington, D.C. He litigates and lobbies to ensure effective implementation of clean air laws and to counter efforts aimed at weakening those laws.

Explore this issue further by accessing **Viewpoints** at www.aw-bc.com/withgott.

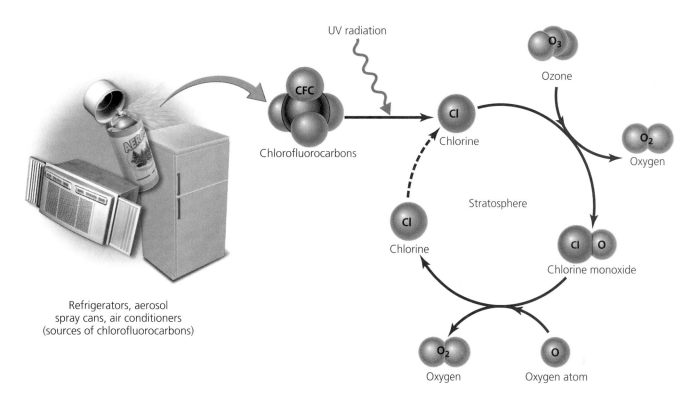

FIGURE 17.15 Chlorofluorocarbons lead to the destruction of ozone molecules in the presence of ultraviolet radiation. A chlorine atom released from a CFC molecule reacts with an ozone molecule, forming one molecule of oxygen gas and one chlorine monoxide (ClO) molecule. The oxygen atom in the ClO molecule will then bind with a stray oxygen atom to form oxygen gas, leaving the chlorine atom to begin the destructive cycle anew. The self-perpetuating nature of this process means that one CFC molecule can destroy a great many ozone molecules.

pinpointed a group of human-made compounds derived from simple hydrocarbons, such as ethane and methane, in which hydrogen atoms are replaced by chlorine, bromine, or fluorine. One class of such compounds, **chlorofluoro-carbons (CFCs),** was being mass-produced by industry at a rate of a million metric tons per year in the early 1970s, and it was growing by 20% a year.

In 1974, atmospheric scientists Sherwood Rowland and Mario Molina showed that CFCs could deplete strato-spheric ozone by releasing chlorine atoms that split ozone molecules, creating from each of them an O_2 molecule and a ClO molecule (Figure 17.15). Three years before Rowland and Molina's study, researcher J. E. McDonald had predicted that ozone loss, by allowing more UV radiation to reach the surface, would result in thousands more skin cancer cases each year. This caught the attention of poli-cymakers, environmentalists, and industry alike (see "The Science behind the Story," ▶ pp. 516–517).

Then in 1985, scientists from the British Antarctic Survey announced that stratospheric ozone levels over Antarctica had declined by 40–60% in the previous decade, leaving a thinned ozone concentration that was

soon dubbed the *ozone hole* (Figure 17.16). Research over the next few years confirmed the link between CFCs and ozone loss in the Antarctic and indicated that deple-tion was also occurring in the Arctic, and perhaps glob-ally. Already concerned about skin cancer, scientists were becoming anxious over the possible effects of increased UV radiation on ecosystems. Research was showing a multitude of effects, including harm to crops and to the productivity of ocean phytoplankton, the base of the marine food chain.

The Montreal Protocol addressed ozone depletion

In light of the science and the ongoing health and ecologi-cal concerns, international efforts to restrict CFC produc-tion finally bore fruit in 1987 with the **Montreal Protocol.** In this treaty, signatory nations (eventually numbering 180) agreed to cut CFC production in half. Five follow-up agreements strengthened the pact, deepening the cuts, ad-vancing timetables for compliance, and addressing related ozone-depleting chemicals. Today the production and use

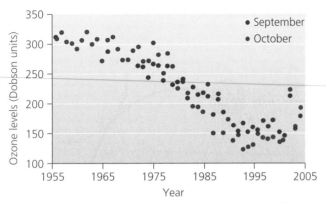

(a) **Monthly mean ozone levels at Halley Bay, Antarctica**

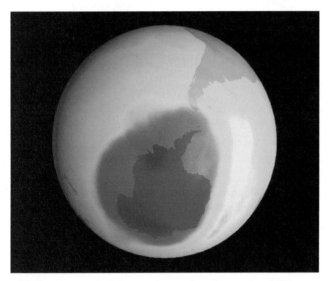

(b) **The "ozone hole" over Antarctica, September 2000**

FIGURE 17.16   The "ozone hole" consists of a region of thinned ozone density in the stratosphere over Antarctica and the southernmost ocean regions. It has reappeared seasonally each September in recent decades. Data from Halley Bay, Antarctica (**a**), show a steady decrease in stratospheric ozone concentrations from the 1960s to 1990. Ozone-depleting CFCs began to be regulated under the Montreal Protocol in 1987, and ozone concentrations stopped declining. Colorized satellite imagery from September 6, 2000 (**b**), shows the "ozone hole" (blue) at its maximal recorded extent to date. Data from British Antarctic Survey.

of ozone-depleting compounds has fallen by 95% since the late 1980s, and scientists can discern the beginnings of long-term recovery of the ozone layer (although much of the 5 billion kg of CFCs emitted into the troposphere has yet to diffuse up into the stratosphere, and CFCs are slow to dissipate or break down). Industry was able to shift to alternative, environmentally safer chemicals, which have largely turned out to be cheaper and more efficient. For these reasons, the Montreal Protocol and its follow-up amendments are widely considered the biggest success story so far in addressing any global environmental problem.

Environmental scientists have attributed this success primarily to two factors:

1. Policymakers engaged industry in helping to solve the problem, and government and industry worked together on developing technological fixes and replacement chemicals. This cooperation reduced the battles that typically erupt between environmentalists and industry.
2. Implementation of the Montreal Protocol after 1987 followed an adaptive management approach (▸ p. 346), altering strategies midstream in response to new scientific data, technological advances, or economic figures.

Because of its success in addressing ozone depletion, the Montreal Protocol is widely seen as a model for international cooperation in addressing other pressing global problems, such as persistent organic pollutants (▸ pp. 426, 428), climate change (▸ pp. 550, 552), and biodiversity loss (▸ pp. 335–336).

Weighing the Issues:
International Cooperation to Solve Global Problems

The Montreal Protocol showed how international collaboration, together with technological advances, can drastically and rapidly address a pressing environmental problem. So far, however, global problems such as organic pollutants, climate change, and biodiversity loss have not seen the same degree of targeted action. Why do you think this is? Besides the effort to halt stratospheric ozone depletion, can you name other success stories in addressing major environmental problems? Are any on the horizon?

Acidic deposition represents another transboundary pollution problem

Just as the problem of stratospheric ozone depletion crosses political boundaries, so does another atmospheric pollution concern—acidic deposition. **Acidic deposition** refers to the deposition of acidic or acid-forming pollutants from the atmosphere onto Earth's surface. This can take place either by precipitation (commonly referred to as *acid rain,* but also including acid snow, sleet, and hail), by fog, by gases, or by the deposition of dry particles. Acidic deposition is one type of **atmospheric deposition,** which refers more broadly to the wet or dry deposition on land of a wide variety of pollutants, including mercury, nitrates, organochlorines, and others.

Primary pollutants   Secondary pollutants

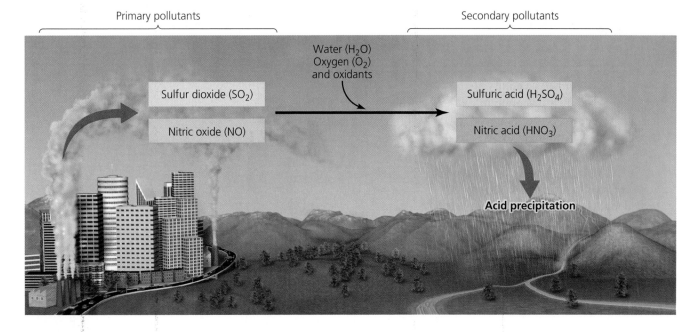

Water (H₂O)
Oxygen (O₂)
and oxidants

Sulfur dioxide (SO₂)   Sulfuric acid (H₂SO₄)

Nitric oxide (NO)   Nitric acid (HNO₃)

Acid precipitation

FIGURE 17.17 Acidic deposition can have consequences many miles downwind from its source. Emissions containing sulfur dioxide and nitric oxide from industries and utilities begin the process. Sulfur dioxide and nitric oxide can be transformed into sulfuric acid and nitric acid through chemical reactions in the atmosphere, and these acidic compounds descend to Earth's surface in rain, snow, fog, and dry deposition.

Acidic deposition originates primarily with sulfur dioxide and nitrogen oxides, pollutants produced largely through fossil fuel combustion by automobiles, electric utilities, and industrial facilities. Once emitted into the troposphere, these pollutants can react with water, oxygen, and oxidants to produce compounds of low pH (▶ p. 98), primarily sulfuric acid and nitric acid. Suspended in the troposphere, droplets of these acids may travel for up to days or weeks, sometimes covering hundreds or thousands of kilometers before falling in precipitation (Figure 17.17).

Acidic deposition can have wide-ranging, cumulative detrimental effects on ecosystems and on our built environment (Table 17.1). Acids can leach basic minerals such as calcium and magnesium from soil, changing soil chemistry and harming plants and soil organisms. Streams, rivers, and lakes may become significantly acidified from runoff. In fact, thousands of lakes in Canada, Scandinavia, the United States, and elsewhere now contain water acidic enough to kill fish. Fish can die when acidic conditions cause toxic aluminum to be more readily accessible to their tissues. Elevated aluminum in the soil hinders water and nutrient uptake by plants. In some regions of the United States, acid fog with a pH of 2.3 (equivalent to vinegar) can envelop forests for extended periods. Some forests in eastern North America have experienced widespread tree die-back from these conditions.

Besides harming trees, acid precipitation also may damage agricultural crops. Moreover, it can erode stone buildings, eat away at cars, and erase the writing from tombstones. Ancient cathedrals in Europe, monuments in

Table 17.1	Effects of Acidic Deposition on Ecosystems in the Northeastern United States

Acidic deposition in northeastern forests has . . .

▶ Accelerated leaching of base cations (ions that counteract acidic deposition) from soil

▶ Allowed sulfur and nitrogen to accumulate in soil

▶ Increased dissolved inorganic aluminum in soil, hindering plant uptake of water and nutrients

▶ Leached calcium from needles of red spruce, leading to tree mortality from wintertime freezing

▶ Increased mortality of sugar maples due to leaching of base cations from soil and leaves

▶ Acidified 41% of Adirondack, New York, lakes and 15% of New England lakes

▶ Lowered lakes' capacity to neutralize further acids

▶ Elevated aluminum levels in surface waters

▶ Reduced species diversity and abundance of aquatic life, and negatively affected entire food webs

Source: Adapted from Driscoll, C.T., et al. 2001. *Acid rain revisited.* Hubbard Brook Research Foundation.

The Science behind the Story

Identifying CFCs as the Main Cause of Ozone Depletion

For half a century after their invention in the 1920s, chlorofluorocarbons (CFCs) were thought to be useful, nontoxic, and environmentally friendly. In the early 1970s, however, scientists became concerned that CFCs could cause long-term damage to the ozone layer. By the late 1980s, evidence for such damage had become strong enough to justify a complete ban on CFC production. In their attempts to understand how CFCs influenced stratospheric ozone, scientists relied on a wide variety of data sources, including historical records, field observations, laboratory experiments, and computer models.

Stratospheric ozone and CFCs had each been the subject of much research before they were linked in the 1970s. Ozone was discovered in 1839, and its presence in the upper atmosphere was first proposed in the 1880s. In 1924, British scientist G. M. B. Dobson built what has become the standard instrument for measuring ozone from the ground. By the 1970s, the Dobson ozone spectrophotometer (see the figure) was being used by a global network of observation stations. However, scientists were unable to establish a clear picture of global trends in ozone concentrations because of

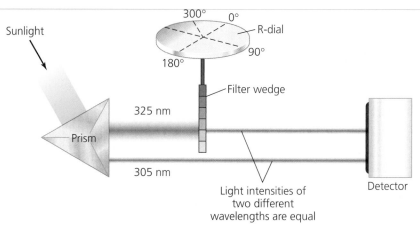

In the Dobson spectrophotometer, UV light passes through a prism that separates wavelengths of 325 nanometers (nm) and 305 nm, and then travels toward the detector. Because ozone absorbs wavelengths of 305 nm but not wavelengths of 325 nm, light that reaches the instrument after passing through the atmosphere contains more light of 325 nm wavelengths. The ratio between the intensities of the two wavelengths of light indicates the amount of ozone in the light's path between the sun and the spectrophotometer. To measure this ratio, the R-dial rotates, causing the filter wedge to block more and more 325-nm light, until the intensities of 325-nm and 305-nm light are equal. At that point, the reading on the R-dial is recorded, and a conversion is used to calculate the atmospheric ozone concentration. Figure adapted from University of Alaska, Fairbanks, http://ozone.gi.alaska.edu.

natural variations in ozone levels and the difficulty of comparing data from different stations.

Research on CFCs also had a long history. First invented in 1928, CFCs were found to be useful as refrigerants, fire extinguishers, and propellants for aerosol spray cans. Starting in the 1960s, CFCs also found wide use as cleaners for electronics and as a part of the process

of manufacturing rigid polystyrene foams. Research on the chemical properties of CFCs showed that they were almost completely inert; that is, they rarely reacted with other chemicals. Therefore, scientists surmised that, at trace levels, CFCs would be harmless to both people and the environment.

However, in June 1974, chemists F. Sherwood Rowland and Mario

Washington, D.C., temples in Asia, and stone statues throughout the world are experiencing billions of dollars of damage as their features gradually wear away (Figure 17.18).

Because the pollutants leading to acid deposition can travel long distances, their effects may be felt far from their sources—a situation that has led to political bickering among the leaders of states and nations. For instance,

much of the pollution from power plants and factories in Pennsylvania, Ohio, and Illinois falls out in states to their east, including New York, Vermont, and New Hampshire, as well as in regions to the north, including Ontario, Quebec, and the maritime provinces of Canada. As Figure 17.19 shows, many regions of greatest acidification are downwind of major source areas of pollution.

Molina published a paper in the journal *Nature,* arguing that the inertness that made CFCs so ideal for industrial purposes could also have disastrous consequences for the ozone layer. More-reactive chemicals are broken down to their constituent atoms in the lower atmosphere. CFCs, in contrast, reach the stratosphere unchanged. Once CFCs reach the stratosphere, intense ultraviolet radiation from the sun breaks them into their constituent chlorine and carbon atoms. Each free chlorine atom, it was calculated, can catalyze the destruction of as many as 100,000 ozone molecules.

Rowland and Molina were the first to assemble a complete picture of the threat posed by CFCs, but they could not have reached their conclusions without the contributions of other scientists. British researcher James Lovelock had developed an instrument to measure extremely low concentrations of atmospheric gases, and American researchers Richard Stolarski and Ralph Cicerone had shown that chlorine atoms can catalyze the destruction of ozone.

Rowland and Molina's finding, which earned them the 1995 Nobel Prize in chemistry, helped spark discussion among scientists, policymakers, and industry leaders over limits on CFC production. As a result, the United States and several other nations banned the use of CFCs in aerosol spray cans in 1979. Other uses continued, however, and by the early 1980s global production of CFCs was increasing.

Then, in 1985, a new finding shocked scientists and spurred the international community to take further action. Scientists at a British research station in Antarctica had been recording ozone concentrations continuously since the 1950s. In May 1985, Joseph Farman and colleagues reported in *Nature* that Antarctic ozone concentrations had been declining dramatically since the 1970s. The decline exceeded even the worst-case predictions.

To determine what was causing the "ozone hole" over Antarctica, expeditions were mounted in 1986 and 1987 to measure trace amounts of atmospheric gases using ground stations and high-altitude balloons and aircraft. Together with other scientists, Dutch scientist Paul Crutzen, who would share the 1995 Nobel prize with Molina and Rowland, analyzed data collected on the expeditions and concluded that the ozone hole resulted from a combination of Antarctic weather conditions and human-made chemicals. In the frigid Antarctic winter, high-altitude clouds, or polar stratospheric clouds, were formed. In the spring, those clouds provided ideal conditions for CFC-derived chlorine and other chemicals to catalyze the destruction of massive amounts of ozone. The problem was exacerbated by the fact that prevailing air currents largely isolated Antarctica's atmosphere from the rest of Earth's atmosphere.

In the following years, scientists used data from ground stations and satellites to show that ozone levels were declining globally. In 1987, those findings helped convince the world's nations to agree on the Montreal Protocol, which aimed to cut CFC production in half by 1998. Within 2 years, however, further scientific evidence and computer modeling showed that more drastic measures would be needed if serious damage to the ozone layer were to be avoided. In 1990, the Montreal Protocol was strengthened to include a complete phaseout of CFCs by 2000. By 1998, the amount of chlorine in the atmosphere appeared to have leveled off, suggesting that the agreements had had the desired effect.

Acid deposition has not been reduced as much as scientists had hoped

Reducing acid precipitation involves reducing amounts of the pollutants that contribute to it. New technology has helped; "scrubbers" that filter pollutants in smokestacks have allowed factories to decrease emissions (▸ pp. 655–656).

As a result of declining emissions of SO_2, average sulfate precipitation in 1996–2000 was 10% lower than in 1990–1994 across the United States and 15% lower in the eastern states. However, because of increasing NO_x emissions, average nitrate precipitation increased nationally by 3% between these time periods.

FIGURE 17.18 Acidic deposition can harm vegetation, alter soil chemistry, affect soil- and forest-dwelling animals, and even eat away at statues and buildings.

A recent report by scientists at New Hampshire's Hubbard Brook research forest, where acidic deposition's effects were first demonstrated in the United States, disputed the notion that the problem of acid deposition is being solved (see "The Science behind the Story," ▸pp. 520–521). Instead, the report said, the effects are worse than first predicted, and the mandates of the 1990 Clean Air Act are not adequate to restore ecosystems in the northeastern United States. At Hubbard Brook, half the soil's calcium content has leached out, leaving the soil less able to neutralize future acid precipitation. This means that forests affected by acidification may take a long time to recover. An additional 80% reduction in sulfur emissions from electric utilities would be needed, the report estimates, to allow New Hampshire streams to recover in 20–25 years. The data on acid deposition show that although there have been many advances in the control of air pollution, more can clearly be done to alleviate outdoor pollution problems. The same can be said for indoor air pollution, a source of human health threats that is less familiar to most of us, but statistically more dangerous.

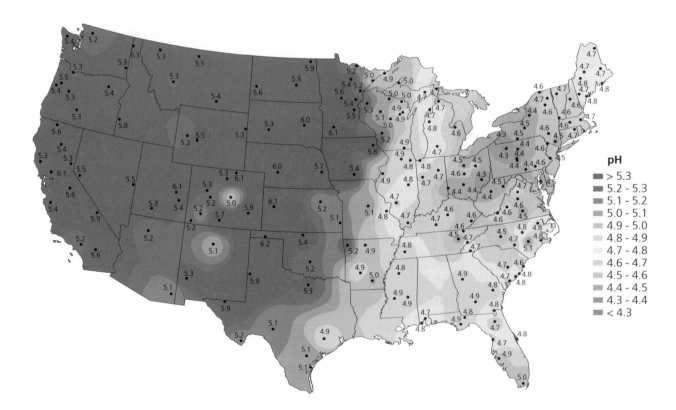

FIGURE 17.19 This map shows pH values for precipitation throughout the United States. Acid precipitation is most serious in parts of the Northeast and Midwest, generally downwind from (roughly east of) areas of heavy industrial development. Data from National Atmospheric Deposition Program. 2005. Hydrogen ion concentration as pH from measurements made at the Central Analytical Laboratory, 2003.

FIGURE 17.20 Indoor air pollution is an under-recognized health threat in both developed nations and developing nations. In the developing world, fires for cooking and heating are often built inside homes, as seen here in a South African kitchen (**a**), exposing family members to particulate matter and carbon monoxide. In most regions of the developing world, indoor air pollution is estimated to cause upwards of 3% of all health risks. In this graph (**b**), disability-adjusted life years (DALYs) indicate the burden of disease in total number of years of healthy life lost, including premature death and disability over a period of time, because of both indoor and outdoor air pollution. Indians suffer most severely from indoor air pollution, with approximately 650,000 years of life lost per million people, followed by sub-Saharan Africans, who suffer roughly 580,000 years loss of life per million people. *Source:* World Bank. 2002. Data from U.N. Development Programme, World Bank, Energy Sector Management Assistance Programme, 2002.

(a) South African family cooking indoors

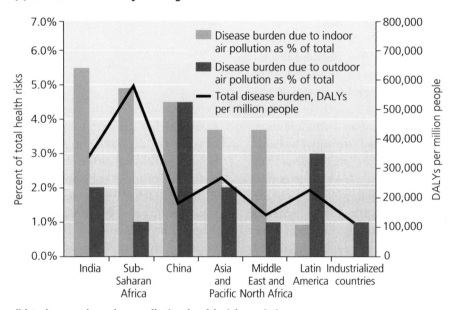

(b) Indoor and outdoor pollution health risk statistics

Indoor Air Pollution

Indoor air generally contains higher concentrations of pollutants than does outdoor air. As a result, the health effects from indoor air pollution in workplaces, schools, and homes outweigh those from outdoor air pollution (Figure 17.20). One estimate, from the U.N. Development Programme in 1998, attributed 2.2 million deaths worldwide to indoor air pollution and 500,000 deaths to outdoor air pollution. Indoor air pollution alone, then, takes roughly 6,000 lives each day.

If the impact of indoor air pollution seems surprising, consider that the average U.S. citizen spends at least 90% of his or her time indoors. Then consider that in the past half century a dizzying array of consumer products has been manufactured and sold, many of which we keep in our homes and offices and have come to play major roles in our daily lives. Many of these products are made of synthetic materials, and, as we saw in Chapter 14, novel synthetic substances are not comprehensively tested for health effects before being brought to market. Products such as insecticides and cleaning fluids can exude volatile chemicals into the air, as can solid materials such as plastics and chemically treated wood products.

In an ironic twist, some attempts to be environmentally prudent during the "energy crisis" of 1973–1974 worsened indoor air pollution in developed countries. To

The Science behind the Story

Acid Rain at Hubbard Brook Research Forest

The effects of acidic deposition are subtle, involving incremental changes in pH levels that take place over long periods of time and affect species with long life cycles, such as trees. For this reason, no single experiment can give us a complete picture of acidic deposition's effects. Nonetheless, one long-term study conducted in the Hubbard Brook Experimental Forest in New Hampshire's White Mountains has been critically important to our understanding of acidic deposition in the United States.

Established by the U.S. Forest Service in 1955, Hubbard Brook was initially devoted to research on hydrology, the study of water flow through forests and streams. In 1963, in collaboration with scientists at Dartmouth University, Hubbard Brook researchers broadened their focus to include a long-term study of nutrient cycling in forest ecosystems. Since then, they have collected and analyzed weekly samples of precipitation. The measurements make up the longest-running North American record of acid precipitation and have helped shape U.S. policy on sulfur and nitrogen emissions.

At Hubbard Brook, only one form of acidic deposition—wet deposition—has been measured regularly. Wet deposition is the

The effects of acidic deposition on trees can be seen in this forest on Mount Mitchell in western North Carolina.

deposition of rain, snow, fog, or sea spray with low pH onto soil, plants, or water. Dry deposition, in contrast, is the deposition of airborne acidic particles. Throughout Hubbard Brook's 3,160 ha (7,800 acres), small plastic collecting funnels, 30 cm (1 ft) in diameter at their openings, channel precipitation into clean bottles, which researchers retrieve and replace each week. Hubbard Brook's laboratory measures acidity and conductivity, which indicates the amount of salts and other electrolytic contaminants dissolved in the water. Concentrations of sulfuric acid, nitrates, ammonia,

and other compounds are measured elsewhere.

By the late 1960s, ecologists Gene Likens, F. Herbert Bormann, and others had found that precipitation at Hubbard Brook was several hundred times more acidic than natural rainwater. By the early 1970s, a number of other studies had corroborated their findings. Together, these studies indicated that precipitation from Pennsylvania to Maine had pH values averaging around 4, and that individual rainstorms showed values as low as 2.1—almost 10,000 times more acidic than ordinary rainwater.

reduce heat loss and improve energy efficiency, building managers sealed off most ventilation in existing buildings, and building designers constructed new buildings with limited ventilation and with windows that did not open. These steps may have saved energy, but they also worsened indoor air pollution by trapping stable, unmixed air—and its pollutants—inside.

Indoor air pollution in the developing world arises from fuelwood burning

Indoor air pollution has the greatest impact in the developing world. Millions of people in developing nations burn wood, charcoal, animal dung, or crop waste inside their homes for cooking and heating with little or no

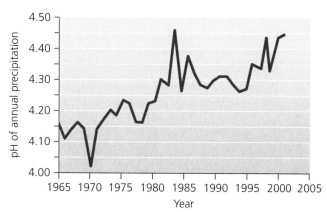

Over the past 40 years, precipitation at the Hubbard Brook Experimental Forest has become slightly less acidic. However, it is still far more acidic than is natural precipitation. Data from Likens, G. E. 2004. *Ecology* 85: 2355–2362.

In 1978, the National Atmospheric Deposition Program was launched to monitor precipitation and dry deposition across the United States. Initially consisting of 22 sites, including Hubbard Brook, the program now comprises more than 200, each of which gathers weekly data on acidic deposition and deposition of other substances. By the late 1980s, this program had produced a nationwide map of pH values. The most severe problems were found to be in the Northeast, where prevailing west-to-east winds were blowing emissions from fossil-fuel-burning power plants in the Midwest. Scientists hypothesized that when sulfur dioxide, nitrogen oxides, and other pollutants arrived in the Northeast, they were absorbed by water droplets in clouds, converted to acidic compounds such as sulfuric acid, and deposited on farms, forests, and cities in the form of rain or snow.

To some extent, the Clean Air Act of 1970 helped reduce acidic deposition in the Northeast. The accompanying figure shows the pH record for an area of Hubbard Brook known as Watershed 6. Between 1965 and 1995, average pH increased slightly, from about 4.15 to about 4.35. In 1990, as a consequence of the Hubbard Brook study and the nationwide research that followed, the Clean Air Act of 1970 was amended to further restrict emissions of sulfur dioxide and other acid-forming compounds. Nonetheless, acidic deposition continues to be a serious problem in the Northeast.

Some of the long-term consequences of acidic deposition are now becoming clear. In 1996, researchers reported that approximately 50% of the calcium and magnesium in Hubbard Brook's soils had been leached out. Meanwhile, acidic deposition had increased the concentration of aluminum in the soil, which can prevent tree roots from absorbing nutrients. The resulting nutrient deficiency slows forest growth and weakens trees, making them more vulnerable to drought and insects. It also reduces the ability of soil and water to neutralize acidity, making the ecosystem increasingly vulnerable to further inputs of acid.

In October 1999, researchers used a helicopter to distribute 50 tons of a calcium-containing mineral called wollastonite over one of Hubbard Brook's watersheds. Their objective was to raise the concentration of base cations to estimated historical levels. Over the next 50 years, scientists plan to evaluate the impact of calcium addition on the watershed's soil, water, and life. By providing a comparison to watersheds in which calcium remains depleted, the results should provide new insights into the consequences of acid rain and the possibilities for reversing its negative effects.

ventilation. In the process, they inhale dangerous amounts of soot and carbon monoxide. In the air of such homes, concentrations of particulate matter are commonly 20 times above U.S. EPA standards, the World Health Organization (WHO) has found. Poverty forces fully half the population and 90% of rural residents of developing countries to heat and cook with indoor fires.

Some people will burn almost any available fuel, even discarded plastic, in indoor fires. In doing so, these people are in effect subjecting themselves to a daily dose of London's smog of 1952. Indoor air pollution from fuelwood burning, the WHO estimates, kills 1.6 million people each year, comprising over 5% of all deaths in some developing nations and 2.7% of the entire global disease burden.

Many people who tend indoor fires are not aware of the health risks. They do not have access to the statistics showing that chemicals and soot released by burning coal, plastic, and other materials indoors can increase risks of pneumonia, bronchitis, allergies, sinus infections, cataracts, asthma, emphysema, heart disease, cancer, and premature death. Many who are aware of the health risks are too poor to have viable alternatives.

Even in the developed world, recognizing indoor air pollution as a problem is still quite novel. Fortunately, scientists have identified the most deadly indoor threats. Particulate matter and chemicals from wood and charcoal smoke are the primary health risks in the developing world. In developed nations, the top risks are cigarette smoke and radon, a naturally occurring radioactive gas.

Tobacco smoke and radon are the most dangerous indoor pollutants in the developed world

The health effects of smoking cigarettes are well known in developed countries, but only recently have scientists quantified the risks of inhaling secondhand smoke. Secondhand smoke, or environmental tobacco smoke, is smoke inhaled by a nonsmoker who is nearby or shares an enclosed airspace with a smoker. Secondhand smoke has been found to cause many of the same problems as directly inhaled cigarette smoke, ranging from irritation of the eyes, nose, and

throat, to exacerbation of asthma and other respiratory ailments, to lung cancer. This hardly seems surprising when one considers that environmental tobacco smoke consists of a brew of over 4,000 chemical compounds, many of which are known or suspected to be toxic or carcinogenic.

Women living with a spouse who smokes have a 24% greater chance of developing lung cancer from secondhand smoke, one study has indicated. Fortunately, the popularity of smoking has declined greatly in the United States and some other nations in recent years. The exposure of young children in the United States has decreased by almost half. A 1998 study found that 20% of young children had been exposed, as compared to 39% in 1986.

After cigarette smoke, radon gas is the second-leading cause of lung cancer in the United States, responsible for an estimated 20,000 deaths per year. Worldwide, the WHO estimates that radon may account for 15% of lung cancer cases. As we saw in Chapter 14 (▶p. 405), radon is a radioactive gas resulting from the natural decay of uranium in soil, rock, or water, which seeps up from the ground and can infiltrate buildings. Radon is colorless and odorless, and it can be impossible to predict where it will occur without knowing details of an area's underlying geology (Figure 17.21). As a result, the only way to determine whether radon is entering a building is to measure radon with a test kit. Testing in 1991 led the EPA to estimate that 6% of U.S. homes exceeded the EPA's maximum recommended level for radon. Since the mid-1980s,

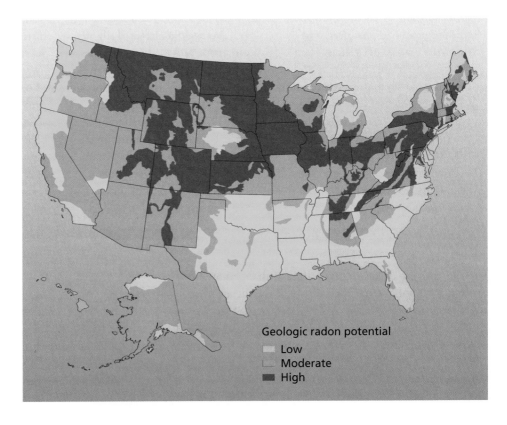

FIGURE 17.21 The risk from radon depends largely on underground geology. This map shows relative levels of risk from radon across the United States, but there is much fine-scale geographic variation from place to place not evident on the map. Testing your home for radon is the surest way to determine whether this colorless, odorless gas could be a problem in your home. Data from U.S. Geological Survey. 1993. Generalized geological radon potential of the United States, 1993.

Geologic radon potential
Low
Moderate
High

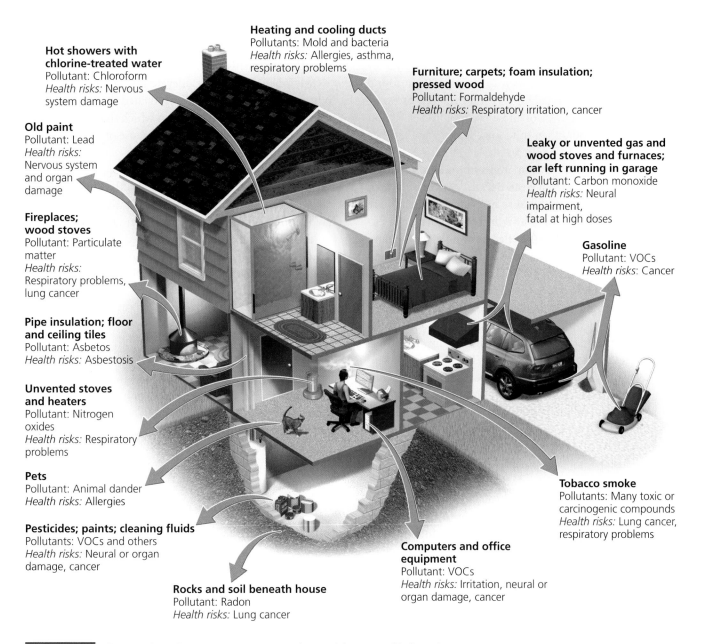

Hot showers with chlorine-treated water
Pollutant: Chloroform
Health risks: Nervous system damage

Heating and cooling ducts
Pollutants: Mold and bacteria
Health risks: Allergies, asthma, respiratory problems

Furniture; carpets; foam insulation; pressed wood
Pollutant: Formaldehyde
Health risks: Respiratory irritation, cancer

Old paint
Pollutant: Lead
Health risks: Nervous system and organ damage

Leaky or unvented gas and wood stoves and furnaces; car left running in garage
Pollutant: Carbon monoxide
Health risks: Neural impairment, fatal at high doses

Fireplaces; wood stoves
Pollutant: Particulate matter
Health risks: Respiratory problems, lung cancer

Gasoline
Pollutant: VOCs
Health risks: Cancer

Pipe insulation; floor and ceiling tiles
Pollutant: Asbetos
Health risks: Asbestosis

Unvented stoves and heaters
Pollutant: Nitrogen oxides
Health risks: Respiratory problems

Pets
Pollutant: Animal dander
Health risks: Allergies

Tobacco smoke
Pollutants: Many toxic or carcinogenic compounds
Health risks: Lung cancer, respiratory problems

Pesticides; paints; cleaning fluids
Pollutants: VOCs and others
Health risks: Neural or organ damage, cancer

Computers and office equipment
Pollutant: VOCs
Health risks: Irritation, neural or organ damage, cancer

Rocks and soil beneath house
Pollutant: Radon
Health risks: Lung cancer

FIGURE 17.22 The typical U.S. home contains a variety of potential sources of indoor air pollution. Shown are some of the most common sources, the major pollutants they emit, and some of the health risks they pose.

18 million U.S. homes have been tested for radon, and 700,000 have undergone radon mitigation. New homes are now being built with radon-resistant features—over a million such homes since 1990.

Many volatile organic compounds pollute indoor air

In our daily lives at home, we are exposed to many indoor air pollutants (Figure 17.22). The most diverse indoor pollutants are VOCs. These airborne carbon-containing compounds are released by everything from plastics to oils to perfumes to paints to cleaning fluids to adhesives to pesticides. VOCs evaporate from furnishings, building materials, color film, carpets, laser printers, fax machines, and sheets of paper. Some products, such as chemically treated furniture, release large amounts of VOCs when new and progressively less as they age. Other items, such as photocopying machines, emit VOCs each time they are used.

Although we are surrounded by products that emit VOCs, they are released in very small amounts. EPA and European surveys have both measured overall levels of VOCs in buildings and have found them to be nearly always less than 0.1 part per million (ppm). This is, however, a substantially greater concentration than is generally found outdoors. When the EPA conducted a survey of U.S. office buildings between 1994 and 1998, it found 34 of the 48 VOCs it tested for, and found at least one type in 81% of buildings. All of these compounds existed at higher levels indoors than outdoors, suggesting that they originated from indoor sources.

The implications for human health of chronic exposure to VOCs are far from clear. Because they exist in such low concentrations and because individuals regularly are exposed to mixtures of many different types, it is extremely difficult to study the effects of any one pollutant. An exception is formaldehyde, which does have clear and known health impacts. This VOC, one of the most commonly synthetically produced chemicals, irritates mucous membranes, induces skin allergies, and causes other ailments. Formaldehyde is used in numerous products, but health complaints have mainly resulted from its leakage from pressed wood and insulation. The use of plywood has decreased in the last decade because of health concerns over formaldehyde.

VOCs also include pesticides, which we examined in Chapters 10 and 14. Three-quarters of U.S. homes use at least one pesticide indoors during an average year, but most are used outdoors. Thus it may seem surprising that the EPA found in a 1990 study that 90% of people's pesticide exposure came from indoor sources. Households that the agency tested had multiple pesticide volatiles in their air, at levels 10 times above levels measured outside. Some of the pesticides found had apparently been used years previously against termites, and then seeped into the houses through floors and walls. DDT, banned 15 years before the study, was found in five of eight homes, probably having been brought in on the soles of occupants' shoes from outdoors.

Living organisms can pollute indoor spaces

Tiny living organisms can be or produce indoor pollutants. In fact, they may be the most widespread source of indoor air pollution in the developed world. Dust mites and animal dander can exacerbate asthma in children. Some fungi, mold, and mildew (in particular, their airborne spores) can cause potentially severe health problems, including allergies, asthma, and other respiratory ailments. Some airborne bacteria can cause infectious disease. One example is the bacterium that causes Legionnaires' disease. Of the estimated 10,000–15,000 annual U.S. cases of Legionnaires' disease, 5–15% are fatal. Heating and cooling systems in buildings make ideal breeding

grounds for microbes, providing moisture, dust, and foam insulation as substrates, as well as air currents to carry the organisms aloft.

Microbes that induce allergic responses are thought to be a major cause of building-related illness, a sickness produced by indoor pollution in which the specific cause may not be identifiable. When the cause of such an illness is a mystery, and when symptoms are general and non-specific, the illness is often called *sick-building syndrome.* The U.S. Occupational Safety and Health Administration (OSHA) has estimated that 30–70 million Americans have suffered ailments due to the environment of the building in which they live. We can reduce the prevalence of sick building syndrome by using low-toxicity building materials and ensuring that buildings are adequately ventilated.

Weighing the Issues:
How Safe Is Your Indoor Environment?

Think about the amount of time you spend indoors. Name the potential indoor air quality hazards in your home, work, or school environment. Are these spaces well-ventilated? What could you do to make the indoor spaces you use safer?

We can reduce indoor air pollution

Using low-toxicity material, monitoring air quality, keeping rooms clean, and providing adequate ventilation are the keys to alleviating indoor air pollution in most situations. In the developed world, we can try to limit our use of plastics and treated wood where possible and to limit our exposure to pesticides, cleaning fluids, and other known toxicants by keeping them in a garage or outdoor shed rather than in the house. The EPA recommends that we test our homes and offices for radon. Because carbon monoxide is so deadly and so hard to detect, many homes are equipped with detectors that sound an alarm if incomplete combustion produces dangerous levels of CO. In addition, keeping rooms and air ducts clean and free of mildew and other biological pollutants will reduce potential irritants and allergens. Finally, it is important to keep our indoor spaces as well ventilated as possible to minimize concentrations of the pollutants among which we live.

Remedies for fuelwood pollution in the developing world include drying wood before burning (which reduces the amount of smoke produced), cooking outside, shifting to less-polluting fuels (such as natural gas), and replacing inefficient fires with cleaner stoves that burn fuel more efficiently. For example, the Chinese government has invested in a program that has placed more fuel-efficient stoves in millions of homes in China. According to WHO studies, this is a relatively cost-efficient means of reducing the health impacts of indoor biomass combustion. Installing hoods, chimneys, or cooking windows can increase ventilation for little cost, alleviating the majority of indoor smoke pollution.

Conclusion

Indoor air pollution is a potentially serious health threat. However, by keeping informed of the latest scientific findings and taking appropriate actions, we as individuals can significantly minimize the risks to our families and ourselves. Outdoor air pollution has been addressed more effectively by government legislation and regulation. In fact, reductions in outdoor air pollution levels in the United States and other developed nations represent some of the greatest strides made in environmental protection to date. Much room for improvement remains, however, particularly in reducing acidic deposition and the photochemical smog resulting from urban congestion. Fortunately, developed nations no longer experience the type of pollution that Londoners suffered in their 1952 killer smog. Nevertheless, avoiding such high pollutant levels in the developing world will continue to pose a challenge as less-wealthy nations industrialize.

REVIEWING OBJECTIVES

You should now be able to:

Describe the composition, structure, and function of Earth's atmosphere

▶ The atmosphere consists of 78% nitrogen gas, 21% oxygen gas, and a variety of permanent and variable gases in minute concentrations. (p. 498)

▶ The atmosphere includes four principal layers: the troposphere, stratosphere, mesosphere, and thermosphere. Temperature and other characteristics vary across these layers. Ozone is concentrated in the stratosphere. (pp. 498–499)

▶ The sun's energy heats the atmosphere, drives air circulation, and helps determine weather, climate, and the seasons. (pp. 500–501)

▶ Weather is a short-term phenomenon, whereas climate is a long-term phenomenon. Fronts, pressure systems, and the interactions among air masses influence weather. (pp. 501–503)

▶ Global convective cells called Hadley, Ferrel, and polar cells create latitudinal climate zones. (pp. 504–505)

Outline the scope of outdoor air pollution and assess potential solutions

▶ Natural sources such as windblown dust, volcanoes, and fires account for much atmospheric pollution, but human activity can exacerbate some of these phenomena. (pp. 505–506)

▶ Human-emitted pollutants include primary and secondary pollutants from point and non-point sources. (p. 506)

▶ To safeguard public health, the U.S. EPA regulates six criteria pollutants (carbon monoxide, lead, nitrogen dioxide, tropospheric ozone, sulfur dioxide, and particulate matter), as well as volatile organic compounds and 188 toxic air pollutants. (pp. 506–509)

▶ Emissions in the United States have decreased substantially since the Clean Air Act of 1970, and ambient air quality is much improved. (pp. 508–509)

▶ Industrial smog like that which blanketed 1952 London is produced by fossil fuel combustion and is still a problem in urban and industrial areas of many developing nations. (pp. 509–510)

▶ Photochemical smog is created by chemical reactions of pollutants in the presence of sunlight. It impairs visibility and human health in urban areas. (pp. 510–511)

Explain stratospheric ozone depletion and identify steps taken to address it

▶ CFCs destroy stratospheric ozone, and thinning ozone concentrations pose dangers to life because they allow more ultraviolet radiation to reach Earth's surface. (pp. 511, 513–514, 516–517)

▶ The Montreal Protocol and its follow-up agreements have proven remarkably successful in reducing emissions of ozone-depleting compounds. (pp. 513–514, 517)

Define acidic deposition and illustrate its consequences

▶ Acidic deposition results when pollutants such as SO_2 and NO react in the atmosphere to produce strong acids that are deposited on Earth's surface. (pp. 514–515)

▶ Acidic deposition may occur a long distance from the source of pollution. (pp. 516, 518)

▶ Water bodies, soils, trees, and ecosystems all experience negative impacts from acidic deposition. (pp. 515–518, 520–521)

Characterize the scope of indoor air pollution and assess potential solutions

▶ Indoor air pollution causes far more deaths and health problems worldwide than outdoor air pollution. (p. 519)

▶ Indoor burning of fuelwood is the developing world's primary indoor air pollution risk. (pp. 519–522)

▶ Tobacco smoke and radon are the deadliest indoor pollutants in the developed world. (pp. 522–523)

▶ Volatile organic compounds and living organisms can pollute indoor air. (pp. 523–524)

▶ Using low-toxicity building materials, keeping spaces clean, monitoring air quality, and maximizing ventilation are some of the steps we can take to reduce indoor air pollution. (pp. 524–525)

TESTING YOUR COMPREHENSION

1. About how thick is Earth's atmosphere? Name one characteristic of each of the four atmospheric layers.

2. Where is the "ozone layer" located? How and why is stratospheric ozone beneficial for people, and tropospheric ozone harmful?

3. How does solar energy influence weather and climate? How do Hadley, Ferrel, and polar cells help to determine long-term climatic patterns and the location of biomes?

4. What factors led to the deadly smog in London in 1952? Describe a thermal inversion.

5. Name three natural sources of outdoor air pollution and three sources caused by human activity.

6. What is the difference between a primary and a secondary pollutant? Give an example of each.

7. What is smog? How is smog formation influenced by the weather? By topography? How does photochemical, or brown-air, smog differ from industrial, or gray-air, smog?

8. How do chlorofluorocarbons (CFCs) deplete stratospheric ozone? Why is this depletion considered a long-term international problem? What was done to address this problem?

9. Why are the effects of acidic deposition often felt in areas far from where the primary pollutants are produced?

10. Name five common sources of indoor pollution. For each, describe one way to reduce one's exposure to this source.

SEEKING SOLUTIONS

1. Consider London's "killer smog" of 1952 and modern urban pollution by photochemical smog. Describe several factors that make it particularly difficult to study causes of air pollution and to develop solutions.

2. How may human activity sometimes exacerbate natural forms of air pollution? Discuss two examples and potential solutions.

3. Describe how and why emissions of major pollutants have been reduced by over 50% in the United States since 1970, despite increases in population and economic activity.

4. International regulatory action has produced reductions in CFCs, but other transboundary pollution issues, including acidic deposition, have not yet been addressed as effectively. What types of actions do you feel are appropriate for pollutants that cross political boundaries?

5. Consider volatile chemicals, such as formaldehyde and other VOCs, that may be emitted at very low levels over long periods of time from manufactured products. What do you think are the best ways to lessen the health impacts of such indoor pollutants?

6. You have just become the head of your county health department, and the EPA has informed you that your county has failed to meet the national ambient air quality standards for ozone, sulfur dioxide, and nitrogen dioxide. Your county is partly rural but is home to a city of 200,000 people and 10 sprawling suburbs. There are several large and aging coal-fired power plants, a number of factories with advanced pollution control technology, and no public transportation system. What steps would you urge the county government to take to meet the air quality standards? Explain how you would prioritize these steps.

INTERPRETING GRAPHS AND DATA

Since the Clean Air Act of 1970, total emissions of carbon monoxide, sulfur dioxide, nitrogen oxides, VOCs, lead, and particulate matter have dropped by over 50% while U.S.

population, energy consumption, and economic productivity have all increased. As you learned from the "Interpreting Graphs and Data" feature in Chapter 3 (▶ pp. 86–87), lead emissions resulted mostly from the combustion of leaded gasoline and dropped precipitously once leaded gasoline was phased out. Consider the other pollutants listed above as you interpret the data in the graph.

1. Relative to 1970, what are the percentage changes by 2004 for each of the five variables graphed above? What are the percentage changes in the per capita values of each variable over this time period?
2. Fossil fuel combustion is the major source of most of the pollutants above. What is the percentage change in aggregate emissions of the six principal pollutants per unit of energy consumed in 2004, compared to 1970?
3. Do you think that additional reductions in emissions are likely to result primarily from changing technology (e.g., hybrid cars, advanced catalytic converters) or from changing behavior (e.g., driving fewer miles per person per year)? Use the data above to support your claim.

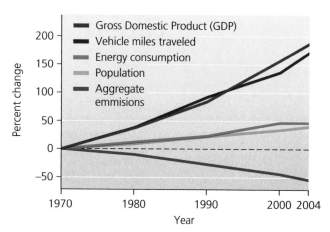

Trends from 1970 to 2004 in economic production, vehicle miles traveled, energy consumption, population, and aggregate emissions of the six principal air pollutants monitored by the EPA. Data from U.S. EPA. 2004. *Air emission trends: Continued progress through 2004.*

CALCULATING ECOLOGICAL FOOTPRINTS

According to EPA data, emissions of nitrogen oxides in the United States in 2000 were 24,899,000 tons, of which 13,251,000 tons were from the transportation sector. Of this amount, 8,150,000 tons came from on-road vehicles, with 5,859,000 tons of this total coming from light-duty cars and trucks. The U.S. Census Bureau estimated the nation's population to be 282,192,162 at mid-year in 2000 and projects that it will reach 300,000,000 by early 2007. Considering these data, calculate the missing values in the table below (1 ton = 2,000 lb).

	Total NO_x emissions (lb)	NO_x emissions due to light-duty vehicles (lb)
You	176.5	41.5
Your class		
Your state		
United States		

Data from U.S. EPA. 2003. *National air quality and emissions trends report 2003*, Appendix A4.

1. By what percentage is the U.S. population projected to increase between 2000 and 2007? Do you think that NO_x emissions will increase, decrease, or remain the same over that period of time? Why? (You may want to refer to Figure 17.11.)
2. If you reduced by half the vehicle miles traveled for which you are responsible, how many pounds of NO_x emissions would you prevent? What percentage of your total NO_x emissions would that be?
3. How might you reduce your vehicle miles traveled by 50%? What other steps could you take to reduce the NO_x emissions for which you are responsible?

Take It Further

Go to www.aw-bc.com/withgott or the student CD-ROM where you'll find:

▶ Suggested answers to end-of-chapter questions
▶ Quizzes, animations, and flashcards to help you study
▶ *Research Navigator*™ database of credible and reliable sources to assist you with your research projects

▶ **GRAPHit!** tutorials to help you master how to interpret graphs
▶ **INVESTIGATEit!** current news articles that link the topics that you study to case studies from your region to around the world

18 Global Climate Change

Muratti Island of the Maldives

Upon successfully completing this chapter, you will be able to:

▶ Describe Earth's climate system and explain the variety of factors influencing global climate

▶ Characterize human influences on the atmosphere and global climate

▶ Delineate modern methods of climate research

▶ Summarize current consequences and potential future impacts of global climate change

▶ Evaluate the scientific, political, and economic debates concerning climate change

▶ Suggest potential responses to climate change

Fishermen of the Maldives

Africa

India

Indian Ocean

Maldives

Central Case: Rising Temperatures and Seas May Take the Maldives Under

"The impact of global warming and climate change can effectively kill us off, make us refugees. . . ."
—ISMAIL SHAFEEU, MINISTER OF ENVIRONMENT, MALDIVES

"More people enjoy health and prosperity now than ever, and a warmer environment . . . should sustain life better than the current one does."
—OIL AND GAS JOURNAL, FEBRUARY 2001

A nation of low-lying coral islands, or atolls, in the Indian Ocean, the Maldives is known for its spectacular tropical setting, colorful coral reefs, and sun-drenched beaches. For visiting tourists it is a paradise, and for 320,000 Maldives residents it is home. But residents and tourists alike now fear that the Maldives could soon be submerged by the rising seas that are accompanying global climate change.

Nearly 80% of the Maldives' land area of 300 km^2 (116 mi^2) lies less than 1 m (39 in.) above sea level. In a nation of 1,190 islands whose highest point is just 2.4 m (8 ft) above sea level, rising seas could be a matter of life or death. The world's oceans rose 10–20 cm (4–8 in.)

during the 20th century as warming temperatures expanded ocean water and as melting icecaps discharged water into the ocean. Current projections are that sea level will rise another 9–88 cm (3.5–35 in.) by the year 2100.

Higher seas are expected to flood large areas of land in the Maldives and to cause salt water to intrude into drinking water supplies. Moreover, if climate change produces larger and more powerful storms, these could worsen flooding and damage the coral reefs that are so crucial to the nation's tourism- and fishing-driven economy. Because of such concerns, the Maldives government has evacuated residents from several of the lowest-lying islands in recent years.

On December 26, 2004, the nation got a taste of what could be in store in the future. The massive *tsunami,* or tidal wave, that devastated coastal areas throughout the Indian Ocean hit the Maldives particularly hard. One hundred people were killed and 20,000 lost their homes, while schools, boats, tourist resorts, hospitals, and transportation and communication infrastructure were destroyed or badly damaged. On a per capita basis, the

Maldives suffered a greater economic shock from the tsunami than any other nation. The World Bank estimates that direct damage in the Maldives totaled $470 million, an astounding 62% of the nation's gross domestic product (GDP). Soil erosion, saltwater contamination of aquifers, and other environmental damage will result in still greater long-term economic losses.

The tsunami was caused *not* by climate change, but by an earthquake. Yet as sea level rises, the damage that such natural events—or ordinary storm waves—can inflict increases considerably. Maldives islanders are not alone in their predicament. Other island nations, from the Galapagos to Fiji to the Seychelles, are also fearing a future in which they may be constantly battling encroaching seawater. Mainland coastal areas of the world, such as the hurricane-battered coasts of Florida and Louisiana, will face similar issues. In one way or another, global climate change seems certain to affect each and every one of us for the remainder of our lifetimes.

Earth's Hospitable Climate

As we learned in Chapter 17, *weather* describes an area's short-term atmospheric conditions (over hours or days), including temperature, moisture content, wind, precipitation, barometric pressure, solar radiation, and other characteristics. *Climate* is an area's long-term pattern of atmospheric conditions. **Global climate change** describes changes in Earth's climate, involving aspects such as temperature, precipitation, and storm frequency and intensity. Although people often use the term *global warming* synonymously in casual conversation, *global warming* refers specifically to an increase in Earth's average surface temperature and thus is only one aspect of global climate change.

Our planet's climate has never been entirely stable and unchanging. However, the climatic changes taking place today are unfolding at an exceedingly rapid rate. Moreover, most scientists agree that human activities, notably fossil fuel combustion and deforestation, are largely responsible for the current modification of Earth's atmosphere and climate. Climatic changes will likely have adverse consequences for ecosystems and for millions of people, including residents of the Maldives, Florida, Louisiana, and other regions. For this reason, increasing numbers of scientists, policymakers, and ordinary citizens are seeking to take action to minimize and mitigate our impacts on the climate system.

The sun and the atmosphere keep Earth warm

Three factors exert more influence on Earth's climate than all others combined. The first is the sun. Without it, Earth would be dark and frozen. The second is the atmosphere. Without it, Earth would be as much as 33 °C (59 °F) colder on average, and temperature differences between night and day would be far greater. The third is the oceans, which shape climate by storing and transporting heat and moisture.

The sun is the source of most of the energy that Earth receives. Earth's atmosphere, clouds, land, ice, and water together reflect about 30% of incoming solar radiation back into space. The remaining 70% is absorbed by molecules in the atmosphere, clouds, or by land, water, or ice at Earth's surface (Figure 18.1).

"Greenhouse gases" warm the lower atmosphere

As Earth's surface absorbs solar radiation, the surface increases in temperature and emits radiation in the infrared portion of the spectrum (▶ pp. 105–106). Some atmospheric gases absorb infrared radiation released from Earth's surface very effectively. These include water vapor, ozone, carbon dioxide (CO_2), nitrous oxide (N_2O), methane (CH_4), and halocarbons. Halocarbons are a diverse group of gases that include chlorofluorocarbons (CFCs; ▶ pp. 513, 516–517) and hydrochlorofluorocarbons (HFCs). Such gases that absorb infrared radiation from Earth's surface are known as **greenhouse gases.** These gases subsequently re-emit infrared energy of slightly different wavelengths, warming the atmosphere (specifically the *troposphere;* ▶ pp. 498–499) and the planet's surface. This warming of the troposphere and Earth's surface is known as the **greenhouse effect.** Despite its wide usage, the term *greenhouse effect* is actually a bit of a misnomer. The greenhouses we use for growing plants hold heat in place by preventing warm air from escaping. Atmospheric greenhouse gases, in contrast, do not trap air, but instead absorb, transform, and radiate heat.

The greenhouse effect is a natural phenomenon, and greenhouse gases (with the exception of the anthropogenic halocarbons) have been present in the atmosphere for billions of years. However, human activities have increased the concentrations of many greenhouse gases in the past 250–300 years, and we have thereby enhanced the greenhouse effect.

Not all greenhouse gases are equally effective in warming the troposphere and surface. *Global warming potential*

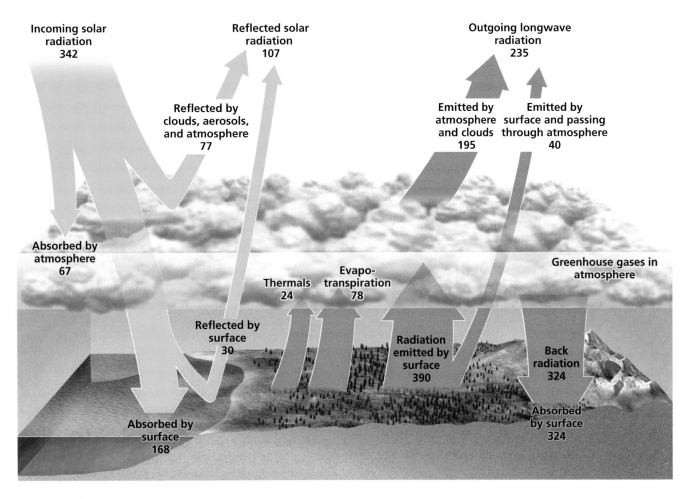

Incoming solar radiation 342

Reflected solar radiation 107

Outgoing longwave radiation 235

Reflected by clouds, aerosols, and atmosphere 77

Emitted by atmosphere and clouds 195

Emitted by surface and passing through atmosphere 40

Absorbed by atmosphere 67

Greenhouse gases in atmosphere

Thermals 24

Evapo-transpiration 78

Reflected by surface 30

Radiation emitted by surface 390

Back radiation 324

Absorbed by surface 168

Absorbed by surface 324

FIGURE 18.1 Earth's climate system is in rough equilibrium; our planet emits about the same amount of energy that it receives from the sun. As greenhouse gases accumulate in the atmosphere, however, they increase the amount of radiation that is emitted from the atmosphere back toward the surface. This illustration shows major pathways of energy flow in watts per square meter. Data from Kiehl, J. T., and K. E. Trenberth. 1997. Earth's annual global mean energy budget. *Bull. Amer. Meteorol. Soc.* 78: 197–208.

refers to the relative ability of one molecule of a given greenhouse gas to contribute to warming. Table 18.1 shows the global warming potential for several greenhouse gases. Values are expressed in relation to carbon dioxide, which is assigned a global warming potential of 1. Thus, a molecule of methane is 23 times as potent as a molecule of carbon dioxide, and a molecule of nitrous oxide is 296 times as potent as a CO_2 molecule.

Carbon dioxide is the primary greenhouse gas

Although carbon dioxide is not the most potent greenhouse gas on a per-molecule basis, its abundance in the atmosphere relative to gases such as methane and nitrous oxide means that it contributes more to the greenhouse

effect. For this reason, changes in the atmospheric concentration of CO_2 are important. Human activities have contributed to a significant increase in the atmospheric concentration of carbon dioxide, from around 280 parts per million (ppm) as recently as the late 1700s to 316 ppm in 1959 to 378 ppm in 2004 (Figure 18.2). The atmospheric CO_2 concentration is now at its highest level in at least 400,000 years, and likely the highest in the last 20 million years. Moreover, it is increasing faster today than at any time in at least 20,000 years.

What has changed since the 1700s to cause the atmospheric concentration of this greenhouse gas to increase so rapidly? As you may recall from our discussion of the carbon cycle in Chapter 7 (▶pp. 195–197), and as we will see further in Chapter 19 (▶pp. 561–562), a great deal of carbon is stored for long periods in the upper layers of

Table 18.1	Global Warming Potentials of Four Greenhouse Gases	
Greenhouse gas	**Relative heat-trapping ability (in CO_2 equivalents)**	
Carbon dioxide	1	
Methane	23	
Nitrous oxide	296	
Hydrochlorofluorocarbon HFC-23	12,000	

Data from Intergovernmental Panel on Climate Change, 2001. *Climate change 2001: The scientific basis.*

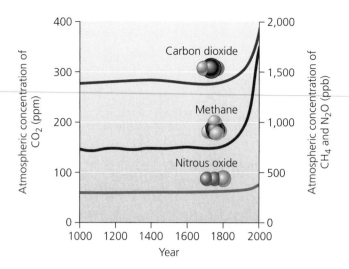

FIGURE 18.2 Global atmospheric concentrations of carbon dioxide (CO_2; top curve), methane (CH_4; middle curve), and nitrous oxide (N_2O; bottom curve) have increased dramatically since 1800. Data from Intergovernmental Panel on Climate Change. 2001. *Third assessment report.*

the lithosphere. The deposition, partial decay, and compression of organic matter (mostly plants) that grew in wetland or marine areas during the Carboniferous Period led to the formation of coal, oil, and natural gas in sediments from that time. In the absence of human activity, these carbon reservoirs would be practically permanent. However, over the past two centuries we have been burning increasing amounts of fossil fuels in our homes, factories, and automobiles. In so doing, we have transferred large amounts of carbon from one reservoir (the underground deposits that stored the carbon for millions of years) to another (the atmosphere). This flux of carbon from lithospheric reservoirs into the atmosphere is the main reason atmospheric carbon dioxide concentrations have increased so dramatically in such a short time.

At the same time, people have cleared and burned forests to make room for crops, pastures, villages, and cities. Forests serve as carbon sinks (▶p. 195), and their removal, especially in areas where they are slow to recover, can reduce the biosphere's ability to absorb carbon dioxide from the atmosphere. Therefore, deforestation has also contributed to increasing atmospheric carbon dioxide concentrations.

Other greenhouse gases add to warming

Carbon dioxide is not the only greenhouse gas increasing in the atmosphere. Methane is also on the rise. We release methane into the atmosphere by tapping into fossil fuel deposits, raising livestock that release methane as a metabolic waste product, disposing of organic matter in landfills, and growing certain types of crops, especially rice. Since 1750, atmospheric methane concentrations have increased by 151% (see Figure 18.2), and the current concentration is the highest in at least 400,000 years.

Nitrous oxide is another greenhouse gas whose atmospheric concentration has increased because of human activities. This gas, a by-product of feedlots, chemical manufacturing plants, auto emissions, and use of nitrogen fertilizers in modern agricultural practices, has risen by 17% since 1750 (see Figure 18.2).

Ozone concentrations in the troposphere have increased by 36% since 1750 as a result of processes described in Chapter 17 (▶pp. 507, 510–511). Halocarbon gases add greatly to the atmosphere's heat-absorbing ability on a per-molecule basis (see Table 18.1). Their overall contribution to global warming, however, has begun to slow because of the Montreal Protocol and subsequent controls (▶pp. 513–514).

Emissions of greenhouse gases from human activity in the United States consist mostly of carbon dioxide. Even after accounting for the greater global warming potential of other gases, carbon dioxide remains the major contributor to warming (Figure 18.3).

Water vapor is the most abundant greenhouse gas in the atmosphere. Its concentration varies, but if tropospheric temperatures continue to increase, the oceans and other water bodies should transfer increasingly more water vapor into the atmosphere. Such a positive feedback mechanism could amplify the greenhouse effect. Alternatively, increased water vapor concentrations could give rise to increased cloudiness, which might, in a negative feedback loop, slow global warming by reflecting more incoming solar radiation back into space. Depending on whether low- or high-elevation clouds resulted, they could either shade and cool Earth (negative feedback) or

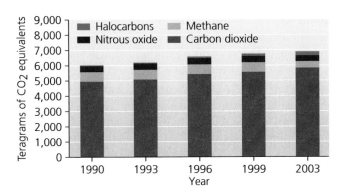

FIGURE 18.3 Emissions of six main greenhouse gases in the United States increased by 13.3% between 1990 and 2003. Even after accounting for the greater global warming potentials of methane, nitrous oxide, and three types of halocarbons, carbon dioxide accounts for nearly 85% of potential warming. Data from U.S. Environmental Protection Agency. 2005. *Inventory of U.S. greenhouse gas emissions and sinks: 1990–2003.* 430-R-05-003. Washington, D.C.: U.S. EPA.

else contribute to warming and accelerate evaporation and further cloud formation (positive feedback). Because of feedback loops (▸pp. 184–185), minor modifications of minor components of the atmosphere can potentially lead to major changes in climate.

Aerosols and other elements may exert a cooling effect on the lower atmosphere

Whereas greenhouse gases exert a warming effect, *aerosols,* microscopic droplets and particles, can have either a warming or cooling effect. Generally speaking, soot aerosols, also known as black carbon aerosols, can cause warming, but most tropospheric aerosols lead to climate cooling. Sulfate aerosols produced by fossil fuel combustion may slow global warming, at least in the short term. When sulfur dioxide enters the atmosphere, it undergoes a number of reactions, some of which lead to acid precipitation (▸p. 515). These reactions, along with volcanic eruptions, also contribute to the formation of a sulfur-rich aerosol haze in the upper atmosphere. Such a haze may reduce the amount of sunlight that reaches and heats Earth's surface. Major volcanic eruptions and the aerosols they release can also exert short-term cooling effects on Earth's climate on the scale of a few years.

The atmosphere is not the only factor that influences climate

Although the atmosphere shapes climate, the amount of energy released by the sun, as well as changes in Earth's rotation and orbit, also affect climate.

Milankovitch cycles During the 1920s, Serbian mathematician Milutin Milankovitch described three kinds of changes in Earth's rotation and orbit around the sun. These variations, now known as **Milankovitch cycles,** result in slight changes in the relative amount of solar radiation reaching Earth's surface at different latitudes over the long term (Figure 18.4). As these cycles proceed, they change the way solar radiation is distributed over Earth's surface. This, in turn, contributes to changes in atmospheric heating and circulation that have triggered climate variation, such as periodic glaciation episodes.

Oceanic circulation The oceans also shape climate. Ocean water exchanges tremendous amounts of heat with the atmosphere, and ocean currents move energy from one place to another. In equatorial regions, such as the area around the Maldives, the oceans receive more heat from the sun and atmosphere than they emit. Near the poles, the oceans emit more heat than they receive. Because cooler water is denser than warmer water, the cooling water at the poles tends to sink, and the warmer

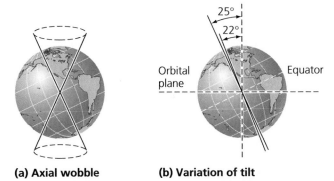

(a) Axial wobble   **(b) Variation of tilt**

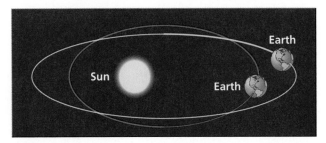

(c) Variation of orbit

FIGURE 18.4 There are three types of Milankovitch cycles. The first is an axial wobble (**a**) that occurs on a 19,000- to 23,000-year cycle. The second is a 3-degree shift in the tilt of Earth's axis (**b**) that occurs on a 41,000-year cycle. The third is a variation in Earth's orbit from almost circular to more elliptical (**c**), which repeats itself every 100,000 years. These variations affect the intensity of solar radiation that reaches portions of Earth at different times, creating long-term changes in global climate.

surface water from the equator moves to take its place. This is one of the principles explaining global oceanic circulation (▸pp. 470–471).

The best-known interaction among ocean circulation, atmospheric circulation, and global climate involves the phenomena named El Niño and La Niña (see "The Science behind the Story," ▸pp. 536–537). **El Niño** conditions are triggered when equatorial winds weaken and allow warm water from the western Pacific to move eastward, preventing cold water from welling up in the eastern Pacific. In **La Niña** events, cold surface waters extend far westward in the equatorial Pacific. Both these events affect temperature and precipitation patterns around the world in complex ways. Scientists are exploring whether globally warming air and sea temperatures may be increasing the frequency and strength of El Niño events.

Another way in which climate and ocean currents interact involves the northern Atlantic Ocean. Here, warm surface water moves from the equator northward, carrying heat to higher latitudes, and keeping Europe much warmer than it would otherwise be. As the surface water of the North Atlantic releases heat energy and cools, it becomes denser and sinks. The deep part of this circulation pattern is called the *North Atlantic Deep Water (NADW)*. Figure 18.5 portrays this circulation pattern as a sort of conveyor belt that moves water and heat from one place to another. Recently, scientists have realized that interrupting the NADW could trigger rapid climate change. If global warming causes large portions of the Greenland Ice

Sheet to melt, freshwater runoff into the North Atlantic would increase. Surface waters would become less dense from such dilution and warming, because warm freshwater is less dense than cold salt water. This could potentially stop the NADW circulation, shutting down the northward flow of warm equatorial water. The entire North Atlantic region, including much of Europe, could cool rapidly as a result.

Methods of Studying Climate Change

To understand how climate is changing today, and to predict future changes, scientists must have a good idea of what climatic conditions were like thousands or millions of years ago. Environmental scientists have developed a number of methods to decipher clues from the past to learn about Earth's climate history.

Proxy indicators tell us about the past

Earth's ice caps and glaciers hold clues about past climate. Over the ages, these huge expanses of snow and ice have accumulated to great depths, preserving within them tiny bubbles of the ancient atmosphere (Figure 18.6). Scientists can examine these trapped air bubbles by drilling into the ice and extracting long columns, or cores. From

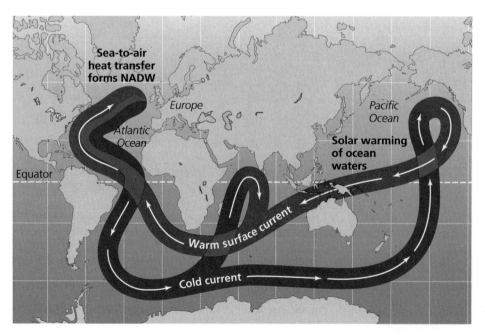

FIGURE 18.5 The atmosphere and the oceans interact. Warm surface currents carry heat from equatorial waters north toward Europe and Greenland, where they release heat into the atmosphere, then cool and sink into the deep ocean. This is how North Atlantic Deep Water (NADW) is formed. Melting of the Greenland Ice Sheet could potentially interrupt the flow of heat from equatorial regions and lead to rapid cooling of the North Atlantic and much of Europe.

(a) Ice core

(b) Micrograph of ice core

FIGURE 18.6 In Greenland and Antarctica, scientists have drilled deep into ancient ice sheets and removed cores of ice like this one **(a),** held by Dr. Gerald Holdsworth of the University of Calgary, to extract information about past climates. Bubbles (black shapes) trapped in the ice **(b)** contain small samples of the ancient atmosphere.

these ice cores, scientists can determine atmospheric composition, greenhouse gas concentrations, temperature trends, snowfall, solar activity, and even (from trapped soot particles) frequency of forest fires.

Scientists also drill cores into beds of sediment beneath bodies of water. Sediments often preserve pollen grains and other remnants from plants that grew in the past, and as we saw with the study of Easter Island (▶ pp. 8–9), the analysis of these materials can reveal a great deal about the history of past vegetation. Because the types of plants that grow in an area depend on the area's climate, knowing what plants occurred in a given location can tell us much about the climate at that time (see "The Science behind the Story," ▶ p. 538). Sources of data such as pollen from sediment cores and air bubbles from ice cores are known as **proxy indicators**, types of indirect evidence that serve as proxies for direct measurement and that indicate the na-

ture of past climate. Other types of proxy indicators include data culled from coral reefs and the tree rings of long-lived trees.

Direct atmospheric sampling tells us about the present

Studying present-day climate is more straightforward, because scientists can directly measure atmospheric conditions. The late Charles Keeling of the Scripps Institution of Oceanography in La Jolla, California, documented trends in atmospheric carbon dioxide concentrations starting in 1958 (Figure 18.7). Keeling collected four air samples from five towers every hour from his monitoring station at the Mauna Loa Observatory in Hawaii. Keeling's data show that atmospheric carbon dioxide concentrations have increased from 315 ppm to 378 ppm since 1958. Today Keeling's colleagues are continuing these measurements, building upon the single best long-term dataset we have of direct atmospheric sampling of any greenhouse gas.

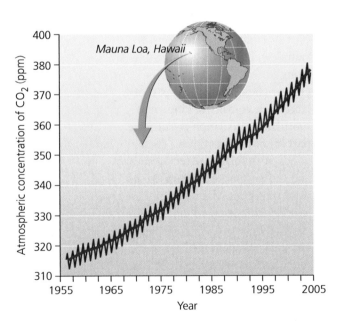

FIGURE 18.7 Atmospheric concentrations of carbon dioxide have risen steeply since 1958, when Charles Keeling began collecting these data at the Mauna Loa Observatory in Hawaii. The jaggedness apparent within the overall upward trend of the so-called "Keeling curve" represents seasonal variation caused by the Northern Hemisphere's vegetation, which absorbs more carbon dioxide during the northern summer, when it is more photosynthetically active. Go to **GRAPHit!** at www.aw-bc.com/withgott or on the student CD-ROM. Data from National Oceanic and Atmospheric Administration, Climate Prediction Center, 2005.

Understanding El Niño and La Niña

The Science behind the Story

Some scientists suspect that the warming of global air and ocean surface temperatures is increasing the frequency of the phenomenon known as El Niño. *El Niño* was originally the name that Spanish-speaking fishermen gave to an unusually warm surface current that sometimes arrived near the Pacific coast of South America around Christmastime. El Niño literally means "the little boy" or "the Christ child," and the current received this name because of its holiday arrival time. El Niño later became the term for the warming of the eastern Pacific that occurs every 2 to 7 years and depresses local fish and bird populations.

Under normal conditions, prevailing winds blow from east to west along the equator in the Pacific Ocean. These winds form part of a large-scale convective loop, or atmospheric circulation pattern; see part (a) of the first figure. These winds push surface waters westward and cause water to "pile up" in the western Pacific. As a result of these winds, water near Indonesia is, at times, about 50 cm (20 in.) higher and approximately 8 °C warmer than water along the coast of South America. The prevailing winds that move surface waters westward also make it possible for deep, cold, nutrient-rich water to form a nutrient-rich upwelling (▸ p. 471) along the coast of Peru.

When prevailing winds weaken, the warm water that has collected in the western Pacific flows "downhill" and eastward toward South America; see part (b) of the first figure. The influx of warm surface water and the

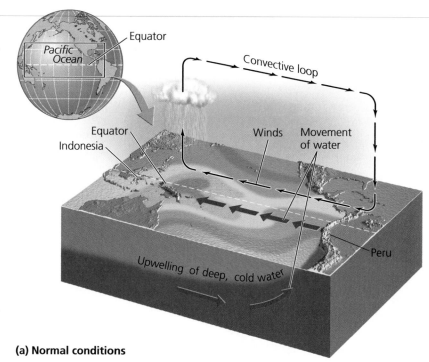

(a) Normal conditions

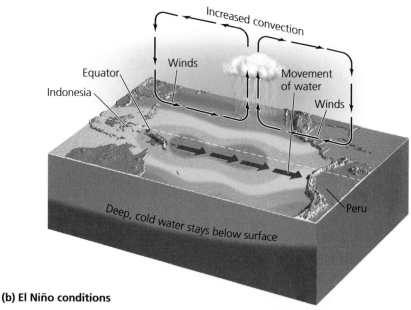

(b) El Niño conditions

In this view, Indonesia is on the left and Peru is on the right. During normal conditions **(a)**, prevailing winds push warm surface waters toward the western Pacific. In this diagram, red and orange water is warmer than blue and green water. The red and orange waters in the west are also at a higher elevation than those in the east. During El Niño conditions **(b)**, the prevailing winds weaken and no longer hold the warm surface waters in the western Pacific. As the warmer water "sloshes" back across the Pacific toward South America, precipitation patterns change.

absence of the normal prevailing winds suppress the upwelling along the coast of Peru, shutting down the delivery of nutrients that support the region's marine life and fisheries. This effect is felt along the entire Pacific Coast of the Americas, and it alters weather patterns around the world. Because warm surface water shapes weather by giving rise to heavy precipitation, its movement across the Pacific can create unusually intense rainstorms and floods in areas that are generally dry, and can cause drought and fire in regions that typically experience wet weather.

Scientists monitor the development of El Niño events with an array of wind- and temperature-sensing buoys, known as the Tropical Atmosphere Ocean Project, or TAO/TRITON, that are anchored across the equatorial Pacific; see part (a) of the second figure. These buoys measure temperature profiles during normal conditions (b) and El Niño events (c). Scientists use such data to determine the extent and severity of El Niño events. Awareness of El Niño conditions can enable governments and individuals to better prepare for extreme weather events and changes in ocean conditions.

La Niña, or "the little girl," is the opposite of El Niño and exerts impacts trending in the opposite direction. It is characterized by the presence of colder-than-normal surface water in the equatorial Pacific Ocean; see part (d) of the second figure. Scientists have made great strides recently in understanding El Niño and La Niña, but a great deal remains to be understood about their causes, their connections

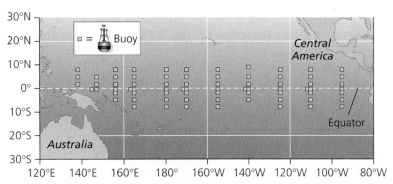

(a) Temperature-sensing buoys: TAO/TRITON array

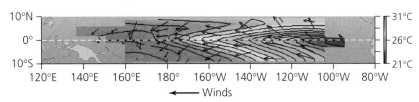

(b) Temperature during normal conditions (Dec. 1993 averages)

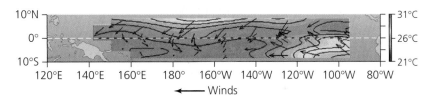

(c) Temperature during El Niño conditions (Dec. 1997 averages)

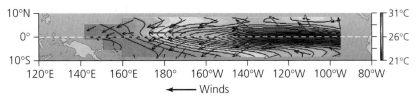

(d) Temperature during La Niña conditions (Dec. 1998 averages)

Scientists gather climate data in the Pacific Ocean by using anchored monitoring stations. These buoys **(a)** are distributed across the Pacific. During normal conditions **(b),** cold water reaches the surface of the eastern Pacific via an upwelling, represented by the blue tongue at the right of the graph. During an El Niño event **(c),** warmer water, shown in orange and red, is present in the eastern Pacific and the cold, blue tongue is absent. During a La Niña event **(d),** colder-than-normal water is present in the equatorial Pacific Ocean. Data from National Oceanic and Atmospheric Administration, Tropical Atmospheric Ocean Project, 2001.

to climate change, and their effects on other environmental systems. They clearly show, however, that the atmosphere, oceans, and regional and global climate are tightly integrated systems.

Scientists Use Pollen to Study Past Climate

Scientists can learn about climate by studying the types of vegetation that grew in particular places in the past. Cores taken from the sediments that collect at the bottoms of lakes and ponds contain pollen and other clues to the plants that lived in those locations at the time each layer was deposited. Tropical plant fossils and pollen tell scientists that warm, wet conditions were present in the past, whereas evidence of cold-loving plants reveal a cooler climate history.

In 1995, J. S. McLachlan of Harvard University and L. B. Brubaker of the University of Washington published a study on the past climate of Washington state's Olympic Peninsula. The researchers found that as climate and hydrological conditions changed, so did the area's vegetation. Their conclusions were based on the characteristics of pollen, fossils, and charcoal that had accumulated over time on the bottom of a small lake and in a cedar swamp, coupled with regional climate data from other sources. Such sediment deposits serve as a timeline. As scientists dig down through

Pollen and larger plant parts preserved in sediments deposited on the bottoms of lakes and ponds over thousands of years can tell scientists about past plant communities and, by extension, about past climates. This sediment core sample was taken from Chesapeake Bay in Maryland.

the layers of sediment, they peer farther back in time.

The research team used a long metal cylinder to pull a 1,811-cm (59.4-ft) core of sediment from the middle of Crocker Lake and a 998-cm (32.7-ft) core of sediment from the middle of the cedar swamp. They wrapped the two cores in plastic and aluminum foil and stored them in large freezers at 4 °C (39.2 °F) to prevent decomposition. The team then began removing 1-mm

thick slices every 10–30 cm (4–12 in.) along the length of the cores and analyzing the pollen and charcoal content of each slice. Pollen is easily transported by the wind and thereby can indicate the types of vegetation that occur over a fairly large area. Pollen, therefore, serves as a clue to regional climate.

The researchers identified between 300 and 1,000 pollen grains to genus and species within each slice. Larger plant fossils, such as those of cones, tree stems, and bark, were also classified. Because these items are not as easily transported as pollen, they are considered to indicate the vegetational and climatological history of smaller, more local, areas. Similarly, fine particles of charcoal, which, like pollen, are easily transported through the atmosphere, tell of the history of forest fires on a regional scale, whereas larger pieces suggest fire occurrence in the immediate area.

In combination with other available information, these clues enabled researchers to piece together a record of the general plant and climate history of the Olympic Peninsula.

Coupled general circulation models help us understand climate change

To understand how climate systems function, and to predict future climate change, scientists attempt to simulate climate processes with sophisticated computer programs. *Coupled general circulation models (CGCMs)* are computer programs that combine what is known about weather patterns, atmospheric circulation, atmosphere-ocean interactions, and feedback mechanisms to simulate climate processes (Figure 18.8). They couple, or combine, climate influences of the atmosphere and oceans in a single simu-

lation. This requires manipulating vast amounts of data and complex mathematical equations—a task not possible until the advent of the supercomputer.

Over a dozen research labs around the world operate CGCMs. The models have been tested to see how closely they simulate known climate patterns when fed the appropriate historical data. In other words, if modelers enter past climate data, and the models produce accurate predictions of current climate, we have reason to believe the models may be realistic and accurate. Figure 18.9 shows temperature results from three such simulations.

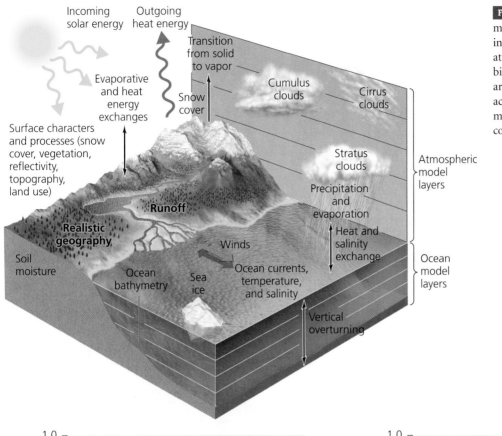

FIGURE 18.8 Modern climate models incorporate many factors, including processes involving the atmosphere, land, oceans, ice, and biosphere. Some of these factors are shown graphically here, but the actual models deal with them as mathematical equations in computer simulations.

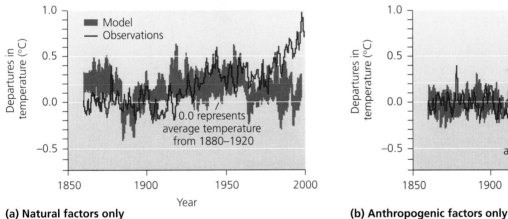

(a) **Natural factors only**

(b) **Anthropogenic factors only**

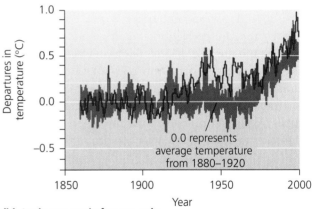

FIGURE 18.9 Scientists can test their climate models by entering climate data from past years and comparing model predictions (blue areas) with actual observed data (red line). Models that incorporate only natural factors (**a**) or only human-induced factors (**b**) do a mediocre job of predicting real climate trends. Models that incorporate both natural and anthropogenic factors (**c**) are most accurate. Data from the Intergovernmental Panel on Climate Change. 2001. *Third assessment report.*

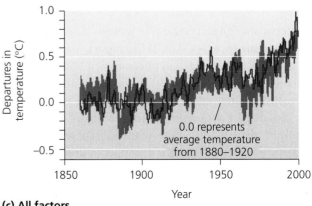

(c) **All factors**

The results in Figure 18.9a are based on natural climate-changing factors alone (such as volcanic activity and variation in solar energy). The results in Figure 18.9b are based on anthropogenic factors only (such as human greenhouse gas emissions and sulfate aerosol emissions). The results in Figure 18.9c are based on natural and anthropogenic factors combined, and this produces the closest match between predictions and actual climate. Results such as those in Figure 18.9c support the notion that both natural and human factors contribute to climate dynamics, and they also indicate that CGCMs can produce reliable predictions. As computing power increases and our ability to glean data from proxy indicators of past climate improves, CGCMs become increasingly reliable.

Climate Change Estimates and Predictions

The most thoroughly reviewed and widely accepted collection of scientific information concerning global climate change is a series of reports issued by the **Intergovernmental Panel on Climate Change (IPCC)**. This international panel of atmospheric scientists, climate experts, and government officials was established in 1988 by the United Nations Environment Programme (UNEP) and the World Meteorological Organization. Its task is to assess information relevant to questions of human-induced climate change. In 2001 the IPCC released its *Third Assessment Report*. This summary of current global trends and probable future trends represents the consensus of atmospheric scientists around the world.

The IPCC report summarizes evidence of recent changes in global climate

The IPCC report's major conclusions concerning recent climate trends address surface temperature; snow and ice cover; rising sea level and warmer oceans; precipitation patterns and intensity; and effects on wildlife, ecosystems, and human societies. The IPCC report is authoritative but, like all science, deals in uncertainties. Its authors have therefore assigned statistical probabilities to each of its conclusions and predictions.

Besides the data on increases in atmospheric concentrations of greenhouse gases that we discussed earlier, the 2001 IPCC report presented a number of findings on how climate change has already influenced the weather, Earth's physical characteristics and processes, the habits of organisms, and our economies. The report concluded, for

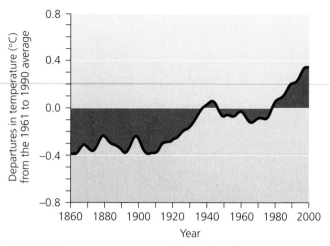

(a) Global temperature measured over the past 140 years

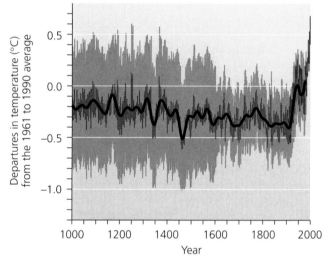

(b) Northern Hemisphere temperature over the past 1,000 years

FIGURE 18.10   Data from thermometers (**a**) show changes in Earth's average surface temperature since 1860. In (**b**), proxy indicators (blue line) and thermometer data (red line) together show average temperature changes in the Northern Hemisphere over the past 1,000 years. The gray-shaded zone represents the 95% confidence range. This record shows that 20th-century warming has eclipsed the magnitude of change during both the "Medieval Warm Period" (10th–14th centuries) and the "Little Ice Age" (15th–19th centuries). Go to **GRAPHit!** at www.aw-bc.com/withgott or on the student CD-ROM. Data from the Intergovernmental Panel on Climate Change, 2001. *Third assessment report.*

instance, that average surface temperatures had increased by 0.6 °C (1.0 °F) during the 20th century (Figure 18.10). It inferred that glaciers, snow cover, and ice caps were melting worldwide. It found that hundreds of species of plants and animals were being forced to shift their

Table 18.2   Major Findings of the IPCC Third Assessment Report, 2001

Weather indicators

▶ Earth's average surface temperature increased 0.6 °C (1.0 °F) during the 20th century (90–99% certainty)

▶ Cold and frost days decreased for nearly all land areas in the 20th century (90–99% certainty)

▶ Continental precipitation increased by 5–10% during the 20th century in the Northern Hemisphere (90–99% certainty), but it decreased in some regions

▶ The 1990s were the warmest decade of the past 1,000 years (66–90% certainty)

▶ The 20th-century Northern Hemisphere temperature increase was the greatest in 1,000 years (66–90% certainty)

▶ Droughts increased in frequency and severity (66–90% certainty)

▶ Nighttime temperatures increased twice as fast as daytime ones (66–90% certainty)

▶ Hot days and heat index increased (66–90% certainty)

▶ Heavy precipitation events increased at northern latitudes (66–90% certainty)

Physical indicators

▶ Average sea level increased 10–20 cm (4–8 in.) during the 20th century

▶ Rivers and lakes in the Northern Hemisphere were covered by ice 2 weeks less from the beginning to the end of the 20th century (90–99% certainty)

▶ Arctic sea ice thinned by 10–40% in recent decades, depending on the season (66–90% certainty)

▶ Mountaintop glaciers retreated widely during the century

▶ Global snow cover decreased by 10% since satellite observations began in the 1960s (90–99% certainty)

▶ Permafrost thawed in many regions

▶ El Niño events became more frequent, persistent, and intense in the past 40 years in the Northern Hemisphere

▶ Growing seasons lengthened 1–4 days per decade in the last 40 years in northern latitudes

Biological indicators

▶ Geographic ranges of many plants, insects, birds, and fish shifted toward the poles and upward in elevation

▶ Plants are flowering earlier, migrating birds are arriving earlier, animals are breeding earlier, and insects are emerging earlier in the Northern Hemisphere

▶ Coral reefs are experiencing bleaching more frequently

Economic indicators

▶ Global economic losses due to weather events rose 10-fold over the past 40 years, partly because of climate factors

Data from Intergovernmental Panel on Climate Change. 2001. *Third assessment report.*

geographic ranges and the timing of their life cycles. These findings and many more were based on the assessment of thousands of scientific studies conducted worldwide over many years, judged by the world's top experts in these fields. Some—but by no means all—of the report's major findings are shown in Table 18.2.

Sea-level rise and other changes interact in complex ways

Few of the processes and impacts examined in the IPCC report function in isolation, because Earth's environmental systems are connected and interact. For instance, warming temperatures are causing glaciers to shrink and disappear in many areas around the world. For the same reason, polar ice shelves that have been intact for millennia are breaking away and melting. As glaciers and ice shelves melt, an increased flow of water into the oceans is causing a rise in sea level. Sea level is also rising because ocean water is warming, and water expands in volume as its temperature increases. In fact, most sea-level rise is calculated to result from thermal expansion of seawater, rather than from runoff from melting glaciers and ice shelves.

Higher sea levels lead to beach erosion, coastal flooding, intrusion of saltwater into aquifers, and other impacts. In 1987, unusually high waves struck the Maldives and triggered a campaign to build a large seawall around Male, the nation's capital city. Known as "The

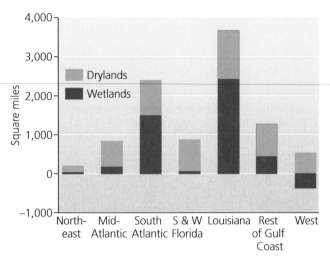

(a) U.S. coastal lands at risk from a 51-cm (20-in.) sea-level rise

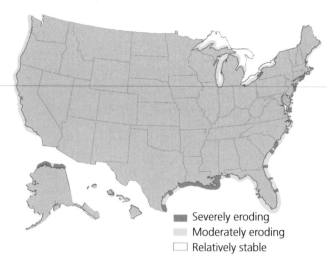

(b) Annual shoreline change

(c) Rescue from floodwaters of Hurricane Katrina

FIGURE 18.11 The United States will lose coastal areas as a result of a rise in sea level. A 51-cm (20-in.) sea level rise would inundate wetlands (red) and drylands (orange) on portions of all coasts **(a)**. Coastal areas around the nation **(b)** are at risk of increased erosion due to sea-level rise. Rescues like this one **(c)** from the floodwaters of Hurricane Katrina's storm surge in coastal Mississippi in 2005 could become more frequent in a future world of higher sea levels. Data (a, b) from National Assessment Synthesis Team, U.S. Global Change Research Program. 2001. *Climate change impacts on the United States: The potential consequences of climate variability and change: Foundation.* Cambridge, U.K.: Cambridge University Press and USGCRP.

Great Wall of Male," the seawall is intended to break up incoming waves and dissipate their energy to prevent the destruction of buildings and roads during storm surges. A *storm surge* is a temporary and localized rise in sea level brought on by the high tides and winds associated with storms. The higher that sea level is to begin with, the further inland a destructive storm surge can reach. "With a mere 1-meter rise [in sea level]," Maldives' President Maumoon Abdul Gayoom warned in 2001, "a storm surge would be catastrophic, and possibly fatal to the nation."

The Maldives is not the only nation with such concerns. In fact, among island nations, the Maldives has fared better than many others. It saw sea level rise about 2.5 mm per year throughout the 1990s, but most Pacific islands are experiencing greater changes. Different regions experience different amounts of sea level change because land elevations may be rising or subsiding naturally, depending on local geological conditions.

The United States is not immune from coastal impacts—and 53% of U.S. residents live within the 17% of the nation's land that lies in coastal areas. Figure 18.11a shows the amounts of wetlands and dry lands in the United States that would be inundated by a 51-cm (20-in.) sea-level rise. Figure 18.11b shows coastal areas in the United States that are eroding because of sea-level rise and other factors. The vulnerability of parts of the nation to coastal flooding due to storm surges became tragically apparent in 2005 when Hurricane Katrina struck New Orleans and the Gulf Coast, followed shortly thereafter by Hurricane Rita. The flooding that devastated the region (Figure 18.11c) could have been even worse had sea level been higher.

The levees surrounding New Orleans that were breached by Katrina's floodwaters are now repaired, but large portions of the city will always remain below sea level. Areas that are now 2.1 m (7 ft) below sea level may be as much as 3.3 m (10 ft) below sea level by 2100, as the land subsides and sea level rises. Just outside the city, marshes of the Mississippi River delta are being lost rapidly, as dams upriver hold back the silt that used to maintain the delta, as land subsides due to the extraction of petroleum deposits, and as rising seas eat away at coastal vegetation. Approximately 2.5 million ha (1 million acres) of Louisiana wetlands have become open water since 1940. Continued loss of these wetlands will mean that New Orleans will have less protection against future storm surges. And coastal Louisiana is not the only U.S. region with these concerns. Houston, Texas, and Charleston, South Carolina are among the major cities most likely to be affected by increased sea level. In southern Florida, coastal wetlands and mangroves are being submerged, and salt water has intruded into aquifers, killing trees and threatening drinking water supplies.

Moreover, the record number of hurricanes and tropical storms in 2005—Katrina and 22 others—left many people wondering if global warming was to blame. Are warmer ocean temperatures spawning more hurricanes, or hurricanes that are more powerful or long-lasting? Scientists are not yet sure of the answers to these questions, but evidence from a number of recent studies indicates that warmer sea surface temperatures are likely increasing the destructive power of storms, and possibly their duration, but are apparently not increasing their frequency.

The IPCC and other groups project future impacts of climate change

Because the consequences of climate change could be substantial, the IPCC and other groups have attempted to predict future climate changes and their potential impacts. One such group is the U.S. Global Change Research Program (USGCRP), created by Congress in 1990 to coordinate federal climate research. In 2000–2001, the USGCRP issued a report highlighting the past and future effects of global climate change on the United States, where annual average temperatures increased by 0.6 °C (1.0 °F) during the 20th century.

The report used CGCMs to develop a series of predictions on impacts of climate change in the United States (Table 18.3). These models, such as the Hadley model designed by British researchers and the Canadian model designed by British Columbian researchers, allowed scientists

Table 18.3 Some Predicted Impacts of Climate Change in the United States
▶ Average U.S. temperatures will increase 3–5 °C (5–9 °F) in the next 100 years
▶ Droughts and flooding will worsen, and snowpack will be reduced
▶ Drought and other factors could decrease crop yields, but longer growing seasons and enhanced CO_2 could increase yields
▶ Water shortages will worsen
▶ Greater temperature extremes will increase health problems and human mortality. Some tropical diseases will spread north into temperate latitudes
▶ Forest growth may increase in the short term, but in the long term, drought, pests, and fire may alter forest ecosystems
▶ Alpine ecosystems and barrier islands will begin disappearing
▶ Southeastern U.S. forests will break up into savanna/grassland/forest mosaics
▶ Northeastern U.S. forests will lose sugar maples
▶ Loss of coastal wetlands and real estate due to sea level rise will continue
▶ Melting permafrost will undermine Alaskan buildings and roads

Adapted from National Assessment Synthesis Team, U.S. Global Change Research Program. 2000. *Climate change impacts on the United States: The potential consequences of climate variability and change: Overview.* Cambridge, U.K.: Cambridge University Press and USGCRP.

to present unique graphical depictions summarizing predicted impacts of climate change on particular geographic areas (Figure 18.12).

Agriculture and forestry Drought and temperature extremes are among the threats that climate change may pose for farms and forests worldwide. Croplands presently near the limits imposed by heat stress and water availability could be pushed beyond their ability to produce food. If average temperatures increase by more than 3–4 °C, most tropical and subtropical areas will likely see decreased crop production, and midlatitude farmlands may also begin to see significant declines.

Conversely, warmer temperatures and longer growing seasons at higher latitudes could potentially increase agricultural productivity there. In fact, of 16 crops in the United States, the USGCRP's agricultural assessment team in 2002 predicted that yields of 13 would increase and only 1 would decrease because of climate change. The overall effect of a warmer climate on agricultural productivity is difficult to predict. Agricultural productivity might remain somewhat stable globally while increasing in some areas and decreasing in others. For this reason,

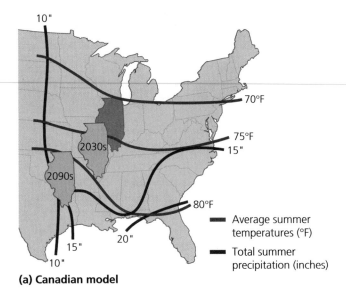

(a) Canadian model

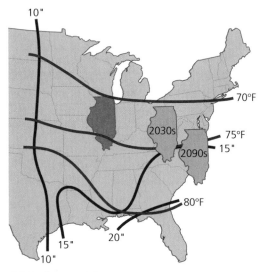

(b) Hadley model

FIGURE 18.12 Using parameters of the Canadian and Hadley models, researchers were able to portray the predicted future climate of the state of Illinois. The Canadian model predicted Illinois to become warmer and drier, attaining in the 2030s a climate similar to Missouri's, and in the 2090s one similar to eastern Oklahoma's. The Hadley model predicted Illinois to become warmer and moister, attaining in the 2030s a climate similar to West Virginia's, and in the 2090s one similar to North Carolina's. Data from National Assessment Synthesis Team, U.S. Global Change Research Program. 2001. *Climate change impacts on the United States: The potential consequences of climate variability and change: Foundation.* Cambridge, U.K.: Cambridge University Press and USGCRP.

some people argue that global warming could benefit their nations, but others point out that it may increase inequities between developed and developing nations.

Research by scientists indicates that climate change could transform U.S. forests. In its 2000–2001 report, the USGCRP predicted that U.S. forests will become more productive, because additional carbon dioxide in the atmosphere will speed rates of photosynthesis. As productivity increases, however, the frequency and intensity of forest fires could increase by 10% or more. Forest communities should in general move northward and upward in elevation. Alpine and subalpine plant communities should become less common as the climate warms, because these mountaintop communities cannot move further upward in elevation. Although some forest types will likely decline, others, including oak-hickory and oak-pine forests, may expand in the eastern United States (Figure 18.13).

Effects on plant communities comprise an important component of climate change, because by drawing in CO_2 for photosynthesis, plants act as sinks for carbon. The widespread regrowth of forests in eastern North America (▸pp. 350, 355) has offset an estimated 25% of U.S. carbon emissions during the past four decades. If climate change increases overall vegetative growth, this could partially mitigate carbon emissions, in a process of negative feedback. However, if climate change decreases overall growth (through drought or fire, for instance), then a positive feedback cycle could increase carbon flux to the atmosphere.

Weighing the Issues:
Agriculture in a Warmer World

Some people maintain that a warmer climate would expand arable lands toward the poles and lead to greater agricultural production globally. In fact, the U.S. Corn Belt is already pushing into Canada. Locate the nations of Russia and Argentina on a world map, and hypothesize how such a poleward shift of agriculture might affect each of these nations. Now hypothesize how it might affect poorer nations near the equator that are already suffering from food shortages and agricultural problems. Thinking back to Chapters 9 and 10, name several factors that could potentially influence crop yields if climate continues changing.

Freshwater and marine systems In regions where climate change increases precipitation and stream flow, erosion and flooding could alter the structure and function of aquatic systems. Where agriculture and other human activities have modified the landscape, such flooding could increase pollution of freshwater ecosystems. In regions where precipitation decreases, lakes, ponds, wetlands, and streams would shrink, affecting the organisms that live in

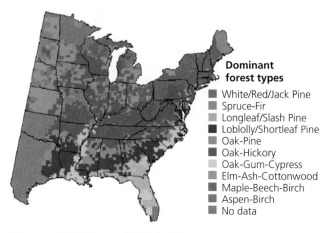

Dominant forest types
- White/Red/Jack Pine
- Spruce-Fir
- Longleaf/Slash Pine
- Loblolly/Shortleaf Pine
- Oak-Pine
- Oak-Hickory
- Oak-Gum-Cypress
- Elm-Ash-Cottonwood
- Maple-Beech-Birch
- Aspen-Birch
- No data

(a) Current distribution (1960–1990)

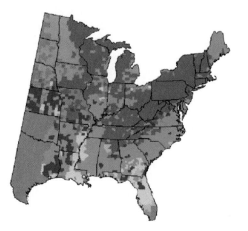

(b) Canadian scenario (2070–2100)

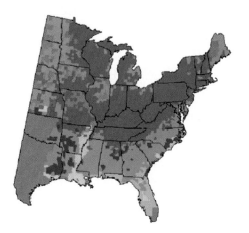

(c) Hadley scenario (2070–2100)

FIGURE 18.13 While some forest types will decline, others, including oak-hickory and oak-pine forests, may expand in the eastern United States. This figure shows predicted forest cover changes based on the Canadian and Hadley models. Data from National Assessment Synthesis Team, U.S. Global Change Research Program. 2001. *Climate change impacts on the United States: The potential consequences of climate variability and change: Foundation.* Cambridge, U.K.: Cambridge University Press and USGCRP.

those habitats, as well as human health and well-being. In 2001, about 1.7 billion people were living in areas with limited water supplies. By 2025, according to the IPCC, this number will likely increase to 5 billion out of a projected world population of 8.4 billion.

The Maldives is likely to suffer from water-related stresses, because its human population is expanding rapidly while rising seas threaten to bring salt water into the nation's wells, just as the 2004 tsunami did. The contamination of groundwater and soils by seawater is particularly threatening to island nations like the Maldives and coastal areas such as the Tampa, Florida, region, which depend on small lenses of freshwater that float atop saline groundwater.

Maldives residents also worry about damage to marine ecosystems, including coral reefs (▶ pp. 474–476). Coral reefs provide habitat for important food fish, reduce wave intensity and protect fragile coastlines from erosion, and provide popular snorkeling and scuba diving sites. Damage to reefs from storm surges would depress tourism and fishing.

Human health As a result of climate change, people could face increased exposure to an array of health problems:

▶ Heat stress resulting from high temperatures and humidity

▶ Respiratory ailments from air pollution, as hotter temperatures promote formation of photochemical smog (▶ pp. 510–511)

▶ Expansion of tropical diseases, such as malaria and dengue fever, into temperate regions

▶ Disease and sanitation problems when floods overcome sewage treatment systems

▶ Injuries and drowning if storms become more frequent or intense

▶ Hunger-related ailments as human population grows and demands on agricultural systems increase

Figure 18.14a shows USGCRP projections regarding the 21st-century July heat index across the United States. One model predicts that the heat index—a product of temperature and humidity—will be 14 °C (25 °F) hotter in much of the southeastern United States. Heat waves can lead to high mortality rates in American cities (Figure 18.14b). In Europe, heat stress killed 35,000 people in August 2003 during a record heat wave.

At the same time, other scientists, including some IPCC contributors, have argued that a warmer world will present fewer diseases and injuries that result from cold weather. The trade-off between an increase in warm-weather ailments and a decrease in cold-weather health problems remains one of the unknowns of global climate change.

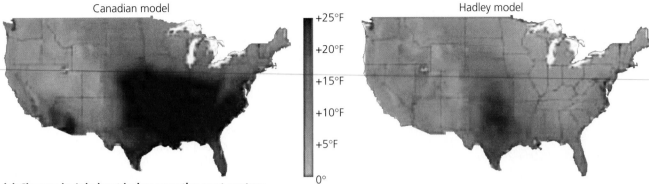

(a) Change in July heat index over the next century

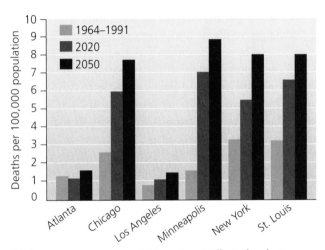

(b) Average summer mortality rates attributed to hot weather episodes

FIGURE 18.14 The Canadian model and the Hadley model provide two projections of the heat index (a product of temperature and humidity) for July across the United States for the upcoming century **(a)**. The Canadian model predicts that the July heat index will be 14 °C (25 °F) hotter than that of much of the present-day southeastern United States. Past and projected future mortality rates attributed to hot weather **(b)** are shown for several U.S. cities. Data from National Assessment Synthesis Team, U.S. Global Change Research Program. 2001. *Climate change impacts on the United States: The potential consequences of climate variability and change: Foundation.* Cambridge, U.K.: Cambridge University Press and USGCRP.

Debate over Climate Change

Virtually all environmental scientists agree that Earth's atmosphere and climate are changing. The great majority of them have concluded that human activity, particularly our emission of greenhouse gases, is the primary reason for this change. Despite this unusually strong scientific consensus, you have no doubt heard a great deal of debate over climate change. To understand why, it is most helpful to break down this debate into components that can be examined separately.

One component involves discussion within the scientific community about the details and mechanisms of climate change and the extent and nature of its likely effects on human welfare and environmental systems. This is part of the normal process of science as researchers gather evidence and test competing hypotheses, trying to get a full and complete picture of the truth. A second arena of debate involves people, primarily nonscientists, who contest the consensus findings and interpretations of the scientific community.

Some of these so-called "greenhouse skeptics" have vested interests (▶ p. 30) in continuing the widespread use of fossil fuels, and some of them have significant sway over policymakers, particularly in the United States. A third arena of debate involves how our societies should respond to climate change. This is a wide-ranging discourse among scientists, economists, business leaders, policymakers, and others.

Scientists agree that climate change is occurring but disagree on many details

Although most scientists have concluded that our greenhouse gas emissions are altering the atmosphere and influencing climate, many details and aspects of climate change science remain uncertain, because climate systems are so complex. Climate scientists have not yet come to firm consensus on the roles played by clouds, water vapor, soot and sulfate aerosols, vegetative carbon sinks, and the oceans and NADW formation. Such uncertainties, coupled with the complexity of feedback mechanisms, make it difficult to predict the future.

Despite these uncertainties, the scientific community now feels that evidence for our role in influencing climate is strong enough that governments should take action to address greenhouse gas emissions. In June 2005, as the leaders of the "G8" industrialized nations met, the national academies of science from 11 nations (Brazil, Canada, China, France, Germany, India, Italy, Japan, Russia, the United Kingdom, and the United States) issued a joint statement urging these political leaders to take action. Such a broad consensus statement from the world's scientists was virtually unprecedented, on any issue. The statement read, in part:

> The scientific understanding of climate change is now sufficiently clear to justify nations taking prompt action. It is vital that all nations identify cost-effective steps that they can take now, to contribute to substantial and long-term reduction in net global greenhouse gas emissions. . . . A lack of full scientific certainty about some aspects of climate change is not a reason for delaying an immediate response that will, at a reasonable cost, prevent dangerous anthropogenic interference with the climate system.

Some challenge the scientific consensus

Despite such clear statements from the scientific community about the risks posed by climate change, many people, primarily nonscientists, have ignored or disputed mainstream scientific findings and interpretations. For instance, just days after the national academies' statement, the *Wall Street Journal* ran an editorial that asked, "What Warming?" and maintained that:

> The scientific case for [climate change] looks weaker all the time. . . . Since [1997], the case for linking fossil fuels to global warming has, if anything, become even more doubtful. The Earth currently does seem to be in a warming period, though how warm and for how long no one knows. In particular, no one knows whether this is unusual or merely something that happens periodically for natural reasons. Most global warming alarms are based on computer simulations that are largely speculative and depend on a multitude of debatable assumptions.

The editorial referenced several scientific studies that contested the data shown in Figure 18.10b, as well as other key findings in climate change science. It did not make clear the extent to which studies demonstrating climate change have earned broad scientific acceptance, whereas studies that deny climate change have not been widely accepted.

The media have played a large role in the public's understanding of climate change, and the U.S. media by tradition try to portray both sides of any issue. However, they often fail to make clear when spokespeople for one side of an issue have stronger evidence or have earned wider support than spokespeople for another side. Author Ross Gelbspan has maintained in two books that the American media have portrayed the climate change debate as much more even and two-sided than it actually is. By giving greenhouse skeptics equal time with mainstream scientists, he and many scientists have argued, the media have amplified the views of greenhouse skeptics out of proportion to their prevalence in the scientific community. Moreover, he documents that the relatively few greenhouse skeptics among scientists are often funded by industries that benefit from fossil fuel use, such as the petroleum and automobile industries.

Industry lobbying has also played a part in making policymakers hesitant to enact policy to reduce carbon emissions. Attempts to improve fuel efficiency for automobiles (▸ p. 582) have repeatedly failed in the U.S. Congress, and attempts to open the Arctic National Wildlife Refuge for oil drilling (▸ pp. 557–559) have repeatedly been made.

Weighing the **Issues:**
Environmental Refugees

Citizens of the Maldives need only look southward to find an omen of their future. The Pacific island nation of Tuvalu, which has been losing 9 cm (3.5 in.) of elevation per decade to rising seas, may become the first casualty of global climate change. Appeals from its 11,000 citizens were heard by New Zealand, which began accepting these environmental refugees in 2003. Do you think this refugee flow casts doubt on the arguments of those who hold that climate change is uncertain? How might the perspective of a Tuvalu resident differ from that of a U.S. oil industry executive, and why? What steps might individual people of Tuvalu or the Maldives or coastal Florida take to protect their ways of life?

How should we respond to climate change?

Even if one accepts that climate change is real and poses significant ecological and economic threats, there is plenty of room for disagreement over appropriate responses. Political and economic debate over how we should respond stems from questions such as:

▸ Would the economic and political costs of reducing greenhouse gas emissions outweigh the costs of unabated emissions and resulting global climate change?

▸ Should industrialized nations, developing nations, or both take responsibility for reducing greenhouse gas emissions?

▸ Should steps to reduce emissions occur voluntarily or as a result of government regulation, political pressure, or economic sanctions?

▸ How should we allocate funds and human resources for reducing emissions and coping with climate change?

Strategies for Reducing Emissions

Since 1990, the generation of electricity, largely through coal combustion, has produced the largest portion (34%, as of 2000) of U.S. greenhouse gas emissions (Figure 18.15). Transportation ranks second at 27%, industry produces 19%, agriculture and residential sources produce 8% each, and commercial sources account for 5%. Tackling climate change will require reducing emissions from these sources, and strategies for doing so involve scientific, technical, political, and economic approaches.

Electricity generation is the largest source of U.S. greenhouse gases

From cooking and heating to the clothes we wear, much of what we own and do depends on electricity. Fossil fuel combustion generates over 70% of U.S. electricity. Coal alone accounts for 56%, along with most of the resultant greenhouse gas emissions. Reducing the volume of fossil fuels we burn to generate electricity would reduce greenhouse gas emissions, as would decreasing electricity consumption. There are two ways to reduce the amount of fossil fuels we use: (1) encouraging conservation and efficiency (▶ pp. 582–583, 585) and (2) switching to renewable energy sources (Chapters 20 and 21).

Conservation and efficiency Conservation and efficiency in energy use can arise from new technologies, such as high-efficiency lightbulbs and appliances, or from individual ethical choices to reduce electricity consumption. In one example of a technological solution, the U.S. Environmental Protection Agency (EPA) promotes energy conservation through its Energy Star Program. The Energy Star Program rates household appliances, lights, windows, fans, office equipment, heating and cooling systems, and appliances for their efficiency in using energy. Following are examples of the energy you can save by choosing Energy Star products:

- ▶ An Energy Star refrigerator can cut your CO_2 emissions by 100 kg (220 lb) annually.
- ▶ An Energy Star washing machine can cut your CO_2 emissions by 200 kg (440 lb) annually.
- ▶ Compact fluorescent lights can reduce the energy you use for lighting by 40%.
- ▶ Energy Star homes use highly efficient construction, duct work, insulation, heating and cooling systems, and windows to reduce energy use by as much as 30%.

FIGURE 18.15 Coal-fired electricity-generating power plants, such as this one in Maryland, are the largest contributors to U.S. greenhouse emissions.

Such technological solutions are popular, and they can be profitable for manufacturers while also saving consumers money. Alternatively, consumers can opt for lifestyle choices rather than technological fixes. For nearly all of human history, people managed without the electrical appliances that most of us take for granted today. It is entirely possible for each of us to simply choose to use fewer greenhouse-gas-producing appliances and technologies or to take practical steps to use electricity more efficiently.

Renewable sources of electricity Technologies that generate electricity without using fossil fuels represent another means of reducing greenhouse gas emissions. These include hydroelectric power (▶ pp. 446–447 and 612–615), geothermal energy (▶ pp. 633–635), photovoltaic cells (▶ pp. 625–626), and wind power (▶ pp. 627–633). We will examine renewable energy sources in more detail in Chapters 20 and 21.

Transportation is the second largest source of U.S. greenhouse gases

Can you imagine life without a car? Most Americans probably can't—a reason why transportation is the second largest source of U.S. greenhouse emissions. One-third of the average American city—including roads, parking lots, garages, and gas stations—is devoted to use by cars. The average American family makes 10 trips by car each day, and governments across the nation spend $200 million per day on road construction and repairs. Registered U.S. automobiles number over 220 million, and this figure is projected to surpass the human population of the country.

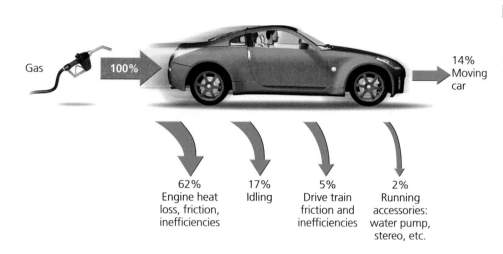

Gas   100%   14% Moving car

62% Engine heat loss, friction, inefficiencies   17% Idling   5% Drive train friction and inefficiencies   2% Running accessories: water pump, stereo, etc.

FIGURE 18.16 Conventional automobiles are extremely inefficient. Almost 85% of useful energy is lost, and only 14% actually moves the car down the road.

Unfortunately, the typical automobile is highly inefficient. Close to 85% of the fuel you pump into your gas tank does something other than move your car down the road. According to the U.S. Department of Energy, only about 13–14% of the fuel energy actually moves the vehicle and its occupants from point A to point B (Figure 18.16). Although more aerodynamic designs, increased engine efficiency, proper maintenance, and improved tire design help to reduce these losses, gasoline-fueled automobiles may always remain somewhat inefficient.

Automotive technology Advancing technology, however, is making possible a number of alternatives to the traditional combustion-engine automobile. These include hybrid vehicles that combine electric motors and gasoline-powered engines for greater efficiency (▶p. 583). They also include fully electric vehicles, alternative fuels such as biodiesel and compressed natural gas (▶pp. 609–610), and hydrogen fuel cells that use oxygen and hydrogen and produce only water as a waste product (▶pp. 637–642).

Driving less, and public transportation Despite these novel options, the high costs of automobile ownership and concerns regarding traffic and the environmental impacts of automobiles are leading many people to make lifestyle choices that reduce their reliance on cars. For example, many people are choosing to live nearer to their place of employment. Many others use buses, subway trains, and other modes of public transportation. Still others bike or walk to work or for their errands (Figure 18.17). Unfortunately, reliable and convenient public transit is not yet available in many U.S. communities. Making automobile-based cities and suburbs more

friendly to pedestrian and bicycle traffic and improving people's access to public transportation stand as major challenges for progressive city and regional planners (▶pp. 388–391).

In a 2002 study, the American Public Transportation Association (APTA) concluded that increasing use of public transportation is the single most effective strategy for conserving energy and reducing environmental pollutants. Already, public transportation in the United States reduces fossil fuel consumption by 855 million gallons of gas (45 million barrels of oil) each year, the APTA has estimated. Yet according to the study, if U.S. residents increased their use of public transportation to the levels of

FIGURE 18.17 Sometimes the most effective solutions are the simplest. By choosing human-powered transportation methods, such as bicycles, we can greatly reduce our individual transportation-related greenhouse gas emissions. An increasing number of people are choosing to live closer to their workplaces and to enjoy the dual benefits of exercise and reduced emissions by walking or cycling to work or school.

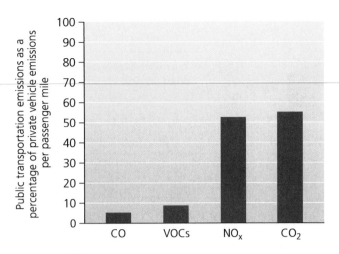

FIGURE 18.18 Relative to travel by private automobile, public transportation reduces almost all types of air pollution, including carbon monoxide (CO), volatile organic compounds (VOCs), nitrous oxides (NOx), and carbon dioxide (CO_2). Data from Shapiro, R. J., et al. 2002. *Conserving energy and preserving the environment: The role of public transportation.* American Public Transit Association.

Canadians (7% of daily travel needs) or Europeans (10% of daily travel needs), enormous amounts of energy and greenhouse emissions would be saved. These savings, the study indicated, could substantially cut air pollution, dependence on imported oil, and the nation's contribution to global climate change (Figure 18.18).

Some international treaties address climate change

In 1992, the United Nations convened the U.N. Conference on Environment and Development Earth Summit in Rio de Janeiro. Nations represented at the Earth Summit signed five documents, including the **U.N. Framework Convention on Climate Change (FCCC).** The FCCC outlined a plan for reducing greenhouse gas emissions to 1990 levels by the year 2000 through a voluntary, nation-by-nation approach.

By the late 1990s, it was apparent that a voluntary approach to slowing greenhouse gas emissions was not likely to succeed. Between 1990 and 2003, for example, U.S. greenhouse emissions (in CO_2 equivalents) increased by 13.3%.

However, some other nations have demonstrated that economic vitality does not require ever-increasing greenhouse gas emissions. For instance, Germany has the third most technologically advanced economy in the world and is a leading producer of iron, steel, coal, chemicals, automobiles, machine tools, electronics, textiles, and other goods—yet it managed between 1990 and 2003 to reduce its greenhouse gas emissions by

18.5%. In the same period, the United Kingdom cut its emissions by 13.0%.

After watching the seas rise and observing the refusal of most industrialized nations to cut their emissions, nations of the developing world—the Maldives among them—helped initiate an effort to create a binding international treaty that would *require* all signatory nations to reduce their greenhouse gas emissions. This effort led to the development of the Kyoto Protocol.

The United States has resisted the Kyoto Protocol

The **Kyoto Protocol** is an outgrowth of the FCCC. Drafted in 1997 in Kyoto, Japan, it mandates signatory nations, by the period 2008–2012, to reduce emissions of six greenhouse gases to levels equal to or lower than those of 1990 (Table 18.4). The treaty was to take effect once nations responsible for 55% of global greenhouse emissions ratified it, and in 2005, the Kyoto Protocol at last came into force after it was ratified by Russia, the 127th nation to ratify it.

The United States, the world's largest emitter of greenhouse gases, has continued to refuse to ratify the Kyoto Protocol. U.S. leaders have called the treaty unfair because it requires industrialized nations to reduce emissions but does not require the same of developing nations, even rapidly industrializing ones such as China and India.

Table 18.4	Emissions Reductions Required and Achieved	
Nation	**Required change,[1] 1990–2008/2012**	**Observed change, 1990–2003[2]**
Russia	0.0%	−38.5%[3]
Germany	−21.0%	−18.5%
United Kingdom	−12.5%	−13.0%
France	0.0%	−1.9%
Italy	−6.5%	+8.3%
Japan	−6.0%	+12.8%
United States	−7.0%	+13.3%
Canada	−6.0%	+24.2%

[1]Percentage decrease in emissions (carbon-equivalents of six greenhouse gases) from 1990 to period 2008–2012, as mandated under Kyoto Protocol
[2]Actual percentage change in emissions (carbon-equivalents of 6 greenhouse gases) from 1990 to 2003. Negative values indicate decreases; positive values indicate increases.
[3]Data through 1999 (most recent available). Russia's substantial decrease was mainly due to economic contraction following the breakup of the Soviet Union.
Data from U.N. Framework Convention on Climate Change, National Greenhouse Gas Inventory Reports, 2005.

1. Calculate the approximate percentage changes in CO_2 emissions from transportation; electricity generation; and residential, commercial, and industrial primary energy use between 1980 and 2003.
2. Between 1980 and 2003, U.S. population increased by 28.4%, and the inflation-adjusted U.S. gross domestic product (GDP) doubled. What quantitative conclusions can you draw from these data about CO_2 emissions per capita? About CO_2 emissions per unit of total economic activity?
3. Imagine you are put in charge of designing a strategy to reduce U.S. emissions of CO_2 from fossil fuel combustion. Based on the data presented here, what approaches would you recommend, and how would you prioritize these? Explain your answers.

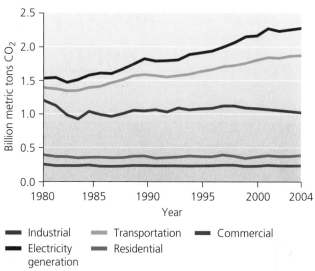

— Industrial   — Transportation   — Commercial
— Electricity   — Residential
   generation

Emissions of CO_2 from fossil fuel combustion by end-use sector in the United States, 1980–2003. Data from U.S. Department of Energy, Energy Information Administration. 2004. *Annual Energy Review 2004,* Report No. DOE/EIA-0384.

CALCULATING ECOLOGICAL FOOTPRINTS

U.S. energy consumption from fossil fuels currently totals 306 gigajoules per person per year. Carbon emitted from fossil fuel combustion is "sequestered" when plants take up the emitted carbon dioxide through photosynthesis and store it as organic matter. Researchers have estimated that for each

100 gigajoules of fossil fuel burned, 1 hectare of ecologically productive land is required for carbon sequestration. Considering these data, calculate the component of the ecological footprint required to sequester carbon from fossil fuel emissions, and record your results in the table.

	Hectares of land to sequester carbon
You	3.06
Your class	
Your state	
United States	

Data from Wackernagel, M., and W. Rees. 1996. *Our ecological footprint: Reducing human impact on the earth.* Gabriola Island, British Columbia, Canada: New Society Publishers.

1. The land area of the United States is about 916 million hectares. What percentage of that land would need to be set aside to sequester all carbon from the fossil fuel consumption of 300 million Americans?
2. What is the environmental fate of carbon dioxide that is released from the combustion of fossil fuels and is not sequestered by plants? Why is this a concern?
3. Name four things you could do to lessen the modification of the global environment caused by your own personal energy consumption.

Take It Further

Go to www.aw-bc.com/withgott or the student CD-ROM where you'll find:

▶ Suggested answers to end-of-chapter questions
▶ Quizzes, animations, and flashcards to help you study
▶ *Research Navigator*™ database of credible and reliable sources to assist you with your research projects

▶ **GRAPHit!** tutorials to help you master how to interpret graphs
▶ **INVESTIGATEit!** current news articles that link the topics that you study to case studies from your region to around the world

19 Fossil Fuels: Energy and Impacts

Oil production facility at Prudhoe Bay, Alaska

Upon successfully completing this chapter, you will be able to:

▶ Survey the energy sources that we use

▶ Describe the nature and origin of coal, and evaluate its extraction and use

▶ Describe the nature and origin of petroleum and evaluate its extraction, use, and future depletion

▶ Describe the nature and origin of natural gas, and evaluate its extraction and use

▶ Outline and assess environmental impacts of fossil fuel use

▶ Evaluate political, social, and economic impacts of fossil fuel use

▶ Specify strategies for conserving energy

Caribou crossing haul road at Prudhoe Bay

Central Case: Oil or Wilderness on Alaska's North Slope?

"The roar alone of road building, drilling, trucks, and generators would pollute the wild music of the Arctic and be as out of place there as it would be in the heart of Yellowstone or the Grand Canyon."
—FORMER PRESIDENT JIMMY CARTER, 2000

"There is absolutely no indication that environmentally responsible exploration will harm the 129,000-member porcupine caribou herd."
—ALASKA SENATOR FRANK MURKOWSKI, 2002

Above the Arctic Circle, at the top of the North American continent, the land drops steeply down from the jagged mountains of Alaska's spectacular Brooks Range and stretches north in a vast, flat expanse of tundra until it meets the icy waters of the Arctic Ocean. Few Americans have been to this remote region, yet it has come to symbolize a struggle between two values in our modern life.

For some U.S. citizens, Alaska's North Slope is one of the last great expanses of wilderness in their sprawling industrialized country—one of the last places humans have left untouched. For these millions of Americans, simply knowing that this wilderness still exists is of tremendous value. For millions of others, this land represents something else entirely—a source of petroleum, the natural resource that, more than any other, fuels our society and shapes our way of life. To these people, it seems wrong to leave such an important resource sitting unused in the ground. Those who advocate drilling for oil here accuse wilderness preservationists of neglecting the country's economic interests, whereas advocates for wilderness argue that drilling will sacrifice the nation's natural heritage for little gain.

Ever since oil was found seeping from the ground in this area a century ago, these two visions for Alaska's North Slope have competed. Now they exist side by side across three regions of this vast swath of land (Figure 19.1). The westernmost portion of the North Slope was set aside in 1923 by the U.S. government as an emergency reserve for petroleum. This parcel of land, the size of Indiana, is today called the National Petroleum

Reserve–Alaska and was intended to remain untapped for oil unless the nation faced an emergency. So far, most of this region's 9.5 million ha (23.5 million acres) remains undeveloped.

East of the National Petroleum Reserve are state lands that experienced widespread development and extraction after oil was discovered at Prudhoe Bay in 1968. Since drilling began in 1977, over 14 billion barrels (1 barrel = 159 L or 42 gal) of crude oil have been extracted from 19 oil fields spread over 160,000 ha (395,000 acres) of this region. The oil is transported across the state of Alaska by the 1,300-km (800-mi) trans-Alaska pipeline south to the port of Valdez, where it is loaded onto tankers.

East of the Prudhoe Bay region lies the Arctic National Wildlife Refuge (ANWR), an area the size of South Carolina consisting of federal lands set aside in 1960 and 1980 mainly to protect wildlife and preserve pristine ecosystems of tundra, mountains, and seacoast. This scenic region is home to 160 nesting bird species, numerous fish and marine mammals, grizzly bears, polar bears, Arctic foxes, timber wolves, musk oxen, and other animals. In most years, thousands of caribou arrive from the south to spend the summer, giving birth to and raising their calves. Because of the vast caribou herd and the other large mammals, ANWR has been called "the Serengeti of North America."

ANWR has been the focus of debate for decades. Advocates of oil drilling have tried to open its lands for development, and proponents of wilderness preservation have fought for its preservation. Scientists, oil industry experts, politicians, environmental groups, citizens, and Alaska residents have all been part of the debate. So have the two Native groups in the area, the Gwich'in and the Inupiat, who disagree over whether the refuge should be opened to oil development. The Gwich'in depend on

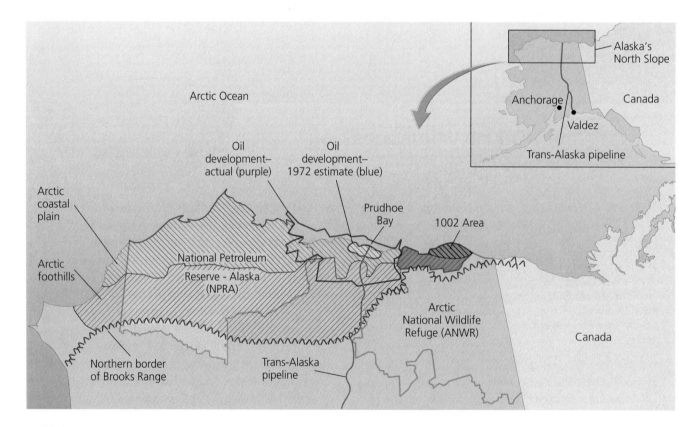

FIGURE 19.1 Alaska's North Slope is the site of both arctic wilderness and oil exploration. In the western portion of this region, the U.S. government established the National Petroleum Reserve–Alaska as an area in which to drill for oil if it is needed in an emergency. It is well explored but not widely developed. To the east of this area, the Prudhoe Bay region is the site of widespread oil extraction, and since 1977 it has produced over 14 billion barrels of oil. Oil development has expanded much farther than experts estimated it would in 1972. Farther east lies the Arctic National Wildlife Refuge, home to untrammeled Arctic wilderness and the focus of debate for years. Proponents of oil extraction and proponents of wilderness preservation have been battling over whether the 1002 Area of the coastal plain north of the Brooks Range should be opened to oil development.

hunting caribou and fear that oil industry activity will reduce caribou herds, whereas the Inupiat see oil extraction as one of the few opportunities for economic development in the area.

In a compromise in 1980, the U.S. Congress put most of the refuge off limits to oil but reserved for future decision-making a 600,000-ha (1.5-million-acre) area of coastal plain. This region, called the 1002 Area (after Section 1002 of the bill that established it), remains undeveloped for oil but can be opened for development by a vote of both houses of Congress. Its unsettled status has made it the center of the oil-versus-wilderness debate, and Congress has been caught between passionate feelings on both sides for a quarter of a century.

In 2005, the Republican-controlled Congress made several efforts to open the refuge to drilling. It added a provision approving drilling to a $450-billion military spending bill that included hurricane relief and other high-priority items. As this book went to press in late December, the House approved the bill but the Senate narrowly blocked it.

Behind the noisy policy debate over ANWR, scientists have attempted to inform the dialogue through research. Geologists have tried to ascertain how much oil lies underneath the refuge, and biologists have tried to predict the potential impacts of oil drilling on Arctic ecosystems. Moreover, scientists and nonscientists alike are debating the relevance of the oil beneath the refuge for the security and prosperity of the nation. We will examine these questions by revisiting Alaska's North Slope as we survey the fossil fuel energy we use to heat and light our homes, power our machinery, and provide the comforts, conveniences, and mobility to which technology has accustomed us.

Sources of Energy

The debate over drilling for oil in the Arctic National Wildlife Refuge is a thoroughly modern debate, pitting the culturally new concept of wilderness preservation against the desire to exploit a resource that has come to guide the world's economy only in the past 150 years. However, people have used—and fought over—energy in one way or another for all of our history.

We use a variety of energy sources

Earth receives energy from several sources, and people have developed many ways to harness the renewable and nonrenewable forms of energy available on our planet (Table 19.1). Most of Earth's energy comes from the sun. We can harness energy from the sun's radiation directly, but solar radiation also makes possible several other energy sources. Sunlight drives the growth of plants, from which we take wood as a fuel source. After their death, plants may impart their stored chemical energy to **fossil fuels** (such as oil, coal, and natural gas), which are highly combustible substances formed from the remains of organisms from past geological ages. Solar radiation also helps drive wind patterns and the hydrologic cycle, making possible other forms of energy, such as wind power and hydroelectric power.

A great deal of energy also emanates from Earth's core, enabling us to harness geothermal power. A much smaller amount of energy results from the gravitational pull of the moon and sun, and we are just beginning to harness the power from the ocean tides that these forces generate.

Table 19.1 **Energy Sources We Use Today**		
Energy source	**Description**	**Type of energy**
Crude oil	Fossil fuel extracted from ground	Nonrenewable
Natural gas	Fossil fuel extracted from ground	Nonrenewable
Coal	Fossil fuel extracted from ground	Nonrenewable
Nuclear energy	Energy from atomic nuclei of uranium mined from ground and processed	Nonrenewable
Biomass energy	Chemical energy from photosynthesis stored in plant matter	Renewable
Hydropower	Energy from running water	Renewable
Solar energy	Energy from sunlight directly	Renewable
Wind energy	Energy from wind	Renewable
Geothermal energy	Earth's internal heat rising from core	Renewable
Tidal and wave energy	Energy from tidal forces and ocean waves	Renewable

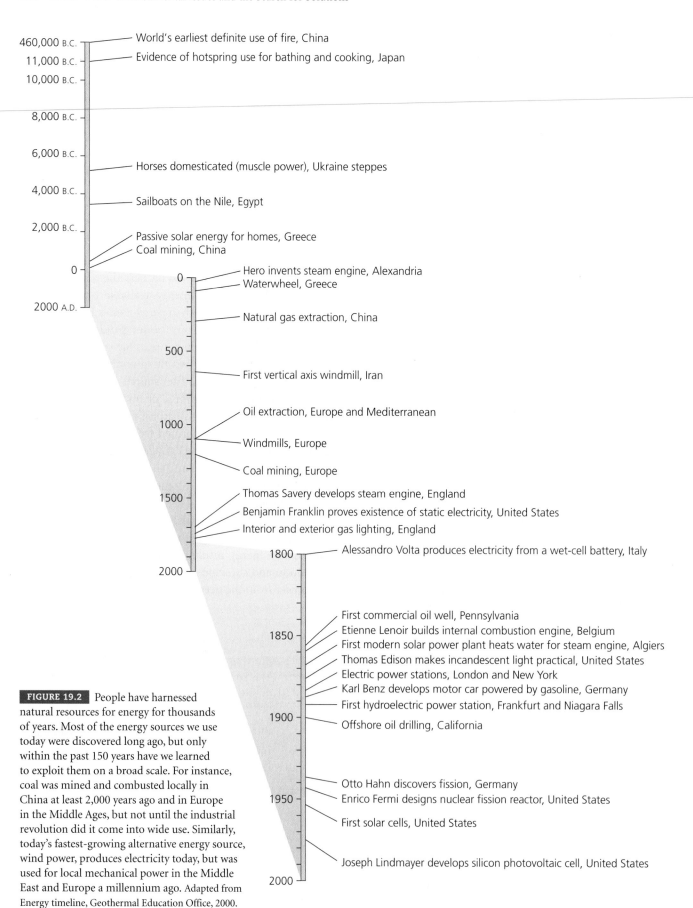

FIGURE 19.2 People have harnessed natural resources for energy for thousands of years. Most of the energy sources we use today were discovered long ago, but only within the past 150 years have we learned to exploit them on a broad scale. For instance, coal was mined and combusted locally in China at least 2,000 years ago and in Europe in the Middle Ages, but not until the industrial revolution did it come into wide use. Similarly, today's fastest-growing alternative energy source, wind power, produces electricity today, but was used for local mechanical power in the Middle East and Europe a millennium ago. Adapted from Energy timeline, Geothermal Education Office, 2000.

An immense amount of energy resides within the bonds among protons and neutrons in atoms, and this energy provides us with nuclear power.

Ever since our ancestors first discovered fire, people have extracted energy from natural resources to cook food and to light and heat dwellings (Figure 19.2). Wood products were our primary sources of energy for heating and cooking. We harnessed animals, wind, and water as energy sources for mechanical work in fields, mills, and granaries. The development of the steam engine in the 18th century allowed us to use concentrated, high-quality energy sources such as coal for a variety of mechanical purposes as economies industrialized. Meanwhile, deforestation in many areas of the world led to dwindling wood supplies, giving many societies incentive to shift to energy sources other than wood.

In the 20th century, fossil fuels became the dominant source of power in industrialized countries and then in developing nations. The high energy content of fossil fuels makes them efficient to burn, ship, and store. Besides providing for transportation, heating, and cooking, these fuels are used to generate electricity, a secondary form of energy that is easier to transfer over long distances and apply to a variety of uses.

As we first noted in Chapter 1 (▸ p. 4), energy sources such as sunlight, geothermal energy, and tidal energy are considered *renewable* because their supplies will not be depleted by our use of them. Other sources, such as timber, are renewable if we do not harvest them at too great a rate. In contrast, energy sources such as oil, coal, and natural gas are considered *nonrenewable,* because at our current rates of consumption we will use up Earth's accessible store of them in a matter of decades to centuries. Nuclear power as currently harnessed through fission of uranium (▸ p. 593) can be considered nonrenewable to the extent that uranium ore is in limited supply.

Although these nonrenewable fuels result from ongoing natural processes, the timescales on which they are created are so long that, once the fuels are depleted, they cannot be replaced in any time span useful to our civilization. It takes a thousand years for the biosphere to generate the amount of organic matter that must be buried to produce a single day's worth of fossil fuels for our society. To replenish the fossil fuels we have depleted so far would take many millions of years. For this reason, and because fossil fuels exert severe environmental impacts, renewable energy sources increasingly are being developed as alternatives to fossil fuels, as we will see in Chapters 20 and 21. Nonetheless, global consumption of the three main fossil fuels has risen steadily for years and is now at its highest level ever (Figure 19.3).

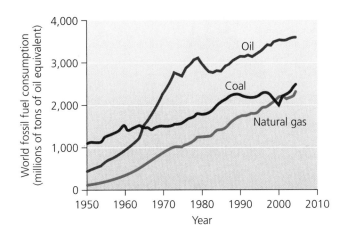

FIGURE 19.3 Global consumption of fossil fuels has risen greatly over the past half century. Natural gas is our fastest-growing fossil fuel today. Oil use rose steeply during the 1960s to overtake coal, and today it remains our leading energy source. Data from Worldwatch Institute. 2005. *Vital signs 2005.*

Fossil fuels are indeed fuels created from "fossils"

The fossil fuels we burn today in our vehicles, homes, industries, and electrical power plants were formed from the tissues of organisms that lived 100–500 million years ago. The energy these fuels contain came originally from the sun and was converted to chemical-bond energy as a result of plants' photosynthesis (▸ pp. 105–106). The chemical energy in these organisms' tissues was then concentrated as these tissues decomposed and their hydrocarbon compounds were altered and compressed (Figure 19.4).

Most organisms that die do not end up as part of a coal, gas, or oil deposit. A tree that falls and decays as a rotting log undergoes mostly **aerobic** decomposition; in the presence of air, bacteria and other organisms that use oxygen break down plant and animal remains into simpler carbon molecules that are recycled through the ecosystem. Fossil fuels are produced only when organic material is broken down in an **anaerobic** environment, one that has little or no oxygen. Such environments include the bottoms of deep lakes, swamps, and shallow seas. Over millions of years, organic matter that accumulated at the bottoms of such water bodies was converted into crude oil, natural gas, and coal. Which fuel formed in any given place depended on the chemical composition of the starting material, the temperatures and pressures to which the material was subjected, the presence or absence of anaerobic decomposers, and the passage of time.

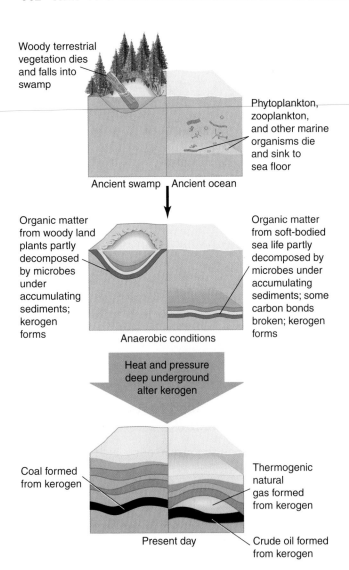

Table 19.2	Nations with Largest Proven Reserves of Fossil Fuels	
Oil (% world reserves)	**Natural gas** (% world reserves)	**Coal** (% world reserves)
Saudi Arabia, 22.1	Russia, 26.7	United States, 27.1
Iran, 11.1	Iran, 15.3	Russia, 17.3
Iraq, 9.7	Qatar, 14.4	China, 12.6
Kuwait, 8.3	Saudi Arabia, 3.8	India, 10.2
United Arab Emirates, 8.2	United Arab Emirates, 3.4	Australia, 8.6
Venezuela, 6.5	United States, 2.9	South Africa, 5.4
Russia, 6.1	Nigeria, 2.8	Ukraine, 3.8
Kazakhstan, 3.3	Algeria, 2.5	Kazakhstan, 3.4
Libya, 3.3	Venezuela, 2.4	Poland, 1.5
Nigeria, 3.0	Iraq, 1.8	Brazil, 1.1

Data from British Petroleum. 2005. *Statistical review of world energy 2005.*

FIGURE 19.4 The fossil fuels we use for energy today consist of the remains of organic material from plants (and to a lesser extent, animals) that died millions of years ago. Their formation begins when organisms die and end up in oxygen-poor conditions, such as when trees fall into lakes and are buried by sediment, or when phytoplankton and zooplankton drift to the seafloor and are buried. Organic matter that undergoes slow anaerobic decomposition deep under sediments forms kerogen. Geothermal heating acts on kerogen to create crude oil and natural gas. Natural gas can also be produced nearer the surface by anaerobic bacterial decomposition of organic matter. Oil and gas come to reside in porous rock layers beneath dense, impervious layers. Coal is formed when plant matter is compacted so tightly that there is little decomposition.

Fossil fuel reserves are unevenly distributed

Fossil fuel deposits are localized and unevenly distributed over Earth's surface, so some regions have substantial reserves of fossil fuels whereas others have very few. How long each nation's fossil fuel reserves will last depends on how much the nation extracts, how much it consumes, and how much it imports from and exports to other nations. Nearly two-thirds of the world's proven reserves of crude oil lie in the Middle East. The Middle East is also rich in natural gas, but Russia contains more than twice as much natural gas as any other country. Russia is also rich in coal, as is China, but the United States possesses more coal than any other nation (Table 19.2).

Developed nations consume more energy than developing nations

Citizens of developed nations generally consume far more energy than do those of developing nations. Per person, the most-industrialized nations use up to 100 times more energy than do the least-industrialized nations (Figure 19.5). Moreover, developed and developing nations tend to apportion their energy use differently. Developing nations devote a greater proportion of energy to subsistence activities, such as food preparation, home heating, and food-growing, whereas industrialized countries use a greater proportion for transportation and industry (Figure 19.6). In addition, people in developing countries often rely on manual or animal energy sources instead of automated ones. For instance, rice farmers in Bali plant rice by hand, but industrial rice growers in California use airplanes. Because industrialized nations rely more on equipment and technology, they use more fossil fuels. In the United States, fossil fuels supply 89% of energy needs. Oil is the most heavily used fossil fuel, constituting 40% of U.S. energy use, compared with 25% for natural gas and 24% for coal.

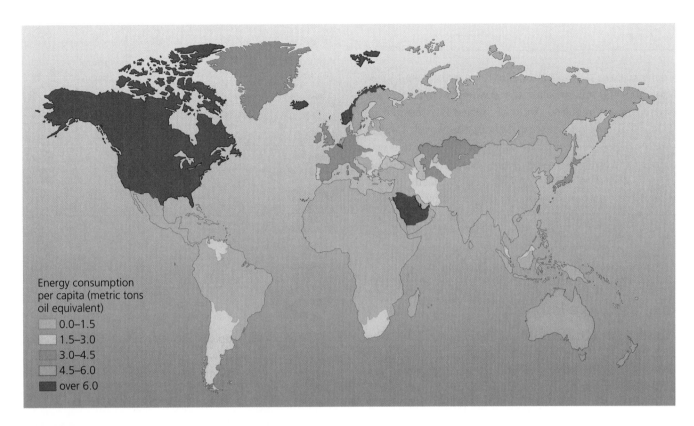

FIGURE 19.5 Regions vary greatly in their consumption of energy per person. People in developed nations consume the most. This map combines all types of energy, standardized to metric tons of "oil equivalent," that is, the amount of fuel needed to produce the energy gained from combusting one metric ton of crude oil. Data from British Petroleum. 2005. *Statistical review of world energy 2005.*

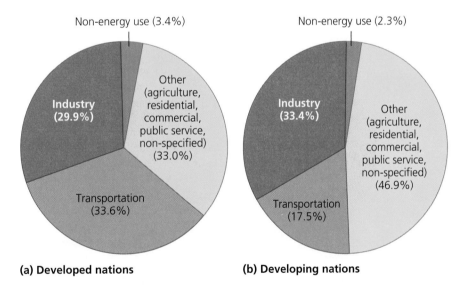

FIGURE 19.6 Developing nations and developed nations show somewhat different profiles of energy use. These data from 2002 show that developed nations of the Organisation for Economic Co-operation and Development (OECD) **(a)** devote nearly twice as much of their energy budgets to transportation as do non-OECD nations **(b),** whereas non-OECD nations devote relatively more energy to agricultural, residential, and other uses. The OECD includes 30 of the most industrialized nations from Europe and North America plus Japan, Korea, Turkey, Australia, and New Zealand. Non-OECD nations include some large, highly industrialized countries, such as Russia, China, Brazil, and Argentina, but consist mostly of smaller and developing nations. Data from International Energy Agency. 2004. *Key world energy statistics 2004.*

Coal

Coal is the world's most abundant fossil fuel. The proliferation 300–400 million years ago of swampy environments where organic material could be buried has resulted in substantial coal deposits throughout the world. **Coal** is organic matter (generally woody plant material) that was compressed under very high pressure to form dense, solid carbon structures (Figure 19.7). Coal typically results when little decomposition takes place because the material cannot be digested or appropriate decomposers are not present. One-quarter of the world's coal is located in the United States, and coal provides one-quarter of the world's commercial energy consumption.

Coal use has a long history

People have used coal longer than any other fossil fuel. The Romans used coal for heating in the second and third centuries in Britain, as have people in parts of China for 2,000–3,000 years. In Arizona, today's Hopi Indians follow ancestral traditions by using coal to fire pottery, cook food, and heat their homes. Once commercial mining began in Europe in the 1700s, coal began to be used widely as a heating source. Coal found an expanded market after the invention of the steam engine, because it was used to boil water to produce steam. Coal-fired steam engines performed work previously done by people or horses, including manufacturing, harvesting, and powering trains and ships. The birth of the steel industry in 1875 increased demand still further because coal fueled the furnaces used to produce steel.

Table 19.3 Top Producers and Consumers of Coal	
Production (% world production)	Consumption (% world consumption)
China, 36.2	China, 34.4
United States, 20.8	United States, 20.3
Australia, 7.3	India, 7.4
India, 6.9	Japan, 4.3
South Africa, 5.0	Russia, 3.8
Russia, 4.7	South Africa, 3.4
Indonesia, 3.0	Germany, 3.1
Poland, 2.6	Poland, 2.1
Germany, 2.0	Australia, 2.0
Kazakhstan, 1.6	South Korea, 1.9

Data from British Petroleum. 2005. *Statistical review of world energy 2005.*

In the 1880s, people began to use coal to generate electricity. In coal-fired power plants, coal combustion converts water to steam, which turns a turbine to create electricity (see "The Science behind the Story," ▶ pp. 566–567). Today coal provides over half the electrical generating capacity of the United States. China and the United States are the primary producers and consumers of coal (Table 19.3).

Coal varies in its qualities

Coal varies from deposit to deposit in many ways, including in water content and the amount of potential energy it contains. Organic material that is broken down anaerobically but remains wet, near the surface, and not well compressed

FIGURE 19.7 Coal forms as ancient plant matter is compacted underground. Scientists categorize coal into several types, depending on the amount of heat, pressure, and moisture involved in its formation. Anthracite coal is formed under greatest pressure, where temperatures are high and moisture content is low. Lignite coal is formed under conditions of much less pressure and heat, but more moisture. Peat is also part of this continuum, representing plant matter that is minimally compacted.

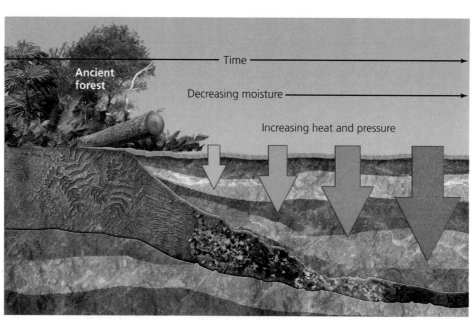

is called **peat.** A kind of precursor stage to coal, peat has been widely used as a fuel in Britain and other locations. As peat decomposes further, as it becomes buried more deeply under sediments, as pressure and heat increase, and as time passes, water is squeezed out of the material, and carbon compounds are packed more tightly together, forming coal. Scientists classify coal into four types: lignite, sub-bituminous, bituminous, and anthracite. Lignite is the least-compressed type of coal, and anthracite is the most-compressed type. The greater the compression, the greater is the energy content per unit volume (see Figure 19.7).

Most coal contains various impurities, including sulfur, mercury, arsenic, and other trace metals, and coal deposits vary in the amount of impurities they contain. For instance, sulfur content varies, depending in part on whether the coal was formed in freshwater or saltwater sediments. Coal in what is today the eastern United States tends to be high in sulfur because it was formed in marine sediments, where sulfur from seawater was present. When high-sulfur coal is burned, it produces sulfate air pollutants, which contribute to industrial smog and acidic deposition (▸pp. 509–510, 514–518). Combustion of coal high in mercury content emits mercury that can bioaccumulate in organisms' tissues, poisoning animals as it moves up food chains. Such pollution problems commonly occur downwind of coal-fired power plants. Scientists and engineers are seeking ways to cleanse coal of its impurities so that it can continue to be used as an energy source while minimizing impact on the environment (see "The Science behind the Story," ▸pp. 566–567).

Coal is mined from the surface and from below ground

We extract coal using two major methods (Figure 19.8). We reach underground deposits with **subsurface mining.** Shafts are dug deep into the ground, and networks of tunnels are dug or blasted out to follow coal seams. The coal is removed systematically and shipped to the surface. When coal deposits are at or near the surface, strip-mining methods are used. In **strip mining,** heavy machinery removes huge amounts of earth to expose and extract the coal. The pits are subsequently refilled with the soil that had been removed. Strip-mining operations can occur on immense scales; in some cases entire mountaintops are lopped off. This environmentally destructive process, called *mountaintop removal,* has become more common recently in the Appalachian Mountains. We will revisit some of coal's environmental impacts later in this chapter. Understanding these impacts is important because society's demand for this relatively abundant fossil fuel may soon rise as supplies of our most-used fossil fuel, oil, decline.

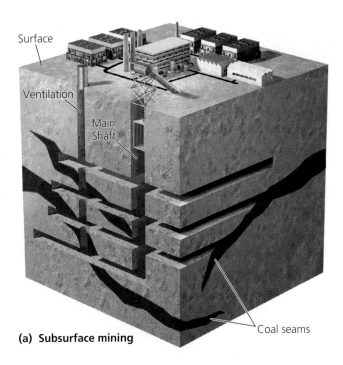

(a) Subsurface mining

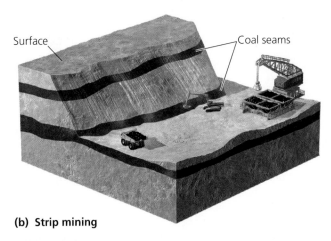

(b) Strip mining

FIGURE 19.8 Coal is mined in two major ways. In subsurface mining **(a),** miners work below ground in shafts and tunnels blasted through the rock; these passageways provide access to underground seams of coal. This type of mining poses dangers and long-term health risks to miners. In strip mining **(b),** soil is removed from the surface, exposing coal seams from which coal is mined. This type of mining can cause substantial environmental impact.

Oil

Oil has been the world's most-used fuel since the 1960s, when it eclipsed coal. It now accounts for 37% of the world's commercial energy consumption. Its use worldwide over the past decade has risen roughly 16%.

The Science behind the Story

How Electricity Is Generated

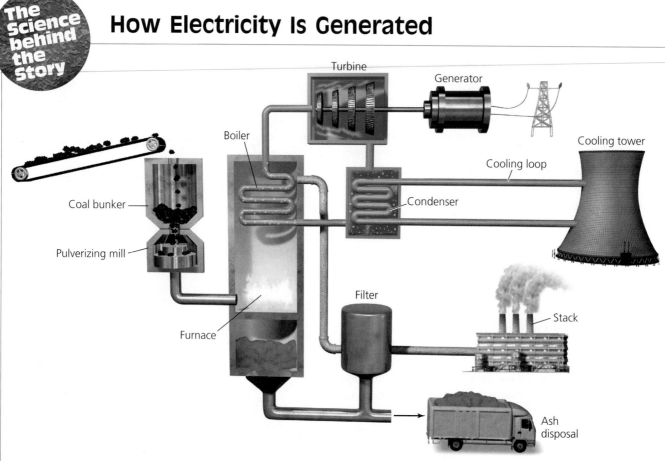

Coal is the primary fuel source used to generate electricity in the United States. Pieces of coal are pulverized and blown into a high-temperature furnace. Heat from the combustion boils water, and the resulting steam turns a turbine, generating electricity by passing magnets past copper coils. The steam is then cooled and condensed in a cooling loop and returned to the furnace. "Clean coal" technologies help filter out pollutants from the combustion process, and toxic ash residue is disposed of in hazardous waste disposal sites.

As the United States tries to balance its growing demand for electricity with rising concerns about environmental and health impacts of coal combustion, power plants continue to rely heavily on coal while scientists work to limit the pollution that use of this fuel creates.

Every new housing subdivision, cell phone, and DVD player means a greater draw on the power grid. To generate more electricity using domestic resources, coal has emerged as a central factor in recent U.S. energy policy.

Coal is used to generate electricity in a process that dates back more than a century (see the figure). Once mined, coal is hauled to power plants, where it is pulverized. The crushed coal is blown into a boiler furnace on a superheated stream of air and burned in a blaze of intense

Heat and pressure underground form petroleum

The sludgelike liquid we know as **crude oil,** or **petroleum,** tends to form within a window of temperature and pressure conditions often found 1.5–3 km (1–2 mi) below the surface. Crude oil is a mixture of hundreds of different types of hydrocarbon molecules characterized by carbon chains of different lengths (▶ pp. 98–99). A chain's length affects its chemical properties, which has consequences for human use, such as whether a given fuel burns cleanly in a car engine. Oil refineries sort the various

heat—typical furnace temperatures often flare at 815 °C (1500 °F). Water circulating around the boiler absorbs the heat and is converted to high-pressure steam. This steam is injected into a **turbine,** a rotary device that converts the kinetic energy of a moving substance such as steam into mechanical energy. The turbine's key components are a drive shaft and a series of fanlike blades attached to the drive shaft. As steam from the boilers exerts pressure on the blades of the turbine, they spin, turning the drive shaft.

The drive shaft is connected to a generator, which also features two main components—a rotor that rotates and a stator that remains stationary. Generators have different designs, but they all make use of the same principle: Moving magnets adjacent to coils of copper wire cause electrons in the copper wires to move, generating alternating electric current. In some generators, the rotor consists of magnets and spins within a stator of coiled copper wire. As the turbine's drive shaft rotates, it causes the rotor to revolve, creating a magnetic field and causing electrons in the stator's copper wires to move, thereby creating an electrical current. This current flows into transmission lines that travel from the power plant out to the customers who use the plant's electricity.

Because coal is the country's largest domestic fossil fuel source but also a major source of pollution, U.S. policymakers have arrived at a dual approach: continuing to rely on coal while trying to make its use less toxic. The national energy legislation passed by Congress in 2005 provided $1.3 billion in tax credits to power plants that use state-of-the-art "clean coal" technologies, and set aside $2.5 billion over 10 years for further research into such technologies. These efforts largely focus on approaches to rid the generation process of toxic chemicals, either before or after the coal is burned. Technologies to do this include *scrubbers,* or materials based on minerals such as calcium or sodium that absorb and remove sulfur dioxide (SO_2) from smokestack emissions. Other approaches use chemical reactions to strip away nitrogen oxides (NO_X), breaking them down into elemental nitrogen and water. Multilayered filtering devices are used to capture tiny ash particles.

Research at the Niles Station power plant in Ohio found a way to remove 95% of SO_2 and 90% of NO_X from the plant's emissions. Researchers first installed smokestacks with efficient filter bags and captured significant amounts of ash and gases. The filters also proved effective at removing dangerous substances such as mercury and selenium. The captured ash and gases were reheated, and small amounts of ammonia were added to chemically convert NO_X into harmless nitrogen gas and water vapor. Further addition of oxygen to SO_2 in the smokestack gas converted the sulfur compounds into sulfuric acid, which the plant bottled and sold for use in fertilizer and manufacturing processes. The gas-scrubbing process worked well enough for the plant to begin using it permanently, and several power plants in Europe adopted the technology.

Some energy analysts and environmental advocates question a policy emphasis on clean coal, however. Coal, they maintain, is an inherently dirty way of generating power and should be replaced outright with cleaner energy sources. Others doubt that the Bush administration's push for clean coal technologies represents a true commitment to cleaner air, because the administration also eased Clean Air Act requirements that would have forced power companies to upgrade their pollution control technology. Without mandates, many power generators have been slow to adopt clean coal technology on their own. However, with the new funding from the 2005 energy legislation, research will press ahead.

hydrocarbons of crude oil, separating those intended for use in gasoline engines from those, such as tar and asphalt, used for other purposes.

The crude oil of Alaska's North Slope was formed when dead plant material (and small amounts of animal material) drifted down through coastal marine waters millions of years ago and was buried in sediments on the ocean floor. These organic remains were then transformed by time, heat, and pressure into the crude oil of today. The shelves of sedimentary rock that now lie beneath Alaska's coastal plain are believed to hold the largest remaining onshore petroleum deposits in North America.

Table 19.4 Top Producers and Consumers of Oil	
Production (% world production)	**Consumption** (% world consumption)
Saudi Arabia, 13.1	United States, 24.9
Russia, 11.9	China, 8.6
United States, 8.5	Japan, 6.4
Iran, 5.2	Russia, 3.4
Mexico, 4.9	Germany, 3.3
China, 4.5	India, 3.2
Venezuela, 4.0	South Korea, 2.8
Canada, 3.8	Canada, 2.6
Norway, 3.9	France, 2.5
United Arab Emirates, 3.3	Italy, 2.4

Data from British Petroleum. 2005. *Statistical review of world energy 2005.*

The age of oil began in the mid-19th century

As long ago as 4,000 B.C., people used solid forms of oil (such as tar and asphalt) from deposits that were easily accessible at Earth's surface. The modern extraction and use of petroleum for energy began in the 1850s. In Pennsylvania, miners drilling for salt occasionally encountered oily rocks instead. At first, entrepreneurs bottled the crude oil from these deposits and sold it as a healing aid, unaware that crude oil is carcinogenic when applied to the skin and poisonous when ingested. Soon, however, a Dartmouth College scholar named George Bissell realized that this "rock oil" could be used to light lamps and lubricate machinery, and in 1854 Bissell started the Pennsylvania Rock Oil Company. By the time the firm developed its drilling technology and struck oil, Bissell was no longer at the company. Instead, Edwin Drake is credited with drilling the world's first oil well, in Titusville, Pennsylvania, in 1859. Over the next 40 years, Pennsylvania's oil fields produced half the world's oil supply and helped establish a fossil-fuel-based economy that would hold sway for decades to come.

Today our global society produces and consumes nearly 750 L (200 gal) of oil each year for every man, woman, and child. The United States consumes fully one-fourth of the world's oil. U.S. oil consumption has increased 16% in the past decade and shows little sign of abating. Table 19.4 shows the top oil-producing and oil-consuming nations.

Petroleum geologists infer the location and size of deposits

Because petroleum forms only under certain conditions, it occurs in isolated deposits. Once geothermal heating separates hydrocarbons from their source material and produces crude oil, this liquid migrates upward through rock pores, sometimes assisted by seismic faulting. It tends to collect in porous layers beneath dense, impermeable layers. Thus, oil deposits are not large black underground lakes, but instead consist of small droplets within holes in porous rock, like a hard sponge full of oil.

Geologists searching for oil (or other fossil fuels) drill rock cores and conduct ground, air, and seismic surveys (Figure 19.9) to map underground rock formations, understand geological history, and predict where fossil fuel deposits might lie. Using such techniques, geologists from the U.S. Geological Survey (USGS) in 1998 estimated, with 95% certainty, the total amount of oil underneath ANWR's 1002 Area to be between 11.6 and 31.5 billion barrels. The geologists' average estimate of 20.7 billion represents their best guess as to the number of barrels of oil the 1002 Area holds.

Some portion of oil will be impossible to extract using current technology and may have to wait for future advances in extraction equipment or methods. Thus, estimates are generally made of "technically recoverable"

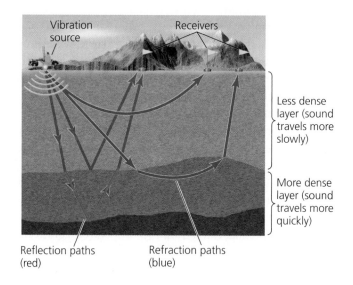

FIGURE 19.9 Petroleum geologists use seismic surveying to locate promising fossil fuel deposits. One method is to create powerful vibrations (by exploding dynamite, thumping the ground with a large weight, or using an electric vibrating machine) at the surface in one location and then measure how long it takes the seismic waves produced to reach receivers at other surface locations. Seismic waves travel more quickly through denser layers, and density differences in the substrate may cause waves to reflect off layers, refract, or bend. Scientists and engineers interpret the patterns of wave reception to infer the densities, thicknesses, and location of underlying geological layers—which in turn provide clues about the location and size of fuel deposits.

amounts of fuels. In its 1998 estimates, the USGS calculated technically recoverable amounts of oil under the 1002 Area to be between 4.3 and 11.8 billion barrels, with an average estimate of 7.7 billion barrels. Because some Native lands and state-owned offshore areas can be developed if the 1002 Area is developed, the USGS also surveyed these areas and estimated that 1.4–4.2 billion additional barrels likely lie beneath them.

However, oil companies will not be willing to extract these entire amounts. Some oil would be so difficult to extract that the expense of doing so would exceed the income the company would receive from the oil's sale. Thus, the amount a company chooses to drill for will be determined by the costs of extraction (and transportation), together with the current price of oil on the world market. Because the price of oil fluctuates, the portion of oil from a given deposit that is "economically recoverable" fluctuates as well. USGS scientists calculated that at a price of $30 per barrel, 3.0–10.4 billion barrels would be economically worthwhile to recover from the 1002 Area. The USGS did not present estimates for today's much higher prices, but as prices climb, economically recoverable amounts approach technically recoverable amounts.

Thus, technology sets a limit on the amount that *can* be extracted, whereas economics determines how much *will* be extracted. The amount of oil, or any other fossil fuel, in a deposit that is technologically and economically feasible to remove under current conditions is termed the **proven recoverable reserve** of that fuel.

We drill to extract oil

Once geologists have identified an oil deposit, an oil company will typically conduct *exploratory drilling*. Holes drilled during this phase are usually small in circumference and descend to great depths. If enough oil is encountered, extraction begins. Just as you would squeeze a sponge to remove its liquid, pressure is required to extract oil from porous rock. Oil is typically already under pressure—from above by rock or trapped gas, from below by groundwater, or internally from natural gas dissolved in the oil. All these forces are held in place by surrounding rock until drilling reaches the deposit, whereupon oil will often rise to the surface of its own accord.

Once pressure is relieved, however, oil becomes more difficult to extract and may need to be pumped out. Even after pumping, a great deal of oil remains stuck to rock surfaces. As much as two-thirds of a deposit may remain in the ground after **primary extraction,** the initial drilling and pumping of available oil (Figure 19.10a). Companies may then begin **secondary extraction,** in which solvents are used or underground rocks are flushed with water or

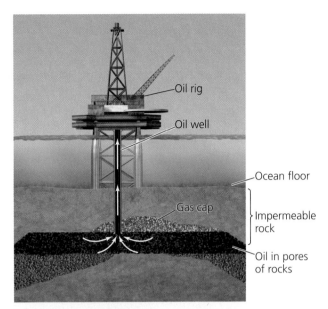

(a) Primary extraction of oil

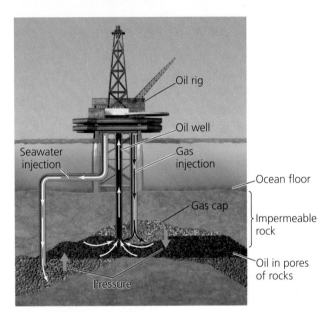

(b) Secondary extraction of oil

FIGURE 19.10 In primary extraction (**a**), oil is drawn up through the well by keeping pressure at the top lower than pressure at the level of the oil deposit. Once the pressure in the deposit drops, however, material must be injected into the deposit to increase the pressure. Thus, secondary extraction (**b**) involves injecting seawater beneath the oil and/or gases just above the oil to force more oil up and out of the deposit.

steam to remove additional oil (Figure 19.10b). At Prudhoe Bay, seawater is piped in from the coast and pumped into wells to flush out remaining oil. Even after secondary extraction, quite a bit of oil can remain; we lack the technology to remove every last drop. Secondary extraction is more expensive than primary extraction, so many U.S. deposits did not undergo secondary extraction when they

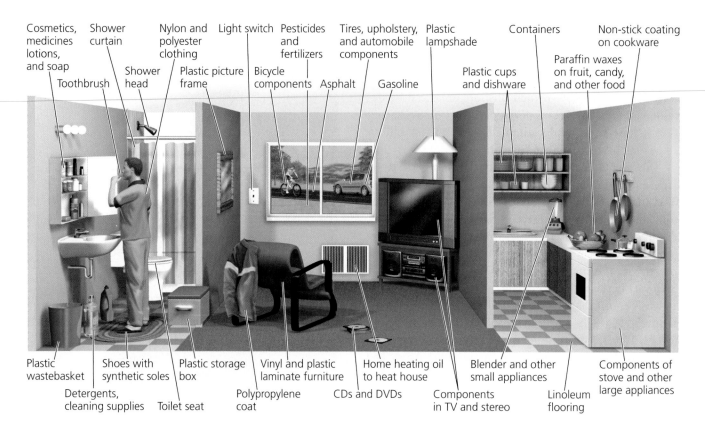

Cosmetics, medicines lotions, and soap | Shower curtain | Nylon and polyester clothing | Light switch | Pesticides and fertilizers | Tires, upholstery, and automobile components | Plastic lampshade | Containers | Non-stick coating on cookware

Toothbrush | Shower head | Plastic picture frame | Bicycle components | Asphalt | Gasoline | Plastic cups and dishware | Paraffin waxes on fruit, candy, and other food

Plastic wastebasket | Shoes with synthetic soles | Plastic storage box | Vinyl and plastic laminate furniture | Home heating oil to heat house | Blender and other small appliances | Components of stove and other large appliances

Detergents, cleaning supplies | Toilet seat | Polypropylene coat | CDs and DVDs | Components in TV and stereo | Linoleum flooring

FIGURE 19.11 Petroleum products are everywhere in our daily lives. The gasoline and other fuels we use for transportation and heating are just a few of the many products we derive from petroleum. These products include many of the fabrics that we wear and most of the plastics that help make up countless items we use every day.

were first drilled because the price of oil was too low to make the procedure economical. When oil prices rose in the 1970s, many drilling sites were reopened for secondary extraction.

Offshore drilling produces much of our oil

Drilling for oil and natural gas takes place not just on land but also in the seafloor on the continental shelves. Offshore drilling has required development of technology that can withstand the forces of wind, waves, and ocean currents. Some drilling platforms are fixed standing platforms built with unusual strength. Others are resilient floating platforms anchored in place above the drilling site. Over 25% of the oil and gas extracted in the United States comes from offshore drilling sites, primarily in the Gulf of Mexico and off the southern California coast. This is why Hurricanes Katrina and Rita in 2005 caused such disruption to U.S. oil supplies. By damaging offshore oil platforms off the coasts of Louisiana and neighboring states and by damaging refineries onshore, the storms interrupted a substantial portion of the nation's oil supply, and prices rose accordingly.

Petroleum products have many uses

Once crude oil is extracted, it is put through refining processes (see "The Science behind the Story," ▸pp. 572–573). Because crude oil is a complex mix of hydrocarbons, we can create many types of petroleum products by separating its various components. Since the 1920s, refining techniques and chemical manufacturing have greatly expanded our uses of petroleum to include a wide array of products and applications, from lubricants to plastics to fabrics to pharmaceuticals. Today, petroleum-based products are all around us in our everyday lives (Figure 19.11).

Because petroleum products have become so central to our lives, many fossil fuel experts today are voicing extreme concern that oil production may soon decline as we continue to deplete the world's oil reserves.

We may have already depleted half our oil reserves

Many scientists and oil industry analysts calculate that we have already extracted half of the world's oil reserves. So far we have used up about 1 trillion barrels of oil, and most estimates hold that an additional 1 trillion barrels, or somewhat more, remain. To estimate how long this remaining oil will last, analysts calculate the **reserves-to-production ratio,** or **R/P ratio,** by dividing the amount of total remaining reserves by the annual rate of production (i.e., extraction and processing). At current levels of production (30 billion barrels globally per year), most analysts estimate that world oil supplies will last about 40 more years.

Many policymakers and industry spokespeople suggest that this means we will have little to worry about for 40 years. However, another group of scientists and analysts insists that we will face a crisis not when the last drop of oil is pumped, but when the rate of production first begins to decline. They point out that when production declines as demand continues to increase (because of rising global population and consumption), we will experience an oil shortage immediately. Because production tends to decline once reserves are depleted halfway, most of these

experts calculate that this crisis will likely begin within the next several years.

To understand the basis of these concerns, we need to turn back the clock to 1956. In that year, Shell Oil geologist M. King Hubbert calculated that U.S. oil production would peak around 1970. His prediction was ridiculed at the time, but it proved to be accurate; U.S. production peaked in that very year and has continued to fall since then (Figure 19.12a). The peak in production came to be known as **Hubbert's peak.** In 1974, Hubbert analyzed data on technology, economics, and geology, to predict that global oil production would peak in 1995. It has kept growing past 1995, but many scientists using newer, better data today predict that at some point in the coming decade, production will begin to decline (Figure 19.12b). Discoveries of new oil fields peaked 30 years ago, and since then we have been extracting and consuming more oil than we have been discovering.

FIGURE 19.12 Because fossil fuels are nonrenewable resources, supplies at some point pass the midway point of their depletion, and annual production begins to decline. U.S. oil production peaked in 1970, just as geologist M. King Hubbert predicted decades previously; this highpoint is referred to as "Hubbert's peak" **(a).** Today many analysts believe global oil production is about to peak. Shown **(b)** is the latest projection, from a 2004 analysis by scientists at the Association for the Study of Peak Oil. Go to **GRAPH**it! at www.aw-bc.com/withgott or on the student CD-ROM. Data from (a) Deffeyes, K. S. 2001. *Hubbert's peak: The impending world oil shortage.* (b) Colin J. Campbell and Association for the Study of Peak Oil, 2004.

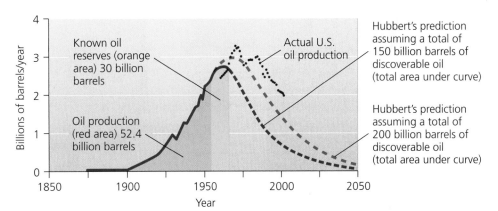

(a) Hubbert's prediction of peak in U.S. oil production, with actual data

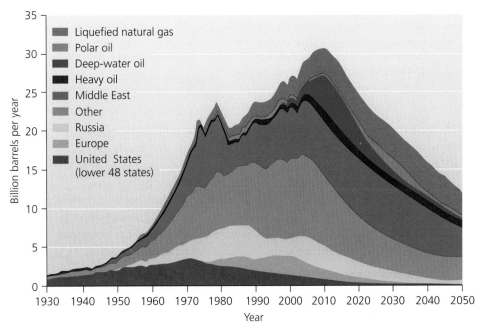

(b) Modern prediction of peak in global oil production

How Crude Oil Is Refined

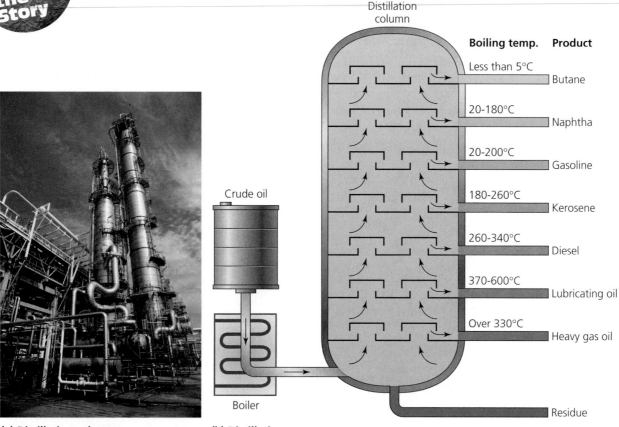

(a) Distillation columns **(b) Distillation process**

Crude oil is shipped to petroleum refineries **(a),** where it is refined into a number of different types of fuel. Crude oil is boiled, causing its many hydrocarbon constituents to volatilize and proceed upward through a distillation column **(b).** Constituents that boil only at the highest temperatures and condense readily once the temperature drops will condense at low levels in the column. Constituents that volatilize readily at lower temperatures will continue rising through the column and condense at higher levels, where temperatures are lower. In this way, heavy oils (generally consisting of long hydrocarbon molecules) are separated from lighter oils (generally those with short hydrocarbon molecules).

Crude oil is a complex mixture of thousands of kinds of hydrocarbon molecules. Through the process of **refining,** these hydrocarbon molecules are separated into classes of different sizes and chemically transformed to create specialized fuels for heating, cooking, and transportation and to create lubricating oils, asphalts, and the precursors of plastics and other petrochemical products. To maximize the production of marketable products while minimizing negative environmental impacts, petroleum engineers have developed a variety of refining techniques.

The first step in processing crude oil is *distillation,* or *fractionation.* This process is based on the fact that different components of crude oil boil at different temperatures. In refineries, the distillation process takes place in tall columns filled with perforated horizontal trays (see the first figure). The columns are cooler at the top than at the bottom. When heated crude oil is introduced into the column, lighter components rise as vapor to the upper trays, condensing into liquid as they cool, while heavier components sink to the lower trays. Light gases, such as butane, boil at less than 32 °C (90 °F), and heavier oils, such as industrial fuel oil, boil only at temperatures above 343 °C (650 °F).

Since the early 20th century, light gasoline, used in automobiles, has been in much higher demand than most other derivatives of crude oil. The demand for high-performance, clean-burning gasoline has also risen. To meet these demands, refiners have developed several techniques to convert heavy hydrocarbons into gasoline.

The general name for processes that convert heavy oil into lighter oil is *cracking*. One of the simplest methods is thermal cracking, in which long-chained molecules are broken into smaller chains by heating in the absence of oxygen. (The oil would ignite if oxygen were present.) Catalytic cracking, a related method, uses catalysts—substances that promote chemical reactions without being consumed by them—to control the cracking process. The result is an increase in the amount of a desired lighter product from a given amount of heavy oil. Today, the most widely used form of cracking is fluidized catalytic cracking, in which a finely powdered catalyst that behaves like a fluid is fed continuously into a reaction chamber with heavy oils. The products of cracking are then fed into a distillation column.

Refiners can also change the chemical composition of oil through a process called *catalytic reforming*. Catalytic reforming uses catalysts to promote chemical reactions that transform certain hydrocarbons that are slightly heavier than gasoline so that they can be blended with gasoline to obtain higher octane ratings. The octane rating reflects the amount of compression gasoline can undergo before it spontaneously ignites. An octane rating of 92 indicates that a gasoline blend is equivalent to a mixture of 92% octane, which spontaneously ignites only at very high compression levels, and 8% heptane, which ignites more easily. Other things being equal, high-compression engines are more powerful than low-compression engines, so gasoline that can withstand high levels of compression—that has a high octane rating—is preferred. Roughly 30–40% of U.S. gasoline is produced using catalytic reforming.

High-octane gasoline can also be produced by combining smaller hydrocarbons to synthesize larger molecules. In this process, called alkylation, molecules such as isobutylene and isobutene, which each have four carbon atoms, are combined in the presence of catalysts to form molecules with eight carbon atoms. The resulting high-octane alkylate can then be blended with lower-octane gasoline.

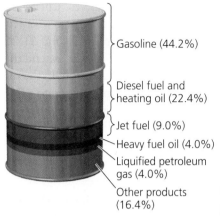

Gasoline (44.2%)

Diesel fuel and heating oil (22.4%)

Jet fuel (9.0%)

Heavy fuel oil (4.0%)

Liquified petroleum gas (4.0%)

Other products (16.4%)

The refining process converts crude oil into a range of petroleum products. Shown are percentages of each major category of product typically produced from a barrel of crude oil. Adapted from U.S. Energy Information Administration.

Besides distilling crude oil and altering the chemical structure of some of its components, refineries also remove contaminants. Sulfur and nitrogen compounds, which can be harmful when released into the atmosphere, are the two most common contaminants in crude oil. Government regulations stemming from legislation such as the Clean Air Act (▶ pp. 506–507) have forced refiners to develop methods of removing such contaminants, particularly sulfur. Some methods successfully remove up to 98% of sulfur.

As a result of all these approaches, each barrel of crude oil is eventually converted into gasoline and a wide variety of other petroleum products (see the second figure).

Predicting an exact date for the coming decline in production is difficult. Nonetheless, it seems certain that Hubbert's peak for global oil will occur and that the divergence of supply and demand could have catastrophic economic consequences. To achieve a sustainable society, we will eventually need to switch to renewable energy sources (Chapters 20 and 21). Energy conservation (▸pp. 582–583, 585) can extend the time we have in which to make this transition. In the meantime, it seems most likely that we may come to rely more heavily on coal and on natural gas.

Weighing the Issues:
The End of Oil

Physicist David Goodstein has calculated that the gap between rising demand and falling supply after world oil production begins to decline may amount to 5% per year. As a result, just 10 years after the production peak we will have only half the oil availability we had at the peak. He worries that we may not be able to modify our infrastructure and institutions fast enough to accommodate other energy sources before the economic impacts of an oil shortage undermine our ability to do so.

Do you think our society could adapt to a 50% decrease in oil availability over 10 years? How do you think we would most likely respond? How do you think we should respond?

Natural Gas

Natural gas is the fastest-growing fossil fuel in use today and provides for one-quarter of global commercial energy consumption. World supplies of natural gas are projected to last perhaps 60 years. This R/P ratio is slightly better than that of oil, but scientists expect that natural gas will likewise experience Hubbert's peak, such that production will decline after half the world's reserves are depleted.

Natural gas is formed in two ways

Natural gas consists primarily of methane, CH_4, and typically includes varying amounts of other volatile hydrocarbons. Natural gas can arise from either of two processes. *Biogenic* gas is created at shallow depths by the anaerobic decomposition of organic matter by bacteria. An example is the "swamp gas" you can sometimes smell when stepping into the muck of a swamp. In contrast, *thermogenic* gas results from compression and heat deep underground. As organic matter is buried more and more deeply under sediments, the pressure exerted by the overlying sediments grows, and temperatures increase. Carbon bonds in the organic matter begin breaking, and the organic matter turns to a substance called *kerogen,* which acts as a source material for both natural gas and crude oil. Further heat and pressure act on the kerogen to degrade complex organic molecules into simpler hydrocarbon molecules. At very deep levels—below about 3 km (1.9 mi)—the high temperatures and pressures tend to form natural gas.

Thermogenic gas may be formed directly, along with crude oil, or it may be formed from crude oil that is altered by heating. Biogenic gas is nearly pure methane, but thermogenic gas contains small amounts of a number of other hydrocarbon gases as well as methane. Most gas extracted commercially is thermogenic and found above deposits of crude oil or seams of coal, so its extraction often accompanies the extraction of those fossil fuels. In Alaska, where natural gas transport remains prohibitively expensive, gas captured during oil drilling is reinjected into the ground for potential future extraction.

One source of biogenic natural gas is the decay process in landfills, and many landfill operators are now capturing this gas to sell as fuel (▸ p. 657). This practice decreases energy waste, can be profitable for the operator, and helps reduce the atmospheric release of methane, a greenhouse gas that contributes to climate change (▸p. 530).

Natural gas was long known but has only recently been widely used in homes

Throughout history, naturally occurring seeps of natural gas would occasionally be ignited by lightning and could be seen burning in parts of what is now Iraq, inspiring the Greek essayist Plutarch around A.D. 100 to describe their "eternal fires." Such natural seepages of gas from underground deposits were used to light some streets and buildings in the United States in the 1800s.

The first commercial extraction of natural gas took place in 1821, but during much of the 19th century its use was localized because technology did not exist to pipe gas safely over long distances. Natural gas was used to fuel streetlamps, but when electric lights replaced most gas lamps in the 1890s, gas companies began marketing gas for heating and cooking. The first long-distance gas pipeline was built in 1891 to carry gas from

Table 19.5 Top Producers and Consumers of Natural Gas	
Production **(% world production)**	**Consumption** **(% world consumption)**
Russia, 21.9	United States, 24.0
United States, 20.2	Russia, 15.0
Canada, 6.8	United Kingdom, 3.6
United Kingdom, 3.6	Canada, 3.3
Iran, 3.2	Iran, 3.2
Algeria, 3.0	Germany, 3.2
Norway, 2.9	Italy, 2.7
Indonesia, 2.7	Japan, 2.7
Netherlands, 2.6	Ukraine, 2.6
Saudi Arabia, 2.4	Saudi Arabia, 2.4

Data from British Petroleum. 2005. *Statistical review of world energy 2005.*

FIGURE 19.13 Horsehead pumps are used to extract natural gas as well as oil. They are a common feature of the landscape in areas such as west Texas. The pumping motion of the machinery draws gas and oil upward from below ground.

deposits in Indiana to homes 195 km (120 mi) away in Chicago. After World War II, wartime improvements in welding and pipe building made gas transport safer and more economical, and during the 1950s and 1960s, thousands of miles of underground pipelines were laid throughout the United States. If laid end to end, our current network of natural gas pipelines would extend to the moon and back twice. Natural gas deposits are greatest in Russia and the United States, and these two nations lead the world in both gas production and gas consumption (Table 19.5).

Natural gas extraction becomes more challenging with time

To access some natural gas deposits, prospectors need only drill an opening, because pressure and low molecular weight drive the gas upward naturally. The first gas fields to be tapped were of this type. Most fields remaining today, however, require that gas be pumped to Earth's surface. In Texas, Kansas, California, and other areas of the United States, it is common to see a device called a horsehead pump (Figure 19.13). This pump moves a rod in and out of a shaft, creating pressure to pull both oil and natural gas to the surface. As with oil and coal, many of the most accessible natural gas reserves have already been exhausted. Thus, much extraction today makes use of sophisticated techniques to break into rock formations and pump gas to the surface. One such "fracturing technique" is to pump salt water under high pressure into the rocks to crack them. Sand or small glass beads are inserted to hold the cracks open once the water is withdrawn.

Other fossil fuels could be used in the future

Although natural gas, crude oil, and coal are the three fossil fuels that power our civilization today, other types of fossil fuels exist. *Oil sands* or *tar sands* are dense, hard, oily substances that can be mined from the ground. *Shale oil* is essentially kerogen, sedimentary rock filled with organic matter that was not buried deeply enough to form oil. *Methane hydrates* occur under the seafloor and were discussed in Chapter 16 (▶ p. 479). These sources are abundant, but technology for extracting usable fuel from them is largely undeveloped, and their extraction will likely remain extremely expensive. Many advocates of sustainability believe it would be a mistake to try to switch to these sources once our conventional fossil fuels become scarce. Such sources will require extensive mining and will emit at least as much carbon dioxide, methane, and other air pollutants as do coal, oil, and gas, so they will further the severe environmental impacts that fossil fuels are already causing.

Environmental Impacts of Fossil Fuel Use

Our society's love affair with fossil fuels and the many petrochemical products we have developed from them has boosted our material standard of living beyond what our ancestors could have dreamed, has eased constraints on travel, and has helped lengthen our life spans. It has had downsides as well, however, including harm to the environment and human health. Concern over these impacts is a prime reason many scientists, environmental advocates, businesspeople, and policymakers are increasingly looking toward renewable sources of energy that exert less impact on natural systems.

Fossil fuel emissions cause pollution and drive climate change

When we burn fossil fuels, we alter certain flux rates in Earth's carbon cycle (▶ pp. 195–197). We essentially take carbon that has been effectively retired into a long-term reservoir underground and release it into the air. This occurs as carbon from within the hydrocarbon molecules of fossil fuels unites with oxygen from the atmosphere during combustion, producing carbon dioxide. Carbon dioxide is a greenhouse gas, and CO_2 released from fossil fuel combustion has been inferred to warm our planet and drive changes in global climate (Chapter 18). Because global climate change may potentially have diverse, severe, and widespread ecological and socioeconomic impacts, carbon dioxide pollution is becoming recognized as the greatest environmental impact of fossil fuel use.

Fossil fuels release more than carbon dioxide when they burn. Methane is a potent greenhouse gas, and other air pollutants resulting from fossil fuel combustion can have serious consequences for human health and the environment. Deposition of mercury and other pollutants from coal-fired power plants is increasingly recognized as a substantial health risk. The burning of oil and coal in our power plants and vehicles releases sulfur dioxide and nitrogen oxides, which contribute to industrial and photochemical smog and to acidic deposition (▶ pp. 509–511, 514–518). Gasoline combustion in automobiles releases pollutants that irritate the nose, throat, and lungs. Some hydrocarbons, such as benzene and toluene, are carcinogenic to laboratory animals and likely also to people. In addition, gases such as hydrogen sulfide can evaporate from crude oil, irritate the eyes and throat, and cause asphyxiation. Crude oil also often contains trace amounts of known poisons, such as lead and arsenic. As a result, people working at drilling operations, refineries, and other jobs that entail frequent exposure to oil or its products can develop serious health problems, including cancer.

Fossil fuels pollute water as well as air. Atmospheric deposition of pollutants exerts many impacts on freshwater ecosystems (▶ p. 515). Moreover, oil from non-point sources, such as industries, homes, automobiles, gas stations, and businesses, runs off roadways and enters rivers and sewage treatment facilities to be eventually discharged into the ocean (▶ pp. 452, 480–481). Although most spilled oil results from these non-point sources, large catastrophic oil spills can have significant impacts on the marine environment. Crude oil's toxicity to most plants and animals frequently leads to high mortality among organisms exposed. This was the case with the *Exxon Valdez* spill in 1989 (Chapter 4), in which oil from Alaska's North Slope, piped to the port of Valdez through the trans-Alaska pipeline, exerted long-term damage to ecosystems and economies in Alaska's Prince William Sound.

Groundwater supplies can also be contaminated with oil, such as when leaks from oil operations penetrate deeply into soil. Of greater concern, as we saw in Chapter 15 (▶ p. 455), thousands of underground storage tanks containing petroleum products have leaked, posing dangers to the public by threatening drinking water supplies.

Coal mining affects the environment

The mining of coal can also have substantial impacts on natural systems and human well-being. Surface strip mining can destroy large swaths of habitat and cause extensive soil erosion. It also can cause chemical runoff into waterways through the process of **acid drainage.** This occurs when sulfide minerals in newly exposed rock surfaces react with oxygen and rainwater to produce sulfuric acid. As the sulfuric acid runs off, it leaches metals from the rocks, many of which are toxic to organisms in high concentrations. Acid drainage is a natural phenomenon, but its rate accelerates greatly when mining exposes many new rock surfaces at once.

Regulations in the United States require mining companies to restore strip-mined land following mining, but impacts are severe and long-lasting just the same. Most other nations exercise less oversight. Mountaintop removal (Figure 19.14) can have even greater impacts than conventional strip mining. When tons of rock and soil are removed from the top of a mountain, it is difficult to keep material from sliding downhill, where immense areas of habitat can be degraded or destroyed and creek beds can be polluted and clogged. Loosening of U.S. government restrictions in 2002 enabled mining companies to legally dump mountaintop rock and soil into valleys and rivers below, regardless of the consequences for ecosystems, wildlife, and local residents.

Whereas mountaintop removal threatens the welfare of nearby residents, subsurface mining raises the greatest health concerns for miners. Underground coal mining is one of our society's most dangerous occupations. Besides risking injury or death from collapsing shafts and tunnels and from dynamite blasts, miners constantly inhale coal dust in the enclosed spaces of mines, which can lead to respiratory diseases, including fatal black lung disease.

FIGURE 19.14 Strip mining in some areas is taking place on massive scales, such that entire mountain peaks are leveled, as at this site in West Virginia. Such "mountaintop removal" can cause enormous amounts of erosion into streams that flow from near the mine into surrounding valleys, affecting ecosystems over large areas, as well as the people who live there.

The costs of alleviating all these health and environmental impacts are high, and the public eventually pays them in an inefficient manner. The reason is that the costs are generally not internalized (▸ pp. 43–44) in the relatively cheap prices of fossil fuels.

Oil extraction can alter the environment

Besides the many impacts summarized above (and discussed in greater detail in other chapters), oil field development has also been shown to have environmental consequences. Drilling activities themselves have fairly minimal impact, but much more than drilling is involved in the development of an oil field (see the photo that opens this chapter, ▸ p. 556). Road networks must be constructed, and many sites may be explored in the course of prospecting. These activities can fragment habitats and can be noisy and disruptive enough to affect wildlife. The extensive infrastructure that must be erected to support a full-scale drilling operation typically includes housing for workers, access roads, transport pipelines, and waste piles for removed soil. Ponds may be constructed for collecting sludge, the toxic leftovers that remain after the useful components of oil have been removed.

Many onshore North American oil reserves are located in arctic or semi-arid areas. With their low rainfall and harsh conditions, plants grow slowly in tundra and semi-desert ecosystems. As a result, even minor changes can have long-lasting repercussions. For example, tundra vegetation at Prudhoe Bay still has not fully recovered from temporary roads last used 30 years ago during the exploratory phase of development.

Whether Prudhoe Bay's oil operations have had a negative impact on the region's caribou is widely debated. Surveys show that the region's summer population of caribou has increased in the 25 years since Prudhoe Bay was developed. Many supporters of drilling have used this trend to argue that oil development does no harm to caribou. Other studies, however, show that female caribou and their calves avoid all parts of the Prudhoe Bay oil complex, including the roads laid down to support it, sometimes detouring miles to do so. These studies also show that the reproductive rate of female caribou in the Prudhoe Bay region is lower than for those in undeveloped areas in Alaska. As a result, although the herd near Prudhoe Bay has increased over the past 25 years, it has not increased as much as have herds in some other parts of Alaska.

Because there is no way of knowing how the particular herd of the Prudhoe Bay region would have performed in the absence of development (that is, there is no control, as there would be for a manipulated experiment), it is difficult to draw conclusions about the actual impacts, if any, of oil development on caribou. In ANWR, the herd that visits the undeveloped 1002 Area rose from 100,000 animals in the 1970s to 178,000 in 1989, and has since declined to 123,000, all for unknown reasons.

FIGURE 19.15 Alaska's North Slope is home to a variety of large mammals, including grizzly bears, polar bears, wolves, arctic foxes, and large herds of caribou. Whether and how oil development may negatively affect these animals are highly controversial issues, and scientific studies are ongoing. The caribou herd near Prudhoe Bay has increased since oil extraction began there, but not by as much as have herds in other parts of Alaska. Grizzly bears such as the ones shown here can sometimes be found near, or even walking atop, the trans-Alaska pipeline.

Many scientists anticipate negative environmental impacts of drilling in ANWR

To predict the possible ecological effects of drilling in ANWR's 1002 Area, scientists have examined the effects of development on arctic vegetation, air quality, water quality, and wildlife, including caribou, grizzly bears, and a variety of bird species in Prudhoe Bay and other Alaska locales where the environment is similar to that of ANWR's coastal plain (Figure 19.15). Scientists have compared different areas, and they have contrasted single areas or populations before and after drilling, as described above with caribou. In addition, scientists have run small-scale manipulative experiments when possible, for example, to study the effects of ice roads and secondary extraction methods such as seawater flushing. In one way or another they have examined the effects of road building, oil pad construction, worker presence, oil spills, accidental fires, trash buildup, permafrost melting, offroad vehicle trails, and dust from roads.

Based on these studies, many scientists anticipate damage to vegetation and wildlife if drilling takes place in ANWR. Vegetation can be killed when saltwater pumped in for flushing deposits is spilled or when plants are buried under gravel pits or roads. Air and water quality can be degraded by fumes from equipment and drilling operations, burning of natural gas associated with oil extraction, sludge ponds, waste pits, and oil spills.

Other scientists, however, contend that drilling operations in ANWR would have little environmental impact.

The roads would be built of ice that will melt in the summer, they point out, and most drilling activity would be confined to the winter, when caribou are elsewhere. Moreover, drilling proponents maintain, Prudhoe Bay is not an appropriate model for ANWR because Prudhoe Bay's development is larger than that projected for ANWR. Furthermore, they say, much of the technology used at Prudhoe Bay is now outdated, and ANWR would be developed with more environmentally sensitive technology and approaches.

Weighing the **Issues:**
The Science of Oil's Arctic Impacts

Evidence from correlative and experimental studies indicates that oil development can harm some arctic plants and animals. However, some scientists say that ice roads, winter drilling, and new technologies will minimize or eliminate such impacts.

Given the evidence and arguments spelled out in the text, suggest several further scientific studies that you believe could be undertaken to help resolve the questions being debated. What evidence would it take to convince you that it is okay to drill, and what evidence would it take to convince you that it is not okay to drill? What would you decide if you had to decide based only on the evidence available now, as policymakers do?

Political, Social, and Economic Impacts of Fossil Fuel Use

The political, social, and economic consequences of fossil fuel use are numerous, varied, and far-reaching. Our discussion focuses on several negative consequences of fossil fuel use and dependence, but it is important to bear in mind that their use has enabled much of the world's population to achieve a higher material standard of living than ever before. It is also important to ask in each case whether switching to more renewable sources of energy would solve existing problems.

Oil supply and prices affect the economies of nations

The prospect of reaching Hubbert's peak for global oil is worrisome because oil is the substance that lubricates the world's economy. Nearly all our modern technologies and services depend somehow on oil. Hurricanes Katrina and Rita and the increased gasoline prices they caused recently served to remind us how much we rely on a steady and ever-increasing supply of petroleum. The hurricanes' economic impact should have come as no surprise, for we have experienced "oil shocks" before, particularly the "energy crisis" of 1973–1974 in the United States. By 1973, with domestic U.S. sources in decline, the nation was importing more and more oil, depending on a constant flow from abroad to keep cars on the road and industries running. Then in 1973, the predominantly Arab nations of the *Organization of Petroleum Exporting Countries (OPEC)* resolved to stop selling oil to the United States. OPEC wished to raise prices by restricting supply and opposed U.S. support of Israel in the Arab-Israeli Yom Kippur War. The embargo created panic in the West and caused oil prices to skyrocket (Figure 19.16), spurring inflation. Short-term oil shortages drove American consumers to wait in long lines at gas pumps.

In response to the embargo, the U.S. government enacted a series of policies designed to reduce reliance on foreign oil. The government called for developing additional domestic sources, such as those on Alaska's North Slope. However, it also called for resuming extraction at sites shut down after primary extraction had ceased being cost-effective, capping the price that domestic producers could charge for oil, importing oil from a greater diversity of nations, funding research into renewable energy sources, and enacting conservation measures we will discuss below. The government also established a stockpile of oil as a short-term buffer against future shortages. Stored deep underground in salt caverns in Louisiana, this is called the *Strategic Petroleum Reserve*. Currently the Reserve contains roughly 700 million barrels of oil; at present rates of U.S. consumption (20.7 million barrels/ day), this equals just over one month's supply.

Nations can become dependent on foreign oil

Putting all of one's eggs in one basket is always a risky strategy. The fact that so many nations' economies are utterly tied to fossil fuels means that those economies are tremendously vulnerable to supplies' becoming suddenly unavailable or extremely costly. Such reliance means that seller nations can potentially control the price of oil, forcing buyer nations to pay more and more as supplies dwindle. With the formation of OPEC in 1960, oil-producing nations tried to exercise even greater power over oil prices by regulating production (although OPEC so far has had only marginal success in this).

In the United States, concern over reliance on foreign oil sources has repeatedly driven the proposal to open

FIGURE 19.16 World oil prices have gyrated greatly over the decades, often because of political and economic events in oil-producing countries. The greatest price hikes in recent times have resulted from wars and unrest in the oil-rich Middle East. Data from U.S. Energy Information Administration.

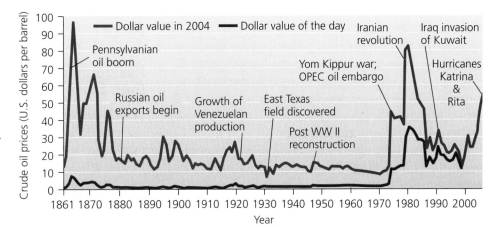

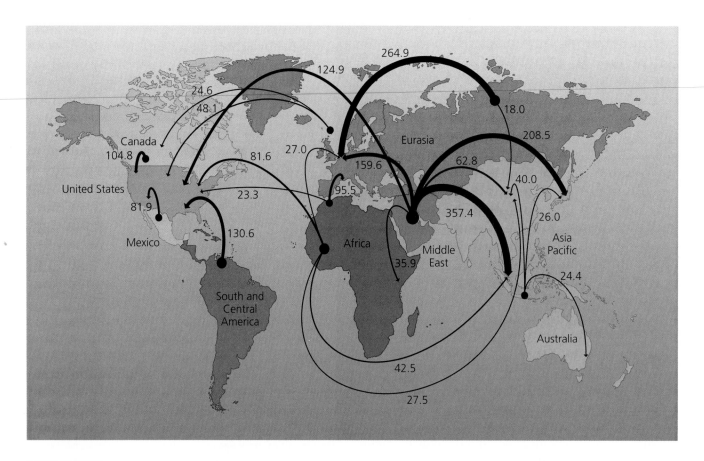

FIGURE 19.17 The global trade in oil is lopsided; relatively few nations account for most exports, and some nations are highly dependent on others for energy. The United States obtains most of its imported oil from Venezuela and Saudi Arabia, followed by Canada, Mexico, Nigeria, and the North Sea. Canada imports some North Sea oil while exporting more to the United States. Numbers in the figure represent millions of metric tons. Data from British Petroleum. 2005. *Statistical review of world energy 2005.*

ANWR to drilling, despite critics' charges that such drilling would likely do little to decrease the nation's dependence. The United States currently imports 65% of its crude oil. With the majority of world oil reserves located in the politically unstable Middle East, crises such as the 1973 embargo are a constant concern for U.S. policymakers. The United States has cultivated a close relationship with Saudi Arabia, the owner of 22% of world oil reserves, despite the fact that that country's political system allows for little of the democracy that U.S. leaders claim to cherish and promote. The world's third-largest holder of oil reserves, at 10%, is Iraq, which is why many people around the world believe the U.S.-led invasion of Iraq in 2003 was motivated primarily to secure access to oil.

To counter dependence on a few major supplier nations, the United States has diversified its sources of petroleum and receives much of it from non–Middle Eastern nations, including Venezuela, Canada, Mexico, and Nigeria. Major trade relations among nations and regions of the world are depicted in Figure 19.17.

As much as U.S. policymakers worry about dependence on foreign fossil fuels, several other nations are far more vulnerable in their energy dependence. Germany, France, South Korea, and Japan stand out as nations that consume far more energy than they produce and thus rely almost entirely on imports for their continued economic well-being. Figure 19.18 contrasts consumption and production for several selected nations.

Residents may or may not benefit from their fossil fuel reserves

The extraction of fossil fuels can be extremely lucrative; many of the world's wealthiest corporations deal in fossil fuel energy or related industries. These industries provide jobs to millions of employees and provide dividends to millions of investors. Development can potentially yield economic benefits for people who live in petroleum-bearing areas, as well. Since the construction of the trans-Alaska pipeline in the 1970s, the state of Alaska has received

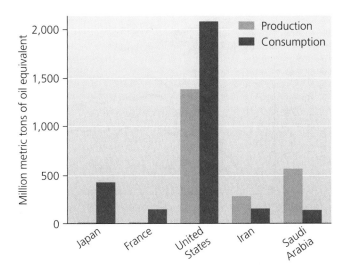

FIGURE 19.18 Japan, France, and the United States are among the nations that consume far more fossil fuel energy than they produce. Iran and Saudi Arabia, like many Middle Eastern nations, produce more fossil fuel energy than they consume and are able to export fuel to high-consumption countries. Data from British Petroleum. 2005. *Statistical review of world energy 2005.*

$60 billion in oil revenues. Alaska's state constitution requires that one-quarter of state oil revenues be placed in the Permanent Fund, which pays yearly dividends to all Alaska residents. Since 1982, each Alaska resident has received annual payouts ranging from $331 to $1,964.

Development of ANWR would add to this fund and create jobs. Some estimates anticipate creation of many thousands of jobs and billions of dollars of income. For this reason, most Alaska residents support oil drilling in ANWR. So do many Inupiat who live on the North Slope, because the income could pay for health care, police and fire protection, and other services that are currently scarce in this remote region.

Alaska's distribution of revenue among its citizenry is unusual, however. In most parts of the world where fossil fuels have been extracted, local residents have not seen great benefits, but instead have frequently suffered. When multinational corporations have extracted oil or gas in developing countries, paying those countries' governments for access, the money often has not trickled down to residents of the regions where the extraction takes place. Moreover, oil-rich developing countries such as Ecuador, Venezuela, and Nigeria tend to have few environmental regulations, and existing regulations may go unenforced if a government does not want to risk losing the large sums of money associated with oil development.

In Nigeria, oil was discovered in 1958 in the territory of the Ogoni, one of Nigeria's native peoples, and the Shell Oil Company moved in to develop oil fields. Although Shell extracted $30 billion of oil from Ogoni land over the years,

the Ogoni still live in poverty, with no running water or electricity. The profits from oil extraction on Ogoni land have gone to Shell and to the military dictatorships of Nigeria. The development resulted in oil spills, noise, and constantly burning gas flares, all of which caused illness among people living nearby. From 1962 until his death in 1995, Ogoni activist and leader Ken Saro-Wiwa worked for fair compensation to the Ogoni for oil extraction and environmental degradation on their land. After years of persecution by the Nigerian government, Saro-Wiwa was arrested in 1994, given a trial widely regarded in the international human rights community as a sham, and put to death by military tribunal.

How will we convert to renewable energy?

Given that fossil fuel supplies are limited and that their use has health, environmental, political, and socioeconomic consequences, the nations of the world have several policy options for guiding future energy use. One option is to commit to using fossil fuels until they are no longer economically practical and to develop other energy sources only after supplies have dwindled. A second option is to fund development of alternative energy sources now and attempt to reduce our reliance on fossil fuels gradually. Third, we could try to end our fossil fuel use as soon as possible and hasten a switch to renewable alternatives.

Weighing the **Issues:**
Will Capitalism Drive a Shift to Renewables?

Many experts say that governments need not take steps to hasten the readiness of renewable energy sources to replace fossil fuels as they decline, because forces of supply and demand in market capitalism will automatically take care of this shift. As fossil fuel supplies dwindle and prices rise, the argument goes, industries and consumers will switch to renewable sources as they become more economical. Do you agree with this outlook? Do you see any possible problems with it? Do you think governments should take steps to encourage the development of renewable energy sources? Why or why not?

In the meantime, it will benefit us to prolong the availability of fossil fuels as we make the transition to renewable sources. We can prolong our access to fossil fuels by instituting measures to conserve energy, primarily through lifestyle changes that reduce energy use and technological advances that improve efficiency.

Energy Conservation

Until our society reaches a point at which we are using solely renewable energy sources, we will face the gradual depletion of nonrenewable sources. In the face of dwindling fossil fuel resources, we will need to find ways to minimize the extent to which we expend energy. **Energy conservation** is the practice of reducing energy use to extend the lifetimes of our nonrenewable energy supplies, be less wasteful, and reduce our environmental impact.

Energy conservation has often been a function of economic need

In the United States, many people first saw the value of conserving energy following the OPEC embargo of 1973–1974. Policies enacted by the U.S. government in response to that event included conservation measures such as a mandated increase in the mile-per-gallon (mpg) fuel efficiency of automobiles and a reduction in the national speed limit to 55 miles per hour (at that time, the most efficient speed to drive a car).

Three decades later, many of the conservation initiatives developed after the 1973 oil crisis have been abandoned. Without high market prices and an immediate threat of shortages, people lack economic motivation to conserve. Government funding for research into alternative energy sources has decreased, speed limits have increased, and recent bills to raise the mandated average fuel efficiency of vehicles have failed in Congress. The average fuel efficiency of new vehicles has fallen from a high of 22.1 mpg in 1988 to 21.0 mpg in 2005 (Figure 19.19). This decrease is due to increased sales of light trucks (averaging 18.2 mpg), including sport-utility vehicles, relative to cars (averaging 24.7 mpg). Transportation accounts for two-thirds of U.S. oil use, and passenger vehicles consume over half this energy. Thus, the failure to improve vehicular fuel economy over the past 20 years, despite the existence of technology to do so, has added greatly to U.S. oil consumption.

Weighing the **Issues:**
More Miles, Less Gas

If you drive an automobile, what gas mileage does it get? How does it compare to the vehicle averages in Figure 19.19? If your vehicle's fuel efficiency were 10 mpg greater, and you drove the same amount, how many gallons of gasoline would you no longer need to purchase each year? How much money would you save? If all U.S. vehicles were mandated to increase fuel efficiency by

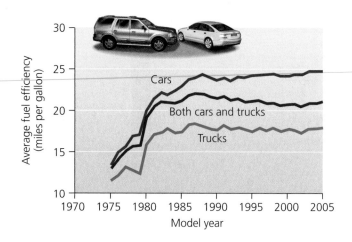

FIGURE 19.19 Fuel efficiency for automobiles in the United States rose dramatically in the late 1970s as a result of legislative mandates, but it has declined slightly since 1988. The decline is due to a lack of further legislation for improved fuel economy and to the increased popularity in recent years of sport-utility vehicles. Data from U.S. Environmental Protection Agency. 2005. *Light-duty automotive technology and fuel economy trends: 1975 through 2005.*

10 mpg, how much gasoline do you think the over 200 million Americans who drive could conserve? What other strategies can you think of to conserve fossil fuels, and how might they compare in effectiveness to a rise in fuel efficiency standards?

Many critics of oil drilling in the Arctic Refuge point out the vast amounts of oil wasted by our fuel-inefficient automobiles. They argue that a small amount of conservation would save the nation far more oil than it would ever obtain from ANWR. Indeed, the USGS's average estimate for recoverable oil in the 1002 Area, 7.7 billion barrels, represents one year's supply for the United States at current consumption rates. Spread over a period of extraction of many years, the proportion of U.S. oil needs that ANWR would fulfill appears strikingly small (Figure 19.20).

Personal choice and increased efficiency are two routes to conservation

Energy conservation can be accomplished in two primary ways. As individuals, we can make conscious choices to reduce our own energy consumption. Examples include driving less, turning off lights when rooms are not being used, turning down thermostats, and cutting back on the use of machines and appliances. For any given individual or business, reducing energy consumption can save money while also helping conserve resources.

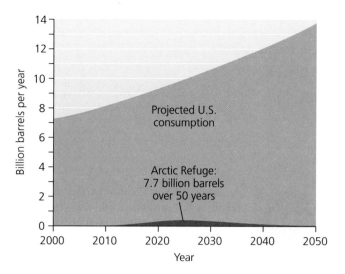

FIGURE 19.20 Opponents of oil drilling in the Arctic National Wildlife Refuge contend that the amount of oil estimated to be recoverable would make only a small contribution toward overall U.S. oil demand. In this graph, the best USGS estimate of oil from ANWR's 1002 Area is shown in red, in the context of total U.S. oil consumption, assuming that current consumption trends are extrapolated into the future and oil production takes place over many years. The actual ANWR contribution, if it comes to pass, would depend greatly on the amount of oil actually present under ANWR, the time it would take to extract it, and future trends in consumption. Adapted from Natural Resources Defense Council. 2002. *Oil and the Arctic National Wildlife Refuge*; and U.S. Geological Survey. 2001. *Arctic National Wildlife Refuge, 1002 Area, petroleum assessment, 1998, including economic analysis.*

We can also conserve energy as a society by making our energy-consuming devices and processes more efficient. In the case of automobiles, we already possess the technology to increase fuel efficiency far above the current average of 21 mpg. We could accomplish such improvement with more efficient gasoline engines, with alternative technology vehicles such as electric/gasoline hybrids (Figure 19.21), or with vehicles that use hydrogen fuel cells (▸ pp. 640, 642).

The efficiency of power plants can be improved through **cogeneration,** in which excess heat produced during the generation of electricity is captured and used to heat workplaces and homes and to produce other kinds of power. Cogeneration can almost double the efficiency of a power plant.

In homes and public buildings, a significant amount of heat is lost in winter and gained in summer because of inadequate insulation (Figure 19.22). Improvements in the design of homes and offices can reduce energy required to heat and cool them. Such design changes can involve the building's location, the color of its roof (light colors keep buildings cooler by reflecting the sun's rays), and its insulation.

Among consumer products, scores of appliances, from refrigerators to lightbulbs, have been reengineered through the years to increase energy efficiency. Even so, there remains room for improvement. Energy-efficient lighting, for example, can reduce energy use by 80%, and new federal standards for energy-efficient appliances have already reduced per-person home electricity use below what it was in the 1970s.

While manufacturers can improve the energy efficiency of appliances, consumers need to "vote with their wallets" by purchasing these energy-efficient appliances. Decisions by consumers to purchase energy-efficient products are crucial in keeping those products commercially available. The U.S. Environmental Protection Agency (EPA) estimates that if all U.S. households purchased energy-efficient appliances, the national annual energy expenditure would be reduced by

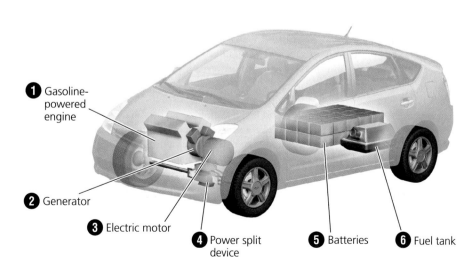

FIGURE 19.21 A hybrid car, such as the Toyota Prius diagrammed here, uses a small, clean, and efficient gasoline-powered engine (1) to produce power that the generator (2) can convert to electricity to drive the electric motor (3). The power split device (4) integrates the engine, generator, and motor, serving as a continuously variable transmission. The car automatically switches between all-electrical power, all-gas power, and a mix of the two, depending on the demands being placed on the engine. Typically, the motor provides power for low-speed city driving and adds extra power on hills. The motor and generator charge a pack of nickel-metal-hydride batteries (5), which can in turn supply power to the motor. Energy for the engine comes from gasoline carried in a typical fuel tank (6).

Drilling for Oil in ANWR

Should we drill for oil in the Arctic National Wildlife Refuge?

ANWR Oil Means Better Living, Not Eco-Apocalypse

Oil development has taken place in the Arctic for over 30 years. Originally there were fears for the environment, just as we are hearing today. The Inupiat feared harm to the caribou and to their lifestyles, while environmentalists claimed apocalypse, and that there were only 2 years of oil.

Well, "2 years" of oil turned into over 28; 3,000 caribou turned into over 32,000; and the Inupiat have turned into the number-one supporters of development. Technology has allowed a 140-acre drill pad to be reduced to 5 acres in size. Improvements in directional drilling now allow us to extract oil 8 miles from drill-point. The EPA, the North Slope Borough, and Alaska Department of Fish and Game monitor activities daily, so one can hardly claim an eco-apocalypse. Mis-information is rife in the ANWR debate.

Furthermore, the NIMBY argument doesn't work with ANWR; the people who live off the 1002 Area want this badly. To them, drilling in the 1002 Area will provide jobs, schools, clinics—a future. They have seen firsthand how their fears of 30 years ago were unfounded. They have seen their communities prosper.

The energy future of the United States should be multi-sourced. Wind turbines, solar panels, and hydrogen cells *all* need oil to build their working parts. None of these sources can produce plastic, bitumen, tires, paints, medicines, or a thousand other oil-based products we use daily. One can scream against oil, yet every one of us uses it every day in every aspect of our lives. We don't know any alternative. We should invest heavily in finding alternatives, but we should not think for a second that we can cut off our oil supply, domestic or foreign, and live as happily as we do now.

Adrian Herrera is a consultant for the lobby group Arctic Power, lobbying Congress to open the 1002 Area of ANWR to responsible oil development. Formerly, Herrera worked as an engineering assistant at Prudhoe Bay, and for an independent ecological research company tasked with monitoring the wildlife and environment in and around the oil fields. He is an Alaskan resident of over 30 years and has worked and lived in the Arctic extensively.

Drilling the Arctic Refuge Is Not a Solution to Our Energy Problems, It's a Distraction

The Arctic National Wildlife Refuge is one of the last unspoiled wild areas in the United States. Its 1.5-million-acre coastal plain is rich in biodiversity, home to nearly 200 species, including polar bears, musk oxen, caribou, and millions of migratory birds.

There is no way to drill in the refuge without permanently harming this unique ecosystem or destroying the culture of the native Gwich'in people, who have depended on caribou for thousands of years. The little oil beneath the refuge is scattered in more than 30 small deposits. To extract it, roads, pipelines, air strips, and other industrial infrastructure would be built across the entire area.

Drilling the Arctic Refuge would do nothing to lower gas prices or lessen our nation's dependence on imported oil. According to the U.S. Geological Survey, the refuge holds less economically recoverable oil than what Americans consume in a year, and it would take 8–10 years for that oil to reach the market. A recent U.S. Energy Department report found that oil from the Arctic Refuge would have little impact on the price of gasoline, lowering gas prices by less than a penny and a half per gallon—in 2025.

If we boosted the fuel economy performance of our cars and trucks just 1 mile per gallon annually over the next 15 years, we would save more than 10 times the oil that could be recovered from the refuge. We have the technology today to accomplish that goal.

The United States has 3 percent of the world's oil reserves but consumes 25 percent of all oil produced each year. We cannot drill our way to lower gas prices. By focusing on efficiency and alternative fuels, we can improve our energy security and preserve the Arctic Refuge for future generations.

Karen Wayland is the Natural Resources Defense Council's legislative director and an adjunct professor at Georgetown University. Dr. Wayland, who holds a dual Ph.D. in geology and resource development, was a legislative fellow for Sen. Harry Reid (D-Nev.) on nuclear waste, water, energy, and Native American issues before joining NRDC's staff.

Explore this issue further by accessing **Viewpoints** at www.aw-bc.com/withgott.

FIGURE 19.22 Many of our homes and offices could be made more energy-efficient. One way to determine how much heat a building is losing is to take a photograph that records energy in the infrared portion of the electromagnetic spectrum (▶ p. 106). In such a photograph, or *thermogram* (shown here), white, yellow, and red signify hot and warm temperatures at the surface of the house, whereas blue and green shades signify cold and cool temperatures. The white, yellow, and red colors indicate areas where heat is escaping.

$200 billion. On an individual consumer's level, studies show that the slightly higher cost of buying energy-efficient washing machines is rapidly offset by savings in water and electricity bills. On the national level, France, Great Britain, and many other developed countries have standards of living equal to that of the United States, but they use much less energy per capita. This disparity indicates that U.S. citizens could significantly reduce their energy consumption without decreasing their quality of life.

Some conservation measures were adopted in the energy bill endorsed by the Bush administration and passed by the U.S. Congress in 2005. Tax credits were offered to consumers who buy hybrid cars or improve energy efficiency in their homes, for example. However, these measures were a relatively minor component, and most of the bill's funding went toward subsidies for nuclear power, cleaner coal technology, ethanol, and wind power, as well as measures to increase production of oil and natural gas. The legislation did not address automotive fuel economy, and as the president signed the bill into law, gas prices rose to their highest inflation-adjusted levels in more than 20 years. Most analysts concluded that overall, the legislation did not notably change the direction of energy policy in the United States.

Both conservation and renewable energy are needed

It is often said that reducing our energy use is equivalent to finding a new oil reserve. Some estimates hold that effective energy conservation in the United States could save 6 million barrels of oil a day. Such a savings would likely far more than offset the energy produced by any oil under the Arctic National Wildlife Refuge while also reducing the negative impacts of fossil fuel extraction and use. Indeed, conserving energy is better than finding a new reserve, because it lessens impacts on the environment while extending our access to fossil fuels.

However, energy conservation does not add to our supply of available fuel. Thus, in the long term, conservation cannot be the sole solution to our energy dilemmas. Regardless of how much we conserve, we will still need energy, and it will need to come from somewhere. The only sustainable way of guaranteeing ourselves a reliable long-term supply of energy is to ensure sufficiently rapid development of renewable energy sources.

Conclusion

Over the past 200 years, fossil fuels have helped us build complex industrialized societies capable of exploring (and exploiting) all parts of the world and even venturing beyond our planet. Today, however, we are approaching a turning point in history: Our production of fossil fuels will begin to decline just as we become increasingly aware of the negative impacts of their use.

We can respond to this new challenge in creative ways, encouraging conservation and developing alternative energy sources. Or we can continue our current dependence on fossil fuels and wait until they near depletion before we try to develop new technologies and ways of life. The path we choose will have far-reaching consequences for human health and well-being, for Earth's climate, and for our environment.

The ongoing debate over the future of the Arctic National Wildlife Refuge is a microcosm of this debate over our energy future. Fortunately, there is not simply a trade-off between benefits of energy for us and harm to the environment, climate, and health. Instead, as evidence builds that renewable energy sources are becoming increasingly feasible and economical, it becomes easier to envision giving up our reliance on fossil fuels and charting a win-win future for humanity and the environment.

You should now be able to:

Survey the energy sources that we use

▶ A variety of renewable and nonrenewable energy sources are available to us. (pp. 559–561)

▶ Since the industrial revolution, nonrenewable fossil fuels—including oil, natural gas, and coal—have become our primary sources of energy. (p. 561)

▶ Fossil fuels are formed very slowly as buried organic matter is chemically transformed by heat, pressure, and/or anaerobic decomposition. (pp. 561–562)

Describe the nature and origin of coal, and evaluate its extraction and use

▶ Coal is our most abundant fossil fuel. It results from organic matter that undergoes compression but little decomposition. (pp. 564–565)

▶ The first fossil fuel to be widely used for heating homes and powering industry, coal is used today principally to generate electricity. (pp. 564, 566–567)

▶ Coal comes in different types and varies in its composition. Combustion of coal that is high in contaminants emits toxic air pollution. (pp. 564–565)

▶ Coal is mined underground and strip-mined from the land surface. (p. 565)

Describe the nature and origin of petroleum and evaluate its extraction, use, and future depletion

▶ Crude oil is a thick liquid mixture of hydrocarbons that is formed underground under certain temperature and pressure conditions. (pp. 566–567)

▶ Scientists locate fossil fuel deposits by analyzing subterranean geology. Geologists estimate total reserves, as well as the technically and economically recoverable portions of those reserves. (pp. 568–569)

▶ Oil drilling often involves primary extraction followed by secondary extraction, in which gas or liquid is injected into the ground to help force up additional oil. (pp. 569–570)

▶ Petroleum-based products, from gasoline to clothing to plastics, are everywhere in our daily lives. (p. 570)

▶ We have nearly depleted half the world's oil. The remainder may last only 40 more years. Once production slows, the gap between rising demand and falling supply may pose immense economic and social challenges for our society. (pp. 570–571, 574)

▶ Components of crude oil are separated in refineries to produce a wide variety of fuel types. (pp. 572–573)

Describe the nature and origin of natural gas, and evaluate its extraction and use

▶ Natural gas consists mostly of methane and can be formed in two ways. (p. 574)

▶ Use of natural gas is growing rapidly. (pp. 574–575)

▶ Natural gas often occurs with oil deposits, is extracted in similar ways, and becomes depleted in similar ways. (pp. 574–575)

Outline and assess environmental impacts of fossil fuel use

▶ Airborne emissions of carbon dioxide from fossil fuel combustion drive global climate change. (p. 576)

▶ Emissions of fuel contaminants and volatile components of oil contribute to air pollution and pose human health risks. (p. 576)

▶ Oil is a major contributor to water pollution. (p. 576)

▶ Strip mining and mountaintop removal can devastate ecosystems locally or regionally, and acid drainage from coal mines pollutes waterways. (pp. 576–577)

▶ Development for oil and gas extraction exerts environmental impacts, although their severity is debated. (pp. 577–578)

Evaluate political, social, and economic impacts of fossil fuel use

▶ Today's societies are so reliant on fossil fuel energy that sudden restrictions in oil supplies can have major economic consequences. (pp. 579–580)

▶ Nations that consume far more fossil fuels than they produce are especially vulnerable to supply restrictions. (pp. 579–581)

▶ People living in areas of fossil fuel extraction do not always benefit from their extraction. (pp. 580–581)

Specify strategies for conserving energy

▶ Increases in automotive fuel efficiency could help us conserve immense amounts of oil. (p. 582)

▶ Energy conservation involves both personal choices and efficient technologies. These two forces interact through the market power of consumer choice. (pp. 582–585)

▶ Conservation helps lengthen our access to fossil fuels and reduce environmental impact, but to build a sustainable society we will also need to shift to renewable energy sources. (p. 585)

TESTING YOUR COMPREHENSION

1. Why are fossil fuels our most prevalent source of energy today? Why are they considered nonrenewable sources of energy?
2. How do developed and developing countries differ in their overall rates of energy consumption and in the ways they use energy?
3. How are fossil fuels formed? How do environmental conditions determine what type of fossil fuel is formed in a given location? Why are fossil fuels often concentrated in localized deposits?
4. How do scientists classify coal? What determines the potential energy contained in coal?
5. Describe how coal is used to generate electricity.
6. How have geologists estimated the total amount of oil beneath the Arctic National Wildlife Refuge (ANWR)

1002 Area? How is this amount different from the "technically recoverable" and "economically recoverable" amounts of oil?
7. How do we create petroleum products? Provide examples of several of these products.
8. Why is natural gas often extracted simultaneously with other fossil fuels? What constraints on its extraction does it share with oil?
9. What are some of the effects of fossil fuel emissions? Compare some of the contrasting views of scientists regarding the environmental impacts of drilling for oil in ANWR.
10. Describe two main approaches to energy conservation, and give a specific example of each.

SEEKING SOLUTIONS

1. Roughly how much oil is left in the world, and how much longer can we expect to use it? How effective were the conservation methods adopted by the United States in response to the "energy crisis" of 1973–1974? Do you think our use of oil and other fossil fuels will give rise to a future crisis? Why or why not? If so, what should we do to avoid such a situation in the future?
2. Compare the effects of coal and oil consumption on the environment. Which process do you think has ultimately been more detrimental to the environment, oil extraction or coal mining, and why? What steps could governments, industries, and individuals take to reduce environmental impacts?
3. Imagine we were living in the 1950s and the United States were facing an oil shortage. Do you think the nation would be debating whether to drill in ANWR? Why

do you think so many people today are concerned about wildlife and about wilderness preservation?
4. If ANWR were located in a developing country, do you think the citizens and policymakers of that country would choose to drill there?
5. If the United States and other developed countries relinquished dependence on foreign oil and on fossil fuels in general, do you think that their economies would benefit or suffer? Might your answer be different for the short term and the long term? What factors come into play in trying to make such a judgment?
6. Contrast the experiences of the Ogoni people of Nigeria with those of the citizens of Alaska. How have they been similar and different? Do you think businesses or governments should take steps to ensure that local people benefit from oil drilling operations? How could they do so?

INTERPRETING GRAPHS AND DATA

The fossil fuels that we burn today were formed long ago from buried organic matter. However, only a small fraction of the original organic carbon remains in the coal, oil, or natural gas that is formed. Thus, it requires approximately 90 metric tons of ancient organic matter—so-called

paleoproduction—to result in just 3.8 L (1 gal) of gasoline. The graph presents estimates of the amount of paleoproduction required to produce the fossil fuels humans have used each year over the past 250 years.

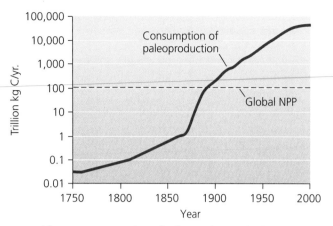

Annual human consumption of paleoproduction by fossil fuel combustion (red line), 1750–2000. The dashed line indicates current annual net primary production (NPP; ▸ p. 192) for the entire planet. Data from Dukes, J. 2003. Burning buried sunshine: Human consumption of ancient solar energy. *Climatic Change* 61: 31–44.

1. Estimate in what year the consumption of paleoproduction represented by our combustion of fossil fuels surpassed Earth's current annual net primary production.
2. In 2000, approximately how many times greater than global net primary production was our consumption of paleoproduction?
3. If on average it takes 7,000 units of paleoproduction to produce 1 unit of fossil fuel, estimate the total carbon content of the fossil fuel consumed in 2000. How does this amount compare to global NPP?

CALCULATING ECOLOGICAL FOOTPRINTS

Wackernagel and Rees calculated the energy component of our ecological footprint by estimating the amount of ecologically productive land required to absorb the carbon released from fossil fuel combustion (see Chapter 18, "Calculating Ecological Footprints," ▸ p. 555). For the average American, this translates into 2.9 of the 5.1 ha of our ecological footprint. Another way to think about our footprint, however, is to estimate how much land would be needed to grow biomass

with an energy content equal to that of the fossil fuel we burn. Assume that you are an average American who burns 287 gigajoules of fossil fuels per year and that average terrestrial net primary productivity can be expressed as 160 megajoules/ha/year. Calculate how many hectares of land it would take to supply our fuel use by present-day photosynthetic production. A gigajoule is 10^9 joules; a megajoule is 10^6 joules.

	Hectares of land for fuel production
You	1,794
Your class	
Your state	
United States	

Data from Wackernagel, M., and W. Rees. 1996. *Our ecological footprint: Reducing human impact on the Earth.* Gabriola Island, British Columbia: New Society Publishers.

1. Compare the energy component of your ecological footprint calculated in this way with the 2.9 ha calculated using the method of Wackernagel and Rees. Explain why results from the two methods differ.
2. Earth's total land area is approximately 1.5×10^{10} ha. Compare this to the hectares of land for fuel production from the table.
3. How large a human population could Earth support at the level of consumption of the average American, if all of Earth's land were devoted to fuel production? Do you consider this realistic? Provide two reasons why or why not.

Take It Further

Go to www.aw-bc.com/withgott or the student CD-ROM where you'll find:

▸ Suggested answers to end-of-chapter questions
▸ Quizzes, animations, and flashcards to help you study
▸ *Research Navigator*™ database of credible and reliable sources to assist you with your research projects

▸ **GRAPHit!** tutorials to help you master how to interpret graphs
▸ **INVESTIGATEit!** current news articles that link the topics that you study to case studies from your region to around the world

Conventional Energy Alternatives

20

Cooling towers of
nuclear power plant,
Shropshire, U.K.

Upon successfully completing this chapter, you will be able to:

▶ Discuss the reasons for seeking alternatives to fossil fuels

▶ Summarize the contributions to world energy supplies of conventional alternatives to fossil fuels

▶ Describe nuclear energy and how it is harnessed

▶ Outline the societal debate over nuclear power

▶ Describe the major sources, scale, and impacts of biomass energy

▶ Describe the scale, methods, and impacts of hydroelectric power

Biomass power plant, Skellefteå, Sweden

Sweden — Ukraine

Atlantic Ocean

Central Case: Sweden's Search for Alternative Energy

"Nowhere has the public debate over nuclear power plants been more severely contested than Sweden."
—WRITER AND ANALYST MICHAEL VALENTI

"If [Sweden] phases out nuclear power, then it will be virtually impossible for the country to keep its climate-change commitments."
—YALE UNIVERSITY ECONOMIST WILLIAM NORDHAUS

On the morning of April 28, 1986, workers at a nuclear power plant in Sweden detected suspiciously high radiation levels. Their concern turned to confusion when they determined that the radioactivity was coming not from their own plant, but through the atmosphere from the direction of the Soviet Union.

They had, in fact, discovered evidence of the disaster at Chernobyl, more than 1,200 km (750 mi) away in what is now the nation of Ukraine. Chernobyl's nuclear reactor had exploded two days earlier, but the Soviet government had not yet admitted it to the world.

As low levels of radioactive fallout rained down on the Swedish countryside in the days ahead, contaminating crops and cows' milk, many Swedes felt more certain than ever about the decision they had made collectively 6 years earlier. In a 1980 referendum, Sweden's electorate had voted to phase out their country's nuclear power program, shutting down all nuclear plants by the year 2010.

But trying to phase out nuclear power has proven difficult. Nuclear power today provides Sweden with one-third of its overall energy supply and nearly half its electricity. If nuclear plants are shut down, something will have to take their place. Aware of the environmental impacts of fossil fuels, Sweden's government and citizens do not favor expanding fossil fuel use. In fact, Sweden is one of the few nations that have managed to decrease use of fossil fuels since the 1970s—and it has done so largely by replacing them with nuclear power.

To fill the gap that would be left by a nuclear phaseout, Sweden's government has promoted research and development of renewable energy sources.

Hydroelectric power from running water was already supplying most of the other half of the nation's electricity, but it could not be expanded much more. The government hoped that energy from biomass sources and wind power could fill the gap.

Sweden has made itself an international leader in renewable energy alternatives, but because renewables have taken longer to develop than hoped, the government has repeatedly postponed the nuclear phaseout. Only one of the 12 reactors operating in 1980 has been shut down so far, and efforts to close a second one have generated sustained controversy.

Proponents of nuclear power say it would be fiscally and socially irresponsible to dismantle the nation's nuclear program without a ready replacement. And environmental advocates worry that if nuclear power is simply replaced by fossil fuel combustion, or if converting to biomass energy means cutting down more forests, the nuclear phaseout would be bad news for the environment. Moreover, Sweden has made international obligations to hold down its carbon emissions under the Kyoto Protocol (▶ pp. 550, 552), so its incentive to keep nuclear power is strong. Nuclear energy is free of atmospheric pollution and seems, to many, the most effective way to minimize carbon emissions in the short term.

In 2003, a poll showed 55% of the Swedish public in favor of maintaining or increasing nuclear power, and 41% in favor of abandoning it. But despite the mixed feelings over nuclear power, Swedes have little desire to return to an energy economy dominated by fossil fuels. A concurrent poll showed that 80% of Swedes supported boosting research on renewable energy sources—a higher percentage than in any other European country.

Alternatives to Fossil Fuels

Fossil fuels helped to drive the industrial revolution and increase our material prosperity. Today's economies are largely powered by fossil fuels; 80% of all primary energy comes from oil, coal, and natural gas (Figure 20.1a), and these three fuels also power two-thirds of the world's electricity generation (Figure 20.1b). However, these nonrenewable energy sources will not last forever. As we saw in Chapter 19, oil production is thought to be peaking, and easily extractable supplies of oil and natural gas may not last half a century more. Moreover, the use of coal, oil, and natural gas entails substantial environmental impacts, as described in Chapters 17, 18, and 19.

For these reasons, most scientists and energy experts, as well as many economists and policymakers, accept that

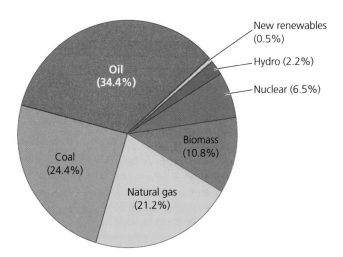

(a) World total primary energy supply

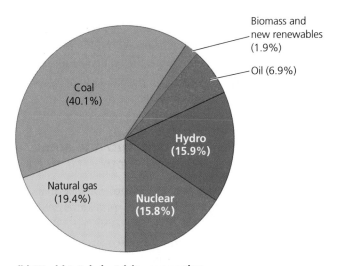

(b) World total electricity generation

FIGURE 20.1 Fossil fuels account for 80% of the world's total supply of primary energy **(a)**. Nuclear and hydroelectric power contribute substantially to global electricity generation, although fossil fuels still power two-thirds of our electricity **(b)**. Data from International Energy Agency. 2005. *Key world energy statistics 2005.* Paris: IEA.

the world's economies will need to shift from fossil fuels to energy sources that are less easily depleted and gentler on our environment. Developing alternatives to fossil fuels has the added benefit of helping to diversify an economy's mix of energy, thus lessening price volatility and dependence on foreign fuel imports.

People have developed a range of alternatives to fossil fuels. Most of these energy sources are renewable and cannot be depleted by use. Most have less impact on the environment than oil, coal, or natural gas. However, at this time most remain more costly than fossil fuels (at least in the short term), and many depend on technologies that are not yet fully developed.

Nuclear power, biomass energy, and hydropower are conventional alternatives

Three alternative energy sources are currently the most developed and most widely used: nuclear energy, hydroelectric power, and energy from biomass. Each of these well-established energy sources plays substantial roles in the energy and electricity budgets of nations today. We can therefore call nuclear energy, hydropower, and biomass energy "conventional" alternatives to fossil fuels.

In many respects, this trio of conventional energy alternatives makes for an eclectic collection. They are generally considered to exert less environmental impact than fossil fuels, but more environmental impact than the "new renewable" alternatives we will discuss in Chapter 21. Yet, as we will see, they each involve a unique and complex mix of benefits and drawbacks for the environment. Although nuclear energy is commonly termed a nonrenewable energy source and hydropower and biomass are generally described as renewable, the reality is more complicated. They are perhaps best viewed as intermediates along a continuum of renewability.

Conventional alternatives provide some of our energy and much of our electricity

Fuelwood and other biomass sources provide 10.8% of the world's primary energy, nuclear energy provides 6.5%, and hydropower provides 2.2%. The less established renewable energy sources account for only 0.5% (see Figure 20.1a). Although their global contributions to overall energy supply are still minor, alternatives to fossil fuels do contribute greatly to our generation of electricity. Nuclear energy and hydropower each account for nearly one-sixth of the world's electricity generation (see Figure 20.1b).

Energy consumption patterns in the United States (Figure 20.2a) are similar to those globally, except that the United States relies less on fuelwood and more on fossil fuels and nuclear power than most other countries. A graph showing changes in the consumption of fossil fuels and conventional alternatives in the United States over the past half century (Figure 20.2b) reveals two things. First, conventional alternatives play minor yet substantial roles in overall energy use. Second, use of conventional alternatives has been growing more slowly than consumption of fossil fuels.

Sweden and some other nations, however, have shown that it is possible to replace fossil fuels gradually with alternative sources. Since 1971, Sweden has decreased its fossil fuel use by 40%, and today nuclear power, biomass, and hydropower together provide Sweden with 60% of its total energy and virtually all of its electricity.

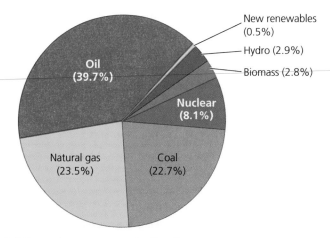

(a) Current energy consumption in the United States by source

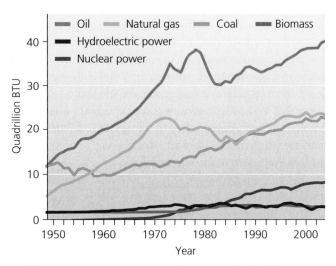

(b) Changes in energy consumption in the United States over the past half-century

FIGURE 20.2 In the United States, fossil fuels account for over 85% of energy consumption, nuclear power for 8.1%, biomass for 2.8%, and hydropower for 2.9% **(a).** U.S. usage of the three fossil fuels has grown faster than biomass or hydropower over the past half century **(b).** Nuclear power grew considerably between 1970 and 2000. Data from Energy Information Administration. 2005. *Annual Energy Review 2004.* U.S. Department of Energy.

Nuclear Power

Nuclear power occupies an odd and conflicted position in our modern debate over energy. It is free of the air pollution produced by fossil fuel combustion, and so has long been put forth as an environmentally friendly alternative to fossil fuels. At the same time, nuclear power's great promise has been clouded by nuclear weaponry, the dilemma of radioactive waste disposal, and the long shadow of Chernobyl and other power plant accidents. As such, public safety concerns and the costs of addressing them have constrained the development and spread of nuclear power in the United States, Sweden, and many other nations.

First developed commercially in the 1950s, nuclear power has expanded 15-fold worldwide since 1970, experiencing most of its growth during the 1970s and 1980s. Of all nations, the United States generates the most electricity from nuclear power—nearly a third of the world's production— and is followed by France and Japan. However, only 20% of U.S. electricity comes from nuclear sources. A number of other nations rely more heavily on nuclear power (Table 20.1). For instance, France and Lithuania each receive roughly four-fifths of their electricity from nuclear sources.

Fission releases nuclear energy

Strictly defined, **nuclear energy** is the energy that holds together protons and neutrons within the nucleus of an atom. We harness this energy by converting it to thermal energy, which can then be used to generate electricity. Several processes can convert the energy within an atom's nucleus into thermal energy, releasing it and making it available for use. Each process involves transforming isotopes (▸ pp. 92–95) of one element into isotopes of other elements, by the addition or loss of neutrons.

Table 20.1 Top Consumers of Nuclear Power			
Nation	Nuclear power consumed[1]	Number of plants[2]	Percentage of electricity generation from nuclear power plants[3]
United States	187.9	104	19
France	101.4	59	78
Japan	64.8	53	23
Germany	37.8	18	28
Russia	32.4	30	16
South Korea	29.6	19	37
Canada	20.5	16	13
Ukraine	19.7	13	45
United Kingdom	18.1	27	22
Sweden	17.3	11	50

Data from International Atomic Energy Agency, British Petroleum, and International Energy Agency.
[1]In million metric tons of oil equivalent, 2004 data.
[2]2003 data.
[3]2003 data.

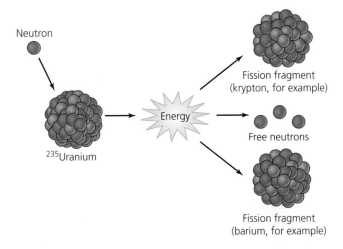

FIGURE 20.3 In nuclear fission, atoms of uranium-235 are bombarded with neutrons. Each collision splits uranium atoms into smaller atoms and releases two or three neutrons, along with energy and radiation. Because the neutrons can continue to split uranium atoms and set in motion a runaway chain reaction, engineers at nuclear plants must absorb excess neutrons with control rods to regulate the rate of the reaction.

The reaction that drives the release of nuclear energy in power plants is **nuclear fission,** the splitting apart of atomic nuclei (Figure 20.3). In fission, the nuclei of large, heavy atoms, such as uranium or plutonium, are bombarded with neutrons. Ordinarily neutrons move too quickly to split nuclei when they collide with them, but if neutrons are slowed down they can break apart nuclei. Each split nucleus emits multiple neutrons, together with substantial heat and radiation. These neutrons (two to three in the case of fissile isotopes of uranium-235) can in turn bombard other uranium-235 (^{235}U) atoms in the vicinity, resulting in a self-sustaining chain reaction.

If not controlled, this chain reaction becomes a runaway process of positive feedback that releases enormous amounts of energy—the same process that creates the explosive power of a nuclear bomb. Inside a nuclear power plant, however, fission is controlled so that only one of the two or three neutrons emitted with each fission event goes on to induce another fission event. In this way, the chain reaction maintains a constant output of energy at a controlled rate, rather than an explosively increasing output.

Nuclear energy comes from processed and enriched uranium

We generate electricity from nuclear power by controlling fission in **nuclear reactors,** facilities contained within nuclear power plants. But this is just one step in a longer process sometimes called the *nuclear fuel cycle*. This

FIGURE 20.4 Enriched uranium fuel is packaged into fuel rods, which are encased in metal and used to power fission inside the cores of nuclear reactors. In this photo, the fuel rods are visible arrayed in a circle within the blue water, at bottom.

process begins when the naturally occurring element uranium is mined from underground deposits, as we saw with the mines on Australian Aboriginal land in Chapter 2.

Uranium is an uncommon mineral, and uranium ore is in finite supply, which is why nuclear power is generally considered a nonrenewable energy source. Uranium is used for nuclear power because it is radioactive. Radioactive isotopes, or **radioisotopes,** emit subatomic particles and high-energy radiation as they decay into lighter radioisotopes, until they become stable isotopes. The isotope uranium-235 decays into a series of daughter isotopes, eventually forming lead-207. Each radioisotope decays at a rate determined by that isotope's **half-life,** the amount of time it takes for one-half of the atoms to give off radiation and decay. Different radioisotopes have very different half-lives, ranging from fractions of a second to billions of years. The half-life of ^{235}U is about 700 million years.

Over 99% of the uranium in nature occurs as the isotope uranium-238. Uranium-235 (with three fewer neutrons) makes up less than 1% of the total. Because ^{238}U does not emit enough neutrons to maintain a chain reaction when fissioned, we use ^{235}U for commercial nuclear power. So, mined uranium ore must be processed to enrich the con-

centration of ^{235}U to at least 3%. The enriched uranium is formed into small pellets of uranium dioxide (UO_2), which are incorporated into large metallic tubes called *fuel rods* (Figure 20.4) that are used in nuclear reactors. After several years in a reactor, enough uranium has decayed so that the fuel loses its ability to generate adequate energy, and it must be replaced with new fuel. In some countries, the spent fuel is reprocessed to recover what usable energy may be left. Most spent fuel, however, is disposed of as radioactive waste.

Fission in reactors generates electricity in nuclear power plants

For fission to begin in a nuclear reactor, the neutrons bombarding uranium are slowed down with a substance called a *moderator,* most often water or graphite. With a moderator allowing fission to proceed, it then becomes necessary to soak up the excess neutrons produced when uranium nuclei divide, so that on average only a single uranium atom from each nucleus goes on to split another nucleus. For this purpose, *control rods,* made of a metallic alloy that absorbs neutrons, are placed into the reactor among the water-bathed fuel rods. Engineers move these control rods into and out of the water to maintain the fission reaction at the desired rate.

All this takes place within the reactor core and is the first step in the electricity-generating process of a nuclear power plant (Figure 20.5). The reactor core is housed within a reactor vessel, and the vessel, steam generator, and associated plumbing are protected within a *containment building.* Containment buildings, with their meter-thick concrete and steel walls, are constructed to prevent leaks of radioactivity due to accidents or natural catastrophes such as earthquakes.

Breeder reactors make better use of fuel, but have raised safety concerns

Using ^{235}U as fuel for conventional fission is only one potential route to harnessing nuclear energy. A similar process, **breeder nuclear fission,** makes use of ^{238}U, which in conventional fission goes unused as a waste product. In breeder fission, the addition of a neutron to ^{238}U forms plutonium (^{239}Pu). When plutonium is bombarded by a neutron, it splits into fission products and releases more neutrons, which convert more of the remaining ^{238}U fuel into ^{239}Pu, continuing the process. Because 99% of all uranium is ^{238}U, breeder fission makes much better use of fuel, generates far more power, and produces far less waste.

However, breeder fission is considerably more dangerous than conventional nuclear fission, because highly reactive liquid sodium, rather than water, is used as a

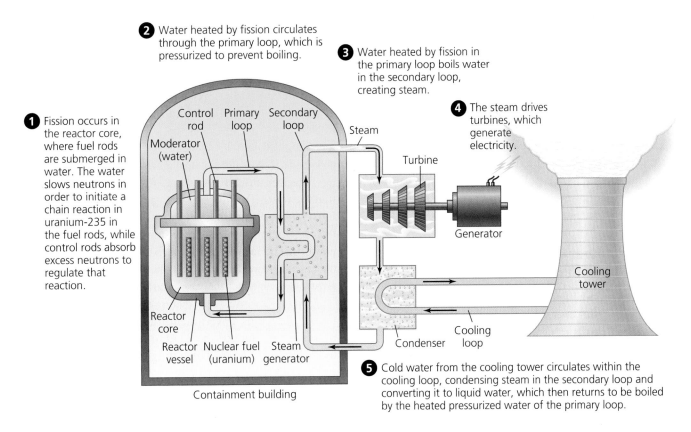

❷ Water heated by fission circulates through the primary loop, which is pressurized to prevent boiling.

❸ Water heated by fission in the primary loop boils water in the secondary loop, creating steam.

❹ The steam drives turbines, which generate electricity.

❶ Fission occurs in the reactor core, where fuel rods are submerged in water. The water slows neutrons in order to initiate a chain reaction in uranium-235 in the fuel rods, while control rods absorb excess neutrons to regulate that reaction.

Control rod Primary loop Secondary loop

Steam

Turbine

Moderator (water)

Generator

Cooling tower

Reactor core

Reactor vessel Nuclear fuel (uranium) Steam generator

Condenser Cooling loop

Containment building

❺ Cold water from the cooling tower circulates within the cooling loop, condensing steam in the secondary loop and converting it to liquid water, which then returns to be boiled by the heated pressurized water of the primary loop.

FIGURE 20.5 In a pressurized light water reactor, the most common type of nuclear reactor, uranium fuel rods are placed in water, which slows neutrons so that fission can occur (1). Control rods that can be moved into and out of the reactor core absorb excess neutrons to regulate the chain reaction. Water heated by fission circulates through the primary loop (2) and warms water in the secondary loop, which turns to steam (3). Steam drives turbines, which generate electricity (4). The steam is then cooled by water from the cooling tower and returns to the containment building (5), to be heated again by heat from the primary loop.

coolant, raising the risk of explosive accidents. Breeder reactors also are more expensive than conventional reactors. Finally, breeder fission can be used to supply plutonium to nuclear weapons programs. As a result, all but a handful of the world's breeder reactors have now been shut down.

Fusion remains a dream

For as long as scientists and engineers have been generating power from nuclear fission, they have been trying to figure out how they might use nuclear fusion instead. **Nuclear fusion**—the process responsible for the immense amount of energy that our sun generates and the force behind hydrogen or thermonuclear bombs—involves forcing together the small nuclei of lightweight elements under extremely high temperature and pressure. The hydrogen isotopes deuterium and tritium can be fused together to create helium, releasing a neutron and a tremendous amount of energy (Figure 20.6).

Overcoming the mutually repulsive forces of protons in a controlled manner is difficult, and fusion requires

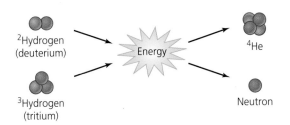

^{2}Hydrogen (deuterium)

^{3}Hydrogen (tritium)

Energy

^{4}He

Neutron

FIGURE 20.6 In nuclear fusion, two small atoms, such as the hydrogen isotopes deuterium and tritium, are fused together, causing the loss of a neutron and the release of energy. So far, however, scientists have not been able to fuse atoms without supplying far more energy than the reaction produces, so this process is not used commercially.

temperatures of many millions of degrees Celsius. Thus, researchers have not yet been able to develop this process for commercial power generation. Despite billions of dollars of funding and decades of research, fusion experiments in the lab still require scientists to input more energy than they produce from the process. Fusion's potentially huge payoffs, though, make many scientists eager to keep trying. If one day we were to find a way to control fusion in a reactor, we could produce vast amounts of energy using water as a fuel. The process would create only low-level radioactive wastes, without pollutant emissions or the risk of dangerous accidents, sabotage, or weapons proliferation. A consortium of nations including Japan, several European countries, and the United States is currently seeking to build a prototype fusion reactor called the International Thermonuclear Experimental Reactor (ITER). Even if this multibillion-dollar effort succeeds, however, power from fusion seems likely to remain many years in the future.

Nuclear power delivers energy more cleanly than fossil fuels

Using conventional fission, nuclear power plants generate electricity without creating air pollution from stack emissions. In contrast, combusting coal, oil, or natural gas emits sulfur dioxide that contributes to acidic deposition, particulate matter that threatens human health, and carbon dioxide and other greenhouse gases that contribute to global climate change. Even considering all the steps involved in building plants and generating power, researchers from the International Atomic Energy Agency (IAEA) have calculated that nuclear power lowers emissions 4–150 times below fossil fuel combustion (see "The Science behind the Story," ▶ pp. 598–599). IAEA scientists estimate that at current global rates of usage, nuclear power helps us avoid

emitting 600 million metric tons of carbon each year, equivalent to 8% of global greenhouse gas emissions.

Nuclear power has additional environmental advantages over fossil fuels, coal in particular. Because uranium generates far more power than coal by weight or volume, less of it needs to be mined, so uranium mining causes less damage to landscapes and generates less solid waste than coal mining. Additionally, in the course of normal operation, nuclear power plants are safer for workers than coal-fired plants.

However, nuclear power also has drawbacks. One is that the waste it produces is radioactive. Radioactive waste must be handled with great care and must be disposed of in a way that minimizes danger to present and future generations. The second main drawback is that if an accident occurs at a power plant, or if a plant is sabotaged, the consequences can potentially be catastrophic.

Given this mix of advantages and disadvantages (Table 20.2), most governments (although not necessarily most citizens) judged the good to outweigh the bad, and today the world has 439 operating nuclear plants.

Weighing the Issues:
Choose your Risk

Given the choice of living next to a nuclear power plant or living next to a coal-fired power plant, which would you choose? What would concern you most about each option?

Table 20.2 Environmental Impacts of Coal-Fired and Nuclear Power		
Type of Impact	**Coal**	**Nuclear**
Land and ecosystem disturbance from mining	*Extensive, on surface or underground*	Less extensive
Greenhouse gas emissions	*Considerable emissions*	None from plant operation; much less than coal over the entire life cycle
Other air pollutants	*Sulfur dioxide, nitrogen oxides, particulate matter, and other pollutants*	No pollutant emissions
Radioactive emissions	No appreciable emissions	*No appreciable emissions during normal operation; possibility of emissions during severe accident*
Occupational health among workers	*More known health problems and fatalities*	Fewer known health problems and fatalities
Health impacts on nearby residents	*Air pollution impairs health*	No appreciable known health impacts under normal operation
Effects of accident or sabotage	No widespread effects	*Potentially catastrophic widespread effects*
Solid waste	*More generated*	Less generated
Radioactive waste	None	*Radioactive waste generated*
Fuel supplies remaining	Should last several hundred more years	Uncertain; supplies could last longer or shorter than coal supplies

Note: More-severe impacts are in *italics*.

FIGURE 20.7 The Three Mile Island nuclear power plant near Harrisburg, Pennsylvania, was the site of the most serious nuclear power plant accident in U.S. history. The partial meltdown here in 1979 was a "near-miss"—radiation was released but was mostly contained, and no health impacts have been confirmed. The incident put the world on notice, however, that a major accident could potentially occur.

Nuclear power poses small risks of large accidents

Although scientists calculate that nuclear power poses fewer chronic health risks than does fossil fuel combustion, the possibility of catastrophic accidents has spawned a great deal of public anxiety over nuclear power. Two events were influential in shaping public opinion about nuclear energy.

The first took place at the **Three Mile Island** plant in Pennsylvania (Figure 20.7), where in 1979 the United States experienced its most serious nuclear power plant accident. Through a combination of mechanical failure and human error, coolant water drained from the reactor vessel, temperatures rose inside the reactor core, and metal surrounding the uranium fuel rods began to melt, releasing radiation. This process is termed a **meltdown,** and it proceeded through half of one reactor core at Three Mile Island. Area residents stood ready to be evacuated as the nation held its breath, but fortunately most radiation remained trapped inside the containment building.

The accident was brought under control within days, the damaged reactor was shut down, and multibillion-dollar cleanup efforts stretched on for years. Three Mile Island is best regarded as a near-miss; the emergency could have been far worse had the meltdown proceeded through the entire stock of uranium fuel, or had the containment building not contained the radiation. Although no significant health effects on residents have been proven

(a) The Chernobyl sarcophagus

(b) Technicians measuring radiation

FIGURE 20.8 The world's worst nuclear power plant accident unfolded in 1986 at Chernobyl, in present-day Ukraine (then part of the Soviet Union). As part of the extensive cleanup operation, the destroyed reactor was encased in a massive concrete sarcophagus **(a)** to contain further radiation leakage. Technicians scoured the landscape surrounding the plant **(b),** measuring radiation levels, removing soil, and scrubbing roads and buildings.

in the years since, the event put safety concerns squarely on the map for U.S. citizens and policymakers.

Chernobyl saw the worst accident yet

In 1986 an explosion at the **Chernobyl** plant in Ukraine (part of the Soviet Union at the time) caused the most severe nuclear power plant accident the world has yet seen (Figure 20.8). Engineers had turned off safety systems to

The Science behind the Story

Assessing Emissions from Power Sources

Combusting coal, oil, or natural gas emits carbon dioxide and other greenhouse gases into the atmosphere, where they contribute to global climate change (Chapter 18). Reducing greenhouse emissions is one of the main reasons so many people want to replace fossil fuels with alternative energy sources.

But determining how different energy alternatives stack up in terms of emissions is a complex process. A number of studies have tried to quantify and compare emission rates of different energy types, but the varying methodologies used have made it hard to synthesize this information into a coherent picture.

Researchers from the International Atomic Energy Agency (IAEA) made an attempt at such a synthesis for the generation of electricity. Experts met at six meetings between 1994 and 1998 and reviewed the existing scientific literature, together with data from industry and government. Their goal was to come up with a range of estimates of greenhouse gas emissions for nuclear energy, each major fossil fuel type, and each major renewable energy source. IAEA scientists Joseph

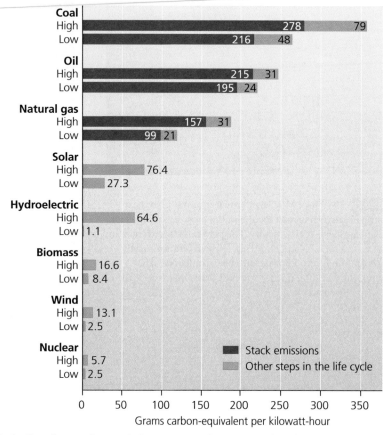

Coal, oil, and natural gas emit far more greenhouse gases than do renewable energy sources and nuclear energy. Red portions of bars represent stack emissions, and orange portions show emissions resulting from other steps in the life cycle. Data from Spadaro, J. V., et al. 2000. Greenhouse gas emissions of electricity generation chains: Assessing the difference. *IAEA Bulletin* 42(2).

Spadaro, Lucille Langlois, and Bruce Hamilton then published the results in the *IAEA Bulletin* in 2000.

The researchers had to decide how much of the total life cycle of electric power production to include

conduct tests, and human error, combined with unsafe reactor design, resulted in explosions that destroyed the reactor and sent clouds of radioactive debris billowing into the atmosphere. For 10 days radiation escaped from the plant, while emergency crews risked their lives (some later died from radiation exposure) putting out fires. Most residents of the surrounding countryside remained at home for these 10 days, exposed to radiation, before the Soviet government belatedly began evacuating more than 100,000 people. In the months and years afterwards, workers erected a gigantic

concrete sarcophagus around the demolished reactor, scrubbed buildings and roads, and removed irradiated materials, but the landscape for at least 30 km (19 mi) around the plant remains contaminated today.

The accident killed 31 people directly and sickened or caused cancer in many more. Exact numbers are uncertain because of inadequate data and the difficulty of determining long-term radiation effects (see "The Science behind the Story," ▸ pp. 604–605). It is widely thought that at least 2,000 people who were children at the time have

in their estimates. Simply comparing the rotation of turbines at a wind farm to the operation of a coal-fired power plant might not be fair, because it would not reveal that greenhouse gases were emitted as a result of manufacturing the turbines, transporting them to the site, and erecting them there. Similarly, because uranium mining is part of the nuclear fuel cycle, and because we use oil-fueled machinery to mine uranium, perhaps emissions from this process should be included in the estimate for nuclear power.

The researchers decided to conduct a "cradle-to-grave" analysis and include all sources of emissions throughout the entire life cycle of each energy source. This included not just power generation, but also the mining of fuel, preparation and transport of fuel, manufacturing of equipment, construction of power plants, disposal of wastes, and decommissioning of plants. They did, however, separate stack emissions from all other sources of emissions in the chain of steps so that these data could be analyzed independently.

Different greenhouse gases were then standardized to a unit of "carbon equivalence" according to their global warming potential (▸ pp. 530–532). For instance, because methane is 21 times as powerful a greenhouse gas as carbon, each unit of methane emitted was counted as 21 units of carbon-equivalence. The researchers then calculated rates of emission per unit of power produced. They presented figures in grams of carbon-equivalent emitted per kilowatt-hour (gC_{eq}/kWh) of electric power produced.

The overall pattern they found was clear: Fossil fuels produce much higher emission rates than renewable energy sources and nuclear energy (see the figure). The highest emission rate for fossil fuels (357 gC_{eq}/kWh for coal) was 4.7 times higher than the highest emission rate for any renewable energy source (76.4 gC_{eq}/kWh for solar power). The lowest emission rate for any fossil fuel (106 gC_{eq}/kWh for natural gas) was nearly 100 times greater than the lowest rate for renewables (1.1 gC_{eq}/kWh for one form of hydropower). Overall, emissions decreased in the following order: coal, oil, natural gas, photovoltaic solar, hydroelectric, biomass, wind, and nuclear.

The majority of fossil fuel emissions were direct stack emissions from power generation, and the amounts due to other steps in the life cycle were roughly comparable to those from some renewable sources.

Within each category, emissions values varied considerably. This was due to many factors, including the type of technology used, geographic location and transport costs, carbon content of the fuel, and the efficiency with which fuel was converted to electricity.

However, technology was expected to improve in the future, creating greater fuel-to-electricity conversion efficiency and therefore lowering emissions rates. Thus, the researchers devised separate emissions estimates for newer technologies expected between 2005 and 2020. These estimates suggested that fossil fuels will improve but still will not approach the cleanliness of nuclear energy and most renewable sources.

Because the IAEA is charged with promoting nuclear energy, critics point out that the agency has clear motivation for conducting a study that shows nuclear power in such a favorable light. However, few experts would quibble with the overall trend in the study's data: Nuclear and renewable energy sources are demonstrably cleaner than fossil fuels.

---

contracted or will contract thyroid cancer from radioactive iodine. Estimates for the total number of cancer cases attributable to Chernobyl, past and future, vary widely, from a few thousand to hundreds of thousands.

Atmospheric currents carried radioactive fallout from Chernobyl across much of the Northern Hemisphere, particularly Ukraine, Belarus, and parts of Russia and Europe (Figure 20.9). Fallout was greatest where rainstorms brought radioisotopes down from the radioactive cloud. Parts of Sweden received high amounts of fallout. The ac-

cident reinforced the Swedish public's fears about nuclear power. A survey taken after the event asked, "Do you think it was good or bad for the country to invest in nuclear energy?" The proportion answering "bad" jumped from 25% before Chernobyl to 47% afterward.

The world has been fortunate not to have experienced another accident on the scale of Chernobyl in the two decades since. There have been smaller-scale incidents; for instance, a 1999 accident at a plant in Tokaimura, Japan, killed two workers and exposed over 400 others to leaked

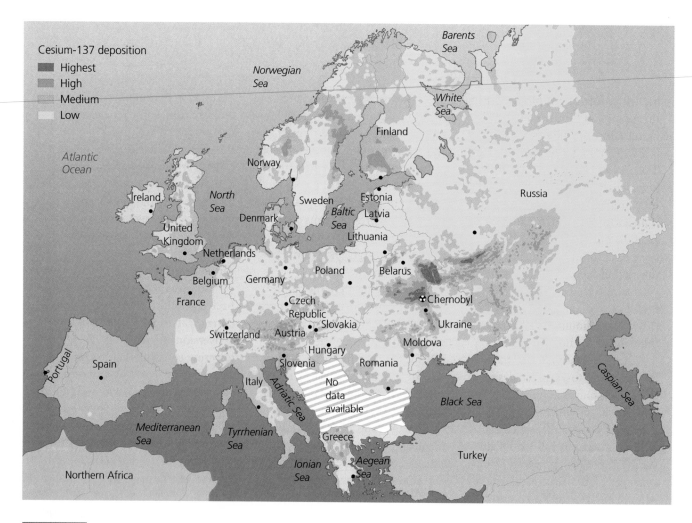

FIGURE 20.9 Radioactive fallout from the Chernobyl disaster was deposited across Europe in complex patterns resulting from atmospheric currents and rainstorms in the days following the accident. Darker colors in the map indicate higher levels of radioactivity. Although Chernobyl produced 100 times more fallout than the U.S. bombs dropped on Hiroshima and Nagasaki in World War II, it was distributed over a much wider area. Thus, levels of contamination in any given place outside of Ukraine, Belarus, and Russia were relatively low; the average European received less than the amount of radiation a person receives naturally in a year. Data from chernobyl.info, Swiss Agency for Development and Cooperation, Bern, 2005.

radiation. And as plants around the world age, they require more maintenance and are therefore less safe. New concerns have also surfaced. The September 11, 2001, terrorist attacks raised fears that similar airplane attacks could be carried out against nuclear plants. Moreover, radioactive material could be stolen from plants and used in terrorist attacks. This possibility is especially worrisome in Russia and other cash-strapped nations of the former Soviet Union, where hundreds of former nuclear sites have gone without adequate security for over a decade. In a cooperative international agreement, the U.S. government, through the "megatons to megawatts" program, has been buying up some of this material and diverting it to peaceful use in power generation.

Waste disposal remains a problem

Even if nuclear power generation could be made completely safe, we would still be left with the conundrum of what to do with spent fuel rods and other radioactive waste. Recall that fission utilizes ^{235}U as fuel, leaving the 97% of uranium that is ^{238}U as waste. This ^{238}U, as well as all irradiated material and equipment that is no longer being used, must be disposed of in a location where radiation will not escape and harm the public. Because the half-lives of uranium, plutonium, and many other radioisotopes are far longer than human lifetimes, this waste will continue emitting dangerous levels of radiation for thousands of years. Thus, radioactive waste must be placed in unusually stable and secure locations where radioactivity will not harm future generations.

Currently, nuclear waste from power generation is being held in temporary storage at nuclear power plants across the United States and the world. Spent fuel rods are sunken in pools of cooling water to minimize radiation leakage (Figure 20.10a). However, the U.S. Department of Energy (DOE) estimates that by 2010, three-fourths of U.S. plants will have no room left for this type of storage. Many plants are now expanding their storage capacity by storing waste in thick casks of steel, lead, and concrete (Figure 20.10b).

In total, U.S. power plants are storing over 49,000 metric tons of radioactive waste, enough to fill a football field to the depth of 3.3 m (10 ft). This waste is held at 125 sites spread across 39 states (Figure 20.11). A 2005 National Academy of Sciences report judged that most of these sites were vulnerable to terrorist attacks. The DOE estimates that over 161 million U.S. citizens live within 125 km (75 mi) of temporarily stored waste.

Because storing waste at numerous dispersed sites creates a large number of potential hazards, nuclear waste managers have long wanted to send all waste to a central repository that can be heavily guarded. In Sweden, that nation's nuclear industry has established a single repository for low-level waste near one power plant and is searching for a single disposal site deep within bedrock for spent fuel rods and other high-level waste. The industry hopes to decide on a site and get it approved by 2008.

In the United States, the multi-year search homed in on Yucca Mountain, a remote site in the desert of southern Nevada, 160 km (100 mi) from Las Vegas (Figure 20.12a, ▶ p. 603). Choice of this site followed extensive study by government scientists (Figure 20.12b), but Nevadans were not happy about the choice and fought against it. In 2002 the site was recommended by the president and approved by Congress, and the Department of Energy is now awaiting approval from the Nuclear Regulatory Commission. If given final approval, Yucca Mountain is expected to begin receiving waste from nuclear reactors, as well as high-level radioactive waste from military installations, in 2010. According to the design, waste would be stored in a network of tunnels 300 m (1,000 ft) underground, yet 300 m (1,000 ft) above the water table (Figure 20.12c) Scientists and policymakers chose the Yucca Mountain site because they determined that:

▶ It is unpopulated, lying 23 km (14 mi) from the nearest year-round residences.
▶ It has stable geology, with minimal risk of earthquakes that could damage the tunnels and release radioactivity.

(a) Wet storage

(b) Dry storage

FIGURE 20.10 Spent uranium fuel rods are currently stored at nuclear power plants and will likely remain at these scattered sites until a central repository for radioactive waste is developed. Spent fuel rods are most often kept in "wet storage" in pools of water **(a)**, which keep them cool and reduce radiation release. Alternatively, the rods may be kept in "dry storage" in thick-walled casks layered with lead, concrete, and steel **(b)**.

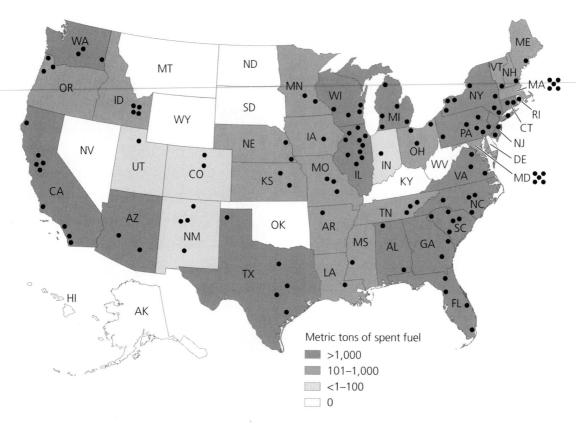

FIGURE 20.11 Nuclear waste from civilian reactors is currently stored at 125 sites in 39 states across the United States. In this map, dots indicate each storage site, and the four shades of color indicate the total amount of waste stored in each state. Data from: Office of Civilian Radioactive Waste Management, U.S. Department of Energy; and Nuclear Energy Institute, Washington, D.C.

▶ Its dry climate should minimize water infiltration through the soil, reducing chances of groundwater contamination.

▶ The water table is deep underground, making groundwater contamination less likely.

▶ The pool of groundwater does not connect with groundwater elsewhere, so that any contamination would be contained.

▶ The location, on federal land, can feasibly be protected from sabotage.

Some scientists, antinuclear activists, and concerned Nevadans have challenged these conclusions. For instance, they argue that earthquakes or volcanic activity could destabilize the site's geology. They also fear that fissures in the mountain's rock could allow rainwater to seep into the caverns.

A greater concern, in many people's minds, is that nuclear waste will need to be transported to Yucca Mountain from the 125 current storage areas and from current and future nuclear plants and military installations. Because this would involve many thousands of shipments by rail and truck across hundreds of public highways through al-

most every state of the union, many people worry that the risk of an accident or of sabotage is unacceptably high.

Weighing the Issues:
How to Store Waste?

Which do you think is a better option—to transport nuclear waste cross-country to a single repository or to store it permanently at numerous power plants and military bases scattered across the nation? Would your opinion be affected if you lived near the repository site? Near a power plant? On a highway route along which waste was transported?

Multiple dilemmas have slowed nuclear power's growth

Dogged by concerns over waste disposal, safety, and expensive cost overruns, nuclear power's growth has slowed. Since the late 1980s, nuclear power has grown by 2.5% per year, about the same rate as electricity generation overall. Public anxiety in the wake of Chernobyl

(a) Yucca Mountain

(b) Scientific testing

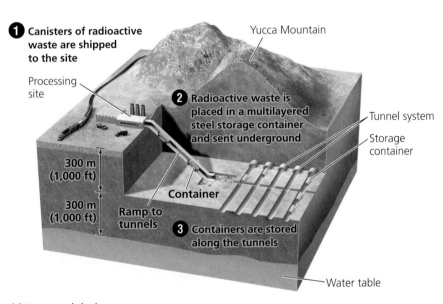

❶ **Canisters of radioactive waste are shipped to the site**

Yucca Mountain

Processing site

❷ **Radioactive waste is placed in a multilayered steel storage container and sent underground**

Tunnel system

Storage container

300 m (1,000 ft)

300 m (1,000 ft)

Container

Ramp to tunnels

❸ **Containers are stored along the tunnels**

Water table

(c) Proposed design

FIGURE 20.12 Yucca Mountain (**a**), in a remote part of Nevada, awaits final approval as the central repository site for all the nuclear waste in the United States. Here (**b**), technicians are testing the effects of extreme heat from radioactive decay on the stability of rock. Waste would be buried in a network of tunnels deep underground yet still high above the water table (**c**).

made utilities less willing to invest in new plants. So did the enormous expense of building, maintaining, operating, and ensuring the safety of nuclear facilities. Almost every nuclear plant has turned out to be more expensive than expected. In addition, plants have aged more quickly than expected because of problems that were underestimated, such as corrosion in coolant pipes. The plants that have been shut down—well over 100 around the world to date—have served on average less than half their expected lifetimes. Moreover, shutting down, or decommissioning, a plant can sometimes be more expensive than the original construction. As a result of these economic issues, electricity from nuclear power today remains more expensive than electricity from coal and

other sources. Governments are still subsidizing nuclear power to keep consumer costs down, but many private investors lost interest long ago.

Nonetheless, nuclear power remains one of the few currently viable alternatives to fossil fuels with which we can generate large amounts of electricity in short order. The International Energy Agency (IEA) predicts nuclear production will peak at 7% of world energy use and 17% of electricity production in 2010, then decline to 5% of world energy use and 9% of electricity production by 2030. The reason is that Asian nations are adding generating capacity, but three-quarters of Western Europe's capacity is scheduled to be retired by 2030. In Western Europe, not a single reactor is under construction today,

Health Impacts of Chernobyl

The Science behind the Story

In the wake of the nuclear power plant accident at Chernobyl in 1986, medical scientists from around the world rushed to study how the release of radiation might affect human health.

Determining long-term health impacts of a discrete event is difficult, so it is not surprising that the hundreds of researchers trying to pin down Chernobyl's impacts sometimes came up with very different conclusions. In an effort to reach some consensus, researchers at the Nuclear Energy Agency (NEA) of the Organization for Economic Co-operation and Development (OECD) reviewed studies through 2002 and issued a report summarizing what scientists had learned in the 16 years since the accident.

Doctors had documented the most severe effects among plant workers and firefighters who battled to contain the incident in its initial hours and days. Medical staff treated and recorded the progress of 237 patients who had been admitted to area hospitals diagnosed with acute radiation sickness (ARS). Radiation destroys cells in the body, and if the destruction outpaces the body's abilities to repair the damage, the person will soon die. Symptoms of ARS include vomiting, fever, diarrhea, thermal burns, mucous membrane damage, and weakening of the immune system by the depletion of white blood cells. In total, 28 (11.8%) of these people died from acute effects soon after the accident, and those who died had had the greatest estimated exposure to radiation.

IAEA scientists in 1990 studied residents of areas highly contaminated with radioactive cesium and

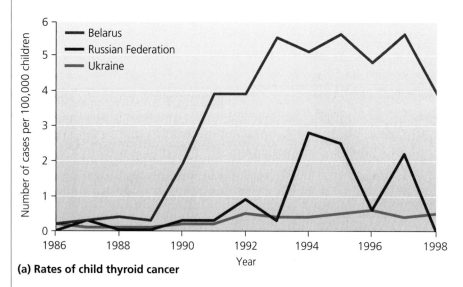

(a) Rates of child thyroid cancer

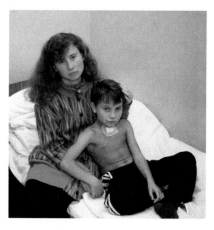

(b) Cancer patient with mother after surgery

The incidence of thyroid cancer **(a)** jumped in Belarus, Ukraine, and southwestern Russia starting 4 years after the Chernobyl accident released high levels of radioactive iodine isotopes. Many babies and young children **(b)** at the time of the accident developed thyroid cancer in later years. Most have undergone treatment and survived. Data (a) from Nuclear Energy Agency, OECD, 2002.

and Germany and Belgium, like Sweden, have declared an intention to phase out nuclear power altogether. However, France is committed to keeping its large share of nuclear-powered electricity. In Asia, Japan is so reliant on oil that it is eager to expand its options in the fastest and easiest way possible. China, India, and South Korea are expanding their nuclear programs to help power their rapidly growing economies. Indeed, Asia hosts 19 of the last 28 nuclear power plants to go into operation and 20 of the 31 plants now under construction.

compared their health with people of the same ages living in uncontaminated settlements nearby. Medical exams of 1,356 people showed no significant differences between the two groups or health abnormalities attributable to radiation exposure. However, the study was criticized for the quality of its data, for its small sample size, and for potential conflict of interest (the IAEA is charged with promoting the nuclear industry). In addition, the study was conducted only 4 years after the accident, before many cancers would be expected to appear.

Nonetheless, studies by the World Health Organization and others have come to similar conclusions. Overall, the NEA summary concluded, there is little evidence for long-term physical health effects resulting from Chernobyl (although psychological and social effects among displaced residents have been substantial). If cancer rates have risen among exposed populations, they have risen so little as to be statistically indistinguishable from normal variation in background levels of cancer.

The one exception is thyroid cancer, for which numerous studies have documented a real and perceptible increase among Chernobyl-area residents, particularly children (because children have large and active thyroid glands). The thyroid gland is where the human body concentrates iodine, and one of the most common radioactive isotopes released early in the disaster was iodine-131 (^{131}I).

Realizing that thyroid cancer induced by radioisotopes of iodine might be a problem, medical workers took measurements of iodine activity from the thyroid glands of hundreds of thousands of people— 60,000 in Russia, 150,000 in Ukraine, and several hundred thousand in Belarus—in the months immediately following the accident. They also measured food contamination and had people fill out questionnaires on their food consumption. These data showed that drinking cows' milk was the main route of exposure to ^{131}I for most people, although fresh vegetables also contributed.

As doctors had feared, in the years following the accident rates of thyroid cancer began rising among children in regions of highest exposure (see the figure). The yearly number of thyroid cancer cases in the 1990s, particularly in Belarus, far exceeded numbers from years before Chernobyl. Multiple studies found linear dose-response relationships (▶ pp. 420–421) in data from Ukraine and Belarus. Fortunately, treatment of thyroid cancer has a high success rate, and as of 2002, only 3 of the 1,036 children

cited in our figure had died of thyroid cancer. By comparing the Chernobyl-region data to background rates elsewhere, researchers calculated that Ukraine would eventually suffer 300 thyroid cancer cases more than normal and that the nearby region of Russia (with a population of 4.3 million) would suffer 349 extra cases, a 3–6% increase above the normal rate.

Critics pointed out that any targeted search tends to turn up more of whatever medical problem is being looked for. But the magnitude of the increase in childhood thyroid cancer was large enough that most experts judge it to be real. The rise in thyroid cancer, the NEA concluded, "should be attributed to the Chernobyl accident until proven otherwise."

Furthermore, thyroid cancer also appears to have risen in adults. Adult cases in Belarus in the 12 years before the accident totaled 1,392, but in the 12 years after Chernobyl totaled 5,449. In the most contaminated regions of Russia, thyroid cancer incidence rose to 11 per 100,000 women and 1.7 per 100,000 men, compared to normal rates of 4 and 1.1 for Russia as a whole. And although rates of childhood cancer may now be falling, rates for adults are still rising. As new cancer cases accumulate in the future, continued research will be needed to measure the full scope of health effects from Chernobyl.

In the United States, the nuclear industry stopped building plants following Three Mile Island, and public opposition scuttled many that were under construction. The $5.5-billion Shoreham Nuclear Power Plant on New York's Long Island was shut down just 2 months after its licensing because officials determined that evacuation would be impossible in this densely populated area should an accident ever occur. Of the 259 U.S. nuclear plants ordered since 1957, nearly half have been cancelled. At its peak in 1990, the United States had 112

Nuclear Power

Can nuclear power help reduce our reliance on fossil fuels? Should we revitalize and expand our nuclear power programs?

Nuclear Power: A Deadly and Needless Energy Source

Nuclear power is deadly and unnecessary. Disasters such as the 1979 Three Mile Island nuclear plant accident, the catastrophic Chernobyl plant explosion in 1986—and worse—are what will happen regularly if the United States and other nations move anew to build nuclear power plants.

The disastrous impacts of nuclear power are acknowledged in government documents. The U.S. Nuclear Regulatory Commission conducted a study in the 1980s, the *Calculation of Reactor Accident Consequences 2.* For the Indian Point 3 plant near New York, it calculated an accident causing 50,000 "peak fatalities," 141,000 "peak early injuries," 13,000 "peak cancer deaths," and $314 billion in property damage in 1980 dollars. The cost of a part of America left uninhabitable for millennia would be nearly $1 trillion today.

Nuclear power is so dangerous that there's a law called the Price-Anderson Act that limits a plant owner's liability for an accident, now $8 billion. If nuclear power is so safe, why the Price-Anderson Act?

The likelihood of an accident is far from "almost impossible," as atomic promoters once claimed. The NRC has conceded a 45% probability of a severe core melt accident every 20 years among the 100 U.S. atomic plants.

And it doesn't take an accident for a nuclear plant to spread radioactivity and contaminate and kill. There are "routine emissions" of radiation at every plant, as well as tons of lethal radioactive waste each plant produces annually, which must be isolated for thousands to millions of years. Moreover, in an age of terrorism, nuclear plants are sitting ducks.

We need not take the colossal risk of atomic power. Safe, clean, sustainable technologies are here today and can unhook us from fossil fuels: wind, solar, geothermal, and hydrogen power, among others. Let's have energy that won't kill us and our children—energy we can live with.

Karl Grossman is a professor at SUNY/College at Old Westbury and an investigative journalist who has authored books and hosted television programs about nuclear power.

Nuclear Power: Essential for Sustainable Global Development

We live in a world that has only *begun* to consume energy. During the next 50 years, as Earth's population expands from 6 billion toward 9 billion, humanity will consume more energy than the combined total used in all previous history.

With carbon emissions threatening human health and the stability of the biosphere, the security of our world requires a massive transformation to clean energy. This crisis is global. Today India and China are gaining rapidly on Europe and America in per capita energy consumption and climate-endangering emissions. Renewables such as solar, wind, and biomass can help. But only nuclear power offers clean energy on a massive scale.

Some "environmental" groups still spread misinformation about nuclear power. But here are the facts:

▶ *Safety.* Using 12,000 reactor-years of experience, nuclear power is the safest large-scale source of energy. The Chernobyl reactor used Soviet technology with no resemblance to today's technology.

▶ *Waste.* Nuclear power extracts enormous energy from tiny amounts of uranium. The small amounts of waste can be safely managed and placed in geological repositories with no long-term environmental harm.

▶ *Proliferation.* Civil nuclear power production does not foster nuclear weapons. Atomic bombs require sophisticated military programs, and worldwide IAEA safeguards prevent illicit diversion of nuclear material.

▶ *Cost.* Nuclear energy is already cost-competitive, and trends point to falling nuclear prices and rising fossil prices. A carbon tax would add to the nuclear advantage.

▶ *Usability.* Nuclear power is operating in countries representing two-thirds of total human population, and usage is expanding.

An informed public debate—focused on facts rather than myths—will demonstrate that nuclear energy is indispensable to sustainable global development.

John Ritch is director general of the World Nuclear Association. From 1994 to 2001, he was U.S. ambassador to the International Atomic Energy Agency and several other U.N. organizations. Previously, he served for 22 years as an advisor to the U.S. Senate Foreign Relations Committee.

Explore this issue further by accessing **Viewpoints** at www.aw-bc.com/withgott.

operable plants; today it has 104. However, the Bush administration advocates expanding U.S. nuclear capacity to decrease reliance on oil imports, and nuclear proponents point out that engineers are planning a new generation of reactors designed to be safer and less expensive.

Weighing the **Issues:**
More Nuclear Power?

Do you think the United States should expand its nuclear power program? Why or why not?

With slow growth predicted for nuclear power, and with fossil fuels in limited supply, where will our growing human population turn for additional energy? Increasingly, people are turning to renewable sources of energy: energy sources that cannot be depleted by our use. Although many renewable sources are still early in their stages of development, two of them—biomass and hydropower—are already well developed and widely used.

Biomass Energy

Biomass energy was the first energy source our species used, and it is still the leading energy source in much of the developing world. *Biomass* (▸ p. 158) consists of the organic material that makes up living organisms. People harness **biomass energy** from many types of plant and animal matter, including wood from trees, charcoal from burned wood, and combustible animal waste products such as cattle manure. Fossil fuels are not considered biomass energy sources because their organic matter has not been part of living organisms for millions of years and has undergone considerable chemical alteration since that time.

Fuelwood and other traditional biomass sources are widely used in the developing world

Over 1 billion people still use wood from trees as their principal power source. In developing nations, especially in rural areas, families gather fuelwood to burn in or near their homes for heating, cooking, and lighting (Figure 20.13). In these nations, fuelwood, charcoal, and manure account for fully 35% of energy use—in the poorest nations, up to 90%.

Fuelwood and other traditional biomass energy sources constitute nearly 80% of all renewable energy used worldwide. Considering what we have learned about the loss of forests (▸ pp. 349–354), however, it is fair to ask whether biomass should truly be considered a renewable resource. In reality, biomass is renewable only if it is not overharvested. At moderate rates of use, trees and other plants can replenish themselves over months to decades. However, when forests are cut too quickly, or when overharvesting leads to soil erosion and forests fail to grow back, then biomass is not effectively replenished.

FIGURE 20.13 Hundreds of millions of people in the developing world rely on fuelwood for heating and cooking. Wood cut from trees remains the major source of biomass energy used today. In theory biomass is renewable, but in practice it may not be if forests are overharvested.

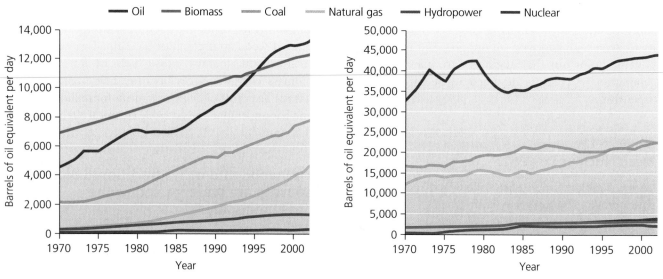

(a) **Energy consumption in developing nations**

(b) **Energy consumption in developed nations in the OECD**

FIGURE 20.14 Energy consumption patterns vary greatly between developing nations (**a**) and developed nations, here represented by nations of the Organisation for Economic Cooperation and Development (OECD) (**b**). Note the large role that biomass (primarily fuelwood) plays in supplying energy to developing countries. Data from Energy Information Administration, U.S. Department of Energy.

The potential for deforestation makes biomass energy less sustainable than other renewable sources, particularly as human population continues to increase.

As developing nations industrialize, fossil fuels are replacing traditional energy sources (Figure 20.14). As a result, biomass use is growing more slowly worldwide than overall energy use.

New biomass sources are being developed in industrialized countries

Besides the fuelwood, charcoal, and manure traditionally used, biomass energy sources in today's world include a number of sources for which innovative uses have recently been developed (Table 20.3). Some of these sources can be burned efficiently in power plants to produce **biopower**, generating electricity in the same way that coal is burned for power (▶ pp. 566–567). Other new biomass sources can be converted into fuels used primarily to power automobiles; these are termed **biofuels.** Because many of these novel biofuels and biopower strategies depend on technologies resulting from a good deal of research and development, they are being developed primarily in wealthier industrialized nations, such as Sweden and the United States.

Many of the new biomass resources are the waste products of preexisting industries or processes. For instance, the forest products industry generates large

amounts of woody debris in logging operations and at sawmills, pulp mills, and paper mills (Figure 20.15). Sweden's efforts to encourage biomass energy have focused largely on using forestry residues. Because so much of the nation is forested and the timber industry is a major part of the national economy, plenty of forestry waste is available. Organic components of waste from municipal landfills also can provide biomass energy, as can animal waste from agricultural feedlots. Residue from agricultural crops (such as stalks, cobs, and husks from corn) could also soon become a major bioenergy source.

Some plants are now specifically grown for the purpose of producing biofuels. These bioenergy crops include various fast-growing grasses, such as bamboo, fescue, and switchgrass; grain and oil-producing crops,

Table 20.3   Major Sources of Biomass Energy
▶ Wood cut from trees (fuelwood)
▶ Charcoal
▶ Manure from domestic animals
▶ Crops grown specifically for biomass energy production
▶ Crop residues (such as corn stalks)
▶ Forestry residues (such as wood waste from logging)
▶ Processing wastes (such as solid or liquid waste from sawmills, pulp mills, and paper mills)
▶ Components of municipal solid waste

FIGURE 20.15 Forestry residues (here, from a Swedish logging operation) are a major source of biomass energy in some regions.

such as corn and soybeans; and fast-growing trees, such as poplar and cottonwood. For biofuels, willow trees are cultivated commercially in Sweden, sugarcane is grown in Brazil, and corn is raised in the United States.

Biofuels can power automobiles

Liquid fuels from biomass sources are helping to power millions of vehicles on today's roads, and some vehicles can run entirely on biofuels. The two primary types of such fuels developed so far are ethanol (for gasoline engines) and biodiesel (for diesel engines).

Ethanol is the alcohol in beer, wine, and liquor. It is produced as a biofuel by fermenting biomass, generally from carbohydrate-rich crops, such as corn, in a process similar to brewing beer. In fermentation, carbohydrates are converted to sugars and then to ethanol. Spurred by the 1990 Clean Air Act amendments and generous government subsidies, ethanol is widely added to gasoline in the United States to reduce automotive emissions. In 2004 in the United States, 14 billion L (3.7 billion gal) of ethanol were produced from corn. This amount has grown each year, as the number of U.S. ethanol plants approaches 100. A number of nations now have vehicles that can run on ethanol (Sweden has dozens of such public buses), and each of the U.S. "big three" automakers is now producing *flexible fuel vehicles* that run on E-85, a mix of 85% ethanol and 15% gasoline. In Brazil, half of all new cars are flexible-fuel cars, and ethanol from sugarcane accounts for 44% of all automotive fuel used.

Biodiesel is produced from vegetable oil, used cooking grease, or animal fat. The oil or fat is mixed with small amounts of ethanol or methanol (wood alcohol) in the

presence of a chemical catalyst. In Europe, where most biodiesel is used, canola oil is often the oil of choice, whereas U.S. biodiesel producers use mostly soybean oil and recycled grease and oil from restaurants. Vehicles with diesel engines can run on 100% biodiesel. In fact, when Rudolf Diesel invented the diesel engine in 1895, he designed it to run on a variety of fuels, and he showcased his invention at the 1900 World's Fair using peanut oil. Since that time, petroleum-based fuel *(petrodiesel)* has been used because it has been cheaper. Because today's diesel engines have been designed to work with petrodiesel, some engine parts wear out more quickly with biodiesel, but one can use both and switch back and forth without making any modifications. Biodiesel can also be mixed with conventional petrodiesel; a 20% biodiesel mix (called B20) is common today.

Biodiesel cuts down on emissions compared with petrodiesel (Figure 20.16). Its fuel economy is almost as good, and it costs just 10–20% more. It is also nontoxic and biodegradable. Increasing numbers of environmentally conscious individuals in North America and Europe are fueling their cars with biodiesel from waste oils, and some buses and recycling trucks are now running on biodiesel. Governments are encouraging its use, too; Minnesota, for instance, mandates that all diesel sold must include a 2% biodiesel component, and already over 40 state and federal fleets are using biodiesel blends.

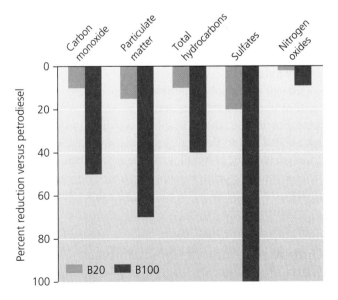

FIGURE 20.16 Burning biodiesel in a diesel engine emits fewer emissions than burning conventional petroleum-based diesel. Shown are the percentage reductions in several major automotive pollutants that one can attain by using B20 (a mix of 20% biodiesel and 80% petrodiesel) and B100 (pure biodiesel). Data from U.S. Environmental Protection Agency.

FIGURE 20.17 Each summer a group of alternative-fuel advocates goes on tour in the B.I.O. Bus, a bus fueled entirely on used vegetable oil from restaurants. Their "Bio Tours" sponsor events across North America that include music, dancing, and seminars on environmental sustainability and alternative fuels. Their motto: "Solar-powered sound, veggie-powered bus."

Some enthusiasts have taken biofuel use a step further. Eliminating the processing step that biodiesel requires, they use straight vegetable oil in their diesel engines. One notable effort is the "B.I.O. Bus," a bus fueled entirely on waste oil from restaurants that a group of students, environmentalists, and artists drives across North America (Figure 20.17). Each summer the group goes on tour with the bus, hosting festive events that combine music and dancing with seminars on environmental sustainability, and spreading the word about nonpetroleum fuels. In the summers of 2003, 2004, and 2005, the bus traveled 25,000 miles.

To run on straight vegetable oil, a diesel engine needs to be modified. Extra parts need to be added, so that there are tanks for both the oil and for petrodiesel, which is often needed to start the engine in cooler weather. Although these parts can be bought for as little as $800, it remains to be seen whether using straight vegetable oil might entail further costs, such as reduced longevity or greater engine maintenance.

Biopower generates electricity from biomass

We can harness biopower in various ways. The most frequently used strategy is simply to burn biomass in the presence of large amounts of air. This can be done on small scales with furnaces and stoves to produce heat for domestic needs, or on large industrial scales to produce district heating and to generate electricity. Power plants built to combust biomass operate in a similar way to those fired by fossil fuels; the combustion heats water, creating steam to turn turbines and generators, thereby generating electricity. Much of the biopower produced so far comes from power plants that generate both electricity and heating through cogeneration (▶p. 583); these plants are often located where they can take advantage of waste material from the forest products industry.

Biomass is also increasingly being combined with coal in existing coal-fired power plants in a process called *co-firing*. Biomass is introduced with coal into a high-efficiency boiler that uses one of several technologies. Up to 15% of the coal can be substituted with biomass, with only minor equipment modification and no appreciable loss of efficiency. Co-firing can be a relatively easy and inexpensive way for fossil-fuel-based utilities to expand their use of renewable energy.

The decomposition of biomass by microbes also produces gas that can be used to generate electricity. The anaerobic bacterial breakdown of waste in landfills produces methane and other components, and this "landfill gas" is now being captured at many solid waste landfills and sold as fuel (▶ p. 657). Methane and other gases can also be produced in a more controlled way in anaerobic digestion facilities. This "biogas" can then be burned in a power plant's boiler to generate electricity.

The process of *gasification* can also provide biopower, as well as biofuels. In this process, biomass is vaporized at extremely high temperatures in the absence of oxygen, creating a gaseous mixture including hydrogen, carbon monoxide, and methane. This mixture can generate electricity when used in power plants to turn a gas turbine to propel a generator. Gas from gasification can also be treated in various ways to produce methanol, synthesize a type of diesel fuel, or isolate hydrogen for use in hydrogen fuel cells (▶ pp. 637–642). An alternative method of heating biomass in the absence of oxygen results in *pyrolysis*, which produces a mix of solids, gases, and liquids. This includes a liquid fuel called pyrolysis oil, which can be burned to generate electricity.

At small scales, farmers and ranchers can buy modular biopower systems that use livestock manure to generate electricity, and small household biodigesters are now providing portable and decentralized energy production. In many developing nations, particularly in Asia, international agencies have distributed hundreds of thousands of efficient biomass-fueled cooking stoves. At large scales, industries such as the forest products industry are using their waste to generate power, farmers are growing crops for biopower, and governments are encouraging biopower. In Sweden, one-sixth of the nation's energy supply now comes from biomass, and biomass provides more fuel for electricity generation than coal, oil, or natural gas. Pulp mill liquors are the main source, but solid wood waste, municipal solid waste, and biogas from digestion are all increasingly being used. In the United States, several dozen biomass-fueled power plants are now operating, and several dozen coal-fired plants are experimenting with co-firing.

Biomass energy brings environmental and economic benefits

Biomass energy has one overarching environmental benefit: It is essentially carbon-neutral, releasing no net carbon into the atmosphere. Although burning biomass emits plenty of carbon, the carbon released is simply the carbon that was pulled from the atmosphere by photosynthesis to create the biomass being burned. That is, because biomass is the product of recent photosynthesis, the carbon released in its combustion is balanced by the carbon taken up in the photosynthesis that created it. Therefore, when biomass sources take the place of fossil fuels, net carbon flux to the atmosphere is reduced. However, this holds only if biomass sources are not overharvested; deforestation will increase carbon flux to the atmosphere because less vegetation means less carbon uptake by plants for photosynthesis.

Biofuels can reduce greenhouse gas emissions that contribute to climate change in additional ways. Capturing landfill gas reduces emissions of methane, a potent greenhouse gas. And biofuels such as ethanol and biodiesel contain oxygen, which when added to gasoline or diesel helps those petroleum fuels to combust more completely, reducing pollution.

Shifting from fossil fuels to biomass energy also can have economic benefits. As a resource, biomass tends to be well spread geographically, so using it should help support rural economies and reduce many nations' dependence on imported fuels. Biomass also tends to be the least expensive type of fuel for burning in power plants, and improved energy efficiency brings lower prices for consumers.

By increasing energy efficiency and recycling waste products, use of biomass energy helps move our industrial systems toward greater sustainability. The U.S. forest products industry now obtains over half its energy by combusting the biomass waste it recycles, including woody waste and liquor from pulp mill processing.

Relative to fossil fuels, biomass also has benefits for human health. By replacing coal in co-firing and direct combustion, biomass reduces emissions of nitrogen oxides and particularly sulfur dioxide. The reason is that wood, unlike coal, contains no appreciable sulfur content. By replacing gasoline and petrodiesel and burning more cleanly, biofuels reduce emissions of various air pollutants.

Biomass energy also brings drawbacks

Biomass energy is not without negative environmental impacts, however. Burning fuelwood and other biomass in traditional ways for cooking and heating leads to health hazards from indoor air pollution (▶ pp. 519–522). In addition, harvesting fuelwood at an unsustainably rapid rate leads to deforestation, soil erosion, and desertification, damaging landscapes, diminishing biodiversity, and impoverishing human societies dependent on an area's resources. In arid regions that are heavily populated and support meager woodlands, fuelwood harvesting can have enormous impacts. Such is the case with many regions of Africa and Asia. In contrast, moister, well-forested areas with lower population densities, such as Sweden, stand less chance of deforestation.

Another drawback of biomass energy is that growing crops specifically to produce biofuels establishes monoculture agriculture, with all its impacts (▶ pp. 281–282), on precious land that might otherwise be used to grow food, be developed for other purposes, or be left as wildlife habitat. Fully 10% of the U.S. corn crop is used to make ethanol, and most U.S. biodiesel is produced from oil from soybeans grown specifically for this purpose (although researchers are studying how to obtain oil from algae, which could take up far less space). Growing

FIGURE 20.18 Ethanol is widely added to gasoline in the United States and some other nations, and 10% of the U.S. corn crop is used to produce ethanol. Ethanol makes for cleaner-burning fuel, but it requires a great deal of industrialized agriculture.

bioenergy crops also requires substantial inputs of energy. We currently operate farm equipment using fossil fuels, and farmers apply petroleum-based pesticides and fertilizers to increase yields. Thus, shifting from gasoline to ethanol for our transportation needs would not eliminate our reliance on fossil fuels. Moreover, growing corn for ethanol (Figure 20.18) yields only small amounts of ethanol per acre of crop grown; 1 bushel of corn creates only 9.4 L (2.5 gal) of ethanol. It is not efficient to grow high-input, high-energy crops merely to reduce them to a few liters of fuel; such a process is less efficient than directly burning biomass. Indeed, even using ethanol as a gasoline additive to reduce emissions has a downside; it lowers fuel economy very slightly, so that more gasoline needs to be used. For all these reasons, many critics have lambasted U.S. subsidies of ethanol, viewing them as effective in gaining politicians votes in politically important farm states, but less effective as a path to sustainable energy use.

On the positive side, crops grown for energy typically receive lower inputs of pesticides and fertilizers than those grown for food. Researchers are currently refining techniques to produce ethanol from the cellulose that gives structure to plant material, and not just from starchy crops like corn. If these techniques can be made widely feasible, then ethanol could be produced primarily from low-quality waste and crop residues, rather than from high-quality crops.

Although biomass energy in industrialized nations currently revolves mostly around a few easily grown crops

and the efficient use of waste products, the U.S. government sees a future of specialized crops that can serve as the basis for a wide variety of fuels and products. Meanwhile, use of traditional fuels in the developing world is expected to increase, and the IEA estimates that in the year 2030, 2.6 billion people will be using traditional fuels for heating and cooking in unsustainable ways. Like nuclear power and like hydropower, biomass energy use involves a complex mix of advantages and disadvantages for the environment and human society.

Weighing the Issues:
Ethanol

Do you think producing and using ethanol from corn or other crops is a good idea? Do the benefits outweigh the drawbacks? Can you suggest ways of using biofuels that would minimize environmental impacts?

Hydroelectric Power

Next to biomass, we draw more renewable energy from the motion of water than from any other resource. In **hydroelectric power,** or **hydropower,** the kinetic energy of moving water is used to turn turbines and generate electricity. We examined hydropower and its environmental impacts in our discussion of freshwater resources in Chapter 15 (▶ pp. 446–449), but we will now take a closer look at hydropower as an energy source.

Modern hydropower uses dams and "run-of-river" approaches

Just as people have long burned fuelwood, we have long harnessed the power of moving water. For instance, waterwheels spun by river water powered mills in past centuries. Today we harness water's kinetic energy in two major ways: with storage impoundments and by using "run-of-river" approaches.

Most of our hydroelectric power today comes from impounding water in reservoirs behind concrete dams that block the flow of river water, and then letting that water pass through the dam. Because immense amounts of water are stored behind dams, this is called the **storage** technique. If you have ever seen Hoover Dam on the Colorado River (▶ p. 432), or any other large dam, you have witnessed an impressive example of the storage

(a) Ice Harbor Dam, Washington

(b) Turbine generator inside MacNary Dam, Columbia River

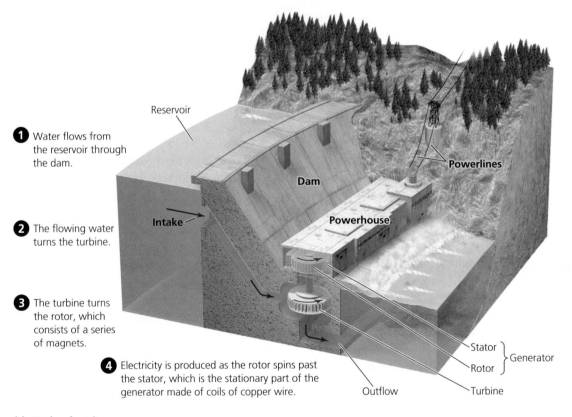

1 Water flows from the reservoir through the dam.

2 The flowing water turns the turbine.

3 The turbine turns the rotor, which consists of a series of magnets.

4 Electricity is produced as the rotor spins past the stator, which is the stationary part of the generator made of coils of copper wire.

Reservoir

Dam

Intake

Powerhouse

Powerlines

Stator ⎫
         ⎬ Generator
Rotor ⎭

Outflow

Turbine

(c) Hydroelectric power

FIGURE 20.19 Large dams, such as the Ice Harbor Dam on the Snake River in Washington (**a**), generate substantial amounts of hydroelectric power. Inside these dams, flowing water is used to turn turbines (**b**) and generate electricity. Water is funneled from the reservoir through a portion of the dam (**c**) to rotate turbines, which turn rotors containing magnets. The spinning rotors generate electricity as their magnets pass coils of copper wire. Electrical current is transmitted away through power lines, and the river's water flows out through the base of the dam.

approach to hydroelectric power. As reservoir water passes through a dam, it turns the blades of turbines, which cause a generator to generate electricity (Figure 20.19). Electricity generated in the powerhouse of a dam is transmitted to the electric grid by transmission lines, while the water is allowed to flow into the riverbed below the dam to continue downriver. The amount of power generated depends on the distance the water falls and the volume of water released. By storing water in reservoirs, dam operators can ensure a steady and predictable supply

FIGURE 20.20 Run-of-river systems divert a portion of a river's water and can be designed in various ways. Some designs involve piping water downhill through a powerhouse and releasing it downriver, and some involve using water as it flows over shallow dams.

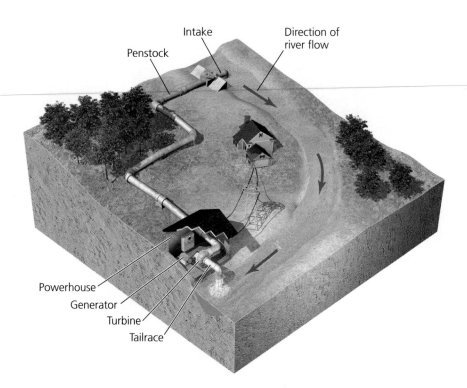

of electricity at all times, even during seasons of naturally low river flow.

An alternative approach is the **run-of-river** approach, in which electricity is generated without greatly disrupting the flow of river water. This approach sacrifices the reliability of water flow across seasons that the storage approach guarantees, but it minimizes many of the environmental impacts of large dams. The run-of-river approach can be followed using several methods, one of which is to divert a portion of a river's flow through a pipe or channel, passing it through a powerhouse and then returning it to the river (Figure 20.20). This can be done with or without a small reservoir that pools water temporarily, and the pipe or channel can be run along the surface or underground. Another method is to flow river water over a dam that is small enough not to impede fish passage, siphoning off water to turn turbines, and then returning the water to the river. Run-of-river systems are particularly useful in areas remote from established electrical grids and in regions without the economic resources to build and maintain large dams.

Hydroelectric power is widely used

Hydropower accounts for 2.2% of the world's primary energy supply and 16.2% of the world's electricity production. For nations with large amounts of river water and the economic resources to build dams, hydroelectric power has been a keystone of their development and wealth. Canada, Brazil, Norway, Austria, Switzerland, and many other nations today obtain large amounts of their energy from hydropower (Table 20.4). Sweden receives 14% of its total energy and nearly half its electricity from hydropower. In the wake of the nation's decision to phase out nuclear power, many people had hoped hydropower

Table 20.4 Top Consumers of Hydropower		
Nation	**Hydropower consumed[1]**	**Percentage of electricity generation from hydropower[2]**
Canada	76.4	57.5
China	74.2	14.9
Brazil	72.4	83.8
United States	59.8	7.5
Russia	40.0	17.2
Norway	24.7	98.9
Japan	22.6	9.9
India	19.0	11.9
Venezuela	16.0	66.0
France	14.8	11.4

Data from British Petroleum and International Energy Agency.
[1]In million metric tons of oil equivalent, 2004 data.
[2]2003 data.

New Renewable Energy Alternatives

21

Iceland's Blue Lagoon
and Svartsengi
geothermal power plant

Upon successfully completing this chapter, you will be able to:

▶ Outline the major sources of renewable energy and assess their potential for growth

▶ Describe solar energy and the ways it is harnessed, and evaluate its advantages and disadvantages

▶ Describe wind energy and the ways it is harnessed, and evaluate its advantages and disadvantages

▶ Describe geothermal energy and the ways it is harnessed, and evaluate its advantages and disadvantages

▶ Describe ocean energy sources and the ways they can be harnessed, and evaluate their advantages and disadvantages

▶ Explain hydrogen fuel cells and assess future options for energy storage and transportation

Iceland's hydrogen buses arrive in Reykjavik

Central Case: Iceland Moves toward a Hydrogen Economy

"I believe that water will one day be employed as fuel, that hydrogen and oxygen which constitute it, used singly or together, will furnish an inexhaustible source of heat and light.... Water will be the coal of the future."
—JULES VERNE, IN THE MYSTERIOUS ISLAND, 1874

"Our long-term vision is of a hydrogen economy."
—ROBERT PURCELL, JR., EXECUTIVE DIRECTOR, GENERAL MOTORS, 2000

The Viking explorers who first set foot centuries ago on the remote island of Iceland were trailblazers. Today the citizens of the nation of Iceland are blazing a bold new path, one they believe the rest of the world may follow. Iceland aims to become the first nation to leave fossil fuels behind and convert to an economy based completely on renewable energy.

Iceland is essentially a hunk of lava the size of Kentucky that has risen out of the North Atlantic from the rift between tectonic plates known as the

Mid-Atlantic Ridge. Most of Iceland's 290,000 people live in the capital city of Reykjavik, leaving much of the island an unpeopled and starkly beautiful landscape of volcanoes, hot springs, and glaciers.

The magma that gave birth to the island heats its groundwater in many places, giving Iceland some of the world's best sources of geothermal energy. Iceland has also dammed some of its many rivers for hydropower. Together these two renewable energy sources provide 73% of the country's energy supply and virtually all its electricity generation. Yet the nation, like most others, also depends on fossil fuels. Oil powers its automobiles, some of its factories, and its economically vital fishing fleet—and together these have given Iceland one of the highest per capita rates of greenhouse gas emission in the world. Because it possesses no fossil fuel reserves, all fossil fuels must be imported—a weak link in an otherwise robust economy that has given its citizens one of the highest per capita incomes in the world.

Enter Bragi Árnason, a University of Iceland professor popularly known as "Dr. Hydrogen." Árnason began arguing in the 1970s that Iceland could achieve independence from fossil fuel imports, boost its economy, and serve as a model to the world by converting from fossil fuels to a renewable energy economy based on hydrogen. He suggested zapping water with Iceland's cheap and renewable electricity in a chemical reaction called electrolysis, splitting the hydrogen from the oxygen and then using the hydrogen to power fuel cells that would produce and store electricity. The process is clean; nothing is combusted, and the only waste product is water.

In the late 1990s Árnason's countrymen began to listen. The nation's leaders decided to embark on a grand experiment to test the efficacy of switching to a "hydrogen economy." By setting an example for the rest of the world to follow, and by getting a head start at producing and exporting hydrogen fuel, these leaders hoped Iceland could become a "Kuwait of the North" and get rich by exporting hydrogen to an energy-hungry world, as Kuwait has done by exporting oil.

The leaders planning the shift to a hydrogen economy sketched a stepped transition in which fossil fuels would be phased out over 30–50 years. Conversion of the Reykjavik bus fleet to run on hydrogen fuel is the first step. After that, Iceland's 180,000 private cars would be powered by fuel cells, and then the fishing fleet would be converted to hydrogen. The final stage would be the export of hydrogen fuel to mainland Europe.

To make this happen, Icelanders in 1999 teamed up with corporate partners looking to develop technology for the future. Auto company Daimler-Chrysler is producing hydrogen buses, oil company Royal Dutch Shell is running hydrogen filling stations to fuel the buses, and Norsk Hydro is providing electrolysis technology.

In 2003, the world's first commercial hydrogen filling station opened in Reykjavik, and three hydrogen-fueled buses began operation. The public-private consortium, *Icelandic New Energy (INE)*, is monitoring the technology's effectiveness and the costs of developing infrastructure. Iceland's citizens are behind the effort; a recent poll showed 93% support among Icelanders.

Meanwhile, Daimler-Chrysler has introduced trios of hydrogen-fueled buses to nine other European cities. Hydrogen buses are also being developed by other companies and run in cities in Europe and throughout the world, from Tokyo to Chicago to Perth to Winnipeg. Hydrogen refueling stations are being demonstrated in Japan, Singapore, and the United States, and fuel-cell passenger automobiles are being tested in Japan and California. A global hydrogen economy could be closer than we suspect.

"New" Renewable Energy Sources

Iceland's bold drive toward a hydrogen economy is one facet of a global move toward renewable energy. Across the world, nations are taking different approaches to figure out how to move away from fossil fuels while ensuring a continued supply of energy for their economies.

In Chapter 20 we explored the two renewable energy sources that are most developed and widely used: biomass energy, the energy from combustion of wood and other plant matter, and hydropower, the energy from running water. These "conventional" alternatives to fossil fuels are renewable energy sources that can be depleted with overuse and that exert some undesirable environmental impacts.

In this chapter we explore a group of alternative energy sources that are often called "new renewables." These include energy from the sun, from wind, from Earth's geothermal heat, and from the movement of ocean water. These energy sources are not truly new. In fact, they are as old as our planet, and people have used them for millennia. They are commonly referred to as "new" because (1) they are not yet used on a wide scale in our modern industrial society, as are fossil fuels and the conventional renewable alternatives; (2) they are harnessed using technologies that are still in a rapid phase of development; and (3) it is widely believed that they will come to play a larger role in our energy use in the future.

The new renewables currently provide little of our power

The new renewable energy sources currently provide only 0.5% of our global primary energy supply. Fossil fuels provide 80% of the world's primary energy, nuclear energy provides 6.5%, and renewable energy sources account for 13.5%, nearly all of which is provided by biomass and hydropower (see Figure 20.1a, ▸ p. 591). The new renewables make a similarly small contribution to our global generation of electricity (see Figure 20.1b, ▸ p. 591). Less than 18% of our electricity comes from renewable energy, and of this amount, hydropower accounts for 90%.

Nations and regions vary in the renewable sources they use. In the United States, most energy supplied by renewable sources comes from hydropower and biomass, in nearly equal proportions. As of 2004, geothermal energy accounted for 5.6%, wind energy for 2.3%, and solar energy for 1.0% (Figure 21.1a). Of electricity generated in the United States from renewables, hydropower accounts

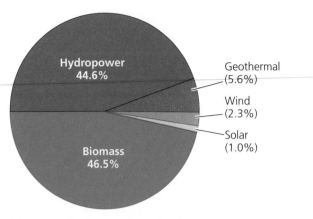

(a) U.S. total primary energy from renewable sources, 2004

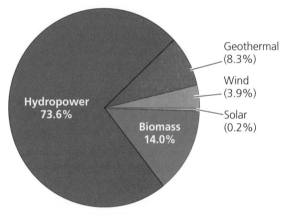

(b) U.S. total electricity generation from renewable sources, 2004

FIGURE 21.1 Only 6% of the total primary energy consumed in the United States each year comes from renewable sources. Of this amount, most derives from hydropower and biomass energy. Geothermal energy accounts for 5.6% of this amount, wind for 2.3%, and solar for 1.0% **(a).** Similarly, only 9% of electricity generated in the United States comes from renewable energy sources, predominantly hydropower and biomass. Geothermal energy accounts for 8.3%, wind for 3.9%, and solar for only 0.2% **(b).** Data from Energy Information Administration, U.S. Department of Energy. 2005. *Annual Energy Review, 2004.*

for nearly 75%, and biomass 14%. As of 2004, geothermal power accounted for 8.3%, wind for 3.9%, and solar for just 0.2% (Figure 21.1b).

The new renewables are growing fast

Although they comprise only a minuscule proportion of our energy budget, the new renewable energy sources are growing at much faster rates than are conventional energy sources. Over the past three decades, solar, wind, geothermal, and ocean energy sources have grown far faster than has the overall primary energy supply (Figure 21.2). Among

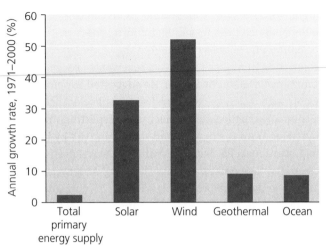

FIGURE 21.2 The "new renewable" energy sources are growing substantially faster than the total primary energy supply. Solar power has grown by 32% each year since 1971, and wind power has grown by 52% each year. Because these sources began from such low starting levels, however, their overall contribution to our energy supply is still small. Go to **GRAPHit!** at www.aw-bc.com/ withgott or on the student CD-ROM. Data from International Energy Agency Statistics, 2002.

renewable sources, the leader in growth is wind power, which has expanded by about 50% *each year* over the past three decades. Because these sources started from such low levels of use, however, it will take them some time to catch up to conventional sources. For instance, the absolute amount of energy added by a 50% increase in wind power is still less than the amount added by just a 1% increase in oil, coal, or natural gas.

The rapid growth of the new renewables has been motivated by concerns over diminishing fossil fuel supplies and the environmental impacts of fossil fuel combustion (Chapter 19). As these concerns build, advances in technology are making it easier and less expensive to harness renewable energy sources. The new renewables promise several benefits over fossil fuels. As they replace fossil fuels, they help alleviate air pollution and the greenhouse gas emissions that drive global climate change (Chapter 18). Unlike fossil fuels, renewable sources are inexhaustible on time scales relevant to human societies. Developing renewables can also help diversify an economy's mix of energy, lowering price volatility and protecting against supply restrictions such as those caused by the 1973 oil embargo or by Hurricane Katrina in 2005 (▶ p. 579). New energy sources also can create new employment opportunities and sources of income and property tax for local communities, especially in rural areas passed over by other economic development. In many rural areas and developing countries, locally based renewable sources may prove cheaper to use for electricity than would extending the electricity grid infrastructure out from cities.

Rapid growth in renewable energy sectors seems likely to continue as population and consumption grow, global energy demand expands, fossil fuel supplies decline, and citizens demand cleaner environments. More governments, utilities, corporations, and consumers are now promoting and using renewable energy sources, and, as a result, the costs of renewables are falling.

The transition cannot be immediate, but it must be soon

If our civilization is to persist in the long term, it will need to shift to renewable energy sources. A key question is whether we will be able to shift soon enough and smoothly enough to avoid widespread war, social unrest, and further damage to the environment. The answer to this question will largely determine the quality of life for all of us in the coming decades.

We cannot switch completely to renewable energy sources overnight, because there are technological and economic barriers. Currently, most renewables lack adequate technological development and lack infrastructure to transfer power on the required scale. However, dramatic improvements in technology and infrastructure in recent years suggest that most of the remaining barriers are political. Renewable energy sources have received far less in subsidies, tax breaks, and other incentives from governments than have conventional sources. By one estimate, of the $150 billion in subsidies the U.S. government provided to nuclear, solar, and wind power in the past half century, the nuclear industry received 96%, the solar industry received 3%, and the wind industry less than 1%. For decades, research and development of renewable sources have suffered from the continuing availability of fossil fuels made inexpensive in part by government policy.

Many corporations in the fossil fuel and automobile industries understand that they will eventually need to switch to renewable sources. They also know that when the time comes, they will need to act fast to stay ahead of their competitors. However, in light of continuing short-term profits and unclear policy signals, companies have not been eager to invest in the transition. Under these circumstances, our best hope may be for a gradual transition from fossil fuels to renewable energy sources, one driven largely by economic supply and demand. However, if the transition proceeds too slowly—if we wait solely for economics to do its work, without government encouragement—

diminishing fossil fuel supplies could outpace our ability to develop new sources, and we could find our economies greatly disrupted, and our environment greatly degraded. Thus, encouraging the speedy development of renewable energy alternatives holds promise for bringing us a vigorous and sustainable energy economy without the environmental impacts of fossil fuels.

Solar Energy

The sun provides energy for almost all biological activity on Earth (▸ pp. 105–106) by converting hydrogen to helium in nuclear fusion (▸ pp. 595–596). Each square meter of Earth's surface receives about 1 kilowatt of solar energy—17 times the energy of a lightbulb. As a result, an average-sized house whose roof is covered in panels that harness solar energy has enough roof area to generate all its power needs. The sun's raw energy is so strong that if only 0.1% of Earth's surface—roughly the combined area of New Mexico and South Dakota—were covered with solar panels, we would have enough solar energy to power all the world's electrical plants. The amount of energy Earth receives from the sun each day, if it could be collected in full for our use, would be enough to power human consumption for 27 years.

Clearly, the potential for using sunlight to meet our energy needs is tremendous. However, all this "free" energy from the sun cannot be harnessed just yet. We are still in the process of developing solar technologies and learning the most effective and cost-efficient ways to put the sun's energy to use.

The most commonly used way to harness solar energy is through **passive solar** energy collection. In this approach, buildings are designed and building materials are chosen to maximize direct absorption of sunlight in winter, even as they keep the interior cool in the heat of summer. This approach contrasts with **active solar** energy collection, which makes use of technological devices to focus, move, or store solar energy.

We tend to think of using solar power as a novel phenomenon, but people have chosen and designed their living sites with passive solar collection in mind for millennia. Moreover, solar energy was first harnessed with active solar technology in 1767, when Swiss scientist Horace de Saussure built a thermal solar collector to heat water and cook food. In 1891, U.S. inventor Clarence Kemp claimed the first commercial patent for a solar-powered water heater. Two California entrepreneurs bought the patent rights and outfitted one-third of the homes in Pasadena, California, with solar water heaters by 1897.

Passive solar heating is simple and effective

One passive solar design technique involves installing low, south-facing windows to maximize sunlight capture in the winter (in the Northern Hemisphere; north-facing windows are used in the Southern Hemisphere). Overhangs block light from above, shading these windows in the summer, when the sun is high in the sky and when cooling, not heating, is desired. Passive solar techniques also include the use of heat-absorbing construction materials. Often called *thermal mass,* these materials absorb heat, store it, and release it later. Thermal mass (of straw, brick, concrete, or other materials) most often makes up floors, roofs, and walls, but also can comprise portable blocks.

Thermal mass may be strategically located to capture sunlight in cold weather and radiate heat in the interior of the building. In warm weather, the mass should be located away from sunlight so that it absorbs warmed air in the interior to cool the building. Passive solar design can also involve planting vegetation in particular locations around a building. By heating buildings in cold weather and cooling them in warm weather, passive solar methods help conserve energy and reduce energy costs.

Active solar energy collection can heat air and water in buildings

One active method for harnessing solar energy involves using *solar panels* or *flat-plate solar collectors,* most often installed on rooftops. These panels generally consist of dark-colored, heat-absorbing metal plates mounted in flat boxes covered with glass panes. Water, air, or antifreeze solutions are run through tubes that pass through the collectors, transferring heat throughout a building. Heated water can be pumped to tanks to store the heat for later use and through pipes designed to release the heat into the building. Such systems have proven especially effective for heating water for residences.

Active solar energy is being used for heating, cooling, and water purification in Gaviotas, a remote town in the high plains of Colombia far from any electrical grid. Engineers have developed inexpensive solar panels that harvest enough solar energy, even under cloudy skies, to boil drinking water for a family of four. They also have designed, built, and installed a unique solar refrigerator in a rural hospital, along with a large-scale solar collector to boil and sterilize water. Their innovations show that solar power does not need to be expensive or confined to regions that are always sunny.

Concentrating solar rays magnifies the energy received

The strength of solar energy can be magnified by gathering sunlight from a wide area and focusing it on a single point. This is the principle behind *solar cookers,* simple portable ovens that use reflectors to focus sunlight onto food and cook it (see Figure 3.13, ▸ p. 71). Such cookers are becoming widespread and proving extremely useful in parts of the developing world.

The principle of concentrating the sun's rays has also been put to work by utilities in large-scale, high-tech approaches to generating electricity. In one approach, mirrors concentrate sunlight onto a receiver atop a tall "power-tower" (Figure 21.3). From the receiver, heat is

FIGURE 21.3 At the Solar Two facility operated by Southern California Edison in the desert of southern California, mirrors are spread across wide expanses of land to concentrate sunlight onto a receiver atop a "power-tower." Heat is then transported through fluid-filled pipes to a steam-driven generator that produces electricity.

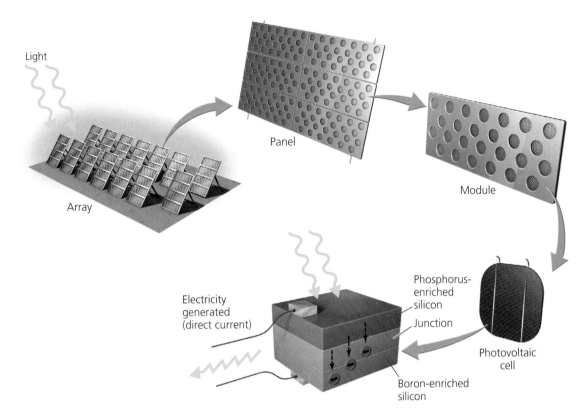

Light

Array

Panel

Module

Electricity
generated
(direct current)

Phosphorus-
enriched
silicon

Junction

Boron-enriched
silicon

Photovoltaic
cell

FIGURE 21.4 A photovoltaic (PV) cell converts sunlight to electrical energy. When sunlight hits a layer of silicon infused with phosphorus, electrons are released and travel toward the layer of silicon laced with boron. This movement of electrons induces an electric current, producing electricity. PV cells are grouped in modules, which can comprise panels, which can be erected in arrays.

transported by fluids (often molten salts) piped to a steam-driven generator to create electricity. These solar power plants can harness light from large mirrors spread across many hectares of land. The world's largest such plant so far—a collaboration among government, industry, and utility companies in the California desert—produces power for 10,000 households. Another approach is the use of solar-trough collection systems, which consist of mirrors that gather sunlight and focus it on oil in troughs. The superheated oil creates steam that drives turbines, as in conventional power plants.

Photovoltaic cells generate electricity directly from sunlight

A more direct approach to producing electricity from sunlight involves photovoltaic (PV) systems. **Photovoltaic (PV) cells** collect sunlight and convert it to electrical energy directly by making use of the *photovoltaic,* or *photoelectric effect,* first proposed in 1839 by French physicist Edmund Becquerel. This effect occurs when light strikes one of a pair of metal plates in a PV cell, causing the release of electrons, which are attracted by electrostatic forces to the op-posing plate. The flow of electrons from one plate to the other creates an electrical current, which can be converted into alternating current (AC) and used for residential and commercial electrical power (Figure 21.4).

The plates of a PV cell are made primarily of silicon, enriched on one side with phosphorus and on the other with boron. Silicon is a semiconductor, so it conducts and controls the flow of electricity. Because of the chemical properties of boron and phosphorus, the phosphorus-enriched side has excess electrons, and the boron-enriched side has fewer electrons. When sunlight strikes the cell surface, it transfers energy and causes electrons to move. When wires connect the two sides, electricity is created as electrons flow from the phosphorus-enriched side to the boron-enriched side. Photovoltaic cells can be connected to batteries that store the accumulated charge until it is needed.

You may be familiar with small PV cells that power your watch or your calculator. Atop the roofs of homes and other buildings, multiple PV cells are arranged in modules. These modules can comprise panels, which can be gathered together in flat arrays. Increasingly, PV roofing tiles are being used instead of these arrays. PV roofing tiles look like normal roofing shingles but generate electricity by the photovoltaic effect. In some remote areas, such as Xcalak, Mexico, PV systems are being used in combination with wind turbines (▶ pp. 627–628) and a diesel generator to power entire villages.

Solar power is little used but fast growing

Although active solar technology dates from the 18th century, it was pushed to the sidelines as fossil fuels gained a stronger foothold in our energy economy. Even as solar technology was being refined for applications ranging from handheld calculators to spacecraft, it was not being developed for the roles that fossil fuels have filled. In recent U.S. history, funding for research and development of solar technology has been erratic. After the 1973 oil embargo, the U.S. Department of Energy funded the installation and testing of over 3,000 PV systems, providing a boost to companies in the solar industry. As oil prices declined, however, so did government support for solar power, and funding for solar has remained far below that for fossil fuels.

Largely because of the lack of investment, solar power currently contributes only a minuscule portion of our energy production. In 2004, solar accounted for only 0.06%—less than 6 parts in 10,000—of the U.S. primary energy supply, and only 0.02% of U.S. electricity generation. However, use of solar energy has grown by nearly one-third annually worldwide since 1971, a growth rate second only to that of wind power. Solar power is proving especially attractive in developing countries, many of which are rich in sun but poor in power infrastructure, and where hundreds of millions of people are still without electricity. Some multinational corporations that built themselves on fossil fuels are now investing in alternative energy as well. BP Solar, British Petroleum's solar energy wing, recently completed $30 million projects in the Philippines and Indonesia, and is working on a $48 million project to supply electricity to 400,000 people in 150 villages.

Sales of PV cells are growing fast—by 25% per year in the United States, for instance, and by 63% annually in Japan, which uses PV roofing tiles widely. Use of solar technology is widely expected to continue increasing as prices fall, technologies improve, and economic incentives are enacted. However, the very small amount of energy currently produced by solar power means that its market share will likely remain small for years or decades to come—unless governments, businesses, and consumers become more motivated by the benefits that solar energy can provide.

Solar power offers many benefits

The fact that the sun will continue burning for another 4–5 billion years makes it practically inexhaustible as an energy source for human civilization. Moreover, the amount of solar energy reaching Earth's surface should be enough to power our civilization once solar technology is adequately developed. Although these overarching benefits of solar energy are clear, the technologies themselves also provide benefits. PV cells and other solar technologies use no fuel, are quiet and safe, contain no moving parts, require little maintenance, and do not even require a turbine or generator to create electricity. An average unit can produce energy for 20–30 years.

Another advantage of solar systems is that they allow for local, decentralized control over power. Homes, businesses, and isolated communities can use solar power to produce their own electricity and may not need to be near a power plant or connected to the grid of a city.

In developing nations, solar cookers enable families to cook food without gathering fuelwood; as a result, they lessen people's daily workload and help reduce deforestation. In locations such as refugee camps, solar cookers are helping relieve social and environmental stress. The low cost of solar cookers—many can be built locally for $2–10 each—has made them available for purchase or donation in many impoverished areas.

In the developed world, most PV systems today are connected to the regional electric grid. As a result, owners of houses with PV systems can sell their excess solar energy to their local power utility through a process called *two-way metering*. The value of the power the consumer sells to the utility is subtracted from the consumer's monthly utility bill.

Finally, a major advantage of solar power over fossil fuels is that it does not pollute the air with greenhouse gas emissions and other air pollutants. The manufacture of photovoltaic cells *does* currently require fossil fuel use, but once it is up and running, a PV system produces no emissions. Consumers may be able to access online calculators to calculate the economic and environmental impacts of installing PV solar cells. For example, a calculator offered by BP Solar estimated that installing a 5-kilowatt PV system in a home in Dallas, Texas, can provide 51% of annual power needs, save $391 per year on energy bills, and prevent the emission of 9,023 lb of carbon dioxide from fossil fuel combustion. Even in overcast Seattle, Washington, a 5-kilowatt system can produce 32% of energy needs, save $336 per year, and prevent the emission of 6,419 lb of carbon dioxide.

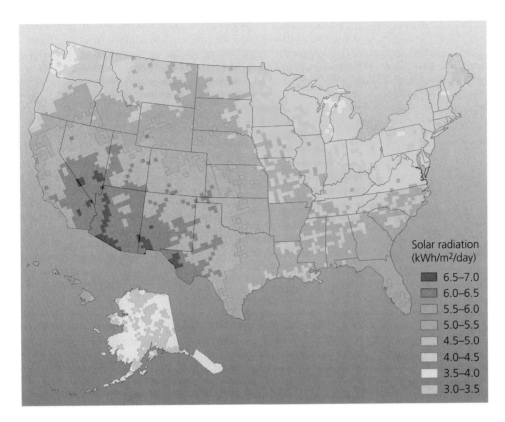

FIGURE 21.5 Because some locations receive more sunlight than others, harnessing solar energy is more profitable in some areas than in others. In the United States, many areas of Alaska and the Pacific Northwest receive only 3–4 kilowatt-hours per square meter per day, whereas most areas of the Southwest receive 6–7 kilowatt-hours per square meter per day. Data from National Renewable Energy Laboratory, U.S. Department of Energy, 2005.

Solar radiation
(kWh/m²/day)

- 6.5–7.0
- 6.0–6.5
- 5.5–6.0
- 5.0–5.5
- 4.5–5.0
- 4.0–4.5
- 3.5–4.0
- 3.0–3.5

Location and cost can be drawbacks for solar power

Solar power currently has two major disadvantages. One is that not all regions are sunny enough to provide adequate power, given current technology. Although Earth as a whole receives vast amounts of sunlight, not every location on Earth does (Figure 21.5). People in cities such as Seattle might find it difficult to harness enough sunlight most of the year to rely on solar power. Daily or seasonal variation in sunlight can also pose problems for stand-alone solar systems if storage capacity in batteries or fuel cells is not adequate or if backup power is not available from a municipal electricity grid.

The primary disadvantage of current solar technology—as with other renewable sources—is the up-front cost of investing in the equipment. The investment cost for solar is higher than that for fossil fuels, and indeed, solar power remains the most expensive way to produce electricity. Proponents of solar power argue that decades of government promotion of fossil fuels and nuclear power—which have received many financial breaks that solar power has not—have made solar power unable to compete. Because the external costs (▸ pp. 43–44) of nonrenewable energy have not been included in market prices, these energy sources have remained relatively cheap. Governments, businesses, and consumers thus have had little economic incentive to switch to solar and other renewables.

However, decreases in price and improvements in energy efficiency of solar technologies so far are encouraging, even in the absence of significant financial commitment from government and industry. At their advent in the 1950s, solar technologies had efficiencies of around 6%, while costing $600 per watt. Recent single-crystal silicon PV cells are showing 15% efficiency commercially and 24% efficiency in lab research, suggesting that future solar technologies may be more efficient than any energy technologies we have today. Solar systems have become much cheaper over the years and now can often pay for themselves in 10–15 years. After that time, they provide energy virtually for free as long as the equipment lasts. With future technological advances, some experts believe that the time to recoup investment could fall to 1–3 years.

Wind Energy

Wind energy can be thought of as an indirect form of solar energy, because it is the sun's differential heating of air masses on Earth that causes wind to blow. We can harness power from wind by using devices called **wind turbines,** mechanical assemblies that convert wind's kinetic energy, or energy of motion, into electrical energy.

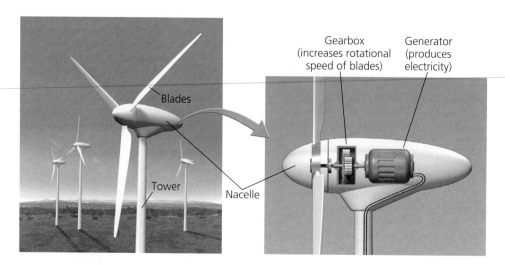

FIGURE 21.6 A wind turbine converts wind's energy of motion into electrical energy. Wind causes the blades of a wind turbine to spin, turning a shaft that extends into the nacelle that is perched atop the tower. Inside the nacelle, a gearbox converts the rotational speed of the blades, which can be up to 20 revolutions per minute (rpm) or more, into much higher rotational speeds (over 1,500 rpm). These high speeds provide adequate motion for a generator inside the nacelle to produce electricity.

Wind has long been used for energy

Today's wind turbines have their historical roots in Europe, where wooden windmills have been used for 800 years. The Netherlands in particular is known for its windmills, whose power has been used to pump water to drain wetlands and irrigate crops, and to grind grain into flour. In each application, wind causes a windmill's blades to turn, driving a shaft connected to several cogs that turn wheels, which either grind grain or pull buckets from a well. In the United States, countless ranches in arid areas of the West and Great Plains feature windmills that draw groundwater up to supply thirsty cattle.

The first wind turbine built to generate electricity was constructed in the late 1800s in Cleveland, Ohio, by inventor Charles Brush, who designed a turbine 17 m (50 ft) tall with 144 rotor blades made of cedar wood. But technology advanced slowly during the 20th century, and it was not until after the 1973 oil embargo that wind energy was funded by governments in North America and Europe. This moderate infusion of funding for research and development boosted technological progress, and the cost of wind power was cut in half in less than 10 years. Today wind power at favorable locations generates electricity for nearly as little cost per kilowatt-hour as do conventional sources, and modern wind turbines appear more like airplane propellers or sleek new helicopters than romantic old Dutch paintings.

Modern wind turbines convert kinetic energy to electrical energy

Wind blowing into a turbine turns the blades of the rotor, which rotate the machinery inside a compartment called a *nacelle,* which sits atop a tall tower (Figure 21.6). Inside the nacelle are a gearbox and a generator, as well as

equipment to monitor and control the turbine's activity. Most of today's towers range from 40 to 100 m (131–328 ft) tall. Higher is generally better, to minimize turbulence (and potential damage) and to maximize wind speed. Most rotors consist of three blades and measure 42–80 m (138–262 ft) across. Turbines are designed to yaw, or rotate back and forth in response to changes in wind direction, ensuring that the motor faces into the wind at all times. Turbines can be erected singly, but they are most often erected in groups called wind parks, or *wind farms.* The world's largest wind farms contain several hundred or thousand turbines spread across the landscape.

Engineers have designed turbines to begin turning at specific wind speeds to harness wind energy as efficiently as possible. Some turbines create low levels of electricity by turning in light breezes. Others are programmed to rotate only in strong winds, operating less frequently but generating large amounts of electricity in short time periods. Slight differences in wind speed can yield substantial differences in power output, for two reasons. First, the energy content of a given amount of wind increases as the square of its velocity; thus if wind velocity doubles, energy quadruples. Second, an increase in wind speed causes more air molecules to pass through the wind turbine per unit time, making power output equal to wind velocity cubed. Thus a doubled wind velocity actually results in an eightfold increase in power output.

Wind power is the fastest-growing energy sector

Like solar energy, wind provides only a minuscule proportion of the world's power needs, but wind power is growing fast—nearly 30% per year globally between 2000 and 2004. Wind provided 3.9% of U.S. renewable electricity

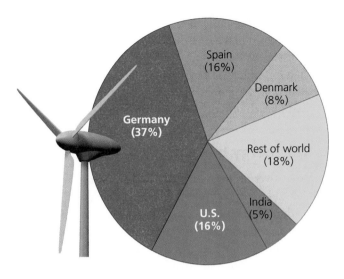

FIGURE 21.7 Most of the world's fast-growing wind power generating capacity is concentrated in a handful of countries. Tiny Denmark obtains the highest percentage of its energy needs from wind, but the larger nations of Germany, the United States, and Spain have so far developed more total wind capacity. Data from Global Wind Energy Council; and American Wind Energy Association. 2005. *Global wind energy market report.* AWEA.

generation in 2004—a small amount but nearly 20 times more than solar power. So far, wind energy production is geographically concentrated; only five nations account for 82% of the world's wind energy output (Figure 21.7). California and Texas account for two-thirds of the wind power generated within the United States. Denmark is a leader in wind power; there, a series of wind farms supplies over 20% of the nation's electricity needs. Ex-

perts agree that wind power's rapid growth will continue, because only a very small portion of this resource is currently being tapped. Meteorological evidence suggests that wind power could be expanded in the United States to meet the electrical needs of the entire country (see "The Science behind the Story," ▶ p. 632).

Offshore sites can be promising

Wind speeds on average are roughly 20% greater over water than over land. There is also less air turbulence over water surfaces than over land surfaces. For these reasons, offshore wind turbines are becoming popular Figure 21.8). Although costs to erect and maintain turbines in water are higher, the stronger, less turbulent winds produce more power and make offshore wind potentially more profitable. Currently, offshore wind farms are limited to shallow water, where towers are sunk into sediments singly or with a tripod configuration to stabilize them. However, in the future towers may be placed on floating pads anchored to the seafloor in deep water. At great distances from land, it may be best to store the generated electricity as hydrogen and then ship or pipe this to land (instead of building submarine cables to carry electricity to shore), but further research is needed on this option.

Denmark erected the first offshore wind farm in 1991. Over the next decade, nine more came into operation across northern Europe, where the North and Baltic Seas offer strong winds. The power output of these farms increased by 43% annually as larger turbines were erected. Several northern European nations are encouraging continued rapid

FIGURE 21.8 More and more wind farms are being developed offshore, because offshore winds tend to be stronger yet less turbulent. Denmark is a world leader in wind power, and much of it comes from offshore turbines. This Danish wind farm is one of several that provide over 20% of the nation's electricity.

growth in the near future. Wind advocates in Iceland are considering developing 240 offshore wind turbines in the nation's waters to meet future electricity demand for its hydrogen economy.

Wind power has many benefits

Like solar power, wind produces no emissions once the necessary equipment is manufactured and installed. As a replacement for fossil fuel combustion in the average U.S. utility generator, the U.S. Environmental Protection Agency has calculated that running a 1-megawatt wind turbine for one year prevents the release of more than 1,500 tons of carbon dioxide, 6.5 tons of sulfur dioxide, 3.2 tons of nitrogen oxides, and 60 lb of mercury. The amount of carbon pollution that all U.S. wind turbines together prevent from entering the atmosphere is greater than the cargo of a 50-car freight train, with each car holding 100 tons of solid carbon, each and every day.

Wind power appears considerably more energy-efficient than conventional power sources. One recent study, which compared the amount of energy that various types of technology produce to the amount they consume, found that wind turbines produce 23 times as much as they consume. For nuclear energy, the ratio was 16:1; for coal it was 11:1; and for natural gas it was 5:1. Wind farms also use less water than do conventional power plants.

Wind turbine technology can be used on many scales, from a single tower for local use to fields of thousands that supply large regions. Small-scale turbine development can help make local areas more self-sufficient, just as solar energy can. For instance, the Rosebud Sioux Tribe of Native Americans in 2003 set up a single turbine on their reservation in South Dakota. The turbine is producing electricity for 220 homes and brings the tribe an estimated $15,000 per year in revenue. Wind resources are rich in this region, and the tribe plans to develop a wind farm nearby in coming years.

Another societal benefit of wind power is that landowners can lease their land for wind development, which provides them extra revenue while also increasing property tax income for rural communities. A single large turbine can bring in $2,000–4,500 in annual royalties while occupying just a quarter-acre of land. Because each turbine takes up only a small area, most of the land can still be used for farming, ranching, or other uses.

Economically, wind energy involves up-front costs for the erection of turbines and the expansion of infrastructure to allow electricity distribution, but over the lifetime of a project it requires only maintenance costs. Unlike fossil fuel power plants, the turbines incur no ongoing fuel costs.

Currently, startup costs of wind farms generally are higher than those of fossil-fuel-driven plants, but wind farms incur fewer expenses once they are up and running. Advancing technology is driving down the costs of wind farm construction; as large wind farms become more efficient, the cost of each unit of electricity produced is dropping.

Wind energy has some downsides

Unlike power sources that can be turned off and on at will, wind is an intermittent resource; we have no control over when wind will occur. This poses little problem, however, if wind is only one of several sources contributing to a utility's power generation. Moreover, several technologies are available to address problems posed by relying on intermittent wind resources. For example, batteries or hydrogen fuel can store energy generated by wind and release it later when needed.

Just as wind varies from time to time, it also varies from place to place. Some areas are simply windier than others. Global wind patterns combine with local topography—mountains, hills, water bodies, forests, cities—to create local wind patterns, and companies study these patterns closely before investing in a wind farm. Meteorological research has given us information with which to judge prime areas for locating wind farms. A map of average wind speeds across the United States (Figure 21.9a) shows that mountainous regions and areas of the Great Plains are best. Based on such information, the young wind power industry has located much of its generating capacity in states with high wind speeds (Figure 21.9b), and is seeking to expand in the Great Plains and mountain states. Provided that wind farms are strategically erected in optimal locations, an estimated 15% of U.S. energy demand could be met using only 43,000 km^2 (16,600 mi^2) of land (with less than 5% of this land area actually occupied by turbines, equipment, and access roads).

Good wind resources, however, are not always near population centers that need the energy. Thus, transmission networks would need to be greatly expanded. Moreover, when wind farms *are* proposed near population centers, local residents often oppose them. Turbines are generally located in exposed, conspicuous sites, and many people object to wind farms for aesthetic reasons, feeling that the structures clutter the landscape. Although polls show wide public approval of wind projects in regions where wind power has already been introduced, new wind projects often elicit the so-called *not-in-my-backyard (NIMBY)* syndrome. For instance, a proposal for North America's first offshore wind farm, in Nantucket Sound between Cape Cod and the islands of Nantucket and Martha's Vineyard, has faced stiff opposition from wealthy

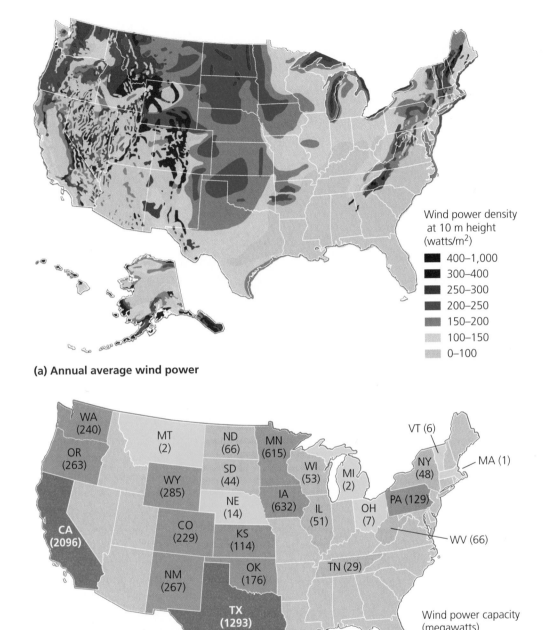

(a) Annual average wind power

Wind power density at 10 m height (watts/m²)
- 400–1,000
- 300–400
- 250–300
- 200–250
- 150–200
- 100–150
- 0–100

(b) Wind generating capacity, 2004

Wind power capacity (megawatts)
- 1,000–2,100
- 100–1,000
- 20–100
- 1–20
- 0

FIGURE 21.9 Wind's capacity to generate power varies according to wind speed. Meteorologists have measured wind speed to calculate the potential generating capacity from wind in different areas. The map in **(a)** shows average wind power in watts per square meter at a height of 10 m (33 ft) above ground across the United States. Such maps are used to help guide placement of wind farms. The development of U.S. wind power so far is summarized in **(b),** which shows the megawatts of generating capacity developed in each state through the end of 2004. *Sources:* (a) Elliott, D. L., et al. 1987. *Wind energy resource atlas of the United States.* Golden, CO: Solar Energy Research Institute; (b) National Renewable Energy Laboratory, U.S. Department of Energy, 2005.

The Science behind the Story

Idaho's Wind Prospectors

Where does wind translate into energy? In Idaho, where resource planners decided that the state's power future lies in generating electricity from wind. Now scores of Idahoans, from small farmers to Native American tribes, have joined in the search for gusts with energy potential.

By handing out wind-measuring devices to interested landowners, Idaho has turned its citizens into "wind prospectors" who pinpoint potential areas for wind farms. Idaho first launched the public wind prospecting program in 2001, after joining several other northwestern states in a regional research effort. People who join the program must collect data on wind speed and direction, share that data with the state, and agree to make it public.

A promising wind farm site requires some infrastructure, such as roads for erecting wind turbines and transmission lines for sending out power the turbines generate. But the single most important factor is the speed and frequency of the wind. Effective commercial wind farms have a steady flow of wind just above ground level, with regular gusts of at least 21 km/hr (13 mi/hr) at a height of about 50 m (164 ft).

Determining whether a site merits further study starts with analyzing existing data. In many parts of the developed world, decades worth of weather information have been compiled into computerized wind maps that indicate general wind conditions. In Idaho, energy planners provide prospectors with starter maps that divide the state into seven wind "classes" and reveal

To determine where to build wind farms, Idaho's wind prospectors use anemometers, which collect and relay wind data. Cup wheels rotate to indicate wind speed, and a vane turns on a vertical axis to reveal wind direction.

which general areas might have enough wind to make a wind farm worthwhile. Areas listed as "Class 3" or higher, with wind speeds of about 23 km (14.3 mi) per hour at 50 m (164 ft) above ground, offer the best possibilities.

Such maps, however, may not provide enough detail about a specific location. A piece of property, for example, may sit in a Class 3 area but be sheltered by a small hill that blocks the wind. Knowing that kind of detail requires site-specific on-the-ground research.

To make such research possible, Idaho loans landowners in areas listed as Class 3 or higher devices called *anemometers* (see the figure), which measure wind speed and direction. The Idaho program uses a common cup anemometer, with an

array of three or four hollow cups set to catch the wind and rotate around a vertical rod. The force of wind on the cups causes them to rotate at a speed proportional to the wind speed; the greater the wind, the faster the cups rotate. Wind direction is measured by a vane that turns on a vertical axis pointing directly into the wind. The cup wheel and wind vane are connected electrically to speed and direction dials, which relay wind data.

More than 80 landowners borrowed anemometers from the state in the Idaho program's first year, and sent data to the state every 60 days for review by energy planners and for subsequent posting online. The studies have generated new funding and wind farm plans in the state. In the fall of 2003, one farm near Idaho Falls won a $500,000 federal grant to help build a 1.5-megawatt wind farm that could supply power for approximately 500 homes.

In eastern Idaho, five anemometers set up on Shoshone-Bannock tribal lands have revealed good prospects for a commercial wind farm on two Native American reservations. The research effort has shown that the lands are "world-class sites" for wind power, according to a state energy official. With average wind speeds in the 29 km/hr (18 mi/hr) range, further study of the tribal lands revealed possible sites for large-scale commercial wind farms, which could mean jobs and revenue for the reservations. Similar wind prospecting programs are now under way on other reservations, as well as in other states, including Utah, Oregon, Virginia, and Missouri.

area residents, even though many of these residents like to think of themselves as environmentalists.

Wind turbines are also known to pose a threat to birds and bats, which can be killed when they fly into the rotating blades. At California's Altamont Pass wind farm, which is located in a region with one of the densest populations of golden eagles in the country, turbines killed many eagles and other raptors during the 1990s. Studies since then at other sites have suggested that bird deaths may be a less severe problem than was initially feared. It has been estimated that roughly one to two birds are killed per turbine per year—far fewer than the millions already being killed annually by television, radio, and cell phone towers, tall buildings, automobiles, domestic cats, pesticides, and other human causes. Bat mortality may be higher, but more research is needed. The key for protecting birds and bats seems to be selecting sites that are not on flyways or in the midst of prime habitat for species that are likely to fly into the blades.

Weighing the Issues:
Wind and NIMBY

If you could choose to get your electricity from a wind farm or a coal-fired power plant, which would you choose? How would you react if the electric utility proposed to build the wind farm that would generate your electricity atop a ridge running in back of your neighborhood, such that the turbines would be clearly visible from your living room window? Would you support or oppose the development? Why? If you would oppose it, where would you suggest the farm be located? Do you think anyone might oppose it in that location?

Geothermal Energy

Geothermal energy is one form of renewable energy that does not originate from the sun. Instead, it is generated from deep within Earth. The radioactive decay of elements amid the extremely high pressures deep in the interior of our planet generates heat that rises to the surface through magma (molten rock, ▶ p. 206) and through fissures and cracks. Where this energy heats groundwater, natural spurts of heated water and steam are sent up from below. Terrestrial geysers and submarine hydrothermal vents (▶ pp. 106–107) are the surface manifestations of these processes. Iceland is built from magma that extruded above the ocean's surface and cooled—magma from the Mid-Atlantic Ridge (▶ pp. 209, 471–472), the area

of volcanic activity along the spreading boundary of two tectonic plates. Because of the geothermal heat in this region, volcanoes and geysers are numerous in Iceland. In fact, the word *geyser* originated from the Icelandic *Geysir,* the name for the island's largest geyser, which recently resumed its periodic eruptions after many years in dormancy.

Geothermal power plants use the energy of naturally heated water to generate power. Rising underground water and steam are harnessed to turn turbines and create electricity. Geothermal energy is renewable in principle (its use does not affect the amount of heat produced in Earth's interior), but the power plants we build to use this energy may not all be capable of operating indefinitely. If a geothermal plant uses heated water at a rate faster than the rate at which groundwater is recharged, the plant will eventually run out of water. This is occurring at The Geysers, in Napa Valley, California, where the first generator was built in 1960. In response, operators have begun injecting municipal wastewater into the ground to replenish the supply. More and more geothermal power plants throughout the world are now injecting water, after it is used, back into aquifers to help maintain pressure and thereby sustain the resource. A second reason geothermal energy may not always be renewable is that patterns of geothermal activity in Earth's crust shift naturally over time, so an area that produces hot groundwater now may not always do so.

Geothermal energy is harnessed for heating and electricity

Geothermal energy can be harnessed directly from geysers at the surface, but most often wells must be drilled down hundreds or thousands of meters toward heated groundwater. Generally, water at temperatures of 150–370 °C (300–700 °F) or more is brought to the surface and converted to steam by lowering the pressure in specialized compartments. The steam is then employed in turning turbines to generate electricity (Figure 21.10).

Hot groundwater can also be used directly for heating homes, offices, and greenhouses; for driving industrial processes; and for drying crops. Iceland heats most of its homes through direct heating with piped hot water. Iceland began putting geothermal energy to use in the 1940s, and today 30 municipal district heating systems and 200 small private rural networks supply heat to 86% of the nation's residences. Other locales are benefiting in similar ways; the Oregon Institute of Technology heats its buildings with geothermal energy for 12–14% of the cost it would take to heat them with

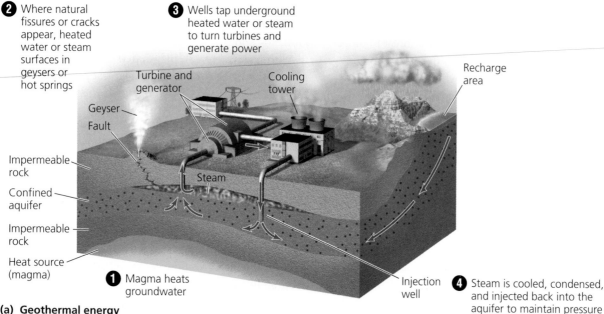

2 Where natural fissures or cracks appear, heated water or steam surfaces in geysers or hot springs

3 Wells tap underground heated water or steam to turn turbines and generate power

Turbine and generator

Cooling tower

Recharge area

Geyser

Fault

Impermeable rock

Confined aquifer

Impermeable rock

Heat source (magma)

Steam

1 Magma heats groundwater

Injection well

4 Steam is cooled, condensed, and injected back into the aquifer to maintain pressure

(a) Geothermal energy

(b) Nesjavellir geothermal power station, Iceland

FIGURE 21.10 With geothermal energy **(a)** magma heats groundwater deep in the earth (1), some of which is let off naturally through surface vents such as geysers (2). Geothermal facilities tap into heated water below ground and channel steam through turbines in buildings to generate electricity (3). After being used, the steam is often condensed and pumped back into the aquifer to maintain pressure (4). At Nesjavellir geothermal power station in Iceland **(b)**, steam is piped from four wells to a condenser at the plant where cold water pumped from lakeshore wells 6 km (3.7 mi) away is heated. The water, heated to 83 °C (181 °F), is sent through an insulated 270-km (170-mi) pipeline to Reykjavik and environs, where residents use it for washing and space heating.

natural gas. Such direct use of naturally heated water is cheap and efficient, but it is feasible only in areas such as Iceland or parts of Oregon, where geothermal energy sources are available and near where the heat must be transported.

Thermal energy from water or solid earth can also be used to drive a heat pump to provide energy. Geothermal *ground source heat pumps* (GSHPs) use thermal energy from near-surface sources of earth and water rather than the deep geothermal heat for which utilities drill. Roughly half a million GSHPs are already used to heat U.S. residences. Compared to conventional electric heating and cooling systems, GSHPs heat spaces 50–70%

more efficiently, cool them 20–40% more efficiently, can reduce electricity use by 25%–60%, and can reduce emissions by up to 72%. These pumps work because soil does not vary in temperature from season to season as much as air does. The pumps heat buildings in the winter by transferring heat from the ground into buildings; they cool buildings in the summer by transferring heat from buildings into the ground. Both types of heat transfer are accomplished by a single network of underground plastic pipes that circulate water. Because heat is simply moved from place to place rather than being produced using outside energy inputs, heat pumps can be highly energy-efficient.

Use of geothermal power is growing

Geothermal energy provides less than 0.5% of total primary energy used worldwide. It provides more power than solar and wind combined, but only a small fraction of the power from hydropower and biomass. Geothermal energy in the United States provides enough power to supply electricity to over 1.4 million homes. At the world's largest geothermal power plants, The Geysers in northern California, generating capacity has declined by more than 50% since 1989 as steam pressure has declined, but The Geysers still provide enough electricity to supply a million residents. Currently Japan, China, and the United States lead the world in use of geothermal power.

Geothermal power has benefits and limitations

Like other renewable sources, geothermal power greatly reduces emissions relative to fossil fuel combustion. Geothermal sources can release variable amounts of gases dissolved in their water, including carbon dioxide, methane, ammonia, and hydrogen sulfide. However, these gases are generally in very small quantities, and it has been estimated that geothermal facilities on average release only one-sixth of the carbon dioxide produced by plants fueled by natural gas. Geothermal facilities using the latest filtering technologies produce even fewer emissions. By one estimate, each megawatt of geothermal power prevents the emission of 7.8 million lb of carbon dioxide emissions and 1,900 lb of other pollutant emissions from gas-fired plants each year.

On the negative side of the ledger, geothermal sources, as we have seen, may not always be truly sustainable. In addition, the water of many hot springs is laced with salts and minerals that corrode equipment and pollute the air. These factors may shorten the lifetime of plants, increase maintenance costs, and add to pollution.

Moreover, use of geothermal energy is limited to areas where the energy can be tapped. Unless technology is developed to penetrate far more deeply into the ground, geothermal energy use will remain more localized than solar, wind, biomass, or hydropower. Places such as Iceland are rich in geothermal sources, but most of the world is not. In the United States, geysers exist in some areas, such as Yellowstone National Park, and hot groundwater and steam exist in various locations in the western part of the country. Nonetheless, many hydrothermal resources remain unexploited around the world, awaiting improved technology and governmental encouragement of their development.

Ocean Energy Sources

The oceans are home to several underexploited energy sources. Each involve continuous natural processes that could potentially provide sustainable energy for our needs. Of the three approaches developed so far, two involve motion and one involves temperature.

We can harness energy from tides and waves

Just as dams on rivers use flowing freshwater to generate hydroelectric power, some scientists, engineers, businesses, and governments are developing ways to use the motion of ocean water to generate electrical power. Two types of kinetic energy show the most promise so far: the energy of wave motion and the energy of tidal motion.

The rising and falling of ocean tides twice each day at coastal sites throughout the world can move large amounts of water past any given coastal point. Differences in height between low and high tides are especially great in long, narrow bays such as Alaska's Cook Inlet or the Bay of Fundy between New Brunswick and Nova Scotia. Such locations are best for harnessing tidal energy, which is accomplished by erecting dams across the outlets of tidal basins. The incoming tide flows through sluices past the dam, and as the outgoing tide passes through the dam, it turns turbines to generate electricity (Figure 21.11). Some designs allow for generating electricity from water moving in both directions. The world's largest tidal generating facility is the La Rance facility in France, which has operated for over 30 years. Smaller facilities now operate in China, Russia, and Canada. Tidal stations release few or no pollutant emissions, but they can have impacts on the ecology of estuaries and tidal basins.

Wave energy could be developed at a greater variety of sites than could tidal energy. The principle is to harness the motion of wind-driven waves at the ocean's surface and convert this mechanical energy into electricity. Many designs for machinery to harness wave energy have been invented, but few have been adequately tested. Some designs are for offshore facilities and involve floating devices that move up and down with the waves. Wave energy is greater at deep-ocean sites, but transmitting the electricity produced to shore would be expensive.

Other designs are for coastal onshore facilities. Some of these designs funnel waves from large areas into narrow channels and elevated reservoirs, from which water is then allowed to flow out, generating electricity as hydroelectric dams do. Other coastal designs use rising and falling

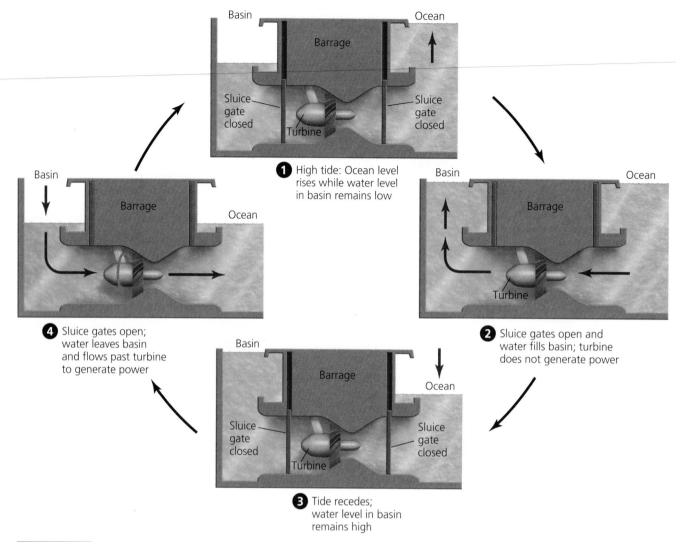

FIGURE 21.11 Energy can be extracted from the movement of the tides at coastal sites where tidal flux is great enough. One way of doing so is involves using bulb turbines in concert with the outgoing tide. At high tide, ocean water is let through the sluice gates, filling an interior basin. At low tide, the basin water is let out into the ocean, spinning turbines to generate electricity.

waves to push air into and out of chambers, turning turbines to generate electricity (Figure 21.12). No commercial wave energy facilities are operating yet, but some have been deployed as demonstration projects in several western European nations.

The ocean stores thermal energy

Besides the motion of tides and waves, other oceanic energy sources we have not yet effectively tapped include the motion of ocean currents, chemical gradients in salinity, and the immense thermal energy contained in the oceans. The concept of **ocean thermal energy conversion (OTEC)**

has been most fully developed. Each day the tropical oceans absorb an amount of solar radiation equivalent to the heat content of 250 billion barrels of oil—enough to provide 20,000 times the electricity used daily in the United States. The ocean's sun-warmed surface is higher in temperature than its deep water, and OTEC approaches are based on this gradient in temperature.

In the *closed cycle* approach, warm surface water is piped into a facility to evaporate chemicals, such as ammonia, that boil at low temperatures. These evaporated gases spin turbines to generate electricity. Cold water piped in from ocean depths then condenses the gases so they can be reused. In the *open cycle* approach, the warm

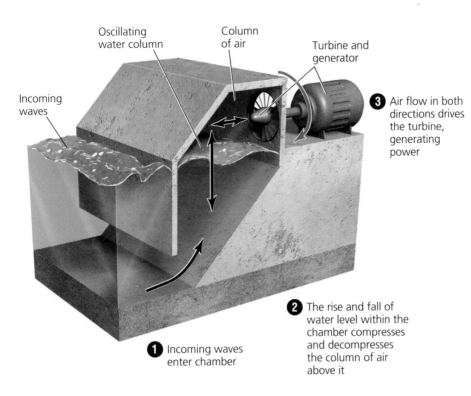

Oscillating water column

Column of air

Turbine and generator

Incoming waves

3 Air flow in both directions drives the turbine, generating power

2 The rise and fall of water level within the chamber compresses and decompresses the column of air above it

1 Incoming waves enter chamber

FIGURE 21.12 Coastal facilities can make use of energy from the motion of ocean waves. As waves are let into and out of a tightly sealed chamber, the air inside is compressed and decompressed, creating air flow that rotates turbines to generate electricity.

surface water is evaporated in a vacuum, and its steam turns the turbines and then is condensed by the cold water. Because the ocean water loses its salts as it evaporates, water can be recovered, condensed, and sold as desalinized freshwater for drinking or agriculture. Research on OTEC systems has been conducted in Hawaii and other locations, but costs remain high, and as of yet no facility is commercially operational.

Weighing the **Issues:**
Your Island's Energy?

Imagine you have been elected the president of an island nation the size of Iceland, and your nation's congress is calling on you to propose a national energy policy. Unlike Iceland, your country is located in equatorial waters. Your geologists do not yet know whether there are fossil fuel deposits or geothermal resources under your land, but your country gets a lot of sunlight and a fair amount of wind, and broad, shallow shelf regions surround its coasts. Your island's population is moderately wealthy but is growing fast, and importing fossil fuels from mainland nations is becoming increasingly expensive.

What approaches would you propose in your energy policy? What specific steps would you urge your congress to fund immediately? What trade relationships would you seek to establish with other countries? What questions would you ask of your economic advisors? What questions would you fund your country's scientists to research?

Hydrogen

All the renewable energy sources we have discussed can be used to generate electricity more cleanly than can fossil fuels. As useful as electricity is to us, however, it cannot be stored easily in large quantities for use when and where it is needed. This is why vehicles rely on fossil fuels for power. The development of fuel cells and hydrogen fuel show promise to store energy conveniently and in considerable quantities and to produce electricity at least as cleanly and efficiently as renewable energy sources.

In the "hydrogen economy" that Iceland's leaders and many energy experts worldwide envision, hydrogen fuel, together with electricity, will serve as the basis for a clean, safe, and efficient energy system. This system will use as a fuel the universe's simplest and most abundant element. In this system, electricity generated from renewable sources that are intermittent, such as wind or solar energy, can be used to produce hydrogen. Fuel cells can then employ hydrogen to produce electrical energy as needed to power vehicles, computers, cell phones, home heating, and countless other applications.

Fuel cell technology has been used since the 1960s in NASA's space flight programs. Basing an energy system on hydrogen could alleviate dependence on foreign fuels and help fight climate change. For these reasons, governments are funding research into hydrogen and fuel cell technology, and automobile companies are investing in research and development to produce vehicles that run on hydrogen.

The Science behind the Story

Algae as a Hydrogen Fuel Source

As scientists search for new ways to generate energy, some are looking past wind farms and solar panels to an unlikely power source—pond scum. Algae are being studied as an innovative way to generate large amounts of hydrogen to move society toward a more sustainable energy future.

Hydrogen's benefits hinge on how hydrogen fuel is produced. Some methods release substantial amounts of carbon dioxide, and other, nonpolluting, processes can be costly. These drawbacks have kept scientists searching for new hydrogen sources.

At the University of California at Berkeley, plant biologist Anastasios Melis thought one possible hydrogen source might be a single-celled aquatic plant known to be a capable, if sporadic, hydrogen producer. The alga *Chlamydomonas reinhardtii* was

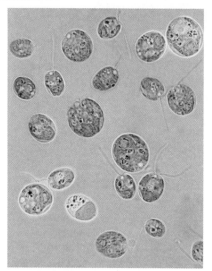

Could green algae such as this provide hydrogen for our energy needs?

known to emit small amounts of hydrogen for brief periods of time when deprived of light.

Melis hypothesized that the alga might be encouraged to produce hydrogen in large amounts. He set

up an experiment with energy experts at the National Renewable Energy Laboratory in Colorado, aiming to develop ways to tweak the alga's basic biological functions so that the plant produced greater quantities of hydrogen.

Green algae, like terrestrial green plants, photosynthesize, drawing in carbon dioxide and water, absorbing energy from light that converts those nutrients into food, and then expelling oxygen as a waste product. Additional nutrients from soil or water, and catalysts called *enzymes* within the plant, keep this process running smoothly. To conduct photosynthesis effectively, *Chlamydomonas reinhardtii* needs the element sulfur as a nutrient. The alga also contains an enzyme called *hydrogenase*, which can trigger the alga to stop producing oxygen as a metabolic by-product and start releasing hydrogen instead.

Hydrogen fuel may be produced from water or from other matter

Hydrogen gas (H_2) does not tend to exist freely on Earth; rather, hydrogen atoms bind to other molecules, becoming incorporated in everything from water to organic molecules. To obtain hydrogen gas for fuel, we must force these substances to release their hydrogen atoms, and this requires an input of energy. Several potential ways of producing hydrogen are being studied (see "The Science behind the Story," above). In **electrolysis,** the process being pursued by Iceland, electricity is input to split hydrogen atoms from the oxygen atoms of water molecules:

$$2H_2O \rightarrow 2H_2 + O_2$$

Electrolysis produces pure hydrogen, and it does so without emitting the carbon- or nitrogen-based pollutants of

fossil fuel combustion. However, whether this strategy for producing hydrogen will cause pollution over its entire life cycle depends on the source of the electricity used for the electrolysis. If coal is burned to create the electricity, then the entire process will not reduce emissions compared with reliance on fossil fuels. If, however, the electricity is produced by some less-polluting renewable source, then hydrogen production by electrolysis would create much less pollution and greenhouse warming than reliance on fossil fuels. The "cleanliness" of a future hydrogen economy in Iceland or anywhere else would, therefore, depend largely on the source of electricity used in electrolysis.

The environmental impact of hydrogen production will also depend on the source material for the hydrogen. Besides water, hydrogen can be obtained from biomass and fossil fuels. Obtaining hydrogen from these sources generally requires less energy input, but results in emissions

Hydrogenase is normally active only after *Chlamydomonas reinhardtii* has been deprived of light. When the alga is deprived of light, the light-dependent reactions of photosynthesis ebb, little oxygen is produced, and hydrogenase is activated. When light returns and the alga begins producing oxygen again, hydrogenase is promptly deactivated, and its associated hydrogen release stops.

Melis's team wanted to activate hydrogenase so that more hydrogen would be produced. But simply keeping the algae in the dark would not escalate hydrogen production because the alga's metabolic functions slowed without light, resulting in small amounts of released hydrogen.

The researchers decided to try limiting the alga's oxygen output another way, by putting it on a sulfur-free, bright-light regimen. The lack of sulfur would hinder photosynthesis, limiting oxygen output enough to activate hydrogenase and trigger hydrogen production. The presence of light would keep the algae metabolically active and releasing large amounts of by-products.

The researchers cultured large quantities of the algae in bottles in labs. Then they deprived the cultures of sulfur but kept the algae exposed to light for long periods of time—in some cases up to 150 hours. After the sustained light exposure, gas and liquids were extracted from the bottles and analyzed.

The analysis supported the team's hypothesis. Without sulfur or photosynthesis, the algae were not producing oxygen. This low-oxygen, or anaerobic, environment had induced hydrogenase, which spurred the algae to begin splitting water molecules and releasing gas. The plants had released amounts of hydrogen that were substantial relative to the size of the algal cultures. Hydrogen also dominated the alga's emissions—in gas collection analysis, approximately 87% of the gas was hydrogen, 1% was carbon dioxide, and the remaining 12% was nitrogen with traces of oxygen. The research teams published their findings in the journal *Plant Physiology* in 2000.

Many questions remain about algae-derived hydrogen, particularly how much fuel can be harvested continuously using this *photobiological* process. Nevertheless, the research results so far are helping to fuel the momentum of a future hydrogen economy.

Within 30 years, some federal energy experts predict that photobiological methods for generating hydrogen could be commonplace—meaning cars on future freeways might just be powered by pond scum.

---

of carbon-based pollutants. For instance, extracting hydrogen from the methane (CH_4) in natural gas entails producing one molecule of the greenhouse gas carbon dioxide for every four molecules of hydrogen gas:

$$CH_4 + 2H_2O \rightarrow 4H_2 + CO_2$$

Thus, whether a hydrogen-based energy system is environmentally cleaner than a fossil fuel system depends on how the hydrogen is extracted.

In addition, some new research suggests that leakage of hydrogen from the production, transport, and use of the gas at Earth's surface could potentially deplete stratospheric ozone and lengthen the atmospheric lifetime of the greenhouse gas methane. Research into these questions is ongoing, because scientists do not want society to switch from fossil fuels to hydrogen without first knowing the possible risks from hydrogen.

Weighing the Issues:
Precaution over Hydrogen?

Some environmental scientists have recently warned that we do not yet know enough about the environmental consequences of replacing fossil fuels with hydrogen fuel. An increase in tropospheric hydrogen gas would deplete hydroxyl (OH) radicals, they hypothesize, possibly leading to stratospheric ozone depletion and global warming from increased concentrations of methane. Some scientists say such effects will be small; others say there could be further effects that are presently unknown. Do you think we should apply the precautionary principle to the development of hydrogen fuel and fuel cells? Or should we embark on pursuing a hydrogen economy before knowing all the scientific answers? What factors inform your view?

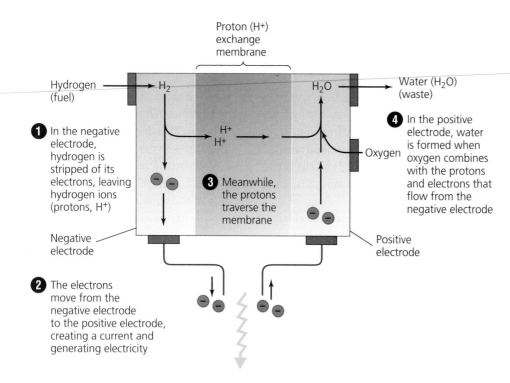

FIGURE 21.13 Hydrogen fuel drives electricity generation in a fuel cell, creating water as a waste product. Atoms of hydrogen are first stripped of their electrons (1). The electrons move from a negative electrode to a positive one, creating a current and generating electricity (2). Meanwhile, the hydrogen ions pass through a proton exchange membrane (3) and combine with oxygen to form water molecules (4).

Fuel cells produce electricity by joining hydrogen and oxygen

Once hydrogen gas has been isolated, it can be used as a fuel to produce electricity within fuel cells. The chemical reaction involved in a fuel cell is simply the reverse of that shown for electrolysis; an oxygen molecule and two hydrogen molecules each split so that their atoms can bind and form two water molecules:

$$2H_2 + O_2 \rightarrow 2H_2O$$

The way this occurs within one common type of fuel cell is shown in Figure 21.13. Hydrogen gas (usually compressed and stored in an attached fuel tank) is allowed into one side of the cell, whose middle consists of two electrodes that sandwich a membrane that only protons (hydrogen ions) can move across. One electrode, helped by a chemical catalyst, strips the hydrogen gas of its electrons, creating two hydrogen ions that begin moving across the membrane. Meanwhile, on the other side of the cell, oxygen molecules from the open air are split into their component atoms along the other electrode. These oxygen ions soon bind to pairs of hydrogen ions traveling across the membrane, forming molecules of water that are expelled as waste, along with heat. While this is occurring, the electrons from the hydrogen atoms have traveled to a device that completes an electric current between the two electrodes. The movement of the hydro-

gen's electrons from one electrode to the other creates the output of electricity.

Hydrogen and fuel cells have many benefits

As a fuel, hydrogen offers a number of benefits. We will never run out of hydrogen; it is the most abundant element in the universe. It can be clean and nontoxic to use, and—depending on the source of the hydrogen and the source of electricity for its extraction—it may produce few greenhouse gases and other pollutants. Pure water and heat may be the only waste products from a hydrogen fuel cell, along with negligible traces of other compounds. In terms of safety for transport and storage, hydrogen can catch fire, but if it is kept under pressure, it is probably no more dangerous than gasoline in tanks.

Hydrogen fuel cells are energy-efficient. Depending on the type of fuel cell, 35% to 70% of the energy released in the reaction can be used. If the system is designed to capture heat as well as electricity, then the energy efficiency of fuel cells can rise to 90%. These rates are comparable or superior to most nonrenewable alternatives.

Fuel cells are also silent and nonpolluting. Unlike batteries (which also produce electricity through chemical reactions), fuel cells will generate electricity whenever

Hydrogen and Renewable Energy

VIEWPOINTS

Is establishing a "hydrogen economy," as Iceland is trying to do, the best way to reduce the use of fossil fuels?

The Role of Renewable Energy for the Hydrogen Economy

Abundant, reliable, and affordable energy is an essential component of a healthy economy. Because hydrogen can be produced from a wide variety of domestically available resources and can be used in heat, power, and fuel applications, it is uniquely positioned to contribute to our growing energy demands, particularly for resource-constrained communities. However, if we are to realize the true benefits of a hydrogen economy, other renewables must play a substantial role in the efficient and affordable production of the hydrogen.

Several renewable options could make a substantial impact in the production of hydrogen: electrolysis powered by wind, photovoltaic, solar-thermal electric, hydropower, and geothermal energy; use of microorganisms and semiconductors to split water; and the thermal and biological conversion of biomass and wastes. Researchers around the globe are working on improving these renewable technologies. As a result, costs continue to drop. Technologies for renewable hydrogen production, coupled with advances in hydrogen production equipment (e.g., electrolyzers) can supply cost-competitive hydrogen and will ultimately play a substantial role in our energy supply.

In addition to the potential supply of affordable hydrogen, these technologies also offer a wide variety of opportunities for developing new centers of economic growth. Most investments in renewable energy are spent on materials and workmanship to build and maintain the facilities, rather than on costly energy imports. Therefore, funds are usually spent regionally and even locally, leading to new jobs and investments in local economies. Because of this synergistic relationship, the shift toward a hydrogen economy will naturally facilitate the advancement of renewable energy. By diversifying our energy supply, we will not only reduce our dependence on imported fuels, but also will benefit from cleaner technologies and investment in our communities.

Susan Hock directs the Electric and Hydrogen Technologies and Systems Center of the National Renewable Energy Laboratory. The center conducts research activities in four areas: distributed power systems integration, hydrogen technologies and systems, geographic information system analysis, and solar measurements and instrumentation.

Is Hydrogen the Answer?

We'll never use the last drop of oil, the last chunk of coal, the last cubic foot of natural gas, or the last pound of uranium. Eventually though, these fossil and nuclear fuels will become too expensive to extract, or politics will make one or more of them unavailable, leaving us to ask how we'll satisfy our voracious appetite in the future.

We should immediately apply all practical energy conservation strategies. Mother Nature is out there making more fossil fuels as we speak, but we don't have time to wait the few million years that will take. The short list of renewables: solar, wind, hydro, biomass, geothermal, waves, tides, and ocean thermal energy conversion. These are all relatively benign and abundant.

An alternative: hydrogen. It can either be burned or electrochemically used in fuel cells to provide useful energy. The by-product or "exhaust" is water. You start with water, get some energy, and end up with water, making it renewable. Another form of hydrogen energy is fusion, hydrogen atoms fusing to form helium plus a lot of energy, the way the sun does it. The catch? It takes about as much energy to extract hydrogen gas from water (by electrolysis) as you get back from your energy conversion device. Until it becomes cheaper (economically and in physical terms), fossil fuels will continue to rule the energy world. The breakthrough may involve using our renewable energy resources to separate hydrogen from other molecules.

Arguably, to reduce our dependence on fossil fuels, the priority list for this country should be:

1. Energy conservation
2. Wind
3. Passive solar
4. Biomass
5. Active solar
6. Hydrogen (chemical)
7. Hydroelectricity
8. Hydrogen (fusion)
9. Others (geothermal, tides, waves, ocean thermal)

Daryl Prigmore has studied energy and the environment since the late 1960s. After receiving bachelor and master of science degrees in mechanical engineering from Colorado State University, he spent 10 years in industry with a company developing solar, geothermal, and low-pollution automotive power systems. He has taught energy science classes for the past 23 years, 20 at the University of Colorado (Colorado Springs).

Explore this issue further by accessing **Viewpoints** at www.aw-bc.com/withgott.

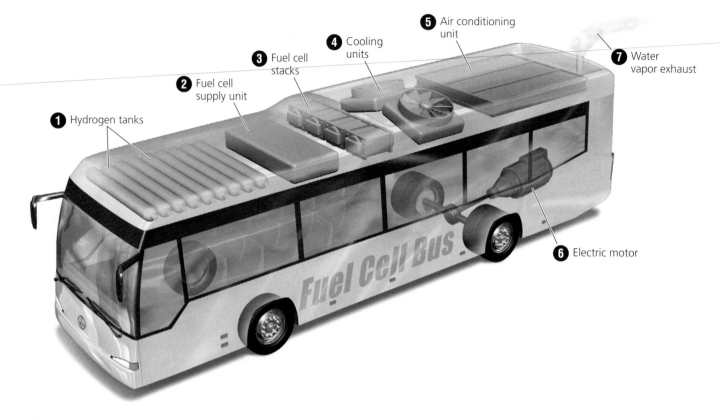

① Hydrogen tanks

② Fuel cell supply unit

③ Fuel cell stacks

④ Cooling units

⑤ Air conditioning unit

⑦ Water vapor exhaust

⑥ Electric motor

FIGURE 21.14 The hydrogen-fueled Citaro buses operating in Reykjavik and other European capitals are designed by Mercedes-Benz and Daimler-Chrysler. Hydrogen is stored in nine fuel tanks (1) The fuel cell supply unit (2) controls the flow of hydrogen, air, and cooling water into the fuel cell stacks (3). Cooling units (4) and the air conditioning unit (5) dissipate waste heat produced by the fuel cells. Electricity generated by the fuel cells is changed from direct current (DC) to alternating current (AC) by an inverter, and it is transmitted to the electric motor (6), which powers the operation of the bus. The vehicle's exhaust (7) consists simply of water vapor.

hydrogen fuel is supplied, without ever needing recharging. For all these reasons, hydrogen fuel cells are being used to power vehicles, including the buses now operating on the streets of Reykjavik and many other European, American, and Asian cities (Figure 21.14).

Conclusion

The coming decline of fossil fuel supplies and the increasing concern over air pollution and global climate change have convinced many people that we will need to shift to renewable energy sources that will not run out and will not pollute. Renewable sources with promise for sustaining our civilization far into the future without greatly degrading our environment include solar energy, wind energy, geothermal energy, and ocean energy sources.

Moreover, by using electricity from renewable sources to produce hydrogen fuel, we may be able to use fuel cells to produce electricity when and where it is needed, helping convert our transportation sector to a nonpolluting, renewable basis.

Most renewable energy sources have been held back for a variety of reasons, including little funding for research and development, and artificially cheap market prices for nonrenewable resources that do not include external costs. Despite this, renewable technologies have progressed far enough to offer hope that we can shift from fossil fuels to renewable energy with a minimum of economic and social disruption. Whether we can also limit environmental impact will depend on how soon and how quickly we make the transition and to what extent we put efficiency and conservation measures into place.

REVIEWING OBJECTIVES

You should now be able to:

Outline the major sources of renewable energy and assess their potential for growth

▶ The "new renewable" energy sources include solar, wind, geothermal, and ocean energy sources. They are not truly "new," but rather are in a stage of rapid development. (pp. 621–622)

▶ The new renewables currently provide far less energy and electricity than the conventional renewables, hydropower and biomass energy—and only a small fraction of the energy and electricity we obtain from fossil fuels. (pp. 621–622)

▶ Use of new renewables is growing quickly, and this growth is expected to continue as people seek to move away from fossil fuels. (pp. 622–623)

Describe solar energy and the ways it is harnessed, and evaluate its advantages and disadvantages

▶ Energy from the sun's radiation can be harnessed using passive methods or by active methods involving powered technology. (pp. 623–624)

▶ Major solar technologies include solar panels, mirrors to concentrate solar rays, and photovoltaic cells. (pp. 624–626)

▶ Solar energy is perpetually renewable, and solar technology creates no emissions and allows for decentralized power. (p. 626)

▶ Solar radiation varies in intensity from place to place and time to time, and harnessing solar energy remains expensive. (p. 627)

Describe wind energy and the ways it is harnessed, and evaluate its advantages and disadvantages

▶ Energy from wind is harnessed using wind turbines mounted on towers. (pp. 627–628)

▶ Turbines are often erected in arrays at wind farms located on land or offshore. Wind farms are developed in locations with optimal wind conditions. (pp. 628–632)

▶ Wind energy is renewable, turbine operation creates no emissions, wind farms can generate economic benefits, and the cost of wind power is nearly competitive with that of electricity generated from fossil fuels. (p. 630)

▶ Wind is an intermittent resource and occurs at adequate strengths only in some locations. Turbines kill some birds and bats, and wind farms can face opposition from local residents. (pp. 630–633)

Describe geothermal energy and the ways it is harnessed, and evaluate its advantages and disadvantages

▶ Energy from radioactive decay in Earth's core rises toward the surface and heats groundwater. Energy from this heated water and steam is harnessed at the surface or by drilling at geothermal power plants. (pp. 633–634)

▶ Use of geothermal energy for direct heating of water, as well as for electricity generation, can be efficient, clean, and renewable. (pp. 633–635)

▶ Geothermal sources occur only in certain areas and may become exhausted if too much water is pumped out without being replenished. (p. 635)

Describe ocean energy sources and the ways they can be harnessed, and evaluate their advantages and disadvantages

▶ Major ocean energy sources include the motion of tides and waves and the thermal heat of ocean water. (pp. 635–637)

▶ Tidal and wave energy is perpetually renewable and holds much promise, but so far technologies have seen only limited development. (pp. 635–637)

Explain hydrogen fuel cells and assess future options for energy storage and transportation

▶ Hydrogen can serve as a fuel to store and transport energy, so that electricity generated by renewable sources can be made portable and used to power vehicles. (p. 637)

▶ Hydrogen can be produced through electrolysis, but it may also be produced by using fossil fuels—in which case its environmental benefits are greatly reduced. (pp. 638–639)

▶ There is some concern that releasing excess hydrogen could have negative impacts on the atmosphere. (p. 639)

▶ Fuel cells create electricity by controlling an interaction between hydrogen and oxygen, and they produce only water as a waste product. (pp. 640, 642)

▶ Hydrogen can be clean, safe, and efficient. Fuel cells are silent, are nonpolluting, and do not need recharging. (pp. 640, 642)

TESTING YOUR COMPREHENSION

1. About how much of our energy now comes from renewable sources? What is the most prevalent form of renewable energy we use? What form of renewable energy is most used to generate electricity?

2. What is causing renewable energy sectors to expand? What renewable source is experiencing the most rapid growth?

3. Describe how passive solar heating works. How does active solar heating work? Give examples of each.

4. Describe the photoelectric effect. Describe a photovoltaic (PV) cell, and explain one way these are used.

5. What are the environmental and economic advantages and disadvantages of solar power?

6. How do modern wind turbines generate electricity? How does wind speed affect the process?

7. What are the environmental and economic benefits of wind power? What are its disadvantages?

8. Define geothermal energy and explain how it is obtained and used. In what ways is it renewable, and in what way is it not renewable?

9. List and describe three approaches to obtaining energy from ocean water.

10. How is hydrogen fuel produced? Is this a clean process? What factors determine the amount of pollutants hydrogen production will emit?

SEEKING SOLUTIONS

1. Why might a hydrogen economy be closer than we think? Why might it instead not come to pass? Do you think water could be "the coal of the future"? Why or why not?

2. For each source of renewable energy discussed in this chapter, what factors are standing in the way of an expedient transition from fossil fuel use?

3. Explain how the use of new renewable energy sources can reduce fossil fuel emissions.

4. Do you think development and implementation of renewable energy resources to replace fossil fuels can be moved forward without great social, economic, and environmental disruption? What steps would need to be taken? Will market forces alone suffice to bring about this transition? Do you think such a shift will be good for the economy?

5. Iceland is giving itself many years to phase in its planned hydrogen economy. Do you think the United

States could transition to a hydrogen economy more quickly, less quickly, or not at all? Why? What steps could the United States take to accelerate such a transition?

6. Imagine you are the CEO of a company that develops wind farms. Your staff is presenting you with three options, listed below, for sites for your next development. Describe at least one likely advantage and at least one likely disadvantage you would expect to encounter with each option. What further information would you like to know before deciding on which to pursue?
   ▶ Option A. A remote rural site in North Dakota
   ▶ Option B. A ridge-top site among the suburbs of Philadelphia
   ▶ Option C. An offshore site off the Florida coast

INTERPRETING GRAPHS AND DATA

Of the new renewable energy alternatives discussed in this chapter, photovoltaic conversion of solar energy is the one that most areas of the United States could most easily adopt. The influx of solar radiation varies with time of day, time of year, and location, so all areas are not equally well

suited. Today's photovoltaic technology is approximately 10% efficient at converting the energy of sunlight into electricity, but new technologies under development may increase that efficiency to as much as 40%.

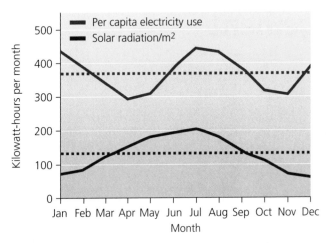

Average per capita residential use of electricity in the United States in 2004 (red line) and average influx of solar radiation per square meter for Topeka, Kansas (blue line). The dashed lines represent the yearly average values for each. Data from Renewable Resource Data Center, National Renewable Energy Laboratory, U.S. Department of Energy (DOE); and Energy Information Administration. 2005. *Annual energy review 2004.* DOE.

1. Given a 10% efficiency for photovoltaic conversion of solar energy, approximately how many square meters of photovoltaic cells would be needed to supply one person's residential electrical needs for a year, based on the yearly average values? How many square meters would be needed if efficiency were improved to 40%?

2. Given the same 10% conversion efficiency, approximately how many square meters of photovoltaic cells would be required to supply one person's residential electrical needs during the month of April? During July? How many square meters would be required to supply the average U.S. household of four people for each of those months?

3. Commercially available photovoltaic systems of this capacity cost approximately $20,000. The average cost of electricity in the United States is approximately 9¢ per kilowatt-hour. At these prices, how long would it take for the PV system to generate $20,000 worth of electricity? Calculate a combination of PV system cost and electricity cost at which the system would pay for itself in 10 years.

CALCULATING ECOLOGICAL FOOTPRINTS

Assume that average per capita residential consumption of electricity is 12 kilowatt-hours per day, that photovoltaic cells have an electrical output of 10% incident solar radiation, and that PV cells cost $800 per square meter. Now refer to Figure 21.5 on p. 627, and estimate the area and cost of the PV cells needed to provide all of the residential electricity used by each group in the table.

	Area of photovoltaic cells	Cost of photovoltaic cells
You	25	$20,000
Your class		
Your state		
United States		

1. What additional information do you need in order to increase the accuracy of your estimates for the areas in the table above?

2. Considering the distribution of solar radiation in the United States, where do you think it will be most feasible to greatly increase the percentage of electricity generated from photovoltaic solar cells?

3. The purchase price of a photovoltaic system is considerable. What other costs and benefits should you consider, in addition to the purchase price, when contemplating "going solar"?

Take It Further

Go to www.aw-bc.com/withgott or the student CD-ROM where you'll find:

▶ Suggested answers to end-of-chapter questions
▶ Quizzes, animations, and flashcards to help you study
▶ *Research Navigator*™ database of credible and reliable sources to assist you with your research projects

▶ **GRAPHit!** tutorials to help you master how to interpret graphs
▶ **INVESTIGATEit!** current news articles that link the topics that you study to case studies from your region to around the world

22 Waste Management

Containers en route to a recycling facility

Upon successfully completing this chapter, you will be able to:

▶ Summarize and compare the types of waste we generate

▶ List the major approaches to managing waste

▶ Delineate the scale of the waste dilemma

▶ Describe conventional waste disposal methods: landfills and incineration

▶ Evaluate approaches for reducing waste: source reduction, reuse, composting, and recycling

▶ Discuss industrial solid waste management and principles of industrial ecology

▶ Assess issues in managing hazardous waste

Fresh Kills Landfill, Staten Island, New York

Central Case: Transforming New York's Fresh Kills Landfill

"An extraterrestrial observer might conclude that conversion of raw materials to wastes is the real purpose of human economic activity."
—GARY GARDNER AND PAYAL SAMPAT, WORLDWATCH INSTITUTE

"Recycling is one of the best environmental success stories of the late 20th century."
—U.S. ENVIRONMENTAL PROTECTION AGENCY

The closure of a landfill is not the kind of event that normally draws politicians and the press, but the Fresh Kills Landfill was no ordinary dump. The largest landfill in the world, Fresh Kills was the primary repository of New York City's garbage for half a century. On March 22, 2001, New York City Mayor Rudolph Giuliani and New York Governor George Pataki were on hand to celebrate as a barge arrived on the western shore of New York City's Staten Island and dumped the final load of 650 tons of trash at Fresh Kills.

The landfill's closure was a welcome event for Staten Island's 450,000 residents, who had long viewed the landfill as a bad-smelling eyesore, health threat, and civic blemish. The 890-ha (2,200-acre) landfill featured six gigantic mounds of trash and soil. The highest, at 69 m (225 ft), was higher than the nearby Statue of Liberty.

New York City had grandiose plans for the site. It planned to transform the old landfill into a world-class public park—a verdant landscape of rolling hills and wetlands teeming with wildlife, and a mecca for recreation for New York's residents. The site certainly had potential. It was two-and-a-half times bigger than Manhattan's Central Park. It was the region's largest remaining complex of saltwater tidal marshes and freshwater creeks and wetlands, which still attracted birds and wildlife. And the mounds offered panoramic views of the Manhattan skyline and the rest of the region. The city sponsored an international competition to select a landscape architecture firm to design plans for the new park.

Meanwhile, with its only landfill closed, New York City began exporting its waste. The city began plans to develop an efficient network of stations to package and transfer the waste and ship it outward by barge and railroad. However, these plans soon fell apart amid neighborhood opposition, economic misjudgments, and accusations of political favoritism and mob influence. New York City instead found itself paying contractors exorbitant prices to haul its garbage away inefficiently, one truckload at a time. In the years following the Fresh Kills closure, trucks full of trash rumbled through neighborhood streets, carrying 12,000 tons of waste each day bound for 26 different landfills and incinerators in New York, New Jersey, Virginia, Pennsylvania, and Ohio. The city sanitation department's budget nearly doubled, and budget woes caused the city to scale back its recycling program. Some New Yorkers suggested reopening Fresh Kills.

The landfill *was* reopened, but not for a reason anyone could have foreseen. After the September 11, 2001, terrorist attacks, the 1.8 million tons of rubble from the collapsed World Trade Center towers, including unrecoverable human remains, was taken by barge to Fresh Kills, where it was sorted and buried. A monument will be erected at the site as part of the new park.

Today, plans for the park are forging ahead. Field Operations, the design firm that won the competition, completed a preliminary master plan in 2005, incorporating suggestions from the public. The plan involves everything from ecological restoration of the wetlands to construction of roads, ball fields, sculptures, and roller-blading rinks. People will be able to bicycle on trails paralleling tidal creeks of the region's largest estuary and reach stunning vistas atop the hills.

This undertaking—one of the largest public works projects in the world—will not be completed overnight. Designers, city officials, and Staten Island residents hope the first portions of the new park will open between 2008 and 2012. In the end, a longtime symbol of waste could be transformed into a world-class center for recreation and urban ecological restoration.

Approaches to Waste Management

As the world's human population rises, and as we produce and consume more material goods, we generate more waste. **Waste** refers to any unwanted material or substance that results from a human activity or process. For management purposes, waste is divided into several main categories. **Municipal solid waste** is nonliquid waste that comes from homes, institutions, and small businesses, whereas **industrial solid waste** includes waste from production of consumer goods, mining, agriculture, and petroleum extraction and refining. **Hazardous waste** refers to solid or liquid waste that is toxic, chemically reactive,

flammable, or corrosive. It can include everything from paint and household cleaners to medical waste to industrial solvents. Another type of waste is *wastewater,* water we use in our households, businesses, industries, or public facilities and drain or flush down our pipes, as well as the polluted runoff from our streets and storm drains. We discussed wastewater in Chapter 15 (▶ pp. 458–461).

We have several aims in managing waste

Waste can degrade water quality, soil quality, and air quality, thereby degrading human health and the environment. Waste is also a measure of inefficiency, so reducing waste can potentially save industry, municipalities, and consumers both money and resources. Waste is also unpleasant aesthetically. For these and other reasons, waste management has become an important pursuit for cities, industries, and individuals.

There are three main components of **waste management:** (1) minimizing the amount of waste we generate, (2) recovering waste materials and finding ways to recycle them, and (3) disposing of waste safely and effectively. Minimizing waste at its source—called *source reduction*—is the preferred approach. There are several ways to reduce the amount of waste that enters the **waste stream,** the flow of waste as it moves from its sources toward disposal destinations (Figure 22.1). Manufacturers can use materials more efficiently. Consumers can buy fewer goods, buy goods with less packaging, and use those goods longer. Reusing goods you already own, purchasing used items, and donating your used items for others also help reduce the amount of material entering the waste stream.

Recovery (*recycling* and *composting*) is widely viewed as the next best strategy in waste management. Recycling involves sending used goods to facilities that extract and reprocess raw materials to manufacture new goods. Newspapers, white paper, cardboard, glass, metal cans, appliances, and some plastic containers have all become increasingly recyclable as new technologies have been developed and as markets for recycled materials have grown. Organic waste can be recovered through *composting*, or biological decomposition. Recycling is not a concept that humans invented; recall that all materials are recycled in ecosystems (▶ pp. 190–191). Recycling is a fundamental feature of the way natural systems function.

Regardless of how effectively we reduce our waste stream, there will likely always be some waste left to dispose of. Disposal methods include burying waste in landfills and burning waste in incinerators. In this chapter we first examine how these approaches are used to manage municipal solid waste, and then we address approaches for managing industrial solid waste and hazardous waste.

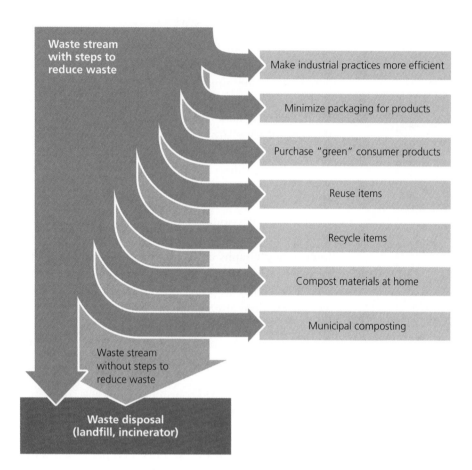

FIGURE 22.1 The most effective way to manage waste is to minimize the amount of material that enters the waste stream. To do this, manufacturers can increase efficiency and consumers can buy "green" products that have minimal packaging or are produced in ways that minimize waste. Individuals can compost food scraps and yard waste at home and can reuse items rather than buying new ones. When we are finished using products, many of us can recycle some materials and compost yard waste through municipal recycling and composting programs. For all remaining waste, waste managers attempt to find disposal methods that minimize impact to human health and environmental quality.

Municipal Solid Waste

Municipal solid waste is waste produced by consumers, public facilities, and small businesses. It is what we commonly refer to as "trash" or "garbage." Everything from paper to food scraps to roadside litter to old appliances and furniture is considered municipal solid waste.

Patterns in the municipal solid waste stream vary from place to place

In the United States, paper, yard debris, food scraps, and plastics are the principal components of municipal solid waste, together accounting for 70% of the waste stream (Figure 22.2). Even after recycling, paper is the largest component of U.S. municipal solid waste. Patterns differ in developing countries; there, food scraps are often the primary contributor to solid waste, and paper makes up a smaller proportion.

Most municipal solid waste comes from packaging and nondurable goods (products meant to be discarded after a short period of use). In addition, consumers throw away old durable goods and outdated equipment as they

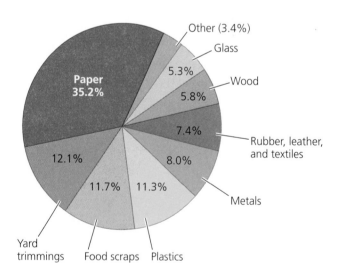

FIGURE 22.2 Paper products comprise the largest component of the municipal solid waste stream in the United States, followed by yard trimmings, food scraps, and plastics. In total, each U.S. citizen generates close to 1 ton of solid waste each year. Data for 2003, from U.S. Environmental Protection Agency, 2005. *Municipal solid waste generation, recycling, and disposal in the United States: Facts and figures for 2003.* EPA530-F-05-003. Washington, D.C.: EPA.

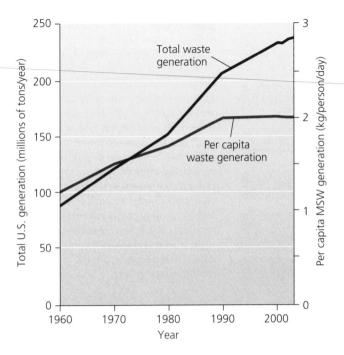

FIGURE 22.3   U.S. waste generation has increased by more than 2.7 times since 1960, and U.S. per-capita waste generation has risen by 66%. Per capita waste generation has leveled off in recent years largely because of recycling and source-reduction efforts. Data from U.S. Environmental Protection Agency, 2005. *Municipal solid waste generation, recycling, and disposal in the United States: Facts and figures for 2003*. EPA530-F-05-003. Washington, D.C.: EPA.

Following the United States in per capita solid waste generation are Canada, with 1.7 kg (3.75 lb) per day, and the Netherlands, with 1.4 kg (3.0 lb) per day. Of developed nations, Germany and Sweden produce the least waste per capita, generating just under 0.9 kg (2.0 lb) per day. Differences among nations result in part from differences in the cost of waste disposal; where disposal is expensive, people have incentive to waste less. The relative wastefulness of the U.S. lifestyle, with its excess packaging and reliance on nondurable goods, has caused critics to label the United States "the throwaway society."

People in developing nations, where consumption is lower, generate considerably less waste. One study (see Interpreting Graphs and Data, ▸ pp. 672–673) indicates that people of high-income nations waste more than twice as much as people of low-income nations. However, wealthier nations also tend to invest more in waste collection and disposal, so they are often better able to manage their waste proliferation and minimize impacts on human health and the environment.

Waste generation is rising in all nations

In the United States since 1960, waste generation has increased (Figure 22.3) by 2.7 times, and per capita waste generation has risen by 66%. Plastics, which came into wide consumer use only after 1970, have accounted for the greatest relative increase in the waste stream during the last several decades (Figure 22.4).

The rising consumption that has long characterized the United States and other wealthy nations is now proceeding rapidly in developing nations. To some extent,

purchase new products. As we acquire more goods, we generate more waste. In 2003, U.S. citizens produced 236 million tons of municipal solid waste, almost 1 ton per person. This means that the average American generates about 2.0 kg (4.4 lb) of trash per day.

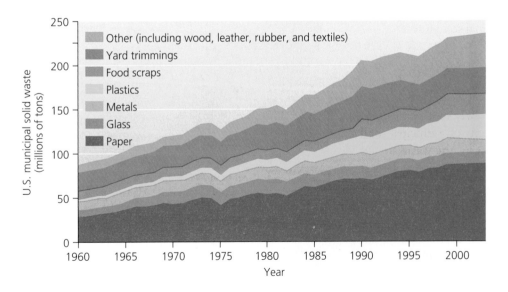

FIGURE 22.4   Amounts of all types of waste have grown in the United States over the past 4 decades, but plastic waste is the only type that has taken up a substantially greater share of the waste stream through time. Data from U.S. Environmental Protection Agency, 2005. *Municipal solid waste generation, recycling, and disposal in the United States: Facts and figures for 2003*. EPA530-F-05-003. Washington, D.C.: EPA.

FIGURE 22.5 Tens of thousands of people used to scavenge each day from the dump at Payatas, outside Manila in the Philippines, finding items for themselves and selling material to junk dealers for 100–200 pesos (U.S. $2–4) per day. That so many people could support themselves this way testifies to the immense amount of usable material needlessly discarded by wealthier portions of the population. The dump was closed in 2000 after an avalanche of trash killed hundreds of people.

this increase reflects rising material standards of living, but an increase in packaging is also to blame. Items made for temporary use, and poor-quality goods designed to be inexpensive, wear out and pile up quickly as trash, littering the landscapes of countries from Mexico to Kenya to Indonesia. Over the past three decades, per capita waste generation rates have more than doubled in Latin American nations and have increased more than fivefold in the Middle East. Like U.S. consumers in the "throwaway society," wealthy consumers in developing nations often discard items that can still be used. At many dumps and landfills in the developing world, in fact, poor people support themselves by selling items they scavenge (Figure 22.5).

In many developed nations, per capita generation rates have leveled off or decreased in recent years. For instance, note in Figure 22.3 that per capita waste production in the United States flattened out during the 1990s. This was due largely to the increased availability of recycling options. Recycling, composting, reduction, and reuse today are taking care of an increasingly larger portion of waste. We will examine these nondisposal approaches to waste management shortly, but let's first assess how we dispose of waste.

Open dumping of the past has given way to improved disposal methods

Historically, people dumped their garbage wherever it suited them. For example, until the mid-19th century, New York City's official method of garbage disposal was to dump it off piers into the East River. As population densities increased, municipalities took on the task of consolidating trash into open dumps at specified locations in order to keep other areas clean. To decrease the volume of trash, these dumps would be burned from time to time. Open dumping and burning still occur throughout much of the world.

As population and consumption rose in developed nations, waste increased and dumps grew larger. At the same time, expanding cities and suburbs forced more people into the vicinity of operating dumps and exposed them to the noxious smoke of dump burning. Reacting to opposition from residents living near dumps, and to the rising awareness of the health and environmental threats posed by unregulated open dumping and burning, many nations improved their methods of waste disposal. Most industrialized nations now bury their waste in lined and covered landfills, and burn their waste in incineration facilities.

In the 1980s in the United States, waste generation increased while incineration was restricted, and recycling was neither economically feasible nor widely popular. As a result, landfill space became limited, and there was much talk of a solid waste "crisis." New York and other East Coast urban areas felt this situation most acutely; Fresh Kills Landfill accepted its all-time annual peak of trash, 29,000 tons, in 1986–1987. Since the late 1980s, however, recovery of materials for recycling has expanded, decreasing the pressure on landfills (Figure 22.6). As of 2003, U.S. waste managers were landfilling 55.4% of municipal solid waste, incinerating 14.0%, and recovering 30.6% for composting and recycling.

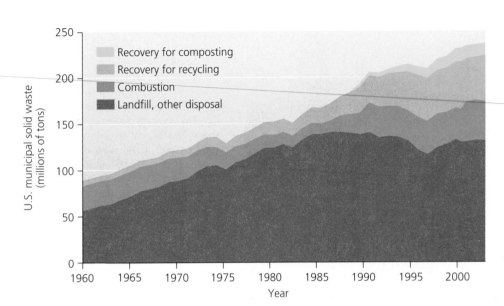

FIGURE 22.6 Since the 1980s, recycling and composting have grown in the United States, allowing a smaller proportion of waste to go to landfills. As of 2003, 55.4% of U.S. municipal solid waste was going to landfills and 14.0% to incinerators, while 30.6% was being recovered for composting and recycling. Go to **GRAPHit!** at www.aw-bc.com/withgott or on the student CD-ROM. Data from U.S. Environmental Protection Agency, 2005. *Municipal solid waste generation, recycling, and disposal in the United States: Facts and figures for 2003.* EPA530-F-05-003. Washington, D.C.: EPA.

Sanitary landfills are regulated by health and environmental guidelines

In modern **sanitary landfills,** waste is buried in the ground or piled up in large, carefully engineered mounds. In contrast to open dumps, sanitary landfills are designed to prevent waste from contaminating the environment and threatening public health (Figure 22.7). Although most municipal landfills in the United States are regulated locally or by the states, they must meet national standards set by the U.S. Environmental Protection Agency (EPA).

Guidelines set forth by the federal **Resource Conservation and Recovery Act (RCRA),** which was enacted in 1976 and amended in 1984, specify how waste should be added to a landfill. In a sanitary landfill, waste is partially decomposed by bacteria and compresses under its own weight to take up less space. Waste is layered along with

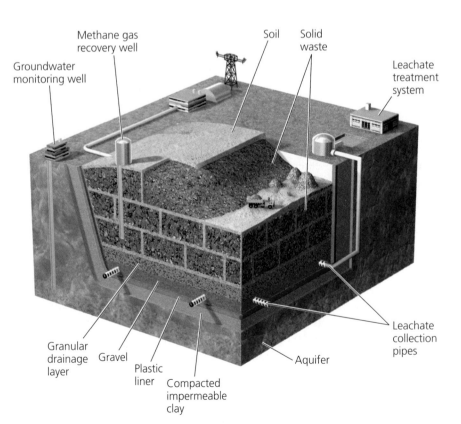

FIGURE 22.7 Sanitary landfills are engineered to prevent waste from contaminating soil and groundwater. Waste is laid in a large depression lined with plastic and impervious clay designed to prevent liquids from leaching out. Pipes of a leachate collection system draw out these liquids from the bottom of the landfill. Waste is layered along with soil until the depression is filled, and continues to be built up until the landfill is capped. Landfill gas produced by anaerobic bacteria may be recovered, and waste managers monitor groundwater for contamination.

soil, which speeds decomposition, reduces odor, and lessens infestation by pests. Limited infiltration of rainwater allows for biodegradation by aerobic and anaerobic bacteria. After a landfill is closed, it is capped with an engineered cover that must be maintained. This cap consists of a hydraulic barrier of plastic that prevents water from seeping down and gas from seeping up; a gravel layer above the hydraulic barrier, which drains water, lessening pressure on the hydraulic barrier; a soil barrier of at least 60 cm (24 in.), which stores water and protects the hydraulic layer from weather extremes; and a topsoil layer of at least 15 cm (6 in.), which encourages plant growth. Vegetation growing in the topsoil helps prevent erosion and returns some soil moisture to the atmosphere by transpiration (▸ p. 204).

Landfill engineers protect against environmental contamination in several ways. U.S. regulations require that landfills be located away from wetlands and earthquake-prone faults, and be at least 6 m (20 ft) above the water table. Regulations require that the bottom and sides of all sanitary landfills be lined with heavy-duty plastic and 60 to 120 cm (2 to 4 ft) of impermeable clay to help prevent contaminants from seeping into aquifers. Regulations also require that area groundwater be monitored regularly for contamination. Because landfills produce **leachate**— liquid that results when substances from the trash dissolve in water as rainwater percolates downward—sanitary landfills have systems to collect and treat it. Typically these systems consist of pipes running from the bottom of the landfill to collection ponds and treatment facilities. Landfill managers are required to maintain leachate collection systems for 30 years after a landfill has closed.

Although it was considered a model for advanced landfill technology at the time of its construction, the Fresh Kills Landfill predated most of the EPA guidelines. As a result, it caused some environmental contamination. However, engineers have retrofitted the landfill with clay liners and a sophisticated leachate collection system. Three of the six mounds have been capped with a "final cover," and the remaining mounds will soon be capped. Indeed, the city's investment of several hundred million dollars in an existing landfill to bring environmental protection measures up to modern standards was unprecedented. Because these safeguards need to be maintained and monitored for 30 years after closure, designs for a public park at Fresh Kills have had to work around these constraints.

Landfills can be transformed after closure

Today thousands of landfills lie abandoned. One reason is that waste managers have closed many smaller landfills

FIGURE 22.8 Old landfills, once properly capped, can serve other purposes. A number of them, such as Cesar Chavez Park in Berkeley, California, shown here, have been developed into areas for human recreation.

and consolidated the trash stream into fewer, much larger, landfills. In 1988 the United States had nearly 8,000 landfills, but today it has fewer than 1,800.

A growing number of cities have been converting closed landfills into public parks (Figure 22.8). The Fresh Kills redevelopment endeavor will be the world's largest landfill conversion project, but it is hardly the first. Such efforts date back at least to 1938, when the site of an ash landfill at Flushing Meadows, in Queens, was redeveloped for the 1939 New York World's Fair. The site subsequently hosted the United Nations and the 1964–1965 World's Fair. Designated a park in 1967, today the site hosts Shea Stadium, the Queens Museum of Art, the New York Hall of Science, and the Queens Botanical Garden, as well as playgrounds, wetlands, and festival events.

Landfills have drawbacks

Despite improvements in liner technology and landfill siting, many experts believe that leachate will eventually escape from even well-lined landfills. Liners can be punctured, and leachate collection systems eventually cease to be maintained. Moreover, landfills are kept dry to reduce leachate, but the bacteria that break down material thrive in wet conditions. Dryness, therefore, slows waste decomposition. In fact, it is surprising how slowly some materials biodegrade when they are tightly compressed in a landfill. Innovative archaeological research has revealed that landfills often contain food that has not decomposed and 40-year-old newspapers that are still legible (see "The Science behind the Story," ▸ pp. 654–655).

Digging Garbage: The Archaeology of Solid Waste

The Science behind the Story

Garbage and *knowledge* are two words rarely put together. But when scientist William Rathje dons trash-flecked clothes and burrows into a city dump, he gleans valuable information about how modern Americans live.

By pulling tons of trash out of disposal sites over the course of decades, Rathje has turned dumpster-diving into a noteworthy field of scientific inquiry that he calls *garbology.* An archaeologist by training who has been called "the Indiana Jones of solid waste," Rathje has brought exacting archaeological techniques to the contents of trash cans.

As a professor at the University of Arizona in the early 1970s, Rathje wanted his students to learn a technique common among archaeologists—sorting through ancient trash mounds to understand the lives of past cultures. With few ancient civilizations or their trash close at hand, however, he arranged for his students to dig through their neighbors' garbage. In 1973, he gave that effort a name, "The Garbage Pro-

"Garbologist" William Rathje has pioneered the study of our culture through the waste we generate.

ject," and began a methodical study of the contents of modern trash.

Rathje asked communities to divert some of their garbage trucks and bring trash to his study teams. With rakes and notebooks, the researchers sorted, weighed, itemized, and analyzed the refuse. They then visited the homes of the people who had generated the trash and asked

residents about their shopping and consumption habits.

Then, in 1987, amid growing debates about how quickly U.S. landfills were filling up, Rathje decided to see what was taking up space in them. The Garbage Project headed to landfills with a truck-mounted bucket auger—a large drill commonly employed by

Another problem is finding suitable areas to locate landfills, because most communities do not want them in their midst. This not-in-my-backyard (NIMBY) reaction (▶ pp. 630, 633) is one reason why New York decided to export its waste and why residents of states receiving that waste are increasingly protesting. As a result of the NIMBY syndrome, landfills are rarely sited in neighborhoods that are home to wealthy and educated people with the political power to keep them out. Instead, they are disproportionately sited in poor and minority communities, as environmental justice advocates have frequently pointed out.

One famous case of long-distance waste transport illustrates the unwillingness of most communities to accept garbage. In Islip, New York, in 1987, the town's landfills were full, prompting town administrators to ship waste by barge to a methane production plant in North Carolina. Prior to the barge's arrival, however, it became known that the shipment was contaminated with 16 bags of medical waste, including syringes, hospital gowns, and diapers. Because of the medical waste, the methane plant rejected the entire load. The barge sat in a North Carolina harbor for 11 days before heading for Louisiana. Louisiana, however, would not permit the barge to dock. The barge

geologists and construction crews to handle everything from excavating soil samples to creating new water wells. Rathje and his researchers dug into landfills around North America, boring as far as 30 m (100 ft) down, often drilling approximately 15 to 20 garbage "wells" at each site, with each well yielding up to 25 tons of trash.

Once excavated, landfill contents were sorted, weighed, and identified. Rathje's teams sometimes froze the trash before they worked with it to make the garbage easier to separate and to limit odor and flies. Smaller bits of trash were put through sieves and sometimes washed with water to make them easier to label. Rathje has excavated at least 21 dumps, including sites in Tucson, San Francisco, Chicago, Phoenix, New York City, and Philadelphia, uncovering a host of interesting facts in the process:

▶ *Not much rot.* Trash doesn't decay much in closed landfills, Rathje has found. In the low-oxygen conditions inside most closed dumps, trash turns into a sort of time capsule. Rathje's teams have found whole hot dogs in most digs, intact pastries that are decades old, and grass clippings that are often still green. Decades-old newspapers are legible and can be used to date layers of trash.

▶ *Paper rules.* Paper-based products make up more than 40% of most landfill content, and construction debris makes up about 20%. Newspapers are often a high-volume item, averaging about 14% of landfill space.

▶ *Plastic packaging no problem.* Rathje says plastic packaging is not the landfill problem many believe it to be. Plastic packaging makes up only about 4.5% of landfill content, and that figure has not increased substantially since the 1970s, Rathje reported in 1997. Fast-food packaging, polystyrene foam, and disposable diapers also aren't a major problem, making up only about 3% of landfill content. If all plastic packaging were to be replaced by containers made of glass, paper, steel, or similar materials, Rathje maintains, the packaging discarded by U.S. households would more than double.

▶ *Poison in small bottles.* Toxic waste comes in all sizes. If nail polish were sold in 55-gallon drums, its chemical composition would make it illegal to throw out in a regular dump. Nail polish, however, is tossed in small bottles. In 1991, after studying Tucson dumps, Rathje calculated that about 350,000 bottles were getting thrown away each year. Luckily, however, the potentially toxic ingredients in nail polish don't always spread far, he found. Paper, diapers, and other nontoxic garbage often absorb toxic materials in landfills and keep the poisons from leaching out.

Through garbology, Rathje has gleaned unique insights into how we can change our often-wasteful habits. Now a consulting professor at Stanford University, Rathje has emerged as a leading expert on how to reduce waste.

traveled toward Mexico, but the Mexican navy prevented it from entering that nation's waters. In the end, the barge traveled 9,700 km (6,000 mi) before eventually returning to New York, where, after several court battles, the waste was finally incinerated at a facility in Queens.

Incinerating trash reduces pressure on landfills

Just as sanitary landfills are an improvement over open dumping, incineration in specially constructed facilities can be an improvement over open-air burning of trash.

Incineration, or combustion, is a controlled process in which mixed garbage is burned at very high temperatures (Figure 22.9). At incineration facilities, waste is generally sorted and metals removed. Metal-free waste is chopped into small pieces to aid combustion and then is burned in a furnace. Incinerating waste reduces its weight by up to 75% and its volume by up to 90%.

However, simply reducing the volume and weight does not rid trash of components that are toxic. The ash remaining after trash is incinerated therefore must be disposed of in hazardous waste landfills (▶ p. 667). Moreover, combustion can create new chemical compounds that can

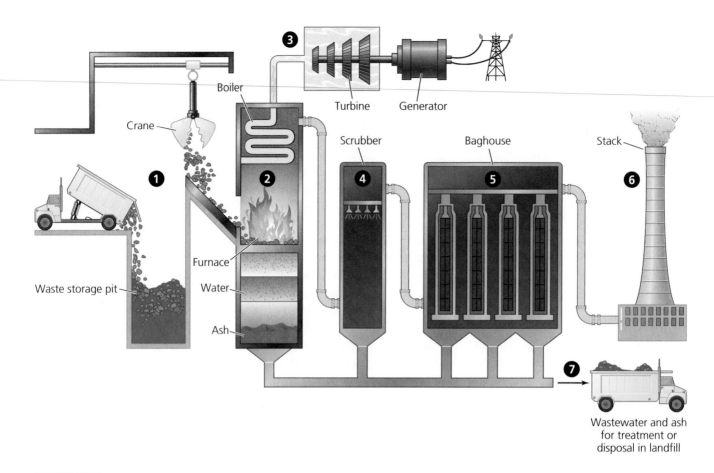

FIGURE 22.9 Incinerators reduce the volume of solid waste by burning it but may emit toxic compounds into the air. Many incinerators are waste-to-energy (WTE) facilities that use the heat of combustion to generate electricity. In a WTE facility, solid waste (1) is burned at extremely high temperatures (2), heating water, which turns to steam. The steam turns a turbine (3), which powers a generator to create electricity. In an incinerator outfitted with pollution control technology, toxic gases produced by combustion are mitigated chemically by a scrubber (4), and airborne particulate matter is filtered physically in a baghouse (5) before air is emitted from the stack (6). Ash remaining from the combustion process is disposed of (7) in a landfill.

be health hazards. When trash is burned, hazardous chemicals—including dioxins, heavy metals, and PCBs (Chapter 14)—can be released into the atmosphere. Such releases caused a backlash against incineration from citizens concerned about health hazards. Most developed nations now regulate incinerator emissions, and some have banned incineration outright.

As a result of real and perceived health threats from incinerator emissions, and of community opposition to these plants, several technologies have been developed to mitigate emissions. *Scrubbers* chemically treat the gases produced in combustion to remove hazardous components and neutralize acidic gases, such as sulfur dioxide and hydrochloric acid, turning them into water and salt. Scrubbers generally do this either by spraying liquids formulated to neutralize the gases or by passing the gases through dry lime. Particulate matter is physically removed from inciner-

ator emissions in a system of huge filters known as a *baghouse.* These tiny particles, called fly ash, often contain some of the worst dioxin and heavy metal pollutants. In addition, burning garbage at especially high temperatures can destroy certain pollutants, such as PCBs. Even all these measures, however, do not fully eliminate toxic emissions.

Weighing the Issues:
Environmental Justice?

Do you know where your trash goes? Where is your landfill or incinerator located? Who lives closest to the facility? Are the people in this neighborhood wealthy, poor, or middle-class? What race or ethnicity are they? Do you know whether the people of this neighborhood protested against the introduction of the landfill or incinerator?

Many incinerators burn waste to create energy

Incineration was initially practiced simply to reduce the volume of waste, but today it often serves to generate electricity as well. Most North American incinerators today are **waste-to-energy** (**WTE**) facilities that use the heat produced by waste combustion to boil water, creating steam that drives electricity generation or that fuels heating systems. When burned, waste generates approximately 35% of the energy generated by burning coal. Over 100 WTE facilities are operating across the United States (mostly in the northeast and south), with a total capacity to process nearly 100,000 tons of waste per day.

Revenues from power generation, however, are usually not enough to offset the considerable financial cost of building and running incineration facilities. Because it can take many years for a WTE facility to become profitable, many companies that build and operate these facilities require communities contracting with them to guarantee the facility a minimum amount of garbage. In a number of cases, such long-term commitments have interfered with communities' later efforts to reduce their waste through recycling and other waste-reduction strategies.

Landfills can produce gas for energy

Combustion in WTE plants is not the only way to gain energy from waste. Deep inside landfills, bacteria decompose waste in an oxygen-deficient environment. This anaerobic decomposition produces *landfill gas,* a mix of gases that consists of roughly half methane (▶ pp. 98–99, 532). Landfill gas can be collected, processed, and used in the same way as natural gas, one of our primary sources of fossil fuel energy (▶ pp. 574–575).

Today more than 330 operational projects collect landfill gas in the United States. Other countries take advantage of this resource as well. In Chile, four facilities in Valparaiso and Santiago supply 40% of the region's demand for natural gas. At Fresh Kills, landfill gas collection wells pull gas upward through a network of pipes by vacuum pressure. Landfill gas collected from Fresh Kills should soon provide enough energy for 25,000 homes. Where gas is not collected for commercial use, it is burned off in flares to reduce smells and greenhouse emissions.

Reducing waste is a better option than disposal

Reducing the amount of material entering the waste stream avoids costs of disposal and recycling, helps conserve resources, minimizes pollution, and can often save consumers and businesses money. Preventing waste generation in this way is known as **source reduction.**

Because much of our waste stream consists of materials used to package goods, manufacturers' choices about packaging have a major effect on the volume of the waste stream. Packaging serves worthwhile purposes—preserving freshness, preventing breakage, protecting against tampering, and providing information—but much packaging is extraneous. Consumers can exercise power by choosing minimally packaged goods, buying unwrapped fruit and vegetables, and buying food in bulk. Consumer preference can give manufacturers incentive to reduce packaging. In addition, manufacturers can use packaging that is more recyclable. They can also reduce the size or weight of goods and materials, as they already have with many items, such as aluminum cans, plastic soft drink bottles, and personal computers.

Increasing the longevity of goods also helps reduce waste. Consumers generally choose goods that last longer, all else being equal. To maximize sales, however, companies often produce short-lived goods that need to be replaced frequently. Thus, increasing the longevity of goods is largely up to the consumer. If demand is great enough, manufacturers will respond.

Weighing the **Issues:**
Reducing Packaging: Is It a Wrap?

Reducing packaging cuts down on the waste stream, but how, when, and how much should we reduce? Packaging can serve very worthwhile purposes, such as safeguarding consumer health and safety. Can you think of three products for which you would not want to see less packaging? Why? Can you name three products for which packaging could easily be reduced without ill effect to the consumer? Would you be any more or less likely to buy these products if they had less packaging?

Reuse is one main strategy for waste reduction

To reduce waste, you can save items to use again, or substitute disposable goods with durable ones. Habits as simple as bringing your own coffee cup to coffee shops or bringing sturdy reusable cloth bags to the grocery store can, over time, have substantial impact. You can also donate unwanted items and shop for used items yourself at yard sales and resale centers. Over 6,000 reuse centers

Table 22.1 Some Everyday Things You Can Do to Reduce and Reuse

▶ Donate used items to charity

▶ Reuse boxes, paper, plastic wrap, plastic containers, aluminum foil, bags, wrapping paper, fabric, packing material, etc.

▶ Rent or borrow items instead of buying them, when possible . . . and lend your items to friends

▶ Buy groceries in bulk

▶ Decline bags at stores when you don't need them

▶ Bring reusable cloth bags shopping

▶ Make double-sided photocopies

▶ Bring your own coffee cup to coffee shops

▶ Pay a bit extra for durable, long-lasting reusable goods rather than disposable ones

▶ Buy rechargeable batteries

▶ Select goods with less packaging

▶ Compost kitchen and yard wastes in a compost bin or worm bin (often available from your community or waste hauler)

▶ Buy clothing and other items at resale stores and garage sales

▶ Use cloth napkins and rags rather than paper napkins and towels

▶ Write to companies to tell them what you think about their packaging and products

▶ When solid waste policy is being debated, let your government representatives know your thoughts

▶ Support organizations that promote waste reduction

exist in the United States, including stores run by organizations that resell donated items, such as Goodwill Industries and the Salvation Army. Besides doing good for the environment, reusing items is often economically advantageous. Used items are quite often every bit as functional as new ones, and much cheaper. Studies from U.S. communities estimate that at least 2–5% of the waste stream consists of reusable items. Table 22.1 presents a sampling of actions that we all can take to reduce the waste we generate.

Composting recovers organic waste

Organic waste, such as yard trimmings and food scraps, can be reduced and recycled by composting. **Composting** is the conversion of organic waste into mulch or humus through natural biological processes of decomposition. The mulch or humus can then be used to enrich soil. Householders can place waste in compost piles, underground pits, or specially constructed containers. As wastes are added, heat from microbial action builds in the interior, and decomposition proceeds. Banana peels, coffee grounds, grass clippings, autumn leaves, and countless other organic items can be con-

verted into rich, high-quality soil, given enough time, through the actions of earthworms, bacteria, soil mites, sow bugs, and other detritivores and decomposers. Home composting is a prime example of how we can live more sustainably by mimicking natural cycles and incorporating them into our daily lives.

Many municipalities are reducing waste through community composting programs—3,800 across the United States at last count. In these programs, food and yard waste are diverted from the waste stream to central composting facilities, where they decompose into mulch that community residents can use for gardens and landscaping. Nearly half of U.S. states now ban yard waste from the municipal waste stream, helping accelerate the drive toward composting. Approximately one-fifth of the U.S. waste stream is made up of materials that can be easily composted. Composting reduces landfill waste, enriches soil and helps it resist erosion, encourages soil biodiversity, makes for healthier plants and more pleasing gardens, and reduces the need for chemical fertilizers.

Recycling consists of three steps

Recycling, too, offers many benefits. **Recycling** consists of collecting materials that can be broken down and reprocessed to manufacture new items. Recycling diverted 55 million tons of materials away from incinerators and landfills in the United States in 2003.

The three basic steps in the recycling loop have given rise to the three elements of the commonly seen recycling symbol (Figure 22.10). The first step is collecting and

1 Collection and processing of recyclable materials by municipalities and businesses

2 Use of recyclables by industry to manufacture new products

3 Consumer purchase of products made from recycled materials

FIGURE 22.10 The familiar recycling symbol consists of three arrows to represent the three components of a sustainable recycling strategy: collection and processing of recyclable materials, use of the materials in making new products, and consumer purchase of these products.

processing used recyclable goods and materials. Communities may designate locations where residents can drop off recyclables or receive money for them. Many of these have now been replaced by the more convenient option of curbside recycling, in which trucks pick up recyclable items in front of houses, usually in conjunction with municipal trash pickup. Curbside recycling has grown rapidly, and its convenience has helped boost recycling rates. Nearly half of all Americans are now served by more than 9,000 curbside recycling programs across all 50 U.S. states.

Items collected are taken to **materials recovery facilities (MRFs),** where workers and machines sort items, using automated processes including magnetic pulleys, optical sensors, water currents, and air classifiers that separate items by weight and size. The facilities clean the materials, shred them, and prepare them for reprocessing. This is the second step in the recycling loop.

Once readied, these materials are used in manufacturing new goods. Newspapers and many other paper products use recycled paper, many glass and metal containers are now made from recycled materials, and some plastic containers are of recycled origin. Some large objects, such as benches and bridges in city parks, are now made from recycled plastics, and glass is sometimes mixed with asphalt (creating "glassphalt") for paving roads and paths. The pages in this textbook are made from recycled paper that is up to 20% post-consumer waste.

If the recycling loop is to function, consumers and businesses must complete the third step in the cycle by purchasing products made from recycled materials. Buying recycled goods provides economic incentive for industries to recycle materials, and for new recycling facilities to open or existing ones to expand. In this arena, individual consumers have power to encourage environmentally friendly options through the free market. Many businesses now advertise their use of recycled materials, a widespread instance of ecolabeling (▶ pp. 50–51). As markets for products made with recycled materials expand, prices continue to fall.

Recycling has grown rapidly and can expand further

The thousands of curbside recycling programs in place today have sprung up only in the last 20 years. Recycling in the United States has risen from 6.4% of the waste stream in 1960 to 23.5% in 2003 (and 30.6% if you include composting), according to EPA data (Figure 22.11). The EPA calls the growth of recycling "one of the best environmental success stories of the late 20th century."

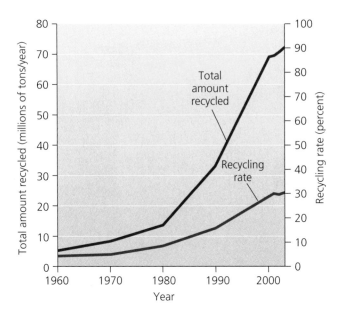

FIGURE 22.11 Recycling has risen sharply in the United States over the past 40 years. Today over 70 million tons of material is recycled, comprising more than 30% of the waste stream. Data from U.S. EPA, 2005.

Recycling rates vary greatly from one product or material type to another and from one location to another. Rates for different types of materials and products range from nearly zero to almost 100% (Table 22.2). Recycling rates among U.S. states also vary greatly, from less than 1% to nearly 50% (Figure 22.12).

Recycling's growth has been propelled in part by economic forces as established businesses see opportunities to save money and as entrepreneurs see opportunities to

Table 22.2 Recovery Rates for Various Materials in the United States	
Material	**Percentage that is recycled or composted**
Auto batteries	93.0
Steel cans	60.0
Yard trimmings	56.3
Paper and paperboard	48.1
Aluminum cans	43.9
Tires	35.6
Plastic milk bottles	31.9
Plastic soft drink containers	25.2
Glass containers	22.0

Data are for 2003, the most recent year available, from U.S. Environmental Protection Agency, Dec. 2005.

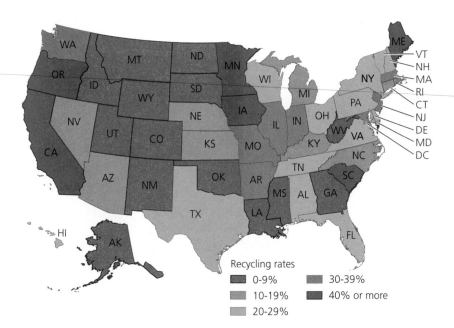

FIGURE 22.12 U.S. states vary greatly in the rates at which their citizens recycle. Data are for 2002 (with earlier data for Alabama, Alaska, and Montana), from Kaufman, S. M., et al., 2004. The state of garbage in America. *BioCycle* 45: 31–41.

Recycling rates

- 0-9%
- 10-19%
- 20-29%
- 30-39%
- 40% or more

start new businesses. However, recycling has also been driven by the desire of municipalities to reduce waste and by the satisfaction people take in recycling. These two forces have driven recycling's rise even though it has often not been financially profitable. In fact, many of the increasingly popular municipal recycling programs are run at an economic loss. The expense required to collect, sort, and process recycled goods is often more than recyclables are worth in the market. Furthermore, the more people recycle, the more glass, paper, and plastic is available to manufacturers for purchase, driving down prices.

Recycling advocates, however, point out that market prices do not take into account external costs (▸ pp. 43–44)—in particular, the environmental and health impacts of *not* recycling. For instance, it has been estimated that globally, recycling saves enough energy to power 6 million households per year. And recycling aluminum cans saves 95% of the energy required to make the same amount of aluminum from mined virgin bauxite, its source material.

As more manufacturers use recycled products, and as more technologies and methods are developed to use recycled materials in new ways, markets should continue to expand, and new business opportunities may arise. We are still at an early stage in the shift from an economy that moves linearly from raw materials to products to waste, to an economy that moves circularly, using waste products as raw materials for new manufacturing processes. The steps we have taken in recycling so far are central to this transition, which many analysts view as key to building a sustainable economy.

Weighing the issues:
Costs of Recycling and Not Recycling

Should recycling programs be subsidized by governments even if they are run at an economic loss? What external costs—costs not reflected in market prices—do you think would be involved in not recycling, say, aluminum cans? Do you feel these costs justify sponsoring recycling programs even when they are not financially self-supporting? Why or why not?

Financial incentives can help address waste

Waste managers have employed economic incentives as tools to reduce the waste stream. The "pay-as-you-throw" approach to garbage collection uses a financial incentive to influence consumer behavior. In these programs, municipalities charge residents for home trash pickup according to the amount of trash they put out. The less waste the household generates, the less the resident has to pay. Over 4,000 of these programs now exist in the United States. Besides reducing waste, they promote equity and fairness; people pay for disposal in accordance with how much they use the service.

"Bottle bills" represent another approach that hinges on financial incentive. Eleven U.S. states have these laws, which allow consumers to return bottles and cans to stores after use and receive a refund—generally 5 cents per bottle or can. The first bottle bills were passed in the 1970s to cut down on litter, but they have also served to decrease the

VIEWPOINTS

Recycling

Will we need to make recycling more economically profitable if it is to continue growing? What, if anything, should we do to encourage recycling? Or should we instead focus on other ways of managing waste?

How to Enhance Recycling?

Certainly recycling needs to be profitable if it is to survive and expand. The question is, how should it be made profitable?

Government subsidies are one common answer; many communities provide small subsidies to keep their recycling programs going. These subsidies will undoubtedly continue because people like municipal recycling programs. But subsidies are not likely to expand much beyond their current, modest level.

In some cases, new technologies are needed. Plastics recycling requires expensive sorting and separation of different plastics to create a high-quality product. With today's technology, recycling of unsorted, mixed plastics yields a low-value product with limited uses. New inventions that improve the sorting process or improve the quality of recycled mixed plastics could make plastics recycling much more profitable.

In other cases, we need new recycling programs and opportunities to keep up with changing lifestyles. Recycling rates are declining for beverage cans and bottles because so many beverages are consumed (and so many containers are discarded) away from home, at parks, beaches, and other public places. A system for recycling in public places could collect the growing quantities of beverage containers, newspapers, and other recyclable materials that are thrown out by people on the go.

Some products should be carefully recycled because it is hazardous to throw them out. Automobile batteries contain large amounts of lead, a toxic substance that should not be tossed in the trash. Some states have laws requiring a deposit on every battery that is sold—which makes it worthwhile to return a dead battery for a refund instead of discarding it. Similar approaches could and should be used with other potentially hazardous products.

Finally, it will never be possible to recycle everything. Along with continuing efforts to expand recycling, we must ensure that there are safe, clean opportunities for disposing of the remaining, nonrecyclable, wastes.

Frank Ackerman is an economist at Tufts University's Global Development and Environment Institute. He has advised the U.S. EPA and state and local agencies on waste management policies. His books include *Why Do We Recycle? Markets, Values, and Public Policies* (Island Press, 1997), and, with Lisa Heinzerling, *Priceless: On Knowing the Price of Everything and the Value of Nothing* (The New Press, 2004).

Recycling: A Mixed Bag

Recycling will continue as long as it is profitable. If it becomes more profitable, we will see more of it. Currently, 55% of all aluminum cans are recycled, a high rate compared to the 30% recycling rate that the Environmental Protection Agency says is average for solid waste. The reason: Aluminum companies can save money by using recycled materials because making cans from bauxite ore is expensive. Paper and cardboard are recycled to a great extent too, partly because cardboard can be made from many kinds of paper.

Plastics aren't recycled nearly as much (about 9%, according to the EPA). One reason is that different kinds of plastic resins can't be mixed. It is expensive for companies to separate the plastics before reprocessing them.

To increase recycling, governments would probably have to force people to recycle and require them to buy products made of recycled materials. To boost the city's recycling rate, Seattle has already made it illegal for residents to put recyclable materials in their regular trash.

This could make sense if recycling always saved resources. But does it?

Not always. The goal of recycling is to save resources, but mandatory recycling often wastes them. Additional trucks must go out into the community, using more energy and adding to air pollution. And reprocessing recyclables is just another form of manufacturing, which inevitably causes some pollution and waste.

Fortunately, there are additional ways to deal with trash. Modern landfills are scientifically engineered to keep garbage dry so that harmful leakage doesn't occur. And there is plenty of space. One widely cited estimate is that the United States could bury all its trash for the next century in a landfill 225 feet deep and 10 miles square.

We should recycle when it makes sense, but we shouldn't be afraid to use other means as well.

Jane S. Shaw is a Senior Fellow of the Property and Environment Research Center (PERC), a nonprofit institute in Bozeman, Montana, dedicated to improving environmental quality through property rights and markets. With Michael Sanera, she is coauthor of *Facts, Not Fear: Teaching Children about the Environment* (Regnery, 1999) and editor of the Greenhaven Press book series, *Critical Thinking about Environmental Issues*.

 Explore this issue further by accessing **Viewpoints** at www.aw-bc.com/withgott.

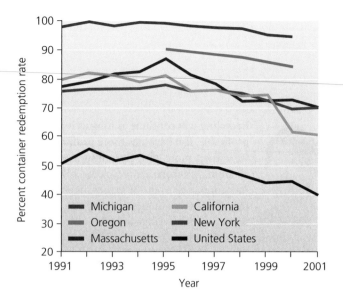

FIGURE 22.13 Data suggest that bottle bills increase recycling rates and that higher redemption amounts boost recycling rates further. The five states with bottle bills shown in this graph all have higher recycling rates than the United States as a whole. Michigan, the only state with a 10-cent deposit, has the highest recycling rate of all. The overall downward trend in rates in all states across the years is likely due to inflationary erosion of the value of deposits. Data from Gitlitz, J., and P. Franklin. 2004. *The 10¢ incentive to recycle.* Arlington, VA.: Container Recycling Institute.

waste stream. In states where they have been enacted, these laws have proved profoundly effective and resoundingly popular; they are recognized as among the most successful state legislation of recent decades (Figure 22.13). It is a testament to the lobbying influence of the beverage industries, which have traditionally opposed passage of bottle bills, that more states do not have such legislation.

States with bottle bills now face two challenges. One is to amend these laws to include new kinds of containers. In New York State, where 80 billion beverage containers have been redeemed since 1982, polls show 70% of the public favoring expanding their bottle bill to include more types of containers, and the legislature is currently considering the issue. The second challenge is to adjust refunds for inflation. In the three decades since Oregon passed the nation's first bottle bill, the value of a nickel has dropped such that today, the refund would need to be 22 cents to reflect the refund's original intended value. Proponents argue that increasing refund amounts will raise return rates, and available data support this view (see Figure 22.13).

One Canadian city showcases the shift from disposal to reduction and recycling

Edmonton, Alberta, has created one of the world's most advanced waste management programs. As recently as 1998, fully 85% of the city's waste was being landfilled, and space was running out. Today, just 35% goes to the new sanitary landfill, while 15% is recycled, and an impressive 50% is composted. Edmonton's citizens are proud of the program, and 81% of them participate in its curbside recycling program.

When Edmonton's residents put out their trash, city trucks take it to their new co-composting plant—at the size of eight football fields, the largest in North America (Figure 22.14a). The waste is dumped on the floor of the facility, and large items, such as furniture, are removed and landfilled. The bulk of the waste is mixed with dried sewage sludge for 1–2 days in five large rotating drums, each the length of six buses. The resulting mix travels on a conveyor to a screen that removes nonbiodegradable items. It is aerated for several weeks in the largest stainless steel building in North America (Figure 22.14b). The mix is then passed

(a) Composting facility, Edmonton, Alberta

FIGURE 22.14 Edmonton, Alberta, boasts one of North America's most successful waste management programs. Edmonton's gigantic composting facility **(a)** is the size of eight football fields. Inside the aeration building **(b),** which is the size of 14 professional hockey rinks, mixtures of solid waste and sewage sludge are exposed to oxygen and composted for 14–21 days.

(b) Aeration building, Edmonton composting facility

through a finer screen and finally is left outside for 4–6 months. The resulting compost—80,000 tons annually—is made available to area farmers and residents. The facility even filters the air it emits with a 1-m (3.3-ft) layer of compost, bark, and wood chips, which eliminates the release of unpleasant odors into the community.

Besides the co-composting facility and the sanitary landfill, Edmonton's program includes a state-of-the-art MRF that handles 30,000–40,000 tons of waste annually, a leachate treatment plant, a research center, public education programs, and a wetland and landfill revegetation program. In addition, 100 pipes collect enough landfill gas to power 4,000 homes, bringing thousands of dollars to the city and helping power the new waste management center. Five area businesses reprocess the city's recycled items. Newsprint and magazines are turned into new newsprint and cellulose insulation, while cardboard and paper are converted into building paper and shingles. Household metal is made into rebar and blades for tractors and graders, and recycled glass is used for reflective paint and signs.

Industrial Solid Waste

Solid waste generated by industry is another major contributor to the waste stream. Each year, U.S. industrial facilities generate about 7.6 billion tons of waste, according to the EPA, about 97% of which is wastewater. Thus, very roughly, 228 million or so tons of solid waste are generated by 60,000 facilities each year—an amount about equal to that of municipal solid waste. In the United States, industrial solid waste is defined as solid waste that is considered neither municipal solid waste nor hazardous waste under the Resource Conservation and Recovery Act.

U.S. waste managers differentiate between municipal and industrial solid waste largely because these categories are regulated differently. Whereas the federal government regulates municipal solid waste, state or local governments regulate industrial solid waste (although with federal guidance). Industrial waste includes more than just waste from factories. It includes waste from everything from manufactured consumer goods to mining activities to petroleum extraction to agricultural waste. Waste is generated at several points along the process from raw materials extraction to manufacturing to sale and distribution (Figure 22.15).

Regulation and economics both influence industrial waste generation

Most methods and strategies of waste disposal, reduction, and recycling by industry are similar to those for munici-

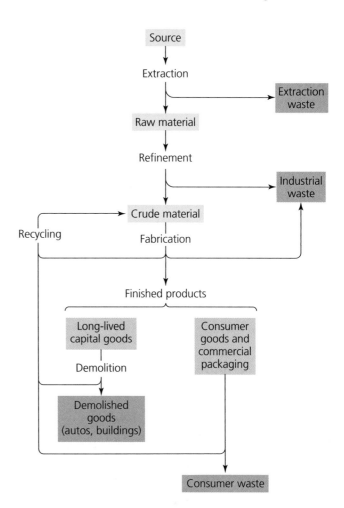

FIGURE 22.15 Industrial and municipal waste is generated at a number of stages throughout the life cycles of products. Waste is first generated when raw materials needed for production are extracted. Further industrial waste is produced as raw materials are processed and as products are manufactured. Waste results from the demolition or disposal of products once they are used by businesses and individuals. At each stage there are often opportunities for efficiency improvements, waste reduction, or recycling.

pal solid waste. For instance, businesses that manage their own waste on site most often dispose of it in landfills, and companies must design and manage their landfills in ways that meet state, local, or tribal guidelines. Other businesses pay to have their waste disposed of at municipal disposal sites. Regulation varies greatly from state to state and area to area, but in most cases, state and local regulation of industrial solid waste is less strict than federal regulation of municipal solid waste. This is one reason that industry continues to suffer public criticism for its waste and pollution impacts. In many areas, industries are not required to have permits, install landfill liners or leachate collection systems, or monitor groundwater for contamination.

Government regulation is not the only factor that influences industrial waste generation. The amount of waste generated by a manufacturing process is one measure of its efficiency, because waste represents resources that are being lost. The less waste produced per unit or volume of product, the more efficient that process is, from a physical standpoint. However, physical efficiency is not always equivalent to economic efficiency. Often it is cheaper for industry to manufacture its products or perform its services quickly but messily. That is, it can be cheaper to generate waste than to avoid generating waste. In such cases, economic efficiency is maximized, but physical efficiency is not. The frequent mismatch between these two types of efficiency is a major reason why the output of industrial waste is so great.

Rising costs of waste disposal, however, enhance the financial incentive to decrease waste and increase physical efficiency. Once either government or the market makes the physically efficient use of raw materials also economically efficient, businesses have financial incentives to reduce their own waste.

Industrial ecology seeks to make industry more sustainable

In an effort to reduce waste, growing numbers of industries today are experimenting with industrial ecology. A holistic approach that integrates principles from engineering, chemistry, ecology, and economics, **industrial ecology** seeks to redesign industrial systems to reduce resource inputs and to minimize physical inefficiency while maximizing economic efficiency. Industrial ecologists would reshape industry so that nearly everything produced in a manufacturing process is used, either within that process or in a different one.

The larger idea behind industrial ecology is that industrial systems should function more like ecological systems, in which almost everything produced is used by some organism, with very little being wasted. Applied to human activity, this principle brings industry closer to the ideal of ecological economists, in which human economies attain sustainability by functioning in a circular fashion rather than a linear one (▸ pp. 45 and 689–690).

Industrial ecologists pursue their goals in several ways. For one, they examine the entire life cycle of a given product—from its origins in raw materials, through its manufacturing, to its use, and finally its disposal—and look for ways to make the process more ecologically efficient. This strategy is called **life-cycle analysis.**

Industrial ecologists also try to identify points at which waste products from one manufacturing process can be used as raw materials for a different process. For instance, used plastic beverage containers cannot be refilled because of the potential for contamination, but they can be shredded and reprocessed to make other plastic items, such as benches, tables, and decks. In addition, industrial ecologists examine industrial processes with an eye toward eliminating environmentally harmful products and materials. Finally, they study the flow of materials through industrial systems to look for ways to create products that are more durable, recyclable, or reusable. Goods that are currently thrown away when they become obsolete, such as computers, automobiles, and some appliances, could be designed to be more easily disassembled, and their component parts reused or recycled.

Attentive businesses are taking advantage of the insights of industrial ecology to reduce waste and save money while lessening their impact on the environment and human health. For example, American Airlines switched from hazardous to nonhazardous materials in its Chicago facility, decreasing its need to secure permits from the EPA. The company also used over 50,000 reusable plastic containers to ship goods, reducing packaging waste by 90%. Its Dallas–Fort Worth headquarters recycled enough aluminum cans and white paper in 5 years to save $205,000. Roughly 3,000 broken baggage containers were recycled into lawn furniture. A program to gather suggestions from employees resulted in over 700 ideas to reduce waste—and 15 of these ideas saved the company over $8 million in the first year of implementation.

Other efforts have gone further. The Swiss Zero Emissions Research and Initiatives (ZERI) Foundation sponsors dozens of innovative projects worldwide that attempt to create goods and services without generating waste. Although most are not fully closed-loop systems, they attempt to approach this ideal. In so doing, they cut down on waste while increasing output and income, and often they generate new jobs as well. One example involves breweries, currently being pursued in Canada, Sweden, Japan, and Namibia (Figure 22.16). Brewers in these projects take waste from the beer-brewing process and use it to fuel a series of other processes. The brewer as a result can make money from bread, mushrooms, pigs, gas, and fish, as well as beer, all while producing little waste.

Hazardous Waste

Solid waste from industrial and municipal sources is a problem largely because of the volumes in which it can accumulate. Hazardous waste is a problem because of its chemical nature. Public awareness of hazardous waste has increased greatly in recent decades, driven by highly publicized instances of toxic contamination at abandoned

FIGURE 22.16 Traditional breweries (**a**) produce only beer while generating much waste, some of which goes toward animal feed. ZERI-sponsored breweries (**b**) use their waste grain to make bread and to farm mushrooms. Waste from the mushroom farming, along with brewery wastewater, goes to feed pigs. The pigs' waste is digested in containers that capture natural gas and collect nutrients used to nourish algae for growing fish in fish farms. The brewer derives income from bread, mushrooms, pigs, gas, and fish, as well as beer.

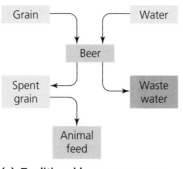

(a) Traditional brewery process

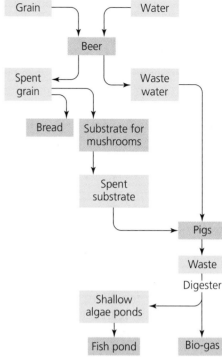

(b) ZERI brewery process

industrial sites. Hazardous wastes are diverse in their chemical composition and may be liquid, solid, or gaseous. By EPA definition, **hazardous waste** is waste that meets one of the following four criteria:

1. *Ignitability*. Substances that easily catch fire (for example, natural gas or alcohol).
2. *Corrosivity*. Substances that corrode metals in storage tanks or equipment.
3. *Reactivity*. Substances that are chemically unstable and readily react with other compounds, often explosively or by producing noxious fumes.
4. *Toxicity*. Substances that harm human health when they are inhaled, are ingested, or contact human skin.

Materials with these characteristics can harm human health and environmental quality. Flammable and explosive materials can cause ecological damage and atmospheric pollution. For instance, fires at large tire dumps in California's Central Valley have exacerbated air pollution and caused highway closures. Toxic wastes in lakes and rivers have caused fish die-offs and closed important domestic fisheries, such as those in Chesapeake Bay.

Hazardous wastes have diverse sources

Industry, mining, households, small businesses, agriculture, utilities, and building demolition are all sources of hazardous waste. Industry produces the largest amounts of hazardous waste, but in most developed nations industrial waste generation and disposal is highly regulated. This regulation has reduced the amount of hazardous waste entering the environment from industrial activities. As a result, households currently are the largest source of unregulated hazardous waste.

Household hazardous waste results from materials commonly used in and around the home that contain toxic, reactive, corrosive, or flammable ingredients. These include

a wide range of items, such as paints, batteries, oils, solvents, cleaning agents, lubricants, and pesticides. U.S. citizens generate 1.6 million tons of household hazardous waste annually, and the average home contains close to 45 kg (100 lb) of it in sheds, basements, closets, and garages.

Although many hazardous substances become less hazardous over time as they degrade chemically, two classes of chemicals are particularly hazardous because their toxicity persists over time: organic compounds and heavy metals.

Organic compounds and heavy metals can be hazardous

In our day-to-day lives, we rely on the capacity of synthetic organic compounds and petroleum-derived compounds to resist bacterial, fungal, and insect activity. Items such as plastic containers, rubber tires, pesticides, solvents, and wood preservatives are useful to us precisely because they resist decomposition. We use these substances to protect our buildings from decay, kill pests that attack crops, and keep stored goods intact. However, the resistance of these compounds to decay is a double-edged sword, for it also makes them persistent pollutants. Many synthetic organic compounds are toxic because they can be readily absorbed through the skin of humans and other animals and can act as mutagens, carcinogens, teratogens, and endocrine disruptors (▶ pp. 409–410).

Heavy metals such as lead, chromium, mercury, arsenic, cadmium, tin, and copper are used widely in industry for wiring, electronics, metal plating, metal fabrication, pigments, and dyes. Heavy metals enter the environment when paints, electronic devices, batteries, and other materials are disposed of improperly. Lead from fishing weights and from hunters' lead shot has accumulated in many rivers, lakes, and forests. In older homes, lead from pipes contaminates drinking water, and lead paint remains a problem, especially for infants. Heavy metals are prone to bioaccumulate (▸ p. 416) when they are fat-soluble and break down slowly. In California's Coast Range, for instance, mercury washed downstream from abandoned mercury mines enters low-elevation lakes and rivers, is consumed by bacteria and invertebrates, and accumulates in increasingly larger quantities up the food chain, poisoning organisms at higher trophic levels and making fish unsafe to eat.

Computers, televisions, VCRs, cell phones, and other electronic devices represent major new sources of potential heavy metal contamination. These products have short lifetimes before people judge them obsolete, and most are discarded after only a few years. The amount of this electronic waste—sometimes called *e-waste*—is growing. In the United States alone, there are well over 300 million television sets, and the National Safety Council has estimated that 500 million computers will be retired between 1997 and 2007, creating several million tons of waste each year. Most e-waste is still disposed of in landfills as conventional solid waste, but recent research suggests that it should instead be treated as hazardous waste (see "The Science behind the Story," ▸ pp. 668–669).

Weighing the **issues:**
Toxic Computers?

The cathode ray tubes in televisions and computer screens can hold up to 5 kg (8 lb) of heavy metals, such as lead and cadmium. These represent the second-largest source of lead in U.S. landfills today, behind car batteries. With a growing number of computer screens being purchased, the transition to high-definition television, and the rapid turnover of computers, what future waste problems might you expect? How do you think these products should be disposed of, and who should be responsible for their disposal?

Several steps precede the disposal of hazardous waste

For many years we discarded hazardous waste without special treatment. In many cases people did not know that certain substances were harmful to human health. In other cases their danger was known or suspected, but it was assumed that the substances would disappear or be sufficiently diluted in the environment. The resurfacing of toxic chemicals years after their burial at Love Canal in upstate New York demonstrated to the public that hazardous waste deserves special attention and treatment.

Today, several methods for disposing of hazardous waste have been developed. All are improvements on the old practice of open dumping, but none is completely satisfactory. A number of steps generally precede disposal. Since the 1980s, many communities have designated sites or special collection days to gather household hazardous waste, or facilities for the exchange and reuse of substances (Figure 22.17). Once consolidated in such sites, the waste is transported for treatment and ultimate disposal.

Under the Resource Conservation and Recovery Act, the EPA sets standards by which states are to manage hazardous waste. RCRA also requires large generators of hazardous waste to obtain permits, and mandates that hazardous materials be tracked "from cradle to grave." As hazardous waste is generated, transported, and disposed of, the producer, carrier, and disposal facility must each report to the EPA the type and amount of material generated; its location, origin, and destination; and the way it is being handled. This process is intended to prevent illegal dumping and to encourage the use of reputable waste carriers and disposal facilities.

FIGURE 22.17 Many communities designate collection sites or collection days for household hazardous waste. Here, workers handle waste from an Earth Day collection event near Los Angeles.

FIGURE 22.18 Unscrupulous individuals or businesses sometimes dump hazardous waste illegally to avoid disposal costs.

Because current U.S. law makes disposing of hazardous waste quite costly, irresponsible companies have sometimes been found guilty of illegally and anonymously dumping waste, creating health risks for residents and financial headaches for local governments forced to deal with the mess (Figure 22.18). However, high costs have also encouraged responsible businesses to heat hazardous waste to transform it into nonhazardous substances prior to disposal.

Many biologically hazardous materials can be broken down by incineration at high temperatures in cement kilns. Some hazardous materials can be treated by exposure to bacteria that break down harmful components and synthesize them into new compounds. Besides bacterial bioremediation, phytoremediation (▸ pp. 92–93) is also used. Various plants have now been bred or engineered to take up specific contaminants from soil, then break down organic contaminants into safer compounds or concentrate heavy metals in their tissues. The plants are eventually harvested and disposed of.

We have three main disposal methods for hazardous waste

Three primary means of hazardous waste disposal have been developed: landfills, surface impoundments, and injection wells. Design and construction standards for landfills that receive hazardous waste are stricter than those for ordinary sanitary landfills. Hazardous waste landfills must have several impervious liners and leachate removal systems, and must be located far from aquifers.

Dumping of hazardous waste in ordinary landfills has long been a problem. In New York City, Fresh Kills largely managed to keep hazardous waste out, but most of the city's older landfills were declared to be hazardous sites because of past toxic waste disposal. Secure landfills for hazardous waste do nothing to lessen the hazards of the materials, but they do help keep these materials isolated from people, wildlife, and ecosystems.

A method for storing liquid hazardous waste, or waste in dissolved form, is in ponds or **surface impoundments.** To create a surface impoundment, a shallow depression is dug and lined with plastic and an impervious material, such as clay. Water containing dilute hazardous waste is placed in the pond and allowed to evaporate, leaving a residue of solid hazardous waste on the bottom (Figure 22.19). This process is repeated until the dry material is removed and transported elsewhere for permanent disposal. Impoundments are not ideal. The underlying layer can crack and leak waste. Some material may

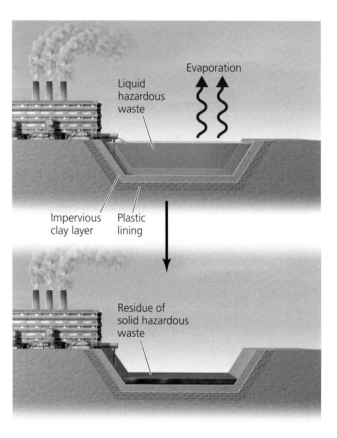

FIGURE 22.19 Surface impoundments are a strategy for temporarily disposing of liquid hazardous waste. The waste, mixed with water, is poured into a shallow depression lined with plastic and clay to prevent leakage. When the water evaporates, leaving a crust of the hazardous substance, new liquid is poured in and the process repeated. This method alone is not satisfactory, because waste can potentially leak, overflow, evaporate, or blow away.

The Science behind the Story

Testing the Toxicity of "E-Waste"

As we began to conduct more of our business, learning, and communication with computers and other electronic devices, many people predicted that our waste, particularly paper waste, would decrease. But instead, the proliferation of computers, printers, VCRs, fax machines, cell phones, and other gadgets has created a substantial new source of waste.

Most of this electronic waste, or "e-waste," is landfilled as conventional solid waste. However, most electronic appliances contain heavy metals that can cause environmental contamination and public health risks. For instance, over 6% of a typical computer is composed of lead.

At the University of Florida, Gainesville, Timothy Townsend's lab was funded by the EPA to determine whether e-waste is toxic enough to be classified as hazardous waste under the Resource Conservation and Recovery Act.

With students and colleagues, Townsend determined in 1999–2000 that cathode ray tubes (CRTs) from computer monitors and color televisions leach an average of 18.5 mg/L of lead, far above the regulatory threshold of 5 mg/L. Following this research, the EPA proposed classifying CRTs as hazardous waste, and several U.S. states banned these items from conventional landfills.

Discarded electronic waste can leach heavy metals and should be considered hazardous waste, researchers say.

Then in 2004, Townsend's lab group completed experiments on 12 other types of electronic devices. To measure their toxicity, Townsend's group used the EPA's standard test, the Toxicity Characteristic Leaching Procedure (TCLP). Designed to mimic the process by which chemicals leach out of solid waste in landfills, this test speeds up the process so that it can be measured in the lab. In the TCLP, waste is ground up into fine pieces and 100 g (3.5 oz) of it is put in a container with 2 L (0.53 gal) of an acidic leaching fluid. The container is rotated for 18 hours, after which the leachate is analyzed for its chemical content.

Researchers look for eight heavy metals—arsenic, barium, cadmium, chromium, lead, mercury, selenium, and silver—and determine for each whether their concentration in the leachate exceeds that allowed by EPA regulations. Of these eight elements, electronic devices contain notable amounts of four: cadmium, chromium, lead, and mercury.

To conduct the standard TCLP, Townsend's team ground up the central processing units (CPUs) of personal computers, creating a mix made up by weight of 15.8% circuit board, 7.5% plastic, 68.2% ferrous metal, 5.4% nonferrous metal, and 3.1% wire and cable. However, grinding up a computer into small bits is no easy task, and it is hard to obtain a sample that accurately represents all components and materials. So the researchers also designed a modified TCLP test in which they placed whole CPUs—with the parts disassembled but not ground up—in a rotating 55-gallon drum full of leaching liquid. Then they tested their 12 types of devices using a combination of the standard and modified TCLP methods.

The team's results are summarized in the figure. Lead was the only heavy metal found to exceed the EPA's regulatory threshold, but this threshold (5 mg/L) was exceeded in the majority of trials. Computer monitors leached the most lead (47.7 mg/L on average), as expected,

evaporate or be blown into surrounding areas. Heavy rainstorms may cause waste to overflow and contaminate nearby areas. For these reasons, surface impoundments are used only for temporary storage.

The third method is intended for long-term disposal. In **deep-well injection,** a well is drilled deep beneath the water table, reaching into porous rock. Once the well has been drilled, wastes are injected into it. The aim is that waste will accumulate in the porous rock and remain deep underground, isolated from groundwater and human contact (Figure 22.20). This idea seems attractive in principle, but in practice wells become

because monitors include the cathode ray tubes already known to be a problem. However, laptops, color TVs, smoke detectors, cell phones, and computer mice also leached high levels of lead. Next came remote controls, VCRs, keyboards, and printers, all of which leached more lead on average than the EPA threshold, and did so in 50% or more of the trials. Whole CPUs and flat panel monitors were the only devices to leach less than 5 mg/L of lead on average, but even these exceeded the threshold more than one-quarter of the time.

The researchers found that items containing more ferrous metals (such as iron) tended to leach less lead. For instance, CPUs contain 68% ferrous metals (compared to only 7% in laptops), and laptops leached seven times as much lead as CPUs. Further experiments confirmed that ferrous metals were chemically reacting with lead and stopping it from leaching.

Townsend says the work suggests that many electronic devices have the potential to be classified as hazardous waste because they frequently surpass the toxicity criterion for lead. However, EPA scientists must decide how to judge results from the modified TCLP methods, and must evaluate other research, before determining whether to alter regulatory standards.

Furthermore, lab tests may or may not accurately reflect what

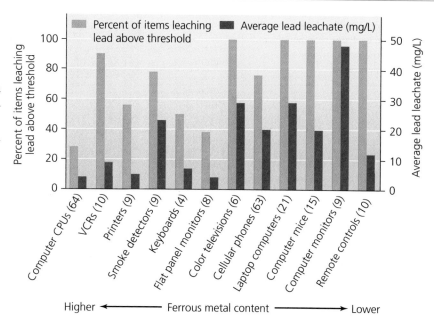

Some proportion of all 12 devices tested exceeded the EPA regulatory standard for lead leachate. Devices with higher ferrous metal content tended to leach less lead. Where both standard and modified TCLPs were used, results are averaged. Data from Townsend, T.G., et al., 2004. RCRA toxicity characterization of computer CPUs and other discarded electronic devices. July 15, 2004, report to the U.S. EPA.

actually happens in landfills. So Townsend is also pursuing research to address this question. His team is filling columns measuring 24 cm (2 ft) wide by 4.9 m (16 ft) long with e-waste and municipal solid waste, burying them in a Florida landfill, and then testing the leachate that results.

As the EPA and more states move toward keeping e-waste out of conventional sanitary landfills, more computers and accessories are being recycled. The devices are taken apart, and parts are either reused or disposed of more safely. Although there are serious concerns about the health risks this may pose to workers doing the disassembly, recycling done responsibly seems likely to be the way of the future.

In many North American cities, businesses, nonprofit organizations, or municipal services now recycle used computers and related devices. So next time you upgrade to a new computer, TV, DVD player, VCR, or cell phone, check out what opportunities may exist in your area to recycle your old ones.

corroded and can leak wastes into soil, allowing them to enter aquifers. Currently the amount of waste disposed of in injection wells is declining, but roughly 34 billion L (9 billion gal) of hazardous waste continue to be placed in U.S. injection wells each year.

Radioactive waste is a special type of hazardous waste

Radioactive waste is particularly dangerous to human health and is persistent in the environment. The dilemma of disposal has dogged the nuclear energy industry and

FIGURE 22.20 A seemingly more satisfactory way of disposing of liquid hazardous waste is to pump it deep underground, in deep-well injection. The well must be drilled below any aquifers, into porous rock separated by impervious clay. The technique is expensive, however, and leakage from the well shaft into groundwater may occur.

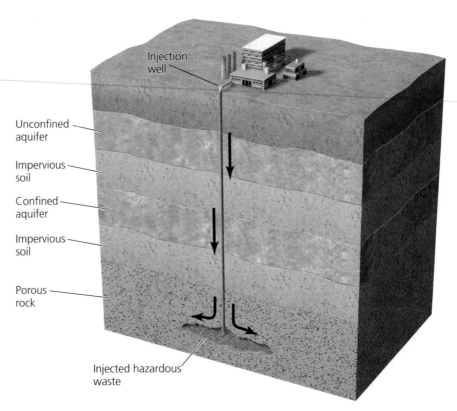

Injection well

Unconfined aquifer

Impervious soil

Confined aquifer

Impervious soil

Porous rock

Injected hazardous waste

the U.S. military for decades. As we saw in our discussion of radioactive waste disposal in Chapter 20 (▶ pp. 600–603), Yucca Mountain in Nevada has recently been approved as the single-site repository for all U.S. nuclear waste.

Currently, a site in the Chihuahuan Desert in southeastern New Mexico serves as a permanent disposal site for radioactive waste. The Waste Isolation Pilot Plant (WIPP) is the world's first underground repository for transuranic waste from nuclear weapons development. The mined caverns holding the waste are located 655 m (2,150 ft) below ground in a huge salt formation thought to be geologically stable. Twenty years in the planning, WIPP became operational in 1999 and will receive thousands of shipments of waste from 23 other locations over the next three decades.

Contaminated sites are being cleaned up, slowly

Many thousands of former military and industrial sites remain contaminated with hazardous waste in the United States, Russia, and virtually every other nation on Earth. For most nations, dealing with these messes is simply too difficult, time-consuming, and expensive. In 1980 the U.S. Congress passed the Comprehensive Environmental Response Compensation and Liability Act (CERCLA). This legislation established a federal program to clean up U.S. sites polluted with hazardous waste from past activities. The EPA administers this cleanup program, called the **Superfund.**

Under EPA auspices, experts identify sites polluted with hazardous chemicals, take action to protect groundwater near these sites, and clean up the pollution. The objective of the act is to charge responsible parties for cleanup of sites (according to the *polluter-pays principle*). For many polluted sites, however, the responsible parties cannot be found or held liable, and in such cases—roughly 1 out of 4 so far—Superfund activities are covered by taxpayers' funds. These funds come from the federal budget ($1.4 billion in 2005), and from a trust fund (which went bankrupt in 2004) established from a tax on chemical raw materials.

Once a Superfund site is identified, EPA scientists evaluate how close the site is to human habitation, whether wastes are currently confined or likely to spread, and whether the site threatens drinking water supplies. Sites that appear harmful are placed on EPA's National Priority List, ranked according to the level of risk to human health that they pose. Cleanup proceeds on a site-by-site basis as funds are available. Throughout the process, the EPA is required to hold public hearings to inform area residents of their findings and to receive feedback.

As of late 2005, 966 of the 1,547 Superfund sites on the National Priority List had been cleaned up. The average cleanup has cost $25 million and has taken 12–15 years. Many sites are contaminated with hazardous chemicals we have no effective way to deal with. In such cases, cleanups simply involve trying to isolate waste from human contact, either by building trenches and clay or concrete barriers around a site or by excavating contaminated material, placing it in industrial-strength containers, and shipping it to a hazardous waste disposal facility. For all these reasons, the current emphasis in the United States and elsewhere is on preventing hazardous waste contamination in the first place.

Conclusion

Our societies have made great strides in addressing our waste problems. Modern methods of waste management are far safer for people and gentler on the environment than past practices of open dumping and open burning. In many countries, recycling and composting efforts are making rapid strides. The United States has gone in a few decades from a country that virtually did not recycle to a nation in which 30% of all solid waste is diverted from disposal. The continuing growth of recycling, driven by market forces, government policy, and consumer behavior, shows potential to further alleviate our waste problems.

Despite these advances, our prodigious consumption habits have created more waste than ever before. Our waste management efforts are marked by a number of difficult dilemmas, including the cleanup of Superfund sites, safe disposal of hazardous and radioactive waste, and frequent local opposition to disposal sites. These dilemmas make clear that the best solution to our waste problem is to reduce our generation of waste. Finding ways to reduce, reuse, and efficiently recycle the materials and goods that we use stands as a key challenge for the new century.

REVIEWING OBJECTIVES

You should now be able to:

Summarize and compare the types of waste we generate

▶ Municipal and industrial solid waste, hazardous waste, and wastewater are major categories of waste. (p. 648)

List the major approaches to managing waste

▶ Source reduction, recovery, and disposal are the three main components of waste management. (pp. 648–649)

Delineate the scale of the waste dilemma

▶ Developed nations generate far more waste than developing nations, but waste everywhere is increasing as a result of growth in population, wealth, and consumption. (pp. 650–651)

Describe conventional waste disposal methods: landfills and incineration

▶ Sanitary landfills are an improvement over open dumping of the past, and extensive efforts guard against contamination of groundwater, air, and soil. Nonetheless, such contamination can occur. (pp. 651–655)

▶ Incinerators with pollution control technology are an improvement over open burning of the past, and these facilities remove the great majority of pollutants. Nonetheless, some pollutants escape, and highly toxic ash needs to be disposed of in landfills. (pp. 655–657)

▶ We can harness energy from landfill gas and electricity generation from incineration. (p. 657)

Evaluate approaches for reducing waste: source reduction, reuse, composting, and recycling

▶ Reducing waste before it is generated is the best waste management approach. Recycling materials comprises the next-best option. (p. 657)

▶ Industry can reduce waste and save money by increasing efficiency. (pp. 657–658)

▶ Consumers can take an array of simple steps to reduce their waste output. (pp. 657–658)

▶ Composting reduces waste while creating organic matter useful for gardening and agriculture. (p. 658)

▶ Recycling has grown in recent years and currently removes 23.5% of the U.S. waste stream. (pp. 658–663)

Discuss industrial solid waste management and principles of industrial ecology

▶ Although regulations differ, industrial waste management practices are similar to those for municipal solid waste. (pp. 663–664)

▶ Industrial ecology urges industrial systems to mimic ecological systems and provides ways for industry to increase its efficiency. (pp. 664–665)

Assess issues in managing hazardous waste

▶ Hazardous waste is regulated and monitored, yet illegal dumping remains a problem. (pp. 664–667)

▶ No fully satisfactory method of disposing of hazardous waste has yet been devised. (pp. 667–670)

▶ Cleanup of hazardous waste sites is a long and expensive process. (pp. 670–671)

TESTING YOUR COMPREHENSION

1. Describe five major methods of managing waste. Why do we practice waste management?

2. Why have some people labeled the United States "the throwaway society"? How much solid waste do Americans generate, and how does this amount compare to that of people from other countries?

3. Name several guidelines by which sanitary landfills are regulated. Describe at least five problems with landfills.

4. Describe the process of incineration or combustion. What happens to the resulting ash? What is one drawback of incineration?

5. What is composting, and how does it help reduce input to the waste stream?

6. What are the three elements of a sustainable process of recycling?

7. What are the goals of industrial ecology?

8. What four criteria are used to define hazardous waste? Why are heavy metals and synthetic organic compounds particularly hazardous?

9. What are the largest sources of hazardous waste? Describe three ways to dispose of hazardous waste.

10. What is the Superfund program? How does it work?

SEEKING SOLUTIONS

1. How much waste do you generate? Look into your waste bin at the end of the day and categorize and measure the waste there. List all other waste you may have generated in other places throughout the day. How much of this waste could you have avoided generating? How much could have been reused or recycled?

2. Some people have criticized current waste management practices as merely moving waste from one medium to another. How might this criticism apply to the methods now in practice? What are some potential solutions?

3. Of the various waste management approaches covered in this chapter, which ones are your community or campus pursuing, and which are they not pursuing? Would you suggest that your community or campus start pursuing any new approaches? If so, which ones, and why?

4. Can manufacturers and businesses benefit from source reduction if consumers were to buy fewer products as a result? How? Given what you know about industrial ecology, what do you think the future of sustainable manufacturing may look like?

5. Imagine you are the head of a major corporation that produces containers for soft drinks and a wide variety of other consumer products. Your company's shareholders are asking that you improve the company's image—while not cutting into profits—by taking steps to reduce waste. What steps would you consider taking?

6. Think of several industries or businesses in your community, as well as the ways these interact with facilities on your campus. Bearing in mind the principles of industrial ecology, can you think of any novel ways that these entities might mutually benefit from one another's services, products, or waste materials? Are there waste products from one business, industry, or campus facility that another might put to good use? Can you design an eco-industrial park that might work in your community or on your campus?

INTERPRETING GRAPHS AND DATA

Using 1990 data from 149 countries, David Beede, an economist at the U.S. Department of Commerce, and David Bloom, a professor of economics at Columbia University, examined global patterns in the generation and management of municipal solid waste (MSW). Beede and Bloom were particularly interested in the relationships among wealth, population size, and per capita generation of MSW. Their results are presented in the accompanying table.

Income category of nation	Total MSW generation		Population size		Percentage of world total GDP	Pounds MSW per capita per day
	Tons/yr (billions)	% of world total	Millions of people	% of world total		
Low	0.658	46.3	3,091	58.5	18.7	1.17
Low-middle	0.160	11.2	629	11.9	9.9	
Upper-middle	0.212	14.9	748	14.2	16.5	
High	0.393	27.6	816	15.4	54.9	
All economies	1.422	100.0	5,284	100.0	100.0	

Data from Beede, D. N. and D. E. Bloom. 1995. The economics of municipal solid waste. *World Bank Research Observer* 10: 113–150.

1. Using the data for total MSW generation and for population size, calculate the pounds of MSW per capita per day for each category of nation.
2. Describe in general terms the relationship between wealth and per capita generation of MSW. Now create a graph that demonstrates this relationship, using the data from the table. Plot the values you calculated for per capita waste generation for each of the four wealth categories of nations. Can you offer at least one possible reason for the trend that you see?
3. Do you think it's possible for wealthy nations to reduce their per capita MSW generation to the rates of poorer nations? Why or why not?

CALCULATING ECOLOGICAL FOOTPRINTS

The 14th annual "State of Garbage in America" survey documents the prodigious ability of U.S. residents to generate municipal solid waste (MSW). According to the survey, on a per capita basis, South Dakotans generate the least MSW (3.72 lb/day), and Kansans the most (9.48 lb/day). The average for the entire country is 7.17 lb MSW per person per day. Compare this number to the data in the "Interpreting Graphs and Data" table. Now calculate the amount of MSW generated in 1 day and in 1 year by each of the groups indicated at each of the rates shown in the accompanying table.

Groups generating municipal solid waste	Per-capita MSW generation rates							
	U.S. average (7.17 lb/day)		South Dakota (3.72 lb/day)		"High-income" countries (2.64 lb/day)		World average (1.47 lb/day)	
	Day	Year	Day	Year	Day	Year	Day	Year
You	7.17	2,617.05						
Your class								
Hometown								
Your state								
United States								
World								

Data sources: Kaufman, S. M. et al. 2004. The state of garbage in America. *BioCycle* 45: 31–41.

1. Suppose your hometown of 50,000 people has just approved construction of a landfill nearby. Estimates are that it will accommodate 1 million tons of MSW. Assuming the landfill is serving only your hometown, for how many years will it accept waste before filling up? How much longer would a landfill of the same capacity serve a town of the same size in another industrialized ("high-income") country?
2. Why do you think U.S. residents generate so much more MSW than people in other "high-income" countries, when standards of living in those countries are comparable?

Take It Further

Go to www.aw-bc.com/withgott or the student CD-ROM where you'll find:

▶ Suggested answers to end-of-chapter questions
▶ Quizzes, animations, and flashcards to help you study
▶ *Research Navigator*™ database of credible and reliable sources to assist you with your research projects

▶ **GRAPHit!** tutorials to help you master how to interpret graphs
▶ **INVESTIGATEit!** current news articles that link the topics that you study to case studies from your region to around the world

23 Sustainable Solutions

Sri Lankan children planting tree seedlings on deforested hillsides around their village

Upon successfully completing this chapter, you will be able to:

▶ List and describe approaches being taken on college and university campuses to promote sustainability

▶ Explain the concept of sustainable development

▶ Discuss how protecting the environment can be compatible with promoting economic welfare

▶ Describe and assess key approaches to designing sustainable solutions

▶ Analyze the roles that consumption, population, and technology play in efforts to achieve sustainability

▶ Explain how time is limited but how human potential to solve problems is tremendous

FIGURE 23.3 In October 2005, 18 teams from colleges and universities across the United States and the world converged on the Mall in Washington, D.C., for the second Solar Decathlon. Each team erected an entire house, of the students' own design, fully powered by solar energy. The University of Colorado won the 2005 competition, edging out Cornell University and California Polytechnic State University. The next Solar Decathlon will be held in fall 2007.

artificial wetland. The treated wastewater is reused in toilets and to irrigate campus plantings. Oberlin's building is set in a landscape of orchards and gardens, with urban agriculture and lawns of grass specially bred to require less chemical care. Students have reforested areas around the building and have restored a wetland. These mimicked natural systems help slow and cleanse stormwater runoff.

The Lewis Center was expensive to construct, and Oberlin failed to incorporate sustainable building practices in later buildings. However, other institutions have adopted sustainable practices in multiple buildings and at less expense, and the movement for "green buildings" continues to expand. One newer example, opened in late 2005, is the Kirsch Center for Environmental Studies at DeAnza College in California. This LEED-certified building is energy-efficient, water-efficient, and climate-responsive; is built with recycled, renewable, and nontoxic materials; is solar-powered; and includes outdoor learning spaces and labs for pollution prevention, energy management, biodiversity, and geographic information systems (GIS; ▸ pp. 352–353).

Efficient energy and water use are vital

Building design is a primary way to reduce the amount—and cost—of energy and water used. In October 2005, teams of students from 18 universities competed in the second-ever *Solar Decathlon*. This remarkable event is sponsored by the U.S. Department of Energy's Office of Energy Efficiency and Renewable Energy, with several co-sponsors. Teams of students from each institution travel

to the National Mall in Washington, D.C., bringing material for the solar-powered homes they have spent months designing. They erect the homes on the Mall, where they stand for 3 weeks (Figure 23.3). The homes are judged on 10 criteria, and prizes are awarded to winners in each category. One of the teams, from the University of Massachusetts at Dartmouth, donated its home to Habitat for Humanity for a Washington, D.C., family to live in. The team is also partnering with Habitat for Humanity to build several other homes with solar technology. The next Solar Decathlon will take place in fall 2007.

Solar energy plays a role on many campuses. The University of Vermont installed the largest solar panel installation in its state to generate part of the campus's power and to demonstrate the feasibility of solar energy. The 500 ft^2 array of panels provides enough electricity for nine desktop computers or 95 energy-efficient light bulbs for 10 hours each day. Other campuses are using wind power for their energy needs. MacAlester College in Minnesota even installed a wind turbine on its campus.

For those students with less inclination to build renewable energy technology themselves, any educational institution can invest in renewable energy by purchasing "green tags" that subsidize wind power and other renewable energy sources. College of the Atlantic in Maine offset 100% of its greenhouse gas emissions by buying green tags for renewable energy.

Students at Lewis and Clark College in Oregon took a dramatic step toward sustainable energy use in 2002, when they made their school the first in the nation to comply with the terms of the Kyoto Protocol (▸ pp. 550, 552).

After conducting an audit of the campus, student leaders reduced greenhouse gas emissions by the percentage required under the Protocol, largely by purchasing carbon offsets from a nonprofit that funds energy efficiency and revegetation projects. Although U.S. leaders have repeatedly refused action to slow greenhouse gas emissions at the national level, citing economic expense, Lewis and Clark students found that becoming Kyoto-compliant on their campus cost only $10 per student per year.

To promote efficiency in water use, sustainability advocates at Manhattanville College in New York persuaded 217 people to reduce the length of their showers one day. On average, people cut their showers short by 30%, saving a total of 5,173 gallons of water. Water-saving technologies, such as living machines and "waterless urinals," also are being installed at a number of campuses. The University of British Columbia in Vancouver, Canada, has had great success in reducing energy and water use. The UBC student body increased by 15% between 1998 and 2003, yet in those 5 years the school reduced energy use by 6%, carbon dioxide emissions by 4%, and water use by 15% (enough water to fill the nearby Vancouver Aquarium 82 times)—all of which saved the university $2 million. Moreover, UBC is sinking $35 million into retrofitting its buildings with energy-efficient upgrades, which should save the school at least $2.5 million annually. In academic buildings, these upgrades should reduce energy use by 20%, water use by 45%, and carbon emissions by 15,000 metric tons each year.

Dining services and campus gardens let students eat sustainably

Campus food service operations can promote sustainable practices by buying organic produce, by purchasing food in bulk or food that involves less packaging, and by composting food scraps. In addition, buying locally grown or produced food supports local economies and cuts down on fuel consumption from long-distance transportation. Some college campuses even have gardens where students can help grow their own food (Figure 23.4).

Sterling College in Vermont provides a model for sustainable campus food service. Many foods are organic, grown by local farmers, or produced by Vermont-based companies; moreover, some foods are grown and breads are baked right on the Sterling campus. Food shipped in is purchased in bulk to reduce packaging. Dish soap is non-petroleum-based and biodegradable, and all kitchen scraps are composted, along with recycled unbleached paper products. By reducing unnecessary equipment use, the college saved money and slashed its electricity use by 23%.

FIGURE 23.4 A number of campuses now include gardens where students can grow organic vegetables that are used for meals in dining halls. Here, a student works in the garden at Middlebury College in Vermont.

Institutional purchasing can be influential

The kinds of purchasing decisions made in dining halls favoring local food, organic food, and biodegradable products are relevant across the entire spectrum of a campus's needs. Ball State's drive to convert to recycled paper is one of many examples in this area. When college and university purchasing departments preferentially buy certified sustainable wood, energy-efficient appliances, goods with less packaging, and other ecolabeled products, they send signals to manufacturers and increase the economic demand for such items.

At Chatham College in Pennsylvania, students chose to honor their school's best-known alumnus, Rachel Carson (▶ pp. 67, 408–409), by seeking to eliminate toxic chemicals used on campus. Administrators agreed to this effort, provided that alternative products to replace the toxic ones worked just as well and were not more expensive. Students brought in the CEO of a company that produced nontoxic cleaning products, who demonstrated to the janitorial staff that his company's products were superior. The university switched to the nontoxic products, which were also cheaper, and proceeded to save $10,000 per year. Chatham students also found a company offering paint without volatile organic compounds and negotiated with it for a free paint job and discounted prices on later purchases. Students also worked with grounds staff to eliminate herbicides and fertilizers used on campus lawns and to find alternative treatments.

Transportation alternatives are many

Many colleges and universities struggle with traffic and parking congestion, sprawl, commuting delays, and pollution from vehicle exhaust. Many are addressing these issues by establishing or expanding bus and shuttle systems; encouraging bicycling, walking, and carpooling; and introducing alternative fuels and vehicles to university fleets. Like Ball State, a number of campuses are using biodiesel fuels, including Hobart and William Smith Colleges in New York, the University of South Carolina, and the University of Vermont. Middlebury College in Vermont has gone farther than all others in this regard— thousands of miles, in fact. Middlebury students began Project Bio Bus, which crisscrosses North America each summer in a biodiesel bus spreading the gospel of this alternative fuel.

Other colleges and universities have also, like Ball State, added alternative vehicles to their fleets (Figure 23.5). Middlebury leases electric vehicles. The State University of New York College of Environmental Science and Forestry in Syracuse acquired six electric vehicles for on-campus use, a gas-electric hybrid car for the fleet, and a delivery van that runs on compressed natural gas. At the same time, it converted its buses to biodiesel.

The University of British Columbia is a leader in transportation efforts. For $20 a month, UBC's "U-Pass" program provides students with biking programs and facilities, expanded campus bus and shuttle services, unlimited use of some city transit systems, rides home in

FIGURE 23.6 Many schools have embarked on habitat restoration projects to beautify their campuses, provide wildlife habitat, restore native plants, and filter water runoff. Here, a student and a volunteer work to maintain a pond in the Cheeseman Environmental Study Area at DeAnza College in California, which showcases a sampling of plant communities native to the region.

emergencies, merchant discounts, and priority parking spaces and ride-matching services for carpoolers. The program has boosted transit ridership to campus by 53% and has decreased single-person car use by 20%.

Campuses are restoring native plants, habitats, and landscapes

No campus sustainability program would be complete without some effort to enhance the campus's natural environment. Along with recycling programs, programs for natural landscaping and habitat restoration make up the most widespread campus sustainability initiatives. Such programs remove invasive species, restore native plant species and communities, improve habitat for wildlife, enhance soil and water quality, reduce pesticide use, and create healthier, more attractive surroundings.

Native plant restoration has proven popular, from desert plants at Arizona State University in Tempe, to prairie plants at University of West Alabama, where a student crew has conducted controlled burns. Some schools, such as California's DeAnza College, feature demonstration gardens of native plant communities (Figure 23.6). Others, such as Loyola College in Maryland, focus on removing invasive species. Still others, such as Ohio State University and the New College of California in San Francisco, have rooftop gardens. At Clemson University, students work with churches and K-12 schools in the community and

FIGURE 23.5 Most buses at the University of California at Davis run on compressed natural gas, and one runs on a novel mixture of natural gas and hydrogen. Several of these low-emission vehicles are shown here lined up in front of Freeborn Hall during a press conference celebrating their purchase.

help them design landscaping plans. At Warren Wilson College in North Carolina, 40 student native plant enthusiasts and the landscaping supervisor used local plants to establish a stock of seeds, built a greenhouse for rearing plants, expanded their arboretum, and started propagating various species of grasses and wildflowers. The landscaping crew uses organic fertilizers and avoids using chemical pesticides. At Northland College in Wisconsin, sustainability advocates wanting to improve biodiversity and water quality have managed to re-landscape half of the campus. They have replaced invasive plants with native ones and have designed and planted a series of meadows that capture stormwater runoff and filter out pollution.

Weighing the Issues:
Sustainability on Your Campus

Are any sustainability efforts being made on your campus? What efforts would you like to see on your campus? Do you foresee any obstacles to these efforts? How could these obstacles be overcome?

Organizations are available to assist campus efforts

Many campus sustainability initiatives are supported by organizations such as University Leaders for a Sustainable Future and the National Wildlife Federation's Campus Ecology program. These organizations act as information clearinghouses for campus sustainability efforts. Each year the NWF program recognizes the most successful campus sustainability initiatives, an award Ball State shared in 2003–2004 along with six other schools. With Ball State's Greening of the Campus conferences and the assistance of these organizations, it is easier than ever to start sustainability efforts on your own campus and obtain the support to carry them through to completion.

Sustainability and Sustainable Development

Efforts toward sustainability on college and university campuses parallel efforts in the world at large. As more people come to appreciate Earth's limited capacity to accommodate our rising population and consumption, they are voicing concern that we will need to modify our behaviors, institutions, and technologies if we wish to sustain our civilization and the natural environment on

which it depends. In the quest for sustainability, the strategies pursued on campuses reflect those pursued in the wider society and also can serve as models.

When people speak of *sustainability,* what precisely do they mean to sustain? Generally they mean to sustain the natural environment, its biodiversity, and its ecological systems in a healthy and functional state—and also to sustain human civilization in a healthy and functional state. The short-term needs of human society and of environmental sustenance are often cast as being in opposition, but environmental scientists recognize that our civilization cannot exist without an intact and functional natural environment. The contributions of biodiversity (▸ pp. 324–328) and ecosystem goods and services (▸ pp. 39–41, 48–51) to human welfare are tremendous. Indeed, they are so fundamental (some would say infinitely valuable, thus literally priceless) that we have long taken them for granted.

Sustainable development involves environmental protection, economic welfare, and social equity

In recent years, people have increasingly drawn the connection between environmental quality and human quality of life. Moreover, we now recognize that very often it is society's poorer people who suffer the most from environmental degradation. This realization has led advocates of environmental protection, advocates of economic development, and advocates of social justice to work together toward common goals. This cooperative approach has given rise to the modern drive for sustainable development.

We first encountered sustainable development in our opening chapter (▸ p. 22), and offered the United Nations' definition: "Development that meets the needs of the present without compromising the ability of future generations to meet their own needs." This definition is taken from the U.N.-sponsored Brundtland Commission (named after its chair, Norwegian prime minister Gro Harlem Brundtland), which produced an influential 1987 report titled *Our Common Future.*

Prior to the Brundtland Report, most people aware of human impact on the environment might have thought "sustainable development" to be an oxymoron—a phrase that contradicts itself. *Development* involves making purposeful changes intended to improve the quality of human life, yet environmental advocates have long pointed out that development often so degrades the natural environment that it threatens the very improvements for human life that were intended. Today many people remain under the impression that protecting the environment is incompatible with serving people's needs. However,

FIGURE 23.7 The 2002 U.N.-sponsored World Summit on Sustainable Development in Johannesburg, South Africa, drew over 10,000 delegates from 200 nations to discuss and set sustainable development goals. In the photo, South African President Thabo Mbeki hugs a boy who performed in the welcoming ceremony. Delegates launched hundreds of initiatives and made millions of dollars of commitments at this conference, but most sustainability proponents said these efforts did not go nearly far enough.

efforts to develop sustainably, undertaken by individuals at all levels—from students on campus to international representatives at the United Nations (Figure 23.7)—are beginning to alter this perception and produce models of success. The question "Can we develop in a sustainable way?" may well be the single most important question in the world today. The myriad actions being taken at campuses such as Ball State's and by governments, businesses, industries, organizations, and individuals across the globe are giving people optimism that developing in sustainable ways is possible.

Environmental protection can enhance economic opportunity

How can protecting environmental quality be good for the economic bottom line? For individuals, businesses, and institutions, reducing resource consumption and waste often saves money—as many colleges and universities discover when they embark on sustainability initiatives. Sometimes savings accrue immediately, and other times an up-front investment results in long-term savings. For society as a whole, attention to environmental quality can also enhance economic opportunity for citizens by providing new types of employment. This contrasts with the common perception that environmental protection measures hurt the economy by costing people

jobs. A prime example is the controversy over logging of old-growth forest in the Pacific Northwest (Chapter 12). Protection for the northern spotted owl (Figure 23.8) under the U.S. Endangered Species Act (▸ pp. 331–334) set limits on timber extraction in the northwestern United States. Proponents of logging rallied opposition to forest protection by claiming that the restrictions cost local loggers their jobs. Some job loss has indeed occurred, but loggers' jobs are far more at risk when timber companies cut trees at unsustainable rates and then leave a region, seeking mature forests elsewhere. As we saw in Chapter 12 (▸ pp. 350–351), this has happened in region after region throughout U.S. history. It is playing out in the Pacific Northwest now, as U.S. timber companies begin moving operations abroad.

The jobs-versus-environment debate frequently overlooks the fact that as some industries decline, others spring up to take their place. As jobs in logging, mining, and manufacturing have decreased in developed nations over the past few decades, jobs have greatly increased in many service occupations and in computer and other high-technology sectors. If we decrease our dependence

FIGURE 23.8 The northern spotted owl (*Strix occidentalis occidentalis*) has become a symbol of the "jobs-versus-environment" debate. This bird of the Pacific Northwest rainforest is considered endangered because of the logging of mature forests. Proponents of logging argue that laws protecting endangered species cause economic harm and job loss. Advocates of endangered species protection argue that unsustainable logging practices pose a larger risk of job loss.

on fossil fuels, experts predict, jobs and investment opportunities will open up in renewable energy sectors, such as wind power and fuel cell technology (Chapter 18).

Moreover, people desire to live in areas that have clean air and water, intact forests, and parks and open space. Environmental protection increases a region's attractiveness, drawing more residents and increasing property values and the tax revenues that help fund social services. As a result, those regions that act to protect their environments are generally the ones that retain and increase their wealth and quality of life.

Thus, environmental protection need not lead to economic stagnation, but instead is likely to enhance economic opportunity. A recent U.S. government review concluded that the economic benefits of environmental regulations greatly exceed their economic costs (see "The Science behind the Story," ▶ pp. 686–687). This connection is also suggested by the fact that both the U.S. and global economies have expanded rapidly in the past 30 years, the very period during which environmental protection measures have proliferated.

What accounts for the perceived economy-versus-environment divide?

If environmental protection and economic development are compatible and even mutually reinforcing, what, then, accounts for the conventional view that we cannot provide for people's needs and protect the environment simultaneously? One proximate explanation lies in the fact that economic development since the industrial revolution has so clearly diminished biodiversity, decreased habitat, and degraded ecological systems. A second proximate explanation lies in the fact that many people view command-and-control environmental policy (▶ p. 71) to pose excessive costs for industry and restrict rights of private citizens.

An ultimate explanation may lie in our species' long history. For the thousands of years that we lived as nomadic hunter-gatherers, population densities were low and consumption and environmental impact were limited. With natural resources in little danger of running out, humans were free to exploit them limitlessly and had little reason to adopt a conservation ethic. The establishment of sedentary agricultural societies beginning about 10,000 years ago, followed by the urbanization that continues today, has increased our impact while also causing us to overlook the connections between our economies and our environments. It is common to hear "humans and the environment" or "people and nature" being set in contrast, as though they were completely separate. Some

philosophers venture to say that the perceived dichotomy between humans and nature is the root of all our environmental problems.

Humans are not separate from the environment

On a day-to-day basis, it is easy to feel disconnected from the natural environment, particularly in developed nations and large cities. We live inside houses; work in shuttered buildings; travel in cars, airplanes, and other enclosed vehicles; and generally know little about the plants and animals around us. Millions of urban citizens have never set foot in an undeveloped area. Just a few centuries or even decades ago, most of the world's people were able to name and describe the habits of the plants and animals that lived nearby. They knew exactly where their food, water, and clothing came from. Today it seems that water comes from the faucet, clothing from the mall, and food from the grocery store. It's little wonder we have lost track of the connections that tie us to the natural environment.

However, this psychological severance, this feeling of isolation, doesn't make our connections to the environment any less real. Consider a thoroughly un-"natural" (yet delicious!) invention of the human species: the banana split (Figure 23.9). Even in this triumph of human creation, seemingly concocted *de novo* at an ice cream shop, each and every element has ties to the resources of the natural environment and exerts environmental impacts. Once we learn to consider where the things we use and value each day actually come from, it becomes easier to see how humans are part of the environment. And once we reestablish this connection, it becomes readily apparent that our own interests are best served by preservation or responsible stewardship of the natural systems around us.

Because what is good for the environment can also be good for people, win-win solutions are very much within reach. We *can* have it both ways, if we learn from what science can teach us, think creatively, and act on our ideas. Ultimately, the only way for our species to "win"—to survive and thrive—is to conduct our activities in ways that sustain the processes and resources of our environment. If the environment "loses" because of our impacts, then so will we.

Weighing the **Issues:**
Unavoidable Impacts?

Ecology teaches us that one species may exclude other species from resources. This has clearly been the case with humanity's recent spread across the planet.

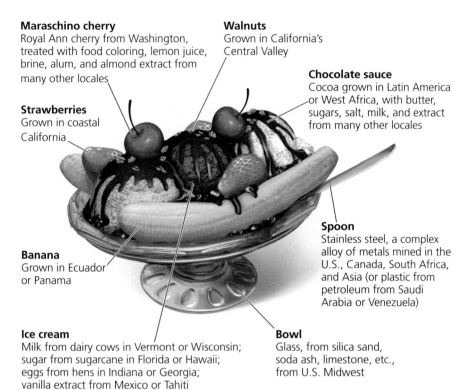

Maraschino cherry
Royal Ann cherry from Washington, treated with food coloring, lemon juice, brine, alum, and almond extract from many other locales

Strawberries
Grown in coastal California

Banana
Grown in Ecuador or Panama

Ice cream
Milk from dairy cows in Vermont or Wisconsin; sugar from sugarcane in Florida or Hawaii; eggs from hens in Indiana or Georgia; vanilla extract from Mexico or Tahiti

Walnuts
Grown in California's Central Valley

Chocolate sauce
Cocoa grown in Latin America or West Africa, with butter, sugars, salt, milk, and extract from many other locales

Spoon
Stainless steel, a complex alloy of metals mined in the U.S., Canada, South Africa, and Asia (or plastic from petroleum from Saudi Arabia or Venezuela)

Bowl
Glass, from silica sand, soda ash, limestone, etc., from U.S. Midwest

FIGURE 23.9 A banana split eaten at an ice cream shop in Tulsa, Denver, or Des Moines consists of ingredients from around the world, whose production has impacts on the environments of many far-away locations. Ice cream requires milk from dairy cows that graze pastures or are raised in feedlots on grain grown in industrial monocultures. Ice cream is sweetened with sugar from sugar beet farms or sugarcane plantations. The banana was shipped thousands of miles by oil-fueled transport from a tropical country, where it grew on a plantation that displaced rainforest and where it was liberally treated with fertilizers and fungicides. Fruits and nuts grown in California's Central Valley were irrigated generously with water piped in from the Sierras. The spoon originated with metal ores mined along with thousands of tons of soil and processed into stainless steel using energy from fossil fuels.

However, in the long scope of Earth's history, biodiversity has increased as organisms have evolved and adapted to environmental changes caused by other organisms. Does our society's development demand some amount of unavoidable harm to other species or to environmental processes? What kinds of impacts might be unavoidable, and what kinds might we be able to avoid?

Strategies for Sustainability

Sustainable solutions to environmental problems are numerous, and we have seen specific examples throughout this book. The challenges lie in being imaginative enough to think of solutions and then being shrewd and dogged enough to overcome whatever political or economic obstacles may lie in the path of their implementation. We will now summarize several broad strategies or approaches that can spawn sustainable solutions (Table 23.2).

We can refine our ideas about economic growth and quality of life

It is conventional among economists and the policymakers who heed their advice to speak of economic growth as an ultimate goal. Many politicians see nurturing an expanding economy as their prime responsibility. Yet economic growth is merely a tool with which we try to attain the real goal of maximizing human happiness. Thus if economic growth depends on an ever-increasing consumption of nonrenewable resources, then we may want to ask ourselves whether attaining happiness truly requires endlessly expanding the size of our economy (▸ pp. 44–47).

Another question we may ask is whether and how to incorporate external costs (▸ pp. 43–44) into the market

Table 23.2 Some Major Approaches to Sustainability
▸ Refine our ideas about economic growth and quality of life
▸ Reduce unnecessary consumption
▸ Limit population growth
▸ Encourage green technologies
▸ Mimic natural systems by promoting closed-loop industrial processes
▸ Think in the long term
▸ Enhance local self-sufficiency and be mindful of globalization
▸ Vote with our wallets
▸ Vote with our ballots
▸ Promote research and education

Assessing Costs and Benefits of Environmental Regulations

The Science behind the Story

Federal regulations that aim to protect environmental quality often result in clearer air or cleaner water. They also usually have financial implications for affected parties. Are such regulations worth the costs to industry, businesses, and consumers? In 2003, a federal study determined that the answer to that question is a definitive yes.

The study, from the White House Office of Management and Budget (OMB), weighed the costs of a range of recently created federal regulations against the economic benefits resulting from the regulations. The OMB examined 107 federal regulations enacted in the United States from 1992 to 2002, many of which dealt with environmental issues.

The study found that the reviewed regulations carried a sizable annual cost, estimated at $36–42 billion. However, their estimated value in terms of public good totaled $146–230 billion. Thus, the economic benefits of these regulations far exceeded their costs. Moreover, of all the regulations, those dealing with environmental protection were found to be especially cost-effective.

Most regulatory benefits were reflected in health and social gains resulting from clean air. Over the studied 10-year period, the benefits of clean-air regulations were found to be 6–10 times greater than the costs of complying with the regulations. The value of reductions in hospitalization and emergency room visits, premature deaths, and lost workdays resulting from improved air quality were estimated to be between $118 billion and $177 billion.

The authors of the OMB review relied on cost-benefit analyses previously made by experts in the EPA and other agencies, scrutinizing them for appropriateness and accuracy before using them. A look at one EPA regulation included in the OMB report—an air pollution regulation called the *Heavy Duty Engine/Diesel Fuel Rule*—shows how the EPA performed its own economic analysis and how the OMB double-checked the EPA's findings.

The diesel regulation aimed to address diesel engines that emit fine soot, sulfur compounds, and nitrogen oxides (NO_x). In the 1990s, the EPA regulation was intended to make diesel trucks and buses run more cleanly by lowering the fuel's sulfur content. However, achieving cleaner diesel fuel would mean substantial—and potentially costly—changes to the way that fuel is made and handled. To determine potential costs, EPA analysts examined how various parties involved with diesel fuel use—from fuel processors to engine manufacturers to vehicle operators—would need to change their processes or equipment to clean up the fuel. For example, a heavy-duty engine redesigned to run more efficiently on cleaner fuel would cost approximately $4,600 more to operate, the analysts calculated. Total research and development costs for emission control were predicted to exceed $600 million. The EPA asked for input from affected parties, scientists, and the public, finally determining that removing sulfur from diesel fuel would be neither cheap nor quick, with total annualized costs of about $4.2 billion by 2030.

EPA analysts then considered the benefits of reduced diesel pollution. Using air quality data from more than 940 smog monitors around the country, the EPA estimated diesel-related air pollution levels with and without the proposed diesel regulation. The analysts examined public

prices of goods and services. Currently, goods and services are priced as though pollution and resource extraction involved no costs to society. If we can make our accounting practices reflect indirect negative consequences and provide a clearer view of the full costs and benefits of any given action or product, then the free market itself can become a force for improving environmental quality, our economy, and our quality of life. If we can incorporate external costs into our price structures, then market capitalism could potentially become the optimal tool for achieving prosperous and sustainable economies. Moreover, implementing green taxes (▶ pp. 72–73) and phasing out harmful subsidies could hasten the shift to sustainability, many economists maintain. The political obstacles to this are considerable, however. Such changes will require educated citizens to push for them and courageous policymakers to implement them.

Estimates of the Total Annual Benefits and Costs of Major Federal Rules, October 1, 1992 to September 30, 2002

Agency	Benefits (millions of 2001 dollars)	Costs (millions of 2001 dollars)
Agriculture	3,094–6,176	1,643–1,672
Education	655–813	361–610
Energy	4,700–4,768	2,472
Health & Human Services	9,129–11,710	3,165–3,334
Housing & Urban Development	551–625	348
Labor	1,804–4,185	1,056
Transportation	6,144–9,456	4,220–6,718
Environmental Protection Agency	120,753–193,163	23,359–26,604
Total	**146,812–230,896**	**36,625–42,813**

Source: U.S. Office of Management and Budget. 2003. *Informing regulatory decisions: 2003 report to Congress on the costs and benefits of federal regulations and unfunded mandates on state, local, and tribal entities.* Washington, D.C.: U.S. OMB.

health statistics, especially data related to respiratory illnesses such as asthma and chronic bronchitis. They then estimated how many cases of disease or premature death might be prevented if people breathed fewer diesel-related pollutants. They determined the economic effects of such health problems by determining the medical, workforce, and social costs. Each premature death was assigned an impact of $6 million, and each case of chronic bronchitis was pegged at $331,000. Many of the estimates were also regionalized to account for different economic conditions in various parts of the country.

In the end, the EPA analysis determined, the benefits of cleaner diesel fuel far exceeded the costs. If sulfur content of diesel were cut from 500 to 15 parts per million, then 2.6 million tons of smog-causing NO_x emissions would be eliminated each year. An estimated 8,300 premature deaths, 5,500 cases of chronic bronchitis, over 360,000 asthma attacks, and about 1.5 mil-

lion lost workdays would be prevented each year. By 2030, the analysis determined, benefits would reach about $70 billion. Following this analysis, the Heavy Duty Engine/Diesel Fuel Rule took effect in early 2001. Requirements are being phased in gradually through 2010.

The OMB, in compiling its own review of federal rules, scrutinized the EPA study to determine whether the EPA estimates made economic sense. The OMB analysts concluded that most of the estimates did, indeed, make sense. The OMB analysts made some minor changes when calculating their own figures. For example, although the EPA said every 1-ton reduction in NO_x emissions represented a benefit of $10,200, the OMB used its own updated figures and determined NO_x reductions to be worth $5,500 per ton. Overall, however, the OMB confirmed the validity of the EPA estimates and used most of the EPA's economic data in the OMB report.

The OMB report, in totaling and highlighting the value of many recent environmental regulations, is now being used to evaluate and support efforts at environmental regulation around the country.

We can consume less

Economic growth is largely driven by consumption, the purchase of material goods and services (and thus the use of resources involved in their manufacture) by consumers (Figure 23.10). Our tendency to believe that more, bigger, and faster are always better is reinforced by advertisers seeking to sell more goods more quickly. Consumption has grown tremendously, with the wealthiest nations leading

the way. The United States, with less than 5% of the world's population, consumes 30% of world energy resources and 40% of total global resources. U.S. houses are larger than ever, sports-utility vehicles are among the most popular automobiles, and many citizens have more material belongings than they know what to do with. We think nothing now of having home computers with high-speed Internet access, let alone the televisions, telephones, refrigerators, and dishwashers that were marvels just decades ago.

FIGURE 23.10 Citizens of the United States consume more than the people of any other nation. Unless we find ways to increase the sustainability of our manufacturing processes, this rate of consumption cannot be sustained in the long run.

Because many of Earth's natural resources are limited and nonrenewable, consumption cannot continue growing forever. Eventually, if we do not shift to sustainable practices of resource use, per capita consumption will drop for rich and poor alike as resources dwindle. Cornucopian critics often scoff at the notion that resources are limited, but we must remember that our perspective in time is limited and that our consumption is taking place within an extraordinarily brief slice of time in the long course of history (Figure 23.11). Our lavishly consumptive lifestyles are a brand-new phenomenon on Earth. We are enjoying the greatest material prosperity in all of human history, but if we do not find ways to make our wealth sustainable, the party may not last much longer.

Fortunately, material consumption is only one way to measure prosperity, and consumption alone does not reflect a person's quality of life. For many people in the United States and other developed nations, the accumulation of innumerable possessions has not brought contentment. Social critics have coined a word for the way material goods often fail to bring happiness to people affluent enough to afford them: **affluenza.** Economic growth is generally equated with "progress," but for many people, true progress consists of an increase in human happiness, not simply growth in material wealth. In the end we are, one would hope, more than simply the sum of what we buy.

We can reduce our consumption while enhancing our quality of life—squeezing more from less—in at least three ways. One way is to improve the technology of materials and the efficiency of manufacturing processes, so that industry produces goods using fewer natural resources. Another way is to develop a sustainable manu-

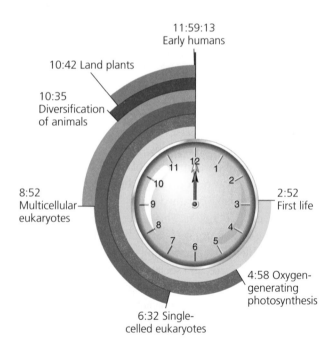

FIGURE 23.11 By viewing Earth's 4.5-billion-year history as a 12-hour clock, we can gain a better understanding of relative timescales across the immense span of geological time. *Homo sapiens,* as a species, has come into existence only during the final 1 or 2 seconds, around 11:59:59. The agricultural and industrial revolutions that have increased our environmental impacts have taken up only a minuscule fraction of a second.

facturing system—one that is circular and based on recycling, in which the waste from a process becomes raw material for input into that process or other processes (▸ pp. 664–665). A third way is for each of us to modify our behavior, attitudes, and lifestyles to minimize consumption. At the outset, such choices may seem like sacrifices, but people who have slowed down the pace of their busy lives and freed themselves of an attachment to excess material possessions say it can feel tremendously liberating.

Population growth must eventually cease

Just as continued growth in consumption is not sustainable, neither is growth in the human population. We have seen (▶ pp. 136–137) that populations may grow exponentially for a time but eventually encounter limiting factors and level off at a carrying capacity. We have increased Earth's carrying capacity for our species with the help of technology, but our population growth cannot continue forever; sooner or later, the human population will stop growing. The question is how: through war, plagues, and famine, or through voluntary means as a result of wealth and education?

The demographic transition (▶ pp. 228–229) that many nations are undergoing provides reason to hope that population sizes will stabilize and begin to fall. This transition is already far along in many developed nations thanks to urbanization, wealth, education, and the empowerment of women. If the demographic transition occurs for today's developing nations, then there is hope that humanity may halt its population growth while creating more prosperous and equitable societies.

Technology can help us toward sustainability

It is largely technology—developed with the agricultural revolution, the industrial revolution, and advances in medicine and health—that has spurred our population increase. Technology has magnified our impacts on Earth's environmental systems. However, technology also can give us ways to reduce environmental impact. Technology is not an independent actor, but is a tool of human agency. The shortsighted use of technology may have gotten us into this mess, but wiser use of environmentally friendly, or "green," technologies can help get us out.

Recall the I=PAT equation (▶ pp. 220–221), which summarizes human environmental impact (I) as the interaction of population (P), consumption or affluence (A), and technology (T). Technology can exert either a positive or negative value in this equation. In recent years, technology has exacerbated environmental impact in developing countries as industrial technologies from the developed world have been exported to poorer nations eager to industrialize. In developed nations, meanwhile, green technologies have begun mitigating

our environmental impact. Catalytic converters on cars have reduced emissions (Figure 23.12), as have scrubbers on industrial smokestacks (▶ p. 656). Recycling technology and advances in wastewater treatment are helping reduce our waste output. Solar, wind, and geothermal energy technologies are producing cleaner renewable energy. Countless technological advances such as these are one reason that people of the United States and western Europe today enjoy cleaner environments, although they consume far more, than people of eastern Europe and rapidly industrializing nations such as China.

Weighing the Issues:
Renewable Energy and Sustainability

Environmental advocates who promote alternative energy sources such as solar, wind, and geothermal power often argue that, given adequate development, green technologies to exploit such renewable sources could provide us clean and limitless energy and thereby help bring a sustainable future. Do you agree? Or do you think abundant renewable energy might simply lead to further unsustainable development and cause further environmental impact? How might a shift to renewable energy sources influence the quest for sustainable development?

Industrial systems can mimic natural systems by recycling and become circular

As industries seek to develop green technologies and sustainable practices, they have available to them an excellent model: nature itself. Natural systems are sustainable; they've been around far longer than we have. As we saw in Chapter 7 and throughout this book, environmental systems tend to operate in cycles consisting of feedback loops and the circular flow of materials. In natural systems, output is recycled into input. In contrast, human manufacturing processes have run on a linear model, in which raw materials are input, go through processing, and create a product, while byproducts and waste are generated and disposed of.

Some forward-thinking industrialists are already making their industrial processes sustainable by transforming linear pathways into circular ones, in which waste is recycled and reused. For instance, several companies now produce carpets with materials that can be retrieved from the consumer when the carpet is worn out, and these materials are then recycled to create new carpeting (recall that

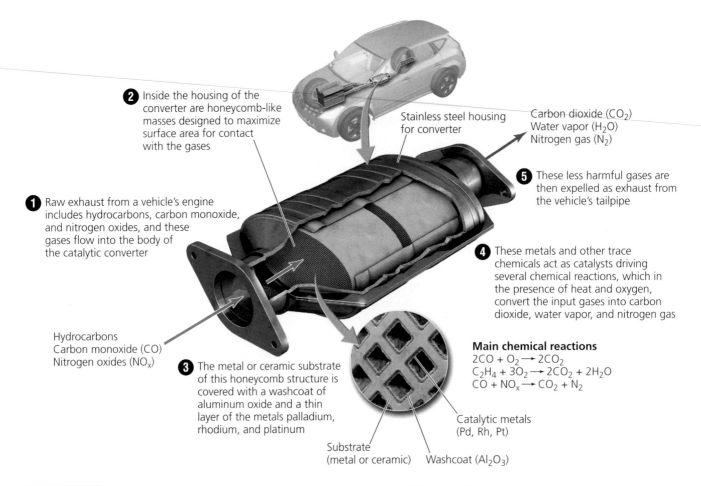

2 Inside the housing of the converter are honeycomb-like masses designed to maximize surface area for contact with the gases

Stainless steel housing for converter

Carbon dioxide (CO_2)
Water vapor (H_2O)
Nitrogen gas (N_2)

5 These less harmful gases are then expelled as exhaust from the vehicle's tailpipe

1 Raw exhaust from a vehicle's engine includes hydrocarbons, carbon monoxide, and nitrogen oxides, and these gases flow into the body of the catalytic converter

4 These metals and other trace chemicals act as catalysts driving several chemical reactions, which in the presence of heat and oxygen, convert the input gases into carbon dioxide, water vapor, and nitrogen gas

Hydrocarbons
Carbon monoxide (CO)
Nitrogen oxides (NO_x)

3 The metal or ceramic substrate of this honeycomb structure is covered with a washcoat of aluminum oxide and a thin layer of the metals palladium, rhodium, and platinum

Main chemical reactions
$$2CO + O_2 \longrightarrow 2CO_2$$
$$C_2H_4 + 3O_2 \longrightarrow 2CO_2 + 2H_2O$$
$$CO + NO_x \longrightarrow CO_2 + N_2$$

Catalytic metals
(Pd, Rh, Pt)

Substrate
(metal or ceramic)

Washcoat (Al_2O_3)

FIGURE 23.12 The catalytic converter is a classic example of a green technology. This device filters air pollutants from vehicle exhaust and has helped bring about the vast improvement of air quality in the United States and other nations.

Oberlin College's green building took advantage of this). Some automobile manufacturers are planning cars that can be disassembled and recycled into new cars. Proponents of this industrial model see little reason why virtually all appliances and other products cannot be recycled, given the right technology. Their ultimate vision is to create industrial processes involving entirely closed loops, generating no waste.

We can base our decisions on long-term thinking

To be sustainable, a solution must work in the long term. Often the best long-term solution is not the best short-term solution, which explains why much of what human societies currently do is not sustainable. Policymakers in democracies often act for short-term good because they compete to produce immediate, positive results so that they will be reelected. This poses a major hurdle for addressing environmental dilemmas, because so many of

these dilemmas are cumulative, worsen gradually, and can be resolved only over a long period. Often the costs for addressing environmental problems are short-term, whereas the benefits are long-term, giving a politician little incentive to tackle the problems. In such a situation, citizen pressure on policymakers becomes especially vital.

Businesses may act according either to long-term or short-term interests. A business committed to operating in a certain community for a long time has incentive to sustain environmental quality. However, a business merely attempting to make a profit and move on has little incentive to invest in environmental protection measures that involve short-term costs.

In 2002, the U.N. Environment Programme conducted analyses to predict the likely effects of a large-scale shift to sustainable development strategies over the 30 years until 2032. The U.N. scientists designed four scenarios that differed in the sustainability of their approaches and then compared projections for a variety of environmental and

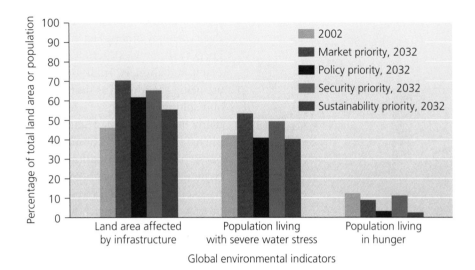

FIGURE 23.13 UNEP scientists projected the status of a number of environmental and socioeconomic indicators in 2032, across four different scenarios. The "markets first" scenario assumed that forces of market capitalism would have the greatest impact on global trends. The "policy first" scenario assumed that governments would take strong action to try to reach social and environmental goals. The "security first" scenario assumed a world of conflict in which security would be governments' highest priority. The "sustainability first" scenario assumed that a sustainable development paradigm would take hold, encouraging sustainable economic development, environmental protection, and social equity. The "sustainability first" approach was consistently predicted to result in the fewest negative impacts. Here it is shown to result in the least land affected by infrastructure, the fewest people living under water stress, and the fewest people living with hunger. Adapted from United Nations Environment Programme, 2002. *Global environment outlook 3 (GEO-3).* Nairobi and London: UNEP and Earthscan Publications.

socioeconomic indicators under the four scenarios. For one indicator after another, the approach that prioritized sustainability was shown to result in the fewest negative impacts and the most positive impacts (Figure 23.13).

A similar review was conducted by scientists with the Millennium Ecosystem Assessment (▸ p. 20) and reported in 2005. These scientists also analyzed four approaches of their own design. Their analysis revealed that the two scenarios involving proactive protection and management of ecosystems best maintained ecosystem services (Figure 23.14a), whether the management was locally based or globally based. It also showed that an approach that prioritized security above all else was the only one that decreased human well-being (Figure 23.14b).

Many wish to promote local self-sufficiency and be mindful of globalization

As our societies become more globally interconnected, diverse impacts, both positive and negative, have resulted. Many proponents of sustainability believe that encouraging local self-sufficiency is an important element of build-

ing sustainable societies. When people are tied closely to the area in which they live, they tend to value the area and seek to sustain its environment and human communities. This line of argument has frequently been made regarding the growing and distribution of food, specifically in encouraging locally based organic or sustainable agriculture. It is estimated that the food Americans eat is transported an average of 2,400 km (1,500 mi) from the place it was grown to the place it is eaten.

Many people who focus their economic activity locally also critique globalization. However, as the ecological economist Herman Daly has pointed out, globalization means different things to different people. People who view it as a positive phenomenon generally accentuate the process by which people of the world's diverse cultures are increasingly communicating with one another and learning about one another's diversity. Books, airplanes, television, and the Internet have made people of every culture more aware of other cultures and thus more likely to respect and celebrate, rather than fear, differences among cultures.

In contrast, people who view globalization in a negative light generally accentuate the homogenization of the

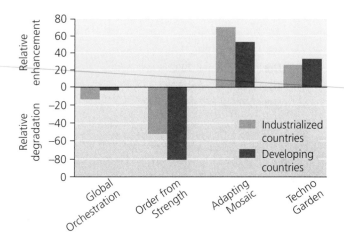

(a) Effect of scenarios on ecosystem services

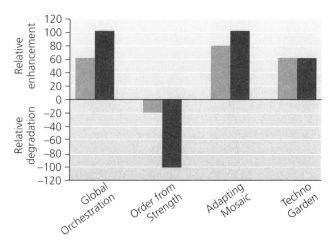

(b) Effects of scenarios on human well-being

FIGURE 23.14 Scientists with the Millennium Ecosystem Assessment analyzed impacts on ecosystem services and human welfare for each of four scenarios for the future. In the "global orchestration" scenario, a globally connected society focuses on trade and economic liberalization and takes a reactive approach to ecosystem problems, but also invests in education and infrastructure and strives to reduce poverty and inequality. In the "order from strength" scenario, nations focus on security in a fragmented world, trading regionally, investing little in public infrastructure, and taking a reactive approach to ecosystem problems. In the "adapting mosaic" scenario, regional watershed-scale ecosystems are the focus of political and economic activity, and local societies support strong institutions and proactively manage ecosystems. In the "technogarden" scenario, a globally connected world relies on green technology, taking a proactive and engineering approach to managing ecosystems and maximizing their delivery of ecosystem services. The assessment revealed that the adapting mosaic and technogarden approaches maintained ecosystem services best **(a),** and that all but the "order from strength" approach promoted human well-being **(b).** Adapted from Millennium Ecosystem Assessment, 2005.

world's cultures, in which a few cultures and worldviews displace many others. For instance, the world's many languages are going extinct more quickly than are species of plants and animals. Traditional ways of life in many areas are being abandoned as more people take up the material and cultural trappings of a few dominant cultures, particularly that of the United States. In recent years, more people have protested this homogenization and the growing power of large multinational corporations. In France, farm activist Jose Bove became a popular hero when he wrecked a McDonald's restaurant with his tractor to protest what many French farmers see as a threat to local French cuisine. In Seattle in 1999, thousands of protesters picketed the meeting of the World Trade Organization, which they viewed as a symbol of the homogenizing effects of Western market capitalism (Figure 23.15).

Daly and others argue that globalization entails a process in which multinational corporations attain greater and greater power over global trade while governments retain less and less. Most opponents of globalization consider businesses less likely than governments to encourage environmental protection, so they feel that globalization will hinder progress toward sustainability. Moreover, the U.S. model of market capitalism promotes a high-consumption lifestyle, which does indeed threaten efforts to advance toward sustainable solutions.

On the positive side, globalization may foster sustainability because Western democracy, as imperfect as it is, serves as a model for many people who live under repressive governments. Open societies allow for entrepreneurship and for the full flowering of creativity in academic research and in business. Millions of minds thinking and searching independently for solutions to problems are more likely to come up with sustainable solutions than the minds of a few who hold authoritarian power.

Weighing the Issues:
Globalization

From your own experience, what advantages and disadvantages do you see in globalization? Have you personally benefited or been hurt by it in any way? In what ways might promoting local self-sufficiency be helpful for the pursuit of global sustainability, and in what ways might it not?

Many people enjoy eating at Vietnamese restaurants in the United States, yet many who protest globalization are quick to criticize the presence of McDonald's restaurants in Vietnam. Do you think this represents a double standard, or are there reasons the two are not comparable?

FIGURE 23.15 Thousands of protesters picketed the World Trade Organization's meeting in Seattle in 1999, criticizing the homogenizing effects of globalization, as well as relaxations in labor and environmental protections brought about by free trade.

Consumers vote with their wallets . . .

A capitalist free-market system driven by consumerism holds at least one great asset for sustainability: Consumers can exercise influence through what they choose to buy. When products produced sustainably are ecolabeled (▸ pp. 50–53), consumers can "vote with their wallets" by preferentially purchasing these products. Consumer choice has helped drive sales of everything from recycled paper to organic produce to "dolphin-safe" tuna. Individuals can multiply their own influence by promoting "green" purchasing habits at the institutions where they are employed or attend school. We have seen how purchasing power at colleges and universities has spurred sales of certified sustainable wood, organic food, energy-efficient appliances, and more. Motivated employees in

businesses and government agencies can often promote change within those institutions by voicing their preferences in purchasing decisions.

. . . as well as with their ballots

Economic decisions are influential, but many of the changes needed to attain sustainable solutions require policymakers to usher them through. Policymakers respond to whoever exerts influence. Corporations and interest groups employ lobbyists to push politicians in one direction or another all the time. Citizens in a democratic republic have the same power, *if* they choose to exercise it. You can exercise this power at the ballot box, by attending public hearings, and by writing letters and making phone calls to policymakers. You may be surprised how little input policymakers receive from the public; sometimes a single letter or phone call can, in fact, make a difference.

Today's major environmental laws came about because citizens pressured their governmental representatives to do something about environmental problems. The raft of legislation enacted in the 1960s and 1970s in the United States and other nations might never have come about had ordinary citizens not stepped up and demanded action from policymakers. We owe it to our children and future generations to be engaged and to act responsibly now so that they have a better world in which to live. The words of anthropologist Margaret Mead are worth repeating: "Never doubt that a small group of thoughtful, committed people can change the world. Indeed, it's the only thing that ever has."

Promoting research and education is vital

None of these approaches will succeed fully if the public is not aware of their importance. An individual's decisions to reduce consumption, purchase ecolabeled products, or vote for candidates who support sustainable approaches will have limited effect unless a great many other people do the same. Individuals can influence large numbers of people by educating others with information and by serving as role models through their actions. The campus sustainability efforts at Ball State University and so many other colleges and universities accomplish both approaches. Moreover, the discipline of environmental science plays a key role in providing information people can use to make wise decisions about environmental problems. By promoting scientific research and by educating the public about environmental science, we can all assist in the pursuit of sustainable solutions.

Sustainability

How can we best pursue a goal of sustainability? What strategies and approaches are most needed and most effective?

GLOBALLY: Implementing Plan B

We are entering a new world. What we do not know is whether it will be a world of decline and collapse or a world of environmental restoration and economic progress.

Sustaining progress depends on restructuring the global economy, shifting from a fossil-fuel-based, automobile-centered, throwaway economy to one based on renewable energy sources, a diverse transportation system, and the comprehensive reuse and recycling of materials.

We can attain such a "Plan B" economy largely by restructuring taxes and subsidies to create a market that tells the ecological truth. Our modern economic prosperity is achieved in part by running up ecological deficits, costs that do not show up on the books, but that someone will eventually pay. Once we calculate the indirect costs of a product or service, we can incorporate them into market prices in the form of a tax, offsetting them with income tax reductions.

Sustaining progress also means eradicating poverty, stabilizing population, and restoring Earth's natural systems. Securing the public outlays needed to reach these goals depends on reordering fiscal priorities. A Plan B budget represents an additional annual expenditure of $161 billion—roughly one-third of the current U.S. military budget, or one-sixth of the global military budget.

Signs of the new economy are emerging all over the world—in the wind farms of Europe, the fast-growing U.S. fleet of gas-electric hybrid cars, the reforested hills of South Korea, the family planning program of Iran, the massive eradication of poverty in China, and the solar rooftops of Japan. What we need to do is doable.

The choice is ours—yours and mine. We can stay with business as usual and preside over an economy that continues to destroy its natural support systems until it destroys itself, or we can adopt Plan B and be the generation that changes direction, moving the world onto a path of sustained progress. The choice will be made by our generation, but it will affect life on Earth for all generations to come.

Lester R. Brown is founder and president of Earth Policy Institute, and founded Worldwatch Institute, serving as its president for 26 years. He helped pioneer the concept of environmentally sustainable development, a concept he uses in his design of an eco-economy. Brown has authored many books and received countless awards, including the 1987 United Nations Environment Prize and the 1994 Blue Planet Prize. In 1995, Marquis *Who's Who*, on the occasion of its 50th edition, selected Lester Brown as one of 50 Great Americans.

ON CAMPUS: A Positive Outlook for Sustainability

Colleges and universities, as centers of learning, have a responsibility to advance goals for global sustainability. Few experiences have as profound an impact on changing individuals and transforming the places where they live as education has.

Courses, research, and actions focused on greening today's campuses will prepare and inspire students to go forward and enter a world in which the environment is inextricably linked to most local and global issues.

The challenges associated with becoming a more sustainable world can be somewhat daunting, but they are not insurmountable. As microcosms of society, institutions of higher education can demonstrate how to achieve goals for sustainability, which have an impact on and are transferable to other sectors of society. Models already exist on many campuses where carbon emissions are reduced, locally grown foods are served in dining halls, building design and construction is certified through LEED standards, and alternative energy is employed in the form of biofuels, wind turbines, and photovoltaics. Not only are colleges and universities recognizing that they have a responsibility, but many are leading the charge for the changes that are taking place.

However, administrators, employees, educators, and students need to understand more about the pathways leading to this change so that we can accelerate its pace. The template for a sustainable future will utilize an integrated, systemic approach, one that creates a shared institutional value, with a key component being the involvement of senior-level leadership. Actions must transcend traditional institutional boundaries and engage a diverse set of individuals, from the trustees to the grounds crew to the students. Collaborations will need to draw on varying types of expertise, resulting in shared outcomes and new networks. Achievements must be celebrated and setbacks should become fuel for reflection. The freedom to explore new ideas and innovation is essential, and these initiatives should leverage even greater change to bring about a more sustainable future.

Nan Jenks-Jay teaches and is director of environmental affairs at Middlebury College. She has created an integrated vision of campus sustainability by advancing Middlebury's exemplary environmental academic program, which received the EPA's Environmental Merit Award and the Vermont Governor's Award for Environmental Excellence. She has written extensively and speaks throughout the United States and abroad on topics related to the environment, sustainability, and higher education.

Explore this issue further by accessing **Viewpoints** at www.aw-bc.com/withgott.

Precious Time

The pace of our lives is getting faster. We do more, consume more, travel more, and often work more. Life's commotion can make it hard to give attention to problems we don't need to deal with on a daily basis. The world's sheer load of environmental dilemmas can feel overwhelming, and even the best intentioned among us may feel we have little time to devote to saving the planet.

However, time is getting short, and impacts on the natural systems we depend on are occurring quickly. Many human impacts continue to become more severe and widespread, including deforestation, overfishing, land clearing, wetland draining, and resource extraction. The window of opportunity for turning some of these trends around is getting short. Even if we can visualize sustainable solutions to our many problems, how can we possibly find the time to implement them before we have done irreparable damage to our environment and our own future?

We need to reach again for the moon

On May 25, 1961, U.S. President John F. Kennedy announced that within the decade the United States would be "landing a man on the moon and returning him safely to the Earth" (Figure 23.16). It was a bold and astonishing statement; the technology to achieve this unprecedented, almost unimaginable, feat did not yet exist. Kennedy's directive had powerful motivation behind it, however. The United States was caught up with the Soviet Union in the Cold War. In this competition for global hegemony, the two nations, held mutually at bay with nuclear weapons, each tried to prove their mettle by other means. The race for dominance in space became the centerpiece of the rivalry. Early victories in the "space race" went to the Soviet Union, which sent into orbit humanity's first satellite, followed it with others, and then sent the first human being into space. The United States, meanwhile, was crashing rockets.

The prospect of "losing" the space race prompted Kennedy's administration to set a national goal and an ambitious timeline to reach it. Congress followed through with funding, NASA performed the science and engineering, and in 1969 astronauts walked on the moon. The United States accomplished this milestone in human history by building public support for a goal and by giving its scientists and engineers the wherewithal to focus on developing technology and strategies to meet the goal.

FIGURE 23.16 U.S. President John F. Kennedy in 1961 called on Congress to fund a space program to send men to the moon and back again before 1970. Addressing our environmental problems and shifting our political, economic, and social institutions to a paradigm of sustainable development will require at least as much vision, resolve, and commitment. The fact that astronauts reached the moon in 1969 demonstrates the power of human ingenuity in meeting a challenge and provides some hope that we may yet be able to meet the bigger challenge of living sustainably on Earth.

The rapid and historic accomplishments of the United States and the Soviet Union during the space race show what societies can accomplish when they focus support for a chosen goal. Similar successes occurred when the United States confronted the Great Depression, threw itself into World War II, and conducted the Marshall Plan after the war to help rebuild western Europe.

Today humanity faces a challenge more important than any previous one—the challenge to achieve sustainability. Attaining sustainability is a larger and more complex process than traveling to the moon. However, it is one to which every single person on Earth can contribute; in which government, industry, and citizens can all cooperate; and toward which governments of all nations can work together. Unlike the space race, there is a real time limit in reaching sustainability. We cannot wait much longer to reduce our impact on the planet; if we delay, the damage may turn permanent. If America was able to reach the moon in a mere 8 years, then certainly humanity can begin down the road to sustainability with comparable speed. Human ingenuity is capable of it; it is just a question of rallying public resolve and engaging our governments, institutions, and entrepreneurs in the race.

We must pass through the environmental bottleneck

Human ingenuity and compassion give us reason to hope that we may achieve sustainability before doing too much damage to our planet and our prospects, but we must be realistic about the challenges that lie ahead. As we decrease the amount of natural capital we can draw on, we give ourselves and the rest of the world's creatures less room to maneuver. Until we implement sustainable solutions, we will be squeezing ourselves through a progressively tighter space, like being forced through the neck of a bottle. The key question for the future of our species and our planet thus becomes whether we can make it safely through this bottleneck. Biologist Edward O. Wilson (▸ p. 328) has written eloquently of this view:

> At best, an environmental bottleneck is coming in the twenty-first century. It will cause the unfolding of a new kind of history driven by environmental change. Or perhaps an unfolding on a global scale of more of the old kind of history, which saw the collapse of regional civilizations, going back to the earliest in history, in northern Mesopotamia, and subsequently Egypt, then the Mayan and many others. . . . People died in large numbers, often horribly. Sometimes they were able to emigrate and displace other people, making them die horribly instead. . . . Somehow humanity must find a way to squeeze through the bottleneck without destroying the environments on which the rest of life depends.

We must think of Earth as an island

We began this book with the vision of Earth as an island, and indeed that is what it is. Islands can be paradise, as Easter Island (▸ pp. 8–9) likely was when the Polynesians first reached it. But if people do not live sustainably on islands, they can turn paradises into desolate graveyards. When Europeans arrived at Easter Is-

land, they witnessed the scene of a civilization that had depleted its resources, degraded its environment, and collapsed as a result. For the few people who remained of the once-mighty culture, life was difficult and unrewarding. They had lost even the knowledge of the history of their ancestors, who had cut trees unsustainably, kicking the base out from beneath their elaborate and prosperous civilization.

As Easter Island's trees disappeared, some individuals must have spoken out for conservation and for finding ways to live sustainably amid dwindling resources. And likely others ignored those calls and went on extracting more than the land could bear, assuming that somehow things would turn out all right. Indeed, whoever cut down the last tree from atop the most remote mountaintop could have looked out across the island and seen that it was the last tree. And yet that person cut it down.

It would be tragic folly to let such a fate occur to our planet as a whole. Yet if human impact on the environment continues growing, it is difficult to see how it can be avoided. By recognizing this, by deciding to shift our individual behavior and our cultural institutions in ways that encourage sustainable practices, and by employing sound science to help us achieve these ends, we may yet be able to live sustainably and happily on our wondrous island, Earth.

Conclusion

In every society facing the dilemma of dwindling resources and environmental degradation, there will be those who raise alarms and those who ignore them. Fortunately, in our global society today we have many thousands of scientists who study Earth's processes and resources. For this reason, we have access to an accumulated knowledge and an ever-developing understanding of our dynamic planet, what it offers us, and what impacts it can bear. The challenge for our global society today, our one-world island of humanity, is to support that science so that we can accurately judge false alarms from real problems and distinguish legitimate concerns from thoughtless denial. This science, this study of Earth and of ourselves, is what offers us hope for our future.

REVIEWING OBJECTIVES

You should now be able to:

List and describe approaches being taken on college and university campuses to promote sustainability

▶ Audits produce baseline data on how much a campus consumes and pollutes. (p. 677)

▶ The most common campus sustainability efforts involve recycling and waste reduction. (pp. 677–678)

▶ Sustainable building design is being pursued on a growing number of campuses. (pp. 678–679)

▶ There are numerous ways to reduce energy and water use, and these efforts often save money. (pp. 679–680)

▶ Providing local food is one way dining services can help in sustainability efforts. (p. 680)

▶ Colleges and universities can favor sustainable products in institutional purchasing. (p. 680)

▶ Institutions can use alternative fuels and vehicles in university fleets and encourage bicycling, walking, and public transportation. (p. 681)

▶ Habitat restoration is one of the most popular campus sustainability activities. (pp. 681–682)

Explain the concept of sustainable development

▶ Sustainable development entails environmental protection, economic development, and socioeconomic equity. (pp. 682–683)

▶ Proponents of sustainable development see no necessary tradeoff between economic development and environmental quality; rather, they feel that the two can enhance one another. (p. 683)

Discuss how protecting the environment can be compatible with promoting economic welfare

▶ Environmental protection and new green technologies and industries can create rich sources of new jobs. (pp. 683–684)

▶ Protecting environmental quality enhances a community's desirability and economy. (p. 684)

Describe and assess key approaches to designing sustainable solutions

▶ There are at least 10 general approaches that can inspire specific sustainable solutions. (pp. 685–693)

Analyze the roles that consumption, population, and technology play in efforts to achieve sustainability

▶ Growth in both population and per capita consumption will likely need to be halted if we are to create a sustainable society. (pp. 687–689)

▶ Technology has traditionally increased our environmental impact, but some new technologies can help reduce environmental impact. (pp. 689–690)

Explain how time is limited but how human potential to solve problems is tremendous

▶ Time for turning around our increasing environmental impacts is running short. (p. 695)

▶ The United States and other nations have met tremendous challenges before, so there is reason to hope that the world will be able to do what is needed to attain a sustainable society. (pp. 695–696)

TESTING YOUR COMPREHENSION

1. In what ways are campus sustainability efforts relevant to sustainability efforts in the broader society?

2. Name one way in which campus sustainability proponents have addressed each of the following areas: (1) recycling and waste reduction; (2) building using green technology; (3) energy and water efficiency; (4) dining services; (5) institutional purchasing; (6) transportation; (7) habitat restoration.

3. What do environmental scientists mean by *sustainable development*?

4. Why is it often thought that economic development and environmental protection are in direct opposition?

5. Describe three ways in which environmental protection can enhance economic well-being.

6. Why are many people now living at the highest level of material prosperity in history? Is this level of consumption sustainable? How can it feel good to consume less?

7. In what ways can technology help us achieve sustainability? How do natural processes provide good models of sustainability for manufacturing? Provide examples.

8. Why do many people feel that local self-sufficiency is important? What consequences of globalization may threaten sustainability? How can open democratic societies help to promote sustainability?

9. Explain Edward O. Wilson's metaphor of the "environmental bottleneck."

10. How can thinking of Earth as an island help prevent us from repeating the mistakes of other civilizations?

SEEKING SOLUTIONS

1. What efforts toward sustainable solutions have been made on your campus? What initiatives have succeeded, and why? What attempts have failed, and why? What lessons can you draw from these successes and failures?

2. What sustainability initiatives would you like to see attempted on your campus? If you were to take the lead in promoting such initiatives, how would you go about it? What obstacles would you expect to face, and how would you deal with them?

3. Choose one item or product that you enjoy, and consider how it came to be. What steps were involved in creating its components, and where did the raw materials come from? How was it manufactured? How was it delivered to you? Think of as many components of the item or product as you can, and determine how each of them was obtained or created.

4. Do you think that we can "have it both ways" by increasing our quality of life through development while also protecting the integrity of the environment? Can what is good for the environment also be good for humans? If so, how? Discuss examples from your course or from other chapters of this book that illustrate possible win-win solutions. Are there cases in which you think win-win solutions will not be possible?

5. Why do many observers find flaws in the conventional outlook that environmental protection necessarily is economically costly? Are you familiar with any cases in your community or at your college that bear on this issue? Describe such a case, and state what lessons you would draw from it.

6. Reflect on the experiences of prior human civilizations and how they came to an end. What is your prognosis for our current human civilization? Do you see a vast world of independent cultures all individually responsible for themselves, or do you consider human civilization to be one great entity? Is Earth an island, or is it a collection of many islands? If we accept that all humans depend on the same environmental systems for sustenance, what resources and strategies do we have to ensure that the actions of a few do not determine the outcome for all and that sustainable solutions are a common global goal?

INTERPRETING GRAPHS AND DATA

An undergraduate class at Penn State University conducted an ecological assessment of one of their biology laboratory buildings in response to the question: "How is this building like an ecosystem?" The result of their assessment was a 52-page report outlining ways to reduce the ecological footprint of the Mueller Laboratory Building in the areas of energy use, water use, communications/computing, furnishings/renovation, maintenance, and food. The students found ways to save an estimated $45,500 per year in the cost of energy alone, for a building that is occupied by 123 scientists and support staff. Their data on the current use and potential savings in the energy component of the ecological footprint are shown in the graph.

1. From the graph, estimate the amount of potential savings for each of the five areas identified in the Mueller Report. Approximately what percentage of the total electrical energy use of the building do these savings together represent?

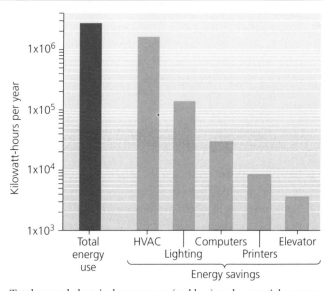

Total annual electrical energy use (red bar) and potential energy savings in five areas (orange bars) at Penn State University's Mueller Laboratory Building. The *y* axis is logarithmic, with each tick mark representing an increase equal to the value indicated just below it on the axis. *Data source:* Penn State Green Destiny Council, 2001. *The Mueller Report: Moving beyond sustainability indicators to sustainability action at Penn State.*

2. What was the approximate cost of electricity in cents per kilowatt-hour used to calculate the savings of $45,500? Do you think this cost may have changed since the report was issued in 2001? How and why?

3. How is the building where you take your course like an ecosystem? How is it not like an ecosystem? Is your building operating sustainably? What improvements do you think could be made to increase its sustainability?

CALCULATING ECOLOGICAL FOOTPRINTS

Of all the choices we make, our transportation choices have some of the greatest repercussions for environmental impact. According to the U.S. Department of Transportation, there are about 136 million passenger cars in the United States (excluding light trucks and SUVs), which travel an average of 12,200 miles per car per year at an average fuel economy of 22 mpg. These figures can be related to the size of our population (P) and the level of affluence (A) in the I = PAT equation (▸ pp. 220–221). The technology (T) term in this case can refer to the automotive technologies that determine fuel economy.

Federal CAFE (Corporate Average Fuel Economy) standards require a manufacturer's fleet of new passenger cars to average 27.5 mpg. With some hybrid gas/electric vehicles you may achieve 60 mpg. Using this information, calculate in the table the impact of choosing more fuel-efficient passenger car technologies. Estimate the number of cars owned by your classmates and by the residents of your hometown.

	Cars	Miles/yr	Gal/year @ 22 mpg	Gal/year @ 27.5 mpg	Gal/year @ 60 mpg	Gal saved @ 27.5 vs. 22 mpg	Gal saved @ 60 vs. 22 mpg
You	1	1.22×10^4	555	444	203	111	352
Your class							
Your hometown							
United States	1.36×10^8						

Data from Bureau of Transportation Statistics. 2005. *National Transportation Statistics 2004.* Washington D.C.: U.S. Department of Transportation.

1. You can reduce your personal environmental impact by choosing different technologies (T), as the table above demonstrates. If more people make these same reductions (P), the total reduction in environmental impact would be larger. What could you choose to do that would alter the value of the other independent variable in the I=PAT equation, namely A, which stands for your affluence, or the amount of resources you use?

2. How many gallons of gasoline would you estimate that you would save per year if you reduced the number of miles you drive per year by 20%? How could you reduce your annual mileage by 20% without affecting your "quality of life?"

3. Suppose the cost of gasoline is $2.50 per U.S. gallon. How much money would you save annually if your vehicle averaged 27.5 mpg instead of 22 mpg? If it averaged 60 mpg? In parts of western Europe, the price of gasoline is triple that in the United States. How much more would your annual driving costs be there at 22 mpg? Would such an additional cost influence your decision as to what type of vehicle to drive?

Take It Further

Go to www.aw-bc.com/withgott or the student CD-ROM where you'll find:

▸ Suggested answers to end-of-chapter questions
▸ Quizzes, animations, and flashcards to help you study
▸ *Research Navigator*™ database of credible and reliable sources to assist you with your research projects

▸ **GRAPHit!** tutorials to help you master how to interpret graphs
▸ **INVESTIGATEit!** current news articles that link the topics that you study to case studies from your region to around the world

Appendix A Some Basics on Graphs

Presenting data in ways that are clear and that help make trends and patterns visually apparent is a vital part of the scientific endeavor. Scientists' primary tool for presenting data and expressing patterns is the graph. Thus, the ability to interpret graphs is a skill that you will want to cultivate early in your study of the sciences. This appendix provides basic information on four of the most common types of graphs—line plots, bar charts, scatter plots, and pie charts—and the rationale for the use of each.

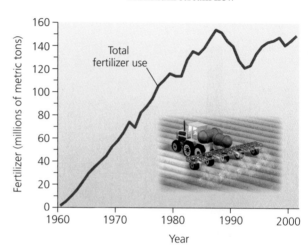

Minimum stream flow

Line Plot

A line plot is drawn when a data set involves a sequence of some kind, such as a sequence through time or across distance (Figure A.1; see ▸ p. 140 and Figure 9.19, ▸ p. 266). Using a line plot allows us to see increasing or decreasing trends in the data. Line plots are appropriate when the variable measured by the *y* axis (the vertical axis) represents continuous numerical data, and when the variable measured by the *x* axis (the horizontal axis) represents either continuous numerical data or sequential categories, such as years.

Global fertilizer use, 1960–2002

FIGURE A.1

One useful technique is to plot two data sets together on the same graph (Figure A.2; see Figure 6.5, ▸ p. 153). This allows us to compare trends in the two data sets to see whether and how they may be related.

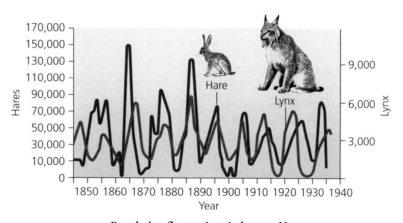

Population fluctuations in hare and lynx

FIGURE A.2

Bar Chart

A bar chart is most often used when one of the variables represents categories rather than numerical values (Figure A.3; see Figure 13.10a, ▸ p. 389). Bar charts allow us to visualize how a variable differs quantitatively among categories.

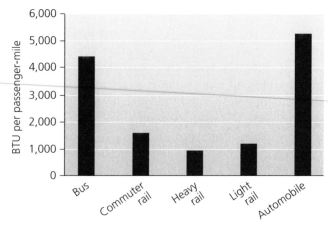

Energy consumption for different modes of transit

FIGURE A.3

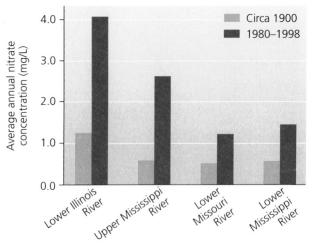

Nitrate concentrations in portions of the Mississippi River watershed

FIGURE A.4

It is often instructive to graph two or more variables together to reveal patterns and relationships (Figure A.4; see Figure 7.4a, ▸ p. 188). Most of the bar charts you will see in this book illustrate several types of information at once in this manner.

Bar charts are usually arrayed so that the bars extend vertically. Sometimes, however, a horizontal orientation may make for a clearer presentation. One special type of horizontally oriented bar chart is the age pyramid used by demographers (Figure A.5; see Figure 8.11, ▸ p. 226). Age categories are displayed on the y axis, with bars representing the population size of each age group varying in width instead of height.

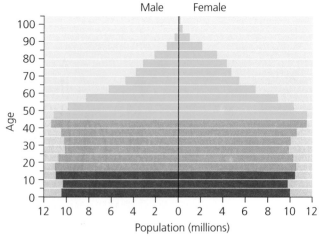

United States age structure, 2005

FIGURE A.5

Scatter Plot

A scatter plot is used most often when there is no sequential aspect to the data, and each data point is independent, having no particular connection to other data points (Figure A.6; see Figure 8.15, ▶ p. 232). Scatter plots allow you to visualize a broad positive or negative correlation between variables on the two axes.

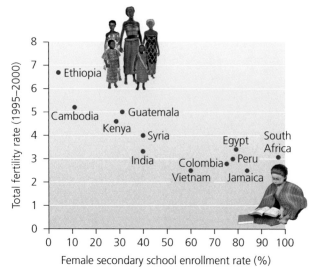

Fertility rate and female education

FIGURE A.6

Pie Chart

A pie chart is used when we wish to compare the proportions of some whole that are taken up by each of several categories (Figure A.7; see Figure 9.2, ▶ p. 247). A pie chart is appropriate when one variable is categorical and one is numerical. Each category is represented visually like a slice from a pie, with the size of the slice reflecting the percentage of the whole that is taken up by that category.

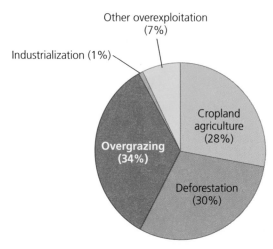

Causes of desertification

FIGURE A.7

But don't stop here. Take advantage of the **GRAPHit!** tutorials on the CD-ROM included with this text, or go to the Withgott/Brennan Companion Website at www.aw-bc.com/withgott. The **GRAPHit!** tutorials allow you to plot your own data, and help you further expand your comprehension of graphs.

Appendix B Metric System

Measurement	Unit and Abbreviation	Metric Equivalent	Metric to English Conversion Factor	English to Metric Conversion Factor
Length	1 kilometer (km)	$= 1,000\ (10^3)$ meters	1 km = 0.62 mile	1 mile = 1.61 km
	1 meter (m)	$= 100\ (10^2)$ centimeters	1 m = 1.09 yards	1 yard = 0.914 m
		$= 1,000$ millimeters	1 m = 3.28 feet	1 foot = 0.305 m
			1 m = 39.37 inches	
	1 centimeter (cm)	$= 0.01\ (10^{-2})$ meter	1 cm = 0.394 inch	1 foot = 30.5 cm
				1 inch = 2.54 cm
	1 millimeter (mm)	$= 0.001\ (10^{-3})$ meter	1 mm = 0.039 inch	
	1 micrometer (μm) [formerly micron (μ)]	$= 0.000001\ (10^{-6})$ meter		
	1 nanometer (nm) [formerly millimicron (mμ)]	$= 0.000000001\ (10^{-9})$ meter		
	1 angstrom (Å)	$= 0.0000000001\ (10^{-10})$ meter		
Area	1 square meter (m²)	$= 10,000$ square centimeters	1 m² = 1.1960 square yards	1 square yard = 0.8361 m²
			1 m² = 10.764 square feet	1 square foot = 0.0929 m²
	1 square centimeter (cm²)	$= 100$ square millimeters	1 cm² = 0.155 square inch	1 square inch = 6.4516 cm²
Mass	1 metric ton (t)	$= 1,000$ kilograms	1 t = 1.103 ton	1 ton = 0.907 t
	1 kilogram (kg)	$= 1,000$ grams	1 kg = 2.205 pounds	1 pound = 0.4536 kg
	1 gram (g)	$= 1,000$ milligrams	1 g = 0.0353 ounce	1 ounce = 28.35 g
			1 g = 15.432 grains	
	1 milligram (mg)	$= 0.001$ gram	1 mg = approx. 0.015 grain	
	1 microgram (μg)	$= 0.000001$ gram		
Volume (solids)	1 cubic meter (m³)	$= 1,000,000$ cubic centimeters	1 m³ = 1.3080 cubic yards	1 cubic yard = 0.7646 m³
			1 m³ = 35.315 cubic feet	1 cubic foot = 0.0283 m³
	1 cubic centimeter (cm³ or cc)	$= 0.000001$ cubic meter $= 1$ milliliter	1 cm³ = 0.0610 cubic inch	1 cubic inch = 16.387 cm³
	1 cubic millimeter (mm³)	$= 0.000000001$ cubic meter		
Volume (liquids and gases)	1 kiloliter (kl or kL)	$= 1,000$ liters	1 kL = 264.17 gallons	1 gallon = 3.785 L
	1 liter (l or L)	$= 1,000$ milliliters	1 L = 0.264 gallons	1 quart = 0.946 L
			1 L = 1.057 quarts	
	1 milliliter (ml or mL)	$= 0.001$ liter	1 ml = 0.034 fluid ounce	1 quart = 946 ml
		$= 1$ cubic centimeter	1 ml = approx. $\frac{1}{4}$ teaspoon	1 pint = 473 ml
			1 ml = approx. 15–16 drops (gtt.)	1 fluid ounce = 29.57 ml
				1 teaspoon = approx. 5 ml
	1 microliter (μl or μL)	$= 0.000001$ liter		
Time	1 second (s)	$= \frac{1}{60}$ minute		
	1 millisecond (ms)	$= 0.001$ second	1 second (s)	$= \frac{1}{60}$ minute
Temperature	Degrees Celsius (°C)	$°F = \frac{9}{5}°C + 32$	$°C = \frac{5}{9}(°F - 32)$	
Energy and Power	1 kilowatt-hour	$= 34,113$ BTUs $= 860,421$ calories		
	1 watt	$= 3.413$ BTU/hr		
		$= 14.34$ calorie/min		
	1 calorie	$=$ the amount of heat necessary to raise the temperature of 1 gram (1cm³) of water 1 degree Celsius		
	1 horsepower	$= 7.457 \times 10^2$ watts		
	1 joule	$= 9.481 \times 10^{-4}$ BTU		
		$= 0.239$ cal		
		$= 2.778 \times 10^{-7}$ kilowatt-hour		
Pressure	1 pound per square inch (psi)	$= 6894.757$ pascal (Pa)		
		$= 0.068045961$ atmosphere (atm)		
		$= 51.71493$ millimeters of mercury (mm hg = Torr)		
		$= 68.94757$ millibars (mbar)		
		$= 68.94757$ (hectopascal hPa)		
		$= 6.894757$ kilopascal (kPa)		
		$= 0.06894757$ bar (bar)		
	1 atmosphere (atm)	$= 101.325$ kilopascal (kPa)		

Appendix C Periodic Table of the Elements

Representative (main group) elements

IA	IIA												IIIA	IVA	VA	VIA	VIIA	VIIIA
1 **H** 1.0079 Hydrogen																		2 **He** 4.003 Helium
3 **Li** 6.941 Lithium	4 **Be** 9.012 Beryllium												5 **B** 10.811 Boron	6 **C** 12.011 Carbon	7 **N** 14.007 Nitrogen	8 **O** 15.999 Oxygen	9 **F** 18.998 Fluorine	10 **Ne** 20.180 Neon
11 **Na** 22.990 Sodium	12 **Mg** 24.305 Magnesium	IIIB	IVB	VB	VIB	VIIB	VIIIB	VIIIB	VIIIB	IB	IIB		13 **Al** 26.982 Aluminum	14 **Si** 28.086 Silicon	15 **P** 30.974 Phosphorus	16 **S** 32.066 Sulfur	17 **Cl** 35.453 Chlorine	18 **Ar** 39.948 Argon
19 **K** 39.098 Potassium	20 **Ca** 40.078 Calcium	21 **Sc** 44.956 Scandium	22 **Ti** 47.88 Titanium	23 **V** 50.942 Vanadium	24 **Cr** 51.996 Chromium	25 **Mn** 54.938 Manganese	26 **Fe** 55.845 Iron	27 **Co** 58.933 Cobalt	28 **Ni** 58.69 Nickel	29 **Cu** 63.546 Copper	30 **Zn** 65.39 Zinc		31 **Ga** 69.723 Gallium	32 **Ge** 72.61 Germanium	33 **As** 74.922 Arsenic	34 **Se** 78.96 Selenium	35 **Br** 79.904 Bromine	36 **Kr** 83.8 Krypton
37 **Rb** 85.468 Rubidium	38 **Sr** 87.62 Strontium	39 **Y** 88.906 Yttrium	40 **Zr** 91.224 Zirconium	41 **Nb** 92.906 Niobium	42 **Mo** 95.94 Molybdenum	43 **Tc** 98 Technetium	44 **Ru** 101.07 Ruthenium	45 **Rh** 102.906 Rhodium	46 **Pd** 106.42 Palladium	47 **Ag** 107.868 Silver	48 **Cd** 112.411 Cadmium		49 **In** 114.82 Indium	50 **Sn** 118.71 Tin	51 **Sb** 121.76 Antimony	52 **Te** 127.60 Tellurium	53 **I** 126.905 Iodine	54 **Xe** 131.29 Xenon
55 **Cs** 132.905 Cesium	56 **Ba** 137.327 Barium	57 **La** 138.906 Lanthanum	72 **Hf** 178.49 Hafnium	73 **Ta** 180.948 Tantalum	74 **W** 183.84 Tungsten	75 **Re** 186.207 Rhenium	76 **Os** 190.23 Osmium	77 **Ir** 192.22 Iridium	78 **Pt** 195.08 Platinum	79 **Au** 196.967 Gold	80 **Hg** 200.59 Mercury		81 **Tl** 204.383 Thallium	82 **Pb** 207.2 Lead	83 **Bi** 208.980 Bismuth	84 **Po** 209 Polonium	85 **At** 210 Astatine	86 **Rn** 222 Radon
87 **Fr** 223 Francium	88 **Ra** 226.025 Radium	89 **Ac** 227.028 Actinium	104 **Rf** 261 Unnilquadium	105 **Db** 262 Unnilpentium	106 **Sg** 263 Unnilhexium	107 **Bh** 262 Unnilseptium	108 **Hs** 265 Unniloctium	109 **Mt** 266 Unnilennium	110 **Uun** 269 Ununnilium	111 **Uuu** 272 Unununium	112 **Uub** 277 Ununbium		114		116			

Transition metals

Rare earth elements

Lanthanides

58 **Ce** 140.115 Cerium	59 **Pr** 140.908 Praseodymium	60 **Nd** 144.24 Neodymium	61 **Pm** 145 Promethium	62 **Sm** 150.36 Samarium	63 **Eu** 151.964 Europium	64 **Gd** 157.25 Gadolinium	65 **Tb** 158.925 Terbium	66 **Dy** 162.5 Dysprosium	67 **Ho** 164.93 Holmium	68 **Er** 167.26 Erbium	69 **Tm** 168.934 Thulium	70 **Yb** 173.04 Ytterbium	71 **Lu** 174.967 Lutetium

Actinides

90 **Th** 232.038 Thorium	91 **Pa** 231.036 Protactinium	92 **U** 238.029 Uranium	93 **Np** 237.048 Neptunium	94 **Pu** 244 Plutonium	95 **Am** 243 Americium	96 **Cm** 247 Curium	97 **Bk** 247 Berkelium	98 **Cf** 251 Californium	99 **Es** 252 Einsteinium	100 **Fm** 257 Fermium	101 **Md** 258 Mendelevium	102 **No** 259 Nobelium	103 **Lr** 262 Lawrencium

The periodic table arranges elements according to atomic number and atomic weight into horizontal rows called periods and vertical columns called groups.

Elements of each group in Class A have similar chemical and physical properties. This reflects the fact that members of a particular group have the same number of valence shell electrons, which is indicated by the group's number. For example, group IA elements have one valence shell electron, group IIA elements have two, and group VA elements have five. In contrast, as you progress across a period from left to right, properties of the elements change, varying from the very metallic properties of groups IA and IIA to the nonmetallic properties of group VIIA to the inert elements (noble gases) in group VIIIA. This reflects changes in the number of valence shell electrons.

Class B elements, or transition elements, are metals, and generally have one or two valence shell electrons. In these elements, some electrons occupy more distant electron shells before the deeper shells are filled.

In this periodic table, elements with symbols printed in black exist as solids under standard conditions (25 °C and 1 atmosphere of pressure), while elements in red exist as gases, and those in dark blue as liquids. Elements with symbols in green do not exist in nature and must be created by some type of nuclear reaction.

Appendix D   Geologic Timescale

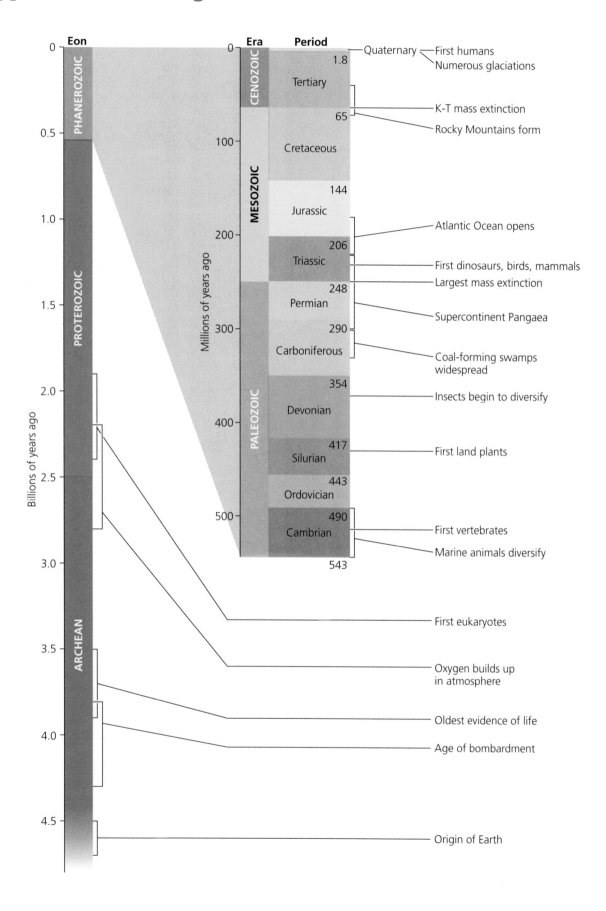

Glossary

abiotic factor Any nonliving component of the *environment*. Compare *biotic factor*.

acidic deposition The deposition of *acidic* or acid-forming pollutants from the *atmosphere* onto Earth's surface by *precipitation*, by fog, by gases, or by the deposition of dry particles.

acid drainage A process in which sulfide minerals in newly exposed rock surfaces react with oxygen and rainwater to produce sulfuric acid, which causes chemical *runoff* as it *leaches* metals from the rocks. Although acid drainage is a natural phenomenon, mining can greatly accelerate its rate by exposing many new rock surfaces at once.

acidic The property of a solution in which the concentration of hydrogen (H^+) *ions* is greater than the concentration of hydroxide (OH^-) ions. Compare *basic*.

active solar energy collection An approach in which technological devices are used to focus, move, or store solar energy. Compare *passive solar energy collection*.

acute exposure Exposure to a *toxicant* occurring in high amounts for short periods of time. Compare *chronic exposure*.

adaptive management The systematic testing of different management approaches to improve methods over time.

adaptive trait (adaptation) A trait that confers greater likelihood that an individual will reproduce.

affluenza Term coined by social critics to describe the failure of material goods to bring happiness to people who have the financial means to afford them.

aerobic Occurring in an *environment* where oxygen is present. For example, the decay of a rotting log proceeds by aerobic decomposition. Compare *anaerobic*.

age distribution The relative numbers of organisms of each age within a *population*. Age distributions can have a strong effect on rates of population growth or decline and are often expressed as a ratio of age classes, consisting of organisms (1) not yet mature enough to reproduce, (2) capable of reproduction, and (3) beyond their reproductive years.

age structure See *age distribution*.

agricultural revolution The shift around 10,000 years ago from a hunter-gatherer lifestyle to an agricultural way of life in which people began to grow their own crops and raise domestic animals. Compare *industrial revolution*.

agriculture The practice of cultivating *soil*, producing crops, and raising livestock for human use and consumption.

A horizon A layer of *soil* found in a typical *soil profile*. It forms the top layer or lies below the *O horizon* (if one exists). It consists of mostly inorganic mineral components such as *weathered* substrate, with some organic matter and humus from above mixed in. The A horizon is often referred to as *topsoil*. Compare *B horizon; C horizon; E horizon; R horizon*.

air pollutants Gases and particulate material added to the atmosphere that can affect *climate* or harm people or other organisms.

air pollution The act of polluting the air, or the condition of being polluted by *air pollutants*.

allergen A *toxicant* that overactivates the immune system, causing an immune response when one is not necessary.

amensalism A relationship between members of different *species* in which one organism is harmed and the other is unaffected. Compare *commensalism*.

anaerobic Occurring in an *environment* that has little or no oxygen. The conversion of organic matter to *fossil fuels (crude oil, coal, natural gas)* at the bottom of a deep lake, swamp, or shallow sea is an example of anaerobic decomposition. Compare *aerobic*.

anthropocentrism A human-centered view of our relationship with the *environment*.

aquaculture The raising of aquatic organisms for food in controlled *environments*.

aquifer An underground water reservoir.

artesian aquifer See *confined aquifer*.

artificial selection *Natural selection* conducted under human direction. Examples include the *selective breeding* of crop plants, pets, and livestock.

atmosphere The thin layer of gases surrounding planet Earth. Compare *biosphere; hydrosphere; lithosphere*.

atmospheric deposition The wet or dry deposition on land of a wide variety of pollutants, including mercury, nitrates, organochlorines, and others. *Acidic deposition* is one type of atmospheric deposition.

atmospheric pressure The weight per unit area produced by a column of air.

atom The smallest component of an *element* that maintains the chemical properties of that element.

autotroph (primary producer) An organism that can use the energy from sunlight to produce its own food. Includes green plants, algae, and cyanobacteria.

Bacillus thuringiensis (**Bt**) A naturally occurring *soil* bacterium that produces a protein that kills many pests, including caterpillars and the larvae of some flies and beetles.

background rate of extinction The average rate of *extinction* that occurred before the appearance of humans. For example, the *fossil record* indicates that for both birds and mammals, one *species* in the world typically became extinct every 500–1,000 years. Compare *mass extinction event*.

basic The property of a solution in which the concentration of hydroxide (OH^-) *ions* is greater than the concentration of hydrogen (H^+) ions. Compare *acidic*.

bedrock The continuous mass of solid rock that makes up Earth's *crust*.

benthic Of, relating to, or living on the bottom of a water body. Compare *pelagic*.

benthic zone The bottom layer of water body. Compare *littoral zone; limnetic zone; profundal zone*.

B horizon The layer of *soil* that lies below the *E horizon* and above the *C horizon*. Minerals that leach out of the E horizon are carried down into the B horizon (or subsoil) and accumulate there. Sometimes called the "zone of accumulation" or "zone of deposition." Compare *A horizon; O horizon; R horizon*.

bioaccumulation The buildup of *toxicants* in the tissues of an animal.

biocentrism A philosophy that ascribes relative values to actions, entities, or properties on the basis of their effects on all living things or on the integrity of the *biotic* realm in general. The biocentrist evaluates an action in terms of its overall impact on living things, including—but not exclusively focusing on—human beings.

biodiesel Diesel fuel produced by mixing vegetable oil, used cooking grease, or animal fat with small amounts of *ethanol* or methanol (wood alcohol) in the presence of a chemical catalyst.

biodiversity (biological diversity) The sum total of all organisms in an area, taking into account the diversity of *species,* their *genes,* their *populations,* and their *communities.*

biodiversity hotspot An area that supports an especially great diversity of *species,* particularly species that are *endemic* to the area.

biofuel Fuel produced from *biomass energy* sources and used primarily to power automobiles.

biogeochemical cycle See *nutrient cycle.*

biological control (biocontrol) The attempt to battle pests and weeds with organisms that prey on or parasitize them, rather than by using *pesticides.*

biological diversity See *biodiversity.*

biomagnification The magnification of the concentration of *toxicants* in an organism caused by its consumption of other organisms in which toxicants have *bioaccumulated.*

biomass energy *Energy* harnessed from plant and animal matter, including wood from trees, charcoal from burned wood, and combustible animal waste products, such as cattle manure. *Fossil fuels* are not considered biomass energy sources because their organic matter has not been part of living organisms for millions of years and has undergone considerable chemical alteration since that time.

biome A major regional complex of similar plant *communities;* a large *ecological* unit defined by its dominant plant type and vegetation structure.

biophilia A hypothetical phenomenon that E. O. Wilson defined as "the connections that human beings subconsciously seek with the rest of life."

biopower The burning of *biomass energy* sources to generate electricity.

bioremediation The attempt to clean up *pollution* by enhancing natural processes of biodegradation by living organisms.

biosphere The sum total of all the planet's living organisms and the *abiotic* portions of the *environment* with which they interact. Compare *atmosphere; hydrosphere; lithosphere.*

biosphere reserve A tract of land with exceptional *biodiversity* that couples preservation with *sustainable development* to benefit local people. Biosphere reserves are designated by UNESCO (the *United Nations* Educational, Scientific, and Cultural Organization) following application by local stakeholders.

biotechnology The material application of biological *science* to create products derived from organisms. The creation of *transgenic* organisms is one type of biotechnology.

biotic factor Any living component of the *environment.* Compare *abiotic factor.*

boreal forest A *biome* of northern coniferous forest that stretches in a broad band across much of Canada, Alaska, Russia, and Scandinavia. Also known as taiga, boreal forest consists of a limited number of *species* of evergreen trees, such as black spruce, that dominate large regions of forests interspersed with occasional bogs and lakes.

breeder nuclear fission A form of *nuclear fission* that uses uranium-238 instead of uranium-235 (which is used in the conventional method to harness *nuclear energy*). Breeder fission makes better use of fuel, generates more power, and produces less waste than does conventional nuclear fission, but it is considerably more dangerous.

Bureau of Land Management (BLM) Federal agency that owns and manages most U.S. rangelands. The BLM is the nation's single largest landowner; its 106 million ha (261 million acres) are spread across 12 western states.

by-catch That portion of a commercial fishing catch consisting of animals caught unintentionally. By-catch kills many thousands of fish, sharks, marine mammals, and birds each year.

capitalist market economy An *economy* in which buyers and sellers interact to determine which *goods* and *services* to produce, how much of them to produce, and how to distribute them. Compare *centrally planned economy.*

captive breeding The practice of capturing members of threatened and endangered *species* so that their young can be bred and raised in controlled *environments* and subsequently reintroduced into the wild.

carbohydrate An *organic compound* consisting of *atoms* of carbon, hydrogen, and oxygen.

carbon cycle A major *nutrient cycle* consisting of the routes that carbon *atoms* take through the nested networks of environmental *systems.*

carcinogen A chemical or type of radiation that causes cancer.

carrying capacity The maximum *population size* that a given *environment* can sustain.

cell The most basic organizational unit of organisms.

cellular respiration The process by which a *cell* uses the chemical reactivity of oxygen to split glucose into its constituent parts, water and carbon dioxide, and thereby release *chemical energy* that can be used to form chemical bonds or to perform other tasks within the cell. Compare *photosynthesis.*

centrally planned economy An *economy* in which a nation's government determines how to allocate resources in a top-down manner. Also called a "state socialist economy." Compare *capitalist market economy.*

chaparral A *biome* consisting mostly of densely thicketed evergreen shrubs occurring in limited small patches. Its "Mediterranean" *climate* of mild, wet winters and warm, dry summers is induced by oceanic influences. In addition to ringing the Mediterranean Sea, chaparral occurs along the coasts of California, Chile, and southern Australia.

chemical energy *Potential energy* held in the bonds between atoms.

chemosynthesis The process by which bacteria in hydrothermal vents use the *chemical energy* of hydrogen sulfide (H_2S) to transform inorganic carbon into *organic compounds.* Compare *photosynthesis.*

Chernobyl Site of a nuclear power plant in Ukraine (then part of the Soviet Union), where in 1986 an explosion caused the most severe *nuclear reactor* accident the world has yet seen. As with *Three Mile Island,* the term is often used to denote the accident itself.

chlorofluorocarbon (CFC) One of a group of human-made *organic compounds* derived from simple *hydrocarbons,* such as ethane and methane, in which hydrogen *atoms* are replaced by chlorine, bromine, or fluorine. CFCs deplete the protective *ozone layer* in the *stratosphere.*

C horizon The layer of *soil* that lies below the *B horizon* and above the *R horizon.* It contains rock particles that are larger and less *weathered* than the layers above. It consists of *parent material* that has been altered only slightly or not at all by the process of *soil* formation. Compare *A horizon; E horizon; O horizon.*

chronic exposure Exposure for long periods of time to a *toxicant* occurring in low amounts. Compare *acute exposure.*

city planning The professional pursuit that attempts to design cities in such a way as to maximize their efficiency, functionality, and beauty.

classical economics Founded by *Adam Smith,* the study of the behavior of buyers and sellers in a free-market *economy.* Holds that individuals acting in their own self-interest may benefit society, provided that their behavior is constrained by the rule of law and by private property rights and operates within competitive markets. See also *neoclassical economics.*

clay *Sediment* consisting of particles less than 0.002 mm in diameter. Compare *sand; silt.*

Clean Air Act of 1970 Revision of prior Congressional *legislation* to control *air pollution* that set stricter standards for air quality, imposed limits on emissions from new stationary and mobile sources, provided new funds for *pollution*-control research, and enabled citizens to sue parties violating the standards.

Clean Air Act of 1990 Congressional *legislation* that strengthened *regulations* pertaining to air quality standards, auto emissions, toxic *air pollution, acidic deposition,* and depletion of the *ozone layer,* while also introducing market-based incentives to reduce *pollution.*

clear-cutting The harvesting of timber by cutting all the trees in an area, leaving only stumps. Although it is the most cost-efficient method, clear-cutting is also the most damaging to the *environment.*

climate The pattern of atmospheric conditions found across large geographic regions over long periods of time. Compare *weather.*

climate diagram (climatograph) A visual representation of a region's average monthly temperature and *precipitation.*

closed system A *system* that is isolated an self-contained. Scientists may treat a system as closed to simplify some question they are investigating, but no natural system is truly closed. Compare *open system.*

coal A *fossil fuel* composed of organic matter that was compressed under very high pressure to form a dense, solid carbon structure.

cogeneration A practice in which the extra heat generated in the production of electricity is captured and put to use heating workplaces and homes, as well as producing other kinds of power.

cold front The boundary where a mass of cold air displaces a mass of warmer air. Compare *warm front.*

command and control An approach to protecting the *environment* that sets strict legal limits and threatens punishment for violations of those limits.

commensalism A relationship between members of different *species* in which one organism benefits and the other is unaffected. Compare *amensalism.*

community A group of *populations* of organisms that live in the same place at the same time.

community-based conservation The practice of engaging local people to protect land and wildlife in their own region.

community ecology The study of the interactions among *species,* from one-to-one interactions to complex interrelationships involving entire *communities.*

competition A relationship in which multiple organisms seek the same limited resource.

composting The conversion of organic *waste* into mulch or humus by encouraging, in a controlled manner, the natural biological processes of decomposition.

compound A *molecule* whose *atoms* are composed of two or more *elements.*

confined (artesian) aquifer A water-bearing, porous layer of rock, *sand,* or gravel that is trapped between an upper and lower layer of less permeable substrate, such as *clay.* The water in a confined aquifer is under pressure because it is trapped between two impermeable layers. Compare *unconfined aquifer.*

conservation biology A scientific discipline devoted to understanding the factors, forces, and processes that influence the loss, protection, and restoration of *biological diversity* within and among *ecosystems.*

conservation district One of many county-based entities created by the Soil Conservation Service (now the Natural Resources Conservation Service) to promote practices that conserve *soil.*

conservation ethic An *ethic* holding that humans should put *natural resources* to use but also have a responsibility to manage them wisely. Compare *preservation ethic.*

consumer See *heterotroph.*

consumptive use *Freshwater* use in which water is removed from a particular *aquifer* or surface water body and is not returned to it. *Irrigation* for *agriculture* is an example of consumptive use. Compare *nonconsumptive use.*

continental shelf The gently sloping underwater edge of a continent, varying in width from 100 m (330 ft) to 1,300 km (800 mi), with an average slope of 1.9 m/km (10 ft/mi).

contingent valuation A technique that uses surveys to determine how much people would be willing to pay to protect a resource or to restore it after damage has been done.

contour farming The practice of plowing furrows sideways across a hillside, perpendicular to its slope, to help prevent the formation of rills and gullies. The technique is so named because the furrows follow the natural contours of the land.

control The portion of an *experiment* in which a *variable* has been left unmanipulated, to serve as a point of comparison with the *treatment.*

controlled burn See *prescribed burn.*

controlled experiment An *experiment* in which the effects of all *variables* are held constant, except the one whose effect is being tested by comparison of *treatment* and *control* conditions.

convective circulation A circular *current* (of air, water, magma, etc.) driven by temperature differences. In the atmosphere, warm air rises into regions of lower *atmospheric pressure*, where it expands and cools and then descends and becomes denser, replacing warm air that is rising. The air picks up heat and moisture near ground level and prepares to rise again, continuing the process.

conventional law International law that arises from conventions, or treaties, that nations agree to enter into. Compare *customary law.*

Convention on Biological Diversity An international treaty that aims to conserve *biodiversity*, use biodiversity in a *sustainable* manner, and ensure the fair distribution of biodiversity's benefits. Although many nations have agreed to the treaty (as of 2005, 188 nations had become parties to it), several others, including the United States, have not.

Convention on International Trade in Endangered Species of Wild Fauna and Flora (CITES) A 1973 treaty facilitated by the *United Nations* that protects endangered *species* by banning the international transport of their body parts.

convergent plate boundary Area where tectonic plates collide. Can result in *subduction* or mountain range formation.

coral reef A mass of calcium carbonate composed of the skeletons of tiny colonial marine organisms called corals.

core The innermost part of the Earth, made up mostly of iron, that lies beneath the *crust* and *mantle.*

Coriolis effect The apparent deflection of north-south air *currents* to a partly east-west direction, caused by the faster spin of regions near the equator than of regions near the poles as a result of Earth's rotation.

correlation A relationship among *variables.*

corridor A passageway of protected land established to allow animals to travel between islands of protected *habitat.*

cost-benefit analysis A method commonly used by *neoclassical economists*, in which estimated costs for a proposed action are totaled and then compared to the sum of benefits estimated to result from the action.

covalent bond A chemical bond in which the uncharged *atoms* in a *molecule* share *electrons*. For example, the uncharged atoms of carbon and oxygen in carbon dioxide form a covalent bond. Compare *ionic bond.*

criteria pollutants Six *air pollutants*—carbon monoxide, sulfur dioxide, nitrogen dioxide, tropospheric ozone, particulate matter, and lead—for which the *Environmental Protection Agency* has established maximum allowable concentrations in ambient outdoor air because of the threats they pose to human health.

cropland Land that humans use to raise plants for food and fiber.

crop rotation The practice of alternating the kind of crop grown in a particular field from one season or year to the next.

crude oil (petroleum) A *fossil fuel* produced by the conversion of *organic compounds* by heat and pressure. Crude oil is a mixture of hundreds of different types of *hydrocarbon* molecules characterized by carbon chains of different length.

crust The lightweight outer layer of the Earth, consisting of rock that floats atop the malleable *mantle,* which in turn surrounds a mostly iron *core.*

culture The overall ensemble of knowledge, beliefs, values, and learned ways of life shared by a group of people.

current The flow of a liquid or gas in a certain direction.

customary law International law that arises from long-standing practices, or customs, held in common by most *cultures.* Compare *conventional law.*

dam Any obstruction placed in a river or stream to block the flow of water so that water can be stored in a reservoir. Dams are built to prevent floods, provide drinking water, facilitate *irrigation*, and generate electricity.

Darwin, Charles (1809–1882) English naturalist who proposed the concept of *natural selection* as a mechanism for *evolution* and as a way to explain the great variety of living things. See also *Wallace, Alfred Russell.*

data Information, generally quantitative information.

deep ecology A philosophy established in the 1970s based on principles of self-realization (the awareness that humans are inseparable from nature) and *biocentric* equality (the precept that all living beings have equal value). Holds that because we are truly inseparable from our *environment*, we must protect all other living things as we would protect ourselves.

deep-well injection A *hazardous waste* disposal method in which a well is drilled deep beneath an area's *water table* into porous rock below an impervious *soil* layer. Wastes are then injected into the well, so that they will be absorbed into the porous rock and remain deep underground, isolated from *groundwater* and human contact. Compare *surface impoundment.*

deforestation The clearing and loss of forests.

demographic transition A theoretical model of economic and cultural change that explains the declining death rates and birth rates that occurred in Western nations as they became industrialized. The model holds that industrialization caused these rates to fall naturally by decreasing mortality and by lessening the need for large families. Parents would thereafter choose to invest in quality of life rather than quantity of children.

demography A *social science* that applies the principles of *population ecology* to the study of statistical change in human *populations.*

denitrifying bacteria Bacteria that convert the nitrates in *soil* or water to gaseous nitrogen and release it back into the *atmosphere.*

density-dependent factor A *limiting factor* whose effects on a *population* increase or decrease depending on the *population density.* Compare *density-independent factor.*

density-independent factor A *limiting factor* whose effects on a *population* are constant regardless of *population density.* Compare *density-dependent factor.*

deoxyribonucleic acid See *DNA.*

dependent variable The *variable* that is affected by manipulation of the *independent variable.*

desalination The removal of salt from seawater.

desert The driest *biome* on Earth, with annual *precipitation* of less than 25 cm. Because deserts have relatively little vegetation to insulate them from temperature extremes, sunlight readily heats them in the daytime, but daytime heat is quickly lost at night, so temperatures vary widely from day to night and in different seasons.

desertification A loss of more than 10% of a land's productivity due to *erosion, soil* compaction, forest removal, *overgrazing,* drought, *salinization, climate* change, depletion of water sources, or other factors. Severe desertification can result in the actual expansion of desert areas or creation of new ones in areas that once supported fertile land.

divergent plate boundary Area where *magma* surging upward to the surface divides tectonic plates and pushes them apart, creating new *crust* as it cools and spreads. A prime example is the Mid-Atlantic ridge. Compare *transform plate boundary* and *convergent plate boundary.*

DNA (deoxyribonucleic acid) A double-stranded *nucleic acid* composed of four nucleotides, each of which contains a sugar (deoxyribose), a phosphate group, and a nitrogenous base. DNA carries the hereditary information for living organisms and is responsible for passing traits from parents to offspring. Compare *RNA.*

dose-response curve A curve that plots the response of test animals to different doses of a *toxicant.* The response is generally quantified by measuring the proportion of animals exhibiting negative effects.

downwelling In the ocean, the flow of warm surface water toward the ocean floor. Downwelling occurs where surface *currents* converge. Compare *upwelling.*

Dust Bowl An area that loses huge amounts of *topsoil* to wind *erosion* as a result of drought and/or human impact; first used to name the region in the North American Great Plains severely affected by drought and topsoil loss in the 1930s. The term is now also used to describe that historical event and others like it.

dynamic equilibrium The state reached when processes within a *system* are moving in opposing directions at equivalent rates so that their effects balance out.

ecocentrism A philosophy that considers actions in terms of their damage or benefit to the integrity of whole ecological *systems,* including both *biotic* and *abiotic* elements. For an ecocentrist, the well-being of an individual organism—human or otherwise—is less important than the long-term well-being of a larger integrated ecological system.

ecofeminism A philosophy holding that the patriarchal (male-dominated) structure of society is a root cause of both social and environmental problems. Ecofeminists hold that a *worldview* traditionally associated with women, which interprets the world in terms of interrelationships and cooperation, is more in tune with nature than a worldview traditionally associated with men, which interprets the world in terms of hierarchies and competition.

ecolabeling The practice of designating on a product's label how the product was grown, harvested, or manufactured, so that consumers buying it are aware of the processes involved and can differentiate between brands that use processes believed to be *environmentally* beneficial (or less harmful than others) and those that do not.

ecological economics A developing school of *economics* that applies the principles of *ecology* and *systems* thinking to the description and analysis of *economies.* Compare *environmental economics; neoclassical economics.*

ecological footprint The cumulative amount of land and water required to provide the raw materials a person or *population* consumes and to dispose of or *recycle* the *waste* that is produced.

ecological restoration Efforts to reverse the effects of human disruption of *ecological systems* and to restore *communities* to their "natural" state.

ecology The *science* that deals with the distribution and abundance of organisms, the interactions among them, and the interactions between organisms and their *abiotic environments.*

economics The study of how we decide to use scarce resources to satisfy the demand for *goods* and *services.*

economy A social *system* that converts resources into *goods* and *services.*

ecosystem All organisms and nonliving entities that occur and interact in a particular area at the same time.

ecosystem ecology The study of how the living and nonliving components of *ecosystems* interact.

ecosystem-based management The attempt to manage the harvesting of resources in ways that minimize impact on the *ecosystems* and ecological processes that provide the resources.

ecosystem service An essential service an *ecosystem* provides that supports life and makes *economic* activity possible. For example, ecosystems naturally purify air and water, cycle *nutrients,* provide for plants to be *pollinated* by animals, and serve as receptacles and *recycling* systems for the *waste* generated by our economic activity.

ecotone A transitional zone where *ecosystems* meet.

ecotourism Visitation of natural areas for tourism and recreation. Most often involves tourism by more-affluent people, which may generate *economic* benefits for less-affluent communities near natural areas and thus provide economic incentives for conservation of natural areas.

ED$_{50}$ (effective dose–50%) The amount of a *toxicant* it takes to affect 50% of a *population* of test animals. Compare *threshold dose; LD$_{50}$.*

E horizon The layer of *soil* that lies below the *A horizon* and above the *B horizon.* The letter "E" stands for "eluviation," meaning "loss," and the E horizon is characterized by the loss of certain minerals through *leaching.* It is sometimes called the "zone of leaching." Compare *C horizon; O horizon; R horizon.*

El Niño The exceptionally strong warming of the eastern Pacific Ocean that occurs every 2 to 7 years and depresses local fish and bird *populations* by altering the marine *food web* in the area. Originally, the name that Spanish-speaking fishermen gave to an unusually warm surface *current* that sometimes arrived near the Pacific coast of South America around Christmas time. Compare *La Niña.*

electrolysis A process in which electrical current is passed through a *compound* to release *ions.* Electrolysis offers one way to produce hydrogen for use as fuel: Electrical current is passed through water, splitting the water *molecules* into hydrogen and oxygen *atoms.*

electron A negatively charged particle that surrounds the nucleus of an *atom.*

element A fundamental type of matter; a chemical substance with a given set of properties, which cannot be broken down into substances with other properties. Chemists currently recognize 92 elements that occur in nature, as well as more than 20 others that have been artificially created.

emergent property A characteristic that is not evident in a *system's* components.

Emerson, Ralph Waldo (1803–1882) American author, poet, and philosopher who espoused transcendentalism, a philosophy that views nature as a direct manifestation of the divine, and who promoted a holistic view of nature among the public.

emigration The departure of individuals from a *population.*

Endangered Species Act (ESA) The primary *legislation,* enacted in 1973, for protecting *biodiversity* in the United States. It forbids the government and private citizens from taking actions (such as developing land) that would destroy endangered *species* or their *habitats,* and it prohibits trade in products made from endangered species.

endemic Native or restricted to a particular geographic region. An endemic species occurs in one area and nowhere else on Earth.

endocrine disruptor A *toxicant* that interferes with the endocrine (hormone) system.

energy conservation The practice of reducing *energy* use as a way of extending the lifetime of our *fossil fuel* supplies, of being less wasteful, and of reducing our impact on the *environment.*

energy An intangible phenomenon that can change the position, physical composition, or temperature of matter.

entropy The degree of disorder in a substance, *system,* or process. See *second law of thermodynamics.*

environment The sum total of our surroundings, including all of the living things and nonliving things with which we interact.

environmental economics A developing school of *economics* that modifies the principles of *neoclassical economics* to address environmental challenges. An environmental economist believes that we can attain *sustainability* within our current economic *systems.* Whereas ecological economists call for revolution, environmental economists call for reform. Compare *ecological economics; neoclassical economics.*

environmental ethics The application of *ethical standards* to environmental questions.

environmental health Environmental factors that influence human health and quality of life and the health of *ecological* systems essential to environmental quality and long-term human well-being.

environmental impact statement (EIS) A report of results from detailed studies that assess the potential effects on the *environment* that would likely result from development projects or other actions undertaken by the government.

environmental justice A movement based on a moral sense of fairness and equality that seeks to expand society's domain of ethical concern from men to women, from humans to nonhumans, from rich to poor, and from majority races and ethnic groups to minority ones.

environmental policy *Public policy* that pertains to human interactions with the *environment.* It generally aims to regulate resource use or reduce *pollution* to promote human welfare and/or protect natural systems.

Environmental Protection Agency (EPA) An administrative agency created by executive order in 1970. The EPA is charged with conducting and evaluating research, monitoring environmental quality, setting standards, enforcing those standards, assisting the states in meeting standards and goals, and educating the public.

environmental science The study of how the natural world works and how humans and the *environment* interact.

environmental studies An academic *environmental science* program that heavily incorporates the social sciences as well as the natural sciences.

environmental toxicology The study of *toxicants* that come from or are discharged into the *environment,* including the study of health effects on humans, other animals, and *ecosystems.*

environmentalism A social movement dedicated to protecting the natural world.

epidemiological study A study that involves large-scale comparisons among groups of people, usually contrasting a group known to have been exposed to some *toxicant* and a group that has not.

equilibrium theory of island biogeography A *theory* that was initially applied to oceanic islands to explain how *species* come to be distributed among them. Since its development, researchers have increasingly applied the theory to islands of *habitat* (patches of one type of habitat isolated within vast "seas" of others). Aspects of the theory include *immigration* and *extinction* rates, the effect of island size, and the effect of distance from the mainland.

erosion The removal of material from one place and its transport to another by the action of wind or water.

estuary An area where a river flows into the ocean, mixing *freshwater* with salt water.

ethanol The alcohol in beer, wine, and liquor, produced as a *biofuel* by fermenting biomass, generally from *carbohydrate*-rich crops such as corn.

ethical standards The criteria that help differentiate right from wrong.

ethics The study of good and bad, right and wrong. The term can also refer to a person's or group's set of moral principles or values.

eukaryote A multicellular organism. The *cells* of eukaryotic organisms consist of a membrane-enclosed nucleus that houses *DNA,* an outer membrane of lipids, and an inner fluid-filled chamber containing *organelles.* Compare *prokaryote.*

European Union (EU) Political and economic organization formed after World War II to promote Europe's economic and social progress. As of 2005, the EU consisted of 25 member nations.

eutrophication The process of *nutrient* enrichment, increased production of organic matter, and subsequent *ecosystem* degradation.

evaporation The conversion of a substance from a liquid to a gaseous form.

even-aged Condition of timber plantations—generally *monocultures* of a single *species*—in which all trees are of the same age. Most *ecologists* view plantations of even-aged stands more as crop *agriculture* than as ecologically functional forests. Compare *uneven-aged.*

evolution Genetically based change in the appearance, functioning, and/or behavior of organisms across generations, often by the process of *natural selection.*

executive branch The branch of the U.S. government that is headed by the president and that includes administrative agencies. Among other powers, the president may approve (enact) or reject (veto) legislation and issue executive orders. Compare *judicial branch; legislative branch.*

experiment An activity designed to test the validity of a *hypothesis* by manipulating *variables.* See *manipulative experiment* and *natural experiment.*

exponential growth The increase of a *population* (or of anything) by a fixed percentage each year.

external cost A negative *externality;* a cost borne by someone not involved in an economic transaction. Examples include harm to citizens from water *pollution* or *air pollution* discharged by nearby factories.

externality A cost or benefit of a transaction that affects people other than the buyer or seller.

extinction The disappearance of an entire *species* from the face of the Earth. Compare *extirpation.*

extirpation The disappearance of a particular *population* from a given area, but not the entire *species* globally. Compare *extinction.*

feedback loop A circular process in which a *system*'s output serves as input to that same system. See *negative feedback loop; positive feedback loop.*

feedlot A huge barn or outdoor pen designed to deliver *energy*-rich food to animals living at extremely high densities. Also called a factory farm or concentrated animal feeding operation (CAFO).

Ferrel cell One of a pair of cells of *convective circulation* between 30° and 60° north and south latitude that influence global *climate* patterns. Compare *Hadley cell; polar cell.*

fertilizer A substance that promotes plant growth by supplying essential *nutrients* such as nitrogen or phosphorus.

first law of thermodynamics Physical law stating that *energy* can change from one form to another but cannot be created or lost. The total energy in the universe remains constant and is said to be conserved.

floodplain The region of land over which a river has historically wandered and periodically floods.

food security An adequate, reliable, and available food supply to all people at all times.

food web A visual representation of feeding interactions within an *ecological community* that shows an array of relationships between organisms at different *trophic levels.*

forestry The professional management of forests.

fossil The remains, impression, or trace of an animal or plant of past geological ages that has been preserved in rock or *sediments.*

fossil fuel A *nonrenewable natural resource,* such as *crude oil, natural gas,* or *coal,* produced by the decomposition and compression of organic matter from ancient life.

fossil record The cumulative body of *fossils* worldwide, which paleontologists study to infer the history of past life on Earth.

free rider A party that fails to invest in controlling *pollution* or carrying out other *environmentally* responsible activities and instead relies on the efforts of other parties to do so. For example, a factory that fails to control its emissions gets a "free ride" on the efforts of other factories that do make the sacrifices necessary to reduce emissions.

freshwater Water that is relatively pure, holding very few dissolved salts.

fundamental niche The full *niche* of a *species.* Compare *realized niche.*

gene A stretch of *DNA* that represents a unit of hereditary information.

gene bank See *seed bank.*

generalist A *species* that can survive in a wide array of *habitats* or use a wide array of resources. Compare *specialist.*

genetically modified (GM) organism An organism that has been *genetically engineered* using a technique called *recombinant DNA* technology.

genetic diversity A measurement of the differences in *DNA* composition among individuals within a given *species.*

genetic engineering Any process scientists use to manipulate an organism's genetic material in the lab by adding, deleting, or changing segments of its *DNA.*

Genuine Progress Indicator (GPI) An *economic* indicator introduced in 1995 that attempts to differentiate between desirable and undesirable economic activity. The GPI accounts for benefits such as volunteerism and for costs such as *environmental* degradation and social upheaval. Compare *Gross Domestic Product (GDP).*

geothermal energy Renewable *energy* that is generated deep within Earth. The radioactive decay of elements amid the extremely high pressures and temperatures at depth generate heat that rises to the surface in magma and through fissures and cracks. Where this energy heats *groundwater,* natural eruptions of heated water and steam are sent up from below.

global climate change Any change in aspects of Earth's *climate,* such as temperature, *precipitation,* and storm intensity. Generally refers today to the current warming trend in global temperatures and associated climatic changes.

good A material commodity manufactured for and bought by individuals and businesses.

greenhouse effect The warming of Earth's surface and *atmosphere* (especially the *troposphere*) caused by the *energy* emitted by *greenhouse gases.*

greenhouse gas A gas that absorbs infrared radiation released by Earth's surface and then warms the surface and *troposphere* by emitting *energy,* thus giving rise to the *greenhouse effect.* Greenhouse gases include carbon dioxide (CO_2), water vapor, ozone (O_3), nitrous oxide (N_2O), halocarbon gases, and methane (CH_4).

green revolution An intensification of the industrialization of *agriculture* in the developing world in the latter half of the 20th century that has dramatically increased crop yields produced per unit area of farmland. Practices include devoting large areas to *monocultures* of crops specially bred for high yields and rapid growth; heavy use of *fertilizers, pesticides,* and *irrigation* water; and sowing and harvesting on the same piece of land more than once per year or per season.

green tax A levy on *environmentally* harmful activities and products aimed at providing a market-based incentive to correct for *market failure.* Compare *subsidy.*

Gross Domestic Product (GDP) The total monetary value of final *goods* and *services* produced in a country each year. The GDP sums all *economic* activity, whether good or bad, and does not account for benefits such as volunteerism or for *external costs* such as *environmental* degradation and social upheaval. Compare *Genuine Progress Indicator (GPI).*

gross primary production The *energy* that results when *autotrophs* convert solar energy (sunlight) to energy of chemical bonds in sugars through *photosynthesis.* Autotrophs use a portion of this production to power their own metabolism, which entails oxidizing *organic compounds* by *cellular respiration.* Compare *net primary production; secondary production.*

groundwater Water held in aquifers underground.

growth rate The net change in a *population*'s size, per 1,000 individuals. Calculated by adding the crude birth rate to the *immigration* rate and then subtracting the crude death rate and the *emigration* rate, each expressed as the number per 1,000 individuals per year.

Haber-Bosch process A process to synthesize ammonia on an industrial scale. Developed by German chemists Fritz Haber and Carl Bosch, the process has enabled humans to double the natural rate of *nitrogen fixation* on Earth and thereby increase *agricultural* productivity, but it has also dramatically altered the *nitrogen cycle.*

habitat The specific *environment* in which an organism lives, including both *biotic* and *abiotic factors.*

habitat use The process by which organisms select and use *habitats* from among the range of options they encounter.

Hadley cell One of a pair of cells of *convective circulation* between the equator and 30° north and south latitude that influence global *climate* patterns. Compare *Ferrel cell; polar cell.*

half-life The amount of time it takes for one-half the atoms of a *radioisotope* to emit radiation and decay. Different radioisotopes have different half-lives, ranging from fractions of a second to billions of years.

harmful algal bloom A *population* explosion of toxic algae caused by excessive *nutrient* concentrations.

hazardous waste *Waste* that is toxic, chemically reactive, flammable, or corrosive. Compare *industrial solid waste; municipal solid waste.*

herbivory The consumption of plants by animals.

heterotroph (consumer) An organism that consumes other organisms. Includes most animals, as well as fungi and microbes that decompose organic matter.

high-pressure system An air mass with elevated *atmospheric pressure,* containing air that descends, typically bringing fair *weather.* Compare *low-pressure system.*

homeostasis The tendency of a *system* to maintain constant or stable internal conditions.

horizon A distinct layer of *soil.* See *A horizon; B horizon; C horizon; E horizon; O horizon; R horizon.*

Hubbert's peak The peak in production of *crude oil* in the United States, which occurred in 1970 just as Shell Oil geologist M. King Hubbert had predicted in 1956.

hydrocarbon An *organic compound* consisting solely of hydrogen and carbon *atoms.*

hydroelectric power (hydropower) The generation of electricity using the *kinetic energy* of moving water.

hydrologic cycle The flow of water—in liquid, gaseous, and solid forms—through our *biotic* and *abiotic environment.*

hydropower See *hydroelectric power.*

hydrosphere All water—salt or fresh, liquid, ice, or vapor—in surface bodies, underground, and in the *atmosphere.* Compare *biosphere; lithosphere.*

hypothesis An educated guess that explains a phenomenon or answers a *scientific* question. Compare *theory.*

hypoxia The condition of extremely low dissolved oxygen concentrations in a body of water.

igneous rock One of the three main categories of rock. Formed from cooling *magma.* Granite and basalt are examples of igneous rock. Compare *metamorphic rock; sedimentary rock.*

immigration The arrival of individuals from outside a *population.*

incineration A controlled process of burning solid waste for disposal in which mixed garbage is combusted at very high temperatures. Compare *sanitary landfill.*

independent variable The *variable* that the scientist manipulates in a *manipulative experiment.*

industrial ecology A holistic approach to industry that integrates principles from engineering, chemistry, *ecology, economics,* and other disciplines and seeks to redesign industrial *systems* in order to reduce resource inputs and minimize inefficiency.

industrial revolution The shift in the mid-1700s from rural life, animal-powered agriculture, and manufacturing by craftsmen to an urban society powered by *fossil fuels* such as *coal* and *crude oil.* Compare *agricultural revolution.*

industrial smog Gray-air smog caused by the incomplete combustion of *coal* or oil when burned. Compare *photochemical smog.*

industrial solid waste Nonliquid *waste* that is not especially hazardous and that comes from production of consumer goods, mining, *petroleum* extraction and *refining,* and *agriculture.* Compare *hazardous waste; municipal solid waste.*

industrial stage The third stage of the *demographic transition* model, characterized by falling birth rates that close the gap with falling death rates and reduce the rate of *population* growth. Compare *pre-industrial stage; post-industrial stage; transitional stage.*

industrialized agriculture A form of *agriculture* that uses large-scale mechanization and *fossil fuel* combustion, enabling farmers to replace horses and oxen with faster and more powerful means of cultivating, harvesting, transporting, and processing crops. Other aspects include *irrigation* and the use of *inorganic fertilizers.* Use of chemical herbicides and *pesticides* reduces *competition* from weeds and *herbivory* by insects. Compare *traditional agriculture.*

infectious disease A disease in which a pathogen attacks a host.

inorganic fertilizer A *fertilizer* that consists of mined or synthetically manufactured mineral supplements. Inorganic fertilizers are generally more susceptible than *organic fertilizers* to *leaching* and *runoff* and may be more likely to cause unintended off-site impacts.

integrated pest management (IPM) The use of multiple techniques in combination to achieve long-term suppression of pests, including *biocontrol,* use of *pesticides,* close monitoring of *populations, habitat* alteration, *crop rotation, transgenic* crops, alternative tillage methods, and mechanical pest removal.

intercropping Planting different types of crops in alternating bands or other spatially mixed arrangements.

interdisciplinary field A field that borrows techniques from several more traditional fields of study and brings together research results from these fields into a broad synthesis.

Intergovernmental Panel on Climate Change (IPCC) An international panel of *atmospheric* scientists, *climate* experts, and government officials established in 1988 by the *United Nations* Environment Programme and the World Meteorological Organization, whose mission is to assess information relevant to questions of human-induced *global climate change.* The IPCC's 2001 *Third Assessment Report* summarizes current and probable future global trends and represents the consensus of atmospheric scientists around the world.

intertidal Of, relating to, or living along shorelines between the highest reach of the highest *tide* and the lowest reach of the lowest tide.

invasive species A *species* that spreads widely and rapidly becomes dominant in a *community,* interfering with the community's normal functioning.

inversion layer In a *temperature inversion,* the band of air in which temperature rises with altitude (instead of falling with altitude, as temperature does normally).

ion An electrically charged *atom* or combination of atoms.

ionic bond A chemical bond in which oppositely charged *ions* are held together by electrical attraction. Compare *covalent bond.*

ionic compound (salt) An association of *ions* that are bonded electrically in an *ionic bond.*

IPAT model A formula that represents how humans' total impact (I) on the *environment* results from the interaction among three factors: *population* (P), affluence (A), and technology (T).

irrigation The artificial provision of water to support *agriculture.*

isotope One of several forms of an *element* having differing numbers of *neutrons* in the nucleus of its *atoms.* Chemically, isotopes of an element behave almost identically, but they have different physical properties because they differ in mass.

judicial branch The branch of the U.S. government, consisting of the Supreme Court and various lower courts, that is charged with interpreting the law. Compare *executive branch; legislative branch.*

kelp Large brown algae or seaweed that can form underwater "forests," providing habitat for marine organisms.

keystone species A *species* that has an especially far-reaching effect on a *community.*

kinetic energy *Energy* of motion. Compare *potential energy.*

K–selected Term denoting a *species* with low biotic potential whose members produce a small number of offspring and take a long time to gestate and raise each of their young, but invest heavily in promoting the survival and growth of these few offspring. *Populations* of K–selected species are generally regulated by *density-dependent factors.* Compare *r–selected.*

Kyoto Protocol An agreement drafted in 1997 that calls for reducing, by 2012, emissions of six *greenhouse gases* to levels lower than their levels in 1990. Although the United States has refused to ratify the protocol, it came into force in 2005 when Russia ratified it, the 127th nation to do so.

La Niña An exceptionally strong cooling of surface water in the equatorial Pacific Ocean that occurs every 2 to 7 years and has widespread climatic consequences. Compare *El Niño.*

land trust Local or regional organization that preserves lands valued by its members. In most cases, land trusts purchase land outright with the aim of preserving it in its natural condition. The Nature Conservancy may be considered the world's largest land trust.

landscape ecology An approach to the study of organisms and their *environments* at the landscape scale, focusing on geographical areas that include multiple *ecosystems.*

lava *Magma* that is released from the *lithosphere* and flows or spatters across Earth's surface.

LD$_{50}$ (lethal dose–50%) The amount of a *toxicant* it takes to kill 50% of a *population* of test animals. Compare *ED$_{50}$; threshold dose.*

leachate Liquids that seep through liners of a *sanitary landfill* and leach into the *soil* underneath.

leaching The process by which solid materials such as minerals are dissolved in a liquid (usually water) and transported to another location.

legislation Statutory law.

legislative branch The branch of the U.S. government that passes laws; it consists of Congress, which includes the House of Representatives and the Senate. Compare *executive branch; judicial branch.*

Leopold, Aldo (1887–1949) American scientist, scholar, philosopher, and author. His book *The Land Ethic* argued that humans should view themselves and the land itself as members of the same *community* and that humans are obligated to treat the land *ethically.*

life-cycle analysis In *industrial ecology,* the examination of the entire life cycle of a given product—from its origins in raw materials, through its manufacturing, to its use, and finally its disposal—in an attempt to identify ways to make the process more *ecologically* efficient.

life expectancy The average number of years that individuals in particular age groups are likely to continue to live.

limiting factor A physical, chemical, or biological characteristic of the *environment* that restrains *population* growth.

limnetic zone In a water body, the layer of open water through which sunlight penetrates. Compare *littoral zone; benthic zone; profundal zone.*

lipid One of a chemically diverse group of *macromolecules* that are classified together because they do not dissolve in water. Lipids include fats, phospholipids, waxes, pigments, and steroids.

lithification The formation of rock through the processes of compaction, binding, and crystallization.

lithosphere The solid part of the Earth, including the rocks, *sediment,* and *soil* at the surface and extending down many miles underground. Compare *atmosphere; biosphere; hydrosphere.*

littoral See *intertidal.*

littoral zone The region ringing the edge of a water body. Compare *benthic zone; limnetic zone; profundal zone.*

loam *Soil* with a relatively even mixture of *clay-, silt-,* and *sand-*sized particles.

lobbying The expenditure of time or money in an attempt to influence an elected official.

logistic growth curve A plot that shows how the initial *exponential growth* of a *population* is slowed and finally brought to a standstill by *limiting factors.*

low-pressure system An air mass in which the air moves toward the low *atmospheric pressure* at the center of the system and spirals upward, typically bringing clouds and *precipitation.* Compare *high-pressure system.*

macromolecule A very large molecule, such as a *protein, nucleic acid, carbohydrate,* or *lipid.*

macronutrient A *nutrient* that organisms require in relatively large amounts. Compare *micronutrient.*

magma Molten, liquid rock.

malnutrition The condition of lacking *nutrients* the body needs, including a complete complement of vitamins and minerals.

Malthus, Thomas (1766–1834) British economist who maintained that increasing human *population* would eventually deplete the available food supply until starvation, war, or disease arose and reduced the population.

mangrove A tree with a unique type of roots that curve upward to obtain oxygen, which is lacking in the mud in which they grow, and serve as stilts to support the tree in changing water levels. Mangrove forests grow on the coastlines of the tropics and subtropics.

manipulative experiment An *experiment* in which the researcher actively chooses and manipulates the *independent variable.* Compare *natural experiment.*

mantle The malleable layer of rock that lies beneath Earth's *crust* and surrounds a mostly iron *core.*

marine protected area (MPA) An area of the ocean set aside to protect marine life from fishing pressures. An MPA may be protected from some human activities but be open to others. Compare *marine reserve.*

marine reserve An area of the ocean designated as a "no-fishing" zone, allowing no extractive activities. Compare *marine protected area.*

marketable emissions permit A permit issued to polluters that allows them to emit a certain fraction of the total amount of *pollution* the government will allow an entire industry to produce. Polluters are then allowed to buy, sell, and trade these permits with other polluters. See also *permit-trading.*

market failure The failure of markets to take into account the *environment*'s positive effects on *economies* (for example, *ecosystem services*) or to reflect the negative effects of economic activity on the environment and thereby on people (*external costs*).

mass extinction event The extinction of a large proportion of the world's *species* in a very short time period due to some extreme and rapid change or catastrophic event. Earth has seen five mass extinction events in the past half-billion years.

materials recovery facility (MRF) A *recycling* facility where items are sorted, cleaned, shredded, and prepared for reprocessing into new items.

maximum sustainable yield The maximal harvest of a particular *renewable natural resource* that can be accomplished while still keeping the resource available for the future.

meltdown The accidental melting of the uranium fuel rods inside the core of a *nuclear reactor,* causing the release of radiation.

metamorphic rock One of the three main categories of rock. Formed by great heat and/or pressure that reshapes crystals within the rock and changes its appearance and physical properties. Common metamorphic rocks include marble and slate. Compare *igneous rock; sedimentary rock.*

methane hydrate An ice-like solid consisting of molecules of methane (CH_4) embedded in a crystal lattice of water molecules. Methane hydrates are being investigated as a potential new source of *energy* from *fossil fuels.*

micronutrient A *nutrient* that organisms require in relatively small amounts. Compare *macronutrient.*

Milankovitch cycle One of three types of variations in Earth's rotation and orbit around the sun that result in slight changes in the relative amount of solar radiation reaching Earth's surface at different latitudes. As the cycles proceed, they change the way solar radiation is distributed over Earth's surface and contribute to changes in *atmospheric* heating and circulation that have triggered the ice ages and other *climate* changes.

Mill, John Stuart (1806–1873) British philosopher who believed that as resources become harder to find and extract, *economic* growth will slow and eventually stabilize into a *steady-state economy.*

molecule A combination of two or more *atoms.*

monoculture The uniform planting of a single crop over a large area. Characterizes *industrialized agriculture.*

Montreal Protocol International treaty ratified in 1987 in which 180 signatory nations agreed to restrict production of *chlorofluorocarbons (CFCs)* in order to forestall stratospheric ozone depletion. Because of its effectiveness in decreasing global CFC emissions, the Montreal Protocol is considered the most successful effort to date in addressing a global *environmental* problem.

Muir, John (1838–1914) Scottish immigrant to the United States who eventually settled in California and made the

Yosemite Valley his wilderness home. Today, he is most strongly associated with the *preservation ethic.* He argued that nature deserved protection for its own inherent values (an *ecocentrist* argument) but also claimed that nature played a large role in human happiness and fulfillment (an *anthropocentrist* argument).

multiple use A principle that has nominally guided management policy for national forests over the past half century. The multiple use principle specifies that the forests be managed for recreation, wildlife habitat, mineral extraction, and various other uses.

municipal solid waste Nonliquid *waste* that is not especially hazardous and that comes from homes, institutions, and small businesses. Compare *hazardous waste; industrial solid waste.*

mutagen A *toxicant* that causes *mutations* in the *DNA* of organisms.

mutation An accidental change in *DNA* that may range in magnitude from the deletion, substitution, or addition of a single nucleotide to a change affecting entire sets of chromosomes. Mutations provide the raw material for evolutionary change.

mutualism A relationship in which all participating organisms benefit from their interaction. Compare *parasitism.*

National Environmental Policy Act (NEPA) A U.S. law enacted on January 1, 1970, that created an agency called the Council on Environmental Quality and required that an *environmental impact statement* be prepared for any major federal action.

national forest Public lands consisting of 191 million acres (more than 8% of the nation's land area) in many tracts spread across all but a few states.

National Forest Management Act *Legislation* passed by the U.S. Congress in 1976, mandating that plans for renewable resource management be drawn up for every national forest. These plans were to be explicitly based on the concepts of *multiple use* and *sustainable development* and be subject to broad public participation.

national park A scenic area set aside for recreation and enjoyment by the public. The national park system today numbers 388 sites totaling 78.8 million acres and includes national historic sites, national recreation areas, national wild and scenic rivers, and other types of areas.

national wildlife refuge An area set aside to serve as a haven for wildlife and also sometimes to encourage hunting, fishing, wildlife observation, photography, environmental education, and other public uses.

natural experiment An *experiment* in which the researcher cannot directly manipulate the *variables* and therefore must observe nature, comparing conditions in which variables differ, and interpret the results. Compare *manipulative experiment.*

natural gas A *fossil fuel* composed primarily of methane (CH_4), produced as a by-product when bacteria decompose organic material under *anaerobic* conditions.

natural rate of population change The rate of change in a *population's* size resulting from birth and death rates alone, excluding migration.

natural resource Any of the various substances and *energy* sources we need in order to survive.

natural science An academic discipline that studies the natural world. Compare *social science.*

natural selection The process by which traits that enhance survival and reproduction are passed on more frequently to future generations of organisms than those that do not, thus altering the *genetic* makeup of populations through time. Natural selection acts on genetic variation and is a primary driver of evolution.

negative feedback loop A *feedback loop* in which output of one type acts as input that moves the *system* in the opposite direction. The input and output essentially neutralize each other's effects, stabilizing the system. Compare *positive feedback loop.*

neoclassical economics A *theory* of *economics* that explains market prices in terms of consumer preferences for units of particular commodities. Buyers desire the lowest possible price, whereas sellers desire the highest possible price. This conflict between buyers and sellers results in a compromise price being reached and the "right" quantity of commodities being bought and sold. Compare *ecological economics; environmental economics.*

net primary production The *energy* or biomass that remains in an ecosystem after *autotrophs* have metabolized enough for their own maintenance through *cellular respiration.* Net primary production is the energy or biomass available for consumption by *heterotrophs.* Compare *gross primary production; secondary production.*

net primary productivity The rate at which *net primary production* is produced. See *productivity; gross primary production; net primary production; secondary production.*

neurotoxin A *toxicant* that assaults the nervous system. Neurotoxins include heavy metals, *pesticides,* and some chemical weapons developed for use in war.

neutron An electrically neutral (uncharged) particle in the nucleus of an *atom.*

new forestry A set of *ecosystem-based management* approaches for harvesting timber that explicitly mimic natural disturbances. For instance, "sloppy clear-cuts" that leave a variety of trees standing mimic the changes a forest might experience if hit by a severe windstorm.

new urbanism A school of thought among architects, planners, and developers that seeks to design neighborhoods in which homes, businesses, schools, and other amenities are within walking distance of one another. In a direct rebuttal to *sprawl,* proponents of new urbanism aim to create functional neighborhoods in which families can meet most of their needs close to home without the use of a car.

niche The functional role of a *species* in a *community.* See *fundamental niche; realized niche.*

nitrification The conversion by bacteria of ammonium ions (NH_4^+) first into nitrite ions (NO_2^-) and then into nitrate ions (NO_3^-).

nitrogen cycle A major *nutrient cycle* consisting of the routes that nitrogen *atoms* take through the nested networks of environmental *systems.*

nitrogen fixation The process by which inert nitrogen gas combines with hydrogen to form ammonium ions (NH_4^+), which are chemically and biologically active and can be taken up by plants.

nitrogen-fixing Term describing bacteria that live in a *mutualistic* relationship with many types of plants and provide *nutrients* to the plants by converting nitrogen to a usable form.

nonconsumptive use *Freshwater* use in which the water from a particular *aquifer* or surface water body either is not removed or is removed only temporarily and then returned. The use of water to generate electricity in hydroelectric *dams* is an example. Compare *consumptive use.*

nonmarket value A value that is not usually included in the price of a *good* or *service.*

non-point source A diffuse source of *pollutants,* often consisting of many small sources. Compare *point source.*

nonrenewable natural resource A *natural resource* that is in limited supply and is formed much more slowly than we use it. Compare *renewable natural resource.*

Northwest Forest Plan A 1994 plan developed by the Clinton administration to allow logging of forests of western Washington, Oregon, and northwestern California with increased protection for *species* and *ecosystems.* The Northwest Forest Plan represented one of the first large-scale applications of *adaptive management.*

nuclear energy The *energy* that holds together *protons* and *neutrons* within the nucleus of an *atom.* Several processes, each of which involves transforming *isotopes* of one *element* into isotopes of other elements, can convert nuclear energy into thermal energy, which is then used to generate electricity. See also *nuclear fission; nuclear reactor.*

nuclear fission The conversion of the *energy* within an *atom's* nucleus to usable thermal energy by splitting apart atomic nuclei. Compare *nuclear fusion.*

nuclear fusion The conversion of the *energy* within an *atom's* nucleus to usable thermal energy by forcing together the small nuclei of lightweight *elements* under extremely high temperature and pressure. Developing a commercially viable method of nuclear fusion remains an elusive goal.

nuclear reactor A facility within a nuclear power plant that initiates and controls the process of *nuclear fission* in order to generate electricity.

nucleic acid A *macromolecule* that directs the production of *proteins.* Includes *DNA* and *RNA.*

nutrient An *element* or *compound* that organisms consume and require for survival.

nutrient cycle The comprehensive set of cyclical pathways by which a given *nutrient* moves through the *environment.*

oceanography The study of the physics, chemistry, biology, and geology of the oceans.

ocean thermal energy conversion (OTEC) A potential *energy* source that involves harnessing the solar radiation absorbed by tropical oceans in the tropics.

O horizon The top layer of *soil* in some *soil profiles,* made up of organic matter, such as decomposing branches, leaves, crop residue, and animal waste. Compare *A horizon; B horizon; C horizon; E horizon; R horizon.*

open system A *system* that exchanges *energy,* matter, and information with other systems. Compare *closed system.*

organelle A structure, such as a ribosome or mitochondrion, inside the *cell* that performs specific functions.

organic agriculture *Agriculture* that uses no synthetic *fertilizers* or *pesticides* but instead relies on biological approaches such as *composting* and *biocontrol.*

organic compound A *compound* made up of carbon *atoms* (and, generally, hydrogen atoms) joined by *covalent bonds* and sometimes including other *elements,* such as nitrogen, oxygen, sulfur, or phosphorus. The unusual ability of carbon to build elaborate molecules has resulted in millions of different organic compounds showing various degrees of complexity.

organic fertilizer A *fertilizer* made up of natural materials (largely the remains or wastes of organisms), including animal manure, crop residues, fresh vegetation, and compost. Compare *inorganic fertilizer.*

overgrazing The consumption by too many animals of plant cover, impeding plant regrowth and the replacement of biomass. Overgrazing can exacerbate damage to *soils,* natural *communities,* and the land's productivity for further grazing.

overnutrition A condition of excessive food intake in which people receive more than their daily caloric needs.

ozone layer A portion of the *stratosphere,* roughly 17–30 km (10–19 mi) above sea level, that contains most of the ozone in the *atmosphere.*

paradigm A dominant philosophical and theoretical framework within a scientific discipline.

parasitism A relationship in which one organism, the parasite, depends on another, the host, for nourishment or some other benefit while simultaneously doing the host harm. Compare *mutualism.*

parent material The base geological material in a particular location.

passive solar energy collection An approach in which buildings are designed and building materials are chosen to maximize their direct absorption of sunlight in winter, even as they keep the interior cool in the summer. Compare *active solar energy collection.*

peat A kind of precursor stage to *coal,* produced when organic material that is broken down by *anaerobic* decomposition remains wet, near the surface, and not well compressed.

peer review The process by which a manuscript submitted for publication in an academic journal is examined by other specialists in the field, who provide comments and criticism (generally anonymously), and judge whether the work merits publication in the journal.

pelagic Of, relating to, or living between the surface and floor of the ocean. Compare *benthic.*

permit-trading The practice of buying and selling government-issued *marketable emissions permits* to conduct environmentally harmful activities. Under such a system, the government determines an acceptable level of *pollution* and then issues permits to pollute. A company receives credit for amounts it does not emit and can then sell this credit to other companies.

pesticide An artificial chemical used to kill insects (insecticide), plants (herbicide), or fungi (fungicide).

petroleum See *crude oil.*

pH A measure of the concentration of hydrogen *ions* in a solution. The pH scale ranges from 0 to 14: A solution with a pH of 7 is neutral; solutions with a pH below 7 are *acidic,* and those with a pH higher than 7 are *basic.* Because the pH scale is logarithmic, each step on the scale represents a tenfold difference in hydrogen ion concentration.

phosphorus cycle A major *nutrient cycle* consisting of the routes that phosphorus *atoms* take through the nested networks of environmental *systems.*

photochemical smog Brown-air smog caused by light-driven reactions of *primary pollutants* with normal atmospheric *compounds* that produce a mix of over 100 different chemicals, ground-level ozone often being the most abundant among them. Compare *industrial smog.*

photosynthesis The process by which *autotrophs* produce their own food. Sunlight powers a series of chemical reactions that convert carbon dioxide and water into sugar (glucose), thus transforming low-quality *energy* from the sun into high-quality energy the organism can use. Compare *cellular respiration.*

photovoltaic (PV) cell A device designed to collect sunlight and convert it to electrical *energy* directly by making use of the *photoelectric effect.*

phylogenetic tree A treelike diagram that represents the history of divergence of *species* or other taxonomic groups of organisms.

Pinchot, Gifford (1865–1946) The first professionally trained American *forester,* Pinchot helped establish the U.S. Forest Service. Today, he is the person most closely associated with the *conservation ethic.*

pioneer species A *species* that arrives earliest, beginning the ecological process of *succession* in a terrestrial or aquatic *community.*

plate tectonics The process by which Earth's surface is shaped by the extremely slow movement of tectonic plates, or sections of *crust.* Earth's surface includes about 15 major tectonic plates. Their interaction gives rise to processes that build mountains, cause earthquakes, and otherwise influence the landscape.

point source A specific spot—such as a factory's smokestacks—where large quantities of *pollutants* are discharged. Compare *non-point source.*

polar cell One of a pair of cells of *convective circulation* between the poles and 60° north and south latitude that influence global *climate* patterns. Compare *Ferrel cell; Hadley cell.*

policy A rule or guideline that directs individual, organizational, or societal behavior.

pollination An interaction in which one organism (for example, bees) transfers pollen (male sex cells) from one flower to the ova (female cells) of another, fertilizing the female flower, which subsequently grows into a fruit.

pollution Any matter or *energy* released into the *environment* that causes undesirable impacts on the health and well-being of humans or other organisms. Pollution can be physical, chemical, or biological, and can affect water, air, or soil.

polymer A chemical *compound* or mixture of compounds consisting of long chains of repeated *molecules.* Some polymers play key roles in the building blocks of life.

population A group of organisms of the same *species* that live in the same area. Species are often composed of multiple populations.

population density The number of individuals within a *population* per unit area. Compare *population size.*

population dispersion See *population distribution.*

population distribution The spatial arrangement of organisms within a particular area.

population ecology Study of the quantitative dynamics of how individuals within a *species* interact with one another—in particular, why *populations* of some species decline while others increase.

population size The number of individual organisms present at a given time.

positive feedback loop A *feedback loop* in which output of one type acts as input that moves the *system* in the same direction. The input and output drive the system further toward one extreme or another. Compare *negative feedback loop.*

post-industrial stage The fourth and final stage of the *demographic transition* model, in which both birth and death rates have fallen to a low level and remain stable there, and *populations* may even decline slightly. Compare *industrial stage; pre-industrial stage; transition stage.*

potential energy *Energy* of position. Compare *kinetic energy.*

precautionary principle The idea that one should not undertake a new action until the ramifications of that action are well understood.

precipitation Water that condenses out of the *atmosphere* and falls to Earth in droplets or crystals.

predation The process in which one *species* (the predator) hunts, tracks, captures, and ultimately kills its prey.

prediction A specific statement, generally arising from a *hypothesis,* that can be tested directly and unequivocally.

pre-industrial stage The first stage of the *demographic transition* model, characterized by conditions that defined most of human history. In pre-industrial societies, both death rates and birth rates are high. Compare *industrial stage; post-industrial stage; transitional stage.*

prescribed (controlled) burns The practice of burning areas of forest or grassland under carefully controlled conditions to improve the health of *ecosystems,* return them to a more natural state, and help prevent uncontrolled catastrophic fires.

preservation ethic An ethic holding that we should protect the natural *environment* in a pristine, unaltered state. Compare *conservation ethic.*

primary extraction The initial drilling and pumping of available *crude oil.* Compare *secondary extraction.*

primary pollutant A hazardous substance, such as soot or carbon monoxide, that is emitted into the *troposphere* in a form that is directly harmful. Compare *secondary pollutant.*

primary succession A stereotypical series of changes as an *ecological community* develops over time, beginning with a lifeless substrate. In terrestrial *systems,* primary succession begins when a bare expanse of rock, *sand,* or *sediment* becomes newly exposed to the atmosphere and *pioneer species* arrive. Compare *secondary succession.*

primary treatment A stage of *wastewater* treatment in which contaminants are physically removed. Wastewater flows into tanks in which sewage solids, grit, and particulate matter settle to the bottom. Greases and oils float to the surface and can be skimmed off. Compare *secondary treatment.*

producer See *autotroph.*

productivity The rate at which plants convert solar *energy* (sunlight) to biomass. *Ecosystems* whose plants convert solar energy to biomass rapidly are said to have high productivity. See *net primary productivity; gross primary production; net primary production.*

profundal zone In a water body, the volume of open water that sunlight does not reach. Compare *littoral zone; benthic zone; limnetic zone.*

prokaryote A typically unicellular organism. The *cells* of prokaryotic organisms lack *organelles* and a nucleus. All bacteria and archaeans are prokaryotes. Compare *eukaryote.*

protein A *macromolecule* made up of long chains of amino acids.

proton A positively charged particle in the nucleus of an *atom.*

proven recoverable reserve The amount of a given *fossil fuel* in a deposit that is technologically and economically feasible to remove under current conditions.

proxy indicator Indirect evidence, such as pollen from *sediment* cores and air bubbles from ice cores, of the *climate* of the past.

public policy *Policy* that is made by governments, including those at the local, state, federal, and international levels; it consists of *legislation, regulations,* orders, incentives, and practices intended to advance societal welfare. See also *environmental policy.*

radioisotopes Radioactive *isotopes* that emit subatomic particles and high-*energy* radiation as they "decay" into progressively lighter isotopes until becoming stable isotopes.

rangeland Land used for grazing livestock.

realized niche The portion of the *fundamental niche* that is fully realized (used) by a *species.*

recombinant DNA *DNA* that has been patched together from the DNA of multiple organisms in an attempt to produce desirable traits (such as rapid growth, disease and pest resistance, or higher nutritional content) in organisms lacking those traits.

recycling The collection of materials that can be broken down and reprocessed to manufacture new items.

Red List An updated list of *species* facing unusually high risks of *extinction.* The list is maintained by the World Conservation Union.

red tide A *harmful algal bloom* consisting of algae that produce reddish pigments that discolor surface waters.

refining Process of separating the *molecules* of the various *hydrocarbons* in *crude oil* into different-sized classes and transforming them into various fuels and other petrochemical products.

regional planning *City planning* done on broader geographic scales, generally involving multiple municipal governments.

regulation A specific rule issued by an administrative agency, based on the more broadly written statutory law passed by Congress and enacted by the president.

regulatory taking The deprivation of a property's owner, by means of a law or *regulation,* of most or all economic uses of that property.

relative humidity The ratio of the water vapor contained in a given volume of air to the maximum amount the air could contain, for a given temperature.

relativist An ethicist who maintains that *ethics* do and should vary with social context. Compare *universalist.*

renewable natural resource A *natural resource* that is virtually unlimited or that is replenished by the *environment* over relatively short periods of hours to weeks to years. Compare *nonrenewable natural resource.*

replacement fertility The *total fertility rate (TFR)* that maintains a stable *population* size.

reserves-to-production ratio (R/P ratio) The total remaining reserves of a *fossil fuel* divided by the annual rate of production (extraction and processing).

resilience The ability of an ecological *community* to change in response to disturbance but later return to its original state. Compare *resistance.*

resistance The ability of an ecological *community* to remain stable in the presence of a disturbance. Compare *resilience.*

resource management Strategic decision making about who should extract resources and in what ways, so that resources are used wisely and not wasted.

resource partitioning The process by which *species* adapt to *competition* by evolving to use slightly different resources, or to use their shared resources in different ways, thus minimizing interference with one another.

Resource Conservation and Recovery Act (RCRA) Congressional *legislation* (enacted in 1976 and amended in 1984) that specifies, among other things, how to manage *sanitary landfills* to protect against environmental contamination.

restoration ecology The study of the historical conditions of *ecological communities* as they existed before humans altered them.

revolving door The movement of powerful officials between the private sector and government agencies.

R horizon The bottommost layer of *soil* in a typical *soil profile.* Also called *bedrock.* Compare *A horizon; B horizon; C horizon; E horizon; O horizon.*

ribonucleic acid See *RNA.*

risk The mathematical probability that some harmful outcome (for instance, injury, death, *environmental* damage, or *economic* loss) will result from a given action, event, or substance.

risk assessment The quantitative measurement of *risk,* together with the comparison of risks involved in different activities or substances.

risk management The process of considering information from scientific *risk assessment* in light of economic, social, and political needs and values, in order to make decisions and design strategies to minimize *risk.*

RNA (ribonucleic acid) A usually single-stranded *nucleic acid* composed of four nucleotides, each of which contains a sugar (ribose), a phosphate group, and a nitrogenous base. RNA carries the hereditary information for living organisms and is responsible for passing traits from parents to offspring. Compare *DNA.*

rock cycle The very slow process in which rocks and the minerals that make them up are heated, melted, cooled, broken, and reassembled, forming *igneous*, *sedimentary*, and *metamorphic* rocks.

r–selected Term denoting a *species* with high biotic potential whose members produce a large number of offspring in a relatively short time but do not care for their young after birth. *Populations* of r–selected species are generally regulated by *density-independent factors*. Compare *K–selected*.

runoff The water from *precipitation* that flows into streams, rivers, lakes, and ponds, and (in many cases) eventually to the ocean.

run-of-river Any of several methods used to generate *hydroelectric power* without greatly disrupting the flow of river water. Run-of-river approaches eliminate much of the *environmental* impact of large *dams*. Compare *storage*.

Ruskin, John (1819–1900) British art critic, poet, and writer who criticized industrialized cities and their *pollution,* and who believed that people no longer appreciated the *environment*'s spiritual or aesthetic benefits.

salinization The buildup of salts in surface *soil* layers.

salt See *ionic compound*.

salt marsh Flat land that is intermittently flooded by the ocean where the *tide* reaches inland. Salt marshes occur along temperate coastlines and are thickly vegetated with grasses, rushes, shrubs, and other herbaceous plants.

salvage logging The removal of dead trees following a natural disturbance. Although it may be economically beneficial, salvage logging can be ecologically destructive, because the dead trees provide food and shelter for a variety of insects and wildlife and because removing timber from recently burned land can cause severe *erosion* and damage to *soil*.

sand *Sediment* consisting of particles 0.005–2.0 mm in diameter. Compare *clay; silt*.

sanitary landfill A site at which solid waste is buried in the ground or piled up in large mounds for disposal, designed to prevent the waste from contaminating the *environment*. Compare *incineration*.

savanna A *biome* characterized by grassland interspersed with clusters of acacias and other trees. Savanna is found across parts of Africa (where it was the ancestral home of our *species*), South America, Australia, India, and other dry tropical regions.

science A systematic process for learning about the world and testing our understanding of it.

scientific method A formalized method for testing ideas with observations that involves several assumptions and a more or less consistent series of interrelated steps.

secondary extraction The extraction of *crude oil* remaining after *primary extraction* by using solvents or by flushing underground rocks with water or steam. Compare *primary extraction*.

secondary pollutant A hazardous substance produced through the reaction of substances added to the *atmosphere* with chemicals normally found in the atmosphere. Compare *primary pollutant*.

secondary production The total biomass that *heterotrophs* generate by consuming *autotrophs*. Compare *gross primary production* and *net primary production*.

secondary succession A stereotypical series of changes as an *ecological community* develops over time, beginning when some event disrupts or dramatically alters an existing community. Compare *primary succession*.

secondary treatment A stage of *wastewater* treatment in which biological means are used to remove contaminants remaining after *primary treatment*. Wastewater is stirred up in the presence of *aerobic* bacteria, which degrade organic pollutants in the water. The wastewater then passes to another settling tank, where remaining solids drift to the bottom. Compare *primary treatment*.

second-growth Term describing trees that have sprouted and grown to partial maturity after virgin timber has been cut.

second law of thermodynamics Physical law stating that the nature of *energy* tends to change from a more-ordered state to a less-ordered state; that is, *entropy* increases.

sediment The eroded remains of rocks.

sedimentary rock One of the three main categories of rock. Formed when dissolved minerals seep through *sediment* layers and act as a kind of glue, crystallizing and binding sediment particles together. Sandstone and shale are examples of sedimentary rock. Compare *igneous rock; metamorphic rock*.

seed bank A storehouse for samples of the world's crop diversity.

septic system A *wastewater* disposal method, common in rural areas, consisting of an underground tank and series of drainpipes. Wastewater runs from the house to the tank, where solids precipitate out. The water proceeds downhill to a drain field of perforated pipes laid horizontally in gravel-filled trenches, where microbes decompose the remaining waste.

service Work done for others as a form of business.

sex ratio The proportion of males to females in a *population*.

shelterbelt A row of trees or other tall perennial plants that are planted along the edges of farm fields to break the wind and thereby minimize wind *erosion*.

silt *Sediment* consisting of particles 0.002–0.005 mm in diameter. Compare *clay; sand*.

sinkhole An area where the ground has given way with little warning as a result of subsidence caused by depletion of water from an *aquifer*.

SLOSS (Single Large or Several Small) dilemma The debate over whether it is better to make reserves large in size and few in number or many in number but small in size.

smart growth A *city planning* concept in which a community's growth is managed in ways that limit *sprawl* and maintain or improve residents' quality of life. It involves guiding the rate, placement, and style of development such that it serves the *environment,* the *economy,* and the community.

Smith, Adam (1723–1790) Scottish philosopher known today as the father of *classical economics*. He believed that when people are free to pursue their own economic self-interest in a competitive marketplace, the marketplace will behave as if guided by "an invisible hand" that ensures that their actions will benefit society as a whole.

social science An academic discipline that studies human interactions and institutions. Compare *natural science*.

soil profile The cross-section of a *soil* as a whole, from the surface to the *bedrock*.

soil A complex plant-supporting *system* consisting of disintegrated rock, organic matter, air, water, *nutrients,* and microorganisms.

source reduction The reduction of the amount of material that enters the *waste stream* to avoid the costs of disposal and *recycling,* help conserve resources, minimize *pollution,* and save consumers and businesses money.

specialist A *species* that can survive only in a narrow range of *habitats* that contain very specific resources. Compare *generalist.*

speciation The process by which new *species* are generated.

species A *population* or group of populations of a particular type of organism, whose members share certain characteristics and can breed freely with one another and produce fertile offspring. Different biologists may have different approaches to diagnosing species boundaries.

species diversity The number and variety of *species* in the world or in a particular region.

sprawl The unrestrained spread of urban or *suburban* development outward from a city center and across the landscape.

steady-state economy An *economy* that does not grow or shrink but remains stable.

storage Technique used to generate *hydroelectric power,* in which large amounts of water are impounded in a reservoir behind a concrete *dam* and then passed through the dam to turn *turbines* that generate electricity. Compare *run-of-river.*

stratosphere The layer of the *atmosphere* above the *troposphere* and below the mesosphere; it extends from 11 km (7 mi) to 50 km (31 mi) above sea level.

strip-mining The use of heavy machinery to remove huge amounts of earth to expose *coal* or minerals, which are mined out directly. Compare *subsurface mining.*

subduction The *plate tectonic* process by which denser ocean *crust* slides beneath lighter continental crust at a *convergent plate boundary.*

subsidy A government incentive (a giveaway of cash or publicly owned resources, or a tax break) intended to encourage a particular activity. Compare *green tax.*

subsistence economy A survival *economy,* one in which people meet most or all of their daily needs directly from nature and do not purchase or trade for most of life's necessities.

subsurface mining Method of mining underground *coal* deposits, in which shafts are dug deeply into the ground and networks of tunnels are dug or blasted out to follow coal seams. Compare *strip-mining.*

suburb A smaller community that rings a city.

succession A stereotypical series of changes in the composition and structure of an *ecological community* through time. See *primary succession; secondary succession.*

Superfund A program administered by the *Environmental Protection Agency* in which experts identify sites polluted with hazardous chemicals, protect *groundwater* near these sites, and clean up the *pollution.*

surface impoundment A *hazardous waste* disposal method in which a shallow depression is dug and lined with impervious material, such as *clay.* Water containing small amounts of hazardous waste is placed in the pond and allowed to evaporate, leaving a residue of solid hazardous waste on the bottom. Compare *deep-well injection.*

survivorship curve A graph that shows how the likelihood of death for members of a *population* varies with age.

sustainability A guiding principle of *environmental science* that requires us to live in such a way as to maintain Earth's systems and its *natural resources* for the foreseeable future.

sustainable agriculture *Agriculture* that does not deplete *soils* faster than they form.

sustainable development Development that satisfies our current needs without compromising the future availability of *natural resources* or our future quality of life.

sustainable forestry certification A form of *ecolabeling* that identifies timber products that have been produced using *sustainable* methods. Several organizations issue such certification.

symbiosis A *parasitic* or *mutualistic* relationship between different *species* of organisms that live in close physical proximity.

synergistic effect An interactive effect (as of *toxicants*) that is more than or different from the simple sum of their constituent effects.

system A network of relationships among a group of parts, elements, or components that interact with and influence one another through the exchange of *energy,* matter, and/or information.

Talloires Declaration A document composed in Talloires, France, in 1990 that commits university leaders to pursue *sustainability* on their campuses. It has been signed by over 300 university presidents and chancellors from more than 40 nations.

temperate deciduous forest A *biome* consisting of midlatitude forests characterized by broad-leafed trees that lose their leaves each fall and remain dormant during winter. These forests occur in areas where *precipitation* is spread relatively evenly throughout the year: much of Europe, eastern China, and eastern North America.

temperate grassland A *biome* whose vegetation is dominated by grasses and features more extreme temperature differences between winter and summer and less *precipitation* than *temperate deciduous forests.*

temperate rainforest A *biome* consisting of tall coniferous trees, cooler and less species-rich than *tropical rainforest* and milder and wetter than *temperate deciduous forest.*

temperature (thermal) inversion A departure from the normal temperature distribution in the *atmosphere,* in which a pocket of relatively cold air occurs near the ground, with warmer air above it. The cold air, denser than the air above it, traps *pollutants* near the ground and causes a buildup of smog.

teratogen A *toxicant* that causes harm to the unborn, resulting in birth defects.

terracing The cutting of level platforms, sometimes with raised edges, into steep hillsides to contain water from *irrigation* and *precipitation.* Terracing transforms slopes into series of steps like a staircase, enabling farmers to cultivate hilly land while minimizing their loss of *soil* to water *erosion.*

theory A widely accepted, well-tested explanation of one or more cause-and-effect relationships that has been extensively validated by a great amount of research. Compare *hypothesis.*

thermal inversion See *temperature inversion.*

Thoreau, Henry David (1817–1862) American transcendental-ist author, poet, and philosopher. His book *Walden,* recording his observations and thoughts while he lived at Walden Pond away from the bustle of urban Massachusetts, remains a classic of American literature.

Three Mile Island Nuclear power plant in Pennsylvania that in 1979 experienced a partial *meltdown.* The term is often using to denote the accident itself, the most serious *nuclear reactor* malfunction that the United States has thus far experienced.

threshold dose The amount of a *toxicant* at which it begins to affect a *population* of test animals. Compare ED_{50}; LD_{50}.

tide The periodic rise and fall of the ocean's height at a given location, caused by the gravitational pull of the moon and sun.

topsoil That portion of the *soil* that is most nutritive for plants and is thus of the most direct importance to *ecosystems* and to *agriculture.* Also known as the *A horizon.*

total fertility rate (TFR) The average number of children born per female member of a *population* during her lifetime.

toxic air pollutant *Air pollutant* that is known to cause cancer, reproductive defects, or neurological, developmental, immune system, or respiratory problems in humans, and/or to cause substantial *ecological* harm by affecting the health of nonhu-man animals and plants. The *Clean Air Act of 1990* identifies 188 toxic air pollutants, ranging from the heavy metal mercury to *volatile organic compounds* such as benzene and methylene chloride.

toxicant A substance that acts as a poison to humans or wildlife.

toxicology The scientific field that examines the effects of poisonous chemicals and other agents on humans and other organisms.

traditional agriculture Biologically powered *agriculture,* in which human and animal muscle power, along with hand tools and simple machines, perform the work of cultivating, harvest-ing, storing, and distributing crops. Compare *industrialized agriculture.*

transform plate boundary Area where two tectonic plates meet and slip and grind alongside one another. For example, the Pacific Plate and the North American Plate rub against each other along California's San Andreas Fault.

transgene A *gene* that has been extracted from the *DNA* of one organism and transferred into the DNA of an organism of an-other *species.*

transgenic Term describing an organism that contains *DNA* from another *species.*

transitional stage The second stage of the *demographic transi-tion* model, which occurs during the transition from the *pre-industrial stage* to the *industrial stage.* It is characterized by declining death rates but continued high birth rates. See also *post-industrial stage.* Compare *industrial stage; post-industrial stage; pre-industrial stage.*

transpiration The release of water vapor by plants through their leaves.

treatment The portion of an *experiment* in which a *variable* has been manipulated in order to test its effect. Compare *control.*

trophic level Rank in the feeding hierarchy of a food chain. Organisms at higher trophic levels consume those at lower trophic levels.

tropical dry forest A *biome* that consists of deciduous trees and occurs at tropical and subtropical latitudes where wet and dry seasons each span about half the year. Widespread in India, Africa, South America, and northern Australia.

tropical rainforest A *biome* characterized by year-round rain and uniformly warm temperatures. Found in Central America, South America, southeast Asia, west Africa, and other tropical regions. Tropical rainforests have dark, damp interiors; lush vegetation; and highly diverse *biotic communities.*

troposphere The bottommost layer of the *atmosphere;* it ex-tends to 11 km (7 mi) above sea level. See also *stratosphere.*

tundra A *biome* that is nearly as dry as *desert* but is located at very high latitudes along the northern edges of Russia, Canada, and Scandinavia. Extremely cold winters with little daylight and moderately cool summers with lengthy days characterize this landscape of lichens and low, scrubby vegetation.

turbine A rotary device that converts the *kinetic energy* of a moving substance, such as steam, into mechanical energy. Used widely in commercial power generation from various types of energy sources.

unconfined aquifer A water-bearing, porous layer of rock, *sand,* or gravel that lies atop a less-permeable substrate. The water in an unconfined aquifer is not under pressure because there is no impermeable upper layer to confine it. Compare *confined aquifer.*

undernutrition A condition of insufficient *nutrition* in which people receive less than 90% of their daily caloric needs.

uneven-aged Term describing stands of trees in timber plantations that are of different ages. Uneven-aged stands more closely approximate a natural forest than do *even-aged* stands.

United Nations (U.N.) Organization founded in 1945 to pro-mote international peace and to cooperate in solving interna-tional economic, social, cultural, and humanitarian problems. Several agencies within it influence *environmental policy,* most notably the United Nations Environment Programme (UNEP), created in 1972.

United Nations Framework Convention on Climate Change (FCCC) International agreement to reduce *greenhouse gas* emissions to 1990 levels by the year 2000, signed by nations represented at the 1992 Earth Summit convened in Rio de Janeiro by the *United Nations.* The FCCC called for a voluntary, nation-by-nation approach, but by the late 1990s it had be-come apparent that it would not succeed. Its imminent failure sparked introduction of the *Kyoto Protocol.*

universalist An *ethicist* who maintains that there exist objective notions of right and wrong that hold across cultures and situa-tions. Compare *relativist.*

upwelling In the ocean, the flow of cold, deep water toward the surface. Upwelling occurs in areas where surface *currents* di-verge. Compare *downwelling.*

urban ecology A scientific field that views cities explicitly as *ecosystems.* Researchers in this field seek to apply the fundamentals of *ecosystem ecology* and *systems* science to urban areas.

urban growth boundary (UGB) In *city planning,* a geographic boundary intended to separate areas desired to be urban from areas desired to remain rural. Development for housing, commerce, and industry are encouraged within urban growth boundaries, but beyond them such development is severely restricted.

urbanization The shift from rural to city and *suburban* living.

variable In an *experiment,* a condition that can change. See *dependent variable* and *independent variable.*

volatile organic compound (VOC) One of a large group of potentially harmful organic chemicals used in industrial processes.

Wallace, Alfred Russell (1823–1913) English naturalist who proposed, independently of *Charles Darwin,* the concept of *natural selection* as a mechanism for *evolution* and as a way to explain the great variety of living things.

warm front The boundary where a mass of warm air displaces a mass of colder air. Compare *cold front.*

waste Any unwanted product that results from a human activity or process.

waste management Strategic decision making to minimize the amount of *waste* generated and to dispose of waste safely and effectively.

waste stream The flow of *waste* as it moves from its sources toward disposal destinations.

waste-to-energy (WTE) facility An incinerator that uses heat from its furnace to boil water to create steam that drives electricity generation or that fuels heating systems.

wastewater Any water that is used in households, businesses, industries, or public facilities and is drained or flushed down pipes, as well as the polluted *runoff* from streets and storm drains.

waterlogging The saturation of *soil* by water, in which the *water table* is raised to the point that water bathes plant roots. Waterlogging deprives roots of access to gases, essentially suffocating them and eventually damaging or killing the plants.

watershed The entire area of land from which water drains into a given river.

water table The upper limit of *groundwater* held in an *aquifer.*

weather The local physical properties of the *troposphere,* such as temperature, pressure, humidity, cloudiness, and wind, over relatively short time periods. Compare *climate.*

weathering The physical, chemical, and biological processes that break down rocks and minerals, turning large particles into smaller particles.

Whitman, Walt (1819–1892) American poet who espoused transcendentalism. See also *Emerson, Ralph Waldo* and *Thoreau, Henry David.*

wilderness area Federal land that is designated off-limits to development of any kind but is open to public recreation, such as hiking, nature study, and other activities that have minimal impact on the land.

wind turbine A mechanical assembly that converts the wind's *kinetic energy,* or energy of motion, into electrical energy.

wise-use movement A loose confederation of individuals and groups that coalesced in the 1980s and 1990s as a response to the increasing success of environmental advocacy. The movement favors extracting more resources from public lands, obtaining greater local control of lands, and obtaining greater motorized recreational access to public lands.

World Bank Institution founded in 1944 that serves as one of the globe's largest sources of funding for *economic* development, including such major projects as *dams, irrigation* infrastructure, and other undertakings.

World Trade Organization (WTO) Organization based in Geneva, Switzerland, that represents multinational corporations and promotes free trade by reducing obstacles to international commerce and enforcing fairness among nations in trading practices.

worldview A way of looking at the world that reflects a person's (or a group's) beliefs about the meaning, purpose, operation, and essence of the world.

zoning The practice of classifying areas for different types of development and land use.

Photo Credits

Selected Sources and References for Further Reading

Chapter 1

Bahn, Paul, and John Flenley. 1992. *Easter Island, Earth island*. Thames and Hudson, London.

Bowler, Peter J. 1993. *The Norton history of the environmental sciences*. W. W. Norton, New York.

Diamond, Jared. 2005. *Collapse: How societies choose to fail or succeed*. Viking, New York.

Ehrlich, Paul. 1968. *The population bomb*. 1997 reprint, Buccaneer Books, Cutchogue, New York.

Esty, Daniel C., et al., 2005. *2005 Environmental sustainability index: Benchmarking national environmental stewardship*. New Haven, Connecticut: Yale Center for Environmental Law and Policy.

Flenley, John, and Paul Bahn. 2003. *The enigmas of Easter Island*. Oxford University Press, New York.

Goudie, Andrew. 2000. *The human impact on the natural environment*, 5th ed. MIT Press, Cambridge, Massachusetts.

Hardin, Garrett. 1968. The tragedy of the commons. *Science* 162: 1243–1248.

Katzner, Donald W. 2001. *Unmeasured information and the methodology of social scientific inquiry*. Kluwer, Boston.

Kuhn, Thomas S. 1962. *The structure of scientific revolutions*, 2nd ed., 1970. University of Chicago Press, Chicago.

Lomborg, Bjorn. 2001. *The skeptical environmentalist: Measuring the real state of the world*. Cambridge University Press, Cambridge.

Malthus, Thomas R. *An essay on the principle of population*. 1983 ed. Penguin USA, New York.

Millennium Ecosystem Assessment. 2005. *Ecosystems and human well-being: General synthesis*. Millennium Ecosystem Assessment and World Resources Institute.

Musser, George. 2005. The climax of humanity. *Scientific American* 293(3): 44–47.

Ponting, Clive. 1991. *A green history of the world: The environment and the collapse of great civilizations*. Penguin Books, New York.

Popper, Karl R. 1959. *The logic of scientific discovery*. Hutchinson, London.

Porteous, Andrew. 2000. *Dictionary of environmental science and technology*, 3rd ed. John Wiley & Sons, Hoboken, New Jersey.

Redman, Charles R. 1999. *Human impact on ancient environments*. University of Arizona Press, Tucson.

Sagan, Carl. 1997. *The demon-haunted world: Science as a candle in the dark*. Ballantine Books, New York.

Schneiderman, Jill S., ed. 2003. *The Earth around us: Maintaining a livable planet*. Perseus Books, New York.

Siever, Raymond. 1968. Science: Observational, experimental, historical. *American Scientist* 56: 70–77.

Valiela, Ivan. 2001. *Doing science: Design, analysis, and communication of scientific research*. Oxford University Press, Oxford.

Van Tilburg, Jo Anne. 1994. *Easter Island: Archaeology, ecology, and culture*. Smithsonian Institution Press, Washington, D.C.

Venetoulis, Jason, et al., 2004. *Ecological footprint of nations 2004*. Redefining Progress, Oakland, California.

Wackernagel, Mathis, and William Rees. 1996. *Our ecological footprint: Reducing human impact on the earth*. New Society Publishers, Gabriola Island, British Columbia, Canada.

World Bank. 2005. *World development indicators 2005*. World Bank, Washington, D.C.

Worldwatch Institute. *State of the world 2005: Redefining global security*. Worldwatch Institute and W. W. Norton, Washington, D.C. and New York.

Worldwatch Institute. *Vital Signs 2005*. Worldwatch Institute and W. W. Norton, Washington, D.C. and New York.

Chapter 2

Balmford, Andrew, et al. 2002. Economic reasons for conserving wild nature. *Science* 297: 950–953.

Barbour, Ian G. 1992. *Ethics in an age of technology*. Harper Collins, San Francisco.

Brown, Lester. 2001. *Eco-economy: Building an economy for the Earth*. Earth Policy Institute and W. W. Norton, New York.

Carson, Richard T., Leanne Wilks, and David Imber. 1994. Valuing the preservation of Australia's Kakadu Conservation Zone. *Oxford Economic Papers* 46: 727–749.

Cole, Luke W., and Sheila R. Foster. 2001. *From the ground up: Environmental racism and the rise of the environmental justice movement*. New York University Press, New York.

Costanza, Robert, et al. 1997. The value of the world's ecosystem services and natural capital. *Nature* 387: 253–260.

Costanza, Robert, et al. 1997. *An introduction to ecological economics*. St. Lucie Press, Boca Raton, Florida.

Daily, Gretchen. 1997. *Nature's services: Societal dependence on natural ecosystems*. Island Press, Washington, D.C.

Daly, Herman E. 1996. *Beyond growth*. Beacon Press, Boston.

Daly, Herman E. 2005. Economics in a full world. *Scientific American* 293(3): 100–107.

Elliot, Robert, and Arran Gare, eds. 1983. *Environmental philosophy: A collection of readings*. Pennsylvania State University Press, University Park.

Field, Barry C., and Martha K. Field. 2001. *Environmental economics*, 3rd ed. McGraw-Hill, New York.

Fox, Stephen. 1985. *The American conservation movement: John Muir and his legacy*. University of Wisconsin Press, Madison.

Gardner, Gary, et al. 2004. The state of consumption today. Pp. 3–23 in *State of the world 2004*. Worldwatch Institute and W. W. Norton, Washington, D.C., and New York.

Gardner, Gary, and Erik Assadourian. 2004. Rethinking the good life. Pp. 164–179 in *State of the world 2004*. Worldwatch Institute and W. W. Norton, Washington, D.C., and New York.

Goodstein, Eban. 1999. *The tradeoff myth: Fact and fiction about jobs and the environment*. Island Press, Washington, D.C.

Goodstein, Eban. 2005. *Economics and the environment*, 4th ed. John Wiley & Sons, Hoboken, New Jersey.

Gundjehmi Aboriginal Corporation. Welcome to the Mirrar site. www.mirrar.net.

Hawken, Paul, Amory Lovins, and L. Hunter Lovins. 1999. *Natural capitalism*. Little, Brown, and Co., Boston.

Kolstad, Charles D. 2000. *Environmental economics*. Oxford University Press, Oxford.

Leopold, Aldo. 1949. *A Sand County almanac, and sketches here and there*. Oxford University Press, New York.

Millennium Ecosystem Assessment. 2005. *Ecosystems and human well-being: Opportunities and challenges for business and industry*. Millennium Ecosystem Assessment and World Resources Institute.

Nash, Roderick F. 1989. *The rights of nature*. University of Wisconsin Press, Madison.

Nash, Roderick F. 1990. *American environmentalism: Readings in conservation history*, 3rd ed. McGraw-Hill, New York.

O'Neill, John O., R. Kerry Turner, and Ian J. Bateman, eds. 2001. *Environmental ethics and philosophy*. Elgar, Cheltenham, U.K.

Pearson, Charles S. 2000. *Economics and the global environment*. Cambridge University Press, Cambridge.

Ricketts, Taylor, et al. 2004. Economic value of tropical forest to coffee production. *Proceedings of the National Academy of Sciences of the USA* 101: 12579–12582.

Sachs, Jeffrey. 2005. Can extreme poverty be eliminated? *Scientific American* 293(3): 56–65.

Singer, Peter, ed. 1993. *A companion to ethics*. Blackwell Publishers, Oxford.

Smith, Adam. 1776. *An inquiry into the nature and causes of the wealth of nations*. 1993 ed., Oxford University Press, Oxford.

Sterba, James P., ed. 1995. *Earth ethics: Environmental ethics, animal rights, and practical applications*. Prentice Hall, Upper Saddle River, New Jersey.

Stone, Christopher D. 1972. Should trees have standing? Towards legal rights for natural objects. *Southern California Law Review* 1972: 450–501.

Tietenberg, Tom. 2003. *Environmental economics and policy*, 4th ed. Addison Wesley, Boston.

Turner, R. Kerry, David Pearce, and Ian Bateman. 1993. *Environmental economics: An elementary introduction*. Johns Hopkins University Press, Baltimore.

Venetoulis, Jason, and Cliff Cobb. 2004. *The genuine progress indicator 1950–2002 (2004 update)*. Redefining Progress, Oakland, California.

Wenz, Peter S. 2001. *Environmental ethics today*. Oxford University Press, Oxford.

White, Lynn. 1967. The historic roots of our ecologic crisis. *Science* 155: 1203–1207.

Chapter 3

Clark, Ray, and Larry Canter. 1997. *Environmental policy and NEPA: Past, present, and future*. St. Lucie Press, Boca Raton, Florida.

Dietz, Thomas, et al. 2003. The struggle to govern the global commons. *Science* 302: 1907–1912.

Fogleman, Valerie M. 1990. *Guide to the National Environmental Policy Act*. Quorum Books, New York.

Fox, Stephen. 1985. *The American conservation movement: John Muir and his legacy*. University of Wisconsin Press, Madison.

French, Hilary. 2000. Environmental treaties gain ground. Pp. 134–135 in *Vital Signs 2000*. Worldwatch Institute and W. W. Norton, Washington D.C., and New York.

Green Scissors, 2004. *Green Scissors 2004: Cutting wasteful and environmentally harmful spending*. Friends of the Earth, Taxpayers for Common Sense, and U.S. Public Interest Research Group.

Herzog, Lawrence A. 1990. *Where north meets south: Cities, space, and politics on the U.S.–Mexico border*. Center for Mexican-American Studies, University of Texas at Austin.

Houck, Oliver, 2003. Tales from a troubled marriage: Science and law in environmental policy. *Science* 302: 1926–1928.

Kraft, Michael E. 2003. *Environmental policy and politics*, 3rd ed. Longman, New York.

Kubasek, Nancy K., and Gary S. Silverman. 2004. *Environmental law*, 5th ed. Prentice Hall, Upper Saddle River, New Jersey.

Myers, Norman, and Jennifer Kent. 2001. *Perverse subsidies: How misused tax dollars harm the environment and the economy*. Island Press, Washington, D.C.

The National Environmental Policy Act of 1969, as amended (Pub. L. 91–190, 42 U.S.C. 4321–4347, January 1, 1970, as amended by Pub. L. 94–52, July 3, 1975, Pub. L. 94–83, August 9, 1975, and Pub. L. 97–258, § 4(b), Sept. 13, 1982). http://ceq.eh.doe.gov/nepa/regs/nepa/nepaeqia.htm.

Shafritz, Jay M. 1993. *The HarperCollins dictionary of American government and politics*. HarperCollins, New York.

Southwest Center for Environmental Research and Policy, and San Diego State University. Tijuana River Watershed Atlas Project. http://geography.sdsu.edu/Research/Projects/TWRP/tjatlas.html

Steel, Brent S., Richard L. Clinton, and Nicholas P. Lovrich. 2002. *Environmental politics and policy*. McGraw-Hill, New York.

Tietenberg, Tom. 2003. *Environmental economics and policy*, 4th ed. Addison Wesley, Boston.

Turner, R. Kerry, David Pearce, and Ian Bateman. 1993. *Environmental economics: An elementary introduction*. Johns Hopkins University Press, Baltimore.

United States Congress. House. H.R. 3378. 2000. The Tijuana River Valley Estuary and Beach Sewage Cleanup Act of 2000.

Vig, Norman J., and Michael E. Kraft, eds. 2002. *Environmental policy: New directions for the twenty-first century*, 5th ed. CQ Press, Congressional Quarterly, Inc., Washington, D.C.

Wilkinson, Charles F. 1992. *Crossing the next meridian: Land, water, and the future of the West*. Island Press, Washington, D.C.

Chapter 4

Alaska Department of Environmental Conservation. 1993. *The Exxon Valdez oil spill: Final report, State of Alaska response*. June, 1993.

Allen, K. C., and D. E. G. Briggs, eds. 1989. *Evolution and the fossil record*. John Wiley & Sons, Hoboken, New Jersey.

Atlas, Ronald M. 1995. Petroleum biodegradation and oil spill bioremediation. *Marine Pollution Bulletin* 31: 178–182.

Atlas, Ronald M., and Carl E. Cerniglia. 1995. Bioremediation of petroleum pollutants. *BioScience* 45: 332–338.

Berry, R. Stephen. 1991. *Understanding energy: Energy, entropy and thermodynamics for every man*. World Scientific Publishing Co.

Bragg, James R., et al. 1994. Effectiveness of bioremediation for the *Exxon Valdez* oil spill. *Nature* 368: 413–418.

Campbell, Neil A., and Jane B. Reece. 2005. *Biology*, 7th ed. Benjamin Cummings, San Francisco.

Fenchel, Tom. 2003. *Origin and early evolution of life*. Oxford University Press, Oxford.

Fortey, Richard. 1998. *Life: A natural history of the first four billion years of life on Earth*. Alfred Knopf, New York.

Gee, Henry. 1999. *In search of deep time: Beyond the fossil record to a new history of life*. Free Press, New York.

Hall, David O., and Krishna Rao. 1999. *Photosynthesis*, 6th ed. Cambridge University Press, Cambridge.

Lancaster, M., 2002. *Green chemistry*. Royal Society of Chemistry, London.

Manahan, Stanley E. 2004. *Environmental chemistry*, 8th ed. Lewis Publishers, CRC Press, Boca Raton, Florida.

McMurry, John E. 2003. *Organic chemistry*, 6th ed. Brooks/Cole, San Francisco.

National Response Team. *NRT fact sheet: Bioremediation in oil spill response*. U.S. EPA, www.epa.gov/oilspill/pdfs/biofact.pdf.

Nealson, Kenneth H. 2003. Harnessing microbial appetites for remediation. *Nature Biotechnology* 21: 243–244.

Ridley, Mark. 2003. *Evolution*, 3rd ed. Blackwell Science, Cambridge, Massachusetts.

United States Environmental Protection Agency. 2003. Oil program. www.epa.gov/oilspill.

Van Dover, Cindy Lee, 2000. *The ecology of deep-sea hydrothermal vents*. Princeton University Press, Princeton.

Van Ness, H.C. 1983. *Understanding thermodynamics*. Dover Publications, Mineola, New York.

Ward, Peter D., and Donald Brownlee. 2000. *Rare Earth: Why complex life is uncommon in the universe*. Copernicus, New York.

Wassenaar, Leonard I., and Keith A. Hobson. 1998. Natal origins of migratory monarch butterflies at wintering colonies in Mexico: New isotopic evidence. *Proceedings of the National Academy of the USA* 95: 15436–15439.

Chapter 5

Alvarez, Luis W., et al. 1980. Extraterrestrial cause for the Cretaceous-Tertiary extinction. *Science* 208: 1095–1108.

Barbour, Michael G., et al. 1998. *Terrestrial plant ecology*, 3rd ed. Benjamin/Cummings, Menlo Park, California.

Begon, Michael, Martin Mortimer, and David J. Thompson. 1996. *Population ecology: A unified study of animals and plants*, 3rd ed. Blackwell Scientific, Oxford.

Breckle, Siegmar-Walter. 1999. *Walter's vegetation of the Earth: The ecological systems of the geo-biosphere*, 4th ed. Springer-Verlag, Berlin, 1999.

Campbell, Neil A., and Jane B. Reece. 2005. *Biology*, 7th ed. Benjamin Cummings, San Francisco.

Clark K. L., et al. 1998. Cloud water and precipitation chemistry in a tropical montane forest, Monteverde, Costa Rica. *Atmospheric Environment* 32: 1595–1603.

Crump, L. Martha, et al. 1992. Apparent decline of the golden toad: Underground or extinct? *Copeia* 1992: 413–420.

Darwin, Charles. 1859. *The origin of species by means of natural selection*. John Murray, London.

Endler, John A. 1986. *Natural selection in the wild*. Monographs in Population Biology 21, Princeton University Press, Princeton.

Freeman, Scott, and Jon C. Herron. 2003. *Evolutionary analysis*, 3rd ed. Prentice Hall, Upper Saddle River, New Jersey.

Futuyma, Douglas J. 2005. *Evolution*. Sinauer Associates, Sunderland, Massachusetts.

Krebs, Charles J. 2001. *Ecology: The experimental analysis of distribution and abundance*, 5th ed. Benjamin Cummings, San Francisco.

Lawton, Robert O., et al. 2001. Climatic impact of tropical lowland deforestation on nearby montane cloud forests. *Science* 294: 584–587.

Molles, Manuel C., Jr. 2005. *Ecology: Concepts and applications*, 3rd ed. McGraw-Hill, Boston.

Nadkarni, Nalini M., and Nathaniel T. Wheelwright, eds. 2000. *Monteverde: Ecology and conservation of a tropical cloud forest*. Oxford University Press, New York.

Pounds, J. Alan. 2001. Climate and amphibian declines. *Nature* 410: 639.

Pounds, J. Alan, et al. 1997. Tests of null models for amphibian declines on a tropical mountain. *Conservation Biology* 11: 1307–1322.

Pounds, J. Alan, and Martha L. Crump. 1994. Amphibian declines and climate disturbance: The case of the golden toad and the harlequin frog. *Conservation Biology* 8: 72–85.

Pounds, J. Alan, Michael P. L. Fogden, and John H. Campbell. 1999. Biological response to climate change on a tropical mountain. *Nature* 398: 611–615.

Powell, James L. 1998. *Night comes to the Cretaceous: Dinosaur extinction and the transformation of modern geology*. W. H. Freeman, New York.

Raup, David M. 1991. *Extinction: Bad genes or bad luck?* W. W. Norton, New York.

Ricklefs, Robert E., and Gary L. Miller. 2000. *Ecology*, 4th ed. W. H. Freeman, New York.

Ricklefs, Robert E., and Dolph Schluter, eds. 1993. *Species diversity in ecological communities*. University of Chicago Press, Chicago.

Savage, Jay M. 1966. An extraordinary new toad (*Bufo*) from Costa Rica. *Revista de Biologia Tropical* 14: 153–167.

Savage, Jay M. 1998. The "brilliant toad" was telling us something. *Christian Science Monitor*, 14 September 1998: 19.

Smith, Thomas M., and Robert L. Smith. 2006. *Elements of ecology*, 6th ed. Benjamin Cummings, San Francisco.

Ward, Peter. 1994. *The end of evolution*. Bantam Books, New York.

Williams, George C. 1966. *Adaptation and natural selection*. Princeton University Press, Princeton.

Wilson, Edward O. 1992. *The diversity of life*. Harvard University Press, Cambridge, Massachusetts.

Whittaker, Robert H., and William A. Niering. 1965. Vegetation of the Santa Catalina Mountains, Arizona: A gradient analysis of the south slope. *Ecology* 46: 429–452.

Chapter 6

Breckle, Siegmar-Walter. 2002. *Walter's vegetation of the Earth: The ecological systems of the geo-biosphere*, 4th ed. Berlin: Springer-Verlag.

Bronstein, Judith L. 1994. Our current understanding of mutualism. *Quarterly Journal of Biology* 69: 31–51.

Chase, Jonathan M., et al., 2002. The interaction between predation and competition: A review and synthesis. *Ecology Letters* 5: 302.

Connell, Joseph H., and Ralph O. Slatyer, 1977. Mechanisms of succession in natural communities. *American Naturalist* 111: 1119–1144.

Drake, John M., and Jonathan M. Bossenbroek. 2004. The potential distribution of zebra mussels in the United States. *BioScience* 54: 931–941.

Estes, J.A., et al. 1998. Killer whale predation on sea otters linking oceanic and nearshore ecosystems. *Science* 282: 473–476.

Ewald, Paul W. 1987. Transmission modes and evolution of the parasitism-mutualism continuum. *Annals of the New York Academy of Sciences* 503: 295–306.

Gurevitch, Jessica, and Dianna K. Padilla. 2004. Are invasive species a major cause of extinctions? *Trends in Ecology and Evolution* 19: 470–474.

Krebs, Charles J. 2001. *Ecology: The experimental analysis of distribution and abundance*, 5th ed. Benjamin Cummings, San Francisco.

Menge, Bruce A., et al. 1994. The keystone species concept: Variation in interaction strength in a rocky intertidal habitat. *Ecological Monographs* 64: 249–286.

Molles, Manuel C. Jr. 2005. *Ecology: Concepts and applications*. 3rd ed. McGraw-Hill, Boston.

Morin, Peter J. 1999. *Community ecology*. Blackwell, London.

Power, Mary E., et al., 1996. Challenges in the quest for keystones. *BioScience* 46: 609–620.

Ricklefs, Robert E., and Gary L. Miller. 2000. *Ecology*, 4th ed. W. H. Freeman and Co., New York.

Shea, Katriona, and Peter Chesson, 2002. Community ecology theory as a framework for biological invasions. *Trends in Ecology and Evolutionary Biology* 17: 170–176.

Sih, Andrew, et al. 1985. Predation, competition, and prey communities: A review of field experiments. *Annual Review of Ecology and Systematics* 16: 269–311.

Smith, Robert L., and Thomas M. Smith. 2001. *Ecology and field biology*, 6th ed. Benjamin Cummings, San Francisco.

Springer, A.M., et al. 2003. Sequential megafaunal collapse in the North Pacific Ocean: An ongoing legacy of industrial whaling? *Proceedings of the National Academy of Sciences of the USA* 100: 12223–12228.

Strayer, David L., et al. 1999. Transformation of freshwater ecosystems by bivalves: A case study of zebra mussels in the Hudson River. *BioScience* 49: 19–27.

Strayer, David L., et al. 2004. Effects of an invasive bivalve (*Dreissena polymorpha*) on fish in the Hudson River estuary. *Canadian Journal of Fisheries and Aquatic Sciences* 61: 924–941.

Thompson, John N. 1999. The evolution of species interactions. *Science* 284: 2116–2118.

Weigel, Marlene, ed., 1999. *Encyclopedia of biomes*. UXL, Farmington Hills, Michigan.

Woodward, Susan L., 2003. *Biomes of Earth: Terrestrial, aquatic, and human-dominated*. Greenwood Publishing, Westport, Connecticut.

Chapter 7

Alling, Abigail, Mark Nelson, and Sally Silverstone. 1993. *Life under glass: The inside story of Biosphere 2*. Biosphere Press.

Capra, Fritjof. 1996. *The web of life: A new scientific understanding of living systems*. Anchor Books Doubleday, New York.

Carpenter, Edward J., and Douglas G. Capone, eds. 1983. *Nitrogen in the marine environment*. Academic Press, New York.

Committee on Environment and Natural Resources, 2000. *An integrated assessment: Hypoxia in the northern Gulf of Mexico*. CENR, National Science and Technology Council, Washington, D.C.

Ferber, Dan. 2004. Dead zone fix not a dead issue. *Science* 305: 1557.

Field, Christopher B., et al., 1998. Primary production of the biosphere: Integrating terrestrial and oceanic components. *Science* 281: 237–240.

Jacobson, Michael, et al. 2000. *Earth system science from biogeochemical cycles to global changes*. Academic Press.

Keller, Edward A. 2004. *Introduction to environmental geology*, 3rd ed. Prentice Hall, Upper Saddle River, New Jersey.

Larsen, Janet. 2004. Dead zones increasing in world's coastal waters. *Eco-economy update #41*, 16 June 2004. Earth Policy Institute, www.earth-policy.org/Updates/Update41.htm.

Mississippi River/Gulf of Mexico Watershed Nutrient Task Force. 2001. *Action plan for reducing, mitigating, and controlling hypoxia in the northern Gulf of Mexico*. Washington, D.C.

Mitsch, William J., et al. 2001. Reducing nitrogen loading to the Gulf of Mexico from the Mississippi River Basin: Strategies to counter a persistent ecological problem. *BioScience* 51: 373–388.

Montgomery, Carla. 2005. *Environmental geology*, 7th ed. McGraw-Hill, New York.

National Oceanic and Atmospheric Administration: National Ocean Service. 2000. Hypoxia in the Gulf of Mexico: Progress toward the completion of an integrated assessment. www.nos.noaa.gov/products/pubs_hypox.html.

National Science and Technology Council, Committee on Environment and Natural Resources. 2003. *An assessment of coastal hypoxia and eutrophication in U.S. waters.* National Science and Technology Council, Washington, D.C.

Rabalais, Nancy N., R. E. Turner, and D. Scavia. 2002. Beyond science into policy: Gulf of Mexico hypoxia and the Mississippi River. *BioScience* 52: 129–142.

Rabalais, Nancy N., R. E. Turner, and W. J. Wiseman, Jr. 2002. Hypoxia in the Gulf of Mexico, a.k.a. "The dead zone." *Annual Review of Ecology and Systematics* 33: 235–263.

Raloff, Janet. 2004. Dead waters: Massive oxygen-starved zones are developing along the world's coasts. *Science News* 165: 360–362. June 5, 2004.

Raloff, Janet. 2004. Limiting dead zones: How to curb river pollution and save the Gulf of Mexico. *Science News* 165: 378–380. June 12, 2004.

Ricklefs, Robert E., and Gary L. Miller. 2000. *Ecology,* 4th ed. W. H. Freeman and Co., New York.

Schlesinger, William H. 1997. *Biogeochemistry: An analysis of global change,* 2nd ed. Academic Press, London.

Skinner, Brian J., and Stephen C. Porter. 2003. *The dynamic earth: An introduction to physical geology,* 5th ed. John Wiley and Sons, Hoboken, New Jersey.

Smith, Robert L., and Thomas M. Smith. 2001. *Ecology and field biology,* 6th ed. Benjamin Cummings, San Francisco.

Stiling, Peter. 2002. *Ecology: Theories and applications,* 4th ed. Prentice Hall, Upper Saddle River, New Jersey.

Takahashi, Taro. 2004. The fate of industrial carbon dioxide. *Science* 305: 352–353.

Turner, R. Eugene, and Nancy N. Rabalais. 2003. Linking landscape and water quality in the Mississippi River Basin for 200 years. *BioScience* 53: 563–572.

Vitousek, Peter M., et al. 1997. Human alteration of the global nitrogen cycle: Sources and consequences. *Ecological Applications* 7: 737–750.

Whittaker, Robert H. 1975. *Communities and ecosystems,* 2nd ed. Macmillan, New York.

Chapter 8

Cohen, Joel E. 1995. *How many people can the Earth support?* W. W. Norton, New York.

Cohen, Joel E. 2003. Human population: The next half century. *Science* 302: 1172–1175.

Cohen, Joel E. 2005. Human population grows up. *Scientific American* 293(3): 48–55.

De Souza, Roger-Mark, et. al., 2003. Critical links: Population, health, and the environment. *Population Bulletin 58(3),* 48 pp. Population Reference Bureau, Washington, D.C.

Eberstadt, Nicholas. 2000. China's population prospects: Problems ahead. *Problems of Post-Communism* 47: 28.

Ehrlich, Paul R., and John P. Holdren. 1971. Impact of population growth: Complacency concerning this component of man's predicament is unjustified and counterproductive. *Science* 171: 1212–1217.

Ehrlich, Paul R., and Anne H. Ehrlich. 1990. The population explosion. Touchstone, New York.

Engelman, Robert, Brian Halweil, and Danielle Nierenberg. 2002. Rethinking population, improving lives. Pp. 127–148 in *State of the world 2002,* Worldwatch Institute and W. W. Norton, Washington D.C., and New York.

Greenhalgh, Susan. 2001. Fresh winds in Beijing: Chinese feminists speak out on the one-child policy and women's lives. *Signs: Journal of Women in Culture & Society* 26: 847–887.

Harrison, Paul, and Fred Pearce, eds. 2000. *AAAS atlas of population & environment.* University of California Press, Berkeley.

Hesketh, Therese, and Wei Xing Zhu, 1997. Health in China: The one child family policy: The good, the bad, and the ugly. *British Medical Journal* 314: 1685.

Holdren, John P. and Ehrlich, Paul R. 1974. Human population and the global environment. *American Scientist* 62: 282–292.

Kane, Penny. 1987. *The second billion: Population and family planning in China.* Penguin Books, Australia, Ringwood, Victoria.

Kane, Penny, and Ching Y. Choi. 1999. China's one child family policy. *British Medical Journal* 319: 992.

Mastny, Lisa. 2005. HIV/AIDS crisis worsening worldwide. Pp. 68–69 in *Vital signs 2005.* Worldwatch Institute and W. W. Norton, Washington, D.C. and New York.

Mastny, Lisa, and Richard P. Cincotta. 2005. Examining the connections between population and security. Pp. 22–41 in *State of the world 2005.* Worldwatch Institute and W. W. Norton, Washington, D.C. and New York.

McDonald, Mia, with Danielle Nierenberg. 2003. Linking population, women, and biodiversity. Pp. 38–61 in *State of the world 2003,* Worldwatch Institute and W. W. Norton, Washington D.C., and New York.

Meadows, Donella, Jørgen Randers, and Dennis Meadows. 2004. *Limits to growth: The 30-year update.* Chelsea Green Publishing Co., White River Junction, Vermont.

Notestein, Frank. 1953. Economic problems of population change. Pp. 13–31 in *Proceedings of the Eighth International Conference of Agricultural Economists.* Oxford University Press, London.

O'Brien, Stephen J., and Michael Dean. 1997. In search of AIDS-resistance genes. *Scientific American* 277: 44–51.

Population Reference Bureau. 2005. *2005 World Population Data Sheet.* Population Reference Bureau, Washington, D.C., and John Wiley & Sons, Hoboken, New Jersey.

Redefining Progress. Programs: Sustainability indicators. www.rprogress.org/newprograms/sustIndi/index.shtml.

Riley, Nancy E. 2004. *China's population: New trends and challenges.* Population Bulletin 59(2), 40 pp. Population Reference Bureau, Washington, D.C.

UNAIDS and World Health Organization. 2005. *AIDS epidemic update: December 2005.* UNAIDS and WHO, New York.

United Nations Economic and Social Commission for Asia and the Pacific. 2005. *2005 ESCAP population data sheet.* UNESCAP, New York.

United Nations Environment Programme. 2003. *Africa environment outlook: Past, present, and future perspectives.* UNEP, New York.

United Nations Population Division. 2004. *World population prospects: The 2004 revision.* UNPD, New York.

United Nations Population Fund. UNFPA, the 2005 World Summit and the millennium development goals. UNFPA. www.unfpa.org/icpd.

United Nations Population Fund. *State of world population 2005.* UNFPA, New York.

United States Census Bureau. www.census.gov.

Wackernagel, Mathis, and William Rees. 1996. *Our ecological footprint: Reducing human impact on the earth.* New Society Publishers, Gabriola Island, British Columbia, Canada.

Chapter 9

Ashman, Mark R., and Geeta Puri. 2002. *Essential soil science: A clear and concise introduction to soil science.* Blackwell Publishing, Malden, Massachusetts.

Brown, Lester R. 2002. World's rangelands deteriorating under mounting pressure. *Eco-Economy Update #6,* 5 February 2002. Earth Policy Institute, www.earth-policy.org/Updates/Update6.htm.

Brown, Lester R. 2004. *Outgrowing the Earth: The food security challenge in an age of falling water tables and rising temperatures.* Earth Policy Institute, Washington, D.C.

Charman, P. E. V., and Brian W. Murphy. 2000. *Soils: Their properties and management,* 2nd ed. Oxford University Press, South Melbourne, Australia.

Curtin, Charles G. 2002. Integration of science and community-based conservation in the Mexico/U.S. borderlands. *Conservation Biology* 16: 880–886.

Diamond, Jared. 1999. *Guns, germs, and steel: The fates of human societies.* W. W. Norton, New York.

Diamond, Jared, and Peter Bellwood. 2003. Farmers and their languages: The first expansions. *Science* 300: 597–603.

Food and Agriculture Organization of the United Nations. 2001. *Conservation agriculture: Case studies in Latin America and Africa. FAO Soils Bulletin No. 78.* FAO, Rome.

Glanz, James. 1995. *Saving our soil: Solutions for sustaining Earth's vital resource.* Johnson Books, Boulder, Colorado.

Goudie, Andrew. 2000. *The human impact on the natural environment,* 5th ed. MIT Press, Cambridge, Massachusetts.

Fox, Stephen. 1985. *The American conservation movement: John Muir and his legacy.* University of Wisconsin Press, Madison.

Halweil, Brian. 2002. Farmland quality deteriorating. Pp. 102–103 in *Vital signs 2002.* Worldwatch Institute and W. W. Norton, Washington D.C., and New York.

Harrison, Paul, and Fred Pearce, eds. 2000. *AAAS atlas of population & environment.* University of California Press, Berkeley.

Jenny, Hans. 1941. *Factors of soil formation: A system of quantitative pedology.* McGraw-Hill, New York.

Larsen, Janet. 2003. Deserts advancing, civilization retreating. *Eco-Economy Update #23,* 27 March 2003. Earth Policy Institute, www.earth-policy.org/Updates/Update23.htm.

Millennium Ecosystem Assessment. 2005. *Ecosystems and human well-being: Desertification synthesis.* Millennium Ecosystem Assessment and World Resources Institute.

Morgan, R. P. C. 2005. *Soil erosion and conservation,* 3rd ed. Blackwell, London.

Natural Resources Conservation Service. 2001. *National resources inventory 2001: Soil erosion.* NRCS, USDA, Washington, D.C.

Natural Resources Conservation Service. Soils. NRCS, USDA. http://soils.usda.gov.

Pieri, Christian, et al. 2002. *No-till farming for sustainable rural development.* Agriculture & Rural Development Working Paper. International Bank for Reconstruction and Development, Washington, D.C.

Pierzynski, Gary M., et al. 2005. *Soils and environmental quality,* 3rd ed. CRC Press, Boca Raton, Florida.

Pretty, Jules, and Rachel Hine. 2001. *Reducing food poverty with sustainable agriculture: A summary of new evidence.* Center for Environment and Society, University of Essex. *Occasional Paper 2001–2.*

Richter, Daniel D. Jr., and Daniel Markewitz. 2001. *Understanding soil change: Soil sustainability over millennia, centuries, and decades.* Cambridge University Press, Cambridge.

Ritchie, Jerry C. 2000. Combining cesium-137 and topographic surveys for measuring soil erosion/deposition patterns in a rapidly accreting area. TEKTRAN, USDA Division of Agricultural Research, January 14, 2000.

Shaxson, T. F. 1999. The roots of sustainability, concepts and practice: Zero tillage in Brazil. *ABLH Newsletter ENABLE; World Association for Soil and Water Conservation (WASWC) Newsletter.*

Soil Science Society of America. 2001. Internet glossary of soil science terms. www.soils.org/sssagloss.

Stocking, M. A. 2003. Tropical soils and food security: The next 50 years. *Science* 302: 1356–1359.

Trimble, Stanley W., and Pierre Crosson. 2000. U.S. soil erosion rates—myth and reality. *Science* 289: 248–250.

Troeh, Frederick R., and Louis M. Thompson. 2004. *Soil and soil fertility,* 6th ed. Blackwell Publishing, London.

Troeh, Frederick R., J. Arthur Hobbs, and Roy L. Donahue 2004. *Soil and water conservation for productivity and environmental protection,* 4th ed. Prentice Hall, Upper Saddle River, New Jersey.

United Nations Convention to Combat Desertification, 2001. *Global alarm: Dust and sandstorms from the world's drylands.* UNCCD and others, Bangkok, Thailand.

United Nations Environment Programme. 2002. Land. Pp. 62–89 in *Global environment outlook 3 (GEO-3).* UNEP and Earthscan Publications, Nairobi and London.

Uri, Noel D. 2001. The environmental implications of soil erosion in the United States. *Environmental Monitoring and Assessment* 66: 293–312.

Wilkinson, Bruce H. 2005. Humans as geologic agents: A deep-time perspective. *Geology* 33: 161–164.

Chapter 10

Bazzaz, Fakhri A. 2001. Plant biology in the future. *Proceedings of the National Academy of the United States of America* 98: 5441–5445.

Brown, Lester R. 2004. *Outgrowing the Earth: The food security challenge in an age of falling water tables and rising temperatures.* Earth Policy Institute, Washington, D.C.

Buchmann, Stephen L., and Gary Paul Nabhan. 1996. *The forgotten pollinators.* Island Press/Shearwater Books, Washington, D.C./Covelo, California.

Commission for Environmental Cooperation. 2004. *Maize and biodiversity: The effects of transgenic maize in Mexico.* CEC Secretariat.

[Correspondence to *Nature,* various authors]. 2002. *Nature* 416: 600–602, and 417: 897–898.

Fedoroff, Nina, and Nancy Marie Brown, 2004. *Mendel in the kitchen: A scientist's view of genetically modified foods.* National Academies Press, Washington, D.C.

Food and Agriculture Organization of the United Nations. 2004. *The state of world fisheries and aquaculture, 2004.* FAO, Rome.

Gardner, Gary, and Brian Halweil. 2000. *Underfed and overfed: The global epidemic of malnutrition.* Worldwatch Paper #150. Worldwatch Institute, Washington, D.C.

Halweil, Brian. 2004. *Eat here: Reclaiming homegrown pleasures in a global supermarket.* Worldwatch Institute, Washington, D.C.

Halweil, Brian. 2005. Aquaculture pushes fish harvest higher. Pp. 26–27 in *Vital signs 2005.* Worldwatch Institute and W. W. Norton, Washington, D.C. and New York.

Halweil, Brian. 2005. Grain harvest and hunger both grow. Pp. 22–23 in *Vital signs 2005.* Worldwatch Institute and W. W. Norton, Washington, D.C., and New York.

Halweil, Brian, and Danielle Niereberg. 2004. Watching what we eat. Pp. 68–95 in *State of the world 2004.* Worldwatch Institute and W. W. Norton, Washington, D.C., and New York.

Harrison, Paul, and Fred Pearce, eds. 2000. *AAAS atlas of population & environment.* University of California Press, Berkeley.

International Food Information Council. 2004. Food biotechnology. IFIC, Washington, D.C. www.ific.org/food/biotechnology/index.cfm.

James, Clive. 2004. *Global status of GM crops, their contribution to sustainability, and future prospects.* International Service for the Acquisition of Agri-biotech Applications.

Kuiper, Harry A. 2000. Risks of the release of transgenic herbicide-resistant plants with respect to humans, animals, and the environment. *Crop Protection* 19: 773.

Liebig, Mark A., and John W. Doran. 1999. Impact of organic production practices on soil quality indicators. *Journal of Environmental Quality* 28: 1601–1609.

Losey, John E., Raynor, Linda S., and Carter, Maureen E. 1999. Transgenic pollen harms monarch larvae. *Nature* 399: 214.

Maeder, Paul, et al. 2002. Soil fertility and biodiversity in organic farming. *Science* 296: 1694–1697.

Mann, Charles C. 2002. Transgene data deemed unconvincing. *Science* 296: 236–237.

Manning, Richard. 2000. *Food's frontier: The next green revolution.* North Point Press, New York.

Miller, Henry I., and Gregory Conko. 2004. *The frankenfood myth: How protest and politics threaten the biotech revolution.* Praeger Publishers, Westport, Connecticut.

Nierenberg, Danielle. 2005. *Happier meals: Rethinking the global meat industry.* Worldwatch Paper #171. Worldwatch Institute, Washington, D.C.

Nierenberg, Danielle. 2005. Meat production and consumption rise. Pp. 24–25 in *Vital signs 2005.* Worldwatch Institute and W. W. Norton, Washington, D.C., and New York.

Nierenberg, Danielle, and Brian Halweil. 2005. Cultivating food security. Pp. 62–79 in *State of the world 2005*. Worldwatch Institute and W. W. Norton, Washington, D.C., and New York.

Nestle, Marion. 2002. *Food politics: How the food industry influences nutrition and health*. University of California–Berkeley Press, Berkeley.

Norris, Robert F., Edward P. Caswell-Chen, and Marcos Kogan. 2002. *Concepts in integrated pest management*. Prentice Hall, Upper Saddle River, New Jersey.

Paoletti, Maurizio G., and David Pimentel. 1996. Genetic engineering in agriculture and the environment: Assessing risks and benefits. *BioScience* 46: 665–673.

Pearce, Fred. 2002. The great Mexican maize scandal. *New Scientist* 174: 14 (15 June 2002).

Pedigo, Larry P., and Marlin E. Rice, 2006. *Entomology and pest management*, 5th ed. Prentice Hall, Upper Saddle River, New Jersey.

Pimentel, David. 1999. Population growth, environmental resources, and the global availability of food. *Social Research,* Spring 1999.

Pinstrup-Andersen, Per, and Ebbe Schioler, 2001. *Seeds of contention: World hunger and the global controversy over GM (genetically modified) crops*. International Food Policy Research Institute, Washington, D.C.

Polak, Paul. 2005. The big potential of small farms. *Scientific American* 293(3): 84–91.

Pringle, Peter. 2003. *Food, Inc.: Mendel to Monsanto—The promises and perils of the biotech harvest*. Simon and Schuster, New York.

Quist, David, and Ignacio H. Chapela. 2001. Transgenic DNA introgressed into traditional maize landraces in Oaxaca, Mexico. *Nature* 414: 541–543.

Ruse, Michael, and David Castle, eds. 2002. *Genetically modified foods: Debating technology*. Prometheus Books, Amherst, New York.

Schmeiser, Percy. Monsanto vs. Schmeiser. www. percyschmeiser.com/

Shiva, Vandana. 2000. *Stolen harvest: The hijacking of the global food supply*. South End Press, Cambridge, Massachusetts.

Smil, Vaclav. 2001. *Feeding the world: A challenge for the twenty-first century*. MIT Press, Cambridge, Massachusetts.

Stewart, C. Neal. 2004. *Genetically modified planet: Environmental impacts of genetically engineered plants*. Oxford University Press, Oxford.

Teitel, Martin, and Kimberly Wilson. 2001. *Genetically engineered food: Changing the nature of nature*. Park Street Press.

The Farm Scale Evaluations of spring-sown genetically modified crops. 2003. A themed issue from *Philosophical Transactions of the Royal Society of London B: Biological Sciences* 358(1439), 29 November 2003.

Tuxill, John. 1999. Appreciating the benefits of plant biodiversity. Pp. 96–114 in *State of the world 1999*, Worldwatch Institute and W. W. Norton, Washington D.C., and New York.

Westra, Lauren. 1998. Biotechnology and transgenics in agriculture and aquaculture: The perspective from ecosystem integrity. *Environmental Values* 7: 79.

Wolfenbarger, L. LaReesa 2000. The ecological risks and benefits of genetically engineered plants. *Science* 290: 2088.

Chapter 11

Balmford, Andrew, et al. 2002. Economic reasons for conserving wild nature. *Science* 297: 950–953.

Barnosky, Anthony D., et al. 2004. Assessing the causes of late Pleistocene extinctions on the continents. *Science* 306: 70–75.

Baskin, Yvonne. 1997. *The work of nature: How the diversity of life sustains us*. Island Press, Washington, D.C.

Bright, Chris. 1998. *Life out of bounds: Bioinvasion in a borderless world*. Worldwatch Institute and W. W. Norton, Washington D.C., and New York.

CITES Secretariat. "Convention on International Trade in Endangered Species of Wild Fauna and Flora. www. cites.org/

Convention on Biological Diversity. www.biodiv.org/

Daily, Gretchen C., ed. 1997. *Nature's services: Societal dependence on natural ecosystems*. Island Press, Washington, D.C.

Ehrenfeld, David W. 1970. *Biological conservation*. International Thomson Publishing, London.

Gaston, Kevin J., and John I. Spicer. 2004. *Biodiversity: An introduction,* 2nd ed. Blackwell, London.

Groom, Martha J., et al. 2005. *Principles of conservation biology,* 3rd ed. Sinauer Associates, Sunderland, Massachusetts.

Groombridge, Brian, and Martin D. Jenkins. 2002. *Global biodiversity: Earth's living resources in the 21st century*. UNEP, World Conservation Monitoring Centre, and Aventis Foundation; World Conservation Press, Cambridge, U.K.

Groombridge, Brian, and Martin D. Jenkins. 2002. *World atlas of biodiversity: Earth's living resources in the 21st century*. University of California Press, Berkeley.

Hanken, James. 1999. Why are there so many new amphibian species when amphibians are declining? *Trends in Ecology and Evolution* 14: 7–8.

Harris, Larry D. 1984. *The fragmented forest: Island biogeography theory and the preservation of biotic diversity*. University of Chicago Press, Chicago.

Harrison, Paul, and Fred Pearce, eds. 2000. *AAAS atlas of population & environment*. University of California Press, Berkeley.

Jenkins, Martin, 2003. Prospects for biodiversity. *Science* 302: 1175–1177.

Louv, Richard. 2005. *Last child in the woods: Saving our children from nature-deficit disorder*. Algonquin Books, Chapel Hill, North Carolina.

MacArthur, Robert H., and Edward O. Wilson. 1967. *The theory of island biogeography*. Princeton University Press, Princeton.

Mackay, Richard. 2002. *The Penguin atlas of endangered species: A worldwide guide to plants and animals*. Penguin, New York.

Maehr, David S., Reed F. Noss, and Jeffrey Larkin, eds. 2001. *Large mammal restoration: Ecological and sociological challenges in the 21st century*. Island Press, Washington, D.C.

Matthiessen, Peter. 2000. *Tigers in the snow*. North Point Press, New York.

Meegaskumbura, Madhava, et al. 2002. Sri Lanka: An amphibian hot spot. *Science* 298: 379.

Millennium Ecosystem Assessment. 2005. *Ecosystems and human well-being: Biodiversity synthesis*. Millennium Ecosystem Assessment and World Resources Institute.

Miquelle, Dale, Howard Quigley, and Maurice Hornocker. 1999. A habitat protection plan for Amur Tiger conservation: A proposal outlining habitat protection measures for the Amur Tiger. Hornocker Wildlife Institute.

Mooney, Harold A. and Richard J. Hobbs, eds. 2000. *Invasive species in a changing world*. Island Press, Washington, D.C.

Newmark, William D. 1987. A land-bridge perspective on mammal extinctions in western North American parks. *Nature* 325: 430.

Pimm, Stuart L., and Clinton Jenkins. 2005. Sustaining the variety of life. *Scientific American* 293(3): 66–73.

Primack, Richard B. 2004. *Essentials of conservation biology,* 3rd ed. Sinauer Associates, Sunderland, Massachusetts.

Quammen, David. 1996. *The song of the dodo: Island biogeography in an age of extinction*. Touchstone, New York.

Relyea, Rick, and Nathan Mills. 2001. Predator-induced stress makes the pesticide carbaryl more deadly to gray treefrog tadpoles. *Proceedings of the National Academy of Sciences, USA* 98: 2491–2496.

Rosenzweig, Michael L. 1995. *Species diversity in space and time*. Cambridge University Press, Cambridge.

Simberloff, Daniel S. 1969. Experimental zoogeography of islands: A model for insular colonization. *Ecology* 50: 296–314.

Simberloff, Daniel. 1998. Flagships, umbrellas, and keystones: Is single-species management passé in the landscape era? *Biological Conservation* 83: 247–257.

Simberloff, Daniel S., and Edward O. Wilson. 1969. Experimental zoogeography of islands: The colonization of empty islands. *Ecology* 50: 278–296.

Simberloff, Daniel S., and Edward O. Wilson. 1970. Experimental zoogeography of islands: A two-year record of colonization. *Ecology* 51: 934–937.

Soulé, Michael E. 1986. *Conservation biology: The science of scarcity and diversity*. Sinauer Associates, Sunderland, Massachusetts.

Takacs, David. 1996. *The idea of biodiversity: Philosophies of paradise.* Johns Hopkins University Press, Baltimore.

United Nations Environment Programme. 2002. Biodiversity. Pp. 120–149 in *Global environment outlook 3 (GEO-3).* UNEP and Earthscan Publications, Nairobi and London.

United Nations Environment Programme. 2003. Sustaining life on Earth: How the Convention on Biological Diversity promotes nature and human well-being. www.biodiv. org/doc/publications/guide.asp.

United States Fish and Wildlife Service. The endangered species act of 1973. Accessible online at http://endangered.fws.gov/esa.html.

Wilson, Edward O. 1984. *Biophilia.* Harvard University Press, Cambridge, Massachusetts.

Wilson, Edward O. 1992. *The diversity of life.* Harvard University Press, Cambridge, Massachusetts.

Wilson, Edward O. 1994. *Naturalist.* Island Press, Shearwater Books, Washington, D.C.

Wilson, Edward O. 2002. *The future of life.* Alfred A. Knopf, New York.

Wilson, Edward O., and Daniel S. Simberloff. 1969. Experimental zoogeography of islands: Defaunation and monitoring techniques. *Ecology* 50: 267–278.

World Conservation Union. 2005. IUCN Red List. www.iucnredlist.org/

Chapter 12

British Columbia Ministry of Forests. Introduction to Silvicultural Systems. www.for.gov.bc.ca/hfd/pubs/SSIntroworkbook/index.htm. British Columbia Ministry of Forests, Victoria, B.C.

Canadian Broadcasting Corporation. 1993. A little place called Clayoquot Sound. CBC broadcast, 13 April 1993. http://archives.cbc.ca/IDC-1-75-679-3918/Science_technology/clearcutting/clip6.

Clary, David. 1986. *Timber and the Forest Service.* University Press of Kansas, Lawrence.

Food and Agriculture Organization of the United Nations. 2005. *Global forest resources assessment.* FAO Forestry, Rome.

Foster, Bryan C., and Peggy Foster. 2002. *Wild logging: A guide to environmentally and economically sustainable forestry.* Mountain Press, Missoula, Montana.

Gardner, Gary. 2005. Forest loss continues. Pp. 92–93 in *Vital signs 2005.* Worldwatch Institute and W. W. Norton, Washington, D.C., and New York.

Harrison, Paul, and Fred Pearce, eds. 2000. *AAAS atlas of population & environment.* University of California Press, Berkeley.

Haynes, Richard W. and Gloria E. Perez, tech. eds. 2001. Northwest Forest Plan research synthesis. *Gen. Tech. Rep. PNW-GTR-498.* USDA Forest Service, Pacific Northwest Research Station, Portland, Oregon.

Jacobs, Lynn. 1991. *Waste of the West: Public lands ranching.* Lynn, Jacobs, Tucson, Arizona.

Landres, Peter, David R. Spildie, and Lloyd P. Queen. 2001. GIS applications to wilderness management: Potential uses and limitations. *Gen. Tech. Rep. RMRS-GTR-80.* USDA Forest Service, Rocky Mountain Research Station, Fort Collins, Colorado.

Myers, Norman, and Jennifer Kent. 2001. *Perverse subsidies: How misused tax dollars harm the environment and the economy.* Island Press, Washington, D.C.

National Forest Management Act of 1976. October 22, 1976 (P.O. 94–588, 90 Stat. 2949, as amended; 16 U.S.C.)

National Round Table on the Environment and the Economy, Environment Canada. Clayoquot Sound Biosphere Reserve. www.nrtee-trnee.ca/eng/programs/Current_Programs/Nature/Case-Studies/Clayoquot-Case-Study-Complete_e.htm.

Natural Resources Canada. 2005. *The state of Canada's forests, 2004-2005.* Natural Resources Canada, Ottawa.

Runte, Alfred. 1979. *National parks and the American experience.* University of Nebraska Press, Lincoln.

Sedjo, Robert A. 2000. *A vision for the US Forest Service.* Resources for the Future, Washington, D.C.

Singh, Ashbindu, et al. 2001. An assessment of the status of the world's remaining closed forests. United Nations Environmental Program, UNEP/DEWA/TR 01–2l, August 2001.

Smith, David M., et al. 1996. *The practice of silviculture: Applied forest ecology,* 9th ed. Wiley, New York.

Smith, W. Brad, et al. 2004. Forest resources of the United States, 2002. *Gen. Tech. Rep.* NC-241, North Central Research Station, USDA Forest Service, St. Paul, Minnesota.

Soulé, Michael E., and John Terborgh, eds. 1999. *Continental conservation.* Island Press, Washington, D.C.

Stegner, Wallace. 1954. *Beyond the hundredth meridian: John Wesley Powell and the second opening of the West.* Houghton Mifflin, Boston.

United Nations Environment Programme. 2002. Forests. Pp. 90–119 in *Global environment outlook 3 (GEO-3).* UNEP and Earthscan Publications, Nairobi and London.

USDA Forest Service. 2001. *U.S. forest facts and historical trends.* FS-696, March 2001.

U.S. National Park Service. 2002. *National Park Service statistical abstract 2002.* NPS Public Use Statistics Office, U.S. Department of the Interior, Denver, Colorado.

Chapter 13

Abbott, Carl. 2001. *Greater Portland: Urban life and landscape in the Pacific Northwest.* University of Pennsylvania Press.

Abbott, Carl. 2002. Planning a sustainable city. Pp. 207–235 in Squires, Gregory D., ed. *Urban sprawl: Causes, consequences, and policy responses.* Urban Institute Press, Washington, D.C.

Beck, Roy, et. al., 2003. *Outsmarting smart growth: Population growth, immigration, and the problem of sprawl.* Center for Immigration Studies, Washington, D.C.

Breuste, Jurgen, et al. 1998. *Urban ecology.* Springer-Verlag.

Cronon, William. 1991. *Nature's metropolis: Chicago and the great West.* W. W. Norton, New York.

Duany, Andres, et al. 2001. *Suburban nation: The rise of sprawl and the decline of the American dream.* North Point Press, New York.

Ewing, Reid, et al. 2002. *Measuring sprawl and its impact.* Smart Growth America.

Ewing, Reid, et al. 2003. Measuring sprawl and its transportation impacts. *Transportation Research Record* 1831: 175–183.

Girardet, Herbert. 2004. *Cities people planet: Livable cities for a sustainable world.* Academy Press.

Hall, Kenneth B. and Gerald A. Porterfield. 2001. *Community by design: New urbanism for suburbs and small communities.* McGraw-Hill, New York.

Horizon International. 2003. Efficient transportation for successful urban planning in Curitiba. www.solutions-site.org/artman/publish/printer_62.shtml.

Jacobs, Jane. 1992. *The death and life of great American cities.* Vintage.

Kalnay, Eugenia, and Ming Cai. 2003. Impact of urbanization and land-use change on climate. *Nature* 423: 528–531.

Kirdar, Uner, ed. 1997. *Cities fit for people.* United Nations, New York.

Litman, Todd. 2004. *Rail transit in America: A comprehensive evaluation of benefits.* Victoria Transport Policy Institute and American Public Transportation Association.

Logan, Michael F. 1995. *Fighting sprawl and city hall.* University of Arizona Press, Tucson.

Metro. www.metro-region.org.

New Urbanism. www.newurbanism.org.

Northwest Environment Watch. 2004. *The Portland exception: A comparison of sprawl, smart growth, and rural land loss in 15 U.S. cities.* Northwest Environment Watch, Seattle.

Portney, Kent. E. 2003. *Taking sustainable cities seriously: Economic development, the environment, and quality of life in American cities (American and comparative environmental policy).* MIT Press, Cambridge, Massachusetts.

Pugh, Cedric, ed. 1996. *Sustainability, the environment, and urbanization.* Earthscan Publications, London.

Sheehan, Molly O'Meara. 2001. *City limits: Putting the brakes on sprawl.* Worldwatch Paper #156. Worldwatch Institute, Washington, D.C.

Sheehan, Molly O'Meara. 2002. What will it take to halt sprawl? *World-Watch* (Jan/Feb 2002): 12–23.

Sprawl City. www.sprawlcity.org.

Stren, R., et al. 1992. *Sustainable cities: Urbanization and the environment in international perspective.* Westview Press, Boulder, Colorado, and San Francisco.

United Nations Environment Programme. 2002. Urban areas. Pp. 240–269 in *Global environment outlook 3 (GEO-3).* UNEP and Earthscan Publications, Nairobi and London.

United States Census Bureau. www.census.gov/

United States Environmental Protection Agency. Smart growth. www.epa.gov/smartgrowth.

Wiewel, Wim, and Jospeh J. Persky., eds. 2002. *Suburban sprawl: Private decisions and public policy.* M. E. Sharpe, Armond, New York.

Chapter 14

Ames, Bruce N., Margie Profet, and Lois Swirsky Gold. 1990. Nature's chemicals and synthetic chemicals: Comparative toxicology. *Proceedings of the National Academy of the USA* 87: 7782–7786.

Bloom, Barry. 2005. Public health in transition. *Scientific American* 293(3): 92–99.

Cagen, S. Z., et al. 1999. Normal reproductive organ development in wistar rats exposed to bisphenol A in the drinking water. *Regulatory Toxicology and Pharmacology* 30: 130–139.

Carlsen, Elisabeth, et al. 1992. Evidence for decreasing quality of semen during past 50 years. *British Medical Journal* 305: 609–613.

Carson, Rachel. 1962. *Silent spring.* Houghton Mifflin, Boston.

Colburn, Theo, Dianne Dumanoski, and John P. Myers. 1996. *Our stolen future.* Penguin USA, New York.

Crain, D. Andrew, and Louis J. Guillette Jr. 1998. Reptiles as models of contaminant-induced endocrine disruption. *Animal Reproduction Science* 53: 77–86.

Guillette, Elizabeth A., et al. 1998. An anthropological approach to the evaluation of preschool children exposed to pesticides in Mexico. *Environmental Health Perspectives* 106: 347–353.

Guillette, Louis J. Jr., et al. 1999. Plasma steroid concentrations and male phallus size in juvenile alligators from seven Florida lakes. *General and Comparative Endocrinology* 116: 356–372.

Guillette, Louis J. Jr., et al. 2000. Alligators and endocrine disrupting contaminants: A current perspective. *American Zoologist* 40: 438–452.

Halweil, Brian. 1999. Sperm counts dropping. Pp. 148–149 in *Vital signs 1999.* Worldwatch Institute and W. W. Norton, Washington, D.C. and New York.

Hayes, Tyrone, et al. 2003. Atrazine-induced hermaphroditism at 0.1 PPB in American leopard frogs (*Rana pipiens*): Laboratory and field evidence. *Environmental Health Perspectives* 111: 568–575.

Hunt, Patricia A., et al. 2003. Bisphenol A exposure causes meiotic aneuploidy in the female mouse. *Current Biology* 13: 546–553.

Kolpin, Dana W., et al. 2002. Pharmaceuticals, hormones, and other organic wastewater contaminants in U.S. streams, 1999–2000: A national reconnaissance. *Environmental Science and Technology* 36: 1202–1211.

Landis, Wayne G., and Ming-Ho Yu. 2004. *Introduction to environmental toxicology,* 3rd ed. Lewis Press, Boca Raton, Florida.

Loewenberg, Samuel. 2003. E.U. starts a chemical reaction. *Science* 300: 405.

Millennium Ecosystem Assessment. 2005. *Ecosystems and human well-being: Health synthesis.* World Health Organization.

Manahan, Stanley E. 2000. *Environmental chemistry,* 7th ed. Lewis Publishers, CRC Press, Boca Raton, Florida.

McGinn, Anne Platt. 2000. *Why poison ourselves? A precautionary approach to synthetic chemicals* Worldwatch Paper #153. Worldwatch Institute, Washington, D.C.

McGinn, Anne Platt. 2002. Reducing our toxic burden. Pp. 75–100 in *State of the world 2002,* Worldwatch Institute and W. W. Norton, Washington, D.C. and New York.

McGinn, Anne Platt. 2003. Combating malaria. Pp. 62–84 in *State of the world 2003,* Worldwatch Institute and W. W. Norton, Washington, D.C. and New York.

Moeller, Dade. 2004. *Environmental health,* 3rd ed. Harvard University Press, 2004.

Nagel, S.C., et al. 1997. Relative binding affinity-serum modified access (RBA-SMA) assay predicts in vivo bioactivity of the xenoestrogens bisphenol A and octylphenol. *Environmental Health Perspectives* 105: 70–76.

National Center for Health Statistics, 2004. *Health, United States, 2004, with chartbook on trends in the health of Americans.* Hyattsville, Maryland.

National Center for Environmental Health; U.S. Centers for Disease Control and Prevention. 2005. *Third national report on human exposure to environmental chemicals.* NCEH Pub. No. 05-0570, Atlanta.

Pirages, Dennis. 2005. Containing infectious disease. Pp. 42–61 in *State of the world 2005.* Worldwatch Institute and W. W. Norton, Washington, D.C., and New York.

Renner, Rebecca. 2002. Conflict brewing over herbicide's link to frog deformities. *Science* 298: 938–939.

Rodricks, Joseph V. 1994. *Calculated risks: Understanding the toxicity of chemicals in our environment.* Cambridge University Press, Cambridge.

Salem, Harry, and Eugene Olajos. 1999. *Toxicology in risk assessment.* CRC Press, Boca Raton, Florida.

Spiteri, I. Daniel, Louis J. Guillette Jr., and D. Andrew Crain. 1999. The functional and structural observations of the neonatal reproductive system of alligators exposed *in ozo* to atrazine, 2,4-D, or estradiol. *Toxicology and Industrial Health* 15: 181–186.

Stancel, George, et al. 2001. "Report of the bisphenol A sub-panel." Chapter 1 in *National Toxicology Program's report of the endocrine disruptors low-dose peer review.* U.S. EPA and NIEHS, NIH.

Stockholm Convention on Persistant Organic Pollutants. www.pops.int/

United States Environmental Protection Agency. 2003. Pesticide registration program. www.epa.gov/pesticides/factsheets/registration.htm.

United States Environmental Protection Agency. 2003. Toxic Substances Control Act. www.epa.gov/region5/defs/html/tsca.htm.

United States Environmental Protection Agency. 2003. *EPA's draft report on the environment.* EPA 600-R-03-050. EPA, Washington, D.C.

Williams, Phillip L., Robert C. James, and Stephen M. Roberts, eds. 2000. *The principles of toxicology: Environmental and industrial applications,* 2nd ed. Wiley-Interscience, New York.

World Health Organization, 2004. *World health report 2004: Changing history.* WHO, Geneva, Switzerland.

Yu, Ming-Ho, 2004. *Environmental toxicology: Biological and health effects of pollutants,* 2nd ed. CRC Press, Boca Raton, Florida.

Chapter 15

American Rivers. 2002. *The ecology of dam removal: A summary of benefits and impacts.* American Rivers, Washington D.C., February 2002.

British Geographical Society and Bangladesh Department of Public Health Engineering. 2001. *Arsenic contamination of groundwater in Bangladesh.* Technical Report WC/00/19, Volume 1: Summary.

De Villiers, Marq, 2000. *Water: The fate of our most precious resource.* Mariner Books.

Gleick, Peter. H. 2003. Global freshwater resources: Soft-path solutions for the 21st century. *Science* 302: 1524–1527.

Gleick, Peter. H., et al. 2004. *The world's water 2004–2005: The biennial report on freshwater resources.* Island Press, Washington D.C.

Harrison, Paul, and Fred Pearce, eds. 2000. *AAAS atlas of population & environment.* University of California Press, Berkeley.

Harvey, Charles F., et al. 2002. Arsenic mobility and groundwater extraction in Bangladesh. *Science* 298: 1602–1606.

Institute of Governmental Studies, University of California, Berkeley. Imperial Valley-San Diego water transfer controversy. www.igs. berkeley.edu/library/htImperialWaterTransfer2003.html.

Jenkins, Matt. 2002. The royal squeeze. *High Country News* 35(1), January 20, 2003.

Jenkins, Matt. 2003. California's water binge skids to a halt. *High Country News* 34(17), September 16, 2002.

Marston, Ed. 2001. Quenching the big thirst. *High Country News* 33(10), May 21, 2001.

Millennium Ecosystem Assessment. 2005. *Ecosystems and human well-being: Wetlands and water synthesis.* Millennium Ecosystem Assessment and World Resources Institute.

Nickson, Ross, et al. 1998. Arsenic poisoning of Bangladesh groundwater. *Nature* 395: 338.

Postel, Sandra. 1999. *Pillar of sand: Can the irrigation miracle last?* W. W. Norton, New York.

Postel, Sandra. 2005. *Liquid assets: The critical need to safeguard freshwater ecosystems.* Worldwatch Paper #170. Worldwatch Institute, Washington, D.C.

Postel, Sandra, and Amy Vickers. 2004. Boosting water productivity. Pp. 46–67 in *State of the world 2004.* Worldwatch Institute and W. W. Norton, Washington, D.C. and New York.

Reisner, Marc. 1986. *Cadillac desert: The American West and its disappearing water.* Viking Penguin, New York.

Sampat, Payal. 2001. Uncovering groundwater pollution. Pp. 21–42 in *State of the world 2001.* Worldwatch Institute and W. W. Norton, Washington, D.C., and New York.

Sibley, George. 1997. A tale of two rivers: The desert empire and the mountain. *High Country News* 29(21), November 10, 1997.

Smith, Lingas, Rahman. 2000. Contamination of drinking water by arsenic in Bangladesh: A public health emergency. *Bulletin of the World Health Organization,* 78(9).

Stone, Richard. 1999. Coming to grips with the Aral Sea's grim legacy. *Science* 284: 30–33.

United Nations Environment Programme. 2002. Freshwater. Pp. 150–179 in *Global environment outlook 3 (GEO-3).* UNEP and Earthscan Publications, Nairobi and London.

United Nations World Water Assessment Programme. 2003. *U.N. world water development report: Water for people, water for life.* Paris, New York, and Oxford, UNESCO and Berghahn Books.

United States Bureau of Reclamation, Lower Colorado Regional Office. www.usbr.gov/lc/region.

United States Environmental Protection Agency. 1998. *Wastewater primer.* EPA 833-K-98-001, Office of Wastewater Management, May 1998.

United States Environmental Protection Agency. 2003. *EPA's draft report on the environment.* EPA 600-R-03-050. EPA, Washington, D.C.

United States Environmental Protection Agency. 2003. *Water on tap: What you need to know.* EPA 816-K-03-007. Office of Water, EPA, Washington, D.C.

Wolf, Aaron T., et al. 2005. Managing water conflict and cooperation. Pp. 80–99 in *State of the world 2005.* Worldwatch Institute and W. W. Norton, Washington, D.C., and New York.

World Health Organization. 2000. *Global water supply and sanitation assessment 2000 report.* WHO, Geneva, Switzerland.

Youth, Howard. 2005. Wetlands drying up. Pp. 90–91 in *Vital signs 2005.* Worldwatch Institute and W. W. Norton, Washington, D.C., and New York.

Chapter 16

Ault, J. S., Bohnsack, J. A. and G.A. Meester. 1998. A retrospective (1979–1996) multi-species assessment of coral reef fish stocks in the Florida Keys. *Fishery Bulletin* 96: 395–414.

Baker, Beth. 1999. First aid for an ailing reef: Research in the Florida Keys National Marine Sanctuary. *BioScience* 49: 173–178.

Bellwood, David R., et al. 2004. Confronting the coral reef crisis. *Nature* 429: 827–833.

Causey, Billy D., Joanne Delaney, and Brian D. Keller. 2001. *The status of coral reefs of the Florida Keys.* Florida Keys National Marine Sanctuary.

[Correspondence to *Science,* various authors]. 2001. *Science* 295: 1233–1235.

Embassy of the People's Republic of China. 2001. Fishing statistics "basically correct," ministry says. Press release, 18 December 2001. http://saup.fisheries.ubc.ca/Media/Chinese_Embassy_18_Dec_2001.pdf.

Food and Agriculture Organization of the United Nations. 2002. Fishery statistics: Reliability and policy implications. FAO Fisheries Department, Rome. www.fao.org/DOCREP/FIELD/006/Y3354M/Y3354M00.HTM.

Food and Agriculture Organization of the United Nations. 2004. *The state of world fisheries and aquaculture, 2004.* FAO, Rome.

Garrison, Tom. 2005. *Oceanography: An invitation to marine science,* 5th ed. Brooks/Cole, San Francisco.

Gell, Fiona R., and Callum M. Roberts. 2003. Benefits beyond boundaries: The fishery effects of marine reserves. *Trends in Ecology and Evolution* 18: 448–455.

Halpern, Benjamin S., and Robert R. Warner. 2002. Marine reserves have rapid and lasting effects. *Ecology Letters* 5: 361–366.

Halpern, Benjamin S., and Robert R. Warner. 2003. Matching marine reserve design to reserve objectives. *Proceedings of the Royal Society of London B:* 270: 1871–1878.

Jackson, Jeremy B. C., et al. 2001. Historical overfishing and the recent collapse of coastal ecosystems. *Science* 293: 629–638.

Larsen, Janet. 2005. Wild fish catch hits limits—Oceanic decline offset by increased fish farming *Eco-economy indicators.* Earth Policy Institute, www.earth-policy.org/Indicators/Fish/2005.htm.

Mastny, Lisa. 2001. World's coral reefs dying off. Pp. 92–93 in *Vital signs 2001.* Worldwatch Institute and W. W. Norton, Washington D.C., and New York.

Myers, Ransom A., and Boris Worm. 2003. Rapid worldwide depletion of predatory fish communities. *Nature* 423: 280–283.

National Academy of Public Administration. 1999. *Protecting Our National Marine Sanctuaries.* Center for the Economy and the Environment, NAPA, Washington, D.C.

National Center for Ecological Analysis and Synthesis (NCEAS) and Communication Partnership for Science and the Sea (COMPASS), sponsors. 2001. *Scientific consensus statement on marine reserves and marine protected areas.* Available online at www.nceas.ucsb.edu/consensus.

National Oceanic and Atmospheric Administration (NOAA). Florida Keys National Marine Sanctuary. www. fknms.nos.noaa.gov.

National Research Council. 2003. *Oil in the sea III: Inputs, fates, and effects.* National Academies Press, Washington, D.C.

Norse, Elliott, and Larry B. Crowder, eds. 2005. *Marine conservation biology: The science of maintaining the sea's biodiversity.* Island Press, Washington, D.C.

Nybakken, James W., and Mark D. Bertness. 2004. *Marine biology: An ecological approach,* 6th ed. Benjamin Cummings, San Francisco.

Palumbi, Stephen. 2003. *Marine reserves: A tool for ecosystem management and conservation.* Pew Oceans Commission.

Pauly, Daniel, et al. 2002. Towards sustainability in world fisheries. *Nature* 418: 689–695.

Pauly, Daniel, et al. 2003. The future for fisheries. *Science* 302: 1359–1361.

Pew Oceans Commission. 2003. *America's living oceans: Charting a course for sea change.* A report to the nation. May 2003. Pew Oceans Commission, Arlington, Virginia.

Pinet, Paul R. 1999. *Invitation to oceanography,* 2nd ed. Jones & Bartlett, Boston.

Roberts, Callum M., et al. 2001. Effects of marine reserves on adjacent fisheries. *Science* 294: 1920–1923.

Sumich, James L., and John F. Morrissey, 2004. *Introduction to the biology of marine life,* 8th ed. Jones & Bartlett, Boston.

Thurman, Harold V., and Alan P. Trujillo, 2004. *Introductory oceanography,* 10th ed. Prentice Hall, Upper Saddle River, New Jersey.

United Nations Environment Programme. 2002. Coastal and marine areas. Pp. 180–209 in *Global environment outlook 3 (GEO-3).* UNEP and Earthscan Publications, Nairobi and London.

United States Commission on Ocean Policy. 2004. *An ocean blueprint for the 21st century.* Final Report. Washington, D.C.

United States Department of Commerce. 1996. *Strategy for stewardship: Florida Keys National Marine Sanctuary final management plan/ environmental impact statement.* 3 vols. Dept. of Commerce, Washington, D.C.

United States Department of Commerce and United States Department of the Interior. Marine protected areas of the United States. www.mpa.gov.

Watson, Reginald, Lillian Pang, and Daniel Pauly. 2001. The marine fisheries of China: Development and reported catches. *Fisheries Centre Research Reports* 9(2). Fisheries Centre, University of British Columbia, Canada.

Watson, Reginald, and Daniel Pauly. 2001. Systematic distortions in world fisheries catch trends. *Nature* 414: 534–536.

Weber, Michael L. 2001. *From abundance to scarcity: A history of U.S. marine fisheries policy.* Island Press, Washington, D.C.

Chapter 17

Ahrens, C. Donald. 2003. *Meteorology today,* 7th ed. Brooks/Cole, San Francisco.

Akimoto, Hajime. 2003. Global air quality and pollution. *Science* 302: 1716–1719.

Bell, Michelle L., and Devra L. Davis. 2001. Reassessment of the lethal London fog of 1952: Novel indicators of acute and chronic consequences of acute exposure to air pollution. *Environmental Health Perspectives* 109(Suppl 3): 389–394.

Bernard, Susan M., et al. 2001. The potential impacts of climate variability and change on air pollution-related health effects in the United States. *Environmental Health Perspectives* 109(Suppl 2): 199–209.

Biscaye, Pierre E., et al. 2000. Eurasian air pollution reaches eastern North America. *Science* 290: 2258–2259.

Boubel, Richard W., et al., eds. 1994. *Fundamentals of air pollution,* 3rd ed. Academic Press, San Diego, California.

Bruce, Nigel, Rogelio Perez-Padilla, and Rachel Albalak. 2000. Indoor air pollution in developing countries: A major environmental and public health challenge. *Bulletin of the World Health Organization* 78: 1078–1092.

Cooper, C. David, and F. C. Alley. 2002. *Air pollution control,* 3rd ed. Waveland Press.

Davis, Devra. 2002. *When smoke ran like water: Tales of environmental deception and the battle against pollution.* Basic Books, New York.

Davis, Devra L., Michelle L. Bell, and Tony Fletcher. 2002. A look back at the London smog of 1952 and the half century since. *Environmental Health Perspectives* 110: A734.

Driscoll, Charles T., et al. 2001. *Acid rain revisited: Advances in scientific understanding since the passage of the 1970 and 1990 Clean Air Act Amendments.* Hubbard Brook Research Foundation. Science Links™ Publication. Vol. 1, no.1.

Ezzati Majid, and Daniel M. Kammen. 2001. Quantifying the effects of exposure to indoor air pollution from biomass combustion on acute respiratory infections in developing countries. *Environmental Health Perspectives* 109: 481–488.

Gardner, Gary. 2005. Air pollution still a problem. Pp. 94–95 in *Vital signs 2005.* Worldwatch Institute and W. W. Norton, Washington, D.C., and New York.

Godish, Thad. 2003. *Air quality,* 4th ed. CRC Press, Boca Raton, Florida.

Hoffman, Matthew J. 2005. *Ozone depletion and climate change: Constructing a global response.* SUNY Press, New York.

Hunt, Andrew, et al. 2003. Toxicologic and epidemiologic clues from the characterization of the 1952 London smog fine particulate matter in archival autopsy lung tissues. *Environmental Health Perspectives* 111: 1209–1214.

Jacobson, Mark Z. 2002. *Atmospheric pollution: History, science, and regulation.* Cambridge University Press, New York.

Kunzli, Nino, et al. 2000. Public-health impact of outdoor and traffic-related air pollution: A European assessment. *Lancet* 356: 795–801.

Lelieveld, Jos, et al. 2001. The Indian Ocean experiment: Widespread air pollution from South and Southeast Asia. *Science* 291: 1031–1036.

Likens, Gene E. 2004. Some perspectives on long-term biogeochemical research from the Hubbard Brook ecosystem study. *Ecology* 85: 2355–2362.

London, Stephanie J., and Isabelle Romieu. 2000. Health costs due to outdoor air pollution by traffic. *Lancet* 356: 782–783.

Molina, Mario J., and F. Sherwood Rowland. 1974. Stratospheric sink for chlorofluoromethanes: Chlorine atom catalyzed destruction of ozone. *Nature* 249: 810–812.

Pal Arya, S. 1998. *Air pollution: Meteorology and dispersion.* Oxford University Press, Oxford.

Parson, Edward A. 2003. *Protecting the ozone layer: Science and strategy.* Oxford University Press, Oxford.

United Nations Environment Programme. Montreal Protocol. http://hq.unep.org/ozone/Treaties_and_Ratification/2B_montreal_ protocol.asp.

United Nations Environment Programme. 2002. Atmosphere. Pp. 210–239 in *Global environment outlook 3 (GEO-3).* UNEP and Earthscan Publications, Nairobi and London.

United States Environmental Protection Agency. 2003. *EPA's draft report on the environment.* EPA 600-R-03-050. Washington, D.C.

United States Environmental Protection Agency. 2003. *Latest findings on national air quality: 2002 status and trends.* EPA 454/K-03-001. Washington, D.C.

Wark, Kenneth, et al., 1997. *Air pollution: Its origin and control,* 3rd ed. Prentice Hall, Upper Saddle River, New Jersey.

World Health Organization. Indoor air pollution. WHO, Geneva, Switzerland. www.who.int/indoorair/en/index.html.

Chapter 18

Alley, Richard B. 2000. *The two-mile time machine: Ice cores, abrupt climate change, and our future.* Princeton University Press, Princeton, New Jersey.

Burroughs, William James. 2001. *Climate change: A multidisciplinary approach.* Cambridge University Press, Cambridge.

Drake, Frances. 2000. *Global warming: The science of climate change.* Oxford University Press, Oxford.

Dunn, Seth. 2001. Decarbonizing the energy economy. Pp. 83–102 in *State of the world 2001.* Worldwatch Institute and W. W. Norton, Washington, D.C., and New York.

Dunn, Seth, and Christopher Flavin. 2002. Moving the climate change agenda forward. Pp. 24–50 in *State of the world 2002.* Worldwatch Institute and W. W. Norton, Washington, D.C., and New York.

Gelbspan, Ross. 1997. *The heat is on: The climate crisis, the cover-up, the prescription.* Perseus Books, New York.

Gelbspan, Ross. 2004. *Boiling point: How politicians, big oil and coal, journalists, and activists are fueling the climate crisis—and what we can do to avert disaster.* Basic Books, New York.

Intergovernmental Panel on Climate Change. 2001. *IPCC third assessment report—Climate change 2001: The scientific basis.* World Meteorological Organization and United Nations Environment Programme.

Intergovernmental Panel on Climate Change. 2001. *IPCC third assessment report—Climate change 2001: Impacts, adaptations, and vulnerability.* World Meteorological Organization and United Nations Environment Programme.

Intergovernmental Panel on Climate Change. 2001. *IPCC third assessment report—Climate change 2001: Mitigation.* World Meteorological Organization and United Nations Environment Programme.

Intergovernmental Panel on Climate Change. 2001. *IPCC third assessment report—Climate change 2001: Synthesis report.* World Meteorological Organization and United Nations Environment Programme.

Intergovernmental Panel on Climate Change. 2001. *Technical Summary of the Working Group 1 report.*

Intergovernmental Panel on Climate Change. www.ipcc.ch/

Kareiva, Peter M., Joel G. Kingsolver, and Raymond B. Huey, eds. 1993. *Biotic interactions and global change.* Sinauer Associates, Sunderland, Massachusetts.

Karl, Thomas R., and Kevin E. Trenberth, 2003. Modern global climate change. *Science* 302: 1719–1723.

Lomborg, Bjorn. 2001. *The skeptical environmentalist: Measuring the real state of the world.* Cambridge University Press, Cambridge.

Mastny, Lisa. 2005. Global ice melting accelerating. Pp. 88–89 in *Vital signs 2005.* Worldwatch Institute and W. W. Norton, Washington, D.C., and New York.

Mayewski, Paul A., and Frank White. 2002. *The ice chronicles: The quest to understand global climate change.* University Press of New England, Hanover, New Hampshire.

McLachlan, Jason S., and Linda B. Brubaker. 1995. Local and regional vegetation change on the northeastern Olympic Peninsula during the Holocene. *Canadian Journal of Botany* 73: 1618–1627.

National Assessment Synthesis Team. 2000. *Climate change impacts on the United States: The potential consequences of climate variability and change.* U.S. Global Change Research Program. Cambridge University Press, Cambridge.

National Research Council, Board on Atmospheric Sciences and Climate, Commission on Geosciences, Environment, and Resources. 1998. *The atmospheric sciences: Entering the twenty-first century.* National Academies Press, Washington, D.C.

National Research Council, Committee on the Science of Climate Change, Division of Earth and Life Studies. 2001. *Climate change science: An analysis of some key questions.* National Academies Press, Washington, D.C.

Nordhaus, William D. 1998. Assessing the economics of climate change: An introduction. In Nordhaus, William D., ed., *Economic and policy issues in climate change.* Resources for the Future Press, Washington D.C.

Parmesan, Camille, and Gary Yohe. 2003. A globally coherent fingerprint of climate change impacts across natural systems. *Nature* 421: 37–42.

Real Climate. www.realclimate.org.

Root, Terry L., et al. 2003. Fingerprints of global warming on wild animals and plants. *Nature* 421: 57–60.

Sawin, Janet L. 2005. Climate change indicators on the rise. Pp. 40–41 in *Vital signs 2005.* Worldwatch Institute and W. W. Norton, Washington, D.C., and New York.

Schneider, Stephen H. and Terry L. Root, eds. 2002. *Wildlife responses to climate change: North American case studies.* Island Press, Washington, D.C.

Seinfeld, John H., and Spyros N. Pandis. 2006. *Atmospheric chemistry and physics,* 2nd ed. Wiley-Interscience, New York.

Shapiro, Robert J., Kevin A. Hassett, and Frank S. Arnold. 2002. *Conserving energy and preserving the environment: The role of public transportation.* American Public Transportation Association, July 2002.

Speth, James Gustave. 2004. *Red sky at morning: America and the crisis of the global environment.* Yale University Press, New Haven, Connecticut.

Stevens, William K. 1999. *The change in the weather: People, weather and the science of climate.* Delta Trade Paperbacks, New York.

Taylor, David. 2003. Small islands threatened by sea level rise. Pp. 84–85 in *Vital signs 2003.* Worldwatch Institute and W. W. Norton, Washington D.C. and New York.

Victor, David G. 2004. *Climate change: Debating America's policy options.* U.S. Council on Foreign Relations Press, Washington, D.C.

United Nations. United Nations Framework Convention on Climate Change. http://unfccc.int/2860.php.

United Nations. Kyoto Protocol. http://unfccc.int/resource/docs/convkp/kpeng.html.

United States Congress. House Committee on Science. 2001. Climate change: The state of the science. Hearing before the Committee on Science, House of Representatives, One Hundred Seventh Congress, first session, 14 March 2001.

Chapter 19

Association for the Study of Peak Oil and Gas. www.peakoil.net/

British Petroleum. 2005. *BP statistical review of world energy.* London, June 2005.

Campbell, Colin J. 1997. *The coming oil crisis.* Multi-Science Publishing Co., Essex, U.K.

Deffeyes, Kenneth S. 2001. *Hubbert's peak: The impending world oil shortage.* Princeton University Press, Princeton, New Jersey.

Deffeyes, Kenneth S. 2005. *Beyond oil: The view from Hubbert's peak.* Farrar, Straus, and Giroux, New York.

Douglas, D. C., P.E. Reynolds, and E. B. Rhode, eds. 2002. *Arctic Refuge coastal plain terrestrial wildlife research summaries. Biological science report.* USGS/BRD/BSR-2002-0001. United States Geological Survey, Washington, D.C.

Dunn, Seth. 2001. Decarbonizing the energy economy. Pp. 83–102 in *State of the world 2001,* Worldwatch Institute and W. W. Norton, Washington, D.C., and New York.

Energy Information Administration, U.S. Department of Energy. www.eia.doe.gov.

Energy Information Administration, U.S. Department of Energy. 1999. *Petroleum: An energy profile, 1999.* DOE/EIA-0545(99).

Energy Information Administration, U.S. Department of Energy. 2005. *Annual energy review 2004.* DOE/EIA-0384(2004). Washington, D.C.

Energy Information Administration, U.S. Department of Energy. 2005. *International energy annual 2003.* Washington, D.C.

Flavin, Christopher. 2005. Fossil fuel use surges. Pp. 30–31 in *Vital signs 2005.* Worldwatch Institute and W. W. Norton, Washington, D.C., and New York.

Freese, Barbara. 2003. *Coal: A human history.* Perseus Books, New York.

Goodstein, David. 2004. *Out of gas.* W. W. Norton, New York.

Holmes, Bob, and Nicola Jones. 2003. Brace yourself for the end of cheap oil. *New Scientist* (August 2, 2003): 9–11.

International Energy Agency. 2005. *Key world energy statistics 2005.* IEA Publications, Paris.

International Energy Agency. 2005. *World energy outlook 2005.* IEA Publications, Paris.

International Energy Agency. 2005. *Resources to reserves: Oil and gas technologies for the energy markets of the future.* IEA Publications, Paris.

Lovins, Amory B., et al. 2004. *Winning the oil endgame: Innovation for profits, jobs, and security.* Rocky Mountain Institute, Snowmass, Colorado.

Lovins, Amory B. 2005. More profit with less carbon. *Scientific American* 293(3): 74–83.

Nellemann, Christian, and Raymond D. Cameron. 1998. Cumulative impacts of an evolving oil-field complex on the distribution of calving caribou. *Canadian Journal of Zoology* 76: 1425–1430.

Pelley, Janet. 2001. Will drilling for oil disrupt the Arctic National Wildlife Refuge? *Environmental Science and Technology* 35: 240–247.

Powell, Stephen G. 1990. Arctic National Wildlife Refuge: How much oil can we expect? *Resources Policy* Sept. 1990: 225–240.

Prugh, Tom, et al. 2005. Changing the oil economy. Pp. 100–121 in *State of the world 2005.* Worldwatch Institute and W. W. Norton, Washington, D.C., and New York.

Ristinen, Robert A., and Jack J. Kraushaar. 1998. *Energy and the environment.* John Wiley and Sons, New York.

Russell, D. E. and P. McNeil. 2005. *Summer ecology of the Porcupine caribou herd.* Porcupine Caribou Management Board, Whitehorse, Yukon.

Sawin, Janet L. 2004. Making better energy choices. Pp. 24–45 in *State of the world 2004.* Worldwatch Institute and W. W. Norton, Washington, D.C. and New York.

Skinner, Brian J., and Stephen C. Porter. 2003. *The dynamic earth: An introduction to physical geology,* 5th ed. John Wiley and Sons, Hoboken, New Jersey.

United States Environmental Protection Agency. 2005. *Light-duty automotive technology and fuel economy trends: 1975 through 2005.* EPA420-R-05-001. EPA Office of Transportation and Air Quality, Washington, D.C.

United States Fish and Wildlife Service. 2001. Potential impacts of proposed oil and gas development on the Arctic Refuge's coastal plain: Historical overview and issues of concern. Web page of the Arctic National Wildlife Refuge, Fairbanks, Alaska. http://arctic.fws.gov/issues1.html.

United States Geological Survey. 2001. *The National Petroleum Reserve-Alaska (NPRA) data archive*. USGS Fact Sheet FS-024-01, March 2001.

United States Geological Survey. 2001. *Arctic National Wildlife Refuge, 1002 Area, petroleum assessment, 1998, including economic analysis*. USGS Fact Sheet FS-028-01, April 2001.

Walker, Donald A. 1997. Arctic Alaskan vegetation disturbance and recovery. Pp. 457–479 in *Disturbance and recovery in Arctic lands*, R.M.M. Crawford, ed. Kluwer Academic Publishers, Dordrecht, Netherlands.

Chapter 20

Aeck, Molly. 2005. Biofuel use growing rapidly. Pp. 38–39 in *Vital signs 2005*. Worldwatch Institute and W. W. Norton, Washington, D.C., and New York.

British Petroleum. 2005. *BP statistical review of world energy*. London, June 2005.

Chandler, David. 2003. America steels itself to take the nuclear plunge. *New Scientist* (August 9, 2003): 10–13.

Energy Information Administration, U.S. Department of Energy. www.eia.doe.gov.

Energy Information Administration. 2005. *Annual energy outlook 2005*. Washington, D.C.

Energy Information Administration, U.S. Department of Energy. 2005. *Annual energy review 2004*. DOE/EIA-0384(2004). Washington, D.C.

Energy Information Administration, U.S. Department of Energy. 2005. *International energy annual 2003*. Washington, D.C.

European Commission/International Atomic Energy Agency/World Health Organization. 1996. One decade after Chernobyl: Summing up the consequences of the accident. Summary of the conference results. Vienna, Austria, 8–12 April, 1996. EC/IAEA/WHO.

International Atomic Energy Agency. 2004. *Annual report 2003*. GC(48)/3. IAEA, Vienna, Austria.

International Atomic Energy Agency. *Nuclear power and sustainable development*. IAEA Information Series 02-01574/FS Series 3/01/E/Rev.1. Vienna, Austria.

International Energy Agency. 2005. *Key world energy statistics 2005*. IEA Publications, Paris.

International Energy Agency. 2005. *World energy outlook 2005*. IEA Publications, Paris.

Klass, Donald L. 2004. Biomass for Renewable Energy and Fuels. In *The Encyclopedia of Energy*, Elsevier.

Lenssen, Nicholas. 2005. Nuclear power rises once more. Pp. 32–33 in *Vital signs 2005*. Worldwatch Institute and W. W. Norton, Washington, D.C., and New York.

Lovins, Amory B., et al. 2004. *Winning the oil endgame: Innovation for profits, jobs, and security*. Rocky Mountain Institute, Snowmass, Colorado.

Murray, Danielle. 2005. Ethanol's potential: Looking beyond corn. *Eco-economy Update #49*, 5 June 2005. Earth Policy Institute, http://www.earth-policy.org/Updates/2005/Update49.htm.

National Renewable Energy Lab, U.S. Department of Energy. www.nrel.gov.

Nuclear Energy Agency, OECD. 2002. *Chernobyl: Assessment of radiological and health impacts. (2002 Update of Chernobyl: Ten Years On)*. OECD, Paris.

Nuclear Energy Agency. 2005. *NEA annual report 2004*. NEA, Organisation for Economic Co-operation and Development. OECD, Paris.

Office of Energy Efficiency and Renewable Energy, U.S. Department of Energy www.eere.energy.gov.

Organisation for Economic Co-operation and Development. 2000. *Business as usual and nuclear power*. OECD Publications, Paris.

REN21 Renewable Energy Policy Network. 2005. *Renewables 2005 global status report*. Worldwatch Institute, Washington, D.C.

Spadaro, Joseph V., Lucille Langlois, and Bruce Hamilton. 2000. Greenhouse gas emissions of electricity generation chains: Assessing the difference. *IAEA Bulletin* 42(2).

Swedish Bioenergy Association (SVEBIO). 2003. *Focus: Bioenergy*. Nos. 1–10. SVEBIO, Stockholm, 2003.

Swedish Energy Agency 2004. *Renewable electricity is the future's electricity*. Swedish Energy Agency, Eskilstuna, Sweden.

Swedish Energy Agency 2004. *Energy in Sweden: Facts and figures 2004*. Swedish Energy Agency, Eskilstuna, Sweden.

Swedish Energy Agency 2004. *The Swedish Energy Agency 2003*. Swedish Energy Agency, Eskilstuna, Sweden.

Swedish Energy Agency 2004. *Energy in Sweden 2003*. Swedish Energy Agency, Eskilstuna, Sweden.

U.N. Food and Agriculture Organization. *Biomass energy in ASEAN member countries*. FAO/ASEAN/EC. FAO Regional Wood Energy Development Programme in Asia, Bangkok, Thailand.

U.S. Environmental Protection Agency. Alternative fuels website. www.epa.gov/otaq/consumer/fuels/altfuels/altfuels.htm.

Chapter 21

American Wind Energy Association. 2005. *Global wind energy market report*. AWEA, Washington, D.C.

Ananthaswamy, Anil. 2003. Reality bites for the dream of a hydrogen economy. *New Scientist,* (November 15, 2003): 6–7.

Arnason, Bragi, and and Thorsteinn I. Sigfusson. 2000. Iceland—a future hydrogen economy. *International Journal of Hydrogen Energy* 25: 389–394.

Ásmundsson, Jón Knútur. 2002. Will fuel cells make Iceland the 'Kuwait of the North?' *World Press Review,* 15 February 2002.

Burkett, Elinor. 2003. A mighty wind. *New York Times magazine*. June 15, 2003.

Chow, Jeffrey, et al. 2003. Energy resources and global development. *Science* 302: 1528–1531.

DaimlerChrysler. 2003. *360 DEGREES/DaimlerChrysler Environmental Report 2003*. DaimlerChrysler AG, Stuttgart, Germany.

Dunn, Seth. 2000. The hydrogen experiment. *WorldWatch* 13: 14–25.

Energy Information Administration, U.S. Department of Energy. www.eia.doe.gov.

Energy Information Administration, U.S. Department of Energy. 2005. *Annual energy review 2004*. DOE/EIA-0384(2004). Washington, D.C.

Energy Information Administration, U.S. Department of Energy. 2005. *International energy annual 2003*. Washington, D.C.

Flavin, Christopher, and Seth Dunn. 1999. A new energy paradigm for the 21st century. *Journal of International Affairs* 53: 167–190.

Hirsch, Tim. 2001. Iceland launches energy revolution. *British Broadcasting Corporation News*, 24 December 2001.

Hydrogen & Fuel Cell Letter. 2003. World's first commercial hydrogen station opens in Iceland. *Hydrogen & Fuel Cell Letter* May 2003.

Idaho Wind Power Working Group for the Idaho Department of Water Resources Energy Division. 2002. *Idaho wind power development strategic plan*. Boise, Idaho.

Idaho Wind Power Working Group for the Idaho Department of Water Resources Energy Division. 2002. *Wind power potential in Idaho by county*. Boise, Idaho.

International Energy Agency. 2002. *Renewables in global energy supply: An IEA fact sheet*. IEA Publications, Paris.

International Energy Agency Renewable Energy Working Party. 2002. *Renewable energy . . . into the mainstream*. SITTARD, The Netherlands.

International Energy Agency. 2005. *Renewables information 2005*. IEA Publications, Paris.

Lovins, Amory B., et al. 2004. *Winning the oil endgame: Innovation for profits, jobs, and security*. Rocky Mountain Institute, Snowmass, Colorado.

Martinot, Eric, et al. 2002. Renewable energy markets in developing countries. *Annual Review of Energy and the Environment* 27: 309–48.

Martinot, Eric, Ryan Wiser, and Jan Hamrin. 2005. *Renewable energy markets and policies in the United States*. Center for Resource Solutions, San Francisco. www.martinot.info/Martinot_et_al_CRS.pdf.

Melis, Anastasios, et al. 2000. Sustained photobiological hydrogen gas production upon reversible inactivation of oxygen evolution in the green alga *Chlamydomonas reinhardtii*. *Plant Physiology* 122: 127–135.

National Renewable Energy Lab, U.S. Department of Energy. www.nrel.gov.

Office of Energy Efficiency and Renewable Energy, U.S. Department of Energy www.eere.energy.gov.

Office of Energy Efficiency and Renewable Energy, U.S. Department of Energy. 2005. *Wind power today: Federal wind program highlights.* DOE/GO-102005-2115. Washington, D.C.

Randerson, James. 2003. The clean green energy dream. *New Scientist* (August 16, 2003): 8–11.

Reeves, Ari, with Fredric Beck. 2003. *Wind energy for electric power: A REPP issue brief.* Renewable Energy Policy Project, Washington, D.C.

REN21 Renewable Energy Policy Network. 2005. *Renewables 2005 global status report.* Worldwatch Institute, Washington, D.C.

Ristinen, Robert A., and Jack J. Kraushaar, 1998. *Energy and the environment.* John Wiley and Sons, New York.

Rocky Mountain Institute webpage. Energy. RMI, Snowmass, Colorado. www.rmi.org/sitepages/pid17.php.

Sawin, Janet. 2004. *Mainstreaming renewable energy in the 21st century.* Worldwatch Paper 169. Worldwatch Institute, Washington, D.C.

Sawin, Janet L. 2005. Global wind growth continues. Pp. 34–35 in *Vital signs 2005.* Worldwatch Institute and W. W. Norton, Washington, D.C. and New York.

Sawin, Janet L. 2005. Solar energy markets booming. Pp. 36–37 in *Vital signs 2005.* Worldwatch Institute and W. W. Norton, Washington, D.C. and New York.

U.S. Department of Energy National Laboratory directors.1997. *Technology opportunities to reduce U.S. greenhouse gas emissions.* DOE, Washington, D.C.

Weisman, Alan. 1998. *Gaviotas: A village to reinvent the world.* Chelsea Green Publishing Co., White River Junction, Vermont.

World Alliance for Decentralized Energy. 2005. *World survey of decentralized energy 2005.* WADE, Edinburgh, Scotland.

Chapter 22

Allen, G. H. and R. A. Gearheart, eds. 1988. *Proceedings of a conference on wetlands for wastewater treatment and resource enhancement.* Humboldt State University, Arcata, California.

Ayres, Robert U., and Leslie W. Ayres. 1996. *Industrial ecology: Towards closing the materials cycle.* Edward Elgar Press, Cheltenham, U.K.

Beede, David N., and David E. Bloom. 1995. The economics of municipal solid waste. *World Bank Research Observer* 10: 113–150.

Diesendorf, Mark, and Clive Hamilton. 1997. *Human ecology, human economy.* Allen and Unwin, St. Leonards.

Edmonton, Alberta, City of. 2003. Waste management. www.edmonton.ca/portal/server.pt/gateway/PTARGS_0_2_104_0_0_35/http%3B/cmsserver/COEWeb/environment+waste+and+recycling/waste/

Energy Information Administration. Municipal solid waste. EIA, U.S. Department of Energy, Washington, D.C. www.eia.doe.gov/cneaf/solar.renewables/page/mswaste/msw.html.

Gitlitz, Jenny, and Pat Franklin. 2004. *The 10-cent incentive to recycle,* 3rd ed. Container Recycling Institute, Arlington, Virginia.

Graedel, Thomas E., and Braden R. Allenby, 2002. *Industrial ecology,* 2nd ed. Prentice Hall, Upper Saddle River, New Jersey.

Integrated Waste Services Association. WTE: About waste-to-energy. www.wte.org/waste.html. IWSA, Washington, D.C.

Kaufman, Scott, et al. 2004. The state of garbage in America. *Biocycle* 45: 31–41.

Lilienfeld, Robert, and William Rathje. 1998. *Use less stuff: Environmental solutions for who we really are.* Ballantine, New York.

Manahan, Stanley E. 1999. *Industrial ecology: Environmental chemistry and hazardous waste.* Lewis Publishers, CRC Press, Boca Raton, Florida.

McDonough, William, and Michael Braungart. 2002. *Cradle to cradle: Remaking the way we make things.* North Point Press, New York.

McGinn, Anne Platt. 2002. Toxic waste largely unseen. Pp. 112–113 in *Vital signs 2002.* Worldwatch Institute and W. W. Norton, Washington D.C., and New York.

New York City Department of Planning. Fresh Kills: Landfill to landscape. www.nyc.gov/html/dcp/html/fkl/ada/about/1_0.html/

New York City Department of Planning. Fresh Kills lifescape. www.nyc.gov/html/dcp/html/fkl/fkl_index.shtml.

New York City Department of Sanitation. 2000. Closing the Fresh Kills landfill. *The DOS Report,* Feb. 2000.

Rathje, William, and Colleen Murphy. 2001. *Rubbish! The archeology of garbage.* University of Arizona Press, March 2001.

Smith, Ronald S. 1998. *Profit centers in industrial ecology.* Quorum Books, Westport.

Socolow, Robert H., et al., eds. 1994. *Industrial ecology and global change.* Cambridge University Press, Cambridge.

United Nations Environment Programme. 2000. *International source book on environmentally sound technologies (ESTs) for municipal solid waste management (MSWM).* UNEP IETC, Osaka, Japan.

United States Environmental Protection Agency. 2005. *Municipal solid waste generation, recycling, and disposal in the United States: Facts and figures for 2003.* EPA530-F-05-003, EPA Office of Solid Waste and Emergency Response.

United States Environmental Protection Agency. Municipal solid waste. www.epa.gov/epaoswer/non-hw/muncpl.

Chapter 23

Bartlett, Peggy, and Geoffrey W. Chase, eds. 2004. *Sustainability on campus: Stories and strategies for change.* MIT Press, Cambridge, Massachusetts.

Brower, Michael, and Warren Leon. 1999. *The consumer's guide to effective environmental choices: Practical advice from the Union of Concerned Scientists.* Three Rivers Press, New York.

Brown, Lester. 2001. *Eco-economy: Building an economy for the Earth.* Earth Policy Institute and W. W. Norton, New York.

Brown, Lester. 2006. *Plan B 2.0: Rescuing a planet under stress and a civilization in trouble.* Earth Policy Institute and W. W. Norton, New York.

Creighton, Sarah Hammond. 1998. *Greening the ivory tower: Improving the environmental track record of universities, colleges, and other institutions.* MIT Press, Cambridge, Massachusetts.

Daly, Herman E. 1996. *Beyond growth.* Beacon Press, Boston.

Dasgupta, Partha, Simon Levin, and Jane Lubchenco. 2000. Economic pathways to ecological sustainability. *BioScience* 50: 339–345.

De Graaf, John, David Wann, and Thomas Naylor. 2002. *Affluenza: The all-consuming epidemic.* Berrett-Koehler Publishers, San Francisco.

Durning, Alan. 1992. *How much is enough? The consumer society and the future of the Earth.* Worldwatch Institute, Washington, D.C.

Erickson, Jon D., and John M. Gowdy. 2002. The strange economics of sustainability. *BioScience* 52: 212.

French, Hilary. 2004. Linking globalization, consumption, and governance. Pp. 144–163 in *State of the world 2004.* Worldwatch Institute and W. W. Norton, Washington, D.C., and New York.

Gardner, Gary. 2001. Accelerating the shift to sustainability. Pp. 189–206 in *State of the world 2001.* Worldwatch Institute and W. W. Norton, Washington, D.C., and New York.

Gardner, Gary, and Erik Assadourian. 2004. Rethinking the good life. Pp. 164–179 in *State of the world 2004.* Worldwatch Institute and W. W. Norton, Washington, D.C., and New York.

Gibbs, W. Wayt. 2005. How should we set priorities? *Scientific American* 293(3): 108–115.

Hawken, Paul. 1994. *The ecology of commerce: A declaration of sustainability.* HarperBusiness, New York.

Keniry, Julian. 1995. *Ecodemia: Campus environmental stewardship at the turn of the 21st century.* National Wildlife Federation, Washington, D.C.

Mastny, Lisa. 2002. Ecolabeling gains ground. Pp. 124–125 in *Vital signs 2002.* Worldwatch Institute and W. W. Norton, Washington D.C., and New York.

McMichael, A. J., et al. 2003. New visions for addressing sustainability. *Science* 302: 1919–1921.

Millennium Ecosystem Assessment. 2005. *Ecosystems and human well-being: General synthesis*. Millennium Ecosystem Assessment and World Resources Institute.

McIntosh, Mary, et al., 2001. *State of the campus environment: A national report card on environmental performance and sustainability in higher education*. National Wildlife Federation Campus Ecology.

Meadows, Donella, Jørgen Randers, and Dennis Meadows. 2004. *Limits to growth: The 30-year update*. Chelsea Green Publ. Co., White River Junction, Vermont.

National Research Council, Board on Sustainable Development. 1999. *Our common journey: A transition toward sustainability*. National Academies Press, Washington, D.C.

National Wildlife Federation. Campus Ecology. www.nwf.org/campusecology.

Office of Management and Budget, Executive Office of the President of the United States, Washington, D.C. 2003. *Informing regulatory decisions: 2003 report to Congress on the costs and benefits of federal and unfunded mandates on state, local, and tribal entities*. Washington, D.C., September 2003.

Sanderson, Eric W., et al. 2002. The human footprint and the last of the wild. *BioScience* 52: 891–904.

Schor, Juliet B., and Betsy Taylor, eds. 2002. *Sustainable planet: Solutions for the twenty-first century*. The Center for a New American Dream. Beacon Press, Boston.

Toor, Will, and Spenser W. Havlick. 2004. *Transportation and sustainable campus communities: Issues, examples, solutions*. Island Press, Washington, D.C.

United Nations. 2002. *Report of the World Summit on Sustainable Development, Johannesburg, South Africa, 26 August–4 September 2002*. United Nations, New York.

United Nations. 2002. *The road from Johannesburg: What was achieved and the way forward*. United Nations, New York.

United Nations Development Programme. 2002. *Human development report 2002*. Oxford University Press, Oxford.

United Nations Division for Sustainable Development. 1990. *Agenda 21*. Accessible online at www.un.org/esa/sustdev/documents/agenda21/index.htm.

United Nations Environment Programme. 2002. Outlook: 2002-2032. Pp. 319–400 in *Global environment outlook 3 (GEO-3)*. UNEP and Earthscan Publications, Nairobi and London.

United States Environmental Protection Agency. 2000. *Regulatory impact analysis: Heavy-duty engine and vehicle standards and highway diesel fuel sulfur control requirements*. EPA420-R-00-026. EPA, Washington, D.C.

University Leaders for a Sustainable Future. www.ulsf.org.

Wackernagel, Mathis, Lillemor Lewan, and Carina Borgström-Hansson. 1999. Evaluating the use of natural capital with the ecological footprint. *Ambio* 28: 604.

Wilson, Edward O. 1998. *Consilience: The unity of knowledge*. Alfred A. Knopf, New York.

World Commission on Environment and Development. 1987. *Our common future*. Oxford University Press, Oxford.

Index